The Science and Technology of I
Packaging

Multilayer Films from Resin and Process to End Use

PLASTICS DESIGN LIBRARY (PDL) PDL HANDBOOK SERIES
Series Editor: Sina Ebnesajjad, PhD (sina@FluoroConsultants.com) President, FluoroConsultants Group, LLC Chadds Ford, PA, United States http://www.FluoroConsultants.com

The **PDL Handbook Series** is aimed at a wide range of engineers and other professionals working in the plastics industry, and related sectors using plastics and adhesives.

PDL is a series of data books, reference works and practical guides covering plastics engineering, applications, processing, and manufacturing, and applied aspects of polymer science, elastomers and adhesives.

Recent titles in the series
Polymers for 3D Printing, Izdebska-Podsiadły (ISBN: 9780128183113)
Applications of Polyurethanes in Medical Devices, Padsalgikar (ISBN: 9780128196731)
Applications of Polymers and Plastics in Medical Devices, Ashter (ISBN: 9780128209806)
Polymer Hybrid Materials and Nanocomposites, Saleh (ISBN: 9780128132944)
Design and Manufacturing of Plastics Products, Pouzada (ISBN: 9780128197752)
Automotive Plastics and Composites, Greene (ISBN: 9780128180082)
The Effect of Long Term Thermal Exposure on Plastics and Elastomers, McKeen (ISBN: 9780323854368)
Introduction to Fluoropolymers, Ebnesajjad (ISBN: 9780128191231)
The Effect of Radiation on Properties of Polymers, McKeen (ISBN: 9780128197295)
Service Life Prediction of Polymers and Coatings, White (ISBN: 9780128183670)
A Practical Guide to Plastics Sustainability, Biron (ISBN: 9780128215395)
Applications of Fluoropolymer Films, Drobny (ISBN: 9780128161289)
Recycling of Flexible Plastic Packaging 1, Niaounakis (ISBN: 9780128163351)
Plasticizers Derived from Post-Consumer PET 1, Langer (ISBN: 9780323462006)
Polylactic Acid 2, Sin (ISBN: 9780128144725)
Durability and Reliability of Polymers and Other Materials in Photovoltaic Modules, Yang, French & Bruckman (ISBN: 9780128115459)
Fluoropolymer Additives 2, Ebnesajjad & Morgan (ISBN: 9780128137840)
The Effect of UV Light and Weather on Plastics and Elastomers 4, McKeen (ISBN: 9780128164570)
PEEK Biomaterials Handbook 2, Kurtz (ISBN: 9780128125243)
Hydraulic Rubber Dam, Thomas et al. (ISBN: 9780128122105)
Electrical Conductivity in Polymer-based Composites, Taherian & Kausar (ISBN: 9780128125410)
Plastics to Energy, Al-Salem (ISBN: 9780128131404)
Recycling of Polyethylene Terephthalate Bottles, Thomas et al. (ISBN: 9780128113615)
Dielectric Polymer Materials for High-Density Energy Storage, Dang (ISBN: 9780128132159)
Thermoplastics and Thermoplastic Composites, Biron (ISBN: 9780081025017)
Recycling of Polyurethane Foams, Thomas et al. (ISBN: 9780323511339)
Introduction to Plastics Engineering, Shrivastava (ISBN: 9780323395007)
Chemical Resistance of Thermosets, Baur, Ruhrberg & Woishnis (ISBN: 9780128144800)
Phthalonitrile Resins and Composites, Derradji, Jun & Wenbin (ISBN: 9780128129661)
The Effect of Sterilization Methods on Plastics and Elastomers, 4e, McKeen (ISBN: 9780128145111)
Polymeric Foams Structure-Property-Performance, Obi (ISBN: 9781455777556)
Technology and Applications of Polymers Derived from Biomass, Ashter (ISBN: 9780323511155)
Fluoropolymer Applications in the Chemical Processing Industries, 2e, Ebnesajjad & Khaladkar (ISBN: 9780323447164)
Reactive Polymers, 3e, Fink (ISBN: 9780128145098)
Service Life Prediction of Polymers and Plastics Exposed to Outdoor Weathering, White, White & Pickett, (ISBN: 9780323497763)
Polylactide Foams, Nofar & Park (ISBN: 9780128139912)
Designing Successful Products with Plastics, Maclean-Blevins (ISBN: 9780323445016)
Waste Management of Marine Plastics Debris, Niaounakis, (ISBN: 9780323443548)
Film Properties of Plastics and Elastomers, 4e, McKeen, (ISBN: 9780128132920)
Anticorrosive Rubber Lining, Chandrasekaran (ISBN: 9780323443715)
Shape-Memory Polymer Device Design Safranski & Griffis, (ISBN: 9780323377973)
A Guide to the Manufacture, Performance, and Potential of Plastics in Agriculture, Orzolek, (ISBN: 9780081021705)
Plastics in Medical Devices for Cardiovascular Applications, Padsalgikar, (ISBN: 9780323358859)
Industrial Applications of Renewable Plastics, Biron (ISBN: 9780323480659)
Permeability Properties of Plastics and Elastomers, 4e, McKeen, (ISBN: 9780323508599)
Expanded PTFE Applications Handbook, Ebnesajjad (ISBN: 9781437778557)
Applied Plastics Engineering Handbook, 2e, Kutz (ISBN: 9780323390408)
Modification of Polymer Properties, Jasso-Gastinel & Kenny (ISBN: 9780323443531)
The Science and Technology of Flexible Packaging, Morris (ISBN: 9780323242738)
Stretch Blow Molding, 3e, Brandau (ISBN: 9780323461771)
Chemical Resistance of Engineering Thermoplastics, Baur, Ruhrberg & Woishnis (ISBN: 9780323473576)
Chemical Resistance of Commodity Thermoplastics, Baur, Ruhrberg & Woishnis (ISBN: 9780323473583)

To submit a new book proposal for the series, or place an order, please contact Brian Guerin, Acquisitions Editor at b.guerin@elsevier.com

The Science and Technology of Flexible Packaging

Multilayer Films from Resin and Process to End Use

Second Edition

Barry A. Morris

William Andrew is an imprint of Elsevier
The Boulevard, Langford Lane, Kidlington, Oxford, OX5 1GB, United Kingdom
50 Hampshire Street, 5th Floor, Cambridge, MA 02139, United States

ISBN: 978-0-323-85435-1

For Information on all William Andrew publications
visit our website at https://www.elsevier.com/books-and-journals

Publisher: Matthew Deands
Acquisition Editor: Brian Guerin
Editorial Project Manager: Clodagh Holland-Borosh
Production Project Manager: Sojan P. Pazhayattil
Cover Designer: Mark Rogers

Typeset by MPS Limited, Chennai, India

Contents

Preface

Since I wrote the first edition of this book in the mid-2010s, a pandemic has accelerated the growth of e-commerce and the world has taken up climate change and plastic waste with greater urgency. These trends have created new packaging needs and driven innovation across the flexible packaging value chain. The goal of this book has been to bridge the gap between "how to" handbooks and high-level "academic" analysis to provide a fundamental understanding of packaging technology that is accessible to the practitioner. Such an approach not only explains why one material may be favored over another for a given application, or how the film manufacturing process impacts packaging performance, but also provides the tools to design new packaging structures for new needs. In essence, it is just the type of book a packaging engineer and material scientist can use as a reference to deal with new challenges.

My goal is to create a book that brings science to the practitioner in a concise and impactful way. Using scientific principles, I explore (and debunk) some of the myths that have persisted in our industry. When discussing key challenges in the packaging industry, I provide a critical review of the scientific literature as well as practical explanations and tips on how to deal with the challenge. Too often packaging engineers fail to seek out the scientific literature. One benefit of this book is easy access to this literature, summarized in a meaningful way. Finally, the book provides insight into why things are done the way they are, which makes the work evergreen: the same principles can be applied to new packaging applications with new materials that are sure to come along in the future.

The second edition updates the first edition with more than 750 new citations and 75 figures and tables. It contains new or expanded sections tailored toward e-commerce (e.g., a new section is added to Chapter 9 on fatigue resistance), metallization, and laminating adhesives. While the principles needed for design for sustainability were covered in the first edition, these have now been organized into a new chapter (Chapter 17) that considers industry challenges around sustainability, life cycle assessment, end-of-life technologies, and design for recycle. Market trends are dynamic and the needs and technologies around sustainable packaging are particularly so. Any chapter can only represent a snapshot in time; while the technology races ahead, the underlying principles remain the same. The concepts brought out in this book will equip the reader with the tools and insight needed to keep up with the changing landscape.

As with the first edition, the second edition is organized into seven parts and two appendices:

1. Introduction
2. Basic processes
3. Material basics
4. Film properties
5. Effect of converting processes on properties
6. End-use considerations and design for sustainability
7. Structure design and modeling
8. Appendices
    A: Writing multilayer film structures
    B: Examples of flexible packaging film structures

The introduction provides a brief history of packaging and its benefits to society. It describes the basic functions of packaging, the packaging value chain and what is important for each player in the value chain. It introduces the use of multilayer films to provide packaging solutions in the most cost-effective manner. It concludes with a discussion of how the major global trends of population growth, changes in population demographics and sustainability drive innovation in packaging.

Part 2 briefly describes the processes involved in the manufacture and use of multilayer films for packaging. Basic differences between technologies and why one may be favored over another for a given application are highlighted. Key unit operations and processing variables are introduced that affect properties which are described in more detail in subsequent chapters.

Part 3 introduces polymers and substrates commonly used in flexible packaging, highlighting important properties, applications, and regulatory issues. Chapters on melt rheology and polymer blending round out the basics.

Part 4 begins the heart of the book, providing detailed descriptions and analysis of the key properties of packaging films from an engineering and scientific perspective. Drawing from personal knowledge/experiments and the scientific and patent literature, the current state of knowledge around these properties is distilled. Each chapter begins with why the property under discussion is important, how to measure it, and typical values. This is followed by a discussion of the science behind how materials influence these properties. Subjects include heat sealing, barrier, physical strength, abuse resistance, adhesion, optical properties, frictional properties, shrinkage, and thermoforming.

Properties of packaging films cannot be thought of in isolation of the process used to make them. Part 5 explores the effect processing has on film properties in tandem with material properties. Three chapters comprise Part 5, the first of which explores the effect of the process on film quality. Flow instabilities and other topics that affect material selection are covered. The second chapter provides examples of how the process directly affects film properties through changes in quench rate and orientation. The final chapter looks at the effect of processing on interlayer adhesion.

How the flexible package is ultimately used—the environment it is subjected to, the types of products being packaged—can influence film properties and design. Part 6 discusses the effect of the end-use environment (temperature, humidity, pressure, and irradiation), packaging–product interactions, and aging on package performance. The cost of the package is a predominant factor in any package design, but the total cost should be considered: the film raw material and converting costs, the packaging line productivity, and waste due to package failure. The cost to the environment is increasingly a consideration in package design. The concept of life cycle assessment is introduced to evaluate the effect of package design on the environment. Such an evaluation should include the environmental cost of producing the package but also the benefits of reducing product waste; often the product being packaged has considerably more environmental impact than the package. Technologies for mechanically and chemically recycling packaging waste are introduced. Concepts for designing packaging for enabling mechanical recycling and composting, along with challenges associated with doing this, are described. A commercial example illustrates some of the challenges of implementing changes from multimaterial packaging to packages for the single-material recycle stream.

The final section brings together the ideas introduced in earlier parts into a concluding section on structure design. Principles of design, analytical methods to determine what structures are currently in use and modeling approaches are covered.

There are also two important appendices. The first is a system for writing multilayer structures, authored by my long-time colleague at DuPont and Dow, Scott Marks. His system provides a coherent method for communicating package design throughout the industry that is used in this book. The second provides some typical packaging structures that have been in use for various applications over the years. The intention is only to provide some examples and not an exhaustive, or even current, list of structures.

Taking on a project like this is a huge endeavor. I have many people to thank, too many to name individually. I want to acknowledge my friends and colleagues at DuPont and Dow who have taught me so much over the years about flexible packaging, and where most of my knowledge has come from. Since the writing of the first edition, DuPont merged with Dow and I had the pleasure of interacting and befriending many new experts on flexible packaging. The bringing together of these two leading material suppliers to the flexible packaging industry gave me new insights that have made their way into the second edition. My long-time association with the Society of Plastics Engineers and TAPPI has also enriched my understanding of polymer processing and film properties.

Finally, I would like to thank my family, Kathy, Elizabeth, and Sara, for their encouragement and for putting up with my long hours of writing and editing.

Barry A. Morris

November 2021

Part I

Why multilayer films?

Chapter 1

Introduction to flexible packaging

Packaging's primary purpose is to protect the product, and it is through this role society derives many benefits such as preventing food and other product waste, enabling economies of scale in the production and distribution of goods, and aiding in the access of healthcare to all parts of the world. Packaging has evolved to meet the changing needs of the value chain; for example, packaging in modern times increasingly serves as a communication platform, providing content information, instructions and a marketing message. Flexible packaging is ideally suited to meet the challenges of the marketplace and is the fastest growing form of packaging. Driving this growth is multilayer film technology that allows specialized layers which provide sealing, barrier or abuse resistance to be combined to meet the packaging requirements in the most cost-effective manner. Adding to this is the variety of manufacturing processes that can be used to assemble these structures. Macro trends involving the global expansion of the middle class, an aging population, new direct consumer distribution channels and the need to mitigate the environmental impact of packaging are driving changes in package design. To meet these challenges, material scientists, process engineers and packaging developers will increasingly need a fundamental understanding of flexible packaging technology.

1.1 History of packaging

The earliest forms of packaging were likely animal skins and other primitive textiles used by hunter-gatherer societies to transport and store food. The development of agriculture with the domestication of grains and other plants for food over 10,000 years ago fostered population growth and the need to store food. This drove the development of new forms of packaging (see Fig. 1.1). Early granaries likely used baskets woven from various plants to store food. The earliest archeological evidence for basket weaving dates back to 10,000 to 12,000 years ago [1], but the technology may have originated even earlier given the poor survivability of these types of artifacts.

Basket weaving is the direct predecessor of pottery since baskets were used as the mold for ceramic vessels before the invention of the pottery wheel. The earliest known ceramic artifact, a statuette of a woman, was found in the modern-day Czech Republic and was dated back to 28,000 BCE (before the common era). The first pottery artifacts appear in China around 18,000–17,000 BCE. From China, pottery spread to the far eastern regions of Russia and Japan, where shards have been dated to 14,000 BCE [2,3]. Starting about 9000 BCE clay-based ceramics were being used to store water and food and their use spread from Asia to the Middle East and Europe [2]. The Egyptians, Greeks and later Romans used pottery vessels called amorpha to transport a variety of food and other products. These two-handled clay pots are about half the size of today's 55-gallon (208L) drum (Fig. 1.2). They have been found on many ship wrecks from these times, indicating they played an important part in early commerce [4,5].

An off-shoot of ceramics is glass; manmade glass dates back to around 2500 BCE or earlier in Mesopotamia [6] and was industrialized in Egypt in 1500 BCE [2]. The blowpipe was invented by the Phoenicians in the third century BCE. The modern era of glass packaging began in 1903 when Michael Owens developed the fully automated bottle-making machine [7].

The development of modern rigid packaging was stimulated by the need to feed large armies. In the late 1700s Napoleon recognized long supply lines and food availability were limiting the mobility of his army and offered a reward for improved packaging. This led Nicholas Appert of France to invent the canning process in 1809. A year later Peter Durand of England received a patent on various forms of packaging and is generally credited with the invention of the tin can [8,9]. Tin was later replaced by steel and aluminum.

Tin foil was marketed for food packaging in the late 1800s but left an off-taste. Aluminum foil was first produced in France and Switzerland in the early 1900s and in the US by 1913. Neher patented a continuous rolling process in 1910. Packaging was an early use. By 1921 the first foil laminated paperboard carton was introduced. The first heat sealing

The Science and Technology of Flexible Packaging. DOI: https://doi.org/10.1016/B978-0-323-85435-1.00008-9

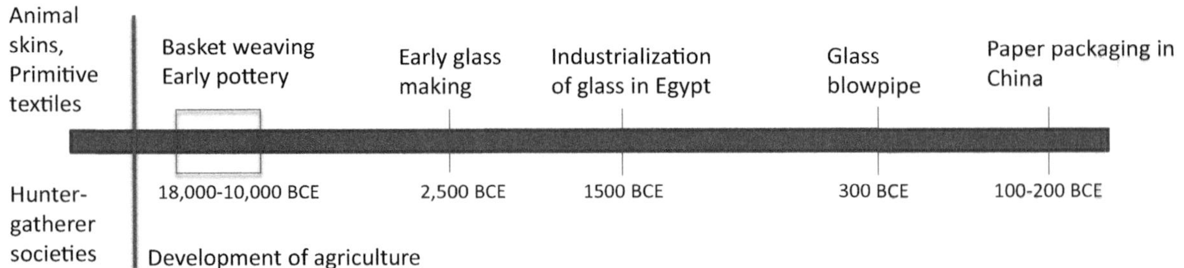

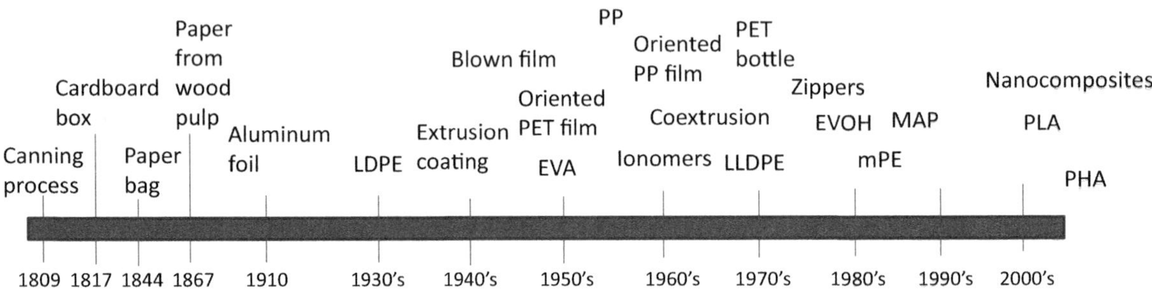

FIGURE 1.1 Milestones in packaging history.

FIGURE 1.2 Amorpha in Turkey. *Reproduced with permission under terms of the GNU Free Documentation License. Photo by Tim Rogers.*

foil was developed in 1938. During World War II, aluminum foil found use as a packaging medium, which helped establish it as a mainstream flexible packaging substrate [10].

Next to animal skins, paper may be the oldest form of flexible packaging [11]. The Chinese used an early form of paper to wrap foods in 100–200 BCE. The technology evolved and was transported to the Middle East, Europe and England over the next 1500 years. Early paper was made from bark, flax fibers or old linen rags. Wood pulp was first used to make paper in 1867. Two other developments in the 1800s contributed to paper's growing use as a packaging medium: the paper making machine and flexographic printing [12]. These developments spurred the commercialization of various packaging forms, such as the paper bag (1844), the cardboard box (1817), corrugated paper (1850s), and the paper carton (1870s) [11].

Paper-based packaging continued to grow well into the mid-1900s, when plastic substrates became widely available. The introduction of cellophane commenced the modern era of flexible packaging. Invented in 1908 by Jacque E. Brandenberger, a Swiss textile engineer, it initially only found limited use for wrapping candy and chocolate. This clear flexible film had one major drawback: it had a very poor moisture vapor barrier, many times worse than wax-coated paper available by this time. After the DuPont Company acquired the US patent rights to manufacture cellophane in the mid-1920s, a young DuPont scientist named William H. Charch came up with a solution: a thin, transparent coating that made the film water tight [13]. This coincided with major changes in the lifestyles of American consumers. Before then, most foods were purchased in open markets or small stores. The modern supermarket was beginning to come onto the scene, emphasizing self-service over customization. The availability of a transparent packaging film transformed the market, allowing the store to prepackage goods in a way that the consumer could see what was being packaged. DuPont heavily advertised its miracle film to consumers for its clarity, protection and convenience: it allowed the shopper to save time by not having to wait for a clerk to serve them [13,14]. Two of these advertisements are reproduced in Figs. 1.3 and 1.4.

By the 1960s cellophane had mostly been replaced by polyvinyl chloride, biaxial oriented polyester (developed in the mid-1950s at DuPont, ICI and Hoechst) [17] or oriented polypropylene (PP) (ICI, 1961) [18]. Metallized film technology, formed by physical or vacuum vapor deposition of a thin layer of aluminum, was first commercialized in the 1930s for Christmas tinsel. It was adapted for packaging films and paper in the 1970s [19].

Polyethylene is the largest volume of plastic used in flexible packaging today, valued for its water barrier, toughness, sealability and transparency. Hans von Peckmann first made polyethylene, although he did it accidentally when

FIGURE 1.3 1948 Cellophane advertisement [15]. *Courtesy of E. I. du Pont de Nemours and Company and Hagley Museum and Library. [15] E. I. du Pont de Nemours and Company, Advertising Department, Modern shoppers prefer food in cellophane: shows what it protects! Protects what it shows! Dpads-1803-00308.tif, Hagley Museum and Library, 1948.*

FIGURE 1.4 1949 Cellophane advertisement [16]. *Courtesy of E. I. du Pont de Nemours and Company and Hagley Museum and Library. E. I. du Pont de Nemours and Company, Advertising Department, New self-service meats make shopping quicker, easier: DuPont Cellophane, Dpads-1803-00320.tif, Hagley Museum and Library, 1949.*

heating diazomethane in a test tube in 1898. The white waxy substance was analyzed by colleagues to contain long molecules which they dubbed polymethylene. These initial experimental forms were not commercially viable. In the 1930s, Imperial Chemical Industries (ICI), an England-based chemical conglomerate, developed the high-pressure free-radical polymerization process for making low-density polyethylene (LDPE). Commercialized in England in 1939, LDPE was used during World War II for radar and cable insulation. DuPont collaborated with ICI to bring production to the U.S for the war effort; production there began in 1943. Union Carbide and Dow Chemical were also producing polyethylene in the US by the early 1950s [20].

Linear low-density polyethylene (LLDPE) and high-density polyethylene (HDPE) were developed in the 1950s. The technology involves using a catalyst and coordination chemistry. There are three basic catalyst chemistries: chromium typically used in slurry and gas phase processes, Ziegler–Natta in gas and solution processes, and metallocene in all processes. The chromium catalyst process was developed at Phillips Petroleum in the 1950s and introduced as Marlex HDPE. Karl Ziegler (Germany) and Guilio Natta (Italy) conducted basic catalyst research in the 1950s; LLDPE using Ziegler–Natta catalysts was commercialized by several companies including Dow and DuPont Canada in the 1970s. Metallocene catalysts were discovered by Walter Kaminsky and Hansjorg Sinn in Germany in 1976 and developed by Dow and ExxonMobil in the 1980s [20].

Throughout the time period between 1950 and 1990 other materials used in packaging were introduced, including ethylene vinyl acetate (EVA), ethylene acid copolymers, ionomers, PP, polyamide 6 (PA6), ethylene vinyl alcohol (EVOH), polyvinylidene chloride (PVDC), polystyrene (PS), and tie resins. New biobased polymers, such as polylactic acid (PLA), polyhydroxyalkanates (PHAs) and starch polymers were commercialized in the 2000s.

As the plastics materials advanced, so did the processes to convert them into flexible packaging. Extrusion coating is a case in point. Following World War II the availability of LDPE for nonwar uses opened up. By 1947, H. P. Smith had developed an early process for coating LDPE onto paper. During this same time period, DuPont, collaborating with the equipment supplier Hartig Engine and Machine Company, developed the process used today involving coating a free film onto paper. DuPont worked closely with St. Regis Company, which was the first to make extrusion-coated products on a commercial scale. International Paper, Sealright and Egan were also early innovators. By 1957, the first LDPE coated milk carton was commercialized. From there, the technology advanced to many other packaging forms [21].

The blown film process also developed with the advent of LDPE. The demand for waterproof packaging films with good low-temperature toughness progressed in the 1940s and 1950s, and LDPE was just the plastic to fit this need. The technology to make blown film was developed through this time period, mostly through activities by film producers that made their own dies and ancillary equipment. Visking Company is reported to be the first company to develop polyethylene extrusion. Most of the technology remained closely guarded company secrets until the late 1950s when resin suppliers began buying up film producers and opening up the technology to grow the market for their resins [22].

Extrusion processes were first used in the mid-1800s to make sheets of gutta-percha, rubber and shellac. Modern looking screw extruders appeared in the mid-1930s in Germany and the United States [23]. As polymer materials were developed so were the sheet and other flat die processes like cast film and extrusion coating.

Combining materials into multilayer flexible packaging has been practiced since the beginning of the modern packaging era. Coatings onto paper, such as wax or clay, were used early on to provide moisture resistance. By the 1940s adhesives were being used to bond sealant layers onto substrates like cellophane and aluminum foil. Coextrusion, where the layers are combined during the film fabrication process, was commercialized for flat die extrusion in the 1960s, followed shortly thereafter by blown film [24,25].

Packaging machinery also evolved with the availability of new materials, from early bag making machines to form-fill-seal machinery. The vertical-form-fill-seal machine was patented by Walter Zwoyer in 1936 [26].

Packaging forms have advanced throughout the years with ever more sophistication. Paper bags were introduced in the 1840s, with glued paper sacks with gussets in the 1870s. Laminated paper cartons with aluminum foil were introduced in 1921. Wax coated milk cartons showed up in the early 1900s [27] and the LDPE coated version in 1957. Modified atmosphere barrier packaging for improving the shelf life of meat products came into use in the 1970s. Zippers appeared in the 1950s but not extensively into flexible packaging until the 1980s. The stand-up pouch gained popularity in the 1990s.

A constant throughout the dynamic growth of multilayer flexible packaging over the past 50 years has been the continued advancement of technology: resin, substrate, process and package form. Packaging materials and forms are continuing to evolve as packaging needs to change. While one or two layers may have been common a few years ago, now 5, 7, and 9 layers are the norm, with some package structures utilizing hundreds of layers. Often there are multiple ways to solve a package need, either by changing layers or the process. For example, a barrier film can be made by laminating a barrier substrate like aluminum foil to a heat seal film, or by co-extruding a barrier resin-like EVOH copolymer with a sealant. Often the choice depends on the secondary needs of the application, the type of equipment the converter has, and increasingly, end-of-life considerations such as recyclability.

1.2 Benefits of packaging

Packaging benefits society by aiding in the distribution of goods [28]. In its simplest form, such as a paper bag, it may be used for transporting or storing goods. Modern packaging, however, has evolved to provide extended protection, abuse resistance, and a platform for communicating benefits and information to the consumer that allows mass merchandizing to support a global economy.

From an engineering perspective, the package protects the product. For a food product, this may mean it helps prevents spoilage or extends the shelf life of the product before the consumer uses it. Barrier and abuse resistance are two primary considerations. Barrier technology may involve the creation of hermetic seals to prevent leakage, blocking light, or providing low permeation to gases such as moisture and oxygen. It also may include preventing food components from migrating into the package or allowing components of the package to migrate into the food. The package may aid in the prevention of harmful microbes from growing by allowing for post packaging sterilization through techniques such as retort, radiation or high-pressure pasteurization.

Packaging abuse comes in many forms. The package may be subjected to slow puncture, as from a sharp product within the package such as a bone or noodle. High-speed impact events may occur during filling, handling and distribution. Vibration during shipping may repeatedly flex the package and cause damage. Cold or hot storage may be a factor. The package may be subjected to internal pressure created by elevation changes during transport. The ability to withstand such abuse must be designed into the package.

From a marketing perspective, the package is an advertisement. In the retail environment, it may be the primary means the brand owner has to inform and entice the consumer. Sparkling graphics can aid in grabbing the attention of the consumer in the retail store and informing the customer why he or she should purchase the product. To this end, the package often must provide a surface for printing. The entire or parts of the package may need to be transparent so consumers can see what they are purchasing. In the direct sales (e-commerce) value chain, the initial section of the product may be more determined by the online ordering process, but good graphics can still be important for repeat sales as the package may still perform an advertising function as it sits in the consumer's pantry. The esthetics and form of the package also play an important role in the opening experience that is important for brand identity and repeat sales.

From an operations perspective, an ideal package has an efficient form factor, allowing for ease of transportation. The shape of the package impacts how many packs can fit on a pallet and on a truck. It also influences what can be put on a shelf—does it stand up on its own or does it need a peg? Can it be stacked one high on the shelf or more than one high? Such forms influence where the product may be displayed in the store.

From a legal perspective, the package performs a service by providing a medium for displaying content. Label laws in many countries require ever more space on the package to the extent that many pharmaceuticals have labels that open out to display all the requisite information. Labeling laws may require a list of ingredients, net weight, dosage and various disclaimers. The brand owner may also want to include contact information, recipe suggestions, instructions on how to recycle, etc. Quick response (QR) codes may be printed that link the consumer to additional information.

The benefits of packaging to society are numerous. Packaging helps reduce world hunger, spurs economic development, enables life-saving measures in healthcare and contributes to sustainability. The UN estimates that the world population will reach 9.7 billion people by 2050 [29], 2 billion more than in 2019. Increasing food production as well as reducing waste will be essential for feeding the world. In 2011, the United Nations Food and Agriculture Organization (FAO) estimated that about 1/3 or 1.3 billion tons of food are wasted each year [30]. A 2013 Institute of Mechanical Engineers study [31] concluded from 30% to 50% of food is wasted. These reports spurred the call for more data and the UN has responded by creating two indices. The first is the Food Loss Index (FLI) prepared by the FAO that includes losses from post-harvest up to but not including retail. Initial estimates for the FLI are around 14%, meaning 14% of food is lost during distribution [32]. Poor distribution and infrastructure are particularly a problem for developing countries. Packaging can play a key role in preserving food throughout the food chain, as has been demonstrated in the developed world. Transferring best practices to the developing world will go a long way to help reduce the potential for future food shortages.

The second index is the Food Waste Index prepared by the UN Environmental Programme. Food waste is defined as food and associated inedible parts (like bones) removed from the human food supply chain in food/grocery retail, food service and households. 931 million tons of food waste was generated in 2019 representing 17% of global food production (taken together with the FLI, 31% of food production is ultimately not consumed). Of the food wasted, 61% occurred in households, 26% in food service and 13% in retail. Household per capital food waste generation is found to be broadly similar across country income groups (developing and developed economies) [33]. This differs from earlier studies [30] that suggest food waste is more prevalent in developed countries whereas food loss is more widespread in developing countries. It points to room for improvement in all economies. Part of the food waste is due to excessive buying by the consumer. Packaging can also play a role here. While single-use packaging seems like a waste, portion control packaging can actually be more sustainable than bulk packaging if it cuts down on the amount of unused food. Advances in barrier and recloseable technologies may also extend shelf life at the consumer level, helping to cut food waste.

Packaging in recent years has often been demonized as part of the sustainability problem. If the whole food chain is considered, however, packaging turns out to be only a small contributor to environmental stress. Büsser, et al. [28] evaluate the impact of packaging on several food products. They find that the impact of packaging is minor compared with the product. For example, for coffee, cultivation, and brewing make up 82%–99% of the nonrenewable energy consumption and greenhouse gas emissions. Packaging only contributes about 1%–5% depending on product package type. The study also considers ozone layer depletion, acidification and eutrophication. These impacts are also not significantly impacted by the package. Similar conclusions are made for the two other food products considered: frozen spinach and butter. For frozen spinach, keeping the product cold throughout distribution, and in particular, at the consumer, is found to be responsible for most of the environmental stress. Butter production overwhelms other factors, including packaging, for environmental impact. Indeed, one can conclude that the package prevents the environmental stress from being even greater due to the prevention of spoilage and damage throughout distribution and use. Section 17.2.2 provides additional examples and discusses a methodology for quantitatively assessing the environmental impact of packaging and product loss.

Although packaging can reduce the environmental impact of the food distribution process, there are still opportunities to reduce environmental stress through material choice, optimizing package converting processes and reduction of packaging. Indeed, the recent trend toward reducing packaging by switching from rigid packaging, such as steel cans or glass jars, to flexible packaging is driven in part by this dynamic (the other driver is lower economic cost). There have been a number of studies trying to compare one package type versus another. Some are of dubious standing, sponsored by special interest groups from one side or another. But intuitively, reducing the amount of material usage should reduce environmental impact. A study by the Flexible Packaging Association [34] shows a substantial reduction in nonrenewable energy consumption and greenhouse gas emissions when switching from rigid or even paper packaging to plastic flexible packaging. For example, switching from glass or metal beverage containers to a plastic stand-up pouch reduces energy consumption by 50% and greenhouse gas emissions by 75%. No doubt this study is motivated by enhancing the status of its members, but nevertheless, reducing the amount of material used in a package makes inherent sense, both from a financial and environmental cost perspective.

Packaging also provides numerous benefits to the health and wellbeing of society [35]. Medical devices packaged in easy-open, tamperproof, sterilizable packaging ensure that instruments are safe for use in the operating room. Packaging enables pharmaceuticals to reach remote areas of the world where cold storage or sanitary conditions do not exist. Emergency relief operations depend heavily on prepackaged goods for fast delivery to areas in need. The health benefits of packaging came to the forefront during the Coronavirus disease (COVID) pandemic as a way to minimize the transfer of the virus.

It is no coincidence that the use of packaging has expanded in the era of mass production and merchandising [36]. Huge business-to-business supply chains have developed involving unfinished parts or goods being transported from place to place. This would not be possible without packaging. Packaging has helped spur economic development in consumer value chains by enabling self-service purchasing by consumers (e-commerce). It has cut the cost of distributing, promoting and ensuring the quality of products, and servicing consumers. Packaging has replaced the sales clerk in many markets by providing information on the label. By taking costs out of the supply chain, the cost of goods sold has decreased, making products more affordable to the mass market.

Packaging does have negative consequences to society and the environment, especially when not used responsibly. These include greenhouse gas emissions during production, possible release of toxins into the environment, the scarring of landscapes from mining pits (aluminum) or clear-cutting (paper), and the accumulation of plastics in our oceans [37–42]. Clearly, more needs to be done to reduce these negative consequences, such as more recycling and better collection of used packaging. Designing packaging to meet societal needs around sustainability (Section 1.5.1) is taken up in detail in Chapter 17. The benefits of packaging, especially flexible packaging, are numerous and generally outweigh the negative consequences.

1.3 Size of market

Flexible packaging has grown faster than other packaging forms over the last several decades; about 19% of the $177 billion total US packaging market is flexible, up from 17% in 2000. It is the second-largest segment after corrugated paperboard, which has also seen recent growth due to e-commerce. The primary driver is cost. Flexible packaging generally uses less material on a weight basis than other packaging forms, resulting in less material usage as well as savings in transportation. Rigid packaging historically held the advantage in terms of reclosability (caps and lids) and high-temperature sterilization, but flexible packaging has closed the gap as new technologies have been introduced.

The size of the flexible packaging market depends on how it is defined. The estimated annual sales for the total US flexible packaging market is $33.6 billion which "includes packaging for retail and institutional food and nonfood, medical and pharmaceutical, industrial materials, shrink and stretch films, retail shopping bags, consumer storage bags, and wraps and trash bags." [43] The Flexible Packaging Association further refines this definition to focus on the segment of the industry that adds significant value usually by providing multiple converting processes such as printing, lamination, coating, extrusion and bag/pouch making. By this definition, retail shopping bags; consumer storage bags; trash bags; and shrink and stretch films used for secondary packaging are excluded. Using this definition, the global flexible packaging market was estimated to be $90.3 billion in 2018 with the following breakdown by region [44]

- North America: 29% ($26.2 billion);
- Eastern Europe: 3% ($2.7 billion);
- Western Europe: 14% ($12.6 billion);
- Middle East and Africa: 5% ($4.5 billion);
- South East Asia and Oceania: 18% ($16.3 billion);
- Central and East Asia: 25% ($22.6 billion);
- Central and South America: 6% ($5.4 billion).

The global growth rate was 4.5% in 2017 and revenue was estimated to be $113 billion by 2023, but this was prior to the COVID pandemic. Regional growth rates generally track GDP with Asia outpacing the rest of the world over the last decade. The growth rate in the US in 2018–19 was 2.2%, slightly lower than the 2.3% change in GDP but on the trendline [43].

The main end-use applications for flexible packaging in the US market include: [43]

- Food (retail and institutional): 59%;
- Retail nonfood: 12%;
- Industrial: 7%;
- Consumer products: 8%;
- Institutional nonfood: 3%;
- Medical and pharmaceutical: 9%.

Again, these estimates come from the Flexible Packaging Association. Other estimates can be found in various market research reports and government reports such as the US Census [45]. Differences in market definitions and the methodologies used to obtain the data can give substantially different results.

Material costs represent on average 56% of sales. On average, pretax profit for converters is 4.4% [46].

1.4 Packaging value chain

The packaging value chain consists of a number of players, as outlined in Fig. 1.5. These include the following:

1. Material suppliers (resins, adhesives, substrates);
2. Film/sheet converters;
3. Film laminators and printers;

Packaging Value Chain

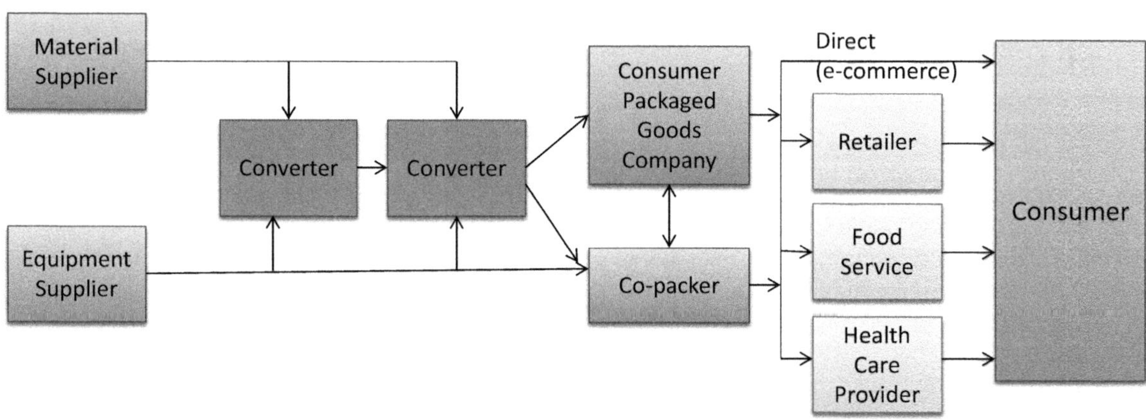

Influencers: Consultants, Media, NGO's, Governments, Brand Consultants, Advertising Agencies, Universities, Trade Associations

Examples of Products or Services Provided

Resin Adhesives Oriented Films Aluminum Foil Paper Ink	Film blowing Film casting Extrusion Coating	Printing Lamination Slitting	Branding Packaging Distribution	Service to Consumer Warehousing and shipping (e-commerce)
Film converting equipment Packaging lines				

FIGURE 1.5 Flexible packaging value chain.

4. Equipment manufacturers (converting and packaging);
5. Packers (including contract packers);
6. Consumer package goods companies;
7. Wholesalers, distributors, warehousers/shippers (e-commerce);
8. Retailers;
9. Consumer.

The material suppliers furnish the raw materials for making the package. These include polymers used by the converter to make films (e.g., polyethylene, PP, polyamide, and EVOH), adhesives and substrates (e.g., paper, aluminum foil, and polymer films). Each of the materials has its own value chain. For example, the resin producer uses monomers such as ethylene to make the polymer. The ethylene, in turn, is generally produced from oil (naphtha-based processes) or natural gas (ethane-based processes). One trend is to move away from fossil fuel-based feedstocks to biosourced or renewable ones. Some examples include:

1. Braskem has introduced grades of polyethylene made from biobased ethanol (from sugarcane).
2. Dow has a partnership with UPM to source feedstocks from byproducts of the paper making process for making polyethylene [47].
3. Coca-Cola uses bottles made from polyester that is partially derived from biosourced monomers.
4. Several companies are developing products from chemical or feedstock recycling of plastic waste where the recycling is broken down to its chemical constituents for reuse (see Chapter 17).

As of this writing, renewable feedstocks are typically more expensive than fossil-derived feedstocks and only a small fraction of the materials used in flexible packaging are biosourced. Similar arguments can be made for new polymers derived from biosources, such as PLA, starch-based polymers and PHAs. New polymers suffer from having to compete against polymers that have been fine-tuned for many years to achieve desired processibility and properties. As with any disruptive technology, it may take time before these polymers gain enough traction to be widely used in flexible packaging. Materials used in flexible packaging, including biobased polymers are described in more detail in Chapter 4.

Film converters process the resins into films, which can be single or multilayer. Common extrusion processes for flexible packaging include blown and cast film, and extrusion coating or lamination. Any of these processes may employ coextrusion technology where more than one layer is combined to make a multilayer structure during the manufacture of the film. Orientation processes such as machine direction, tenter frame or double-bubble may also be used to provide greater strength, dimensional stability or shrink. These processes are described in Chapter 2.

Film lamination and printing are often performed by the converter who makes the film but may be done in a separate operation at another plant or company. Lamination involves combining films or substrates, typically using either an adhesive lamination or extrusion lamination process. In adhesive lamination, solvent-based, water-based or liquid solventless adhesives are used to bind the substrates together. In extrusion lamination, molten polymer is used as the adhesive. Thermal lamination is a third process less practiced in the packaging industry. One example is a thermal lamination of a barrier film to a single-layer tray to improve the barrier performance of the tray. The film is brought in through a secondary unwind and nipped to the sheet just after it is made and is still hot.

Equipment suppliers provide film fabrication, lamination, coating, printing, winding and unwinding, and packaging equipment to the industry. The packaging equipment suppliers are probably the most diverse given the myriad of packaging forms on the market today. These include everything from stick packs, to pillow pouches, to stand up pouches, to juice boxes, to composite cans to bag-in-box, to trays with lids, etc. There are many ways to package a product and many ways to produce a given package form.

Packers put the product into the package, often on high-speed packaging lines, seal it up and send it off for distribution. At one time the packer was synonymous with the brand owner. The brand owner did its own packing, often at multiple locations that made sense for local distribution. Increasingly, independent or contract packers have set up shop. Initially, they handled the overflow from the brand owners. They have since evolved to provide services for private labels, or even brand owners who have decided not to invest in their own lines.

Consumer packaging goods (CPG) companies or brand owners control the development, production, distribution, and marketing and branding of the product. The CPGs may be local, regional, national or international in scope. They may choose to contract some of these functions out to co-packers or marketing service companies. Increasingly, retailers have created their own "private" labels that compete with traditional brand owners.

In the traditional "brick and mortar" retail value chain, the CPG sells its product through a retailer where the consumer physically comes to purchase the product. There may be distributors and wholesalers between the CPG and the retailer. Increasingly, the CPG or retailer may sell all or part of their product online through one or more e-commerce business models. These models include the "delivered basket" where the consumer orders online and the basket of goods is directly shipped to the consumer. Another is "click and collect" where the consumer orders online but picks up the basket of goods at a physical store. Bulk shipments on pallets to physical retail stores generally are less abusive to the package than individual delivery. It is estimated that direct shipments result in three times more touchpoints from the time the product is packaged to when it arrives at the consumer versus traditional physical retail business models [48]. This has implications for package design, as is discussed in Chapter 9.

The consumer is the ultimate customer for consumer goods and many food products. Consumer preferences and purchasing decisions in the end determine which products will be manufactured and packaged.

In addition to the consumer goods industry, the food service and healthcare industries are important users of flexible packaging. Packaging structures for food service are often similar to that in retail, although the size or form of the package may differ. Also, labeling and graphics may be less critical for food service than retail. Medical devices, such as surgical instruments, are packaged in sterilizable flexible packaging. Blister packaging is often used for pharmaceuticals.

The electronics, automotive and other industries use flexible packaging during internal manufacturing operations as parts are moved from place to place, sometimes in complex international supply chains.

Throughout the value chain, there are influencers that impact packaging decisions. These influencers include consultants, package design companies, brand and advertising consultants, co-packers, trade associations, trade magazines and other media outlets, universities, nongovernment organizations (NGOs) and government (especially regulators). NGOs and government agencies are increasingly influencing change in the packaging industry to enhance sustainability and reduce waste.

1.5 Needs along the value chain

Packaging requirements can be quantified into what is critical to quality (CTQ). Various players along the value chain express CTQs differently. The consumer drives the decisions of the brand owner or end-user, which is chiefly concerned with the manufacture, marketing, distribution and safety of the product. The end-user CTQs are expressed in terms of the package form and features that can provide for the consumer's needs. Converters are generally responsible for designing and making the film for the package; their needs center on the materials and processes that can satisfy the end-user CTQs. Material and equipment suppliers focus on satisfying the needs of the converter. Overriding all of these requirements are societal needs around sustainability and product safety.

1.5.1 Society needs and sustainability

There are basic needs for which individual consumers may not have sufficient knowledge to make informed decisions or be willing to pay for but nevertheless benefit society as a whole. Examples include product safety and environmental impact. Product safety considerations are typically dealt with through government regulations, such as food contact packaging regulations. These are discussed in chapter 4.

Environmental impact and sustainability have many facets and are taken up in earnest in chapter 17. There are two broad policy frameworks for ensuring sustainability of packaging, Sustainability Materials Management (SMM) and Circular Economy (CE). SMM is espoused by the US Environmental Protection Agency and supports the efficient use, reuse and recycling of materials. It focuses on minimizing material usage and carbon impact [49]. CE proposes rethinking our linear packaging value chain that results in the single-use of many packaging materials. It envisions a future economy that embraces recycling, drastically reducing leakage and plastics into the natural environment, and ultimately decoupling plastics and packaging from fossil feedstocks [42]. CE has been embraced by the European Union, China and many global brand owners, who have announced aggressive packaging recycling goals.

The drive toward a more sustainable economy takes on different forms in different countries and regions and affects package design and end-of-life in many ways. Examples include:

1. Extended producer responsibility fees act as a tax on packaging in some regions such as Europe. This drives the reduction of material usage and recycling.
2. Some countries/regions have put forth packaging recycle content goals. Many brand owners have also set their own goals for recycle content in their packaging. Postconsumer plastic waste is especially challenging for incorporation into flexible packaging due to quality, performance and safety considerations. For example, food safety regulations

have limited the use of postconsumer recycling in food packaging. The value chain is starting to respond with better control over sourcing, new cleaning technologies, and the development of polymer compatibilizers and performance boosters.

3. The European Union Strategy for Plastics has an objective that by 2030 all plastic packaging will be reusable or easily recyclable in a cost-effective manner. Many brand owners have put forth recyclability goals. This is driving innovation in package design, such as moving away from difficult to recycle multimaterial structures toward structures that are compatible with single-material recycle streams such as the polyethylene stream. Standards have been introduced by the recycling industry to control what goes into these mono-plastic streams [50–53]. This is leading to the development of new materials, such as tougher grades of polyethylene and barrier coatings that make up for the performance features of the eliminated layers but are compatible with the recycle stream.
4. New technologies and infrastructure for the recycling of packaging are being developed. This includes processes to separate multimaterial layers of multilayer packaging for applications where monomaterial packaging will not suffice. It also includes investment in chemical recycling where the plastic is broken down to its chemical constituents that can subsequently be reused to make new plastics.
5. Biobased materials, such as paper, are getting renewed attention. Routes to using bio-based monomers to make conventional polymers such as polyesters and polyoefins are being developed. New biobased materials such as starch-based polymers, PLA and PHA are also being considered for some applications.
6. Compostable materials such as paper and compostable polymers such as PLA are being considered, especially for closed-loop venues where the packaging can be collected and taken to an industrial composting facility.

Many of these technologies are still in development and add cost; it is not clear which will win out in the long run. Sustainability and end-of-life considerations for flexible packaging have become important drivers for package innovation and will continue to be in the future. More in-depth discussion of these considerations and their implication for package design can be found in Chapter 17.

1.5.2 Consumer needs

What do consumers expect from the package? First and foremost they are looking for protection. For a food product, they want freshness and an assurance of safety. They want the product to be free of contamination, be the correct weight and not crushed inside the package. They do not want the product to leak from the package, whether it is a liquid or powder.

Information is a second need. What is the product? This may be in the form of a description or picture, or if the package is transparent, a window to the product itself. In this age of self-service, the package may also need to provide guidance as to why to buy this product over others (new flavor, new size, etc.). A list of ingredients, net weight, "use by" dates, and instructions for use may be other information expected on the package. A recycle symbol designating which recycle stream it can become part of is another feature that many consumers look for.

Increasingly, the consumer is looking for the package to provide convenience, such as easy-open features, reclose-ability, or shaped for on-the-go consumption.

1.5.3 End-user needs

The end-user is a term used to describe the players in the value chain that use the film to package the product. This may be the CPG or brand owner or the packer. The end-user is concerned with translating the consumer needs to package function. An example for a food company is given in Table 1.1. Here the following needs are translated to packaging requirements:

1. Preserving the freshness of the product may involve achieving certain barrier requirements, such as the amount of oxygen that may permeate into the package over the intended shelf-life of the product. If the barrier requirement can be quantified, the materials and their thickness can be specified to meet the target. A light barrier may also be a requirement, which can be achieved through incorporating opaque layers such as aluminum foil or metallized films, adding pigments such as titanium dioxide (TiO_2) to a plastic layer or by printing. Moisture, CO_2, and ethylene barrier or transmission may also be important for ensuring freshness. Fresh produce, for example, benefits from being packaged in films with high gas transmission to allow the produce to "breath." Micro-perforation, a process which imparts small holes into a film, is often used to achieve such permeation.

TABLE 1.1 Example of packaging needs of a food company based on consumer needs.

Consumer needs	Translates to
Freshness	Barrier and flavor retention properties
Not contaminated	Heat seal properties, off odors, abuse resistance
Correct weight	Heat seal properties, abuse resistance
Product not crushed	Stiffness, durability
Information	Print quality

2. Freshness may also involve preventing off-flavors from migrating from the package to the product, or flavors or active ingredients in the product from migrating out.
3. Preventing the product from being contaminated and achieving barrier performance involves establishing good heat seals. Sometimes the product itself, if it is powdery or greasy, may contaminate the seal area, making sealing even more demanding. The choice of the sealant can be critical in these situations.
4. The seal must also be able to withstand the extreme environments it may be subjected to during shipping and handling. High altitudes may put pressure on the seal as the package expands due to the thinning of the air surrounding it. Extreme cold, as in blast freezers, or hot temperatures during filling, sterilization or shipping may affect seal durability.
5. Off-odors permeating through the package or from the package itself can also contaminate the product.
6. Maintaining the correct weight involves preventing the package from leaking, which relates back to seal and barrier performance, as well as puncture resistance.
7. The stiffness, puncture resistance, and impact toughness of the package all contribute to preventing the product from being crushed during transportation and distribution.
8. Fulfilling the consumer's needs for information may require good printing quality, which starts with having a good surface for printing. Also, it may be important for the consumer to see the product, so providing transparency or a window in the package may be needed.

   The end-user has a number of other needs, such as
9. Processability of the packaging film on the filling equipment. This requires the film to have the right stiffness and frictional properties to be easily handled. Seal performance will contribute to packaging line speed. The packaging film must also be easy to cut and withstand the temperature of the filling process.
10. Compliance with food packaging regulations, such as the US FDA or the European food laws.
11. Ability to withstand post-filling operations, such as sterilization, pasteurization or freezing.
12. Meet commitments around sustainability: recycle content, the ability of the package to be recycled, use of renewable materials, etc. This translates to materials selection that meets all the rest of the package performance requirements in addition to the sustainability goals.

1.5.4 Converter needs

The flexible packaging converter purchases base films and resins to produce the film roll stock for the end-user. Cost is a key driver throughout the value chain. The focus at this stage of the chain is providing function at the lowest cost. Frequently this involves combining more than one material into a multilayer film. Often there are several ways to put together a packaging structure for a given application. The converter must weigh the options, based on the functional requirements, process assets they may have in place (or wish to invest in) and cost.

One consideration is whether lamination or coextrusion should be used to make the multiple layer structure. Lamination involves combining two films with an adhesive or polymer melt (extrusion lamination). In coextrusion, layers are combined while in the melt during the film converting process. Like lamination, coextrusion brings together desired properties of different materials into one structure. It eliminates solvent handling and emissions from using solvent-based lamination adhesives, although water-based and solventless adhesives are available for adhesive lamination. Its main advantage over-lamination is that it eliminates process steps: multiple materials are combined in one pass, which saves cost and time.

Coextrusion is inherently a more flexible process than mono-layer extrusion but is more complex: more extruders, more complicated flow geometries, greater potential for flow issues, higher cost of capital investment and higher scrap costs if the waste generated during the process cannot be recycled. These factors all must be weighed by the converter when determining what process to use.

The converter must translate the needs of the end-user into a cost-effective functional packaging film, using a combination of resin and process technology. An example is given in Table 1.2. Here the end-user needs are translated into material and structure design requirements:

1. Barrier requirements are met by the selection of materials. Often there is a choice. For example, for high oxygen barrier, the converter may choose between a number of materials, such as aluminum foil, metallized films, AlOx or SiOx coated film, PVDC, EVOH, and polyamide. The cost, the ability to meet the other requirements of the package such as transparency or recyclability, and compatibility with the converter's process will dictate which to use. For example, if transparency is required, then aluminum foil or metallized films would not be considered. See Chapter 8.
2. The barrier is also influenced by the structure design. The permeation of plastic films is typically inversely proportional to the thickness, so increasing thickness improves the barrier, but increases cost. Splitting barrier layers into two or more layers may be beneficial. Protecting barrier layers such as aluminum foil from stress cracking during use may also be an effective strategy.
3. The bending stiffness of the film is affected by the stiffness (modulus) of each of the layers, their thickness and their position within the structure. Stiffer layers near the outside of the film have a greater effect on increasing stiffness than near the center. As structures are downsized to meet cost or sustainability targets, these design features become increasingly important to maintain the stiffness, strength and durability of the package, as well as for consumer acceptance (perception of quality). See Chapter 9.
4. The chemistry of the sealant polymer affects sealability, particularly when dealing with powdery or greasy contamination in the seal area. See Chapter 7.
5. The durability of the packaging film when exposed to puncture, impact, scratch and other threats is influenced by the durability of the layers within the package. The adhesion between the layers may also play a role. Controlled delamination during an impact may help dissipate energy and prevent failure of the package. See Chapters 9 and 10.
6. Packaging line speed is a function of how fast the package will seal. This is determined by how fast the heat arrives at the interface (film thickness) and the characteristics of the sealant. Sealants with low heat seal initiation temperature and high hot tack strength allow for faster line speeds. See Chapter 7.
7. The slip properties of the film are determined by the film coefficient of friction of the requirements, and are often controlled by the use of additives to the polymer. See Chapter 12.
8. The ability of the film to withstand post-filling steps such as sterilization, retort, or blast freezing will depend on the temperature ranges of the materials. Retort, for example, typically involves exposing the filled package to

TABLE 1.2 Translating end-user needs by the converter.

End-user needs	Translates to
Barrier requirements	Selection of materials and layer thickness
Stiffness	Selection of materials and layer thickness
Heat seal performance	Characteristics of heat seal material including seal around contamination and grease resistance
Durability	Toughness of materials and interlayer adhesion
Packing Line speeds	Characteristics of heat seal material including seal initiation temperature and hot tack
Slip properties	COF, additive packages
Postfilling steps	Material temperature ranges
Recyclability or recycle content	Selection of materials
Compliance with regulations	Selection of materials; presence of functional barrier

121°C for 30 minutes or more under pressure. Materials that melt below this temperature may not be suitable. See Chapters 8 and 16.

9. Designing a packaging film structure for sustainability depends on the goal of the end-user. Reducing climate change impact may involve enhancing the performance of the package to improve the shelf-life of the product since product waste may contribute more to greenhouse gas emissions than the package. Increasing package material circularity and reducing packaging waste in the environment may involve designing the package to be included in a monomaterial collection stream such as all-polyethylene. To meet barrier and toughness requirements, the use of orientation, special grades of polymer and compatibilizers may be needed. Incorporating postconsumer packaging recycle into the structure may require using blends with virgin polymers to add back strength and toughness. Moving from fossil-based polymers to natural or biobased ones may involve material substitutions, such as increased use of paper/paperboard or bio-based polymers. These concepts are discussed in Chapter 17.
10. Many applications for flexible packages involve the packaging of food or medicine that are regulated by government agencies for safety. The materials used in these packaging applications must comply with these regulations. As a result, the materials available to the package engineer are only a subset of the materials that may be available for other applications. Material selection and testing of the final structure for compliance are important. In some applications, placing a functional barrier layer between the noncompliant material and the product may allow for compliance. Regulatory compliance is briefly discussed in Chapter 4.

Addressing these needs is explored in-depth in this book along with the scientific principles behind them. As needs change, these principles can be used to form the basis of new solutions.

1.6 Assembling a package: benefits of multiple layers

The simplest packaging structure consists of a single material or layer. There are a number of packaging forms and applications where this works well, such as glass jars, metal cans (although cans are usually coated with an epoxy, so even seemingly single layer packages may be more than one layer), polyethylene terephthlate (PET) bottles, and polyethylene (PE) bags. Each of these has its limitations, however, that prevent it from being used more broadly. The glass jar is heavy and breakable; metal cans are not transparent and so cannot show the product; PET bottles lack sufficient oxygen barrier for many products and work best for liquids, and PE bags lack oxygen barrier and may be limited in the quality of printing that may be applied.

Adding an additional layer gains some functionality, often in a way that reduces overall cost (even after including the additional processing cost of making a more complex structure) over what would be needed to make a single layer perform the same function. Or more often, the multilayer structure allows the package to perform a combination of functions not possible with a single layer.

Obviously, a two-layer structure consists of an inner (product side) and an outer layer. In a typical two-layer flexible packaging structure the inner layer provides sealability and the outer layer abuse resistance, printability or barrier. Some examples are shown in Table 1.3. Common ethylene-based sealant layers include EVA, LDPE, polyethylene plastomers and ionomers. PP and polyester-based sealants may be found in oriented PP and PET films, respectively. Sealants provide low-temperature sealability for fast packaging line speeds. They also may contribute to barrier properties. For example, ionomers have exceptional oil and grease resistance. By themselves, however, sealants often lack the stiffness, structural integrity and abuse resistance needed for the application. This is the role of the outside layer, which also may bring additional functionality. HDPE provides a moisture barrier and is often used in packaging dry foods such as cereal and crackers. Polyamide (PA) may be used where exceptional thermoformability may be needed, such as in brick packages for cheese. Oriented PET film provides an excellent surface for printing. Paper is a low-cost substrate often used in pouches and bags. With two layers, greater functionality can be achieved than either layer alone.

TABLE 1.3 Examples of the functionality of two-layer flexible packaging structures.

Inner layer function	Outer Layer function	Example structure	Example application
Low-temperature sealing, easy open	Stiffness, moisture barrier	(HDPE-EVA), (HDPE-Ionomer)	Cereal or cracker liner (bag in box)
Sealing	Stiffness, low cost	Paper/LDPE	Dry soup, spice pouches

Three layers provide even more versatility in package design. The primary function of the inner layer is still sealability. The outer layer again provides abuse resistance, heat resistance (during sealing), stiffness, structural integrity, a surface for printing and in some cases moisture barrier. The same substrates as discussed in the two-layer examples are typically used as the outer layer in three-layer structures. The center layer may provide a variety of functions. PVDC or metallization layers may be used to provide an oxygen barrier. Many films are printed and laminated to a sealant layer with the print side facing inward, protected from outside abuse by the film layer. This is known as reverse printing. Adhesives may be used in the center layer to allow for the lamination of two dissimilar materials. Recycling can be incorporated into the center layer for some applications. Pigments such as TiO_2 may be added to a core layer to provide a light barrier. Examples of such structures are shown in Table 1.4. One example is a medical forming web. Here an ionomer, traditionally used as a sealant in flexible packaging, finds use as the core providing outstanding thermoformabilty and puncture resistance. Less expensive EVA or PE layers are used as the skin to bond to the Tyvek spun-bond polyethylene top layer of the package. An alternative structure for this application is (PA6-tie-PE) where the PA6 provides the thermoformability and puncture resistance.

Examples of extending the concept of multiple layers to five are shown in Table 1.5. Again the innermost layer is typically a sealant layer such as PE, EVA or ionomer. Layer 2 is often an adhesive layer followed by a barrier layer, another adhesive layer and the abuse resistance layer. The examples shown here are all made by coextrusion. The

TABLE 1.4 Examples of three-layer flexible packaging structures.

Inner layer function	Center layer function	Outer layer function	Example structure	Example application
Sealing	Print	Temperature and abuse resistance, printability	OPET/print/EVA	Cheese packaging
Sealing	O_2 barrier	Temperature and abuse resistance	OPET/PVDC-adhesive/EVA	Meat packaging
Sealing	Adhesion	Temperature and abuse resistance	PA/adh/mPE, OPET/adh/EVA	Cheese packaging
Sealing	Recycle	Temperature and abuse resistance	(HDPE/recycle/LLDPE)	Bags
Sealing	Light barrier	Temperature and abuse resistance, moisture barrier	(HDPE-HDPE + TiO_2-EVA)	Snack package
Sealing	Thermoforming and puncture resistance	Sealing	(EVA-Ionomer-EVA)	Medical device packaging forming web
Sealing	Adhesion	Thermoforming and puncture resistance	(PA6-tie-PE)	Medical device packaging forming web

TABLE 1.5 Examples of five-layer flexible packaging structures.

Layer 1	Layer 2	Layer 3	Layer 4	Layer 5	Example structures	Example application
Sealing	Adhesion	Barrier	Adhesion	Abuse resistance, Moisture Barrier	(HDPE-tie-PA6-tie-EVA), (HDPE-tie-EVOH-tie-Ionomer)	Barrier snack package
Sealing	Adhesion	Barrier	Adhesion	Abuse resistance, Thermoforming	(PA6-tie-EVOH-tie-Ionomer)	Bacon package
Sealing	Adhesion	Thermo-forming	Barrier	Abuse resistance, Thermoforming	(PA6-EVOH-PA6-tie-Ionomer)	Meat forming web
Sealing	Adhesion	Moisture barrier	Moisture barrier	Moisture barrier	(HDPE-HDPE-HDPE-HDPE-EVA)	Snack package

barrier polymers, such as PA 6 and EVOH, do not bond to most sealant polymers. The industry has developed tie resins to bond these materials together during coextrusion, as explained in Chapter 10. The first example is an extension of the two-layer (HDPE-sealant) snack package film discussed earlier. Five layers allow the incorporation of a PA 6 or EVOH layer to provide oxygen and an aroma barrier. HDPE provides the moisture barrier.

The next two structures in Table 1.5 represent barrier-forming webs often used in meat packaging. The ionomer sealant provides excellent sealability, especially in the presence of grease. PA is combined with EVOH to help with thermoformability as well as abuse resistance.

The final structure in Table 1.5 highlights one of the advantages five or more layers bring to operational flexibility. Here a two-layer snack package structure is made with five layers. The HDPE thickness is determined by the amount of moisture barrier needed for the application, and often is greater than what a two-layer film process is designed to run. By putting the HDPE in multiple layers, the converter can easily make the desired structure. Indeed, operational flexibility is one of the primary motivations for the converter to purchase seven and even nine-layer film lines. The operational flexibility gained with additional layers may also improve the control of the layer distribution within the film. Without such flexibility, the converter may resort to bumping up the thickness of an expensive barrier layer to ensure the film meets specifications. Better process control may allow the layer to be downgauged.

The capability of incorporating more than five layers also allows additional structures to be made. Some examples of such structures are shown in Table 1.6. Additional layers may also allow easier incorporation of post-industrial or postconsumer recycling into the structure.

More examples of multilayer structures can be found in Appendices A and B.

1.7 Packaging trends

Worldwide interest in reducing climate change and leakage of used packaging into the environment, especially marine environments, is a major driver for packaging innovation today. This driver is part of a larger set of macro trends around population growth and demographics that is motivating advances in package design. Table 1.7 outlines key macro trends and how these influence consumer needs, brand owner behavior and packaging design.

The first major trend is the growing middle class; it is estimated that 2 billion more people will enter the middle class in the next 10 years [54]. Most of this will be in the developing world, with China and India each expected to have 1 billion in the middle class by 2040. With this comes additional purchasing power and improved living standards that affect buying behavior. This is expected to drive several consumer needs that will influence brand owner and retailer decisions, and packaging innovation:

1. The first is the trend toward greater health, safety and wellness. Consumers want a healthier diet, particularly as consumers in the developing world rise out of poverty into the growing middle class. This brings with it the challenge of food safety and the role packaging plays. The implication for the brand owner is the need to improve product quality. This translates to packages and package technology that ensure safety and freshness, such as the use of quality sealants to prevent leakers, high barrier resins or coatings, and in-package sterilization.
2. The growth of online shopping is part of a greater trend toward convenience. E-commerce involves many more touch-points during delivery to the consumer, which drives the need for packaging innovation around durability. The emergence of an intelligent value chain with the "internet of things" may allow automated ordering and tracking of packaging compositions for ease of recycling.

TABLE 1.6 Examples of structures with greater than five layers.

Sample structure	Application
PE/Paperboard/PE/Al/(ACR-LDPE)	Aseptic pouch
(PP-regrind-tie-EVOH-tie-PP)	Retort bowl
OPET/print/adh/met-OPP/adh/(LLDPE-LLDPE-LLDPE)	Stand up pouch
(LDPE-ACR)/ACR/Al/ACR/(ACR-LDPE)	Toothpaste tube
(PA6-tie-PA6-EVOH-PA6-tie-LLDPE), (PA6-tie-PA6-EVOH-PA6-tie-Ionomer)	Meat forming web

ACR, Acid copolymer resin.

TABLE 1.7 Packaging drivers and trends.

Macro trends	Consumer need	Brand owner/retailer implications	Packaging implications
Growing middle class (global)	Health, safety, wellness	Produce quality products	Package design to ensure product safety and freshness
	Convenience: e-commerce	Online ordering; customer shipping	More durable packaging
	Reduce offering overload	Differentiate for share	Enhanced shelf impact and experiential features
	Affordability/value	Minimize cost by increasing productivity; maximize profits	Lightweight structures
Aging population	Health, safety, wellness	Produce quality products	Package design to ensure product safety and freshness
	Convenience, simplicity	Improve experience	Easy-open solutions
Sustainability	Sustainability	People, planet, profits	Reduce, recycle, renewable, protect

3. The consumer is inundated with choices, especially in the developed world. The brand owner and retailer must continually innovate to gain share. This leads to new package designs that have shelf impact and give the consumer a hands-on experience. Such experiences increasingly include interactions with smartphones that may require changes in graphics and other technology.
4. Finally, the consumer is always looking for value. This drives the brand owners to greater productivity to minimize cost. One consequence of this is the light-weighting of packaging structures. A Flexible Packaging Association white paper illustrates the progress than has been made on this front with several examples, including [55]:
   a. Over a 6-year period the weight of a national brand hot dog package was reduced by 30%.
   b. A candy bar flexible package was reduced by 60% in a 6-year period.
   c. Cereal bag liners and stretch films have been reduced by using a tough ionomer film.
   d. Coextrusion technology has allowed cheese packages to be 30% lighter without sacrificing barrier performance.

The second macro trend recognizes the changing population demographics. People are living longer and in many parts of the world, birth rates are low. This is resulting in unprecedented aging of the general population. The UN projects that in 2030 the total world population will reach 8.5 billion with roughly 1 billion people 65 years and older (11.7%). By 2050 the number of people in the 65 + age group will reach almost 1.5 billion or 15.9% of the total population [56]. Some of the implications of this include:

1. Like the growing middle class, aging will increase the value of health, safety, and wellness.
2. Consumers in general value simplicity and convenience, but the aging population values these even more. Brand owners strive to improve the consumer experience with easy open and recloseable features, packages shaped to be held in the hand and microwaveable packaging.

The third macro trend is sustainability. Packaging has a role in reducing climate change by helping prevent food and other product waste, but designing packaging for circularity is becoming increasingly important: changing package format for improved shipping and handling efficiencies, reducing material usage through light-weighting, use of mono-materials to facilitate mechanical recycling and incorporating materials made from renewable resources (bio-based or from recycle). An example of the first is the transition to square 1-gallon (3.8 L) plastic milk jugs from round ones in the US market at large discount stores like Walmart and Costco [57]. The design eliminates the need for a shipping crate since the jugs are stackable. This saves labor by 50% and cuts water use (for cleaning the crates) by 60%–70%. Because of the form factor, more gallons fit on a truck and crates do not need to be shipped back. They also save space in the store. Other products have followed suit, such as the switch from round to square yogurt cups. Chapter 17 discusses approaches to designing packaging for ease of recycling and other sustainability topics.

We are in a time of rapidly changing demands on packaging, especially in the area of sustainability. Satisfying these trends will require continued advances in flexible packaging technology: new materials, refinements of converting processes, new packaging structures, optimization of packaging equipment and better ways to recycle and mitigate end-of-life challenges. The remainder of this book introduces flexible packaging technology with a focus on the scientific principles that underlie these technologies. Understanding the science will enable material scientists and packaging engineers to rapidly develop new materials and package designs that satisfy new packaging needs.

References

[1] C. Scarre, Smithsonian Timelines of the Ancient World, 1st American ed, Dorling Kindersley, London, 1993.

[2] The American Ceramic Society, A brief history of ceramics and glass. <https://ceramics.org/about/what-are-engineered-ceramics-and-glass/brief-history-of-ceramics-and-glass> , n.d. (accessed 07.08.21).

[3] A. Lawler, World's oldest pottery? Science. <https://www.sciencemag.org/news/2009/06/worlds-oldest-pottery>, 2009 (accessed 07.08.21).

[4] National Geographic News, Ancient Greek wreck found in Black Sea.<http://news.nationalgeographic.com/news/2003/01/0110_030113_black-sea_2.html>, 2013 (accessed 18.05).

[5] National Geographic Society, Ancient shipwrecks of the Black Sea, video. <https://www.nationalgeographic.org/media/ancient-shipwrecks-black-sea/>, nd (accessed 07.08.21).

[6] Corning Museum of Glass, The Origins of Glassmaking. <https://www.cmog.org/article/origins-glassmaking>, 2011 (accessed 07.08.21).

[7] American Society of Mechanical Engineers, ASME designates the Owens "AR' bottle machine as an international historic engineering landmark. <http://files.asme.org/ASMEORG/Communities/History/Landmarks/5612.pdf>, 1983 (accessed 07.08.21).

[8] S.J. Risch, Food packaging history and innovations, J. Agric. Food Chem. (2009) 8089–8092.

[9] K.R. Berger, A brief history of packaging, University of Florida IFAS Extension, ABE321.<https://ufdc.ufl.edu/IR00001524/00001>, 2002 (accessed 07.08.21).

[10] The Aluminum Association, Foil and packaging.<https://www.aluminum.org/product-markets/foil-packaging>, n.d. (accessed 07.08.21).

[11] P. Hook, J.E. Heimlich, A history of packaging, Ohio State University Extension Fact Sheet, Community Development, 700 Ackerman Road, Suite 235, Columbus, OH 43202-1578, CDFS-133. <http://ohioline.osu.edu/cd-fact/0133.html>, n.d. (accessed 07.08.21).

[12] D. Twede, The origins of paper based packaging,", The Future of Marketing's Past, Proceedings of the 12th Conference on Historical Analysis and Research in Marketing, CHARM 2005, Long Beach, California, 2005, pp. 288–300. April 28- March 1, 2005.

[13] A. Kinnane, DuPont: From the Banks of the Brandywine to Miracles of Science, E.I. Du Pont de Nemours and Co, 2002.

[14] C. Bernat, Supermarket packaging: the shift from glass to aluminum to plastic, Atl (2012)25 January, viewed 07 Aug 2021, from. Available from: http://www.theatlantic.com/health/archive/2012/01/supermarket-packaging-the-shift-from-glass-to-aluminum-to-plastic/251875/?single_page = true.

[15] E. I. du Pont de Nemours and Company, Advertising Department, Modern shoppers prefer food in cellophane: shows what it protects! Protects what it shows! Dpads-1803-00308.tif, Hagley Museum and Library, 1948.

[16] E. I. du Pont de Nemours and Company, Advertising Department, New self-service meats make shopping quicker, easier: DuPont Cellophane, Dpads-1803-00320.tif, Hagley Museum and Library, 1949.

[17] Wikipedia, BoPET. <http://en.wikipedia.org/wiki/BoPET#History> 2021 (accessed 07.08.21).

[18] Wikipedia, Innovia Films, Ltd. <http://en.wikipedia.org/wiki/Innovia_Films_Ltd>, 2021 (accessed 07.08.21).

[19] J.F. Hanlon, R.J. Kelsey, H.E. Forcinio, Handbook of Package Engineering, third ed., CRC Press, 1998, p. 117.

[20] M. Demirors, The history of polyethylene, in: E. Strom, S. Rasmussen (Eds.), 100 + Years of Plastics. Leo Baekeland and Beyond, American Chemical Society, Washington, DC, 2011, pp. 115–138. ACS Symposium Series.

[21] T. Bezigian (Ed.), Extrusion Coating Manual, fourth ed., TAPPI Press, Atlanta, GA, 1999, pp. 1–2.

[22] W.C. Andrews, K.J. Williams, in: G.A. Giles, D.R. Bain (Eds.), Technology of Plastics Packaging for the Consumer Market, Sheffield Academic Press, Oxford, 2001, p. 160.

[23] BCC Research, History of plastic sheet. <http://www.bccresearch.com/blog/report-archives/history-plastic-sheet.html> 2013 (accessed 18.05.13).

[24] D. Djordjevic, Coextrusion, Rapra Rev Rep 6 (2) (1992). Rapra Publishing.

[25] T.D. Frey, J.A. Albright, Coextrusions for flexible packaging, in: Marilyn Baker (Ed.), Wiley Encyclopedia of Packaging Technology, John Wiley and Sons, New York, 1986, p. 199.

[26] Vijay, Flexible packaging - part 1 (VFFS machines history, current market and future direction), FoodBevHub. <https://www.foodbevhub.com.au/post/60-flexible-packaging-part-1-vffs-machines-history-current-market-and-future-direction/>, 2020 (accessed 07.08.21).

[27] E. Chertoff, The surprising history of the milk carton, The Atlantic <https://www.theatlantic.com/national/archive/2012/08/the-surprising-history-of-the-milk-carton/260587/>, 2012 (accessed 07.08.21).

[28] S. Büsser, R. Steiner, N. Jungbluth, LCA of packed food products: the function of flexible packaging, ESU-services Ltd. commissioned by Flexible Packaging Europe, Düsseldorf, DE and Uster, CH, 2008.

[29] United Nations, Department of Economic and Social Affairs, Population Division, Population Estimates and Projections Section, UN World Population Prospects, the 2019 Revision. <https://population.un.org/wpp/Publications/Files/WPP2019_Highlights.pdf>, 2019 (accessed 07.08.21).

[30] J. Gustavavsson, C. Cederberg, U. Sonesson, R. van Otterdijk, A. Meybreck, Global food loses and food waste: Extent, causes and prevention, Food and Agriculture Organization of the United Nations, Rome, 2011. Study conducted for the International Congress, SAVE FOOD! at Interpack2011, Düsseldorf, Germany.

[31] Institution of Mechanical Engineers, Global food. Waste not, want not. <https://www.imeche.org/policy-and-press/reports/detail/global-food-waste-not-want-not>, 2013 (accessed 03.08.21).

[32] Food and Agricultural Organization of the United Nations, Food loss and food waste. <http://www.fao.org/food-loss-and-food-waste/flw-data>, 2021 (accessed 03.08.21).

[33] H. Forbes, T. Quested, C. O'Connor, Food wastes index report 2021, UN Environmental Programme. <https://www.unep.org/resources/report/unep-food-waste-index-report-2021>, 2021 (accessed 03.08.21).

[34] Flexible Packaging Association, Flexible Packaging, Less Resources, Less Footprint, More Value, 3rd ed., 2009.

[35] Plastics Industry Association, Plastics 101. <https://thisisplastics.com/topics/plastics-101/>, n.d. (accessed 08.08.21).

[36] S. Anthony, A public service announcement for packaging?" Packaging World.<https://www.packworld.com/issues/workforce/blog/13361874/a-public-service-announcement-for-packaging>, 2013 (accessed 08.08.21).

[37] National Geographic Magazine, We depend on plastic. Now we are drowning in it, National Geographic Society, June 2018.

[38] N.A. Welden, A. Lusher, Microplastics: from origin to impacts, in: T.M. Letcher (Ed.), 2020, Plastic Waste and Recycling: Environmental Impact, Societal Issues, Prevention, and Solutions, Academic Press, 2020, pp. 223–249.

[39] R. Hurley, A. Horton, A. Lusher, L. Nizzetto, Plastic waste in the terrestrial environment, in: T.M. Letcher (Ed.), 2020, Plastic Waste and Recycling: Environmental Impact, Societal Issues, Prevention, and Solutions, Academic Press, 2020, pp. 163–193.

[40] C.R. Cook, R.U. Halden, Ecological and health issues of plastic waste, in: T.M. Letcher (Ed.), 2020, Plastic Waste and Recycling: Environmental Impact, Societal Issues, Prevention, and Solutions, Academic Press, 2020, pp. 513–527.

[41] E. Kosior, I. Crescenzi, Solutions to the plastic waste problem on land and in the oceans, in: T.M. Letcher (Ed.), 2020, Plastic Waste and Recycling: Environmental Impact, Societal Issues, Prevention, and Solutions, Academic Press, 2020, pp. 415–446.

[42] Ellen Macarthur Foundation, The new plastics economy: rethinking the future of plastics & catalysing action. <https://www.ellenmacarthurfoundation.org/publications/the-new-plastics-economy-rethinking-the-future-of-plastics-catalysing-action>, 2017 (accessed 27.07.20).

[43] Flexible Packaging Association, Flexible packaging facts and figures, as of June 2019. <https://www.flexpack.org/facts-and-figures>, nd (accessed 28.07.21).

[44] Flexible Packaging, FPA publishes '2020 state of the flexible packaging industry report. <https://www.flexpackmag.com/articles/90681-fpa-publishes-2020-state-of-the-flexible-packaging-industry-report>, 2020 (accessed 28.07.21).

[45] U.S. Census Bureau, North American Industry Classification System. <https://www.census.gov/naics/>, 2021 (accessed 24.05.22).

[46] J. Homan, European and Global Markets for flexible packaging, FPA Annual Conference, Arizona. <https://www.flexpack.org/MEMONL/mo_presentations/2012_am/Jan_Homan.pdf>, 2012 (accessed 18.05.13).

[47] Dow Chemical Company, Dow and UPM partner to produce plastics made with renewable feedstock, Press Release. <https://corporate.dow.com/en-us/news/press-releases/dow-and-upm-partner-to-produce-plastics-made-with-renewable-feedstock.htmlSmithers>, 2019 (accessed 26.12.20). Pira, Versatility and Innovation to Drive Global Flexible Packaging Market. <https://www.smitherspira.com/versatility-and-innovation-to-drive-global-flexible-packaging-market.aspx> (accessed 18.05.13).

[48] Amcor, Stand out and deliver. <https://www.amcor.com/insights/educational-resources/north-america-ecommerce-ebook>, 2020 (accessed 30.08.20).

[49] Michael Niaounakis, Recycling of Flexible Packaging, Elsevier, 2020.

[50] Association of Plastics Recyclers, APR Design® Guidance. <https://plasticsrecycling.org/staging/apr-design-guide>, 2020 (accessed 18.01.21).

[51] CEFLEX, Designing for a circular economy, Recyclability of polyolefin-based flexible packaging. <https://guidelines.ceflex.eu/resources/>, 2020 (accessed 23.01.20).

[52] Plastics Recyclers Europe, "Design for recycle," <https://www.plasticsrecyclers.eu/design-recycling>, n.d. (accessed 07.08.21).

[53] RecyClass, Making plastics packaging circular. <https://recyclass.eu>, 2019 (accessed 08.08.21).

[54] F. Chambon, E. Lernoux, A look at future trends in the flexible packaging industry, Packaging Europe, <https://packagingeurope.com/future-trends-in-the-flexible-packaging-industry/>, 2021 (accessed 28.07.21).

[55] Flexible Packaging Association, Flexible packaging lightweighting strategies and innovation. http://www.flexpack.org/sustainable_packaging/FPA_Lightweighting_White_Paper.pdf, 2011 (accessed 10.06.13).

[56] United Nations, World population prospects 2019, UN Department of Economic and Social Affairs, Population Dynamics. <https://population.un.org/wpp/Graphs/DemographicProfiles/Line/900>, 2021 (accessed 07.08.21).

[57] Rosenbloom, 2008]S. Rosenbloom, Solution or mess? A milk jug for a green earth, N Y Times (2008) June 30, viewed 10 June 2013, from, http://www.nytimes.com/2008/06/30/business/30milk.html?pagewanted = all

Part II

Basic processes

Chapter 2

Converting processes for flexible packaging

The purpose of the converting process is to transform raw materials in the form of polymer pellets or film substrates (including paper and foil) into a film structure for use in packaging. Converting processes are used to produce films and to physically or esthetically modify them. This often involves more than one process or step that may include film formation, orientation, surface treatment, metallization, lamination, coating, printing, winding and slitting. The major converting processes are listed in Table 2.1 along with their corresponding starting materials and the finished products they produce. Often there is more than one way to make a functional structure, the choice of which depends on a number of factors such as run-length and existing assets of the converter. This chapter briefly describes these processes, their key components and processing parameters that affect the properties of the final product.

2.1 Extrusion

Extrusion is a common component of all polymer film converting processes. The extruder melts the polymer pellets and pumps them through a die, which forms the film or coating. Almost all film and coating lines use single screw extruders, shown schematically in Fig. 2.1. More sophisticated extruders, such as twin-screw extruders and kneaders, are used in the polymer processing industry where good mixing and control of the heat history are required for sensitive materials, such as in the production of additive masterbatches. These are rarely used in film extrusion since single screw extruders typically provide sufficient mixing at a much lower cost for packaging converting operations.

Polymer pellets are fed into the rear of the extruder through a feed hopper. Typically, the rate of rotation of the extruder screw underneath the hopper determines the rate of feed and output of the extruder. This is known as flood feeding. In some operations, separate feeders on top of the hopper "starve feed" the extruder so that screw speed and feed rate are independently controlled.

The output of a single screw extruder increases with the diameter of the screw; as the diameter increases so does the volume of material that can be conveyed. Idealistically, the output volume increases with the cube of the diameter from simple geometric considerations. However, this is generally unachievable since other factors such as heat transfer also influence output. Typical rates can be found in Table 2.2.

The key performance requirements for the extruder are to fully melt the polymer, to produce a uniform melt temperature at the die exit (which minimizes film or coating gauge variability), to generate sufficient pressure to sustain flow through the downstream process (typically transfer pipes, screen packs, feedblock, and die), and to provide steady output. The important factors affecting performance are the screw design, barrel temperatures and screw speed. The function of the screw can be divided into three operations: solids conveying; compression and melting; and metering or pumping. Each operation occurs in about one-third of the extruder length, as shown in Fig. 2.2.

Solids conveying is achieved by the frictional differences between the polymer pellets against the barrel and screw root as the screw rotates. The static friction of the pellets against the barrel needs to be greater than the sliding friction at the screw root, otherwise, the pellets will simply rotate with the screw. An analogy is the rotation of a screw and nut—if the nut is not held in place it will rotate with the screw as it is tightened. (Another analogy for understanding the rotational direction of the screw is to consider a drill—a forward drill pulls material toward you and a reverse drill pushes material away.) Screw cooling and barrel heating are sometimes used to create differential friction for conveying. Adding grooves to the barrel in the feed zone can increase the solids conveying for some polymers by 20%–40% [2], but care must be taken when using this approach. Some polymers may be too hard; the resultant pressure build-up may cause damage or even break the screw. This is especially true for retrofitted smooth bore systems that have not been designed for the high pressure and torque required with grooves [3]. Some polymers may be too soft—if the polymer melts in the grooves, the effect of the grooves

The Science and Technology of Flexible Packaging. DOI: https://doi.org/10.1016/B978-0-323-85435-1.00009-0

TABLE 2.1 Major converting processes.

Unit operation	Processes	Raw materials	Finished product
Film formation	Blown Film Cast film	Polymer pellets	Single or multilayer film
Coating or Lamination	Extrusion Coating Extrusion Lamination Adhesive Lamination Liquid coating	Film, paper or foil substrate(s) Polymer, adhesive, or coating formulation	Multilayer film or laminate
Film orientation	Tenter Frame Double Bubble Machine Direction Orienter	Polymer or sheet	Oriented film
Metallization	Vacuum metallization	Film substrate and metal	Metallized film
Printing	Gravure Flexographic	Film Ink	Printed film
Slitting	Slitter	Film	Film

FIGURE 2.1 Schematic diagram of a single screw extruder.

TABLE 2.2 Typical output and specific output for a 30:1 length to diameter ratio (L/D) single screw extruder running at 100 rpm [1].

Screw diameter		Output		Specific output	
inches	mm	lb/h	kg/h	lb/h/rpm	kg/h/rpm
0.75	19	8	3.6	0.08	0.036
1.0	25	13	5.9	0.13	0.059
1.25	32	30	13.6	0.30	0.14
1.5	38	50	22.7	0.50	0.23
2.0	51	120	54.5	1.20	0.55
2.5	63.5	250	114	2.50	1.14
3.5	89	600	273	6.00	2.73
4.5	114	1200	545	12.0	5.45
6.0	152	2500	1140	25.0	11.4
8.0	203	5200	2360	52.0	23.6

Source: J. Frankland, Get smarter on extrusion sizes, Plastics Technology, July (2012) 22.

may be negated and the grooves can become a source of hold-up and polymer degradation. Light grooves can be a good compromise for some polymers [4]. Discussion with the extruder manufacturer and the polymer supplier prior to running a new polymer on a grooved barrel extruder is advised to avoid potential problems.

As the solid pellets are conveyed forward, they are compressed. In what is typically referred to as a "standard" screw design, this is achieved by slowly reducing the height of the screw channel. The ratio of the final screw channel

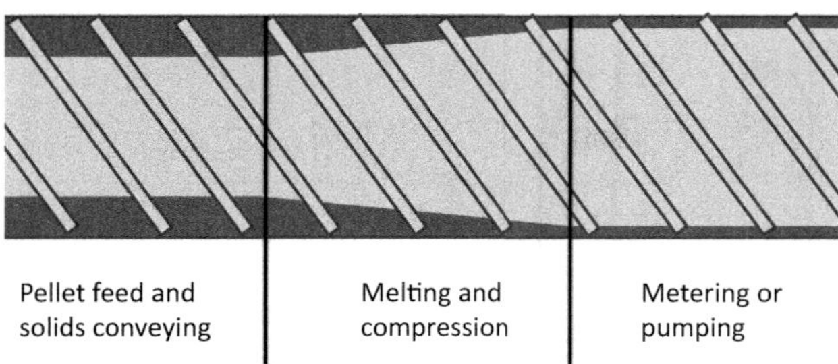

FIGURE 2.2 Operational design of a screw.

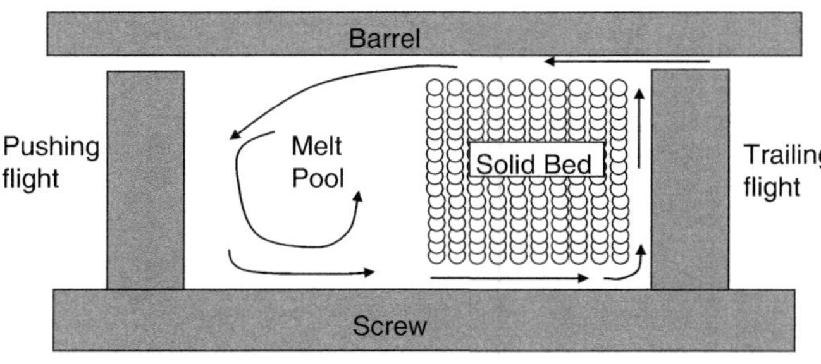

FIGURE 2.3 Schematic representation of melting in a single screw extruder [5]. *Reproduced from B. A. Morris, Polymer blending for packaging applications, in Ebnesajjad, S., (ed.), Plastic Films in Food Packaging, Materials, Technology and Applications, Elsevier, Amsterdam, 2013, pp. 311–343.*

depth (metering section) to the initial (solids conveying section) screw channel depth is known as the compression ratio. This ratio typically ranges from about 2:1 to 3:1 and compression takes place over approximately the second third of the screw length. As the compression takes place, friction and heat transferred from the barrel to the solid bed begin to melt the polymer. Forces compact the solid bed at the trailing flight and the melt forms a pool in front of it. The polymer melt pool increases in width as the polymer travels down the barrel length (Fig. 2.3). Ideally, the polymer is entirely melted at the end of the compression or melting zone.

Frictional forces tend to dominate the melting process in single screw extruders. To speed up this process and increase output, many modern screw designs introduce a second flight that segregates the melt pool from the solid bed. These designs are known as barrier screws with the barrier being the second flight that forms a barrier between the solid bed and melt pool. Forcing the separation of the melt from the solid bed increases the frictional heat on the solid bed, accelerating its melting. There are many different variations on this general scheme, including changing the width or depth of the melt pool along the screw length. An example of varying the width is provided in Fig. 2.4. One of the challenges is the transition to and from the barrier section of the screw. The pros and cons of each approach are discussed in the general literature [6–8].

The final section of the screw builds pressure in the melt to pump it through the die. The metering section output is a function of screw speed, screw geometry, pressure drop and polymer viscosity. As the screw rotates, drag is introduced between the stationary barrel and the rotating surface of the screw, propelling the melt forward. There is also a pressure-driven component to the flow, and depending on the pressure profile near the extruder exit, this may be positive or negative. A simple model for the drag and pressure-driven flow is given by Eqs. (2.1) and (2.2) [9]. Fig. 2.5 shows the definition of variables in these equations. Here several assumptions have been made including Newtonian flow (viscosity is independent of shear rate), fully developed flow and no leakage across the flight tips. Despite these assumptions, the sum of Eqs. (2.1) and (2.2) usually comes close to predicting the output of the metering section. These can be used as a starting point for troubleshooting [10].

$$Q_{\text{drag}} = \frac{1}{2}F_d\pi^2D^2h\left(1-\frac{ne}{t}\right)\sin\varphi\cos\varphi \tag{2.1}$$

$$Q_{\text{pressure}} = \frac{F_p\pi Dh^3\left(1-\frac{ne}{t}\right)\sin^2\varphi\Delta P}{12\mu L} \tag{2.2}$$

where Q_{drag} is the drag flow, Q_{pressure} is the pressure-driven flow, F_d is the channel correction factor $= 0.140(\text{h/w})^2 - 0.645(\text{h/w}) + 1$, F_p is the channel correction factor $= 0.162(\text{h/w})^2 - 0.742\ \text{h/w} + 1$, D is the screw diameter, N is the screw speed (rpm), h is the screw meter section channel depth, n is the number of flights on the screw, e is the thickness of flight, t is the flight lead (pitch), ϕ is the flight helix angle, μ is the viscosity of melt calculated at shear rate $= \pi\text{ND/h}$, and L is the length of the metering section being investigated.

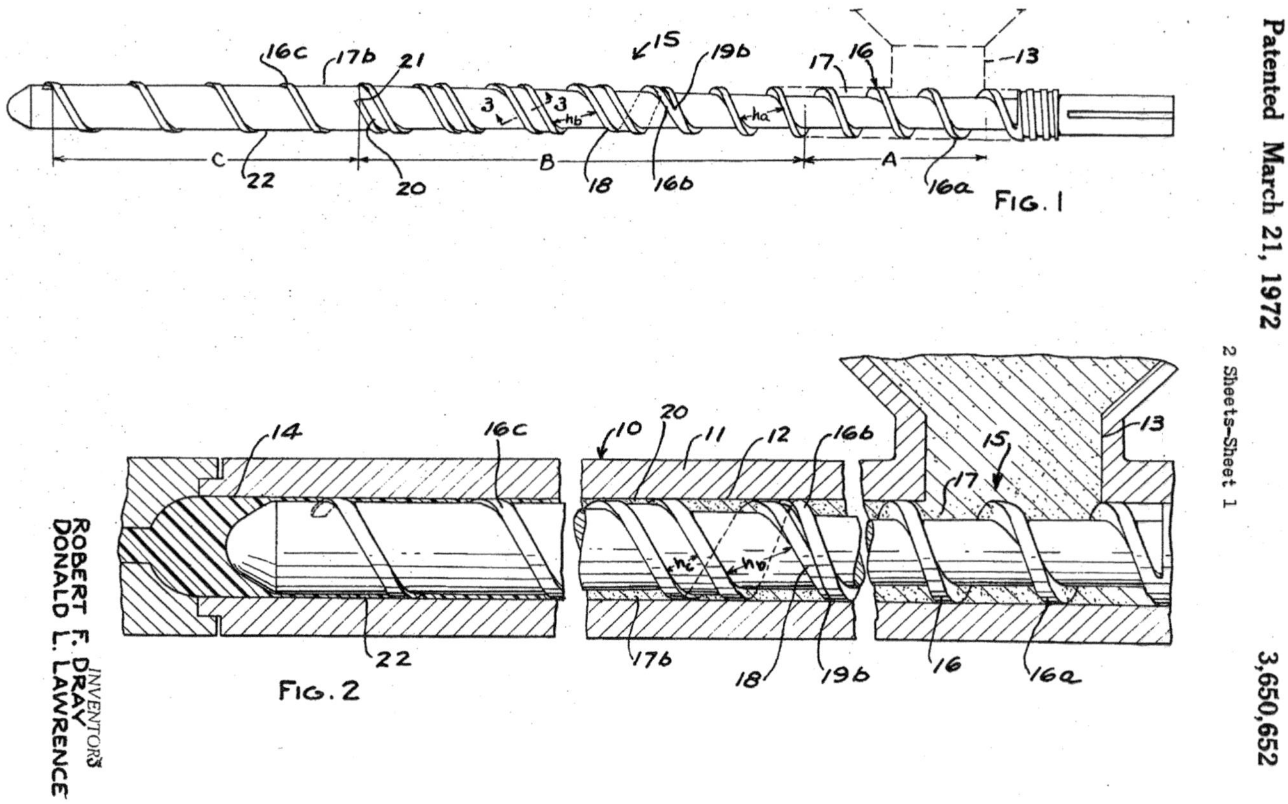

FIGURE 2.4 Example of a barrier screw design where the width of the secondary channel is varied along the melting zone. *Reproduced from R.F. Dray, D.L. Lawrence, Apparatus for extruding plastic material, US Patent 3650652 A, 1972.*

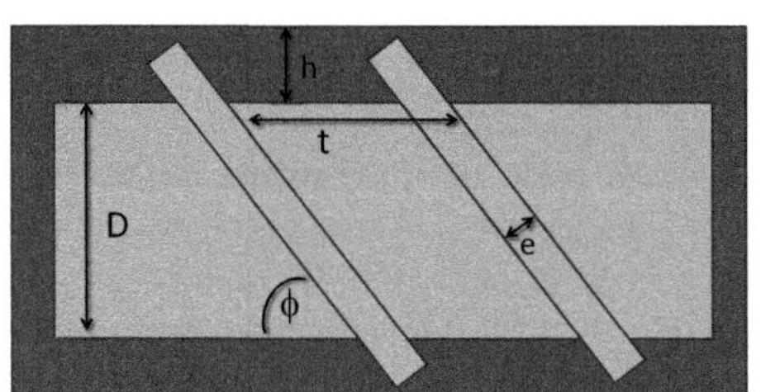

FIGURE 2.5 Metering section of a single screw extruder.

A good screw design will keep the pressure-driven component of flow to about 10% or less of the total flow. If the pressure-driven component becomes too high, small fluctuations in pressure can result in large changes in output (surging), which is undesirable. In situations where precise control of output is required, a separate melt pump may be used at the end of the extruder to decouple pumping from the other functions of the extruder. Melt pumps are more often found on sheet lines than film lines.

Achieving temperature uniformity at the extruder exit is critical for good gauge control coming out of the die. To reduce temperature variability in the extruder, often times a mixing element is incorporated into the screw design, typically as part of the metering section. There are a number of different types of mixing sections, three of which are shown in Fig. 2.6. Inserting pins is one of the easiest ways to redistribute the flow, but stagnant areas behind the pins setup the potential for polymer degredation and gel formation. More sophisticated designs, such as the Saxton mixer [11], eliminate the dead zones while dividing up the flow. In the Maddock mixer, the polymer flows down a fluted section and over a small clearance between the screw and barrel; this introduces high shear stresses to break up agglomerates. Alternatively, a static mixer may be incorporated into the transfer pipe between the extruder and die.

The temperature of the melt is determined by external heat transfer through the extruder barrel and by the amount of internal heat generated by the polymer flow through viscous energy dissipation. The barrel of the

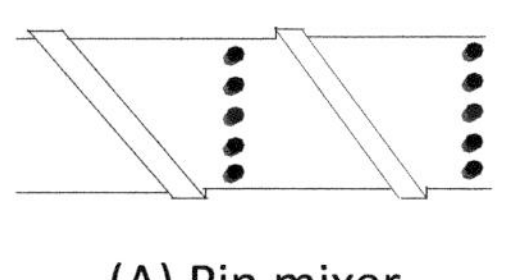

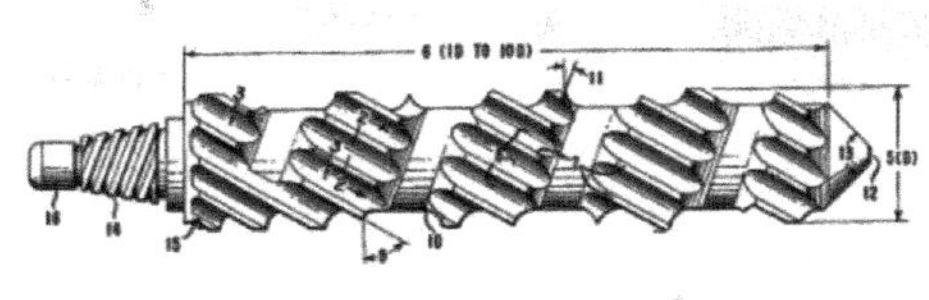

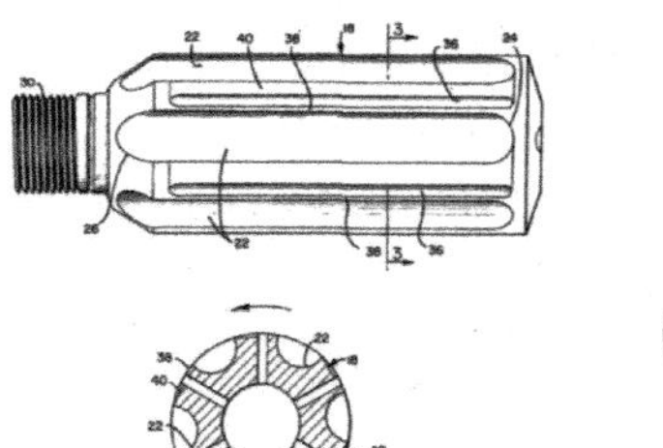

FIGURE 2.6 Common mixing elements used in single screw extrusion: (A) pin mixer, (B) Saxton mixer [11], (C) Maddock mixer [12].

extruder is typically jacketed with heaters and coolers (either water or air cooling). As the size of the extruder is increased, however, the effectiveness of the barrel heating/cooling is diminished by the reduction of surface area to polymer volume and the polymer temperature is dominated by its internal frictional heat. The frictional heat or energy dissipation from polymer flow is a function of the shear rate times the shear stress, or the viscosity times the shear rate squared. The typical shear rate in the extruder ranges from about 10 to 100/second. The shear rate increases with decreasing channel depth and increasing screw speed. Processes requiring higher temperatures such as extrusion coating may operate with higher screw speed or shallower channels than processes trying to minimize temperatures.

The viscosity of the polymer, the screw design and operating conditions are typically optimized for the application. In the blown film process, high viscosity polymers and operating at low temperatures help achieve sufficient melt strength and bubble stability. Blown film operations often use screws with deep flights, run at low screw speed and utilize barrel cooling to achieve low melt temperatures. In extrusion coating and lamination, high temperatures and lower molecular weight polymers are used to reduce the viscosity of the melt for better wetting of the substrate, a first step to achieving adhesion. High temperature and shear also promote oxidation of the melt curtain for improved adhesion of some polymers to polar substrates. In addition to operating at high screw speed and utilizing a screw with shallow channel depths, an extrusion coating extruder may be equipped with a back pressure valve to further increase shear and residence time for temperature build-up.

Further reading on extrusion

1. Tadmor, Z. and Klein, I., 1970, *Engineering principles of plasticating extrusion*, Van Nostrand Reinhold Co., New York.
2. Rauwendaal, C., 2014, *Polymer Extrusion*, 5th ed., Hanser Publications, Cincinnati, OH.
3. Vlachopoulos, J. and Wagner, J. R. (eds.), 2001, *The SPE guide on extrusion technology and troubleshooting*, The Society of Plastics Engineers, Brookfield, CT.
4. Campbell, G. A. and Spalding, M. A., 2010, *Troubleshooting & analysis of single-screw extrusion, Hanser Gardner Publications*, New York.
5. Chung, C., 2019, *Extrusion of Polymers: Theory and Practice*, 3rd ed., Hanser Publications, Cincinnati, OH.
6. Tadmor, Z. and Gogos, C., 2006, *Principles of polymer processing, 2nd ed.*, John Wiley & Sons, Hoboken, NJ
7. Lafleur. P. G. and Vergnes, B., 2014, *Polymer Extrusion*, John Wiley & Sons, Hoboken, NH.
8. del Pilar Noriega E., M. and Rauwendaal, C., 2019, *Troubleshooting the Extrusion Process: A Systematic Approach to Solving Plastic Extrusion Problems*, 3rd ed., Hanser Publications, Cincinnati, OH.
9. Giles, Jr., H. F., Wagner, Jr., J. R. and Mount, III, E. M., 2014, *Extrusion: The definitive processing guide and handbook*, 2nd ed., William Andrew (Elsevier), Waltham, MA.
10. Griff, A. L., *The plastics extrusion operating manual*, self-published: https://www.griffex.com/manu.htm.

2.2 Film converting

2.2.1 Blown film

The blown film process converts polymer resin into a plastic film. Polymer pellets are fed into an extruder, melted, and pumped out a tubular die (Fig. 2.7). In a traditional air quenched process, air is forced into the middle of the tube, expanding the tube in the hoop direction as it is accelerated in the machine direction by rotating nip rollers. Typically the tube is positioned to travel up a vertical tower where a nip roll pinches off and flattens (or folds) the film to lay flat prior to winding. The expansion of the tube occurs in a semimolten state while cooling air is blown across the outside of the bubble, resulting in a film having an orientation in both the machine and transverse (hoop) directions. In addition to outer bubble air cooling, larger lines also circulate air inside the bubble to increase the cooling rate and productivity.

In a water quench system, the tube is stretched while molten around a mandrel or guide and then rapidly quenched in a water bath. The resulting film typically has less orientation and crystallinity than an air quenched film, providing higher transparency, less stiffness and deeper and more uniform thermoformability. Water quenched film can also be more efficiently oriented in a subsequent in-line orientation process such as machine orientation or double or triple bubble process.

2.2.1.1 Key components

2.2.1.1.1 Extruder

One or more extruders melt and pump the polymer through the blown film die. See Section 2.1 for details.

2.2.1.1.2 Tubular die

The die distributes the melt into a tube with uniform flow in the axial direction at the die exit. This minimizes gauge variation in the final film. The objective is to streamline flow without die lines. Doing this without excessive residence time and pressure is also important. The key factors affecting performance are uniform temperature entering the die, die design as it affects the flow geometry, die gap and materials of construction. Spiral mandrel designs (Fig. 2.8) are typically used to distribute the flow more evenly and to avoid die lines. Stacked or pancake die designs have also found use in coextrusion to help thermally isolate polymers of different viscosities or thermal stability and minimize residence time.

At the center of the tubular die exit there is an inlet for air to be blown into the tube to prevent it from collapsing and to create the bubble (the air is entrapped by the nip at the top of the tower). The air pressure inside the bubble determines the amount of expansion, the final bubble diameter and its lay-flat (one half the final circumference of the bubble).

2.2.1.1.3 Air ring and inner bubble cooling

The air ring blows air on the outside of the bubble, cooling the melt and solidifying it into a film. Key parameters include the airflow rate, distribution and temperature. Poor bubble stability can result in nonuniform film thickness, line

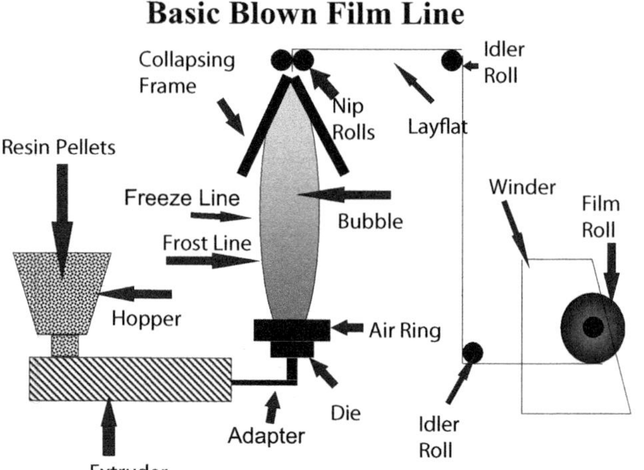

FIGURE 2.7 Blown film process. *Reproduced with permission from The Dow Chemical Company.*

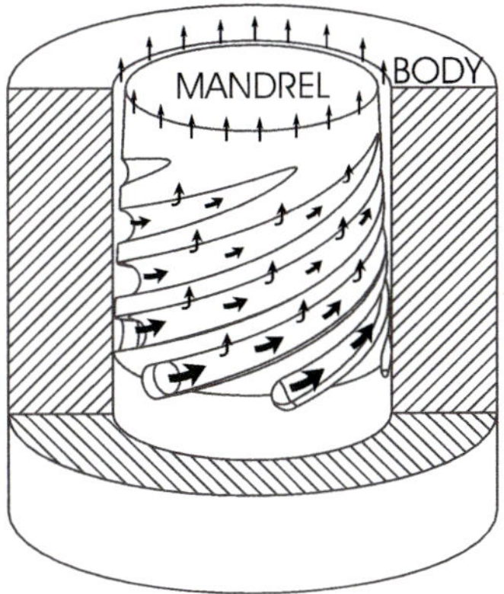

FIGURE 2.8 Schematic representation of cylindrical spiral mandrel die design. *Reproduced from J. Perdikoulias, Multilayer blown (tubular) film dies, in Wagner, J. (ed.), Multilayer Flexible Packaging, 2nd (ed.), Elsevier, Oxford, 2016, pp. 123–128.*

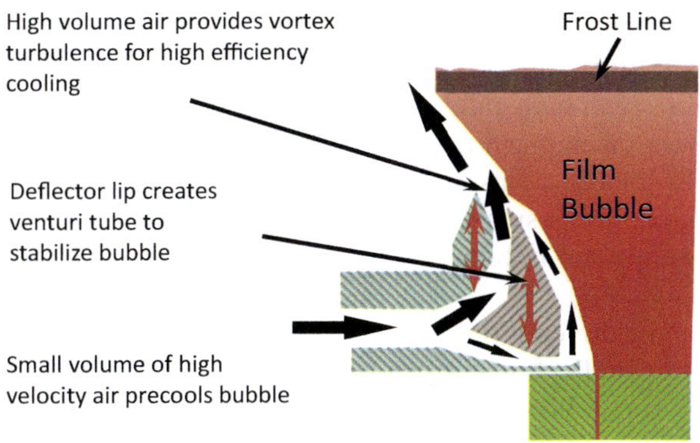

FIGURE 2.9 Schematic representation of a dual lip air ring. *Reproduced with permission from The Dow Chemical Company.*

breaks and limited output. Therefore the design of the flow manifold is critical, as a good design can help stabilize the bubble [14,15]. Bubble stability is described in more detail in Section 13.4.3. Most air rings today also have a dual lip to better control the airflow and create a venturi tube that helps seat the bubble and hold it in place. A schematic of the dual lip air ring is shown in Fig. 2.9.

Cooling by blowing air on the outside of the bubble becomes less effective as the line size gets bigger. On larger lines, chilled air is also circulated inside the bubble and continuously replenished to help cool the bubble more rapidly. This feature is known as inner bubble cooling (IBC) and allows an increase in line speed and output. The chosen quench rate influences several film properties, including crystallinity, tear strength, impact strength, transparency and haze.

2.2.1.1.4 Collapsing frame and nip

At the nip at the top of the tower the bubble transitions from a tube to its lay-flat with the help of the collapsing frame. The design of the collapsing frame is critical to preventing the formation of wrinkles during this transition [16]. The speed of the driven nip roll (haul-off roll) determines the final haul-off speed and the amount of draw in the machine direction. The collapsing frame and nip rolls are typically rotated to help randomize bands of nonuniform gauge on the wound roll. Early versions of this technology rotated the die rather than the haul-off.

2.2.1.1.5 Gauge control

A sensor is typically used to measure the gauge variation across the lay-flat during haul-off. Algorithms have been developed that allow these measurements to be used for automatic gauge correction [17,18]. On lines with automatic die adjustment capability, a signal is sent back to a controller that adjusts the thickness uniformity around the circumference of the die. One approach is to selectively heat or remove heat from bolts along the circumference to flex a membrane in the die lip [19]. Another approach is to locally adjust the quench air speed along the air ring [20].

There are many types of in-line gauge monitoring technologies. Historically gamma backscatter or capacitance techniques have been used to measure the total thickness of the bubble inline [21] and transmission techniques such as beta, gamma or X-ray after the film has been laid flat. Beta gauges emit beta particles (high-speed electrons) from a radioactive source. The beta particles pass through the film and are detected by a sensor on the other side. The amount of particles that transmit through the film is proportional to the mass of the film. Gamma gauges are similar except that the radioactive source emits gamma rays or high-energy photons. Gamma-ray transmission is highly dependent on the atomic number of the material it passes through, so in some cases, discrimination can be made between layers in a multilayer film. X-rays and gamma rays are essentially the same, differing by their origin: gamma rays arise from a nuclear source whereas X-rays come from high-speed electrons [22,23].

Newer advances in technology, such as near-infrared (IR) spectroscopy, full-spectrum IR, and optical probes [24], are starting to make possible the in-line measurement of interior layers, such as barrier layers. Coupled with feedback control, these developments may significantly reduce layer-to-layer variation in coextrusion.

2.2.1.1.6 Wind-up

The wind-up is a critical piece of equipment in any film converting process. There are several different types of winders in the market in order to accommodate a wide variety of possible substrates, from elastic/extensible films to heavy paperboard. Winder types fit into three general categories: center, center-surface and two-drum surface winders (see Fig. 2.10).

In a center winder, torque is directly applied through the wind-up roll. This is the simplest method and the best for many materials, such as elastic materials and flexible packaging films with high caliper variation. Generally, the tension is tapered to allow for higher tension at the start of the roll to prevent dishing and telescoping, two problems that occur when the edges of the film on each wrap do not line up properly. In some systems, the tapering is simply accomplished by maintaining a constant tension and relying on the natural reduction in tension as the diameter increases (with each successive wrap). Some versions of center winders have a nondriven nip roll to remove entrapped air during winding to increase roll hardness.

Surface-center winders have both a driven center wind-up roll and a driven nip. These are the most versatile winders since they utilize all the mechanisms of winding: tension, nip and torque. Each is independently controlled. The nip roll isolates the winding tension from the process tension, allowing the center drive to apply more torque at the beginning of the roll to ensure a tight start to prevent telescoping and dishing

Pure surface winders are also known as two-drum winders. Two externally driven nip rolls control the speed of winding, while a third nip roll provides additional nip pressure during startup. These winders find use in the industry due to their consistency of quality roll generation when producing films of varying elasticity, size or film thickness.

Further reading on winding

1. Lynch, R., "The basics of web handling,"; Roisum, D. R. and Smith, R. D., "Winding," and Smith, R. D., "Continuous unwinding and splicing technology and equipment," 2017, in Durling, W., ed., *Extrusion Coating Manual, 5th ed.*, TAPPI Press, Atlanta.
2. Frye, K. G., 1991, *Winding*, TAPPI Press, Atlanta

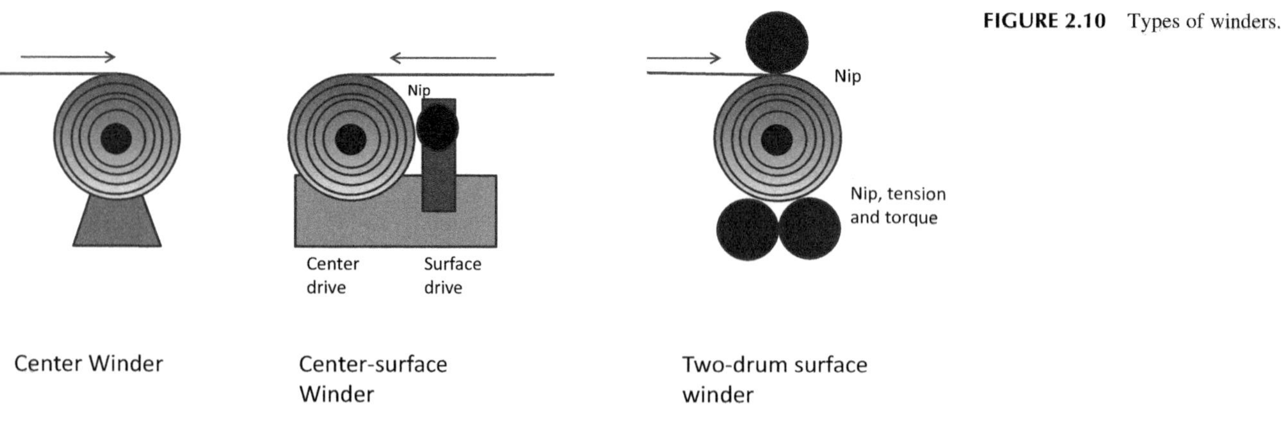

FIGURE 2.10 Types of winders.

3. Roisum, D. R., 1994, *The mechanics of winding*, TAPPI Press, Atlanta.
4. Roisum, D. R, 1996, *The mechanics of rollers*, TAPPI Press, Atlanta.
5. Roisum, D. R., 1998, *The mechanics of web handling*, TAPPI Press, Atlanta.
6. Smith, R. D., 2014, *The ultimate roll and web defect troubleshooting guide*, 3rd ed., TAPPI press, Atlanta.

2.2.1.2 Coextrusion blown film

In coextrusion, multiple layers are combined in the melt during the film fabrication process. There are two general approaches for joining layers. One is to feed different polymer melts from multiple extruders into a combining operation prior to forming the final film shape. The hardware that performs the combining operation is known as a feedblock. Feedblocks are commonly used in flat die extrusions such as coextrusion cast film, sheet and extrusion coating. The second approach is to form the final film shape first and then combine the melt streams. This is the approach used in coextrusion blown film.

In a conventional coextrusion blown film die, each layer has its own spiral mandrel distribution system, as described previously in Fig. 2.8. However, they are now nestled together within a larger die as shown schematically in Fig. 2.11. A disadvantage of this system is that the outer layers have an increasingly larger flow diameter and residence time. In the late 1980s, Brampton Engineering introduced stacked die technology where a series of flat "pancakes" that distribute flow horizontally rather than vertically are stacked together vertically. This allows a flat spiral distribution to be used, as illustrated in Fig. 2.12. A further cross-section view of the stacking geometry is shown in Fig. 2.13. Advantages of the stacked die configuration are reduced residence time and the ability to thermally isolate layers, which can be important for coextruding thermally sensitive resins with high melt point polymers. By thermally isolating the layers, the thermally sensitive polymer can be run at colder temperatures to avoid thermal degradation. Most coextrusion blown film dies in the market today use a variation on these two basic designs [13].

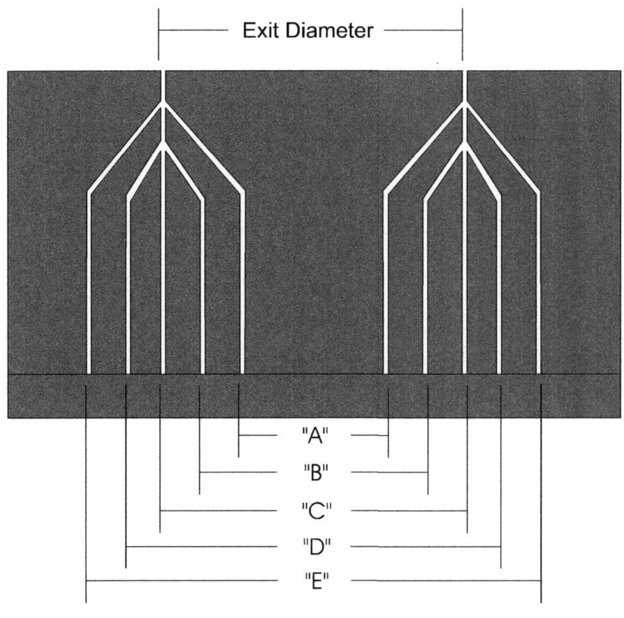

FIGURE 2.11 Cross-sectional schematic of a conventional cylindrical spiral mandrel coextrusion die with five nestled layers. *Reproduced from J. Perdikoulias, Multilayer blown (tubular) film dies, in Wagner, J. (ed.), Multilayer Flexible Packaging, 2nd (ed.), Elsevier, Oxford, 2016, pp. 123–128.*

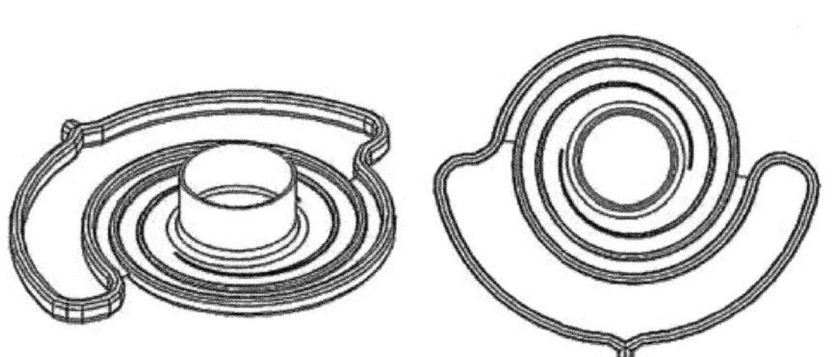

FIGURE 2.12 A flat spiral distribution system with two spirals. *Reproduced from J. Perdikoulias, Multilayer blown (tubular) film dies, in Wagner, J. (ed.), Multilayer Flexible Packaging, 2nd (ed.), Elsevier, Oxford, 2016, pp. 123–128.*

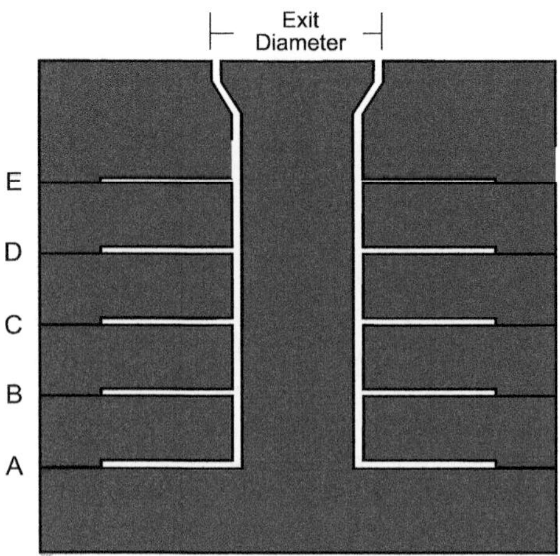

FIGURE 2.13 Cross-sectional schematic of a stacked type of coextrusion die. *Reproduced from J. Perdikoulias, Multilayer blown (tubular) film dies, in Wagner, J. (ed.), Multilayer Flexible Packaging, 2nd (ed.), Elsevier, Oxford, 2016, pp. 123–128*

2.2.1.3 *Stress–strain factors affecting film orientation and properties*

The properties of blown film are influenced by the degree of crystallization and amount of orientation imparted during the fabrication process. These in turn are determined by the stress–strain history: the amount of stretching, the rate of stretching, and the temperature at which it occurs. The stress–strain history results from a complex interplay between cooling, take-up speed and bubble shape. This section defines key terms, relationships and calculations that describe stress and strain in the blown film process. Chapters 14 and 15 provide specific examples of how these relate to film mechanical properties, such as stiffness, impact resistance and tear strength, and interlayer adhesion.

The amount of stretching, also known as strain or elongation, in the machine direction is characterized by the take-up ratio (TUR), sometimes called the draw-down ratio (DDR), which is the ratio of the take-up speed to the velocity of the melt as it exits the die. The take-up speed is the velocity of the bubble at the nip and is determined by the speed of the driven nip roll.

$$\mathrm{TUR} = \mathrm{DDR} = v_{\mathrm{f}}/v_0 \tag{2.3}$$

where TUR is the take-up ratio, DDR is the draw-down ratio, v_{f} is the final film velocity or take-up speed, and v_0 is the initial velocity or velocity of the melt as it exits the die.

Here the subscript "0" is used for the initial state and "f" for the final state. Stretching in the cross or hoop direction is characterized by the blow-up ratio (BUR), which is defined as the final film diameter divided by the die diameter:

$$\mathrm{BUR} = D_{\mathrm{f}}/D_0 \tag{2.4}$$

where BUR is the blow-up ratio, D_{f} is the final tube diameter, and D_0 is the die diameter.

Since the velocity of the polymer melt at the die exit is not typically measured, Eq. (2.3) is not in a readily useable form. A mass balance approach can be used to derive a more useful form of Eq. (2.3). This is possible since the mass flow rate at the die exit is equal to the product of the velocity, area of the die opening and the density of the polymer melt:

$$\dot{M} = v_0 \pi D_0 X_0 \rho_0 \tag{2.5}$$

where $\dot{M}$ is the mass flow rate, v_0 is the velocity of polymer at die exit, D_0 is the die diameter, X_0 is the die gap, and ρ_o is the density of polymer melt (at die exit).

A mass balance at the nip roll yields the following:

$$\dot{M} = v_f \pi D_f X_f \rho_f \tag{2.6}$$

where X_f is the final film thickness and ρ_f is the density of polymer at the nip roll (solid).

Here the die gap is assumed to be much less than the die diameter ($X_0 << D_0$) so that the area simplifies to the circumference times the die gap. Similarly, it is assumed that $X_{\mathrm{f}} << D_{\mathrm{f}}$.

Equating Eqs. (2.5) and (2.6), the following relationship for TUR (DDR) is obtained:

$$\mathrm{TUR} = \frac{1}{\mathrm{BUR}} \frac{X_o}{X_f} \frac{\rho_o}{\rho_f}. \tag{2.7}$$

For many polyolefins, the ratio ρ_0/ρ_f ranges from about 0.7 to 0.9. Ethylene vinyl alcohol and polyamide 6 have ratios around 0.85−0.9, as do many amorphous polymers like polystyrene. Many references ignore the density ratio, but the error can be significant depending on the accuracy needed.

The amount of stretching only tells part of the story. The force imparted by the process is what orients the polymer chains and influences properties. The force per area is known as stress. For an elastic solid, stress is proportional to the amount of stretching (elongation) through a material parameter called the modulus:

$$\sigma_{\text{elastic solid}} = G\varepsilon \tag{2.8}$$

where $\sigma_{\text{elastic solid}}$ is the stress, G is the elastic modulus, and ε is the strain or elongation.

The modulus is related to stiffness and is a function of temperature; as the bubble cools the modulus increases. The increasing modulus provides melt strength that allows the bubble to sustain stretching. In monolayer blown film, much of the stretching occurs near the freezing point of the polymer where the modulus and stress are highest. In coextrusion, the maximum amount of stretching occurs when the first polymer starts to solidify. The other layers in the coextrusion may still be molten. As a result, the stress on these layers and their properties may differ from what they would be if they were made into monolayer films. This is described in detail in Chapter 14.

Polymers are known to be viscoelastic (see Chapter 5 for further discussion on the rheology of polymers), meaning they behave like elastic solids as well as viscous fluids. For a viscous fluid, the stress is proportional to the rate of strain (also known as the rate of elongation or rate of stretching) through the viscosity:

$$\sigma_{\text{viscous fluid}} = \eta_e\dot{\varepsilon} \tag{2.9}$$

where $\sigma_{\text{viscous fluid}}$ is the stress, η_e is the elongational viscosity, and $\dot{\varepsilon}$ is the rate of strain or elongation. Therefore, the stress for most polymers will be a function of both the strain (amount of stretching) and the rate of strain.

The rate of strain in the machine direction depends on the change in velocity of the polymer as it travels up the bubble. Similarly, the rate of strain in the circumferential or transverse direction is related to the change in bubble diameter. These in turn are a function of the cooling rate; as the polymer cools it gains melt strength that is needed to sustain the stretching of the bubble. A faster cooling rate generally results in a higher rate of strain or stretching. A characteristic process time can be defined as the time it takes the polymer melt to travel from the die exit to the frost line; the process time is inversely related to the rate of strain as is shown below. The frost line is an indicator of when the polymer is starting to freeze. In some films, the bubble becomes hazy at this point as it begins to crystallize. In other films, the frost line can be determined by where the bubble stops changing shape. The cooling rate can be monitored by an IR thermometer. Butler [25] has defined several transitions as shown in Fig. 2.14: the crystalline-line height (CLH) where crystallization first begins, the frost line height (FLH) where the change in the bubble shape ends, the plateau line height where the temperature plateau ends and the freeze-line height (FZH) where primary crystallization ends.

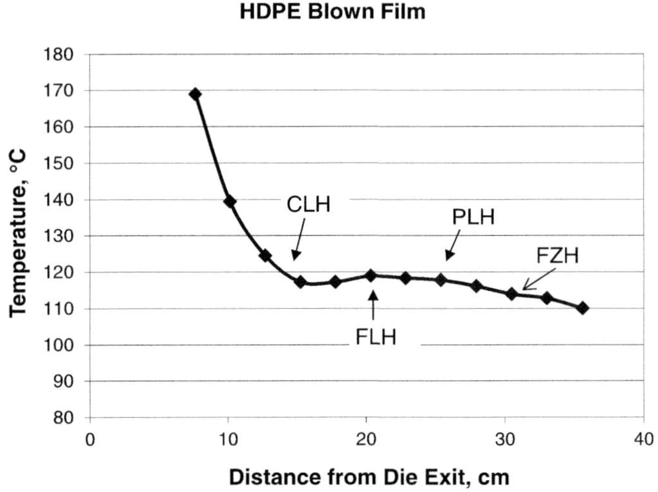

FIGURE 2.14 Definition of transitions in blown film [26]. *Reproduced with permission from B. A. Morris, Effect of process and material parameters on interlayer peel strength in coextrusion coating, film casting and film blowing, J. Plast. Film Sheeting, 26 (2010) 343−376.*

The velocity (v) of an element of polymer within the extruded tube or web is equal to its change in position (z) with time (t):

$$v = \frac{dz}{dt}. \tag{2.10}$$

The instantaneous rate of strain is equal to the change in velocity with respect to distance; the strain is found by integrating the instantaneous rate of strain over time:

$$\dot{\varepsilon}(t) = \frac{dv}{dz} = \frac{1}{v}\frac{dv}{dt}. \tag{2.11}$$

$$\varepsilon(t) = \int_0^t \dot{\varepsilon} dt = \int_0^t \frac{1}{v}\frac{dv}{dt} dt = \ln\left(\frac{v}{v_o}\right). \tag{2.12}$$

At the frost line height, the bubble diameter and velocity cease to change and stretching of the polymer ends; the bubble diameter and velocity reach their final values (and are designated with the subscript f). The process time, t_f, is defined as the time it takes the polymer to travel from the die exit to the frost line. Here the subscript f is used to designate values at the frost line. t_f represents the time over which the bubble is stretched. Using Eq. (2.12), the total amount of strain imparted in the bubble is found by integrating the rate of strain from time zero to t_f:

$$\varepsilon_{\text{tot}} = \ln\left(\frac{v_f}{v_o}\right) = \ln(\text{DDR}). \tag{2.13}$$

A typical velocity profile in blown film is shown in Fig. 2.15. The change in velocity from the die exit to the frost line is nearly linear. Assuming a linear velocity profile and integrating Eq. (2.10) yields the following equation for the process time:

$$t_f = \frac{h_f}{v_f - v_o} \ln\left(\frac{v_f}{v_o}\right) \tag{2.14}$$

where h_f is the frost line height.

Using a mass balance, Eq. (2.14) can be recast into more easily measured parameters:

$$t_f = \left[\frac{h_f}{v_f\left(1 - \frac{[\text{BUR}]X_f \rho_f}{X_0 \rho_0}\right)}\right] \ln\left(\frac{X_0 \rho_0}{(\text{BUR})X_f \rho_f}\right). \tag{2.15}$$

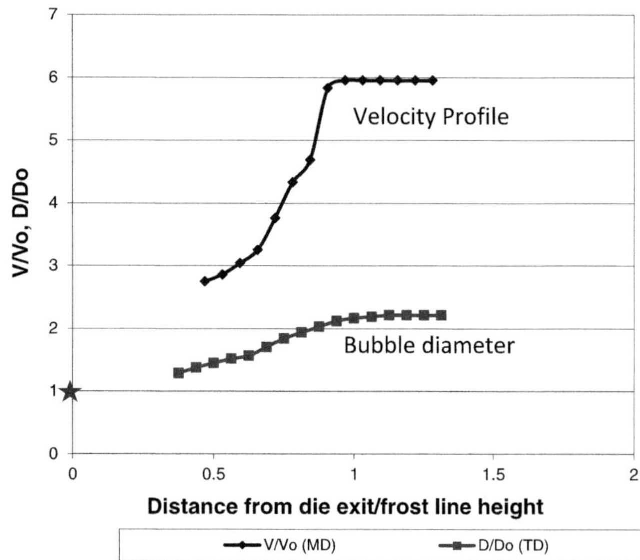

FIGURE 2.15 Machine direction velocity (V) and bubble diameter (D) profiles in blown film. Taken from video camera marker experiments on a 2-layer film. The x-axis is the vertical distance from the die exit divided by the frost line height. Data collection from $x = 0$ to 0.5 is blocked by the air ring. The value at $x = 0$ is 1 by definition. *Reproduced with permission from B. A. Morris, Effect of process and material parameters on interlayer peel strength in coextrusion coating, film casting and film blowing, J. Plast. Film Sheeting, 26 (2010) 343–376.*

In monolayer blown film, stress is high near the front line since the modulus, rate of strain and viscosity are all near their maximum. As a result, substantial orientation is imparted in both the machine and transverse directions that influence film properties. Patel et al. [27], Babel and Campbell [28], and others have proposed models for how these parameters affect mechanical and other properties of monolayer blown film. These are explored in more detail in Chapters 14 and 15, along with their extension to coextrusion.

2.2.1.4 Scale-up

Scale-up factors are important for predicting process conditions that will duplicate film properties made on different size film lines. Many of the concepts introduced in the previous section are useful scale-up factors since film properties are related to the stress–strain history. Kanai et al. [29] use a thin-membrane approximation of Pearson and Petrie [30–32] to derive scale-up factors for blown film. Shirodkar et al. [33] propose using process time (Eq. (2.14)) as a scale-up factor and show that dart impact and tear resistance correlate well with process time. Careful examination of Kanai's scaling parameters shows that when BUR, TUR and h_f/D_f are held constant, so too is process time.

Butler et al. [34] have also proposed scale-up factors based on membrane models and a heat transfer analysis. The output in Kanai's scaling model increases with the die gap times the square of the die diameter ($X_0D_0^2$) [35], which means that the die gap needs to be changed to achieve a consistent stress value on the polymer flow in the die at both scales of manufacturing. As a consequence, if the die gap is held constant the final film thickness must be changed to maintain constant properties. This is not practical in most situations. Butler et al. propose scaling the output linearly with die diameter. They explicitly use the process time in their scaling model, proposing a Film Fabrication Number (known as the Deborah number in the Rheology literature) that relates the time scale of the process (Eq. (2.14)) to the relaxation time of the polymer. A comparison of the two approaches can be found in Table 2.3.

Further reading on blown film

1. Macnamra, Jr., J. F., ed., 2020, *Film extrusion manual, 3rd ed.*, TAPPI Press, Atlanta.
2. Kanai, T. and Campbell ,G. A., 2011, Film processing, Hanser Publications.
3. Giles, Jr., H. F., Mount III, E. M. and Wagner, Jr., J. R. (eds), 2007, *Extrusion: the definitive process guide and handbook*, William Andrews, pp. 453-463.
4. Perdikoulias, J., 2010, "Multilayer blown (tubular) film dies," in Wagner, J. (ed.), *Multilayer flexible packaging, technology and applications for the food, personal care and over-the-counter pharmaceutical industries*, Elsevier, Oxford, pp.97-102.
5. Butler, T. I., 2001, "Blown film extrusion," in Vlachopoulos, J. and Wagner, J. R. (eds.), *The SPE guide on extrusion technology and troubleshooting*, The Society of Plastics Engineers, Brookfield, CT.

TABLE 2.3 Blown film scale-up models.

Parameter	Kanai	Butler
BUR	Constant	Constant
DDR or TUR	Constant	Constant
T_{melt}	Constant	Constant
h_f/D_f	Constant	h_f adjusted to keep t_f constant
λ/t_f		Constant
v_0	D_0	Constant
v_f	D_0	Constant
X_0	D_0	Constant
X_f	X_0	Constant
h_f	D_0	Adjusted to keep t_f constant
Q	$X_0D_0^2$	D_0

h_f, frost line height, λ, relaxation time of polymer, Q, volumetric throughput.

2.2.2 Cast film

The cast film process involves extruding the polymer through a flat die, drawing it down to the final thickness by pinning the melt extrudate onto a rotating cold roll and rapidly solidifying it (Fig. 2.16). In the cast film process, the polymer is stretched in the molten state in the machine direction in the air gap between the die exit and the chill roll and very quickly quenched. This is quite different from the blown film process previously discussed, where the polymer is stretched near its freezing point in both the machine and transverse directions while being more slowly quenched. As a result, cast film tends to have lower crystallinity, higher clarity, greater flexibility and less balanced orientation in the machine and transverse directions than films of a similar composition made on the blown film process.

2.2.2.1 Key components

2.2.2.1.1 Extruder

One or more extruders melt and pump the polymer through the cast film die. See Section 2.1 for details.

2.2.2.1.2 Die

The application requires that the die distribute the melt into a flat sheet. This is achieved by designing the die to produce a uniform flow at the die exit across the width of the die. Key factors influencing performance are the flow channel geometry including the manifold design and pressure distribution. The flow should be streamlined to minimize residence time and prevent any regions where flow may stagnate and contribute to polymer degradation. T-slot dies use a wide channel perpendicular to the die entrance to distribute the flow across the width of the die. Pressure across a land region regulates the flow. (See Fig. 2.17A.) These are more commonly used in extrusion coating. So-called "coat-hanger" dies (Fig. 2.17B) use a manifold to create an equal pressure drop across the width of the die. This helps create uniform flow out the die. Most modern cast film dies use some type of coat-hanger geometry.

The width of the melt extrudate can be narrowed by attaching deckles which restrict the flow at the die edges. Early versions of this technology were externally applied to the die. Modern dies have internal deckles with rods and pins that can be slid to adjust the flow at the edges. Internal deckles are illustrated in Fig. 2.18 [36].

2.2.2.1.3 Chill roll and pinning technology

After the melt curtain exits the flat die, it is drawn down to its final thickness and quenched by pinning it to a fast moving cold roll. The film thickness is determined by the die exit velocity (which is fixed by the die dimensions and extruder output) and the line speed, which is controlled by the chill roll speed. The ratio of the line speed to the die exit velocity is known as the draw-down ratio (DDR, see Eq. 2.16).

Two types of pinning technology are common, the vacuum box and the air knife. These are shown in Fig. 2.16. As their names suggest, the vacuum box draws the melt curtain onto the roll with a vacuum whereas the air knife blows air across the curtain to force it onto the roll. The chill roll is usually chrome plated and internally cooled with chilled water. The chill roll temperature is an important operating parameter for quenching. If the chill roll temperature is too hot, the polymer may stick to the roll. Therefore, a matted surface is sometimes used to help prevent sticking.

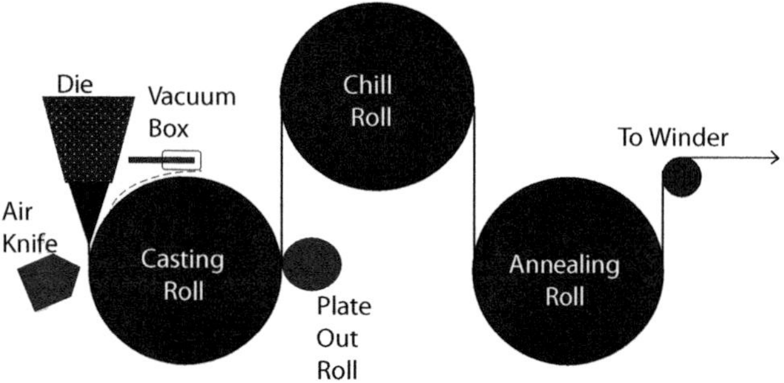

FIGURE 2.16 Cast film process. *Reproduced with permission from The Dow Chemical Company.*

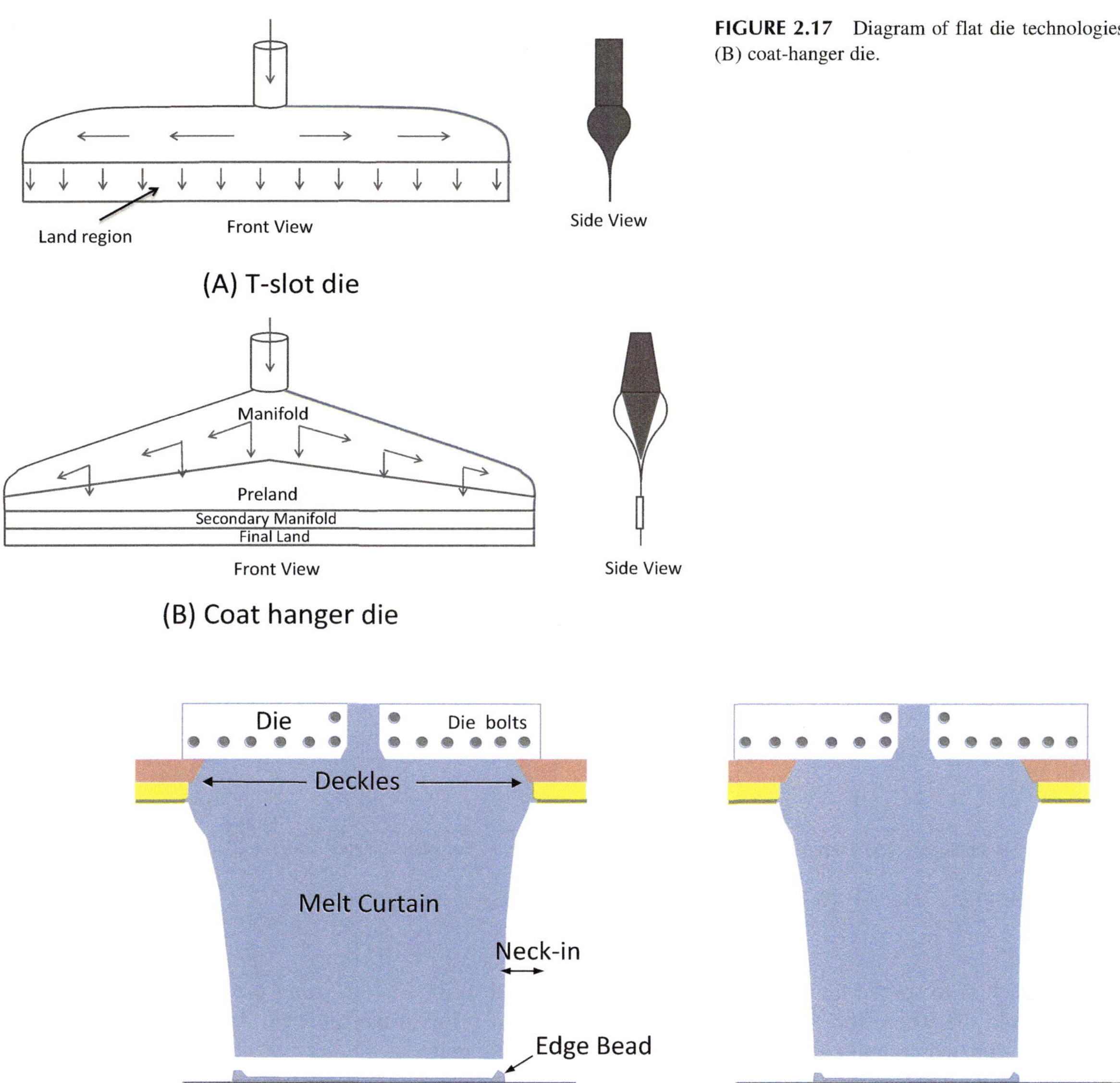

FIGURE 2.17 Diagram of flat die technologies: (A) T-slot die, (B) coat-hanger die.

FIGURE 2.18 Illustration of internal deckled die. Deckles have been inserted further into the die channel on the right to narrow the width of the melt curtain. *Based on S. Iuliano, Extrusion coating dies, in W. Durling (Ed.), Extrusion Coating Manual, 5th ed., TAPPI Press, Atlanta, GA, 2017, pp. 31–35.*

2.2.2.1.4 Wind-up and other components

Wind-up technology similar to that described in Section 2.2.1.1 applies here. Automatic gauge control is often employed in cast film. A common setup utilizes differentially heated die bolts spaced approximately 2.5 cm (1-inch) apart across the width of the die. Gauge variation across the film width is continuously measured and fed back to the automatic gauge control system, which adjusts the heat on the die bolts to control the local die gap and flow [37].

2.2.2.2 Coextrusion

Two approaches are used to combine layers in flat die coextrusion, whether the process is cast film, cast sheet or extrusion coating: feedblock technology and multimanifold dies. These are illustrated in Fig. 2.19. In the first, each extruder feeds the polymer melt into a combining device known as a feedblock. The function of the feedblock is to combine the polymer melt streams in a way that results in a uniform layer distribution prior to entering the die. To help achieve uniform layer distribution, several strategies can be employed, including vanes and laminar plates, as shown in Fig. 2.20. Most feedblocks allow flexibility in layer ratio and the number of layers through the use of specific plugs, pins and other devices.

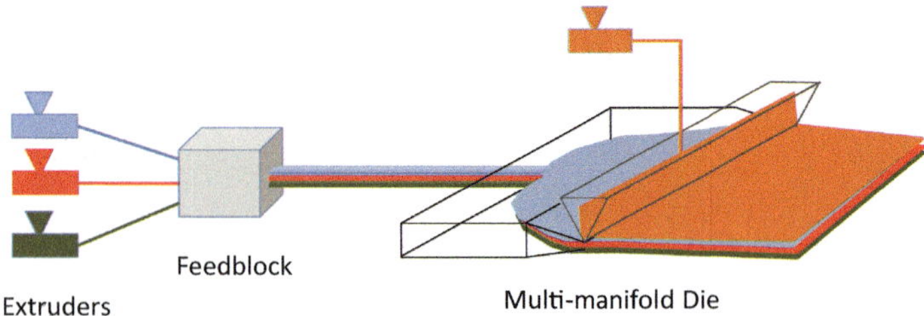

FIGURE 2.19 Approaches to combining layers in flat die coextrusion. Here the first three layers are combined using a feedblock, spread to the final width in the die and then joined with the fourth layer in a multimanifold die.

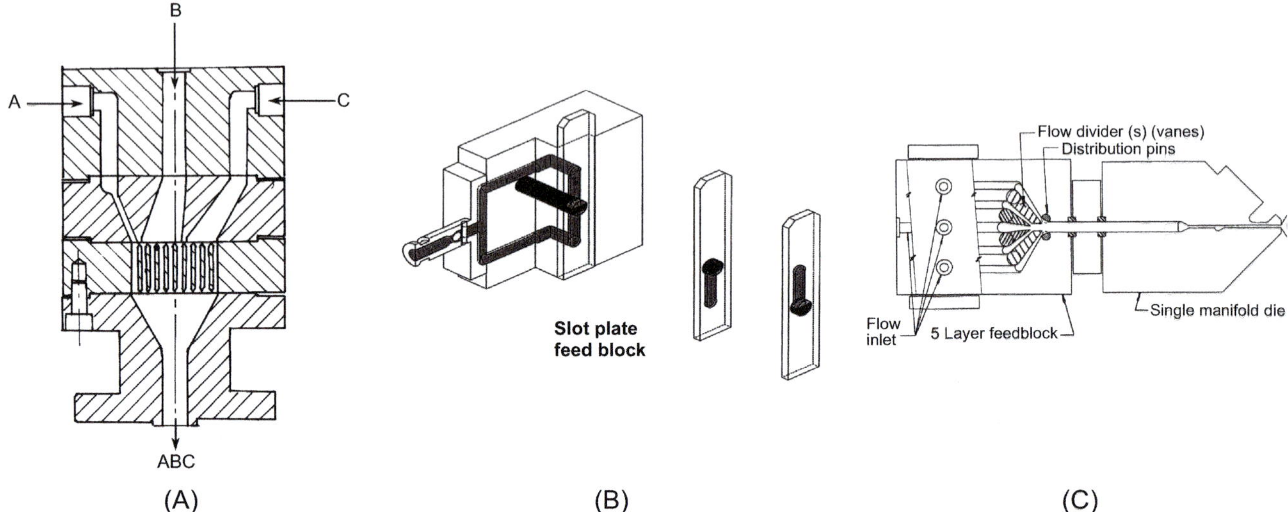

FIGURE 2.20 Feedblock technologies. (A) Laminar plates (Dow design), (B) modular slot plate (Welex design), (C) five-layer feedblock with adjustable vanes and distribution pins (Cloeren design). *Reproduced with permission from E.M. Mount, III, Coextrusion equipment for multilayer flat films and sheets, in: J. Wagner (Ed.), Multilayer Flexible Packaging, 2nd ed., Elsevier, Oxford, 2016, pp. 99–122.*

The second approach is to spread the polymer melt to its final width prior to combining the flow. Typically this is done in a multimanifold die where the melt streams are combined just prior to the die exit. A variation on this is to employ more than one die and combine the melt streams just after the die exit.

Feedblock technology is generally less expensive and more versatile than multimanifold die technology. Multimanifold dies are typically designed for a specific structure. Adding more layers, for example, becomes difficult. Multimanifold dies are less sensitive to flow instabilities and allow more diverse materials to be combined. Flow instability is covered in more detail in Chapter 13. Also multimanifold dies allow polymers with different melt temperatures to be combined with minimal heat transfer between them. This may be important when extruding a thermally sensitive polymer with a polymer with a high melting point; the thermally sensitive polymer can be processed at a lower temperature to avoid thermal degradation. Multimanifold dies may be combined with feedblocks to further increase versatility.

2.2.2.3 *Key factors affecting properties*

The crystallinity and orientation of the polymer chains in cast film are influenced by the quench rate, imposed strain and the rate of strain of the process. These in turn affect film properties such as haze, stiffness, thermoformability and tear strength. The DDR characterizes the strain or orientation in the machine direction:

$$\mathrm{DDR} = \frac{v_f}{v_0} \tag{2.16}$$

where v_f is the line speed (final speed of the film coming off the chill roll) and v_0 is the velocity of polymer at the die exit.

A simple mass balance equating the mass flow rate at the die exit with the film at the chill roll yields:

$$w_0 X_0 v_0 \rho_0 = w_f X_f v_f \rho_f \tag{2.17}$$

where w_0 is the width of the die opening (deckled), w_f is the width of the melt curtain upon contact with the chill roll, X_O is the die gap, X_f is the final film thickness, ρ_0 is the melt density, and ρ_f is the solid density.

Substituting Eq. (2.17) into (2.16):

$$\text{DDR} = \frac{w_0 X_0 \rho_0}{w_f X_f \rho_f}. \tag{2.18}$$

Neck-in is the tendency for the melt curtain to narrow in width as it is drawn down to its final thickness in the air gap between the die exit and the chill roll (see Fig. 2.18). A simple definition is $(w_0 - w_f)/2$ (the factor 2 in the denominator arises from the convention that neck-in is the change in width at each edge). If neck-in is small, then $w_0 \cong w_f$ and Eq. (2.18) simplifies to

$$\text{DDR} = \frac{X_0 \rho_0}{X_f \rho_f}. \tag{2.19}$$

For a given polymer, the density ratio (ρ_0/ρ_f) is constant, so that the DDR is proportional to the die gap and inversely proportional to the final film thickness: DDR $\propto X_0/X_f$. For a constant die gap, the thinner the film, the more draw-down, and the greater the orientation.

The quench rate in cast film is very rapid, on the order of a few milliseconds, since the melt curtain is pinned directly against the chill roll. Rapid quench rates generally result in lower levels of crystallinity and density (for crystalline polymers). Some of the consequences of lower crystallinity are described in Section 2.2.2.4.

Another important parameter is the time in the air gap (TIAG). As is shown in later chapters, time in the air gap is important for adhesion and orientation since it is inversely related to the rate of strain. For cast film and extrusion coating, the polymer velocity in the air gap fits well to a Newtonian fluid [38] (see Fig. 2.21):

$$\frac{v}{v_o} = \text{DDR}^{\left(\frac{z}{L}\right)} \tag{2.20}$$

where z is the vertical distance from the die exit and L is the length of air gap (distance between the die exit and the chill roll).

It then follows from Eqs. (2.10) to (2.12) that

$$\varepsilon = \ln\left(\frac{v}{v_o}\right) = \frac{z}{L}\ln(\text{DDR}) \tag{2.21}$$

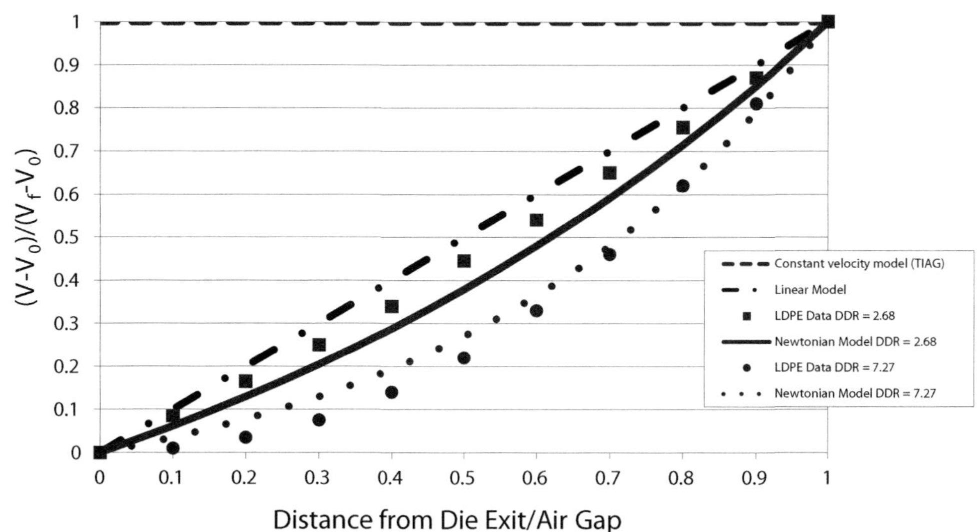

FIGURE 2.21 Velocity profile of polymer melt in air gap of cast film/extrusion coating. *Reproduced with permission from B. A. Morris, Understanding why adhesion in extrusion coating decreases with diminishing coating thickness, J Plastic Film & Sheeting, 24 (2008) 53–88. Data for LDPE melt taken from K. Canning, B. Bian, A. Co, Film casting of a low density polyethylene melt, in Proceedings of ANTEC 1999, Annual Technical Conference –Society of Plastic Engineers, Vol. 57, 1999, pp. 386–390 [39]. Constant model assumes* $v = v_f$ $(t_f = TIAG = L/v_f)$. *Linear model assumes* $v = v_o + (v_f - v_o)z/L$. *Newtonian model is shown in Eq. (2.20).*

$$\dot{\varepsilon} = \frac{v_f}{L} \mathrm{DDR}^{\left(\frac{z}{L}-1\right)} \ln(\mathrm{DDR}) \tag{2.22}$$

$$t_f = \frac{L}{v_f}\left[\frac{\mathrm{DDR}-1}{\ln(\mathrm{DDR})}\right] \tag{2.23}$$

where ε is the strain, $\dot{\varepsilon}$ is the rate of strain, and t_f is the process time.

Many practitioners simplify Eq. (2.23) by ignoring the term in brackets. This value is typically known as the TIAG [40]:

$$\mathrm{TIAG} = \frac{L}{v_f}. \tag{2.24}$$

In English units, this can be written:

$$\mathrm{TIAG(ms)} = \frac{L[\text{inches}]}{v_f\left[\frac{ft}{\min}\right]} \times 5000. \tag{2.25}$$

In metric units:

$$\mathrm{TIAG(ms)} = \frac{L[\text{mm}]}{v_f\left[\frac{m}{\min}\right]} \times 60. \tag{2.26}$$

Eq. (2.24) assumes that the polymer instantaneously accelerates to the final line speed as it exits the die. While not physically correct, TIAG does prove useful for explaining adhesion and other data, especially in extrusion coating. Eq. (2.23) should be used if changes are being made in DDR.

The time in the air gap for cast film is typically fairly short since the die exit is positioned close to the chill roll (L is small).

2.2.2.4 Comparison of blown and cast film

In the blown film process, the melt is simultaneously cooled and stretched in both the machine and transverse directions over a relatively long postdie quench time (on the order of 1 second or more). Most of the stretching takes place near the freezing point. In the cast film process, the polymer is stretched primarily in the machine direction before the polymer is quenched, but over a much shorter process time (on the order of 0.1 second or less). The result is that blown film typically has more crystallinity and is more balanced in orientation between the machine and transverse directions. Greater crystallinity produces a more dense film, higher haze, less transparency and higher barrier properties. Cast film is typically softer and more transparent. It also may thermoform more easily into deeper cavities while maintaining uniform thickness distribution due to lower crystallinity.

In addition to these property differences, there are a number of operational differences, as summarized in Table 2.4. These include [41–43]:

- Blown film typically has greater flexibility in width (by changing BUR or using die lip inserts), produces less trim waste and is easier to transition. As a result, blown film is favored for short runs.
- Gauge control is typically better for cast film. Variation of +/−2% or less is common for cast film, versus +/−5% to 10% for blown film.
- Cast film lines typically run at a higher output than blown film. Blown: 5–24 + lb/circumferential inch (0.9–4.3 + kg/cm) of die; cast: 15–30 + lb/in (2.7–5.4 +) of die width.
- Blown film lines typically require less capital investment than cast film lines.
- Blown film lines typically use resin with higher viscosity (lower melt index) and higher melt strength than cast film lines. Processing temperatures are also typically lower on blown film lines to help achieve good melt strength, which improves bubble stability.

Further reading on cast film

1. Macnamra, Jr., J. F., ed., 2020, *Film extrusion manual, 3rd ed.*, TAPPI Press, Atlanta.
2. Ivey, J. 2001, "Cast film and sheet extrusion," in Vlachopoulos, J. and Wagner, J. R. (eds.), The SPE guide on extrusion technology and troubleshooting, The Society of Plastics Engineers, Brookfield, CT, pp. 12-1–12-12.

TABLE 2.4 Comparison of blown and cast film.

Attribute	Blown Film	Cast Film
Film Properties		
Crystallinity (density)	Higher	Lower
Haze	Higher	Lower
Gloss	Lower	Higher
Barrier (moisture or gas)	Higher	Lower
Stiffness	Higher	Softer
Thermoformability	Shallower	Deeper
Orientation	MD and TD	MD
Operation		
Flexibility in width	Greater	Less
Gauge variability	More	Less
Trim waste	Less	Higher
Output per film width	Lower	Higher
Flexibility for short runs	Greater	Less
Initial capital investment	Lower	Higher

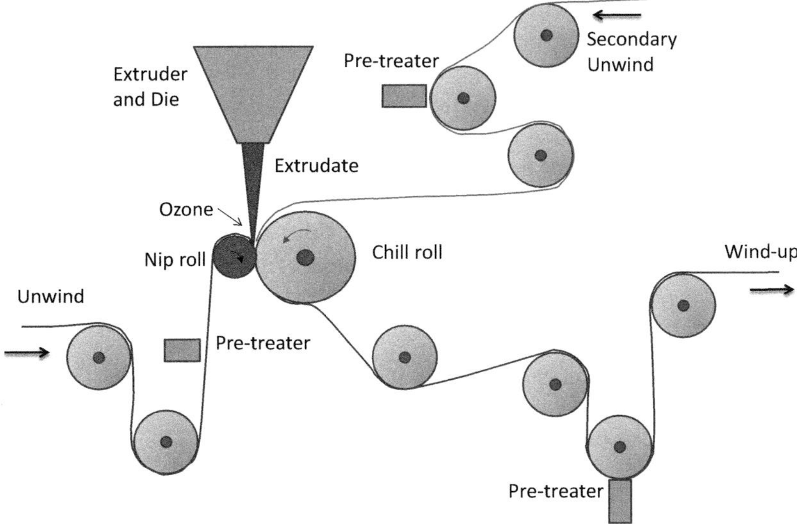

FIGURE 2.22 Extrusion coating and lamination process.

2.3 Coating and lamination

2.3.1 Extrusion coating and lamination

In the extrusion coating process (Fig. 2.22), the solid substrate, such as paper, foil, film or fabric, is continuously unwound from a roll, sometimes primed or treated to aid in adhesion, coated with a molten polymer, passed through a nip to contact a cold roll, cooled and wound again.

The extrusion lamination process uses the same equipment as in extrusion coating. Instead of coating a single substrate, an additional unwind station is used to feed a second substrate into the nip region on the chill roll side. The molten polymer is laid in between to bond the substrates together, creating a multilayer laminate.

The key components of the extrusion coating/lamination process are similar to those used in cast film.

2.3.1.1 Key components

2.3.1.1.1 Extruder

The extruder melts and pumps the polymer through the flat die. Details on the extrusion process can be found in Section 2.1. Extrusion coating generally involves extruding the polymer at high temperatures (e.g., 320°C for LDPE) in order to achieve adhesion. To attain such temperatures, shallow screw channel depths are often employed to increase shear rate and viscous heating. A restrictor valve at the end of the extruder may also be used to help generate temperature and pressure.

2.3.1.1.2 Die

A flat die similar to that used in cast film is utilized. See Section 2.2.2.1 for details. As the polymer is drawn down from the die exit to the chill roll, it has a tendency to narrow in width. This is known as neck-in. As the melt curtain necks-in, the edges become thicker (known as edge bead). These phenomena are illustrated in Fig. 2.18. Modern "edge bead reduction" dies have adjustable flags or rods that fine-tune the flow path near the edge of the die exit. This thins out the flow to help compensate for beading. This is illustrated in Fig. 2.23. Edge bead reduction dies are T-shaped (Fig. 2.17A) so that shims can go all the way to the base of the manifold and slide in and out without there being a gap. Evening out the flow near the final land, or using a longer land, moderates variations in flow to smoothen the web and prevent die lines.

2.3.1.1.3 Substrate unwind and preparation

The substrate is unwound from a roll with proper tension control. Depending on the structure and application, the surface of the substrate may need to be physically or chemically modified prior to coating to improve adhesion. A number of technologies may be used for in-line surface preparation, such as

- Flame treatment [44,45]: This is often used to burn off oils and other impurities on the surface of aluminum foil, or to remove raised fibers, flash off moisture, remove loose starch and preheat paper or paperboard.
- Corona or plasma treatment [44–51]: Ionization of various gases may be employed to change the surface chemistry of the substrate and remove impurities and/or to promote adhesion to the coating polymer.
- Chemical priming [52]: Chemicals may be applied to the substrate to promote adhesion to the coating polymer. One commonly used primer is polyethylene imide (PEI). Others include polyurethanes and aqueous dispersions of ethylene acrylic acid copolymers. Priming is usually accompanied by in-line drying, depending on the chemistry employed. See Section 10.4.1.1 for more details.
- Combinations of these technologies: For example, corona treatment may be used prior to the application of the chemical primer to ensure the primer wets the surface. Generally, the surface tension of the substrate must be 10 mN/m higher than the primer to ensure good wetting and coating. Corona or plasma treatment is often used to raise the surface tension of the substrate. See Section 10.3.1.

Further discussion of the effect of surface treatment on adhesion can be found in Chapter 10. Treatment may also be used to modify the polymer melt curtain. For example, ozone is used to promote oxidation of LDPE to improve its adhesion to polar substrates such as aluminum foil. Details can be found in Section 15.1.2.1.

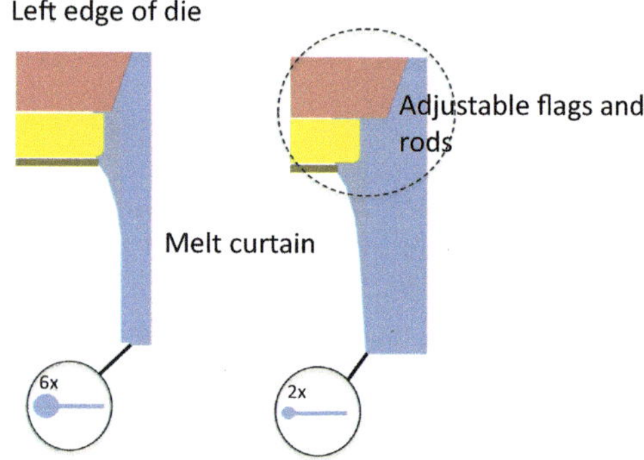

FIGURE 2.23 Illustration of edge bead reduction technology. Flags and rods can be adjusted to thin out the flow near the edge of the melt curtain to reduce edge bead. *Based on S. Iuliano, Extrusion coating dies, in W. Durling (Ed.), Extrusion Coating Manual, 5th ed., TAPPI Press, Atlanta, GA, 2017, pp. 31–35.*

2.3.1.1.4 Chill and nip roll

The molten polymer impinges onto the substrate and subsequently comes into contact with the chill roll (or the second substrate that is in contact with the chill roll in extrusion lamination). The offset between when the polymer contacts the substrate and the chill roll is a parameter used to promote adhesion as long as the hot polymer does not burn through on the substrate: the offset can be more substantial with thermally resistant substrates like paper. A rubber nip roll applies pressure to help ensure contact between the polymer and substrate. The nip pressure also helps press the molten polymer into porous or fibrous substrates for better adhesion. Heat is transferred to the chill roll to solidify the polymer.

Chilled water or other cooling media cools the chill roll. The roll is typically chrome plated to help prevent the polymer from sticking to it. It may also be matted to aid in the release of the polymer. The polymer may contain a release agent to help prevent chill roll sticking as well.

Time and pressure in the nip are important for adhesion, particularly to porous substrates such as paper. The nip pressure is at a maximum at the centerline of the nip region as shown in Fig. 2.24. Ideally, the polymer should still be molten and able to flow at the midpoint so that the pressure can be most effectively utilized. The polymer should be solidified by the time it exits the nip to avoid chill roll sticking and other issues [53]. The time in the nip is determined by the length of the nip region divided by the line speed and can be increased by choosing a softer nip roll. The resulting increase in the deformation of the nip, however, reduces the maximum nip pressure as shown in Fig. 2.24. There are tradeoffs between roll hardness, line speed, nip force and cooling rate that are explored in more detail in Section 15.1.1.3.1.

2.3.1.1.5 Wind-up

Like blown and cast film, wind-up is a key processing step for extrusion coating and lamination. See Section 2.2.1.1 for a discussion on winder technology.

2.3.1.2 Coextrusion

The coextrusion process for extrusion coating and lamination is the same as for coextrusion cast film (Section 2.2.2.2). Both feedblock and multimanifold die technology are commercially employed in coextrusion coating/lamination. One of the advantages of coextrusion is the ability to tailor the adhesive polymer used to bond to the substrate(s).

2.3.1.3 Key parameters affecting properties

Unlike cast or blown film, the physical properties of multilayer films made by extrusion coating or lamination are primarily determined by the substrates rather than the extrudate or process. Extrusion coating is used to apply an attribute to a substrate that it does not have, such as barrier, sealing performance, strength or tear resistance, or print protection. Extrusion lamination is used to bring together substrates with different functionalities; the lamination polymer provides adhesion. but may add additional functionality such as barrier or opacity. Sealing is not an attribute of the laminating polymer since it is not on the surface, but can be an important functionality for polymer coatings.

Adhesion is influenced by the extrusion melt temperature at the die exit, polymer viscosity, process time or TIAG, DDR, chill roll temperature and time in the nip. In general, higher processing temperature increases adhesion to the substrate, through better wetting and, when present, chemical interaction. Lower viscosity increases wetting. Increasing process time (Eq. (2.23)) or TIAG (Eq. 2.24) increases the time for oxidation of LDPE. Oxidation creates small amounts of acids, aldehydes and ketones that help improve adhesion to polar substrates such as aluminum foil,

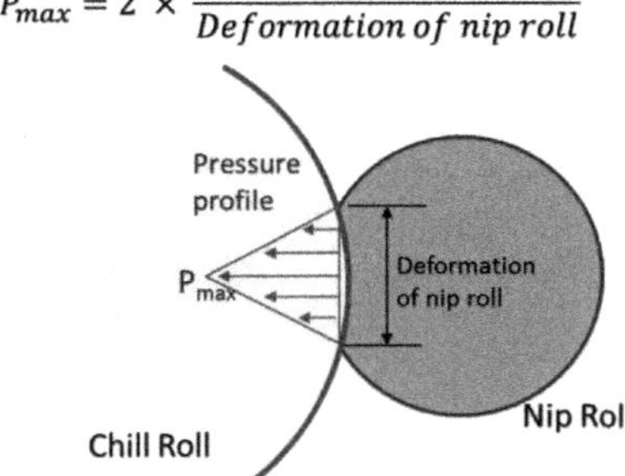

FIGURE 2.24 Pressure distribution in the nip region. *Reproduced with permission from The Dow Chemical Company.*

metallized films, printed films and paper. Increasing process time also reduces the rate of strain (Eq. (2.22)), which reduces stress and can increase peel strength performance [54]. These effects are explored more deeply in Chapter 15.

Increasing TIAG has its limits. TIAG can be increased by either increasing the air gap (L) or reducing line speed. Reducing line speed decreases productivity. Increasing air gap leads to more neck-in. Neck-in can limit coating width, and cause a thickening of the coating near the edges (edge bead). The choice of polymer can influence neck-in: greater melt strength reduces neck-in. Also, die design can help reduce edge beads as described in Section 2.3.1.1 and Fig. 2.23.

Extrusion temperature also has its limitations. Polymers must be run below their auto-ignition temperature. But even before this temperature is reached, excessive temperatures can lead to thermal stability issues (such as crosslinking that leads to gel), generation of off-odor species and poor melt strength (greater neck-in). High temperature can also negatively affect heat seal and hot tack performance through crosslinking [55,56] or partial degradation of the coating [57].

Polyethylene is the most commonly used coating or laminating polymer. It relies on oxidation to bond to polar substrates. The PE melt curtain can be treated with ozone to promote oxidation and often enables faster line speeds [58,59]. Specialty polymers, such as ethylene-acid copolymers [60], can also be used to overcome these processing limitations since they have inherent functional groups that do not rely on oxidation. See Section 10.4 for a discussion of polymers for extrusion coating/lamination. These specialty copolymers may also provide greater resistance to delamination in the presence of aggressive food products, as described in Section 16.2.1.2.

High line speed and draw down can lead to web instabilities known as edge weave and draw resonance (see Section 13.4.2.2). The line speed or DDR where this occurs is influenced by the polymer flow characteristics (rheology).

Once the hot polymer contacts the substrate it has only a few milliseconds to wet the surface before it solidifies. The nip pressure aids wetting. The balance between time in the nip and quench rate can influence adhesion. For paper substrates, the nip pressure helps drive the molten polymer into the pores to achieve physical bonding around the fibers. If the polymer freezes too soon, poor penetration into the paper may result in poor adhesion [61]. The effect of processing on adhesion is further explored in Chapter 15.

Further reading on exrtrusion coating and lamination

1. Bezigian, T. (ed.), 1999, *Extrusion coating manual, 4th ed.*, TAPPI Press, Atlanta.
2. Durling, W., ed., *Extrusion Coating Manual, 5th ed.*, TAPPI Press, Atlanta, Ga.
3. Cooper, J., 2001, "Extrusion Coating," in Vlachopoulos, J. and Wagner, J. R. (eds.), The SPE guide on extrusion technology and troubleshooting, The Society of Plastics Engineers, Brookfield, CT, pp. 13-1-13-16.
4. Gregory, B. H., 2005, *Extrusion coating: A process manual*, Trafford Publishing, Victoria, B.C.

2.3.2 Adhesive lamination

Adhesive lamination is a method for bonding dissimilar substrates together in a continuous roll-to-roll process. The technique involves applying a liquid adhesive to one or both substrates followed by subsequent drying or curing by heat, UV light, moisture or e-beam radiation. In dry lamination (Fig. 2.25), the adhesive is applied to one substrate and dried prior to the second substrate being introduced. The dried adhesive is activated by a heated nip roll. In wet lamination, the adhesive is applied to one or both substrates just prior to the combining of the two substrates via a nip. This is followed by drying. At least one of the substrates must be porous to allow the water or solvent to evaporate.

The adhesive may be solvent-based, water-based, or "solventless." Solvent and water-based adhesives require drying whereas solventless adhesives are low viscosity liquids such as polyurethanes or isocyanates that are applied to one or both substrates and subsequently cured. The chemistry of laminating adhesives is described in more detail in Section 10.5.

2.3.2.1 Key components

2.3.2.1.1 Unwind and surface preparation

The substrates are unwound and typically pretreated with flame, corona, ozone or plasma to clean and prepare the surface for the application of the adhesive. The substrate generally needs surface energy that is 10 mN/m greater than the adhesive to achieve good wetting. The treatment removes surface oils and contaminants as well as increases the surface energy.

2.3.2.1.2 Adhesive application

The adhesive is typically applied via a gravure cylinder or other coating method (see Section 2.3.4). This involves a secondary cylinder that picks up the adhesive from a bath and transfers it to the applicator roll.

(A)

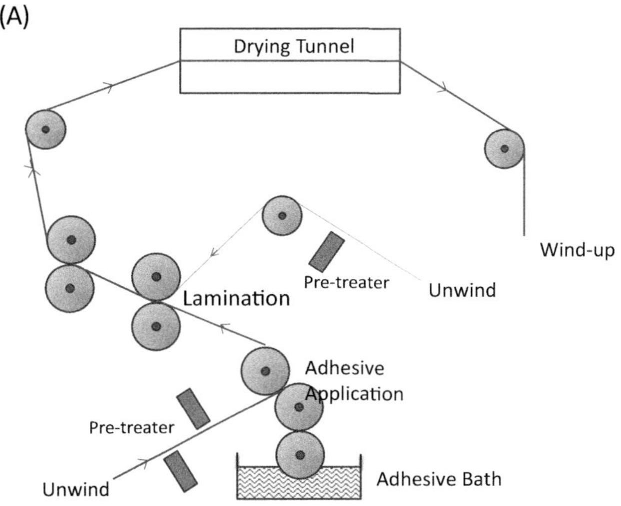

(B)

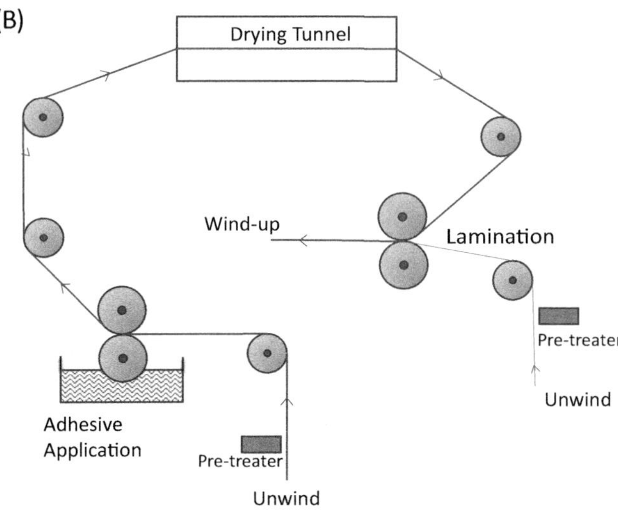

FIGURE 2.25 Adhesive lamination processes: A) wet lamination process B) dry lamination process.

2.3.2.1.3 Curing or drying

For solvent and water-based adhesives, drying is usually accomplished with heaters (such as IR heaters) or hot airflow. For cured adhesives, the curing method and time depend on the functional chemistry of the adhesive. For example, UV adhesives are cured by UV lamps.

2.3.2.1.4 Wind-up

Proper tension control is important during lamination to prevent curl, wrinkling, dishing, telescoping and other potential problems. See Section 2.2.1.1 for more details on winding.

2.3.2.2 Key factors affecting properties

Substrate surface chemistry, adhesive chemistry, coat weight, curing conditions and tension control are critical parameters affecting mechanical and processing performance. The adhesive must wet the surface of the substrate as a prerequisite to bonding. Contaminants on the substrate surface or a poor match in surface energies with the adhesive can contribute to poor wettability. Adhesion fundamentals are explored in Chapter 10.

The coat weight is a function of the adhesive viscosity, line speed and solids content. Coat weight can affect peel strength and other properties such as barrier performance.

Drying involves evaporation of the solvent or water and additionally curing with some adhesive chemistries. Drying must be accomplished within the residence time of the drying tunnel; drying time is a function of the evaporation rate, the solids content of the adhesive, line speed and drying temperature. The thermal stability of the substrate, however, needs to be considered as a limitation. Oriented films, for example, sag, distort or shrink when they get too hot so care must be taken to not exceed the temperature limit. Incomplete curing can leave unreacted monomers or oligomers that can migrate out of the adhesive and affect sealing or other film properties.

Tension control is important during lamination to prevent stresses from building up, which can lead to delamination. Proper tension also helps avoid problems such as telescoping that contribute to defective rolls.

2.3.3 Comparison of extrusion and adhesive lamination

The selection of extrusion or adhesive lamination to make a packaging film structure often comes down to what equipment the converter has in place. Table 2.5 summarizes some of the differences. Extrusion coating generally uses less expensive materials but takes more time and material to setup the process. Thus adhesive lamination is often favored for short runs. Adhesive lamination is also favored in situations where its curing chemistry brings advantage, such as temperature resistance for retort applications [62–64].

2.3.4 Liquid coating

Low viscosity coatings that are solvent, aqueous or solventless based are used in packaging for a variety of reasons, including for providing adhesion or barrier, or for the modification of surface properties such as abuse resistance, anti-stick (antiblock) or release, heat resistance and lubricity. Following application, they are dried or cured. Examples include lacquers, barrier coatings, hot melt adhesive coatings, heat seal coatings and cold seal coatings.

2.3.4.1 Key components

The typical components found on a coating line are shown in Fig. 2.26. Modern coating lines often are designed with multiple coating stations to coat both sides of the substrate or to lay down multiple layers. To add operational flexibility the coating stations may be modular so they can be swapped out with different coating processes depending on the coating needs.

2.3.4.1.1 Unwinder

Like other roll-to-roll processes already discussed, unwinding and winding is an important part of any coating line. Proper tension control is critical.

TABLE 2.5 Comparison of advantages of extrusion and adhesive lamination.

Extrusion lamination	Adhesive lamination
No solvents	Short job changes
Lamination of rough or porous substrates possible (e.g., paper)	Potentially lower energy use
Good flexibility of laminate	Very cost-effective at $<5000\ m^2$
Good puncture resistance	Lower capital investment
Good environmental stress crack resistance	Wider opportunity for high bond strength
Cost-effective at long ($>5000\ m^2$) runs	No heat deformation of substrates
Immediately ready to convert into packages (no cure time)	
No residual monomers	
Low cost in resin handling and disposal	

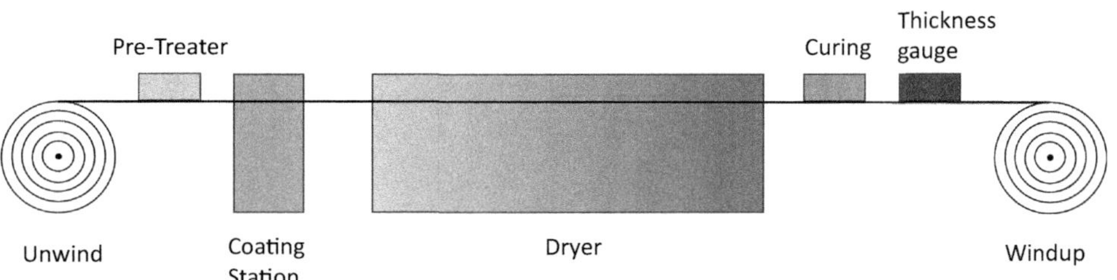

FIGURE 2.26 Key components of liquid coating process.

2.3.4.1.2 Substrate preparation (pretreater)

The substrate (plastic film, paper, etc.) may be treated prior to coating to remove contaminants and increase the surface energy. Flame, corona or plasma is most often used. See Section 2.3.1.1 for details.

2.3.4.1.3 Coater

There are many types of coaters used today. As described by Gutoff and Cohen [65], they fall into four categories based on the method used to control the coating weight or thickness:

- Self-metered;
- Doctored;
- Premetered;
- Hybrid.

In the self-metered method, the coating bead or meniscus conditions control the coating weight. There are two types of self-metering methods. In one type, the solution viscosity and line speed control the solution volume that is applied; increasing these proportionally increases coating weight. Dip, meniscus and forward roll coating are examples of this method. In the second type, additional rolls are added that create a nip. The coating bead is split in the nip with part of the solution remaining on the substrate and the rest returning to the coating pan. A reverse roll coater is an example of this method.

The doctored method involves a two-part process. The coating solution is applied to the substrate by any suitable method. A doctoring device, such as a rod, knife or blade, is used to remove excess solution to control the final coating weight.

In the premetered method, all the coating solution is applied to the substrate—there is no recirculation. The flow rate is controlled by a metering pump or extruder and the thickness is simply the flow rate divided by the substrate line speed. This method provides coatings with good uniformity (typically $\pm 2\%$), coverage that is independent of viscosity, and multilayer coating capability. Elimination of recirculation additionally aids in preventing bubbles and contamination.

The hybrid methods control coating weight by combining more than one method. For example, in the gravure method, an engraved cylinder picks up the solution and any excess solution is removed by a doctoring blade. The solution transfers from the engraved cells of the cylinder to the substrate. Coating weight is primarily controlled by the engraved cell volume and the fraction transferred, and secondarily by the doctoring blade design and pressure of the gravure cylinder against the substrate and impression roll. Table 2.6 provides a summary of many coating methods in use today.

2.3.4.1.4 Drying or curing

Following the coater, the solution is either dried (aqueous and solvent-based coatings) or cured. Hot air, IR heaters, hot rolls, or some combination may be used. UV, e-beam, heat and in some cases, moisture are used to cure some coatings, depending on the chemistry.

2.3.4.1.5 Online measurement

The coating weight is often measured online for quality control. Several methods are described in Section 2.2.1.1.

TABLE 2.6 Description of coating methods.

Name	Type	Description	Thickness determined by	Operating parameters				Historical use
				Viscosity (mPa-s)	Wet thickness (μm)	Line Speed (m/s)	Coverage uniformity (%)	
Dip	Self-Metered	Web dipped into coating solution	• Fluid properties • Coating speed • Withdrawal angle	20–2000	10–200	0.5–7.5	± 10	Continuous, irregular shaped objects such as fibers
Meniscus or Bead-Roll	Self-Metered	Web contacts solution in pan and transfers by surface tension	• Viscosity • Solution surface tension • Coating speed	• Low		0.1–0.2		Adhesives and photographic coatings
Kiss	Self-Metered	Unsupported web contacts with an applicator roll	• Viscosity • Kiss roll speed • Coating speed	50–1000	5–75	0.5–5.5	± 10	Applying excess coating prior to metering device
Forward Roll	Self-Metered	Web passes between applicator roll and backup roll each rotating in the direction of the web		20–2000	10–200	0.5–7.5	± 10	Not used much due to poorer quality and operating than reverse roll.
Reverse Roll	Self-Metered	Web passes between applicator roll rotating in opposite direction and backup roll. A reverse turning metering roll removes excess fluid from the applicator roll. Fluid remaining on the applicator roll transfers to the web.	• Metering and applicator roll gap • Metering roll to applicator roll speed ratio • Applicator roll speed capillary number	200–50000	14–450	0.1–8.5	± 2–10	Widely used, versatile method able to handle low and high viscosity fluids over a wide coating speed range.
Mayer Rod or Wire-Wound Rod	Doctored	A wire-wound rod removes excess fluid from the moving substrate	• Fluid rheology • Substrate speed • Web tension • Rotation direction	50–1000	4–80	0.05–5	± 7–10	Low solids, low viscosity fluids
Air knife (metering mode)	Doctored	Web is dipped into fluid against a backup roll and excess coating is removed by a focused air jet	• Coating line speed • Viscosity • Air knife characteristics	1–50	1–200	0.2–2	± 5	Versatile and once widely used but newer methods are more precise and less expensive. Photographic coatings

TABLE 2.6 (Continued)

Name	Type	Description	Thickness determined by	Operating parameters				Historical use
				Viscosity (mPa-s)	Wet thickness (μm)	Line Speed (m/s)	Coverage uniformity (%)	
Air Knife (Squeegee mode)	Doctored	Air knife operates at high pressure and coating speeds. Most of the coating is blown off the web.	Used for porous webs such as paper. Coating is absorbed into the pores.	5–500	Depends on pore depth	0.6–10	± 5	Paper coatings
Knife	Doctored	Two-part applicator where the coating is applied to by any of many methods and a stationary knife wipes off the excess fluid.	• Knife configuration • Knife pressure • Gap between knife and substrate • Hydrodynamic forces • Tends to level rough surfaces rather than create uniform coating	50–40000	10–750	1.7–25	± 10	Simple, easy to operate and maintain.
Blade	Doctored	Similar in configuration to knife coating. Knives are rigid and thick, blades thin and may be flexible	• Blade thickness • Blade angle • Force pushing blade against the substrate • Tends to level rough surfaces rather than create uniform coating	50–40000	10–750	1.7–25	± 10	Pigmented coatings
Slot Die	Premetered	The coating fluid is pumped at a controlled rate to a coating die positioned close to the substrate. Fluid exits the die lips and forms a bead on the surface of the substrate.	• Flow rate per unit width divided by coating speed • Uniformity is determined by the die design	15–20000	10–250	0.02–8.5	± 2	Allows simultaneous coating of multiple layers
Extrusion	Premetered	Like a slot die but a high viscosity fluid is pumped via an extruder. Die may have internal manifolds to control flow uniformity.	• Extruder speed • Line speed	50000–300000	13–525	0.6–9	± 5	Often used in food packaging. Allows coextrusion of multiple layers.
Slide	Premetered	A series of rectangular plates with flow distribution channels is	• Flow rate per unit width	1–500	25–250	0.1–5	± 2	Photographic films and paper

(Continued)

TABLE 2.6 (Continued)

Name	Type	Description	Thickness determined by	Operating parameters				Historical use
				Viscosity (mPa-s)	Wet thickness (μm)	Line Speed (m/s)	Coverage uniformity (%)	
		used to simultaneously coat multiple layers.	divided by coating speed					
Curtain (standard)	Premetered	A weir or die directs a curtain of solution flow onto the substrate. Curtain is wider than the substrate and excess is returned to the coating reservoir.	• Not strictly premetered since flow overcoats edge of substrate • Need high flow (and therefore line speed) to get stable curtain	150–2000	25–250	1.5–6.5	± 5	Widely used for coating adhesives, paints, food products (e.g., chocolate) and electronic coatings
Curtain (precision)	Premetered	Curtain is narrower than the substrate. Multilayer coating is possible and neck-in is controlled by edge rods.	• Flow rate per unit width divided by coating speed • Need high flow (and therefore line speed) to get stable curtain	5–500	5–500	1.5–15	± 2	Originally for photographic coatings, now pressure sensitive adhesives
Gravure (direct)	Hybrid	A gravure cylinder engraved with small patterns picks up the fluid from a reservoir pan. A doctor blade removes excess fluid. Transfer to the substrate occurs upon contact with gravure roll, typically backed by an impression roll.	• Coating configuration (e.g., forward, reverse, offset, chamber) • Pressure • Gravure pattern	1–500	3–65	0.1–12	± 2	Long runs and thin coatings with low viscosity fluids
Microgravure	Hybrid	A gravure process using small diameter gravure roll and no impression roll. Coating is laid down by kiss coating.	• Gravure pattern	1–4000	8–80	0.005–1.7	± 2	Low coat weight on thin gauge substrates. Imaging, electronics, packaging, batteries and other specialty applications

Source: Data from E.B. Gutoff, E.D. Cohen, Water- and solvent-based coating technology, in J.R. Wagner Jr. (Ed.), Multilayer Flexible Packaging, 2nd ed.. Elsevier, Cambridge, MA, 2016, pp. 205–234.

2.3.4.1.6 Wind-up

As with any roll-to-roll process, proper tension control during wind-up is critical for preventing roll problems such as telescoping and blocking. See Section 2.2.1.1 for more details on winding.

Further reading on liquid coating

1. Gutoff, E. B. and Cohen, E. D., 2016, "Water- and solvent-based coating technology," in Wagner, Jr., J. R., *Multilayer flexible packaging, 2nd ed.*, Elsevier, Cambridge, MA, pp. 205-234.
2. Gutoff, E. B. and Cohen, E. D., 2006, *Coating and drying defects. Troubleshooting operating problems, 2nd ed.*, John Wiley & Sons, Inc., Hoboken, NJ.
3. Gutoff, E. B. and Cohen, E. D., 1992, *Modern coating and drying technology*, VCH Publishers, New York.

2.3.5 Comparison of liquid and extrusion coating

Like lamination, selecting between extrusion and liquid coating often comes down to what equipment the converter has. In general, liquid coating has advantages for shorter runs and has greater flexibility in the chemistries used. Extrusion coating allows thicker coating weight, which can be important for coating porous substrates like paper to achieve a continuous coating. On the other hand, thin coatings may be desired for ease of recycling or repulping. The advantages of each technology are summarized in Table 2.7.

2.4 Orientation

Biaxial orientated polypropylene (BOPP), biaxial orientated polyester (BOPET), and biaxial orientated polyamide (BOPA) are often used as substrates in multilayer laminations or as stand-alone films. Biaxially oriented polyethylene (BOPE) is starting to make inroads—first in Asia in 2017, then in Europe and the rest of the world. This is driven by the call for more sustainable and recyclable (in the existing PE recycling stream) structures for flexible packaging (see Chapter 17). With this, OPE is increasingly replacing OPET, OPA, and OPP for printing and barrier substrates in flexible packaging. Oriented polylactic acid (OPLA) and polystyrene (OPS) films have also found use in packaging. Orientated coextruded films (EVA-PVDC-EVA) are used as shrink bags for primal meat packaging. Machine direction oriented (MDO) films are used in strapping and tapes and more recently in hygienic films and synthetic papers [66]. Coextruded MDO structures are starting to find wider use although challenges remain [66,67]. MDO films based on PE are complementing BOPE films for flexible packaging redesign. Transverse oriented films are used as shrink sleeves.

TABLE 2.7 Comparison of advantages of extrusion and liquid coating.

Extrusion coating	Liquid coating
No solvents	Short job changes
Continuous coating of rough or porous substrates possible (e.g., paper)	Potentially lower energy use
Thicker coatings possible (improved caulking for heat sealing)	Very cost-effective at $<5000\ m^2$
Barrier and other functionality can be augmented by coextrusion	Lower capital investment
Improves flex crack resistance of substrates like aluminum foil	Wider selection of chemistries or functionality possible
Cost-effective at long ($>5000\ m^2$) runs	Wide selection of coating technologies
Immediately ready to convert into packages (no cure time)	No heat deformation of substrates
No residual monomers	Can be patterned so that full surface is not coated (e.g., coating just in the seal area)
Low cost in resin handling and disposal	

The effect of orientation on properties and applications are explored in more detail in Chapter 11. In general, orienting a film increases tensile strength and barrier properties while decreasing haze and elongation at break. These films are flatter and thinner than conventional films.

There are three main types of commercial processes for producing oriented film: MDO; tenter frame for transverse, machine or biaxial orientation; and double or triple bubble for biaxial orientation. These are depicted schematically in Figs. 2.27–2.29.

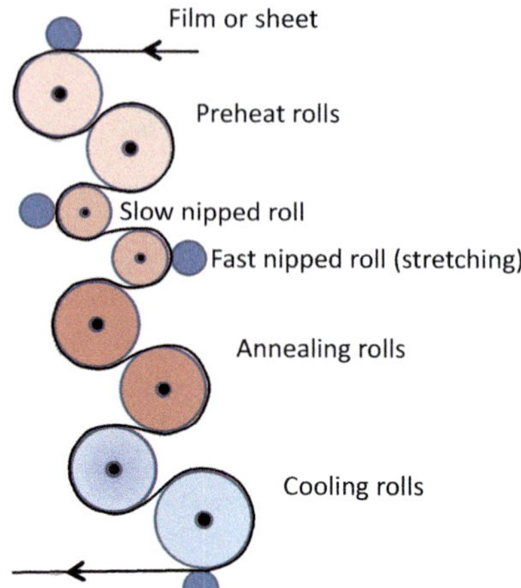

FIGURE 2.27 The machine orientation process.

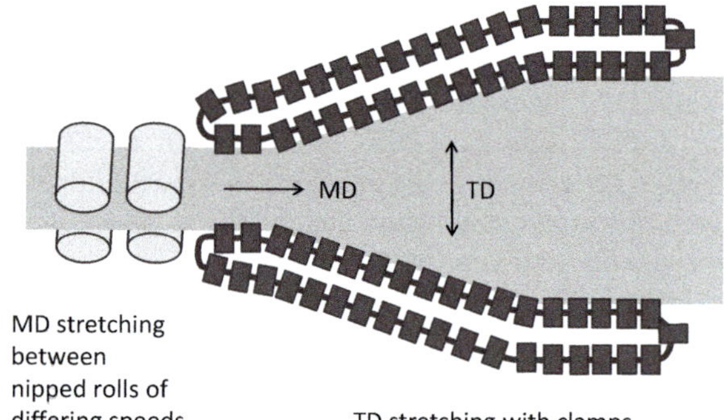

FIGURE 2.28 The tenter frame process.

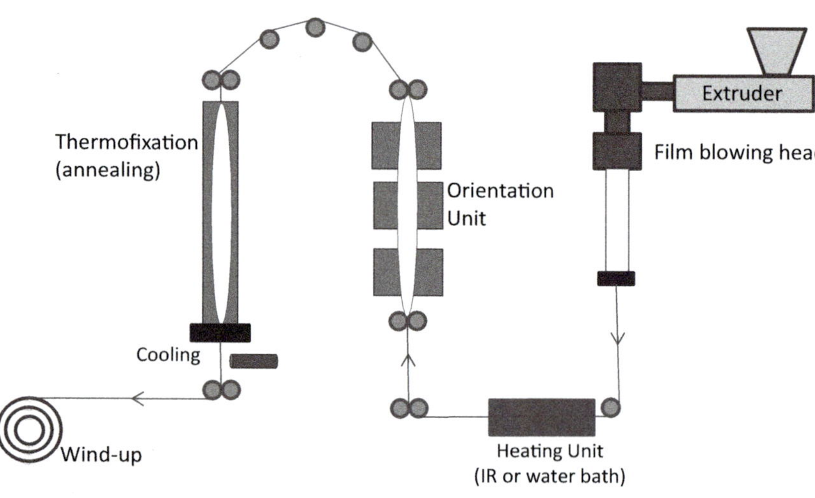

FIGURE 2.29 The double bubble or triple bubble process.

2.4.1 Machine direction orientation

MDO machines consist of a set of heated rolls, often 12 or more in series. The first few rolls preheat the film to the stretching temperature. This feeds into a set of two nipped rolls operating at different speeds. The differential speed determines the amount of stretching or orientation. This is followed by heated annealing rolls to lock in the orientation. Finally, the film is cooled to room temperature on a set of cooling rolls [66].

The orientation temperature, draw ratio (ratio of the speeds of the nipped rolls), rate of stretching (difference in nipped roll speeds divided by the distance between the nipped rolls) and annealing temperature are all important parameters influencing the final properties. Getting these right for a given polymer is often the challenge in making a good film from this process. Splitting and brittleness are two potential outcomes if not done correctly [67].

In addition to being used as a stand-alone process, MDO lines are sometimes incorporated in-line with a blown or cast film line. In addition, the machine direction orientation process is typically used as the first step of a sequential tenter frame biaxial orientation process.

2.4.2 Tenter frame

In a sequential tenter frame process, the polymer sheet is stretched in the machine and transverse directions in separate serial stages. Typically the sheet is stretched in the machine direction first, often using heated rolls operating at different speeds as described for the MDO process. The film then goes into the transverse stretching oven where heaters raise the temperature of the sheet to its transverse stretching temperature. Clamps are attached to the side of the film that are on a track that accelerates the sheet in the transverse direction. Following orientation, the film undergoes a thermofixation or annealing step to improve dimensional stability where it is exposed to a temperature that is typically above the orientation temperature. The film remains clamped during annealing in order to prevent shrinkage and to lock the orientation in place. Polymers have a memory and subsequent heating of the film below the thermofixation temperature will preserve the orientation (in a shrink film, the film is heated above its thermofixation temperature to initiate the shrinkage).

The stretching in the machine and transverse directions may also be done simultaneously in order to achieve a more balanced orientation in the two directions. Some polymers are resistant to stretching in series but still can be stretched in a simultaneous process. Key variables affecting performance include the stretching temperature, annealing temperature, the draw ratio in each direction, the rate of stretching and for biaxial orientation, the sequence of the stretching (sequential or simultaneous).

2.4.3 Double and triple bubble

In the double bubble process, the polymer is first extruded as a tube into a water bath. The water rapidly quenches the polymer, reducing its crystallinity compared to conventional air-quench blown film. The lower level of crystallinity helps in the subsequent orientation process.

Following quenching, the tube is reheated to the orientation temperature using a hot water bath or IR heater. This two-step process of quenching and reheating the tube helps control the crystallinity as well as produces a more uniform temperature for orientation. The tube is expanded by blowing air inside the tube, similar to a conventional blown film process but from the rear of the process. The difference is that in the double bubble process the tube is oriented at lower temperatures so that the polymer is in a semisolid state, producing higher stresses during orientation. Stretch ratios are also typically much higher than in the conventional blown film process. As a result, double bubble films have more orientation than conventional blown films. Melt strength is also not a factor as it is in the blown film process.

The oriented film may undergo a thermofixation step in a third bubble, where the tube is reheated to the annealing temperature. This helps control shrinkage in a shrink film or improve dimensional stability in a nonshrink film.

Important factors influencing performance include the orientation and thermofixation temperatures, the BUR (transverse orientation stretch ratio) and the TUR (machine direction stretch ratio).

The double bubble process produces a film with more uniform properties in the machine and transverse directions than a sequential tenter frame process. Tenter frame processes have a much higher output than double bubble processes and are the predominant method used to make BOPP, BOPET and BOPE film. Double bubble processes are often used to make shrink film/bags and sausage casings. More recently they have found use in the production of coextruded barrier films [68,69]. The tenter frame process is also capable of making multilayer barrier films.

The technology of film orientation is taken up in more detail in Chapter 11.

Further reading on film orientation

1. DeMeuse, M. T., 2011, *Biaxial stretching of film, Principles and applications*, Woodhead Publishing, Oxford.
2. Breil, J., 2013, "Biaxially oriented films for packaging applications," in Ebnesajjad, S., ed., *Plastic films in food packaging, Materials, technology and applications*, Elsevier, Amsterdam, pp. 53-70.
3. Hatfield, E., 2010, "Machine direction oriented film technology," in Wagner, J. (ed.), *Multilayer flexible packaging, technology and applications for the food, personal care and over-the-counter pharmaceutical industries*, Elsevier, Oxford, pp. 113-118.

2.5 Metallization

Vacuum metallization is a physical vapor deposition process where thin coatings are deposited onto a substrate by thermal evaporation and condensation of metals and ceramics. Vacuum metallization is used to deposit a thin (typically 3–40 nm) coating of aluminum onto a plastic film, paper, or other substrate to provide decoration (graphics) or barrier to light, moisture and gases. Modern metallizers are roll-to-roll processes that can run at speeds up to 1000 m/minute with widths of 4.45 m [70]. Oriented polyester and polypropylene are the most often used substrate films, but polyamide (nylon), polyethylene and paper are also metallized.

Similar technology is used to deposit ceramic or inorganic coatings of aluminum oxide (AlO*x*) or silicon oxide (SiO*x*) for producing transparent coatings that provide moisture and gas barrier. ("*x*" is used here since the exact stoichiometry is generally unknown or variable.) Table 2.8 summarizes several technologies commercially used to produce these coatings. Packages made from these transparent films pass through traditional metal detectors, allow the packaged product to be seen, are microwaveable, and are easier to sort in recycle collection streams than metallized films which often get rejected by metal detectors at recycling facilities. The cost of inorganic coatings continues to be higher than metallized films.

2.5.1 Key components

There are many different technologies and configurations used for metallization but the basic design is shown in Fig. 2.30. The film is placed into a vacuum chamber where the aluminum (or ceramic) is vaporized and then condensed onto the film surface.

2.5.1.1 Vacuum chamber

The deposition process takes place inside a large vacuum chamber. Vacuum is used to prevent the metal or inorganic vapors from colliding with any gas prior to depositing onto the substrate.

2.5.1.2 Heating source and evaporant vessel

In the most common system used for metallizing packaging films, aluminum wire is fed to a "boat" which heats the metal above its melting point using a resistive heating source. Inductive heating using an externally applied magnetic field or direct heating using an electron beam are also sometimes used. The pros and cons of these systems are summarized in Table 2.9.

TABLE 2.8 Inorganic or ceramic coating technologies [71].

Process	Advantages	Disadvantages
Reactive evaporation of AlOx	• Low material cost • No yellowing • High productivity • Low investment	• Process difficult to control • Layer sensitive to elongation • Medium barrier properties
Plasma chemical vapor deposition (SiOx or hydrocarbons)	• Low material costs • No yellowing	• High investment • Low productivity
Electron beam evaporation of SiOx	• Low material costs • No yellowing • High productivity • Good barrier • High mechanical resistance	• High investment

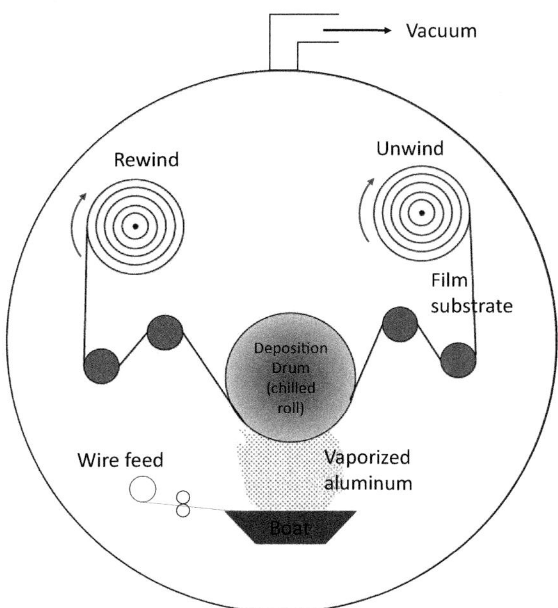

FIGURE 2.30 Vacuum metallization process. *Adapted from C.A. Bishop, E.M. Mount III, Vacuum metallizing for flexible packaging, in J.R. Wagner, Jr. (Ed.), Multilayer Flexible Packaging, 2nd ed., Elsevier, Cambridge, MA, 2016, pp. 235–255.*

TABLE 2.9 Heating sources for vacuum deposition processes.

Heating process	Typically used for	Pros	Cons
Resistive heating	Metallization for packaging films	• Continuously wire fed	• Spitting may produce uneven coating if the heating is nonuniform
Induction heating	Inorganic (SiOx)	• Constant evaporation rate, good thickness control • No splashes/spitting • High yield, low running cost	• Thought to be difficult to operate • Can not be wire fed • Poor economics
Electron beam	Plasma activated evaporation of Al; plasma enhanced chemical vapor deposition	• Simpler process than resistive • Can directly evaporate ceramics (e.g., silica to produce SiOx coating) • Versatile configurations	• Higher cost

Source: Based on T. Glaw, Transparent inorganic barrier films, TAPPI European PLACE Conference, 2007.

In reactive evaporative deposition of AlO*x*, the AlO*x* is produced in the process. Aluminum is thermally heated like in the metallization process and oxygen is injected into the aluminum vapor prior to its deposition onto the film substrate. Plasma may be introduced to assist the process by energizing the AlO*x* molecules for better adhesion to the substrate.

2.5.1.3 Pre and posttreatment

The substrate may be treated prior to deposition to clean the surface of low molecular weight contaminants and increase the surface energy. Flame, corona, or plasma (atmospheric and vacuum) are most often used.

2.5.1.4 Deposition roll and substrate handling

The vaporized aluminum or inorganic molecules condense onto the colder substrate. In most newer machines the substrate film is supported against a cold roll, as shown in Fig. 2.30, for improved productivity. Unwinders, winders and tension control rolls are critical for controlling the tension and dimensional stability of the film during deposition.

2.5.2 Key factors affecting properties

The critical quality properties of metallized films depend on the application but usually involve appearance, optical density, metal adhesion to the substrate film, and barrier properties. Metallized films are typically used as components of packaging structures and are further converted in lamination (extrusion or adhesive) or coating processes. The durability of the metallized coating through these processes is often a key consideration. Some of the sources of poor packaging conversion yield and rate include scratches, lanes and metal pick-off affecting appearance; wrinkles, hard wind and telescoping in the rolls during roll formation; poor adhesion; damage to the frail coating during subsequent converting processes that reduce barrier performance; and film property variation [72].

Several factors affect the quality of the coating and the performance of the film. These include substrate film quality, the evaporation and deposition process, and web handling during metallization and through subsequent converting processes nullN.

2.5.2.1 Substrate quality

A clean, high-energy surface with little or no dimensional distortion during the deposition process is critical for good coating performance. The substrate type influences dimensional stability and strength. The high modulus of oriented polyethylene terephthalate (OPET) and oriented polypropylene (OPP) films contribute to their predominant use as substrate films for metallized films for packaging. Other substrate factors include film gauge uniformity and the use of antiblock to control unwinding. Antiblock additives are inorganic particles such as silica or talc that create a bumpy film surface that limit contact with other surfaces (see Chapter 12). Surface topography can affect coating quality. A rough surface may also scratch and damage the coating during subsequent handling in converting processes.

The surface chemistry or surface energy of the substrate influences metal adhesion and the nucleation and growth of crystals are important for barrier performance. High surface energy favors improved wetting and denser coatings with tighter grain boundaries and better barriers. OPET has high surface energy favorable for metallization and generally provides better barrier performance than polypropylene.

The surface of polymer films is often treated with plasma or other means. Plasma creates new polymer chain ends through chain scission at the surface that enhances adhesion. However, too much treatment causes low molecular weight chains to form that can be detrimental to adhesion. The treatment also removes contaminants such as oligomers, slip agents and other additives that may be at the surface. The treatment level should be optimized to provide good cleaning of the surface without too much chain scission [73]. Plasma pretreatment of OPP is thought to enhance nucleation and growth of an AlO_x coating by incorporating functional groups on the surface of the film [70].

Several investigators have explored coextruding polypropylene (PP) with a skin layer that favors higher adhesion of the metallization to the film substrate and better barrier performance. Some of the materials considered for the skin layer include propylene-ethylene copolymers (with 3.5% ethylene), propylene-butene copolymers, ethylene-propylene-butene terpolymers, maleic anhydride grafted PP, high-density polyethylene, linear low-density polyethylene, ethylene vinyl alcohol, amorphous polyamide copolyesters and sulfonated copolyester [74–79].

Surface imperfections on the film can cause defects in the coating. Debris on the surface can be coated and later flake off leaving behind an uncoated area. Struller et al. [70] report that volatiles can come off during thermofixation of oriented PP that cause divets on the surface that result in defects during deposition. Surface contamination from fillers, additives or residual products from the polymerization process can form a weak boundary layer and poor adhesion. In-line treatment of the film helps clean the surface of contaminants. Vacuum plasma treatment is well suited since it removes low molecular weight species [72].

2.5.2.2 Evaporation and deposition

Several elements make up this critical part of the process which influences the uniformity of the coating:

- The vacuum level may change within the production cycle and should be leveled out prior to deposition. Consistency of the vacuum level is needed for uniform deposition of the metal atoms without prior collisions for reliable adhesion and barrier performance.
- The heating source (see Table 2.9) and boat conditions affect the quality of the coating. Many metallizer machines require sophisticated control systems and multiple "boats" to achieve a uniform coating across the width of the film. A nonuniform evaporation rate can result in "spitting" or "splashing" of the molten aluminum in the boat that results in defects in the coating [80].

- Mount [81] studies the effect of metallizer line speed and deposition roll temperature on the adhesion and barrier properties of metallized OPP after it was converted into a three-layer laminate: OPET-met/LDPE/OPP. He measures several properties of the laminate including oxygen barrier, moisture barrier, metal pick-off, green bond to LDPE and aged bond to LDPE. Only the aged bond strength to LDPE is significantly impacted by the metallizer process parameters, with substantial interaction between line speed and deposition roll temperature. At low line speed, increasing deposition roll temperature decreases aged bond strength. The opposite is true for high line speed. Mount postulates that at high line speed, the film surface is exposed to the aluminum vapors for a shorter amount of time, and the nucleation and crystallization of the coating must be faster to increase the bond strength.
- For standard aluminum coatings, a low amount of oxide will form between the metallization layer and substrate and a larger layer on the surface of the metallization (growing up to 3–4.5 nm in thickness over a year) due to oxygen in the environment. In reactive thermal vaporization of aluminum with oxygen to produce AlOx coatings, the AlO*x* stoichiometry depends on the feed ratio of O_2 to Al. An oxygen rich or lean coating can be formed. If the O_2 to Al ratio is too high, the structure becomes porous and barrier performance is lost. If the ratio is too low, the coating becomes grayish and nontransparent [82].
- Plasma or other chemical alterations of the metal or oxide vapor may improve adhesion and other performance parameters. A hybrid physical and chemical coating process has been developed for aluminum coatings that promote the adhesion of the first aluminum particles to contact the substrate through chelation [83]. Trassl et al. [84] report that a plasma-assisted reactive evaporation process for AlO*x* coatings has a much wider process window and requires a less demanding control system for stable coatings over wide widths than a standard reactive evaporation process. It also creates a higher density coating for high barrier performance with improved mechanical stability during and after the metallization process.
- Struller et al. [70]. find that pre and posttreatment of AlO*x* coated OPP reduces oxygen and moisture vapor permeability. Pretreatment is thought to clean the surface and nucleate crystallization, as described above. Posttreatment is thought to help densify the AlO*x* coating.

2.5.2.3 Web handling

Tension control and roll hardness are important, both in the deposition process and for the handling of the coated film afterwards. Tension should be controlled so that the film elongates no more than 1% for winding; tension affects wrinkling, width changes (elongation in the machine direction creates dimensional changes in the transverse direction) and roll hardness (through recoverable strain) [61]. The judicious use of antiblock and slip additives can help with preventing damage to the coating during unwinding. Damage can occur during the subsequent converting process. Metallized films are typically less susceptible to damage than ceramic coatings. Organic coating technology has been developed that can be deposited via vacuum deposition on a second roll within the vacuum chamber that helps protect AlOx coatings. This technology is reported to [84–86]:

- Improve oxygen barrier of Al or AlO*x* coating by about 2x,
- Protect the barrier during winding and other converting operations,
- Allow the possibility of printing on the AlO*x* coating (with a top coat).

Metal or ceramic coatings theoretically have very high barrier properties but in the end-use, the performance is always less than theoretical. This is due to defects from the metallization and converting processes causing pinholes in the coating, as well as pinholes formed from flexing the film in use (see Chapter 8). The more the metallized film is handled, the greater the number of pinholes. The use of protective coatings and burying the metallization layer within laminations help mitigate the creation of pinholes and improve flex crack resistance. Converting methods can also impact performance. For example, Migliorini et al. [75]. describe how the metal coating is more prone to crazing and cracking from heat introduced during extrusion lamination if the metallized film is introduced on the secondary unwind. In this configuration, the film is in contact with the rubber nip roll rather than the cold chill roll when the molten polymer contacts the substrate. This allows the heat to distort the underlying base film which crazes the coating.

Further reading on metallization processes

1. Bishop, C. A. and Mount, III, E. M., 2016, "Vacuum metallizing for flexible packaging," in Wagner, Jr., J. R., ed., *Mutlilayer Flexible Packaging*, 2nd *ed.*, Elsevier, Cambridge, MA, pp. 235-255.
2. Bakish, R., 2009, "Metallizing, Vacuum," in Yam, K. L,2009, *Wiley Encyclopedia of Packaging Technology (3rd Edition)*, John Wiley & Sons. p. 742-750.

3. Bishop, C. A., 2015, *Vacuum Deposition onto Webs, Films and Foils (3rd Edition)*. Elsevier.
4. Ataya, V., 2020, "Metallizing," in Macnamra, Jr., J. F., ed., 2020, *Film extrusion manual, 3rd ed.*, TAPPI Press, Atlanta.
5. Bishop, C. A. and Mount III, E. M., 2012, *Metallizing Technical Reference*, The Association of International Metallizers, Coaters and Laminators.

2.6 Printing

There are two general types of a high-speed roll-to-roll printing used in packaging: gravure and flexographic. Historically Europe favored gravure and the US flexographic, but both have gained popularity across regions.

2.6.1 Gravure printing

In gravure printing (Fig. 2.31), an engraved metal roll is used to directly transfer the ink to the film substrate. The depressions in the engraved surface hold the ink and the ridges or un-engraved portions of the roll are the nonprinted areas. The roll is immersed in a bath of ink where the etched regions pick-up the ink. A blade removes the excess ink prior to contact with the film. An impression cylinder on the backside of the film forces contact of the film with the depressed, ink-filled regions of the gravure roll, transferring the ink to the film. The film is passed through a dryer to drive off the solvent or water. The gravure press often contains many printing stations in-line, each for a different color of ink, or for applying cold seal or barrier emulsions.

The gravure rolls are chemically, mechanically or laser etched to form the pattern of small cells of depressions that hold the ink. The rolls are hard and long-lasting. One of the advantages of this process is that the rolls are durable, allowing for accurate reproduction of a design over many prints at high speeds. The initial costs are high to produce the print rolls but for long runs the gravure process is quite cost-effective.

2.6.2 Flexographic printing

In flexographic printing (Fig. 2.32), the pattern is made into plates, typically from a photopolymer. The plates are attached to a roll. Ink is transferred from a bath to an anilox roll, which contains millions of tiny divots to take up the ink. The anilox roll then contacts the printing roll, transferring the ink to the surface of the plates. The printing roll then

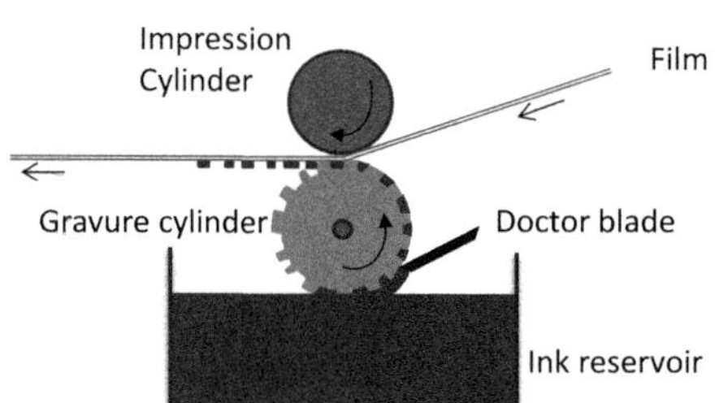

FIGURE 2.31 The gravure printing process [87]. *Based on Lentz, J., Printing, in: Bakker, M. (ed.), The Wiley Encyclopedia of Packaging Technology, John Wiley & Sons, New York, 1986, pp.554–561.*

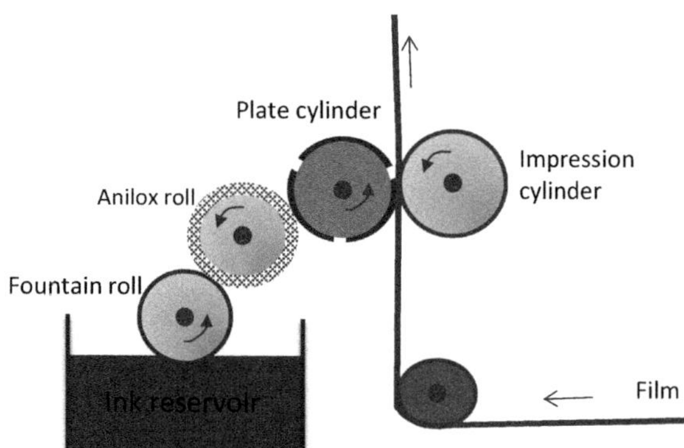

FIGURE 2.32 The flexographic printing process [87]. *Based on Lentz, J., Printing, in: Bakker, M. (ed.), The Wiley Encyclopedia of Packaging Technology, John Wiley & Sons, New York, 1986, pp.554–561*

contacts the surface of the film. While gravure printing relies on impressing the film into the cavities of the roll where the ink resides, the ink on the flexographic plates is on the ridges of the pattern.

The plates are relatively inexpensive, can be made quickly and changed out rapidly, which favors flexographic printing for short runs. The quality and productivity of flexographic printing have advanced so that choosing between flexographic and gravure printing is often difficult.

Further reading on film printing processes

1. Lentz, J., 1986, "Printing," in Bakker, M. (ed.), *The Wiley encyclopedia of packaging technology*, John Wiley & Sons, New York, pp.554-561.
2. Kirwin, M. and Strawbridge, J. W., 2003, "Plastics in Food Packaging," in Coles, R., McDowell, D. and Kirwin, M. J., *Food Packaging Technology*, Blackwell Publishing Ltd, Oxford, pp. 211-212.

References

[1] J. Frankland, Get smarter on extrusion sizes, Plast. Technol. (2012) 22. July.
[2] J. Schut, Understanding grooved feed, Plast. Technol. (2000)April, viewed 29 May 2013, from. Available from: http://www.ptonline.com/articles/understanding-grooved-feed.
[3] C. Rauwendaal, Polymer Extrusion, Hanser Publishers, Munich, 1986, pp. 246–256.
[4] W. Hellmuth, Light grooved feed throat extrusion – versatile, efficient, and retrofitable, in Proceedings of the International Polyolefins Conference 2010, Society of Plastics Engineers, Houston, TX, February 21–24, 2010.
[5] B.A. Morris, Polymer blending for packaging applications, in: S. Ebnesajjad (Ed.), Plastic Films in Food Packaging, Materials, Technology and Applications, Elsevier, Amsterdam, 2013, pp. 311–343.
[6] C.I. Chung, Scientifically designed barrier screw, SPE ANTEC Conference, 2013.
[7] G.A. Campbell, M.A. Spalding, Troubleshooting & Analysis Of Single-Screw Extrusion, Hanser Gardner Publications, New York, 2010, pp. 627–633.
[8] C. Rauwendaal, Polymer Extrusion, 4th (ed.), Carl Hanser Verlag, Munich, 2001, pp. 477–490.
[9] W.A. Kramer, E.L. Steward, Single-screw extruders, in: J. Vlachopoulos, J.R. Wagner (Eds.), The SPE Guide on Extrusion Technology and Troubleshooting, The Society of Plastics Engineers, Brookfield, CT, 2001, pp. 1–9.
[10] G.A. Campbell, M.A. Spalding, Troubleshooting & Analysis Of Single-Screw Extrusion, Hanser Gardner Publications, New York, 2010, pp. 597–624.
[11] R. Saxton, Extruder mixing screw, U.S. Patent 3006029, 1964.
[12] B. Maddock, Mixing of thermoplastic materials, U.S. Patent 3730492A, 1973.
[13] J. Perdikoulias, Multilayer blown (tubular) film dies, in: J. Wagner (Ed.), Multilayer Flexible Packaging, 2nd (ed.), Elsevier, Oxford, 2016, pp. 123–128.
[14] P. Waller, What to do when the bubble won't behave, PlasticsTechnology (2002)viewed 19 June 2013, from. Available from: http://www.ptonline.com/articles/what-to-do-when-the-bubble-won't-behave.
[15] T. Butler, Blown film bubble instability induced by fabrication conditions, SPE ANTEC Tech. Pap. (2000) 156–164.
[16] P. Waller, Blown film processes and troubleshooting, the ultimate quality control tool, 2009 Flex Packaging Summit (2009)TAPPI PLACE, viewed on 15 June, 2013, from. Available from: http://www.tappi.org/content/events/09PLACESY/Course_Papers/waller.pdf.
[17] W. Hellmuth, Novel gauging technology for contact-sensitive blown films, SPE ANTEC Conference, 2005, pp. 2902.
[18] M. Saini, M. Nisoli, Method for measuring the thickness profile of a flattened tubular film, European Patent EP1027368B1, 2007.
[19] H.G. Gross, Control of the thickness distribution of blown film by changing the flow channel gap of the die over the circumference, SPE ANTEC Conference, 2008.
[20] H. Tamber, Blown film cooling systems, in: J.F. Macnamara Jr. (Ed.), Film Extrusion Manual: Process, Materials, Properties, 3rd ed., TAPPI Press, Atlanta, GA, 2020, p. 148.
[21] D. Wright, Latest developments in on-line measurement and control of multilayer blown films: full spectrum infrared (FSIR) and air ring, in TAPPI PLACE Conference, Portsmouth, VA, 2008.
[22] I.B. Bensen, Online thickness measurement of single layers and co-extruded plastic films using infrared absorption, Pop, Plast, & Packaging 40 (6) (1995) 49–56.
[23] B. Embleton, S. Axelrod, On-line product quality measurements for continuous sheet processes, in Annual Technical Conference - Society of Plastics Engineers, 58th(Vol. 1), 2000, pp. 357–361.
[24] T. Blalock, S. Heveron-Smith, Multi-layer film measurement using dual-light interferometry, in TAPPI PLACE Conference, Indianapolis, IN, 2004.
[25] T.I. Butler, R. Patel, S.Y. Lai, J. Spuria, Blown film frost line-freeze line interactions, in TAPPI Polymers, Laminations & Coatings Conference, 1993, pp. 13–25.

[26] B.A. Morris, Effect of process and material parameters on interlayer peel strength in coextrusion coating, film casting and film blowing, J. Plast. Film Sheeting 26 (2010) 343–376.

[27] R.M. Patel, T.I. Butler, J.L. Walton, G.W. Knight, Investigation of processing-structure-properties relationships in polyethylene blown films, Polym Eng Sci 34 (19) (1994) 1506–1514.

[28] A.K. Babel, G.A. Campbell, Correlating the plastic strain with the properties of the low density polyethylene blown film, J Plast. Film Sheeting 9 (1993) 246–258.

[29] T. Kanai, M. Kimura, Y. Asano, Studies on scale-up of tubular film extrusion, in SPE ANTEC Conference, 1986, pp. 912–916.

[30] J.R.A. Pearson, C.J.S. Petrie, The flow of a tubular film. Part 1. Formal mathematical representation, J. Fluid. Mech. 40 (1970) 1.

[31] J.R.A. Pearson, C.J.S. Petrie, The flow of a tubular film. Part 2. Interpretation of the model and discussion of solutions, J. Fluid. Mech. 42 (1970) 609–625.

[32] J.R.A. Pearson, C.J.S. Petrie, A fluid-mechanical analysis of the film-blowing process, Plast. Polym. 38 (1970) 85.

[33] P. Shirodkar, V. Firdaus, H. Fruitwals, Scale-up of LLDPE blown film, Plast. Eng. (1994) 27. Dec.

[34] T. I. Butler, S. Y. Lai, R. Patel, 1994, "Scale-up factors effecting crystallization in polyolefin blown film," TAPPI Polymers, Laminations & Coatings Conference, pp. 289–304.

[35] A.M. Sukhadia, The use of bubble kinematics in the scale up and processing-property analysis of the blown film process, SPE ANTEC Conf (1994) 202–210.

[36] T. Bezigian, A survey of current bead reduction die technology, in: T. Bezigian (Ed.), Extrusion Coating Manual, 4th (ed.), TAPPI Press, Atlanta, 1999, pp. 55–63.

[37] Kirtley, Coat weight/thickness measurement and control, in: T. Bezigian (Ed.), Extrusion Coating Manual, 4th (ed.), TAPPI Press, Atlanta, 1999, pp. 117–118.

[38] B.A. Morris, Understanding why adhesion in extrusion coating decreases with diminishing coating thickness, J. Plast. Film Sheeting 24 (2008) 53–88.

[39] K. Canning, B. Bian, A. Co, Film casting of a low density polyethylene melt, in: Proceedings of ANTEC 1999, Annual Technical Conference –Society of Plastic Engineers, Vol. 57, 1999, pp. 386–390.

[40] V. Antonov, A Soutar, Foil adhesion with copolymers: time in the air gap, in TAPPI 1991 Polymers, Laminations & Coatings Conference Proceedings, TAPPI PRESS, Atlanta, 1991, pp. 553.

[41] B. Prinsen, Why choose blown, cast or lamination processes to achieve barrier properties, in TAPPI PLACE, 2009 Flexible Packaging Summit. <http://www.tappi.org/content/events/09PLACESY/Course_Papers/prinsen.pdf>, 2009 (accessed 08.06.13).

[42] H.F. Giles Jr., E.M. Mount III, J.R. Wagner Jr., Extrusion: The Definitive Process Guide and Handbook, William Andrews Publishing, 2007, pp. 460–462.

[43] R. Knittel, Blown film vs the cast film process, SPE Extrusion Div (1995)Viewed on 2 June, 2013 from. Available from: http://www.extrusion-wiki.com/wiki/(X(1)S(0bpdnb45tivsdq450plakjus))/CC-V22-3-B.ashx.

[44] E. Laiho, T. Ylanen, Flame, corona, ozone – do we need all pretreatments in extrusion coating? in: T. Bezigian (Ed.), Extrusion Coating Manual, 4th (ed.), TAPPI Press, Atlanta, 1999, pp. 89–97.

[45] R. Wolf, Surface treatment, in: J.F. Macnamara Jr (Ed.), Film Extrusion Manual, 4th ed., TAPPI Press, Atlanta, 2020, pp. 157–172.

[46] T. Gilbertson, M. Plantier, Corona surface treatment, in: W. Durling (Ed.), Extrusion Coating Manual, 5th (ed.), TAPPI Press, Atlanta, Ga, 2017.

[47] D.A. Markgraf, Corona treatment enhanced adhesion for extrusion coating, in: T. Bezigian (Ed.), Extrusion Coating Manual, 4th (ed.), TAPPI Press, Atlanta, 1999, pp. 65–74.

[48] R. Wolf, Atmospheric-pressure plasmas, in: J.F. Macnamra Jr. (Ed.), Film Extrusion Manual, 3rd ed., TAPPI Press, Atlanta, 2020, pp. 173–186.

[49] R. Wolf, Plasma surface treatment, in: W. Durling (Ed.), Extrusion Coating Manual, 5th (ed.), TAPPI Press, Atlanta, Ga, 2017.

[50] J. DiGiacomo, Flame plasma surface treatment, in: W. Durling (Ed.), Extrusion Coating Manual, 5th (ed.), TAPPI Press, Atlanta, Ga, 2017.

[51] J.D. DiGiacomo, Flame plasma surface treatment, in: T. Bezigian (Ed.), Extrusion Coating Manual, 4th (ed.), TAPPI Press, Atlanta, Ga, 1999, pp. 121–130.

[52] D.J. Bentley Jr., F.M. Singer, Chemical primers to enhance adhesion and other properties, in: T. Bezigian (Ed.), Extrusion Coating Manual, 4th (ed.), TAPPI Press, Atlanta, Ga, 1999, pp. 99–108.

[53] B.A. Morris, Adhesion in extrusion lamination and coating, in: W. Durling (Ed.), Extrusion Coating Manual, 5th (ed.), TAPPI Press, Atlanta, Ga, 2017, pp. 163–177.

[54] B.A. Morris, N. Suzuki, The case against oxidation as a primary factor for bonding acid copolymers to foil, in TAPPI Polymers, Laminations & Coatings Conference, 2000.

[55] P.E. Waghorn, The control of oxidation and odour in extrusion coating, Polythene Extrusion Coaters Group, Chicago, 1963.

[56] V.G. Crolius, W.E. Ebeling, R.C. Parsons, The effect of processing variables on the adhesion strength of polyethylene-coated aluminum foil, TAPPI 45 (5) (1962) 351–356.

[57] G. Schubert, O. Plassmann, S. Husch, Hot tack of extrusion coatings on aluminum foil and paper – Does the process always lead to the same result? in TAPPI PLACE Conference, 2010.

[58] T. Gilbertson, M. Plantier, Ozone surface treatment, in: W. Durling (Ed.), Extrusion Coating Manual, 5th (ed.), TAPPI Press, Atlanta, Ga, 2017.

[59] P.B. Sherman, Ozonation of polymer melt for improved adhesion, in: T. Bezigian (Ed.), Extrusion Coating Manual, 4th (ed.), TAPPI Press, Atlanta, Ga, 1999, pp. 75–87.

[60] J.D. Vansant, S.B. Marks, B.A. Morris, "Acid copolymers for extrusion lamination and coating,", in: W. Durling (Ed.), Extrusion Coating Manual, 5th (ed.), TAPPI Press, Atlanta, Ga, 2016.

[61] Y. Trouilhet, Morris, Prediction of temperature profiles across coating and substrate in the nip, in TAPPI Polymers, Laminations and Coatings Conference, 1999.

[62] E.M. Petrie, Laminating adhesives for flexible packaging, Omnexus 4 Adhesives & Sealants. <http://www.packagingconnections.com/downloads/download195_0.pdf> (accessed 11.06.13).

[63] R. Wolf, A technology decision – adhesive lamination or extrusion coating/lamination? TAPPI PLACE Conference, April, Albuquerque, NM <http://www.tappi.org/content/events/10PLACE/papers/wolf1.pdf>, 2010 (accessed 11.06.13).

[64] H. Schetschok, E. Kupsch, Adhesive lamination vs. extrusion lamination, Trends and technology update, in TAPPI European PLACE Conference, May, Bregenz, Austria. <http://www.tappi.org/Downloads/Conference-Papers/2011/2011-13th-TAPPI-European-PLACE-Conference/11EURO30.aspx>, 2011 (accessed 11.06.13).

[65] E.B. Gutoff, E.D. Cohen, Water- and solvent-based coating technology, in: J.R. Wagner (Ed.), Multilayer Flexible Packaging, 2nd ed., Elsevier, Cambridge, MA, 2016, pp. 205–234.

[66] R. Breese, Benefits of machine direction orientation (MDO) film in flexible packaging applications, Converting Q (2011) 32–36.

[67] J.H. Shut, MDO films: lots of promise, big challenges, Plast Technol (2005)February 2005, viewed 18 June 2013, on. Available from: http://www.ptonline.com/articles/mdo-films-lots-of-promise-big-challenges.

[68] F. Alexy, Replacing aluminum in high barrier packaging with Triple Bubble® blown film extrusion, in: Multilayer Packaging Films 2012 Conference, 2012.

[69] E. Adolfo, All encompassing extrusion technology for producing a wide spectrum of simultaneously biooriented films, in: 2018 SPE ANTEC Conference, 2018.

[70] C.F. Struller, P.J. Kelly, N.J. Copeland, Aluminum oxide barrier coatings on polymer films for food packaging applications, Surface Coatings Technol., 241 (2014) 130–137. Also 56th Annual Technical Conference Proceedings, Society of Vacuum Coaters, Providence, RI, April 20–25, 2013.

[71] T. Glaw, Transparent inorganic barrier films, TAPPI European PLACE Conference, 2007.

[72] C.A. Bishop, E.M. Mount, Vacuum metallizing for flexible packaging, in: J.R. Wagner Jr. (Ed.), Multilayer Flexible Packaging, 2nd ed., Elsevier, Cambridge, MA, 2016, pp. 235–255.

[73] C.A. Bishop, Vacuum Deposition onto Webs, Films and Foils, 3rd (Ed.), Elsevier, 2015Retrieved from. Available from: https://app.knovel.com/hotlink/toc/id:kpVDWFFE11/vacuum-deposition-onto/vacuum-deposition-onto.

[74] E.M. Mount, III, A.J. Benedict, Film having metallizable surface, in European Patent application EP9030227A1, 1990.

[75] R.A. Migliorini, S.J. Pellingra, K.A. Sheppard, Metallized multilayer film, in U.S. Patent 6863964B2, 2002.

[76] L.A. Parr, R.G. Peet, Multi-layer film with grafted sydiotactic polypropylene adhesion promoting layer, U.S. Patent 6703134B1, 2000.

[77] C.R. Hart, Metallized films having an adherent copolyester coating, U.S. Patent 4,971,863, 1990.

[78] R. Balloni, J.K. Keung, E.M. Mount III, Process for manufacturing a metallized polyolefin film and resulting film; improved adhesion using isotactic polymer, U.S. Patent 4888237, 1989.

[79] F. All, G. Duncan, Oriented polypropylene film substrate and method of manufacture, U.S. Patent 4,345,005, 1982.

[80] C.A. Bishop, Is it possible to predict a film's barrier performance before vacuum metallizing, Converting Quarterly <http://www.convertingquarterly.com/vacuum-web-coating/is-it-possible-to-predict-a-films-barrier-performance-before-vacuum-metallizing-part-1-of-3>, 2020 (accessed 30.05.20).

[81] E.M. Mount, III, Metallizer process settings impact on lamination bond strength, in Society of Vacuum Metallizers, 52nd Annual Technical Conference Proceedings, Santa Clara, CA, May 9–14, 2009, pp. 716–719.

[82] N. Ahmed, Review and evaluation of AlOx clear barrier production processes, idvac. <http://www.idvac.co.uk/review-and-evaluation-of-alox-clear-barrier-production-processes/>, 2014 (accessed 31.05.20).

[83] Converting Quarterly, BOBST Alubond®: breakthrough in high metal adhesion.<http://www.convertingquarterly.com/industry-news1/bobst-alubondr-breakthrough-in-high-metal-adhesion>, 2018 (accessed 31.05.20).

[84] R. Trassl, G. Hoffmann, A. Wolff, Transparent barrier coatings for environmental-friendly packaging applications, in Web Coating & Handling Conference <https://www.aimcal.org/uploads/4/6/6/9/46695933/trassl_abs.pdf>, 2012 (accessed 31.05.20).

[85] Knowfort Technologies BV, Freshure® coatings, Environmentally friendly transparent barrier coatings. <http://www.knowfort.com/files/knowfort_freshure_leaflet.pdf> (accessed 31.05.20).

[86] S. Jahromi, Advances in protection of aluminum oxide using inline vacuum deposited organic top coats, in AIMCAL Web Coating and Handling Conference Europe, 2016.

[87] J. Lentz, Printing, in: M. Bakker (Ed.), The Wiley Encyclopedia of Packaging Technology, John Wiley & Sons, New York, 1986, pp. 554–561.

Chapter 3

Flexible packaging equipment

Packaging equipment transforms the flexible film into a package form, fills it with a product, seals it to prevent leaks, and prepares it for shipping. This chapter briefly describes packaging equipment commonly used in primary flexible packaging, as outlined in Table 3.1. Secondary and tertiary packaging, such as pallet wrapping, are not covered. There are many forms of packaging equipment available on the market today. The purpose of this chapter is not to describe these in great detail, but to point out some of the key features and applications. The chapter begins with a discussion of the unit operations common to many types of packaging equipment: registration, thermoforming, filling, gas flush/modified atmosphere packaging (MAP) and sealing.

3.1 Unit operations

Flexible packaging equipment comes in a wide variety of designs but most share one or more common unit operations including registration, filling and sealing. This section explores the factors that affect the performance of these operations. Additionally, thermoforming and modified atmosphere operations are considered. Information provided here provides a framework for later sections describing specific types of packaging equipment

3.1.1 Registration

Registration involves the controlled movement of the film through the machine while various operations are performed. Critical to quality is the movement of the film without wrinkling or distortion. Factors influencing performance include the frictional properties of the film, film stiffness, and machine design. The frictional properties of the film are generally characterized by the coefficient of friction (COF), as measured by ASTM D1894. The COF is defined as "the ratio of the force required to move one surface over another over the total force applied normal to the surfaces" [1]. Several types of COF can be measured, as detailed in Section 12.1.2.1. Static COF involves the friction between a weighted sled and the film surface as the sled is *initially* dragged across the surface (i.e., it is a start-up value). The kinetic COF is the measured frictional force after the force has reached a steady state. Both may be important in film registration. COF can also be measured against different surfaces. Film-to-film COF involves measuring the frictional drag on the film against itself. Film-to-metal COF involves measuring the drag of the film against a metal surface. Depending on the process, both may be important for registration.

The COF of barefoot plastic films can often be quite high, near 1.0. (1 is the practical limit of the COF measurement: values exceeding 1 are not meaningfully differentiated from 1). Often values of around 0.2–0.3 are desired for high slip applications. To achieve this, additives may be used to impart a more slippery surface on the film. These slip additives are typically fatty acid amides that bloom to the surface over time, so care must be taken to allow sufficient time for the slip agent to reach the surface before measuring COF. A more detailed discussion on the factors that influence film COF and the use of slip agents is provided in Chapter 12.

Blocking the film on the roll can be another issue. Blocking is the resistance of the film to unwind. It is particularly an issue with soft or tacky plastics, or if the film is too tightly wound. "Antiblocking" additives are often added to the plastic to reduce blocking. These are typically inorganic materials such as diatomaceous earth, silica, or talc. These make the surface bumpier, reducing the contact between the layers of film in the roll.

Film stiffness is important because if a film is too "flimsy" it may be too difficult to handle, resulting in wrinkles and other problems. If the film is too stiff, it may not bend around sections of the machinery. Both the tensile stiffness (as measured by a tensile or Young's modulus) and the bending stiffness are important for registration. Tensile stiffness is needed to resist stretching as the film is pulled through the process. Bending stiffness correlates with ease of

The Science and Technology of Flexible Packaging. DOI: https://doi.org/10.1016/B978-0-323-85435-1.00016-8

TABLE 3.1 Examples of packaging equipment and their uses.

Equipment/category	Package form	Examples of products packaged
Vertical form fill seal	Pouches Pillow Packs	Chips Cereal and other bag-in-box Cookies Nonfood: resin, fertilizer
Horizontal form fill seal (flow wrappers)	Flow wrappers	Candy bars Cookies Snack crackers/cookies Medical products Disposable razors
Horizontal thermoform fill seal	Tray with lid, shingle packs	Processed meat Cheese Fish Medical items
Tray sealer	Tray, cup or bowl with lid	Microwavable meals Ovenable meals Other dual oven food items Dairy products
Vertical pouch	Pouches Sachets	Condiments Shampoo/soap sachets Water Oil Sauces
Vertical pouch	Stick packs	Powder drink mixes Pharmaceuticals
Horizontal pouch	Pouches	Dry soups Mixed ingredient food products Jello Cake mixes
Stand-up pouch	Pouches	Frozen vegetables Snacks Candy Liquids
Vacuum packaging	Shrink bags Horizontal FFS with vacuum Vacuum skin packaging	Primal meat Fresh meat Fish
Blister packaging	Carded blister packs	Pharmaceuticals Retail merchandize
Skin packaging	Carded skin packs	Retail merchandize
Clamshell	Clamshell	Retail merchandize

handling. For single-layer films, the bending stiffness increases with the thickness cubed. For multilayer films, the bending stiffness is a complex relationship between the thickness, placement, and tensile modulus of each of the layers. Modeling of this relationship is described in Chapter 9.

Film strength and equipment design also play a role. Proper tension control and keeping the drag force below the yield strength of the film are important to prevent the film from stretching.

One of the concepts for designing packaging structures for recycling (see Chapter 17) is to replace multimaterial laminates with single material structures, such as all-PE structures. The replacement structures often have lower stiffness and thermal resistance than the structures they are intended to replace, such as laminates with oriented polyester film. This presents challenges with registration.

3.1.2 Thermoforming

Thermoforming involves heating the film to the *forming temperature* and shaping it into the form of a cavity. This typically entails drawing a vacuum to pull the film into the cavity, using a plug to push the film into the cavity, or both. During the forming process, the film is drawn down in thickness in proportion to the additional area created.

Uniform thickness and layer distribution after forming as well as minimizing spring back are critical to quality. Factors that influence performance include the thermoforming temperature, plug assist design, film design/construction, draw ratio, the heating process, and time.

Most plastic materials have a range of temperatures over which they can be readily thermoformed. Since thermoforming involves biaxially stretching of the polymer, many of the same considerations for material selection and temperature settings as for oriented films apply to thermoforming. (Film orientation processes are introduced in Chapter 2 and Chapter 11 discusses both thermoforming and orientation in more detail.) Forming is typically done below the highest melting point of the layers in the structure. Otherwise, the film may sag too much during the process, resulting in operational issues. Forming in the solid-state also helps impart more orientation to the formed package, providing better strength. But too much orientation or residual stress may result in spring back, where the final shape of the package is not within the desired dimensional tolerances due to relaxation of stresses created during forming.

High forming temperatures may distort one or more of the layers as some of the layers may be molten at that temperature. This may particularly be true for the sealant layer, which is often designed with a low melting point to achieve low-temperature sealability and high packaging line speeds. Since the sealant is an outer layer of the film, care must be taken that the molten sealant does not stick to the plug assist or mold.

When designing a multilayer film structure for thermoforming, the individual thermoforming temperature ranges of each of the layers must overlap so that a common forming temperature can be found [2]. Sometimes materials may have to be substituted to achieve a workable system.

The stretch or draw ratio and draw rate may affect formability. Some materials do not draw well. For example, standard grades of ethylene vinyl alcohol (EVOH) have limited drawing performance. At high draw ratios, they become distorted and discontinuous. Using grades tailored for thermoforming helps reduce this problem. Early on this meant using grades with higher ethylene content, but in more recent years the EVOH suppliers have come out with special thermoforming grades with lower ethylene content (and higher barrier performance) [3,4]. Blends have also proven effective. Chou and Deak [5] show that adding a minor amount of amorphous polyamide to EVOH can significantly improve deep draw thermoformability without significant impact on barrier properties. Similarly, Fetell [6] blended in a minor amount of ionomer to improve formability. Adjustments to the thermoforming process may also be effective [7].

The thermoforming process and key resin properties, such as strain hardening, are described in further detail in Chapter 11.

3.1.3 Filling

The filling operation involves dispensing an accurate number or volume of product into the package. There are two general methods to account for the product being packaged: by count or by weight (or volume). When dosing a specific count or number of items into the package, three steps typically take place: representation, detection and product handling [8].

The objects must first be detected, typically by an electric signal, so that they can be accounted for in a control system. Detection is done by optical or nonoptical methods. A typical optical method uses a photosensor to detect the object. Nonoptical methods use other means to "touch" the article. One method is to channel the objects into an escapement device that is designed so that the product exactly fits into a receiving device. This involves a physical touch of the article. Electrical or magnetic fields are also used for detection.

During product handling, the objects are typically channeled into a single file during conveying. This facilitates counting. The articles are then accumulated and dispensed into the package.

Dry or liquid products are filled by weight or volume. The goal is to fill the container or package with a specific weight or volume, as denoted on the package. Label laws often require that the package contain the amount indicated. Variability in filling or measuring inevitably occurs, generally requiring that the package be overfilled to ensure that the law is met. The capability of the filling system to meet a target filling weight can be described by statistics. Typically the variability follows a normal probability distribution and the standard deviation can be used to characterize the variability or error. Guidelines, such as +/− 3 standard deviations (99.73%) can be established to ensure compliance with laws. The specification of a filling system will be dictated by its cost and ability to minimize filling error.

Volumetric systems are less expensive and sufficient for constant density dry products. They are also easier to maintain than weight systems. Weight systems are generally more accurate and allow for record-keeping. Their cost has come down in recent years. There are many types of dry goods feeders, including augur, gravity-fed, vibratory, belt, screw, vibration bin discharge, cascade filling and vacuum filling. Each has its advantages and disadvantages [9]. There are several weighing systems on the market, the details of which are beyond the scope of this book.

In the packaging process, there are many approaches to accurately deliver the target volume or weight to the package. Some systems fill volumetrically to a point, weigh the partially filled package and fill with a weight system to reach the goal. Others use loss in weight throughout. Multiple feeders and feeder types may be used on a single packaging line. Selection of the feeder must take into account production rate, product cost, packaging type and cost, machine efficiency, maintenance cost, equipment cost, operating cost, and layout [9].

3.1.4 Gas flush/modified atmosphere packaging

Oxygen in the headspace or entrained with the product during filling can contribute to food spoilage. Often the headspace will be flushed with an inert gas such as nitrogen or an atmosphere different from air (20.9% oxygen, 79% nitrogen, and 0.04% carbon dioxide) during filling or prior to sealing. This is known as modified atmosphere packaging (MAP) and has been in practice since the 1970s as a way to extend the shelf life of food products.

Oxygen accelerates oxidation and microbial growth. Replacing the oxygen retards oxidation, reduces enzymatic reactions, and decreases loss of product water through respiration. Gas flushing involves introducing a continuous gas stream to flush air out of the package. This technique only gets the O_2 down to 2%–5% and is not sufficient for oxygen-sensitive foods, but allows high packing rates. In compensated vacuum gas flushing, the headspace is evacuated by pulling a vacuum and then flushed with a modified atmosphere gas. This gives lower residual oxygen than gas flushing and is sufficient for oxygen-sensitive foods [10]. Vacuum packaging is a special type of MAP where the oxygen is removed from the package without replacing it with a modified atmosphere.

A simple gas flush of N_2 may be used, or other gases or mixtures of gases may be introduced. Table 3.2 gives a list of commonly used gases and the reasons for their use. It should be noted that completely excluding the oxygen can lead to anaerobic microbial growth, which can be deleterious as well. An example of a harmful anaerobic bacterium is *Clostridium botulinum.*

MAP may be used in combination with active packaging, such as oxygen and carbon dioxide scavengers, and ethanol and carbon dioxide emitters. The specific gas mixture used depends on the product. Typically it involves some combination of nitrogen, oxygen and carbon dioxide. For example, a high oxygen environment is used to preserve the color of fresh red meat, CO_2 and N_2 help prevent mold in bakery items, and N_2 is used to prevent rancidity in dried goods. Other examples are given in Table 3.3.

TABLE 3.2 Gases used in modified atmosphere packaging [11].

Gas	Some common uses in modified atmosphere packaging
Oxygen	Maintain red color in fresh red meat
Nitrogen	Inert gas to replace O_2 or other gas
Carbon dioxide	Acidification of water provides direct suppression of microbial growth (> 10%–15% CO_2); suppresses respiration of some fresh produce and can suppress ripening
Carbon monoxide	Suppresses microbial growth at low concentrations. Also helps red meat retain its red color longer. Toxic at higher concentrations so a challenge to handle
Sulfur dioxide	Helps control mold growth and microbial growth in some foods (e.g., wines, fruit juices, etc.) Some people are hypersensitive to it so under scrutiny
Ethanol	Antifungal activity; retain flavor of tomatoes
Argon	Some reports of antimicrobial effect

Source: Adapted from information in J. Brandenburg, Modified atmosphere packaging, in: K.L. Yam (Ed.), Wiley Encyclopedia of Packaging Technology, third ed., John Wiley & Sons, 2009, pp. 787–794.

TABLE 3.3 Some modified atmospheres used in food packaging.

Food type	Modified atmosphere	Comment
Raw red meat	20%–30% CO_2, 70%–80% O_2	Red color stability, microbial inhibition
Raw poultry	25%–50% CO_2, 50%–75% N_2 100% CO_2 (bulk pkg)	CO_2 inhibits microbial growth
Cooked, cured, processed meat	25%–50% CO_2, 50%–75% N_2	Inhibits oxidation. Food poisoning from anaerobic bacteria is a concern with oxygen free atmosphere
Fish	CO_2/N_2	Inhibits aerobic bacteria food poisoning, but anaerobic bacteria is a concern. Some add O_2. Typically stored at 3°C for less than 10 days
Fresh produce	Low O_2/high CO_2	Inhibits respiration. Composition of gas specific to different types of produce
Dairy	N_2 or N_2/CO_2	Inhibits oxidative rancidity in powdered milk, cheeses and fat spreads

Source: Adapted from M. Mullan, D. McDowell, Modified atmosphere packaging, in: R. Coles, D. McDowell, M.J. Kirwan (Eds.), Food Packaging Technology, Blackwell Publishing and CRC Press, Oxford, England, 2003, pp. 303–339.

In fresh produce packaging, respiration of the product changes the atmosphere within the package with time. For the full effect of MAP in these applications, the film must be permeable. For highly respiring foods, such as mushrooms, beansprouts, leeks, herbs, peas, and broccoli, requiring very high permeation, micro-perforated films are often used.

In MAP, the package contents are not sterilized, as in retort or other sterilization techniques. MAP only slows the degradation mechanisms. It is typically used in combination with refrigeration. In some cases, introducing a modified atmosphere that inhibits one type of spoilage mechanism may introduce another, or potentially introduce conditions that may cause concerns for food safety. For example, minimizing oxygen in the atmosphere may delay oxidation, rancidity, and growth of aerobic microbes, but favor the growth of harmful anaerobic microbes such as botulism. Refrigeration, limiting shelf life, and the use of preservatives may be required to maintain food safety.

3.1.5 Heat sealing

Heat sealing involves the fusing of films together under heat and pressure. It is important for preventing the contents of the package from leaking out. Sealing may involve sealing films to themselves, as in pouches, or other films, as in horizontal-thermoform fill seal packages, or to rigid substrates, as in lidding for trays or cups. Often, a separate sealant layer is designed into a multilayer film to achieve heat sealability to prevent distortion of other layers of the film, achieve high packaging line speeds, or to seal in the presence of product contamination. Some of the commonly used sealant resins are described in Chapters 4 and 7. A discussion of the science of heat sealing, modeling, and easy-open technology is given in Chapter 7. Here, the types of heat sealing and the basic factors that influence the quality of heat sealing are discussed.

3.1.5.1 Types of heat sealing

3.1.5.1.1 Bar sealing

Bar sealing is the most commonly used method of sealing. Heated jaws are clamped down on the outside surface of the two films being joined. Heat transfers from the jaws through the film, melting the interface. The jaws are removed and the seal area is allowed to cool. One or both jaws may be heated. The surface of the jaws may be coated with a fluoropolymer or another nonstick surface to prevent sticking to the outside of the film. The jaws may have a serrated design with ridges to obtain a higher seal strength. This technique is used on many laminated and coextruded film structures.

3.1.5.1.2 Impulse sealing

In impulse sealing the bar is covered with silicone rubber and a nichrome ribbon over one or both surfaces. The ribbon is covered with a fluoropolymer impregnated cloth. Heating and cooling steps occur while the bars are engaged. Electric current is sent through the ribbon to heat the surface. Cooling occurs with circulating water in the closed position.

Impulse sealing is slower than hot bar sealing. It is used on tacky materials and unsupported films.

3.1.5.1.3 Hot wire or hot knife sealing

Hot wire or hot knife sealing is typically used on unsupported films. The film is cut through with the hot wire forming a bead. The method is often used on shrink films but does not result in a reliable hermetic seal.

3.1.5.1.4 Hot air/gas sealing

In hot air or gas sealing, heated gas is passed over both surfaces to be sealed. When the surfaces are sufficiently heated, the gas stream is removed and the surfaces are joined between chilled jaws. This technique is typically used to seal heavy or thick packaging materials, such as coated paperboard milk cartons and toothpaste tubes.

3.1.5.1.5 Pneumatic sealing

Pneumatic sealing is used for carded skin packaging (see Section 3.2.6.3). An object is placed on a porous or perforated paperboard. A heated film is draped over the object and board. Air is withdrawn with a vacuum through the board, collapsing the film against the object and board. The residual heat in the film creates the seal.

3.1.5.1.6 Ultrasonic sealing

In ultrasonic sealing, a sonic horn is used to impart sound waves to "hammer" or "rub" the films together at high frequency. The friction at the interface generates heat to create the weld. This technique is good for oriented films where direct heat may distort the film. It also works well for thick films where it takes too long to transfer the heat through the film in hot bar sealing. Other advantages over hot bar sealing include the ability to seal through contamination, less heat transferred to the product, lower operating expense (since energy is only consumed during the short sealing cycle, not continuously to heat the bars), and the potential for smaller material footprint since the seal area is typically smaller than with a conventional hot bar seal. Disadvantages include the potential for squeezing out the sealant and substantially higher capital expense [12–17].

3.1.5.1.7 Induction sealing

Induction sealing uses alternating current applied to a metal to generate the heat for sealing. Typically, this is used in lidding films containing a layer of aluminum for sealing to the mouth of filled containers. Heat is generated by the electrical resistance of the foil.

3.1.5.1.8 Radiofrequency sealing

In radiofrequency (RF) sealing a dielectric (nonconductive) material is heated by the transfer of energy from a high-energy electromagnetic field (RF). Polar groups within the plastic try to reorient themselves to the alternating field (typically above 27 million times/s or 27 MHz). The internal friction generates heat to melt and fuse the polymer. Pressure is applied to assist in the sealing. The quality of the seal is a function of the RF power, sealing time and pressure. The sealant polymer must have a high degree of polarity (high dielectric constant) to respond to the RF field. Of the materials used in flexible packaging, flexible polyvinyl chloride (PVC) is best suited for this method. Ethylene vinyl acetate and ethylene butyl acrylate copolymers with high levels of vinyl acetate or butyl acrylate also have limited application for RF sealing.

Further reading on types of heat sealing

- Morris, B. A., Darby, D., 2009, "Sealing, heat," in Yam, L.K. (ed.), *Wiley Encyclopedia of Packaging Technology*, John Wiley & Sons, pp. 1089–1096.
- Hishinuma, K., 2009, *Heat Sealing Technology and Engineering for Packaging: Principles and Applications*, DEStech Publications Inc., Lancaster, PA.

3.1.5.2 Types of seals

There are two possible configurations used when sealing a film to itself: fin and lap seals. The most common is a conventional fin seal where the film is folded such that the inside surfaces of two different regions of the film are sealed together. In multilayer films, this means the sealant layer is sealing to itself; there are no issues with poor compatibility or adhesion.

In the lap seal, the inside surface is sealed to the outside surface of the film. With a single-layer film, there is no issue with compatibility. But with multilayer films, an adequate sealant must be selected to ensure it seals well to the outside layer of the film.

In horizontal thermoform fill seal processes, care should be taken to ensure that the sealant layers of the top and bottom webs are compatible. For best results, it is recommended to use the same sealant in each film. In lidding applications, the lidding film sealant is often comprised of different chemistry than the container to which it is sealed. Here the sealant must be chosen to wet and adhere to the container surface.

3.1.5.3 Seal crimper and die design

In hot bar sealing, the leading and trailing end seals are typically sealed using a heated mechanical crimper assembly. The crimper assembly consists of a pair of dies or crimpers that have a pattern of grooves. The grooves on the top and bottom crimpers are aligned so that the peaks on the top crimper line up with the valleys of the bottom crimper. The geometry of these groove patterns, including the angle of the groove, whether the peaks are truncated (to prevent tearing of the film), depth and size of the grooves, and orientation of the grooves are important. Orientation of the grooves or serrations is usually described in relation to the longitudinal dimension of the die or bar, as shown in Fig. 3.1. Horizontal serrations are parallel to the longitudinal dimension. In a typical end seal on a vertical form fill seal (VFFS) or horizontal form fill seal (HFFS) packaging line, horizontal serrations are perpendicular to the machine direction of the film and perpendicular to the direction of pull as the package is opened. Vertical serrations are oriented perpendicular to the longitudinal dimension of the seal bar. Diagonal serrations of almost any angle can be found and the design of these patterns is often more art than science.

Lako Tool [18] discusses problems associated with crimp design. The crimp pattern can affect the packaging line speed. Horizontal designs allow for higher line speeds than vertical designs. They show data generated by the Mobil Chemical Company on several polypropylene films. Horizontal serrations allow a 43%–67% increase in line speed over vertical serrations at a constant seal temperature. The horizontal serration pattern also gives a wider packaging speed window (the low end of the line speed window is determined by overheating and distortion of the seal area and the high end by poor seal strength).

Theller [19] discusses serration design. Serration angle, land length and other aspects of die geometry can be important. By varying the geometrical configuration of the serration, it is possible to improve hot tack and reduce channel leakers. Theller found for a (HDPE–EVA + PB) film structure sealed to itself (EVA sealant to EVA sealant), vertical serrations produced higher heat seal strength than horizontal serrations or flat bars with no serrations (horizontal serrations and flat bars gave the same seal strength). They attributed this not to changes in molecular diffusion at the sealant interface but to orientation effects with respect to the direction of stress. Vertical serrations act as corrugations that are perpendicular to the peel front (parallel to the direction of pull) as the sample is peeled apart (see Fig. 3.1 bottom right). The serrations act as reinforcements, requiring extra force to flatten or stretch out the corrugation as the front moves through. Horizontal serrations are parallel to the peel front and lack this reinforcing effect.

This phenomenon of high seal strength with vertical serrations is not apparent when Theller seals a more flexible, thinner LDPE film. Indeed, the vertical serrations give slightly lower seal strengths than the horizontal serrations or flat die. Theller proposes a model to explain the results. The serrations create a corrugation in the seal area. The sealing force is greatest on the slanted sides of the serration and lowest at the peaks and valleys since the tool is generally rounded off to prevent breakage of the film. This results in periodic sealed and lightly sealed regions. When the serrations are vertical the measured force is the sum of the fully sealed and lightly sealed regions, resulting in a lower seal value. Since the seal test measures the maximum seal strength, horizontal serrations give the fully sealed value and is equal to that of the flat bar.

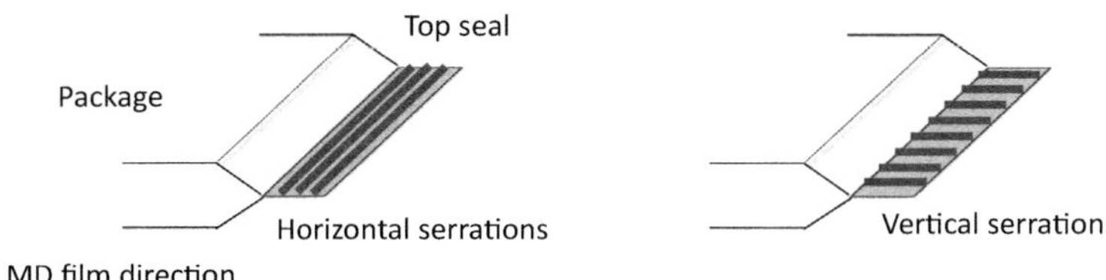

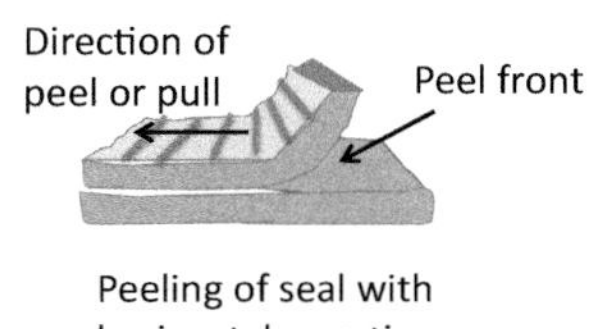

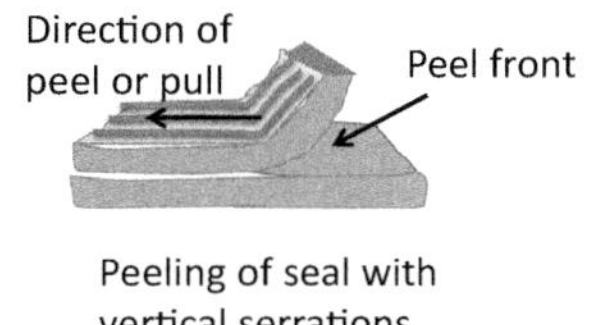

FIGURE 3.1 Illustration of the direction of serrations and peel front. Top left: horizontal serrations in top seal. Top right: vertical serrations in top seal. Bottom left: Location of peel front and direction of pull/peel for a seal with horizontal serrations. Bottom right: location of peel front and direction of peel for seal with vertical serrations.

Vertical serration designs can result in leakers. Bower [20] describes a horizontal design that is claimed to provide more shear flow and reduce leakers. The key feature of the design is interlocking crimp patterns with at least one male "land" truncated to a lower height than the corresponding groove on the opposite seal bar. Other patents describe variations on crimper designs [21–27].

3.1.5.4 Factors influencing heat seal performance

The primary function of the seal is to prevent the leaking of the contents of the package to the environment, and ingress of the environment (such as oxygen and moisture) into the package. Factors influencing hot bar sealing performance include:

- sealing properties of the film
- packaging line speed
- seal bar temperature
- uniformity of temperature across the seal bar
- uniformity of pressure across the seal bar
- package design (e.g., gussets)
- weight of the product (especially for VFFS lines)
- contamination in the seal area.

The sealing properties of multilayer film are explored in more detail in Chapter 7. The temperature range over which a quality seal can be obtained is an important consideration. If the temperature is too low, the seal strength may be variable. If it is too high, the film may distort or tear. The optimum seal temperature range is a function of the sealing time (known as the dwell time), film thickness, and the thermal and physical properties of the film. Time and temperature are inversely related. As is described in Chapter 7, sufficient time and temperature are needed for the molten polymer to fuse and build strength at the sealant interface. In a typical heat seal curve (Fig. 3.2), the seal strength as a function of seal bar temperature at constant dwell time and seal bar pressure is plotted. Seal strength is zero at low temperatures until at some point the seal strength begins to increase. This point is known as the heat seal initiation temperature (HSIT). At higher temperature the seal strength levels off to the ultimate seal strength. The point where it starts to level off is known as the plateau initiation temperature. The HSIT and plateau initiation temperature are related to the melting behavior of the polymer [28].

Thickness influences the time it takes the heat to reach the interface in hot bar sealing. The transfer of heat from the outside surface to the interface is roughly proportional to the square of the thickness. As a result, at a given dwell time, the HSIT may be higher for thicker films [29].

The upper end of the sealing temperature range is influenced by the physical and thermal properties of the film. Shrinkage and distortion of the film may occur at elevated temperatures. Often at the ultimate seal strength, the seal

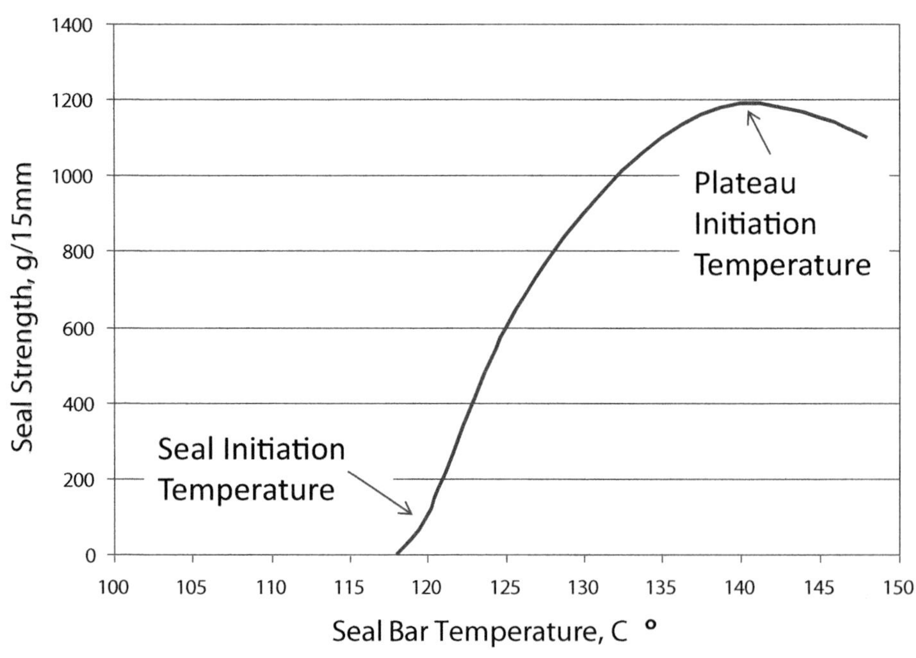

FIGURE 3.2 Heat seal curve. Center portion of the curve represents the heat seal window or temperature range for optimum sealing. *Reproduced with permission from The Dow Chemical Company.*

strength exceeds the tear resistance of the film. The tear resistance may be weakened as the result of some of the film having been squeezed out during the heat seal process at high temperatures and pressures.

Increasing packaging line speed reduces the dwell time for sealing. The exact relationship between line speed and dwell time depends on the fraction of the machine cycle time devoted to sealing. With registration, cutting and other operations, the sealing maybe 50% or less of the cycle time. For example, if the machine is running at 60 packs per minute and half the cycle time is devoted to sealing, the dwell time is 60 packs/min × min/60 s × 0.5 = 0.5 second. If there are multiple sealing steps within the cycle then the calculation is adjusted accordingly.

The seal bar temperature may vary throughout the day as the packaging machine is slowed down or sped up. The temperature may also not be uniform across the width of the bar. This is shown in a study by DuPont where measurements were made on several commercial-scale packaging lines at their technical service laboratory. Four different pieces of equipment are studied using thermocouples and a data logger to simultaneously record the temperatures at three places across the seal bar. The temperature setpoints are chosen to represent the normal operating range of the machine. The temperatures are recorded long enough to capture drift from the on/off of the heating cycle. The trials are done in a static state with no film. Results from the study are given in Fig. 3.3. Variation from the setpoint is as high as 54°C.

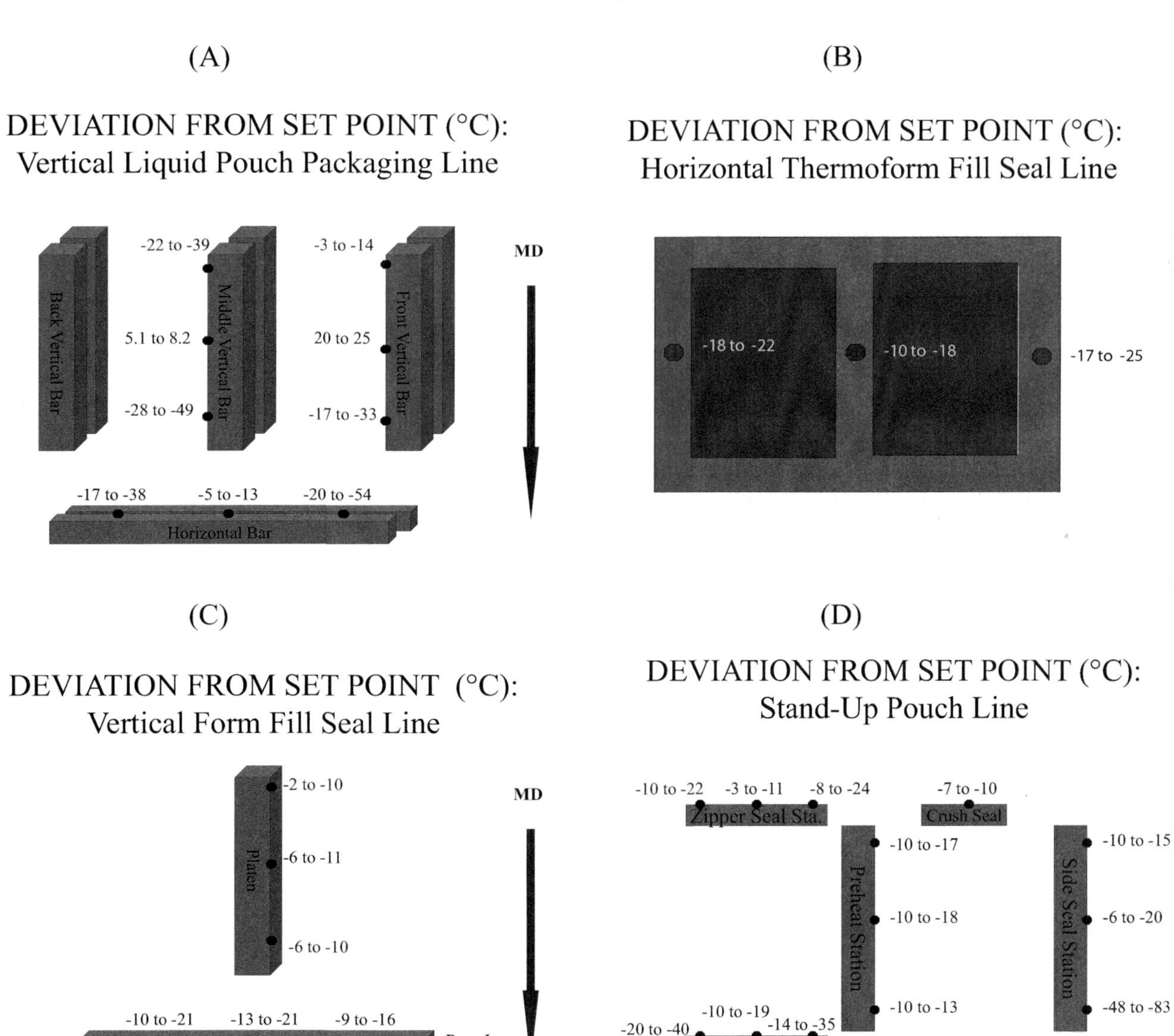

FIGURE 3.3 Measurements of temperature variation (deviation from setpoint) across heat seal bars on commercial-scale packaging lines: (A) vertical liquid pouch line, (B) horizontal thermoform fill seal line, (C) vertical form fill seal line, (D) stand-up pouch line. *Reproduced with permission from The Dow Chemical Company.*

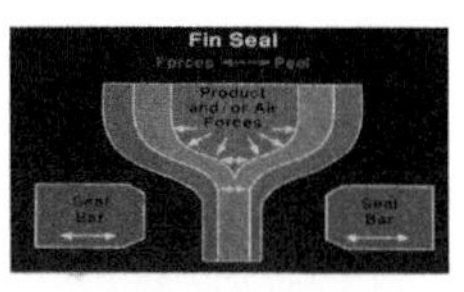

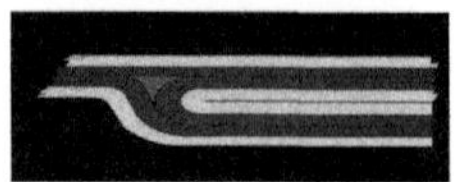

FIGURE 3.4 Importance of hot tack. *Reproduced with permission from The Dow Chemical Company.*

Some of the possible causes for such variation include maintenance and cleanliness of the machine, calibration of the machine's sensors, controlling thermocouple position (e.g., is it properly situated in the thermocouple well?), and machine design considerations. Advances have likely been made in temperature control since this study was made, but it underscores the amount of variation that is possible.

Pressure is another factor in seal performance. Sufficient pressure is needed for the surface of the two films to be in intimate contact for fusion to occur. Too much pressure, however, can result in squeezing out of the sealant from the seal area, causing poor seal strength and tearing. This is covered in more detail in Chapter 7. Variation in pressure across the seal bar can also result in seal issues. Misalignment of the seal jaws, bowing of long seal bars and crimp patterns are possible causes of pressure variation.

Gussets or folds in the seal area are a particular challenge for hot bar sealing. The folded areas are thicker than the rest of the seal area, resulting in longer times needed to obtain a good seal. Gussets may also affect the pressure distribution along the seal bar, with the gusseted areas experiencing high pressure and the other areas' lower pressures.

Gussets also tend to spring back after sealing while the seal area is still warm and the sealant has not had time to solidify. The resistance of the sealant to opening forces such as spring back in HFFS or the weight of the package in VFFS is known as a hot tack (see Fig. 3.4). It is also a factor for sealing around particulate contamination in the seal area (Section 7.4.1.5). Hot tack is an important consideration in selecting a sealant and is dealt with in considerable detail in Chapter 7. In general, polymers with greater melt strength have higher hot tack performance.

Contamination of the seal area with the powdery or greasy product during filling may affect seal performance. Equipment design can help reduce the potential for contamination, but in industrial environments, contamination is always a possibility. The film may build-up a static charge during handling that attracts powder contamination. Static charge mitigation is important for preventing contaminated seals. Optimum sealant selection may also help reduce sealing issues due to contamination. There is no generally accepted laboratory method for comparing sealing performance in the presence of contamination. This topic is discussed in greater detail in Chapter 7.

3.2 Brief description of packaging equipment

3.2.1 Pouch machines

Pouches come in a wide variety of forms, such as pillow packs, stick packs, gusseted packs, and stand-up pouches. Pouch styles are differentiated by their sealing. Some general types are depicted in Fig. 3.5. Pillow packs have straight top and bottom seals or end seals with a long seal up the side. The side seal maybe a lap seal, where the outside surface of the film seals to the inside surface of the same film, or a fin seal where the inside seals to the inside. In a gusseted end seal, the film is folded along the sides to provide a more rectangular profile. Three and four-sided seal pouches are other special forms. Stand-up pouches generally have gussets at the bottom to provide a flat surface, a fin or lap seal along the side and a flat seal on top.

Pouch machines come in a wide variety of designs but generally fit into two categories: VFFS or HFFS machines.

3.2.1.1 Vertical form fill seal

As pictured in Fig. 3.6, in a common form of a VFFS line a flexible packaging film is shaped over a mandrel (also referred to as a forming collar) into a tube. The bottom and side are sealed, the product is dropped into the tube from

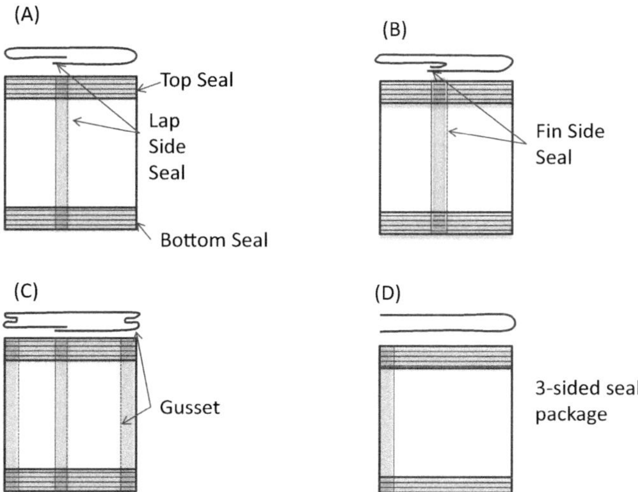

FIGURE 3.5 Types of seals found on pouches (A) lap seal, (B) fin seal, (C) gusseted seal, (D) 3-sided seal. Top insets show top view and arrangement of film for sealing. Gray areas designate seal regions [30]. *Adapted with permission from Moyer, G., Form/fill/seal, vertical, in: M. Bakker (Ed.), Wiley Encyclopedia of Packaging Technology, John Wiley & Sons, New York, 1986, pp. 367–369.*

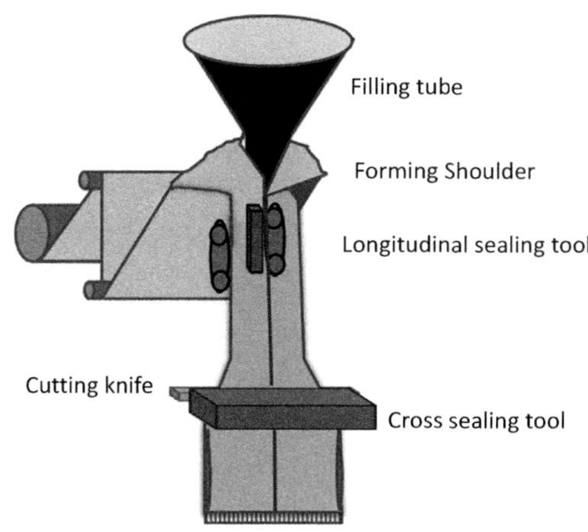

FIGURE 3.6 Vertical form fill seal line [30]. *Adapted with permission from G. Moyer, Form/fill/seal, vertical, in: M. Bakker (Ed.), Wiley Encyclopedia of Packaging Technology, John Wiley & Sons, New York, 1986, pp. 367–369.*

above, and the top is sealed (at the same time as the bottom seal of the next package) and cut. The sealed pouch drops onto a conveyor belt for secondary packaging and shipping.

Packaging speeds of up to 100 bags per minute are common. Bag sizes range from 13 to 64 cm (5–25 in.) for food packaging, and even longer for nonfood packaging. Typical products packaged on these lines include chips and other snacks, cereal, cookies (that are dump filled) and bagged candy. Often these packages form the inner package of a "bag and box" package structure, where the filled bag is stuffed into a cardboard box used as a secondary package. The outer package provides some degree of protection, but more importantly, provides a good form factor for sitting on a shelf as well as a print surface. VFFS is also often used for nonfood packaging, such as for polymer resins and fertilizer.

An important consideration for VFFS is a strong bottom seal that can withstand the weight of the product during filling. The seal is still molten at this point in the process. The ability of the seal to withstand an opening force before it solidifies is known as hot tack, an important sealing property that is covered in detail in Chapter 7.

The side seal may be a fin or lap seal, or in the case of a 3- or 4-sided seal, a conventional hot bar seal. The 3- or 4-side seal pouch or sachet is often used for condiments, shampoo and soap sachets, oil and sauces. Sachet machines can often run faster than pillow pack machines since the product does not drop as far.

Stick packs are long and thin and are often used for powdered drink mixes and pharmaceuticals.

3.2.1.2 Horizontal form fill seal

Various types of packages can be made on HFFS machines, including flow wraps, 3 or 4- sided seal pouches and stand-up pouches. In a horizontal flow wrap machine, the film is formed around a mandrel (forming box) in the horizontal dimension, filled with the product, sealed and cut (Fig. 3.7). For a 0.2 m (8-in.) package, these high-speed lines

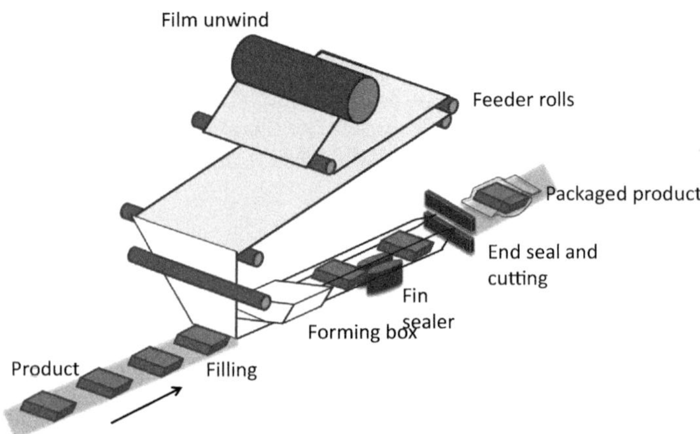

FIGURE 3.7 Horizontal flow wrap line [31]. *Adapted from A.A. Hughes, Food packaging machinery, in: Ebnesajjad (Ed.), Plastic Films Food Packaging, Elvesier, 2013, pp. 210.*

run up to 81 m (3200 in.) per minute, or 400 packages/min. Package size can vary from 0.13 to 0.4 m (5–16 in.) long and 0.013–0.18 m (0.5–7 in.) wide. There may be multiple lanes to increase the number of packages per minute. The filling may be automated or manual.

Some of the products typically packaged on flow wrapper lines include candy bars, cookies, crackers, medical products and disposable razors.

Three- or four-sided seal horizontal pouch machines are used to package a variety of dry soups, spices, cake mixes, and mixed ingredient food products. Lines can run at high speed (200 packs/minute or higher) with multiple fillers and one or two webs. A variety of sized pouches can be made from small (5 × 5 cm or 2 × 2 in.) to large (20 × 25 cm or 8 × 10 in.).

The use of stand-up pouches (SUP) has seen substantial growth over the last 20 years. The rollsock or film used in SUPs is often stiffer and thicker than other flexible packaging. It typically comprises of laminate structures of two or three plies to ensure stiffness (for the stand-up function), barrier, quality printing and sealing performance. Much of the early growth of this package form came from the replacement of metal cans, where stand-up pouches offer the following advantages:

- customizable sizes/shapes
- reduced raw material costs
- reduced storage costs
- reduced freight costs
- less material usage
- greater area for printing and better graphics.

Stand-up pouches still suffer from lower filling speeds versus cans. In recent years, stand-up pouches have replaced other forms of flexible packaging, mainly for their superior graphics and ability to stand-up and be seen (billboard effect) on the grocery shelf.

3.2.2 Horizontal thermoform fill seal

In a horizontal thermoform fill seal packaging line (Fig. 3.8), a bottom forming web is registered into a heating and forming station where a cavity is thermoformed into the web, typically with a vacuum and plug assist. The film then registers to a filling station where the product is laid into the cavity, either manually or automatically. A second "top" web is then laid down on top of the filled cavity, air is evacuated, and if necessary gases are added. (In a technology known as MAP, the gas environment inside the package is adjusted to extend the shelf life of the product. See Section 3.1.4.) The top and bottom web are then heat-sealed together. Often multiple lanes are thermoformed and filled in parallel. In a final step, the individual packages are cut from the web.

The bottom forming web is often different in composition and thickness than the top web. In particular, the bottom web is often thicker to accommodate the reduction in wall thickness during thermoforming. The bottom web may be flexible or semirigid with thicknesses up to about 35 mils (0.89 mm).

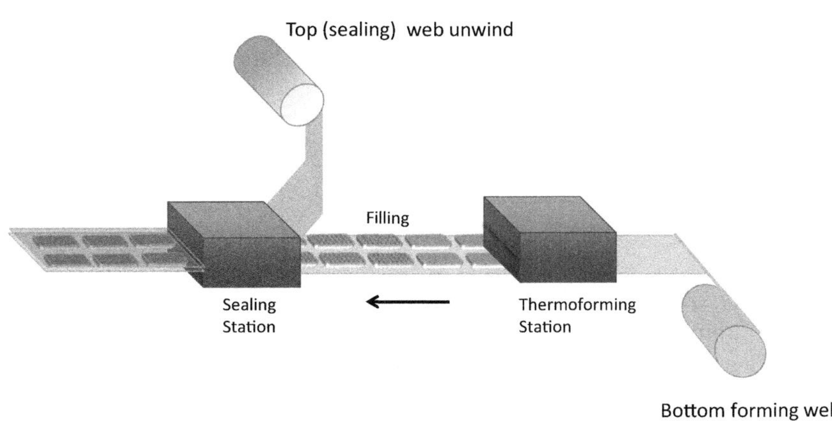

FIGURE 3.8 Horizontal thermoform fill seal machine.

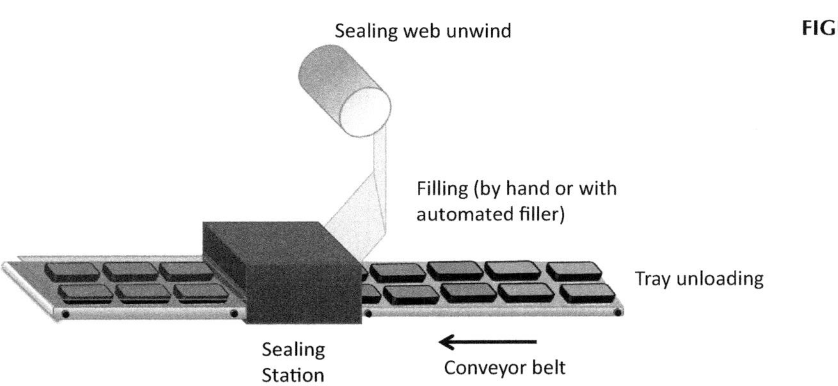

FIGURE 3.9 Tray sealing line.

Typical products packaged on these lines include block cheese, processed meat, fish, and medical devices. For medical device packaging, the top web is often a porous paper or nonwoven, such as DuPont's Tyvek, to facilitate sterilization by ethylene oxide.

3.2.3 Tray sealer

In a tray sealer (Fig. 3.9), preformed trays are filled with product and covered with a lidding film. The headspace air is evacuated and gas flushed (if necessary), and the lidding film is sealed to the tray. Machines are typically designed to handle a wide variety of tray materials such as paperboard, metal, and plastic (PET, PP, and PS are the most common).

Tray sealing machines are versatile since they can handle various bowls and cups in addition to trays. They are used for a variety of packaging applications, such as microwavable and dual ovenable meals, MAP for fresh meat, and dairy goods.

3.2.4 Vacuum packaging

Vacuum packaging is a version of MAP where the gases are evacuated and the film forms or collapses around the product. It is often used for the packaging of fresh meat products. There are four main methods:

1. Heat-shrinking a flexible bag around the product, such as primal meat.
2. Placing a preformed plastic bag around the product and evacuating the air with vacuum.
3. Using a HFFS process with vacuum.
4. Forming a film over the product where the product acts as the "forming mold" in a process known as vacuum skin packaging (VSP).

A film with good gas barrier is typically used to preserve the vacuum inside the package. VSP has caught on, particularly in Europe. In VSP the product is placed on a film or tray. The sealing film is heated and draped over the product as vacuum evacuates the atmosphere surrounding the product. The softened film is pulled down by the vacuum and takes the shape of the product like a second skin. For fresh meat packaging, VSP offers several advantages over MAP trays:

- Provides a transparent "skin" around the product,
- Helps retain the juices in the product and extends shelf life,
- Uses less packaging material.

3.2.5 Aseptic packaging

In aseptic packaging, the food and packaging materials are sterilized and maintained in a sterile environment throughout the packaging process. The rollstock is typically a multilayer barrier film or laminate. As it enters the machine it is sterilized by gamma radiation or hydrogen peroxide. The film is formed, filled, and sealed.

Some products packaged with aseptic packaging include juices and dairy products. Individual children's juice drinks are often packaged in this way, such as the Tetra Brick (Tetra Pak) introduced in the 1980s.

3.2.6 Primary retail display packaging

There are three main types of primary flexible packaging used for display of nonfood merchandize in the retail environment: clamshell, blister, and carded skin packaging. Shrink and collation films are often used for secondary packaging. Usage of stand-up pouches (previously discussed in Section 3.2.1.2) as primary packaging for retail products has grown but these packages generally are not transparent and do not put the product on display.

3.2.6.1 Clamshell packaging

Clamshell packaging consists of a clear thermoformed sheet, usually hinged, with a top and bottom portion. The product is placed in the cavities formed in the sheet and the sheet is welded in place.

Clamshell packaging has become a major form of packaging for retail goods, such as hardware goods, automotive parts and supplies, and consumer goods and household items. PVC and polyethylene terephthalate (PET) are two common materials used for retail clamshell packaging due to their combination of clarity, toughness, and low cost.

Graphics are typically provided by inserting a printed paper card within the clamshell.

3.2.6.2 Blister packaging

Blister packaging is a form of retail packaging where the product sits in a plastic blister on top of a paperboard. There are four main components to blister packaging: a preformed plastic blister, the paperboard card, a heat seal coating, and printing inks [32].

The preformed plastic blister is made of clear plastic such as PVC or PET. The type of plastic depends on many factors including the size of the product, whether the product has sharp or pointed edges, impact resistance needs, aging requirements, and cost.

The paperboard provides stiffness and a surface for heat seal and graphics. Its size and thickness depend on the product packaged and the stresses that the package will see during its use.

A heat seal coating is applied to the board to seal the blister in place. The coating must be compatible with the board (or ink) and blister.

The printing ink provides graphics and must bond well to the substrates.

There are several types of blister packages as shown in Fig. 3.10. In the conventional design, the blister fits over the product and its flange is heat-sealed to the board. In a fold-over card design, the board is folded so that the blister flange sits below the top portion of the board and above the bottom portion. The bottom and top portions of the board are sealed together away from the blister flange. A hole in the top portion is punched out to allow the blister flange to protrude. In the hinged blister/fold-over card variation, the blister folds over the product and is set in place by a sealed hinged blister. In a sandwich or double card design the blister is sandwiched and sealed between two boards with no hinge.

In a typical assembly operation, the blister is filled with the product, the paperboard is set on top and heat-sealed in place.

Blister packaging is also used in pharmaceutical packaging where aluminum foil and barrier films, such as Aclar polychlorotrifluoroethylene, are often used instead of paperboard and PVC to achieve good moisture barrier.

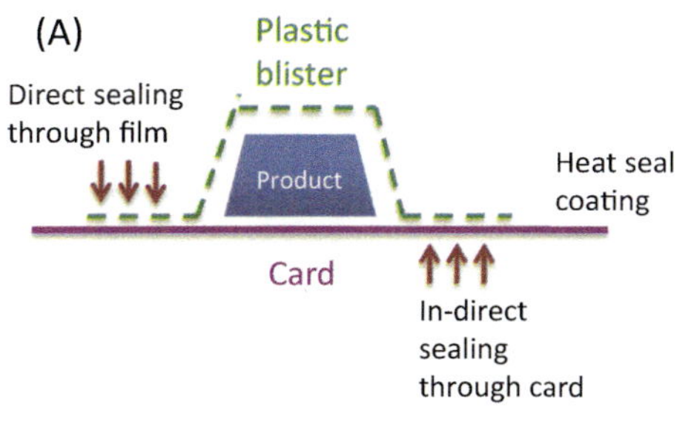

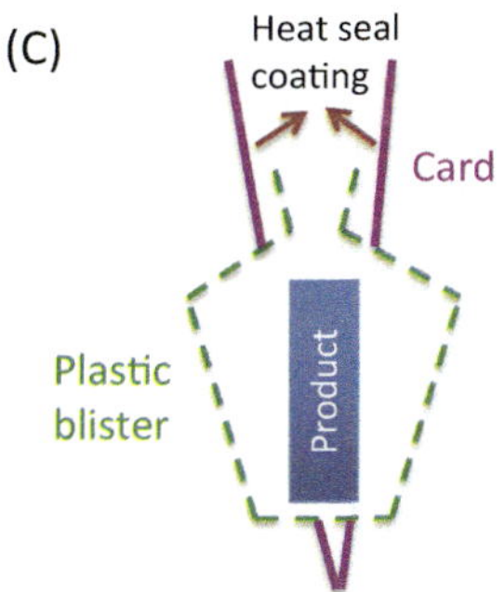

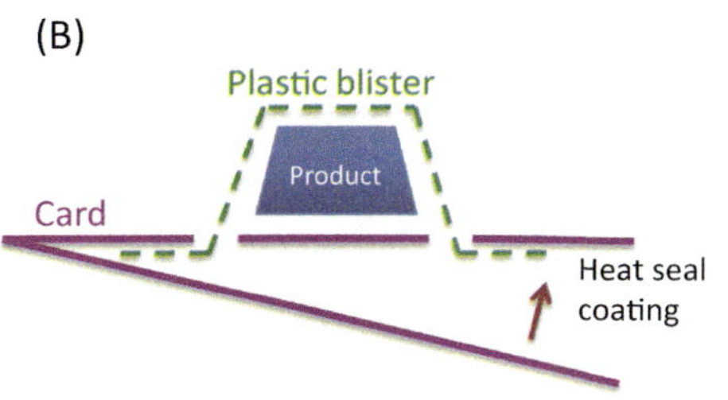

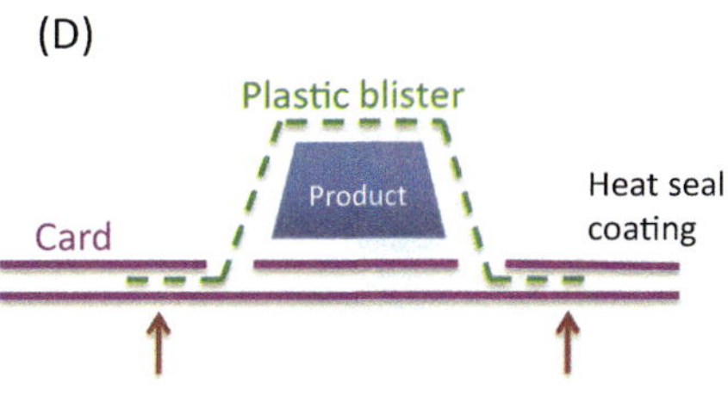

FIGURE 3.10 Forms of retail blister packaging: (A) conventional surface seal, (B) fold-over design, (C) hinged blister/fold-over card variation, (D) double card design. *Based on J.M. Gresher, Carded packaging, in: M. Bakker (Ed.), Wiley Encyclopedia of Packaging Technology, John Wiley & Sons, New York, 1986, pp. 124–132.*

3.2.6.3 *Skin packaging*

In skin packaging, the product serves as the mold over which a heated plastic film (skin) is draped. Vacuum is drawn through a porous board to pull the film against the product and board, conforming it to the shape of the product and creating a heat seal. Its application in food packaging is discussed in Section 3.2.4.

Retail skin packaging has four main components: plastic film, porous paperboard, heat seal coating, and ink. The plastic film is typically low-density polyethylene (LDPE), PVC or ionomer. PVC is clear and inexpensive. LDPE is inexpensive, but lacks the clarity of PVC or ionomer. Ionomers provide better abrasion and puncture resistance, superior draw-ability for high profile items, excellent resistance to board curl, better gloss, cold crack resistance, faster cycle time, and are cost-effective due to their higher yield per unit weight.

The paperboard must be porous enough for the vacuum to draw the film down around the product. For this reason, solid bleached sulfate board is often used. Clay coatings are generally not utilized since these reduce the porosity of the board. (Clay coatings sometimes are applied to paperboard to smooth the surface for high-quality printing and graphics.) In some cases the board may be perforated to achieve the desired porosity.

The board is coated with a heat seal coating, typically applied as a latex or water-based dispersion. Ethylene-acrylic acid copolymer dispersions are often used for good compatibility with LDPE and ionomer, as well as good sealability.

Inks are used for printing the board. The ink needs to be compatible with both the board and the heat seal coating.

In the skin packaging machine (Fig. 3.11), product is placed on the printed heat seal coated paperboard. The plastic film is held in a frame above the product, heated and draped over the product, while vacuum is pulled through the porous paperboard from below. The residual heat in the film activates the heat seal coating, forming a bond between the film and board.

3.2.6.4 *Comparison of retail display packaging types*

Consumer products companies are generally looking for packages that meet the following requirements:

- Transparent so that the consumer can see the product,
- Tamper evident or resistant to pilferage (to reduce in-store losses to theft),
- Easy to display (typically on a peg or shelf),
- Ability to have graphics to inform the consumer about the benefits of the product.

Choosing between the package forms is not always a clear choice. Table 3.4 gives some advantages and disadvantages for each type.

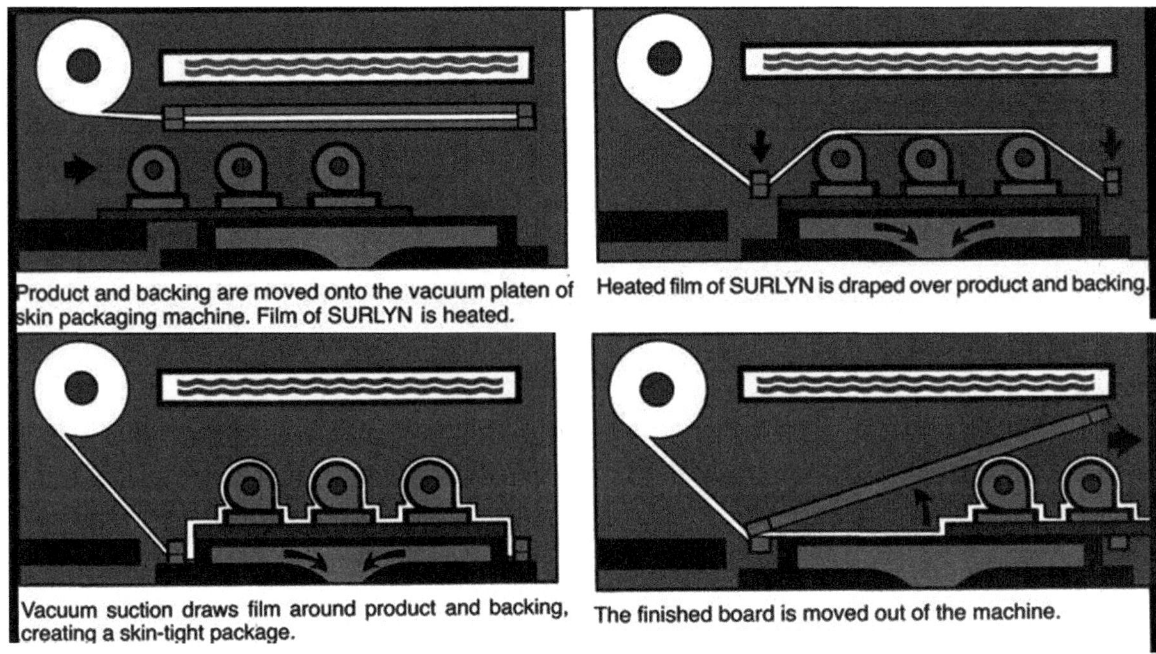

FIGURE 3.11 Skin packaging process. *Reproduced with permission from E. I. du Pont de Nemours and Company.*

TABLE 3.4 Comparison of retail packaging types.

Attribute	Clamshell	Blister	Skin
Cost	Highest material costs	May have advantage over Skin Packaging if product is much smaller than the card	May have advantage over Blister Packaging if product is about the same size as the card
High speed/ automation	Good	Good	Typically packed on smaller, manual machines; more labor intensive
Transparency, ability to see the product	Good	Good	Good. Allows the consumer to interact with the product
Graphics/ printing	Achieved with paper inserts; generally low "billboard space"	Best graphics: Paperboard may be clay coated to achieve excellent graphic quality; large "billboard space"	Largest "billboard space," Clay coating not used due to its effect on porosity
Retail space	Typically bulky, taking up more retail space	Easy for pegged display	Easy for pegged display
Ease of opening	Difficult to open (good for stopping in-store theft, bad for consumers)	Easier to open than clamshell	Can be designed for easy open or difficult to open
Other	Perception of higher quality; Easily allows packaging multiple items in single package		Difficult to package round items; Holds products in place during distribution

Further reading on packaging equipment

- Yam, K. L., 2009, *Wiley Encyclopedia of Packaging Technology*, John Wiley & Sons.
- Paine, F. A., Paine, H. Y., 1992, *A Handbook of Food Packaging*, Springer, pp. 97–158.

References

[1] ASTM International, D1894-11: Standard test method for static and kinetic coefficients of friction of plastic film and sheeting, 2011.

[2] Eval Americas, Thermoforming of Eval® resin containing structures, Technical Bulletin No. 150, n.d. Available from: <http://www.eval.eu/media/36418/ea%20-%20technical%20bulletin%20no%20150.pdf> (accessed 13.06.20).

[3] Eval-Americas, EVAL™ products, n.d. Available from: <http://www.eval-americas.com/us/about-eval/eval-products.aspx> (accessed 13.06.20).

[4] G. Nippon, Grade lineup, n.d. Available from: <http://www.soarnol.com/eng/grade/index.html> (accessed 13.06.20).

[5] R.T. Chou, G.I. Deak, (DuPont) Blends of ethylene vinyl alcohol copolymer and amorphous polyamide, and multilayer containers made, U.S. Patent 4,990,562, 1991.

[6] A.I. Fetell, Nylon-containing ionomer improves the performance of ethylene vinyl alcohol (EVOH) copolymers in multilayer structures, in: TAPPI Polymers, Laminations & Coatings Conference, 1997, pp. 503–505.

[7] W.A. Oberle, (DuPont), Process for thermoforming EVOH barrier sheets and products relating hereto, European patent EP0699125 B1, 1998.

[8] D. Madison, Filling machinery, by count, in: M. Bakker (Ed.), Wiley Encyclopedia of Packaging Technology, John Wiley & Sons, New York, 1986, pp. 294–295.

[9] R.F. Bardsley, Filling machinery, dry-product, in: M. Bakker (Ed.), Wiley Encyclopedia of Packaging Technology, John Wiley & Sons, New York, 1986, pp. 295–300.

[10] M. Mullan, D. McDowell, Modified atmosphere packaging, in: R. Coles, D. McDowell, M.J. Kirwan (Eds.), Food Packaging Technology, Blackwell Publishing and CRC Press, Oxford, England, 2003, pp. 303–339.

[11] J. Brandenburg, Modified atmosphere packaging, in: K.L. Yam (Ed.), Wiley Encyclopedia of Packaging Technology, 3rd Edition, John Wiley & Sons, 2009, pp. 787–794.

[12] D. Considine, Ultransonic Packaging Sealing - Technology and Applications Review, Herrmann Ultrasonics, Bartlett, IL, 2007. March 22, 2007.

[13] S. Bach, N. Bunk, J.-P. Majschak, Effect of laminate structure and morphology on fracture mechanics and seam strength of sealed seams, in: TAPPI European PLACE Conference, Dresdon, Germany, 2013.

[14] J. Dun, S. Bensasson, K. Thürling, Polyolefins for ultrasonic sealing, in: TAPPI European PLACE Conference, Dresdon, Germany, 2013.

[15] K. Thurling, S. Bach, J.-P. Maschak, M. Nase, Influence of material parameters on the ultrasonic sealing behavior of polymer films for packaging, in: 14th International Scientific Conference "Polymer Materials", Halle, PT67, 2010.

[16] K.S. Suresh, M.R. Rani, K. Prakasan, R. Rudramoorthy, Modeling of temperature distribution in ultrasonic welding of thermoplastics for various joint designs, J Mater Process Technol 186 (2007) 138–146.

[17] N. Tateishi, T.H. North, R.T. Woodhams, Ultrasonic welding using tie-layer materials. Part I: Analysis of process operation, Polym Eng Sci 32 (9) (1992) 600–611.

[18] Lako Tool, Solutions, n.d. Available from: <http://www.lakotool.com/solutions.htm> (accessed 14.08.13).

[19] H.W. Theller, Heat sealability of flexible web materials in hot-bar sealing applications, J Plastic Film & Sheeting 5 (1) (1989) 66–93.

[20] W.B. Bower, Heatseal die, Crown Zellerbach Corporation, U.S. Patent 4,582,555, 1986.

[21] L.M. Montano, L.E. Smith, Split crimper for heat sealing packaging material, Lako Tool and Manufacturing, U.S. Patent 8,387,344B2, 2013.

[22] R.L. Knauf, D.F. Menzner, Seal for plastics capable of thermoplastic sealing and means for effecting such sealing, Marathon Cheese Corporation, U.S. Patent 4, 001, 075, 1977.

[23] L.E. Smith, Crimper bar set for forming hermetically sealed packages, U.S. Patent 4, 807,426, 1987.

[24] L.E. Smith, Crimper assembly for sealing overlapping portions of a sheet of packaging material, Lako Tool & Manufacturing, Inc., U.S. Patent 6,230,781B1, 2001.

[25] D.J. Stevens, Sealing assembly, Cryovac Australia Pty. Ltd., U.S. Patent 6,877,542 B2, 2005.

[26] L.M. Montano, L.E. Smith, Split crimper for heat seal packaging material, Lako Tool and Manufacturing, U.S. patent 7,941,991 B2, 2011.

[27] L.M. Montano, L.E. Smith, Crimper assembly for sealing overlapping portions of a sheet of packaging material, Lako Tool and Manufacturing, U.S. Patent 6,736,182 B2, 2004.

[28] F.C. Stehling, P. Meka, Heat sealing of semicrystalline polymer films. II. Effect of melting distribution on heat-sealing behavior of polyolefins, J. Appl. Poly. Sci 51 (1994) 105–109.

[29] B.A. Morris, Predicting the heat seal performance of ionomer films, J Plastic Film & Sheeting 18 (July) (2002) 157–167.

[30] G. Moyer, Form/fill/seal, vertical, in: M. Bakker (Ed.), Wiley Encyclopedia of Packaging Technology, John Wiley & Sons, New York, 1986, pp. 367–369.

[31] A.A. Hughes, Food packaging machinery, in: Ebnesajjad (Ed.), Plastic Films Food Packaging, Elvesier, 2013, p. 210.

[32] J.M. Gresher, Carded packaging, in: M. Bakker (Ed.), Wiley Encyclopedia of Packaging Technology, John Wiley & Sons, New York, 1986, pp. 124–132.

Part III

Material basics

Chapter 4

Commonly used plastics and substrates in flexible packaging

Material selection is the key to packaging performance. This chapter introduces some of the commonly used materials in a flexible package and their important characteristics, beginning with a discussion of the basic functions they must provide: structural integrity, barrier, sealing, adhesion and esthetics. A brief introduction to polymer chemistry as it relates to package performance is provided followed by brief descriptions of polymers and substrates. The polymer resins include polyolefins, polyamides, barrier resins, polyesters, and bio-based polymers. Additives that enhance the performance of these resins are reviewed. Substrates include polypropylene, polyethylene terephthalate (PET), aluminum foil, metallized films, cellophane, and paper/paperboard. A final section describes regulatory considerations such as compliance with food contact regulations.

Plastics have become an integral part of many flexible package structures. Plastics are very large molecules called polymers that can be molded or shaped into films and other articles. Plastics is an engineering or application-driven term whereas polymers is a more scientifically based term. All plastics are polymers but not all polymers are plastics. Natural materials like cotton, wool, DNA and proteins are polymers but not considered plastics. This book will ignore the nonplastic types of polymers and use the terms plastics and polymers interchangeably. The packaging industry also uses the term resin to describe plastics used in packaging. While some industries use the term "resin" to mean low molecular weight (MW) polymers such as tackifiers or epoxies, this book will use the term to mean high MW plastics.

4.1 Polymer resin and substrate function

As introduced in Chapter 1, the basic purpose of a flexible package is to provide protection, communication and promotion. Material attributes that help provide this functionality may include such properties as mechanical strength, stiffness, puncture resistance, tear resistance, optical transparency, barrier, sealing, adhesion, esthetics and printability. A specific combination of materials is selected to meet the functional requirements in the most economical manner, and increasingly, in the most sustainable way. Often materials used in flexible packaging provide more than one function. For example, high-density polyethylene (HDPE) in a cereal liner bag-in-box structure provides stiffness, puncture resistance, and moisture barrier. As more layers are incorporated into flexible packaging, the contribution of each of the layers has become more specialized.

4.1.1 Structural integrity

Structural integrity involves ensuring the package has enough strength to hold up during the packaging process and the end-use, has sufficient stiffness to be easily handled on the packaging line and in some cases for the end-use (such as for stand-up pouches), and survive abuse. Some of the specific properties that are used to gauge whether the material provides structural integrity are tensile strength, yield strength, elongation at break, modulus, tear strength, impact strength, and puncture resistance.

Stiff materials such as paper, oriented polypropylene (OPP), or oriented polyester (OPET), are generally used for structural integrity.

The temperature the package will be exposed to during its use is a factor in selecting materials for structural integrity since many properties depend on temperature. High temperatures may melt or distort the polymer or substrate, resulting in poor performance. Low temperatures can also be a problem, especially when it comes to toughness. Many materials become brittle at low temperatures.

Material choice and structure design considerations for structural integrity is covered in detail in Chapter 9.

The Science and Technology of Flexible Packaging. DOI: https://doi.org/10.1016/B978-0-323-85435-1.00017-X

4.1.2 Barrier

Moisture, oxygen, carbon dioxide, and light barrier are the four most common barriers needs in flexible packaging. Oil and grease barrier is another need for some applications. The amount of barrier required depends on the product being packaged and the shelf-life requirements. A gas barrier is generally characterized by the permeability of the gas through the film. Permeability is inversely related to barrier; the higher the permeability the lower the barrier performance. The gas barrier of many resins and substrates is influenced by environmental factors such as temperature and relative humidity (RH). It is important, therefore, to select materials that have the required barrier performance under the conditions of the end-use.

Highly crystalline polyolefins such as HDPE and polypropylene (PP) are often the most cost-effective moisture barrier materials. Polyvinylidene chloride (PVDC), cyclic olefin copolymers (COCs), and chlorotrifluoroethylene (CTFE) are used for more specialized applications. Aluminum foil and metalized film are also used for a moisture barrier.

Polar polymers such as ethylene vinyl alcohol (EVOH), polyvinyl alcohol (PVOH), and PVDC provide outstanding oxygen barrier performance. Aluminum foil and metalized film also provide an exceptional oxygen barrier. Polyamide 6 (PA6) and PET provide a moderate barrier to oxygen.

Light barrier is typically provided by adding opacifiers such as titanium dioxide (TiO_2) into one or more layers, through metallization, by incorporating aluminum foil or paper, or by printing.

Oil and grease resistance is generally improved with higher crystallinity and/or polarity of the polymer. HDPE provides better oil resistance than LDPE due to its higher crystallinity. Polar polymers such as PA, PET, and PVDC also provide excellent oil resistance.

Barrier needs and technology are covered in detail in Chapter 8.

4.1.3 Sealing

The sealant layer is the innermost layer of the flexible packaging structure. Its chief function is to provide a hermetic seal to keep the contents of the package in and contamination out. For high-speed packaging operations, a low seal initiation temperature is desired to achieve short sealing times. Heat is applied to the outside surface of the film and transfers inward, melting the sealant. The seal initiation temperature is the minimum temperature at which seal strength is obtained (see Chapter 7 for details.) The melting point of the polymer is a primary determinant of the seal initiation temperature; above the melting point, the polymer chains at the interface have the mobility to interdiffusion and build strength during the sealing process. Therefore polymers with low melting points generally make good sealants. As a class, the ethylene copolymers such as ethylene vinyl acetate (EVA) copolymer, low-density polyethylene (LDPE), linear low-density polyethylene (LLDPE), polyethylene plastomers and ionomers have low melting points and provide low seal initiation temperature.

Other factors involved in selecting the sealant resin are hot tack (the ability of the seal to withstand opening forces prior to solidification), seal through or around a product that may contaminate the seal area (such as greasy food or powders), and the ability to flow or calk voids formed near folds or around contaminant particles. The sealant may also contribute to other functions of the package such as optical clarity, puncture resistance and stiffness.

Equipment aspects of sealing are described in Chapter 3 and material and package design aspects are covered in Chapter 7.

4.1.4 Adhesion

Many of the resins and substrates used in multilayer flexible packaging do not bond well to each other. Adhesives are used to tie the layers together so that the package behaves as a single entity. There are generally two types of adhesives used in flexible packaging: liquid adhesives used in adhesive lamination and polymer adhesives used in coextrusion or extrusion coating and lamination. The liquid adhesives use water or organic solvents or are "solventless" adhesives that are designed to bond films together during lamination. They require heat after application to remove the solvent or water, or to provide curing (although some adhesives cure by exposure to UV light or other means).

The polymer adhesives are applied in the molten state and require cooling after application. Ethylene copolymers such as EVA, ethylene methyl acrylate (EMA) copolymer, ethylene acrylic acid copolymer, and special maleated grades of PE are often used as tie resins. The polar comonomer helps bond to polar substrates whereas the nonpolar ethylene component of the copolymer helps tie to nonpolar substrates such as PE. The converting process may also impart polarity and functionality. LDPE is often run at elevated temperatures (~320°C) in extrusion coating to promote the formation of acids, aldehydes, and ketones that help bond to aluminum foil.

Details on adhesion technology and the effect of processing on adhesion are provided in Chapters 10 and 15, respectively.

4.1.5 Esthetics

Optical clarity and transparency are important for some packaging applications. For these, the substrates need to be selected to have good transparency in addition to performing their primary function. For example, a clear barrier polymer such as EVOH or PVDC would be favored over an opaque barrier such as a metallized film for a transparent package.

Printability is another factor in selecting substrates. For good printing, paper may be calendered to provide a smoother surface and/or treated with flame, corona or plasma discharge to clean the surface and increase surface energy. Many polymer film substrates are also corona or plasma treated to increase surface energy prior to printing. OPET is often favored in applications requiring a stable print surface.

4.1.6 Typical properties

Table 4.1 provides some typical properties and test methods that characterize the functionality of various resins and substrates used in flexible packaging. These are just a starting point, as many applications have more targeted performance tests that have been developed to determine if a packaging structure will meet the needs of the end-use. Most of these properties are material properties. A key consideration is translating material properties into a set of specifications for packaging performance. For example, what tensile strength or impact resistance test values are needed for the package to survive the packaging and the transportation process? Often, material specifications are just a starting point. In the ultimate assessment, the structure is prototyped and tested on an actual packaging line and through distributrion channels to simulate the abuse it may be exposed to during packaging, transportation, distribution, and retail.

Table 4.2 lists some of the common resins used in flexible packaging and the functions they most often provide. Table 4.3 provides a range of typical properties for several frequently used resins and film substrates in flexible packaging. These data were taken from a number of sources. A common problem with comparing material properties from multiple sources is that differences in test methods and the method of fabricating the samples may have a significant impact on the measured values. For example, films made on the cast and blown film processes will likely have differences in machine and transverse direction physical properties. Even with samples prepared on the same type of fabrication process, differences in processing conditions may impact the properties. One should be careful about comparing property data from different sources.

4.2 Commonly used plastics in flexible packaging

This section provides a brief introduction to frequently used plastics in flexible packaging, highlighting their chemistry, key properties and typical applications.

4.2.1 Polymer chemistry as it relates to packaging performance

The term polymer comes from the Greek "poly" and "mer", meaning many mers or repeat units. Polymers are long-chain molecules made up of numerous reoccurring units or groups of atoms. A number of aspects of polymer chemistry and technology are introduced in this section that lay the groundwork for understanding the chemistry of the polymers used in packaging.

4.2.1.1 Molecular weight

The MW of a polymer is directly related to the number of repeat units on the polymer chain. The longer the chain, the higher the MW. The polymers used in flexible packaging generally have a high MW. Polymers with low MW tend to be brittle since individual chains may be easily pulled away from the others. Polymers gain strength once they are long enough to become entangled. The entanglement MW is a characteristic of the polymer that varies with chemistry. For the most part, polymers with an MW of greater than 1000 repeat units are used in packaging [21].

TABLE 4.1 Commonly measured properties and test methods for flexible packaging films.

Property	ASTM method	Why it is important
Density	D792	Enters into cost (per area), needed for layer control
Melt flow rate or melt index	D1238	Indicates flowability in a polymer process; inversely related to viscosity and molecular weight; typically used for polyolefins
Viscosity (melt or solution)	Various	Indicates processability; related to molecular weight
Secant or tensile modulus	D882	Helps determine how stiff the package will be; also related to strength, integrity, impact resistance, and puncture resistance
Tensile strength	D882	Important for gauging the strength of the package
Elongation at break	D882	Related to toughness
Tear strength	D1922 (Elmendorf) D1004 (Graves)	Notched or un-notched tear resistance. Related to toughness. Also important for some easy open applications
Melt temperature (semicrystalline polymers)	D3418	Related to seal initiation temperature; determines extrusion temperature
Glass transition temperature (Tg)	E1356 (DSC) E1640 (DMA)	Material becomes brittle below Tg and may lead to failure; Tg defines lower use temperature
Freezing temperature	D3418	Affects how fast the film quenches in converting processes
Vicat softening temperature	D1525	An indication of temperature resistance and resistance to deformation
Impact strength	D3420 (Spencer) D1709 (Dart Drop)	High speed impact toughness
Puncture resistance	F1306, D2582	Slow speed puncture resistance
Clarity, haze, gloss	D1746 (clarity) D1003 (haze) D2547 (gloss)	Optical properties
Oxygen permeation value	F1927, D3985, F2622	Inversely related to oxygen barrier
Moisture vapor transmission rate (MVTR)	F1249, E96	Inversely related to moisture barrier

4.2.1.2 *Molecular weight distribution*

An inherent consequence of the polymerization process is that not all the polymer chains are the same length. There are a number of ways to characterize molecular weight distribution (MWD), the two most common being the number average molecular weight (M_n) and the weight average molecular weight (M_w). Mathematically, these are given by

$$M_n = \sum_{i=1}^{\infty} n_i M_i = \frac{\sum_{i=1}^{\infty} N_i M_i}{\sum_{i=1}^{\infty} N_i} \tag{4.1}$$

$$M_w = \sum_{i=1}^{\infty} w_i M_i = \frac{\sum_{i=1}^{\infty} N_i M_i^2}{\sum_{i=1}^{\infty} N_i M_i} \tag{4.2}$$

TABLE 4.2 Function of commonly used plastics in flexible packaging.

Resin	Sealant	Tie-layer adhesive	Moisture barrier	Oxygen barrier	Structural layer
LDPE	X	X			X
LLDPE	X				X
HDPE			X		X
PP	X		X		X
PS					X
PET	x				X
PVC					X
EVA	X	X			X
Ionomer	X	X			
EMA	X	X			
ACR (EAA or EMAA)	X	X			
Tie resin		X			
COC			X		X
PA				X	X
EVOH				X	
PVDC			X	X	
PLA	X				X

Note: Abbreviations can be found in Appendix A.

where M_n is the number average molecular weight, M_w is the weight average molecular weight, N_i is the number of molecules of chain length i, M_i is the molecular weight of molecules with chain i, n_i is the number fraction of molecules of chain length $i\left(=\frac{N_i}{\sum_{i=1}^{\infty} N_i}\right)$, and w_i is the weight fraction of molecules of chain length $i\left(=\frac{N_i M_i}{\sum_{i=1}^{\infty} N_i M_i}\right)$.

M_n is highly influenced by the short-chain molecules, whereas M_w is more heavily influenced by the longer chains. In general, $M_n < M_w$ unless all the chains are the same length. The use of M_n over M_w or other representations depends on the property; some properties correlate well with M_n, others with M_w or other variations of average MW. Modern techniques to measure MW and MWD typically employ gel permeation chromatography (GPC), which involves dissolving the polymer and detecting the size via a number of sensors, which taken together can provide information on M_n and M_w.

The ratio of M_w to M_n is known as the polydispersity index, PDI:

$$\mathrm{PDI} = \frac{M_w}{M_n} \tag{4.3}$$

Polymers such as PA or PET made by condensation polymerization have a theoretical PDI of 2. Polyolefins, made via a free radical or coordination catalyst polymerization process, generally have a higher PDI.

MWD influences the flow or rheological properties of the polymer. In general, broad MWD polymers tend to be easier to melt process. Their viscosities tend to decrease with increasing shear rate, which has some advantages in a number of polymer processes. This is discussed in more detail in Chapter 5.

4.2.1.3 Crystallinity

In some polymers, the chains organize into ordered structures known as crystals. Steric hindrances and other considerations prevent all the chains from participating in the crystals; hence this class of polymers is known as semicrystalline

TABLE 4.3 Typical property values of commonly used resins in flexible packaging.

Property	Method	Units	LDPE	LLDPE	m-LLDPE	mPE plastomer	EVA (18% VA)	Ionomer	HDPE	BOPE	Cast PP	OPP	BOPET	BOPLA	PA6	EVOH (32 m% ethy)
Density	ASTM D792/ D1505	g/cc	0.92–0.93	0.91–0.93	0.91–0.93	0.9–0.915	0.94	0.95	0.96	0.93	0.9	0.9	1.39	1.24	1.15	1.18
Tensile Strength at Break (MD/TD)	ASTM D882	MPa	26/22	50/35	53/41	60/45	31	24–37/ 23–41	80/37	60/100	45/38	140/ 240	200/235	62/90	90–120	34
Elongation at Break (MD/TD)	ASTM D882	%	130/540	570/850	500/700	600/600	530	300–500/ 300–450	420/4	260/ 100	650/700	180/60	116/91	180/ 100	300–900	15
Secant Modulus (MD/TD)	ASTM D882	MPa	240/290	190/220	192/201	92/92	48	150–480	1100/ 1600	350/ 510	480/450	1720/ 3100	4900/ 5100	3340/ 3780	600–1000	4500
Elmendorf Tear (MD/TD)	ASTM 1923	g/micron	10/2.7	11/30	8/17	8/20	7	0.1–1.5	18/1.6	23/11	0.8/3.5		1.2/1.2	0.6/0.5	2/5	0.5/0.6
Dart Impact strength	ASTM D1709A	g/micron	3.1	4	22	>16		6–14	< 1.5	25	6					
Melting Point	DSC	°C	110–113	121	119	99	89	92–100	135	125	145–161	161	254		220	183
Vicat	ASTM D1525	°C	96	94	102	86	69	65–79	125–129					57 (Tg)		172
MVTR, 38 C, 90% RH		g-mil/m2-day	15–20	12–18	12–18	18–23	110	18–24	6–12		12.4	6	20	466	70	75.6
OPV, 23 C, 45% RH		cc-mil/m2-d	8000	8500	8500	8000–12000	15000	5200–8200	3000		2790	2300	70	1245	280	0.3
References			[1,2]	[3,4]	[4,5]	[4,6]	[7]	[8]	[4,9]	[10,11]	[12]	[13,14]	[1,15,16]	[17,18]	[19]	[20]

polymers. The noncrystalline regions are known as amorphous regions. Common semicrystalline polymers include polyethylene (PE), PP, PA6, EVOH, and PET.

Some polymers such as polystyrene (PS) are amorphous; they lack any crystalline regions.

Semicrystalline polymers have melting and freezing points, although these generally occur over a broader temperature range than low MW molecules. The breadth of the melting point range impacts heat seal and other properties. For example, the melting point of conventional LLDPE is often reported at 121°C, which represents the peak of the melting endotherm in a differential scanning calorimeter measurement. However, the crystals do not completely melt until around 130°C. It is this temperature that corresponds to the ultimate seal strength [22].

The long-chain nature of polymers also impacts the rate at which the polymer freezes from the melt. Some polymers form crystals relatively slowly so that the fabrication process can have a substantial impact on final properties. An example is PET. Its slow rate of crystallization is taken advantage of when making film and sheet. In the APET (amorphous polyester) process, the film/sheet is rapidly quenched so that the PET remains in the amorphous state (polymer melts are amorphous since all the crystals have been melted). The resulting sheet is transparent and pliable but has poor temperature resistance. Applications include thermoformed containers for refrigerated foods. The amorphous sheet can be stretched into biaxial oriented film [biaxial oriented polyethylene terephthalate (BOPET)]. During the stretching process, the polymer chains align and organize into crystals, producing a transparent, tough semicrystalline film. In the crystalline PET (CPET) process, a nucleating agent is added to the PET so that when it is quenched into a sheet it crystallizes. The crystals form large structures called spherulites that diffract light. The resulting sheet is opaque but has good temperature stability and heat resistance. Trays made from CPET are used for packaging dual ovenable foods, which can be heated up by the consumer in a microwave or convection oven. A consequence of the large crystals is poor low-temperature toughness; ethylene copolymers are often added for freezer applications to improve impact resistance [23–26].

The above example shows that the presence and morphology of the crystals affects properties. Some of the typical properties influenced by crystallinity include:

- Stiffness. The mobility of the polymer chains is impeded by the presence of the crystals, increasing the stiffness or modulus.
- Melting point. Higher levels of crystallinity often result in higher melting point and larger heats of fusion.
- Barrier. The crystals may impede the transport of liquids or gases through the polymer, reducing permeability.
- Optical properties. The crystals often are large enough to diffract light, causing a reduction in transparency and an increase in a haze.

PE is a semicrystalline polymer. The amount of crystallinity is controlled by the level of short-chain branching and is characterized by density. Density is inversely proportional to a specific volume. Polymer with a higher level of crystallinity has a lower volume per unit mass or higher density since the chains are packed closer together in the crystalline phase than in the amorphous phase. HDPE has a crystallinity of around 50%–60% and low-density PE (LDPE) around 20%–30%. Compared with LDPE, HDPE has a higher moisture barrier [lower moisture vapor transmission rate (MVTR)], a higher melting point (135°C vs 110°C), higher modulus [760 MPa (110 kpsi) vs 140–210 MPa (20–30 kpsi)] and higher haze.

The size, shape, and organization of the crystals, known as crystalline morphology, influence properties. The differences in crystalline morphology in OPET and CPET are discussed above. Another example is PP. Standard PP forms large spherulites when made into sheet or film. Nucleating agents are often added to PP to improve its clarity. The nucleating agent forms many sites upon which the crystals can initiate as the melt is cooled. This not only increases the speed of crystallization but because so many additional sites have been created, the resulting size of the growing crystals is smaller. The smaller crystals diffract less light. The result is clarified PP with low haze.

Quench rates and orientation during film fabrication can affect the amount of crystals and their morphology. In general, a faster quench from the melt to solid results in a lower level of crystallinity. Cast film often has lower haze than blown film due to the faster quench rate in the cast film process. Orientation also influences the morphology of the crystals. The crystals in PE, for example, have been observed to change from a twisted ribbon to a row nucleation morphology with low and high orientation processes [27].

A second class of polymers is amorphous polymers which have no crystallinity. An example is PS. The characteristic temperature for amorphous polymers is the glass transition temperature (Tg). Above Tg the polymer is said to be in its rubbery state; the polymer is tough but less stiff and may creep. Below Tg the polymer is in its glassy state; it is much stiffer but can be brittle.

4.2.1.4 Polymerization process

The polymerization process affects molecular characteristics such as MW, MWD, and branching. The three main processes for manufacturing polymers used in flexible packaging are briefly described below.

4.2.1.4.1 Free radical

Free radical or addition polymerization involves a chain reaction with unsaturated monomers, such as vinyl molecules, forming long polymer chains. Vinyl molecules have a double bond of the type $CH_2 = CHX$ and $CH_2 = CXY$ where X and Y represent a number of possible chemical moieties, such as a hydrogen atom (ethylene), a methyl group (propylene), or a benzene ring (styrene). Not all vinyl molecules polymerize into high MW polymers. Homopolymers of ethylene (resulting in LDPE), various copolymers of ethylene and vinyl monomers such as vinyl acetate, and polystyrene are the most important examples of free radical polymers used in flexible packaging.

Vinyl free radical polymerization involves several steps:

- **Free radical generation:** Typically, thermal or photochemical degradation of peroxide is used to generate intermediates known as free radicals, which result in very reactive unpaired electrons.
- **Initiation:** The free radicals combine with the vinyl molecule to add to the double bond, creating a new radical. The regeneration of the free radical is the hallmark of a chain reaction.
- **Propagation:** The chain radical adds successive monomers to grow the polymer chain.
- **Termination:** Propagation could go on indefinitely until all the monomer is consumed but for the fact that some free radicals combine to form a paired-electron covalent bond and a loss of free radical activity. Typically, the reaction is run so that only a small amount of free radicals are created to reduce the probability of termination.
- **Chain transfer:** The free radical can be transferred to other species. This may be used to control MW (by creating additional polymer molecules for each radical chain initiated). It also leads to the formation of branches, which has a substantial impact on MWD and properties.

Because of the nature of the chain addition process, the MWD of free radical polymerized polymers can be broad.

4.2.1.4.2 Coordination

Coordination or stereospecific polymerization involves directing the monomers in their approach to the growing polymer chain. Most often coordination polymerization is conducted using a catalyst introduced into a solution, slurry, or supported gas phase (fluidized bed) reactor. The technology came into existence with the work of Ziegler and Natta in the 1950s, resulting in the highly successful commercialization of linear polyethylene (both HDPE and LLDPE) and PP.

Ziegler–Natta catalysts are formed by complexes of alkyls of metals from groups I–III with halides and other derivatives from the transition metals of groups IV–VIII in the periodic table [28]. Solid and supported catalysts based on titanium represent an important class, which are typically used with a co-catalyst such as organoaluminum. More recently, metallocene catalysts, representing a combination of mono and *bis*-metallocene complexes of titanium, zirconium, or hafnium have been developed. Complexes of transition metals with multidentate oxygen and nitrogen-based ligands represent another class of these catalysts [29].

Chromium based catalysts are another type of commercially important coordination catalysts used to make HDPE [30].

The mechanisms of these catalysts are beyond the scope of this introductory chapter. In brief, these catalysts control the approach and regularity of the growing polymer chain through the formation of complexes. Advances have led to a number of technologies, including heterogeneous catalysts with multi-modal comonomer or MW distributions, sterospecificity (especially for PP), and narrow comonomer and MW distributions. The consequences of these catalysts on the properties of polyethylene and PP are reviewed later in this chapter.

4.2.1.4.3 Condensation

Condensation polymerization generally involves stepwise reactions that result in the loss of a small molecule such as water at each reaction step. Many polymers are produced via condensation polymerization but the two most important for packaging applications are polyester and polyamide (nylon).

For example, PET can be synthesized by reacting terephthalic acid and ethylene glycol, liberating water, as shown in the following equation:

$$n\ C_6H_4(CO_2H)_2 + n\ HOCH_2CH_2OH \rightarrow [(CO)C_6H_4(CO_2CH_2CH_2O)]_n + 2n\ H_2O \quad (4.4)$$

PET can also be polymerized by reacting dimethyl terephthalate and ethylene glycol, liberating methanol.

Following the work of Flory, a statistical analysis shows that the distribution of polymers made via condensation polymerization has the following relationship [31,32]:

$$\frac{M_w}{M_n} = 1 + p \tag{4.5}$$

where p is the extent of reaction.

For high MW polymers, the extent of reaction is large and approaches a value of 1. The PDI approaches 2.0, which has been observed experimentally. PET and nylon generally have narrow MWD, which poses some processing issues that are discussed in Chapter 5.

4.2.1.5 Molecular forces

The nature of the chemical bonds that hold polymers together influences their properties. This section reviews the main types of molecular forces acting on polymers used in packaging.

4.2.1.5.1 Primary bonds

The covalent bond is the primary type of molecular force linking atoms within the polymer chain. Covalent bonds form when one or more pairs of valence electrons are shared between atoms. Bonds between carbon atoms are covalent bonds.

Ionic bonds involve an electrostatic attraction between oppositely charged ions, typically a metal salt and an electron donor. Sodium chloride is an example of small molecule containing ionic bonds. Ionic bonding does not usually come into play for polymers used in packaging, except for a special type of polymer called an ionomer. Ethylene based ionomers are used in flexible packaging. These are ethylene methacrylic or acrylic acid copolymers that are partially neutralized with metal cations such as zinc or sodium. The ionic interactions form a crosslink-like structure that is melt reversible. Such polymers have an attractive balance of properties, especially for heat seal applications.

4.2.1.5.2 Secondary bonds

Bonds not associated with filling the outer electron shells are generally known as secondary forces. These are considerably weaker than the primary bonds, but because they may occur in large numbers, especially between polymer chains, they can have a considerable impact on the properties of polymers.

Dipole forces exist when atoms within the molecule carry equal and opposite charges, forming a dipole moment. These molecules are said to be polar. At short distances, these forces provide an attractive force. The strength of these forces depends on the alignment of the dipoles, which can be disrupted by thermal motion. Hence these forces are strongly dependent upon temperature.

A polar molecule can induce a temporary dipole moment in a nearby nonpolar molecule. This interaction between polar and induced dipoles is called an induction force. The strength of this force is dependent on the polarizability of the molecule and is independent of temperature.

Dispersion forces exist between nonpolar molecules. They arise from temporary perturbations in the electron clouds surrounding neighboring molecules, creating time-varying dipole moments. They are independent of temperature and are the main secondary force acting on polymers unless strong dipoles are present.

Hydrogen bonding is ionic in character. It typically involves an interaction between hydrogen present in an acid group with a basic oxygen atom (such as in carbonyls, ethers, or hydroxyls) or nitrogen in amines or amides. They influence the properties of polar molecules like polyamide and many proteins. Properties may be susceptible to moisture.

4.2.1.5.3 Effect of bonding on polymer properties

The strength of a polymer's intermolecular forces impacts its physical properties. (Intermolecular forces refer to interactions between polymer chains, as opposed to intramolecular forces which involve bonding within the polymer chain.) Polymers with weak intermolecular forces are more flexible and elastomeric in nature. Strong interactions increase stiffness and strength. Polar interactions contribute to the high stiffness and strength of polyesters and polyamides. Polyethylene, which only has dispersive intermolecular forces, is more flexible. Within the polyethylene family, higher density increases stiffness. Higher density is the result of greater crystallinity, which involves closely ordered structures and interaction between chains.

The types of bonds and their strength also influence surface energies and bonding at interfaces between polymers or at surfaces (inks, etc.). Polar polymers are more likely to have higher surface energies for bonding to inks. As is described in Chapter 10, matching polarities is one strategy to improve bonding between polymers at an interface.

4.2.1.6 Branching and isomerism

Beyond MW, molecular weight distribution and the number and type of chemical bonds, there are other structural features of polymers that influence their properties. One of these is branching. Branching comes in two types: short chain and long-chain branching (LCB). Short-chain branching typically involves less than about 10 carbon atoms. In polyethylene, the presence of short-chain branching arises from backbiting during free radical polymerization (LDPE) or the deliberate incorporation of short-chain comonomers such as butene, hexene, and octene (LLDPE). The presence of these short chains disrupts the ability of the primary chains to form ordered structures or crystals. Increasing the amount of short-chain branching reduces the density of polyethylene.

LCB involves the formation of side branches off the main chain that rival in length to the main chain. In polyethylene, free radical polymerization is conducive to the creation of long branches. LDPE is often referred to as branched PE whereas PE produced via Ziegler–Natta or metallocene catalysts are referred to as linear polyethylene (even if short chains are present). Some single-site catalysts allow for an occasional long chain to form [33–35].

LCB can have a substantial influence on the flow properties of polymers. Branched polymers are typically more shear thinning (viscosity decreases with increasing shear rate) and have greater melt strength than their linear counterparts. These translate into improved melt processibility which is described in greater detail in Chapter 5.

Polymers with a side group in the repeat structure can have different properties depending on the orientation of the side group. An example is PP, which has a methyl side group. When the side groups are all arranged on the same side of the polymer chain the polymer is referred to as isotactic (see Fig. 4.1). If the methyl group alternates from side to side, it is called syndiotactic. If the side groups are randomly oriented, the polymer is atactic. Isotactic PP is highly crystalline, exhibits high stiffness and strength, and is the most important type used commercially. Atactic PP is amorphous and rubbery in nature.

Copolymers have more than one repeat unit. The distribution of the copolymer repeat units within the chain influences properties. This distribution can range from purely random to purely alternating to blocks (Fig. 4.2). In block copolymers, the comonomers are arranged in groups. Examples of block copolymers are propylene-ethylene block copolymers where the blocks are EP rubber and various styrene copolymers such as styrene-ethylene-butene-styrene blocks. Alternating comonomer distributions are rather rare. One example is an alternating ethylene-carbon monoxide

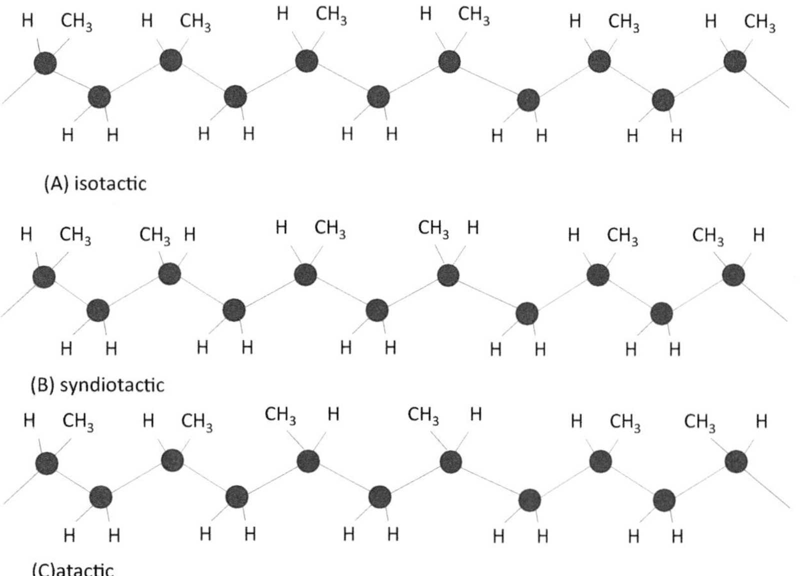

FIGURE 4.1 Isomerism of polypropylene: (A) isotactic, (B) syndiotactic, and (C) atactic.

AAAAAABBBBBBBBBBBBBBBAAAAAAA

(A) block copolymer

ABABABABABABABABABABABABABA

(B) alternating copolymer

AAABAAAABAABAAAABABBAAABA

(C) random copolymer

FIGURE 4.2 Copolymer types where (A) and (B) represent different comonomers or repeat units. a) Block copolymer, b) alternating copolymer, c) random copolymer.

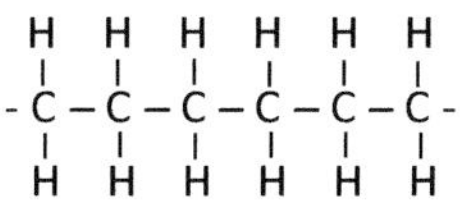

```
 H   H   H   H   H   H
 |   |   |   |   |   |
-C - C - C - C - C - C-
 |   |   |   |   |   |
 H   H   H   H   H   H
```

FIGURE 4.3 Chemical structure of polyethylene.

(ECO) copolymer [36]. Random copolymers include many ethylene copolymers made by free radical polymerization, such as EVA copolymers.

Manipulation of molecular architecture is often achieved through catalyst design. For example, the introduction of single-site catalysts for producing LLDPE and PP copolymers has allowed resin manufactures to better control the MWD and comonomer distributions, resulting in unique structures. Examples include plastomers (uniform MWD and comonomer distribution) and Dow Inc.'s olefin block copolymers.

4.2.2 Polyethylene

Polyethylene is one of the most commonly used resins in flexible packaging due to its low cost, outstanding toughness and flexibility. Polyethylene homopolymers have one of the simplest chemical structures of any polymer; they consist simply of carbon and hydrogen atoms, as illustrated in Fig. 4.3. Yet polyethylene comes in a number of varieties, characterized by differences in crystallinity (as measured by density), long and short-chain branching, MWD, comonomer, and comonomer distribution. Historically, PE was available in three general classes: LDPE, LLDPE, and HDPE. With the advancement of single-site catalyst technology and reactor design, the LLDPE family of resins, in particular, has fractured into a number of different product types. Free radical polymerization has also yielded a variety of ethylene copolymers that are suitable for packaging applications. Several general references are available on polyethylene and ethylene copolymers, including Mcnamara [37], Durling [38], and Spalding and Chatterjee [39].

4.2.2.1 Low-density polyethylene

LDPE was developed by ICI in the 1930s and was the first type of PE to be commercialized. It is made by free radical polymerization on high-pressure autoclaves and tubular reactors. (An autoclave is a stirred tank reactor.) As a characteristic of the free radical polymerization process, LDPE contains long chaing branching (LCB), which provides better melt processibility than LLDPE. The LCB contributes to greater shear thinning behavior (see Chapter 5), which results in lower pressure during extrusion. The LCB also contributes to increased blown film bubble stability and less neck-in during extrusion coating. LDPE generally has lower ultimate seal strength, dart impact resistance, and transverse direction tear strength than LLDPE.

LDPE has a typical density between about 0.92 and 0.94 g/cc. LDPE is a homopolymer; density or crystallinity is controlled by the amount of short-chain branching with arises from the free radical polymerization process. Density is controlled over a moderate range through changes in reactor synthesis conditions, namely reactor temperature and pressure.

There are subtle differences between LDPE produced on tubular and autoclave reactors. Historically, the tubular reactors produced a somewhat narrower MWD that is well suited to blown film applications, whereas autoclave polymers have more LCB and tend to be favored for extrusion coating due to the lower neck-in of the melt curtain. These distinctions have become blurred as technology has evolved. Differences in synthesis conditions and other considerations also play a role in how a particular grade of LDPE will process on a given converting process.

Some of the uses for LDPE in flexible packaging include:

- Sealant or tie resin.
- Adhesive or sealant in extrusion coating onto foil, metallized film or paperboard.
- Blending additive for LLDPE in blown film to improve processing (increase bubble stability to improve output) and reduce haze.

4.2.2.2 Linear low-density polyethylene

LLDPE, as its name implies, consists primarily of linear chains with no or little LCB. It is made via a coordination catalyst process. The original catalysts were developed by Ziegler and Natta and grades made with this technology are referred to as such. Ziegler–Natta LLDPE was commercialized in the 1970s. Chrome-based catalysts were developed by Phillips Petroleum in the 1950s and are used to produce LLDPE and HDPE with broad MWD. In the 1990s, LLDPE made with single site or metallocene catalysts were commercialized by Dow Chemical and ExxonMobil. These catalysts allow better control over MWD and comonomer distribution.

The density of LLDPE typically ranges from 0.885 to 0.94 g/cc and is controlled by introducing a comonomer during polymerization. Three comonomers are used commercially: butene, hexene and octene. The catalysts only tolerate nonpolar comonomers; currently, copolymers of ethylene and polar comonomers are only commercially made using the free radical (LDPE) polymerization process. The increasing comonomer content of LLDPE increases the number of short chains. Short chains disrupt crystallinity and lower the density. The size of the comonomer molecule determines the length of the short-chain branch: butene produces a 2-carbon branch, hexene a 4-carbon branch and octene a 6-carbon branch. The length of the short-chain branch influences properties. For example, films made from butene LLDPE generally have lower toughness than hexene or octene LLDPE [40,41].

Three types of processes are used commercially to make LLDPE:

- **Gas-phase fluidized bed reactor:** Ethylene is introduced in the gas phase and as the polyethylene polymerizes it precipitates as a solid. The temperature is kept below 115°C to maintain the precipitate as a solid powder and the pressure is typically around 2.4 MPa (350 psi).
- **Solution:** The raw materials and polymer are in the liquid state in this process. The ethylene, hydrogen, comonomers and catalyst are dissolved in a hydrocarbon fluid. The reactor temperature is >150°C and pressure typically around approximately 3.4 MPa (500 psi). The polymer solution goes through a series of separators to remove the polymer and recycle the gases. Make-up ethylene helps cool the reactor.
- **Slurry loop process:** The raw materials are in the liquid state and the polymer is in the solid state in this process. Ethylene, hydrogen, and comonomers are dissolved in a hydrocarbon fluid. The reactor temperature is kept below 90°C to keep the polymer solid. Pressure is typically <2 MPa (300 psi). The polymer is separated from the fluid in a flash tank and the ethylene is recovered.

Each process is capable of producing a broad range of resin densities and MWs. A single catalyst or multiple catalysts, as well as multiple reactors in series, can be used to produce a wide variety of MWDs. Various additives to deactivate the catalyst and improve stability are added at the end of the process. Due to the differences in synthesis conditions and additives, there are often subtle differences between grades from different suppliers of nominally the same viscosity and density.

Important manufacturing parameters that influence the properties of LLDPE include:

- **Comonomer content:** Higher levels of butene, hexene, or octene reduce the crystallinity and density. Lower density results in greater flexibility (lower modulus) and greater toughness.
- **Comonomer type:** Hexene and octene LLDPE generally have superior physical properties compared to butene-LLDPE.
- **Catalyst type:** Chrome-based catalysts generally produce LLDPE with a very broad MWD but are more often used to produce high MW HDPE. Ziegler–Natta catalysts produce a heterogeneous MWD and comonomer distribution, with more comonomer on low MW chains. Single-site catalysts produce a more controlled MWD and comonomer distribution. The difference in chain structure is shown schematically in Fig. 4.4.

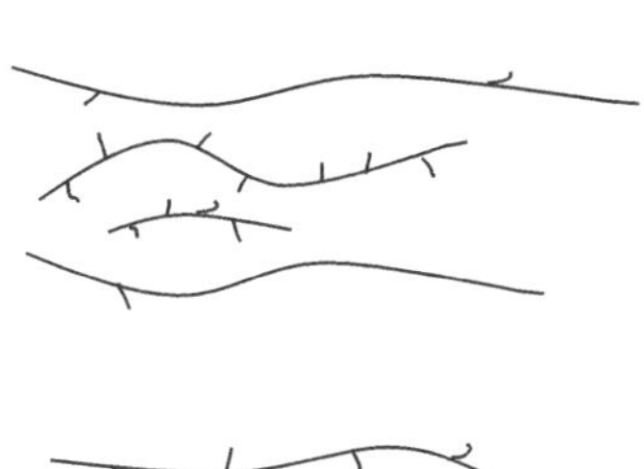

FIGURE 4.4 Differences in chain structure of A) Ziegler–Natta and B) metallocene-catalyzed LLDPE.

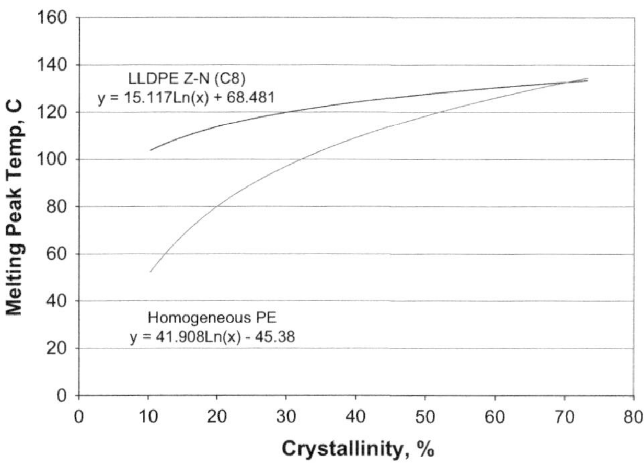

FIGURE 4.5 Peak melting temperature for homogeneous PE (single site catalyst) is compared to heterogenous (Ziegler–Natta) LLDPE octene copolymers from a solution process. *Reproduced with permission from T.I. Butler, "PE processes," in J. Wagner, (Ed.), Multilayer Flexible Packaging, Technology and Applications for the Food, Personal Care and Over-The-Counter Pharmaceutical Industries, Elsevier, Oxford, 2010, pp. 15–30.*

- **Production process:** The process (gas, solution, or slurry), number of catalysts, and number of reactors in series influence MWD and comonomer composition.

  The comonomer (or short-chain branching) distribution and MWD affects a number of properties of LLDPE. The PDI ranges from about 2 to 30 across all LLDPE product families. Ziegler–Natta catalysts produce a broad MWD with the shorter chains more likely to have a higher amount of comonomer. A fraction of the long chains will have little or no comonomer or short chain branching and will be HDPE-like. As a consequence, as the comonomer content is increased, the peak melting point of the polymer remains at about 125°C, despite a lower level of crystallinity (see Fig. 4.5); the heat of fusion or the amount of energy needed to melt the polymer decreases but the melting point is nearly constant. This influences sealing properties.

  Metallocene catalysts allow the MW and comonomer distributions to be more tightly controlled. The class of metallocene-catalyzed LLDPE and very low-density polyethylene (VLDPE) known as plastomers have a narrow MWD and more uniform comonomer distribution. As a result, as the comonomer content is increased, the peak melting point decreases since the polymer does not contain the HDPE-like component (long chains with little comonomer content) found in Ziegler–Natta LLDPE. Benefits of the narrow MWD and comonomer distribution over Ziegler–Natta LLDPE include:
- Fewer extractables, allowing VLDPE resins with higher comonomer content and lower density to comply with food contact regulations and be more easily handled (due to a reduction of the tacky low MW component).
- Higher impact strength and puncture toughness.
- Lower melting point and seal initiation temperature.
- Higher hot tack strength.
- Greater clarity.

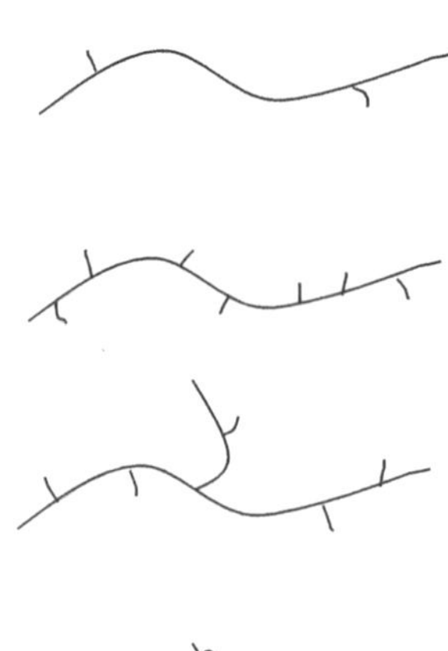

FIGURE 4.6 Chain structure differences between polyethylene types. A) HDPE, B) LLDPE, C) LDPE, D) ethylene copolymer.

The primary difference between LDPE and LLDPE lies in the molecular architecture. LDPE typically has a broad MWD and long-chain branches. The chain structure of the different polyethylene types is illustrated in Fig. 4.6. Some of the property differences include the following:

- LLDPE typically draws down more readily, allowing thinner layers in extrusion coating.
- LLDPE orients more easily but shrinks less after orientation than LDPE. LDPE is often blended into LLDPE to increase shrink in shrink film applications.
- LDPE has greater melt strength, resulting in better bubble stability in blown film, improved edge stability in cast film/extrusion coating, and lower neck-in in cast film and extrusion coating and lamination.
- The viscosity of LLDPE typically decreases less with increasing shear rate than LDPE resulting in higher extrusion pressures and power requirements. LDPE is often blended into LLDPE at about 20% to help with bubble stability and extruder power. Early generation metallocene LLDPE was even more challenging to process than Z-N LLDPE, but newer generations have started to overcome these limitations by introducing a limited amount of LCB [42–44].
- LLDPE is generally tougher with a higher tear strength and dart impact strength than LDPE. New generations of metallocene LLDPE have an improved stiffness/toughness balance [45–48].
- Ziegler–Natta LLDPE has a higher melting temperature than LDPE (126°C vs 110°C), resulting in a higher seal initiation temperature.
- Depending on its comonomer content, metallocene LLDPE may have a lower melting point than LDPE, resulting in lower seal initiation temperature.

Some of the uses for LLDPE in flexible packaging include:

- Sealants;
- Components of tie resins;
- Bulking layers;
- Structural layers.

4.2.2.3 High-density polyethylene

HDPE is a linear polyethylene with no or little comonomer content and density in the range of 0.95–0.97 g/cc. It is made via a coordination catalyst similar to the LLDPE process. Ziegler–Natta, chrome-based and metallocene (single site) catalysts are all used to make versions of HDPE.

HDPE has the highest level of crystallinity of the polyethylenes, typically around 50%–60%. The high crystallinity provides property benefits and tradeoffs compared with LLDPE and LDPE:

- Greater stiffness (modulus in the range of 690–1030 MPa or 100–150 kpsi);
- Higher melting temperature (135°C);
- High moisture barrier (low MVTR);
- Excellent grease and oil resistance;
- Lower clarity, higher haze.

In multilayer flexible packaging HDPE is typically used as a structural layer and for its moisture barrier.

Special grades of HDPE with ultrahigh MW have been developed that are often used for monolayer bags for their superior strength and downgaugability. Suppliers have optimized the processibility of these grades by using multiple reactors or catalysts to introduce bimodal MWDs.

New grades have also been developed in recent years with improved moisture vapor barrier using nucleating agents and other additives, bimodal MWDs, and single-site catalyst technology [49–51].

4.2.2.4 Ethylene copolymers (of polar comonomers)

While LLDPE is a copolymer of ethylene and a nonpolar alpha-olefin comonomer that acts as a short-chain branch to disrupt crystallinity, there is another class of ethylene copolymers involving polar comonomers. These random copolymers are made on a high-pressure free radical process (LDPE process). The introduction of the polar comonomer:

- lowers the crystallinity (but typically increases the density due to the higher MW of the comonomer; unlike LDPE or LLDPE, density is not a good indicator of crystallinity in ethylene copolymers);
- lowers the melting point, reducing the seal initiation temperature;
- increases the polarity of the polymer, increasing adhesion to some substrates;
- for some monomers (e.g., acrylic or methacrylic acid) adds a reactive functional group that provides adhesion to some substrates;
- increases toughness;
- reduces stiffness and increases flexibility;
- increases optical transparency and reduces haze.

Typical polar comonomers used in flexible packaging are given in Table 4.4. The comonomer content typically ranges from about 3 to 30 wt.%. The higher the comonomer content, the lower the crystallinity. At too high comonomer content, the polymer becomes tacky and difficult to handle.

Ethylene copolymers are made in the high-pressure LDPE process and thus have similar molecular architecture: short-chain branching and LCB (see Fig. 4.6) and broad MWD. Consequently, they offer the same processing advantages as LDPE. The lower pressure LLDPE process has yet to be adopted for commercially producing polar

TABLE 4.4 Typical comonomers used in ethylene copolymers in flexible packaging.

Monomer	Typical wt. %	Chemical structure	Typical use
Vinyl acetate (VA)	6%–18% (sealant) 20%–28% (adhesive)	$H_2C=C(H)-O-C(=O)-CH_3$	Sealant Adhesive
Methyl acrylate (MA)	6%–20%	$H_2C=C-C(=O)-O-CH_3$	Sealant Adhesive Blending component
Butyl acrylate (BA)	6%–30%	$H_2C=C-C(=O)-O-CH_2-CH_2-CH_2-CH_3$	Low-temperature toughness Blending component
Acrylic acid	3%–9%	$H_2C=C(H)-C(=O)-OH$	Tie resin (to aluminum) Sealant
Methacrylic acid	4%–12%	$H_2C=C(CH_3)-C(=O)-OH$	Tie resin (to aluminum) Sealant

ethylene copolymers; the polarity of the monomers tends to poison the catalyst and causes poor operational efficiencies.

Ionomers are a special class of ethylene copolymer. These are copolymers of ethylene and methacrylic or acrylic acid that are partially neutralized with a metal salt. Their chemical structure is shown schematically in Fig. 4.7. The ionic interactions between the acid groups and metal salt form clusters within the polyethylene matrix, as shown in Fig. 4.8. Like PE, ionomers have amorphous and crystalline regions. The presence of the ionic clusters enhances this morphology and gives rise to a unique balance of properties well suited for flexible packaging applications:

- In addition to covalent carbon-carbon bonds, and secondary dispersive bonds found in PE, ionomers have hydrogen bonding between acid groups and ionic bonding between the acid groups and metal salts. These additional forces act like crosslinks in the solid state, providing outstanding strength and toughness.
- The presence of the ionic clusters further impedes the movement of the polymer chains in the solid state, giving rise to higher stiffness than other ethylene copolymers of the same comonomer content or crystallinity. This is illustrated in Fig. 4.9.
- The formation of ionic interactions reduces the amount of primary crystallization. As a result, ionomers have exceptional clarity and low haze.
- The ionic domains are polar in nature, imparting excellent oil and grease resistance.
- The ionic "crosslinks" are thermally reversible so that ionomers are melt-processible. Like LDPE, they contain long-chain branches that contribute to their overall processibility. But the ionic interactions, while weakened, persist into the melt state and impart exceptional melt strength. As a result, ionomers have outstanding bubble stability, thermoformability and hot tack.

Ethylene copolymers are utilized in a wide variety of flexible packaging applications, primarily as sealants and adhesives or components of tie resins. EVA is used in the highest volumes, primarily as a sealant or adhesive resin. Acrylate copolymers, acid copolymers and ionomers are specialty polymers used in demanding applications. Within the acrylate copolymer family, EMA is most often used in packaging. EMA has similar properties as EVA but can be

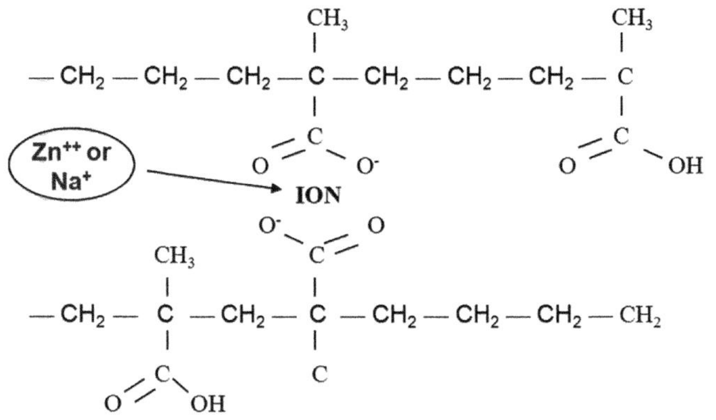

CH_3 CH_3
—CH_2—CH_2—CH_2—C—CH_2—CH_2—CH_2—C
C C
O O⁻ O OH
Zn^{++} or Na^+ → ION
O⁻ O
CH_3 C
—CH_2— C —CH_2— C —CH_2—CH_2—CH_2—CH_2
C C
O OH

FIGURE 4.7 Chemical structure of an ionomer.

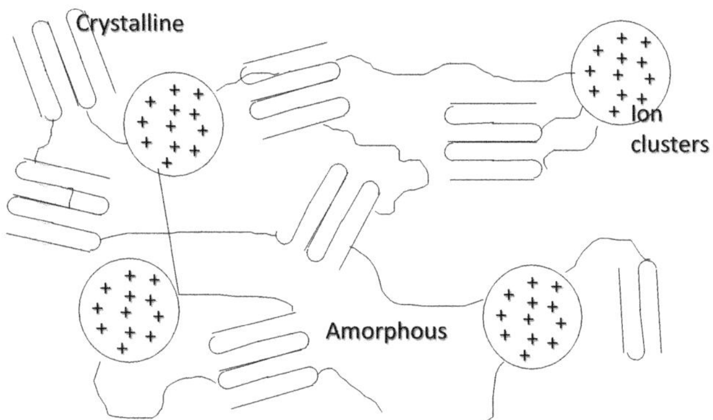

FIGURE 4.8 Ionomer morphology. *Reproduced with permission from The Dow Chemical Company.*

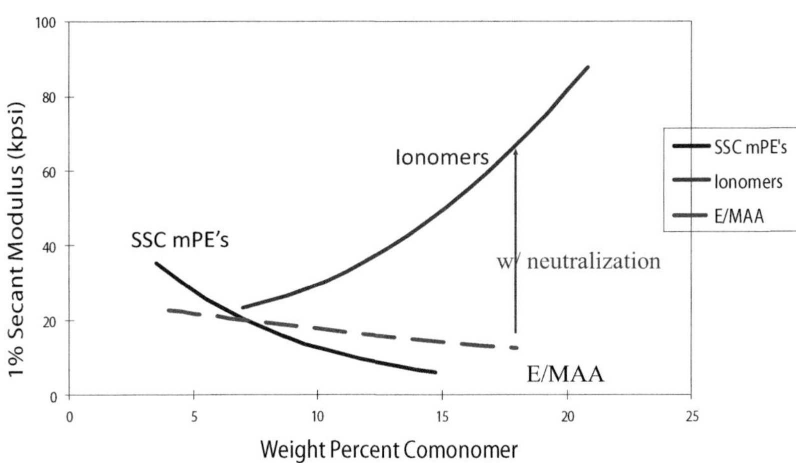

FIGURE 4.9 Effect of comonomer content on modulus (stiffness) for several ethylene copolymers [52]. *Based on data from J. R. de Garavilla, "Ionomer, acid copolymer, and metallocene polyethylene resins: a comparative assessment of sealant performance," TAPPI J., 78(6) (1995) 191–203. Reproduced with permission from The Dow Chemical Company.*

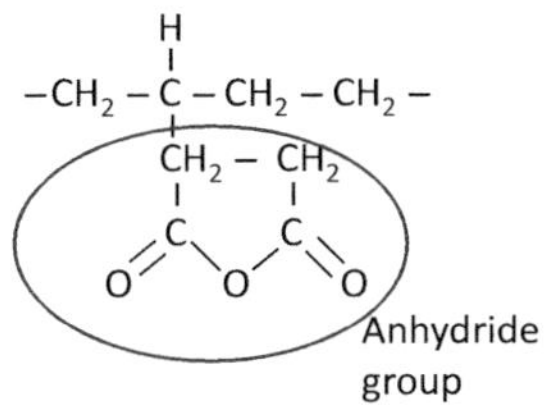

FIGURE 4.10 Chemical structure of maleic anhydride-grafted polyethylene.

extruded at higher temperatures, which is important for some applications. Ethylene butyl acrylate is mostly used as a modifier to improve the toughness of other polymers. Butyl acrylate comes in several isomers with n-BA being the one most often used.

4.2.2.5 Maleic anhydride-grafted polyethylene

Maleic anhydride is free-radically grafted onto polyethylene to impart chemical functionality. The chemistry is illustrated in Fig. 4.10. Many different types of polyethylene and ethylene copolymers have been grafted and are available commercially, with HDPE, LLDPE and PP being the most common. The grafting reaction is a post-synthesis reaction, typically done in a reactive melt extrusion process or by dissolving the polyolefin and reacting in solution. The level of anhydride functionality ranges from about 0.02 to 2 wt.%.

In flexible packaging, maleic anhydride-grafted PE and PP are typically used as components of tie resins (see Sections 4.2.10 and 10.4). The anhydride functionality bonds to the amine end groups on PA6 and to the OH groups on EVOH. It is also used as a compatibilizer for recycling multilayer packaging structures. Maleic anhydride-grafted PE is used in nonpackaging applications such as for toughening and compatibilizing engineering polymers.

4.2.3 Polypropylene

PP finds wide use in flexible packaging for its strength and high melting point. It comes in a variety of forms, depending on its tacticity, crystallinity, MW, MWD, and through the introduction of comonomers.

PP is derived from propylene (also called propene or methyl ethylene). A methyl group extends from the polymer chain on every other carbon atom, as shown in Fig. 4.11. The position of this methyl group relative to its neighbors, known as tacticity, has a profound effect on properties. Only the isotactic form (where the methylene groups are aligned on the same side of the chain) is commercially important for film applications.

PP was initially made via a high-pressure free radical polymerization process but the technology produced a low MW oligomer that was not suited for flexible packaging [53]. In the 1950s and 1960s coordinated catalyst technology

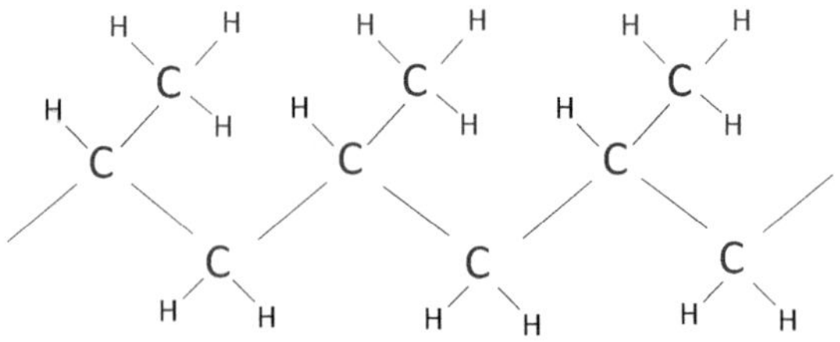

FIGURE 4.11 Chemical structure of isotactic polypropylene. Methyl side groups are highlighted in red (gray in print version).

was developed by Ziegler and Natta that produced high MW, predominately isotactic PP that was commercially viable and is used today [54]. More recently, metallocene catalyst technology has been developed for PP.

PP is produced on slurry, solution, and gas-phase processes similar to those described in Section 4.2.2.2 for LLDPE and HDPE.

Homopolymer isotactic PP is one of the most commonly used types of PP in flexible packaging. Its melting point of 161°C makes it suitable for retort sterilization and other high-temperature end-uses. It has high stiffness (typically around 1030 MPa or 150 kpsi) and strength. It is typically used as a structural layer in rigid packaging (blow molded bottles and injection molded or thermoformed tubs or trays), oriented or cast film, or coextrusion blown film.

Random copolymers of propylene with ethylene and butene using Ziegler–Natta catalysts have lower melting points (ranging from 135°C to 155°C) and are often used as sealant layers in multilayer cast or BOPP films. In these copolymers, the comonomer comprises up to about 6 wt.% of the total monomer content and is randomly distributed along the polymer chain. Block copolymers have higher levels of comonomer, with blocks of ethylene-propylene rubber phases embedded in the PP matrix. Block copolymers are not typically used in flexible packaging.

Similar to PE, the development of metallocene catalysts has allowed tighter control over the MWD, comonomer distribution, and tacticity of PP. A new generation of metallocene propylene-ethylene or other alpha-olefin random copolymers have been introduced with ethylene contents up to 20 wt.% [55–57]. These have melting points ranging from 50°C to 135°C, depending on the comonomer content, making them suitable sealants for mono-material all-PP flexible packaging structures.

PP by itself is subject to oxidative and thermal instability and, because of its crystalline structure, has a high level of haze. Antioxidants, thermal stabilizers, and clarifying agents are often added by the manufacturer to improve stability and enhance optical clarity.

PP films are used alone or as a substrate for lamination in multilayer flexible packaging. These films may be cast or biaxially oriented, in a double bubble or tenter frame process (see Section 2.4). BOPP films are often multilayer films themselves, made via coextrusion, with surface layers for heat sealing (random copolymer PP) or metallization. PP is typically used as a substrate for its strength, moisture barrier and high-temperature stability. Orientation of PP improves clarity, strength and moisture barrier performance. Packaging of snacks and other dry foods is a major application for BOPP films.

4.2.4 Specialty polyolefins

4.2.4.1 Polybutene-1

Polybutene-1 is a polymer derived from butene. Its primary application in flexible packaging is as a blending resin for PE, EVA, or ionomers to create easy-open seals. This technology is described in detail in Section 7.7.3.

4.2.4.2 Cyclic olefin copolymers

COCs are polymers of ethylene and norborene. The structure is given in Fig. 4.12. These are amorphous polymers available with a variety of glass transition temperatures, ranging from about 30°C to 180°C. They are stiff, transparent, and have a low MVTR. They find application in pharmaceutical packaging for their moisture barrier and as blending resins with polyolefins for downgauging and enhancing moisture barrier [58,59].

4.2.5 Polyamides

Polyamide or nylon was invented by Wallace Carothers at DuPont in the 1930s. Polyamides come in two types: those produced from the condensation reaction of diacids and diamines and those produced from the addition polymerization of ring compounds that contain both acid and amine ends. An example of the former is PA66, which is derived from

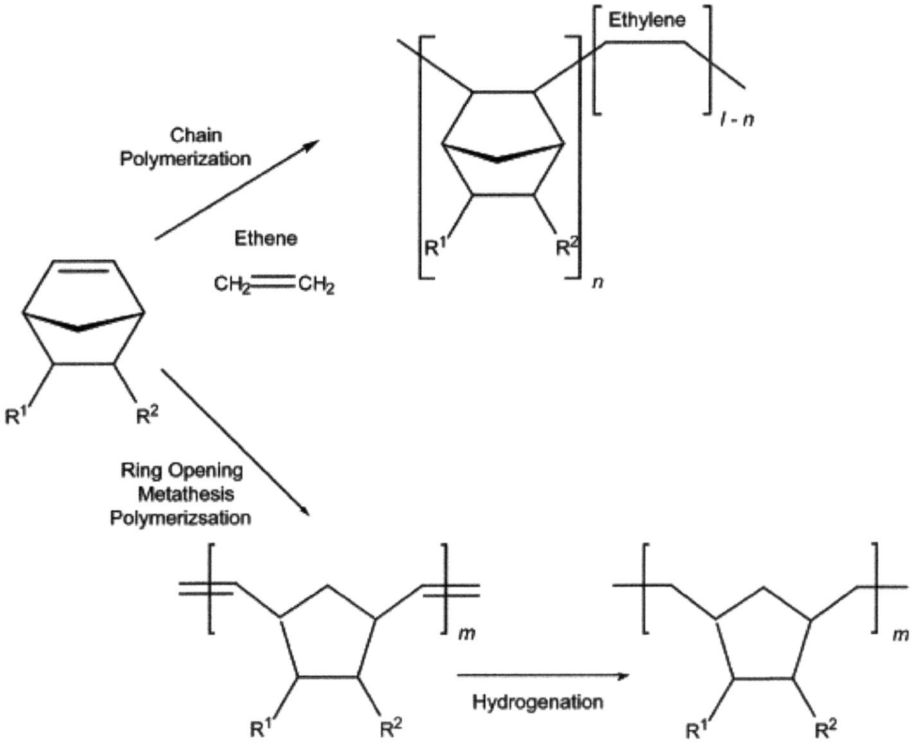

FIGURE 4.12 Chemical structure of cyclic olefin copolymers [60]. *Reproduced with permission from J.Y. Shin, J.Y. Park, C. Liu, J. He, S.C Kim, "Chemical structure and physical properties of cyclic olefin copolymers," Pure Appl. Chem. 77, (2005) 801–814.*

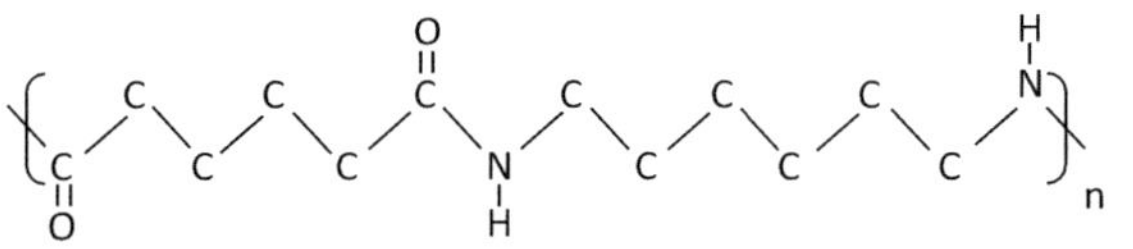

FIGURE 4.13 Chemical structure of PA66.

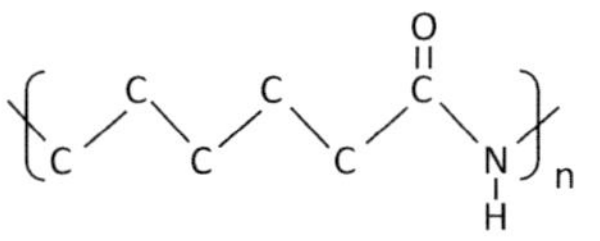

FIGURE 4.14 Chemical structure of PA6.

the reaction of adipic acid (6 carbons) with hexamethylene diamine (6 carbons), as shown in Fig. 4.13. The nomenclature for PA66 comes from the number of carbon atoms from the diamine (first 6) and diacid (second 6). Other PA types follow the same convention. For example, PA610 is derived from hexamethylene diamine (6) and sebacic acid (10).

PA6 is an example of the second type of polyamide. It is polymerized from ring-opening caprolactam (6 carbons). Its chemical structure is shown in Fig. 4.14.

The two most important nylon types commercially are PA6 and PA66, with PA6 predominant in packaging. PA6 provides abuse resistance, strength, tear resistance, temperature resistance, clarity, thermoformability and moderate oxygen barrier. Oriented PA6 is used as a substrate for some flexible packaging applications, but more often PA6 is found in coextruded cast or blown film structures. Meat and cheese packaging are two applications that often utilize one or more layers of PA6. Here, EVOH is often sandwiched between two layers of PA6 to improve the thermoformability of the structure.

PA66 is an engineering plastic more often used as fibers and for injection molded parts, but sometimes is used in flexible packaging for outstanding temperature resistance. Its melting point of 255°C (vs 230°C for PA6) makes it more difficult to process than PA6 in many coextrusion processes, especially with thermally sensitive polymers like EVOH. PA66 is sometimes used in high-temperature applications such as boil-in-bag packaging.

The longer chain polyamides (referring to the number of carbons in the repeat structure) such as PA610, 612, and 12 are specialty polyamides that are generally not used in packaging. As the chain length increases, the polymer becomes more linear and polyolefin-like (less polar). PA6 is highly polar and absorbs up to 2.8% water (at 23°C, 50% RH) [61]. The longer chain polyamides absorb less water (which is an advantage for injection molded parts) but their oxygen barrier suffers.

There are three specialty polyamides that bear mentioning for flexible packaging:

1. **PA6/66:** This is a copolymer of 6 and 66 that has a lower melting point than PA6 (196°C vs 230°C). It is used in coextrusion applications where the melting or thermoforming temperature of PA6 is too high.
2. **PA 6I/6T:** This is an amorphous polyamide derived from hexamethylenediamine and a combination of aromatic diacids: isophthalic acid (6I) and terephthalic acid (6T). It is manufactured by DuPont (Selar PA) and EMS (Grivory). In flexible packaging it is used primarily as a blending resin in PA6: 10%–20% amorphous nylon improves the clarity, oxygen barrier, extrusion processability, and thermoformability of PA6 [62].
3. **MXD6:** This semiaromatic polyamide is derived from m-xylenediamine and adipic acid. Mitsubishi Gas is the predominant supplier. It has the best oxygen barrier performance of any of the commercial polyamides. While not typically used in flexible packaging, it is used in PET injection blow molded bottles as part of an oxygen scavenging technology (with iron) or as a separate oxygen barrier layer.

Several general references are available on polyamides for packaging, including Djordjevic [63] and Saez and Hamilton [64].

4.2.6 Polyvinyl alcohol

Polyvinyl alcohol is a water-soluble polymer with an excellent oxygen barrier at low RH. It is made by first polymerizing vinyl acetate and then hydrolyzing the acetate groups to –OH groups, as shown in Fig. 4.15. There are three general types of PVOH: fully hydrolyzed (98% or higher), partially hydrolyzed (typically 87%–89% hydrolyzed), and intermediate (89%–98% hydrolysis). A number of copolymers have also been commercialized, the most prominent EVOH described in the next section.

PVOH is a challenge to melt process since its degradation temperature is close to its melting point of 230°C. Air Products introduced a copolymer with about 5 mol% comonomer that is reported to be melt-processable and still water-soluble [65]. Solution casting has proved to be a commercially viable process for making water-soluble films from PVOH [66].

Water-soluble films for hospital laundry bags, detergent bags/pods and other applications are the leading application for PVOH in flexible packaging. The temperature at which PVOH dissolves can be moderated by the level of hydrolysis: partially hydrolyzed grades can be dissolved in cold water whereas fully hydrolyzed grades only dissolve in hot

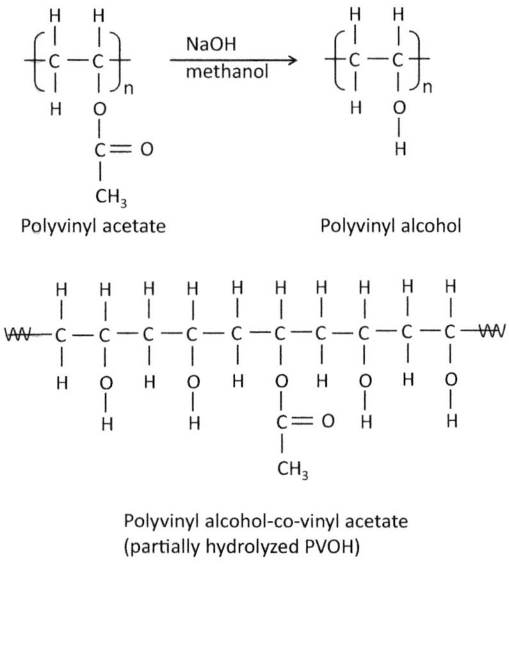

NaOH / methanol

Polyvinyl acetate

Polyvinyl alcohol

Polyvinyl alcohol-co-vinyl acetate (partially hydrolyzed PVOH)

FIGURE 4.15 Chemical structure of polyvinyl alcohol.

water. Early versions of the technology were limited by a technical challenge: the ingredients being packaged can catalyze further hydrolysis of partially hydrolyzed PVOH. Over time the temperature at which the PVOH will dissolve increases. Various modification technologies have been developed to solve this problem [67].

A good general reference on polyvinyl alcohol is Finch [68].

4.2.7 Ethylene vinyl alcohol

EVOH is one of the high oxygen barrier resins used in flexible packaging. It is derived from EVA copolymers where the vinyl acetate is completely hydrolyzed to hydroxyl (−OH) groups. Unlike EVA copolymers described in Section 4.2.2.4, these EVA copolymers have much higher levels of vinyl acetate, in the range of 56−73 mol% (EVA used as sealants typically contain about 5 mol% vinyl acetate). EVOH is a random copolymer with a chemical structure that is a combination of ethylene and vinyl alcohol units as shown in Fig. 4.16.

EVOH is characterized by its mole percent ethylene and melt flow. Grades with ethylene contents ranging from 27 to 48 mol% are available commercially. The ethylene content is a tradeoff between ease of processing and barrier performance. As the mole percent ethylene is decreased, the oxygen transmission rate (at low RH) decreases but processing becomes more difficult. An easy way to remember these effects is that the higher the ethylene content, the more like polyethylene and less like PVOH the molecule is.

The barrier performance of EVOH is highly influenced by the RH of the environment. This is explored in more detail in Chapter 8. Fig. 4.17 shows the relationship between oxygen transmission and RH. Above an RH of about 70%, the transmission rate increases rapidly. The effect of RH on the oxygen transmission rate is more pronounced with lower ethylene content.

EVOH comes in a variety of flow rates suitable for processing on blown and cast film and extrusion coating processes. EVOH is transparent, stiff and highly crystalline, and while it is an excellent oxygen barrier, it has a relatively

```
 H   H   H   H   H   H   H   H   H   H   H   H
 |   |   |   |   |   |   |   |   |   |   |   |
-C - C - C - C - C - C - C - C - C - C - C - C-
 |   |   |   |   |   |   |   |   |   |   |   |
 H   O   H   O   H   H   H   H   H   O   H   H
     |       |                       |
     H       H                       H
 Vinyl alcohol    Ethylene
```

FIGURE 4.16 Chemical structure of ethylene vinyl alcohol.

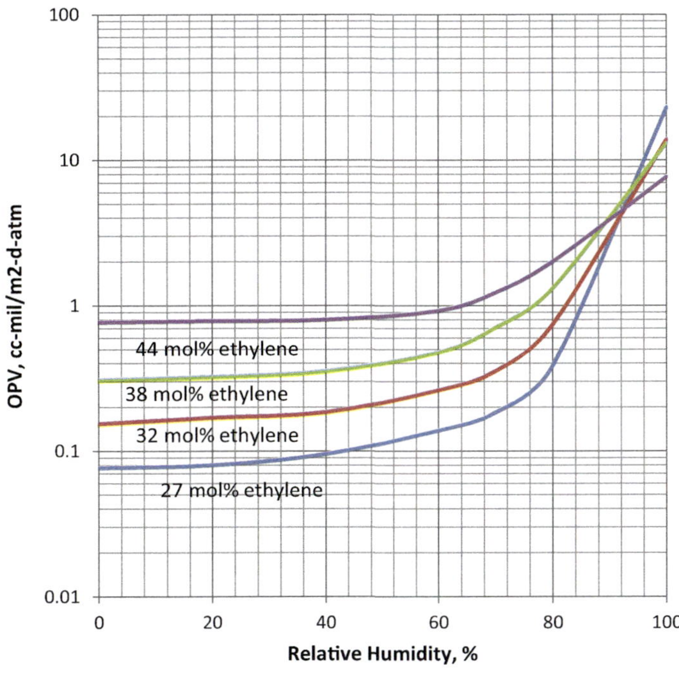

FIGURE 4.17 Oxygen transmission rate of ethylene vinyl alcohol as a function of relative humidity [69]. *Data from Kuraray, Technical bulletin No. 110: Gas barrier properties of resins. <http://www.eval-americas.com/media/36916/tb_no_110.pdf> (accessed 28.06.20).*

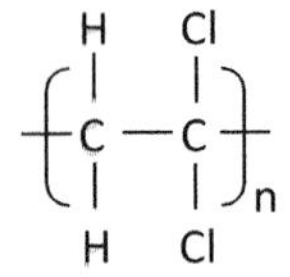

FIGURE 4.18 Chemical structure of polyvinylidene chloride.

high moisture vapor transmission rate. EVOH is used primarily in coextruded structures for both rigid and flexible packaging. Because of its high crystallinity, it can be difficult to thermoform or orient. Several technologies have been developed to overcome these deficiencies, including:

- Sandwiching the EVOH layer between layers of PA6.
- Blending in amorphous nylon [70–72] or ionomers [73–75].
- Development of specialty grades by EVOH resin producers [76–80].
- Splitting the EVOH layer into two or more thinner layers [81]. New microlayer technology reduces the crystallinity of EVOH, making the film more flexible and orientable [81,82]. See Section 8.5.2.
- Using a rapid quench cast film or water quench blown film process, which suppresses crystallinity [83,84].

EVOH has become a ubiquitous barrier resin in transparent coextruded flexible packaging. It can be found in many applications where long shelf-life is needed. Examples include processed meat and cheese packaging; instant coffee machine cups; lidding films; high barrier formable films for fresh pasta; barrier cereal liners; thermoformed cups and trays; extrusion coated paper for soups, coffee and tea; and toothpaste tubes.

4.2.8 Polyvinylidene chloride

PVDC is a clear barrier polymer made from vinylidene dichloride monomer (see Fig. 4.18). Commercialized by Dow Chemical in the 1940s, PVDC is available as a powder for extrusion or solvent solutions, as an aqueous latex or emulsion, and as a solvent coating. It provides a unique combination of oxygen, moisture, oil and grease resistance.

PVDC is sold commercially as a copolymer with other monomers such as vinyl chloride, acrylic or acrylates, and nitrile. The homopolymer decomposes near its melting point making melt processing a challenge. The comonomers reduce the crystallinity, lowering its melting point and allowing it to be melt processed. The typical vinylidene dichloride level is 70%–90%. Lower levels of vinylidene chloride decrease the melting point but also reduce the barrier performance. Even with the incorporation of comonomers, PVDC is difficult to melt process into a film, generally requiring temperature isolation and special materials for construction. Recent advances in equipment design, such as temperature isolation and encapsulation, and new product offerings have expanded the application potential for PVDC [85–87].

PVDC brings a distinctive combination of performance attributes, including [86]:

- Moisture barrier;
- Oxygen barrier;
- Aroma barrier;
- Grease and oil resistance;
- Good seal performance;
- Excellent transparency and gloss;
- Good antifog properties;
- Excellent printability;
- Good scratch and abrasion resistance;
- Flex crack resistance.

Other high-performance barrier polymers are sensitive to moisture or RH. PVDC uniquely has good moisture and oxygen barrier that does not significantly change with increasing RH.

PVDC is often coated as an aqueous latex or solvent coating onto BOPP or BOPET substrates (see Section 4.3.1), or as part of an adhesive formulation for adhesive lamination. Coated substrates are used in dry food packaging, retort pouches, transparent barrier stand-up pouches and pharmaceutical blister packs, to name a few applications. Coextrusions of (EVA-PVDC-EVA) have long been used for barrier shrink bags for meat packaging.

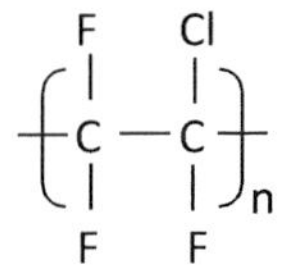

FIGURE 4.19 Chemical structure of poly-chloro-tri-fluoro-ethylene.

4.2.9 Poly-chloro tri-fluoro-ethylene

The chemical structure of poly-chloro tri-fluoro-ethylene (PCTFE) is given in Fig. 4.19. PCTFE film is sold commercially under the Aclar tradename by Honeywell. It has the best moisture barrier of any transparent thermoplastic film. It is often laminated with PET, PVC, or other substrates. Its largest application is for pharmaceutical blister packaging [88].

4.2.10 Tie resins and adhesives

Tie resins are adhesive polymers used to bond layers together in coextrusion or extrusion coating/lamination. Tie resins and adhesives are covered in more detail in Chapter 10. A few examples are highlighted here:

- LDPE is the workhorse resin used to bond to aluminum foil, paper, and other substrates in extrusion lamination. When extruded at temperatures above 300°C, oxidation occurs, producing polar groups such as aldehydes, acids, and ketones. These provide a moderate level of adhesion to polar substrates [89,90].
- For more demanding applications for bonding to aluminum foil, especially when product resistance is needed, acid copolymers are used [91].
- EVA and EMA are used to bond to PVDC, PET, and other polar substrates in coextrusion film and coating applications [92].
- Functionalized polyolefins, especially maleic anhydride-modified LLDPE, are used as tie layers to bond polyolefins to PA or EVOH in coextrusions [93–95]. See Section 10.4.

Adhesives are solvent-borne, aqueous, or two-part cured liquids that bond substrates together in adhesive lamination. Primers may also be applied to substrates to promote adhesion. Curing occurs through heat, UV or e-beam radiation. Many chemistries are used as adhesives, including polyether imide, isocyanates, and polyurethanes. Some commonly used examples include [96,97]:

- Solvent-borne two-part urethane made up of polyester and or isocyanate.
- Water-borne two-part urethane using urethane/water-dispersible isocyanate, or polyfunctional aziridine/urethane.
- 100% solids two-part urethane with polyol/isocyanate prepolymer.
- Acrylic emulsions, either uncured or two-part with a water-dispersible isocyanate (forming acrylic urethane upon curing).
- UV curable adhesive based on acrylic or cationic epoxy.

Section 10.5 describes adhesive chemistry in more detail, including advantages and disadvantages.

4.2.11 Polyester

The most commonly used polyester in flexible packaging is PET. The chemistry of this polymer is described in Section 4.2.1.4 [see Eq. (4.4)]. Other versions of polyester include:

- **Polybutylene terephthalate (PBT):** A rigid polymer used primarily in engineering plastics applications. This polymer is made from butylene glycol and terephthalic acid.
- **Polytrimethyl terephthalate (PTT):** A rigid polymer used mostly for carpet fibers and some apparel applications. This polymer is made from 1,3-propane diol and terephthalic acid. A bio-based production route to making 1,3-propane diol was commercialized by DuPont and Tate & Lyle, making this polymer more economical. So far it has not found use in flexible packaging.
- **Glycol-modified PET (PETG):** In this copolymer, some of the ethylene glycol is replaced by cyclohexane dimethanol (CHDM). Because CHDM is larger than ethylene glycol, its addition reduces the crystallinity of PET and lowers its melting point. It is a clear, transparent amorphous polymer that has not found general use in flexible packaging. One potential application is as a sealant with outstanding oil resistance. Other glycol-modified polyesters include

PCT (terephthalic acid + CHDM) and PCTG (terephthalic acid + ethylene glycol + CHDM). These are typically used in transparent injection molded parts.
- **Polyethylene naphthlate (PEN):** This polymer, made from ethylene glycol and naphthalene-2,6-dicarboxylate, has a higher barrier to oxygen than PET. Its high cost to performance ratio has limited its use in flexible packaging.
- **Polyethylenefuran (PEF) and polytrimethyl furan (PTF):** These are 100% bio-based polymers using 2,5-furandicarboxylic acid. PEF has been commercialized by Avantium and PTF is under development by DuPont [98]. Touted for their improved CO_2 and O_2 barrier performance, they are targeted primarily at rigid bottles for carbonated beverages.

Most PET on the market today is derived from petrochemical feedstocks. A number of new developments are underway to commercialize partially and ultimately fully bio-based versions of PET. In 2009, Coca-Cola introduced its PlantBottle [99], made from bio-based ethylene glycol, which makes up about 30% of PET. By 2019 30% of their North American packaging volume and 7% of global volume was using this technology, which has also been licensed to other companies [100]. Pepisco announced its intent to develop a 100% bio-based bottle in 2011 [101] and joined the NaturALL Alliance (founded by Danone, Nestle and others) in 2018. The Alliance is striving to make bottles 75% bio-based by 2020 and 95% by the end of 2022 [102]. Other biopolyesters are discussed in Section 4.2.14. The role bio-based polymers may play in sustainable package design is taken up in Chapter 17.

The two largest applications for PET in packaging are injection stretch blow molded bottles and biaxially oriented film. While blow molded bottles lie outside the scope of this book, it should be noted that these bottles have undergone significant downgauging over the last several years. Some bottles have such thin walls they might be considered flexible.

OPET or BOPET film (both acrynoms are used in the industry for biaxial oriented PET) is typically produced on a tenter frame process (see Chapter 3). PET crystallizes during orientation into small crystals, producing a strong, stiff, heat stable, transparent film. OPET has excellent abuse resistance and is a dimensionally stable substrate for printing. It also provides heat resistance during heat sealing. It has good oxygen barrier but can be coated with PVDC or PVOH, or metallized with aluminum to provide an outstanding barrier. It is also used as a substrate for AlO_x and SiO_x coatings for transparent barrier applications. By itself it has poor heat sealability; PET film is typically laminated to a sealant film or coated with a sealant.

As with any film substrate, dimensional stability and surface chemistry are important for subsequent converting and packaging processing. The dimensional stability of OPET is typically controlled by an annealing step after orientation; the oriented film is heat-treated to prevent shrinkage. The surface chemistry may be modified by various treatments (corona, plasma, etc.) or by coatings (PVDC, primers, lacquers, heat seal coatings, etc.). These treatments or coatings may be applied by either the film manufacturer or a converter.

OPET film is used as a laminating layer in a variety of packaging applications including lidding films (structural layer), cheese packaging (abuse resistance) and stand-up pouches (printing surface and abuse resistance).

PET and PETG have been used in some coextrusion films, including multilayer oriented films (see Section 11.2.4.4). PET is extrusion coated for dual ovenable paperboard trays for frozen food packaging (paperboard/primer/PET) and lidding for dual ovenable trays (OPET/PET) [98]. Two other packaging forms for PET are briefly discussed in Section 4.2.1.3. These are APET and CPET:

- APET (amorphous polyester) takes advantage of the slow rate of crystallization of PET. A thin sheet is cast and quickly quenched against a cold roll. The resulting mostly amorphous (<10% crystallinity) sheet is transparent and easily thermoformed. APET is used for clamshells, trays, and other packaging formats for refrigerated foods or retail packaging.
- CPET (crystalline polyester) is formulated with a nucleating agent and typically a polymer modifier to improve low-temperature toughness. Trays are produced via a two-stage thermoforming process that allows sufficient time for crystallization (typically 20%–25% crystallinity). The result is a temperature-resistant, opaque tray or tub that is typically used for dual oven use (conventional and microwave oven).

While not usually considered flexible packaging, APET and CPET trays are an integral part of many packages involving flexible lidding films. The chemistry and crystallinity of these trays play an important role in the adhesion and design of the lidding film structure.

4.2.12 Polystyrene

The chemical structure of polystyrene is shown in Fig. 4.20. The presence of the bulky phenyl group (benzene ring) prevents PS from crystallizing: PS is an amorphous polymer with a glass transition temperature of 100°C.

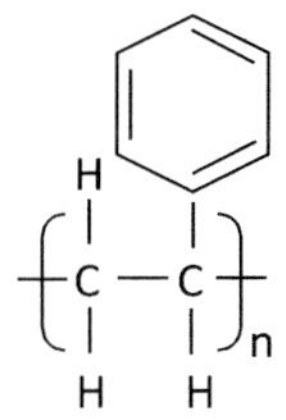

FIGURE 4.20 Chemical structure of polystyrene.

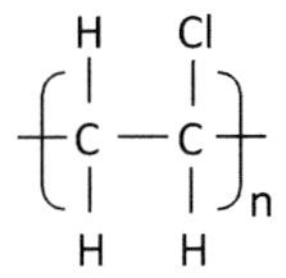

FIGURE 4.21 Chemical structure of polyvinyl chloride.

At room temperature, it is a stiff polymer that is often used in thermoformed cups and trays. It can also be easily foamed and historically has been used in foamed clamshells for quick-service restaurants and other applications. PS can also be oriented into films for various applications, although the clear film tends to be brittle.

General-purpose PS is subject to crazing that leads to splitting and brittleness. To overcome these deficiencies, polybutadiene rubber is introduced during the polymerization process to make high impact polystyrene (HIPS). The rubber particles stop the cracks from propagating, improving toughness.

4.2.13 Polyvinylchloride

Polyvinyl chloride is a widely used polymer with the chemical structure shown in Fig. 4.21. It comes in rigid and flexible varieties. Flexible PVC contains plasticizers. Some plasticizers in the phthalate family have become controversial in recent years and the industry has moved to replace them.

Flexible PVC can be made into a low cost, transparent flexible film suitable for wraps. It has a high oxygen transmission rate. One application in the US has been for in-store packaging of beef cuts and ground beef on foamed PS trays. The high oxygen transmission rate helps retain the red color that is desired by many consumers.

4.2.14 Bio-based polymers

Bio-based polymers are those polymers derived fully or partly from bio-based resources such as plants or bacteria. Bio-based should not be confused with biodegradable or compostable, two end-of-life scenarios. Some bio-based polymers are biodegradable or compostable but others are not, as shown in Table 4.5. Likewise, some petroleum-based polymers are biodegradable or compostable. Bio-based simply refers to the origin of the monomers or polymers.

A resurgence in the development and introduction of bio-based polymers occurred in the mid-2000s during a time of high petroleum costs and concern over impending shortages. Resourcing polymers from plants was thought to be a good strategy to even out the cyclic pricing of some commodity petroleum-based polymers and ensure long-term supply. Bio-based polymers were also seen as a way to contribute to lower greenhouse gas emissions and this is a key driver today. The Ellen MacArthur Foundation and World Economic Forum have set forth a vision for a sustainable plastic future involving a circular economy where plastic packaging is reused or recycled to eliminate leakage into the environment. One aspect of this low-carbon vision is the decoupling of virgin plastics from fossil feedstocks by using biosourced polymers [103]. Many consumer products companies have adopted this vision. An alternative to biobased monomers for completing the circular economy is to use monomers derived from chemical recycling of post consumer plastic waste. Recycling technology is described in Chapter 17.

For the most part, bio-based polymers are more expensive than their petroleum-based counterparts, but costs may come down with scale as they gain greater adoption. Many biosourced polymers use food crops such as corn or sugar cane as the starting point, which may put pressure on scarce land and water resources. Alternative sources are being explored, such as waste oils from the wood pulp industry [104].

TABLE 4.5 Classification of biosourced, petroleum-sourced and biodegradable polymers used in packaging.

Polymer	Abbreviation	Petroleum-based	Bio-based	Biodegradable or compostable
Polyethylene	PE	Yes	Available	No
Polypropylene	PP	Yes	Available	No
Polystyrene	PS	Yes	No	No
Polyvinylchloride	PVC	Yes	Available	No
Polyethylene terephthalate	PET	Yes	Partly bio-based available	No
Ethylene vinyl alcohol	EVOH	Yes	No	No
Polyvinylidene chloride	PVDC	Yes	No	No
Polyamide 6 (nylon 6)	PA6	Yes	No	No
Polybutylene adipate terephthalate	PBAT	Yes	No	Yes
Polyvinyl alcohol	PVOH	Yes	No	Yes
Polyglycolic acid	PGA	Yes	No	Yes
Polycaprolactone	PCL	Yes	No	Yes
Polybutylene succinate	PBS	Partly	Partly	Yes
Polyamide 11	PA11	No	Yes	No
Polylactic acid	PLA	No	Yes	Yes
Polyhydroxyalkanoate	PHA	No	Yes	Yes
Starch		No	Yes	Yes
Cellulose-based		No	Yes	Yes

There are three general approaches for making bio-based polymers:

1. Use natural bio-based polymers with some modification. An example is starch-based polymers.
2. Make monomers from a biosource, such as fermentation of sugars, followed by a conventional polymerization process to make the polymer. Examples include bio-polyethylene and polylactic acid (PLA).
3. Produce new polymers directly from bacteria. An example is polyhydroxyalkanoate (PHA).

These polymers can further be classified according to ease of adoption: those that are simply bio-derived versions of polymers long used in flexible packaging (such as bio-based PE and PET) and those that are new polymers or new to flexible packaging (PLA, PHA, starch-based polymers). The former enjoys the advantage of benefiting from many years of product evolution to meet the needs of the application.

4.2.14.1 *Biosourced monomer based polymers*

Braskem began making polyethylene from ethylene derived from sugar cane in 2010. The ethylene monomer is derived from ethanol, which in turn comes from the fermentation of sugar. Once the ethylene is made and purified, it is polymerized in the same process as conventional polyethylene. The advantage of this approach is that the properties of the polymer are the same as the petroleum-based analog. The bio-based polymer takes advantage of the years of innovation and customizing that have taken place in polyethylene technology to tailor properties for flexible packaging. Several other PE manufacturers have announced plans to use waste oils in their feedstock streams [104–106]. For example, Dow is partnering with DPM to use renewable naphtha made from crude tall oil that is a byproduct of the wood pulping process in some of their PE plants. So far, these efforts represent just a small part of the $100 billion PE market.

As described in Section 4.2.11, several partly biosourced polyester resins are in various stages of development where one or more monomers are substituted with biosourced ones.

The main disadvantage to using bio-based monomers to replace petroleum-based ones has been cost. The bio-based monomers are typically more expensive than the monomers they replace. A more successful approach was taken by DuPont in the development of its polytrimethylene terephthlate resin (PTT). A key monomer for the production of PTT is 1,3-propane diol, which is very costly to produce from petroleum-based streams. DuPont and its development partner Tate & Lyle developed a bio-based route to making 1,3-propane diol that is considerably less expensive. Once their production facility came on line, the existing petroleum-based process was shuttered. The resulting PTT is 37% bio-based. The largest application for PTT is for carpet fibers [107]. Currently, there are no flexible packaging applications for PTT, but such a strategy of developing bio-based routes to monomers that are less expensive than their petroleum counterparts will likely be more successful in the marketplace.

4.2.14.2 New bio-based polymers

4.2.14.2.1 Polylactic acid

PLA is a high strength, high modulus polymer made from lactic acid. Bacterial fermentation is used to make lactic acid, which is then converted to the polymer as shown in Fig. 4.22. It is fully bio-derived and is compostable in an industrial facility.

Due to the chiral nature of lactic acid, several forms of PLA exist. Table 4.6 shows the effect of L and D monomers on the thermal properties of PLA: the melting point decreases with increasing amounts of D. When made into a clear, transparent sheet for thermoformed trays and other packaging applications, it is in its amorphous state (PLA is slow to crystallize and crystallinity takes away from optical clarity). Because of its low Tg, PLA sheet is typically used for refrigerated foods. PLA sheet is also brittle; various additives have been developed to help improve toughness.

PLA also can be made into an oriented film. Typical properties of PLA biaxially oriented film are shown in Table 4.3. It has physical properties (tensile strength, % elongation and modulus) approaching that of OPET, but about an order of magnitude higher oxygen and moisture vapor transmission rates [1,18,108].

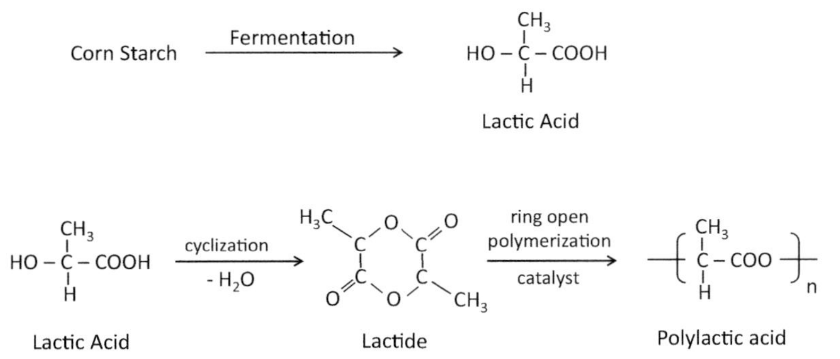

FIGURE 4.22 Chemical structure of polylactic acid.

TABLE 4.6 Thermal properties of polylactic acid copolymers [108].

Copolymer ratio L:D monomer	Glass transition (Tg), °C	Melting temperature (Tm), °C
100:0	63	178
95:5	59	164
90:10	56	150
85:15	56	140
80:20	56	125

Source: Data from R.P. Babu, K. O'Conner, R. Seeram, Current progress on bio-based polymers and their future trends, Prog. Biomater. 2 (1) (2013). Available from: <http://www.progressbiomaterials.com/content/2/1/8>.

4.2.14.2.2 Polyhydroxyalkanoates

Polyhydroxyalkanoates (PHAs) are a family of biodegradable polyesters produced by bacterial fermentation. A variety of biological feedstocks, including cellulosics, vegetable oils, and organic waste, may be used depending on the specific PHA. Microbes are engineered to produce particular PHA chemistries during the fermentation process. The PHA accumulates within the microbe cells, which is harvested by concentrating and drying the cells, followed by chemical extraction and purification. Hundreds of PHA monomers have been identified. Poly-3-hydroxybutyrate (PHB) is one of the simplest and earliest to be developed and has physical properties similar to PS and PP. It has several shortcomings, however, including a slow rate of crystallization, a narrow processing window, and a tendency to creep. Various copolymers of poly-3-hydroxybutric acid and other monomers have been developed to overcome these deficiencies, including poly(3-hydroxybutyrate-co-3-hydroxyvalerate) or PHVB. PHVB has better toughness and increased tensile strength than PHB. The chemical structures of PHB and PHVB are shown in Fig. 4.23.

PHAs can be processed into films and sheets and have properties that may be suitable for flexible packaging. Physical properties can be tuned for the application by varying the chemical composition or through blending with other biopolymers and the use of various additives [109,110]. Oxygen and moisture permeation is lower than most biopolymers but a factor of five or more higher than PET [110]. A commercial scale 50,000T capacity plant was built by Metabolix and Archer Daniels Midland in early 2010 only to be shuttered within a couple of years due to lack of market demand. Cost compared with polyolefins and other nonrenewable sourced polymers have limited their adoption in the past. Renewed interest in PHA's for their biodegradability, even in the ocean, has attracted several companies to build pilot and small scale (5–8 kT) manufacturing facilities [111]. PHAs degrade in the environment without the need for industrial composting like PLA. Target applications include replacing polyolefins and PET in single-use plastic applications such as straws, shopping bags, and food packaging [109,111].

4.2.14.2.3 Polybutylene succinate

Polybutylene succinate (PBS) is an aliphatic polyester with properties somewhat similar to PET and polyethylene. It is synthesized from succinic acid and 1,4-butanediol. Bio routes to succinic acid have been commercialized and several companies have announced development activities for bio-based butanediol. 1,4-Butanediol has been produced commercially from hydrocarbon feedstocks for many years.

PBS is a semicrystalline polymer with a Tg of around −32°C and a Tm of around 115°C. The thermal, mechanical, and crystallization behavior is similar to PE. It is tougher than PLA, and more flexible (lower modulus). It has relatively poor flex crack resistance and so is often blended with other polymers such as PLA or starch.

4.2.14.2.4 Starch polymers

Starch is a bio-based polymer that is the end product of plant photosynthesis. It is a natural carbohydrate that can be obtained commercially from a variety of sources including corn, rice, wheat, and potatoes. Starch contains linear polysaccharide amylose and highly branched polysaccharide amylopectin.

Starch alone does not melt process well and becomes brittle with age due to a phenomenon known as retrogradation. Several strategies have been developed to create "thermoplastic" starches [112,113,114,115]:

- Destructuring or gelatinizing starch by exposing the starch to heat and pressure and blending it with other polymers [116].
- Compounding with thermoplastic polymers such as PE as a filler. As such it does not bring much functionality other than a lower cost.

(A)

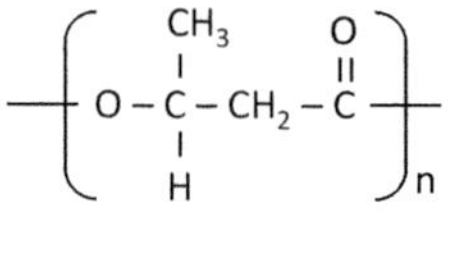

$$-\!\left[O-\underset{H}{\overset{CH_3}{C}}-CH_2-\overset{O}{\overset{\|}{C}}\right]_n\!-$$

(B)

$$-\!\left[O-\underset{H}{\overset{CH_3}{C}}-CH_2-\overset{O}{\overset{\|}{C}}-O-\underset{H}{\overset{CH_2CH_3}{C}}-CH_2-\overset{O}{\overset{\|}{C}}\right]_n\!-$$

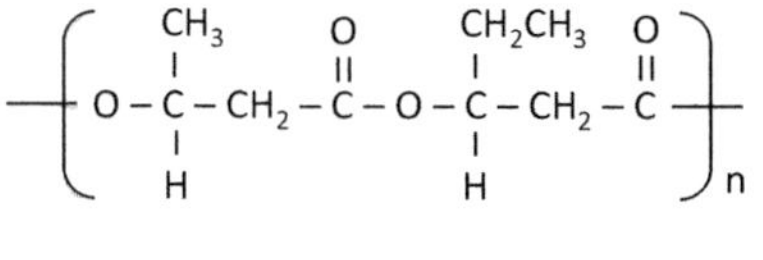

FIGURE 4.23 Chemical structure of (A) PVB and (B) PHVB.

- Compounding with a water-soluble polymer such as PVOH, plasticizers (typically a polyol such as glycerol), and water [117,118].
- Chemical modification of starch, such as esterification, etherification, and epoxidation [119–121].
- Compounding with a biodegradable polymer such as aliphatic polyesters (PLA, PBAT, PCL), typically adding polyol plasticizers and water [122–125].
- Compounding with a nonbiodegradable polymer such as PE [126–128].

Thermoplastic starches and hybrids are typically used in biodegradable/compostable garbage and grocery bags and various agricultural films. Their use in traditional flexible packaging so far has been limited.

4.2.15 Biodegradable or compostable polymers

Biodegradability and compostability both refer to the breakdown of plastic by microorganisms. These terms refer to end-of-life scenarios and not to the origin of carbon. As shown in Table 4.5, some biosourced polymers are biodegradable or compostable, as are some petroleum-based ones. Not all biosourced polymers are biodegradable or compostable. Biodegradability is the more general term with few standards and therefore can be misleading. Compostability refers to a specific process for forming a compost (a soil additive or nutrient) that involves human management [129]. There are two general types of composting processes: home and industrial. End-of-life claims around compostability generally refer to industrial composting, which refers to commercial operations that generally use more aggressive conditions than home. Plastics that can be composted in an industrial facility often cannot be composted in a home operation. Industrial composting is covered by standards, EN13432 and ASTM D 6400 or 6868, which prescribe the time limit for decomposition:

- Fragmentation and loss of visibility in final compost in 3 months: mass of residues $<10\%$ of initial;
- Biodegradation into CO_2: 90% within 6 months;
- Absence of negative effects on the composting process;
- Heavy metal content below allowable levels.

There are no standards for home composting of plastics. For biodegradability, there is a European standard for biodegradability in soil, EN 17033, but no international standards exist for water or ocean degradation. Plastics that degrade in soil or compost facilities may behave differently in water or other environments.

The infrastructure for commercial composting is limited in many countries. Claims around packaging compostability have no real value if composting is not available. Most commercial composting of plastic packaging today is from close-looped systems, such as sports stadiums, amusement parks, or large events such as the Olympic games. Here the collection, sorting, and distribution to composting centers can be controlled.

PLA, starch-based polymers, poly butylene adipate co-terephthalate (PBAT), and their blends are the largest volume of compostable polymers used in flexible packaging today. Interest in PHA is growing but volumes are still small. PBS and PVOH are also biodegradable. PLA, starch polymers, PBS, and PVOH are covered in Sections 4.2.14 and 4.2.6. PBAT is described below.

Additives are available for petroleum-based polymers such as polyethylene that promote the breakdown of the polymer over time. They typically act through oxidation catalysts or enzymatic activity to fragment the polymer into microscopic pieces that are more amendable for biological activity. They are controversial for a number of reasons, including their contribution to littering (products labeled as biodegradable are more likely to be littered by the consumer); fragmentation into "microplastics" that may take years to fully degrade, absorb toxic moieties and be harmful to the environment; and contribution of fossil carbon dioxide to the atmosphere if they do break down aerobically (causing a net increase in atmospheric carbon dioxide) and methane, an even more powerful greenhouse gas than carbon dioxide, if they break down anaerobically (such as in a landfill). The rate of degradation slows when in deep water oceans and so the additives may not mitigate marine pollution. The jury is still out on these additives with industry groups pitted against one another making conflicting claims [130–132].

4.2.15.1 Poly butylene adipate co-terephthalate

PBAT is a petroleum-based polymer that is industrially compostable. Some suppliers have announced their intention to develop bio-based routes to one or more of the monomers. The polymer has LLDPE-like properties and is used in compostable bag applications. It is also used in blends with PLA and other compostable polymers.

4.2.16 Additives

Polymers by themselves often do not have all the properties needed for a given application. Additives have been developed by the industry to address issues in processing, handling and stability, as well as provide product features such as improved flexibility. Some additives may be incorporated during the manufacturing process to aid in polymerization or isolation. These may include catalysts, emulsifiers, and processing lubricants. In addition, reaction byproducts and residual monomers may be present. These are not the subject of this section, except to note that some additives are used to counter these residual components. The additives described here are purposely incorporated into the polymer by the resin supplier or the film converter to tailor properties. This section provides a brief introduction. The reader is referred to the references for further details. Good overviews of additives for flexible packaging can be found in Refs. [133], [134], and [135].

In general, there are three classes of additives: stabilizers, modifiers, and fillers.

4.2.16.1 Stabilizers

Stabilizers include acid scavengers and other agents to neutralize residual catalysts or monomers left from the resin manufacturing process, as well as antioxidants, hydrolysis retardation agents, UV light stabilizers, and UV blockers to prevent degradation with time.

Deactivation of catalysts and co-catalysts may result in acid residues such as hydrogen chloride (HCl). Residual HCl may result in corrosion issues and interfere with other additives. Acid scavengers are added to neutralize these acids. Examples include zinc stearate, calcium stearate, zinc oxide, and dihydrotalcite [133].

Thermo and thermo-oxidative degradation is of concern for polyolefins, especially during processing. Oxidative degradation over long service times is also important for durable goods, but not as much for flexible packaging given the short shelf life of most packaging applications. Degradation proceeds along the pathways described in Fig. 4.24. Thermal energy from processing, UV light, catalyst residues, and other factors create alkyl radicals (R ·). This in turn sets off the auto formation of additional free radicals and radical precursors, including peroxy radicals (ROO), hydroperoxide (ROOH), alkoxy radicals (RO ·) and hydroxyl radicals (· OH). Here R refers to the main polymer chain. The creation of these free radicals results in crosslinking or scission of the polymer chain, leading to gels and other film defects (crosslinking) or embrittlement (scission).

There are two predominant chemistries used commercially to short circuit the degradation cycle: hindered phenols and phosphites. As depicted in Fig. 4.24, the phenolic antioxidants attack the degradation cycle by reacting with the RO · and · OH free radicals to yield inactive ROH and H_2O byproducts. They also intercept the ROO · free radicals. The phosphite antioxidants react with hydroperoxides to yield inactive ROH. The use of one type of antioxidant over another often depends on the particular polymer being stabilized and the application. Phenolic antioxidants are frequently used to help stabilize the polymer both during processing and in the end-use. Phosphite antioxidants tend to help more during polymer processing. Often a combination of phenolic and phosphite antioxidants is used. Typically the antioxidant system is added to the polymer by the resin supplier during their manufacturing process. Some of the commonly used antioxidants in polyolefins are described in Table 4.7.

Polyesters undergo hydrolysis, which causes chain scission and embrittlement. Drying is the most important method of preventing hydrolysis during processing, but epoxide, carbodiimide and phosphorous acid ester additives may be

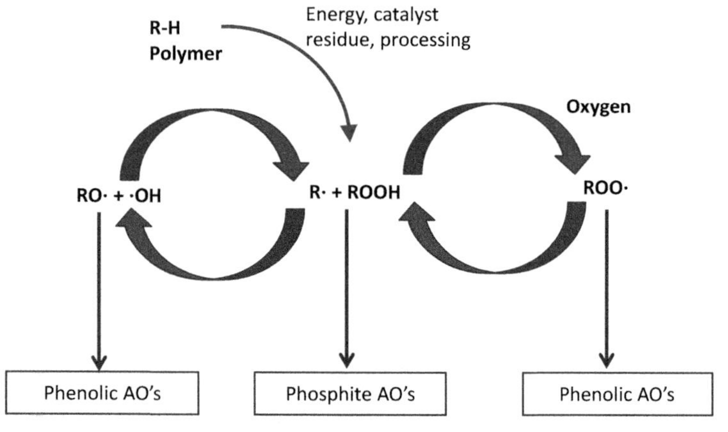

FIGURE 4.24 Auto-oxidative degradation of polyolefins and antioxidant intervention mechanisms [136]. *Adapted from F. Stricker, M. Horton, New stabilizer solutions for polyolefin film grades, in: European TAPPI PLACE Conference, 2003. <http://www.tappi.org/content/enewsletters/eplace/2004/8-1stricker.pdf> (accessed 28.06.20).*

TABLE 4.7 Commonly used antioxidants for stabilizing polyolefins.

Chemical name	Chemical structure	Grade	Supplier
Octadecyl-3-(3,5-di-tert-butyl-4-hydroxyphenyl) propionate	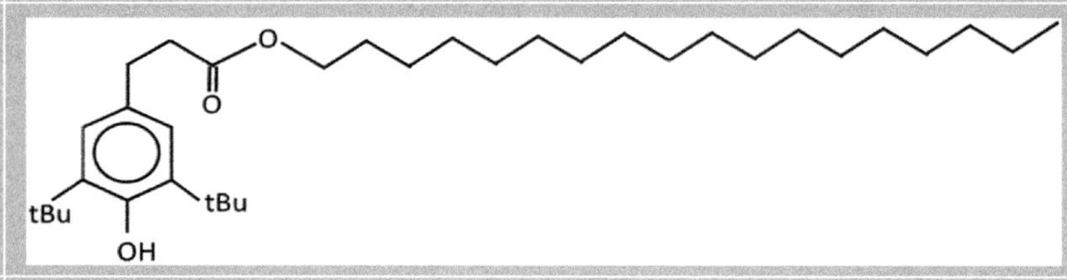	Irganox 1076 Anox PP18 Songnox 1076 Plaox 1076 Antioxidant 1076	BASF Addivant Songwon Polymate North Wanxing Chemical
Tetrakis (methylene-3-(3,5-di-tert-butyl-4-hydroxyphenyl)propionate) methane	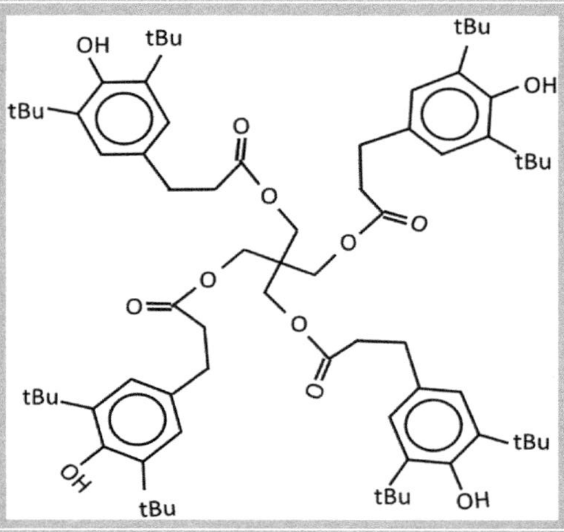	Irganox 1010 Anox 20 Songnox 1010 Antioxidant 1010 Plaox 1010	BASF Addivant Songwon North Wanxing Chemical Polymate
1, 3, 5-Trimethyl-2,4,6-tris-(3,5-di-tert-butyl-4-hydroxybenzyl)benzene	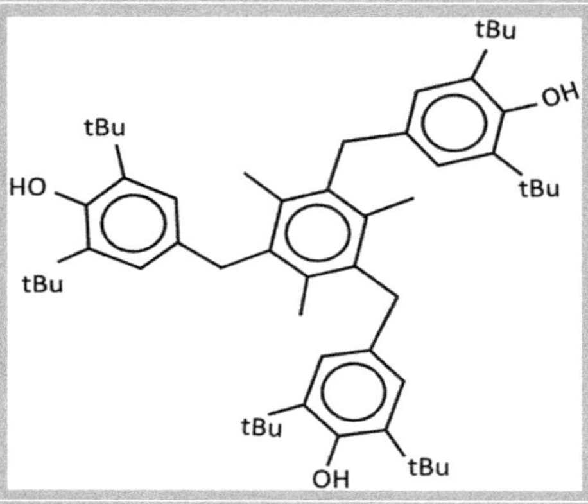	Ethanox 330 Irganox 1330 Anox 330 Sunox 555	SI Group BASF Addivant Sunshow

(*Continued*)

TABLE 4.7 (Continued)

Chemical name	Chemical structure	Grade	Supplier
Tris(2,4-di-tert-butlyphenhyl) phosphite	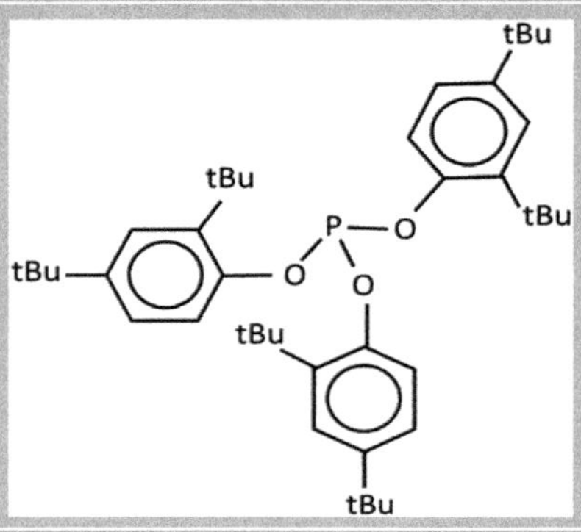	Irgafas 168 Songnox 1680 Antioxidant 168 Plaox 168	BASF Songwon North Wanxing Chemical Polymate
Butylated hydroxytoluene	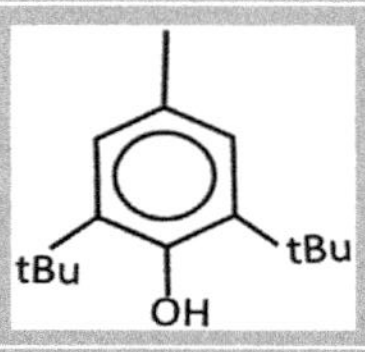	BHT	
4-Ethyl-2,6-di-tert-butylphenol	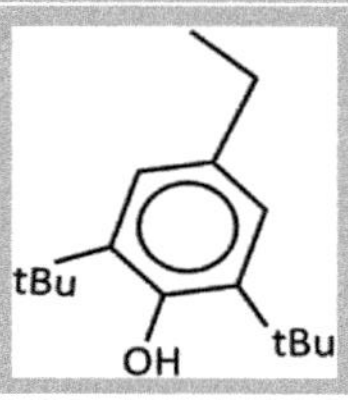	BHEP	

used to help retard hydrolysis. Polyamides also undergo hydrolysis and thermal-oxidative degradation. Copper salts, aromatic amines, and phenolic antioxidants are used to help stabilize polyamides [134].

UV stabilization is typically used for durable goods that are exposed outdoors for long periods of time. For packaging, the film itself will likely be stable enough without additives for the short duration of the use of the package. UV stabilizers and blockers, however, may be incorporated to help protect the product from UV exposure. The most effective way to block UV light is the use of opacifiers, such as metallization and pigments, for example, TiO_2. For transparent packaging, UV blockers such as hindered amine light stabilizers (HALS) and hydrobenzoates may be used. Antioxidants also help in stabilizing the film during UV exposure.

Discoloration of plastic films is known to occur as a result of complex interaction amongst additives and between additives and byproducts in the resin. For example, discoloration of EVA encapsulant films in photovoltaic modules has been attributed to antioxidants, UV stabilizers, and peroxide byproducts [137,138].

Most packaging films do not contain such additives, but overoxidation of phenolic antioxidants has caused discoloration issues, particularly in polyolefin films. Under harsh processing conditions or exposure to gaseous oxides of nitrogen (NO_x) from incomplete combustion of fossil fuels, a variety of colors from yellow to red and orange have been observed. The latter phenomenon is known as gas fading or pinking and often is associated with exposure of films to NO_x from propane-powered forklift trucks in warehouses. The color forms near the surface of the plastic, and near the edges of the rolls. Color formation increases with exposure time to NO_x and the amount of phenolic antioxidant in the resin. Some phosphites may also break down to low MW phenols when exposed to heat and humidity, with subsequent color formation from the same pathway as for primary hindered phenol antioxidants [133]. Additionally, some pigments, such as some grades of TiO_2, may contribute to the problem [139,140].

The mechanism for the color formation is thought to be the formation of quinone species, such as stilbenaquinone [140]. Several techniques are used to try to control the problem [141,142]:

- Modifying the ratio of phosphite to phenolic antioxidant (toward more phosphite). Phosphites do not have the same functionality as phenolic antioxidants and so this approach can only be taken so far. Also, care must be taken in selecting the phosphite portion of the antioxidant system to ensure phenols are not created through hydrolysis reactions.
- Using isocyanurate-type phenolic antioxidants that do not readily form quinone structures.
- Avoiding blending resins with different antioxidant chemistries/formulations that may upset the balance of phosphite to phenolic ratio.
- Improving ventilation in the warehouse and covering the film rolls to prevent NOx from entering.
- Adding acid neutralizers to buffer the pH.
- Using other antioxidant chemistries, such as hydroxylamine, tocopherols (vitamin E), or HALS [143–147].

The last approach has been well studied in the literature but has been slow to be adopted except for special grades developed for applications that are most sensitive to the issue. For the most part, yellowing is not an issue and standard phenolic antioxidants are satisfactory.

4.2.16.2 Modifiers

4.2.16.2.1 Pigments and opacifiers

Pigments and mineral fillers may be added to the polymer for light barrier or esthetics. For white films with an excellent light barrier, titanium dioxide (TiO_2) is often used. The amount of light barrier or opacity depends on the type of pigment (refractive index, particle size), loading, how well it is dispersed, and thickness of the film [148]. TiO_2 in its raw form is highly reactive. Different coatings are applied by the manufacturer to make it suitable for use in plastics. The coating type may impact polymer stability as well as how easily it can be dispersed into the polymer matrix. Some polymers may be impacted more than others for a given grade of TiO_2. Consultation with the TiO_2 producer or masterbatch supplier is recommended.

Calcium carbonate ($CaCO_3$) is another inorganic often added to help opacify a film. Calcium carbonate brings other benefits to PE film, such as improved output and toughness, believed to be due to changes in heat conductivity and the crystalline morphology of the polyethylene [149,150]. In the extrusion coating of PE, calcium carbonate is reported to increase output, adhesion, and printability [151,152].

The addition of mineral fillers can be used to make white oriented film through the creation of small air voids during orientation [153,154]. The particle size and dispersion of the cavitation agent are critical. Using such films may be less expensive than printing for creating a white background.

4.2.16.2.2 Slip agents

A film needs a certain degree of slipperiness to run along rolls and other metal surfaces in high-speed packaging lines. The degree of slipperiness is measured as the coefficient of friction (COF). A higher COF means the film is tackier. Many sealant polymers, such as EVA or LLDPE, have COFs greater than about 1 and are too tacky to run on high-speed lines. Slip agents have been developed to bring the COF of these resins down to the operable range of about 0.1–0.3.

Slip agents typically belong to a class of chemicals known as fatty acid amides [155]. They are added to the polymer either by the resin manufacturer or by the converter just prior to extruding into the film. Fatty acid amides are generally soluble in the polymer melt, but as the polymer is quenched into the solid film, the fatty acid amide is driven away from the crystallizing regions to the surface of the film. Once on the surface, they act as soaps to provide "slip." The effectiveness of a given fatty acid amide depends on the size and shape of the slip molecule, how much has bloomed to the surface and the bloom rate. Generally, once the surface of the film has a monolayer of slip molecules on it, the ultimate COF is achieved, which depends on the structure of the molecule. The bloom rate depends on the specific interactions between the polymer and slip agent. In some cases, it may take a few minutes but in others, it may take days. The net effect is that a slip agent that works well in one polymer may not work as well in a second. Higher concentrations may be required to achieve the same COF in the same amount of time for LLDPE than LDPE, for example. Adding too much slip, however, can lead to problems with adhesion and esthetic effects (chalking). A more in-depth discussion of slip agents and blooming rates is provided in Chapter 12.

Some of the more commonly used slip agents for polyolefins are described in Table 4.8.

4.2.16.2.3 Antiblock

Blocking refers to when a roll of film becomes difficult to unwind due to the film surfaces binding to one another. Some of the factors that influence blocking include:

- Softness and tackiness of the polymer: softer/tacky polymers tend to be more prone to blocking than harder polymers.
- Crystallinity of the polymer: higher crystallinity often reduces blocking of polyolefins.
- Glass transition temperature: amorphous polymers below their glass transition temperature are also less prone to blocking.
- Roll tension: high roll tension may increase blocking.
- Temperature: higher ambient temperatures may make blocking worse.

In addition to selecting the right polymer and controlling roll tension and temperature, antiblock additives are used to prevent blocking. There are two general types of antiblock agents: organic and inorganic. Examples of organic additives include some varieties of fatty acid amides, such as stearate, which helps reduce the blocking of PE films. Like fatty acid amides used for slip agents, these organic additives bloom to the surface. The morphology of the fatty acid structure formed at the surface acts to reduce the blocking of the films; some fatty acid amides act as antiblock agents and others as slip agents depending on their chemical structure.

Examples of inorganic antiblock additives include diatomaceous earth, silica, and talc. These are dispersed throughout the bulk of the polymer. They are thought to reduce blocking by creating small bumps at the surface of the film. The small bumps reduce the surface contact between the films in the roll, reducing friction and blocking. The rougher surface also reduces friction at the surface. Often inorganic antiblock is added in conjunction with a fatty acid amide slip agent. The synergy between the two types of additives typically acts to reduce the COF more efficiently than either one alone.

The particle size of the inorganic additive is important. If it is too large, it may create holes in the film rather than bumps. Optical clarity may also suffer, depending on the concentration. If the particle size is too small, the size of the bumps may be too small to prevent blocking. More detail on additive technology for preventing blocking can be found in Chapter 12.

Another form of blocking is in the blown film process where the collapsing tube may block in the nip at the top of the cooling tower. Sealant layers with low melting points, such as EVAs and plastomers, are more prone to this problem. Some of the same strategies used to control blocking in the roll are often tried for mitigating this type of blocking.

4.2.16.2.4 Chill roll release

In cast film and extrusion coating, the melt curtain contacts a cold roll and is quenched. The polymer may stick to the roll as it is drawn off. Factors that influence chill roll sticking include:

- The metallurgy of the roll surface. Chrome plating is often used to help prevent sticking.
- The roughness of the roll surface. The polymer often releases easier with a matted roll than a smooth one. Matting the roll surface, however, can affect the optical clarity of the film/coating.

TABLE 4.8 Commonly used slip agents for polyolefins.

Slip agent	Structure	Carbons	Saturation	Uses
Erucamide		22	Mono unsaturated	COF: PE, EVA
Oleamide		18	Mono unsaturated	COF: PE
Stearamide		18	Saturated	COF: EVA, ionomer; Antiblock: PE
Behenamide		22	Saturated	Chill roll release and antiblock
n-oleylpalmitamide		34	Mono unsaturated	COF: Ionomer

- Chill roll temperature.
- Extrudate temperature.
- Polymer properties: some polymers are tackier or adhere more strongly to the metal surface than others.
- Use of additives.

Additives for aiding in releasing the polymer from the chill roll must be fast-acting. Typically fatty acid amides are used. For example, behenamide provides good chill roll release properties for acid copolymers and ionomers.

4.2.16.2.5 Lubricants

External lubricants serve to decrease frictional forces between the polymer and metal surfaces, or when compounding with fillers, between the filler and polymer. They are generally not used in polyolefins for film applications. They are more prevalent for processing PVC or PA, particularly in injection molding. Some examples of external lubricants include polyethylene waxes, oxidized polyethylene waxes, paraffins, metal soaps (such as zinc stearate), esters, amides, and fatty acids [156].

4.2.16.2.6 Plasticizers

Plasticizers or internal lubricants act to reduce the bulk viscosity and increase the flexibility of a polymer. They are typically semisoluble in the polymer, opening up the polymer chain with the soluble portion and providing intermolecular lubrication with the less soluble portion [156,157]. Plasticizers also act to lower the glass transition temperature of polymers such as PVC. This makes the polymer more flexible, particularly at cold temperatures.

The most widely used plasticized polymer is flexible PVC [157]. The addition of a plasticizer completely transforms PVC from a very rigid polymer used in pipes to a flexible material suitable for flexible films. Most other polymers used in flexible packaging do not use plasticizers. Phthalate esters have been the predominant plasticizer used in PVC, the most common being di-2-ethylhexyl phthalate (DEHP), also known as dioctyl phthalate (DOP). Some types of phthalate plasticizers have come under scrutiny due to concerns over their toxicity and potential to be endocine disruptors; outright bans of some phthalate plasticizers are in various stages of consideration by regulatory bodies around the world. In response, the industry has shifted away from these plasticizers to alternatives for applications involving human health.

Water acts as a plasticizer for PA6. Some nylon films are "moisturized" in a post-fabrication step to help improve low-temperature toughness and puncture resistance [158,159].

4.2.16.2.7 Antistats

Plastics are naturally nonconducting and are prone to static build-up. This becomes an issue with high-speed wind-up of films, unwinding of films for packaging operations, and in the end-use. Dust attraction, an electric discharge that may damage electronic components, and electric shock are examples of end-user issues.

In addition to equipment modifications such as static dissipators on unwinders, three methods are typically used to reduce static buildup: addition of antistatic additives, addition of conductive fillers such as carbon black, and coating the surface with an antistatic agent. Fillers are often used in thick-walled parts where opacity is not an issue. In films, additives are more often used and can be divided into two categories, blooming and permanent [160].

Blooming additives, such as glycerol mono stearate, diffuse to the surface during processing. They are typically added at 1%–2% and are low in cost to use. Usually, they act as surfactants that are incompatible with the polymer matrix. Upon migrating to the surface of the film, they attract moisture. The charge is transferred from the polymer to the water layer on the surface for dissipation. Examples of blooming typing antistats are ethoxylated amines, glycerides, ammonia salts, phosphates and sulfonamides [101]. Most do not work well at low RH and wear off with time. They can also cause delamination and adhesion issues.

Permanent additives are inherently dissipative or conductive polymers that are typically blended with polymers at 20%–30% levels. They usually work at low RH and do not wear off with time. Polyether ester amides (such as Pebax) are one group of permanent antistatic polymer additives. They are relatively expensive to use, can cause some esthetic issues in processing since they are often not compatible with polyolefins, and may yellow with aging. Potassium based ionomers have also been introduced as permanent antistat additives for polyolefin films [160,161]. They are more compatible with polyolefins, resulting in better esthetics.

Two methods are typically used to measure the antistatic properties of films requiring antistatic performance: surface resistivity and static charge decay time. Surface resistivity is evaluated by measuring the potential difference between set distances on the surface of the film when a charge is applied, as per ASTM D257. Polyolefins typically have surface resistivities of 10^{14} ohms/square or higher. Applications requiring antistat additives generally need values of around 10^{12} ohms/square or less. Static decay time is measured per US Federal Test Standard 101 C, Test Method 4046. The initial charge is +/−5000 V. Static decay times of less than 10 seconds, and preferable less than 2 seconds are generally required. The RH and temperature are two key parameters that affect these measurements and should be specified.

4.2.16.2.8 Process aids

Polymer processing aids are typically used to reduce melt fracture of polymers, especially linear polyethylene. Melt fracture is a type of flow instability that begins as a roughening of the surface (shark skin) and at higher output can lead to severe distortion of the polymer. PPAs help extend the critical shear rate or throughput for the onset of melt fracture, allowing higher line speeds. The most common PPAs are based on fluoropolymers or fluoroelastomers [162]. They are inherently incompatible with the LLDPE matrix and coat out onto metal surfaces. The PPA forms a lubricating boundary layer at the surface of the flow channel. Advances in the understanding of the mechanisms and optimal compositions have occurred over the years [163−166], which has resulted in lower concentrations of additives needed. PPAs have also been found to help reduce die lip build-up and gel. The process is dynamic and the PPA must be continuously replaced. Various strategies are used, including adding the PPA to the resin during manufacture and pre-coating the die and flow channel surfaces by introducing a PPA masterbatch at the beginning of film production. Melt fracture and die lip buildup and their mitigation are further discussed in Chapter 13.

4.2.16.2.9 Nucleating and clarifying agents

Nucleating and clarifying agents help seed the formation of crystals in semicrystalline polymers, producing more crystals that are smaller and more uniform in size. The smaller crystals refract less light and reduce haze and increase transparency. Crystallization occurs at a higher temperature than it would normally, which provides some processing advantages.

When semicrystalline polymers cool from the melt, lamella organizes into macrostructures called spherulites. These spherulites grow until they impinge on neighboring spherulites. Large spherulites can result in high levels of haze and affect physical properties. There are three types of nucleation: homogeneous, orientation-induced, and heterogeneous. Homogenous nucleation occurs naturally as the polymer is supercooled below its crystallization temperature. Orientation-induced nucleation occurs when molecules align during orientation, typically during film fabrication. Heterogeneous nucleation involves the introduction of a foreign phase or surface, typically the nucleating agent, for the crystal to grow on. By introducing the nucleation agent, more crystals form at a higher temperature than would otherwise occur with pure homogeneous nucleation. More crystals result in small sizes as they impinge upon each other sooner, as illustrated in Fig. 4.25.

Nucleation generally results in two types of advantages: better optical quality (lower haze and higher transparency) and crystallization at higher temperatures. The latter can have advantages in productivity, since quenching or cooling is often a limiting step of the film fabrication process. In injection molding, nucleating agents help reduce cycle time by allowing the mold to be opened sooner.

PP has been the most common polyolefin to nucleate. A subset of nucleating agents known as clarifying agents are added to PP to improve its optical clarity. PP is particularly amenable to nucleating agents due to its inherent slow rate of primary crystallization.

Polyethylene nucleation has been more difficult. Milliken introduced a nucleating agent for linear polyethylene [167]. Addition of the nucleating agent increases the peak crystallization temperature and reduces the crystallization half time. An increase in gloss and a reduction in haze have been measured for several types of linear polyethylene blown films. Improvements in physical properties, hot tack and barrier are also noted.

Nucleating agents are typically small inorganic particles with melting points above that of the polymer. As the polymer melt cools, the solid particle forms the site around which the crystal forms. The most effective nucleating agents for PP are derivatives of benzoic acid. The most widely used is sodium benzoate. Kaolin and talc are also commonly used [168−170].

Clarifying agents are typically organic molecules. Clarifying agents are nucleating agents, but not all nucleating agents are clarifying agents. Inorganic nucleating agents may actually increase haze. The most widely used clarifying agents for PP are derivatives of dibenzylidene sorbitol (DBS). Clarifying agents melt in the same temperature range as the PP. As the polymer cools from the melt, the clarifying agent solidifies, initiating crystallization of the PP at a higher

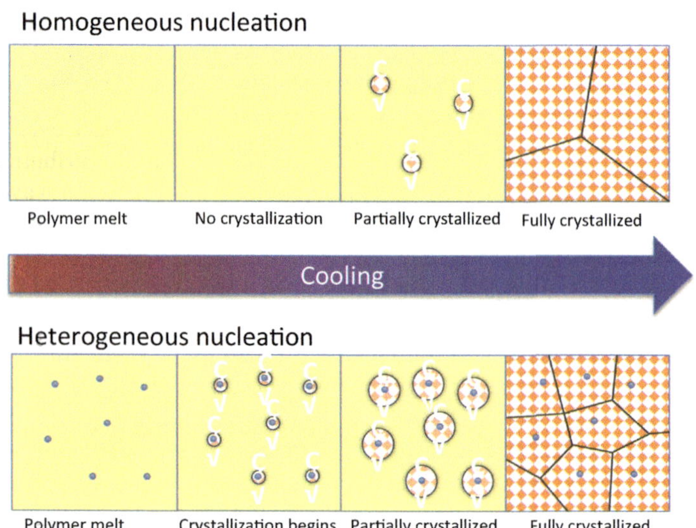

FIGURE 4.25 Illustration of homogeneous (top) and heterogeneous (bottom) nucleation of crystallization.

temperature than it would with just homogeneous crystallization. Unlike inorganic nucleating agents, clarifying agents are transparent, aiding in the reduction of haze.

Care must be taken in the processing of clarified PP; high temperatures can degrade the clarifying agent, resulting in color and organoleptic issues.

4.2.16.2.10 Antifog

In certain types of packaging, particularly for refrigerated food items, condensation may form on the inside surface of the film. Tiny droplets can form that detract from the clarity and esthetics of the package. Antifog agents have been developed that change the surface energy of the film, allowing the water droplets to better wet the film. These types of additives are typically blended in with the resin and bloom to the surface with time. Examples include ethoxylated alkylphenols, polyethoxylated oleic acid esters, and sorbitan esters of fatty acids [133].

4.2.16.2.11 Tackifiers

Tackifiers may be added to change the cling, tack, or adhesive properties of a film. Often used in hot melt and pressure-sensitive adhesive formulations, tackifiers modify the viscous properties of polymers and elastomers, aiding in flow and tack. A common cling agent for stretch wraps is polyisobutylene. For hot melts, pressure-sensitive adhesives and other adhesive polymers, there are hundreds of tackifiers to choose from. They are conveniently categorized by their basic raw material feed streams [171]:

- Rosin resins are derived from wood rosin or gum rosin or the byproducts of the paper making process (tall oil rosin).
- Hydrocarbon resins are made from petroleum-based feedstocks that are either aromatic, aliphatic, or mixtures.
- Terpene resins are derived from terpene feedstocks such as wood sources or citrus fruit. They have broad capability and impart adhesive properties to a wide range of polymer types but are limited in supply and high in cost.

4.2.16.2.12 Tougheners

Some polymers are inherently brittle. Modifiers have been developed to toughen the polymers. Although most often used for rigid applications, the basic concept is to add a rubbery component to the polymer. When designed properly, the rubbery phase helps stop cracks from propagating through the polymer, increasing its toughness. Some examples include

- Copolymerization of butadiene with styrene (HIPS) to stop crazing of polystyrene.
- Blending in ethylene-propylene rubber or ethylene-propylene diene monomer rubber. (EPDM) into PP, or creating impact block copolymers of PP and rubber with multiple reactors.
- Compounding nylon with maleic anhydride-modified rubber to form supertough nylon [172].

- Adding epoxy functionalized ethylene copolymers to CPET to improve cold temperature toughness [23,26].
- Adding EMA to APET to improve refrigeration temperature toughness [23].
- Adding rubbery polymers to PLA to reduce brittleness [173].

The rubber modifier must be compatible with the polymer matrix to ensure good adhesion during impact events. Special compounding steps and chemical functionality is often used to obtain the optimum domain size and distribution of the modifier in the polymer, adding to the cost. In many cases, the size of the rubbery phase is greater than the wavelength of light, resulting in haze or even opacity. For these reasons, tougheners are not often used in flexible packaging.

4.2.16.2.13 Compatibilizers

When combining polymers with other polymers or inorganic fillers, good interfacial bonding is important to ensure mechanical integrity between the phases. Most polymer blends are immiscible (as is shown in Chapter 6), resulting in discreet and continuous phase morphology. The minor or discreet phase forms droplets or ribbons within the major or continuous phase. Compatiblizers act to improve the adhesion between the phases, or between the polymer and an inorganic filler. This can result in higher strength and integrity.

In multilayer flexible packaging, the individual layers are not always compatible with each other. Tie layers are used to bond the layers together. More and more often, scrap film is recycled (postindustrial or postconsumer mechanical recycling). Sometimes compatibilizers are added to the recycle melt stream to improve the properties of the blend. An example is the use of maleic anhydride-modified polyolefins for compatibilizing PA or EVOH with PE or PP [174,175]. Chapter 17 provides some additional examples of compatibilizers for plastic recycle streams, or incorporating compatibilizers into a film structure to ensure compatibility in postrecycle processing.

4.2.16.3 Fillers

Fillers are typically inorganic materials such as calcium carbonate, talc, mica, or glass fibers. Except for calcium carbonate, already discussed in Section 4.2.15.2.1, fillers are not typically used in flexible packaging. They increase stiffness at the expense of flexibility and clarity.

4.2.16.4 General considerations

The combination of two or more additives may be synergistic or antagonistic. For example, combining an inorganic antiblocking agent with a fatty acid amide slip agent can result in lower or higher COF than either one alone. Additives also may impact other properties such as adhesion or optical clarity. When using more than one blooming additive there may be competition at the surface with less than desired results. Care must be taken when selecting additives to test for all contingencies.

In coextrusion, migrating additives may diffuse to adjacent layers or interfaces as well as the free surface where it is intended. Often additional additives must be put into the adjacent layers to account for this. An example is the use of slip additives to control COF, which is discussed in more detail in Chapter 12.

4.3 Commonly used substrates in flexible packaging

Premade film, foils, and paper are commonly used as components of multilayer flexible packaging structures. This section highlights some of the commercially important substrates.

4.3.1 Oriented films

Oriented plastic films are often used in multilayer laminates in flexible packaging for their strength, stiffness, printing surface, heat resistance, and barrier performance. A machine direction orienter, tenter frame, or double/triple bubble process (see Section 4.2.4) orients the polymer in the semisolid state, increasing mechanical strength, stiffness, dimensional stability, barrier and transparency. BOPP and BOPET are most often used, with polyamide (BONy or BOPA) used in some specialty applications. Some of the characteristics and applications for these films include:

- **BOPP:** Described in Section 4.2.3, BOPP has high gloss and moisture barrier. It is the most economical of the three, with lower cost and higher yield. With a density of 0.90 g/cc (vs 1.4 g/cc for PET), a 17.8 μm (0.70 mil) BOPP film yields 6% more per kg of film than 12.2 μm (0.48 mil) BOPET film. It is used in many dry foods applications such as chip bags.

- **BOPET:** Described in Section 4.2.11, BOPET offers high stiffness and strength, good heat resistance for heat sealing, an excellent printing surface and a good balance of moisture and oxygen barrier. It performs well at low thickness and is used in stand-up pouches, cheese packaging and many other applications.
- **BONy:** Oriented polyamide is the most expensive of the three. It provides good oxygen barrier and outstanding puncture and tear resistance.

Oriented PLA films are also available that are bio-based, compostable, and approach the mechanical properties of BOPET film. Typical properties of oriented films are included in Table 4.3.

Oriented films may be made of multiple layers, with a core layer for strength and skin layers for adhesion or sealing. They also are often laminated with a heat seal film or other substrates and printed. For improved barrier performance they may be coated with PVOH or PVDC or metallized.

Oriented polyethylene films have traditionally been used as shrink films. With the trend toward replacing complex multimaterial, multilayer laminates with structures that are mono-material or compatible enough to be recycled in a mono-material recycle stream, the development of dimensionally stable BOPE is underway. The BOPE can be reverse printed or metallized and replace BOPET in laminates with PE-based sealant films so that the packages can be placed in the all-PE recycle stream. PE is challenging to orient on existing lines designed for PP due to differences in crystallization, crystalline morphology, and orientation behavior. Lin et al. [10,176], report on the development of a novel PE grade that allows stretching $5\times$ in the machine direction and $9\times$ in the transverse on a tenter frame machine. They report the oriented film has higher dart impact and puncture strength, higher modulus, easier tear and better optical properties than standard blown film of the same density (0.926 g/cc). A laminate with BOPE compared with an incumbent laminate with BOPA film has equivalent performance. BOPE provides a good balance of performance attributes but still has limitations in heat resistance and stiffness. Compared with BOPP or BOPET, the use of sealant resins with a lower heat seal initiation temperature is generally required to prevent the heat from the sealing jaws from distorting the BOPE film. These and other challenges for design for recycling are taken up in Chapter 17.

4.3.2 Vacuum metallized films with aluminum or inorganic oxides (SiO_x and AlO_x)

Metallized film is used to provide moisture, oxygen, and light barrier and for esthetics. A variety of substrate films for metallization are used in flexible packaging, including PET, PP, PA, and PE. Of these metallized oriented PET and PP are most common. The metallization process is described in Chapter 2.

The thickness of the metallization layer ranges from about 3 to 40 nm. As the thickness increases, the optical density (opacity) and gas barrier performance increase. For oxygen and moisture barrier applications, typically around 30 nm are needed.

Metallized films are used in many packaging applications, including:

- Bags for snack foods such as potato chips (met-BOPP for light barrier).
- Coffee pouches (met-BOPET or BONy).
- Candy wrappers (met-BOPP or met-BOPET).
- Stand-up pouches (met-BOPP or met-BOPET).
- Meat or cheese packages (met-BOPET).

The metallized layer is typically turned inward in a laminate structure to protect the thin metal layer. Some example structures include: BOPP/print/adhesive/met-BOPP and print/BOPET-met/adhesive/PE sealant film.

Metallized film has replaced aluminum foil in some barrier packaging applications. While in theory foil has superior barrier properties, handling and flex cracking can reduce its performance. An economic analysis by Lush and Ferrari shows that metallized films may provide a lower total cost in many applications when taking into account price per square meter, handling and processing [177].

Transparent barrier coatings of SiO_x and AlO_x onto film substrates are also available. These provide moisture and oxygen barrier but not opacity. They are useful for applications not requiring a light barrier, where the product is meant to be seen, or when aluminum coatings interfere with online packaging metal detectors used to detect possible contamination in the product. Metal detectors may also interfere with the sorting process for the recycling of postconsumer packaging containing metallized film. Electron beam deposition or plasma-enhanced chemical vapor deposition processes are typically used to apply these inorganic coatings (see Section 2.5). At present these coatings are about two to three times more expensive than aluminum metallization [178].

4.3.3 Aluminum foil

Aluminum foil has long been used as a barrier layer in flexible packaging. The production of aluminum foil begins with a series of cold roll steps where the thickness of an aluminum ingot is reduced in half with each step. In the final step below 60 microns, the foil is doubled over and sprayed with oil (for release). The foils are rolled together to the final thickness, separated, re-rolled, slit into the desired width and annealed. Annealing is a critical step to ensure the quality of the foil. The annealing

- removes the oil,
- creates a stronger, more uniform oxide crystalline structure on the surface,
- creates a wettable surface,
- provides a more homogenous surface, and
- controls oxide crystal growth to prevent sticking between foil layers in the roll.

A number of quality issues can arise from the manufacturing process, including [179]:

- While most of the oil is removed during annealing, traces will remain. Residual oil is more likely to be near the center of the roll than the edges, causing differences in wettability across the width of the foil.
- There may be differences in wettability between the oil and nonoil sides of the foil.
- Less oxygen ingresses to the center of the roll during annealing; this may result in differences in oxide structure and properties from the edges to the center. Oxides form in two layers with the layer in contact with the foil providing barrier and the outer layer providing adhesion (surface properties). The nature and thickness of these layers change during annealing.
- Precipitates such as AlFeSi and Al_3Fe may form at the oxide surface causing defects that may affect quality. While pure aluminum may have the best barrier (defect-free), alloys and precipitates are needed to achieve the strength requirements for packaging applications. There may be a tradeoff between strength and defects.

Aluminum foil is available in a variety of thicknesses, from about 6 microns and higher. Aluminum foil is often the most expensive layer in a multilayer flexible package, so the thinnest possible thickness is used. Since in theory aluminum foil is an impenetrable barrier, the minimum thickness is determined by defects that may occur during the foil manufacturing process, during handling by the converter and on the packaging line, and in the end-use. Pinholes and flex cracking often lower the barrier performance of foil-based packaging [180].

Aluminum foil is typically coated with a lacquer or water-based latex, or laminated to a polymer film via adhesive or extrusion coating or lamination. Vinyl based lacquer-coated aluminum foils are used in many lidding applications, particularly for dairy products such as yogurt. Laminations are used in many applications such as pharmaceutical packaging, aseptic liquid packaging and condiment sachets.

4.3.4 Cellophane

Cellophane is a transparent film made from cellulose that has low permeability to oxygen, moisture, oil, grease, and bacteria. As described in Chapter 1, its introduction in the early 20th century allowed for the mass production and promotion of food items and changed the retail environment in the US and elsewhere. It has largely been displaced by less expensive petrochemical-based films, such as BOPET.

Cellophane is made from a rather complex process. Cellulose from wood or other sources is dissolved in alkali and carbon disulfide to form a viscose solution. The viscose is extruded through a slit into a bath of sulfuric acid and sodium sulfate to reconvert the viscose into cellulose. The film passes through a series of baths to remove excess sulfate and add softeners or plasticizers. Various mixtures of glycols, such as propylene, ethylene, or triethylene glycol, are typically used as a plasticizer to keep the film from getting too brittle for packaging applications. After the film is dried, it is typically coated. The uncoated film has few packaging applications as it loses too much water and becomes brittle, or absorbs too much water and loses its natural gas barrier properties. Nitrocellulose-based waterproof coatings were first developed by DuPont in the 1930s and are still used today. Other coatings have been developed, including ones based on PVDC, which provide additional functionality, such as heat resistance to heat seal jaws during heat sealing. Varying the plasticizer and coating technology allows the properties of cellophane to be tailored to a wide variety of application needs [181].

Cellophane is inherently stiff with a secant modulus in the range of 5200 MPa (750 kpsi), higher than BOPET. It has high temperature resistance (typically up to 177°C), excellent moisture and oxygen barrier (depending on the coating), and oil and grease resistance.

Cellophane lost ground to lower cost petrochemical-based films such as BOPET and BOPP in the 1980s and 1990s. It has made somewhat of a resurgence in recent times due to the fact it is biosourced, compostable, and biodegradable. Its sustainability record is clouded by its energy-intensive manufacturing process, the potential negative impact of some of the chemicals used, and its inability to be recycled. Significant progress in recent years, however, has been made by leading manufacturers in reducing their environmental footprint [182].

4.3.5 Paper and paperboard

Paper has been used as a packaging substrate for over two millennia. The main ingredient for making paper today is wood pulp. The properties of the paper vary depending on the type of wood and pulping process. Paper produced from softwood pulp (from conifer trees) has good strength and tear resistance, but is porous and rough. Hardwood pulp (from deciduous trees) is suitable for making smooth paper for writing and printing. Often a blend is used [183].

Mechanical and chemical processes have been developed to make wood pulp. The mechanical process produces a weak paper that is not suitable for packaging. The kraft (sulfate) process for chemically pulping paper is the dominant process used today. It produces higher yields and stronger paper, and the chemicals are more easily recovered than competing processes.

Unbleached paper is generally stronger, stiffer, and more course than bleach counterparts, but bleach paper is used for more applications.

While there is no official delineation between paper and paperboard, in general the cut-off is around 300 microns (12 mils): paper is below 300 microns thick and paperboard above. Basis weight (weight per area) is a standard measurement used to specify papers. A standard ream of paper in the US is based on 500 sheets of 24 inch × 36 inch paper or 3000 ft^2 (279 m^2). The number of sheets and paper size that make up a ream may differ by region or application, so it is important to pay attention to what is meant by a ream. The basis weight is often reported in lb/ream, kg/ream or g/m^2 (gsm). The conversion factor to grams per square meter (gsm) is 1.6725 gsm/lb/ream for a ream of 3000 ft^2.

Typical packaging papers range from about 30 to 150 gsm (18–90 lb/ream) but may be as low as 17 gsm (10 lb/ream) and as high as 335 gsm (200 lb/ream). Paperboard generally starts at 250 gsm (150 lb/ream) but typically is in the range of 200–600 gsm (120–360 lb/ream) for packaging. Density can vary from about 2.2 to 4.4 g/cm^3 and so a wide range of thickness and strength is available.

Another classification is course versus fine papers. Course packaging papers are made from unbleached kraft softwood pulps. Fine papers are typically made from bleached pulps and used in demanding printing and barrier applications.

Some of the properties that differentiate one paper or paperboard from another include [184]:

- Paper specifications:
  - Basis weight,
  - Specific volume,
  - Thickness,
  - Relative humidity,
  - Porosity.
- Optical properties:
  - Opacity,
  - Whiteness,
  - Gloss.
- Surface properties:
  - Smoothness,
  - Roughness,
  - Surface energy or contact angle.
- Strength:
  - Tensile strength,
  - Elongation at break,
  - Tear strength,
  - Stiffness.
- Absorption:
  - Moisture,

- Water vapor,
- Ink,
- Oil and grease.

TAPPI, ISO, and DIN test methods have been developed for these properties.
Good general references on paper and paperboard include Sikora [185], Attwood [186], and Riley [187].

4.3.5.1 Common paper types used in packaging

4.3.5.1.1 Kraft

Kraft comes from the German word meaning strong. Kraft paper is made from at least 80% sulfate wood pulp. It is course and exceptionally strong, making it well suited for a packaging substrate. It is sometimes made with a rough surface to prevent bags from sliding off pallets. It accepts print from letterpress, flexography, and offset processes. It finds use in multiwall bags, shipping sacks, tape, sacks, envelopes, and wraps.

4.3.5.1.2 Bleached

Bleached paper is used in applications that place a high value on printing, writing, or special functionality (such as a barrier). It is generally not as strong as kraft paper but can be made with a good balance of strength and printability. Bleached pulps are generally white, bright, and soft. They are susceptible to chemicals used to impart special functionality. The whiteness helps in printing, which can be enhanced by clay coating on one or both sides.

4.3.5.1.3 Specialty papers

"Greaseproof paper" is manufactured to have a pore-free structure to inhibit grease from penetrating. It is aided by the fact that cellulose is hydrophilic and so resists hydrophobic materials like grease. Chemical or mechanical refinements, as well as calendaring, are used to produce the structure.

Grease resistant paper resists staining from oil and grease. This is achieved by a variety of methods, such as treatment with fluorochemicals or applying various coatings such as wax or PVDC. Oelophobic chemicals have been developed to impart both grease and water resistance. Examples of these chemicals include alkylsuccinic anhydrides and alkylketene dimers. These may be added during the papermaking process or as sizing in an inline coating process. Specialty coatings have been developed for specific chemical resistance.

Grease and water resistance paper has many applications, including quick-service restaurant food packaging, pet food packaging and meat wraps.

Coated paper generally refers to clay-coated paper. Clay is often used to provide a smooth, glossy surface for printing. Other coatings include extrusion coatings and aqueous coatings for barrier and heat seal applications.

4.3.5.2 Paperboard for packaging

Paperboard comes in a variety of single and multi-ply structures. Single-ply paperboard from bleached wood pulp is used where cleanliness and purity are important, together with good strength and printability. Multi-ply is used over a broader range of applications since quality outer plies can be combined with less expensive middle plies. They may be combined with plastic coatings or be used in laminations to extend their application range.

Examples of paperboard use in flexible packaging include polyethylene coated milk cartons, composite cans, and laminations with foil and polyethylene for aseptic juice boxes.

4.4 Material specifications

The manufacture of any material produces variability in composition and properties. Suppliers use a variety of tools to control this variation within a set of sales specifications. The sales specifications are set according to the needs of the users and the capabilities of the manufacturing unit. Sales specifications set too tight may result in low yields and poor economics for the supplier. Specifications set too wide may cause problems with the customer, resulting in loss of business. Hence, materials suppliers are continuously striving to improve their quality control (QC) to satisfy customer needs at the highest yields. Some of the tools used to achieve QC include statistical process control, inline measuring of key properties, inline monitoring of important process variables, established procedures, and standard operating conditions.

The nature of the manufacturing process determines the type and frequency of measurements for QC. A continuous film process, for example, may have an inline gauge for total thickness as well as video inspection for defects; these measurements are made on a continuous basis. Other tests, such as heat seal or burst strength are destructive tests and can only be made periodically, perhaps on samples taken at the end of each roll. A resin producer may have a continuous process but rely on monitoring and control of key process variables while taking samples periodically to measure properties for release to sales. The frequency of the sampling can be determined by statistics and depends on the variability of the process, run rate, and other considerations.

Many properties used for QC are designed to ensure the composition is under control and may not be directly related to the end-use. For example, melt flow index is often used as a QC measurement for polyolefin production. Melt flow index is a measure of how much polymer melt flows through a standard orifice using a standard weight, time, and temperature. It is inversely related to viscosity and MW, but the relationship is not exact. The deficiencies of relying on melt flow index for polymer processing are discussed in Chapter 5. Here it is important to note that melt flow index is used as a surrogate for MW control. The higher the melt flow index, the lower the MW for a given polymer type. But other factors influence the melt flow index as well, such as melting point, branching, and MWD. Hence, two polymers with the same melt flow index from two different manufacturers may have very different flow characteristics in a film converting process. For a given manufacturer and product, however, maintaining the melt flow index within a specified range is a good way to ensure the product is consistent from lot to lot. It is also important to note that typically melt flow index is not the only parameter that is being tested or measured to ensure the process is under control. With experience, the manufacturer may have developed relationships between key process parameters such as reactor temperature or pressure and MW. The combination of ensuring the process is under control as well as periodic testing is the bedrock of modern manufacturing.

Often properties that a manufacturer specifies differ from scientific measurements that may be conducted in a research environment. Returning to the example of melt flow index, in a laboratory environment, GPC is typically used to characterize the MW and MWD. This test is time-consuming and not very robust for a manufacturing environment. The melt flow index test can be done in 10 minutes for the quick turnaround times needed for QC. Similarly, nuclear magnetic resonance spectroscopy may be the best scientific technique to determine molecular structure, but Fourier transform infrared spectroscopy (FTIR) is often used because it is faster. Or a simple density test with a density gradient column may be sufficient to ensure the comonomer content is within spec for a LLDPE. Cost, the turnaround time for test results, and robustness in a manufacturing environment are all considered when developing QC tests and specifications.

For QC tests, the manufacturer will often make standards to correlate the QC test with a known laboratory test. From this data, the manufacturer can set reasonable specifications. Developing such tests is not always easy and may require substantial time and effort for new products. For example, FTIR is often used to measure the composition of the comonomer in copolymers, but the FTIR spectrum from different comonomers may interfere and obscure the presence of others or change the quantitative relationship between peak size and composition. Subtractive techniques may help, where a baseline spectrum is subtracted from the measured spectrum to eliminate interference. This may introduce other errors, however. The importance of developing QC tests for new products should not be overlooked during the development process.

Properties that are specified by the manufacturer may also differ from what may be reported on a product datasheet. Product specifications are meant to ensure the process is under control and the product is consistent from lot to lot. Product data sheets often list "typical" properties. These are properties that are reasonable to expect from the product but may not be measured on every lot or used to determine if the product is fit for sale. Data sheet properties are meant to provide a starting point for product use and design.

Some typical properties that may be specified by resin manufacturers are given in Table 4.9. Generally, a resin supplier will specify four types of characteristics: rheology, composition, appearance, and additives. Rheology is the measurement of how the polymer flows. For PE, the melt flow index is the most commonly specified measurement of flow. Melt flow index is the weight in grams that flows from a specified orifice in 10 minutes under a weight of 2160 g at a temperature of 190°C. Details are provided in ASTM D1238 and ISO 1183. Other weights and temperatures may be specified, and the industry convention is to call this measurement melt flow rate. For example, for PP, 230°C is typically used. Intrinsic and inherent viscosities are solution viscosities and are typically used for polyamides and polyesters.

For the purposes of QC, rheology is used as a tool to ensure the MW is within specification. The viscosity is related to M_W raised to the 3.4 power and so small changes in MW will have a large impact on viscosity. Viscosity is also impacted by MWD and branching. Some resin producers use inline measurements of viscosity that correlate to melt flow index or other specified rheological measurements.

TABLE 4.9 Typical properties specified by manufacturers.

Category	Test	Comment
Rheology	Melt Flow Index Melt Flow Rate Intrinsic Viscosity	These measurements are surrogates for MW
Composition	Density FTIR	For PE, density is an indication of comonomer content; FTIR and other spectroscopy techniques are used for measuring comonomer content
Appearance/ contamination	Gel/fisheyes Black specs	Typically a count per standard area of film or weight of pellets (black specks), or a pass/fail vs. a standard
Additives	Additive concentration	Some suppliers may provide data on measured amounts of additives, such as slip agents
Films	Thickness Surface energy Defects Optical properties	These are some of the properties that may be specified
Paper	Basis weight Thickness Density or specific volume Smoothness Porosity Absorption (water, etc.)	These are some of the properties that may be specified

For copolymers, some sort of measurement of composition is typically specified. For LLDPE, the amount of butene, hexene, or octene is monotonically related to the polymer density. Therefore density is measured and specified. For an ethylene copolymer such as EVA, the amount of VA is specified. FTIR is often used to measure the comonomer content of the polymer.

Film appearance is important in flexible packaging. Most polymers sold into the flexible packaging market will have a specification around film defects that may originate with the polymer manufacturing process, such as imperfections or "gel" above a specified size and black specs. The resin manufacturer will extrude a sample into a film and measure defects per unit area. Visual standards may be used to rate film quality but in recent years most of this has been automated with cameras and digital defect counters.

Some resins may contain additives for specific functionality, such as slip and antiblock additives. The amount of the additive may be specified by some manufacturers, rather than a property such as COF. COF will depend on the slip additive concentration, thickness of the film, presence of adjacent layers in the film, and temperature. Converters often add to the amount of slip in the product for a given application. Hence, they need to know how much is in the product to begin with.

Substrates such as film, foil, and paper have their own set of specifications. Thickness uniformity is one of the most important properties. Surface characteristics maybe another. For a film, the surface energy or dyne level may be specified. A foil may be classified by the class of finish (A, B, etc.). Film defects and optical properties may be specified. For paper, properties such as basis weight, porosity, and smoothness may be specified.

4.5 Regulatory considerations

Compliance with appropriate government regulations for food, medical, and other packaging applications is an important consideration for material selection and structure design. The materials and additives that comply with food and medical packaging regulations are only a small subset of available materials and thus regulations may limit the choice. There are also materials, or components of materials, that may not be regulated but nevertheless be unsuitable due to public perception. Recent examples of chemicals that have come under public scrutiny due to possible health concerns include some types phthalate plasticizers used in PVC, Bisphenol A (BPA) found in polycarbonate and epoxies, and perfluorooctanoic acid (PFOA), and perfluorooctanesulfonic acid (PFOS) fluorochemicals.

This introductory chapter will not go into detail about regulatory compliance. Good references on the subject include Ebnesajjad [188] and Knight and Creighton [189]. Since food packaging is the largest application of flexible packaging, the focus is on describing the process for determining food packaging compliance in the US and in Europe. The rest of the world often follows one of these, but other countries may have their own local regulations. Special considerations for healthcare packaging are also briefly described. Communication with regulatory agencies, the study of the complete regulations and consulting sound legal advice is recommended for anyone ascertaining the compliance status of an actual package.

4.5.1 Food packaging regulatory compliance in the US

In the US every component of a food package must comply with the appropriate regulation for food packaging. For multilayer packaging, this means each layer must comply and if a layer comprises more than one material (such as a blend), each component must comply. The individual substance that may reasonably be expected to migrate to the food in its intended use in the food contact material must be covered by one of the following:

- A regulation listed in Title 21 Code of Federal Regulations (CFRs);
- A prior sanction letter;
- Criteria for generally considered safe status (GRAS);
- A threshold of regulation (TOR) exemption;
- An effective food contact notification (FCN).

An important difference between the US and Europe is that the US Food and Drug Administration (FDA) generally considers each polymer as a different material and subject to regulation, even if it is comprised of monomers used in already approved materials. In Europe, the regulatory process is built upon the status of the monomers; generally, if the monomer is on the approved list then any polymer made from that monomer is approved.

It is the responsibility of the manufacturer of the food packaging article to ensure that their article complies with the regulations. The regulations may set limitations on the use of materials depending on the application, such as temperature in use and the type of food being packaged. Extraction tests may be required to ensure the article meets the limitations set forth by the regulations. The regulations provide guidance for the type of testing required.

When determining whether a component used in a food package complies with the regulations, there are several places to begin:

- Consult the component manufacturer's product datasheet, which may have a statement regarding FDA compliance.
- Consult 21 CFR 174–179 to see if the component is appropriately regulated as an indirect food additive for the intended use.
- Consult 21 CFR 182–186 to determine if the component is GRAS or consult the GRAS notices.
- Consult 21 CFR 181 to see if the use of the component has been sanctioned for food contact.
- Consult the listing of Threshold of Regulation Exemptions.
- Consult the listing of Effective Food Contact Substance Notifications.

Be aware that FCNs are for a specific manufacturer for a specific substance and conditions of intended use (see Section 409(h)(2)(C) of the Federal Food, Drug, and Cosmetic Act). They may not be used for similar substances or different manufacturers. FCNs are proprietary to the manufacturer and must be obtained from that manufacturer.

Section 21 CFR 174–179 contains the list of indirect food additives for food packaging. It can be further broken down as follows:

- 21 CFR 175: Adhesives and Components of Coatings;
- 21 CFR 176: Paper and Paperboard Components;
- 21 CFR 177: Polymers;
- 21 CFR 178: Adjuvants, Production Aids, and Sanitizers;
- 21 CFR 179: Irradiation in the Production, Processing, and Handling of Food.

The regulations can be found on the FDA website: http://www.fda.gov/.

If one or more of the components in the article are not in the above lists, this does not mean the material is inherently unsafe. It simply means no one has taken the material through the appropriate regulatory approval process. The options for getting a new material approved include submitting a Threshold of Regulation Exemption Request, submitting a Food Contact Substance Notification, or satisfying the criteria for GRAS. Consultation with attorneys who specialize in food regulations is often advisable.

4.5.2 Food packaging regulatory compliance in the European Union

In May 2011, the European Union Plastics Regulation (No. 10/2011) went into effect in all the European Union states, replacing previously existing national laws. This regulation was published in the Journal of the European Union on January 15, 2011 [190]. The complete regulation is beyond the scope of this chapter, but some highlights are selected below. Note that this regulation is for plastics; other substrates such as foil or paper are covered by other regulations.

- For synthetic macromolecules (which includes polymers and plastics), the starting monomers, as well as the macromolecules, should be assessed for risk.
- In macromolecules produced by natural means, such as fermentation, the final product may contain non or incompletely reacted substances. The final product should be assessed for its health risk.
- Directive 2002/72/EC, which the Plastics Regulation supersedes, had separate lists of authorized monomers for additives and polymers. These have been merged into the new regulation.
- Polymers may be used as the main structural component of plastics, as well as additives. For the case of polymeric additives, if the monomers or starting substances are the same as a structural polymer, and the monomers have already been authorized, the additive is considered authorized. If not, an assessment of the additive is necessary.
- Substances used to manufacture polymers may contain impurities. During manufacture, reaction and degradation byproducts may form. Main impurities and byproducts, if they are relevant, should be considered in the assessment, but all impurities and byproducts need not be considered.
- Aids to the manufacture of polymers, such as catalysts, chain transfer agents, chain extending agents, and chain stopping agents, when used in minute amounts and not intended to be in the final product are not subject to the authorization procedure.
- The release of substances from food contact materials and articles should not bring about unacceptable changes to the food. The regulation sets specific migration limits (SML) for substances based on toxicity data. According to good manufacturing practices, the upper limit for migration of all substances (overall migration limit) from the article to the food is 10 mg/dm^2. This is based on 60 mg/kg of food. For packaging of foods intended for infants and children, whose consumption per body weight is higher, special provisions should be set.
- The regulation sets forth migration testing protocols. Simulants are used to represent different food categories being packaged, such as 3% acidic acid solution for acetic foods. Time and temperature conditions for the testing protocols are established for the different food categories.
- The regulations allow for the concept of a functional barrier in multilayer food packaging. If a layer prevents the migration of substances, then nonauthorized substances may be used in layers on the nonfood side of the barrier layer, provided the migration is below a specified detection limit. The maximum limit is 0.01 mg/kg of food for the nonauthorized substances. Substances that are mutagenic, carcinogenic or toxic to reproduction, as well as nano-engineered substances, are not covered by the functional barrier concept.
- For multi-material layers within a multilayer package, each material should comply, and the concept of the functional barrier also applies.
- Coatings, printing inks, and adhesives are not yet covered by specific EU legislation. These may be covered by individual country regulations within the EU.
- This regulation is only for new materials. After the materials have been used the regulation no longer applies as they may have been contaminated with other materials such as food products. A separate regulation (EU No. 282/2008) has been created for the use of postconsumer mechanical recycled plastics in food packaging where the recycling process must be authorized. Polymers made from monomers generated from chemical recycling (chemically breaking down recycled plastics to the constituents) or unused film scrap generated by the film producer and not exposed to food products (postindustrial recycling) are covered by the virgin plastic regulation (EU No. 10/2011). Also, if the recycling is behind a functional barrier then EU No. 10/2011 applies.

At the risk of being too general, the basic requirements for establishing whether an article is authorized for use in contact with food are as follows:

- The substances used to manufacture plastics used in the layers and as components of multicomponent layers must be on the authorized Union list. The list contains monomers or other starting substances, additives excluding colorants, polymer production aids excluding solvents, and macromolecules obtained from microbial fermentation.

- Substances used in the manufacture of plastic layers in plastic materials and articles are subject to specific and overall migration limits, as set out in the regulations. The specific migration limits (SMLs) are expressed in mg of substance per kg of food (mg/kg). For substances with no SML or other restrictions, a generic SML of 60 mg/kg is used. The plastic article is subject to total or overall migration limits as well.
- In multilayer packaging, the functional barrier concept may be invoked, as described above.
- Formulas are given in the regulation for testing and computing limits for packaging.

The complete regulation should be consulted for a given application. This brief description is solely intended to give the reader an appreciation of the process and is not a substitution for reading the complete regulation.

4.5.3 Heathcare packaging

For packaging of pharmaceuticals, medical devices and other medical products, the package materials and structures must meet the appropriate regulations and expectations of the governing bodies and standards organizations. These include European Union Directives, US Federal Drug Administration and International Conference on Harmonization regulations and guidances, European regulations and standards from groups such as the US Pharmacopeial Convention, Japanese Pharmacopeia, European Pharmacopeia, and the World Health Organization. The issues that must be addressed by the material and package suppliers include the potential for migratory species in the flexible packaging structure, tolerance of the package to downstream processes such as sterilization, reliability of supply, notification of change, and requests for plant audits by the film converters and end-users.

Even after meeting all of the relevant regulations, there is inherently an increased amount of liability and risk in these applications. Companies that can provide high-quality packaging for improved healthcare, however, are meeting an important societal need that will only grow as the human population increases and more people gain access to quality healthcare.

References

[1] DuPont Teijin Films, Oxygen and water vapour barrier properties of flexible packaging films. <http://usa.dupontteijinfilms.com/informationcenter/downloads/Oxygen_And_Walter_Vapour_Barrier_Properties_of_Flex_Pack_Films.pdf>, n.d. (accessed 28.06.20).

[2] ExxonMobil Chemical, LDPE LD 071. <https://www.exxonmobilchemical.com/en/resources/product-data-sheets/Polyethylene/Exxonmobil-LDPE>, n.d. (accessed 28.06.20).

[3] ExxonMobil Chemical, LLDPE LL 1001. <https://www.exxonmobilchemical.com/en/productselector#/grade-detail/150000000546> n.d. (accessed 28.06.20).

[4] T.I. Butler, B.A. Morris, PE based multilayer film structures, in: J. Wagner (Ed.), Multilayer Flexible Packaging, Technology and Applications For The Food, Personal Care and Over-The-Counter Pharmaceutical Industries, Elsevier, Oxford, 2010, p. 211.

[5] ExxonMobil Chemical, Exceed mLLDPE exceed 1018CA. <http://www.b2bpolymers.com/TDS/Exceed_1018CA.pdf>, n.d. (accessed 28.06.20).

[6] The Dow Chemical Company, Affinity™ PL 1880G polyolefin plastomer. <https://www.dow.com/en-us/document-viewer.html?ramdomVar=2022656873883621l0&docPath=/content/dam/dcc/documents/en-us/productdatasheet/400-2/400-00071423en-01-affinity-pl-1880g-polyolefin-plastomer-tds.pdf>, n.d. (accessed 28.06.20).

[7] DuPont, Elvax® ethylene vinyl acetate resins, packaging grade selection guide, 1992.

[8] DuPont, Surlyn® ionomer resin selector guide, 1994.

[9] ExxonMobil Chemical, HDPE HD 7845.30. <https://www.exxonmobilchemical.com/en/productselector#/grade-detail/150000100677>, n.d. (accessed 28.06.20).

[10] Y. Lin, J. Xu, J. Pan, X.-B. Yun, M. Demirors, Biaxially oriented polyethylene (BOPE) films fabricated via tenter frame process and applications thereof, SPE ANTEC Conf. (2018).

[11] Guangdong Decro Film New Materials Co., Ltd., Decro® typical properties: co-extruded oriented down-gauge PE film, 2019.

[12] Blueridge Films, Inc., Cast PP film properties. <https://www.blueridgefilms.com/cast-film>, n.d. (accessed 28.06.20).

[13] Inteplast Group, Ltd., AmTopp – EM BOPP. <https://calculator.inteplast.us/calculator-bopp/>, n.d. (accessed 28.06.20).

[14] Impex Global, LLC, Biaxially oriented polypropylene film (BOPP). <https://www.impexglobalfilms.com/bopp.html>, n.d. (accessed 28.06.20).

[15] DuPont-Teijin Films, Mylar polyester film, physical-thermal properties. <http://usa.dupontteijinfilms.com/informationcenter/downloads/Physical_And_Thermal_Properties.pdf>, n.d. (accessed 28.06.20).

[16] R. Breeze, Economic benefits of utilizing MDO film in flexible packaging, in: TAPPI PLACE Conference, 2007. Available from: <http://www.tappi.org/content/events/07place/papers/breese.pdf> (accessed 28.06.20).

[17] Bi-Ax International, Inc, Evlon compostable plastic film data sheet: co-extruded plain polylactide film, 2019.

[18] Natureworks, Natureworks PLA polymer 4032D biaxially oriented films – high heat. <https://www.natureworksllc.com/~/media/Files/NatureWorks/Technical-Documents/Technical-Data-Sheets/TechnicalDataSheet_4032D_films_pdf.pdf>, n.d. (accessed 28.06.20).

[19] L.W. McKeen, Film Properties of Plastics and Elastomers, 3rd ed., Elsevier, Oxford, 2012.

[20] Kuraray, EVAL EVOH resins. http://www.evalevoh.com/us/eval-properties.aspx>, n.d. (accessed 21.01.14).

[21] J.M. Dealy, K.F. Wissbrun, Melt rheology and its role in plastics processing, Theory and applications, Chapman & Hall, New York, 1995, p. 370.

[22] F.C. Stehling, P. Meka, Heat sealing of semicrystalline polymer films. II. Effect of melting distribution on heat-sealing behavior of polyolefins, J. Appl. Polym. Sci. 51 (1994) 105–119.

[23] J. Scheirs, Additives for the modification of poly(ethylene terephthalate) to produce engineering-grade polymers, in: J. Scheirs, T. Long (Eds.), Modern Polyesters: Chemistry and Technology of Polyesters and Copolyesters, 2004. Available from: http://dx.doi.org/10.1002/0470090685.ch14.

[24] R.P. Saltman, Partially grafted thermoplastic compositions, U.S. Patent 5091478, 1992.

[25] L.S. McTaggart, Thermoformed polyester articles having impact resistance and high temperature stability, US Patent 3960807, 1976.

[26] E.J. Deyrup, Toughened thermoplastic polyester compositions, US Patent 4753980, 1988.

[27] A. Keller, M.J. Machin, Oriented crystallization in polymers, J. Macromol. Sci. (Phys.) B1 (1967) 41–91.

[28] F.W. Billmeyer Jr., Textbook of Polymer Science, 2nd edition, Wiley-Interscience, New York, 1971, p. 320.

[29] R. Hoff, Preparation of polymer-supported metallocene catalysts, Handbook of Transition Metal Polymerization Catalysts, 2nd Edition, John Wiley & Sons, 2018.

[30] M. Demirors, The history of polyethylene, in: E. Strom, S. Rasmussen (Eds.), 100 + Years of Plastics. Leo Baekeland and Beyond, American Chemical Society, Washington, DC, 2011, pp. 115–138. ACS Symposium Series.

[31] F.W. Billmeyer Jr., Textbook of Polymer Science, 2nd edition, Wiley-Interscience, New York, 1971, pp. 255–279.

[32] H. Fogler, M. Gurman, R7.1 polymerization, Elements website. <http://www.umich.edu/~elements/07chap/html/polymerization.pdf>, 2008 (accessed 28.06.20).

[33] S.Y. Lai, J.R. Wilson, G.W. Knight, J.C. Stevens, Elastic substantially linear olefin polymers, US Patent 5272236, 1993.

[34] J.B.P. Soares, A.E. Hamielec, Semicrystalline polyolefins – narrow MWD and long chain branching: best of both worlds, SPE ANTEC Conf. (1996) 2334–2341.

[35] Y.S. Kim, C.I. Chung, S.Y. Lai, K.S. Hyun, Melt rheological and thermodynamic properties of polyethylene homopolymers and poly(ethylene/α-olefin) copolymers with respect to molecular composition and structure, J. Appl. Polym. Sci. 59 (1996) 125–137.

[36] E. Drent, J.A.M. Van Broekhoven, M.J. Doyle, Efficient palladium catalysts for the copolymerization of carbon monoxide with olefins to produce perfectly alternating polyketones, J. Organomet. Chem. 417 (1991) 235–251.

[37] J.F. Macnamra, Jr., (Ed.), Film extrusion manual, 3rd ed., TAPPI Press, Atlanta, 2020.

[38] W. Durling, (Ed.), Extrusion Coating Manual, 5th ed., TAPPI Press, Atlanta, Ga, 2017.

[39] M. Spalding, A. Chatterjee (Eds.), Handbook of Industrial Polyethylene and Technology, Wiley – Scrivener Publishing, Hoboken, NJ, 2017.

[40] P. Gupta, G.L. Wilkes, A.M. Sukhadia, R.K. Krishnaswamy, M.J. Lamborn, S.M. Wharry, et al., Does the length of the short chain branch affect the mechanical properties of linear low density polyethylenes? An investigation based on films of copolymers of ethylene/1-butene, ethylene/1-hexend and ethylene/1-octene synthesized by a single site metallocene catalyst, Polymer 46 (2005) 8819–8837.

[41] T.I. Butler, PE processes, in: J. Wagner (Ed.), Multilayer Flexible Packaging, Technology and Applications For The Food, Personal Care nd Over-The-Counter Pharmaceutical Industries, Elsevier, Oxford, 2010, pp. 15–30.

[42] C.-T. Lue, Easy processing metallocene polyethylene, in: SPE ANTEC Conference, 1998.

[43] P. Tas, L. Nguyen, S. Marshall, M. Kashanian, Bubble stability and maximum output of SURPASS sLLDPE resin in blown film, in: TAPPI PLACE Division Conference, New Orleans, LA, September 2004, pp. 320–334.

[44] D. Falla, Superior processability and economics of single site LLDPE vs. Conventional metallocene LLDPE blown films, in: SPE ANTEC Conference, 2013.

[45] Packaging World, Dow: high-performance film resins. New INNATE family of resins provides stiffness/toughness balance with improved processing and sustainability profiles. <https://www.packworld.com/design/flexible-packaging/product/13368996/dow-highperformance-film-resins>, 2015 (accessed 21.06.20).

[46] Packaging World, ExxonMobil Chemical Company: resin for tough, stiff film. <https://www.packworld.com/design/flexible-packaging/product/13346749/exxonmobil-chemical-company-resin-for-tough-stiff-film>, 2010 (accessed 21.06.20).

[47] Packaging World, ExxonMobil: polymers for liquid and food packaging. <https://www.packworld.com/home/product/13370428/exxonmobil-polymers-for-liquid-and-food-packaging>, 2016 (accessed 21.06.20).

[48] Packaging World, ExxonMobil: new polymer grade. <https://www.packworld.com/home/product/13375113/exxonmobil-new-polymer-grade>, 2018 (accessed 21.06.20).

[49] N. Aubee, T. Chuang, D. Checknita, P. Chisholm, P. Lam, S. Marshall, et al., U.S. Patent Application US20080118749A1, 2008.

[50] O. Lightbody, P. Chisholm, N. Aubee, T. Tikuisis, Barrier properties of HDPE film, US Patent Application US 2014/0309351A1, 2014.

[51] L. Sherman, High-moisture barrier sHDPE gets increasing play in flexible food packaging, Plast. Technol. (2015), accessed online on 21 June 2020. Available from: https://www.ptonline.com/articles/high-moisture-barrier-shdpe-gets-increasing-play-in-flexible-food-packaging.

[52] J.R. de Garavilla, Ionomer, acid copolymer, and metallocene polyethylene resins: A comparative assessment of sealant performance, TAPPI J. 78 (6) (1995) 191–203.

[53] A. Calhoun, Polypropylene, in: J. Wagner (Ed.), Multilayer Flexible Packaging, Technology and Applications For The Food, Personal Care and Over-The-Counter Pharmaceutical Industries, Elsevier, Oxford, 2010, pp. 31–36.

[54] E.P. Moore Jr., Polypropylene Handbook, Hanser Publishers, New York, 1996, pp. 6–9.
[55] ExxonMobil, Vistamaxx™ performance polymers Low sealing temperatures in cast PP films for fast packaging line speeds. <https://www.exxonmobilchemical.com/en/library/library-detail/11978/great_southeast_case_story_en>, 2014 (accessed 21.06.20).
[56] S. Srinivas, S. Datta, G. Racine, Novel propylene-based specialty elastomers - structure and properties, in: SPE ANTEC Conference, 2004.
[57] D. Sexton, P. Ansems, L. Hazlitt, Versify™ plastomers and elastomers: resin technology, performance and applications, in: Originally Presented at SPE International Polyolefins Conference, 2004. Available from: <https://pdfs.semanticscholar.org/bddc/ed6e8f2c625adde5446b77a1-b06e20e56a8f.pdf>, n.d. (accessed 21.06.20).
[58] R. Jester, Heat seal characteristics of cyclic olefin copolymer/polyethylene blends, TAPPI PLACE Conference, Boston, MA, 2002, pp. 1308–1334.
[59] R. Jester, Cyclic olefin copolymer enhances polyolefin blends for film packaging, Plastics Technology. <http://www.ptonline.com/articles/cyclic-olefin-copolymer-enhanced-polyolefin-blends-for-film-packaging>, 2011 (accessed 28.06.20).
[60] J.Y. Shin, J.Y. Park, C. Liu, J. He, S.C. Kim, Chemical structure and physical properties of cyclic olefin copolymers,"IUPAC Technical Report Pure Appl. Chem. 77 (2005) 801–814.
[61] S. Gerbig, Polyamide moisture absorption. Relative dimensional change of various nylon products, UL IDES. <http://articles.ides.com/polyamide-moisture-absorption.asp>, n.d. (accessed 09.01.14).
[62] DuPont, DuPont™ Selar® PA 3426 blends with nylon 6. <http://www2.dupont.com/Selar/en_US/assets/downloads/selar_pa3426_nylon_-blends.pdf>, n.d. (accessed 09.01.14).
[63] D. Djordjevic, Guide to nylon chemistry, equipment & processing (coating and casting), In: TAPPI Polymers, Laminations & Coatings Conference, 1997, pp. 341–382.
[64] S. Saez, J. Hamilton, Nylons, in: J.F. Macnamra, Jr., (Ed.), Film extrusion manual, 3rd ed., TAPPI Press, Atlanta, 2020.
[65] R.K. Tubbs, Polyvinyl alcohol acrylate and methacrylate copolymers, in: C.A. Finch (Ed.), Polyvinyl Alcohol – Developments, Second Edition, John Wiley & Sons, Ltd., England, 1992, p. 399. 1992.
[66] G.W. Brant, Polyvinyl alcohol casting solution, U.S. Patent 2491642 A, 1949.
[67] M. Masuda, Recent advances in polyvinyl alcohol film, in: C.A. Finch (Ed.), Polyvinyl Alcohol – Developments, Second Edition, John Wiley & Sons, Ltd., England, 1992, pp. 403–431. 1992.
[68] C.A. Finch, (Ed.), Polyvinyl Alcohol – Developments, 2nd ed., John Wiley & Sons, Ltd. England, 1992.
[69] Kuraray, Technical bulletin No. 110: Gas barrier properties of resins. <http://www.eval-americas.com/media/36916/tb_no_110.pdf>, n.d. (accessed 28.06.20).
[70] R.T. Chou, G.I. Deak, Blends of ethylene-vinyl alcohol copolymer and amorphous polyamide, and multilayer containers made therefrom, EP Patent Application 380123 A1, 1990.
[71] D.E. Ofstein, Blends of amorphous nylon and ethylene-vinyl alcohol copolymer for oxygen barrier containers, US Patent 500302, 1991.
[72] R. Chou, I.-H. Lee, The behavior of ethylene vinyl alcohol amorphous nylon blends in solid-phase thermoforming, J. Plastic Film. & Sheeting 13 (1997) 74–93.
[73] A.I. Fetell, Ethylene-vinyl alcohol copolymer blends with balanced properties for laminates, WO 9709380 Patent Application, 1996.
[74] A.I. Fetel, Nylon-containing ionomer improves the performance of ethylene vinyl alcohol (EVOH) copolymers in multilayer structures, 1997 TAPPI Polymers, Laminations & Coat. Conf. Proc. (1997) 503–505 (1997).
[75] E. Gimenez, J.M. Lagaron, L. Cabedo, R. Gavara, J.J. Saura, Study of the thermoformability of ethylene-vinyl alcohol copolymer based barrier blends of interest in food packaging applications, J. Appl. Polym. Sci. 91 (2004) 3851–3855.
[76] Eval Americas, EVAL™ EVOH barrier resins and monolayer film. <http://www.eval-americas.com/media/124484/01_eval_-_euen.pdf>, n.d. (accessed 28.06.20).
[77] Mitsuibishi Chemical, Soarnol™ specialty grades. <http://www.soarnol.com/eng/grade/index.html>, n.d. (accessed 28.06.20).
[78] M. Ikko, H. Onishi, T. Yamamoto, K. Ninomiya, Study on orientable EVOH, in: European TAPPI PLACE Conference, 2009. Available from: <http://www.tappi.org/content/enewsletters/eplace/2004/10-3matsui.pdf> (accessed 28.06.20).
[79] G. Nippon, Soarnol® grade line up. <http://www.soarnol.com/eng/grade/index.html>, 2013 (accessed 28.06.20).
[80] E. Chow, High Barrier Forming Films with EVOH, TAPPI/CETEA VII International Conference on Flexible Packaging, Brazil, 2012.
[81] E.D. Medlock, Nanolayer film structures with ethylene vinyl alcohol resins, in: TAPPI PLACE Conference, 2012.
[82] P.C. Lee, J. Dooley, J. Robacki, S. Jenkins, R. Wrisley, Flex barrier property enhancement in film structures using microlayer coextrusion technology, in: SPE ANTEC Conference, 2013.
[83] K. Xiao, I-H. Lee, R. Armstrong, Comparison of water-quench versus air-quench blown film processes – part I: flat film properties, in: SPE ANTEC Conference, vol. 70, 2012, pp. 972–977.
[84] K. Xiao, I-H. Lee, R. Armstrong, Comparison of water-quench versus air-quench blown film processes – part II: thermoformability, in: SPE ANTEC Conference, vol. 70, 2012, pp. 978–983.
[85] V. Renard, Saran PVDC barrier resin concepts for demanding coextrusion and lamination applications, in: TAPPI PLACE European Conference, 2005. Available from: <http://www.tappi.org/content/enewsletters/eplace/2004/13-1Renard.pdf> (accessed 28.06.20).
[86] K. Paisley, PVDC – new developments, new opportunities, in: TAPPI PLACE Conference, 2007.
[87] T.R. Mueller, PVdC – past and current barrier material, in: TAPPI PLACE Conference, 2007.
[88] Honeywell, Honeywell Aclar®. <https://www.packagingcomposites-honeywell.com/aclar/>, n.d. (accessed 28.06.20).
[89] G.D. Jerdee, Polyethylene for extrusion laminating and coating, in: W. Durling (Ed.), Extrusion Coating Manual, 5th ed., TAPPI Press, Atlanta, Ga, 2017.

[90] B. Morris, Adhesion in extrusion laminating and coating, in: W. Durling (Ed.), Extrusion Coating Manual, 5th ed., TAPPI Press, Atlanta, Ga, 2017.

[91] J. Vansant, B. Morris, S. Marks, Acid copolymers for extrusion laminating and coating, in: W. Durling (Ed.), Extrusion Coating Manual, 5th ed., TAPPI Press, Atlanta, 2017, pp. 235–239.

[92] I-H. Lee, Bonding 'unjoinable' polymers, DuPont. <http://www2.dupont.com/Packaging_Resins/en_US/assets/downloads/white_papers/Bonding_Unjoinable_Polymers_Lee_92011.pdf>, 2011 (accessed 23.01.14).

[93] S.R. Tanny, Tie layer adhesives, in: T.I. Butler (Ed.), Film Extrusion Manual. Process, Materials, Properties, 2nd ed., TAPPI Press, Atlanta, 2005, pp. 403–412.

[94] B. Morris, I-H. Lee, "Tie polymers," in: J.F. Macnamra, Jr. (Ed.), Film Extrusion Manual, third ed., TAPPI Press, Atlanta, 2020.

[95] B.A. Morris, Factors influencing adhesion in coextruded structures, TAPPI J. 75 (8) (1992) 107–110.

[96] T.E. Rolando, *Flexible packaging – adhesives, coatings and processes: Rappa Review Reports, report 122*, Rappa Technology, Ltd., 2000.

[97] L. Jopko, Flexible packaging adhesives – the basics, in: TAPPI PLACE Conference, 2004.

[98] Torradas, J., "Polyester," in: J.F., Macnamra, Jr., (Ed.), Film Extrusion Manual, third ed., TAPPI Press, Atlanta, 2020.

[99] Plastics News, Coke's PlantBottle use swells globally. <http://www.plasticsnews.com/article/20130607/NEWS/130609933/cokes-plantbottle-use-swells-globally>, 2013 (accessed 28.06.20).

[100] Coca-Cola, Inc., Coca-Cola expands access to PlantBottle IP. <https://www.coca-colacompany.com/news/coca-cola-expands-access-to-plant-bottle-ip>, 2019 (accessed 24.06.20).

[101] Pepsi to launch 100% biobased PET bottle, Plastics Today, March 16, 2011. Available from: <http://www.plasticstoday.com/articles/pepsi-launch-100-biobased-pet-bottle-03179> (accessed 28.06.20).

[102] A. Barrett, Pepsico goes for biobased bottles, Bioplastics N. (2018), accessed online on 24 June 2020. Available from: https://bioplasticsnews.com/2018/09/10/pepsico-goes-for-bioplastic-bottles/.

[103] Ellen Macarthur Foundation, The new plastics economy: rethinking the future of plastics & catalysing action. <https://www.ellenmacarthur-foundation.org/publications/the-new-plastics-economy-rethinking-the-future-of-plastics-catalysing-action>, 2017 (accessed 27.06.20).

[104] The Dow Chemical Company, Dow and UPM partner to produce plastics made with renewable feedstock, Press release. <https://corporate.dow.com/en-us/news/press-releases/dow-and-upm-partner-to-produce-plastics-made-with-renewable-feedstock.html>, 2019 (accessed 27.06.20).

[105] W. Beacham, SABIC takes first steps into renewable polyolefins, ICIS (2014), accessed online on 27 June 2020. Available from: https://www.icis.com/explore/resources/news/2014/06/10/9789706/sabic-takes-first-steps-into-renewable-polyolefins/.

[106] LyondellBasell, LyondellBasell and Neste announce commercial-scale production of bio-based plastic from renewable materials, Press release. <https://www.lyondellbasell.com/en/news-events/products--technology-news/lyondellbasell-and-neste-announce-commercial-scale-production-of-bio-based-plastic-from-renewable-materials/>, 2019 (accessed 27.06.20).

[107] H. Liu, Y. Xu, Z. Zheng, D. Liu, 1,3-Propane diol and its copolymers: research, development and industrialization, Biotechol J. 5 (11) (2010) 1137–1148.

[108] R.P. Babu, K. O'Conner, R. Seeram, Current progress on bio-based polymers and their future trends, Prog. Biomater. (2013)2:8, viewed on 28 June 2020. Available from: http://www.progressbiomaterials.com/content/2/1/8.

[109] R. Lingle, PHA bioplastics a 'tunable' solution for convenience food packaging, Plastics Today, 2018. Available from: <https://www.plasticstoday.com/packaging/pha-bioplastics-tunable-solution-convenience-food-packaging/157388153458558> (accessed 21.06.20).

[110] E. Bugnicourt, P. Cinelli, A. Lazzeri, V. Alvarez, Polyhydroxyalkanoate (PHA): review of synthesis, characteristics, processing and potential applications in packaging, eXPRESS Polym. Lett. 8 (11) (2014) 791–808, accessed online on 21 June 2020. Available from: https://ri.conicet.gov.ar/bitstream/handle/11336/4563/CONICET_Digital_Nro.5438_A.pdf?sequence.

[111] A. Tullo, PHA: a biopolymer whose time has finally come, Chem. & Eng. N. 97 (53) (2019), accessed online on 21 June 2020. Available from: https://cen.acs.org/business/biobased-chemicals/PHA-biopolymer-whose-time-finally/97/i35.

[112] A. Wojtowicz, L.P.B.M. Janssen, L. Moscicki, Blends of natural and synthetic polymers, in: L.P.B.M. Janssen, L. Moscicki (Eds.), Thermoplastic Starch, Wiley-VCH Verlag GmbH & Co, 2009, pp. 35–53.

[113] A. Mohammadi Nafchi, M. Moradpour, M. Saeidi, A. Alias, Thermoplastic starches: properties, challenges, and prospects, Starch/Staerke 65 (2013) 61–72.

[114] Y. Zhang, C. Rempel, Q. Liu, Thermoplastic starch processing and characteristics-a review, Crit. Rev. Food Sci. Nutr. 54 (10) (2014) 1353–1370.

[115] L. Averous, Biodegradable multiphase systems based on plasticized starch: a review, J. Macromol. Sci. Part C – Polym. Rev. C44 (2004) 231–274.

[116] G. Lay, J. Rehm, R.F. Stepto, M. Thoma, J-P. Sachetto, D.J. Lentz, et al., Polymer compositions containing destructured starch, US Patent 5,095,054, 1992.

[117] J.M. Mayer, M.J. Hepfinger, E.A. Welsh, Method of producing biodegradable starch-based product from unprocessed raw materials, US Patent 5322866, 1994.

[118] L. Yu, S. Coombs, G. Bruce, Y. Christie, Biodegradable polymer, US Patent 7326743B2, 2008.

[119] T. Kiyotsuna, H. Minematsu, Carbonated starch derivative and its production method, Japanese Patent JP2004359737A, 2003.

[120] K. Neelam, S. Vijay, S. Lalit, Various techniques for the modification of starch and the applications of its derivatives, Int. Res. J. Pharm. 3 (5) (2012) 25–31.

[121] D. Demirgoz, C. Elvira, J.F. Mano, A.M. Cunha, E. Piskin, R.L. Reis, Chemical modification of starch based biodegradable polymeric blends: effects on water uptake, degradation behaviour and mechanical properties, Polym. Degrad. Stab. 70 (2) (2000) 161–170.

[122] J. Loercks, W. Pommeranz, H. Schmidt, R. Timmermann, E. Grigat, W. Schulz-Schlitte, Thermoplastic starch utilizing a biodegradable polymer as melting aid, US Patent 6472497, 2002.

[123] C. Bastioli, V. Bellotti, A. Montino, Polymers of hydrophobic nature, filled with starch complexes, US Patent 6962950, 2005.

[124] C. Bastioli, V. Bellotti, G. Tredici, I. Guanella, R. Lombi, Complexed starch containing compositions having high mechanical properties, US Patent 6,348,524 B2, 2002.

[125] C. Bastioli, V. Bellotti, G.D. Cella, L.D. Giudice, S. Montino, G. Perego, Biodegradable polymeric compositions comprising starch and a thermoplastic polymer, US Patent 7,176,251 B1, 2007.

[126] V. Flaris, N. Alexander, S. Bagrodia, Morphological characterization of novel bio-based hybrid resin, SPE ANTEC Conf. Proc. (2008) 803–806.

[127] B. Raj, U.S. K, Siddaramaiah, "Low density polyethylene/starch blend films for food packaging applications, Adv. Polym. Technol. 23 (2004) 32–45. Available from: https://doi.org/10.1002/adv.10068.

[128] B.D. Favis, F. Rodriquez, B.A. Ramsay, Polymer compositions containing thermoplastic starch, US Patent 6.605,657 B1, 2003.

[129] A. Barrett, What is the difference between biodegradable, compostable and OXO Degradable? BioPlasticsNews (2019), accessed online on 27 June 2020. Available from: https://bioplasticsnews.com/2019/04/13/what-is-the-difference-between-biodegradable-compostable-and-oxo-degradable/.

[130] T. Clark, Plastics: establishing the path to zero waste, a pragmatic approach to sustainable management of plastic materials, Enso Plast. (2015) viewed 27 June 2020. Available from: http://www.ensoplastics.com/download/Plastics_EstablishingthePathtoZeroWaste.pdf.

[131] Sustainable Packaging Coalition, The SPC position against biodegradability additives for petroleumbased plastics. <http://www.sustainable-packaging.org/content/?type = 5&id = position-against-biodegradability-additives-for-petroleum-based-plastics>, 2015 (accessed 02.12.15).

[132] P.J. Kershaw, Biodegradable plastics & marine litter: misconceptions, concerns and impacts on marine environments, United Nations Environment Program. <https://www.unenvironment.org/resources/report/biodegradable-plastics-and-marine-litter-misconceptions-concerns-and-impacts>, 2015 (accessed 27.06.20).

[133] T. Finnegan, R. King, III, "Additives for film products," in: J.F., Macnamra, Jr., (Ed.), Film Extrusion Manual, third ed., TAPPI Press, Atlanta, 2020.

[134] K. Keck-Antoine, E. Lievens, J. Bayer, J. Mara, D.-S. Jung, S.-L. Jung, Additives to design and improve the performance of multilayer flexible packaging, in: J. Wagner (Ed.), Multilayer Flexible Packaging, 2nd ed., Elsevier, Oxford, 2016, pp. 53–76.

[135] H. Zweifel, R. Maier, M. Schiller, Plastics Additives Handbook, 6 ed., Hanser Plublications, Cincinnati, 2009.

[136] F. Stricker, M. Horton, New stabilizer solutions for polyolefin film grades, in: European TAPPI PLACE Conference, 2003. <http://www.tappi.org/content/enewsletters/eplace/2004/8-1stricker.pdf> (accessed 28.06.20).

[137] M. Ezrin, G. Lavigne, P. Klemchuk, W. Holley, S. Agro, J.G.L. Thomas, et al., Discoloration of EVA encapsulant in photovoltaic cells, in: SPE ANTEC Conference, 1995, pp. 3957–3960.

[138] F.J. Pern, Ethylene vinyl acetate (EVA) encapsulants for photovoltaic modules: Degradation and discoloration mechanisms and formulation modifications for improved photostability, Die Angew. Makromolekulare Chem. 252 (1997) 195–216.

[139] ExxonMobil Chemical, Tip. technology, discoloration (2003).

[140] T. Tikuisis, L. Ellis, S. Chisholm, Improving the gas-fading performance of polyolefin film resins by addressing the over-oxidation of phenolic antioxidants, in: TAPPI PLACE Conference, 2005.

[141] Y. Ye, R.E. King III, Phenol vs. phenol free stabilization and the impact of selected grades of TiO_2: can we make non-discolored white LLDPE film by adding phenol-free masterbatch to phenolic AO stabilized resin? in: TAPPI PLACE Conference, 2007.

[142] R. King, Additives: polyolefins discoloration, Plast. Technol. Mag. (2017). Available from: https://www.ptonline.com/articles/additives-polyolefin-discoloration.

[143] R.E. King III, Stabilization of a clarified gamma irradiated controlled rheology polypropylene random copolymer, in: SPE Polyolefins RETEC Conference, Houston, pp. 479–504, 1999.

[144] R.E. King III, Discoloration resistant polyolefin films, in: TAPPI Polymers, Laminations, Adhesive, Coatings and Extrusions Conference, San Diego, August 2001.

[145] R.E. King III, Discoloration resistant polyolefin films – part 2, in: TAPPI Polymers, Laminations, Adhesive, Coatings and Extrusions Conference, Boston, September 2002.

[146] R.E. King III, Discoloration resistant polyolefin films – part 3, in: TAPPI Polymers, Laminations, Adhesive, Coatings and Extrusions Conference, Orlando, August, 2003.

[147] R.E. King III, Discoloration resistant LLDPE: an overview of stabilization issues & opportunities, in: SPE Polyolefins RETEC, Houston, TX, 2006.

[148] Chemours, Polymers, light and the science of TiO_2. <https://www.tipure.com/en-/media/files/tipure/legacy/polymers-light-science-tio2.pdf>, 2018 (accessed 28.06.20).

[149] M.D. Roussel, A.R. Guy, L.G. Shaw, J.E. Cara, The use of calcium carbonate in polyolefins offers significant improvement in productivity, in: TAPPI PLACE Conference, 2005.

[150] F.A. Ruiz, Film physical property effects on low-level calcium carbonate addition to LLDPE, in: TAPPI PLACE Conference, 2004.

[151] A.R. Guy, Calcium carbonates for polyolefin extrusion coating applications, in: TAPPI Polymers, Laminations, and Coatings Conference Proceedings, Indianapolis, IN, 2004.

[152] F.A. Ruiz, Mineral enhancement of extrusion coated polyethylene, TAPPI Polymers, in: Laminations & Coatings Conference Proceedings, TAPPI Press, pp. 89–95 1994.

[153] E.M. Mount III, Control of dispersed polymer domain size formation in melt processing, in: TAPPI PLACE Conference, 2007. Available from: <http://www.tappi.org/content/events/07place/papers/mount.pdf> (accessed 28.06.20).

[154] D-C. Kong, J.A. Larter, E.M. Mount III,, Opaque polymeric films cavitated with PBT and polycarbonate, US Patent 6500533 B1, 2002.

[155] A. Maltby, R. Marquis, "Slip agents," in: J.F., Macnamra, Jr., (Ed.), Film Extrusion Manual, third ed., TAPPI Press, Atlanta, 2020.

[156] Struktol Company of America, LLC, Plastics additives: lubricants and homogenizing additives for thermoplastic compounding. <https://www.struktol.com/products/plastic-additives/>, n.d. (accessed 28.06.20).

[157] D.F. Cadogan, C.J. Howick, Plasticizers, Kirk-Othmer Encyclopedia of Chemical Technology, John Wiley & Sons, Wiley Online Library, 2000.

[158] E. Noon, Fundamentals of nylon for flexible packaging, in: TAPPI PLACE Conference, 2004.

[159] A. Mollison, Films from nylon blends, in: Canadian Patent CA1222086 A1, 1987.

[160] N. Maki, S. Nakano, H. Sasaki, Development of a packaging material using non-bleed-type antistatic ionomer, Packaging Technol. Sci. 17 (2004) 249–256.

[161] B.A. Morris, New developments in ionomer technology for film applications, J. Plastic Film. & Sheeting 23 (April) (2007) 97–108.

[162] P.S. Blatz, Extrudable composition consisting of a polyolefin and a fluorocarbon polymer, US Patent 3,125,547, 1964.

[163] C. Lavallee, Advances in polymer processing additives, In: TAPPI PLACE Conference, 2005. Available from: <http://www.tappi.org/content/enewsletters/eplace/2006/09-5Lavallee.pdf> (accessed 28.06.20).

[164] M.G. Meillon, D. Morgan, D. Bigio, K. Migler, S. Oriani, a description of the fluoroelastomer coating evolution as a polymer processing aid, SPE ANTEC Tech. Pap. (2005).

[165] S.R. Oriani, Optimizing process aid performance by controlling fluoropolymer particle size, J. Plastic Film. Sheeting 21 (July) (2005) 179–198.

[166] D. Bigio, Fluoropolymer process aids: the effect of processing conditions, in: SPE ANTEC Conference, 2009, pp. 1547–1556.

[167] D.L. Dotson, A novel nucleating agent for polyethylene, in: TAPPI PLACE Conference, 2007. Available from: <http://www.tappi.org/content/events/07place/papers/dotson.pdf> (accessed 28.06.20).

[168] N.D. Lyondell, Equistar technical tip: additives: nucleating and clarifying agents. <http://www.lyondellbasell.com/techlit/techlit/Tech%20Topics/General/Nucleating%20and%20Clarifying%20Agents.pdf>, n.d. (accessed 09.02.14).

[169] N.D. SpecialChem, Nucleating agents & clarifiers selection tips for polypropylene. <https://polymer-additives.specialchem.com/selection-guide/nucleating-agents-selection-for-polypropylene>, n.d. (accessed 28.06.20).

[170] K. Hoffman, G. Huber, D. Mader, Nucleating and clarifying agents for polyolefins, Macromol. Symposia 176 (2001) 83–92.

[171] N.D. Eastman, Tackifier center: resources for formulation. <http://www.eastman.com/Markets/Tackifier_Center/Pages/introduction.aspx>, n.d. (accessed 28.06.20).

[172] B.N. Epstein, Tough thermoplastic nylon compositions, U.S. Patent 4,174,353A, 1979.

[173] Natureworks, Technology focus report: toughened PLA. <http://www.natureworksllc.com/~/media/Technical_Resources/Properties_Documents/PropertiesDocument_Toughened-Ingeo_pdf.pdf>, 2007 (accessed 28.06.20).

[174] L.B. Weaver, S.F. Dalla Valle, M.P. Franca, Improving mechanical properties of polyethylene and polypropylene recycled streams using polyoefin elastomers and functionalized polyolefins, in: SPE ANTEC Conference, 2011.

[175] J.M. Torradas, Polymer modifiers in flexible packaging, in: TAPPI PLACE Conference, 2012.

[176] Y. Lin, J. Alaboson, J. Wang, K. Hausmann, J. Xu, J. Pan, et al., Biaxially oriented polyethylene (BOPE) films fabricated via tenter frame process and applications thereof, in: AIMCAL Conference, 2018. Available from: <https://www.aimcal.org/uploads/4/6/6/9/46695933/alaboson_-presentation.pdf> (accessed 29.06.20).

[177] J. Lush, D. Ferrari, An economic value model for converters comparing the total cost of metallized film vs. foil in flexible packages, in: TAPPI PLACE Conference, 2006.

[178] C.A. Bishop, E.M. Mount III, Vacuum metallizing for flexible packaging, in: J. Wagner (Ed.), Multilayer Flexible Packaging, Technology and Applications For The Food, Personal Care and Over-The-Counter Pharmaceutical Industries, Elsevier, Oxford, 2010, pp. 185–202.

[179] G. Shubert, Adhesion of aluminum foil to coatings – stick with it, in: European TAPPI PLACE Conference, 2004.

[180] L. Murray, The impact of foil pinholes and flex cracks on the moisture and oxygen barrier of flexible packaging, in: TAPPI PLACE Conference, 2005.

[181] C.C. Taylor, Cellophane, in: M. Bakker (Ed.), The Wiley Encyclopedia of Packaging Technology, John Wiley & Sons, New York, 1986, pp. 159–163.

[182] Innovia, Natureflex next generation packaging, Sustainability and life cycle performance. <http://www.innoviafilms.com/InnoviaFilms-Kentico5/media/PDFs/MARKET%20BROCHURES/NATUREFLEX/Natureflex-Prod-Life-Cycle-Assess.pdf>, n.d. (accessed 27.12.13). Note: Innovia sold its cellophane business to Futamura in 2016: http://www.futamuragroup.com/en/divisions/cellulose-films/products>.

[183] L.D. Shackford, A comparison of pulping and bleaching of kraft softwood and eucalyptus pulps, in: 36th International Pulp and Paper Congress and Exhibition, Sao Paulo, Brazil, October 13–16, 2003. Available from: <http://www.celso-foelkel.com.br/artigos/outros/A%20Comparison%20of%20Pulping%20and%20Bleaching%20of%20Kraft%20Softwood%20and%20Eucalyptus%20Pulps.pdf> (accessed 28.06.20).

[184] SAPPI Fine Paper, Paper standards and measurements. <https://fdocuments.in/document/paper-standards-and-measurements.html>, 2007 (accessed 28.06.20).

[185] M. Sikora, Paper, in: K. Yam (Ed.), Wiley Encyclopedia of Packaging Technology, 3rd ed., John Wiley & Sons, New York, 2009, pp. 908–912 (updated by staff for 3rd ed).

[186] B.W. Attwood, Paperboard, in: K. Yan, (Ed.), Wiley Encyclopedia of Packaging Technology, 3rd ed., John Wiley & Sons, New York, 2009, pp. 913–920.

[187] A. Riley, Paper and paperboard packaging, In: A. Emblem, H. Emblem, (Eds.), Packaging Technology - Fundamentals, Materials and Processes, Elsevier, 2012, pp. 178–239.

[188] S. Ebnesajjad, A survey of regulatory aspects of food packaging, in: S. Ebnesajjad (Ed.), Plastic Films In Food Packaging, Materials, Technology and Applications, Elsevier, Amsterdam, 2013, pp. 345–388.

[189] D.J. Knight, L.A. Creighton, *Regulation of food packaging in Europe and the US: Rappa review reports, Rappa Technology Limited*, 2004.

[190] The Official Journal of European Union, Commission Regulation (EU) No 10/2011of 14 January 2011on plastic materials and articles intended to come into contact with food. <https://eur-lex.europa.eu/LexUriServ/LexUriServ.do?uri=OJ:L:2011:012:0001:0089:en:PDF>, 2011 (accessed 25.01.14).

Chapter 5

Rheology of polymer melts

The term "rheology" is derived from the Greek word *rheo* or flow. It is the study of how materials flow, or more specifically, how materials deform under an applied force. Polymers exhibit many complex and interesting flow behaviors that impact extrusion, coextrusion, film fabrication, coating, heat sealing and other aspects of flexible packaging.

5.1 Basic terms

In engineering terms, materials deform or flow when a force is applied. The amount of deformation is known as *strain* and how fast the deformation occurs is known as the *rate of strain*. The applied force per unit area is the *stress*.

There are two fundamental modes of strain: shear and elongation. Fig. 5.1 illustrates a simple shear flow. Here a fluid is sandwiched between two parallel plates. The bottom plate is stationary and a tangential force F is applied to the top plate. The fluid is sheared as the top plate moves. Fig. 5.2 illustrates an elongational flow where the fluid is stretched upon application of the force. Some examples of shear flow include pressure-driven flow down a circular pipe or rectangular channel, as in a transfer pipe or die, and much of the flow in a single-screw extruder. Examples of elongational flow include the drawdown of the melt after it exits the die in the blown, cast film and extrusion coating processes, film orientation in a tenter frame or double bubble process, and certain types of mixing devices in extrusion. Most polymer processes have a combination of shear and elongational flow and are known as *rheologically complex flows*. Such complex flows are difficult to study and model, but insight can be gained from the analysis of simple shear and elongational flow.

In Fig. 5.1, the force F is known as the shear force, which acts over the area (A) of the top plate (normal to flow). F/A is the *shear stress*. The rate of strain or shear rate is equal to the change in velocity divided by the separation distance between the two plates, h. Since the bottom plate is stationary, the change in velocity is simply U, the velocity of the top plate. The Greek letters sigma (σ) [or in some texts, tau (τ)] and gamma dot ($\dot{\gamma}$) are used to designate the shear stress and shear rate, respectively. The viscosity is a material property that describes the resistance to deformation and is equal to the shear stress divided by the shear rate:

$$\eta = \frac{\sigma}{\dot{\gamma}} = \frac{F/A}{U/h} \tag{5.1}$$

where η is the shear viscosity.

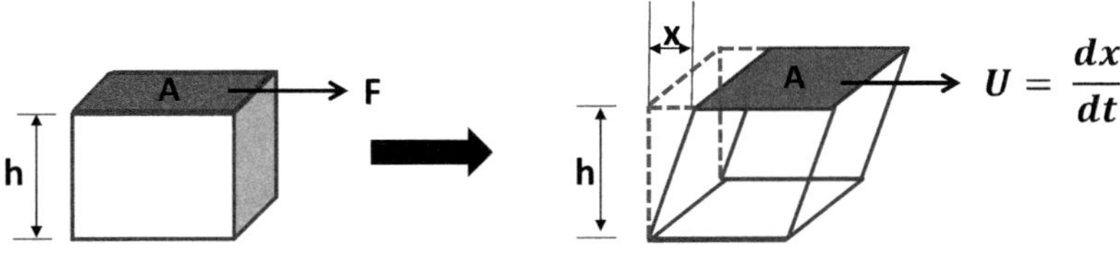

FIGURE 5.1 Simple shear flow. F, force, A, area, U, velocity.

Shear Stress

$$\sigma = \frac{F}{A}$$

Shear Strain

$$\gamma = \frac{x}{h}$$

Shear Rate

$$\dot{\gamma} = \frac{d\gamma}{dt} = \frac{1}{h}\frac{dx}{dt} = \frac{U}{h}$$

The Science and Technology of Flexible Packaging. DOI: https://doi.org/10.1016/B978-0-323-85435-1.00010-7

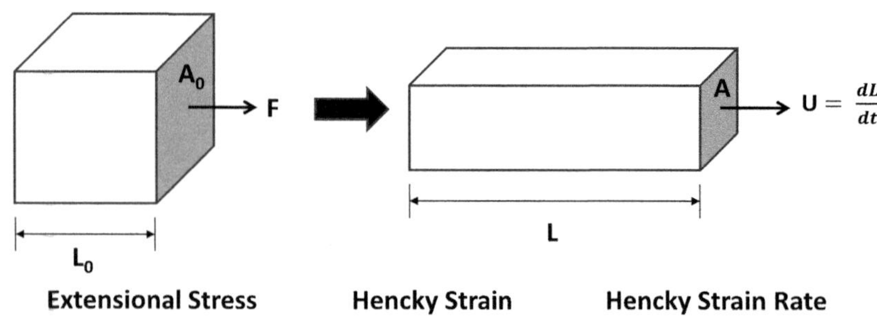

FIGURE 5.2 Extensional flow. *F*, force, *A*, area, *U*, velocity.

TABLE 5.1 Typical units for stress and viscosity and their conversion factors.

Force per area (stress)			Viscosity		
Unit	Multiply by	New unit	Unit	Multiply by	New unit
Newton/m^2	1	Pascal (Pa)	centipoise	0.001	Pa-s
lb$_f$/in^2 (PSI)	6896	Pa	poise	0.1	Pa-s
lb$_f$/ft^2	47.88	Pa	lb$_f$-s/in^2	6895	Pa-s
dyne/m^2	0.1	Pa	lb$_f$-s/ft^2	47.88	Pa-s
atmosphere	1.01E + 05	Pa			
kg$_f$/cm^2	9807	Pa			

Similarly, for elongational flow, tensile or elongational stress and rate of strain are defined, as illustrated in Fig. 5.2. The logarithm of the strain is used when the deformation is in increments. The *elongational or extensional viscosity* is the tensile stress divided by the strain rate:

$$\eta_E = \frac{\sigma_E}{\dot{\varepsilon}} \tag{5.2}$$

where η_E is the extensional viscosity (in some texts a lower case e is used: η_e), σ_E is the extensional stress, and $\dot{\varepsilon}$ is the rate of strain.

The extensional viscosity is sometimes associated with what is called the melt strength of the polymer. Melt strength is a subjective term that also may be related to elasticity and other properties of the polymer. Section 5.3.6 describes methods for measuring elongational viscosity and melt strength.

The stress is typically given in units of pounds force per square inch (psi) or in Newtons per square meter [Pascals (Pa)]. Shear rate has units of s^{-1}. Viscosity has units of Pa-s. Other units are given in Table 5.1. The viscosity of water and other small molecules are on the order of 10^{-2} Pa-s, whereas most polymer melts have viscosities in the range of $10-10^5$ Pa-s. Table 5.2 gives an order of magnitude value for the viscosities of several common materials that put the viscosity of polymers into perspective.

If the shear viscosity of a fluid does not change with the shear rate, the fluid is known as a *Newtonian* fluid. Most polymers are *non-Newtonian*. A typical shear viscosity versus shear rate curve is given in Fig. 5.3. A log–log plot is normally used since the viscosity spans several decades over the shear rate range of interest in polymer processing. At very low shear rates, the viscosity curve is nearly horizontal; the viscosity is independent of the shear rate. This is known as the Newtonian plateau. The value of the viscosity in the plateau region is called the zero shear viscosity, η_o. At higher shear rates, a key characteristic of many polymers is a decrease in viscosity with increasing shear rate, a phenomenon known as *shear-thinning* behavior. On the log–log plot, the change in viscosity in this region is nearly linear. The viscosity in this region can be fit to a power-law model:

$$\eta = m\dot{\gamma}^{n-1}. \tag{5.3}$$

TABLE 5.2 Viscosity of some common materials at room temperature [1].

Liquid	Approximate viscosity (Pa-s)
Glass	10^{40}
Molten glass (500°C)	10^{12}
Bitumen	10^{8}
Molten polymers	10^{3}
Liquid honey	10
Glycerol	1
Olive oil	10^{-1}
Water	10^{-3}
Air	10^{-5}

Source: Adapted from H.A. Barnes, J.F. Hutton, K. Walters, An Introduction to Rheology, 1989, Elsevier, Amsterdam, pp. 13.

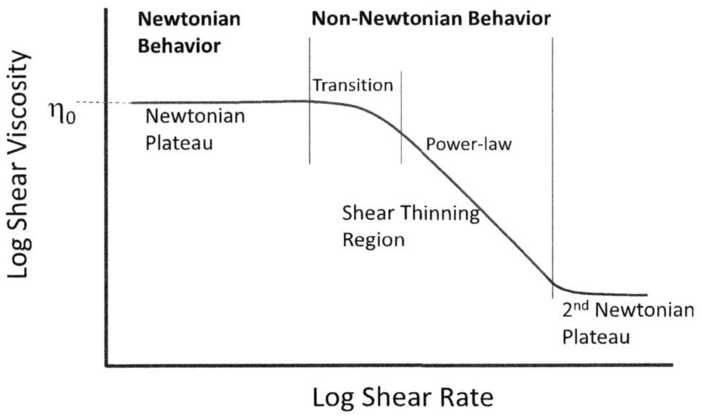

FIGURE 5.3 Typical polymer shear viscosity versus shear rate curve.

Here n is the power-law exponent and m is the consistency index. The value of n may be between 0 and 1. The closer n is to 1, the more Newtonian-like the fluid is. Fluids with lower values of n are said to be more shear thinning. The value of m can be obtained by looking at where the viscosity crosses the shear rate $= 1\ s^{-1}$ line. Factors that affect the values of m and n are discussed in Section 5.4.

At very high shear rates, the viscosity levels off again. This region is known as the second Newtonian plateau. The second Newtonian region is rarely reached in film extrusion but sometimes becomes important at the very high shear rates experienced in injection molding.

A Newtonian model (constant viscosity) can be used to fit the first part of the curve and the power-law model for the second. Other models have been developed to fit the whole shear viscosity versus shear rate curve. Two popular models are the Cross model and the Carreau–Yasuda models. These are described in Table 5.3.

Similarly, in extensional flow, some polymers exhibit non-Newtonian behavior. An increase in elongational viscosity with an increasing rate of strain is known as *strain hardening*. Conversely, a decrease in elongational viscosity with an increasing rate of strain is known as *strain softening*. These are illustrated in Fig. 5.4. for polyethylene.

If the material acts like a solid, the strain rather than the rate of strain is the appropriate value to describe how the material is deformed. For an elastic solid, the application of applied stress instantaneously results in a strain. The *modulus* is the material parameter that describes the resistance of a solid to deformation and is defined as the ratio of the applied stress to the resulting strain:

$$E = \frac{\sigma_E}{\epsilon} \text{ or } G = \frac{\sigma}{\gamma} \tag{5.4}$$

where E is Young's modulus, ε is the elongational strain, G is the shear modulus, and γ is the shear strain.

TABLE 5.3 Curve fitting models for shear viscosity.

Model	Name
$\eta = \eta_0$	Newtonian model
$\eta = m\dot{\gamma}^{n-1}$ m = consistency index n = power-law exponent	Power-law model
$\eta = \eta_0(1+(\lambda\dot{\gamma})^a)^{\frac{n-1}{a}}$ η_0 = zero shear viscosity λ, a and n are fitted parameters	Carreau–Yasuda model
$\eta = \frac{\eta_0}{1+(\lambda\dot{\gamma})^{1-n}}$ η_0 = zero shear viscosity λ and n are fitted parameters	Cross model

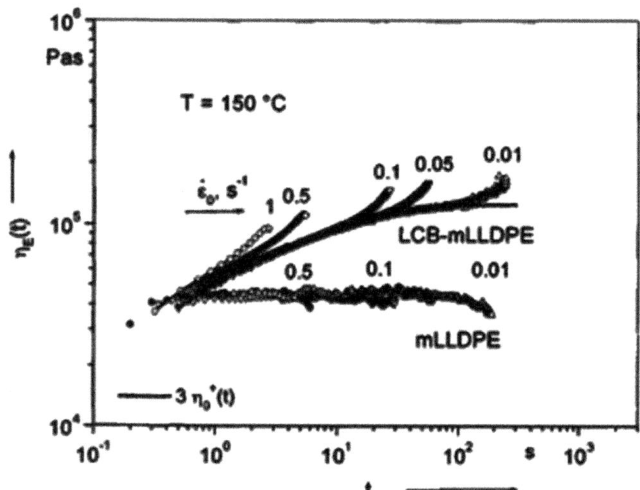

FIGURE 5.4 Examples of elongational viscosity behavior. Time-dependent elongational viscosity (see Section 5.3.6 for definition) of a linear and a slightly long-chain branched metallocene linear low-density polyethylene (mLLDPE) of similar molecular mass distributions. The linear mLLDPE exhibits strain-softening behavior whereas the slightly branched mLLDPE shows strain hardening behavior [2]. *Reproduced with permission from C. Gabriel, H. Munstedt, Strain hardening of various polyolefins in uniaxial elongational flow, J. Rheology, 47(3) (2003) 619–630.*

Small molecules such as water in the liquid state are known as viscous liquids. Their deformation and flow can be characterized solely by viscosity. Simple elastic solids, known as Hookean solids, on the other hand, deform but do not flow under applied stress. Their deformation can be characterized by the modulus. Polymers typically fall in between viscous fluids and elastic solids; they are known as *viscoelastic* fluids. A characteristic relaxation time can be defined for a polymer that is related to how fast it responds to stress. Relaxation experiments and measurement of relaxation time are described in Section 5.5.1. To determine whether a polymer acts like a fluid, solid or both, the relaxation time can be compared with the characteristic time scale for the process. A dimensionless number known as the Deborah number (De) is defined as the polymer relaxation time divided by the processing time:

$$\mathrm{De} = \frac{\lambda}{t_f} \tag{5.5}$$

where De is the Deborah number, λ is the relaxation time of the polymer, and t_f is the characteristic process time.

If De >> 1, the polymer will behave as an elastic solid. Conversely, if De << 1, it behaves like a viscous fluid.

The viscoelastic nature of a polymer arises from its molecular structure; polymers are long-chain molecules with long relaxation times compared to small molecules like water. The viscoelasticity of polymer melts gives rise to *a normal force or stress*, as illustrated in Fig. 5.5. The normal forces act in the direction perpendicular to flow and are responsible for phenomena such as the Weissenberg effect (where a viscoelastic fluid rises up a stirring rod when agitated), secondary or ciculatory flow patterns in polymer flow, and the swelling of polymer melts as they exit a die.

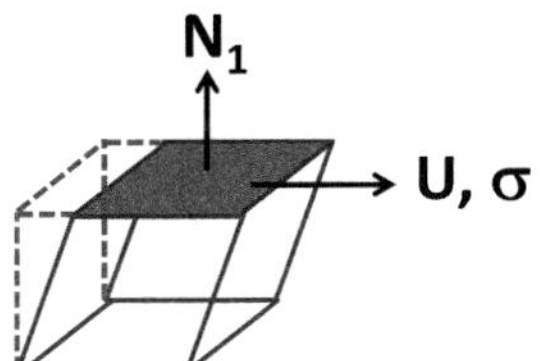

FIGURE 5.5 Illustration of normal force N_1 exerted on two parallel plates by a viscoelastic fluid in shear flow. The top plate is moved at velocity U and the bottom plate is stationary. A shear stress, σ, is applied in the direction of flow.

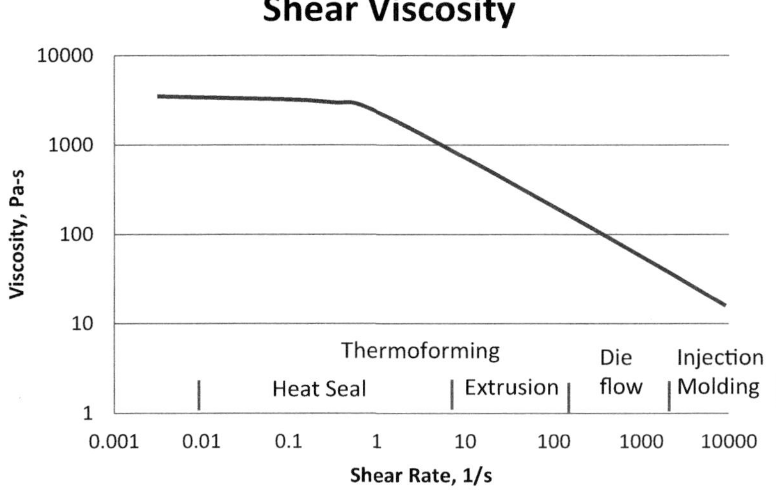

FIGURE 5.6 Typical shear rate ranges for commonly used processes in flexible packaging.

5.2 Importance of rheology in flexible packaging

The rheology of polymer melts plays a key role in many aspects of flexible packaging, including the extrusion and fabrication of films, coating and lamination to make final film structures, thermoforming, and sealing during the packaging process.

The specific property of importance depends on the type of flow and nature of the process. An example of shear flow is given in Fig. 5.6 where the shear rate ranges for commonly used processes in flexible packaging are highlighted. Different regions of the shear viscosity versus shear rate curve are central to the different processes. For example, the shear-thinning region is critical for extrusion and flow in the die during film fabrication whereas the lower shear rate region is more important for heat sealing and thermoforming.

Another role rheology plays is in characterizing materials. Rheological measurements are sensitive to small changes in molecular architecture and are useful tools for new polymer development and for troubleshooting problems related to film fabrication and packaging performance.

This section provides a brief overview of some of the rheological phenomenon that impact polymer processing and end-use performance. Many of these examples are discussed in greater detail in later chapters.

5.2.1 Extrusion and film fabrication

The shear-thinning behavior of polymer melts is important for extrusion. The extruder power or torque needed to pump the molten polymer through the extruder and die is related to the viscosity. Greater shear-thinning lowers the viscosity and power requirement. When linear low-density polyethylene (LLDPE) was first introduced, film fabricators struggled with processing it on their lines designed to process low-density polyethylene (LDPE). LLDPE is less shear thinning, resulting in higher viscosity at high shear rates than LDPE of the same molecular weight. Because of this, die pressures became excessive and extruder power requirements exceeded that of the existing equipment. Processors had to slow their lines down to process it. Adding a small amount of LDPE to LLDPE substantially increases the shear-thinning behavior and melt strength, allowing higher outputs [3–5].

The viscosity is also critical for melting the polymer in the extruder. Heat conduction from the heated extruder barrel is generally not sufficient to melt the polymer, particularly for large lines running at high output. Most single-screw extruders rely on the heat generated as the polymer is sheared to assist in melting the polymer. This is known as viscous heat generation. The amount of heat that is generated is related to the product of the shear rate and shear stress:

$$D_{\text{visc}} = \sigma\dot{\gamma} = \eta\dot{\gamma}^2 \tag{5.6}$$

where $D_{\text{visc}} =$ rate of viscous heat dissipation. Hence the viscosity plays a significant role in generating heat for melting. The adiabatic average temperature rise for polymer melts in an extruder is about 3°C for every 1000 psi (6.9 MPa) pressure drop, but local temperature rise near the wall can be greater since both the shear stress and shear rate are high in this region.

Many applications in flexible packaging involve polymer blends, either with other polymers or with inorganic fillers or pigments. As described in detail in Chapter 6, when mixing polymers, matching their viscosities often provides the best results. When dispersing pigments or inorganic fillers, shear and elongational stresses act to break up the particles for a more uniform distribution.

The flow of polymer melts is described by the fundamental equations of motion coupled with a "constitutive equation" that describes the material relationship between stress and deformation. Newtonian and power-law fluids, for example, are two simple constitutive equations for viscous fluids. Section 5.5 discusses more universal constitutive equations for viscoelastic fluids. Solving these equations for general polymer processing requires complex numerical simulation. Analytical solutions, however, have been obtained for several simple geometries and constitutive relationships that provide insight into many flow situations. Solutions for Newtonian and power-law fluids in a simple drag and pressure-driven flow are given in Table 5.4. The flow in the metering section of a single-screw extruder, for example, is dominated by drag flow that is a form of simple shear flow shown in Fig. 5.1. In a simplified representation of the flow in the extruder, the screw flight is unwound into a rectangular channel, the screw is assumed to be stationary and the barrel rotates. The top plate in Fig. 5.1 represents the moving barrel.

The flow in transfer pipes and the die is pressure-driven flow. The velocity profile in the channel is parabolic for a Newtonian fluid with a maximum at the centerline of the channel and zero at the channel wall. The classic assumption in fluid mechanics is to assume "no-slip" at the channel wall, meaning the fluid will take on the same velocity as the wall. There are situations where polymers exhibit some slip. The reader is directed to Vlachopoulis and Strutt [7] for details. Note that the shear rate and shear stress is zero along the centerline and maximum at the walls, and that the shear stress is a linear function of the distance from the centerline. For shear-thinning fluids ($n < 1$), the parabola becomes flattened as n decreases. The flow is said to be more "plug-like." This is illustrated in Fig. 5.7.

The elasticity of the polymer can give rise to normal stresses that produce recirculation flows in the corners of rectangular channels. An illustration of such a flow is shown in Fig. 5.8. These are also known as secondary flows and generally are smaller in magnitude than the main flow. These flows can produce stagnant areas in the extruder and die, leading to long hold-up times and polymer degradation, and contributing to film imperfections such as die lip build-up and gel. In addition, elasticity and secondary flows may contribute to nonuniform film thickness and layer distribution in coextrusion. These phenomena are explored in greater depth in Chapter 13.

Elasticity also contributes to the swelling of the polymer as it exits a die. This phenomenon, known as *melt swell*, is illustrated in Fig. 5.9. A melt swell ratio can be defined as the ratio of the maximum diameter of the melt just after exiting the orifice to the orifice diameter. While the primary mechanism involves the release of normal stresses at the die exit, other factors are also involved [7]. Newtonian fluids, which have no elasticity, exhibit some swell, with a melt swell ratio of 1.13, due to streamlining rearrangement at the die exit. Some non-Newtonian polymers exhibit ratios of up to 4. Melt swell typically decreases as the die length increases, suggesting a memory effect. Melt swell may contribute to unwanted deposits on the die lips (known as die drool or die lip build-up) during film fabrication but is typically more of a problem for profile extrusion, where the final shape of the article may be affected by the amount of swell.

Melt fracture or extrudate distortion is an extreme example of a rheological phenomenon that affects productivity and quality. For linear polymers like LLDPE, as the throughput through a die is increased, at some point the surface takes on a rough appearance. This is known as sharkskin and typically occurs at a critical stress level of about 0.1–0.5 MPa [7]. It is generally thought that sharkskin is the result of rupturing of the melt as it accelerates near the die exit [8,9]. Addition of polymer processing aides, such a fluoroelastomers, which act to promote slip at the wall, increase the critical stress level and delay the onset of sharkskin to higher throughput. Other proposed techniques to delay sharkskin include heating the die lips to reduce the viscosity and stress, flaring the exit angle of the die and using special coatings on the metal surface [10,11].

Further increases in throughput may eventually lead to a flow regime where the extrudate is severely distorted as it exits the die (see Fig. 5.10) and wild pressure fluctuations occur. This is known as gross melt fracture or extrudate distortion. Such distortion results in poor appearance, nonuniform thickness and other problems. Several mechanisms have been

TABLE 5.4 Equations for simple shear flows [6].

Flow	Equation	Picture of flow
Drag flow between flat plates (Couette flow)	$v_x = \left(\frac{V_0}{B}\right) y$ $v_{avg} = \frac{V_0}{2}$ $Q = \frac{V_0 B w}{2}$ $\sigma_{yx} = \text{constant}$ $\dot{\gamma} = \frac{V_0}{B}$	Bottom plate is stationary
Pressure-driven flow in circular channel (Poisseuille flow)	*Newtonian fluid* $v_z = v_{max}\left[1 - \left(\frac{r}{R}\right)^2\right]$ $v_{max} = \frac{R^2}{4\eta}\left(\frac{\Delta P}{L}\right)$ $v_{avg} = \frac{v_{max}}{2}$ $Q = \frac{\pi R^4}{8\eta}\left(\frac{\Delta P}{L}\right)$ $\Delta P = \frac{8Q\eta L}{\pi R^4}$ $\sigma_{rz} = -\frac{\Delta P}{2L} r$ $\sigma_w = -\frac{\Delta P}{2L} R$ $\dot{\gamma} = -\frac{2v_{max}}{R}\left(\frac{r}{R}\right)$ $\dot{\gamma}_w = -\frac{2v_{max}}{R} = \frac{4Q}{\pi R^3}$ *Power-law fluid* $v_z = v_{max}\left[1 - \left(\frac{r}{R}\right)^{\frac{n+1}{n}}\right]$ $v_{max} = \frac{n}{n+1}\left[\frac{R^{n+1}}{2m}\left(\frac{\Delta P}{L}\right)\right]^{\frac{1}{n}}$ $v_{avg} = \frac{n+1}{3n+1} v_{max}$ $Q = \pi \frac{n}{3n+1}\left[\frac{1}{2m}\left(\frac{\Delta P}{L}\right)\right]^{\frac{1}{n}} R^{\frac{1}{n}+3}$ $\Delta P = 2mR^{-(3n+1)} L\left[\left(3 + \frac{1}{n}\right)\frac{Q}{\pi}\right]^n$ $\sigma_{rz} = -\frac{\Delta P}{2L} r$ $\sigma_w = -\frac{\Delta P}{2L} R$ $\dot{\gamma} = -\frac{n+1}{n}\frac{v_{max}}{R}\left(\frac{r}{R}\right)^{\frac{1}{n}}$ $\dot{\gamma}_w = -\frac{n+1}{n}\frac{v_{max}}{R} = \left(\frac{3n+1}{4n}\right)\frac{4Q}{\pi R^3}$	

(Continued)

TABLE 5.4 (Continued)

Flow	Equation	Picture of flow
Pressure-driven flow for a rectangular channel	*Newtonian fluid* $v_x = v_{\max}\left[1-\left(\frac{y}{b}\right)^2\right]$ $v_{\max} = \frac{1}{2}\left[\frac{b^2}{\eta}\left(\frac{\Delta P}{L}\right)\right]$ $v_{avg} = \frac{2}{3}v_{\max}$ $Q = \frac{2}{3}\frac{1}{\eta}\frac{\Delta P}{L}b^3 w$ $\Delta P = \frac{3\eta L}{2b^3}\frac{Q}{w}$ $\sigma_{yx} = -\frac{\Delta P}{L}y$ $\sigma_w = -\frac{\Delta P}{L}b$ $\dot{\gamma} = -\frac{2v_{\max}}{b}\left(\frac{y}{b}\right)$ $\dot{\gamma}_w = -\frac{2v_{\max}}{b} = \frac{3Q}{2wb^2}$ *Power-law fluid* $v_x = v_{\max}\left[1-\left(\frac{y}{b}\right)^{\frac{n+1}{n}}\right]$ $v_{\max} = \frac{n}{n+1}\left[\frac{b^{n+1}}{m}\left(\frac{\Delta P}{L}\right)\right]^{\frac{1}{n}}$ $v_{avg} = \frac{n+1}{2n+1}v_{\max}$ $Q = \frac{2n}{2n+1}\left[\frac{1}{m}\frac{\Delta P}{L}\right]^{\frac{1}{n}}b^{\frac{1}{n}+2}w$ $\Delta P = \frac{mL}{b^{2n+1}}\left[\left(1+\frac{1}{2n}\right)\frac{Q}{w}\right]^n$ $\sigma_{yx} = -\frac{\Delta P}{L}y$ $\sigma_w = -\frac{\Delta P}{L}b$ $\dot{\gamma} = -\frac{n+1}{n}\frac{v_{\max}}{b}\left(\frac{y}{b}\right)^{\frac{1}{n}}$ $\dot{\gamma}_w = -\frac{n+1}{n}\frac{v_{\max}}{b} = \frac{2n+1}{n}\frac{Q}{2wb^2}$	
Single-screw extruder, metering section • Newtonian fluid • 1st term drag flow • 2nd term pressure-driven flow	$Q = \frac{1}{2}\pi^2 D^2 HN\sin\theta\cos\theta - \frac{\pi DH^3}{12\mu}\sin^2\theta\frac{\Delta P}{L}$ where Q = volumetric throughput D = screw diameter H = channel depth N = speed of rotation θ = helix angle μ = viscosity ΔP = pressure drop in metering section L = length of metering section	

Source: Adapted from J. Vlachopoulos, Introduction to Polymer Processing, Second ed., 1994, McMaster University.

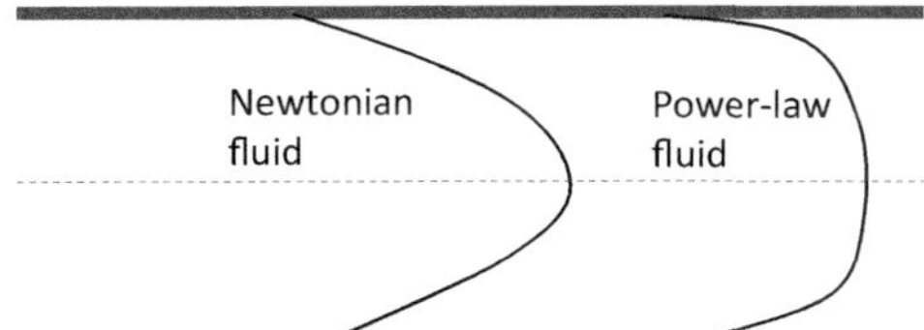

FIGURE 5.7 Velocity profile of pressure-driven flow in a circular channel. The velocity of a Newtonian fluid is parabolic. A power-law fluid has more plug-like flow, depending on the value of the power-law exponent.

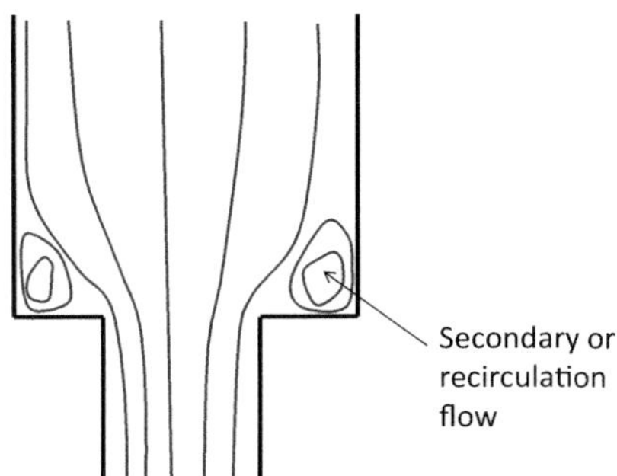

FIGURE 5.8 Example of recirculation secondary flows in the corners of a sudden contraction in a convergent flow channel. Lines represent streamlines or pathways of the flow through the channel.

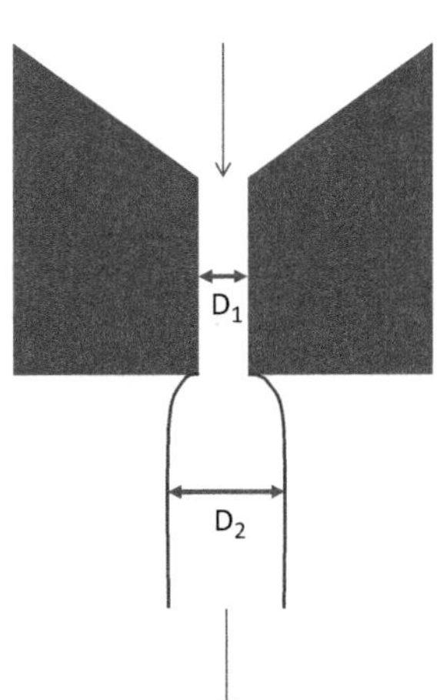

FIGURE 5.9 Illustration of melt swell. The solid region represents the solid mass of the die. The flow channel has a diameter D_1 and the maximum melt diameter after the die exit is D_2.

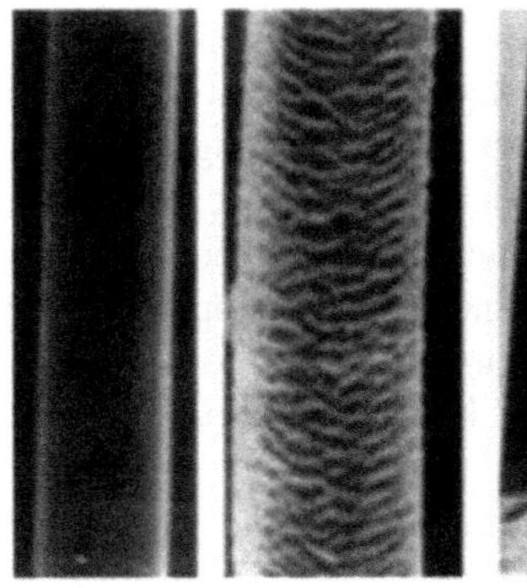
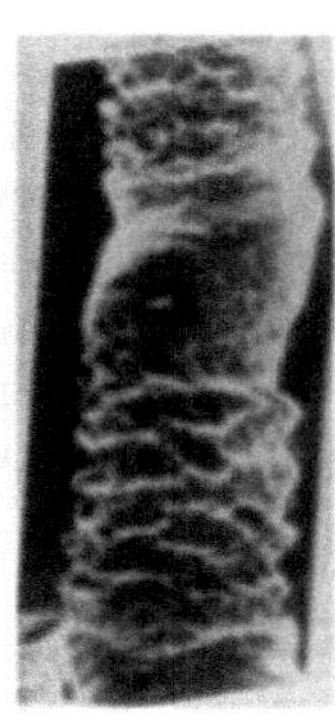
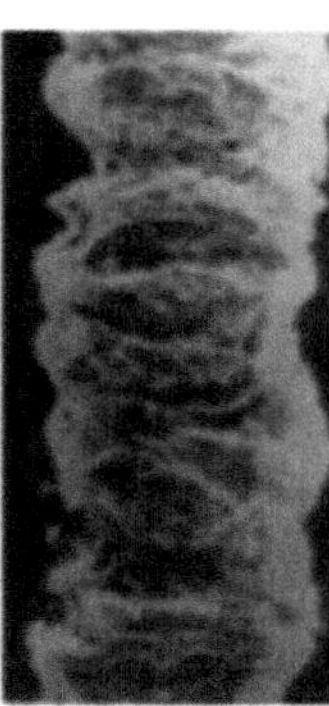

FIGURE 5.10 Linear low-density polyethylene extrudates obtained from a capillary at apparent shear rates of 37, 112, 750, and 2250 s^{-1}. *Reproduced with permission from J. Vlachopoulos, D. Strutt, Rheology of molten polymers, in: J.R. Wagner, Jr. (Ed.), Multilayer Flexible Packaging, second ed., 2016, Elsevier, Amsterdam, pp. 77–96.*

proposed, including instabilities in the entry flow region (related to the secondary flows described earlier), elastic instabilities, and a slip-stick phenomenon at the die wall. It is likely more than one mechanism may be involved [12–18].

These flow instabilities are pursued further in Chapter 13.

In the coextrusion process, two or more polymers are combined in the melt. This can introduce additional flow instabilities related to the rheology of the polymers that are discussed in detail in Chapter 13. Three examples are illustrated in Fig. 5.11. Fig. 5.11A shows that the polymers may reorganize as they flow together down the channel after

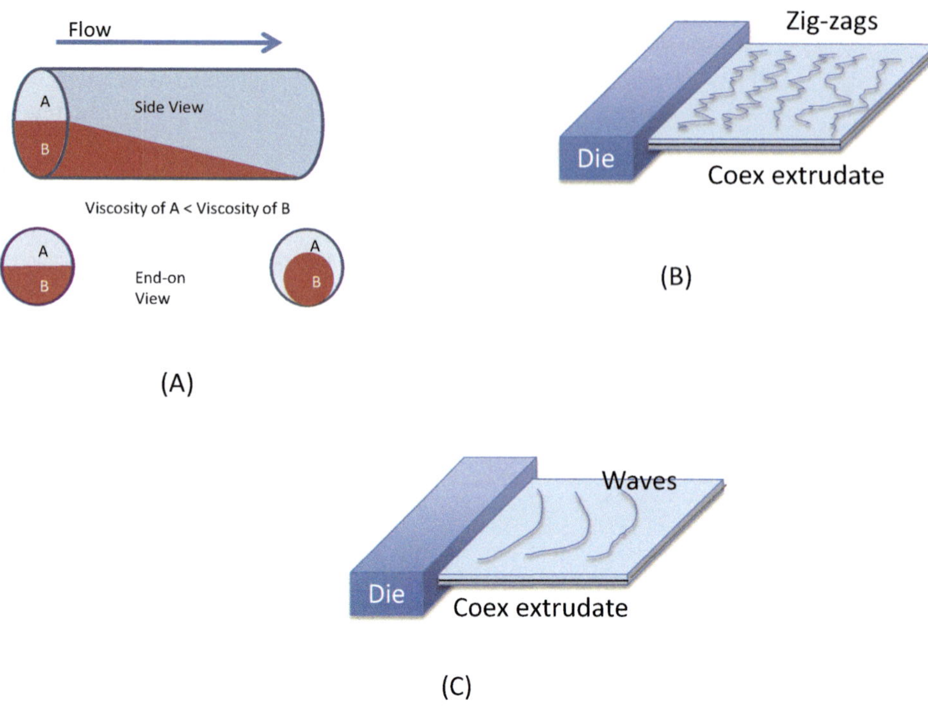

FIGURE 5.11 Examples of flow instabilities in coextrusion: (A) layer rearrangement due to viscosity mismatch, (B) zig-zag flow instability, and (C) wave flow instability.

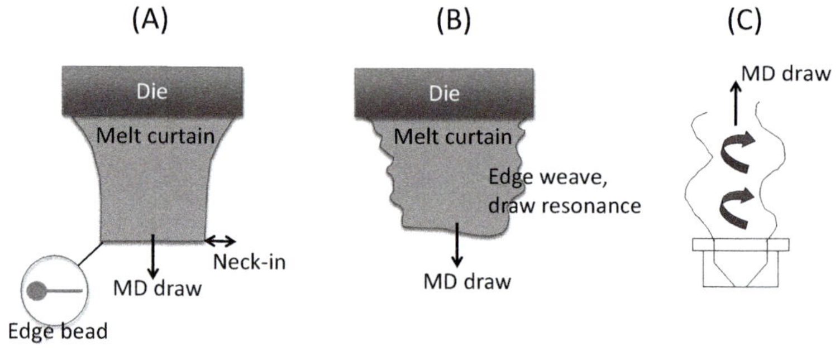

FIGURE 5.12 Examples of flow phenomena during film fabrication: (A) neck-in of width of melt curtain in cast film and extrusion coating, (B) edge weave and draw resonance in high-speed cast film or extrusion coating, and (C) helical bubble instability in blown film.

they have been combined. If the polymers have different viscosities, the material with the lowest viscosity tends to migrate toward the high shear rate region near the channel walls. The driving force for this behavior is the lowering of the overall energy of the system. This reorganization or encapsulation can lead to layer nonuniformity. Matching viscosities of the polymers and contouring the channel dimensions are two approaches that are used to counter this effect [19–21].

In coextrusion flow, if the interfacial stress at one of the layer interfaces in the die land region exceeds a critical value (typically around 40–80 kPa), a chevron pattern may develop in the film. This is known as zig-zap instability and is illustrated in Fig. 5.11B. Steps that reduce the interfacial stress, such as raising the temperature (to lower viscosity), increasing the thickness of the skin layer (to move the interface further away from the high-stress region near the die wall), and reducing the skin layer viscosity or reducing output, help alleviate this type of instability [22–24].

Wood grain and parabola-like patterns may develop in the coextruded film or sheet, an indication of "wave" type instability (see Fig. 5.11C). This instability is believed to originate as the layers are first brought together in the feedblock or die. Differences in velocity, acceleration and even elasticity are thought to be involved. Changes in die or feedblock design, resin types and film structure may be needed to alleviate this type of instability [25–27].

Once the polymer melt exits the die, elongational flow predominates in film fabrication and extrusion coating/lamination processes. The polymer is drawn down in thickness by impinging it onto a fast-moving roll in cast film and extrusion coating, or in the case of blown film, by the combined action of blowing air inside the bubble to expand it laterally and accelerating the speed in the machine direction using nipped rolls. Several flow phenomena and instabilities (illustrated in Fig. 5.12) may arise that are influenced by the rheology of the polymer:

- In extrusion coating and cast film, the width of the melt curtain decreases as it is drawn down in thickness (Fig. 5.12A). This is known as neck-in. Neck-in reduces the width of the web, which can result in productivity issues. It also results in greater film or coating thickness near the edges, known as edge bead. Modern dies use inserts at the die edge to thin down the curtain before it exits the die to compensate (see Section 2.3.1.1). Increasing the melt strength of the polymer also helps. For example, LDPE with long-chain branching (LCB) typically has less neck-in than LLDPE, which has no LCB. Ionomers and acid copolymers, which contain LCB, as well as strong intermolecular forces (hydrogen bonding or ionic interactions) that contribute to melting strength, have even less neck-in. Broadening the molecular weight distribution (MWD) of PE typically reduces neck-in. Too little neck-in, however, can result in not enough of an edge-bead, which can lead to edge tear and line breaks [28,29].
- Drawing down the melt curtain at high speed can result in edge instability or draw resonance in extrusion coating or cast film (Fig. 5.12B). Also known as edge weave, this can result in line breaks or nonuniform coatings/films. The onset of the instability is related to polymer rheology. Newtonian fluids have a critical draw ratio for the onset of draw resonance of 20.1. The draw ratio (DDR) is the ratio of the final line speed to the velocity of the polymer as it exits the die (Eq. (2.16)). For non-Newtonian fluids, the critical DDR may be higher or lower. Linear polymers, such as LLDPE, tend to draw down more readily than branched polymers. But when it comes to draw resonance, branched PE is more stable than LLDPE and stability increases with increasing MWD and LCB [15,30–33].
- In blown film, various modes of bubble instability may occur, one of which is illustrated in Fig. 5.12C [31,34–36]. These instabilities result in nonuniform thickness and ultimately lower line speeds. Increasing melt strength and strain hardening behavior improves bubble stability [3,31,34–43]. Linear polyethylene is known to have poorer bubble stability than branched polyethylene. Since the introduction of LLDPE and metallocene LLDPE, blown film technology has evolved to better handle these materials, such as increasing the die gap and using better cooling systems. Resin suppliers have tried to tweak the catalyst technology to improve melt strength by manipulating the chain architecture (MWD and LCB). Processors often blend a small amount of LDPE into LLDPE to improve runnability and productivity, even though this negatively impacts film mechanical properties.

These flow issues and instabilities are taken up in more detail in Chapter 13.

5.2.2 Packaging equipment operations and end-use

Polymer rheology affects many aspects of flexible packaging operations and end-use. One example is thermoforming. During forming, the polymer is heated and stretched into a cavity. Strain hardening is an important polymer characteristic to ensure uniform stretching and thickness in the final package. If an imperfection begins to occur where one region of the film is drawn down more than another, strain hardening will increase its resistance to deformation. This self-correcting mechanism helps keep the thickness uniform throughout the forming process [43,44]. Chapter 11 discusses thermoforming in detail.

In heat sealing, if the sealing pressure or temperature is too high, the sealant may flow too much, resulting in the sealant being squeezed out from the seal area. This can lead to poor seal strength. Depending on the geometry of the heat seal jaw, the shear rate during sealing may be quite small. Polymers with higher zero shear viscosity may have fewer squeeze out problems [45,46]. Conversely, some amount of flow may be favorable for good sealing performance. Voids can form in wrinkles, around solid particles contaminating the seal region and in folded regions such as gussets. Flow into these voids is needed to "caulk" these regions to ensure a hermetic seal [47]. See Chapter 7 for more information.

After a heat seal is made but has not yet solidified, it may be subjected to opening forces such as the weight of the product in a vertical form fill seal packaging line. The resistance to seal failure from these opening forces is known as a hot tack. Hot tack strength is a function of the sealing temperature and generally decreases with increasing temperature as the melt strength of the polymer decreases. The sealing temperature range over which the hot tack strength exceeds the opening forces is an important operational parameter. (See Chapter 7 for details.)

The force required to pull a seal apart or to separate layers within a multilayer film structure is called the seal or peel force. The tabs that separate during the peel are known as the peel arms. The peel force is a complex function of the energies to 1) irreversibly elongate the peel arm, 2) bend the peel arm and 3) create a new surface area (this energy is called the fracture energy). The fracture energy is related to the bonding at the interface but is also influenced by local viscoelastic energy losses at the peel front as the specimen is pulled apart. As shown in Chapter 10, this viscoelastic energy contribution can be a significant portion of the measured peel strength.

The stress–strain behavior of the film impacts the strength of the package. While typically outside the realm of melt rheology, factors such as modulus, elongation at yield, elongation at break, and tensile strength are important material properties that involve the deformation of the polymer. An example is the creep resistance of a pouch hanging on a peg in a store. The seal must not stretch and fail under the weight of the product. The change in strain under applied stress is known as creep and is a fundamental rheological property of the material. Similarly, tear strength and impact resistance of the film is influenced by the viscoleastic properties of the polymer.

5.2.3 Polymer characterization

Rheological measurements are very sensitive to changes in polymer structure and molecular architecture and often compliment other analytical techniques to characterize polymers. This can be useful during the development of new polymers, improving the properties of existing polymers, or troubleshooting quality issues.

An example is characterizing small changes in MWD or branching. Gel permeation chromatography (GPC) is a commonly used method for measuring MWD. Some rheological measurements, such as shear viscosity, are very sensitive to the high molecular weight (MW) fraction in the polymer. For high MW polymers above the entanglement MW, the shear viscosity is proportional to the MW raised to the 3.4 power. Small changes in MW can result in large changes in the viscosity and complement the findings of GPC.

At small deformations and slow rates of deformation rheological properties are typically linear. The rheological properties of polymers under such conditions are known as *linear viscoelasticity*. At larger deformations and fast rates of deformation, the rheological response becomes nonlinear. For characterizing molecular architecture (such as MWD), phase behavior, and polymer blends, linear viscoelasticity measurements are most useful. Methods such as dynamic rheology measurements (see Section 5.3.5), creep and recovery measurements (Section 5.5.2), and stress relaxation (Section 5.5.1) experiments are often used. Deformation and rates of deformation during polymer processing, however, are typically in the nonlinear regime. Methods used to characterize fundamental polymer properties do not often translate to the nonlinear behavior found in film fabrication. Different tools are used to measure properties in this regime, such as the capillary rheometer.

5.3 Rheological measurements

5.3.1 Melt flow index or melt flow rate

The most commonly used rheological measurement for polyolefins is the melt flow rate (MFR). It is a measure of the mass of polymer that flows from an orifice under a given weight at a prescribed temperature. The melt flow indexer (the name of the instrument) consists of a heated tube, a piston for pushing the melted polymer, a weight that lies on top of the piston, and an orifice at the end of the tube. The dimensions of the orifice are specified in ASTM D1238 (ISO 1183) and fixed by the design of the instrument. The polymer pellets are introduced into the tube and the piston is inserted to remove air from above the melting resin. After a prescribed amount of time to allow the polymer to melt and come to temperature, the weight is placed on top of the piston. The polymer that flows from the orifice in 10 minutes is weighed. The result is reported in grams per 10 minutes. The weight and temperature must also be reported. For polyethylene, the standard temperature and weight is 190°C and 2160 g and the measurement is often called the melt flow index (MFI). Higher temperatures are used for resins with higher melting points. For example, PP is typically measured at 230°C. MFR is the more general term whereas MFI is typically used for the specific conditions of 190°C /2160 g, but this is not universally followed. The test conditions should always be reported to eliminate any confusion.

The MFR of a resin is inversely related to viscosity and molecular weight; high MFR resins have lower MW and viscosity. The exact relationship between viscosity and MFR cannot be determined for several reasons:

- The shear rate is not controlled during the test. The stress (weight) is fixed rather than the shear rate (as is the case for a capillary rheometer). The shear rate will vary with the MFR. A crude calculation can be made for the shear rate in the oriface under the standard weight of 2160 g:

$$\dot{\gamma} = \frac{1838}{\rho} \text{MFR} \tag{5.7}$$

  where ρ = the melt density of the polymer. Assuming ρ = 766 kg/m^3 for a typical PE melt, $\dot{\gamma}$ [s^{-1}] = 2.4 × MFR.
- The orifice length to diameter ratio (*L/D*) is zero and entrance effects may result in significant errors. Corrections for capillary rheometry results are described in Section 5.3.3. These corrections cannot be made on the melt indexer.

TABLE 5.5 Melt flow index ranges of typical polyethylene grades by process.

Process	Typical MFI range for PE, g/10 min (190°C/2160 g)
Blown film	1 – 5
Blown film (high stock UMW-HDPE)	0.01 – 1
Cast film	2 – 7
Extrusion coating	4[a]–15

[a]This range is typical for grades of LDPE made on an autoclave reactor. Grades made on a tubular reactor for extrusion coating may be lower in MFI.

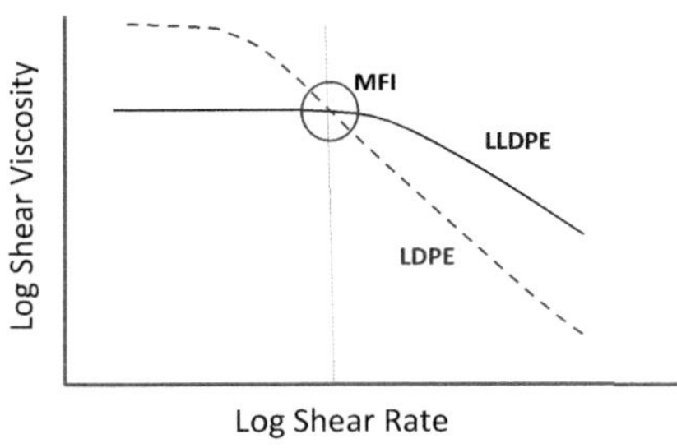

FIGURE 5.13 Comparison of viscosity versus shear rate curves for two types of polyethylene with the same metal flow index. Dashed curve is LDPE and solid curve LLDPE.

- Because of the short *L/D*, the flow in the melt flow indexer is a combination of shear and elongational flow [48]. Capillary or cone and plate rheometry, on the other hand, involves simple shear flow without any complications from the elongational flow. It is challenging to separate the shear component of the MFR measurement to compare with the shear viscosity measured in the rheometers.

Resin producers use the MFR test as a quality control test. For polymers of similar compositions made in the same process, the MFR is a good control test. Over the years the MFR has been adopted by the converting industry as a surrogate for viscosity. As long as it is only used for general guidance, MFR is suitable for this purpose. Table 5.5 gives the typical MFI range for various processes. It is important to note, however, that resins of the same MFI can behave substantially different in a polymer process. This is because the MFI only captures one point on the viscosity versus the shear rate curve. Differences in shear-thinning behavior, or other rheological phenomena, are not captured by the MFI. This is illustrated in Fig. 5.13 where the viscosity curves for an LLDPE and LDPE for the same MFI are shown. The LLDPE is less shear-thinning and will extrude at a higher pressure than the LDPE. This is not captured by the MFI. Even within the same polymer family, grades with the same MFR can have significantly different processing behavior. For example, Clevenhag and Ovelby [49] screen many different grades of LDPE from different manufacturers for suitability in a high-speed extrusion coating process. Even though all the grades have MFI's in the typical range for extrusion coating, this is not enough to predict performance. They develop other screening measurements that predict suitability much better.

Measuring MFR at higher weights has been used to try to capture the shear-thinning behavior of polymer melts. For example, a 10 kg weight (at 190°C) is sometimes used for polyethylene. The short hand for this test is I10 (with I2 used to designate the standard MFI test). While two data points are better than one, the method is not a good substitution for using the full shear viscosity versus shear rate curve for polymer processing issues.

5.3.2 Solution viscosity

Manufacturers of polyethylene terephthalate (PET) and polyamide 6 (PA 6) report a solution viscosity as a means of describing MW. The polymer is dissolved in a solvent at low concentrations and the viscosity is measured either using a glass capillary viscometer (the time for the solution to flow through the capillary is related to the viscosity) or a rotational viscometer (such as a Brookfield viscometer). The solution viscosity is compared with the viscosity of the solvent; the ratio of the solution viscosity to the solvent viscosity is defined as the relative viscosity, η_r:

$$\eta_r = \eta/\eta_s \tag{5.8}$$

where η is the solution viscosity and η_s is the solvent viscosity.

The inherent viscosity, η_{inh}, is the natural logarithm of the relative viscosity divided by the concentration:

$$\eta_{\text{inh}} = (\ln \eta_r)/C \tag{5.9}$$

where C is the concentration in g/100 mL in the US and g/mL in Europe. Finally, the intrinsic viscosity, $[\eta]$, is defined as the limit of the inherent viscosity as the concentration approaches zero:

$$[\eta] = \lim_{C \to 0} (\eta_{\text{inh}}). \tag{5.10}$$

The inherent viscosity is found by measuring the solution viscosity at several concentrations, plotting η_{inh} versus C and extrapolating to zero concentration.

The inherent viscosity or IV is directly related to MW by the Mark–Houwink equation:

$$[\eta] = KM_v^a \tag{5.11}$$

where K and a are a constants that depend on the polymer-solvent system, and M_v is the viscosity average molecular weight.

Values for K and a are tabulated in the literature [50–52]. The viscosity average molecular weight is related to the MWD in the following way:

$$M_v = \left(\sum_i w_i M_i^a \right)^{1/a} \tag{5.12}$$

where w_i is the weight fraction of chain length M_i and M_i is the molecular weight of chain with i repeat units.

M_v is typically about 80%–90% of weight average molecular weight (M_w) and so the following approximation is often used:

$$[\eta] = kM_w^a \tag{5.13}$$

[M_w is defined by Eq. (4.2)].

The PET industry typically uses IV, tested according to ASTM D4603. The polyester is dissolved in a solvent of 60% phenol/40% tetrachlorethane. Bottle grades of PET have an IV of 0.70–0.85 dL/g. The viscosity of polyamide is typically reported as the relative viscosity. Formic acid, sulfuric acid and m-creosol are commonly used solvents. Refer to ASTM D789 and D4878.

The IV and relative viscosity relate directly to the molecular weight of the polymer and are reported mostly for product design and quality control purposes. The brittleness of PET, for example, is related to molecular weight and IV. For polymer processing, however, these values merely represent a benchmark. Processing conditions, particularly the presence of moisture during processing, can significantly affect the melt viscosity. Furthermore, much like the MFR, these values only represent a portion of the viscosity versus shear rate curve. Shear-thinning and other behaviors are not captured by the solution viscosity. For that, the melt viscosity must be measured.

5.3.3 Capillary melt shear viscometry

Probably the most common viscometer used in industry for polymer melts is the capillary viscometer or rheometer. It is easy to operate, relatively inexpensive and captures viscosity data over a shear rate range of about 10–10,000 s^{-1}, which covers most polymer film and coating processes. Polymer pellets or powder are charged to a heated metal tube and allowed to melt. A piston pushes the polymer melt through a capillary die at a constant speed (shear rate). The pressure drop is measured, which is related to the stress. The ratio of the stress to shear rate is viscosity. By taking measurements at several piston speeds, the viscosity over a range of shear rates is obtained.

Capillary viscometry differs from the MFR test in several important ways:

- Capillary viscometry controls the speed (shear rate) and measures the stress; the melt flow indexer uses a constant force (stress) and measures a flow rate. For reasons pointed out in Section 5.3.1, it is difficult to convert melt flow indexer data into a viscosity versus shear rate curve.
- Capillary viscometry allows the user to generate data over a wide shear rate range. The melt flow indexer generally captures data at only one stress and the shear rate varies with the molecular weight of the polymer being tested (see Eq. (5.7)).
- Capillary viscometry uses long dies that reduce entrance effects. The data can be corrected for entrance effect errors (see below). The melt flow indexer uses a short *L/D* orifice that introduces entrance effect errors that cannot be corrected.

5.3.3.1 Basic equations

The calculations for the capillary rheometer use equations that describe fully developed pressure-driven flow in a circular tube for a Newtonian fluid, and then correct for non-Newtonian and entrance effects. As given in Table 5.4, the shear rate at the wall is related to the velocity:

$$\dot{\gamma}_w = \frac{2}{R} V_{\max} \tag{5.14}$$

where $\dot{\gamma}_w$ is the shear rate at the wall of the die, R is the radius of die, and $V_{\max}$ is the maximum velocity in the channel.

It can be shown that

$$V_{\max} = 2V_{\text{avg}} \tag{5.15}$$

and, therefore,

$$\dot{\gamma}_w = \frac{4}{R} V_{\text{avg}}. \tag{5.16}$$

The average velocity, V_{avg}, is simply the piston speed. The volumetric flow rate, Q, is also related to the average velocity:

$$Q = V_{\text{avg}} \pi R^2. \tag{5.17}$$

Therefore, in terms of Q, the shear rate at the wall is

$$\dot{\gamma}_w = \frac{4Q}{\pi R^3}. \tag{5.18}$$

The shear stress is determined by the pressure drop:

$$\sigma = \frac{\Delta P}{2L} r. \tag{5.19}$$

where ΔP is the pressure drop, L is the length of the die, and r is the radial distance from the centerline.

Eq. (5.19) shows that the shear stress varies linearly with r and is greatest at the wall:

$$\sigma_w = \frac{\Delta P}{2L} R. \tag{5.20}$$

The shear rate and shear stress are calculated using Eqs. (5.18) and (5.20), knowing the dimensions of the die and the piston speed and measuring the pressure drop. Without any corrections, these are known as the apparent shear rate and apparent shear stress. The apparent viscosity, η_A, is simply the ratio of the apparent shear stress to the apparent shear rate:

$$\eta_A = \frac{\sigma_w}{\dot{\gamma}_w}. \tag{5.21}$$

Often this is how the data will be reported from a resin supplier. The corrections described in the next section may change the absolute results by 20% or more and are important for scientific experiments but are often not necessary to gain useful insight for polymer processing.

5.3.3.2 Corrections to capillary rheology data

There are two corrections to capillary rheology data that are often applied. The first deals with the error associated with assuming the fluid is Newtonian and the second with entrance flow effects.

Even if the true nature of the flow is unknown, an elegant solution has been derived, often attributed to Rabinowitch, that allows the shear rate to be corrected for non-Newtonian effects. The procedure involves obtaining pressure drop data for a variety of flow rates, die lengths and die radii and plotting $[4Q/\pi R^3]$ versus $[(\Delta P)R/2L]$. All should fall on the same line on a double log–log plot, the slope of which is b. The true shear rate at the wall is related to b:

$$\dot{\gamma}_w = \left(\frac{3+b}{4}\right)\dot{\gamma}_A \tag{5.22}$$

where

$$b = \frac{d(\log\dot{\gamma}_A)}{d(\log\tau_w)} = \text{slope of plot}. \quad (5.23)$$

For a power-law fluid, $b = 1/n$ and

$$\dot{\gamma}_w = \frac{3n+1}{4n}\frac{4Q}{\pi R^3}. \quad (5.24)$$

The pressure drop (ΔP) is measured from the reservoir on top of the die to the die exit, as shown in Fig. 5.14. This pressure drop includes contributions from the entrance to the capillary or die, the drop across the die itself, and the pressure drop at the exit of the die:

$$\Delta P = P_d - P_a = \Delta P_{\text{entrance}} + \Delta P_{\text{die}} + \Delta P_{\text{exit}} \quad (5.25)$$

where P_d is the driving pressure, P_a is the ambient pressure at the die exit, $\Delta P_{\text{entrance}}$ is the pressure drop in the entrance region to the die, ΔP_{die} is the pressure drop through the die, and ΔP_{exit} is the pressure drop at the exit of the die.

The exit pressure drop is generally small but the entrance pressure drop can be significant. The entrance pressure drop is attributed to elastic effects since the value for a Newtonian fluid is small. Bagley proposed a method to correct for the entrance and exit effects by measuring the driving pressure at several values of L/R (by using several dies of constant radius but varying lengths). Plotting P_d versus L/R for a given piston speed (constant Q) yields a line that can be extrapolated back to $P_d = 0$ (here P_d is assumed to be the gauge pressure, $P_d - P_a$). The distance from this intercept and $L/R = 0$ is known as the Bagley end correction, e:

$$\sigma_w = \frac{P_d}{2\left(\frac{L}{R} + e\right)}. \quad (5.26)$$

The value of e for a Newtonian fluid has been shown to be 1.15 but can be near 20 for polymer melts near the onset of sharkskin [7]. More typically, the entrance effect may result in a 10%–30% change in the calculated shear stress. As the L/R of the die is increased, the relative effect of the entrance pressure is diminished. Above L/R of about 50 the entrance effect becomes negligible.

A simple way to correct for entrance effects is to use a 2-bore capillary rheometer. This device has two bores that run off the same drive (and at the same piston speed). One bore has a die with a long L/R (typically 32) and the other one has an L/R of 0, as shown in Fig. 5.15. Studies have shown that the P_a measured with the $L/R = 0$ die is equivalent to the entrance pressure drop [53], and

$$\sigma_w = \frac{\left(P_d^L - P_d^0\right)R}{2L} \quad (5.27)$$

where P_d^L is the driving pressure measured in the bore with the long die and P_d^0 is the driving pressure measured in the bore with the $L/R = 0$ die.

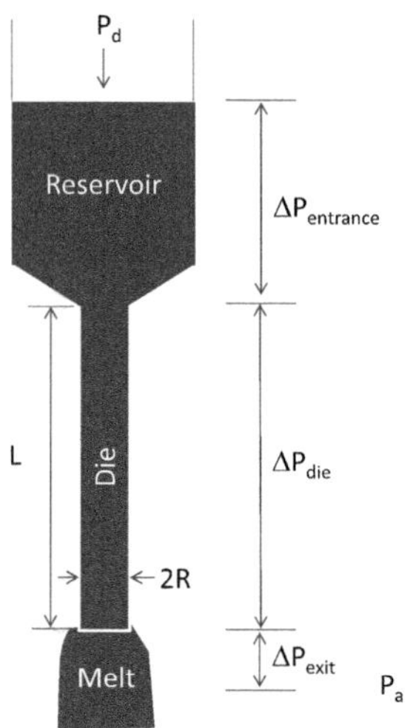

FIGURE 5.14 Contributions to pressure drop in a capillary viscometer.

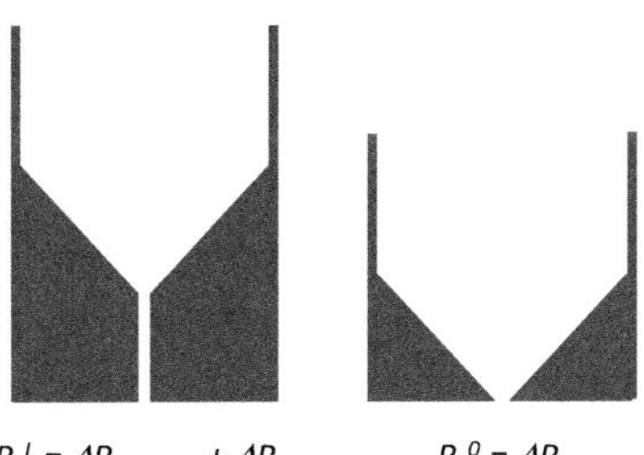

FIGURE 5.15 Dual-bore capillary rheometer.

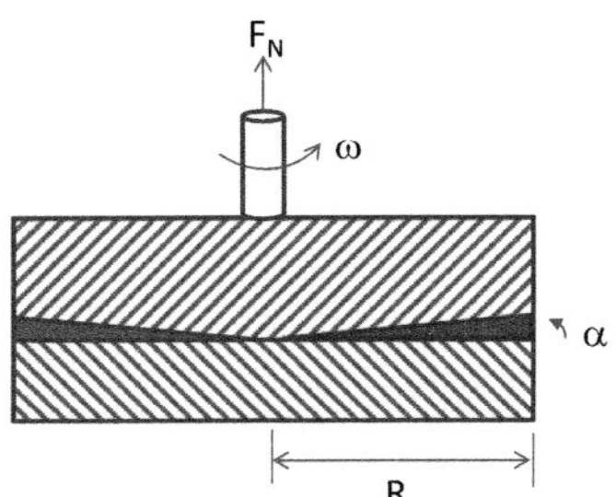

FIGURE 5.16 Cone and plate viscometer.

5.3.4 Cone and plate viscometer

The cone and plate viscometer, shown schematically in Fig. 5.16, is a type of rotational viscometer. The angular speed and torque determine the shear rate and shear stress, as given in Eqs. (5.28) and (5.29):

$$\dot{\gamma} = \frac{\omega}{\alpha} \tag{5.28}$$

$$\sigma = \frac{3M}{2\pi R^3} \tag{5.29}$$

where ω is the rate of rotation, α is the cone angle (typically less than 5 degrees), M is the torque, and R is the radius.

Rotational viscometers are typically used over a shear rate range of 10^{-2} to 5 s^{-1}. At higher rotational speeds, flow instabilities and deviations from the simple shear flow assumed in Eqs. (5.28) and (5.29) occur.

5.3.5 Dynamic rheology measurements

Dynamic measurements are used to study the linear viscoelastic behavior of polymer melts. Linear behavior occurs at low levels of deformation. From the linear response information about the structure and behavior of the polymer can be deduced. Nonlinear behavior, which often occurs in polymer processing, is more intractable to scientific analysis and prediction.

Dynamic measurements are conducted by imposing a sinusoidal strain on the polymer using a parallel plate or cone and plate rotational viscometer. A perfectly elastic material responds in phase with the imposed sinusoidal strain γ_o sinωt. This is because the stress is directly proportional to the strain:

$$\sigma = G\gamma = G\gamma_o \sin\omega t. \tag{5.30}$$

A perfectly viscous material responds 90 degrees out of phase since stress is proportional to the rate of strain rather than the strain:

$$\sigma = \eta\dot{\gamma} \tag{5.31}$$

$$\sigma = \eta\frac{d\gamma}{dt} = \eta\frac{d(\gamma_o \sin\omega t)}{dt} = \eta\omega\gamma_o \cos\omega t = \eta\omega\gamma_o \sin(\omega t + 90^\circ). \tag{5.32}$$

Viscoelastic materials have both elastic and viscous behavior and measuring the in-phase and out-of-phase behavior provides useful insight into molecular structure and performance. The in-phase modulus, G', is known as the *storage modulus* and represents the elastic component. The out-of-phase modulus, G'', is known as the *loss modulus* and represents the viscous component:

$$G'(\omega) = \frac{\text{in-phase stress}}{\text{maximum strain}} \tag{5.33}$$

$$G''(\omega) = \frac{\text{out-of-phase stress}}{\text{maximum strain}} \tag{5.34}$$

where ω typically ranges from 0.01 to 10^3 rads/s. Furthermore, in-phase and out-of-phase viscosities are defined as

$$\eta'(\omega) = \frac{G''}{\omega} \tag{5.35}$$

$$\eta''(\omega) = \frac{G'}{\omega}. \tag{5.36}$$

The magnitude of the viscosity is known as the complex viscosity, η^*:

$$\eta^* = \sqrt{\eta'^2 + \eta''^2}. \tag{5.37}$$

An empirical relationship known as the Cox–Merz rule allows complex viscosity at a given frequency to be equated with steady-state viscosity at a shear rate corresponding to the same frequency:

$$\eta(\dot{\gamma}) = \eta^*(\omega). \tag{5.38}$$

The Cox–Merz rule holds for most but not all polymers. This rule allows the viscosity to be measured over a broad shear rate range by combining the results of multiple measurements. While the steady-state cone and plate viscometer goes up to about 5 s^{-1}, when operated in the dynamic mode, frequencies of up to about 500 s^{-1} can be achieved. This puts the data comfortably in the lower range of capillary viscometers, which can extend the range even further.

The dynamic modulus and viscosity of many polymers also obey time-temperature superposition. By operating the viscometer at different temperatures, the curves are shifted horizontally and vertically to create master rheology curves that span eight decades, typically from 10^{-4} to 10^4 s^{-1}.

The elastic or storage modulus has been shown to be related to the first normal stress difference (a measure of the normal forces or elasticity of the polymer) in the limit as the frequency approaches zero:

$$2G' = N_1 as\ \omega \rightarrow 0. \tag{5.39}$$

where N_1 is the first normal stress difference. Since N_1 is difficult to measure by other means, dynamic measurements are often used.

5.3.6 Extensional measurements

In the linear viscoelasticity regime of small deformations and rates of deformation, it can be shown that the extensional viscosity is three times the shear viscosity:

$$\eta_E = 3\eta. \tag{5.40}$$

Thus, shear measurements can be used to predict extensional properties at small deformations. For large or rapid deformations, there is no such relationship between extension and shear.

Extensional viscosity measurements are notoriously difficult to conduct in a controlled scientific manner. Stretching a polymer above its softening point poses challenges in clamping, temperature control and achieving sufficient strain and strain rates to be meaningful for polymer processing [54]. One method pioneered by Meissner uses rotary clamps to stretch a polymer ribbon in a hot oil bath [55,56]. Münstedt [57] has also developed devices. Besides the practical challenges mentioned above, these techniques are limited to strain rates of less than about 1 s^{-1}, making it difficult to study nonlinear behavior typical of polymer processing.

In more recent years, Sentamanat [58–60] has introduced an attachment to a rotational viscometer for extensional measurements. The device employs a dual drum wind-up mechanism to achieve rates of strain of up to 20 s^{-1}. It too is limited in its use. For example, a minimum zero shear viscosity of 10,000 MPa is recommended. Hence the device is restricted to higher MW polymers and lower temperatures. Test times are also very short at high rates of strain and only transient measurements can be made. Andrade et al. [61] review the pros and cons of several extensional rheometer designs and propose an updated version of the Meissner design that is said to achieve higher strain rates than the Sentamanat design.

These devices measure the start-up stress, usually expressed as the shear stress growth coefficient, $\eta_E{}^{+}(t)$. In start-up flow, the material is initially at rest and subjected to a steady extension at a rate $\dot{\varepsilon}$. The shear stress growth coefficient is defined as

$$\eta_E^{+}(t,\dot{\varepsilon}) \overset{\text{def}}{=} \frac{\sigma_E(t,\dot{\varepsilon})}{\dot{\varepsilon}}. \tag{5.41}$$

At long times, the shear stress growth coefficient is assumed to approach a steady-state elongational viscosity:

$$\lim_{t\to\infty} \eta_E^{+}(t,\dot{\varepsilon}) = \eta_E(\dot{\varepsilon}). \tag{5.42}$$

The shear stress growth coefficient is typically plotted versus time (see, e.g., Fig. 5.4). The relationship 3η is often superimposed on the plot so that deviations from linear viscoelasticity are obvious.

To characterize extensional behavior at the high rates of strain encountered during polymer processing, a pressure-driven converging flow measurement is sometimes used [62]. This method, often called the Cogswell method, uses measurements of the entrance pressure in a capillary rheometer (see Section 5.3.3.2). A dual-bore capillary rheometer is often used. This method computes an average steady-state elongational viscosity:

$$\dot{\varepsilon}_{\text{avg}} = \frac{4\dot{\gamma}^2\eta}{3(n+1)\Delta P_{\text{entrance}}} \tag{5.43}$$

$$\eta_{E_{\text{avg}}}(\dot{\varepsilon}) = \frac{9(n+1)^2(\Delta P_{\text{entrance}})^2}{32\dot{\gamma}^2\eta} \tag{5.44}$$

where η is the shear viscosity, n is the power-law index, and $\Delta P_{\text{entrance}}$ is the pressure drop in the entrance region to the die (as given by Eq. (5.25)).

Cogswell makes several simplifying assumptions in his analysis that may or may not hold true for a given polymer melt. Several studies have tried to relate this method to the more rigorous stretching flow measurements with varying degrees of success [37,63–67].

Perhaps the most popular industrial test for elongational flow is the extrudate drawing or melt strength test. Often known as the Rheotens test, this method uses the filament extruded from a capillary rheometer. The filament is drawn down by a set of motor-driven rolls. The vertical force or torque is measured. This is illustrated in Fig. 5.17. The method typically involves ramping up the speed of the rolls until the filament breaks. The force generally steadies out to a constant value and is recorded as the "melt strength or tension." The speed at break is converted into a maximum draw rate. This method simulates several industrially important processes, such as fiber spinning, extrusion coating, film casting and blowing (although film and coating processes are biaxial extension processes). Because the temperature and strain are not held constant, the results do not correspond to well-defined rheological properties. Nevertheless, the method is useful for measuring differences in extensional behavior of materials and at strains and rates of strain representative of polymer processing. Several studies have used the technique for this purpose [3,4,37,42,68–72].

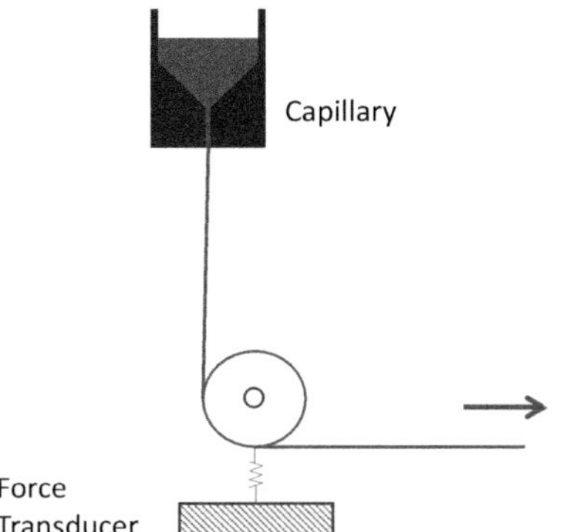

FIGURE 5.17 Extrudate drawing or melt strength test.

5.4 Factors influencing polymer rheology

5.4.1 Strain and rate of strain

It has previously been shown that the viscoelastic properties of polymer melts are a function of strain and rate of strain. The shear viscosity of most polymers is constant at low shear rates but decreases at higher shear rates (shear-thinning behavior) as shown in Fig. 5.3. Shear-thinning behavior is beneficial for many polymer processes. For example, in blown film, it helps lower the pressure, allowing for higher throughput.

Similarly, polymers may be strain hardening or softening in extensional flows (see Fig. 5.4). Such behavior has been found to influence the processibility of polymers, such as drawability in cast film, bubble stability in blown film and achieving uniform thickness during thermoforming.

Some of the molecular factors that influence these behaviors are explored in Section 5.4.5.

5.4.2 Temperature

Increasing the temperature generally decreases the viscosity, as illustrated in Fig. 5.18. On a molecular level, increasing the temperature provides more energy for the molecules to move, increasing the ability of the long chains to reptate or slide past one another. Increasing the temperature has a similar effect as reducing the average molecular weight (see Section 5.4.5.1):

- The zero shear viscosity decreases.
- The critical value of the shear rate that delineates the transition between the Newtonian region and the power-law region increases.
- The slope of the power-law region (the power-law index, n) remains essentially unchanged.

For polymers that follow time-temperature superposition, the effect of temperature can be computed using a shift factor. "Reduced" variables are introduced to collapse viscosity versus shear rate curves onto a single master curve. From the master curve, the viscosity at any temperature and shear rate can then be computed:

$$\eta_R\left(\dot{\gamma}_r, T_0\right) = \frac{\eta(\dot{\gamma}, T)}{a_T} \tag{5.45}$$

$$\dot{\gamma}_r = a_T \dot{\gamma} \tag{5.46}$$

where η_R is the reduced viscosity, T_0 is the reference temperature, and T is the temperature.

$$\mathrm{a_T} = \text{shift factor} = \frac{\eta_0(T)}{\eta_0(T_0)} \tag{5.47}$$

For amorphous polymers, when $T > T_g + 100$ K, the shift factor typically follows an Arrhenius-like or exponential dependence:

$$a_T = \exp\left[\frac{E_a}{R}\left[\frac{1}{T} - \frac{1}{T_0}\right]\right] \tag{5.48}$$

where E_a is the activation energy of flow, R is the ideal gas constant, T is the temperature (K), and T_0 is the reference temperature (K).

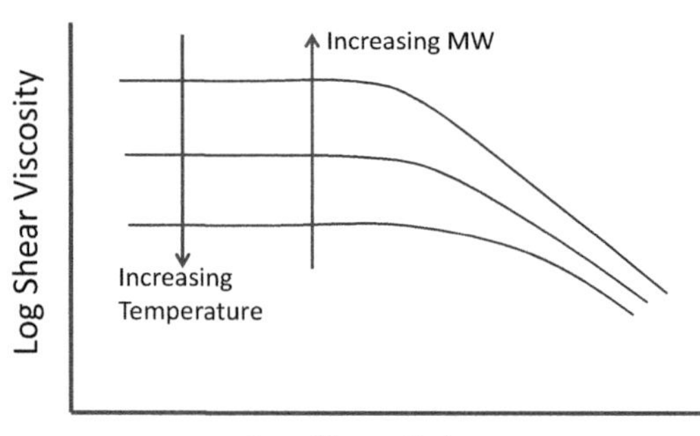

FIGURE 5.18 Effect of temperature and molecular weight on shear viscosity (η) of polymer melts.

TABLE 5.6 Viscosity change with temperature for some common polymer melts.

Polymer	Reference temperature (°C)	Change in viscosity
Polyethylene	190	1.4%/°C
Polystyrene	190	3.4%/°C
PA 6	280	1.4%/°C
PET	280	2.8%/°C

Courtesy of E. I. du Pont de Nemours and Company.

For $T_g < T < T_g + 100$ K, the so-called WLF equation typically holds:

$$\log a_T = \frac{-c_1^0(T - T_0)}{c_2^0 + (T - T_0)} \tag{5.49}$$

where c_1^0 and c_2^0 are constants.

For semi-crystalline polymers, typically an Arrhenius relationship (Eq. (5.48)) is used for temperatures above the crystalline melting point ($T > T_m$).

The change in viscosity per degree of temperature change for a few polymers is given in Table 5.6. For common polyolefin grades, an increase in temperature of 10°C decreases the viscosity by about 15%.

5.4.3 Pressure

Increasing pressure increases the viscosity. Generally, an exponential model can be used:

$$\eta_0(P) = \eta_0(P_0)\exp(\alpha P) \tag{5.50}$$

where P is the pressure, P_0 is the reference pressure, and α is a constant.

A typical value of α for polymer melts is 2×10^{-8} Pa^{-1} (1.4×10^{-4} psi^{-1}). For a 6.9 MPa (1000 psi) pressure rise, the viscosity increases by about 15%.

5.4.4 Time

Changes in viscosity with time generally indicate a chemical reaction that is altering the MW of the polymer. Examples include crosslinking (increasing viscosity), reactions at interfaces (increasing viscosity) and chain scission (decreasing viscosity). For example, the MW of PET resin is very sensitive to moisture. During extrusion, moisture may cause chain scission, reducing MW and viscosity and causing the polymer to become brittle in the solid-state.

5.4.5 Molecular factors

The size (MW), size distribution (MWD), architecture (branching) and interchain interactions affect the ability of long-chain polymers to flow.

5.4.5.1 Molecular weight

Increasing the MW generally increases the viscosity as the polymer chains become more entangled and have less mobility to flow. This is illustrated in Fig. 5.18. Similar to decreasing the temperature, increasing MW:

- increases zero shear viscosity,
- decreases the critical shear rate where the Newtonian region ends and the power-law region begins, and
- has little effect on the slope of the power-law region.

Below the critical M_w [weight average molecular weight, Eq. (4.2)] for entanglement (M_c), the zero shear viscosity is proportional to the M_w:

$$\eta_0 = K_1 M_w \text{ for } M_w < M_c. \tag{5.51}$$

Above M_c, the effect of MW is much stronger:

$$\eta_0 = K_2 M_w^{3.4} \text{ for } M_w > M_c \tag{5.52}$$

Here K_1 and K_2 are constants that depend on the polymer and temperature. M_c typically ranges from about 5000 to 30,000 g/mol and most commercial polymers used in flexible packaging are well above their M_c. For regions outside the Newtonian plateau, the power-law exponent in Eq. 5.52 is typically less than 3.4.

Elastic or normal forces are very sensitive to M_w, as shown in Eq. (5.53):

$$N_{1,0} \propto M_w^{7.0} \tag{5.53}$$

where $N_{1,0}$ is the zero shear rate first normal stress difference.

Because G' is proportional to N_1 (Eq. (5.39)), dynamic measurements are a good tool for probing the presence of high MW components.

5.4.5.2 *Molecular weight distribution*

As illustrated in Fig. 5.19, broader MWD or higher polydispersivity (M_w/M_n) generally:

- decreases the shear rate at which shear-thinning begins,
- gives a more gradual transition region (from the Newtonian region to the shear-thinning or power-law region), and
- decreases the viscosity at shear rates beyond the Newtonian region.

G' is particularly sensitive to changes in MWD. $N_{1,0}$ is proportional to high order moments of the MWD that are heavily influenced by the highest MW chains in the polymer.

5.4.5.3 *Branching*

Short-chain branching generally has little effect on the shape of the shear viscosity versus shear rate curve; adding short-chain branches may reduce the viscosity across the shear rate range if the melting point of the polymer is significantly reduced due to lower crystallinity. Increasing the length of the short-chain (e.g., replacing butene with octene in LLDPE) may have some effect but MW and MWD effects often dominate.

LCB increases the zero shear viscosity and gives more shear-thinning behavior (steeper slope in the power-law region). This leads to lower viscosity at high shear rates. The combination of high viscosity at low shear rates and low viscosity at high shear rates has advantages in many polymer processes, as described earlier. The effect of branching on the shear viscosity versus shear rate curve is illustrated in Fig. 5.20.

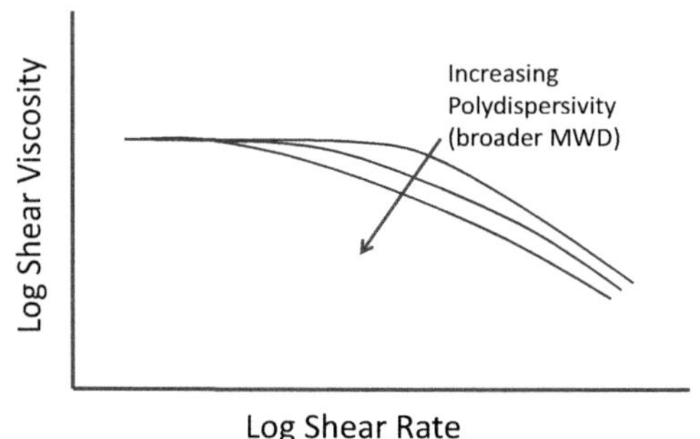

FIGURE 5.19 Effect of molecular weight distribution on shear viscosity of polymer melts.

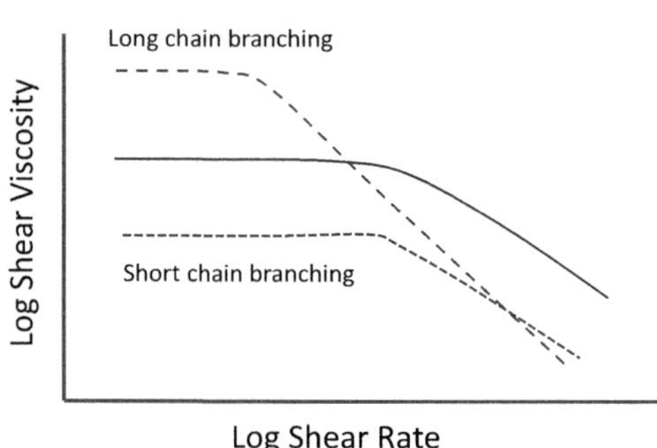

FIGURE 5.20 Effect of branching on shear viscosity of polymer melts.

Here it is instructional to compare LDPE (short-chain branching and LCB) and LLDPE (short-chain branching but little or no LCB). Decreasing the density of LLDPE (increasing short-chain branching) generally has little impact on viscosity. With LDPE, decreasing the density typically results in a greater amount of short and long-chain branches, and so the impact on viscosity can be more significant. Determining the effect of LCB on the viscosity of LDPE is challenging since MWD and LCB effects are often confounded in MWD measurements.

LCB has a substantial effect on extensional viscosity. LDPE is strain hardening while LLDPE is typically strained softening (see Fig. 5.4). Several studies of metallocene LLDPE show that introducing a small amount of LCB changes the extensional viscosity behavior from strain-softening to hardening [2,73].

5.4.5.4 Chemical interactions

The presence of specific inter- (between molecular chains) and intra-chain (within a chain) interactions can influence the rheology. A commercially important example in flexible packaging is ionomers. Introduced in Chapter 4, ionomers are partially neutralized acid copolymers. They are essentially LDPE but with acid-acid interactions (hydrogen bonding) and ionic bonding. Like LDPE, they contain LCB. The hydrogen and ionic bonding tend to accent the effects of the branches. Compared with LDPE, ionomers typically have higher activation energy to flow (their viscosity is more sensitive to changes in temperature), higher zero shear viscosity (for a given MFI), more shear-thinning behavior, greater strain hardening and higher melt strength [74,75].

5.4.5.5 Fillers and additives

Inorganic fillers tend to increase the viscosity of polymer melts. The size, shape and concentration of the filler influence the rheological behavior. This is discussed in greater detail in Chapter 6. Additives such as plasticizers often decrease the viscosity. The specific effect will depend on the nature of the additive, its miscibility with the polymer, and its concentration.

5.5 Relaxation, creep and constitutive equations

The Newtonian, power-law, Cross, and Carreau–Yasuda viscosity models introduced in Table 5.3 relate the shear stress to the rate of strain for purely viscous fluids. For viscoelastic material responses, models that account for both viscous and elastic behavior are needed. It is often helpful to turn to mechanical analogies to represent these behaviors, as shown in Fig. 5.21. An ideal viscous fluid is represented by a dashpot, a mechanical damper that comprises a piston inside a cylinder with a viscous fluid. The viscous friction of the fluid resists the force exerted on the piston. It deforms slowly (not instantaneous), its rate of deformation is related to the imposed stress through the viscosity and when the imposed stress is removed, it does not return to its original length. An ideal elastic or Hookean solid is represented by a spring. The force exerted on a spring is proportional to the displacement through the spring constant, analogous to the stress being proportional to the strain through the material modulus. Like a spring, ideal elastic solids deform instantaneously, and when the stress is removed, return to their original length. (An electrical analogy can also be used, where a capacitor represents the viscous fluid and a resistor the elastic solid.)

By combining these ideal elements, relationships between stress and strain known as constitutive equations can be developed. Constitutive equations are needed to solve the equations of motion and can be very complex for polymer flow. This approach is illustrated with two experiments that represent two of the extremes of viscoelasticity: stress relaxation and creep/recovery.

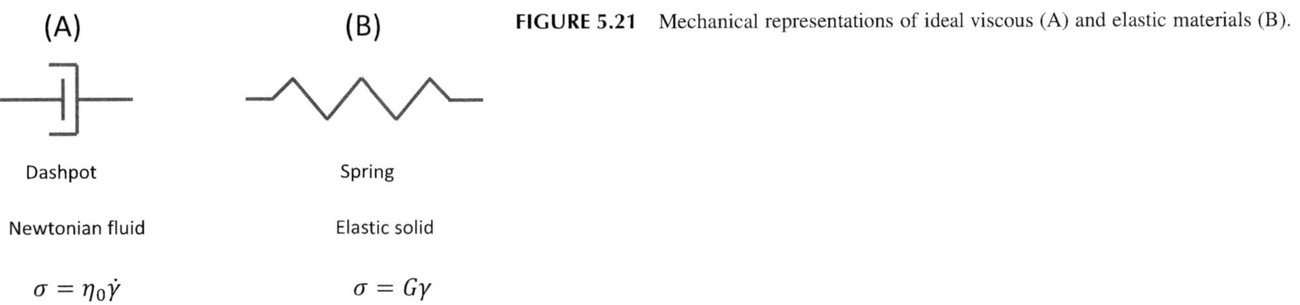

FIGURE 5.21 Mechanical representations of ideal viscous (A) and elastic materials (B).

5.5.1 Stress relaxation

In a stress relaxation experiment, the polymer is subjected to a constant strain at time zero (Fig. 5.22A) and the resultant stress is measured as a function of time. The response typically looks like Fig. 5.22B where the stress initially spikes up and slowly decreases or relaxes. In the ideal case, the response can be represented as a dashpot and spring in series, as shown in Fig. 5.22C. The displacement initially stretches the spring, resulting in stress. Over time, the force recedes as the dashpot moves. Mathematically, the force in both the spring and dashpot are equal. If Eq. (5.30) for the elastic solid is differentiated and combined with Eq. (5.31) for a viscous fluid, the following equation for the shear rate is obtained:

$$\frac{\dot{\sigma}}{G} + \frac{\sigma}{\eta} = \dot{\gamma} \tag{5.54}$$

or

$$\sigma + \lambda\dot{\sigma} = \dot{\gamma}. \tag{5.55}$$

The solution to Eq. 5.55 is

$$\sigma = \sigma_0 e^{-t/\lambda} \tag{5.56}$$

where

$$\lambda = \frac{\eta}{G} = \text{relaxation time.} \tag{5.57}$$

Eq. (5.55) is known as the Maxwell model. The solution results in an exponential decay of the stress with time, similar to what is shown in Fig. 5.22. The parameter λ has units of time and can be thought of as a characteristic relaxation time. In real polymer responses, a single relaxation time is not sufficient. Instead, a spectrum of relaxation times best fit the data. It can be shown that λ is the longest relaxation time. A value for λ from dynamic rheology measurements can be obtained by letting $G = G''$ (the loss modulus) and using Eq. (5.36) for η in the limit as the frequency approaches zero:

$$\lambda = \frac{\eta}{G} = \frac{G'}{G''\omega} \text{ for } \omega \to 0. \tag{5.58}$$

In practice, λ is taken as the reciprical of the frequency ω_c where the G' and G'' curves cross over (see Fig. 5.23) or $G' = G''$.

Higher values of λ indicate greater elastic effects. In the context of the Deborah number introduced in Eq. (5.4), longer relaxation time means the polymer chains cannot respond (by relaxing or flowing) as much to the imposed strain during the time scale of the process and therefore behave more solid-like. This also is apparent in the dynamic rheology

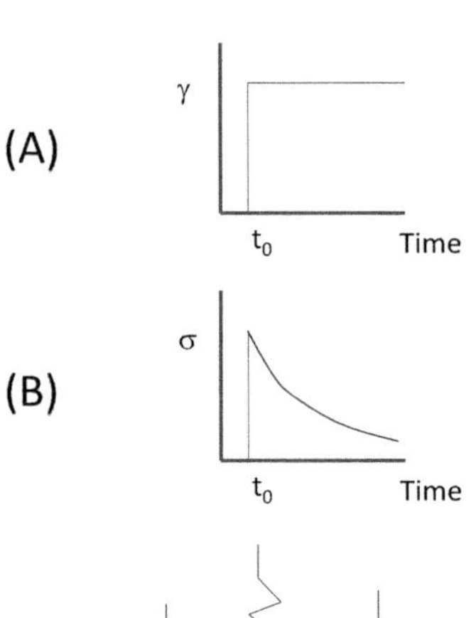

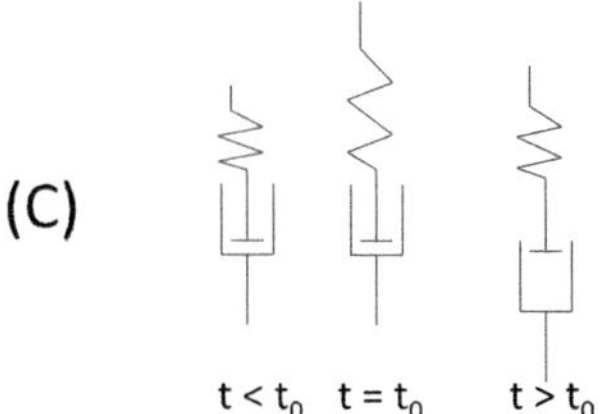

FIGURE 5.22 Stress relaxation experiment. (A) Imposed strain, (B) stress response, and (C) 2-element mechanical Maxwell model.

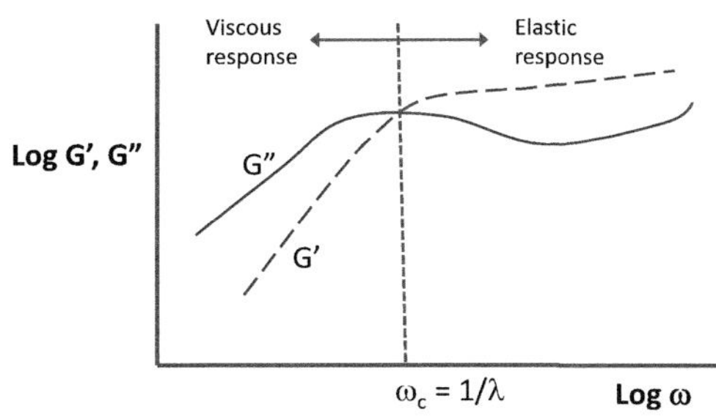

FIGURE 5.23 Relaxation time from dynamic modulus data.

data illustrated in Fig. 5.23. At frequencies or shear rates below the cross over frequency ($1/\lambda$), times scales ($1/\omega$) are much longer than λ and viscous effects dominate ($G'' > G'$). At frequencies or shear rates above the cross over frequency, elastic effects dominate ($G' > G''$).

Some investigators use the λ parameter from the Cross equation (Table 5.3) for a characteristic relaxation time because it has units of time. This is generally a poor choice since it is not supported by theory or experimental evidence [7].

5.5.2 Creep/recovery

In the second type of experiment, the polymer is subjected to an applied load or stress at time zero and the strain is measured as a function of time (Fig. 5.24A). The ideal response is shown in Fig. 5.24B. The mechanical analogy is a dashpot and spring in parallel, as shown in Fig. 5.24C. The stress on each element is

$$\sigma_1 = G\gamma \tag{5.59}$$

$$\sigma_2 = \eta\dot{\gamma}. \tag{5.60}$$

For elements in parallel, the stress is additive:

$$\sigma = G\gamma + \eta\dot{\gamma}. \tag{5.61}$$

Solving for the strain

$$\gamma = \frac{\sigma}{G}\left(1 - e^{-t/\lambda}\right) \tag{5.62}$$

where $\lambda = \eta/G$ is the retardation time. This is known as the Voigt model. Retardation time refers to the delay in the response to the imposed stress. As λ approaches zero, the material behaves like an ideal elastic solid.

The Maxwell model can only handle viscoelastic fluid and the Voigt model viscoelastic solid behaviors. For more complex behavior, more elements are added. An example is given in Fig. 5.25 for a general creep response curve. Here a three-element model in series is used:

- ideal elastic solid (spring)
- retarded elasticity (spring and dashpot in parallel)
- ideal viscous fluid (dashpot)

At time t_1 the stress is imposed. Prior to t_1 the polymer is in equilibrium. At t_1 the spring stretches instantaneously, but the retarded elastic and viscous components remain stationary. This corresponds to the rapid rise of the strain at t_1. Between t_1 and t_2 the strain increases as the dashpots increase in length. At t_2 the stress is removed and the spring element responds instantaneously. Over time, the spring in the retarded elastic model relaxes, lowering the overall stress.

The elements of the model correspond to the physical changes occurring in the polymer, as shown in Fig. 5.25. Region (a) represents the bending of bond angles and lengths and (b) the elastic distension of secondary bonds. This is analogous to the stretching of the spring. Region (c) involves short-range chain segment displacement, (d) uncoiling of the chain and (e) chain-chain slip. These factors are responsible for the retarded elastic and ideal viscous behavior. Once the stress is removed at t_2 the strain from bending of the covalent bond angles and lengths and the ideal elastic distension of the secondary bonds recover instantly. The short-range chain segment displacement and uncoiling recover with time, but the chain-chain slip is permanent (as represented by the extended dashpot in the model) and is responsible for the unrecoverable strain.

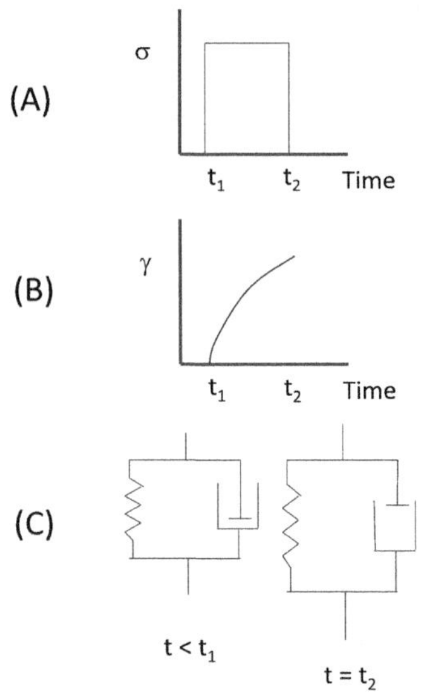

FIGURE 5.24 Creep experiment. (A) Imposed stress, (B) strain response, and (C) 2-element mechanical Voigt model.

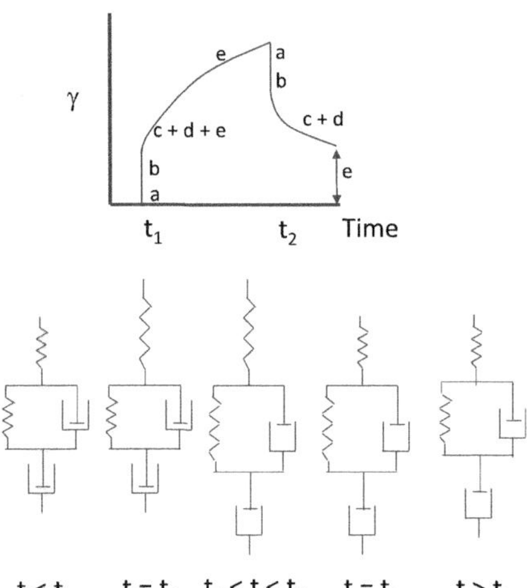

FIGURE 5.25 General creep behavior and the three-element Maxwell–Voigt model.

Constitutive equations can be very complicated with complex derivatives and many parameters. For example, one of the most powerful constitutive equations is the K-BKZ model, which uses over two dozen experimentally fit parameters. Other models are based on molecular motions such as the reptation models of De Gennes [76] and Doi and Edwards [77]. Many help in broadening our understanding of molecular motion but are too cumbersome or limiting in their assumptions to be of practical use in predicting polymer flow behavior. Indeed, the scientific literature is filled with papers on numerical simulation of viscoelastic fluids, yet one should use caution when interpreting the results. Most of the simulations use limiting assumptions because of the complexity of the behavior of these materials.

References

[1] H.A. Barnes, J.F. Hutton, K. Walters, An Introduction to Rheology, Elsevier, Amsterdam, 1989, p. 13.

[2] C. Gabriel, H. Munstedt, Strain hardening of various polyolefins in uniaxial elongational flow, J. Rheology 47 (3) (2003) 619–630.

[3] A. Ghijsels, J.J.S.M. Ente, J. Raadsen, Melt strength behavior of PE and its relation to bubble stability in film blowing, Intern. Polym. Process. 5 (4) (1990) 284–286.

[4] A. Ghijsels, J.J.S.M. Ente, J. Raadsen, Melt strength behavior of polyethylene blends, Int. Polym. Process. 7 (1) (1992) 44–50.

[5] C. Cunningham, N. Savargaonkar, L. Kardos, T. Karjala, New low density polyethylene products for improved throughput, in: TAPPI PLACE Conference, 2012, Seattle, WA.

[6] J. Vlachopoulos, Introduction to Polymer Processing, 2nd ed., McMaster University, 1994.

[7] J. Vlachopoulos, D. Strutt, Rheology of molten polymers, in: J.R. Wagner Jr. (Ed.), Multilayer Flexible Packaging, 2nd ed., Elsevier, Amsterdam, 2016, pp. 77–96.
[8] R. Rutgers, M.J. Mackley, The correlation of experimental surface instabilities with numerically predicted exit surface stress concentrations and melt strength for linear low density polyethylene, J. Rheol. 44 (2000) 1319.
[9] M.M. Denn, Extrusion instabilities and wall slip, Ann. Rev. Fluid Mech. 33 (2001) 265.
[10] K. Chaloupkova, M. Zatloukal, Theoretical and experimental analysis of the die drool phenomenon for metallocene LLDPE, Polym. Engr Sci. 47 (2007) 871–881.
[11] J. Musil, M. Zatloukal, Historical review of die drool phenomenon in plastics extrusion, Polym. Rev. 54 (2014) 139–184.
[12] A.W. Ramamurthy, Wall slip in viscous fluids and influence of materials of construction, J. Rheol. 30 (1986) 337–357.
[13] X. Yang, H. Ishida, S.-Q. Wang, Wall slip and absence of interfacial flow instabilities in capillary flow of various polymer melts, J. Rheol. 42 (1998) 63–80.
[14] A.M. Sukhadia, Sharkskin melt fracture behavior of medium and high molecular weight polyethylenes in film blowing, in: SPE ANTEC Conference, 1997, pp. 1059–1066.
[15] C.J.S. Petrie, M.M. Denn, Instabilities in polymer processing, AICHE J. 22 (1976) 209–236.
[16] E.B. Bagley, H.P. Schreiber, Effect of die entrance geometry on polymer melt fracture and extrudate distortion, Trans. Soc. Rheol., V. (1961) 341–353.
[17] J. Vlachopoulos, M. Alam, Critical stress and recoverable shear for polymer melt fracture, Polym. Eng. Sci. 12 (1972) 184–192.
[18] J.L. White, Critique on flow patterns in polymer fluids at the entrance of a die and instabilities leading to extrudate distortion, Appl. Polym. Symp. 20 (1973) 155–174.
[19] J. Lee, J.S. Allen, M. Gupta, Observations from liquid-liquid encapsulation in coextrusion of inelastic Newtonian fluids, Annu. Technical Conf. Soc. Plast. Engineers, 69th 2 (2011) 1360–1366.
[20] P. Yue, C. Zhou, J. Dooley, J. Feng, Elastic encapsulation in biocomponent stratified flow of viscoelastic fluids, J. Rheol. 52 (2008) 1027–1042.
[21] C.D. Han, Bicomponent coextrusion of molten polymers, J. Appl. Polym. Sci. 17 (1973) 1289–1303.
[22] W.J. Schrenk, N.L. Bradley, T. Alfrey, H. Maack, Interfacial flow instability in multilayer coextrusion, Polym. Eng. Sci. 18 (1978) 620.
[23] C.D. Han, R. Shetty, Studies on multilayer film coextrusion II. Interfacial instability in flat film coextrusion, Polym. Eng. Sci. 18 (1978) 180–186.
[24] M. Zatloukal, W. Kopytko, J. Vlcek, P. Saha, Investigation of zig-zag type of interfacial instabilities in coextrusion, in: SPE ANTEC Conference, 2005, pp. 101–105.
[25] R. Ramanathan, R. Shanker, T. Rehg, S. Jons, D.L. Headley, W.J. Schrenk, "Wave" pattern instability in multilayer coextrusion – an experimental investigation, in: SPE ANTEC Conference, 1996, pp. 224–228.
[26] M. Zatloukal, J. Vlcek, C. Tzoganakis, P. Saha, Viscoelastic stress calculation in multilayer coextrusion dies: die design and extensional viscosity effects on the onset of "wave" interfacial instabilities, Polym. Engr Sci. 42 (2002) 1520–1533.
[27] M. Martyn, R. Spares, P.D. Coates, M. Zatloukal, Imaging and analysis of wave type interfacial instability in the coextrusion of low-density polyethylene melts, J. Non-Newtonian Fluid Mech. 156 (2009) 150–164.
[28] T. Dobroth, L. Erwin, Causes of edge bead in cast films, Polym. Eng. Sci. 26 (1986) 462–467.
[29] K. Kanning, A. Co, Edge effects in film casting, J. Plastic Film. & Sheeting 16 (2000) 188–203.
[30] M.J. Bortner, P.J. Doerpinghaus, D.G. Baird, Effects of sparse long chain branching on the spinning stability of LLDPEs, Intern. Polym. Process. 19 (2004) 236–243.
[31] W. Minoshima, J.L. White, Instability phenomena in tubular film, and melt spinning of rheologically characterized high-density, low-density and linear low-density polyethylenes, J. Non-Newtonian Fluid Mech. 19 (1986) 275–302.
[32] O. Zavinska, J. Claracq, S. van Eijndhoven, Non-isothermal film casting: determination of draw resonance, J. Non-Newtonian Fluid Mech. 151 (2008) 21–29.
[33] A. Co, Draw resonance in film casting, Chemical Industries (Dekker), 103, 2005, pp. 287–319 (Polymer Processing Instabilities).
[34] T.I. Butler, Blown film bubble instability induced by fabrication conditions, TAPPI Polymers, Laminations & Coat. Conf. 2 (1999) 815–828.
[35] J. Laffargue, L. Parent, P.G. Lafleur, P.J. Carreau, Y. Demay, J.F. Agassant, Investigation of bubble instabilities in film blowing process, Intern. Polym. Process. 17 (2002) 347–353.
[36] P. Tas, L. Nguyen, S. Marshall, M. Kashanian, Bubble stability and maximum output of surpass sLLDPE resin in blown film, TAPPI PLACE. Conf. (2004) 320–334.
[37] H.M. Laun, H. Schuch, Transient elongational viscosities and drawability of polymer melts, J. Rheol. 33 (1989) 119.
[38] G.J. Field, P. Micic, S.N. Bhattacharya, Melt strength and bubble stability of LLDPE/LDPE blends, Polym. Int. 48 (1999) 461–466.
[39] H. Munstedt, T. Steffl, A. Malmberg, Correlation between rheological behavior in uniaxial elongation and film-blowing properties of various polyethylenes, Rheol. Acta 45 (2005) 14–22.
[40] J. Meissner, Basic parameters, melt rheology, processing and end-use properties of three similar low density polyethylene samples, Pure Appl. Chem. 42 (1975) 551–612.
[41] H.H. Winter, A collaborative study on the relation between film blowing performance and rheological properties of two low-density and two high-density polyethylene samples, Pure Appl. Chem. 55 (1983) 943–976.
[42] J.N.D. Vlachopoulos, Extensional viscosity and melt strength and their role in film blowing, SPE Extrusion Division Consultants Corner, 2015.
[43] S. Kurzbeck, H. Munstedt, Elongational properties of polymer melts and correlations with their processing behavior in blown film and thermoforming, in: Progress and Trends in Rheology V, Proceedings of the European Rheology Conference, 1998, 5 h Potoroz, Slovenia, pp. 402–403.

[44] T. Spence, D. Hylton, Rheological studies of commercial thermoforming materials, SPE ANTEC Conf. (1998) 681–689.
[45] B.A. Morris, J.M. Scherer, Squeeze-out of sealant during hot bar sealing: modeling insights, SPE ANTEC Conf. (2013).
[46] B.A. Morris, J.M. Scherer, Preventing squeeze-out of sealant during hot bar sealing: modeling and experimental insights, SPE ANTEC Conf. (2014).
[47] M. Kapur, M. Bilgen, P. O'Connell, Importance of application representative sealant test methods, 2014 SPE ANTEC Conference, 2014, p. 1412.
[48] D.C. Rohlfing, J. Janzen, What's happening in the melt-flow plastometer: The role of elongational viscosity, SPE ANTEC Conf. (1997) 1010–1014.
[49] P.-A. Clevenhag, C. Oveby, Rheological indicators to predict the extrusion coating performance of LDPE, TAPPI PLACE. Conf. (2003).
[50] H.L. Wagner, The Mark-Houwink-Sakurada equation for the viscosity of linear polyethylene, J. Phys. Chem. Ref. Data 14 (1985) 611–616.
[51] M. Kurata, W.H. Stockmayer, Intrinsic viscosities and unperturbed dimensions of long chain molecules, Adv. Polym. Sci. 3 (1963) 196.
[52] M. Kurata, Y. Tsunashima, M. Iwama, K. Kamada, in: J. Brandrup, E.H. Imnmergut (Eds.), Chapter IV in Polymer Handbook, 2nd ed., John Wiley & Sons, New York, 1975.
[53] J.M. Dealy, K.F. Wissbrun, Melt rheology and its role in plastics processing, Theory and Applications, Chapman & Hall, 1995, p. 322.
[54] J.M. Dealy, K.F. Wissbrun, Melt rheology and its role in plastics processing, Theory and Applications, Chapman & Hall, 1995, p. 241.
[55] J. Meissner, Rheometer zur Untersuchung der deformationsmechanischen Eigenschaften von Kunststoff-Schmelzen unter definierter Zugbeanspruchung, Rheol. Acta 8 (1969) 78.
[56] J. Meissner, Dehnungsverhalten von Polyäthylen-Schmelzen, Rheol. Acta 10 (1971) 230.
[57] H. Munstedt, New universal extensional rheometer for polymer melts. measurements on a polystyrene sample, J. Rheol. 23 (1979) 421–436.
[58] M. L. Sentmanat, Dual windup extensional rheometer, US Patent 6,578,413, 2003.
[59] M.L. Sentmanat, Miniature universal testing platform: from extensional melt rheology to solid-state deformation behavior, Rheologica Acta 43 (2004) 657–669.
[60] M.L. Sentmanat, E.B. Muliawan, S.G. Hatzikiriakos, Fingerprinting the processing behavior of polyethylenes from transient extensional flow and peel experiments in the melt state, Rheologica Acta 44 (2005) 1–15.
[61] R.J. Andrade, P. Harris, J.M. Maia, High strain extensional rheometry of polymer melts: Revisiting and improving the Meissner design, J. Rheol. 58 (2014) 869–890.
[62] F.N. Cogswell, Converging flow of polymer melts in extrusion dies, Polym. Eng. Sci. 12 (1972) 64–73.
[63] B.H. Bersted, Refinement of the converging flow method of measuring extensional viscosity in polymers, Polym. Eng. Sci. 33 (1993) 1079–1083.
[64] A.D. Gotsis, Q. Ke, Comparison of three methods to measure the elongational viscosity of polymer melts: entry flow, fiber spinning and uniaxial elongation, SPE ANTEC Conf. (1999).
[65] D.M. Binding, An approximate analysis for contraction and converting flows, J. Non-Newtonian Fluid Mech. 27 (1988) 173–189.
[66] D.M. Binding, Further considerations of axisymmetric constriction flows, J. Non-Newtonian Fluid Mech. 41 (1991) 27–42.
[67] A.D. Gotsis, A. Odriozola, The relevance of entry flow measurement for the estimation of extensional viscosity of polymer melts, Rheologica Acta 37 (1998) 430–437.
[68] S.K. Goyal, The influence of polymer structure on melt strength behavior of PE resins, Plast. Eng. (1995) 25. Feb 1995.
[69] F.P. La Manta, D. Acierno, Influence of molecular structure on the melt strength and extensibility of polyethylenes, Polym. Eng. Sci. 25 (5) (1985) 279–283.
[70] D. Acierno, D. Curto, F.P. La Manta, A. Valenza, Flow properties of low density/linear low density polyethylenes, Polym. Eng. Sci. 26 (1) (1986) 28–33.
[71] M.H. Wagner, A. Bernnat, The rheology of the rheotens test, J. Rheol. 42 (4) (1998) 917–928.
[72] V. Rauschenberger, H.M. Laun, A recursive model for Rheotens tests, J. Rheol. 41 (3) (1997) 719–737.
[73] P.J. Doerpinghaus, D.G. Baird, Separating the effects of sparse long-chain branching on rheology from those due to molecular weight in polyethylene, J. Rheol. 47 (3) (2003) 717–736.
[74] T. Takahashi, J. Watanabe, K. Minagawa, K. Koyama, Effect of ionic interaction on elongational viscosity of ethylene-based ionomer melts, Polymer 35 (26) (1994) 5722–5728.
[75] N.K. Tierney, R.A. Register, Ion hopping in ethylene-methacrylic acid ionomer melts as probed by rheometry and cation diffusion measurements, Macromolecules 35 (2002) 2358–2364.
[76] P.G.D. Gennes, Scaling Concepts in Polymer Physics, Cornell University Press, Ithaca, NY, 1979.
[77] M. Doi, S.F. Edwards, The Theory of Polymer Dynamics, Oxford University Press, New York, 1986.

Further reading

[1] W.W. Graessley, Viscoelasticity and flow in polymer melts and concentrated solutions, in: J.E. Mark (Ed.), Physical Properties of Polymers, third ed., Cambridge University Press, 2004.
[2] J.D. Ferry, Viscoelastic Properties of Polymers, third ed., Wiley, New York, 1980.
[3] J.M. Dealy, J. Wang, Melt Rheology and its Applications in the Plastics Industry, Springer, 2013.
[4] C.W. Macosko, Rheology: Principles, Measurements, and Applications, Wiley-VCH, New York, 1994.

Chapter 6

Polymer blending for packaging applications

6.1 Introduction

The blending of polymers is becoming increasingly important in packaging applications to enhance properties, improve processing or lower cost. Tailoring surface properties, such as coefficient of friction (COF), adding color, promoting adhesion, increasing output, improving stability, enhancing barrier performance and obtaining easy-opening features, are just a few of the attributes that can be achieved by blending.

The simplest blends can be made by mixing ingredients in the extruder used to convert the resin into a film or coating. For more complex blends, specialized screw designs or customized compounding equipment may be required to achieve the desired properties. These machines incorporate various mixing modes, such as flow rearrangement (distributive mixing) or high stress levels to break up particles (dispersive mixing). Complex shear and elongational flow fields may also be used to obtain optimum mixing.

The final blend properties will depend not only on the flow and stress history, which is process dependent, but also on the polymers' thermal, rheological and interfacial properties. Most polymer blends are immiscible, resulting in the minor component forming a separate dispersed phase or domain within the major component. The major component forms a continuous phase or matrix. In some cases the phases are co-continuous, forming an interpenetrating network. The phase size and shape are known as the blend morphology. Blend morphology has a profound effect on the final properties and is the subject of many studies. Morphology is influenced by many factors, including:

- Interfacial tension (thermodynamics).
- Dispersed to continuous phase viscosity ratio.
- Elasticity of each phase.
- Minor component concentration.
- Mixing and melting order.
- Stress and flow history.

A basic understanding of these complex relationships between polymer properties and processing can aid in optimizing blends for a given application.

This chapter highlights some fundamentals of blending polymers for packaging applications. The literature is too broad for a detailed review of polymer blending and alloying. The goal is to familiarize the reader with those aspects of polymer blending that are important for in-line mixing of resins and other ingredients during film converting. Some blend requirements, however, are too demanding for in-line mixing and are best left to a resin manufacturer or compounder. Some of the technology that may be used to design and produce such blends is also reviewed. Dispersion of rigid or inorganic particles in polymers, nanocomposites, rheology and characterization are also briefly discussed.

6.2 Why blend?

Even with the flexibility of controlling properties by introducing specific functional layers within a multilayer film, blending can be critical for meeting package requirements. Blending may be needed to make the polymer stable enough to extrude or have the right surface properties after extrusion. Blending may be a way to tailor specific properties into a layer, such as a barrier or heat seal performance. The resin manufacturer often adds polymer additives, such as antioxidants, catalyst killers and various processing aids, to the polymer. But the film manufacturer may add these and other additives, such as slip, antiblock, antifog, antistatic agents or processing aids such as fluoroelastomers for reducing

The Science and Technology of Flexible Packaging. DOI: https://doi.org/10.1016/B978-0-323-85435-1.00006-5

sharkskin and melt fracture. Typically, these additives comprise less than 1% of the final composition. They are usually in the form of a powder or liquid and therefore difficult to feed into single-screw extruders usually used in film converting operations. Because of this, and their low concentrations, they are typically first made into a masterbatch, a highly concentrated blend of the additive with a carrier resin. Masterbatches are usually produced by companies with specialized compounding equipment. The film manufacturer blends the masterbatch into the resin at the extruder feed hopper used to make the film or sheet.

Pigments such as titanium dioxide (TiO_2) and various colorants are typically added as a masterbatch. Since the additive or pigment is well dispersed in the masterbatch, blending can be done in a single-screw extruder without special screw designs or compounding equipment.

Blending may be used to reduce the resin cost. For example, a metallocene polyethylene (mPE) plastomer may be diluted with standard linear low-density polyethylene (LLDPE) or low-density polyethylene (LDPE) to lower cost. Recycle may be incorporated back into the film structure. One form of recycling is in-house scrap or postindustrial recycle (PIR), which may be ground up and introduced into the extruder hopper as "regrind." Depending on the composition of the PIR, a compatibilizer may be added to ensure good adhesion between the phases or to control the blend morphology. Ethylene copolymers and anhydride modified polyolefins often make good compatibilizers due to their nonpolar and polar functionality, and the ability to incorporate reactive groups to promote specific interactions with the recycle components [1].

Postconsumer recycle (PCR) may be added to the film. Virgin polymer is often blended with the PCR to enhance the mechanical properties. Other additives may be needed to improve thermal stability, control odor or enhance compatibility. These may be added as masterbatches. Compatibilizers may also be added to multimaterial, multilayer film structures to make them recyclable in a mono-material recycle stream. For example, Öner-Deliormanlı et al. [2]. describe a novel compatibilizer by Dow that is added to PE layers in a multilayer barrier film containing PE and ethylene vinyl alcohol (EVOH) so it can be placed in the "all PE" recycle stream by the consumer. In this design for recycling approach, which is explored in detail in Chapter 17, the compatibilizer helps control the domain size of EVOH once the package is recycled so that the presence of the EVOH does not reduce optical or mechanical performance.

Blending may also help improve resin processability. Two grades of the same resin with differing flow properties (such as melt flow index) may be blended together to achieve the proper flow for a given process. LDPE is typically blended into LLDPE to help increase output by reducing extruder pressure and torque and improving melt strength and bubble stability [3,4]. Other examples where blending improves processing include blending amorphous polyamide (nylon) into polyamide 6 (PA 6) to increase extruder output and adding amorphous PA or ionomer to EVOH to improve thermoformability [5,6].

A polymer deficient in one property is often blended with another one to enhance that property. Blending in cyclic polyolefins (COCs), for example, can enhance LLDPE stiffness [7]. Soft polymers are often blended into harder polymers to improve the toughness. Examples include blending ethylene vinyl acetate (EVA) into LLDPE, mPE into LLDPE and ethylene-propylene-diene-monomer rubber (EPDM) or mPE into polypropylene (PP) [8].

In the engineering polymer world, rubber is added to polyamide (PA) to improve low-temperature toughness [9]. Adding polyethylene terephthalate (PET) to polycarbonate (PC) lowers cost and improves chemical resistance and processability. Blends of PC and acrylonitrile-butadiene-styrene (ABS) have a lower cost than PC and higher heat deflection temperature and toughness than ABS [10].

Barrier properties may be enhanced by blending. High-density polyethylene (HDPE) is blended into LLDPE or LDPE to improve moisture barrier performance. Amorphous polyamide is blended into PA 6 to improve the oxygen barrier at high relative humidity [11]. DuPont invented a laminar barrier technology where PA is blended into HDPE forming large platelets that impede the flow of species trying to migrate through. The laminar morphology is accomplished by choosing specific resins and processing conditions [12–15].

Adhesion may be promoted by blending together several resins. Adding ethylene methyl acrylate (EMA) copolymer or EVA to PE can improve adhesion to certain inks. Maleic anhydride (MAH) modified polyolefins are blended with PE or EVA to improve adhesion to PA or EVOH in coextrusion. An acid-based additive has been developed for enhancing the adhesion of LDPE to aluminum foil in extrusion coating [16].

Additives may also help control adhesion during the heat seal process. "Contaminants" are often blended into sealant resins to achieve easy openability. Examples include blends of polybutene-1 (PB) with LDPE, LLDPE, EVA or ionomers and EVA with ionomers [17–20]. The blend morphology and phase compatibility is important for these applications. For example, in the blends containing PB, the PB is the minor phase. It forms spherical particles that are stretched into fibers and ribbons during the film fabrication process. The poor compatibility between the PB and matrix resin results in failure along these fibers and ribbons near the sealant interface, lowering the seal strength.

6.3 Blending processes

Blending requires that the ingredients be brought together in the right proportions and then homogeneously mixed. In polymer blending, the former is important because there is little back mixing in most continuous mixing devices used by the industry. This is particularly true of the single-screw extruder, which is the predominant device used by the film converting industry. For simple blends, such as those adding masterbatches, the pellets are premixed together and fed into the extruder hopper. In more sophisticated compounding devices, such as twin-screw extruders, ingredients can be added at various stages along the length of the device.

In all cases, the ingredients must be transported to the line. Modern material handling systems allow material to be conveyed remotely from rail cars, silos or large boxes, typically using pneumatic systems.

Once the polymer enters the extruder or other mixing device, it is melted and the mixing mechanism depends on the specific device employed. Mixing in single-screw extruders is the focus of this section. More sophisticated devices are briefly introduced and compared and contrasted with single-screw extruders. While specialty compounding devices, such as twin-screw extruders, kneaders and continuous mixers, have historically been used by the resin manufacturer or toll compounder and not by film converters, they are beginning to be used as in-line compounders for some large-scale film operations [21].

6.3.1 Pellet premixing

Pellet mixers fit into two general types: off-line batch mixers and in-line feeders/mixers. Batch mixers can be as simple as a cement or tumbler mixer. The ingredients are weighed and poured into the mixer, blended and then transported to the extruder hopper. Batch mixers generally are less expensive and occupy less space than in-line mixers. They can be used to feed more than one extruder. Also, the ingredients' weight can be accurately measured, depending on the scale being used. On the other hand, they can be labor intensive, leave no automatic records and open up the chance for human weighing errors and the possibility of pellet segregation during transport. Human weighing errors and record-keeping can be dealt with by using automatic weighing systems, as described below.

In-line mixers generally are positioned above the extruder hopper. Each ingredient is fed by individual feeders. Fig. 6.1 shows a typical pellet blender. The ingredient feeders meter out a specific volume or mass over time, either by the constant speed of an auger (volumetric feeders) or by an auger whose speed is controlled by the feed hopper loss in weight (gravimetric feeder). Volumetric feeders are less expensive but need to be manually calibrated for each ingredient since differences in density, bulk density and compressibility affect the feed rate. Gravimetric feeders have become the most commonly used feeders. Their calibration is usually much simpler than volumetric feeders. For example, the feeder may measure the weight loss in 30 seconds of operation at 10% of the maximum feeder speed to compute a feed factor that takes into account the variation due to bulk density, density, etc. The whole calibration process is usually done electronically and requires little operator interaction. Since the feed hopper weight is monitored by the computer control system, gravimetric feeders allow the individual feed rates to be recorded. Alarms can be set to ensure each ingredient is being fed to the extruder. Care must be taken when the feed hopper is refilled. Various control schemes are used to ensure feed continuity as the weigh cell is recalibrated, such as temporarily going into volumetric mode. Understanding how the gravimetric feeder handles refilling is helpful in ensuring a trouble-free operation.

In-line mixers generally take up more space and are more expensive than batch mixers. They allow, however, for changing ingredient proportions during processing and can record the actual ingredient weights entering the extruder, which can be important for process control. They are also generally less labor intensive.

The ingredients in premixed pellet blends may segregate if they differ in density, size or shape and are transported over a long distance. Keeping the transport distance as short as possible and avoiding mixing powders with pellets can minimize segregation.

Using properly sized feeders and augers is important to ensure ingredients are controlled to the correct proportions. An oversized feeder will likely have less precision and accuracy. There are also practical limits to pellet blending. A 1% pellet blend corresponds to blending one pellet in every 100. When possible, it is best to strive for ingredients of 5% or higher for greater control and accuracy.

Care must be taken to clean thoroughly the feeders, mixers and transport lines when changing over products. One wrong pellet in a product can cause film imperfections and other quality issues.

The film producer incurs extra risk when blending since the blend properties cannot be directly measured or controlled, especially if the blend is a thin layer in a multilayer film. Pre-made (melt) blends from a resin manufacture or compounder, however, can be tested for properties to ensure they meet specifications. The error associated with pellet

FIGURE 6.1 Pellet blender. *Courtesy of AEC.*

mixing is directly related to the precision and accuracy of the pellet premixing system, how well it is maintained; and procedures that have been adopted to ensure the ingredients never run dry. Gravimetric pellet mixers typically have precisions ranging from about 0.1% to 1%, depending on hopper, feeder and auger size.

6.3.2 Melt blending

Once the ingredients have been fed in the correct proportions into the hopper of the mixing device, typically an extruder, the polymers and/or additives must be homogeneously blended together. This requires that the polymers be in the molten state. The mixing device melts the polymers, provides a means for mixing, and generates pressure for subsequent operations such as making film (when mixing in-line) or pelletizing (when compounding in a separate step). In some operations where pressure generation requirements are high or flow and pressure fluctuations need to be minimized, a separate pressure-generating device such as a gear pump or single-screw extruder is inserted in-line. Separating pressure generation from mixing also helps to control the temperature when processing thermally sensitive materials.

Mixing is described as either distributive or dispersive and is illustrated in Fig. 6.2. In distributive mixing, the polymer is rearranged by deformation. Separation and rearrangement of flow and kneading are two examples of distributive mixing. In dispersive mixing, particles are broken up and dispersed within the polymer matrix. Shear stress is important for overcoming the yield stress of the material. Dispersing pigment particles in a masterbatch is an example of dispersive mixing.

The single-screw extruder is the most commonly used device for film production. It efficiently melts the polymer and generates pressure for extruding the polymer through a flat or annular die. It is suitable for many blending

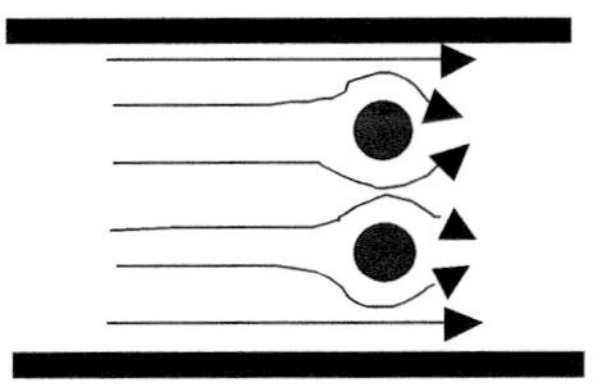

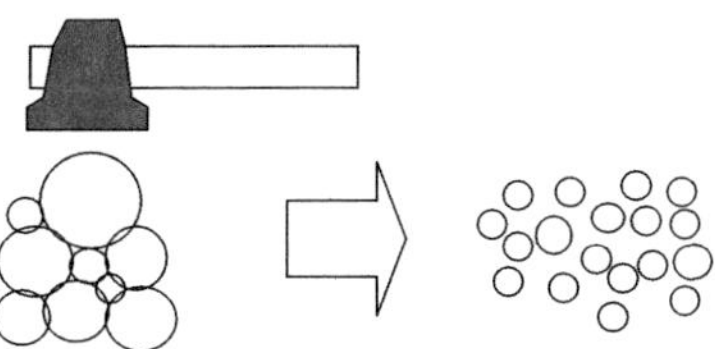

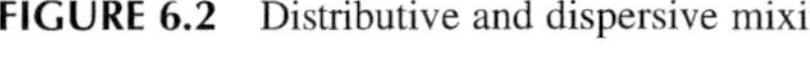

FIGURE 6.2 Distributive and dispersive mixing.

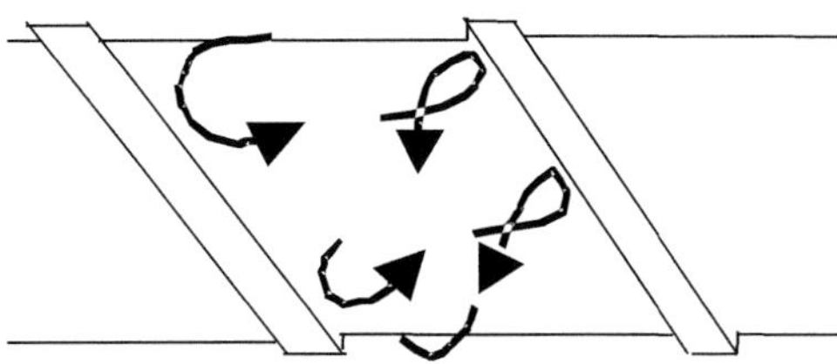

FIGURE 6.3 Illustration of circulatory flow patterns in the metering section of a single-screw extruder [27]. *Adapted from S. Middleman, Fundamentals of Polymer Processing, McGraw-Hill Book Company, New York, 1977, pp. 154.*

applications but has its limitations, many of which can be overcome by optimizing the screw design. Single-screw extruders rely on the difference in friction between the solid polymer pellets and the barrel and screw surfaces to propel the pellets forward. Slippery ingredients may impede the pellet flow and cause surging and other unwanted effects. Single-screw extruders are typically flood fed, meaning the throughput is determined by the extruder screw speed, not how fast the ingredients are fed to the feed hopper. (In flood feeding, the feed hopper is kept partially filled with a reservoir of ingredients that fill (or "flood") the screw channel below by gravity as the extruder screw is rotated.) In more sophisticated compounding devices, such as twin-screw extruders, the throughput is decoupled from the screw speed. The output in these devices is determined by the feeder rate and the screw speed can be increased or decreased to change the mixing intensity and energy input. Single-screw extruders are also generally not well equipped to handle powder (due to potential problems with segregation in the feed hopper and nonuniform melting) and liquid feeds.

The melting mechanism for pellet blends with different melt points in a single-screw extruder is not well understood. The standard melting model for uniformly melting pellets envisions a solid bed compressed against the trailing edge of the screw channel accompanied by an ever-lengthening melt pool that results from the frictional heat near the barrel surface [22–24] (see Fig. 6.23). Disruptions in the solid bed can result in poor melt quality; unmelted particles may exit the extruder, plugging screen packs or appearing in the final film as "gel" particles [25,26]. A disruption in the solid bed can cause surging. Phase inversions and other aspects of blending of polymers that have different melt temperatures may cause disruptions in the solid bed and are not well documented in the literature. The low melt point polymer may also act as a lubricant, impeding the melting of the high melt point component since viscous heat generation from the friction between the polymer and barrel wall is one of the primary avenues for polymer melting.

Conversely, Morris and Kirschbaum [28] find that adding 1%–2% of a low melting ethylene copolymer to polylactic acid (PLA) lowers power consumption by up to 30%, which may lead to higher output on extrusion lines limited by power. Baird et al. [29]. use a highly instrumented extruder to show that melting of the additive alters the frictional characteristics in the solids conveying section of the extruder as well as the melting behavior in the melting zone. This results in less viscous energy dissipation and a more energy-efficient process.

In a standard screw, the polymer melt is subject to nonuniform temperature and flow histories. There are circulatory flow patterns in the metering section of the screw (Fig. 6.3). These patterns aid distributive mixing but, since they are nonuniform, they may lead to uneven mixing. The shear stress and flow rates vary across the flow channel (Fig. 6.4). As a consequence, the temperature may vary across the flow channel. Nonuniformity in the melt temperature at the extruder exit can cause film thickness variability. Mixing elements incorporated into the screw design or at the extruder exit (such as static mixers) help homogenize the melt temperature for proper control of film thickness. These same mixing devices also help blend polymers. Mixing elements that have been utilized over the years include pins, restriction rings and more specialized designs such as the pineapple, Dulmage, Saxton and Maddock mixers [30]. Mixing and temperature nonuniformities increase as the extruder diameter increases; as a result, screw design becomes even more important as scale increases.

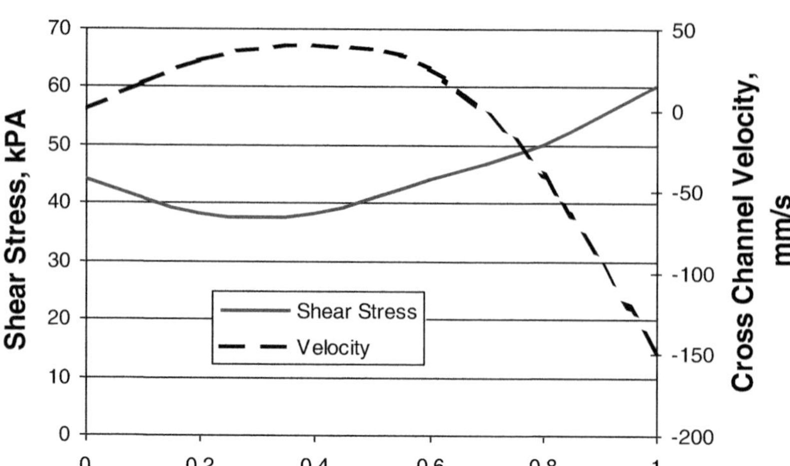

FIGURE 6.4 Example of stress and velocity distribution of polyethylene in the cross channel direction of the metering section of a single-screw extruder. Calculations were done using commercial software from Polydynamics, Inc.

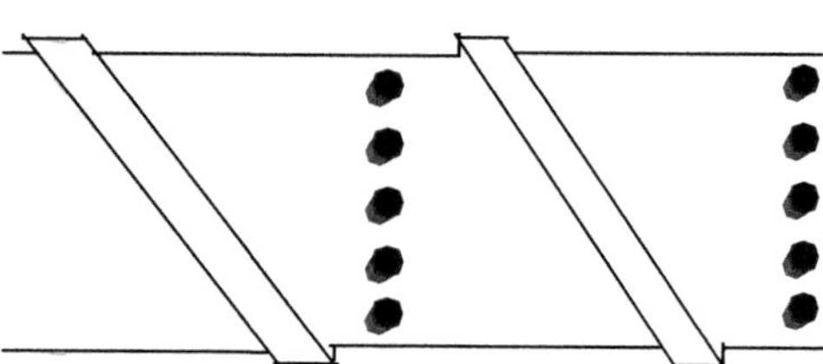

FIGURE 6.5 Pin mixing section.

Some mixing elements are characterized as distributive mixing elements—they achieve mixing by disrupting the flow. One of the easiest ways to do this is by inserting pins (Fig. 6.5), or other elements that impede and redistribute the flow. Pins can cause problems, however, as stagnant flow zones are created behind the pins, setting up the potential for gel formation with some thermally sensitive polymers. More sophisticated designs, such as the Saxton mixer [31] (Fig. 6.6), eliminate the dead zones while dividing up the flow. Other elements are dispersive in nature, such as the well-known Maddock mixer (Fig. 6.7). In this design, the polymer flows down a fluted section and over a small clearance between the screw and barrel, which introduces high shear stresses to break up agglomerates. Its actual effectiveness at dispersive mixing is suspect, particularly compared to specialized compounding devices, such as twin-screw extruders, but it is often used for multipurpose mixing.

In recent years, several new high-performance screw designs have been developed that try to improve the melting and mixing efficiency while increasing output. These screws generally work by having the polymer flow through a series of relatively tight clearances. Any remaining solid material is both entrapped and given time to melt or subjected to high shear or elongational flow fields that aid in melting and mixing [30,32]. The high-performance section replaces the whole metering section of the screw. Some of the more commonly used high-performance screws include the energy transfer [33], variable barrier energy transfer [34,35], double wave [36,37], stratablend [38], unimix [39] and CRD [40] screws. An example of a variable barrier energy transfer screw element is shown in Fig. 6.8.

Although optimizing the screw design allows many polymer blends to be made on single-screw extruders, in some cases, specialty compounding devices may be necessary. This is especially true when intensive dispersive mixing, reaction or devolatilization is required. There are several compounder designs on the market. For general purpose compounding, co-rotating twin-screw extruders are often employed [41,42]. Kneaders provide good distributive and dispersive mixing through an interaction of oscillating action of the rotation of their single screw and pins along the barrel. They easily allow the introduction of liquid feeds and minimize temperature buildup, which is important for thermally sensitive polymers. They are generally only used in specialty compounding because their output is not high compared to other equipment and they must be coupled with a pressure-generating device such as a gear pump or single-screw extruder to smooth out the oscillations. Tangential (nonintermeshing) counter-rotating twin-screw

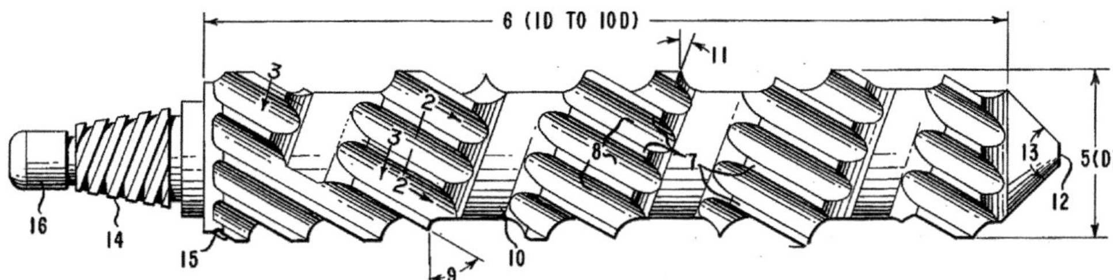

FIGURE 6.6 Saxton mixing section. *Reproduced from R. Saxton, Extruder mixer screw, US patent 30060, 1961.*

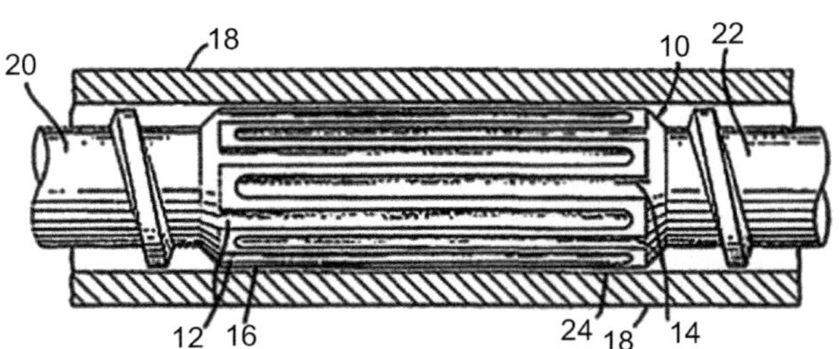

FIGURE 6.7 Maddock type mixing element. *Reproduced from G. Le Roy, US patent 3486192A, 1969.*

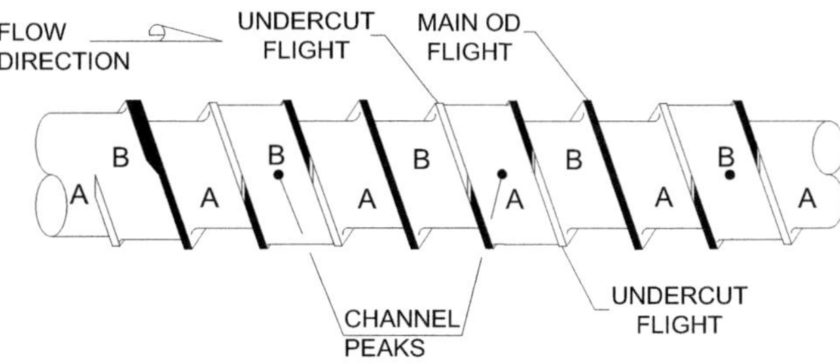

FIGURE 6.8 Illustration of a variable barrier energy transfer screw. *Courtesy of Jeff Myers of BARR, Inc.*

extruders are often used for reactive extrusion and devolatilization since they can have long L/Ds (length to diameter ratios). Conical or parallel intermeshing counter-rotating twin screws are often used in PVC compounding and have good pumping characteristics. The Farrel continuous mixer operates like a continuous Banbury mixer and is useful for making masterbatches with high filler loadings. Planetary mixers and other multiscrew devices are also available for specialty applications.

Of the specialty compounding devices, the co-rotating twin-screw extruder is the most often used. Co-rotating twin-screw extruders offer considerable versatility in design and operation. As described earlier, feeding is independent of extruder speed, which allows the mixing and energy intensity to be independently varied. Multiple feeding ports are common and screw designs can be changed from job to job to tailor the process. The screws are built from modular elements, which can be changed to suit the application. These elements come in several geometries with different mixing intensities. They provide good distributive and dispersive mixing and impart both shear and elongational flow fields. Fig. 6.9 shows some twin-screw modular elements. Some twin-screw extruders have screws that are self-wiping, which results in a narrow residence time distribution for better mixing control. One drawback is that they are poor melt pumps. Because of this, temperatures can be difficult to control at the extruder exit. Twin-screw extruders are not used alone for in-line compounding on a film line. They are adapted for better melt pumping by coupling the twin-screw extruder with a single-screw extruder or gear pump [21].

Twin-screw extruders and other specialty compounding devices are considerably more expensive than single-screw extruders.

6.4 Physics of blending

The properties of a polymer blend are influenced by specific interactions between the molecules (thermodynamics) and their response to deformation (rheology). The thermodynamics determine whether the blend forms a single phase

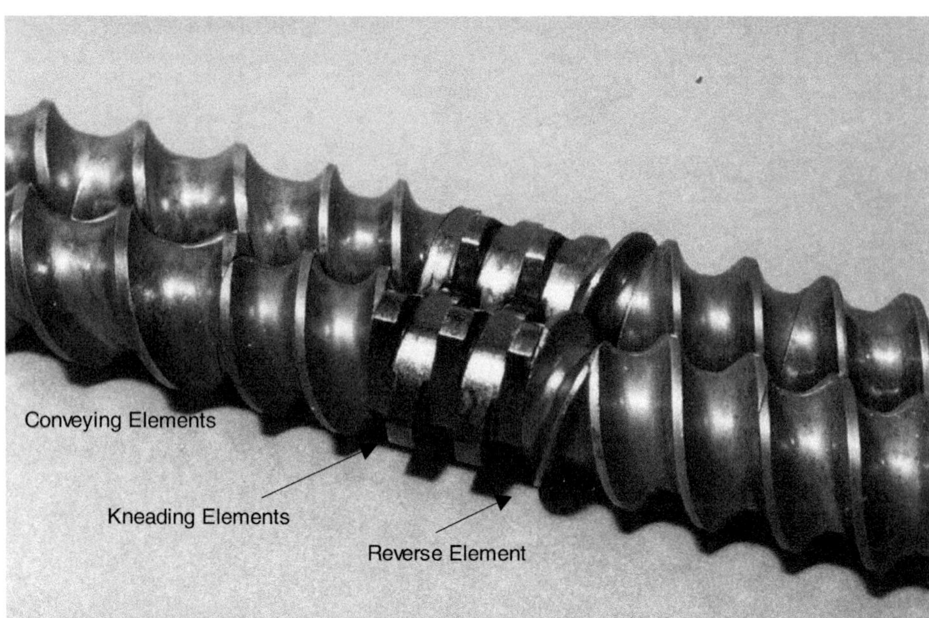

FIGURE 6.9 Twin-screw extruder modular elements.

(miscible) or multiple phases (immiscible). Miscible blends typically follow the rule of mixtures, namely the blend properties are directly proportional to the component ratio. Blends of different grades of the same polymer, such as two grades of LDPE, are typically miscible. An example is an additive masterbatch whose carrier resin is of the same type as the polymer it is being blended with. Indeed, it is generally important to select a masterbatch carrier resin that is miscible or at least highly compatible with the blending polymer to avoid significant changes in properties.

Most polymer blends of different polymers, however, are not miscible. There are only a few commercially important miscible blends on the market today and most are not used in flexible packaging. For example, SABIC's Noryl, which is a blend of polyphenylene oxide (PPO) and high impact polystyrene (HIPS), is an engineering polymer.

How does one determine whether a blend is miscible? One method is to look at transparency, either by microscopy or light scattering. An immiscible blend forms separate domains within the polymer matrix that may diffract or scatter light if they are large enough. Another technique is to measure the glass transition temperature (Tg) using differential scanning calorimetry (DSC) or other thermal analyses. A miscible blend will have a single Tg, typically between that of the components. Both techniques use a fairly large sample or probe size, which can at times be misleading. Sometimes blends appear to be miscible because the probe size is too large. Other techniques with smaller probe sizes are X-ray and neutron scattering and various spectroscopy techniques, such as infrared (IR) and nuclear magnetic resonance (NMR).

Occasionally, the word "compatibility" is used to describe the degree to which polymers interact. Miscibility can be thought of as the state of maximum compatibility. Compatibility is a subjective term and is not well defined. Miscibility has a specific definition—two polymers are miscible if they form a single phase over their entire composition range at a given temperature. In some contexts, two polymers may be miscible over a specified composition range. Furthermore, the molecular weight of the polymers may change their miscibility.

Immiscible blends, by definition, form multiple phases. In the simplest case, a two-component blend, the minor phase forms domains within a continuous major component matrix. The domain sizes, shapes and distribution are known as the morphology. Fig. 6.10 shows some immiscible blend morphologies. The morphology is influenced by the concentration, thermodynamics, component rheology and the flow and stress history during mixing and processing.

The morphology is critically important to the final blend properties, which often do not follow the rule of mixtures (Fig. 6.11). In many cases, blends are designed to create a specific morphology to achieve certain property gains. An example is super-tough PA or nylon, which is a PA66 (or PA 6)-rubber blend [9]. The rubber must achieve a certain domain size (typically between 0.1 and 1 micron) in order to stop cracks from propagating during impact [45–48]. Another example is Selar RB laminar technology developed by DuPont. Here PA is dispersed in HDPE so that the PA phase forms platelets parallel to the surface. The platelets create a tortuous path for diffusion of small molecules, resulting in improved barrier performance [12].

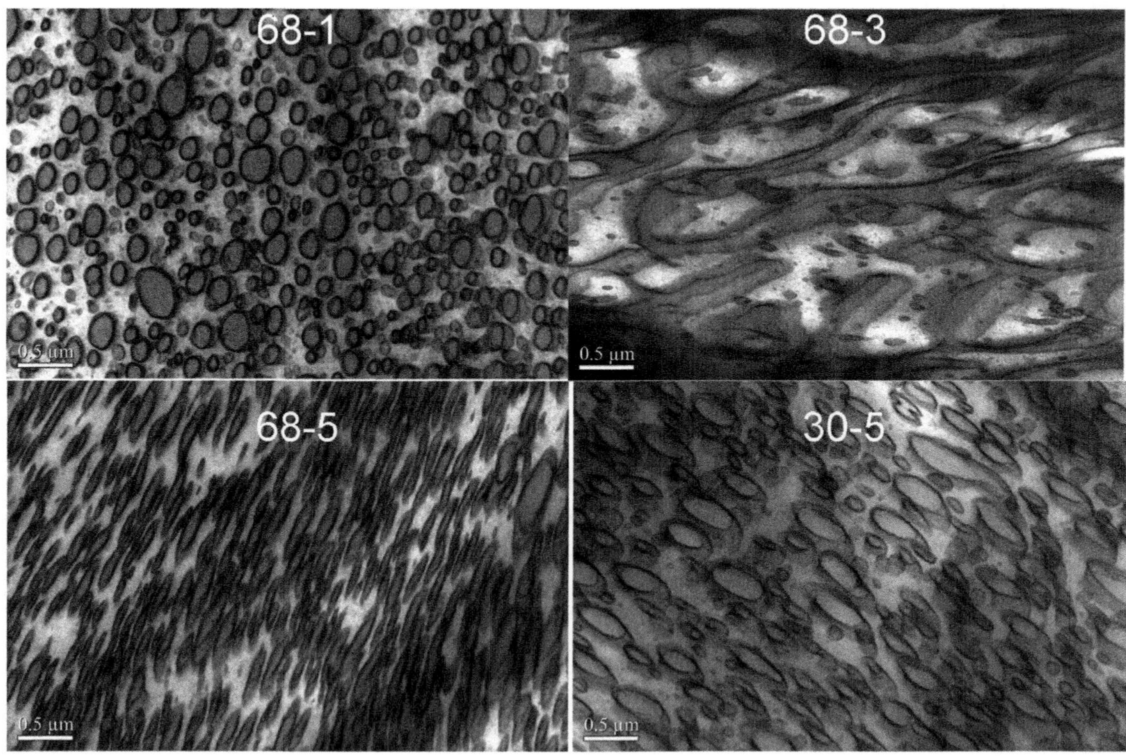

FIGURE 6.10 Transmission electron micrographs of immiscible polyethylene–styrene polymer blends of varying chemistries [43,44]. *Courtesy of Barbara Wood and I-Hwa Lee, DuPont. Further examples of TEM's of immiscible blends can be found in B.A. Wood, Transmission electron microscopy of detection of rubber modifiers, in: 161st Technical Meeting of the Rubber Division of the American Chemical Society, Savannah, GA, April 29-May 1, 2002, Paper 45. B.A; Wood, Transmission electron microscopy of polymer blends, in: K. Finlayson (Ed.), Advances in Polymer Blends and Alloys Technology, Vol 3, Technomic Publishing Co., Inc, Lancaster, PA, 1992.*

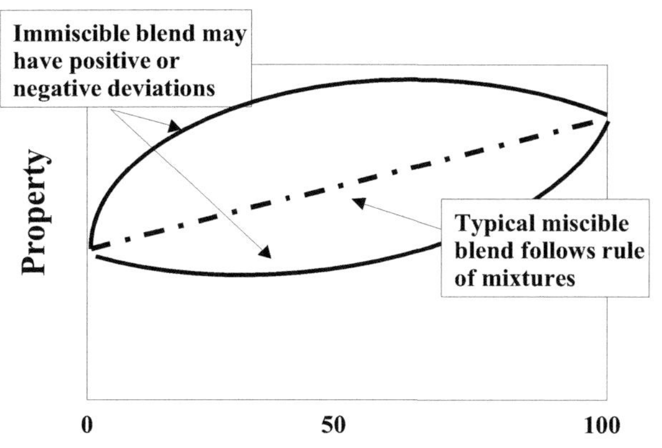

FIGURE 6.11 Properties versus percent of polymer A in polymer blends.

Once a specific morphology has formed, it may change with further processing. The domains may coalesce or be stretched through orientation (see Fig. 6.10). Compatibilizers are frequently used to stabilize the blend morphology. Compatibilizer technology and its role in morphology development and stabilization is described in Section 6.4.2.

There is a third type of blend system, namely melt-miscible blends that are miscible in the melt state but phase separate in the solid state. An example is polyoxymethylene/polylactide (POM/PLA) [49]. From an ease of processing perspective, this phenomenon may be useful for obtaining a fine dispersion of one component into another during melt blending.

6.4.1 Thermodynamics

When two materials are mixed together, there must be a decrease in the free energy for them to transform into a single material or miscible blend. This can be expressed mathematically as:

$$\Delta G_m < 0 \text{ for miscibility} \tag{6.1}$$

where ΔG_m is the Gibbs free energy of mixing. The free energy includes enthalpy and entropic contributions:

$$\Delta G_m = \Delta H_m - T\Delta S_m \tag{6.2}$$

where ΔH_m is the enthalpy of mixing, T is the temperature, and ΔS_m is the entropy of mixing.

For most polymer blends, ΔH_m is positive and ΔS_m is nearly zero. Thus, it is rare that polymer blends are miscible. Coleman et al. [50]. derive the following relationship for ΔG_m based on the work of Flory and Huggins:

$$\frac{\Delta G_m}{RT} = \left[\frac{\Phi_A}{N_A} ln\Phi_A + \frac{\Phi_B}{N_B} \mathrm{In}\Phi_B\right] + \chi\Phi_A\Phi_B + \left(\frac{\Delta G_H}{RT}\right) \tag{6.3}$$

where ΔG_m is the Gibbs free energy of mixing, R is the ideal gas constant, T is the temperature (K), Φ_A is the volume fraction of polymer A, Φ_B is the volume fraction of polymer B, N_A is the degree of polymerization of polymer A, N_B is the degree of polymerization of polymer B, χ is the interaction parameter, and ΔG_H is the free energy of specific interaction between polymers, including hydrogen bonding.

The first term on the right-hand side of Eq. (6.3) arises from combinatory entropy. Since N, which is related to molecular weight, is large for polymers, this term is nearly zero. The second term, $\chi\Phi_A\Phi_B$, also arises from entropy and is always positive. The interaction parameter, χ, is defined by

$$\chi = \frac{V_{ref}}{RT}[\delta_A - \delta_B]^2 \tag{6.4}$$

where V_{ref} is the reference volume, δ_A is the solubility parameter for polymer A, and δ_B is the solubility parameter for polymer B.

The final term, $\Delta G_H/RT$, is negative when specific interactions are present.

Eqs. (6.3) and (6.4) show that compatibility is improved by matching solubility parameters and miscibility is achieved only when specific interactions are present. Nonpolar polymer blends, such as PP–PE blends, have no specific interactions beyond weak dispersive forces. It will be shown that even though their solubility parameters are nearly equal, they are not miscible.

Eq. (6.4) introduces the solubility parameter. The solubility parameter is a measure of molecular attractiveness and arises from the energy per unit volume required to remove a molecule from a liquid or solid. This is known as the cohesive energy density and is illustrated in Fig. 6.12. The cohesive energy density is a function of the forces that hold the material together. The solubility parameter is the square root of the cohesive energy density. It contains contributions from both nonpolar (dispersive forces) and polar (dipole–dipole and hydrogen bonding) interactions.

Comparing solubility parameters is a way to quantify the "like dissolves like" principle of chemistry. For two polymers to be miscible, the difference in solubility parameters should be between 0.1 and 3 Hildebrands, depending on their interaction strength. Coleman et al. [50] define a critical solubility parameter difference, $\Delta\delta_c$, below which the polymers may be miscible. As shown in Table 6.1, the value of $\Delta\delta_c$ depends on what interactive forces are present.

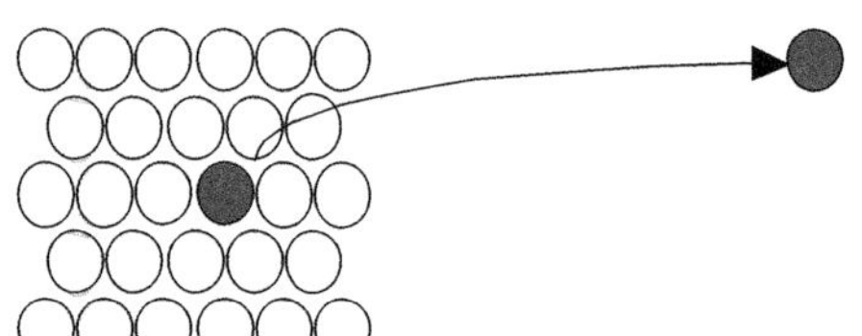

FIGURE 6.12 Illustration of cohesive energy density.

TABLE 6.1 Critical solubility parameter difference upper limit.

Specific interactions involved	Polymer blend examples	$\Delta\delta_{critical}$ $(cal/cm^3)^{1/2}$
Dispersive forces only	Polybutadiene–polyethylene (PBD-PE)	< 0.1
Dipole–dipole	Polymethylmethacrylate–polyethylene oxide (PMMA-PEO)	0.5
Weak	Polyvinyl chloride–butadiene acrylonitrile copolymer (PVC-BAN)	1.0
Weak to moderate	Polystyrene acrylonitrile–polymethylmethacrylate (SAN-PMMA)	1.5
Moderate	Polycarbonate-polyesters	2.0
Moderate to strong	Nylon-polyethylene oxide (Nylon-PEO)	2.5
Strong	Poly vinyl phenol–poly vinyl acetate (PVPh-PVAc)	3.0
Very strong	Polymethacrylic acid-polyethylene oxide (PMMA-PEO)	> 3.0

Source: From M. Coleman, J. Graf, P. Painter, Specific Interactions and the Miscibility Of Polymer Blends, Techonomic Publishing, Lancaster, PA, 1991.

TABLE 6.2 Solubility parameters for some polymers used in packaging.

Polymer	Solubility parameter $cal^{1/2}/cm^{3/2}$ (Hildebrands)	Source
PE (polyethylene)	7.7–8.4	Van Krevelen [51]
PP (polypropylene)	8.2–9.2	Van Krevelen
PS (polystyrene)	8.5–9.3	Van Krevelen
PVC (poly vinyl chloride)	9.4–10.8	Van Krevelen
PVDC (poly vinylidene chloride)	9.9–12.2	Van Krevelen
PVOH (polyvinyl alcohol)	12.6–14.3	Van Krevelen
EVOH (44mol% ethylene) (ethylene vinyl alcohol)	17	Kuraray literature [52]
EVOH (32mol% ethylene)	19	Kuraay literature
Polyamide 6 (Nylon 6)	12.6	Kuraray literature
EVA(9%VA) (ethylene vinyl acetate)	8.1	DuPont Calculation [53]
EVA (25%VA)	8.2	DuPont Calculation
EMA (20%MA) (ethylene methyl acrylate)	8.3	DuPont Calculation
PET (polyethylene terephthalate)	10.3	Wu [54]

Table 6.2 lists solubility parameter values for some polymers used in packaging films. Returning to the PE–PP blend example, the values from the table show that the difference in their solubility parameters is less than 1 ($\Delta\delta < 1$ Hildebrands). However, since only nonpolar dispersive forces are present, the critical solubility parameter difference ($\Delta\delta_c$) is less than 0.1. Thus, these polymers are not miscible.

It should be noted that there is some controversy around solubility parameters in the literature. They are exact for polymers with only physical interactions. There are errors associated with trying to extend the concept to polymer systems involving hydrogen bonding and other polar interactions and with the indirect methods used experimentally to measure them. These errors typically result in a broader range of values than is useful for predicting miscibility, hence matching solubility parameters is not an absolute condition for the miscibility of polymers. They are useful guides for compatibility, however, and since most polymer blends are not miscible, this may be their greatest strength. The solubility parameter difference has been related to the interphase thickness between immiscible polymers [see Eq. (10.21)].

Immiscible polymers with a solubility parameter difference of about 0.5 Hildebrands or less may still build up enough strength at the interface for good mechanical properties [55].

Returning to the PP–PE blend example, PE is often used to modify the properties of PP. For example, ethylene–propylene rubber (EPR) is blended with PP to improve PP impact toughness. The EPR forms a separate phase with sufficient adhesion to the PP that the properties are enhanced, as would be predicted by their close solubility parameters.

LDPE–LLDPE blends are perhaps the most commercially important blends used in flexible packaging applications. A distinguishing difference between LDPE and LLDPE is long-chain branching (LCB) in LDPE, which contributes to its relative processing ease. Typically, about 10%–30% LDPE is added to LLDPE to improve melt strength and bubble stability in blown film production. Blends with greater than 60% LDPE are used for improved optics and shrink properties for collation shrink films and masking films. Hussein et al. [56–59]. find that LDPE and LLDPE are not miscible over their entire composition range and describe a number of nuances:

- Miscibility is favored for LLDPE-rich blends.
- The miscibility range increases by lowering the LLDPE molecular weight (MW) and by increasing the short-chain branch length (replacing the butene comonomer with octene).
- At the same molecular weight and short chain branch content, a Zeigler–Natta LLDPE is more miscible with LDPE than metallocene LLDPE (m-LLDPE), which has a narrow molecular weight and comonomer distribution.
- At the same molecular weight distribution (MWD) and molecular weight (MW), an m-LLDPE with higher branch content is more miscible with LDPE than an m-LLDPE with lower branch content.
- Comonomer type does not have an effect on miscibility of m-LLDPE with LDPE.

Patel et al. [4] review the LLDPE-LDPE blend literature and conduct experiments to better understand the mechanisms behind the change in properties. Fig. 6.13 shows the substantial deviation from the rule of mixtures for melt strength (measured by the Rheotens method described in Section 5.3.6) and Elmendorf tear measured in the machine direction. The max/min in these properties at around 60% LDPE had not been well studied prior to their work. By comparing properties of blown film (with some orientation) with compression-molded films (no or little orientation) they are able to isolate the effects of orientation:

- The positive deviation from the rule of mixtures at 60%–80% LDPE is explained by invoking the model of Bersted et al. [60,61] which shows that low shear viscosity increases with increasing LCB (higher entanglements), but reaches a maximum and decreases at high LCB (pure LDPE) due to molecular coil size reduction. Adding a minor amount of LLDPE expands the coil size, leading to an increase in melt strength.

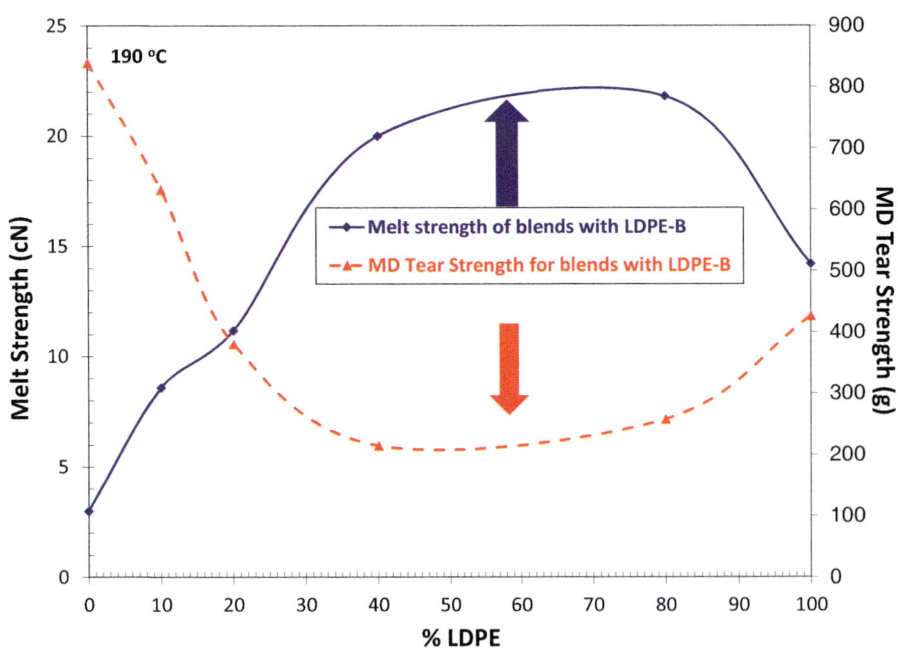

FIGURE 6.13 Tear strength and melt strrength properties of LDPE–LLDPE blends. LLDPE is an octene LLDPE resin made with a Ziegler-Natta catalyst, 0.920 g/cc density and 0.95 g/10 min MI (190 C/2160 g). LDPE-B has a 0.923 g/cc density and 0.64 g/10 min MI and is made on a tubular reactor. *Reproduced with permission from R. Patel, T. Karjala, N. Savargaonkar, P. Salibi, L. Liu, J. Plastic Film and Sheeting, 35(4) (2019) 401–421.*

- Tear strength of the compression-molded films decreases linearly with increasing % LDPE in the blend, following the rule of mixtures. This is attributed to a reduction in tie chain concentration. Tie chains link the crystalline domains within the polyethylene and help build strength. Machine direction (MD) tear strength of blown film shows a significant negative deviation from the rule of mixtures as shown in Fig. 6.13. The deviation is almost a mirror image of the melt strength behavior with a minimum at around 40% LDPE. Conversely, cross direction (CD) tear strength exhibits a positive deviation from the rule of mixtures with a maximum of around 20%–40% LDPE. The authors conclude that orientation is responsible for the deviation from the rule of mixtures in the tear response in blown film. Small x-ray scattering results correlate with the tear behavior. This is also consistent with previous investigations showing blow-up ratio and draw-down ratio influence the properties of these blends [62].
- Dart impact and puncture resistance of the blown and compression-molded films mostly follow the rule of mixtures. Both properties decrease with increasing LDPE content, which is attributed to the reduction in tie chain concentration. There is an increase in the puncture resistance at about 10% LDPE which is believed to be due to higher gloss (less friction with the impact probe).

One can add specific interactions to promote miscibility. Coleman et al. [50]. have developed software that uses group contribution theory to predict solubility parameters of polymers that, in many cases, agrees well with experimental results. This is illustrated in Fig. 6.14. Here, vinyl acetate (VA) and styrene (St) contents are varied in an ethylene vinyl acetate copolymer–styrene vinyl phenol (EVA-StVPh) copolymer blend. Increasing the VA and the vinyl phenol (VPh) content promotes polar interactions (ΔG_H) that drive ΔG_m below zero [Eq. (6.3)]. The experimental data are represented by the open and closed circles: the open circles signify miscible blends and the closed circles immiscible blends. The line demarking the region of miscibility comes from Eq. (6.3), using the software to predict the solubility parameters with specific interactions. The miscibility region depends on the comonomer percent in each polymer.

A practical note about compatibility and miscibility concerns masterbatches. The carrier resin in the masterbatch should be compatible and, ideally miscible, with the resin in which it is being blended. This way, the carrier resin will not change the letdown resin properties. Poor compatibility can lead to poor optical properties, reduced barrier performance and, for sealants, poor seal performance. Benkreira and Britton [63] find that better dispersion occurs when the carrier resin is lower in viscosity and has a lower melting point than the host polymer, for reasons that will become clearer in the next section.

Solubility parameters may also prove useful for predicting miscibility of low molecular weight additives in polymers.

6.4.2 Morphology development in immiscible blends

To understand how morphology develops in a polymer blend, the breakup of a single droplet suspended in a fluid during flow is first considered. The droplet is held together by interfacial tension, which arises from a nonuniform force distribution acting on the molecules at the interface (Fig. 6.15). Inside a material, a molecule is bound to its neighbors by attractive forces related to the cohesive energy density. At an interface, however, molecules are only partly surrounded by their own kind. The material across the interface may exert different attractive forces. The difference in these attractive forces gives rise to interfacial tension. The more alike the materials are, the lower the interfacial tension and the smaller the driving force holding the droplet together. If the droplet has a radius a, then this holding force, $F_{interfacial}$, is proportional to Γa, where Γ is the interfacial tension.

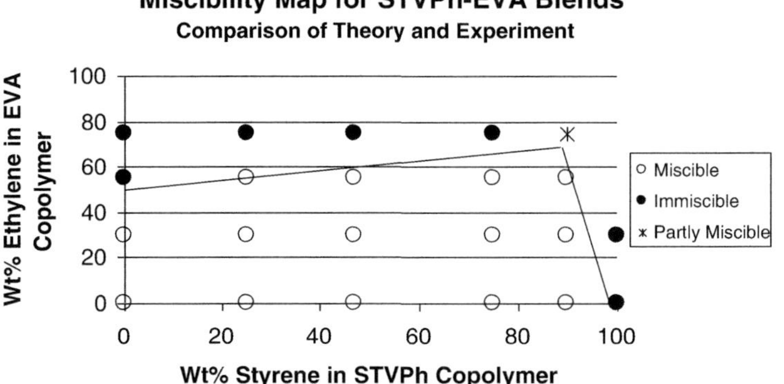

FIGURE 6.14 Example of using specific interactions between molecules to achieve miscible blends. *Adapted from M. Coleman, J. Graf, P. Painter, Specific Interactions and the Miscibility of Polymer Blends, Techonomic Publishing, Lancaster, PA, 1991.*

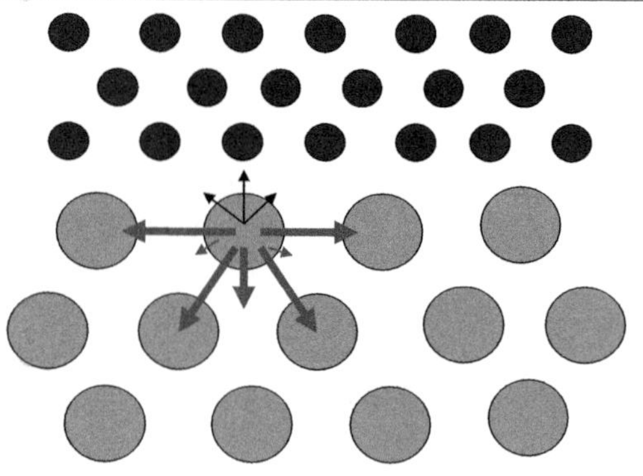

FIGURE 6.15 Illustration of the origins of interfacial tension.

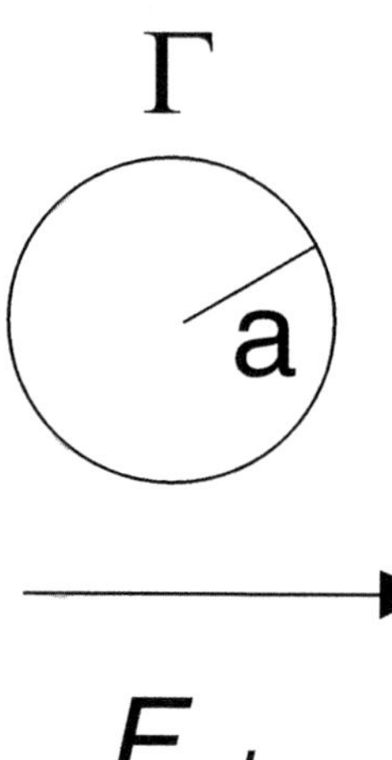

FIGURE 6.16 Balance of drag and interfacial forces on a spherical droplet in shear flow.

The interfacial tension can be related to solubility parameters introduced earlier. Wu [54] shows that the interfacial tension is related to the surface tensions of the two polymers:

$$\Gamma_{12} = \Gamma_1 + \Gamma_2 - 2\phi\sqrt{\Gamma_1\Gamma_2} \tag{6.5}$$

where Γ_{12} is the interfacial tension between polymers 1 and 2, Γ_1 is the surface tension of polymer 1, Γ_2 is the surface tension of polymer 2, and φ is the interaction parameter. This is known as the Good and Girifalco equation. Wu tabulates the interaction parameter, φ, for several polymer combinations. They range from 0.79 to 0.98 for the polymers considered. The surface tension can be related to the solubility parameter:

$$\Gamma_1 = 0.2575\frac{\delta_1^2}{\sqrt[3]{\rho_1}} \tag{6.6}$$

where Γ_1 is the surface tension (dynes/cm), ρ_1 is the density of polymer (g/mL), and δ_1 is the polymer 1 solubility parameter $(\text{cal/mL})^{1/2}$.

Combining Eqs. (6.5) and (6.6), recognizing polymer densities are around 1 g/mL and assuming $\varphi = 1$, the interfacial tension is found to be directly proportional to the square of the difference in solubility parameters:

$$\Gamma_{12} \cong 0.26(\delta_1 - \delta_2)^2. \tag{6.7}$$

Note in this analysis the effect of temperature is ignored.

As shown in Fig. 6.16, the flow field exerts a drag force that acts to break up the droplet. For shear flow, the drag force is equal to the viscosity of the continuous phase, η_c, times the shear rate, $\dot{\gamma}$, acting over the area of the particle ($\approx a^2$):

$$F_{\text{drag}} \sim \eta_c \dot{\gamma} a^2.$$

The dimensionless capillary number, Ca, is defined as the ratio of the drag force to the interfacial force:

$$\text{Ca} = \frac{F_{\text{drag}}}{F_{\text{interfacial}}} = \frac{\eta_c \dot{\gamma} D_d}{2\Gamma_{12}} \tag{6.8}$$

whereD_d is the droplet diameter ($=2a$).

When Ca exceeds a critical value, $\text{Ca}_{\text{critical}}$, the droplet breaks up because the drag force exceeds the force holding the droplet together. This hydrodynamic instability was first proposed by G. I. Taylor [64]. Taylor found that $\text{Ca}_{\text{critical}}$ for a Newtonian droplet imbedded in a Newtonian fluid is a function of the viscosity ratio (the ratio of the droplet viscosity to the continuous phase viscosity): η_d/η_c. This has since been confirmed by Grace [65] and others and is illustrated in Fig. 6.17. In a shear flow, $\text{Ca}_{\text{critical}}$ reaches a minimum when the viscosity ratio is 1. When the viscosity ratio exceeds about 3.5, the droplet cannot be broken up in a shear flow, as indicated by the rapid rise in $\text{Ca}_{\text{critical}}$. The bottom curve in Fig. 6.17 shows the relationship for $\text{Ca}_{\text{critical}}$ in elongational flow. Elongational flow is more effective for breaking up droplets than shear flow; $\text{Ca}_{\text{critical}}$ is lower and droplets can be broken up even at high viscosity ratios.

For transient flow, a different mechanism for drop breakup has been proposed by Tomotika [66], based on Rayleigh's instability theory. Here, the droplet becomes an elongated ellipse or cylinder that, upon cessation of flow, breaks up into small droplets due to capillary disturbances, provided that the wavelength of these disturbances is greater than $2\pi a$ [67,68].

The single droplet analysis provides considerable insight into the dispersion of polymer blends (Fig. 6.18). Typically, the minor component domain size (D_d) of the blend is desired to be small. For example, for good clarity, the dispersed phase should be less than the wavelength of light, about 0.3 μm. For toughening, small soft rubber domains help prevent cracks from propagating. Eq. (6.8) shows that the droplet size can be decreased by increasing the continuous phase viscosity, increasing the shear rate or by decreasing the interfacial tension. The droplet size is also reduced by matching the polymer viscosities (a good guideline is to choose a viscosity ratio between 0.01 and 2) and using a mixing device that imparts elongational flow. The interfacial tension can be minimized by reducing the difference in solubility parameters [Eq. (6.7)] or by introducing specific interactions.

Compatibilizers are sometimes used to reduce the interfacial tension between polymers. Block or random copolymers often make good compatibilizers since they can be designed to contain two functionalities, each compatible with one of the polymers being blended. An example is using styrene–ethylene–butadiene–styrene copolymer (SEBS) to compatibilize high-density polyethylene (HDPE) and polystyrene (PS) blends [69]. Another example is the use of olefin block copolymers to compatibilize PP and PE, for toughening or recycle applications [45,70,71]. A second approach is to add functional groups to the compatibilizer that react with one of the polymers. An example is using an ionomer to compatibilize PA (polyamide or nylon) and PE blends. The acid groups in the ionomer react with the PA amine end

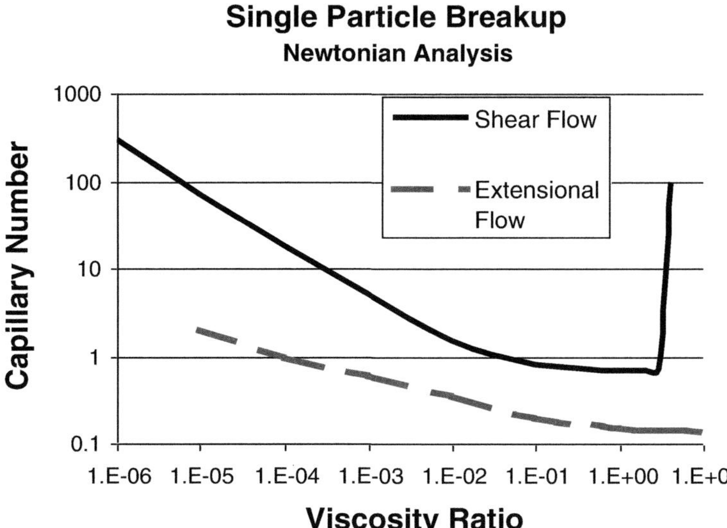

FIGURE 6.17 Critical Capillary Number for the breakup of Newtonian droplet in shear and elongational flow. *Data from H.P. Grace, Dispersion in high viscosity immiscible fluid systems and application of static mixers as dispersion devices in such systems, Chem. Eng. Commun. (14) (1982) 225–277.*

Ways to Reduce Dispersed Phase Particle Size

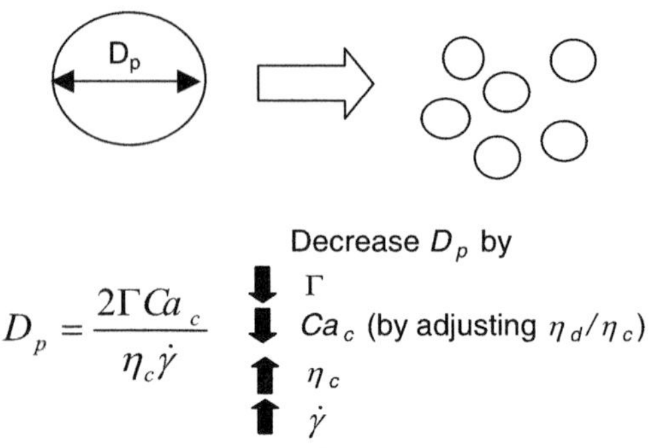

FIGURE 6.18 Learnings from single droplet breakup mechanics.

groups and the ethylene backbone is compatible with PE [72]. Another example is the use of maleated polyolefins to compatiblize PE- or PP-recycle streams containing PA or EVOH [1,73].

While the breakup of a single Newtonian droplet in a Newtonian fluid is well understood, there are several difficulties in extending the analysis to polymer blends. The first is that polymer melts are typically non-Newtonian in their flow behavior: non-Newtonian fluids have viscosities that vary with shear rate and often exhibit elastic effects. Several investigators have studied droplet breakup for non-Newtonian fluids as summarized in Table 6.3. This phenomenon, however, is still not as well understood as the Newtonian case [74–79]. One aspect of non-Newtonian behavior is the polymer melt elasticity, characterized by the first and second normal stress differences. (See Chapter 5 for an introduction to melt elasticity or consult a rheology text such as Dealy and Wang [80] for a broader discussion.) Some investigators have used the dynamic storage modulus, G', obtained from dynamic mechanical analysis (DMA), as a way to characterize elasticity. G' is proportional to the first normal stress difference, but only at low shear rates. At higher shear rates typical of polymer processing, G' may underestimate the polymer elasticity. In general, it has been found that when the droplet is more elastic than the continuous phase, it is more difficult to disperse. An elastic force aids in holding the droplet together, resulting in larger $Ca_{critical}$ values and larger droplet sizes. Conversely, when the continuous phase is more elastic than the droplet, it is easier to break up the droplet because the matrix resin elasticity adds to the drag force on the droplet to break it up. The experimental studies agree mostly with this generalization but with many exceptions; quantification and modeling of this behavior in flow regimes typical of polymer processing is still in its infancy [74].

Experimental studies with non-Newtonian fluids have also revealed different droplet breakup mechanisms. Fortelny and Juza [74] list seven different mechanisms in their review of the experimental studies (see Fig. 6.19). A Newtonian droplet immersed in a Newtonian fluid breaks up via Taylor [64] and Tomotika (Raleigh) [66] instabilities. Under some conditions, non-Newtonian fluids also break up in this manner, but other mechanisms have also been observed. In some cases, the droplet flattens in the flow direction and becomes elongated perpendicular to the flow [87,89]. The ends of highly elongated particles find themselves in different planes with respect to flow and are torn apart by the velocity differences. At very high viscosity ratios, greater than the cut-off of 3.5 for Newtonian droplets, Mighri and Huneault [89] observe breakup via attrition or erosion at the droplet surface. Other mechanisms include elongation of the droplet in the flow direction with pinch-off of the tips into smaller droplets (tip streaming), parallel breakup and the formation of large and satellite droplets. The ability to predict the mechanism that will prevail is limited, as is the ability to predict final droplet size even in the relatively simple case of a single droplet in a complex fluid.

As the droplet concentration increases, the break-up mechanism becomes more complex. Investigators [68,74,110–114] have found that when the polymer concentration exceeds 0.5%–1%, the droplet size after shearing is much greater than that predicted by the single droplet experiments. As the concentration increases, the probability that droplets will collide and coalesce increases. The coalescence kinetics are not well understood and are thought to be critically important to the final blend morphology. Utracki and Shi [68] observe that the same factors that enhance droplet breakup, namely high shear rates and reduced droplet viscosity, favor coalescence. Compatibilizers are shown to suppress coalescence [112,113,115–118]. One proposed mechanism is that compatibilizers act to inhibit coalescence by providing a protective shell around the droplet (Fig. 6.20) [115,116]. The shell acts to repulse other droplets. Thus, dispersion stabilization may be more important to the droplet size than reducing interfacial tension. Indeed, Mighri and Huneault [89] find that an EPR–PP droplet–matrix system with very low interfacial tension breaks up in a similar

TABLE 6.3 with caption: Summary of studies of viscoelastic droplet deformation in a viscoelastic matrix under shear or elongational flow.

Source	Experimental or theoretical	Droplet	Matrix	Flow	Result
Flumerflet [81]	Experimental	Newtonian	Viscoelastic	Shear	Minimum droplet size and critical shear rate for breakup increases with matrix elasticity
Tavgac [82]	Experimental	Viscoelastic	Viscoelastic	Shear	Effect of elastic matrix depends on the viscosity ratio:when ratio is small, matrix elasticity stabilizes the dropletwhen ratio is high elasticity helps break up the droplets.
Gauthier [83]	Experimental	Viscoelastic	Newtonian	Shear	Small viscosity ratio: similar to Newtonian drop in Newtonian matrix.
					High viscosity ratio: $Ca_c > Ca_{c\text{-}Newtonian}$
Parabodh and Stroeve (in Utracki and Shi) [68]	Experimental	Viscoelastic	Newtonian	Shear	For viscosity ratio < 0.5, droplet viscoelasticity has a stabilizing effect. For viscosity ratio > 0.5 droplet viscoelasticity has a destabilizing effect.
DeBruijn [84]	Experimental	Viscoelastic	Newtonian	Shear	$Ca_{critical}$ for elastic droplet is slightly higher than for Newtonian droplet
Elmendorp and Maalcke [85]	Experimental	Viscoelastic	Viscoelastic	Shear	Viscoelastic drop deformation in Newtonian matrix decreases with increasing drop elasticity. Newtonian drop deformation in viscoelastic matrix increases with increasing matrix elasticity.
Varanasi [86]	Experimental	Viscoelastic	Newtonian	Shear	At any viscosity ratio $Ca_{critical}$ increases with increasing droplet elasticity.
Levitt et al. [87]	Experimental	Viscoelastic	Viscoelastic	Shear	For high elastic matrix the droplet widened in the direction perpendicular to flow. The width of the flattened drops depends on the differences in second normal stress differences between the phases.
Mighri et al. [88]	Experimental	Viscoelastic	Viscoelastic	Shear	For high matrix elasticity ($k' < 0.37$), the elastic drop deformation in an elastic matrix is higher than for the Newtonian case with same viscosity ratio and interfacial tension. In some cases, droplet widening in the direction perpendicular to flow is observed. For $k' > 0.37$, elastic drops deform less than the Newtonian case. The critical shear rate for droplet breakup increases with increasing k'. For $k' < 4$, $Ca_{critical}$ increases rapidly with k'; for $k' > 4$, $Ca_{critical}$ levels off at 1.75.
Mighri and Huneault [89]	Experimental	Viscoelastic	Viscoelastic	Shear	Studied model systems and PS/PE system with viscosity ratio between 5 and 20 under relatively high shear rate ($1-20\ s^{-1}$) in shear flow. In PS/PE system, the PS drop widens in the direction perpendicular to flow which contributes to droplet breakup (the ends of the highly elongated droplet are in different flow planes). At high shear attrition from the surface was the breakup mechanism. An EPR/PP system has similar

(Continued)

TABLE 6.3 (Continued)

Source	Experimental or theoretical	Droplet	Matrix	Flow	Result
					breakup mechanisms as the PS/PE system despite a 10x difference in interfacial tension.
Lerdwijitjarud et al. [90]	Experimental	Viscoelastic	Newtonian	Shear	Viscosity ratio controlled to around 1.0. Shear rates up to 5 s^{-1} used. Deformation decreases for isolated droplets at constant *Ca* as the droplet elasticity increases. $Ca_{critical}$ increases with increasing droplet elasticity. 10% blends also studied.
Li and Sundararaj [91]	Experimental	Viscoelastic	Newtonian	Shear	Studied deformation of a viscoelastic droplet in Newtonian matrix at high Ca in Couette flow. If the elastic contribution is weaker than the viscous, the droplet deforms in the flow direction. If the elastic contribution is stronger, it deforms in the vorticity direction. Bigger drops have more deformation than smaller drops.
Tanpaiboonkul et al. [92]	Experimental	Viscoelastic	Viscoelastic	Shear	PS droplet in HDPE matrix. Transient effects: initially droplet exhibits oscillation in the flow direction but eventually stretches preferentially in vorticity direction. Oscillation amplitude decreases with increasing droplet elasticity and oscillation period depends primarily on, and increases with, viscosity ratio. At steady state, the droplet length in the vorticity direction increases with increasing capillary number, viscosity ratio and droplet elasticity, and if Ca is above a critical value and elasticity is high enough, break up into smaller droplets.
Sibillo et al. [93]	Experimental	Newtonian	Viscoelastic	Shear	Matrix elasticity inhibits droplet breakup. Follows theory of Greco, 2002.
Lin et al. [94]	Experimental	Viscoelastic	Viscoelastic	Shear	PS droplet in PE matrix with and without a compatiblizer. Breakup is by elongation in vorticity direction. Compatibilizer does not delay breakup except if interfacial crosslinking reaction occurs.
Lin and Sundararaj [95]	Experimental	Viscoelastic	Viscoelastic	Shear	PC droplet in PE matrix forms sheet during initial shear, then forms thin thread or sheet structure elongated in the flow direction. Sheet formation and breakup determined by viscosity and stress ratios and Deborah number.
Lin et al. [96]	Experimental	Viscoelastic	Viscoelastic	Shear	PC droplet in PE matrix in Couette flow. Breakup by 2 mechanisms observed: (1) erosion and (2) elongation perpendicular to flow.
Han and Funatsu [97]	Experimental	Viscoelastic	Viscoelastic	Elongational	Viscoelastic dropletss are more stable than Newtonian droplets in both Newtonian and viscoelastic matrices.

(Continued)

TABLE 6.3 (Continued)

Source	Experimental or theoretical	Droplet	Matrix	Flow	Result
Milliken and Leal [98]	Experimental	Viscoelastic	Newtonian	Elongational	Viscoelastic droplets with viscosity ratios less than 1 have smaller deformation and $Ca_{critical}$ than Newtonian droplets. For viscosity ratio >1 the viscoelastic droplet deformation behavior is similar to Newtonian droplets.
Delaby et al. [99]	Experimental	Viscoelastic	Viscoelastic	Elongational	For negligible interfacial tension, viscoelastic droplets deform less than the surrounding media when the viscosity ratio is less than 1 and more when the viscosity ratio is greater than 1.
Miejer and Janssen [100]	Experimental	Not disclosed	Not disclosed	Elongational	At small viscosity ratios, the droplet deformation in planar elongational flow resembles that for the matrix.
Chin and Han [101]	Experimental	Viscoelastic	Viscoelastic	Elongational	Higher droplet elasticity results in less deformation compared to the matrix.
Shanker et al. [102]	Experimental	Viscoelastic	Viscoelastic	Elongational	Higher droplet elasticity results in less deformation compared to the matrix.
Mighri et al. [103]	Experimental	Viscoelastic	Viscoelastic	Elongational	Droplet deformation increases with increasing matrix elasticity. Defined k' as first normal stress difference ratio divided by the viscosity ratio. When $k' < 0.2$, the matrix elasticity has a greater effect on deformation than the drop elasticity. The opposite is true with $k' > 0.2$.
Tretheway and Leal [104]	Experimental	Newtonian	Viscoelastic	Elongational	Droplet is more pointed at the ends and $Ca_{critical}$ decreases in steady flow as elasticity of matrix fluid is increased. Also examines unsteady flow and relaxation.
Van Oene [105]	Theoretical	Viscoelastic	Viscoelastic	Shear	Droplet elasticity greater than the matrix stabilizes the droplet; Matrix elasticity greater than the droplet destabilizes the droplet.
Mukherjee and Sarkar [106]	Theoretical	Viscoelastic	Newtonian	Shear	Modeled Oldroyd B fluid deformation in Newtonian matrix under steady shear flow. For viscosity ratio $<=1$, increasing Deborah number reduces deformation due to increase in inhibiting normal stresses inside the droplet. At higher viscosity ratio, the change in deformation changes with De. At first, deformation decreases with increasing De but above a critical De it increases. Strain rate inside the droplet is reduced at higher viscosity ratios, reducing the inhibiting viscoelastic stress.
Greco [107]	Theoretical	Viscoelastic	Viscoelastic	Shear and Elongational	Assumes the materials are simple second-order fluids, which includes elastic effects for slow flows. Uses a perturbation method to analyze the first normal stress difference effect on droplet shape for small droplet deformations.

(Continued)

TABLE 6.3 (Continued)

Source	Experimental or theoretical	Droplet	Matrix	Flow	Result
Maffettone and Greco [108]	Theoretical	Viscoelastic	Viscoelastic	Shear and Elongational	Develops a phenomenological model for the dynamics of a droplet immersed in an immiscible fluid. Assumes the droplet is ellipsoidal. Finds that elastic droplets deform less than Newtonian droplets. Elastic droplets in Newtonian matrix under shear are forbidden to breakup at lower viscosity values than the Newtonian case. In elongational flow, the droplet breakup up is easier when the matrix is elastic and more difficult when the droplet is elastic.
Chen et al. [109]	Theoretical	Viscoelastic	Viscoelastic	Shear	Model for breakup of PC or PE droplet in PE melt. An erosion mechanism is predicted similar to previous experimental results.

manner as a PS–PE system with an order of magnitude greater interfacial tension. They suggest that interfacial instabilities may not be required for non-Newtonian droplets to break apart.

Juza and Fortelny [119] show that elasticity of the matrix polymer acts to inhibit coalescence. Addition of nanoparticles has also been found to suppress coalescence [120,121].

As the minor component concentration is increased, the morphology may change. At high enough concentrations, the minor component becomes the continuous phase. Prior to this phase inversion, there may be a concentration range where both polymers have continuity – called the co-continuous region. These phase transitions are known to be complex functions of the concentration, viscosity ratio, elasticity ratio and interfacial tension (compatibilizers) [122–131]. An example of the transition between phase morphologies is shown in Fig. 6.21 for HDPE/PS blends studied by Bourry and Favis [69,122]. At low PS levels, the PS forms droplets in the HDPE matrix. As the PS concentration is increased, fibers are formed. At about 70% PS, the morphology becomes co-continuous; the HDPE and PS phases form an interpenetrating network structure where both phases are continuous. Higher PS levels results in HDPE droplets in a PS matrix. The addition of a SEBS compatibilizer decreases the domain size across the entire concentration range (open squares in the figure).

Another concern in applying single droplet studies directly to polymer blends is uncertainties about the flow field in the mixing device and how to calculate the appropriate viscosity ratio. Polymer mixing devices, such as extruders, have complex flow fields involving both shear and elongational flow. The flow fields are typically nonuniform; the shear rate varies across the screw channel (see Fig. 6.4). Further complicating matters are temperature changes along the extruder. Given these complexities, Lyngaae-Jorgensen [132] proposes measuring the $Ca_{critical}$ vs viscosity ratio curve for the given device.

One must also consider the fact that the polymer melt viscosity varies with temperature and shear rate. Generally, there are two schools of thought on calculating the viscosity ratio. Some argue that the zero shear viscosities should be used. Others suggest using the viscosity at a representative shear rate for the process. However, it is stress, not shear rate, that drives dispersion and droplet breakup. Stress is determined by the flow field and is continuous across an interface, whereas the shear rate is discontinuous. A better method is to plot the viscosity as a function of stress (stress is equal to the viscosity times the shear rate) and compare the blend components at a representative stress for the process. This is illustrated in Fig. 6.22. Here, polymers A and B are the minor and major components, respectively. At a constant shear rate equal to 100 s^{-1}, the viscosity ratio is 1.6. But, at constant stress, the ratio is 2.6, very close to the 3.5 cut-off for shear flow.

The polymer temperature changes along the extruder as the polymer becomes molten and is conveyed to the exit. Huneault et al. [133]. show that, for a PS/HDPE blend, the viscosity ratio varies by six orders of magnitude depending on the temperature (Fig. 6.23). Computing the viscosity ratio just at the final extrusion temperature may be misleading.

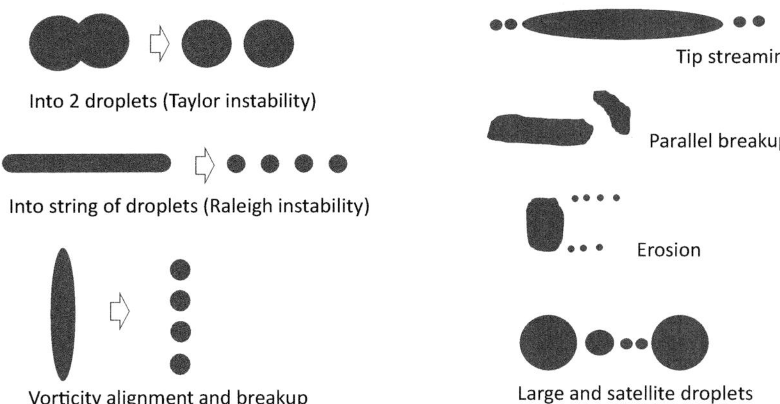

FIGURE 6.19 Single droplet breakup mechanisms in shear flow. *Based on I. Fortelný, J. Jůza, Description of the droplet size evolution in flowing immiscible polymer blends, Polymers (11) (2019) 761.*

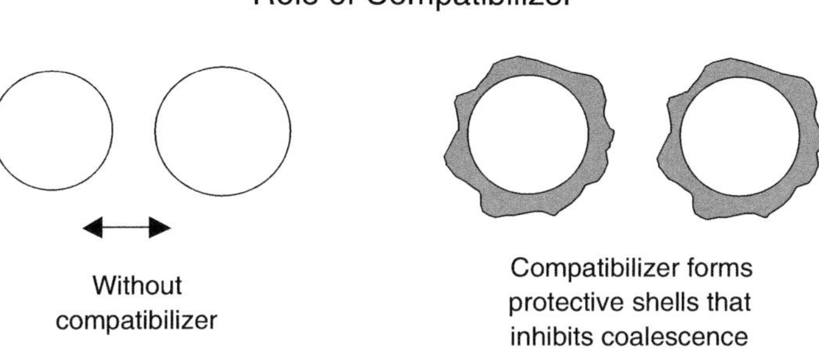

FIGURE 6.20 Role of compatibilizers in reducing coalescence in polymer.

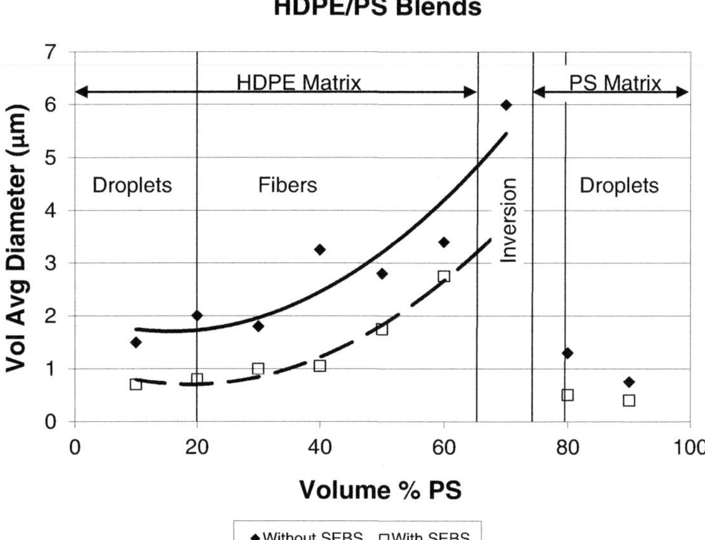

FIGURE 6.21 Blend morphology of PS-HDPE blends as a function of concentration. *Based on data from D. Bourry, B. Favis, Co-continuity and phase inversion in HDPE/PS blends: the role of interfacial modification, SPE ANTEC Conf., pp. 2001.*

They find that, for this blend, the mixing is much better than what they would have expected just by looking at the viscosity ratio at the final temperature (200°C). Notice in Fig. 6.23 that the viscosity ratio is sometimes less than 1 and other times greater than 1. This indicates that, for the two polymers, the viscosity temperature dependence is not the same, another complication in computing the viscosity ratio.

So far in this discussion, the polymers have been assumed to both be molten at the time they are mixed. This is generally not the case as mixing can begin during melting. Ghosh et al. [134] study the softening/mixing of two similar amorphous polymers and find that laminar sheets or striations form that subsequently break up into droplets. Lindt and

Calculating Viscosity Ratio: Constant Shear Rate or Shear Stress

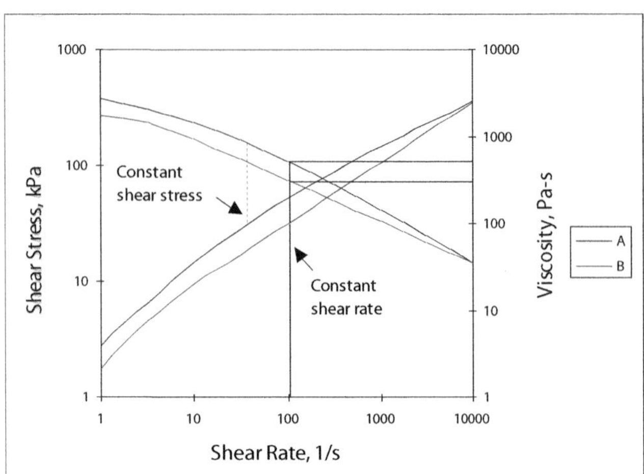

	A	B	A/B
Constant Shear Rate	520	320	1.6
Constant Shear Stress	840	320	2.6

FIGURE 6.22 Calculating viscosity ratio.

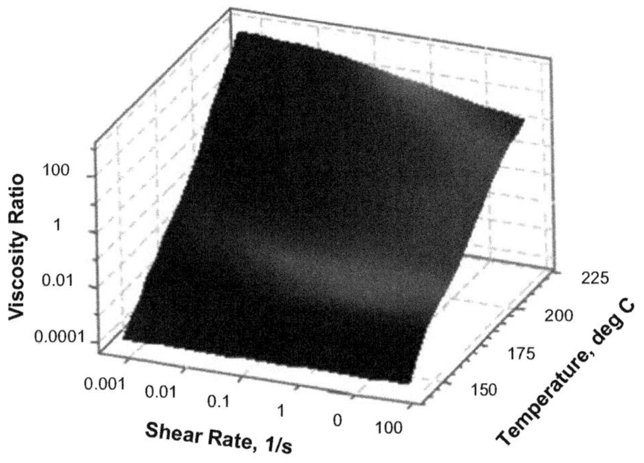

FIGURE 6.23 HDPE-PS blend viscosity ratio variation as a function of shear rate and temperature. *Based on M. Huneault, M. Champagne, L. Daigneault, M. Dumoulin, Blend morphology development during compounding in a twin-screw extruder, SPE ANTEC Tech. Pap., 1995, pp. 2020.*

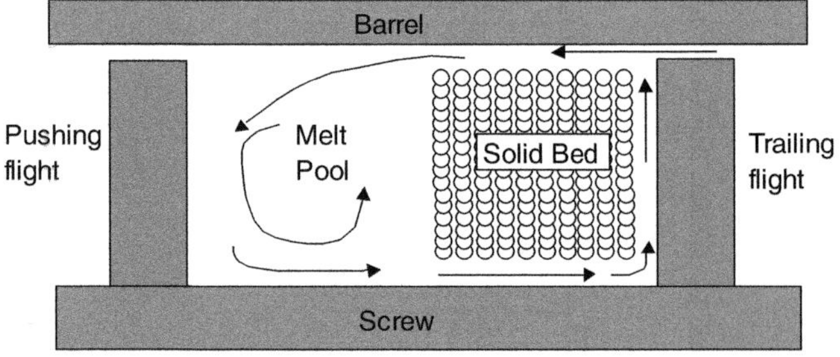

FIGURE 6.24 Single-screw extrusion solid bed melting model.

Ghosh [135] find that the striations come from the single-screw extruder melting process. A thin molten layer forms between the pellet solid bed and the barrel or screw surface caused by frictional heat and heat transferred from the barrel (see Fig. 6.24 for the solid bed melting model). The high stress and deformation rates in this region transform the molten polymer into laminar morphologies as the pellets melt. The domain size decreases from a few millimeters (pellets) to about 50 to 100 μm due to the melting process. Their measurements, as well as calculations, confirm these striations and abrupt domain thickness changes for blends of both rheologically similar and dissimilar resins. Huang et al.

[136,137] study morphology development of blends of two semicrystalline polymers (PP as the major component and PA6 as the minor) by sampling along the length of an extruder. The PP melts first and the PA6 forms striations followed by breakup into droplets; the nature of the striations and droplet formation depends on the viscosity ratio.

Benkreira et al. [138]. find that, when mixing a masterbatch into a host polymer, most of the mixing occurs during melting and little thereafter. They propose a laminar mixing model where the creation and stretching of striations during melting leads to a reduction in dimensions. Scott and Macosko [139] and Sandararaj et al. [140] describe how component melting and softening in a batch mixer leads to domain size reduction. They find sheets and ribbons are formed first, which become unstable due to flow and interfacial tension effects. Holes form, then a lace-like structure followed by irregular shaped particles and finally spheres (Fig. 6.25). Breakup is driven by interfacial instabilities. Burch and Scott [141,142] observe similar behavior for miscible polymers during the initial mixing stages and propose instabilities due to dynamic interfacial tension as the cause. Willemse et al. [143] find that sheets or striations form during single-screw extrusion and that the final droplet size depends on the striation dimensions at the start of breakup rather than the capillary number. The sheet to droplet breakup during shear is very effective at dispersing the minor phase, more so than the elongated droplet to subdroplet mechanism found in single drop experiments.

In a single-screw extruder, Tyagi and Ghosh [144] follow a PP–EVA (70/30 wt.%) blend as it develops its morphology using freeze experiments pioneered by Maddock [22]. In these "screw-pull" experiments, after the extruder has reached a steady state, the screw is stopped, and the polymer is quickly quenched. The screw is pulled and samples from the screw channels are analyzed. Tyagi and Ghosh find that striations form in the feed zone. The dimensions quickly diminish along the channel length. In the compression zone, the shear and elongation increase, causing the striations to break up. The droplets that form are an order of magnitude smaller in size than the striation thickness prior to breakup. The breakup into droplets requires a step-up in shear and stretching. They conclude that, for an extruder to be an effective mixer, periodic flow reorientations should occur. This concept is used in high-performance screw designs where the polymer flows through tight clearances.

Domingues et al. [145] develop a model for the morphology development along a single-screw extruder for a two-component blend, taking into account stretching, breakup and coalescence in the melting and metering sections. Since a two-material melting model does not exist, they use the primary component in a conventional single-screw extruder melting model and insert droplets of the secondary component within this melt. A Newtonian droplet breakup model based on Taylor and Grace is assumed. To validate the model, they conduct screw-pull experiments with a blend of 90% HDPE/10% PP. Better dispersion (smaller droplets) occurs at lower screw speed (which provides longer residence time for droplet breakup). This trend agrees with the model results. They consider this encouraging and point out the errors in the model: model assumes initial droplets of uniform size, not the real distribution; the model's simplified flow has longer flow paths and residence time; and parameters are taken from literature on similar materials.

Next they model different screw geometries and conduct a parametric study. The viscosity ratio showed complex effects with 0.01 giving smaller droplets than a ratio of 1, but 10^{-4} gives larger droplets. The model predicts long fiber-like droplets in some cases, consistent with what is seen in some of the experiments. Higher interfacial tension results in faster dispersion but the droplet size plateaus in the metering zone; lower interfacial tension slows dispersion and droplets continue to decrease in size in the metering section, resulting in smaller droplets. Extrusion conditions tend to affect dispersion through changes in residence time and shear rate.

Wilczynski et al. [146,147] study melting of polymer blends in a starve-fed single-screw extruder. While most single-screw extruders are run flood fed, the differences in behavior may prove enlightening. Melting of single polymers in the starve-fed extruder is found to proceed in two stages. At first the pellets do not fill the flights and heat

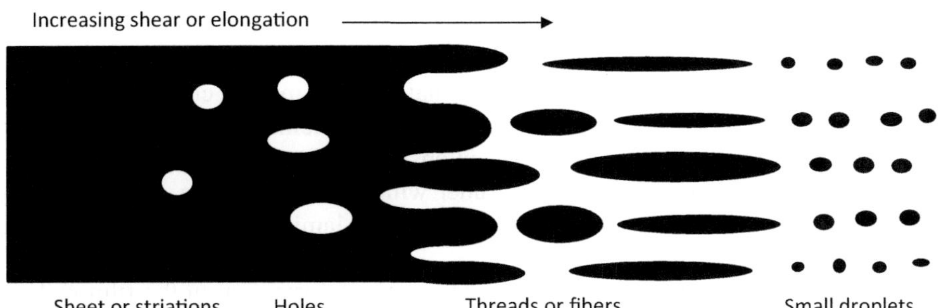

FIGURE 6.25 Morphology development during melting and subsequent shear in an extruder. Striations or sheets form during melting. Shear and stress cause holes to form, followed by threads and particle breakup. This process leads to large changes in the minor phase morphology.

conduction from the barrel wall is dominant. As the pellets travel down the channel the flights become filled. In this region, solid pellets are suspended in the previously melted polymer and further melting occurs primarily by viscous dissipation. Using the screw-pull technique, these investigators observe the melting of 85/15 blends of HDPE/PS, PP/PMMA and PP/PS. For HDPE/PS blends, HDPE melts first with PS granules dispersed in the PE. The PS granules slowly decrease in size as they travel down the channel. For PP/PMMA blends, PP melts first. Particles of unmelted PP and PMMA can be seen in the molten PP at first. PP/PS blends behave similarly. With flood feeding, a solid bedforms at the pushing flight (in accordance with the Tadmor melting model—see Fig. 6.24) that comprises of unmelted pellets of both components with molten polymer blend at the active flight.

Li et al. [148] find that the minor phase melts faster during mixing when a compatibilizer is present. They attribute this to faster heat transfer between the phases due to their intimate contact. Faster melting can affect the final blend morphology.

Shih et al. [149] study the solid–melt transition for many polymer blends in a batch mixer equipped with a glass window. They find that blends go through four stages from solid to melt:

- *Elastic solid pellets*: Below the glass transition temperature (Tg) or melting point (Tm) the solid pellets are observed sliding in the mixer. Torque is low.
- *Deformed solid pellets*: As the temperature increases, some components begin to soften and deform. The torque starts to rise.
- *Transition material*: This takes on several different forms including:
  - a fluid with suspended solid particles
  - fractured or semifluid material
  - a dough-like material.

  This is the transition zone between solid and liquid. Different blends take on different transition phases. The temperature increases sharply and then remains constant near the polymer melt temperature (Tm) if it is crystalline. Torque rises sharply.
- *Viscoelastic material*: The typical liquid appearance is observed. Temperature rises to about 20°C–50°C above the matrix resin Tm or Tg and the torque decreases.

During the transition phase, they observe several phase transitions that aid in the mixing process. Shih [150,151] and others [131,152,153] find that phase inversions are to be expected when the minor component melts first. The molten minor component surrounds the still solid major component. As the major component melts, it becomes a continuous phase. This inversion is accompanied by a spike in the mixer torque and dramatically reduces the minor component phase size. They also describe how phase inversion can explain some unique morphologies, such as major component regions embedded within the minor component domains.

The phase inversion onset is a function of the volume fraction ratio, viscosity ratio, elasticity ratio, and surface interactions (surface tension or compatibility, the introduction of compatibilizers, etc.) with a number of models proposed in the literature [122–131]. A simple expression accounting for the primary volume fraction and viscosity effects is given by: [69,142]

$$\frac{\eta_1}{\eta_2} \sim \frac{\Phi_1}{\Phi_2} \tag{6.9}$$

where Φ is the volume fraction.

Two polymers with the same viscosity will undergo a phase inversion when the minor component reaches about 50% of the blend composition. If the minor component viscosity is substantially less than the major component, the minor component can become the continuous phase. This is the case when the minor component melts first. As the major component melts and its viscosity decreases, then conditions no longer favor the minor component as the continuous phase and the phase inversion occurs.

As described in Section 6.3.2, mixing elements may be incorporated into the screw design in the metering section of the screw. These may further impact the morphology. Huang et al. [137] show that a fluted mixing element produces a finer and more uniform dispersed phase in a PP-PA6 blend, whereas a pineapple mixing element is more conducive to producing a thin laminar morphology. Huang et al. [15,154] find that shear intensity and viscosity ratio play a role in creating laminar morphologies of a HDPE-PA6 blend in a single-screw extruder with a convergent die. For a viscosity ratio of < 2, spherical droplets were formed. For viscosity ratios between 2 and 7, laminar morphology could be obtained by controlling the shear intensity of the metering zone.

Finally, the morphology of the dispersed phase may change as the blend undergoes deformation in the film making or coating process downstream of the extruder. As the polymer is drawn down to its final shape and thickness, stress and orientation may elongate the domains. Several morphologies have been described for immiscible blends in blown film, including elongated particles, such as fibrils and ribbons [155–159]. The factors influencing this morphology are covered in Chapter 14.

6.5 Dispersion of rigid particles and nanocomposites

Dispersing inorganic and other particles in polymers is not well studied. As White [41] explains, this may be because breaking up agglomerates is hard to quantify, particularly since most fillers have a particle size distribution. Generally, it is thought that high stresses are needed to achieve dispersive mixing. Such stresses are typically not found in a single-screw extruder. Coupling agents, lubricants and other additives may be employed to aid in dispersion. Coupling agents are similar to compatibilizers but are typically lower in molecular weight. They function by interacting with the functional groups on the surface of the inorganic particles to provide better adhesion between the particle and polymer. Given these complexities, most mineral-filled polymers are manufactured in specialized compounding equipment as described earlier. In packaging applications, blending fillers is generally limited to letting down masterbatches. Here, the filler, pigment or additive has been predispersed and simple distributive mixing is all that is required.

One noteworthy aspect of dispersing rigid particles into polymers is particle attrition. Fillers, such as glass fibers, mica and clay, may have long aspect ratios that are important for the blend properties. For rod-like particles, the aspect ratio is defined as the length divided by the diameter. Too much dispersive mixing can reduce their aspect ratio, degrading performance. Filler abrading the mixing device surfaces can also cause excessive wear. Adding mineral filler in a downstream stage where the polymer is molten rather than in the first stage with the solid pellets can significantly reduce abrasion and attrition.

A key question when dealing with pigments is how do you know if good mixing has been achieved? One commonly used method is to compare a sample against a visual standard. Spectroscopic techniques have also been used, as well as microscopic image analyses [138]. Various indices, such as striation thickness, variance in minor component concentration and segregation scales, have been used, as discussed by McKelvy [160]. More recently, computer image analysis (red-green-blue correlations) has been used. Translating such color analysis to a quantitative distributive mixing index has been difficult. Alemaskin et al. [161,162] develop an index by employing Shannon entropy. They use both computer simulation and experiments to verify the method. They simulate mixing two ABS colors in a conventional single-screw extruder metering section using a numerical particle tracer analysis. The results were compared with extrusion experiments under similar conditions. Here, the extruder is stopped, rapidly quenched and the screw pulled to obtain samples along the extruder length. The color homogeneity evolution is measured using digital computer imaging. Color homogeneity indices based on entropic considerations are computed for both the simulations and digital images and they qualitatively agree. Such an approach can potentially be used for scale-up, control and process design optimization.

A polymer filled with nanoparticles is a special case. These are typically blends of polymer with inorganic particles that have nanometer dimensions (less than 0.1 μm). At levels between 3 and 5 wt.%, they have been found to be effective in improving stiffness, mechanical strength, barrier, electrical conductivity, and flame retardancy. Although nanotechnology is still in its infancy, two nanofillers have received the most attention, carbon nanotubes and clay. Carbon nanotubes have been found to impart good electrical properties for shielding and other potential specialty electronics applications. Because of their cost carbon nanotube composites are not suited for packaging applications.

Of greater interest are clay nanocomposites [163–169]. Many studies have been done with montmorillonite clay, initially at Toyota. In its natural state, each montmorillonite particle is an agglomeration of many layers of nanosized platelets. The platelet length and width range from a few tenths of a micron to about 1.5 μm. Their thickness is only about 1 nm and is the reason they are considered nanomaterials. Their high aspect ratios give them their unique properties.

The montmorillonite platelets are hydrophilic and are held together at a distance of about 3.5 Å by attractive forces. Clay suppliers add surfactants to the clay to promote interplatelet expansion to about 20 Å. To obtain the best properties, the platelets must be completely separated from each other. This complete dispersion into the polymer matrix is known as exfoliation and is illustrated in Fig. 6.26. There are two typical ways to accomplish this. One is to introduce the clay during the polymerization process. The monomer is absorbed into the spaces between platelets. As the monomer polymerizes the platelets separate. This process is only amenable to certain polymers where the clay can be introduced during polymerization. Most often it has been applied to PA.

The second method is to compound the clay particle with an already polymerized polymer and rely on the stress generated during mixing to separate the platelets. The surfactant selection is critically important; it must help turn the

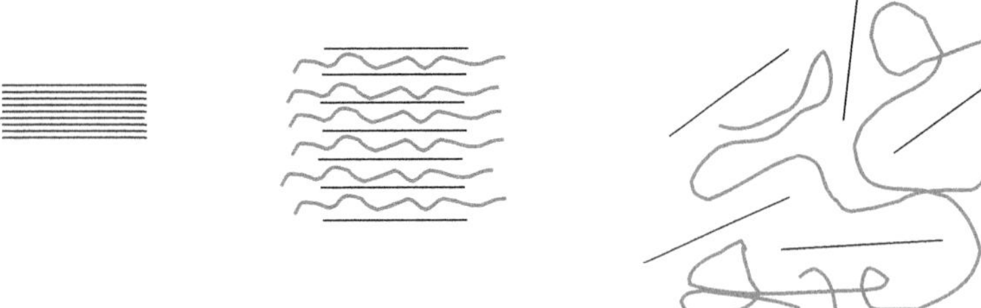

FIGURE 6.26 Illustration of states of clay nanocomposites.

hydrophilic environment between the platelets into one that is compatible with the monomer or polymer. In some cases, compatibilizers are used to aid in the exfoliation. For example, in PP or HDPE–montmorillonite nanocomposites, maleic anhydride grafted PP or HDPE is often added as a compatibilizer [170]. Nevertheless, so far it has proven difficult to fully exfoliate montmorillonite in a compounding process and is the focus of continued R&D. The first approach of polymerizing in the presence of the clay has been more successful.

Adding nanosized particles to the polymer restrains the molecule's movement, more so than in conventional fillers because of the extremely high surface area that is generated. The restrained motion enhances stiffness and heat deflection temperature. Because the individual platelets are small, toughness and optical properties do not suffer. Of particular importance for packaging applications is the potential for increased barrier performance, especially the oxygen barrier. The orientation of the platelets is flexible and their longer dimensions line up in the extrusion and orientation direction. The resulting platelet network structure creates a more tortuous path for gas molecules to diffuse through. Reductions in gas permeability of 60%–80% for PA or EVOH nanocomposites have been reported [163,169].

For most packaging applications, the reduction in gas permeation so far has not been enough to justify the additional cost. Color is another issue with natural clays. Better barrier performance has been achieved in coatings, which may prove useful for recyclable packaging [171]. Questions around cost, performance, safety and regulatory compliance (the US and European Union have not yet developed regulations for use of nanocomposites for food packaging) have slowed the adoption of nanocomposites in flexible packaging. Recent research has focused on the use of nanocomposites in biopolymers to overcome their deficiencies in mechanical, barrier and processing properties. Good interaction between the nanofiller and the biopolymer help with dispersion. Nanofillers may also bring other functionality to a package such as antimicrobial, sensing and scavenging [172].

Nanoparticles have been used to change or stabilize the morphology of immiscible polymer blends. For example, Bose et al. [173] find multiwalled carbon nanotubes (MWCNTs) compatiblize co-continuous blends of PA6 and ABS. Tiawai and Paul [174] find that organo nanoclay particles stabilize and reduce the dispersed phase size of PP-PS blends. The nanoparticle shape (aspect ratio) is found to influence which phase the particle ends up in and its ability to stabilize the interface [175].

6.6 Rheology of polymer blends

Utracki [176] provides a good review of the rheology of polymer–polymer blends. Miscible blends are easier to characterize than immiscible ones. Often a log-mean rule is used to estimate the miscible blend viscosity, although there may be some positive deviation:

$$\log(\eta_{\text{blend}}) = X_A \log(\eta_A) + X_B \log(\eta_B) log(\eta_{blend}) = X_A log(\eta_A) + X_B log(\eta_B) \quad (6.10)$$

where X_A is the weight fraction of polymer A in the blend, X_B is the weight fraction of polymer B in the blend, η_{blend} is the viscosity of blend (also works with melt flow rate), η_A is the viscosity of polymer A, and η_B is the viscosity of polymer B.

Eq. (6.10) is useful for estimating the viscosity or melt flow index (MFI) for two grades of the same polymer. For example, a 70–30 wt.%, LDPE1–LDPE2 blend where the LDPE1 MFI is 4 and LDPE2 MFI is 12 g/10 min, yields:

$$\log(MFI_{blend}) = 0.7\log(4) + 0.3\log(12)$$
$$MFI_{blend} = 5.6\frac{g}{10\text{ min.}}$$

For immiscible blends, Lyngaae-Jorgensen [132] points out that steady state data, such as capillary rheometry data, look superficially like homogeneous polymers. Unless they have been compatibilized [176], however, their behavior may be complex. These blends typically do not follow time-temperature superposition and mixing rules since the individual phases may behave differently. Their transient behavior and die swell may also be complicated. The immiscible blend morphology is the driving force for this complex behavior.

Reactive polymer blends are a special case. Typically, the reactive blend viscosity increases due to the formation of cross-links that effectively increase the polymer molecular weight. An example is blending maleic anhydride modified PE with PA. The PA viscosity can increase by an order of magnitude. The viscosity rise is related to the interfacial reaction as well as the amount of new interfacial area created; as the domain size of the dispersed PE phase decreases, more interfacial area is created and viscosity increases.

Some additives can cause gels and other problems when blended with a polymer. For example, the acid groups in acid copolymers and ionomers can react with a coupling agent used in some TiO_2 grades, causing gel and other problems. Most masterbatch suppliers have learned how to deal with this through the design of the masterbatch. On the other hand, adding wax, tackifiers and other low molecular weight additives to polymers will reduce their viscosity.

How rigid fillers effect viscosity can also be complex. For very dilute solutions of rigid spheres in a Newtonian fluid, Albert Einstein derived the following result:

$$\eta = \eta_f(1 + 2.5\varphi) \tag{6.11}$$

where η is the viscosity of the suspension, η_f is the viscosity of the suspending fluid, ϕ is the volume fraction of the rigid spheres.

The assumptions for this equation break down quickly for filled polymers: the filler loading is not typically dilute, the filler is not spherical and the polymer matrix is non-Newtonian. Dealy and Wissbrun [177] describe the following complexities:

- non-Newtonian effects caused by particle concentration and aspect ratio,
- buoyancy effects,
- particle migration and agglomeration.

A number of equations have been proposed, including the following by Maron and Pierce [178]

$$\eta = \frac{\eta_f}{[1-(\varphi/A)]^2} \tag{6.12}$$

Here, A is a constant that is a function of the filler. Dealy and Wissbrun propose that A can be thought of as the maximum filler packing fraction. For example, the value for A for spheres is 0.68, close to the theoretical maximum packing fraction of 0.637. The value for A decreases with the increasing filler aspect ratio (0.44 and 0.16 for aspect ratios equal to 8 and 30, respectively). The value of A for filler with a nonuniform particle size distribution is greater than that for uniform particle size. A nonuniform filler size distribution, therefore, can reduce the concentrated suspension viscosity.

There are other complications. One is the viscosity to choose when dealing with polymers, since η_f varies with shear rate. Dealy and Wissbrun point out that plotting the viscosity vs shear stress rather than shear rate, as noted earlier for polymer–polymer blends, gives the best results. At low shear rates, the filled polymer may exhibit yield stress, below which the material behaves like a solid (and does not flow). The yield stress increases with filler concentration.

6.7 Characterizing polymer blends

The properties of polymer blends depend on their composition and morphology. While miscible polymer blends may follow the rule of mixtures where properties vary linearly with composition, immiscible blends typically do not. Several analytical techniques are helpful for verifying the composition of a blend and determining its morphology. These are summarized in Table 6.4.

For polymer blends with different chemistries, spectroscopy techniques such as Fourier transform infrared (FTIR) spectroscopy or Raman spectroscopy may be helpful for determining the composition. These techniques can be used to determine the presence of different chemical moieties associated with each polymer. Creating standards with different blend ratios may help in establishing a baseline for quantitative analysis.

DSC measures thermal transitions such as the glass transition tempertature (Tg) and melting point (Tm). A blend of miscible amorphous polymers will have a single Tg that falls between the Tg of the components. For blends of immiscible polymers of differing Tg or Tm, the height of the DSC endotherm peaks may help quantify the blend composition. DMA may also be used to measure the Tg of polymer blends.

For blends of polymers of similar composition, a combination of techniques may be helpful. For example, blends of LLDPE and LDPE may have overlapping Tm endotherms in a DSC analysis. The presence of the long-chain branches from the LDPE component can sometimes be determined by various liquid chromatography techniques used to characterize molecular weight distribution. Rheological measurements are also useful for teasing out the presence and amount of LCB (see Chapter 5).

Various elemental analysis techniques may be helpful for verifying the presence and amount of inorganic components in a blend. These include atomic absorption and inductively coupled plasma techniques. For total inorganics, an ash test is useful.

TABLE 6.4 Some common characterization techniques for polymer blends.

Characterization	Technique	What it measures	When to use
Composition	Spectroscopy techniques such as Fourier Transform Infrared Spectroscopy (FTIR) and Raman Spectroscopy	Chemical composition of components.	Especially useful when the components are chemically distinct.
Composition	Differential Scanning Calorimetry (DSC)	Melting points; glass transition temperature (Tg)	Useful if the polymers have different melting points. Miscible amorphous polymer blends will have a single Tg between the components.
Composition	Dynamic Mechanical Analysis (DMA)	Modulus vs. temperature (Tg)	Miscible amorphous polymer blends will have a single Tg between the components.
Composition	Ash test	Inorganic content	Can be used to determine the total amount of inorganic content.
Composition	Atomic Absorption Spectroscopy	Elemental analysis	Inorganic content.
Composition	Inductively Coupled Plasma (ICP) spectroscopy techniques	Elemental/ metals analysis	Inorganic content.
Blend morphology	Optical Microscopy	Size and shape of phases	When phases differ in optics.
Blend Morphology	Electron Microscopy (scanning or transmission)	Size and shape of phases	Useful for small domain sizes
Blend Morphology	Atomic Force Microscopy (AFM)	Size and shape of phases	Useful when phases differ in hardness or modulus
Rheological measurements	Melt flow rate; melt viscosity; dynamic viscoelastic measurements (see Section 6.6 and Chapter 5)	Flow or viscoelastic properties	May be useful for quality control if standards are known. May provide insight into morphology/ composition when combined with other techniques.
Physical or optical properties	Tensile tests, haze, transparency, color, etc.	Physical or optical property of blend	May be useful for quality control if standards are known.

Microscopy techniques are typically used to discern the phase morphology of immiscible polymer blends or to assess the dispersion of inorganics in polymers. These include optical, electron and atomic force microscopy (AFM). A contrast between phases is important for these techniques. Various stains and lighting conditions (such as polarized light) may help. With AFM, differences in modulus or hardness can be used to quantify the phase size and shape. As discussed in Section 6.5 various microscopy and spectroscopy techniques, along with computer image analysis, may be used to characterize the quality of pigment dispersions. Microscopy techniques are also helpful for measuring the attrition of particles during polymer processing.

Finally, from a practical and quality control perspective, various rheology or physical properties may be measured. Often these can be related to composition and morphology through comparison with standards of known composition and morphology.

6.8 Conclusion

Blending is an important aspect of polymer/property design for many packaging applications. Achieving a consistent quality blend with the desired properties requires proper attention to both process and product design. The polymer blend properties are a complex function of the stress history imposed by the process and the rheological and thermodynamic properties of the components. Based on experience and science, several guidelines are proposed for reducing these complexities in practice:

- Careful control of the blend components fed to the mixing device feed hopper is important since most continuous mixers, such as extruders, do not have very good back mixing.
- All things being equal, target at least 4%–5% or more for the minor component for better accuracy and control.
- Single-screw extruders can be used for many blending applications where only simple distributive mixing is required. The limitations of single-screw extruders can be mitigated to some extent by using barrier screws and specialized mixing sections. For more demanding applications, high-performance screws or ultimately specialty compounding devices, such as twin-screw extruders, may be required.
- When using a masterbatch concentrate, make sure the masterbatch carrier resin is compatible with the resin it is being blended into. Also, viscosity differences between the masterbatch and the matrix resin need to be considered; in some cases using a low viscosity materbatch carrier resin can help.
- Very few polymer blends are miscible. For miscible blends, the viscosity can be approximated using the log-mean rule [Eq. (6.10)].
- The immiscible blend morphology influences the properties, which generally deviate from the mixing rule. Some examples include using small domains of a soft resin to improve toughness without detracting from optical clarity and laminar morphologies to help improve barrier performance.
- The immiscible blend morphology is influenced by:
  - the component concentrations,
  - stress history imposed by the mixing device,
  - the viscosity and mutual attractiveness (thermodynamics) of the components.
- The minor component domain size of an immiscible blend can be reduced by:
  - increasing the stress (by increasing the viscosity of the matrix phase and the shear rate or rate of elongation during mixing),
  - matching component polymer viscosities (a viscosity ratio in the range of 0.01–2 is preferred, particularly for shear flow),
  - lowering the interfacial tension by matching solubility parameters or using compatabilizers,
  - minimizing coalescence through minimizing minor component concentration and use of compatibilizers.
- Compatibilizers reduce interfacial tension and coalescence thus reducing domain size. They also reduce the minor phase melt time, which can affect domain size.
- Elongational flow is generally more efficient than shear flow for dispersing the minor component of an immiscible blend. The flow in single-screw extruders is typically dominated by shear flows unless special mixing elements are used. Both shear and elongational flow can be found in twin-screw extruders.
- Most mixing in a single-screw extruder occurs as the polymers melt, forming sheets or striations that break up into holes and droplets as the shear and elongation are increased along the extruder. This is because the shear stresses are highest in the initial part of the extruder. Increased shear and stretching in the compression zone or high-performance metering zones (such as energy transfer screws) promote sheet breakup.

- Dispersive mixing may be required to break up agglomerates when blending mineral fillers or pigments with polymers. For this reason, these blends are often made as masterbatches using specialized compounding equipment and let down using a single-screw extruder by the film manufacturer.
- Spectroscopy (especially FTIR), thermal analysis (e.g., DSC) and microscopy techniques are useful for characterizing polymer blends. Rheology, physical and optical properties may also aid in quality control.

References

[1] H. Karlheinz, Compatibilizers, polymeric (recycling of multilayer structures), in: J.C. Salamone (Ed.), Polymeric Materials Encyclopedia, Vol 2, CRC Press, 1996, pp. 1364–1377.
[2] D. Öner-Deliormanli, B. Walther, E. Garcia-Meitin, S. Kodjie, Enhancing value of barrier film recycle stream with a novel recycle compatibilizer, in: TAPPI PLACE Conference, 2014.
[3] A. Ghijsels, J.J.S.M. Ente, J. Raadsen, Melt strength behavior of polyethylene blends, Int. Polym. Process. 7 (1) (1992) 44–50.
[4] R. Patel, T. Karjala, N. Savargaonkar, P. Salibi, L. Liu, Fundamentals of structure-property relationships in blown films of LLDPE/LDPE blends, J. Plastic Film. Sheeting 35 (4) (2019) 401–421.
[5] A.I. Fetel, Nylon-containing ionomer improves the performance of ethylene vinyl alcohol (EVOH) copolymers in multilayer structures, TAPPI Polymers, Laminations & Coat. Conf. Proc. (1997) 503–505.
[6] A.I. Fetel, Amorphous nylon improves the performance of EVOH copolymers in multilayer structures, Future-Pak (1977).
[7] R. Jester, Heat seal characteristics of cyclic olefin copolymer/polyethylene blends, TAPPI PLACE Conference (2002) 1308–1334. Boston, MA.
[8] G. Qiu, P. Zhang, J. Pan, Y. Xie, S. Zhoa, Mechanical properties and morphology of polypropylene toughened with metallocene polymerized polyethylene elastomers, Polym-Plastics Technol. Eng. 42 (1) (2003) 33–44.
[9] B.N. Epstein, Tough thermoplastic nylon compositions, US patent 4174358, 1979.
[10] Encyclopedia of Polymer Science and Technology. 2nd edn., Wiley & Sons, New York, 1987.
[11] DuPont, DuPont Selar® PA blends with nylon 6. <http://www2.dupont.com/Selar/en_US/assets/downloads/selar_pa3426_nylon_blends.pdf>, n.d. (accessed 06.04.14).
[12] P.M. Subramanian, Permeability barriers by controlled morphology of polymer blends, Poly. Eng. Sci. 25 (1985) 483–487.
[13] M.R. Kamal, H. Garmabi, S. Hozhabr, L. Arghyris, The development of laminar morphology during extrusion of polymer blends, Can. Polym. Eng. Sci. 35 (1995) 41–51.
[14] H. Garmabi, M.R. Kamal, Improved barrier and mechanical properties of laminar polymer blends, J. Plastic. Film. Sheeting 15 (2) (1999) 120–130.
[15] H.-X. Huang, Y.-F. Huang, S.-L. Yang, Developing laminar morphology in higher-viscosity ratio polyblends by controlling flow fields in a single-screw extruder, Polym. Int. 54 (2005) 65–69.
[16] B.A. Morris, Improving the adhesion of LDPE to aluminum foil through blending, SPE ANTEC Tech. Pap. (2004) 1101–1105.
[17] A.M. Soutar, Polymer combinations for peel/seal applications, TAPPI Semin. Notes, Coextrusion (1986) 31–43.
[18] S. Pirtle, L. Mergenhagen, J. Wooster, R. Cotton, The effects of resin selection and extrusion variables on peelable seal performance, TAPPI PLACE. Conf. (2004).
[19] C.C. Hwo, Polybutylene blends as easy open seal coats for flexible packaging and lidding, J. Plastic Film. & Sheeting 3 (4) (1987) 245–260.
[20] L.K. Mergenhagen, T.J. Pope, C.A. Timper, Polyolefin plastomer-based peelable seal formulations, TAPPI Polymers, Laminations & Coat. Conf. 1 (1996) 63–71.
[21] C. Martin, Evolution of direct extrusion using high speed twin screw extruders.", SPE ANTEC Technical Pap. (2002).
[22] B.H. Maddock, A visual analysis of flow and mixing in extruder screws, SPE J (1959) 383–389. May.
[23] Z. Tadmor, I. Duvdevani, I. Klein, Melting in plasticating extruders theory and experiments, Poly. Eng. Sci. 7 (1967) 198.
[24] E. Mount, A review of melting mechanisms in single screw extrusion, SPE ANTEC Conf. (2005) 304–310.
[25] M.A. Spalding, K.S. Hyun, Troubleshooting mixing problems in single-screw extruders, SPE ANTEC Tech. Pap. (2003) 229.
[26] G.A. Campbell, M.A. Spalding, Analyzing and troubleshooting single-screw extruders,", Hanser Publications, Munich, 2013.
[27] S. Middleman, Fundamentals of Polymer Processing, McGraw-Hill Book Company, New York, 1977, p. 154.
[28] B.A. Morris, D. Kirschbaum, Modification of PLA for improved processing and properties, Bioplastics Process. Conf. (2007). Charlotte, NC.
[29] J.C. Baird, J.P. Christiano, B.A. Morris, An extrusion study: examination of the improved processing characteristics of a PLA impact modified blend, SPE ANTEC Conf. (2009). Chicago, IL.
[30] C. Rauwendaal, Polymer Extrusion, Hanser, New York, 1986.
[31] R. Saxton, Extruder mixer screw, US patent 3006029, 1961.
[32] J. Schut, Single-screw compounding is learning new tricks, Plastics Technology, 2019. <https://www.ptonline.com/articles/single-screw-compounding-is-learning-new-tricks> (accessed 11.07.20).
[33] S.A. Somers, M.A. Spalding, J. Dooley, K.S. Hyun, An experimental study of the flows in an energy transfer screw, SPE ANTEC Tech. Pap. (2002) 307.
[34] T.A. Hogan, M.A. Spalding, E.K. Kim, R. Barr, J. Meyers, Performance analysis of a variable barrier energy transfer screw, SPE ANTEC Tech. Pap. (2003) 92.

[35] J.A. Myers, R.A. Barr, Improved screw design for maximum conductive melting, SPE ANTEC Tech. Pap. (2002) 154.
[36] G.A. Kruder, W.N. Calland, Fundamentals of wave screw extrusion, SPE ANTEC Tech. Pap. (1990) 74.
[37] P. Fan, J. Vlachopoulos, N. Smith, H. Sheth, Computer simulation of melt flow in wave screws, SPE ANTEC Tech. Pap. (1998) 97–101.
[38] S.A. Somers, M.A. Spalding, K.R. Hughes, J.D. Frankland, Performance of a Stratablend mixing screw for single-screw extrusion, SPE ANTEC Tech. Pap. (1998) 272.
[39] J.A. Myers, Feedscrew for injection molding and extrusion, US Patent 5,342,125, 1994.
[40] C. Rauwendaal, G. Ponzielli, Performance characteristics of elongational mixing screws, SPE ANTEC Tech. Pap. (2004) 178–182.
[41] J.L. White, Twin Screw Extrusion, Technology and Principles, Hanser Publications, Munich, 1990.
[42] J.E. Curry, Practical aspects of processing blends, in: M.J. Folkes, P.S. Hope (Eds.), Polymer Blends and Alloys, Blackie Academic & Professional, London, 1993.
[43] B.A. Wood, Transmission electron microscopy of detection of rubber modifiers, in: 161st Technical Meeting of the Rubber Division of the American Chemical Society, Savannah, GA, April 29-May 1, 2002, Paper 45.
[44] B.A. Wood, Transmission electron microscopy of polymer blends, in: K. Finlayson (Ed.), Advances in Polymer Blends and Alloys Technology, Vol 3, Technomic Publishing Co., Inc, Lancaster, PA, 1992.
[45] J. Munro, G. Wu, J. Huang, D. Zhang, S. Bawiskar, B. Morris, A review of impact modification technologies for different thermoplastics using ethylene copolymers, SPE Virtual ANTEC Conf. (2020).
[46] S. Wu, Phase structure and adhesion in polymer blends: a criterion for rubber toughening,", Polymer 26 (1985) 1855–1863.
[47] C.B. Bucknall, D.R. Paul, Notched impact behavior of polymer blends: part 1: new model for particle size dependence, Polymer 50 (2009) 5539–5548.
[48] C.B. Bucknall, D.R. Paul, Notched impact behavior of polymer blends: part 2: dependence of critical particle size on rubber particle volume fraction, Polymer 54 (2013) 320–329.
[49] H. Ohme, S. Kumazawa, J. Kumaki, Resin composition and molded article, film and fiber each comprising the same, US Patent application 2007/0260019A1, 2007.
[50] Coleman, M., Graf, J., Painter, P., 1991, Specific Interactions and the Miscibility of Polymer Blends, Techonomic Publishing, Lancaster, PA.
[51] D.W. Van Krevelen, Properties of Polymers: Their Estimation and Correlation with Chemical Structure, Elsevier Scientific Publishing Co., Amsterdam, 1976.
[52] Kuraray, Eval properties <http://www.eval-americas.com/us/eval-properties.aspx>, n.d. (accessed 11.07.20).
[53] E. McBride, DuPont Intern. memo. (1993).
[54] S. Wu, Polymer Interface and Adhesion, Marcel Dekker, Inc, New York, 1982.
[55] D.J. Walsh, personal. Commun. (2005).
[56] I.A. Hussein, M.C. Williams, Rheological study of heterogeneities in melt blends of ZN-LLDPE and LDPE: influence of Mw and comonomer type, and implications for miscibility, Rheol. Acta 43 (6) (2004) 602–614.
[57] I.A. Hussein, M.C. Williams, Rheological study of the miscibility of LLDPE//LDPE blends and the influence of Tmix, Can. Polym. Eng. Sci. 41 (4) (2001) 696–701.
[58] T. Hameed, I.A. Hussein, Rheological study of the influences of Mw and comonomer type on the miscibility of m-LLDPE and LDPE blends, Polymer 43 (25) (2002) 6911–6929.
[59] I.A. Hussein, T. Hameed, B. Sharkh, K. Mezghani, Miscibility of hexane-LLDPE and LDPE blends: influence of branch content and composition distribution, Polymer 44 (2003) 4665–4672.
[60] B.H. Bersted, J.D. Slee, C.A. Richter, Prediction of rheological behavior of branched polyethylene from molecular structure, J. Appl. Polym. Sci. 26 (1981) 1001.
[61] B.H. Bersted, On the effects of very low levels of long chain branching on rheological behavior in polyethylene, J. Appl. Polym. Sci. 30 (1985) 3751.
[62] U. Yilmazer, Effects of thc processing conditions and blending with linear low-density polyethylene on the properties of low-density polyethylene films, J. Appl. Polym. Sci. 42 (1991) 2379–2384.
[63] H. Benkreira, R. Britton, The optimization of masterbatch formulations for use in single screw machines, Intern. Polym. Process. 9 (1994) 205–210.
[64] G.I. Taylor, The formation of emulsions in definable fields of flow, Proc. Roy. Soc, Ser. A 146 (1934) 501.
[65] H.P. Grace, Dispersion in high viscosity immiscible fluid systems and application of static mixers as dispersion devices in such systems, Chem. Eng. Commun. 14 (1982) 225–277.
[66] S. Tomotika, On the stability of a cylindrical thread of a viscous liquid surrounded by another viscous fluid, Proc. Roy. Soc, Ser. A 150 (1935) 322.
[67] J. Kang, D. Bigio, T. Smith, Study of breakup mechanisms in cavity flow, SPE ANTEC Conf. (1995) 1988–1993.
[68] L.A. Utracki, Z. Shi, Development of polymer blend morphology during compounding in a twin-screw extruder. Part I: droplet dispersion and coalescence – a review," Poly, Eng. Sci. 32 (1992) 1824–1833.
[69] D. Bourry, B. Favis, Co-continuity and phase inversion in HDPE/PS blends: the role of interfacial modification, SPE ANTEC Conf. (1995) 2001.
[70] J. Eagan, J. Xu, R. Di Girolamo, C. Thurber, W. Macosko, A. LaPointe, et al., Combining polyethylene and polypropylene: enhanced performance with PE/iPP multiblock polymers, Science 355 (2017) 814–816.

[71] Y. Zeng, D. Abebe, A. Singh, L. Effler, Y. Hu, M. Kapur, et al., Compatibilizing polypropylene (PP) contaminated polyethylene (PE) waste streams in film applications, SPE Int. Polyolefins Conf. (2020).
[72] W. Sinthavathavorn, M. Nithitanakul, R. Magaraphan, B.P. Grady, Blends of polyamide 6 with low-density polyethylene compatibilized with ethylene-methacrylic acid based copolymer ionomers: effect of neutralizing cations, J. Appl. Polym. Sci. 107 (2008) 3090–3098.
[73] J. Torradas, R.O. Pimentel, D.M. Dean, Additives to improve regrind utilization and recycling of high barrier blow molded containers, SPE ANTEC Conf. (2013).
[74] I. Fortelný, J. Jůza, Description of the droplet size evolution in flowing immiscible polymer blends, Polymers 11 (2019) 761.
[75] H. Stone, Dynamics of drop deformation and breakup in viscous fluids, Annu. Rev. Fluid Mech. 26 (1994) 65–102.
[76] S. Guido, Shear-induced droplet deformation: Effects of confined geometry and viscoelasticity, Curr. Opin. Colloid & Interface Sci. 16 (2011) 61–70.
[77] S. Guido, V. Preziosi, Droplet deformation under confined Poiseuille flow, Adv. Colloid Interface Sci. 161 (2010) 89–101.
[78] S. Guido, F. Greco, Dynamics of a liquid drop in a flowing immiscible liquid, Rheology Rev. (2004) 99–142.
[79] P. Van Puyvelde, P. Moldenaers, Rheology and morphology development in immiscible polymer blends, Rheology Rev. (2005) 101–145.
[80] J.M. Dealy, J. Wang, Melt Rheology and Its Applications in the Plastics Industry, Springer, 2013.
[81] R.W. Flumerfelt, Drop break-up in simple shear fields of viscoelastic fluids, Ind. Eng. Chem. Fundam. 11 (1972) 312–318.
[82] T. Tavgav, Drop deformation and breakup in simple shear fields, PhD Thesis (Chemical Engineering), University of Houston, Houston, Texas, 1972.
[83] F. Gauthier, H. Goldsmith, S. Mason, Particle motions in non-Newtonian media. II: Poiseuille flow, Trans. Soc. Rheol. 15 (1971) 297–330.
[84] R. De Bruijn, Deformation and breakup of drops in simple shear flow (Ph.D. thesis), Eindhoven University of Technology, Eindhoven, The Netherlands, 1989.
[85] J. Elmendorf, R. Maalcke, A study on polymer blending microrheology: Part I, Poly Engr Sci. 25 (1985) 1041–1047.
[86] P. Varanasi, M. Ryan, P. Stroeve, Experimental study on the breakup of model viscoelastic drops in uniform shear flow, Ind. Eng. Chem. Res. 33 (1994) 1858–1866.
[87] C.L. Levitt, C.W. Macosko, S. Pearson, Influence of normal stress difference on polymer drop deformation,", Poly. Eng. Sci. 36 (1996) 1647–1655.
[88] F. Mighri, P. Carreau, A. Ajji, Influence of elastic properties on drop deformation and breakup in shear flow, J. Rheol. 42 (1998) 1477–1490.
[89] F. Mighri, M.A. Huneault, Drop deformation and break-up mechanisms in viscoelastic model fluid systems and polymer blends, Can. J. Chem. Eng. 80 (2002) 1028–1035.
[90] W. Lerdwijitjarud, A. Sirivat, R. Larson, Influence of dispersed-phase elasticity on steady-state deformation and break-up of droplets in simple shearing flow of immiscible polymer blends, J. Rheol. 48 (2004) 843–862.
[91] H.-P. Li, U. Sundararaj, Experimental investigation of viscoelastic drop deformation in Newtonian matrix at high capillary number under simple shear flow, J. Non-Newtonian Fluid Mech. 165 (2010) 1219–1227.
[92] P. Tanpaiboonkul, W. Lerdwijitjarud, A. Sirivat, R. Larson, Transient and steady-state deformations and breakup of dispersed-phase droplets of immiscible polymer blends in steady shear flow, Polymer 48 (2007) 3822–3835.
[93] V. Sibillo, M. Simeone, S. Guido, Break-up of a Newtonian drop in a viscoelastic matrix under simple shear flow, Rheologica Acta 43 (2004) 449–456.
[94] B. Lin, F. Mighri, M.A. Huneault, U. Sundararaj, Effect of premade compatibilizer and reactive polymers on polystyrene drop deformation and breakup in simple shear, Macromolecules 38 (2005) 5609–5616.
[95] B. Lin, U. Sundararaj, Sheet formation during drop deformation and breakup in polyethylene/polycarbonate systems sheared between parallel plates, Polymer 45 (2004) 7605–7613.
[96] B. Lin, U. Sundararaj, F. Mighri, M.A. Huneault, Erosion and breakup of polymer drops under simple shear in high viscosity ratio systems, Polym. Engr. Sci. 43 (2003) 891–904.
[97] C.D. Han, K. Funatsu, An experimental study of droplet deformation and break-up in pressure-driven flows through converging and diverging channels, J. Rheol. 22 (1978) 113–133.
[98] W. Milliken, L. Leal, Deformation and breakup of viscoelastic drops in planar extensional flows, J. Non-Newtonian Fluid Mech. 40 (1991) 255–379.
[99] I. Delaby, B. Ernst, R. Muller, Droplet deformation in polymer blends during uniaxial elongational flow: influence of viscosity ratio for large capillary numbers, J. Rheol. 38 (1994) 1705–1720.
[100] H.E.H. Meijer, J.M.H. Janssen, Mixing of immiscible liquids, in: Manas-Zloczower, Z. Tadmor (Eds.), Mixing and Compouding of Polymers, Hanser Publishers, Germany, 1994, pp. 85–147.
[101] H. Chin, C. Han, Studies on droplet deformation and breakup: droplet deformation in extensional flow, J. Rheol. 23 (1979) 557–590.
[102] R. Shanker, R. Spradling, H. Pham, S. Namhata, C. Bosnyak, Effect of viscoelasticity on transient drop dispersion, in: SPE ANTEC Conference, 1996, pp. 2596–2600.
[103] F. Mighri, A. Ajji, P. Carreau, Influence of elastic properties on drop deformation in elongational flow, J. Rheol. 41 (1997) 1183–1201.
[104] D.C. Tretheway, L.G. Leal, Deformation and relaxation of Newtonian drops in planar extensional flows of a Boger fluid, J. Non-Newtonian Fluid Mech. 99 (2001) 81–108.
[105] Oene, H.J. Van, Modes of dispersion of viscoelastic fluids in flow, J. Coll. Inter. Sci. 40 (1978) 448.
[106] S. Mukherjee, K. Sarkar, Effects of viscosity ratio on deformation of a viscoelastic drop in a Newtonian matrix under steady shear, J. Non-Newtonian Fluid Mech. 160 (2009) 104–112.

[107] F. Greco, Drop deformation for non-Newtonian fluids in slow flows, J. Non-Newtonian Fluid Mech. 107 (2002) 111−131.
[108] P. Maffettone, F. Greco, Ellipsoidal drop model for single drop dynamics with non-Newtonian fluids, J. Rheol. 48 (2004) 83−100.
[109] H. Chen, U. Sundararaj, K. Nandakumar, Modeling of polymer melting, drop deformation, and breakup under shear flow, Polym. Engr. Sci. 44 (2004) 1258−1266.
[110] C. Macasko, Morphology development and control in immiscible polymer blends, Macromol. Symp. 149 (2000) 171−184.
[111] J.J. Elmendorp, Van der Vegt, A.K., A study on polymer blending microrheology: Part IV. The influence of coalescence on blend morphology origination, Polym. Eng. Sci. 26 (1986) 1332−1338.
[112] S.P. Lyu, F.S. Bates, C.W. Macosko, Coalescence in polymer blends during shearing, AIChE J. 46 (2000) 229−238.
[113] E. Van Hemelrijck, P. Van Puyvelde, S. Velankar, C.W. Macosko, P. Moldenaers, Interfacial elasticity and coalescence suppression in compatibilized polymer blends, J. Rheology 48 (2004) 143−158.
[114] I. Fortelny, A. Zivny, Extensional flow induced coalescence in polymer blends, Rheologica Acta 42 (2003) 454−461.
[115] S. Uttandaraman, C.W. Macosko, Drop breakup and coalescence in polymer blends: the effects of concentration and compatibilization, Macromolecules 28 (1995) 2647−2657.
[116] S. Milner, H. Xi, How copolymers promote mixing of immiscible homopolymers, J. Rheol. 40 (1996) 663−687.
[117] A.J. Ramic, J.C. Stehlin, S.D. Hudson, A.M. Jamieson, I. Manas-Zloczower, Influence of block copolymer on droplet breakup and coalescence in model immiscible polymer blends, Macromolecules 33 (2000) 371−374.
[118] L.G. Leal, Droplet coalescence and breakup with application to polymer blending, J. Cent. South. Univ. Technol. 14 (2007) 1−5.
[119] J. Juza, I. Fortelny, Flow induced coalescence in polymer blends, Chem. & Chem. Technol. 7 (2013) 53−60.
[120] J. Vermant, G. Cioccolo, K. Golapan Nair, P. Moldenaers, Coalescence suppression in model immiscible polymer blends by nano-sized colloidal particles, Rheologica Acta 43 (2004) 529−538.
[121] S. Vandebril, J. Vermant, P. Moldenaers, Efficiently suppressing coalescence in polymer blends using nanoparticles: Role of interfacial rheology, Soft Matter 6 (2010) 3353−3362.
[122] D. Bourry, B.D. Favis, Co-continuity and phase inversion in HDPE/PS blends: influence of interfacial modification and elasticity, J. Polym. Science, Part. B: Polym. Phys. 36 (1998) 1889−1899.
[123] P. Martin, P.J. Carreau, B.D. Favis, R. Jerome, Investigating the morphology/rheology interrelationships in immiscible polymer blends, J. Rheology 44 (2000) 569−583.
[124] S. Steinmann, W. Gronski, C. Friedrich, Cocontinuous polymer blends: influence of viscosity and elasticity ratios of the constituent polymers on phase inversion, Polymer 42 (2001) 6619−6629.
[125] S. Steinmann, W. Gronski, C. Friedrich, Quantitative rheological evaluation of phase inversion in two-phase polymer blends with cocontinuous morphology, Rheologica Acta 41 (2002) 77−86.
[126] D. Wu, J. Zhang, M. Zhang, W. Zhou, D. Lin, The co-continuous morphology of biocompatible ethylene-vinyl acetate copolymers/poly (ε-caprolactone) blend: effect of viscosity ratio and vinyl acetate content, Colloid Polym. Sci. 289 (2011) 1683−1694.
[127] W. You, W. Yu, Control of the dispersed-to-continuous transition in polymer blends by viscoelastic asymmetry, Polymer 134 (2018) 254−262.
[128] N. Marin, B.D. Favis, Co-continuous morphology development in partially miscible PMMA/PC blends, Polymer 43 (2002) 4723−4731.
[129] G.-W. Pu, Y.-W. Luo, A.-N. Wang, B.-G. Li, Tuning polymer blends to cocontinuous morphology by asymmetric diblock copolymers as the surfactants, Macromolecules 44 (2011) 2934−2943.
[130] N. Kitayama, H. Keskkula, D.R. Paul, Reactive compatibilization of nylon 6/styrene−acrylonitrile copolymer blends. Part 1, Phase Invers. Behavior," Polym. (2000) 8041−8052.
[131] P. Potschke, D.R. Paul, Formation of co-continuous structures in melt-mixed immiscible polymer blends, J. Macromol. Sci., Polym. Rev. C43 (2003) 87−141.
[132] J. Lyngaae-Jorgensen, Rheology of polymer blends, in: M.J. Folkes, P.S. Hope (Eds.), Polymer Blends and Alloys, Blackie Academic & Professional, London, 1993.
[133] M. Huneault, M. Champagne, L. Daigneault, M. Dumoulin, Blend morphology development during compounding in a twin-screw extruder, SPE ANTEC Tech. Pap. (1995) 2020.
[134] A.K. Ghosh, S. Ranganathan, J.T. Lindt, S. Lorek, Blend morphology development in a single screw extruder, SPE ANTEC Tech. Pap. (1991) 232−236.
[135] J.T. Lindt, A.K. Ghosh, Fluid mechanics of the formation of polymer blends. Part I: Formation of laminar structures, Polym. Eng. Sci. 32 (1992) 1802−1813.
[136] H.-X. Huang, Y.-F. Huang, X.-J. Li, Detecting blend morphology development during melt blending along an extruder, Polym. Test. 26 (2007) 770−778.
[137] H.-X. Huang, G. Jiang, X.-J. Li, Development of polymer blend morphology along an extruder with different screw geometries, Intern. Polym. Process. 23 (2008) 47−54.
[138] H. Benkreira, R. Shales, M. Edwards, Mixing on melting in single screw extrusion, Intern. Polym. Process. 7 (1992) 126−131.
[139] C. Scott, C. Macosko, Morphology development during the initial stages of polymer-polymer blending, Polymer 36 (3) (1995) 461−470.
[140] U. Sundararaj, Y. Dori, C. Macosko, Sheet formation in immiscible polymer blends: model experiments on initial blend morphology, Polymer 36 (1995) 1957−1968.
[141] H. Burch, C. Scott, Evolution of structure in the softening/melting regime of miscible polymer mixing,", Polymer. Eng. Sci. 41 (6) (2001) 1038−1048.

[142] H.E. Burch, C.E. Scott, Effect of viscosity ratio on structure evolution in miscible polymer blends, Polymer 42 (2001) 7313–7325.

[143] R.C. Willemse, E. Ramaker, J. van Dam, A. Posthuma de Beor, Morphology development in immiscible polymer blends: initial blend morphology and phase dimensions, Polymer 40 (1999) 6651–6659.

[144] S. Tyagi, A.K. Ghosh, Morphology development during blending of immiscible polymers in screw extruders, Polymer. Eng. Sci. 42 (2002) 1309–1321.

[145] N. Domingues, A. Gaspar-Cunha, J.A. Covas, Estimation of the morphology development of immiscible liquid–liquid systems during single screw extrusion, Polym. Engr. Sci. 50 (2010) 2194–2204.

[146] K.J. Wilczynski, A. Lewandowski, K. Wilczynski, Experimental study of melting of polymer blends in a starve fed single screw extruder, Polym. Eng. Sci. 56 (2016) 1349–1356.

[147] K.J. Wilczynski, A. Nastaj, K. Wilczynski, A computer model for starve-fed single-screw extrusion of polymer blends, Adv. Polym. Techn. 37 (2018) 2142–2151.

[148] H. Li, G.-H. Hu, J.A. Sousa, Morphology development of immiscible polymer blends during melt blending: effects of interfacial agents on the liquid-solid interfacial heat transfer, J. Polym. Sci. Part. B: Polm. Phys. 37 (1999) 3368–3384.

[149] C. Shih, D. Tynan, D. Denelsbeck, Rheological properties of multicomponent polymer systems undergoing melting or softening during compounding, Poly. Eng. Sci. 31 (23) (1991) 1670–1673.

[150] C. Shih, Fundamentals of polymer compounding: The phase inversion mechanism during mixing of polymer blends, Adv. Polymer. Tech. 11 (3) (1992) 223–226.

[151] C-K. Shih, Morphology development during polymer blending: recent advances and further examinations of the phase inversion mechanism, in: SPE ANTEC Conference Proceedings, 55th ed., Vol 2, 1997, pp. 2588–2590.

[152] R. Ratnagiri, C. Scott, Phase inversion during compounding of polycaprolactone/polyethylene blends, in: SPE ANTEC Conference, 1996, pp. 59.

[153] U. Sundararaj, C. Macosko, C.-K. Shih, Evidence for inversion of phase continuity during morphology development in polymer blending, Poly. Eng. Sci. 36 (13) (1996) 1769–1781.

[154] H.-X. Huang, HDPE/PA-6 blends: influence of screw shear intensity and HDPE melt viscosity on phase morphology development, J. Mater. Sci. 40 (2005) 1777–1779.

[155] C. David, M. Trojan, R. Jacobs, M. Piens, Morphology of polyethylene blends: I. From spheres to fibrils and extended co-continuous phases in blends of polyethylene with styrene-isoprene-styrene triblock copolymers, Polymer 32 (3) (1991) 510.

[156] C. David, M. Getlichermann, M. Trojan, Morphology of polyethylene blends. II: formation of a range of morphologies in blends of polyethylene with increasing content of styrene-isoprene-styrene triblock copolymers during extrusion-blowing, Poly. Eng. Sci. 32 (1) (1992) 6.

[157] M. Getlichermann, C. David, Morphology of polyethylene blends: III. Blend films of polyethylene with styrene-butadiene or styrene-isoprene copolymers obtained by extrusion-blowing, Polymer 35 (12) (1994) 2542.

[158] R. Ronzalez-Nunez, B. Favis, R. Carreau, Factors influencing the formation of elongated morphologies in immiscible polymer blends during melt processing, Poly. Eng. Sci. 33 (13) (1993) 851.

[159] S.-Y. Lee, S.-C. Kim, Morphology and oxygen barrier properties of LLDPE/EVOH blends, Intern. Polym. Processing, XI (1996) 238.

[160] J.M. McKelvy, Polymer Processing, John Wiley, New York, 1967.

[161] K. Alemaskin, M. Camesasca, I. Manas-Zloczower, et al., "Entropic mixing characterization in a single screw extruder," SPE ANTEC Conference, 2004, pp. 167–172.

[162] K. Alemaski, I. Manas-Zloczower, M. Kaufman, Color mixing in single screw extruder: simulation & experimental validation, in: SPE ANTEC Conference, 2005, pp. 255–259.

[163] T. Lan, Nanocomposite materials for packaging applications, in: SPE ANTEC Conference, 2007.

[164] R. Leaversuch, Nanocomposites broaden roles in automotive, barrier packaging, Plast. Technol. (Oct 2001) 64.

[165] L. Sherman, Chasing nanocomposites, Plast. Technol. (November 2004) 56.

[166] E. Manias, K. Strawhecker, L. Wu, V. Kuppa, Concurrent enhancement of various materials properties in polymer/clay nanocomposites, Proc. Am. Chem. Soc. Compos. 16 (2001) 236–246.

[167] S. Dahman, Melt processed polymer-clay nanocomposites: properties and applications, SPE ANTEC Conf. 58 (2) (2000) 2407–2411.

[168] K. Akkapeddi, E. Socci, T. Kraft, J. Facinelli, D. Worley, A new family of barrier nylons based on nanocomposites and oxygen scavengers, SPE ANTEC Conf. 61 (3) (2003) 3845–3848.

[169] M. Krook, M. Hedenqvist, Polymer-clay nanocomposities: one way to improve the barrier properties of polymers used in packaging, TAPPI PLACE. Conf. (2002) 194–197.

[170] C. Lotti, C.S. Isaac, M.C. Branciforti, R.M.V. Alves, S. Liberman, R.E.S. Bretas, Rheological, mechanical and transport properties of blown films of high density polyethylene nanocomposites, Eur. Polym. J. 44 (2008) 1346–1357.

[171] J.-Y. Huang, J. Limqueco, Y.Y. Chieng, X. Li, W. Zhou, Performance evaluation of a novel food packaging material based on clay/polyvinyl alcohol nanocomposite, Innovative Food Sci. & Emerg. Technol. 43 (2017) 216–222.

[172] K. Fatyeyeva, C. Chappey, S. Marais, 13 - Biopolymer/clay nanocomposites as the high barrier packaging material: recent advances, in: A.M. Grumezescu (Ed.), Food Packaging, Academic Press, 2017, pp. 425–463.

[173] S. Bose, A.R. Bhattacharyya, A.P. Bondre, A.R. Kulkarni, P. Potschke, Rheology, electrical conductivity, and the phase behavior of cocontinuous PA6/ABS blends with MWNT: correlating the aspect ratio of MWNT with the percolation threshold, J. Polym, Sci., Part. B: Polym. Phys. 46 (2008) 1619–1631.

[174] R.R. Tiawari, D.R. Paul, Effect of organoclay on the morphology, phase stability and mechanical properties of polypropylene/polystyrene blends, Polymer 52 (2011) 1141–1154.
[175] A. Goldel, A. Marmur, G.R. Kasaliwal, P. Potschke, G. Heinrich, Shape-dependent localization of carbon nanotubes and carbon black in an immiscible polymer blend during melt mixing, Macromolecules 44 (2011) 6094–6102.
[176] L.A. Utracki, Polymer Alloys and Blends: Thermodynamics and Rheology, Hanser Publishers, New York, 1989.
[177] J. Dealy, K. Wissbrun, Melt Rheology and Its Role in Plastics Processing: Theory and Applications, Chapman & Hall, London, 1995, pp. 392–400.
[178] S. Marion, P. Pierce, Application of ree-eyring generalized flow theory to suspensions of spherical particles, J. Coll. Sci. 11 (1956) 80.

Part IV

Film properties

This section provides detailed descriptions and analysis of the key properties of packaging films from an engineering and science perspective. Each chapter begins with why the property under discussion is important, how to measure it, and typical values. This is followed by a discussion of the science behind how materials influence these properties. The effect of the converting process and end-use conditions on these properties is covered in more detail in Parts V and VI.

Chapter 7

Heat sealing in flexible packaging

Heat sealing is one of the primary unit operations of flexible packaging. Different methods of heat sealing are covered in Section 3.1.5. In this chapter, heat sealing is described from a science and engineering perspective: why it is important, how to measure it, and factors that influence performance, as well as the scientific fundamentals of heat sealing with an eye toward understanding how to troubleshoot heat seal issues. The technology of easy-open seals and reclosable technology is highlighted. Hot bar heat sealing, the predominant method used in flexible packaging, is the focus but ultrasonic sealing (US) is introduced. The chapter ends with a comparison of sealant attributes that may be important for selecting one sealant over another for a specific application.

7.1 Why it is important

Heat sealing in flexible packaging involves the fusion of one film with another or to a rigid container to create a hermetic seal. In hot bar sealing, the predominant method used in flexible packaging, heat and pressure from hot bars contacted with the film provides the driving force for fusion. An example of film-to-film sealing is the creation of a pouch or bag where the film is sealed to itself. An example of film-to-container sealing is a lidding film used to seal a tub or tray. Sealing helps keep product in the package until the consumer or user opens it and protects the product from the outside environment. A good heat seal helps the package provide one of its primary functions of protecting the product, as described in Chapter 1. For the packaging of food, the seal helps ensure product freshness.

In multilayer flexible packaging the innermost layer (the outside film layer closest to the product) provides the sealing functionality. Specialized sealant materials have been developed to optimize the sealing function, which often warrants its own layer. This is because polymer characteristics that provide good sealing may be suboptimum for other key attributes of the package, such as barrier, stiffness and strength. Other layers can provide these functions. The sealant layer may provide a number of features specific to the sealing function, including:

- Low seal initiation temperature for high-speed packaging operations and to prevent the distortion of other layers of the film.
- The ability to seal around or through product contamination in the seal area.
- Low extractables and minimum effect on food quality (e.g., flavor scalping or impartation).
- Tamper-proof or tamper evidence (especially for medical packaging).
- Seal integrity (strength and hermeticity).
- Ease of opening (specific seal strength).

Resistance to premature opening during the packaging operation (hot tack strength).

Even in multilayer packaging the sealant layer often is called upon to provide other attributes to the package beyond sealing. These may include:

- Transparency,
- Toughness (impact resistance, tear resistance, puncture resistance),
- Oil or chemical resistance,
- Moisture resistance,
- Abrasion resistance,
- Stiffness,
- Thermoformability,
- Adhesion,
- Recloseability.

The Science and Technology of Flexible Packaging. DOI: https://doi.org/10.1016/B978-0-323-85435-1.00007-7

Of particular importance is the ultimate seal strength, which determines the ease of opening the package, hermeticity, which relates to package integrity, and hot tack. Hot tack is the ability to resist opening forces during packing while the seal is still molten. This is critical since it can take several seconds before the seal area cools and the seal solidifies. During this time, the seal may be stressed by several external forces, as illustrated in Fig. 7.1. In vertical form fill seal operations, the weight of the product puts a burden on the seal immediately after filling. In horizontal filling operations, the gusseted areas tend to spring back, putting stress on the seal.

A sealant requires a balance of properties to ensure good performance:

- Low melting point to allow the molecular chains enough mobility to diffuse and build strength across the interface, yet high enough crystallinity to help aid in the build-up of that strength.
- Sufficient flow to fill or caulk voids in the seal area, but high enough viscosity to avoid being squeezed out from the seal area during sealing.
- Low molecular weight to speed diffusion across the interface, but high enough molecular weight and branching to provide melt strength for hot tack and solid state strength for seal strength.
- High polarity to provide good oil and grease resistance, yet low polarity to bond to adjacent layers.

Typical sealants used today include low-density polyethylene (LDPE), linear LDPE (LLDPE), metallocene LLDPE (including lower density plastomers), ethylene vinyl acetate (EVA) copolymers, acid copolymer resins (ACRs) and ionomers. Random polypropylene (PP) copolymers and polyesters are also used in some applications. Heat seal coatings based on EVA, ACR, acrylics, polyester, PP or polyvinylidene chloride (PVDC) are also used.

7.2 How to measure

7.2.1 Seal strength of packages: the peel test

Seal strength is a measure of the force required to pull open a package or a sealed film. In the lab, this is typically measured using a peel test. The peel test is generally performed on a tensile tester. The specimen is cut into strips, typically 1-inch (25.4-mm), 25-mm or 15-mm wide. The tabs or ends of the strip that have not been sealed are placed into the jaws of the tensile tester. These tabs are known as the peel arms. The jaws are separated at a fixed rate, typically in the range of 200–300 mm/minute (8–12 inches/minute), and the force is measured as a function of time. A schematic of the peel test is shown in Fig. 7.2. The peel force typically goes through a maximum as the peel is initiated and then settles down to a steady plateau value, as shown in Fig. 7.3. The maximum or plateau value may be reported as the seal strength, whichever is more appropriate for the application. The maximum value is most often used [1].

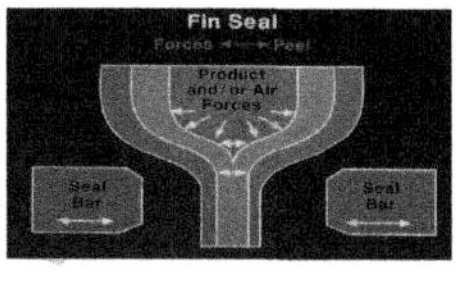

A. Internal package pressure

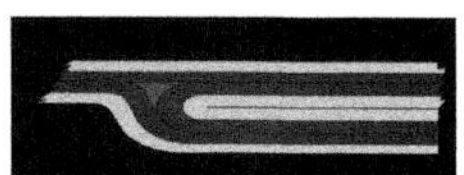

B. Holding shut wrinkles and folds

FIGURE 7.1 Examples of why hot tack is important. (A) Internal pressure from the weight of the product or air in a vertical form fill seal operation. (B) Holding shut wrinkles and folds in a horizontal form fill seal operation. *Reproduced with permission from The Dow Chemical Company.*

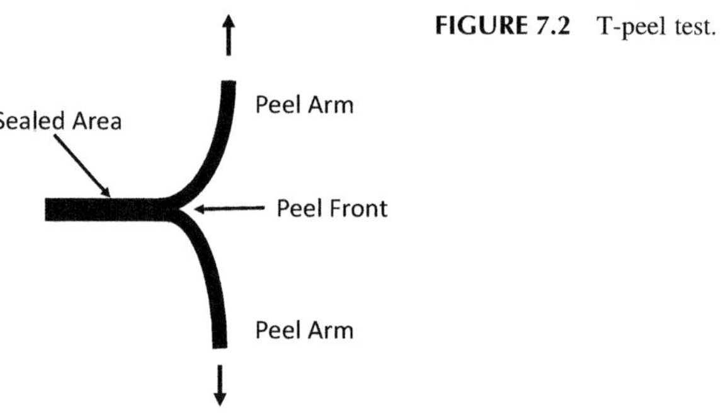

FIGURE 7.2 T-peel test.

The angle, speed, and temperature of the pull can all be important and affect the measurement result. As is described in more detail in Section 7.5.7, the peel test is a measure of the energy to pull the specimen apart which involves bending and stretching the peel arm and the creation of a new interfacial area. The angle of pull directly affects the bending energy, but this contribution is often small. More importantly, the angle can influence the concentration of stresses at the peel front, which can affect peel strength. In most controlled peel tests in the lab, the angle is fixed at 180 degrees in either a "T-peel" configuration (Fig. 7.2) or pulled completely back on itself. These typically give different results.

The speed of the pull affects the viscoelastic properties of the film, which can have a significant effect on the measured seal strength. As the seal is pulled apart, local resistance to deformation near the peel front can impact the energy to create new surfaces. This gives rise to greater resistance to peel. Tear strength and other properties can also play a role.

Temperature also impacts seal strength through its effect on the viscoelastic energy of deformation near the peel front as well as bulk properties of the peel arm such as elongation and stiffness.

The result of the heat seal measurement is typically reported as the seal strength or force divided by the width of the specimen. Some of the typical units and their conversions are given in Table 7.1. It is also helpful to record how the

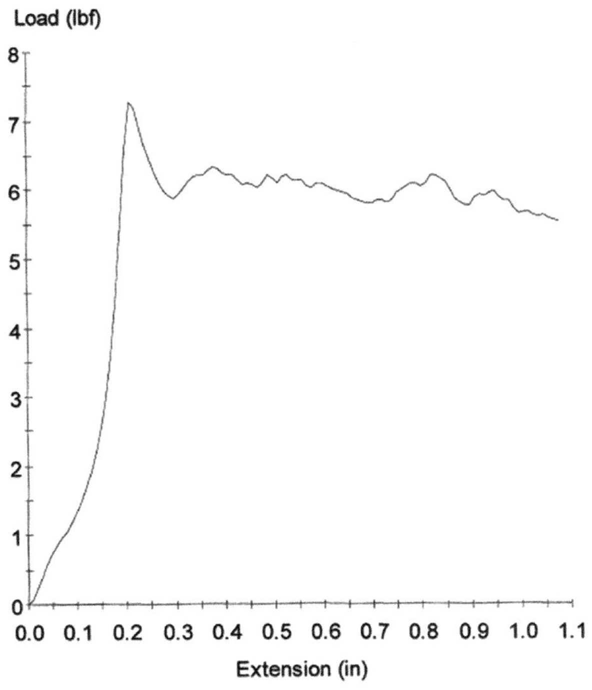

FIGURE 7.3 Example of peel force (load) measurement during a peel test. Extension refers to the amount of separation of the peel arms in the tensile tester.

TABLE 7.1 Some common units and their conversion factors for peel strength

	Multiply current value in row units by factor to get column units					
	lb/in	g/25 mm	g/15 mm	N/15 mm	N/25 mm	N/m
lb/in	1	454	757	7.42	4.45	178
g/25 mm	0.0022	1	1.67	0.0163	0.0098	0.392
g/15 mm	0.00132	0.599	1	0.0098	0.00587	0.235
N/15 mm	0.135	61.3	102	1	0.600	24
N/25 mm	0.225	102	170	1.67	1	40
N/m	0.00562	2.55	4.25	0.0417	0.025	1

specimen fails with descriptive language such as "peel followed by delamination," "peel followed by tear," "elongation," and "cohesive failure." These descriptors can help troubleshoot sealing problems (Section 7.10) and provide insight into variability in results. Often variability is due to changes in the failure mode.

7.2.2 Laboratory heat seal tests

Heat seal tests are often conducted in the lab under controlled conditions to compare materials and film structures. Several ASTM methods describe heat seal or peel tests, including F88, F2029, and F2824 [1–3]. The heat seal test generally involves the following steps:

- Conditioning the films for a prescribed amount of time at a fixed temperature and relative humidity (typically 23°C and 50% RH).
- Cutting strips to a standard width, typically 25.4 or 15 mm.
- Sealing the films on a laboratory hot bar sealer under a fixed set of conditions:
  - Bar width,
  - Sealing temperature,
  - Sealing time (known as the dwell time),
  - Sealing pressure,
  - Top and/or bottom bar heated.
- Conditioning the sealed films for a prescribed amount of time at a fixed temperature and relative humidity. For lab testing, the time is often 24 hours at 23°C and 50% RH.
- Measuring the heat seal strength, as described in Section 7.2.1. Typically, five or more specimens are measured for a given sample, or enough to ensure the accuracy of the measurements.

  Conditioning of the samples prior to heat sealing is important as the performance of some materials and films change with moisture and other effects of aging (see Section 7.10.2.5). For materials with suspected aging effects, conducting experiments with different aging times and conditions may be warranted.

  There is no standard set of heat seal settings and these are known to vary from one investigator to the next. Some typical ranges are given below:
- **Seal bar width:** 0.5–1 inch (12.7–25.4 mm), with 15–25.4 mm the most common.
- **Sealing temperature:** Usually varied over a range to construct a heat seal curve of seal strength versus sealing temperature.
- **Dwell time:** 0.25–3 seconds, with 0.5 seconds being most common. This is often set to match the sealing time on a packaging line. For example, a packaging machine operating at 60 packs/minute (1/second) with half the cycle time devoted to sealing would have a dwell time of 0.5 seconds.
- **Sealing pressure:** 20–70 psi (0.14–0.48 MPa). This is generally set at the minimum needed to ensure good contact.
- **One or both bars heated:** This is up to the investigator and should be noted.

Some heat sealers will have a Teflon film or tape covering the seal bar or the investigator will insert a high-temperature film [such as Teflon or polyethylene terephthlate (PET)] between the seal bar and the sample to prevent the specimen from sticking to the seal bar during sealing. This should be noted since it affects the heat transfer between the heated bar and the film.

Conditioning the sealed specimens after heat sealing is important to ensure the seal has cooled to room temperature and crystallization has occurred. Cooling can take several seconds, much longer than the heat-up portion of the heat seal process. This is because during heat-up the hot bar is in direct contact with the film surface whereas during cool-down the film may only be in contact with air. Heat transfer is much more efficient for direct contact than it is in air. The polymer may undergo various changes over the first 24 hours, such as secondary crystallization and moisture absorption. Hence, waiting 24 hours before conducting the peel test is more representative of what the seal strength may be in a package. Some seals are known to age over several weeks, especially some types of peel-seal or easy-open seals (see Sections 7.7.3 and 7.10.2.5). Longer conditioning times may be needed to study these systems.

7.2.3 Laboratory hot tack measurement

A number of home-grown hot tack testing techniques were developed in the early days of flexible packaging, several of which are reviewed by Soutar [4,5] and Brodie [6]. One device was developed by the Dow Chemical Company [7],

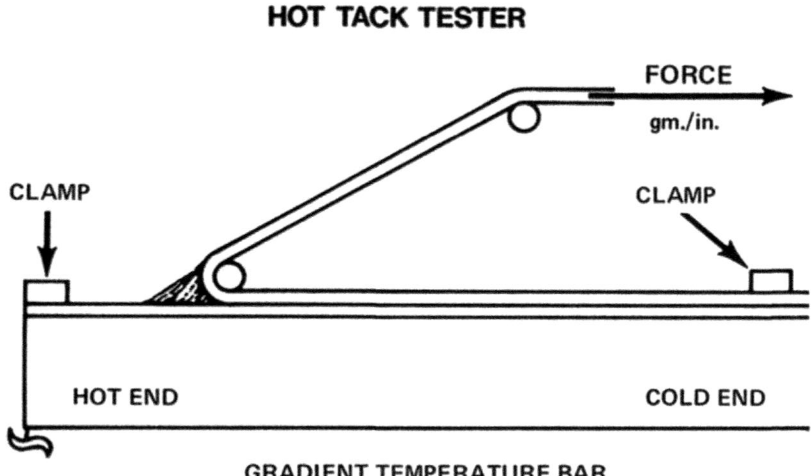

FIGURE 7.4 Dow chemical hot tack tester (c. the 1960s). *From Hot tack strength. Surlyn update, vol. 1, issue 4, 1970, DuPont. Reproduced with permission from The Dow Chemical Company.*

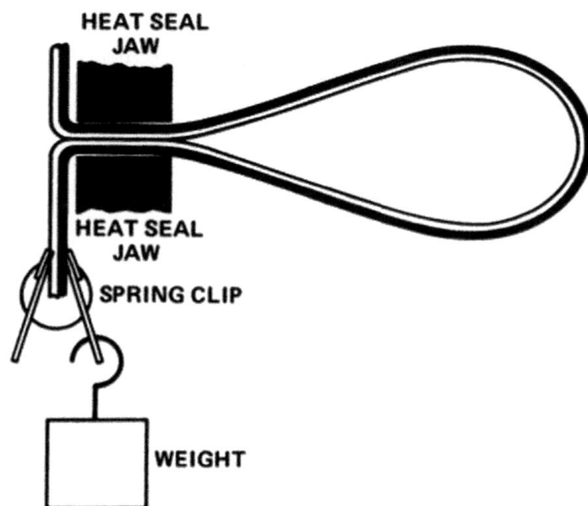

FIGURE 7.5 Deadweight hot tack test. *From Hot tack strength. Surlyn update, vol. 1, issue 4, 1970, DuPont. Reproduced with permission from The Dow Chemical Company.*

shown in Fig. 7.4 [8]. It measures the peel strength of a sample with an imposed temperature gradient. A sample of the film, sealant side up, is clamped onto a heated bar that is hot on one end and cold on the other. A second film sample, sealant side down, is placed on top of the first and clamped at the cold end. After the temperature has time to equilibrate, the bonded films are pulled apart from the hot end. The peel force is measured and plotted versus the distance of pull, which corresponds to a temperature profile measured by thermocouples along the length of the bar. The film requires a temperature-resistant outside layer for the clamps to hold onto and was used primarily for the testing of polyethylene coated paper in the late 1960s.

Dead weight tests were developed by DuPont to showcase their sealant resins at tradeshows [8]. A clip is fastened to one peel arm to which a weight is attached. As the seal jaws are released, the seal is exposed to the force of the weight, as shown in Fig. 7.5. Soutar [4,5] suggests that this technique is mostly a measure of the creep behavior of the polymer.

Frito-Lay developed a version of a dead weight test. A frame is used to control the distance and angle of the film extending from a standard bar heat sealer. A weight is hung from the frame so that when the seal jaws are released, a force is applied to the hot seal. The distance the seal has opened is measured and if it is above a critical value (typically 3/16 inch), the test is deemed a failure at that temperature and weight. If the seal creep is less than the critical value, the test is repeated with a higher weight [6]. Hot tack is reported as the maximum weight prior to failure for a given temperature.

The DuPont Spring Test was used well into the 1990s. Spring steel is cut in the shape of a tensile bar; the width of the bar at its narrowest point and thickness determine the spring force. The spring is bent into a U shape and inserted into the structure to be tested, as shown in Fig. 7.6. The ends of the film are placed into a laboratory heat sealer. The films are sealed under standard conditions. DuPont used a dwell time of 3 seconds to ensure the seal was complete and to accurately measure the interfacial temperature with a thermocouple. As the seal jaws are released, the spring immediately exerts a force on the molten seal. The sample is allowed to cool down and the separation of the seal to the nearest 0.1-inch is measured. Different springs representing different forces are used to measure several seal separations ranging from 10% to 90% separation. The results are plotted and interpolated to find the force required for 20% separation, which is deemed the hot tack strength for that sealing temperature. The process is repeated for different sealing temperatures to generate a hot tack curve.

In 1979, The Swedish Packaging Research Institute introduced the Packforsk hot tack testing device [4,9], an automated device that measures the force to pull the specimen apart. Later known as the DTC hot tack tester, this device (Fig. 7.7) consists of two heated sealing jaws, clamps to hold the test sample with the lower clamp dropping to pull the seal apart, a transducer to measure the force, and controls for setting the dwell time, temperature, pressure of the heat seal, peel speed and the delay time between when the seal is completed and when the force is measured. A sample with high heat resistant outer layers, such as paper or PET, is generally required to prevent distortion during the measurement. Various versions of this basic design have come out over the years, some of which are cam-driven and others motor-driven [10,11]. These quickly became the preferred method of measuring hot tack.

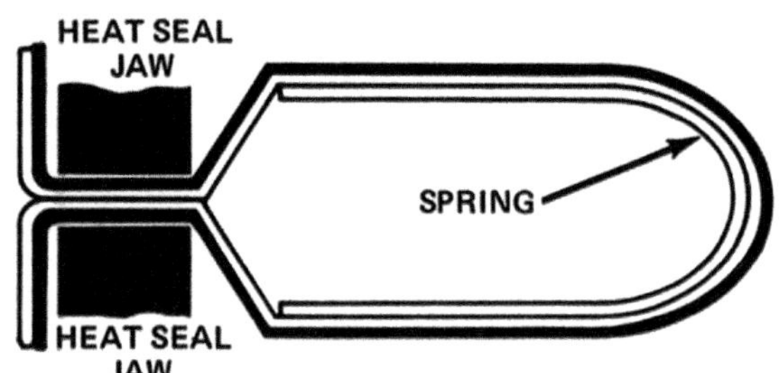

FIGURE 7.6 DuPont spring test for hot tack. *From Hot tack strength. Surlyn update, vol. 1, issue 4, 1970, DuPont. Reproduced with permission from The Dow Chemical Company.*

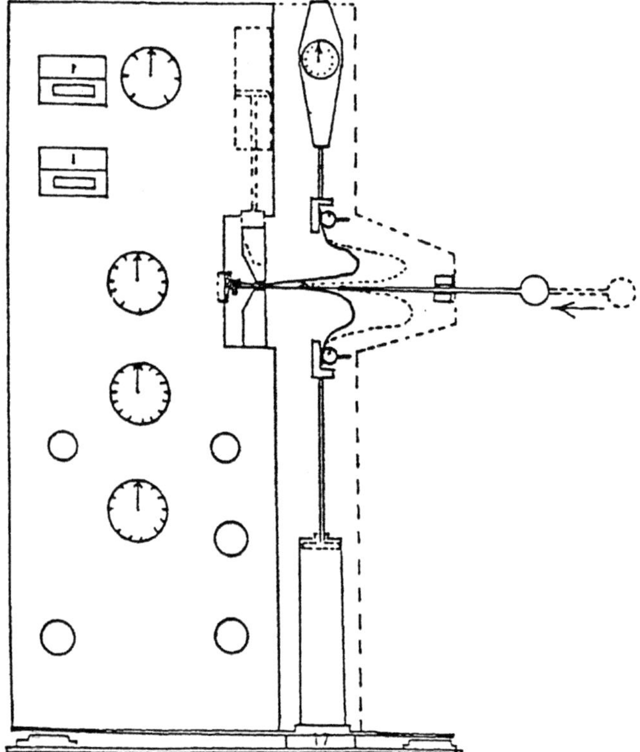

FIGURE 7.7 Packforsk hot tack tester. *Courtesy The Dow Chemical Company.*

Soutar [4,5] discusses the pros and cons of the DuPont Spring Test versus automated tests such as the Packforsk test. The spring test exerts the separation force immediately after the seal jaws are released, closely representing what typically occurs on a packaging line. The automated tests usually have a delay of 0.05–2 seconds following the seal before the force is measured. Cooling during this delay time was registered as a concern. Both methods are quantifiable, although the automated tests tend to only record the maximum force prior to full separation, whereas the spring method gives a wealth of information regarding the force for partial ("mooning") to full separation of the seal. (An automated device introduced by Theller [11] removes this negative.) Theller [11] suggests that the speed of pull in automated hot tack testers is not representative of the dynamics of packaging operations. Indeed, the speed is two orders of magnitude greater than the heat seal test. The spring method is less expensive in terms of initial capital outlay, but does require more time (to conduct each test, as well as to calibrate the springs prior to each test) and is subject to variability due to operator experience. Hence, the chief advantages of the automated hot tack devices are their speed and reproducibility.

ASTM F1921 [12] provides guidance around measuring hot tack but does not specify the equipment or conditions. Key variables for the automated hot tack testers include:

- Seal bar width,
- Seal dwell time,
- Sealing pressure,
- Delay time,
- Pull speed.

Falla and Li [10] survey the literature of major resin suppliers and lists a range of typical values used for these parameters. The seal bar width is 5 mm and the seal time is typically 0.5 seconds. Pressure ranges from 0.21 to 0.69 MPa. Delay time is 0.02–0.5 seconds and pull speed ranges from 33 to 500 mm/s.

Berger et al. [13] find that sample width, temperature and pressure are key factors influencing hot tack measurements. They use a Theller automated device. The method for preparing and cutting the sample is not found to be significant. Falla and Li [10] compare a 45 degrees cam-driven with a 90 degrees motor-driven hot tack tester for measuring hot tack of polyethylene films and conclude the motor-driven device gives more consistent and reliable results. Close examination of their results, however, suggests that differences in delay time may have influenced the results. Falla and Li go on to study factors that influence the variability of the motor-driven device:

- **Hand-cut versus punch die-cut specimens:** They find significantly more variation in the specimen width with the hand-cut specimens. Whether this translates into more variation in hot tack strength is not discussed. Berger et al. [13] examine the shape of the cut (straight vs sinusoidal) and find it has no statistically significant effect on the hot tack strength.
- **Teflon tape:** They recommend the use of Teflon tape on the seal bars to prevent sticking to an all polyolefin film.
- **Film tension:** The amount of slack in the film after the sample is loaded into the seal jaws is varied. Using a slight tension gives better consistency in the hot tack measurements.
- **Pressure across seal bar:** They recommend testing the uniformity of the pressure across the seal bar using a pressure indicating sensor film.
- **Seal time:** They cite literature sources that show the hot tack plateaus at around 0.5 seconds for polyolefin films and use this throughout their work. Hocking [9] shows that the onset of the plateau depends on the structure of the film being tested, with a thicker paper/LDPE/Al/LDPE film taking longer than a thinner Al/LDPE structure.
- **Peel speed:** They test pull speeds ranging from 100 to 500 mm/s. As the pull speed is decreased, the hot tack initiation temperature (HTIT) (which they define as the temperature at which 1 N/25 mm is achieved) and the maximum hot tack strength decrease. They find a pull rate of 200 mm/s results in the most differentiation between the different films tested.
- **Delay time:** Falla and Li use a delay time of 0.5 seconds. Hocking shows data varying delay time from 0.1 to 3 seconds for several structures. Hot tack strength is shown to increase and then level off with increasing delay time. He attributes the initial rise to the cooling of the interface, increasing the viscosity of the sealant. The leveling off is explained as radiant heat deforming the film, resulting in nonuniform peeling and lower values. Hocking states that the delay time in production packing machines is of the order of 0.1 seconds and they use this time for routine testing. Berger et al. [13] show data for delay times (or cooling time on the Theller instrument) of 0, 0.125, and 0.25 seconds for a (HDPE-EVA) film. Hot tack strength increased monotonically from 0 to 0.25 seconds. Soutar [4] questions the use of long delay times, citing cooling as a concern.

- **Sealing pressure:** Hocking recommends a sealing pressure of 0.1–0.5 MPa, which is typical of most heat sealing operations. Falla and Li use a pressure of 0.28 MPa (40 psi). Berger et al. use a range of 0.69–2.1 MPa.

7.2.4 Relating lab results to packaging line studies

Care must be taken when comparing hot tack or heat seal results from the literature. The film structure (thickness, layers), heat seal test parameters and hot tack test parameters and even preparation methods are likely to be different from one investigator to another. For example, Schubert et al. [14] find that foil containing structures have higher hot tack than nonfoil structures with the same sealant. They also show that the extrusion temperature in extrusion coating has a substantial impact on hot tack performance.

Even when looking at results from a single study comparing different sealants one should be cautious. The investigator may pick test conditions that are optimum for one type of sealant but not for another, skewing the results. For example, ionomers give excellent hot tack results under conditions of long seal times and very short delay times whereas metallocene plastomers give good results for shorter seal time but long delay times. Neither condition may be representative of a commercial packaging line. Studies on monolayer films may be useful for screening sealants, but in actual multilayer structures, interlayer adhesion and the tensile properties of the other layers may impact sealing performance. Indeed, laboratory heat seal and hot tack tests should be regarded as just a first step in qualifying materials. Ultimately, the structure needs to be tested on the actual packaging line. Lab tests do not always correlate with packaging line experience. Several examples are given by Wooster and Mergenhagen [15], Bilgen et al. [16] and Kapur et al. [17] where sealant films with the best hot tack performance in the lab do not provide the best hermeticity on a vertical form film seal packaging line. Factors such as caulkability may also play a role on the real packaging line.

Analysis of failures can be instrumental in finding the underlying cause of sealing problems. This subject is taken up in detail in Section 7.10. The ultimate test is the measure of leaker rates on actual package lines. Standard test methods for measuring packaging leakers are described in Table 7.2 [18]. A common method is to select sampling from the run and test for hermicity by immersing the sealed package in water under a vacuum and looking for bubbles. The location of the bubbles can be examined to try to determine the cause of the failure. An even more sensitive test is to inject helium into the sealed package and use a helium detector along the seal to find channel leakers [19]. In-line methods using gas detection, differential pressure, ultrasonic waves or infrared radiation are commercially available [19–24].

Kapur et al. [17] construct hermeticity maps of the operational window of packaging structures. Fig. 7.8 shows an example comparing two films run on a vertical form fill seal line. Hermeticity (measured with a bubble emission leaker test) is plotted versus sealing dwell time (based on the packaging rate of the line) and seal bar temperature. The maps are useful for assessing the breadth of the operating window, which involves many aspects of sealing including heat seal strength, hot tack strength, and caulking (flow) in gusset regions and other challenging areas in complex ways that

TABLE 7.2 Test methods for detecting package leakers

Method	Standard	Description
Gas Leak Test	ASTM F2391-05(2016)	Helium or CO_2 is sealed or injected into the package and leaks detected with a mass spectrometer. Indirectly measures leaks. Small leaks can be detected. In-line processes have been automated.
Ultrasonic Seal Test	ASTM E1002-11(2018)	Uses ultrasonic energy to detect location of leaks or contamination in seal area. Nondestructive, rapid and detects small leaks.
Vacuum Decay Test	ASTM F2338-09(2013)	Sealed package is placed in vacuum and pressure measured versus time. Rate of loss of vacuum is used to estimate size/number of leaks. Nondestructive but does not show where the leaks are located.
Bubble Emission Test	ASTM D3078-02(2013)	Sealed package is typically placed in a water bath under vacuum. Pressure differential provides driving force for leaks, which can be seen as bubbles. Rapid and allows location of leaks but only detects large leaks. Destructive.
Dye Penetration Test	ASTM F3039-15	A dye is place on one side of the seal and the opposite side is examined for penetration. Rapid but is destructive and only detects large leaks.

are not captured in a laboratory ASTM test. Their drawback is that they require large quantities of film and run time on a packaging line and are specific to a particular package format.

7.3 Typical values

7.3.1 Heat seal curves

A plot of seal strength versus seal bar temperature is known as the heat seal curve. It provides information about the performance and operating window of the film. An example is shown in Fig. 7.9. At low seal bar temperature, this is no tack or seal strength. As the temperature is increased, the heat seal reaches the heat seal initiation temperature (HSIT) where the seal strength rises rapidly. The actual definition of the onset of sealing varies in the literature and throughout the industry. Some define HSIT as the temperature at which the seal strength first becomes nonzero. Others define it as the temperature at which the seal strength reaches an arbitrary seal strength threshold, such as 100 g/inch or 1 N/25 mm. Because of the rapid rise in seal strength in this region, both approaches give similar results.

As the seal bar temperature is increased further, the seal strength will reach a plateau. The temperature at which the plateau begins is known as the plateau initiation temperature, T_{pi}. The seal strength at the plateau is known as the

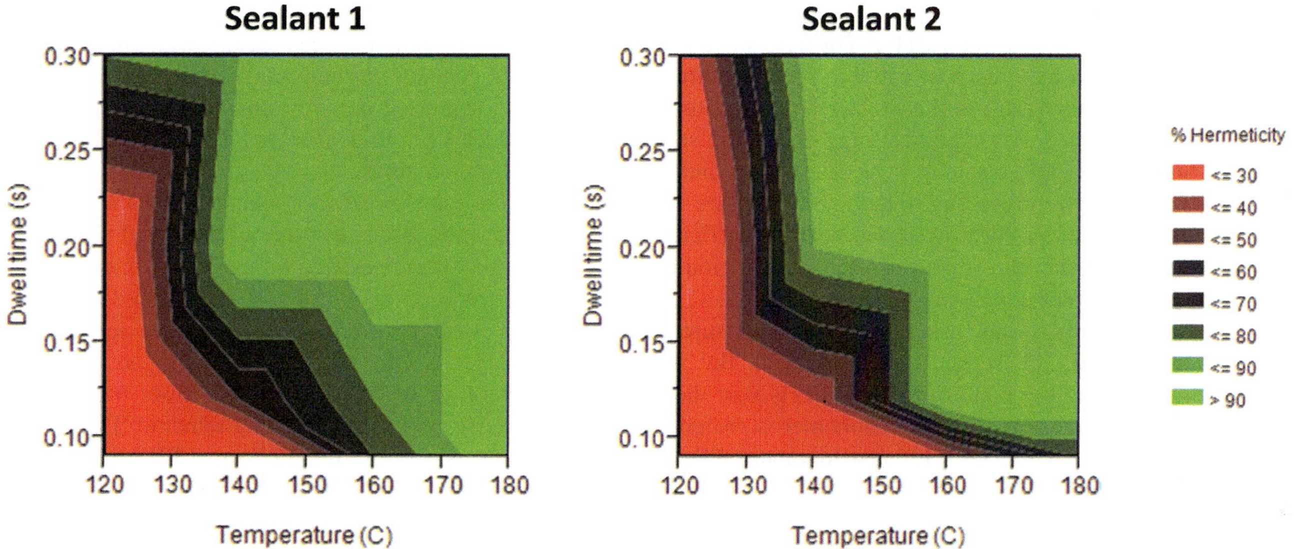

FIGURE 7.8 Example of a hermeticity map for two films [17]. *Reproduced from M. Kapur, M. Bilgen, P. O'Connell, Importance of application representative sealant test methods, in: SPE ANTEC, 2014, pp. 1393–1399, with permission from The Dow Chemical Company.*

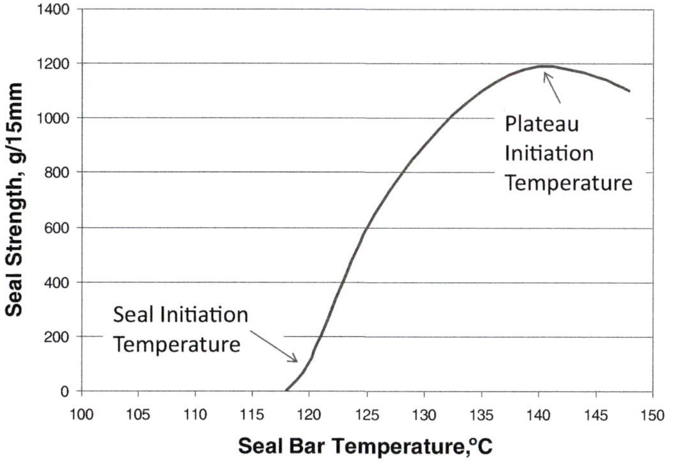

FIGURE 7.9 Typical heat seal curve.

ultimate seal strength. Ultimate seal strength values vary with sealant type, film structure and the film fabrication process (orientation, etc.). Tear strength, tensile strength and heat distortion often limit the value of the ultimate seal strength, which may depend on other layers in the structure [25]. Typical values are in the range of 400–1200 g/25 mm for easy-open seals and above 1200 g/25 mm for regular hermetic seals.

Li et al. [26] find that the ultimate seal strength of an mPE (0.918 g/cc density) monolayer blown film (25 μm thick) is considerably less than the ultimate tensile strength of the unsealed film. Wide-angle X-ray scattering reveals an increase in crystallinity during the heat sealing process and the orientation of the crystalline lamellae changes to a more isotropic state. This has also been observed by other investigators [27–32]. Tensile tests of film subjected to heat histories mimicking the heat seal process (exposing the film to temperatures above the melting point and cooling back down) show a 50% drop in tensile strength due to less strain hardening behavior. They propose that during the heat sealing process the region adjacent to the seal area sees enough heat to melt out the orientation in the film, leaving more isotropic orientation, less strain hardening and lower tensile strength. They hypothesize that this is responsible for the ultimate seal strength being lower than the ultimate tensile strength of an unsealed film. The region within the seal is double in thickness and so does not significantly contribute to this weakening. Regions outside the immediate adjacency do not see enough heat to melt out the orientation. Yamada et al. [33] also find a weakening of the film area adjacent to the seal area limits the seal strength of PP films.

The practical heat seal temperature window is defined by the HSIT on the lower end and by distortion and other limitations on the upper end. This is illustrated in Fig. 7.10. Section 7.5 shows that HSIT is related to the melting point of the sealant. Operating near the HSIT is not wise since in this region the seal strength varies greatly with small changes in temperature. This may result in variable seal strength, poor sealing and channel leakers. At the upper end, the excessive temperature may result in the sealant becoming too fluid and being squeezed out from the seal area, resulting in poor seal performance. Also, the heat may distort the structural layers, such as high-density polyethylene (HDPE), oriented PP (OPP), or PET, resulting in poor esthetics and functionality. High temperatures can also result in the seal bar cutting through the film resulting in tearing and failure. The middle region is the sweet spot where the seal strength has reached its maximum performance level, there is sufficient flow to caulk voids in gussets and wrinkles and seal around contamination, and the film is not distorted. This region is known as the heat seal operating temperature window.

To ensure a good operational temperature window, the sealant and structural layers should be matched for performance. For example, when using a high-temperature structural layer such as oriented polyethylene terephthalate (OPET) film, an LLDPE sealant may work well. But if the structural layer is OPP film, LLDPE may not be a good choice since the melting point of the LLDPE is near the distortion temperature of the OPP. Here a lower melting sealant, such as EVA, plastomer, ACR or ionomer is preferred to extend the operational temperature window. This is illustrated in Fig. 7.11 for LLDPE/PET and Ionomer/OPP. The OPP structure is 19-μm OPP/12.7-μm LPDE/8.9-μm met-OPET/(12.7-μm LDPE-25-μm ionomer) and the upper heat seal temperature limit is a result of the distortion of the

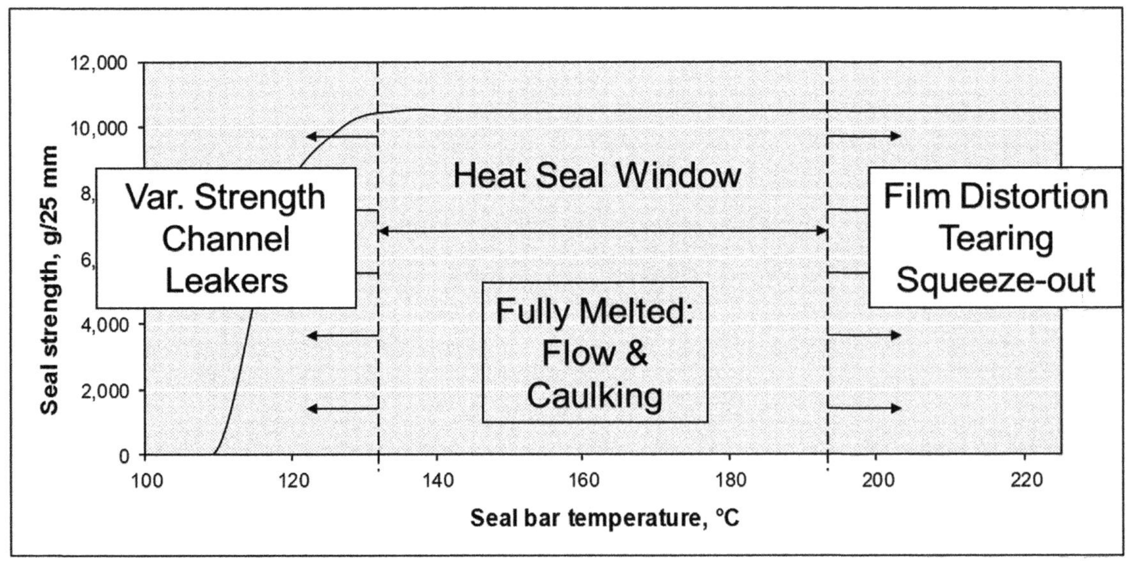

FIGURE 7.10 Illustration of heat seal temperature window. *Reproduced with permission of The Dow Chemical Company.*

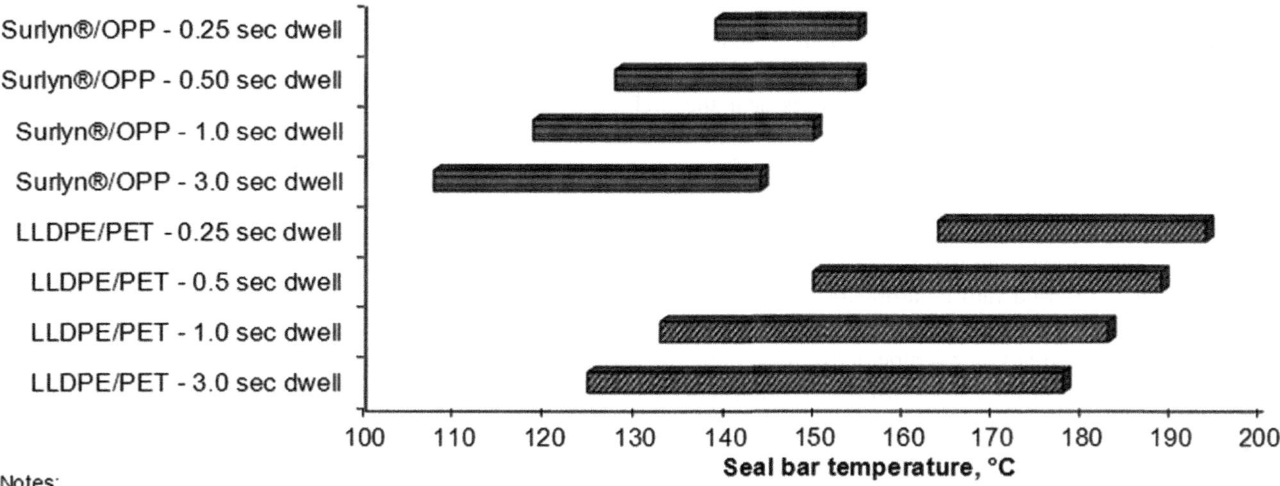

Notes:
1) OPP/Surlyn - (75 ga OPP/0.5 mil LPDE/35 ga met PET/0.5 mil LDPE/1.5 mil Surlyn coex)
Upper temperature limit is set by distortion due to melting of the OPP.
2) PET/LLDPE - (48 ga PET/0.7 mil LDPE/35 ga met PET/0.7 mil LDPE/1.5 mil coex (LDPE/LLDPE))
Upper temperature limit is set by distortion due to melting of the LDPE inside the structure.

FIGURE 7.11 Heat seal window of some packaging film structures. *Reproduced with permission of The Dow Chemical Company.*

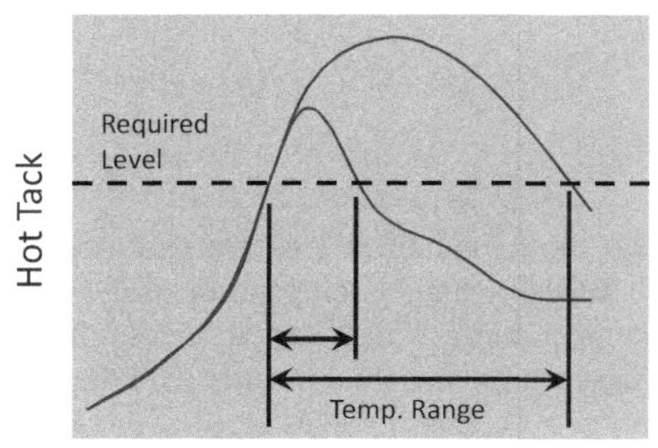

FIGURE 7.12 Hot tack versus seal temperature curves for two films illustrating differences in the breadth of the hot tack temperature window. *Reproduced with permission of The Dow Chemical Company.*

OPP. The PET structure is 12-μm OPET/17.8-μm LDPE/8.9-μm met-OPET/17.8-μm LDPE/(12.7-μm LDPE-25-μm LLDPE). Here the upper-temperature limit is much higher due to the temperature resistance of the outside OPET layer. Since both film structures have about the same thickness, the lower seal temperature limit is governed by the sealant layer. Especially at short dwell times, there is very little overlap in the seal temperature windows of the two structures, indicating the LLDPE sealant would not be suitable for the OPP structure since its HSIT would be higher than the distortion temperature (for 0.25 seconds dwell).

Fig. 7.11 shows the effect of four different sealing dwell times for each structure. As the seal dwell time is decreased, the seal initiation temperature increases. This is due to less time available for heat to travel through the film; the temperature at the sealant-sealant interface is reduced and the seal bar temperature must be increased to compensate. Thus, the seal temperature window is a function of both the design of the film structure and the sealing conditions.

7.3.2 Hot tack curves

A plot of hot tack strength versus seal bar temperature is known as the hot tack curve. An example is given in Fig. 7.12. Like the heat seal curve, the hot tack curve has an initiation region of rapidly rising hot tack strength, the onset of which is known as the HTIT. This is due to the build-up of strength at the interface and corresponds to the same region in the heat seal curve. The HTIT and HSIT temperature may differ due to differences in sealing conditions (dwell time, etc.) as well as the nature of the adhesion at the interface (melt adhesion for hot tack and solid-state adhesion for heat seal). Unlike the heat seal curve that plateaus, the hot tack curve typically goes through a maximum followed by a decrease in hot tack as the temperature is increased further. This is the result of a reduction in melt strength of thermoplastic polymers with increasing temperature.

In addition to the HTIT there are three other parameters of note. One is the temperature at which the maximum hot tack occurs. For most polyolefins, this occurs near the melting point of the polymer. The second is the peak or maximum hot tack strength. The third is the breadth of the hot tack curve, or hot tack temperature window. This is generally determined by a given hot tack strength and may be related to the force exerted on the seal during packing. For example, in a vertical form fill seal operation, the weight of the product acts as an opening force exerted on the molten seal. Gusset areas exert opening forces due to the numerous folds and the spring-back nature of polymer films. Once the package is filled, it may see forces due to handling or secondary packaging. For example, cereal is often packaged in a bag liner that is inserted into a cardboard box. The filled bag is stuffed into the box shortly after sealing; the stuffing force can be significant, and the seal may still be molten.

The breadth of the hot tack temperature range is perhaps more important than the peak hot tack strength since it provides information regarding the operational window. Changes in packaging machine line speed, nonuniformity of the temperature across the seal bar and other operational factors favor the use of structures with broad temperature windows to ensure good seal performance.

While the hot tack curve provides useful information, it is difficult to quantitatively translate the measurement to actual field experience. One reason is that calculating the opening force can be challenging, especially for a gusset. Also, most hot tack measurements today test to failure, whereas partial failure or "mooning" of seals may be what is important from a consumer perspective. The old DuPont spring method, now mostly out of favor, produced results for 20% of seal area failure. The maximum force that is measured today may not relate to actual field issues. Still, the hot tack curve is a useful means to compare different sealants and structures.

7.4 Factors that influence heat seal performance

7.4.1 Heat seal operation

7.4.1.1 Dwell time

Since heat sealing is a melt fusion process, time and temperature are the two most fundamental parameters affecting seal performance. Time and temperature are interrelated. For shorter dwell times, a higher temperature is typically needed to achieve the same performance. Conversely, for lower sealing temperatures, longer dwell times may be required. The sealing dwell time on a packaging machine is the time the sealing bars are in actual contact with the film. The dwell time is determined by the line speed and the fraction of the registration cycle that involves sealing. For example, if a machine is packaging 60 packages/minute and half the cycle time is devoted to heat sealing, the dwell time is 60 packages/minutes $\times$ minute/60 seconds $\times$ $1/2 = 0.5$ seconds.

Given that the dwell time is fixed by the machine parameters and line speed and that it takes a finite amount of time for heat to conduct through the thickness of the film once the seal bars contact the outer surface, the need for low seal initiation temperature sealants becomes apparent. The lower the HSIT, the faster the line speed can be for a given film structure.

7.4.1.2 Temperature

Temperature is important during the sealing operation as well as in the end-use. Time and temperature impact the melting of the sealant during sealing. This is critical for wetting, diffusion and flow (caulking). Too much temperature, however, can lead to distortion of the film and squeeze-out of the sealant. Oriented films, such as OPP and PET, are prone to shrinkage when the temperature is raised above their thermal setting temperature. Other films may melt and shrink. The viscosity of a polymer melt decreases with increasing temperature. Squeeze-out occurs when the sealant becomes too fluid and is pushed away from the seal area by the pressure on the seal. Some flow is desirable to fill voids and seal around contamination, but too much flow results in thinning down the sealant layer and loss of strength. Squeeze-out is further discussed in Section 7.6.4.

Nonuniform temperature across the seal bar can cause variability in seal strength and lead to leakers. Without proper control and maintenance, the temperature differential can be quite large. Even in a laboratory environment, significant temperature differences were found across the seal bars on packaging equipment, as is described in Section 3.1.5.4 and pictured in Fig. 3.3. The temperature may also vary throughout the day as a packaging line is slowed and sped up for various reasons.

The temperature the package is exposed to during its use can also influence seal strength. High temperatures experienced during hot fill or retort sterilization operations can put stress on the bond that has formed during sealing, especially if the exposure temperature exceeds the melting point of the sealant (and hot tack becomes an important predictor of package performance). Cold temperatures, such as blast freezing, may embrittle the film and cause seal failures.

Even the temperature at which the seal strength is measured can impact the result. Temperature affects the stiffness and other viscoelastic properties of the sealant and other layers in the structure, which can affect the dynamics of the peel test. See Section 7.5.7.

7.4.1.3 Pressure

Pressure is required to bring the films into intimate contact to allow bonding to take place. Too much pressure, however, can result in a squeeze-out of the sealant, especially at elevated temperatures. A general guideline is to use just enough pressure to ensure good contact and no more. Operators sometimes mistakenly increase the sealing temperature and pressure to try to resolve a sealing issue. This often results in worse seals rather than better ones.

7.4.1.4 Seal bar design

The crimper design and its effect on heat sealing are described in Section 3.1.5.3. Lengthening the seal (wider bars) is one approach to increasing hot tack as the separation distance is longer before seal failure. Crimping may also help. Wider bars also help prevent squeeze-out. Wider seals, however, require more film, which can result in higher cost.

Covering the seal bar with Teflon or other high-temperature resistant material can help prevent the hot bar from sticking to the film. This may be necessary if the outside layer of the film is not temperature resistant.

Some sealing operations heat just one bar whereas others heat both bars. This depends on the design of the packaging line. For example, in a lidding operation, it makes sense to only heat the film side since the substrate may be too thick for the heat to reach the interface within the time frame of the sealing operation.

7.4.1.5 Product contamination

There is no industry-accepted standard method for measuring how a sealant will perform in the presence of product contamination. Most references in the literature focus on food-based contaminants, both solid and liquid. Liquid contaminants such as various oils are often painted onto the seal area with a brush. These usually cover the entire seal area and generally do not represent the type of splattering that may occur on a packaging line. Solid contaminants include various powders, such as flour meal. These are often dusted on, again covering the entire seal area. It should not be surprising that most studies show that the presence of product contamination increases the seal initiation temperature, sometimes substantially. In some cases, the ultimate seal strength is also reduced.

Mihindukulasuriya and Lim [34] model the effect of water contamination in the seal area on heat transfer and melting of the sealant interface (see Section 7.6.1). They find the interfacial temperature is only decreased by about 4°C. The water fills in microvoids at the interface and increases heat transfer but acts as a heat sink. These two effects balance out. They hypothesize that an oil contaminant would wet a PE sealant more completely and may further reduce the resistance to heat flow. Heat transfer alone is not responsible for the poorer seal performance when liquid contaminants are present.

Hahm [35] took a different approach, studying the effect of a liquid marinade sauce on sealing. She postulates that the liquid contaminant is squeezed across the seal surface during sealing, resulting in leakers and channels across the entire seal width. Sealants that are able to encapsulate a liquid droplet should perform better. Using films of OPET/sealant (12/51 μm), she puts a droplet of controlled size (2.5 mm) onto the seal area and heat seals the films at 40 psi (0.28 MPa) pressure and 1 seconds dwell time. The sealants include LLDPE, EVA, mPE plastomer, and ionomer (zinc and sodium). She measures the final droplet size and describes the seal. Each sealant is found to have an optimum sealing temperature. At low temperatures, the sealant is not fluid enough to flow around the drop. At high temperatures, the marinade and sealant are too low in viscosity. The ionomers perform best, containing the droplet and preventing its expansion across the seal. The ionomer seals tightly even after venting steam. All the other sealants leave channels. Some examples from her study are shown in Fig. 7.13.

Ward and Li [36] consider sealing around solid contamination by simulating the contaminant with a metal wire. They are inspired by two ASTM methods, F2095 describing a pressure decay leak test for flexible packaging where changes in pressure are used to calculate the effective diameter of the leaker and F1929 for medical packaging where a tungsten wire is placed in the seal area to create a standard leak. They place a copper wire across the width of the seal area and seal it in place, creating a channel leaker. The size of the leak is estimated using the pressure decay method.

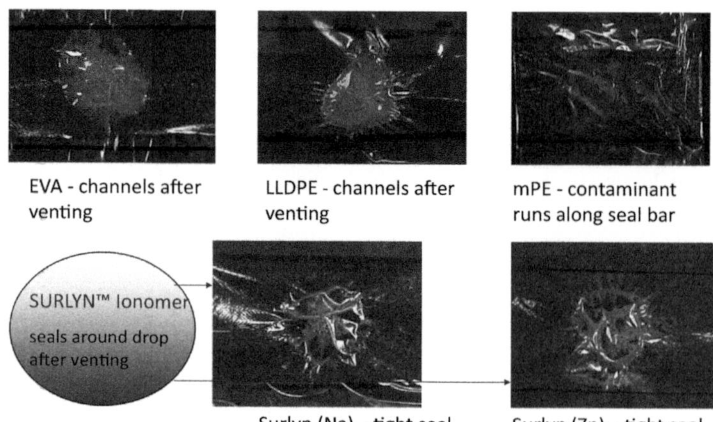

FIGURE 7.13 Examples of sealing in the presence of a liquid marinade sauce [35]. *From D. M. Hahm, Sealing study with meat marinades on seal surface, in: Am. Meat Inst. Conf., October 1999. Reproduced with permission from The Dow Chemical Company.*

A smaller calculated diameter indicates better caulkability or seal around the wire. Using a 9-layer coex film structure, (PA-tie-LLDPE-tie-PA-tie-LLDPE-LLDPE-sealant), they test 30 different sealant resins, including various LLDPE resins and plastomers, EVAs and an ionomer. The ionomer gave the best results followed by several plastomers. The leaker size decreases with:

- Increasing wire diameter. This was unexpected. They think it may be due to increased pressure near the edge of the wire, resulting in more flow of the sealant.
- Decreasing zero shear viscosity of the sealant (biggest effect).
- Increasing sealant hot tack (next biggest).
- Decreasing sealant density (of LLDPE).
- Narrower temperature transition from high-viscosity semisolid to low viscosity fluid of the sealant.
- Increasing sealant secant modulus.

They postulate there are three regimes important for caulking and sealing around solid contamination:

1. During the heat seal, where low viscosity is important for flow around the contaminant.
2. During cooling of the seal, where the hot tack is critical.
3. After the seal has cooled down, where a high modulus helps resist pressures near the solid contaminant.

Ionomers provide a unique combination of outstanding hot tack, relatively high modulus (for a sealant resin) and flowability.

Still, many of the test methods are unsatisfying and seem to be subject to manipulation. Most of the studies have been conducted by resin manufacturers and show their resin to be better than a competitive resin [35–40]. Competing manufactures counter with studies showing their product performs best. Some general concepts can be considered:

- When dealing with a powder product, it helps to minimize static attraction between the sealant and the product. This helps prevent contamination from reaching the seal area. For example, DuPont recommends an amide (slip agent) containing ionomer for situations where static is an issue [41].
- The chemical affinity of the sealant to the contaminant can be important, as is shown in the Hahm study. Ionomers and ACRs are more polar than PE and may be better suited for containing oily contaminants.
- The presence of long-range intermolecular attractive forces such as hydrogen bonding and ionic interactions may aid in sealing through contamination, helping the sealant to lock onto the opposing sealant surface. These types of forces are found in acid copolymers and ionomers.
- The rheology of the sealant may be important, especially for sealing around and encapsulating solid contaminants. Linear polyethylene often has more flowability during sealing than branched PE. This is supported by Ward and Li's study. Sadeghi and Ajji [42], however, show that greater shear thinning behavior can lead to better caulking, which would favor branched polymers like LDPE, EVA, ACR and ionomers.
- Hot tack can help the sealant encapsulate the particle or droplet [36,43].
- Increasing seal bar pressure can help push contaminants out of the seal area or assist in flow around them [34]. Too much pressure can result in problems as previously discussed.

- Ultrasonic sealing works well for sealing through some bulk product contamination. The vibration of the horn helps push contaminants away from the seal area [44]. See Section 7.9.

7.4.2 Film fabrication and package design

7.4.2.1 Film thickness

In hot bar sealing, the time for heat to conduct through the thickness of a film is related to the square of the thickness. The thicker the film structure, the longer the dwell time needs to be for the sealant interface to reach the bar temperature. For some high-speed processes, the dwell time is too short and the thickness of the film too great for the interfacial temperature to actually reach the bar temperature. In these situations the bar temperature is set above the desired interfacial temperature to achieve sufficient performance. The limit to this approach is the dimensional stability of the film structure; if the sealing temperature is too high, the film may get distorted and other sealing methods may need to be used.

Fig. 7.14 shows an example of the effect of thickness on the heat seal curve. Here the sealant is the same ionomer resin. In the thin structure, the sealant is a monolayer film whereas in the thick structure the sealant is laminated to paper. The dwell time is constant at 0.5 seconds. The HSIT of the thick structure is about 25°C higher than the thin structure. Section 7.6 on modeling of heat seal performance revisits this example in more detail.

The thickness of the film structure may affect the measured seal strength by changing the bending stiffness during the peel, which may affect the peel strength directly through a change in the energy to bend the peel arm and indirectly by changing the local angle of failure and stress concentration. The thickness of the sealant also affects the amount of local viscoelastic energy dissipation as new interfacial area is created during the peel test. These factors are taken up in greater detail in Section 7.5.7.

7.4.2.2 Sealant resin

As is described in more detail in Section 7.5, heat sealing involves melting and fusion of the sealant at the interface. Semicrystalline polymers that are commonly used as sealants do not have a single melting point like small molecules do. Rather they melt over a temperature range. A common method used to measure melting behavior is by differential scanning calorimetry (DSC), which measures the heat flux while heating or cooling a sample compared to a reference. Primary transitions such as melting and crystallization show up as peaks. A typical DSC curve plotting heat flux versus temperature is shown in Fig. 7.15. The melting point typically reported in a technical data sheet is the value at the peak of melting. Some of the crystals will melt well before the peak temperature, allowing for some tack. Others will melt after the peak, especially those associated with longer chains. Stehling and Meka [45] conducted a comprehensive study of the melting behavior of polyolefins used as sealants and correlated their DSC melting behavior with heat seal strength. They found that the temperature of the onset of the plateau or ultimate seal strength (T_{pi}) correlates with the temperature at which all the crystals have melted and the HSIT corresponds to a certain fraction of melting. Hence the melting point of the sealant is a key parameter that can be used to predict seal initiation temperatures. Table 7.3 gives the melting points of some common sealant resins. In general, the lower the melting point, the lower the HSIT.

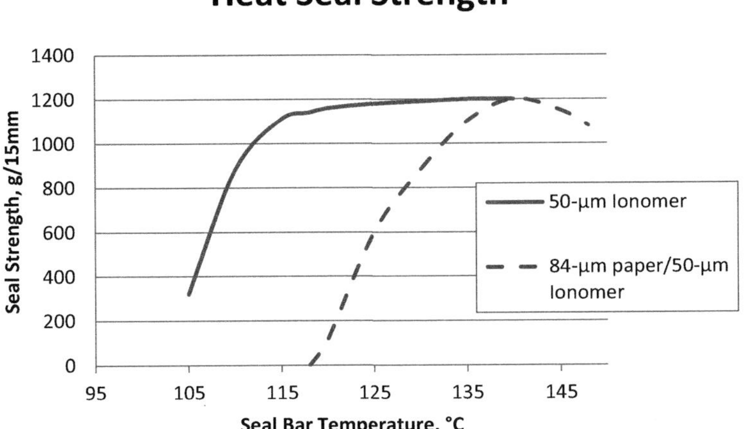

FIGURE 7.14 Heat seal curves for two structures of different thickness. Sealing conditions: 0.5 s dwell time, 40 psi (0.28 MPa) seal pressure, 25 mm bar width, top bar heated only. *Reproduced with permission from The Dow Chemical Company.*

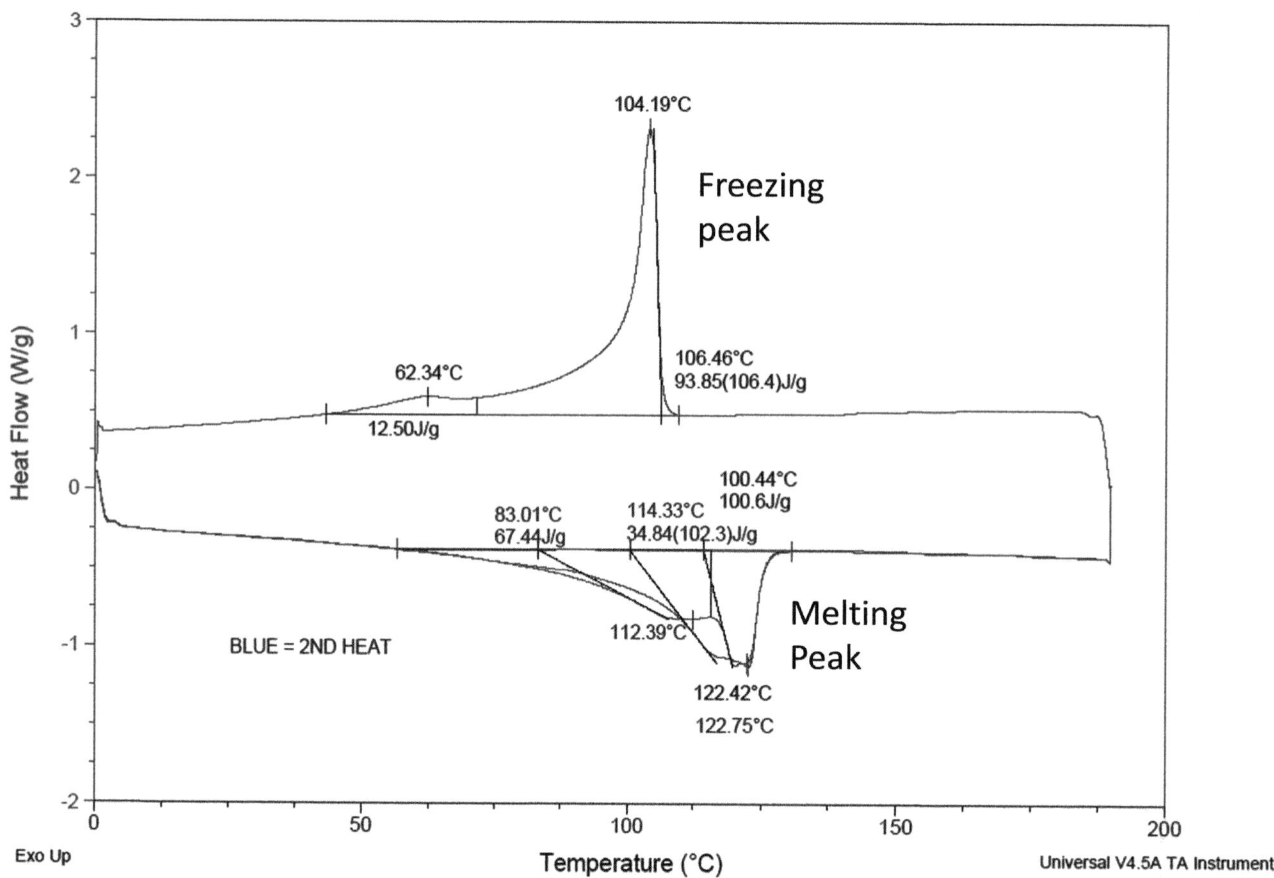

FIGURE 7.15 Differential scanning calorimetry curve for polyethylene. *Courtesy The Dow Chemical Company.*

As described in Section 4.2.2.2 one of the factors affecting melting behavior of LLDPE is their molecular weight distribution (MWD) and comonomer distribution. Single-site or metallocene catalyzed LLDPE has a much narrower MWD and comonomer distribution than their Ziegler—Natta (Z—N) counterparts. This results in lower peak melting points as well as narrower temperature ranges over which melting takes place. A comparison of the melting points of the two technologies is given in Fig. 4.5.

The molecular architecture also influences the ultimate seal strength. Linear polyethylene (LLDPE, HDPE) generally has higher ultimate seal strength than polyethylene and ethylene copolymers containing long-chain branching (LCB) (LDPE, EVA, ionomers, etc.). This is likely due to differences in ultimate tensile strength and tear strength behavior.

7.4.2.3 Substrate adhesion

For lidding and other applications where the sealing film must bond to a different substrate (as opposed to heat sealing of a pouch where the sealant seals to the same material), the affinity or adhesion of the sealant to the substrate becomes important. In many lidding applications, the tray or cup is made of a polymer with a high melting point, such as PET, or even a nonpolymer material such as paperboard or aluminum. The substrate may not melt during the sealing operation and the development of seal strength relies on the sealant wetting and adhering to the surface of the substrate. A number of factors play a role (see Chapter 10) including the ability of the sealant to flow, tackiness, and polarity.

7.4.2.4 Package design

The film structure ideally suited for hot bar sealing has an outer layer that is temperature resistant and an innermost layer that has a low seal initiation temperature. This provides the greatest sealing temperature window for operations, allowing, for example, higher seal bar temperatures to be used to achieve strong seals at shorter dwell times. This permits faster cycle times and line speeds.

TABLE 7.3 Differential scanning calorimetry peak melting temperature for common sealants

Polymer	DSC peak melting temperature (°C)
EVA	
4% VA	101–105
7.5% VA	98
12% VA	95
18% VA	89
EMA	
6% MA	102
18% MA	81
23% MA	75
27% MA	64
Acid Copolymer	
7% AA	102
9% AA	97
4% MAA	109
9% MAA	103
12% MAA	97
Ionomer	88–100
PE	
HDPE	135
MDPE	126–130
LLDPE (Ziegler–Natta)	118–125
LDPE	98–115
mLLDPE, 0.910–0.920 g/cc	100–110
Polyolefin plastomer, 0.885–0.910 g/cc	80–100
Polyolefin elastomer, 0.87–0.88 g/cc	60
PP	
Homopolymer	161
Random copolymer	145–155

Gussets or folds in the film in the corners of pouches are prone to sealing problems due to the difference in thickness compared with the rest of the seal area. This presents challenges in pressure distribution and the transfer of heat within the many interfaces within the gusset region. The folds provide spring-back forces that can lead to seal failure. The folds also create void regions where the sealant must flow or "caulk" to prevent leakers.

Hot bar sealing of lidding films onto thick trays or cups can be a challenge; the thick flange of the tray can act as a heat sink and reduce the temperature of the interface.

As flexible packaging structures become more sophisticated with more and more layers, care must be taken to ensure the sealing performance is maintained. Factors such as interlayer adhesion and the hot tack performance of layers next to the sealant become important, especially as the sealant layer is downgauged. This is brought out in a study by Vansant [43] that examines the effect of coextrusion on the hot tack performance of an ionomer, a sealant known for its broad hot tack window. Vansant coextrusion coats three layers onto 50-μm aluminum foil: Al/(ACR-B-

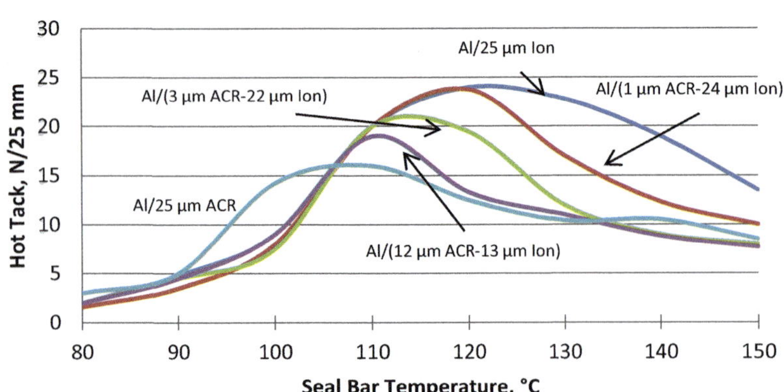

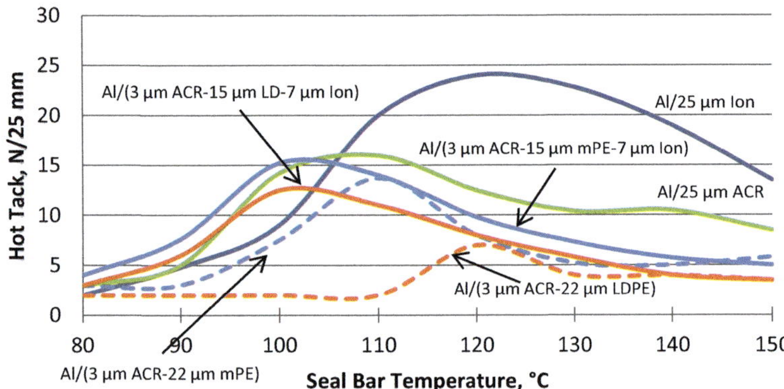

FIGURE 7.16 Hot tack of coextruded structures. Top graph is acid copolymer resin (ACR)-Ionomer coextrusion coatings onto Al foil: Al/(ACR-Ionomer). Bottom graph is ACR-LDPE-Ionomer coextrusion coatings: Al/(ACR-LDPE-Ionomer) [43]. *From J. D. Vansant, The effect of coextrusion on the hot-tack performance of a common ionomer resin, in: TAPPI PLC Conf., 1998, pp. 743–754. Reproduced with permission from The Dow Chemical Company.*

Ionomer) and measures hot tack and heat seal performance. The goal is to understand the effect of downgauging the sealant layer by adding a less expensive bulking layer (layer B). The ACR is chosen to ensure good adhesion to the aluminum foil. Three different resins for the B layer are studied: the ACR, a LDPE and a metallocene plastomer (mPE). For controls, all ACR (ACR-ACR-ACR) and ionomer (Ion-Ion-Ion) coatings are also run. The total thickness of the coating is 25-μm and the ratio of the B to ionomer layer is varied.

Selected hot tack results are shown in Fig. 7.16. The ratio of ionomer to B layer and composition of the B layer has a substantial impact on the performance:

- The 100% ionomer control gives the highest hot tack over a broad temperature range.
- The (ACR-ACR-Ion) structure approaches the hot tack performance of the ionomer when the ACR layer is thin compared to the ionomer. Thinner layers of ionomer result in poorer hot tack. Vansant attributes this to cohesive or internal failure within the ACR. The thinner the ionomer layer, the easier it is for a crack to break through this layer and reach the ACR, lowering hot tack.
- With LDPE as the core layer, the hot tack is good until the LDPE reaches its melting point. Below Tm (around 120°C) interfacial adhesion seems to dominate; failure occurs at the LDPE-ionomer interface. Above 120°C enough LDPE is melted to transfer failure to within the LDPE layer and the hot tack approaches that of LDPE.
- With mPE as the core, the hot tack resembles that of the ACR alone. The melting point of the mPE is similar to that of the ionomer, increasing the likelihood that the core layer will flow with the ionomer layer to help seal and caulk wrinkled and gusseted areas.

Vansant points out that this work involves measuring hot tack on a lab tester that measures the force to completely pull the specimen apart. This may be too severe for some sealing situations. In some circumstances, such as seal around contamination and preventing channel leakers in gussets and wrinkled areas, "surface hot tack" may be all that is needed to obtain good seal integrity. These applications may be able to tolerate a sublayer that has poorer hot tack than the surface sealant layer.

Ward and Crowe [46] use a one-dimensional heat conduction model to show how placing a layer next to the sealant layer that has higher thermal conductivity conducts the heat away from the sealant faster after sealing. They hypothesize this will help broaden the hot tack temperature window of the package by enhancing melt strength. Hot tack measurements on experimental film structures containing layers of a medium-density polyethylene (MDPE) with a higher thermal conductivity bear this out.

7.4.2.5 Additives, treatments, and converting

Tackifiers and plasticizers are sometimes added to polymers to promote adhesion. The former may alter the surface chemistry and change the viscoelasticity of the sealant. Plasticizers generally help increase the mobility of rigid polymer chains, aiding in diffusion. They may also act as lubricants, reducing adhesion by decreasing intermolecular slippage.

Low molecular weight fatty acid amides are often added to control the coefficient of friction (COF) of the film. These "slip agents" bloom to the surface to create a low friction coating. They typically do not interfere with heat seal performance unless too much is used, which results in a "chalking" effect on the surface. At normal levels, they are thought to revert back into the polymer as the sealant melts.

Various lubricants and processing aids such as silicone oils can interfere with seal performance by acting as contaminants or weak boundary layers at the interface.

Inert fillers such as calcium carbonate or talc may reduce seal performance by reducing the amount of polymer-polymer contact and increasing the viscosity, which affects spreading, wetting and caulking.

Farley and Meka [47] find that corona treatment increases HSIT and reduces the ultimate seal strength of LLDPE. Corona treatment is usually used to promote adhesion (to ink or coatings) and is typically not applied to the sealant side of the film. Some treatments may inadvertently expose the sealant side in what is called back side treatment. In these cases, sealing may be an issue. Interestingly, at very low levels of treatment, Farley and Meka find that seal strength is enhanced. The improvement is negated if a slip agent is added. At higher levels of treatment, the HSIT increases by 5°C–17°C. They compare the effect with crosslinking the surface with ebeam or chemical crosslinking and conclude that the negative effect of corona treatment on heat seal is due to crosslinking of the PE.

Richards [48] evaluates the effect of corona treatment and processing conditions on heat seal and hot tack of LDPE extrusion coated onto paper. Increasing corona treatment level increases HSIT but also increases the ultimate seal strength (contrary to what Farley and Meka find for LLDPE). Hot tack strength decreases with increasing treatment level; the untreated control has 50% higher hot tack than treated samples. Richards also finds increasing the time in the air gap (32–58 ms) or increasing the coating temperature (from 307°C to 327°C) on the extrusion coating line is detrimental to hot tack. Lower temperature reduces adhesion to the paper, which can be overcome by ozone treatment of the melt curtain in the air gap (to promote oxidation). Crosslinking of PE can occur with excessive thermal history in the film-fabrication or coating process. This is especially true in high-temperature processes such as extrusion coating and cast film.

7.5 Science of heat sealing

Heat sealing involves the formation of an adhesively bonded interface through melting, flowing, wetting, and either diffusion of polymer chains across the interface or the presence of attractive chemical interactions. A successful heat seal allows mechanical work to be transferred across the sealed interface. The resulting seal typically either has a diffuse or a sharp interface. The diffuse interface is the result of the substantial movement of molecular segments across the interface over time. This type of seal typically occurs in film-to-film heat seal applications involving a common sealant on each side of the interface. To be consistent with the literature the term "autoadhesion" is used here to describe this type of sealing. It is also sometimes called "self-adhesion." With sufficient chain penetration across the interface, the interface becomes diffuse and given enough time disappears (and becomes part of the bulk polymer). Long-range chemical interactions such as covalent, ionic or electrostatic forces are not necessary; nonpolar polymers such as polyethylene that lack chemical functionality for such long-range forces work well for autoadhesion. While their secondary intermolecular forces are individually weak, they become numerous as penetration proceeds. Entanglement of their high molecular weight chains and crystallization further build strength across the interface.

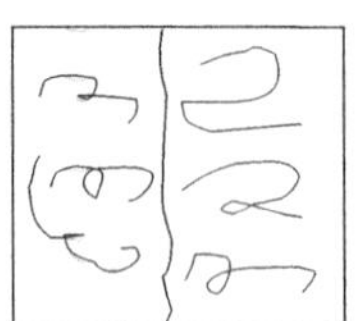

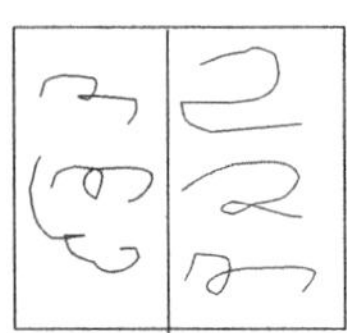

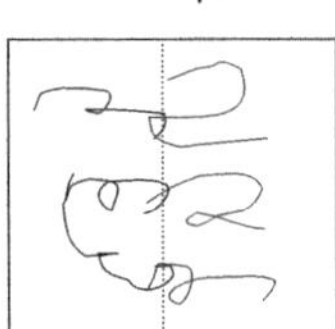

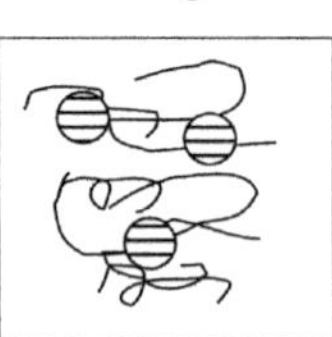

FIGURE 7.17 Mechanisms of self-adhesion during heat seal. *HSIT*: heat seal initiation temperature; T_{pi}: plateau initiation temperature; T_c: crystallization temperature [45]. *Adapted from F. C. Stehling, P. Meka, Heat sealing of semicrystalline polymer films. II. Effect of melting distribution on heat-sealing behavior of polyolefins, J. Appl. Polym. Sci. 51 (1994) 105–119.*

Sharp interfaces often occur in lidding applications where the sealant film bonds to a rigid tray, cup or bowl. Here the tray material may not melt during the sealing process. The sealant melts and wets the surface of the tray material but cannot penetrate the surface of the tray substrate that is still crystalline or glassy. The seal forms almost instantaneously due to intermolecular attractive forces across the interface and are strongly influenced by the chemical nature of the two surfaces.

This section focuses primarily on the diffusion mechanism of heat sealing since sealing films to themselves (autoadhesion) predominates in flexible packaging. The bonding of dissimilar substrates is described in Section 7.7.4 (lidding films) and adhesion more generally in Chapter 10.

The modern view of autoadhesion during hot bar sealing is shown schematically in Fig. 7.17. The two films are brought together and heat and pressure is applied from the outside. Initially the surfaces of the films are rough on a microscopic scale and little molecular contact is made. As the heat transfers through the film and begins to melt (or soften in the case of amorphous polymers) the interface, the surfaces of the films become smoother, allowing more intimate contact and wetting. The partial melting also produces tack. Further melting of the sealant results in molecular motion and the start of diffusion across the interface. This accounts for the sharp rise in seal strength with increasing temperature in the seal initiation region of the heat seal curve. Interfacial strength increases with increasing molecular penetration, and upon cooling with crystallization. Seal strength increases with increasing temperature and diffusion until the ultimate seal strength is reached. A plateau region follows where the seal strength exceeds the tensile or tear strength of the film or regions adjacent to the seal area [26]. At even higher temperatures the film distorts, tears or enough sealant is squeezed out of the seal area, limiting the seal performance.

Stehling and Meka [45] evaluate many types of polyolefins and find seal initiation occurs when the temperature reaches the point where, on average, 77% of the polymer is amorphous, as determined by DSC (see Fig. 7.15). This accounts for differences in the initial crystallinity of the polymer; a sealant with a lower level of crystallinity, to begin with, will reach the point of 77% amorphous at a lower temperature. Indeed, semicrystalline polymers with low levels of crystallinity, and particularly polyethylene and ethylene copolymers, have long been used as sealants for their low seal initiation temperatures. Mazzola et al. [49] report similar findings for propylene-ethylene copolymers where the HSIT corresponded to 40% melting of the crystalline domains.

Stehling and Meka, and van Alsten [50] find that that the onset of the plateau region of the heat seal curve corresponds with the temperature at which all the crystals have melted (the temperature where the melting peak completely disappears on the DSC curve). Van Alsten finds this also corresponds to the peak hot tack temperature for polyethylene. Najarzadeh and Ajji [51] show that the correlation of HSIT and T_{pi} with amorphous fraction in DSC measurements depends on the dwell time (the time the hot bars are in contact with the film); longer dwell time decreases the amount of melting needed to build strength at the interface. The crystals that melt at the highest temperature are associated with the longest molecules in the polymer. It is only after the longest chains are free from the confines of the crystals that they are able to migrate across the interface. The longest chains are needed to build maximum strength at the interface [52].

The diffusion and strength build-up at the interface is influenced by the molecular forces present in the polymer. The strength and reach of these forces determine how much molecular penetration is needed to achieve the ultimate seal strength. With enough time and temperature, the interphase region will disappear, leaving a continuous bulk polymer. Further strength builds as the polymer crystallizes and the interpenetrating chains participate in the crystalline structure.

7.5.1 Early theories of autoadhesion

By the early 1960s, the diffusion model of autoadhesion of high molecular weight polymers was gaining acceptance as the primary mechanism for heat sealing. McKelvey [53] describes heat sealing along the lines depicted in Fig. 7.17. Voyutskii [54] reviewed the literature on autoadhesion in his monogram in 1963, arguing that the diffusion model best explains the experimental results of the day. While most of his examples do not involve heat sealing and packaging, they provide the basis for our understanding today. Two other early theories no longer favored are briefly described here.

One is the adsorption theory where adhesion is thought to be due to secondary interactions between the surface molecules of the adhesive and substrate; the nature and magnitude of the chemical interaction are unimportant since secondary interactions are present in all polymers. Secondary bonds, also called intermolecular dispersive forces or van der Waals forces, are described in Section 4.2.1.5.2. These are weaker bonds that do not include covalent, ionic or electrostatic forces that make up the polymer chain. Wetting was thought to be the key. The stages of adhesion according to this theory include migration of the polymer chain to the surface followed by the establishment of an adsorption equilibrium that brings the contact between the molecules of the adhesive and substrate to less than 5 Å needed for secondary forces to be important. This theory may play a role in the adhesion of polymers to solid substrates such as glass but early proponents may have fallen victim to thermodynamic arguments showing wetting alone gives a very high work of adhesion. Wu [55] discounts this since the work of adhesion due to wetting is an equilibrium calculation for reversible fracture. Almost all adhesion applications, however, involve irreversible fracture and large viscoelastic deformation of the polymer. The calculated ideal adhesion due to wetting is often two orders of magnitude larger than actual measurements due to the presence of defects and stress concentration. For sharp interfaces, secondary forces alone are not sufficient to overcome these defects. For diffuse interfaces, where the polymer segments have migrated sufficiently across the interface, secondary forces can provide good adhesion. This gives credence to the diffusion theory for autoadhesion.

The electrical theory of adhesion was proposed by Deryagin and Krotova [56–59] and is based on the phenomena of two dielectrics. The adhesive substrate is compared with a condenser and the double electric layer produced at the interface of two polymers is analogous to the plates of the condenser. When stripping the adhesive from the substrate a potential difference develops that increases with separation distance. Some observations of electron emissions during stripping were observed that gave credence to the theory. Voyutskii gives a number of arguments against this theory including that it does not explain adhesion between nonpolar polymers or conductive polymers and calculates that any electric contribution is dominated by molecular diffusion.

7.5.2 Wetting

There are three primary factors that influence bonding at the interface:

1. Wetting;
2. Diffusion;
3. Interfacial bond strength and ordering (crystallinity).

The wetting of solid substrates is dealt with in more detail in Chapter 10. The thermodynamic driving force for wetting can be described by a spreading coefficient:

$$\lambda_{12} = \gamma_2 - \gamma_1 - \gamma_{12} \tag{7.1}$$

where

λ_{12} = spreading coefficient of substance 1 onto substance 2
γ_1 = surface tension of substance 1 (adhesive)
γ_2 = surface tension or energy of substance 2 (adherend)
γ_{12} = interfacial tension of substances 1 and 2.

For spontaneous wetting, $\lambda_{12} \geq 0$. For autoadhesion, the two substances are the same. Therefore, $\gamma_1 = \gamma_2$ and $\gamma_{12} = 0$ so that $\lambda_{12} = 0$. For the case of dissimilar sealant and substrate (e.g., lidding applications) matching solubility parameters, first introduced in Chapter 6, to minimize interfacial tension can help improve wetting and adhesion.

In autoadhesion, wetting is not instantaneous. The surface is rough, at least on a micros[illegible]ic scale, and sufficient time (and temperature) is needed to smooth out the surface for wetting. The development of [illegible] proceeds as wetted pockets randomly form and radiate outwardly as the polymer melts and flows. The [illegible] area, $\Phi(t)$, can be modeled as a two-dimensional nucleation and growth process [59]

$$\Phi(t) = 1 - \exp\left(-kt^m\right)$$

where

Φ = fraction of interface wetted
t = time
k, m = constants.

The application of pressure helps ensure interfacial contact and heat transfer for melting of the sealants, a[illegible] [illegible]quent flow and wetting. The time dependence of viscous flow for wetting and diffusion is similar since they depend on the same molecular dynamic process. Hence, it is difficult to isolate the importance of each mechanism. Some investigators have proposed that wetting alone is sufficient for adhesion (as in the adsorption theory described in Section 7.5.1) but the preponderance of evidence suggests diffusion is also important for developing strength at the interface [55,59].

For nonpolar sealants like polyethylene, wetting and diffusion are the primary mechanisms for adhesion development. These molecules contain no polar or chemical functionality and rely on very short-range secondary (dispersive) forces for their cohesive strength. Dispersive forces are relatively weak by themselves. Significant penetration and interdiffusion of molecular segments across the interface are required for enough of these secondary bonds to form to build strength. The rate of diffusion is thus critical and is influenced by molecular architecture, thermal motion (temperature) and time. Chain entanglement and crystal formation also serve to build seal strength.

With polar sealants, additional forces come into play such as hydrogen bonding and ionic interactions. These forces may hinder the diffusion rate, but because they are longer-range forces, cohesive strength can build with less penetration than for sealants relying solely on secondary interactions.

Polar sealants are also ideally suited to wet and adhere to dissimilar substrates found in lidding applications, especially if the substrate is polar in nature, such as PET. Here a sharp interface is typically created and dispersion forces alone are not sufficient for good adhesion development. This topic is further explored in Section 7.5.4.

7.5.3 Diffusion at the interface

Diffusion involves the movement of molecular segments. The transfer of whole chains across the interface has the effect of erasing the interface and transforming interfacial properties into bulk ones. Such complete movement and erasure of the interface is difficult in the time scales of heat seal operations, and may not be necessary. Ends or small segments can move more readily across the interface and Wool [60] shows that these are responsible for strength development. To build strength, sufficient penetration is required for the creation of molecular bridges and entanglements, and possibly co-crystallization. For polar molecules longer-range attractive forces may also be important.

The polymer needs to be in the melt state (or rubbery state for amorphous polymers) so that the molecules have enough thermal energy for movement and are not constrained by the crystalline domains. As described earlier, Stehling and Meka [45] found that seal initiation only takes place once most of the polymer is in the amorphous or melt state.

The modern theory of molecular dynamics, developed by de Gennes [61], Rouse [62], Doi and Edwards [63,64], among others, has brought theoretical grounding to the proposals of Voyutskii and others. Wool [60] nicely summarizes the current thinking. In the bulk amorphous polymer, the chains are constrained by the entangled network structure. These constraints confine the movement of the chain to a tube-like region defined by the instantaneous location of the entanglements and the chain. The chain can wiggle within the tube due to thermal fluctuations (Brownian motion). This results in random movement of the chain along the contour of the tube. As the head of the chain extends out from the tube by a chain segment length, the tail of the tube disappears, and a new tube is formed. In this way, the chain "snakes" its way through the network structure. This is known as reptation theory, coined and first proposed by de Gennes and refined by Doi and Edwards.

Wool [60] defines five time regions associated with diffusion that occur in order:

1. Short-range Fickian diffusion of monomers. By monomers, Wool is referring to the diffusion of individual repeat units on the polymer chain.
2. Rouse relaxation of entanglements. (P. E. Rouse [62] developed the molecular dynamic model for dilute solutions of polymers from which the reptation model for concentrated polymer solutions or melts is based.)
3. Rouse relaxation of the whole chain.
4. Reptation.
5. Long-range Fickian diffusion of the polymer chains.

There are three important time scales for diffusion, again defined by Wool:

1. τ_e, Rouse relaxation time between entanglements, characteristic of the wiggling motion
2. τ_{RO}, Rouse relaxation time of the whole chain, characteristic of the movement along the tube
3. T_r, reptation time, also known as the tube renewal time. This is the time required for the chain to escape its initial tube. It is the characteristic relaxation time for the chain and refers to 70% escaping (complete escape occurs at $1.94T_r$). At $t > T_r$ the interfacial region disappears and behaves like the bulk.

The important time for heat sealing is $\tau_{RO} < t < T_r$. In this region, chain segments can diffuse across the interface and strength develops. Times less than τ_{RO} may be important for bonding of dissimilar surfaces. T_r is typically much larger than τ_{RO} since T_r scales as M^3 (where M = molecular weight) and τ_{RO} as M^2 and the molecular weight for sealant polymers is large (on the order of 100,000 g/mol).

Wool computes the time dependence of the average segment penetration distance across the interface as

$$\langle l \rangle \approx 2\left(Dt/\pi\right)^{1/2} \tag{7.3}$$

where

$\langle l \rangle$ = average minor segment length as it escapes the tube,
D = diffusion coefficient.

Eq. (7.3) is valid up to the point where t approaches T_r (where $T_r \cong L^2/(2D)$) and $M > M_e$. Here L is the tube length from the reptation model and M_e is the entanglement molecular weight. The diffusion coefficient relates the diffusive flux to the concentration gradient (Fick's first law):

$$J = -D\frac{\partial c}{\partial x} \tag{7.4}$$

where

J = diffusion flux (amount of chain segments per unit area per time),
c = concentration,
x = dimension of diffusion (perpendicular to interface, typically in the thickness dimension).

The average penetration distance at an interface, $\mathscr{X}$ (defined for only one side of the interface; the total interface thickness is 2 $\mathscr{X}$), is found to scale with the square root of the average segment penetration distance:

$$\mathscr{X} \approx \langle l \rangle^{\frac{1}{2}}. \tag{7.5}$$

Scaling relationships such as the ones presented in this section are helpful for understanding relationships between physical properties and fundamental molecular and physical parameters. Most of these molecular dynamic models, such as the Rouse model, are too difficult to use for estimating the exact relationships but nevertheless provide key insight into relationships.

The equilibrium penetration distance is found to be $0.8R_g$, where R_g is the radius of gyration of the polymer chain (a characteristic length scale of the chain in its relaxed coiled state) and scales as $M^{1/2}$. The scaling law for the interface thickness becomes

$$\mathscr{X}(t) = \mathscr{X}_\infty\left(t/T_r\right)^{1/4} \sim t^{1/4}M^{-1/4} \text{ at } t < T_r \tag{7.6}$$

$$\mathscr{X}(t) = \mathscr{X}_\infty\left(t/T_r\right)^{1/2} \sim t^{1/2}M^{-1} \text{ at} t > T_r \tag{7.7}$$

where

$\mathscr{X} =$ average interpenetration distance (half of interfacial thickness),
$\mathscr{X}_\infty =$ equilibrium monomer interdiffusion distance at T_r,
$T_r =$ reptation time,
$M =$ molecular weight of polymer chain,
$t =$ time.

At T_r the scaling relationship switches from reptation to Fickian diffusion, where $\mathscr{X}^2 \sim Dt$.

Wool presents scaling laws for a number of mechanical properties at the interface. Most important for this discussion is the fracture energy, the energy to create a new interfacial area, which is related to the seal strength (see Section 7.5.7). The fracture energy is found to scale with the minor segment penetration distance across the interface, $\langle l \rangle$:

$$G_a \approx t^{1/2} M^{-1/2} \tag{7.8}$$

where

$G_a =$ fracture energy,
$t =$ time,
$M =$ molecular weight.

Wu [65] reports that experimental data shows the fracture strength often develops as $t^{1/4}$ rather than $t^{1/2}$, and that the $t^{1/4}$ behavior is predicted by kinetic theory. This suggests that G_a may scale with $\mathscr{X}$ rather than with $\langle l \rangle$. Others have associated the $t^{1/4}$ dependence for the case of healing of previously fractured surfaces where the chain ends have become segregated [51]. Heat seal results for PE films have been shown to follow a $t^{1/2}$ relationship, as shown in Fig. 7.18. This is taken up further in Section 7.5.3.2.2.

The molecular weight dependence given in Eq. (7.8) is suspect since the mechanics of polymer welds, particularly heat seals, are highly complex, involving more than the deformation of the first few hundred angstroms at the interface. Experimental studies indeed show a complex relationship between molecular weight and peel strength, underscoring the competing effects of reduced diffusion and increased strength when the molecular weight is increased.

7.5.3.1 Effect of time, temperature, and pressure of the heat seal process on diffusion

Seal dwell time, seal bar temperature and pressure are controlled parameters which influence the heat history at the interface and the extent of diffusion during the heat seal process. As shown in Eq. (7.6), the amount of molecular diffusion or penetration increases with the fourth root of time up to the reptation time, after which Fickian center-to-mass diffusion best describes the molecular motion ($\mathscr{X} \sim Dt^{1/2}$).

The diffusion coefficient increases with increasing temperature. An Arrhenius relationship is often used to describe this relationship:

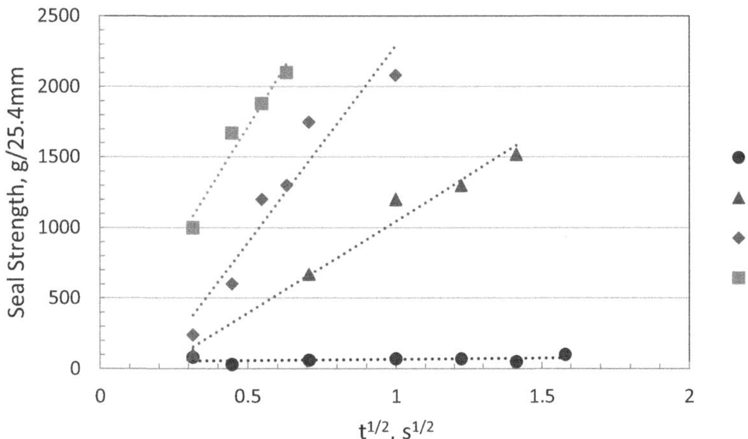

FIGURE 7.18 Effect of dwell time on development of seal strength for an LLDPE film (50 μm cast film, hexene Z-N LLDPE with density 0.918 g/cc, 2 g/10 min MFI). Dashed lines are linear trend lines for data at different seal bar temperature [51]. *Data from Z. Najarzadeh, A. Ajji, A novel approach toward the effect of seal process parameters on final seal strength and microstructure of LLDPE, J. Adhes. Sci. Techn. 28(16) (2014) 1592–1609.*

$$D = D_0 \exp\left(\frac{-E_d}{RT}\right) \tag{7.9}$$

where

D = diffusion coefficient,
D_0 = constant,
E_d = activation energy for diffusion,
R = ideal gas constant,
T = temperature (K).

Wu [65] tabulates values of E_d for several polymers (4−15 kcal/mol) and comments that the activation energies for diffusion and viscous flow are in the same range, making it difficult to distinguish between these two mechanisms.

The temperature at the interface can be modeled by simple heat conduction models described more fully in Section 7.6. Meka and Stehling [66] develop a finite element model that agrees well with experimental measurements with micro thermocouples. Morris [67] uses a finite difference algorithm to compute temperature profiles at the sealant interface for different thickness films. For the thicker structure investigated, the interface temperature never reaches the seal bar temperature in the time scale of the heat seal operation. He couples the heat conduction model with a model for molecular diffusion to predict penetration and strength at the interface (see Section 7.6 for details). Most models assume heat transfer occurs only in the thickness dimension, but lateral heat transfer may also be important, especially for aluminum foil containing laminates and for understanding what happens near the edges of the seal area [26,68,69]. Aluminum has a thermal conductivity two orders of magnitude higher than plastics and heat may travel faster laterally within the foil layer than across the thickness of the laminate. As described in Section 7.3.1, Li et al. show that loss of orientation and strength due to melting at the edges of the seal area may be responsible for the heat seal plateau [26].

Meka and Stehling [66] show that beyond a small amount of pressure to ensure good contact, additional pressure has little impact on the heat seal of polyethylene films. Yuan et al. [25] show similar behavior for a PET/LLDPE laminate. Najarzadeh and Ajji [51] map the effect of pressure on heat sealing an LLDPE film. At very low pressures, HSIT decreases with increasing pressure as the surfaces come into close enough contact to allow wetting and diffusion. Following the initial effect on HSIT, pressure has little impact on heat seal performance until very high pressures are reached. In theory, high pressure may reduce the volume available for segmental motion (depending on the compressibility of the polymer) [70]. This may decrease the diffusion coefficient. In practice, at high pressure the sealant flows and is squeezed out of the seal area before this occurs: the seal jaw pressure needs to be large enough to ensure good contact, but not too large as to cause excessive flow from the seal area.

7.5.3.2 Effect of polymer properties on diffusion

7.5.3.2.1 Thermal properties

The polymer chains must be mobile for diffusion to take place. In semicrystalline polymers, the crystalline structure constrains the movement of the molecules. Enough thermal energy must be transferred to the polymer to at least partially melt the crystals for appreciable movement to take place. As described earlier, studies have shown that the initiation region of the heat seal versus temperature curve occurs when part of the crystals have melted and the onset of the plateau region when the crystals have fully melted [45,46,50,51]. Amorphous polymers must be well above their glass transition temperature for effective diffusion. The thermal conductivity and heat capacity of the materials in the film structure influence how fast the thermal energy gets to the sealant interface. Layers between the seal bar and the sealant with low conductivity or high heat capacity may slow heat from reaching the interface and slow initial diffusion and strength development. They may also slow subsequent cooling after the seal jaws are released and influence hot tack behavior [36,71,72].

For many ethylene copolymers, the crystalline melting point (Tm) decreases with increasing comonomer content. This is true for metallocene LLDPE, EVA, ethylene methyl acrylate (EMA), ethylene acrylic acid (EAA), ethylene methacrylic acid (EMAA) and ionomers. One exception among the commonly used sealant resins is standard LLDPE made with Z−N catalysts. These are heterogeneous in MWD and comonomer distribution. The longer chains in these polymers tend to have little comonomer content, even as the overall comonomer content is increased. The peak Tm is determined by the crystallization of the longest chains. As a result, the peak Tm of Z-N LLDPE remains almost constant as total comonomer content is increased, even though the amount of crystallinity decreases (see Fig 4.5).

7.5.3.2.2 Molecular architecture

Eqs. (7.6)–(7.8) show that as the molecular weight increases, the amount of chain diffusion decreases. The reptation model shows that the interfacial thickness scales with M^{-1} (Eq. 7.6) and the center of mass diffusion scales with M^{-2} [64]. This agrees with experimental evidence.

LCB also slows diffusion. The branching entangles the polymer network resulting in less chain movement. Of the common ethylene-based sealant resins, LDPE, EVA, EMA, EAA, EMAA and ionomer contain LCB. The linear polyethylenes (LLDPE, mPE plastomers) do not. Some metallocene-catalyzed plastomers have a sparse amount of LCB that provides improved bubble stability in blown film. The diffusion coefficients for these materials lie somewhere between LLDPE and LDPE.

The molecular weight and comonomer distribution of LLDPE have a significant effect on diffusion and the build-up of seal strength. Schuman et al. [73,74] use layer multiplier technology (see Section 8.5.2) to produce film structures with 32 alternating layers of LLDPE and HDPE with an average layer thickness of 17 and 25 microns, respectively. They anneal the films at 200°C for up to 10,000 minutes, periodically running a DSC scan. Since these two polymers are miscible, as the interface disappears through interdiffusion the two melting peaks merge into a single peak. By following this change with annealing time, they calculate the rate of diffusion. They find that the rate of disappearance of the layers (interface) depends on the MWD and especially the longer chain molecules. The boundaries between the layers remain stationary during the characteristic diffusion time of the main components and only move at long times associated with the diffusion of the high molecular weight fraction.

Mueller et al. [52]. study the heat seal behavior of a heterogeneous Z–N LLDPE with a polydispersivity (Mw/Mn) of 5.1. The resin is made into a blown film and heat-sealed over a temperature range of 100°C–130°C for 1–100,000 seconds. DSC shows that complete melting takes place at 130°C. Long dwell times are used to characterize diffusion since the short time scales of the commercial heat seal process is of the same order as the time for the interface to reach the seal bar temperature. Over the temperature range of 100°C–125°C, and especially near 115°C, there is a strong dependence of temperature and sealing time on heat seal strength. The development of seal strength is found to increase as $t^{1/2}$ in agreement with Wool's scaling analysis. Wool [75] suggests that to build strength in a polymer melt, the chain segments form bridges across the interface (entanglements). For melts, the minimum number of bridges required is three and the scaling goes as $t^{1/2}$. Mueller et al. find this to be true for semicrystalline LLDPE polymers with the presence of crystals as an anchoring point.

Mueller et al. find that the seal times needed to reach the ultimate seal strength are always longer than the characteristic diffusion time based on theory. They consider several possibilities for this behavior. They conclude that the change in diffusion coefficient with temperature is not enough to explain the results. They also find that in the narrow range of 113°C–117 °C where the greatest change in seal strength takes place, the amorphous fraction only changes from 70% to 75%. They conclude that this is not likely a major factor. The MWD and comonomer distribution of this heterogenous polymer does provide a reasonable explanation, however. These polymers have a broad melting point range with the shorter MW, high comonomer content chains forming less perfect crystals with lower melting points. They propose that at temperatures below 115°C, only the crystals with the low MW fraction have melted. While these chains are free to diffuse across the interface, they make poor tie molecules and form weak crystals upon cooling and recrystallization. Higher temperatures and longer times are required to melt the crystals formed by the longer, less branched chains. It is only when these chains diffuse across the interface that the seal strength can approach its ultimate value. This results in a considerable increase in the entanglement time.

Qureshi et al. [76] study the hot self-adhesion of three Z–N LLDPE resins that are polydisperse in MWD and heterogeneous in comonomer distribution. Fifty-micron monolayer blown films are laminated to OPET: OPET/adh/LLDPE. Sealing is done within a controlled temperature chamber on a tensile tester so that the peel strength is measured at the sealing temperature (the temperatures are well below the distortion temperature of the OPET so that the films can be grasped by the tensile tester even if the LLDPE is completely molten). The seal time is varied from 0.5 to 25 seconds.

The measured peel strength increases with total time (seal time plus the peel time) and curves produced at different seal times overlap. This shows that diffusion continues during the peel (when the polymers are in the melt). This observation comes up again with respect to the delay time in standard hot tack measurements (Section 7.6.3). Peel strength increases with $t^{1/2}$ at short contact times but saturates or levels out at longer times (when the time-independent value is reached). The time for saturation is T_r in Wool's analysis: the time for the chain to completely diffuse out of the tube and when the interphase region begins to behave like the bulk. As described in the development of Eq. (7.6), T_r is related to the time for the polymer to diffuse the distance R_g, the radius of gyration of the polymer chain. Qureshi et al.

calculate T_r and find it to be two orders of magnitude less than the measured time to reach the ultimate seal strength. This is similar to the findings of Mueller et al.

Qureshi et al. explain the discrepancy by the presence of an inert surface layer made up of predominately low MW, high comonomer content chains. This layer is approximately 100 nm thick. Both enthalpic and entropic forces drive this segregation of chains. One is flow-induced segregation due to higher shear rates at the wall of the flow channels, favoring low molecular weight (lower viscosity) chains to move toward the wall. Another driving force favors the amorphous fraction (high comonomer content) to be near the surface [77,78]. This is consistent with the high comonomer content chains being near the surface since the comonomers in LLDPE act as short-chain branches that disrupt crystallinity. In a second paper, Qureshi et al. [79]. present atomic force microscopy (AFM) evidence showing the presence of this amorphous surface layer.

As soon as the two films are joined in the melt, the low MW, high comonomer chains at the surface quickly diffuse but cannot generate good adhesion in the melt (they disentangle at the same rate as they entangle). But once the surfaces are joined, they become part of the bulk and the thermodynamics for segregation no longer exist. Homogenization of the interphase by mutual interdiffusion brings long chains to the interface. This interdiffusion process has a $t^{1/2}$ dependence. Not only do the long chains have lower diffusion coefficients than the short chains, they have a longer distance to travel, moving through the former amorphous surface layer, before diffusing across the interface.

In a follow-up paper, Qureshi et al. [79] compare the three Z–N LLDPE's with a metallocene hexene-LLDPE. All have a similar molecular weight (MW) but the mPE has a narrower MWD and comonomer distribution. The seal strength of the Z–N LLDPE increases with $t^{1/2}$ whereas the seal strength of the mPE shows no dependence on time (t^0). This indicates the mPE diffusion rate is much faster than the Z–N LLDPE. This is consistent with the narrower MWD of the mPE and its lack of the high MW fraction contained in the Z–N LLDPE. Furthermore, AFM shows no presence of an amorphous surface layer on the mPE blown film as is found with Z-N LLDPE.

Qureshi et al. [79] also study blends of two octene mPE resins with two different densities, 0.87 and 0.91 g/cc. These densities correspond to different comonomer contents, with the 0.87 g/cc grade having considerably more comonomer than the 0.91 g/cc grade. When they blend 5%–20% of the 0.87 g/cc grade into the 0.91 g/cc grade they observe heterogeneous behavior: $t^{1/2}$ dependence. At blends greater than 20% the behavior reverts back to the t^0 dependence observed for the hexene mPE.

Sadeghi and Ajji [42] characterize three metallocene PE plastomers. Rheology and gel permeation chromatography results show that two contain a sparse amount of LCB whereas the third is linear (no LCB). The relaxation spectrum for the linear polymer is narrower and shifted to lower relaxation times than the two mPE resins with a small amount of LCB. The diffusion coefficient is inversely related to the longest relaxation time, known as the terminal relaxation time (see Eq. 7.21). The longer relaxation time of the sparsely branched mPE's suggest they will take longer for molecular segment diffusion and strength to build at the seal interface. Indeed, hot tack for the linear mPE is substantially higher than for the sparsely branched mPE's, which they attribute to increased diffusion, penetration and entanglement.

Boutrouka et al. [80,81] consider the sealing of PP, important for OPP films and retort packaging applications. They compare homopolymer PP (hPP), conventional random copolymer PP (rPP), a metallocene propylene-ethylene copolymer elastomer (ExxonMobil Vistamaxx) and their blends. The elastomer has 10.5% ethylene, is amorphous (no melting point), and has a much narrower MWD than the hPP or rPP. The elastomer has a 70°C lower HSIT than the rPP, which they attribute to its lack of crystallinity and narrow MWD, both of which ease diffusion across the interface. Adding elastomer to the rPP reduces its HSIT. The elastomer has a lower ultimate seal strength than the hPP or rPP, however. The ultimate seal strength of the blends is found to increase with increasing crystallinity. They conclude that seal strength is influenced by the co-crystallization of polymer chains at the interface.

7.5.3.2.3 Chemistry

The rigidity of the polymer chain influences diffusion rates. Polymers that contain aromatic groups typically are stiffer and slower to diffuse. The presence of bulky side groups can also impede the movement of the chains. Polar forces such as hydrogen bonding and specific chemical interactions also generally increase rigidity and change the diffusion behavior [54,82].

Polymer crosslinking also impedes molecular mobility and hinders diffusion.

For bonding of dissimilar polymers, compatibility is important. Small chain segments diffuse across the interface but this may not be enough to build strength [83]. A significant number of bridges are required; the formation of such bridges is not thermodynamically favored for incompatible polymers. Chemical means to promote adhesion at an interface when the polymers are not compatible are described in Chapter 10.

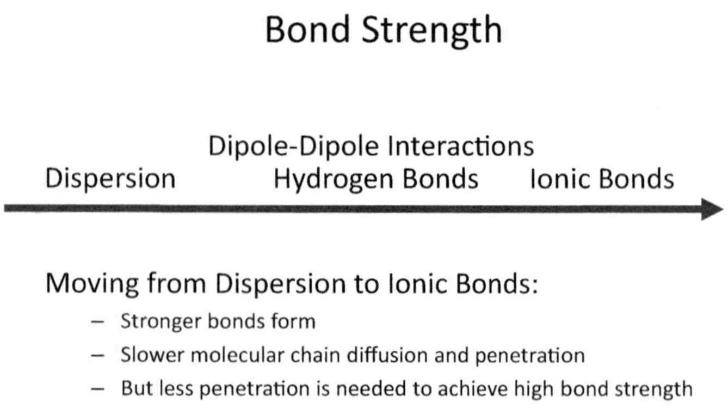

FIGURE 7.19 Continuum of polymer intra and inter molecular forces.

TABLE 7.4 Types of molecular forces present in some common sealant resins.

Sealant	Dispersion	Dipole–dipole	Hydrogen bonding	Ionic
LLDPE, mPE, LDPE	X			
PP	X			
EVA	X	X		
EMA, EEA, EBA	X	X		
EAA, EMAA	X		X	
Ionomer	X		X	X

7.5.4 Interfacial bond strength and crystallization

Once sufficient wetting and diffusion take place, the strength of the interface builds as the polymer chains bridge across the interface and come close enough for intermolecular forces to take hold. The adhesive strength of the interface is determined by the nature of the molecular forces that are present. For common sealants used in flexible packaging, these range from nonpolar dispersion forces to polar forces (dipole–dipole, hydrogen bonding and ionic interactions). These forces can be thought of as a continuum, as shown in Fig. 7.19. Moving from left to right, the molecular forces increase in strength and stronger bonds form. These same forces can impede the movement of the chains resulting in slower diffusion and less penetration, but less penetration is needed to build strength.

The types of intermolecular forces present in some of the commonly used sealant resins are given in Table 7.4. For nonpolar molecules like PE, only weak dispersion forces are present. The relatively weak dispersion forces are compensated by their number. A high level of penetration is needed to essentially erase the interface and achieve the strength of the bulk.

The polar ethylene copolymers such as EVA and the acrylate copolymers contain relatively polar side groups. These are particularly helpful in wetting and bonding to dissimilar substrates such as PET or PP for lidding applications. In autoadhesion, the polar groups likely help in establishing bond strength. It is difficult to separate out this effect from the influence the comonomer has on lowering the melting point, which allows the chains to diffuse at lower temperatures and provides low HSIT.

Ionomers and their acid copolymer precursors have hydrogen bonding between the oxygen atom in one molecule and the hydrogen atom in another. These are stronger than dispersion forces and act over a longer distance. In addition to helping with autoadhesion, the acid groups bond well to aluminum foil and enhance melt strength.

Ionomer sealant bonds include dispersion, hydrogen bonding and ionic interactions. As described in Chapter 4, ionic bonding provides many of the favorable attributes of ionomers for flexible packaging: high melt strength and hot tack, low seal initiation temperature (by suppressing cyrstallinity), oil and grease resistance, toughness and transparency. Van Alsten [50] shows that the ultimate heat seal strength occurs only after the polymer has fully melted. He and others [84] have shown that the metal ions diffuse several orders of magnitude faster than the polymer chains. Hence, van Alsten concludes that the development of strength at the seal interface involves the diffusion of large chains, similar to what

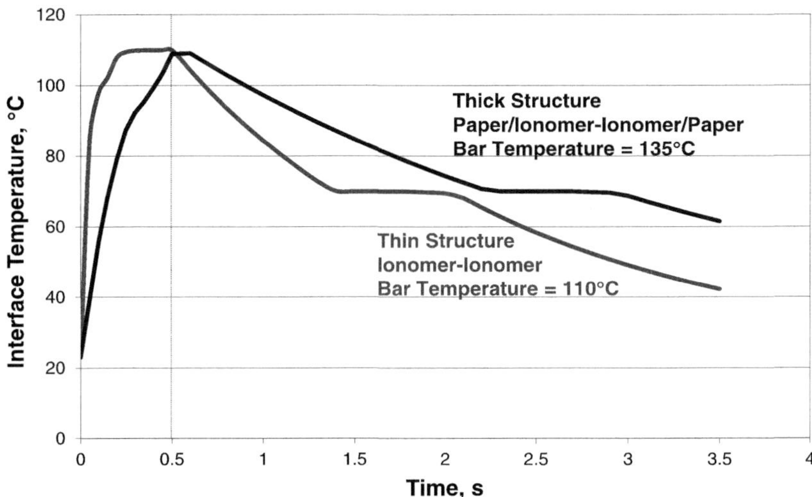

FIGURE 7.20 Model calculation of temperature of the interface during hot bar sealing for the two film structures shown in Fig. 7.14. Each film is sealed at its respective plateau initiation temperature (110°C and 135°C for the thin and thick structure) for 0.5 s [67]. *Reproduced with permission from B.A. Morris, Predicting the heat seal of ionomer films, J. Plastic Film, Sheeting 18 (2002) 157–167.*

happens in polyethylene. Morris [67], however, shows that ionomers only need penetration on the order of about $0.1R_g$ to reach the ultimate seal strength, ten times less than that typically needed for polyethylene.

In the heat seal process, once the seal bars are removed, the interface begins to cool. Whereas heating of the interface takes on the order of a second or less, cooling takes several seconds or more. This is because heat transfer from direct contact with the hot bar is much faster than convective cooling from the air. A typical temperature profile is shown in Fig. 7.20 based on the modeling of Morris [67], which is described in Section 7.6. The long times at temperatures above the freezing point (around 70°C for this example) underscore the importance of hot tack.

As the interface cools, the molecular mobility decreases, reducing the rate of diffusion. For semicrystalline polymers, the chains organize into crystalline structures as the temperature drops to the crystallization temperature. The ability of sealants to crystallize at the interface contributes to interfacial strength.

The influence of crystallization at the interface of two different immiscible semicrystalline polymers is complex. The shrinkage of the first polymer to crystallize may cause an influx of the other polymer across the interface if the second polymer is still molten. As it solidifies, this may create mechanical interlocking between the two polymers and add to the interfacial strength. The interfacial strength is dependent on a number of factors including the compatibility of the two polymers, their crystallization temperatures, their crystallization rates, and the rate of quenching. The nature of the influxes and the effect of co-crystallization of incompatible polymers are still not well understood [85].

7.5.5 Hot tack

Hot tack is a function of interfacial bonding and melt strength. The hot tack versus sealing temperature curve, introduced in Fig. 7.12, goes through a maximum due to the competing effects of temperature on chain diffusion and melt strength: diffusion increases with increasing temperature and melt strength decreases. The maximum or peak hot tack temperature is typically near where the heat seal strength reaches its plateau value. These may not exactly be the same due to differences in measurement. In the heat seal test, there may be significant diffusion that occurs during the initial stages of cooling which shifts the curve to the left. In the hot tack test, long delay times may mimic this effect but may not be representative of the actual sealing operation.

The initial stages of the hot tack test are similar to the heat seal test. The films are brought together with pressure but initially are not in intimate contact due to the microscopic roughness of the surfaces. As the heat from the seal bars transfers through the film, the interface melts and smooths out, creating greater contact. The polymer chain segments become mobile, wet the surface, diffuse across the interface and build strength through entanglements and molecular interactions. As described in Section 7.2.3, most commercial hot tack testers today have a programmed delay time between when the seal bars are released and the hot seal is pulled apart. During the delay, further chain diffusion may occur as well as cooling (which may increase melt strength). Following the delay, the force to separate the films is measured, typically at much higher pull rates than those used in the heat seal test. This force is not only a function of the hot adhesion at the interface but also the resistance to elongational stress, which is related to the melt strength or elongational viscosity.

Diffusion is described in detail in Section 7.5.3. This discussion pertains to hot tack as well. One difference is that the amount of chain penetration, bridging and entanglement needed to build hot adhesion strength may be greater than for cold adhesion (heat seal). This is because the hot tack measurement is done while the sealant is molten, whereas in the heat seal measurement the sealant has solidified and crystallized. Hot adhesion is also more heavily influenced by strong molecular forces than cold adhesion because of the lack of crystallinity in hot adhesion. The presence of stronger bonds such as hydrogen and ionic bonding not only helps with interfacial adhesion but also increases the melt strength.

Melt strength is related to elongational viscosity and decreases with increasing temperature, typically in an exponential manner. Several factors influence melt strength including molecular forces and chain architecture. In general, the presence of longer-range molecular forces increases melt strength and hot tack. The strength and range of these forces are in the following order: dispersion $<$ hydrogen bonding $<$ ionic or covalent bonds. Crosslinking, for example, increases melt strength but runs the risk of rendering the polymer difficult to seal since this also impedes diffusion. Ionomers have outstanding melt strength due to the presence of ionic interactions which act like crosslinks but still allow for flow and sealing.

The molecular architecture also has a substantial impact on melt strength as described in Chapter 5. Increasing MW, introducing LCB and broadening the MWD tend to increase melt strength. Just as strong molecular forces impede chain movement and diffusion, so to do these factors. A recurring theme in hot tack is that factors that help the left side of the curve often negatively affect the right side of the curve (and vice versa) making sealants with broad hot tack temperature windows rare.

Some of the resin design strategies for improving hot tack are shown in Fig. 7.21. These include:

- Broadening the hot tack temperature window by lowering the sealant melting point (to lower the HTIT) and increasing the sealant melt strength (through changes in chemistry or architecture). The melting point for ethylene copolymers can often be lowered by increasing the comonomer content.
- Increasing the peak hot tack strength by increasing the melt strength, increasing the hot adhesion between the layers (by increasing chain diffusion and entanglement or introducing long-range molecular forces) or inducing crystallization.
- Decreasing the peak hot tack temperature by lowering the sealant melting point or increasing the rate of diffusion (lower MW, narrower MWD, less LCB, etc.).

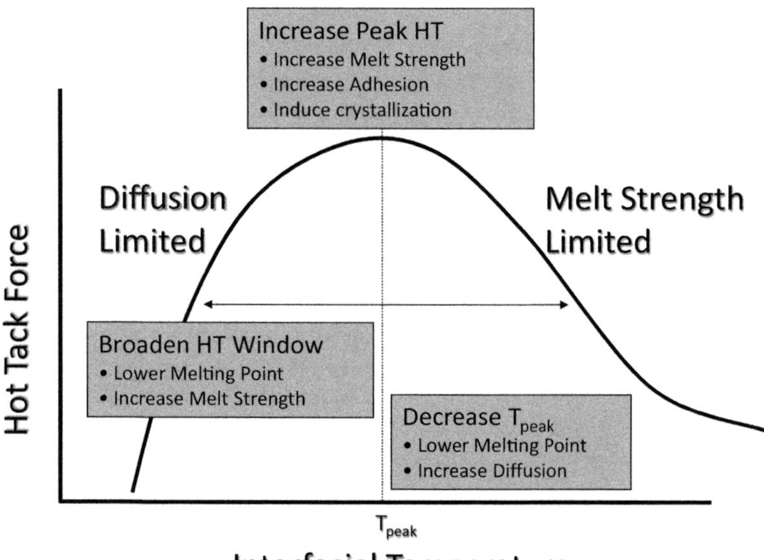

FIGURE 7.21 Resin design strategies for hot tack performance.

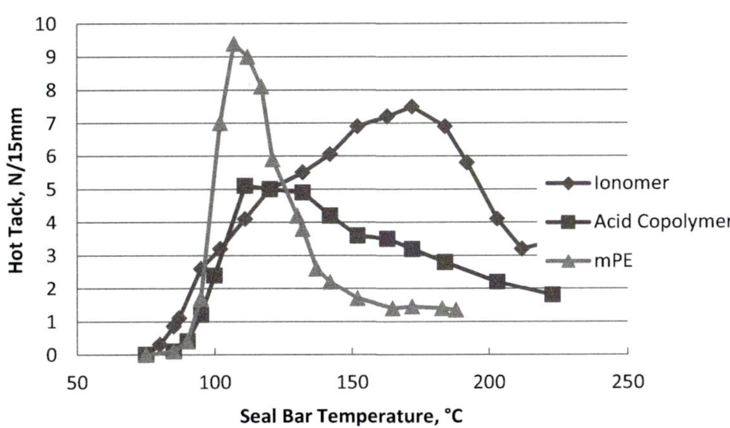

FIGURE 7.22 Comparison of hot tack performance of different ethylene-based sealants. Packforst hot tack tester, 0.5 s dwell time, 0.1 s delay time. *Data from The Dow Chemical Company.*

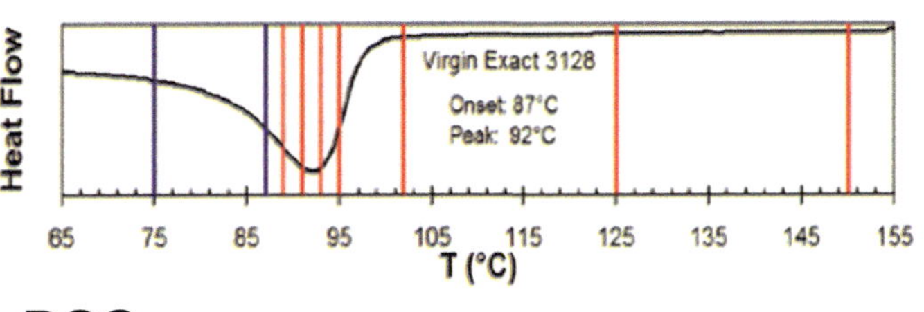

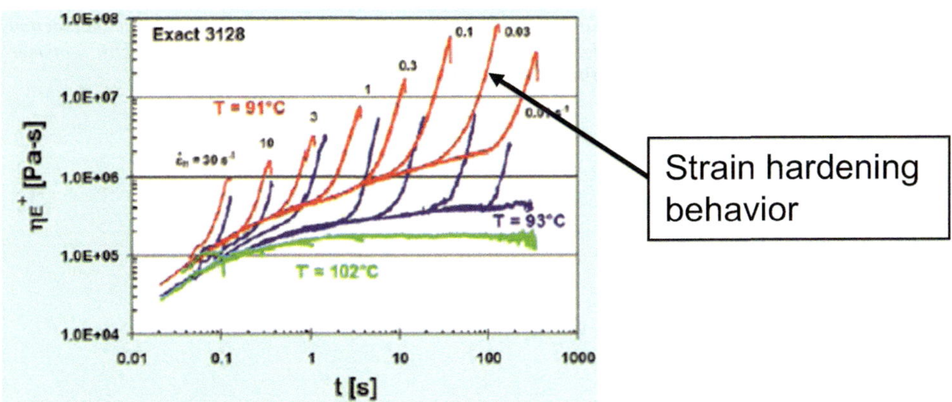

FIGURE 7.23 DSC (top) and transient elongational viscosity (bottom) of a mPE [88]. *Reproduced with permission from M. Sentmanat, S.G. Hatzikiriakos, Flow induced crystallization of polymers in extension, in: SPE ANTEC Conf., 2010, pp. 51–55.*

Some typical hot tack curves for PE and ethylene copolymers are shown in Fig. 7.22. The two families of ethylene-based sealants with the best hot tack performance are the metallocene or single-site catalyst plastomers and ionomers. They achieve their performance in substantially different ways.

The mPE plastomers typically have very high hot tack strength over a narrow temperature range. The peak hot tack temperature occurs at the temperature where the final crystals melt [39,50]. This exceptionally high hot tack may due at least in part to strain-induced crystallinity. Normally the crystallization rate is zero at the melting point (the crystallization temperature is typically considerably lower than the melting point). Extensional flow causes the chains to line up in the direction of extension, enabling crystallization. Metallocene PE is more likely to undergo strain-induced crystallization than Ziegler–Natta PE due to its more uniform comonomer and chain size distribution [86,87]. The comonomers act as short chain branches and do not participate in the crystals. In heterogeneous Z-N LLDPE, the shorter chains have a high comonomer content that is not conducive to crystallization. The longer chains have lower comonomer content but are more constrained and less likely to form crystals in the time frame of the heat seal operation or hot tack test.

Sentmanat and Hatzikiriakos [88] use uniaxial extensional rheometry to study the behavior of a metallocene plastomer with a peak melting point of 92°C, the onset of melting at 87°C, density of 0.900 g/cm^3, Mw/Mn of 2.0, MFI of 1.2 g/10 minutes, MW of 80 kg/mol and no LCB. Their study focuses on a narrow temperature range near the peak melting point where the maximum hot tack is typically observed. In the solid-state (< 87°C) the polymer exhibits classic plastic yielding. At 95°C and above there is no strain hardening, typical of LLDPE melts. When the temperature ranges between 89°C and 93°C the polymer is partially molten and exhibits strain hardening behavior, as shown by the steeply rising transient elongational viscosity in Fig. 7.23. DSC of rapidly quenched samples shows evidence of an increase in crystallinity in the samples stretched at 91°C or 93°C. This strain-induced crystallinity may explain the exceptional hot tack performance of these resins. Note that under quiescent conditions, the mPE has a crystallization temperature of 73°C, nowhere near the melting point.

Qureshi et al. [76] observe an overshoot in hot tack strength at the beginning of the hot tack measurement. They propose this is due to strain-induced crystallization since the strain rate at initiation is higher than the steady strain rate. While they observe overshoot at high peel rates, none is observed at low peel rates, which supports the hypothesis. In a second study, Qureshi et al. [79] compare metallocene and Z–N catalyzed (heterogeneous) polyethylene of similar densities and molecular weights. At high peel speeds (500 m/minute), the metallocene LLDPE gives substantially higher peel strength than the heterogeneous LLDPE. As described in Section 7.5.3.2.2 they conduct their measurements in a heated chamber so the peel strength is essentially a measure of hot tack. At 50 m/minute there is no difference in hot tack between the two polymers. They attribute the change in behavior to strain-induced crystallization in the mPE at the higher peel rate.

The scientific explanation for the high peak hot tack strength of mPE reveals how laboratory hot tack measurements may not represent actual performance in a commercial packaging operation. Most commercial hot tack testers use a

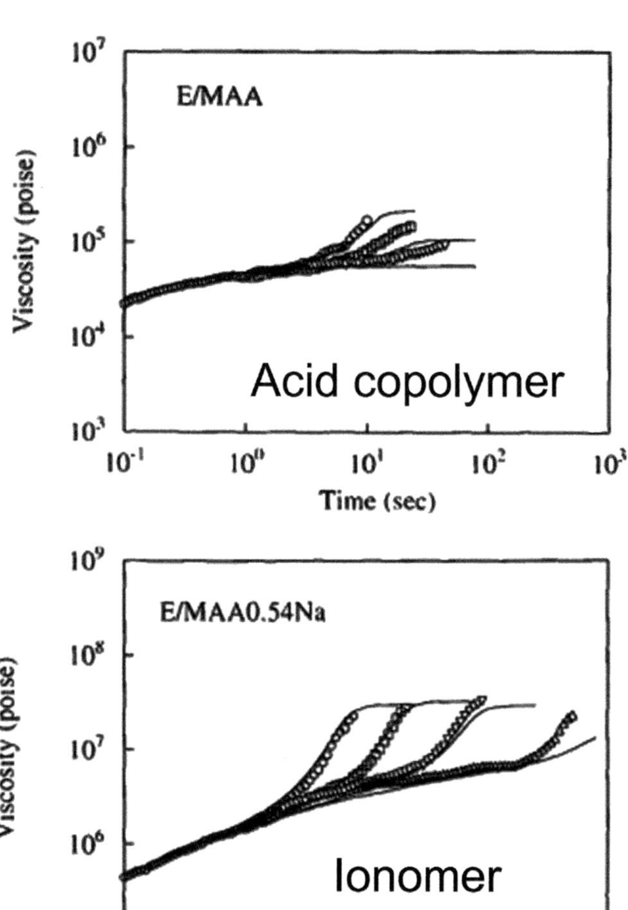

FIGURE 7.24 Transient elongation viscosity of an ionomer and acid copolymer [89]. *Reproduced with permission from T. Takahashi, J. Watanabe, K. Minagawa, K. Koyama, Effect of ionic interaction on elongational viscosity of ethylene-based ionomer melts, Polymer 35 (1994) 5722–5728.*

high peel rate that is more characteristic of the registration speed of the packaging line than the pulling force acting on the hot seal. High peel rates favor strain-induced crystallization. Many users of hot tack testers set the delay time for 0.1–0.5 seconds whereas the opening force in many applications occurs almost instantaneously. While the heat transfer is slow during the delay time there is nevertheless a small amount of cooling that takes place (see Fig. 7.20 for example). Early work with standard grades of polyethylene shows that delay time has a minimal effect on the hot tack curve [9], but these materials are not sensitive to strain-induced crystallization. The work of Sentmanat and Hatzikiriakos shows that for the newer metallocene LLDPEs and plastomers, even small changes in temperature can greatly increase the potential for strain-induced crystallization.

For acid copolymers and ionomers, the presence of LCB, hydrogen bonding, and for the case of ionomers, ionic interactions, impart strain hardening behavior in the melt. This is depicted in Fig. 7.24. The amount of strain hardening increases in the following order: LDPE $<$ ACR $<$ ionomers. The strain hardening arises from chain entanglements and molecular interactions rather than from strain-induced crystallization. As a result, the strain hardening behavior of ACR and ionomers persists over a very broad temperature range. The measurements shown in Fig. 7.22 are conducted at 140°C, well above the peak melting point of around 95°C.

mPE is characterized by a sharp melting peak over a narrow temperature range, rapid self-diffusion across the interface, strain-induced crystallization within a few degrees of the melting point and a reduction in melt strength with further temperature increases. Ionomers are slower to diffuse across the interface due to their broader melting point range, long-chain branches and strong ionic intermolecular forces. Because of the longer-range forces present, chain penetration at the interface is not as critical. Once the ultimate hot tack strength is achieved, its magnitude persists over a much broader temperature range than mPE due to its outstanding melt strength.

7.5.6 Caulking

Caulking is a term used to describe flow in the sealant area to fill voids and channels. When the films are first brought together, microscopic surface roughness prevents wetting. As the interface heats up, flow smooths out the surface roughness ensuring good contact for bonding. Further flow may be required to seal around solid contamination and fill voids in folded regions such as gussets or when wrinkles are present. An example of a gusset area is shown in Fig. 7.1B. Caulking is a balancing act; too much flow (and too little hot tack) may cause mooning of the seal where the weight of the product partially opens up the seal. Excessive flow can also result in squeeze-out of the sealant from the seal area and loss of seal integrity.

Modeling approaches for squeeze flow and caulking are described in Section 7.6.4. The shear rate in the heat seal process is generally on the order of $0.1-10s^{-1}$. The zero shear viscosity of the sealant is often a good benchmark to use for comparing resins. For comparisons within the same polymer family, the melt flow index (MFI) is a good surrogate. Sealants with higher MFI often caulk better. For example, high MFI EVA sealants are typically used for packaging of blocks of cheese because of their good caulking. These are extrusion coated onto OPET film. Extrusion coating and cast film resins often have higher MFI and better caulking than blown film resins due to the inherent viscosity requirements of the processes.

When comparing across sealant families, it is best to look at the full viscosity versus shear rate curves. As described in Chapter 5, linear polyethylene typically has a lower zero shear viscosity than branched polyethylene (LDPE) at the same melt flow index. As a result, LLDPE will typically caulk better than LDPE at temperatures near their melting points. Standard rheology curves do not tell the whole story, however. Viscosity is typically measured at temperatures well above the melting point of the polymer. The viscosity versus temperature curve can be extrapolated down to temperatures near the melting point using common modeling techniques such as an Arrhenius relationship (see Chapter 5), but such relationships fail to hold in the semimolten state during melting. Konaganti et al. [71,72] use oscillatory measurements (storage and loss modulus, G' and G'') in the solid-state through melting to compute the complex viscosity (see Eqs. 5.35–5.37). By coupling these measurements with a heat conduction model for the temperature at the seal interface, they compute the viscosity as a function of time and temperature during sealing. This relates to interfacial smoothing and the initiation of heat seal and hot tack strength as well as the potential for good caulkability.

The importance of caulking in actual packaging line operations is inferred by Bilgen et al. [16] and Kapur et al. [17]. They compare the performance of sealants in various multilayer film structures on a vertical form fill seal line. They produce operational maps of dwell time and temperature (see Fig. 7.8) using bubble emission hermeticity tests to delineate performance. In a side-by-side comparison of LLDPE-type sealants they find several cases where the sealant providing the broadest operating window has the same or even somewhat poorer hot tack performance in a standard ASTM hot tack test. They attribute this to better caulkability.

Caulking and hot tack both contribute to reducing leakers and mooning of flexible packaging. Section 7.4.1.5 describes a study by Ward and Li [36] where a metal wire is inserted into the seal area to simulate a contaminant particle. This test also provides insight into caulking since it estimates the size of the void in the region around the wire left after sealing. They conclude that a combination of flow during sealing (as indicated by zero shear viscosity), hot tack to prevent spring back during cooling after the seal bars are removed, and stiffness in the solid-state provides the best hermeticity. An ionomer and several plastomers provide the best results in their experiment. One word of caution concerning the Ward and Li study: they report a zero shear viscosity of 3300 Pa-s at 190°C for the ionomer in their investigation. This is the lowest value of all the sealants, and it contributes to their correlation showing low viscosity is important for caulking. Typically, this grade would be expected to have a zero shear viscosity of around 12,000 Pa-s at 190°C [90–92]. Also, the viscosity of ionomers increases much faster with decreasing temperature than LLDPE. As a result, the actual zero shear viscosity of the ionomer at the seal temperature of 130°C is likely among the highest of the sealants they tested. This suggests that it may be the hot tack and modulus of the ionomer that is responsible for its outstanding performance in their test rather than superior caulking. Caulking, however, is likely to be a major factor for good performance of the plastomers.

7.5.7 Fracture mechanics

The measurement of seal strength, whether in a controlled manner in the lab, or in an uncontrolled manner by the consumer, is essentially a peel test. The peel test, introduced in Section 7.2.2, measures the energy or force needed to pull the specimen apart. During the test, a peel front moves through the specimen that can be thought of as a crack propagating through the length of the seal. The engineering discipline that studies crack propagation and failure of materials is known as fracture mechanics. A fracture mechanics analysis of the peel test helps provide insight into the factors that influence it.

A detailed analysis of the peel test is developed in Chapter 10. Here basic concepts are introduced and a discussion of the insights the analysis provides for heat seal strength is provided.

The peel force is not simply the bond strength; many other factors are involved. As the peel arms are separated and pulled apart, they may stretch irreversibly depending on the stiffness and yield strength of the materials in the peel arm. The peel arms are bent to an imposed angle, typically 180° for a T-peel test, although the actual local angle may differ depending on the stiffness of the arms. New interfacial area is created as the peel front advances. The measured force is a function of the energy required to bend the peel arms (E_b), the energy to irreversibly stretch the peel arms (E_e) and the energy to create new interfacial area, known as the fracture energy (G_a):

$$\text{Peel Strength} = f(E_b, E_e, G_a) \tag{7.10}$$

Often E_b and E_a are small compared to G_a for flexible packaging. So far the bond strength has not been explicitly mentioned. This is introduced through G_a. G_a is a function of the work of adhesion (W_a) times the local viscoelastic energy dissipation at the peel front (Φ):

$$G_a = W_a(1 + \Phi(R, T)) \tag{7.11}$$

W_a includes bonding due to wetting, diffusion and chemical interaction. The local viscoelastic energy dissipation, Φ, can be significant compared with the bond strength and have a substantial impact on the measured peel strength. Φ is a function of the pull rate (R) and temperature (T).

Insights from fracture mechanics into factors that affect seal strength include:

- The peel strength is rate dependent. The pull speed can significantly change the measured seal strength. This is a consequence of Eq. (7.11).
- The peel strength is temperature-dependent, through changes in Φ and in the bulk properties of the film. For example, an increase in temperature typically decreases the modulus, which affects bending angles and irreversible elongation.
- Changing the angle of pull changes seal strength through several effects. The imposed peel angle has a direct effect on E_b. Changing the angle will likely change the actual angle at the interface; this may change how stress is concentrated and impact the separation dynamics.
- Changing the modulus of the sealant changes the local viscoelastic energy dissipation, the bending energy and the energy to irreversibly elongate the peel arm, each of which may affect seal strength.

- Increasing sealant thickness may increase seal strength. Fracture mechanics predicts that increasing thickness will increase peel strength up to a point and then level off. At the front of the crack tip there is a region of local deformation. At very small sealant thickness, the size of the deformation zone is constrained by the thickness of the sealant. As the thickness increases, the size of the deformation zone increases; peel strength is raised through an increase in Φ. At some critical thickness, the zone of deformation reaches its natural size and is no longer constrained by the thickness of sealant. At thicknesses greater than the critical thickness, no further increase in fracture energy occurs. See Section 10.2.3.3 for further details.
- Other layers in the film structure besides the sealant can influence the measured peel strength. Poor interlayer adhesion can reduce seal strength, as described in Section 7.7 on peelable seal technology. Elongation and stretching of the peel arm during testing can lead to higher peel strength values; all the layers in the structure impact elongation. The stiffness of the layers can also influence the actual angle at the interface.

An example of the last point is illustrated by work done by DuPont researchers several years ago. They developed special copolymers for bonding to aluminum foil. They made a series of polymer compositions that had the same chemical functionality (for bonding to the foil surface) but differed in modulus. Their screening test was to make heat seals of the polymer sandwiched between two layers of aluminum foil: Al/film/Al. They found that the peel strength increased substantially as they decreased the modulus of the film. The actual application involved extrusion coating the polymer onto the foil with the final structure: Al/coating. When they conducted trials on a pilot coating line, they found the change in modulus had no effect on the peel strength between the coating and foil. After some investigation, they concluded that the difference was due to the local angle at the interface. In the Al/film/Al structure, the local angle is controlled by the stiff foil layers and does not change when the modulus of the film is decreased. The observed increase in peel strength is attributed to greater local viscoelastic deformation (Φ). For the Al/coating structure, as the modulus of the coating was decreased, the local angle of pull also increased, concentrating more of the stress on the interface and counteracting the effect of greater Φ. This example also highlights the need to conduct laboratory tests using structures that mimic the end use.

One final example is used to illustrate the concepts of fracture mechanics and the influence of factors beyond bond strength on peel strength. Suppose a person walks down the street on a hot summer day and steps on some old chewing gum. The gum is very difficult to remove from the shoe. It stretches as it is pulled. If the shoe is put in a freezer and the gum is allowed to cool down, it becomes easy to remove. The gum no longer stretches. The bond strength of the gum to the shoe is the same for the two cases. What has changed is the elongational properties of the gum. It is the total energy, not just the bond strength that determines how hard it is to remove the gum.

7.6 Modeling heat seal and hot tack

7.6.1 Seal interface temperature

Understanding how the sealant interface temperature heats and cools during the sealing process is important for predicting melting, diffusion, cooling and recrystallization that control strength development. Several investigators have modeled the flow of heat to and from the seal interface. Yuan et al. [25] use heat flow rate to correlate with the dwell time or temperature needed. Heat flow rate is the initial temperature difference between the seal bar and film (the driving force for heat flow) times the thermal conductance (a measured value) of the laminate – laminates with greater heat flow rate seal at shorter dwell times.

The unsteady-state heat conduction equation, Eq. (7.12), describes the change in temperature across the film as a function of time. Most investigators assume that the conduction of heat is one-dimensional in the thickness dimension since the film thickness is typically much smaller than the seal area width and length. A few investigators, such as Bach [69], develop a two-dimensional model. This can be important when one or more layers conduct heat in the lateral direction much more quickly than other layers conduct heat in the thickness dimension, such as when aluminum foil is in the structure. A two-dimensional model may also provide insight into what happens in the region immediately adjacent to the seal area. Li et al. [26] show that melting of LLDPE blown film in this region leads to a loss of orientation and strength upon recrystallization. Cooling from lateral transfer of heat after sealing may also influence hot tack. Here the focus is on summarizing the one-dimensional modeling studies from the literature.

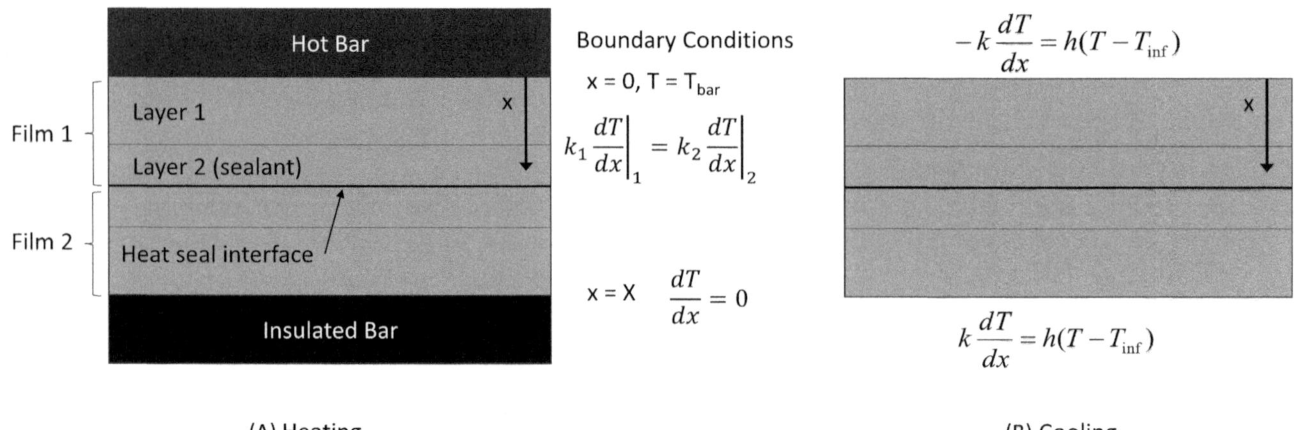

FIGURE 7.25 Illustration of boundary conditions for modeling heat conduction in hot bar sealing: (A) heat up portion of process with hot bar in direct contact with upper film an insulated rubber bar in contact with the lower film, and (B) cool-down in air after bars are released.

Unsteady state heat conduction in the thickness dimension (x) is given by the following equation:

$$\frac{\partial T}{\partial t} = \alpha \frac{\partial^2 T}{\partial x^2} \tag{7.12}$$

where

T = temperature,
t = time,
x = distance in the thickness dimension,
α = thermal diffusivity = $\rho k/C_p$,
ρ = density,
k = thermal conductivity,
C_p = heat capacity.

Initial and boundary conditions must be specified to solve this equation.

7.6.1.1 Initial condition

At $t = 0$, the films are assumed to be at ambient temperature.

For models considering cooling in addition to heat-up, the results from the heat-up model can be used as the initial conditions for the cool down:

At $t = t_{\text{dwell}}$, T = temperature at end of heat-up.

7.6.1.2 Boundary conditions

At the external surfaces, $x = 0$ and X (X is the thickness of two films being sealed), three types of boundary conditions are typically allowed (see Fig. 7.25):

• Type 1: Temperature is specified at the surface (e.g., seal bar is in direct contact with the packaging film with no resistance to heat transfer).

$$\text{At} x = 0 \text{ or } X, T = T_1 \tag{7.13}$$

• Type 2: The surface is insulated from heat loss or gain (e.g., film is in contact with an insulated surface). In mathematical terms, this means the change in temperature is zero.

$$\text{At} x = 0 \text{ or } X, \frac{dT}{dx} = 0 \tag{7.14}$$

- Type 3: The surface of the film is being heated or cooled by heat convection with an outside medium such as air. Here it is convenient to introduce a heat-transfer coefficient, h. The boundary conditions can be written as follows:

$$\text{At } x = 0, \ -k\frac{dT}{dx} = h(T - T_{\text{inf}}) \tag{7.15}$$

$$\text{At } x = X, \ k\frac{dT}{dx} = h(T - T_{\text{inf}}) \tag{7.16}$$

Here T_{inf} is the temperature of the cooling or heating medium far away from the surface. The type 3 boundary condition is the most general of the three. When $h = 0$, the second boundary condition is obtained. Letting h approach infinity yields the first boundary condition. The type 3 boundary condition can also be used to simulate resistance to heat transfer between the hot bar and film due to incomplete contact. Other approaches for modeling resistance to heat transfer at interfaces are described below.

For multiple layers, the heat flux at each interface is equal. For example, at the interface between layers 1 and 2:

$$k_1\frac{dT}{dx}\bigg|_1 = k_2\frac{dT}{dx}\bigg|_2. \tag{7.17}$$

McKelvy [53] was one of the early users of this modeling approach for predicting bonding performance. Meka and Stehling [66] used finite element analysis (FEA) to solve Eq. (7.12) for a monolayer PE film. They achieve good agreement between their model results and measurements at the sealant interface using a thin wire thermocouple for the heat-up part of the sealing process. The agreement deviates at high seal temperatures, which they attribute to the model incorrectly accounting for the phase change and to distortion and squeeze-out of the sealant at the elevated temperatures. Morris [67] describes a finite difference model that handles multiple layers and deals with phase changes by incorporating changes in enthalpy at a single melting or freezing temperature. He models both the heat-up and cooling part of the process. Moffitt [93] uses an FEA approach and also accounts for the enthalpy change during phase transitions for a monolayer film. He ignores cooling. These investigators ignore changes in thermal conductivity, specific heat and density with temperature.

Konaganti et al. [71,72] and Ward and Crowe [46] use DSC to measure the specific heat of individual layers in a multilayer film as a function of temperature. The specific heat includes the enthalpy change due to the phase change between melt and solid. They incorporate these data into their FEA model. Konagati et al. also make rheological characterizations (modulus and complex viscosity) using small amplitude oscillatory measurements through the phase change. By coupling the heat conduction model with the viscosity measurements, they compute the change in viscosity at the interface as a function of sealing time. Comparing different sealant resins allows them to propose a rheological fingerprint for LLDPE resins with good caulkability and low HSIT. They also model the cooling part of the process and find the seal area becomes hardened within about 0.1 seconds after the seal bars are released. Ward and Crowe apply the model to multilayer film structures to enhance hot tack. By placing layers next to the sealant layer with higher thermal conductivity, the model predicts the sealant interface will cool more rapidly after the seal bars are released. For their test structure, they use layers of MDPE in the core of the film, which has a higher conductivity than the LLDPE sealant. The model is validated with hot tack measurements that show a broadened hot tack operating window with the new structure.

Mihindukulasuriya and Lim [34] use a one-dimensional FEA heat conduction model to study the effect of a liquid contaminant layer at the seal interface. They use an apparent specific heat versus temperature profile to account for

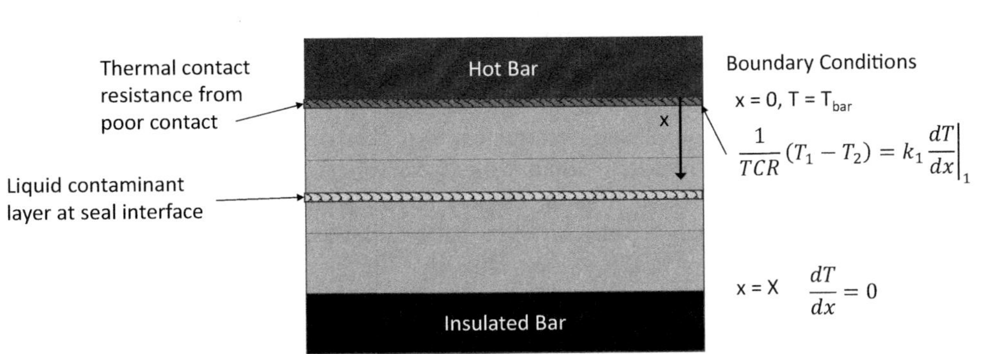

FIGURE 7.26 Concept of thermal contact resistance boundary condition.

polymer melting over a wide temperature range. To account for imperfect heat transfer between layers, they introduce a thermal contact resistance (TCR):

$$\mathrm{TCR} = \Delta T/q \tag{7.18}$$

where

ΔT = temperature difference between two contacting bodies,
q = heat flow rate or flux.

This is illustrated in Fig. 7.26. TCR is a function of the surface characteristics (roughness, presence of pores, etc.), properties of the contaminant (if present), and pressure. These authors determine TCR empirically by fitting the model predictions with experimental measurements using a thermocouple. TCR is applied to the model through the boundary conditions between layers by adjusting Eq. (7.17) to include the heat flux of an intermediary resistance layer:

$$q = \frac{1}{\mathrm{TCR}}(T_1 - T_2) = k_1 \left.\frac{dT}{dx}\right|_1 \tag{7.19}$$

where

q = heat flux,
T_1 = temperature at interface on layer 1 side,
T_2 = temperature at interface on layer 2 side.

One should note that measuring the temperature at the seal interface using a thin wire thermocouple in this and other studies may introduce inaccuracies due to the size of the thermocouple. Even though thin wires are used (typically around 50–75 μm), the size is of the same order of magnitude as the sealant layer thickness. Therefore, a significant amount of distortion of the film near the thermocouple likely occurs.

Mihindukulasuriya and Lim apply their model to the sealing of two 60-μm LLDPE films. A layer of Kapton polyimide film is placed between the heated seal bar and the LLDPE film (generally used to prevent the bar from sticking to the LLDPE). The bottom layer of LLDPE is against an unheated rubber bar and is assumed to be insulated. A thin layer of water is introduced in the model between the two LLDPE films to study the effect of contamination in the seal area. TCRs are used to model the imperfect heat transfer between the Kapton and LLDPE, the two LLDPE films, and when the liquid contaminant is present, between the contaminant and LLDPE layers. The model shows only a small decrease in the interfacial temperature when the water contaminant is present, and this is borne out in experimental validation with milk in the seal area. The results show that the water contaminant reduces the thermal resistance at the interface, presumably by filling in air voids due to micro surface roughness. This increases the heat transfer. The water, however, acts as a heat sink. These two effects nearly balance out. The authors postulate that an oil-based contaminant may wet the PE even better than water and further reduce the thermal resistance.

Aghkand et al. [94] analyze a 3-layer coextruded cast film: (PA6 – tie – mLLDPE). The total thickness is 100 μm with layer ratios 40/10/50. One film uses 130 μm of mLLDPE (keeping the PA and tie thickness constant) to evaluate the effect of sealant thickness. Three fast response thermocouples (75 μm) are placed in different locations to show that the heat transfer is primarily in the thickness dimension, validating the use of a one-dimensional model. They use a scanning electron microscope to show that little squeeze-out occurs under the conditions of their experiments, allowing them to ignore dimensional changes in their FEA model. The thermal parameters are assumed to vary with temperature and are fit to experimental data for density, thermal conductivity and specific heat. They use apparent specific heat measurements from DSC to account for enthalpy changes during melting. Perfect heat transfer is assumed between the layers in the film since they were made by coextrusion. Since the PA surface of the film has microroughness and does not melt during the sealing process, they follow the approach of Mihindukulasuriya and Lim and introduce a TCR between the seal bar and the PA surface. Rather than relying on empirical determination of the TCR from thermocouple measurements, they develop a model for TCR based on plastic deformation that uses surface roughness (from AFM measurements), contact pressure and surface micro-hardness as input parameters. The model simulations agree well with their experimental thermocouple measurements for interfacial temperature as a function of dwell time. They remark that the input parameters for their TCR model will be difficult to measure for industrial applications.

Aghkand et al. compare an mLLDPE sealant with an HDPE sealant. The time for the sealant interface to reach the seal bar temperature is the same at temperatures below Tm. At 140°C and 160°C, above Tm, the HDPE sealant interface takes longer to reach the seal bar temperature than for mLLDPE. This is due to the larger heat of fusion for HDPE because of its higher crystallinity; energy is absorbed as the crystals melt. Another factor is the change in thermal conductivity with melting. For mLLDPE thermal conductivity is found to increase but for HDPE it decreases, slowing the heat to the interface.

7.6.2 Heat seal

De Gennes, Wool and others have developed the theoretical framework for understanding diffusion and scaling of strength build-up at the interface, as described in Section 7.5.3. Predictive modeling based on this framework, however, remains challenging [93]. For example, Wool gives a relationship for the penetration distance as a function of time and diffusion coefficient, but ambiguities regarding the value of the diffusion coefficient make this approach impractical. The reptation-based models are still lacking; while describing a powerlaw dependence of the weld strength on time and MW, they are incomplete, failing for example to account for the thermal history. On the other hand, purely empirical-based models lack a scientific basis. Hybrid models, combining scientific and empirical modeling, is one approach, such as that employed by Ezekoye [93] for the weld strength of polymers.

Moffitt [95–97] develops a mathematical model for heat sealing of linear, semicrystalline polymers such as LLDPE based on reptation theory. The model includes transient heat transfer with phase change to accomplish amorphous contact, interfacial wetting kinetics, and chain interdiffusion via reptation. The kinetics of wetting is described by the model of Wool and O'Conner [98] and De Gennes reptation scaling concepts are used. The model does not include any contributions to the seal strength from cooling or crystallization.

The model predictions are compared with the experimental results reported by Stehling and Meka [45]. The model shows that interfacial melting, wetting and chain interdiffusion are all important. The wetting and interdiffusion times are similar in magnitude; the inclusion of noninstantaneous wetting is needed to get good agreement with the experimental results. The model also shows good agreement with Stehling and Meka on the amount of melting (amorphous fraction) needed for seal initiation. Moffitt constructs curves of relative heat seal strength vs MFI for a given sealing dwell time and temperature. He finds the optimum MFI (190°C /2160 g) is in the range of 2–10 g/10 minutes for dwell times of 0.5–1 seconds. This range of MFI's is what is close to what is typically used in heat seal applications in the industry.

Morris [67] develops a heat seal model for ionomers that includes heat transfer. Most laboratory experiments on seal performance are done under somewhat long dwell times using standard structures. Changing the design of the film structure or the dwell time can significantly alter the transfer of heat to the interface. This can have a substantial impact on plateau initiation temperature, T_{pi}. The thicker the structure or shorter the dwell time, the higher the value of T_{pi} since less heat gets transferred to the interface. An example is shown in Fig. 7.14. Here a thin (50-μm ionomer) and thick (84-μm paper/50-μm ionomer) film structure with the same ionomer sealant is compared. (The ionomer in this example is a sodium ionomer with an MFI of 1.3 g/10 minutes.) The thin and thick structures have T_{pi}'s of 110°C and 135°C, respectively. Both values are considerably above the peak melting point of the ionomer (95°C).

Presumably, T_{pi} is directly related to a critical penetration distance of chain-segment diffusion across the seal interface associated with the ultimate seal strength. If the penetration distance of one structure can be related to T_{pi}, the T_{pi} for any other structure should be able to be computed from heat transfer modeling. An obvious explanation for the difference between the curves in Fig. 7.14 is that it takes a longer time for heat to transfer through the thicker structure. This necessitates the use of higher seal bar temperatures to achieve equivalent molecular penetration during the short duration of the heat seal test. To determine if heat transfer considerations alone are sufficient to make predictive calculations, Morris uses a heat transfer model to compute the temperature of the interfaces as a function of time, as described in Section 7.6.1. The heat conduction equation simplifies to one dimension (in the direction perpendicular to the surface or across the thickness of the film) since the thickness is so much thinner than the lateral dimensions of the film. The model handles multiple layers by assuming constant heat flux at each interface. An enthalpy approach is used to take into account the latent heat of fusion as the polymer changes from solid to melt [99]. Initially, the film is at room temperature. For boundary conditions, he assumes that one surface is at the seal bar temperature (T_{pi}) and the other surface is insulated (rubber bar). This computation is run for a sealing time of 0.5 seconds after which the boundary conditions are changed to reflect cooling in air (by using a convective heat transfer coefficient).

The results are given in Fig. 7.20. Here structure 1 (thin) has a bar temperature of 110°C and structure 2 (thick) has a bar temperature of 135°C. These are the plateau initiation temperatures from Fig. 7.14. If heat transfer considerations alone can be used to predict equal performance, the curves in Fig. 7.20 should be equivalent. The results show

- Considering only the heat-up part of the process when the seal bars are in contact with the film, the heat transfer curves are not equivalent. The interface of the thin structure reaches its T_{pi} (110°C) in 0.2 seconds whereas the interface of the thick structure barely reaches 110°C (and never 135°C) during the 0.5 seconds dwell time.
- Cooling of the thick structure is much slower than cooling of the thin structure.
- If the whole process from heat-up to cool-down is considered, then the heat history of the interface may be equivalent. But this is difficult to determine from this calculation. A better way to quantify the equivalency of these two curves is needed.

These results suggest that the molecular chain-segment penetration across the interface needs to be included in the model. Morris assumes the penetration can be modeled as Fickian diffusion in a semiinfinite slab:

$$W = 4\sqrt{Dt} \tag{7.20}$$

where

W = penetration distance,
D = diffusion coefficient,
t = time.

The penetration distance is assumed to be proportional to the interfacial strength. In Section 7.5.3, it is found that the diffusion is not Fickian up until the reptation time and that penetration of chain segments is actually proportional to $t^{1/4}$. The fracture energy (G_a), however, scales as $t^{1/2}$ (Eq. 7.8), and so the approach is valid.

Rather than computing a diffusion coefficient, Morris uses an approach originated by Register [100] for modeling hot tack (Section 7.6.3) involving the terminal relaxation time. The terminal relaxation time, t_d, is the time it takes a polymer chain to diffuse a distance characteristic of its own size in the melt. The characteristic size is the radius of gyration R_g, the size of the coiled chain. Assuming a Fickian diffusion model, the penetration is related to the square root of the diffusion coefficient times the time (Eq. 7.20), thus

$$R_g = 4\sqrt{Dt_d} \tag{7.21}$$

where

R_g = radius of gyration or characteristic size of the coiled chain,
D = center of mass diffusion coefficient,
t_d = terminal relaxation time.

The terminal relaxation time can be determined from rheological measurements. It is related to the product of the recoverable compliance and zero shear viscosity:

$$t_d = J_e^0 \eta_0 \tag{7.22}$$

where

J_e^0 = recoverable compliance (strain/stress),
η_0 = zero shear viscosity.

Vanhoorne and Register [101] and Tierney and Register [84,102] measure the zero shear viscosity of several ionomers using a constant stress cone and plate rheometer with shear rates ranging from 10^{-5} to $1\ s^{-1}$. The recoverable compliance is obtained from creep relaxation experiments on the cone and plate rheometer. They compute the terminal relaxation time and relate it to the compositional variables of the ionomer such as acid level, cation and percent neutralization.

Combining Eqs. (7.20) and (7.21) yields an expression for the dimensionless penetration distance, W^*:

$$W^* = \frac{W}{R_g} = \sqrt{t/t_d} \tag{7.23}$$

An Arrhenius equation similar to Eq. (7.9) is used to fit the temperature dependence of the terminal relaxation time:

$$t_d = A\exp\left[\frac{Ea}{R}\left(\frac{1}{T_i} - \frac{1}{408}\right)\right] \tag{7.24}$$

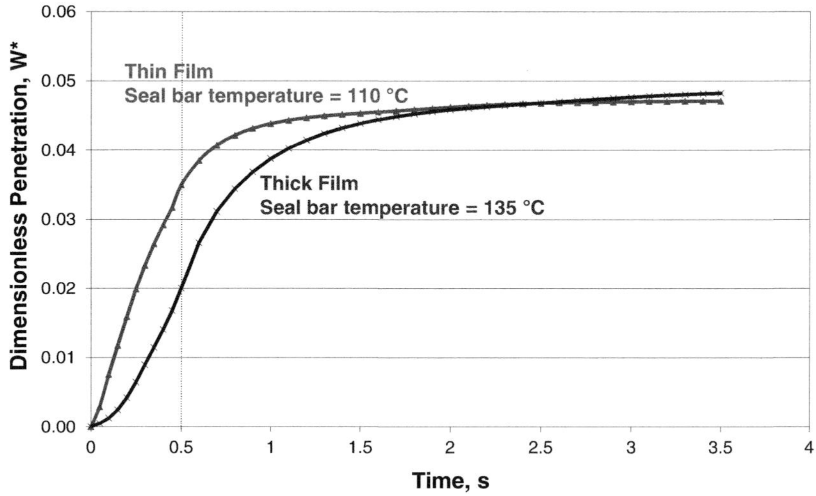

FIGURE 7.27 Model calculation of dimensionless penetration during heat sealing for two structures from Fig. 7.14 [67]. *Reproduced with permission from B. A. Morris, Predicting the heat seal of ionomer films, J. Plastic Film, Sheeting 18 (2002) 157–167.*

where

t_d = terminal relaxation time,
A = factor related to polymer chemistry and architecture,
E_a = activation energy,
R = ideal gas constant,
T_i = temperature [K] at the interface at increment time i.

The heat transfer model uses a finite difference algorithm where the time is divided into small increments for the purposes of the computation. Using Eq. (7.24), t_d is computed for a given time increment. By summing over all past t_d, the average t_d over that time period is obtained:

$$\frac{1}{t_d^{\text{avg}}} = \frac{1}{n}\sum_{i=1}^{n}\frac{1}{t_d^i} \quad (7.25)$$

The dimensionless penetration distance becomes:

$$W^* = \sqrt{t/t_d^{\text{avg}}} \quad (7.26)$$

Results of the W^* computation for the thin and thick structures at their respective T_{pi}'s are given in Fig. 7.27. Both structures now have the same penetration, as would be expected from the heat seal results. It is encouraging that model results using terminal relaxation times derived from independent measurements give excellent results when applied to this sealing application. Key learning's from the analysis include:

- The results suggest that W^* can be used as a parameter for predicting heat seal performance. As W^* increases, the seal strength should increase due to increasing chain bridging and entanglement at the interface. At some point, the increase in strength levels off, either because the interface has been erased, or the solid-state yield strength exceeds the tear strength. The penetration at this point is identified as the critical dimensionless penetration, Wc^*, and is directly related to T_{pi}. Once Wc^* is known for a given sealant structure, the model can be used to predict whether similar structures of differing thickness, heat seal bar temperature and dwell time will be on the seal strength plateau by comparing W^* with Wc^*.
- Both the heat-up and cool-down phases of the process need to be included. During the heat-up phase the thin structure reached about 70% of the ultimate penetration and the thick structure about 40%. A significant amount of penetration occurs during the cooling-phase.
- Penetration of only a small fraction (around 5%) of the coiled size of the molecular chain (R_g) is needed for the ionomer to achieve maximum seal strength. The scaling laws described earlier suggest for PE to achieve maximum peel strength its chains need to diffuse a distance on the order of R_g. The ionic interactions in ionomers act over longer distances than the dispersive forces in PE, suggesting that less entanglement is necessary for the ionomers.

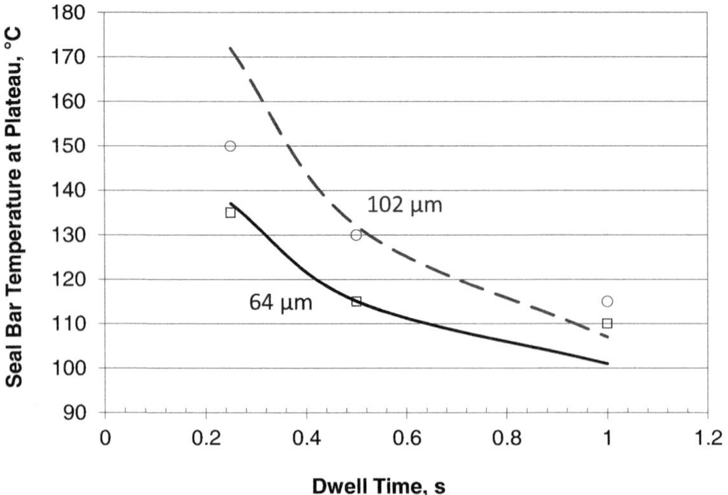

FIGURE 7.28 Comparison experimental measurement and model prediction of heat seal plateau initiation temperature. Structure is 13-mm OPET/ionomer with two different thicknesses of ionomer film. Symbols are experimental data points and solid lines model results for $Wc^* = 0.081$ [67]. *Reproduced with permission from B. A. Morris, Predicting the heat seal of ionomer films, J. Plastic Film, Sheeting 18 (2002) 157–167.*

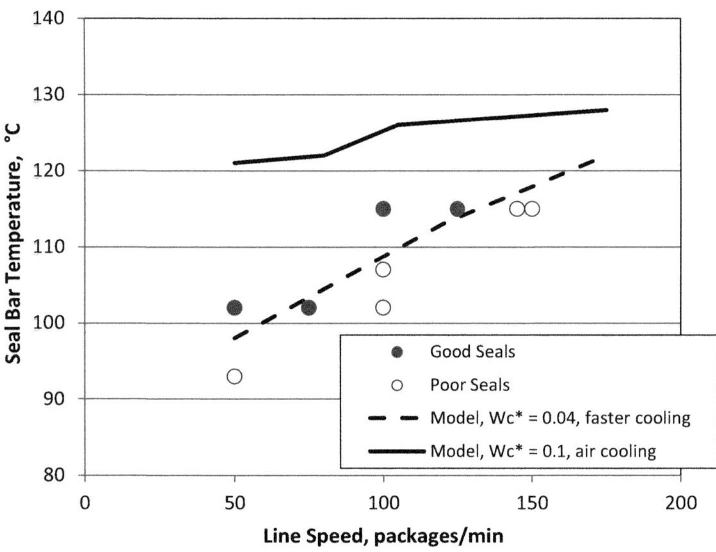

FIGURE 7.29 Results of horizontal form fill seal trial. Open circles represent experimental results of poor seals, closed circles good seals. Solid lines are model predictions. Packaging structure is 13-μm OPET/25-μm Ionomer2 [67]. *Reproduced with permission from B. A. Morris, Predicting the heat seal of ionomer films, J. Plastic Film, Sheeting 18 (2002) 157–167.*

- The model has limitations. In particular, the penetration at temperatures near the melting point may not be accurately portrayed by the Arrhenius relationship given in Eq. (7.24). Also, the computed value of the penetration depends on the assumptions made in the heat transfer model. The ability to translate results from a lab experiment to a high-speed packaging line will depend on how accurately the heat seal conditions are portrayed.

Morris tests the model by comparing the results with laboratory heat seal measurements on two cast films of ionomer 63-μm and 102-μm thick. The films are heat-sealed at dwell times ranging from 0.25 to 1.0 seconds (at 0.28 MPa pressure, 25-mm seal bar width, top bar heated only). 13-μm films of OPET are placed on top of the films to protect the seal bar. The measured value of T_{pi} for each dwell time is plotted in Fig. 7.28 along with results from the model. The model predictions agree well with the data for both films, confirming the utility of the model. At long dwell times, thickness has little impact on T_{pi}. At short dwell times, thicker films require much higher seal bar temperatures to achieve the ultimate heat seal strength.

Morris also compares model predictions with the results of a trial run on a Doboy Super Mustang horizontal packaging line. Such lines package candy bars, crackers, cookies, baseball cards and other items at speeds up to about 300 packages per minute. In the trial, the quality of the seal at several seal bar temperatures at line speeds ranging from 50 to 150 packages per minute are evaluated without packaging any product. The seal area on the front and back seals comprises 26-mm out of the total packaging length of 190-mm. Thus, the dwell time ranges from 0.168 to 0.056 seconds. The packaging structure is 13-μm OPET/25-μm Ionomer2, a zinc ionomer with a MFI of 5 g/10 minutes.

In Fig. 7.29, the open circles represent weak seals and closed circles good seals. The modeling of laboratory heat seal data gives a value for Wc^* of 0.1. Using this value for Wc^* overpredicts the transition from good to bad seals, as shown by the solid curve in Fig. 7.29. A better fit is found using a value of $Wc^* = 0.04$, suggesting either the temperature on the packaging line is reading too low or the heat seal plateau does not have to be reached in order to achieve good seals. Also, the slope of the transition curve between good and bad seals is too shallow if it is assumed the seal cools in air, the standard assumption used in the model. It is likely that the seal area cools faster because the bottom of the film comes into contact with a cold surface after the heat seal has been made. Assuming a faster cooling rate achieves better agreement between the model and experimental results.

The results point to how laboratory heat seal experiments can help predict seal performance on high-speed packaging lines, but at the same time underscore several limitations. These include the proper selection of heat transfer parameters and what constitutes a good heat seal (plateau strength or otherwise). In practice, data from an existing film structure run on the high-speed line can be used to determine Wc* and the heat transfer coefficients, rather than relying on laboratory data. Once these parameters are determined for a given line, the model should be useful for predicting the performance of other structures and line speeds on that line.

7.6.3 Hot tack

Register [100,103], in some unpublished work supported by DuPont's ethylene copolymer business (a business now part of Dow), develops a model for hot tack of ionomers. There are two fundamental issues for optimizing hot tack performance:

- What is the ultimate hot tack (at long contact times at a given seal temperature after interdiffusion has completely erased the interface)? This should be related to bulk elongational viscosity or melt strength.
- How long does it take to develop this ultimate hot tack at a given sealing temperature? This should relate to polymer diffusivity, D.

The goal of the work is to establish what rheological properties will yield maximum hot tack within the constraints of the process (sealing temperature (T_s) and dwell or contact time (t_c)).

Monolayer blown film from a series of ionomers made from the same acid copolymer base resin (MW and acid level) at different neutralization levels are laminated to paper and the hot tack measured over a range of 1–15 seconds contact (dwell) time and 120°C–220°C seal bar temperature. Register plots hot tack force versus contact time for several temperatures, as shown in Fig. 7.30. The plateau hot tack strength is the "ultimate hot tack force." The measurement process exerts a baseline force that is subtracted out, yielding the net ultimate hot tack (NUHT). This is what is used in the analysis.

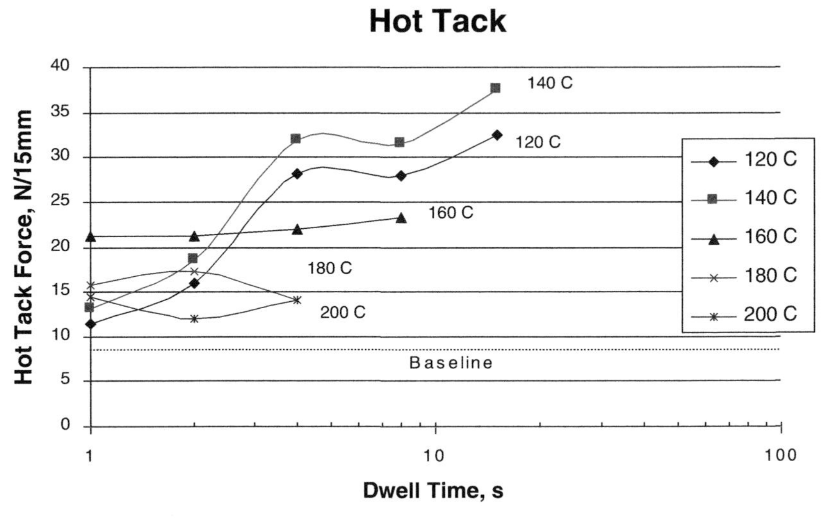

FIGURE 7.30 Hot tack force as a function of seal contact time (dwell time) and seal bar temperature for one ionomer [100]. *From R. A. Register, Optimizing hot tack in packaging sealing operations, Present, 1999, DuPont. Reproduced with permission from The Dow Chemical Company.*

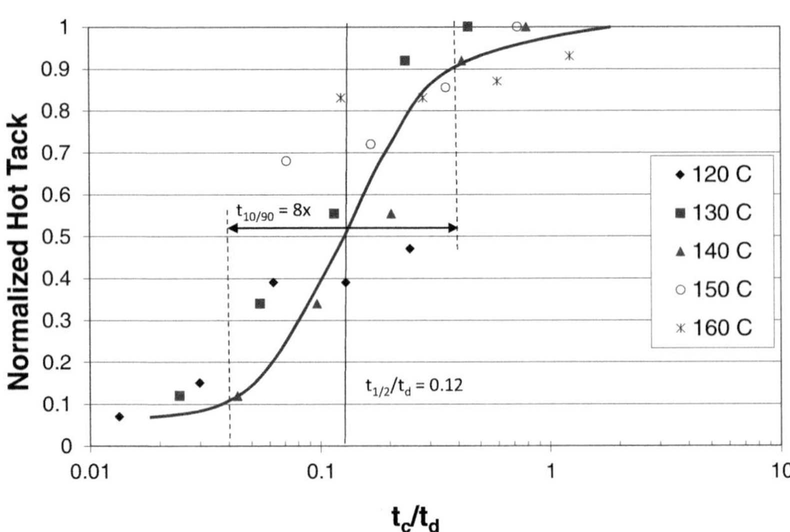

FIGURE 7.31 Example of a master hot tack curve for an ionomer [100]. *From R. A. Register, Optimizing hot tack in packaging sealing operations, Present, 1999, DuPont. Reproduced with permission from The Dow Chemical Company.*

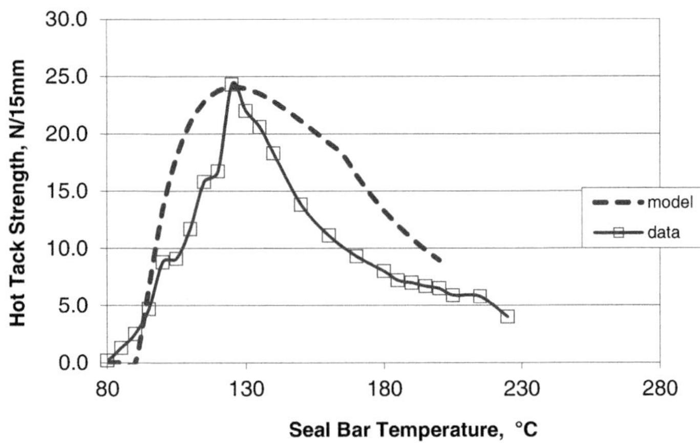

FIGURE 7.32 Comparison of hot tack model with experimental data for an ionomer film at long seal dwell time. 3.3-mil kraft paper/2-mil Ionomer, 3-s dwell. *Reproduced with permission from The Dow Chemical Company.*

An Arrhenius plot of NUHT versus $1/T$ gives unique curves for each ionomer. The slope is related to the activation energy and differs from the activation energy for shear flow (Eq. 5.48) by more than a factor of two: 45 versus 100 kJ/mol for zinc and 35 versus 80 kJ/mol for sodium ionomers for hot tack versus shear flow, respectively. This means that with increasing seal temperatures, ionomers retain their hot tack better than they retain their shear viscosity. The difference is likely due to differences in deformation. It is suggested that NUHT may correlate better with elongational viscosity or melt strength.

Register creates a master hot tack versus temperature curve by scaling the hot tack at a given temperature by NUHT and t_c by t_d. (Here t_d is the terminal relaxation time, the time for the polymer to diffuse one random coil distance. The determination of t_d from rheological measurements is described in Section 7.6.2.) One such master curve is shown in Fig. 7.31. Note that the x-axis is related to W^{*2} per Eq. (7.23). The data are expected to collapse onto a single sigmoidal shaped curve with the center at $t_c/t_d < 1$ since ionomers do not need to diffuse a whole coil size to achieve ultimate interfacial strength. Indeed, the data show t_c/t_d is centered around 0.12 for this grade of ionomer. The data also show that 80% of the hot tack develops in a narrow range of t_c/t_d of less than a decade (8x to go from 10% to 90% of the ultimate hot tack). Since the midpoint is around 0.12, 80% of the NUHT is achieved at $t_c/t_d = 0.3$. If $t_c << 0.3t_d$ then no hot tack develops. If $t_c >> 0.3t_d$ all the hot tack develops (Normalized NUHT = 1).

From a polymer design perspective, the above relationship can be flipped to suit the sealing process. The dwell or contact time is dictated by the packaging line speed. Thus, to achieve good hot tack, the rheology of the ionomer should

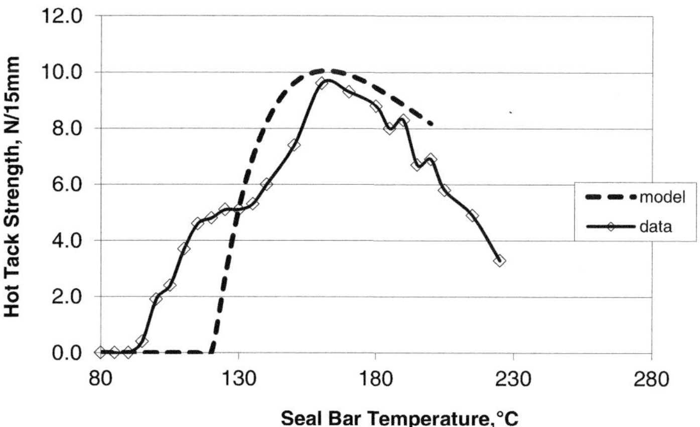

FIGURE 7.33 Comparison of hot tack model with experimental data for an ionomer film at short dwell time. 3.3-mil kraft paper/2-mil Ionomer, 0.5-s dwell. *Reproduced with permission from The Dow Chemical Company.*

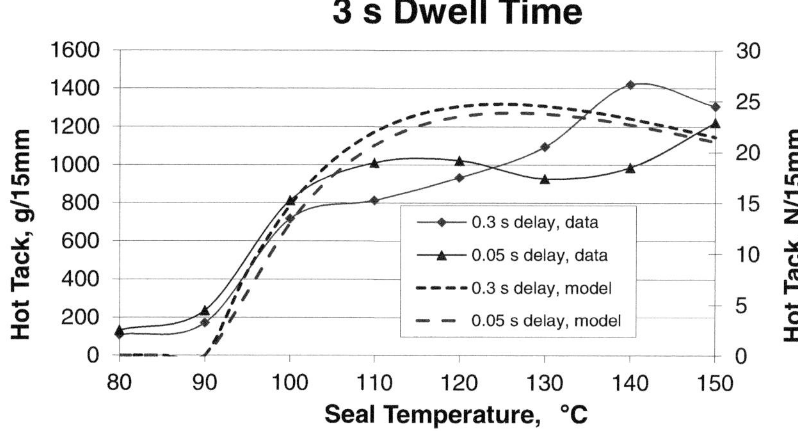

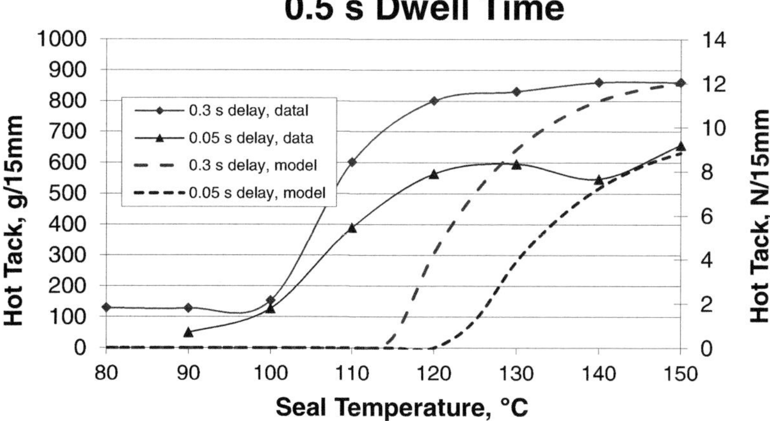

FIGURE 7.34 Effect of delay time on hot tack of an ionomer. 3.3-mil kraft paper/2-mil Ionomer. Top: 3 s dwell time. Bottom: 0.5 s dwell time. *Reproduced with permission from the Dow Chemical Company.*

be adjusted so that $t_d = 3t_c$ to $5t_c$. For ionomers, this can be done through tailoring the amount of neutralization. Register compares his model's prediction for peak hot tack temperature at a given dwell time with data on grades of ionomer used commercially for heat seal applications. The model shows good agreement.

Moffitt uses his model to determine a design range for the MFI of LLDPE for optimum heat seal at a given dwell time (Section 7.6.2). The same can be done with Register's hot tack model. As shown in Eq. (7.22), t_d is related to the zero shear viscosity, which in turn generally correlates with the inverse of the MFI. The result predicts that ionomers with MFIs in the range of 1– 10 g/10 minutes have good hot tack, which is what is used commercially. This is also in

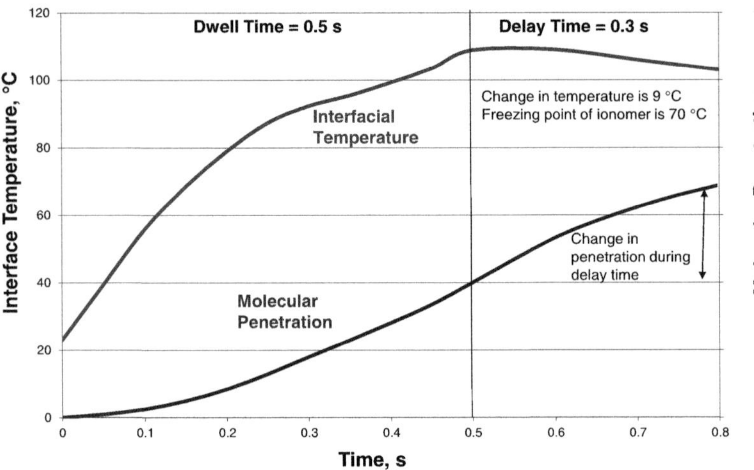

FIGURE 7.35 Model prediction of temperature and penetration profiles of an ionomer sealant. *Reproduced with permission from the Dow Chemical Company.*

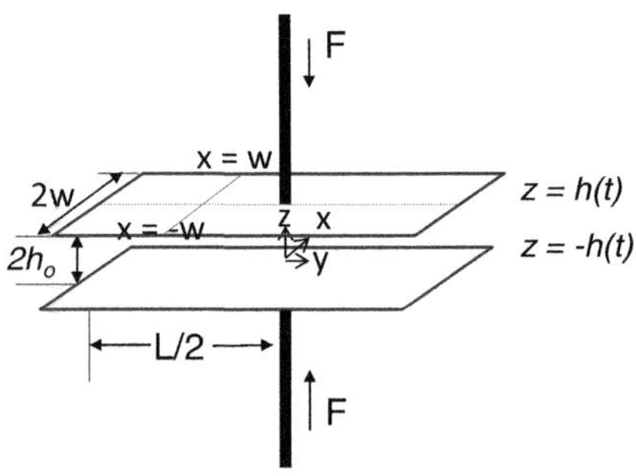

FIGURE 7.36 Squeeze flow between flat plates [90]. *Reproduced from B.A. Morris, J.M. Scherer, Squeeze-out of sealant during hot bar sealing: Modeling insights, in: SPE ANTEC Conf., 2013, with permission from The Dow Chemical Company.*

agreement with Moffitt's study of LLDPE. While these models may merely validate what has long been known in the industry about suitable sealants, they provide the theoretical basis for this performance. Such scientific understanding may allow the development of other chemistries with good heat seal performance. It should be noted that while the concept of a NUHT master curve likely holds for polyethylene sealants, the location of t_c/t_d where 80% of NUHT is achieved is expected to shift to higher values (~ 1) due to the presence of only short-range dispersion forces. The terminal relaxation time for PE, especially linear PE, is substantially smaller than for ionomers; while more penetration across the interface is needed to build strength, the rate of diffusion is much faster for linear PE than ionomers.

In some unpublished work, Morris extends the hot tack model of Register by coupling it with his heat conduction model (Section 7.6.1). This allows for the prediction of the whole hot tack curve. Results of the model compared with experimental measurements by J. R. de Garavilla at DuPont are shown in Figs. 7.32 and 7.33 at two different sealing dwell times. The model correctly predicts the shape of the curve but has difficulty predicting the onset of hot tack at short dwell times.

The model can be used to gain insight into the effect of delay time on the hot tack of ionomers. Fig. 7.34 shows that the influence of delay time is insignificant at long dwell times, with good agreement between experiment and model prediction. At short dwell times, a substantial increase in hot tack is measured as the delay time is increased. The model correctly captures this difference but fails to predict the onset of hot tack. Nevertheless, the model can be used to discern the underlying mechanism for higher hot tack at longer delay times at short dwell times. Fig. 7.35 shows the model predictions for the interfacial temperature and molecular penetration (W^*) profiles during the heat seal and subsequent delay time of the hot tack experiment. Many have hypothesized that cooling of the seal area is responsible for the increased hot tack at long delay times. These results show that the cooling is insufficient to significantly affect the strength of the interface; the temperature only decreases by 9°C and is well above the freezing point of the ionomer. Also, ionomers do not undergo strain-induced crystallization at these temperatures. Furthermore, the model reveals that substantial molecular penetration occurs during the delay. This is likely the source of the increased hot tack. At longer dwell times sufficient penetration occurs during sealing so that the influence of delay time is minimal.

Recall in Section 7.5.5 that mPE plastomers may undergo strain-induced crystallization near their melting point. A 9°C reduction in interfacial temperature, therefore, may have a substantial impact on hot tack performance for these materials. It is likely both cooling and penetration influence the impact of delay time on the hot tack performance of mPE.

7.6.4 Squeeze flow

In Section 7.3.1 the concept of the heat seal temperature window is introduced. Excessive flow of the sealant may occur at the high end of the temperature range [66]. Morris and Scherer [90–92] model this behavior by adapting an approach originally developed for predicting the squeeze flow between flat disks found in the rheology literature [104–111]. Their work is summarized here. While the focus is on squeeze-out, the work also provides insights into caulking and flow within the seal area (Section 7.5.6).

7.6.4.1 Model development

The film area under the sealing bars can be represented by two parallel flat plates under a constant applied force (Fig. 7.36). A rectangular coordinate system is defined, originating at the center point between the plates with z in the direction of plate separation, x in the width direction and y in the length direction. The following assumptions are made:

- The initial film thickness (h_o) is much less than the seal area half width ($h_o << w$), so the flow is mostly in the x-direction.
- Seal length is much greater than the width ($L >> 2w$), so the flow in the y-direction is ignored.
- Inertia and gravity effects are negligible.
- Flow is quasisteady state.
- Flow is Newtonian.
- No slip at the surface of the plates.
- Process is isothermal.

Details of the model derivation are found in Morris and Scherer [90,92]. The equations of continuity and motion are solved with the following results:

$$v_x = \frac{3}{2}\frac{(-\dot{h})}{h}x\left(1 - \left(\frac{z}{h}\right)^2\right) \quad (7.27)$$

$$v_z = \frac{3}{2}\dot{h}\left(\frac{z}{h} - \frac{1}{3}\left(\frac{z}{h}\right)^3\right) \quad (7.28)$$

$$p - p_a = \frac{3}{2}\frac{(-\dot{h})}{h^3}\mu w^2\left(1 - \left(\frac{x}{w}\right)^2\right) \quad (7.29)$$

where

v_x = velocity in x-direction,
v_z = velocity in z-direction,
p = pressure,
μ = viscosity,
h = half distance between moving plates,
w = half width of the plates,
p_a = atmospheric pressure.

The change in separation between the plates is found to be [90]

$$\frac{1}{h^2} - \frac{1}{h_o^2} = \frac{2Pt}{w^2\mu} \quad (7.30)$$

This is the key result that is used to model squeeze-out during hot bar sealing.

An expression equivalent to Eq. (7.30) for circular disks is [104]

$$\frac{1}{h^2} - \frac{1}{h_o^2} = \frac{16Pt}{3R^2\mu} \tag{7.31}$$

where

R = radius of the circular disk,
P = $F/\pi R^2$.

This is a form of Stephan's equation and may be useful for analyzing the squeeze flow near the long edge of the seal bar (in the rectangular plate analysis the flow is assumed to be far away from the length edge of the bar). Bird, Armstrong and Hassager [104] also derive expressions for squeezing flow between disks for powerlaw fluids (Scott's equation [106]) and for fluids following the Criminale–Ericksen–Filbey constitutive equation.

The viscosity must be known to use the model. In the derivation, the viscosity is assumed to be independent of shear rate (Newtonian), yet the viscosity of most polymers decreases with increasing shear rate. This is accounted for by choosing a representative viscosity at a shear rate characteristic of the sealing process.

Morris and Scherer [90,92] derive the following expression for average shear rate near the edge of the plates:

$$\dot{\gamma}_{avg} = \frac{3Ph}{2w\mu}(x = \pm w, \quad w >> h). \tag{7.32}$$

Table 7.5 gives some typical values for the variables and their effect on the average shear rate given by Eq. (7.32).

From a mass balance, the amount of polymer squeezed out from the plates is related to the negative of the change in separation distance. The percent squeeze-out is defined as the volume of polymer squeezed out from the seal area divided by the total volume of polymer:

$$\%\text{Squeeze} - \text{Out} = 100(h_o - h)/h_o. \tag{7.33}$$

TABLE 7.5 Typical shear rate values near the edge of the seal bar as computed by Eq 7.32 [90]. P is the pressure, h_0 the initial sealant layer thickness, h the final sealant layer thickness, % squeeze out is defined by Eq. 7.33, w is the half width of the seal bar, t is sealing time and viscosity is viscosity of the sealant.

P	h_o	h	Squeeze out	w	t	Viscosity	Shear rate
Pa	mm	mm	%	mm	s	Pa-s	s^{-1}
275000	0.025	0.025	0	6.35	1	20000	0.08
275000	0.025	0.025	0.55	6.35	1	2000	0.81
620000	0.025	0.020	21.5	1.25	1	2000	7.5

Source: B.A. Morris, J.M. Scherer, Squeeze-out of sealant during hot bar sealing: Modeling insights, in: SPE ANTEC Conf., 2013. Reproduced with permission from The Dow Chemical Company.

TABLE 7.6 Description of sealant polymers used in squeeze flow study [91].

Resin	Grade	Manufacturer	Manufacturing Process	Density g/cc	MFI (190°C/2160 g) g/10 min
Ionomer	SURLYN 1601	Dow	Free Radical	0.95	1.3
Plastomer	AFFINITY PL1881G	Dow	Single site catalyst	0.906	1

Source: B.A. Morris, J.M. Scherer, Preventing squeeze-out of sealant during hot bar sealing: modeling and experimental insights, in: SPE ANTEC Conf., 2014, pp. 1452–1459. Reproduced with permission from The Dow Chemical Company.

7.6.4.2 Experimental results

Morris and Scherer conduct experiments to compare with the model predictions. Films with the structure (PA-tie-PA-EVOH-PA-tie-sealant) with (25/8/13/8/13/8/25 vol%) and total thickness 100-μm are produced on a 9-layer blown film line. Two sealants described in Table 7.6 are used. The adjacent tie layer is BYNEL 41E687, an anhydride-modified LLDPE-based resin with an MFI of 1.7 g/10 minutes and manufactured by Dow.

The sealants, an ionomer and a metallocene polyethylene plastomer, are chosen to study the effect of molecular architecture on performance. Ionomers have LCB, broad MWD and ionic interactions. Plastomers are substantially linear polymers with a nearly uniform molecular weight and comonomer distribution.

The viscosity of the sealant resins is measured using a TA Instruments ARES LS2 controlled strain parallel plate rheometer. The frequency is varied from 0.01 to 100 radians per second at 130°C, 150°C, and 170°C. Results are found to follow time-temperature superposition. The master curve is fit to a Carreau–Yasuda model:

$$\eta_r = \frac{\eta_o}{\left[1+(\lambda\dot{\gamma}_r)^a\right]^{(1-n)/a}} \tag{7.34}$$

where

$\eta_r =$ reduced viscosity ($= b_T\eta/a_T$),
$\eta =$ viscosity,
$a_T =$ horizontal shift factor,
$b_T =$ vertical shift factor,
$\dot{\gamma}_r =$ reduced shear rate ($a_T\,\dot{\gamma}$),
$\dot{\gamma} =$ shear rate,
$\eta_o, \lambda, n, a =$ model parameters.

The shift factors follow an Arrhenius relationship:

$$a_T = \exp\left[\frac{E_a}{R}\left(\frac{1}{T} - \frac{1}{T_o}\right)\right] \tag{7.35}$$

$$b_T = \exp\left[\frac{E_b}{R}\left(\frac{1}{T} - \frac{1}{T_o}\right)\right] \tag{7.36}$$

where

E_a and $E_b =$ activation energies,
$T =$ temperature (K),
$T_o =$ reference temperature (K).

The rheology model parameters for the sealants are given in Table 7.7. Fig. 7.37 shows the viscosity versus shear rate data. The shape and temperature dependence of the viscosity curves differ due to the differences in molecular architecture and chemistry. The ionomer sealant has a higher low shear rate viscosity which decreases substantially as the

TABLE 7.7 Carreau–Yasuda rheology model parameters for the two sealant polymers used in the squeeze-out study.

Parameter	Units	Ionomer	Plastomer
η_o	Pa-s	774000	37700
λ	s	4.339	6.895
n		0.419	0.589
a		0.530	0.977
T_o	K	423.15	423.15
E_a	kJ/mol	98	59
E_b	kJ/mol	5.87	10.13

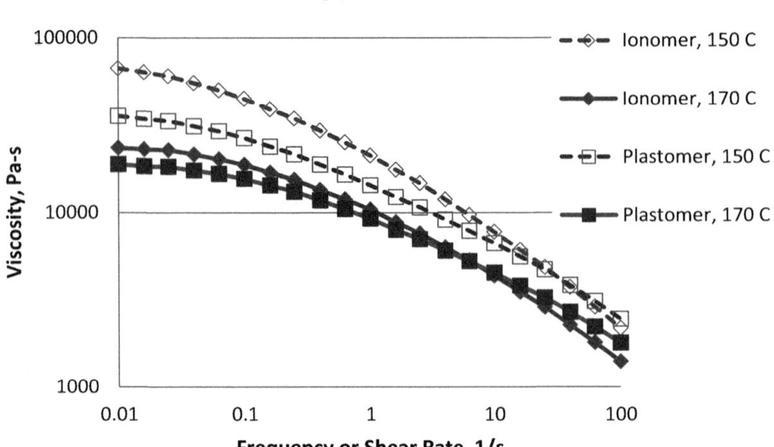

FIGURE 7.37 Complex viscosity of sealant resins, as measured by a parallel plate rheometer [91]. *Reproduced from B.A. Morris, J.M. Scherer, Preventing squeeze-out of sealant during hot bar sealing: modeling and experimental insights, in: SPE ANTEC Conf., 2014, pp. 1452–1459, with permission from The Dow Chemical Company.*

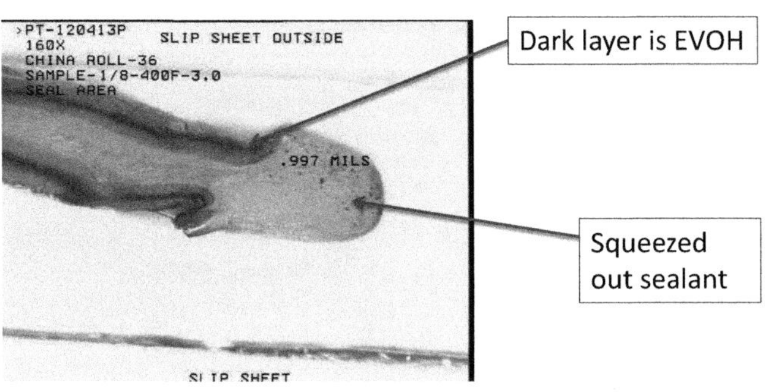

FIGURE 7.38 Photo of squeeze out of sealant taken from an optical microscope. Seal conditions are 204°C seal bar temperature, 3 s dwell, 0.345 MPa (50 psi) bar pressure, 3.2 mm bar width [90]. *Reproduced from B.A. Morris, J.M. Scherer, Squeeze-out of sealant during hot bar sealing: Modeling insights, in: SPE ANTEC Conf., 2013, with permission from The Dow Chemical Company.*

shear rate is increased. This shear-thinning behavior is due to the presence of LCB and ionic interactions. The viscosity decreases more rapidly with an increase in temperature than the linear plastomer, as indicated by the ionomers higher *Ea*. Consequently, as the temperature is increased, the low shear rate viscosity of the ionomer and plastomer become closer in value. At even higher temperatures, the low shear rate viscosity of the ionomer will cross over that of the plastomer and be the lower viscosity polymer.

Heat seals are made on a Sentinel heat sealer at several temperatures. Pressure, dwell time and seal bar width are held constant at 0.345 MPa (50 psi), 3 seconds, and 12.7 mm (0.5-inch), respectively. Both top and bottom seal bars are heated. The sealed samples are allowed to cool to room temperature, cross-sectioned and examined under an optical microscope.

The experimental procedure involves cutting a cross-section from the seal edge of each sample. Next, the samples, with a polyester slip sheet placed between the films to allow for a more defined squeeze-out area, are placed into the holder of the microscope and viewed at 40–63X magnification. Pictures and measurements of the squeeze widths are taken under the microscope. An example of a severe squeeze-out from the edge of the seal area is shown in Fig. 7.38. The darkest layer is EVOH and the two dark layers next to the EVOH are PA6. The region between the dark layers is the tie and sealant.

Calculated % squeeze-out results from the microscopy results are given in Table 7.8 and plotted in Fig. 7.39. They show the ionomer having a very low squeeze-out until around 200°C. This is consistent with the model predictions discussed below. The plastomer has a higher amount of squeeze-out until the ionomer crosses over at around 200°C.

7.6.4.3 Model results

Eq. (7.30) is used to explore the effect of heat seal parameters and sealant rheology on squeeze-out. The film structure used in the experiments is used as the basis, where the initial separation distance, h_o, is taken to be the thickness of the sealant and adjacent tie resin. The next layer in, a PA layer, is assumed to not melt or flow during the heat seal process since it has a much higher melting point than the sealing temperature.

TABLE 7.8 Percentage of sealant squeeze-out from experimental results. Seal conditions: 0.345 MPa, 3 s dwell, and 12.7 mm bar width, two-side heated seal bars [91].

Seal	Squeeze-out	
Temperature (°C)	Plastomer	Ionomer
121	6%	7%
149	12%	7%
177	10%	6%
204	16%	33%

Source: Reproduced from B.A. Morris, J.M. Scherer, Preventing squeeze-out of sealant during hot bar sealing: modeling and experimental insights, in: SPE ANTEC Conf., 2014, pp. 1452–1459, with permission from The Dow Chemical Company.

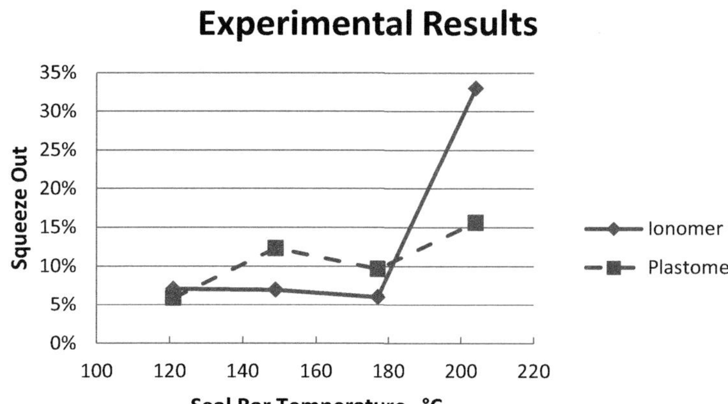

FIGURE 7.39 Experimental results of squeeze-out plotted versus seal bar temperature [91]. *Reproduced from B.A. Morris, J.M. Scherer, Preventing squeeze-out of sealant during hot bar sealing: modeling and experimental insights, in: SPE ANTEC Conf., 2014, pp. 1452–1459, with permission from The Dow Chemical Company.*

To use Eq. (7.30) some of the parameters must be defined with respect to the experiment. The values for P and w are straight-forward: the seal bar pressure and half-width of the seal bar, respectively. The time parameter, t, is taken as the sealing process dwell time. As is discussed in Section 7.6.4.4 there is a time lag for the temperature of the sealant to reach the bar temperature. The time lag will depend on the film thickness and whether one or both bars are heated. The time lag can be predicted with standard heat transfer models (Section 7.6.1) but is ignored in these calculations.

Eqs. (7.34)–(7.36) are used to predict the viscosity of the fluid. For simplicity the rheology of the tie resin is ignored and only the viscosity of the sealant is used in the calculations. Given that the thickness of the sealant layer is typically much larger than the tie layer, this is thought to be a good assumption.

The viscosity is a function of temperature and shear rate for a given sealant. The temperature is assumed to be the seal bar temperature (subject to the time lag limitation described above). The average shear rate at the edge of the seal bar is used for the shear rate (Eq. 7.32). Eq. (7.32) shows that the shear rate is a function of the final separation distance, h, is computed. An iterative process is used. An initial value for the shear rate is assumed (typically $1s^{-1}$), the viscosity is computed using the rheology model for the sealant (Eqs. 7.34–7.36), h computed from Eq. (7.30), and the average shear rate computed with Eq. (7.32). This value for the shear rate is then compared with the assumed shear rate. If they do not agree, the newly computed value for the shear rate is used to repeat the calculation. Usually, the two shear rates agree after three iterations. Typical shear rates range from about 0.01 to $10s^{-1}$, depending on the amount of squeeze-out.

The results of a factor analysis are shown in Figs. 7.40–7.44 for the ionomer sealant, where each of the input parameters to Eq. (7.30) is varied one at a time. The results show:

- Squeeze-out increases almost linearly with increasing pressure (Fig. 7.40).
- As expected, squeeze-out increases nearly linearly with dwell time (Fig. 7.41). Some nonlinearly is introduced due to the change in shear rate and its effect on the viscosity.

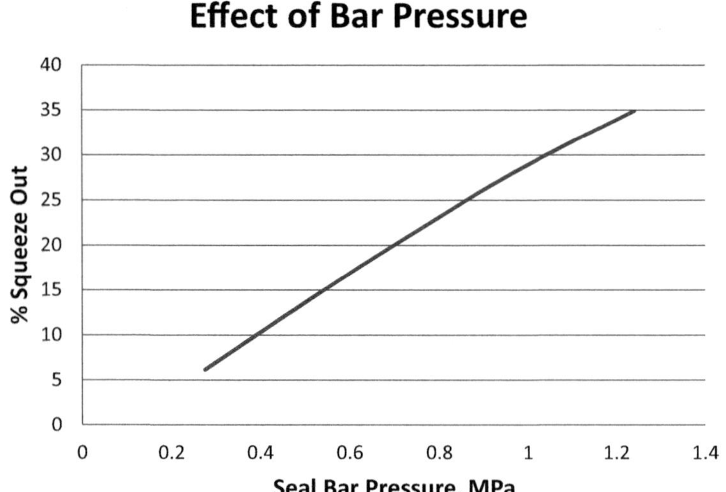

FIGURE 7.40 Model prediction (Eq. 7.30) of % squeeze-out of an ionomer as a function of sealing pressure. Seal temperature = 175°C, bar width $(2w) = 2.5$ mm, dwell time = 1 s, sealant thickness = 0.05 mm [91]. *Reproduced from B.A. Morris, J.M. Scherer, Preventing squeeze-out of sealant during hot bar sealing: modeling and experimental insights, in: SPE ANTEC Conf., 2014, pp. 1452–1459, with permission from The Dow Chemical Company.*

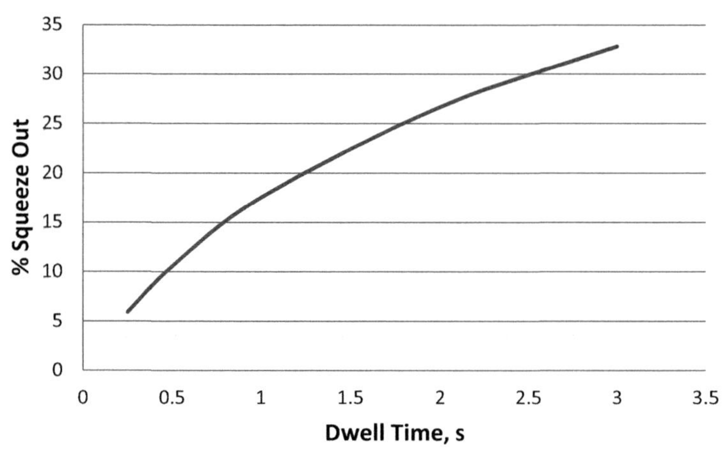

FIGURE 7.41 Model prediction (Eq. 7.30) of % squeeze-out of an ionomer sealant as a function of sealing dwell time. Seal temperature = 175°C, bar width $(2w) = 2.5$ mm, seal bar pressure = 0.6 MPa, sealant thickness = 0.05 mm [91]. *Reproduced from B.A. Morris, J.M. Scherer, Preventing squeeze-out of sealant during hot bar sealing: modeling and experimental insights, in: SPE ANTEC Conf., 2014, pp. 1452–1459, with permission from The Dow Chemical Company.*

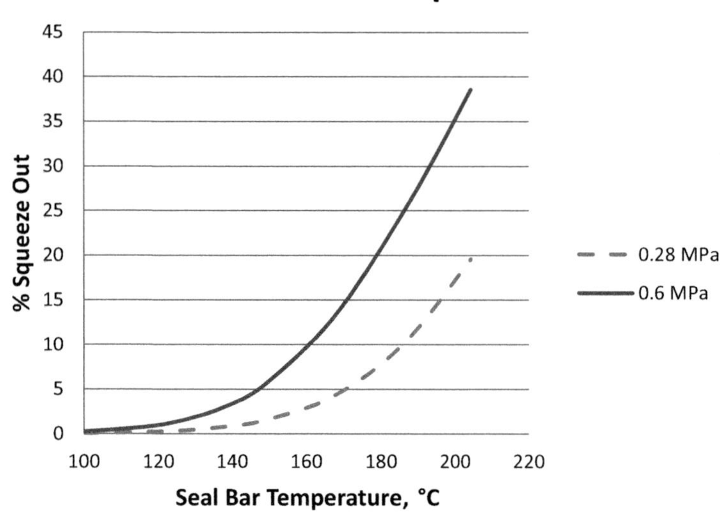

FIGURE 7.42 Model prediction (Eq. 7.30) of % squeeze-out of an ionomer sealant as a function of sealing temperature at two sealing bar pressures. Sealing dwell time = 1 s, bar width $(2w) = 2.5$ mm, sealant thickness = 0.05 mm [91]. *Reproduced from B.A. Morris, J.M. Scherer, Preventing squeeze-out of sealant during hot bar sealing: modeling and experimental insights, in: SPE ANTEC Conf., 2014, pp. 1452–1459, with permission from The Dow Chemical Company.*

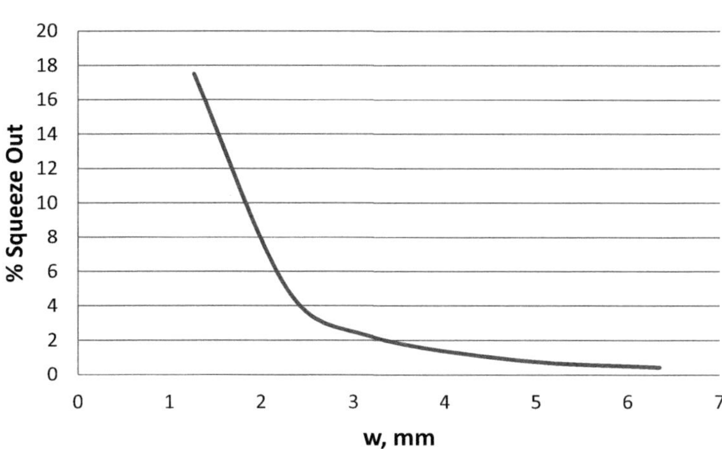

FIGURE 7.43 Model prediction (Eq. 7.30) of % squeeze-out of an ionomer sealant as a function of sealing bar half width (w). Seal temperature = 175°C, seal bar pressure = 0.6 MPa, dealing dwell time = 1 s, sealant thickness = 0.05 mm [91]. *Reproduced from B.A. Morris, J.M. Scherer, Preventing squeeze-out of sealant during hot bar sealing: modeling and experimental insights, in: SPE ANTEC Conf., 2014, pp. 1452–1459, with permission from The Dow Chemical Company.*

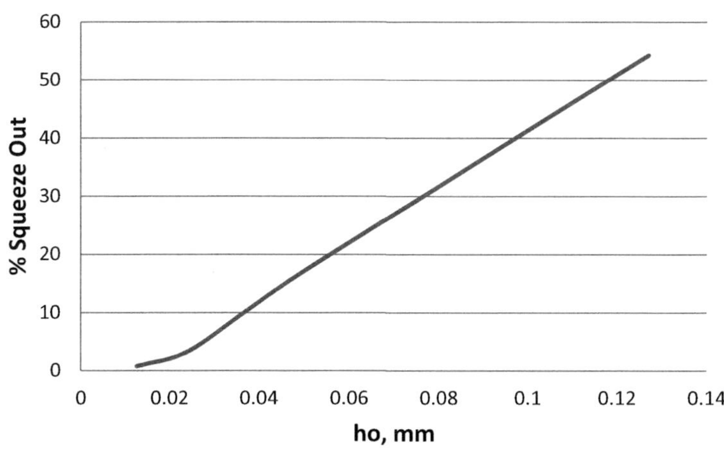

FIGURE 7.44 Model prediction (Eq. 7.30) of % squeeze-out of an ionomer sealant as a function of initial sealant thickness (h_o). Seal temperature = 175°C, seal bar pressure = 0.6 MPa, dealing dwell time = 1 s, seal bar width ($2w$) = 2.5 mm [91]. *Reproduced from B.A. Morris, J.M. Scherer, Preventing squeeze-out of sealant during hot bar sealing: modeling and experimental insights, in: SPE ANTEC Conf., 2014, pp. 1452–1459, with permission from The Dow Chemical Company.*

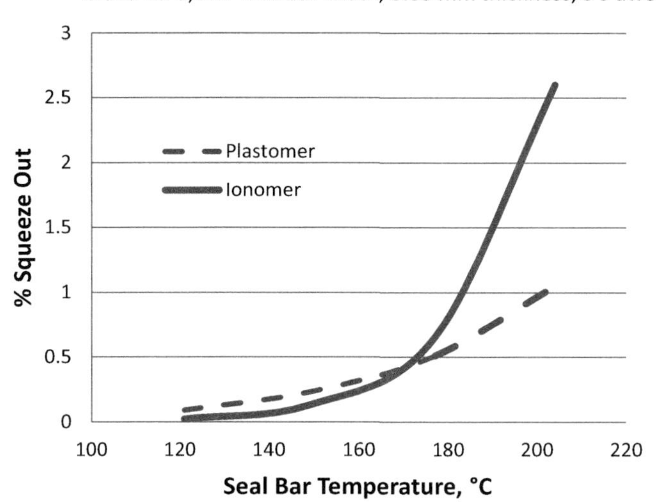

FIGURE 7.45 Model prediction (Eq. 7.30) of % squeeze-out of different sealants a as a function of seal bar temperature. Sealing dwell time = 3 s, bar width ($2w$) = 12.7 mm, sealant thickness = 0.05 mm [91]. *Reproduced from B.A. Morris, J.M. Scherer, Preventing squeeze-out of sealant during hot bar sealing: modeling and experimental insights, in: SPE ANTEC Conf., 2014, pp. 1452–1459, with permission from The Dow Chemical Company.*

- As seal temperature increases (Fig. 7.42), the amount of squeeze-out increases exponentially. This is due to the exponential decrease in viscosity with increasing temperature (Eqs. 7.35 and 7.36).
- As the seal bar width increases, squeeze-out decreases (Fig. 7.43); more squeeze-out is expected in seal designs involving thin serrations, which can concentrate the force on the seal.
- As sealant thickness (h_o) is increased, squeeze-out increases (Fig. 7.44).

Fig. 7.45 highlights the effect of the sealant type on squeeze-out. The sealant type has the greatest impact on the temperature dependence of the squeeze-out due to the differences in rheology (Fig. 7.37). The ionomer has the least amount of squeeze-out at low sealing temperatures, where it is most often used. This is consistent with its very high viscosity at low shear rates at these temperatures. The plastomer also has low squeeze-out, although not as low as the ionomer. The amount of squeeze-out of the ionomer increases with temperature faster than the plastomer, so that at about 170°C the two curves cross-over. This again is consistent with the higher activation energy (*Ea*) of the ionomer.

7.6.4.4 Insights and limitations

The model parameters used to construct Fig. 7.45 were chosen to simulate the experimental conditions in Fig. 7.39. Comparing the two figures, the model correctly predicts the experimental trends. In both cases, the ionomer has lower squeeze-out than the plastomer, despite having a higher MFI, until the temperature reaches 170°C–200°C. At that point the squeeze-out of the ionomer begins to take off.

While the model predicts the correct trends, the actual values differ by about an order of magnitude. More work needs to be done to bring the model predictions in line with the experimental observations. Several factors may be involved, including:

- Errors in the experimental technique and measurements.
- Non-Newtonian flow effects. The model assumes Newtonian flow and quasisteady state. At the low shear rates characteristic of the heat seal process many polymers are near the Newtonian plateau region of the viscosity versus shear rate curve. For these polymers, the Newtonian assumption may be a good representation. Other polymers, such as the ionomer used in this analysis, undergo a transition from the Newtonian plateau to the power-law region over the range of shear rates found in heat sealing. One consequence is that the velocity in the x-direction will be less parabolic with respect to z and more like plug flow since the viscosity will be lower in the higher shear rate region near $z = h$. Future work should help determine if the iterative approach used here to estimate the viscosity is sufficient to get good agreement with experimental results.
- Heat transfer effects. The model assumes the temperature is constant, yet it takes time for the heat to transfer through the film and bring the sealant up to the seal bar temperature [66]. The heat-up time will take longer with thicker films [67]. Lower temperatures during part of the sealing time will result in higher viscosities and less squeeze-out than calculated by the model. Shortening the sealing time or lowering the effective temperature may compensate for this in the model.
- The temperature gradient in the z-direction during heat-up will also affect the velocity profile. Since the heat is applied at the outside surface of the film, the outside portion of the sealant ($z = h$) will be hotter than the sealant-

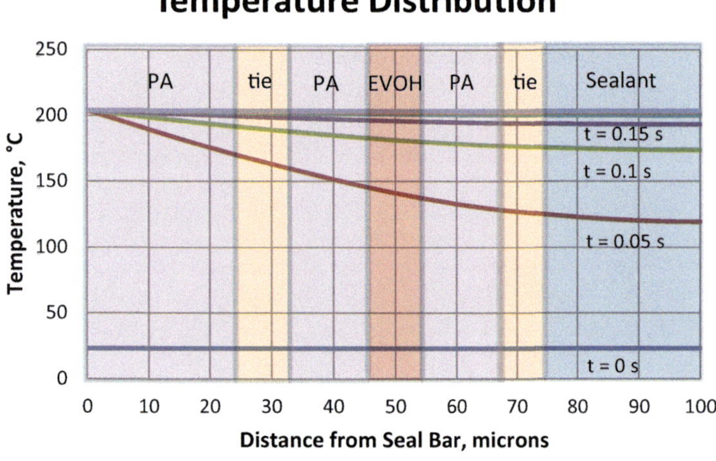

FIGURE 7.46 Heat transfer analysis of the seal area during squeeze flow. Results from the solution of the unsteady state heat conduction equation in one dimension for 100-micron thick film with bar temperature of 204°C. *Reproduced with permission of the Dow Chemical Company.*

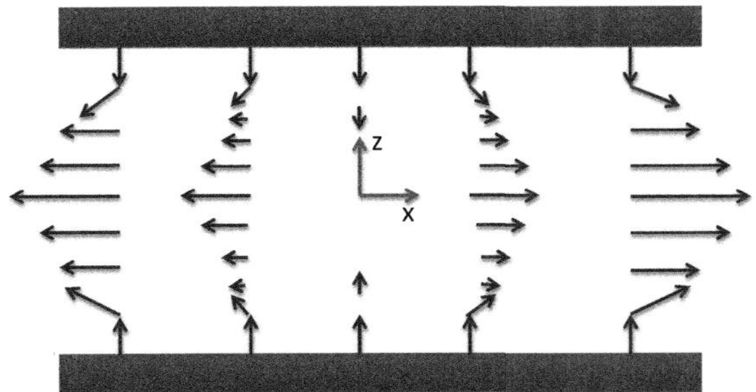

FIGURE 7.47 Illustration of flow patterns in the heat seal area during squeeze from based on Eqs. (7.19) and (7.20). Note the flow patterns in the z and x directions are not at scale. *Reproduced with permission of The Dow Chemical Company.*

sealant interface ($z = 0$) until the temperature comes to equilibrium. This lowers the viscosity at the outside of the sealant, resulting in more plug-like flow (similar to the effect of non-Newtonian flow behavior for the higher shear rates near $z = h$). The heat profile at different snap shots in time, as computed using a heat conduction model, is shown in Fig. 7.46. For the film structures used in these experiments, the model computations show it takes about 0.15 seconds for the sealant interface to reach the seal bar temperature. (This model ignores any resistance between the seal bar and film and therefore the time may actually be somewhat longer.)

Even without quantitative agreement between the model and experimental results, the model is useful for better understanding the factors that increase the chances for squeeze-out problems during heat seal. These include:

- Higher temperature;
- Higher pressure;
- Longer time;
- Lower sealant viscosity at low shear rates;
- Narrow seal width;
- Thicker sealant layers.

The development of the model also provides insight into the flow patterns within the seal area, which are important not only for squeeze-out but also caulking. Eqs. (7.27) and (7.28) describe the velocity in the x and z directions. These are illustrated in Fig. 7.47. Starting at the center ($x = 0$), the velocity in the z-direction is small at the surface of the plate and decreases to zero at the centerline between the plates. The velocity in the x-direction increases as x increases, reaching a maximum at the bar edge ($x = w$). The flow in the x-direction is parabolic, similar to pressure-driven flow through a rectangular channel or pipe.

The model confirms what is known in the field: squeeze-out should not be a problem if the seal temperature is controlled within the operating temperature range of the sealant. Only if the temperature or pressure is set too high will a significant squeeze-out occur. In the case of the plastomer and ionomer, both provide very low squeeze-out over the typical seal temperature range for which they are designed to be used.

One consequence of the model predictions is that squeeze-out may be a particular challenge in ultrasonic sealing (US). There the heat is generated at the sealant-sealant interface through friction. This can create high temperatures in a very short amount of time. Squeeze-out may occur if the horn pressure is too high. Further discussion of US is provided in Section 7.9.

The factors that influence squeeze-out should affect caulking in the same way. Caulking in the seal area will increase with increasing dwell time, bar pressure, bar temperature, sealant thickness and decreasing bar width. One of the challenges with sealing pouches with gussets or folds is caulking voids that form in the transition from the folded to nonfolded regions of the seal area. The pressure and thickness is lower in this region, and as as this work shows, this reduces squeeze flow.The rheology of the sealant, especially at low shear rates, is critical for filling these voids.

7.7 Easy open seal technology

7.7.1 Introduction

The definition of "easy open" depends on the application but generally refers to seal strength in the range of about 250–700 g/15 mm (400–1200 g/25 mm). Above 700 g/15 mm the seal is known as a lock-up seal. Below about 250 g/15 mm the seal may fail during handling and shipping (e.g., altitude changes during shipping are known to put stress

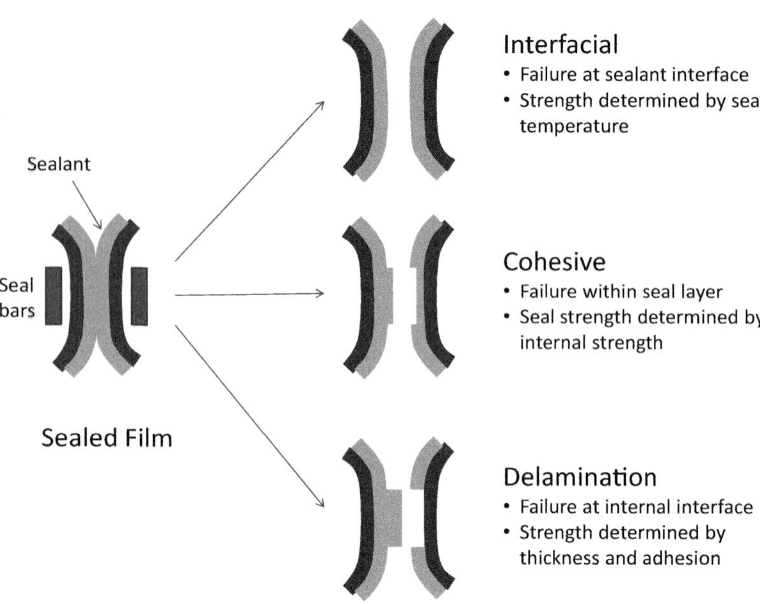

FIGURE 7.48 Mechanisms of easy open seals. *Reproduced with permission from The Dow Chemical Company.*

on seals). Film-to-film seals within the easy open range are often called peel seals (or seal peels). Most film-to-film seals are lock-up unless specific technology has been introduced to control the seal strength within the peel seal range.

There are three main methods used to control seal strength, as illustrated in Fig. 7.48:

- **Interfacial or adhesive:** Here there is a clean failure along with the sealant-sealant or sealant (lidding)-substrate interface. Interfacial failure often occurs in standard heat seals at low sealing temperatures and is difficult to control since it typically occurs on the fast-rising portion of the heat seal versus the temperature curve. Typically, cohesive or delamination technology is used for sealant-to-sealant peel-seal applications. The interfacial mechanism is often used when sealing to a dissimilar substrate as in a lidding application. Here the adhesion to the substrate is primarily due to wetting and polar interactions and not interdiffusion (if the substrate does not melt during sealing). The sealant can be cleanly peeled away from the substrate.
- **Cohesive:** Failure occurs within a layer, usually the sealant layer. This is typically accomplished by "contaminating" the sealant with an incompatible polymer. The incompatible polymer forms a separate domain within the sealant resin, which is the continuous phase. At the sealant-to-sealant interface, the sealant bonds to the sealant on the other side, providing a hermetic seal. As the peel arms are pulled apart, the stress separates the sealant from the contaminating polymer since this is the weakest interface, propagating a crack that opens the seal. The size and shape of contaminating polymer domains, their position relative to the interface, and their adhesion to the sealant resin influence the seal strength. Compatibility, mixing, dispersion and orientation all play a role.
- **Delamination:** Failure occurs at an interface adjacent to the seal layer in a multilayer film. Here the seal-to-seal interfacial strength is greater than the sealant to adjacent layer bond strength. The adjacent layer or the sealant itself is often modified to control the adhesion between the layers. During the peel, an initial burst or force is need for the peel front to propagate through the sealant layer to reach the adjacent interface. This initial opening force is influenced by the tear strength of the sealant and can be controlled by changing the thickness of the sealant layer.
- **Combinations are possible:** For example, Hayden [112] describes a peel seal system where failure occurs cohesively in the layer adjacent to the sealant: Sealant/cohesive peel layer/backing layer. During the peel, the peel front propagates through the sealant like the delamination method, but instead of relying on the adhesion between the sealant and the adjacent layer the failure is within the adjacent layer.

Creating an easy tear feature is another method to provide ease of opening of a flexible package. Since this is not necessarily a function of the sealant or sealant technology, it will not be covered in this chapter. Tear properties are discussed in Chapter 9.

7.7.2 Examples

The interfacial peel seal mechanism is most often used in lidding applications, which are taken up in Section 7.7.4.

There are numerous examples of cohesive peel seals involving blends of incompatible polymers. Many involve blends of polybutene-1 (PB-1) with polyolefins or ethylene copolymers. These include blends of PB-1 with EVA or LDPE [113–120], EMA [118], tie resins [118], EAA [119], ionomers [121,122] and mPE [123]. Since PB-1 blends are so common, this technology is discussed in greater detail in Section 7.7.3. Other blends include EAA with EVA [119], ionomer with EVA [113], ionomer with PP [124,125], and mPE with PP [119]. Typically, the minor component is added in the range of 10%–30% but higher levels may be needed if the compatibility between the two polymers is good.

One delamination type peel seal involves the use of ionomers in combination with HDPE. This technology is often used in snack packages where the HDPE provides moisture barrier and strength. The ionomer chemistry can be tailored to control the adhesion to HDPE, either directly through the synthesis of the polymer or by blending zinc and sodium-based ionomers. Sodium ionomers have a lower affinity toward HPDE than zinc ionomers and the blend can be used to adjust the adhesion. Earlier versions of this technology used blends of ionomer and EVA to control the adhesion to HDPE [113,126].

Falla [127] describes a commercially available formulated sealant based on an ethylene-propylene copolymer that fails interfacially at low seal temperatures, but by delamination to HDPE at higher seal temperatures. He cautions that the bond strength at the HDPE-sealant interface is high and care should be taken to provide just enough heat to seal the package to avoid high seal strength. He compares this sealant with an ionomer and several PB-1 blends in a (HDPE-sealant) blown film structure, where the HDPE is a new high moisture barrier grade. His standard PB-1 formulation comprises 55% LDPE, 30% low melting LLDPE and 15% PB-1, although different levels of LDPE and PB-1 were tested. The ionomer structure gives the best physical properties, the broadest peel seal temperature window and the lowest seal strength. The EP sealant provides the highest hot tack. Both the ionomer and EP sealant provide suitable peel seal performance in combination with the new HDPE grade. The seal strength of the PB-1 blends is found to decrease with increasing LDPE or PB-1 content.

Busche and Bostian [128] describe a delamination system involving a sealant layer and controlled adhesion of the sealant layer to a tie layer adjacent to the sealant layer. The adhesion of the tie layer to a third layer is stronger than the adhesion to the sealant so that failure occurs along the tie-sealant interface. The adhesion of the tie layer to the sealant may be controlled by adding PB-1 to the tie layer. Hwo [115] teaches that PB-1 blends really fail cohesively near the interface, so the concepts of Busche and Bostian [128] and Hadyen [112] are similar: the sealant is unaltered so it can perform its sealing function, but failure is programed to fail either by delamination or cohesively within the adjacent layer.

Delamination and cohesive peel seal technologies each have their pros and cons. Delamination peel seals allow for the use of unaltered sealants; adding incompatible polymers may affect sealing and other properties important to the application. Delamination peel seals typically have a broader temperature window for the easy-open seal strength range. This is because the seal strength is determined by the adhesion between the sealant and adjacent layer rather than the complex morphology and melting behavior of the blend. Delamination seals also typically do not change with time. Some peel seals, particularly those containing PB-1, are prone to vary with age due to changes that take place in the crystalline form of the PB-1 [115]. Cohesive peel seals tend to turn white (stress whitening) as they are peeled, providing tamper evidence. Cohesive peel seals also can be designed to be less "stringy"; in the delamination peel seal, the initial tearing through of the sealant layer to reach the adjacent interface can leave some polymer strings behind. Cohesive seals are not altogether immune to stringiness, however. Improper control of the morphology of the discontinuous phase can lead to stringiness [119,120].

A variation on peel seal technology is the peel-lock seal. For example, a pouch may require lockup seals around three sides and an easy-open seal on top. Alternatively, a frangible seal may be desired in the interior of a pouch to provide easy open between two compartments. These concepts are shown in Fig. 7.49. Here it is desirable to have a sealant that is a hybrid between a peel seal and lock-up, as shown in Fig. 7.50. At low sealing temperatures, the seal has low seal strength over a wide temperature range. At high seal temperature, the seal strength increases to the lock-up range. Such seals are typically formed by blending different polymers or heat sealing dissimilar surfaces [129]. Standard cohesive type peel seal blends do not lock-up at high seal temperatures and special technologies have been developed for this purpose, such as ionomer/propylene copolymer blends [87,88]. Frangible seals can also be assisted by the design of the seal geometry. For example, DuPont used finite element modeling to optimize the seal geometry for a dual compartment pouch [130].

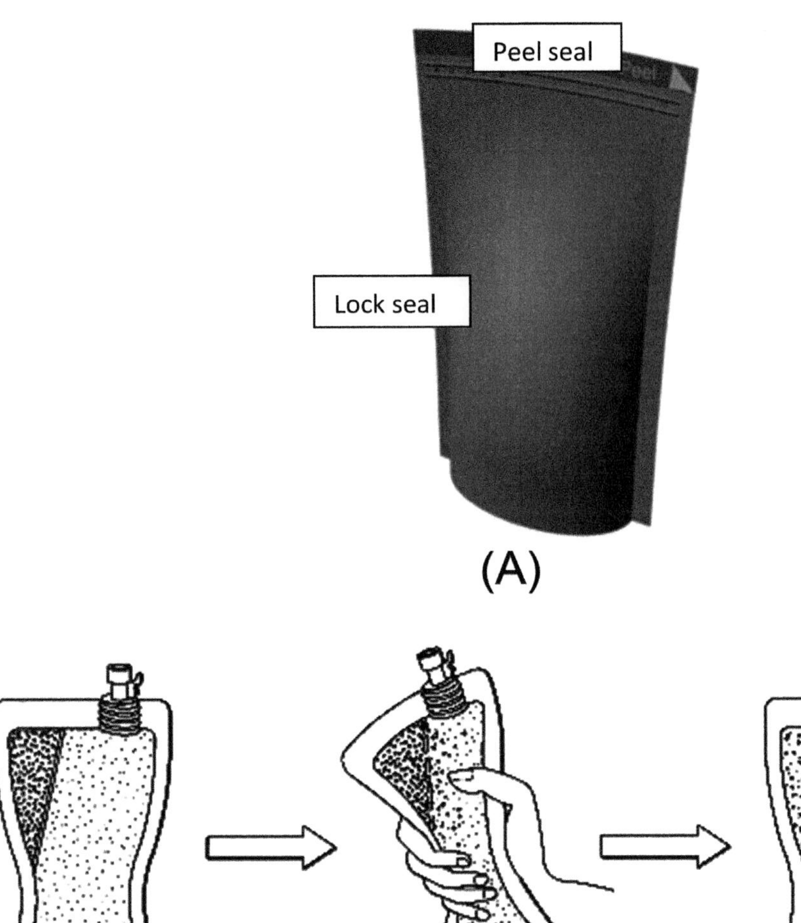

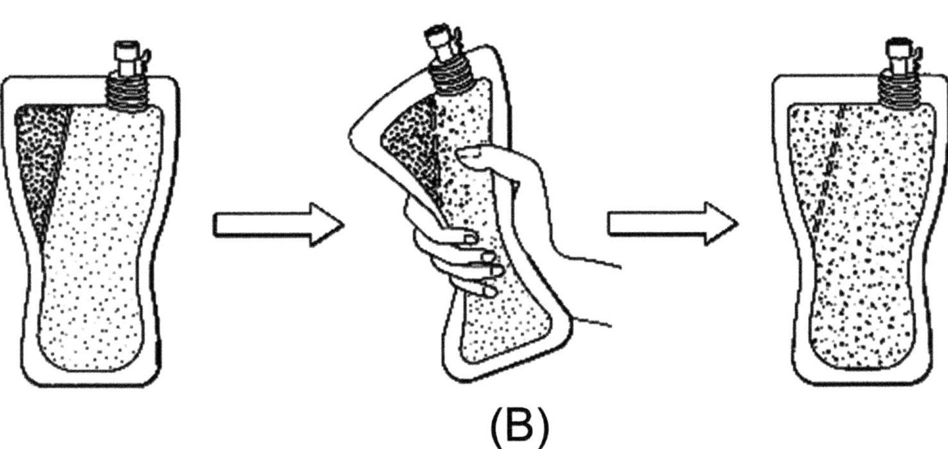

FIGURE 7.49 Peel-lock seal packaging concepts. (A) Single compartment pouch with top peelable seal and side lock-up seals. (B) Dual compartment pouch with interior frangible seal for mixing compartments. *(A) Reproduced with permission from the Dow Chemical Company. (B) Reproduced from R.A. Bourque, et al., Multiple compartment pouch and beverage container with frangible seal, in: US Patent 7306095 B1, 2007.*

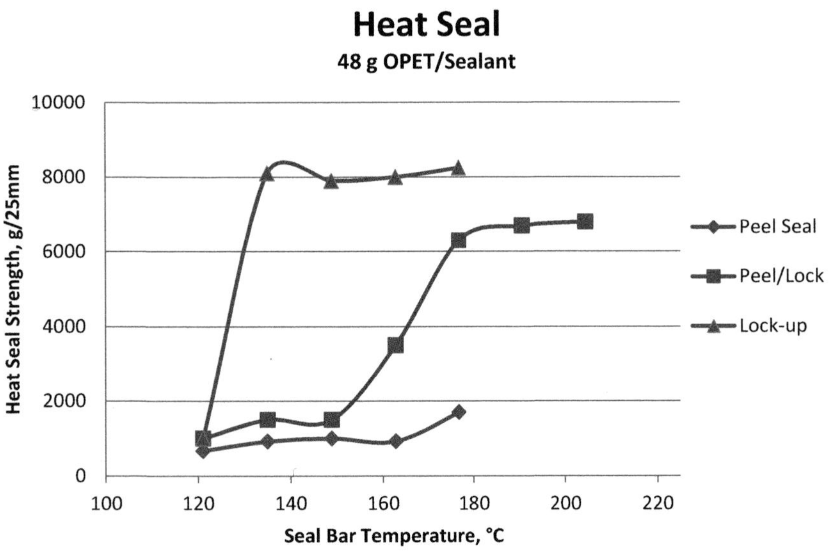

FIGURE 7.50 Examples of lock-up, peel seal and peel lock heat seal curves. *Reproduced with permission from the Dow Chemical Company.*

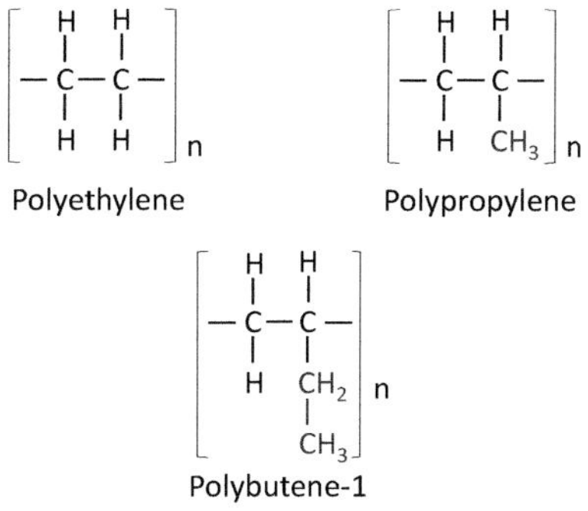

Polyethylene

Polypropylene

Polybutene-1

FIGURE 7.51 Comparison of the chemical structures of polyethylene, polypropylene and polybutene-1.

7.7.3 Polybutene-1 blends

Since blends with polybutene-1 (PB-1) are one of the most commonly used and well-documented technologies for cohesive-failure peel seals, they are considered here in detail. Many of the issues encountered with this system are also prevalent in other blend systems.

The chemical structure of Polybutene-1 is similar to isotactic PP, with a butyl group (2-carbon) protruding from the polymer backbone instead of a methyl group (see Fig. 7.51). The extra polarity makes it less compatible with most types of polyethylene but it adheres well to PP. It has a melting temperature similar to PE. Homopolymers and copolymers with alpha-olefins are available.

When dispersed in a PE, EVA or ionomer matrix, the poor adhesion between the PB and matrix creates poor cohesive strength that results in an easy-open seal. Some advantages of PB-based easy-open systems include: [117,120]

- Broad seal temperature range.
- Consistent and reproducible peel force.
- Tailorable seal strength.
- Initiation seal force similar to the propagation seal force.
- Processable on conventional film fabrication equipment.
- Little effect on hot tack.

Concentration, compatibility and dispersion are the key factors influencing seal strength. These factors control the interfacial area between the dispersed PB-1 phase and the matrix polymer, which in turn influence the seal strength. Generally, the higher the concentration of PB-1 the lower the seal strength as more weak interfacial area is created. Typical concentrations are 8–20 wt.% for LDPE and EVA [115] and higher for LLDPE and mPE (around 35% according to Mergenhagen et al. [123]). If the concentration is too low, the seal becomes stringy upon opening.

The order of increasing compatibility is: Ionomer < EVA (VA < 18%) < LDPE < HDPE < LLDPE < mPE. Resins with good compatibility, such as mPE, typically require much higher concentrations of PB-1 to achieve peelable seal strengths. Higher compatibility increases the interaction between the PB and matrix, increasing interfacial adhesion and cohesive strength. The PB copolymers tend to have greater compatibility with PE. Increasing VA content of EVA beyond about 18 wt.% increases seal strength of PB-EVA blends. In theory, increasing VA content should reduce the compatibility with PB-1 since the EVA becomes more polar. Increasing VA content also decreases the crystallinity, however, which may be responsible for the rise in peel strength. Most EVA sealants contain 18% VA or less so this effect is not a hindrance for using EVA-PB peel seal blends.

Three-component blends are also used, such as 15% PB-1, 30% mLLDPE and 55% LDPE [120,127]. The mLLDPE provides low HSIT and improved hot tack and the LDPE helps control the cohesive strength to produce low seal strength; greater levels of LDPE or PB-1 reduce the seal strength.

The dispersion of the PB-1 in the sealant is critical, influencing domain size and distribution. All things being equal, a smaller domain size results in lower and more uniform seal strengths since more weak interfacial area is created. Good dispersion also helps prevent delamination to adjacent layers and stringiness of the seal during peeling. As discussed in Chapter 6, however, the domain size is influenced by interfacial adhesion; greater adhesion tends to produce smaller domain size. Therefore, the effect of domain size can be complex.

The film fabrication process can influence the PB-1 domain size and shape. High shear rates near the die wall can result in smaller domain sizes and lower seal strength. Orientation during film blowing or casting can result in elongated domains. De Clippelier [116] suggests that greater orientation may be better (lower seal strength) whereas Pirtle

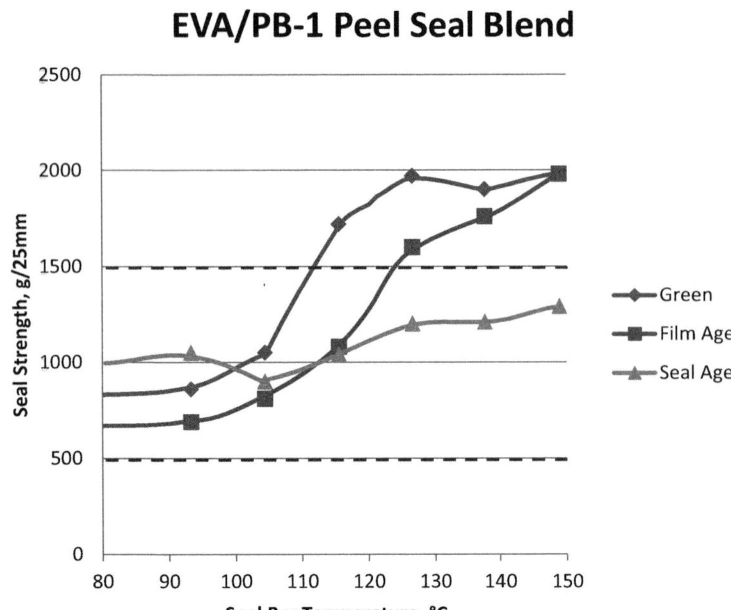

FIGURE 7.52 Heat seal curves for an 80% EVA (18% VA), 20% PB-1 blend. Film structure is (56-μm HDPE—8-μm sealant) blown film. The three curves show the effect of aging. "Green" measurements are for heat seals made and measured within a day of making the film. "Film Aged" represents film stored at 22°C (72°F), 50% RH for 14 days before sealing and measuring. "Seals Aged" designates film sealed within a day of making the film but the seal strength was measured after the sealed film had been stored at 22°C, 50% RH for 14 days. Dashed horizontal lines represent the typical range for easy open seals. *Reproduced with permission of the Dow Chemical Company.*

[119] suggests the opposite, that a more oval shape is preferred. Pirtle suggests increasing the frost line height, lowering the draw down ratio and maintaining a layer distribution consistent with the die design to minimize stress and orientation to obtain the optimum morphology. It is likely that these factors are important but may vary in the way they affect the seal strength depending on specific circumstances. The effect of blown film processing on dispersed phase morphology is explored in more detail in Section 14.3.

Matching the viscosities of the PB-1 and matrix resin is important for minimizing the domain size. PB-1 is more shear-thinning than many of the polymers it is blended with. Therefore, simply matching melt indexes is not sufficient – careful attention to the whole rheology curve is a better approach.

Cast film generally gives higher seal strength than blown film with the same composition. Higher levels of PB-1 are added to cast film peel-seal blends to compensate. Hwo [115] attributes the increased seal strength in cast film to the PB-1 being in a different crystalline form and more amorphous (due to the rapid quench rate) than in blown film. This form of PB-1 has a lower melting point and stronger adhesion to PE.

Pasquali [120] recommends the thickness of the sealant layer be 15 μm (no less than 10 μm) and the adjacent layer thicker to avoid problems with sealing through the sealant layer, delamination and stringiness.

The angle of the pull during the peel test affects the mechanism of failure and seal strength. Nase et al. [131,132] find that the mode of failure changes with peel angle. Interlaminar failure occurs parallel to the interface whereas translaminar failure occurs perpendicular to the interface (propagating through the thickness of the sealant). For angles less than 120 degrees failure was interlaminar with a minimum peel strength occurring at 120 degrees. For angles 140–180 degrees failure was translaminar and higher peel strengths were measured.

Seal strength of blends of PE or EVA with PB-1 ages down with time after the film has been fabricated (prior to sealing) as well as after heat sealing (see Fig. 7.52). Reductions in seal strength of 40% or more have been reported [115,117]. PB-1 has four crystalline forms and some of these forms can change from one to another over time. These crystalline forms have different adhesion characteristics that affect seal strength. Hwo [115], for example, describes the transition of Type II crystals to Type I crystals and correlates this transition with the age-down in seal strength. The user of PB-1 blend peel seal formulations needs to establish correlations for seal strength measured off the packaging line with seal strength when the package is opened by the consumer, which may be several weeks later.

One method to reduce or eliminate the age-down issue is to pick a matrix resin that requires less PB-1 addition to achieve a peel seal. The lower the level of addition, the less affect aging has on the overall performance of the film. Ionomers are one such sealant. Brodie [122] shows that less than 10% PB-1 can be added to some grades of ionomer to achieve a cohesive peel seal. This is considerably less than the amount required for LDPE or EVA blends (15%–20%), and much less LLDPE (20%–25%) or mPE (25%–35%) blends.

7.7.4 Lidding films

There are two general types of sealing technologies used for lidding applications: sealant coatings and sealant resins. Both are designed to bond to the tray or cup stock, typically made from PP, PET, PS, PE, polyvinyl chloride (PVC), paperboard or aluminum.

7.7.4.1 Heat seal coatings

A heat seal coating is essentially water- or solvent-based adhesive that is applied to a flexible substrate and dried to form a nonblocking coating. The adhesive is activated with heat and pressure to provide a peelable adhesive bond to another rigid or flexible substrate. Heal seal coatings are used in many applications including lidding for food packaging (such as for yogurt cups), retail skin packaging and healthcare (medical device packaging, blister packaging, etc.). The coating is typically applied using coating techniques (see Section 2.3.4) such as air knife, gravure, and flexographic coating processes.

The properties of heat seal coatings must meet the needs of both the converter and the end-use. For the converter to apply the coating to the substrate properties such as viscosity, percent solids, shelf stability, pH, color, density and adhesion to the primary and secondary substrates are important. Specific end-use needs include activation temperature, block resistance, heat seal strength, hot tack strength, resistance to heat or cold (e.g., microwave oven or retort conditions) and regulatory compliance. Given these varying needs, heat seal coatings are typically formulated products that may contain several components such as [133]

- Polymers;
- Tackifiers and additives;
- Lubricants, waxes and slip modifiers;
- Antioxidants and fillers;
- Antiblock and heat stabilizing agents;
- Solvents and diluents;
- Water, dispersants, pH control additives, defoamers and surfactants.

EVA based formulations have long been used as heat seal coatings. Gel lacquer is one of the oldest technologies. This solvent-based technology is supplied as a gel that requires heating to 38°C (100 °F) or higher for application. Solvents such as toluene also are added. Despite the challenges, they provide many useful benefits, including:

- Bonding to a broad range of substrates, including PET, PP, PS and PVC.
- Good product resistance.
- Ability to be used in hot fill food and autoclave medical applications.
- Relatively low activation temperature.

Newer formulations with EVA have been developed that can be applied at ambient temperatures. These formulations tend to be more specific for the substrates they bond to but have better product resistance. Drawbacks include more expensive solvents and lower percent solids.

Another category of solvent-based heat seal coatings consists of true solutions at room temperature. Several chemistries are available for specific applications:

- **Vinyl/acrylics:** These formulations have better chemical resistance than EVA coatings, excellent clarity and heat resistance (for hot fill and autoclave). They do not work for PET or polyolefin substrates and require higher heat seal temperatures (121°C–177°C or 250–350 °F).
- **Polyesters:** These high-temperature resistant coatings can be used for retort or ovenable applications. They are coated onto OPET or aluminum foil and bond to CPET, APET, PVC and PVDC coated PS. These coatings are easy to apply, have excellent clarity, can be modified with fillers to provide specific seal strengths, have excellent chemical resistance to grease, oil and fats, and have high hot tack strength and temperature resistance. They are higher priced than other coatings and generally only adhere to PET or foil films.
- **PP dispersions:** These are used for foil and PP containers, typically in retort pouches, medical device packages and aseptic pouches requiring excellent product or chemical resistance. They have specific adhesion to aluminum foil and require high processing temperatures to drive off the solvent.

Water-based coatings are typically made from EVA, EMA and EAA chemistries. They are aqueous dispersions formulated for broad use. Benefits of these technologies include [133]

- No volatile organic compounds;
- Minimal health and safety issues;
- Lower seal temperatures;
- Adhesion to a wide range of film substrates;
- Seals to a wide range of cup substrates including PP, PS, PVC, and PET;
- Formulated for peel seal and hot tack characteristics.

Their main drawbacks derive from being water-based; the coatings are susceptible to freezing, and may have drying and foaming issues.

Cushing and Cooper [134] compare the hot tack performance of EAA, acrylic, and PVDC based aqueous coatings applied to a primed BOPP substrate. Hot tack of the film sealed to itself is measured. The EAA coating provides the best hot tack over a broad temperature range with a low seal initiation temperature. The acrylic coating bonds are said to be versatile and seal well to themselves and to other polymers. The hot tack and seal initiation temperature in this comparison, however, is inferior to the EAA. PVDC provides good oxygen and moisture barrier but can also be used as a sealant. It shows higher seal initiation temperature, lower hot tack and requires higher coat weight to achieve similar bond strength. It is also known to yellow with time. One of the drawbacks of aqueous heat seal coatings is that they typically do not bond well to film substrates (EAA does bond well to aluminum foil and paper). A primer is typically applied to the film prior to coating the sealant. This requires an extra processing step or a tandem coating line. Cushing and Cooper show the results of a new aqueous coating that does not require a primer and has similar hot tack performance as the EAA coating with a primer.

Heat seal coatings are typically applied at 1–2 g/m^2 coat weight (dry) whereas extruded sealants are much thicker (10–16 g/m^2 for extrusion coatings and 25 g/m^2 for laminated films). Thin coatings have a disadvantage in applications that require caulking (such as sealing around contamination or sealing near folds of pouches) but Cushing and Cooper show that thin coatings can have significantly lower seal initiation temperature than thicker extruded sealants. This is presumably related to heat transfer (see Section 7.6.2) and chemistry.

Another advantage of heat seal coatings over heat seal resins is that they can be pattern coated. This allows the coating to be applied just in the seal area of the container. This minimizes food exposure and the potential for contamination, reduces cost and avoids membraning of the seal layer during opening.

Lacquers and other heat seal coatings often fail by rupture, which Soutar [113] points out, can be an advantage.

7.7.4.2 Peelable seal resins

Peelable seal resins are applied to lidding substrate films, foils or paper using an extrusion process rather than a liquid coating process. Typical processes include extrusion coating and adhesive or extrusion lamination of films that contain the peelable seal resin. No drying or removal of liquids are involved (except if adhesive lamination is used).

There are many requirements for extrudable lidding structures including [135]

- Easy-open peeling;
- Reliable sealing over a wide temperature range;
- Versatility for use with a variety of substrates;
- Compliance with regulations;
- Good machinability
- Good processability at the extrusion stage.

Like heat seal coatings, heat seal resins are typically formulated to meet the specific requirements of the application. Many factors influence sealing force including:

Sealant composition. Many lidding sealant resins are EVA or EMA based. These copolymers have a good balance of polarity and softness to bond to a number of substrates. Soutar [73] describes a blend of an EVA-based tie resin and an ionomer for bonding to PVC. Dow offers a variety of EMA and EVA based single-bag (preblended) lidding sealants in its APPEEL product line [136]. Pascal [135] describes an EMA based composition. EVA or EMA by themselves often lack the versatility to bond to a number of different substrates or provide the right level of seal strength. They are typically modified with other resins, tackifiers, softeners or elastomers to achieve the desired performance [135].

Substrate chemistry. Eq. (7.1) shows that the substrate surface energy must be greater than the adhesive for good wetting. This is often accomplished by surface treatment such as corona treatment. Eq. (7.1) also shows that the interfacial energy should be small. In Chapter 6, the concept of the solubility parameter was introduced. Eq. (6.7) shows that the interfacial energy or tension is minimized when the solubility parameters closely match. One of the reasons EVA and EMA work well as lidding sealants is that their solubility parameters more closely match that of common substrates such as PET or PVC than PE or PP [137].

The geometry of the seal area. Pascal [135] describes how geometry can influence the initiation force independent of the sealant characteristics. One influence is the concentration of stress, as was highlighted in Section 7.7.2 for the two-compartment pouch.

Peel angle and speed. As described in Section 7.5.7, the angle and speed in any peel test influences the measured peel force. Care should be taken to mimic the end-use when running peel tests in the lab for screening lidding sealant candidates.

The thickness of the sealant layer, especially when flow or caulking may be needed to fill in voids or to seal around contamination. Heat seal resins typically have a much higher thickness than heat seal coatings and caulk better.

Thermal conductivity and thickness of the lidding structure. Thinner structures seal faster (or seal at lower seal temperatures) than thicker ones since heat is transferred more quickly to the interface. Foil structures also generally seal faster than nonfoil structures because aluminum has a higher thermal conductivity than paper or polymer substrates.

Plasticity of the sealant and structure. Pascal [135] provides an example of three structures with the same sealant and tie resin bonding to a PS substrate: (1) Al/tie/sealant, (2) Paper/tie/OPET/tie/sealant, and (3) OPET/tie/sealant. The structures provide significantly different seal strength in the order of $1 > 2 > 3$. He attributes this to differences in the peel dynamics (fracture mechanics) related to the stiffness of the structure.

Soutar [113] compares field studies with lab results and shows that lab heat seals are a good starting point for screening lidding sealant candidates. Pascal [135] describes a lab method for studying the initiation and propagation of the seal strength.

Some of the advantages of extrudable lidding sealants over lacquers and other sealant coatings include [138]:

- A well-designed extrudable sealant resin is very versatile, providing peelable seal functionality with multiple substrates. This may allow the converter or the brand owner to use the same resin or lidding film for many applications and reduce the number of products in their inventory.
- Extrudable sealants can be used with multiple converting processes, such as extrusion coating and coextrusion blown or cast film.
- Extrudable sealants can be converted with higher coat weights (thickness) that provide better caulking and improved sealing in the presence of contaminants such as milk solids.
- Some versions of extrudable sealants have better product resistance.
- Extrudable sealants may have lower seal initiation temperatures.
- Extrudable sealants are solvent-free.

As described in the proceeding section, some of the advantages of heat seal coatings are their low coat weight and ability to be pattern coated just in the seal area.

7.8 Reclosable seal technologies

Some of the methods used to create a reclosable flexible package include:

- Zippers.
- Pressure-sensitive strips that the consumer can apply after rolling up the end of the opened package to keep it from unrolling.
- Pressure-sensitive adhesives (PSA) incorporated into the body of package that are exposed when the package is opened.

Zipper technology has been around for several decades and is probably the most extensively used technique for incorporating a reclosable feature into a flexible package. Zippers have been incorporated into a wide variety of package forms, adapted to many different packaging machine formats and used across most food types from shredded cheese to deli meats and snacks. They come in a variety of styles such as press-to-close and sliders. The most widely used press-to-close zippers typically consist of integral profiles and flanges made by profile extrusion [139].

Zippers are normally added to the package on the packaging line while the pouch is being formed. The zipper is a separate entity usually in the form of a tape that is heat-sealed to the package. The zipper material must be compatible with the sealant on the package film for proper bonding.

Some recent innovations in zipper technology include: [139]

- Incorporation of easy-to-open technology with reclosable zippers. Previously zipper packages often required scissors to get into. Combining an easy tear (perforation, tear notch, laser scoring, etc.) with a hermetic seal that is easy to open (Section 7.7) has become more common, enhancing the consumer experience.
- Double profile zippers and multialign closures [140] have made reclosing the package easier, reducing challenges with aligning the zipper profiles. These new profile designs also allow the incorporation of tactile or auditory signals. For example, the consumer may hear a snapping sound or receive a tactile signal when the zipper has been successfully closed.

The use of PSA in flexible packaging has grown in recent years but the concept dates back at least until the 1990s. Straus et al. [141] describe a reclosable flexible package with a tacky adhesive layer and a thin skin layer on a substrate such as polyester film. The thin skin (sealant) layer fractures when opened, exposing the adhesive layer to form a resealable seal surface. The tacky adhesive is preferably a pressure-sensitive adhesive and typically contains 40%–80% of a thermoplastic elastomer (such as styrene butadiene, polyurethane or EVA), 20 to 60% tackifier and up to 15% of other components.

Timm and Bublitz [142] describe a multilayer film with the structure: Substrate/PSA/cold-seal//cold-seal/substrate. Adhesion of the cold seal to the PSA and substrate is stronger than the adhesion of the PSA to the substrate so that when the seal is pulled apart, the PSA is exposed for resealing.

Exner and Dagestad [143] disclose structures involving a cold seal under a sealing layer. Selective adhesion allows the cold seal to be partially exposed after opening, allowing for reclosability.

Trouilhet and Roulin [144] and Haedt et al. [145] describe polyester seal layers in contact with a PSA layer that are exposed upon opening.

Cruz [146] describes a multilayer film with the structure: Tie layer (anhydride-modified polyolefin)/EVOH/PSA. The interface between EVOH and PSA provides a peelable/reclosable seal.

Ranganathan et al. [147] disclose a tray with a liner comprising a sealant that is at least 80% ionomer and a laminated lidding film sealed to the tray liner. The lidding film consists of a substrate film and a coextruded film. The substrate film contains a high modulus thermoplastic (BOPET, BOPP, or BOPA) outer layer and a barrier inner layer. The coextruded film has the structure (Sealant-PSA) where the sealant is a blend of 80% ionomer and 20% polyolefin. The barrier layer of the substrate film is laminated to the PSA layer of the coextruded film, and the lidding film is sealed to the tray liner. When the package is opened, at least part of the PSA is exposed and can be reclosed by adhering to part of the sealant layer of the coextruded film that remains sealed to the tray.

Trouilhet [148] finds that an ionomer and PSA can be coextrusion coated onto a variety of substrates including BOPP-met, Paper, Al-foil, and BOPET/tie/Al. The seal strength can be adjusted from 5 to 20 N/15 mm by changing the PSA layer thickness. The resulting package can be reclosed up to 10 times with steady seal strength of 1.5 N/15 mm. The ionomer works well since it has a low elongation at break to allow for easy tear through the sealant to expose the PSA during opening. The ionomer also brings other known attributes as a sealant such as low HSIT, oil resistance, seal through contamination and stiffness.

Moehlenbrock and Ranganathan [149] describe a discreet laminate base strip containing PSA that can be incorporated onto a pouch using conventional packaging lines to provide a reclosable feature.

These are just a sampling of the many patents and publications in the field. Most involve an extrudable pressure sensitive adhesive disclosed by Bardiot et al. [150].

7.9 Ultrasonic sealing

US involves the use of a horn to focus sound waves that impart an oscillatory force on the films (or film and substrate), which generates frictional heat to melt and seal the interface. Some of the reported advantages of US over hot bar or conductive sealing (CS) include:

- Lower operating expense since less energy is consumed during sealing (the ultrasonic horn is only engaged during the sealing process, whereas energy to heat the bars in CS is continuous).
- The ability to seal heat sensitive products. In US the heat is generated from the heat seal interface and transfers outward rather than from the outside of the film inward, reducing the amount of heat transferred to the product.

- Smaller film footprint since the seal area is typically narrower for US.
- The ability to seal through contamination. The sonic force pushes many powders and other products out of the seal area during sealing [151]. Bach et al. [44], however, note that US only has a marginal advantage with bulk materials such as wheat flour and CS is better for other materials in their study.
- The ability to seal oriented films without distorting them.
- The ability to seal poor conducting films such as paperboard, or thick films in a shorter amount of time, increasing packaging line speeds.

Despite these advantages, US is not widely used in flexible packaging. Some of the disadvantages that have kept it from overtaking CS include:

- Higher capital expense.
- Higher maintenance cost.
- Inability to seal some geometries, such as fin seals. US is mostly designed for flat surfaces.
- Difficulty in maintaining barrier due to squeeze-out.
- Lower seal strength [44,152,153].

Several studies that have been conducted on ultrasonic welding of thermoplastic parts are relevant to heat sealing [154–156]. These studies reveal that:

- Amorphous polymers generally weld better than semicrystalline polymers.
- The temperature at the interface rises quickly which is attributed to asperities at the surface that concentrate the stress. The interfacial temperature levels off above Tg or Tm.
- The temperature of the bulk also rises, although not as fast as the interface, which is attributed to viscous dissipation of the bulk molecules from the oscillations rather than to conduction of heat from the interface. The rise in the temperature in the bulk is too fast for conduction to be the major factor.
- When a tie layer or sealant is used, substantial squeeze-out can occur from the seal area once the bulk temperature of the sealant reaches its melting point.

Bach et al. [44,152,157] study the differences between US and CS by sealing the same film structure with the two processes. Seal seam strength is often lower for US than for CS. In one example with a commercial PA/PE laminate, the seal strength of the US is only 40% of CS. Several differences between the two processes that influence this behavior are described:

- Pressure is more important for US than for CS. In CS, only sufficient pressure to ensure the films are in contact is needed. For US, higher pressures help transmit the vibrations to the interface for melting.

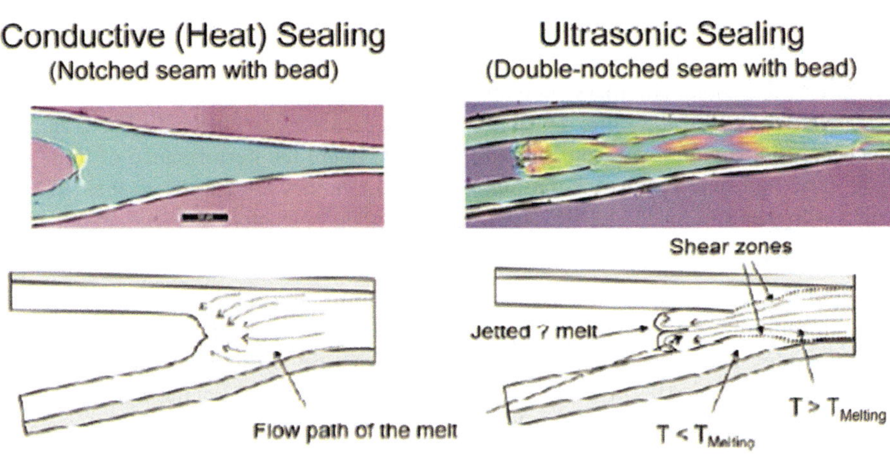

FIGURE 7.53 Picture of area near seal edge in ultrasonic and conduction sealing [109]. *Reproduced with permission from S. D.R. Wilson, Squeezing flow of a Bingham material, J. Non-Newton Fluid Mech. 47 (1993) 211–219.*

- The temperature is generated from the outside in for CS and from the inside out for US. This creates a more uniform temperature profile in the seal area for CS. Hot spots were observed in US. Also, cooling is more rapid near the edges of the seal area in US than for CS. This impacts flow and crystallization.
- Melt is squeezed out from the seal area in both processes, but in US a melt bubble with a double notch at the edge of the seam forms (see Fig. 7.53). Infrared thermal images and transmission electron microscopy images show a higher level of orientation in the seam area for US. As a result, a different crystalline orientation structure forms; the inter-lamellar chains are stretched and molecular diffusion is constrained.
- For the US seals, failure is often along the morphological interface where the flow has occurred. This depends on the seal force. At low force, no squeeze-out is observed but the seal strength is weak. Failure occurs along the seal to seal interface, presumably due to insufficient melting and diffusion. At high seal force, the interface heats up to well above the melting point and a significant amount of squeeze-out occurs. Failure is along the flow interface and seal strength is reduced. Optimum seal strength occurs with a small amount of squeeze-out. For strong bonds to form, $T >> Tm$, but this sets up conditions for squeeze flow and failure along flow lines.
- The double notch at the edge of the US seal area also tends to concentrate the stress during peeling and reduces seal strength.
- In CS, failure is clean along the sealant interface at low temperatures. At high temperatures, failure occurs by breakage (tearing) along the seam side. If squeeze-out occurs, the profile is concave rather than the convex melt bubble found in US and likely has little effect on the seal strength other than thinning of the sealant in the seal area.

Van Dun et al. [153] compare different polyethylene sealants with the goal to optimize seal performance in US. They measure seal performance of 50-μm sealant films laminated to OPET or OPP film. In one study they compare various linear polyethylenes of different densities. The ultimate seal strength develops at lower seal temperature as the density of the LLDPE is decreased. This is analogous to lower seal initiation temperature in hot bar sealing. Reducing the density also increases the ultimate seal strength. Thus, an ULDPE outperforms a LLDPE.

Van Dun et al. relate this behavior to the linear elastic behavior of the polymers. The viscoelastic heat generation for small cyclic deformations is given by Eq. (7.37):

$$\dot{Q} = \pi f \gamma^2 G\} \tag{7.37}$$

where

$\dot{Q} =$ heat generation,
$f =$ frequency,
$\gamma =$ strain amplitude,
$G'' =$ loss modulus.

The horn frequency is 20 kHz. Rheology measurements at this frequency are difficult so they use time-temperature superposition to translate this to measurements at 10 Hz. They assume 20 kHz at 25°C is equivalent to 10 Hz at 10°C. Plotting G'' at these conditions for the different sealants, they find G'' decreases with decreasing density, in the opposite direction for increased heat generation according to Eq. (7.37). With lower density polyethylene, however, the applied sealing force produces a larger deformation, which increases the strain amplitude. Since the heat generation is a function of the strain amplitude squared, this overwhelms the effect of decreasing loss modulus. At higher densities, apparently, there is not enough heat generation to fully melt the crystals, resulting in lower seal strength.

They also compare several other commercial sealant polymers including EAA, mPE plastomer, ionomer and LDPE resins. The plastomer performs the same as the ULDPE of the same density. The EAA performs exceptionally well, and this corresponds to its higher G''. The ionomer and LDPE give lower seal strength. They compare the performance of these sealants with their hot bar (CS) seal strength. The US seal strengths for the plastomer and ionomer are similar to their CS seal strength. For the EAA, the US seal strength is higher and for the LDPE it is lower than their respective CS seal strengths.

In other studies, Van Dun et al. find:

- A propylene-ethylene copolymer (VERSIFY from Dow Chemical) outperforms the ULDPE and LLLDPE.
- A ULDPE outperforms an LLDPE when sealing a 4-layer structure (to simulate a gusset region).
- A coextruded structure gives a more complex performance than a monolayer structure with the same sealant. The coextruded structure (40 μm [LLDPE + LDPE] – 10 μm sealant) is compared with a 50-μm monolayer sealant film. At low sealing force, the seal strengths are the same, but as the sealing force is increased, the monolayer outperforms the coextruded film. Layers outside of the seal participate in the heat generation, deformation and flow.

To date, only a few studies have been published elucidating the factors that influence sealant selection and optimization for US. This technology remains a ripe area for further research and development.

7.10 Failure analysis and troubleshooting

The performance of heat seals does not always conform to expectations. Many factors may be involved. By understanding the symptoms and their root causes, the packaging professional can identify what corrective actions to take.

7.10.1 Types of seal failures

The first step in a failure investigation is to understand how the seal is failing. This will help determine what the root cause may be. There are three types of failure:

1. Interfacial;
2. Delamination;
3. Cohesive.

It is not a coincidence that these failure modes are the same as for creating peelable seals described in Section 7.7.1. In peelable seals, the failure mode is carefully and purposely controlled. In the failure of heat seals, these modes may be undesired and the cause of poor performance.

A seal fails interfacially if it cleanly peels at the sealant-sealant or sealant-substrate. It is as if the seal was never made or was only tacked together. As described below, failure in this mode may point to too low sealing temperature or time, or contamination.

In delamination, the failure occurs along the interface between adjacent layers. The seal-to-seal or seal-to-substrate strength is greater than the interlayer adhesion so that when the sealed film is pulled apart, the peel front tears through the sealant layer and failures at the adjacent interface. This may suggest a poorly cured adhesive, a problem with a tie layer, or perhaps chemical attack of the bond strength by a product ingredient.

Cohesive failure occurs when the seal fails within the seal layer (not at the interface). This is an indication of poor cohesion that may be the result of poor mixing, migration of additives to near the surface or polymer degradation near the seal surface.

7.10.2 Common causes of failure

In this section, several root causes of seal failures are described along with their symptoms and actions that can be taken to eliminate the cause. Table 7.9 provides this information in the reverse form, listing the symptoms followed by the causes and corrective action. The reader is referred to Section 7.4 for further discussion on factors that affect seal performance.

7.10.2.1 Sealing variables

The three primary variables in the hot bar sealing process are time, temperature and pressure. Improper settings or control over these variables can negatively impact seal performance.

If the sealing temperature is too high, the film may shrink, causing wrinkles and channel leakers. High temperatures may distort one or more layers and cause loss of orientation. In extreme conditions, the seal bar will burn through the film, causing holes in the seal area.

If the sealing temperature is too low, not enough heat will transfer to the sealant interface for bonding. The film will delaminate interfacially.

Solutions to these issues involve first checking to make sure the bar temperature is set to a temperature appropriate for the film structure. The film structure may have to be redesigned. For example, if the sealant seal-initiation temperature is greater than the distortion temperature of one of the layers, either the sealant or the other layer will need to be changed. The film thickness must also be taken into account. The bar temperature set point should be higher for a thicker film with the same sealant and seal time to compensate for the longer time it will take for the heat to transfer to the sealant interface.

If the set temperature is OK, then the next step is to check to make sure the actual bar temperature is within a few degrees of the set point. There may be a hot or cold spot from a broken heater or the thermocouples controlling the temperatures may be in the wrong location to properly control the temperature.

TABLE 7.9 Troubleshooting guide to heat seal issues.

Symptom	Root cause	Action
• Shrinkage or wrinkles in entire structure • Distortion of outside layer • Loss of orientation • "Burn-through"	• Seal temperature too high • Seal time too long	• Reduce temperature set point • Check thermocouple placement • Check heating elements • Reduce seal time
• Interfacial failure	• Seal temperature too low • Not enough contact pressure • Seal time too short	• Increase bar temperature set point • Check thermocouple placement • Check heating elements • Increase pressure • Increase seal time
• Squeezing sealant out of seal area	• Seal temperature too high • Seal pressure too high • Inconsistent pressure • Misaligned crimp teeth	• Reduce temperature set point • Check thermocouple placement • Check heating elements • Reduce seal pressure • Check pressure profile with carbon paper impressions • Align crimp teeth
• Cohesive failure	• Poor mixing during extrusion • Poor purging during extrusion • Too much release additive, slip, lubricant, etc. • Carrier resin for additive masterbatch incompatible with sealant • Degradation of polymer during extrusion • Gel	• Optimize extrusion process • "Disco" purge procedure (vary rpm during purging – see Section 13.1) • Practice good purge procedures • Reduce additive level • Check composition of carrier resins for additive masterbatches • Eliminate dead zones and hot spots in extruder
• Poor seal strength	• Product contamination in seal area	• Look for evidence of dust, powder, oils, etc. in seal area • Eliminate static charge that attracts powdery contaminates to the seal area • Select sealant optimized for sealing through the type of contamination encountered • Adjust packaging line operation to minimize contamination of seal area
• Poor seal strength	• Substances transferred to seal areas	• Check for source of contaminants from contact with other layers in roll (e.g., back side transfer) • Check for sources of known trouble makers: silicones, stearates, slip additives, release sprays, residual solvent
• Peelable seal blend too high or low in seal strength	• Poor mixing of incompatible polymers • Too much stress or orientation	• Adjust level of polymers in blend • Adjust mixing in extruder • Adjust orientation
• Channel leakers	• Poor caulking of wrinkles, gusset areas • Poor hot tack • Contamination	• Select sealant for better flow and/or hot tack • Check for sources of contamination • Note: Higher pressure may not improve caulking and just thin out sealant layer.
• Seal strength changes with time	• Product ingredient degrades bonding • Low end-use temperature embrittles sealant • Low atmospheric pressure during shipping	• Select sealant that resists product migration and loss of adhesion • Select sealant for low temperature toughness
• Weak surface layer or cohesive skin layer failure	• Corona treatment • Back-side corona treatment exposing sealant surface	• Check dyne levels for evidence of treatment
• Interlayer delamination	• Incomplete cure of adhesive	• Check for insufficient drying (too much residual solvent) • Check for incorrect recipe • Increase curing time
• Partial seal failure	• "Mooning" due to lack of hot tack strength • Hot seals pulling apart due to film sticking to seal bars or high COF	• Select sealant with higher hot tack • Select outside surface layer with higher temperature resistance or protect seal bars with Teflon • Decrease COF of film

If the sealing pressure is too high, the sealant may be squeezed from the seal area. Examining a cross sectioned sample of the seal area near the seal edge may reveal this. Solutions include reducing the pressure set point, checking to make sure the pressure is uniform across the seal bar by making impressions using pressure sensitive paper (or carbon paper) and ensuring the crimper teeth are aligned. If the pressure is too low, not enough contact is being made and the seal will fail interfacially. In this case, try increasing the pressure.

If the dwell time is too short, the seal will not be made and it will fail interfacially. If the dwell time is too long, there may be degradation of the polymer in one or more layers and distortion. Time and temperature go hand in hand; increasing time is equivalent to increasing temperature.

7.10.2.2 Contamination

There are many potential sources of contamination that may affect seal performance. One involves the introduction of foreign matter during extrusion. The addition of an incompatible polymer may reduce seal strength by creating a second phase with weak adhesion between the phases, the same concept purposely used to create peelable seals. Additives or contaminants may migrate to the surface and form a weak boundary layer. These types of issues may be caused by a variety of problems including poor purging; addition of too much slip, release agent, lubricant, filler, antiblock, etc.; degradation of the polymer; and creation of gel that cannot be melted or sealed.

Corrective action may include choosing the carrier resin for an additive masterbatch to be compatible with the sealant resin so that an inadvertent peel seal is not created. Using proper purge procedures such as varying the screw speed randomly up and down during purging can help dislodge hung up resin. See Section 13.1 for details. Streamlining dead spots in the extruder, transition pipe and die may also help.

Product contamination is another source of seal issues. Dust and powders attracted by static charge to the seal area are a challenge. Examples include cereal, powdered products, starch, coffee and shredded cheese. Oil and greasy foods such as meat are another challenge. Liquids are sometimes purposely sealed through to eliminate headspace gases. Corrective action includes taking steps to prevent or discharge static charge and selecting sealants that excel at sealing through the type of contamination being encountered.

Foreign substances may be transferred to the sealant surface from contact with the outside layer in the roll (known as backside transfer), from additives migrating from adjacent layers in the film or from various processing aids used in the manufacturing area [158,159]. These may include silicones, stearates, slip agents, release sprays, and residual solvents. Silicones and fluorocarbon sprays are particularly problematic for sealing and should be avoided.

7.10.2.3 Film converting

Surface treatments, such as corona, flame, ozone or plasma treatment, are generally detrimental to heat sealing performance. These are used primarily to increase the surface energy of films to improve adhesion to coatings and inks. They promote oxidation of the surface and their effectiveness can be evaluated by measuring the surface energy using dyne solutions or other surface techniques. Oxidation may reduce the molecular weight of the surface molecules, decreasing the strength of the interface. In other cases, oxidation may promote crosslinking at the surface, which impedes molecular diffusion during heat sealing and reduces seal strength. Overtreatment can create a skin layer that readily delaminates from the bulk.

Surface treatment of the seal layer is normally avoided for these reasons. But backside treatment can inadvertently treat the front side of the film (the sealant) if the treatment station is not working properly. When treatment is suspected as a cause for poor heat seal performance several corrective actions can be taken, including

- Measure the surface energy of the sealant surface to see if it is higher than expected. This may be a tip-off for backside treatment.
- Try turning off the treatment to see if this improves the seal strength.
- Carefully examine seals to see if some of the symptoms of overtreatment are present.
- Check to make sure the treater is not causing the backside (sealant side) to be treated with the front side.

Another potential problem is the incomplete curing of solvent or water-based adhesives. Interlayer delamination may occur during the seal strength measurement due to too much residual solvent, using the incorrect formulation, or not providing enough curing time.

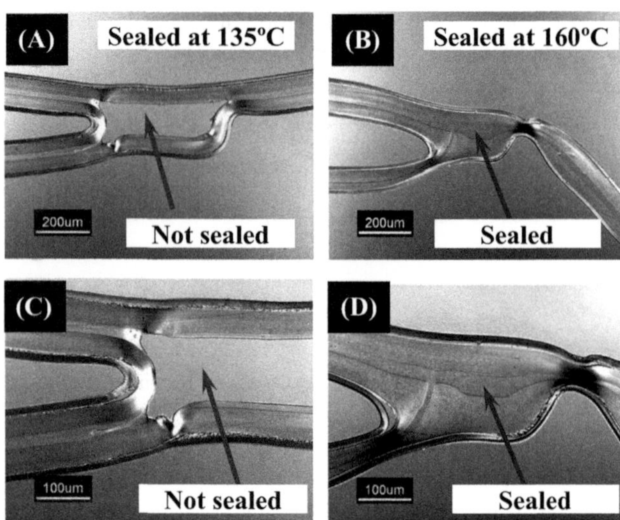

FIGURE 7.54 Photos taken under cross polarized light with an optical microscope of the cross section of a seal area with folds or gussets. The seals in pictures a) and c) were formed at 135°C and leaked. The seals in pictures b) and d) formed at 160°C show complete caulking of the voids near the fold and were hermetic. *Source: E.I. Garcia-Meitin, C.Y. Lai, P. O'Connel, Development of sealants for flexible packaging using light microscropy, SPE ANTEC Conference, 2015. Reproduced with permission of the Dow Chemical Company.*

7.10.2.4 Package design: channel leakers and gussets

Channel leakers are capillaries that run across the width of the seal area creating a pathway for leakage of product from the package or ingress of gasses from the external environment. These often form from wrinkles or around product contamination. A contributor to channel leakers is not having enough seal resin, or a seal resin with the right flow properties, to flow into the channels to caulk them shut.

Gussets are multi-ply regions of the seal area, such as where the film is folded or where a fin seal meets the end seal (see Fig. 7.1B). These areas can be particularly challenging to seal since there is often a gap between the single and multi-ply regions. Like channel leakers, selecting a sealant that can flow into and caulk the gap is important. Fig. 7.54 shows an example of a seal with a hole that was not caulked by the sealant.

Increasing the hot tack of the sealant can also help prevent channel leakers and improve seal integrity in the regions near gussets.The resilience of the film in the gusset or wrinkle acts to open up the hole after sealing if hot tack is not sufficient.

7.10.2.5 Aging effects

Seal strength may age up or down with time. In some situations, the film may change prior to sealing in a way that affects its ability to be sealed. Physical changes, such as changes in crystallinity or absorbed moisture, may take place in the roll during storage. This may be influenced by storage conditions such as temperature and relative humidity. Many semicrystalline polymers undergo secondary crystallization in the first few days or weeks after fabrication into a film. This is a process where even at room temperature the molecular chains may be able to form imperfect ordered structures (hence the name secondary crystals) over time. This may cause shrinkage in one or more layers and put stress on interfaces, causing changes in adhesion and seal strength. Crystallization increases stiffness, which may impact bending and stretching of the peel arm during peeling. The presence of crystals may also increase the time or temperature needed to melt the polymer for sealing.

Other aging effects may occur after the seal has been made while the package is stored. Filling, storage and end-use conditions should be considered. For example, hot fill or retort may distort the film or melt the seal, weakening the seal. Internal interlayer bonds may also weaken over time, which may affect seal strength. Changes in the crystal form may also affect seal performance. The example of the different forms of PB-1 and its effect on peelable seals is discussed in Section 7.7.3.

A special form of aging involves product migration. Some components of the product being packaged may diffuse through the sealant layer and attack the adhesion to an adjacent layer. A well-known example is acetic acid in condiments such as ketchup and salad dressings. These products have been packaged in sachets and pouches that use aluminum foil as the barrier and LDPE as the sealant. The adhesion of LDPE to aluminum foil relies on the formation of oxidative byproducts such as acid groups during extrusion coating. The acid groups bond to the oxides on the aluminum surface. But these bonds are rather weak and susceptible to attack by the acetic acid that readily migrates through the PE. A solution is to use an acid copolymer tie layer. The high acid level in these copolymers provides better adhesion that is more resistant to attack from the acetic acid, as well as slows down the diffusion of the acetic acid through the layer.

Aging and product migration effects are discussed in greater detail in Chapter 16.

7.10.2.6 Environmental conditions

There are many environmental factors that the package may be subjected to during packaging, sterilization or the end-use that can affect seal strength. Examples include:

- Too high temperature. Retort or hot fill conditions may exceed the melting point of the sealant or other layers in the structures, weakening seal strength or interfacial adhesion.
- Too low temperature. The seal area may become embrittled if the temperature gets too low. Blast freezing can be problematic for some package structures.
- Low atmospheric pressure found during shipment of packages over mountains can cause the interior of gas flushed or modified atmosphere packaging to expand and stress the seal.
- High internal pressure, such as cooked-in deli meats during cooking. The combination of high pressure and temperature puts stress on the seal.

7.10.2.7 Low hot tack strength

Partial seal failure, often referred to as mooning, results from the seal being pushed open by the product due to low hot tack strength. On horizontal form fill seal lines this can be the result of high line speeds exerting forces from registration and product filling. In vertical form fill seal operations, the product falling on the newly created bottom seal imparts an opening force. The weight and drop height of the product are key factors. Lowering the seal temperature generally increases hot tack strength but this can only be safely done if the sealing temperature is well above the onset of the seal plateau region. Otherwise, the seal strength may be affected. Equipment modifications may also help with reducing opening forces or cooling the seal during packing. Short of this, switching to a sealant with better hot tack is the best way to deal with this problem. As described in Section 7.4, interlayer adhesion and the hot tack of layers adjacent to the seal layer may also impact hot tack; adjustments to the film structure may help improve hot tack.

Hot seals can also be pulled apart from the jaw sticking on the outside of the film. Coating the seal bar with Teflon can help, but the thickness of the Teflon layer must be taken into account when considering the time needed for the heat to transfer to the sealant interface. Higher seal bar temperatures may be needed to compensate. Changing the film structure so that the outside layer has greater temperature resistance can also help.

High COF of the film/metal surfaces of the packaging line can act to pull hot seals apart. Fatty acid amide slip agents are often less effective at the elevated temperatures present just after sealing. These agents act by blooming to the surface. At elevated temperatures they are more soluble in the polymer; some of the slip may migrate back into the polymer leaving less on the surface to control COF. Using nonmigratory slip agents or adding additional migratory slip agents may help. Adding too much migratory slip agent, however, can cause other problems such as chalking of the surface. Changing the outside layer to a more temperature resistant one may help. Increasing the hot tack strength of the sealant also can help.

7.10.3 Analytical techniques

There are several analytical tools that can be used to help identify the cause of heat seal problems. These include

- Microscopy. This can be very useful for identifying the location and type of the failure. Optical and electron microscopy may be worthwhile.
- Fourier transform infrared (FTIR) analysis. This is useful for making sure the correct sealant is being used and for identifying the composition of any obvious contaminants that may be in the seal area.
- Electron spectroscopy for chemical analysis (ESCA) (also known as X-ray photoelectron spectroscopy) and time of flight secondary ion mass spectroscopy (TOF-SIMS). These are two complementary techniques for identifying chemical and molecular species near the surface of films. They have a much finer resolution than FTIR. Where the measurement size of FTIR is in microns, the effective measurement size for these techniques is in nanometers ($<$ 100 nm lateral resolution and $<$ 1 nm depth resolution). Thus, they are useful for identifying trace contaminants near the surface that may affect heat seal performance. ESCA uses electrons and TOF-SIMS larger ions to bombard the surface under vacuum. Detectors sense the atoms and molecules that are knocked off from the bombardment.

- Differential scanning calorimeter (DSC). This technique can determine the melting points and crystallinity of the sealant. It may also be useful for detecting changes in crystallinity with age. More sophisticated techniques such as wide and small angle x-ray scattering can provide more detail on crystalline structure and orientation, which may prove useful in identifying problems.
- Thermal mechanical analysis (TMA) and AFM. Van Ness [159] shows how TMA can be used to detect crosslinking at the surface that may occur during film fabrication or extrusion coating. AFM is a more modern technique that is more sensitive and versatile. It can be used to detect soft and hard regions on the surface of the film, which may be helpful for understanding seal strength variability.

7.10.4 Troubleshooting check list

The following checklist is useful for systematically troubleshooting heat seal problems (DuPont, 2016):

1. How did the seal fail?
   - Interfacial, delamination, cohesive
2. Are the sealing conditions optimized?
   - Time, temperature, pressure
3. Compare problem material with a control
   - Use lab testing for comparison
4. Are correct surfaces being sealed?
   - FTIR for identification
   - Roll wound incorrectly may put seal surface on outside
5. Can seal be made if surface is cleaned with a solvent (mineral spirits, alcohol)?
   - Removes possible contamination
6. Check for surface treatment
   - Dyne level measurement
7. Does material seal on a different packaging line?
   - Check condition of seal bars/jaws
8. Has anything changed?
   - Product
   - Film
   - Equipment
   - Product storage/transportation
9. Call suppliers! Provide as much information as possible, including samples of good and bad seals.

7.11 Selecting sealant resins

Many factors go into selecting a sealant. Some have been discussed in preceding sections and involve sealing performance such as seal initiation temperature, ultimate seal strength, hot tack, sealing temperature range, hermeticity, seal through contamination and caulking. Often, however, the sealant contributes to other features of the package such as clarity, puncture resistance, abrasion resistance, tear strength, moisture barrier and oil and grease resistance. The sealant may also need to contribute to the ease of making the film structure, such as having melt strength for bubble stability in blown film and minimizing neck-in in extrusion coating. Thermal resistance may be important during converting as well as in the end use (as in cook-in or pasteurization processes).

At the same time, the sealant needs to be the most cost-effective solution. Cost, however, is not just the price of the resin, but involves ease of processing into a film (and its effect on yields), packaging line efficiencies (line speed) and package quality (leaker rates). Sometimes a higher priced sealant will yield a lower cost package when the total cost is considered. Total cost modeling is discussed in Chapter 16.

The sealant is part of a packaging structure that increasingly is designed to be recycled after use. This may influence the selection of the chemistry of the sealant, such as using a PE-based sealant in a package structure that can be recycled in a PE recycle stream. This may also require the use of higher performance sealants; for example, using a sealant with lower seal initiation temperature to compensate for low temperature resistant structural layers in the packaging structure. Design for recycle is taken up in detail in Chapter 17.

TABLE 7.10 Description of some common ethylene-based sealant resin families.

Resin	LDPE	LLDPE	VLDPE	mLLDPE	Plastomers	EVA	ACR	Ionomer
Name	Low Density Polyethylene	Linear Low Density Polyethylene	Very Low Density Polyethylene	Metallocene Polyethylene	Metallocene Plastomer	Ethylene Vinyl Acetate	Acid Copolymer Resin	Acid Copolymer Resin partially neutralized with a metal cation
General Characteristics as a Sealant	Low cost, general purpose sealant with good processing in extrusion coating and blown film due to its high melt strength (via long-chain branching). Excellent clarity in blown film. Moderate adhesion to aluminum foil in extrusion coating.	Low cost, general purpose sealant. Better toughness than LDPE in film applications. More difficult to process than LDPE.	Lower melting point range, broader hot tack, lower seal initiation temperature than LLDPE.	Premium sealant. Lower melting point, better clarity and greater toughness than LLDPE. More difficult to process. Low seal initiation temperature and high hot tack. Excellent organoleptics.	Premium sealant. Very low seal initiation temperature, high hot tack, good caulking, low extractables and organoleptics, high OTR and MVTR, good toughness. Like mLLDPE, less shear thinning and more difficult to process than LLDPE.	Premium sealant. Lower melting point, better clarity and greater toughness than LDPE. Low seal initiation temperature. High shear thinning for ease of process.	Specialty sealant,. Lowermelting point, better clarity and toughness than LDPE. Excellent adhesion to metal in extrusion coating. Low heat seal initiation and broad hot tack window.	Specialty sealant. Low melt point, excellent clarity, low seal initiation temperature, very broad hot tack temperature window, excellent oil and grease resistance, excellent optical properties, good adhesion to metal, excellent formability, excellent abrasion resistance and depending on the grade higher stiffness than ACR, mPE and EVA's.
Polymer-ization Process	High Pressure, Free Radical Initiated	Low Pressure (1000 psi), catalyst (Ziegler–Natta)	Low Pressure (1000 psi), catalyst (Ziegler–Natta)	Low Pressure (1000 psi), catalyst (metallocene)	Low Pressure (1000 psi), catalyst (metallocene)	High Pressure, Free Radical Initiated	High Pressure, Free Radical Initiated	High Pressure, Free Radical Initiated
Comonomer	None	butene, hexene, octene	butene, hexene, octene	butene, hexene, octene	hexene, octene	vinyl acetate	acrylic acid (AA) or meth-acrylic acid (MAA)	acrylic acid (AA) or methacrylic acid (MAA)
MWD	Very Broad	Broad	Broad	Narrow	Narrow	Very Broad	Very Broad	Very Broad
Comonomer Distribution	None	Broad	Broad	Narrow	Narrow	Random	Random	Random
Long-Chain Branching	Yes	No	No	No or Limited	No or Limited	Yes	Yes	Yes
Density (g/cc)	0.915–0.93	0.915–0.935	0.89–0.91	0.910–0.920	0.885–0.910	0.94–0.95	0.94–0.95	0.94–0.95
Typical Melting Point, °C	108	123	115–123	100–112	80–105	87 (18% VA)	97–103	87–100
Modulus, kpsi (MPa)	30 (207)	30–40 (207–275)	10–20 (69–138)	17–30 (120–207)	7–17 (50–120)	7 (48)	20 (138)	7–70 (48–480)

TABLE 7.11 Summary of qualitative assessment of common ethylene-based sealant polymers.

ATTRIBUTE	LDPE	LLDPE	VLDPE	mLLDPE	Plastomer	EVA	ACR	Ionomer
Low Seal Initiation Temperature	♦♦	♦	♦♦	♦♦♦♦	♦♦♦♦♦	♦♦♦♦	♦♦♦	♦♦♦♦
Ultimate Seal Strength	♦♦♦	♦♦♦♦♦	♦♦♦♦♦	♦♦♦♦♦	♦♦♦♦♦	♦♦♦	♦♦♦	♦♦♦
Peak Hot Tack Strength	♦	♦♦	♦♦	♦♦♦♦♦	♦♦♦♦	♦	♦♦♦	♦♦♦♦
Hot Tack Temperature Window	♦	♦♦♦	♦♦♦	♦♦	♦♦♦	♦	♦♦♦♦	♦♦♦♦♦
Caulkability and Hermeticity	♦♦	♦	♦♦♦	♦♦♦♦	♦♦♦♦♦	♦♦♦	♦♦♦♦	♦♦♦♦♦
Seal Through Contamination	♦♦	♦♦	♦♦	♦♦♦♦	♦♦♦♦	♦♦	♦♦♦♦	♦♦♦♦♦
Grease and Oil Resistance	♦♦♦	♦♦♦	♦	♦	♦	♦	♦♦♦♦	♦♦♦♦♦
Adhesion to PE	♦♦♦♦♦	♦♦♦♦♦	♦♦♦♦♦	♦♦♦♦♦	♦♦♦♦♦	♦♦♦♦♦	♦♦♦♦	♦♦♦
Adhesion to Al Foil	♦♦						♦♦♦♦♦	♦♦♦♦
Low MVTR	♦♦♦	♦♦♦♦	♦♦	♦♦	♦♦	♦♦	♦♦	♦♦
Transparency	♦♦	♦	♦♦	♦♦♦	♦♦♦♦	♦♦♦♦♦	♦♦♦♦	♦♦♦♦♦
Melt Strength	♦♦♦	♦♦	♦	♦	♦	♦♦♦	♦♦♦♦	♦♦♦♦♦
Tear Strength	♦♦♦	♦♦♦♦♦	♦♦♦♦♦	♦♦♦♦	♦♦♦♦	♦♦♦♦	♦♦♦	♦♦
Abrasion Resistance	♦♦♦	♦♦♦	♦♦	♦♦	♦♦	♦♦	♦♦♦♦	♦♦♦♦♦
Puncture Resistance	♦♦♦	♦♦♦	♦♦	♦♦	♦♦	♦♦	♦♦	♦♦♦♦♦
Stiffness	♦♦♦	♦♦♦♦	♦	♦	♦	♦	♦♦	♦♦♦♦♦

The greater the number of diamonds, the better the performance is for a given attribute.

Polyethylene and ethylene copolymers are the most widely used sealant resins in multilayer flexible packaging. In the sections that follow, qualitative rankings for several important attributes are provided for a representative set of ethylene-based sealant resins. These resin families are described in detail in Table 7.10. The rankings are summarized in Table 7.11. These rankings represent qualitative comparisons based on the author's experience and should only be used as a rough guide. The performance of individual grades within a polymer family may vary greatly. Blends of different resins are often used to improve processability and performance or reduce cost. The fabrication process will influence properties through changes in orientation, stress and crystallinity. This guide should only be used as a starting point for discussions with resin suppliers on specific applications. Chapter 4 introduces these resins and several references provide relevant properties and applications [160–162].

For this discussion, PP-based sealants are ignored. It is simply noted that random copolymers of propylene and ethylene (rPP) are commonly used as the sealant layer in oriented PP films. The rPP provides good adhesion to the homopolymer structural layer of OPP and melts at 10°C–20°C lower so provides an adequate seal temperature window. It also has excellent high temperature resistance, making it suitable for retort applications. More recently, metallocene propylene-ethylene copolymers and elastomers such as Versify from Dow Chemical and Vistamaxx from ExxonMobil have found use as sealants (or as a blend component for sealants) in PP structures. These have even lower seal initiation temperature than traditional propylene random copolymers [80,81,163].

7.11.1 Seal initiation temperature

HSIT generally follows the melting point of the polymer. Metallocene polyethylene melts over a narrow temperature range, which may provide an advantage over EVA and other low melting polymers for high-speed packaging operations. The order of decreasing seal initiation temperature is:

LLDPE > VLDPE > LDPE > ACR, Ionomer > EVA, mLLDPE > Plastomer

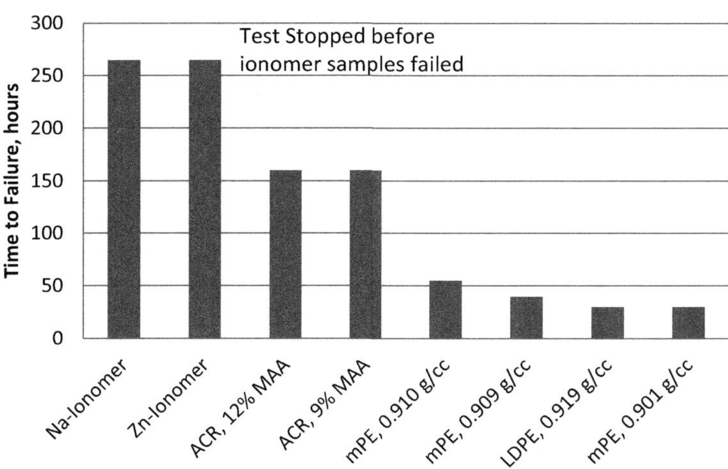

FIGURE 7.55 Oil resistance of some common sealant resins. Penetration of corn oil through 50-µm film at 40°C [39]. *Adapted from J.R. De Garavilla, Ionomer, acid copolymer, and metallocene polyethylene resins: a comparative assessment of sealant performance, TAPPI J. 78(6) (1995) 191–203. Reproduced with permission from the Dow Chemical Company.*

7.11.2 Ultimate seal strength

Linear polymers have higher ultimate seal strength than those with LCB:

mPE, VLDPE, LLDPE > EVA, Ionomer, ACR, LDPE.

7.11.3 Peak hot tack strength

There is considerable variability amongst grades within each sealant family on this attribute. Also, blends of different families are often used (e.g., blends of plastomer with LLDPE or LDPE) which further muddles the ranking of these materials.

mLLDPE > Plastomer > Ionomer > ACR > VLDPE > LLDPE > EVA > LDPE.

7.11.4 Breadth of hot tack temperature window

Ionomers generally have high hot tack over a very wide temperature range due to their outstanding melt strength arising from ionic interactions. Some individual grades of plastomers and mLLDPE also exhibit broad temperature windows.

Ionomer > ACR > VLDPE > LLDPE > Plastomer > mLLDPE > EVA > LDPE

7.11.5 Caulking and hermetiticy

As described in Section 7.5.6 caulking is related to the viscosity at low shear rates. Viscosity is influenced by molecular weight, LCB and crystallinity. Linear polymers generally have lower viscosity at low shear rates than branched polymers. Within the linear PE sealant families, plastomers exhibit flow at the lowest temperatures due to their low and distinct melting point. VLDPE starts to melt at low temperatures but due to its broad MWD and comonomer distribution does not finish melting and exhibit considerable flow until much higher temperatures. EVA is known for good caulkability and this may be due to the ability to select grades with higher MFI for a given process due to their good processability. Ionomers generally exhibit outstanding hermetiticy over a wide operating window due to their good flow and hot tack.

Ionomer, plastomer > mLLDPE, ACR > EVA > VLDPE > LDPE > LLDPE

7.11.6 Seal through or around contamination

There is a lot of disagreement in the literature on the ranking of the sealants on this attribute, as is discussed in Section 7.4.1.5. Ionomers and ACRs have historically been used in applications involving sealing of greasy products such as processed meats. PE and EVA still find their way into many meat applications, and so the suitability of a given sealant depends on the severity and type of contamination. Seal around contamination also involves caulkability and hot tack considerations.

7.11.7 Grease and oil resistance

The oil resistance of polyethylene is a strong function of its density; the higher the density, the greater the crystallinity and the better the oil resistance. PE sealants typically have low crystallinity and poor oil resistance. Ionomers and ACRs, because of the polar nature of their comonomers, have excellent oil resistance:

Ionomer > ACR > LLDPE, LDPE > EVA, VLDPE, mLLDPE > plastomer.

Fig. 7.55 compares the oil resistance of several sealant resins. These results were obtained using the ground glass technique where the time for corn oil to penetrate through the film is measured. The test is accelerated by using a temperature of 40°C. Even though the test was stopped before the penetration through the ionomer films occurred, the results clearly show the superior performance of the ionomers versus the polyethylene sealants. The lowest density polyethylene resins with the lowest seal initiation temperatures provide little resistance to oil.

7.11.8 Adhesion to polyethylene

Adhesion to PE is important in many coextruded flexible packaging films. The adjacent layer to the sealant layer is often a PE-based tie resin or a PE structural or bulking layer. Good adhesion helps improve heat seal strength and hot tack. Lower adhesion can be an advantage for the design of delamination peelable seals.

mPE, LLDPE, VLDPE, EVA, LDPE > ACR > Ionomer.

7.11.9 Adhesion to aluminum foil

Adhesion to foil is important for some extrusion coating applications. The order of performance, highest to lowest:

ACR > Ionomer > LDPE.

7.11.10 Moisture vapor transmission rate

The higher the crystallinity, the lower is the moisture vapor transmission rate (MVTR) and the better the moisture barrier. The order of increasing MVTR:

LLDPE < LDPE < VLDPE, mPE, EVA, ACR, Ionomer.

7.11.11 Oxygen barrier

None of the polyethylene or ethylene copolymer sealants have good oxygen barrier performance compared with high barrier polymers such as EVOH, PVDC or even medium barrier polymers such as PA6 or PET. For high oxygen transmission for fresh produce packaging and other breathable applications, polyolefin elastomers, plastomers, and high VA EVA work best.

7.11.12 Clarity and transparency (low haze)

Clarity and haze are influenced by how the film is fabricated, especially how fast it is quenched. Cast films generally have lower haze than blown films due to the faster rate of cooling in the cast film process. All things being equal, the resin order from highest clarity to lowest is:

EVA, Ionomer > ACR > plastomer > mLLDPE > VLDPE > LDPE > LLDPE.

7.11.13 Melt strength

Melt strength is important for hot tack, bubble stability, thermoforming and low neck-in (extrusion coating). The order of melt strength from the highest to lowest is:

Ionomer > ACR > LDPE, EVA > LLDPE > VLDPE > mLLDPE, plastomer.

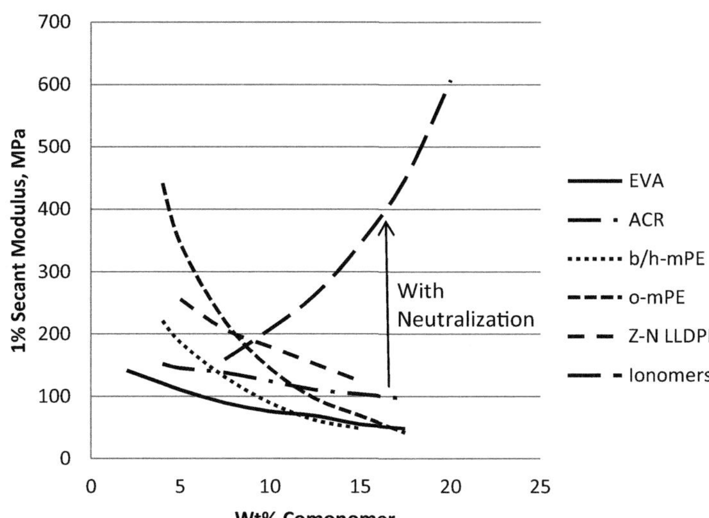

FIGURE 7.56 Film stiffness (secant modulus) as a function of comonomer content for several classes of ethylene-based sealant resins [39]. *Adapted from J.R. De Garavilla, Ionomer, acid copolymer, and metallocene polyethylene resins: a comparative assessment of sealant performance, TAPPI J. 78(6) (1995) 191–203. Reproduced with permission from the Dow Chemical Company.*

7.11.14 Tear strength

Blown film of linear PE has high tear strength in the TD but low tear strength in the MD. The opposite is true for branched PE and ethylene copolymers. The general order for TD tear strength is:

LLDPE, VLDPE, mLLDPE, plastomer > EVA > LDPE > ACR > Ionomer

7.11.15 Abrasion resistance

Abrasion resistance depends on the type of abrasion. Standard laboratory abrasion tests with a "sandpaper" like abrader often gum up soft sealant resins and the results are not very useful. Ionomers perform very well on scratch tests.

Ionomer > ACR > LDPE, LLDPE > mLLDPE, plastomer, EVA, VLDPE

7.11.16 Puncture and impact resistance

Puncture resistance is important for protecting products with sharp features such as bones or noodles. The stiffness and toughness of ionomer provide good slow puncture resistance:

Ionomer > LDPE, LLDPE > ACR, mLLDPE, plastomer, EVA, VLDPE.
Linear PE generally performs well in higher speed impact tests such as dart impact.
mLLDPE, plastomer > LLDPE, ionomer > EVA, ACR, LDPE.

7.11.17 Stiffness

The stiffness or modulus of the sealant can have a disproportionate effect on the bending stiffness of a package, allowing the structure to be downgauged [164]. A model of bending stiffness is described in Chapter 9. The melting point and seal initiation temperature of polyethylene plastomers and ethylene copolymers such as EVA and ACR decrease with increasing comonomer content. This is due to the reduction in cystallinity. The crystals constrain the movement of polymer chains. Fewer crystals result in more chain movement and lower stiffness. Therefore, most premium sealant resins have a low modulus as illustrated in Fig. 7.56. An exception is ionomers. The ionic interactions in ionomers constrain molecular movement, raising the stiffness.

The order of stiffness from highest to lowest is:

Ionomer > LLDPE > LDPE > ACR > VLDPE, EVA, mLLDPE, plastomer.

Resin suppliers can supply comparisons and help in selecting grades from their portfolio. In Chapter 18 the development of tools to help select resins is described. One such tool was developed by DuPont [165] that allows the user to input attributes and assign important factors for a given application. The tool then assesses which family of sealant resins provide the most benefit based on the attributes chosen.

References

[1] ASTM International, Standard Test Method for Seal Strength of Flexible Barrier Materials, (F88/F88M-15), 2015.

[2] ASTM International, Standard Test Method for Making Heatseals for Determination of Heatsealabilitiy of Flexible Webs as Measured by Seal Strength, (F2029-16), 2016.

[3] ASTM International, Standard Test Method for Mechanical Seal Strength Testing for Round Cups and Bowl Containers With Flexible Peelable Lids, (F2824-20), 2020.

[4] A.M. Soutar, Hot testing: or is it?, in: TAPPI Polymers, Laminations and Coatings Conference, 1994, p. 383.

[5] A.M. Soutar, Hot tack testing: a review, J. Plastic Film. Sheeting 12 (1996) 304–334.

[6] V. Brodie, III, Hot tack of sealant resins, in: TAPPI Polymers, Laminations And Coatings Conference, 1986, pp. 187–196.

[7] B.A. Cooper, W.H. Wharton, Adhesion testing method and apparatus, in: U.S. Patent 3412606, 1968.

[8] DuPont Literature, Hot tack strength, Surlyn® Update 1 (4) (1970).

[9] C. Hocking, Latest findings on hot tack show worth of test unit, Package Eng. (1983).

[10] D. Falla, M. Li, Reliable hot tack testing of polyolefin films, in: SPE ANTEC Conference, 2014.

[11] H.W. Theller, Hot tack tester, in: U.S. Patent 5847284, 1998.

[12] ASTM International, Standard Test Methods for Hot Seal Strength (Hot Tack) of Thermoplastic Polymers and Blends Comprising the Sealing Surfaces of Flexible Webs, (F1921/F1921M-18), 2018.

[13] K.R. Berger, M. Knoelke, B. Welt, Influence of sample preparation and instrument settings on hot tack measurements of thin sealant films, J. Plastic Film. Sheeting 20 (2004) 55–63.

[14] G. Schubert, O. Plassmann, S. Husch, Hot tack of extrusion coatings on aluminum foil and paper – does the process always lead to the same result? TAPPI PLACE. Conf. (2010).

[15] J.J. Wooster, L.K. Mergenhagen, Assessing real-life sealant performance in flexible packaging, in: TAPPI Polymers, Laminations & Coatings Conference, 2000, pp. 617–630.

[16] M. Bilgen, J. Van dun, R. Patel, J. Garnett, A. Bafna, Insuring seal integrity and broad operating window, TAPPI PLACE. Conf. (2012).

[17] M. Kapur, M. Bilgen, P. O'Connell, Importance of application representative sealant test methods, SPE ANTEC (2014) 1393–1399.

[18] C.L. Sand, Testing the integrity of package seals, Food Technol. Mag. 73 (7) (2019)accessed online on 26 July 2020 at. Available from: https://www.ift.org/news-and-publications/food-technology-magazine/issues/2019/july/columns/packaging-seals-tests-standards.

[19] E.I. Du Pont de Nemours, Co, Helium leak detection for quantifying low seal leak rates in flexible and rigid packages, Res. Discl. 313 (1990) 372.

[20] M.A. Pascall, J. Richtsmeier, J. Riemer, B. Farahbakhsh, Non-destructive packaging seal strength analysis and leak detection using ultrasonic imaging, Packaging Technol. & Sci. 15 (6) (2002) 275–285.

[21] Qipack, 100% in-line leak detection, Packaging World Mag. (2012)accessed online on 26 July 2020 at. Available from: https://www.packworld.com/machinery/primary-packaging/press-release/13360781/100-inline-leak-detection.

[22] Ametek Mocon, LeakMatic II. <https://www.ametekmocon.com/products/packageleakdetectors/leakmatic-ii>, n.d. (accessed 26.07.20).

[23] Witt, Inline Package Leak Detection: LEAK-MASTER MAPMAX. <https://www.wittgas.com/products/package-leak-detectors/inline-mapmax.html>, n.d. (accessed 26.07.20).

[24] Oxipack, Fully automatic leak testing. <https://www.oxipack.com/in-line-testing/>, n.d. (accessed 26.07.20).

[25] C.S. Yuan, A. Hassan, M.I.H. Ghazali, A.F. Ismail, Heat sealability of laminated films with LLDPE and LDPE as the sealant materials in bar sealing application, J. Appl. Polym. Sci. 104 (2007) 3736–3745.

[26] D. Li, S. Schrader, Q. Qian, L. Ge, Strength and failure of sealed polyethylene films, SPE FlexPackCon Conf. (2020).

[27] L.C. Nicastro, J.S. Paik, R.W. Keown, A.B. Metzner, Change in crystallinity during heat sealing of cast polypropylene film, J. Plastic Film. & Sheeting 9 (2) (1993) 159–167.

[28] K. Miyata, T. Abe, H. Tokanai, A. Nishioka, G. Murasawa, T. Koda, Effect of crystallization at film interface in heat sealing process, Annu. Technical Conf. Soc. Plast. Eng. Vol. 2 (2011) 1467–1471. 69th.

[29] K. Miyata, Y. Ichikawa, M. Noguchi, A. Nishioka, T. Koda, G. Murasawa, The effect of casting condition on heat seal property of high density polyethylene film, SPE ANTEC Conf. (2010).

[30] K. Miyata, T. Ozama, A. Nishioka, T. Koda, G. Murasawa, The effect of molecular weight on heat seal property for high density polyethylene film, SPE ANTEC Conf. (2010).

[31] K. Miyata, T. Ozama, K. Katsuno, A. Nishioka, T. Koda, G. Murasawa, Effect of molecular structure on heat seal properties for high density polyethylene film, SPE ANTEC Conf. (2010).

[32] K. Miyata, T. Hiyama, K. Yamada, Effect of LLDPE content on heat seal properties for HDPE/LLDPE film, SPE ANTEC Conf. (2011).

[33] K. Yamada, R. Konishi, Y. Hashimoto, Y. Hashimoto, T. Tsujii, K. Miyata, Effect of heater surface shape for heat-sealed parts on oriented polypropylene/cast polypropylene (OPP/CPP) film, SPE ANTEC Conf. (2014).

[34] S. Mihindukulasuriya, L.-T. Lim, Heat sealing of LLDPE films: heat transfer modeling with liquid presence at film-film interface, J. Food Eng. 116 (2) (2013) 532–540.

[35] D.M. Hahm, Sealing study with meat marinades on seal surface, Am. Meat Inst. Conf. (1999). October 1999.

[36] D. Ward, M. Li, Seal through contamination and 'caulkability' an evaluation of sealants' ability to encapsulate contaminates in the seal area, TAPPI PLACE. Conf. (2016).

[37] P. Mesnil, J. Arnaunt, R. Halle, N. Rohse, Seal through contamination performance of metallocene plastomers, TAPPI PLC Conf. (2000).

[38] J. Cooper, A.P. Moreno, Metallocene copolymers for extrusion coating applications designed specifically for exceptional ultimate hot tack and high temperature hot tack strength, TAPPI PLC Conf. (2000).

[39] J.R. De Garavilla, Ionomer, acid copolymer, and metallocene polyethylene resins: a comparative assessment of sealant performance, TAPPI J. 78 (6) (1995) 191–203.

[40] K. Hausmann, All you want to know about hot-tack, DuPont Lit. (1997).

[41] DuPont literature, Sealing through product residue,, Surlyn® Update 1 (6) (1970).

[42] F. Sadeghi, A. Ajji, Application of single site catalyst metallocene polyethylenes in extruded films: Effect of molecular structure on sealability, flexural cracking and mechanical properties, Can. J. Chem. Engg. 9999 (2014) 1–8 (published online).

[43] J.D. Vansant, The effect of coextrusion on the hot-tack performance of a common ionomer resin, TAPPI PLC Conf. (1998) 743–754.

[44] S. Bach, K. Thurling, J.-P. Majschak, Ultrasonic sealing of flexible packaging films – principle and characteristics of an alternative sealing method, Packaging Technol. Sci. 25 (2011) 233–248.

[45] F.C. Stehling, P. Meka, Heat sealing of semicrystalline polymer films. II. Effect of melting distribution on heat-sealing behavior of polyolefins, J. Appl. Polym. Sci. 51 (1994) 105–119.

[46] D. Ward, J. Crowe, Heat transfer modelling in multilayer films used for flexible packaging, SPE ANTEC Conf. (2018).

[47] J.M. Farley, P. Meka, Heat sealing of semicrystalline polymer films. III. Effect of corona discharge treatment of LLDPE, J. Appl. Polym. Sci. 51 (1994) 121–131.

[48] S. Richards, The effects of surface treatment on heat seal and hot tack, TAPPI Int. Flex. Packaging Extrusion Div. Conf. (2018).

[49] N. Mazzola, C.A. Cáceres, M.P. França, S.V. Canevarolo, Correlation between thermal behavior of a sealant and heat sealing of polyolefin films, Polym. Test. 31 (2012) 870–875.

[50] J.G. Van Alsten, Ionic and chain interdiffusion and interfacial strength development in ionomers of poly(ethylene-co-methacrylic acid), Macromolecules 29 (1996) 2163–2168.

[51] Z. Najarzadeh, A. Ajji, A novel approach toward the effect of seal process parameters on final seal strength and microstructure of LLDPE, J. Adhes. Sci. Techn. 28 (16) (2014) 1592–1609.

[52] C. Mueller, G. Capaccio, A. Hiltner, E. Baer, Heat sealing of LLDPE: relationships to melting and interdiffusion, J. Appl. Polym. Sci. 70 (1998) 2021–2030.

[53] J.M. McKelvey, Bonding operations, Polymer Processing, John Wiley & Sons, New York, 1962, pp. 356–371.

[54] S.S. Voyutskii, Autoadhesion and Adhesion of High Polymers, Interscience Publishers, New York, 1963.

[55] S. Wu, Polym. interface Adhes, Marcel Deker, New York, 1982, p. 338.

[56] B.W. Deryagin, N.A. Krotova, Adhesion, Izd. Akad. Nauk, M-L (1949) (cited in Voyutskii, 1963).

[57] B.W. Deryagin, N.A. Krotova, Doklady Akad. Nauk SSR 61 (1948) 849 (cited in Voyutskii, 1963).

[58] B.W. Deryagin, N.A. Krotova, Usphekhi Fiz. Nauk 36 (1948) 387 (cited in Voyutskii, 1963).

[59] R.P. Wool, Polymer Interfaces, Structure and Strength, Hanser Publishers, New York, 1995, p. 278.

[60] R.P. Wool, Polymer Interfaces, Structure and Strength, Hanser Publishers, New York, 1995, pp. 40–82.

[61] P.G. de Gennes, Scaling Concepts in Polymer Physics, Cornell University Press, 1979.

[62] P.E. Rouse, A theory of linear viscoelastic properties of dilute solutions of coiling polymers, J. Chem. Phys. 21 (1953) 1272.

[63] M. Doi, S.F. Edwards, The Theory of Polymer Dynamics, Oxford University Press, 1986.

[64] M. Doi, Introduction to Polymer Physics, Clarendon Press, Oxford, 1996, p. 97.

[65] S. Wu, Polymer Interface Adhesions, Marcel Deker, New York, 1982, pp. 388–395.

[66] P. Meka, F. Stehling, Heat sealing of semicrystalline polymer films. I. Calculation and measurement of interfacial temperatures: effect of process variables on seal properties, J. Appl. Polym. Sci. 51 (1994) 89–103.

[67] B.A. Morris, Predicting the heat seal of ionomer films, J. Plastic Film. Sheeting 18 (2002) 157–167.

[68] G. Schubert, S. Hüsch, What does hot tack mean? TAPPI Eur. PLACE. Conf. (2011).

[69] S. Bach, Coupled finite-element modelling of heat sealing and ultrasonic sealing processes for multilayer packaging films, 2011 Eur. TAPPI PLACE. Conf. (2011).

[70] R.P. Wool, Polymer Interfaces, Structure and Strength, Hanser Publishers, New York, 1995, p. 286.

[71] V.K. Konaganti, S. Sadeghi, S.K. Goyal, Modeling & simulation of heat sealing process in multilayer polyethylene films, AMI Polyethyl. Films Conf. (2019).

[72] V.K. Konaganti, S. Sadeghi, S.K. Goyal, Thermo-rheological modeling and simulation of heat sealing process for multi-layer flexible packaging applications, SPE ANTEC Conf. (2018).

[73] T. Schuman, E.V. Stepanov, S. Nazarenko, G. Capaccio, A. Hiltner, E. Baer, Interdiffusion of linear and branched polyethylene in microlayers studied via melting behavior, Macromolecules 31 (1998) 4551–4561.

[74] T. Schuman, S. Nazarenko, E.V. Stepanov, S.N. Maganov, A. Hiltner, E. Baer, Solid state structure and melting behavior of interdiffused polyethylenes in microlayers, Polymer 40 (1999) 7373–7385.

[75] R.P. Wool, Polymer Interfaces, Structure and Strength, Hanser Publishers, New York, 1995, p. 225.

[76] N.Z. Qureshi, E.V. Stepanov, G. Capaccio, A. Hiltner, E. Baer, Self-adhesion of polyethylene in the melt. I. Heterogeneous copolymers, Macromolecules 34 (2001) 1358–1364.

[77] F. Scheffold, A. Budkowski, U. Steiner, E. Eiser, J. Klein, Fetters, et al., Surface phase behavior in binary polymer mixtures. II. Surface enrichment from polyolefin blends, J. Chem. Phys. 104 (1996) 8795–8806.

[78] P. Brant, A. Karim, J.F. Douglas, F.S. Bates, Surface composition of amorphous and crystallizable polyethylene blends as measured by static SIMS, Macromolecules 29 (1996) 5628–5634.
[79] N.Z. Qureshi, M. Rogunova, E.V. Stepanov, G. Capaccio, A. Hiltner, E. Baer, Self-adhesion of polyethylene in the melt. 2. Comparison of heterogeneous and homogeneous copolymers, Macromolecules 34 (2001) 3007–3017.
[80] R. Boutrouka, S.H. Tabatabaei, A. Ajji, Enhancement of heat seal properties of polypropylene films by elastomer incorporation, Int. Polym. Process. 33 (3) (2018) 322–326.
[81] R. Boutrouka, S.H. Tabatabaei, A. Ajji, Effect of molecular structure on the heat seal performance of polypropylene films, SPE ANTEC (2014).
[82] M. Bulacu, E. van der Giessen, Effect of bending and torsion rigidity on self-diffusion in polymer melts: a molecular-dynamics study, J. Chem. Phys. 123 (2005) 114901.
[83] R.P. Wool, Polymer Interfaces, Structure and Strength, Hanser Publishers, New York, 1995, pp. 336–339.
[84] N.K. Tierney, R.A. Register, Ion hopping in ethylene-methacrylic acid ionomer melts as probed by rheometry and cation diffusion measurements, Macromolecules 35 (2002) 2358–2364.
[85] R.P. Wool, Polymer Interfaces, Structure and Strength, Hanser Publishers, New York, 1995, p. 395.
[86] F. Bustos, P. Cassagnau, R. Fulchiron, Effect of molecular architecture on quiescent and shear-induced crystallization of polyethylene, J. Polym. Science, Part. B: Polym. Phys. 44 (11) (2006) 1597–1607.
[87] M. Derakhshandeh, S.G. Hatzikiriakos, Flow-induced crystallization of high-density polyethylene: the effects of shear and uniaxial extension, Rheol. Acta 51 (2012) 315–327.
[88] M. Sentmanat, S.G. Hatzikiriakos, Flow induced crystallization of polymers in extension, SPE ANTEC Conf. (2010) 51–55.
[89] T. Takahashi, J. Watanabe, K. Minagawa, K. Koyama, Effect of ionic interaction on elongational viscosity of ethylene-based ionomer melts, Polymer 35 (1994) 5722–5728.
[90] B.A. Morris, J.M. Scherer, Squeeze-out of sealant during hot bar sealing: Modeling insights, SPE ANTEC Conf. (2013).
[91] B.A. Morris, J.M. Scherer, Preventing squeeze-out of sealant during hot bar sealing: modeling and experimental insights, SPE ANTEC Conf. (2014) 1452–1459.
[92] B.A. Morris, J.M. Scherer, Modeling and experimental analysis of squeeze flow of sealant during hot bar sealing and methods of preventing squeeze-out, J. Plastic Film. & Sheeting 32 (2016) 34–55.
[93] O.A. Ezekoye, C.D. Lowman, M.T. Fahey, A.G. Hulme-Lowe, Polymer weld strength predictions using a thermal and polymer chain diffusion analysis, Polym. Engg Sci. 38 (1998) 976–991.
[94] Z.K. Aghkand, E. Jalali Dil, A. Ajji, C. Dubois, Simulation of heat transfer in heat sealing of multilayer polymeric films: effect of process parameters and material properties, Ind. & Eng. Chem. Res. 57 (43) (2018) 14571–14582.
[95] R.D. Moffitt, A mathematical model for the heat sealing of linear, semicrystalline polymers, SPE ANTEC Conf. (2006) 1027–1032.
[96] R.D. Moffitt, Industrial applications of heat sealing modeling, SPE ANTEC Conf. (2007) 2312–2317.
[97] R.D. Moffitt, Heat seal design and optimization, Polym. Prepr. 48 (2007) 735–736.
[98] R.P. Wool, K.M. O'Conner, A theory of crack healing in polymers, J. Appl. Phys. 52 (1981) 5953.
[99] N. Shamsundar, E. Rooz, Numerical methods for moving boundary problems, in: W.J. Minkowycz, et al. (Eds.), Handbook of Numerical Heat Transfer, John Wiley & Sons, New York, 1988.
[100] R.A. Register, Optimizing hot tack in packaging sealing operations, Present. DuPont (1999).
[101] P. Vanhoorne, R. Register, Low shear melt rheology of partially neutralized ethylene methacrylic acid ionomers, Macromolecules 29 (2) (1996) 598–604.
[102] N.K. Tierney, R.A. Register, The role of excess acid groups in the dynamics of ethylene-methacrylic acid ionomer melts, Macromolecules 35 (2002) 6284–6290.
[103] R.A. Register, Letters to J. W. Paul, DuPont, 1996–1997.
[104] R.B. Bird, R.C. Armstrong, O. Hassager, 2nd (Ed.), Dynamics of Polymeric Liquids, Vol 1 Fluid Mechanics, 189, John Wiley & Sons, New York, 1987, p. 20. 539.
[105] J. Stefan, Versuche über die scheinbara adhäsion, Akad. Wiss. Math. Nat. Wien. 69 (1874) 713–735.
[106] J.R. Scott, Theory and application of parallel-plate plastometer, Trans. Inst. Rubber Ind. 7 (1931) 169.
[107] G. Dai, R.B. Bird, Radial flow of Bingham fluid between two fixed circular disks, J. Non-Newton Fluid 8 (1981) 349–355.
[108] T.J. Pratt, D.F. James, The viscous squeeze-film behavior of shear-thinning fluids, J. Non-Newton Fluid Mech. 27 (1988) 27–46.
[109] S.D.R. Wilson, Squeezing flow of a Bingham material, J. Non-Newton Fluid Mech. 47 (1993) 211–219.
[110] J. Engmann, C. Servais, A.S. Burbidge, Squeeze flow theory and applications to rheometry: a review, J. Non-Newton Fluid Mech. 132 (2005) 1–27.
[111] E. Mitsoulis, Flows of viscoelastic materials: models and computations, Rheology Rev. (2007) 135–178.
[112] A. Hayden, A novel peel-seal system, TAPPI PLACE. Conf. (2004).
[113] A.M. Soutar, Polymer combinations for peel/seal applications, 1986 Coextrusion Conf. (1986) 31–43.
[114] C.C. Hwo, Peel seals, in: US Patent 4882229 A, 1989.
[115] C.C. Hwo, Polybutylene blends as easy open seal coats for flexible packaging and lidding, J. Plastic Film. Sheeting 3 (1987) 245–260.
[116] J. De Clippeleir, PE/PB blends for film applications, Specialty Plastic Films '97, Zurich, Switzerland, 1997.

[117] J. De Clippeleir, S. Hirata, I. Roucourt, Easy-opening packaging technology. A technical perspective from Basell Polyolefins, TAPPI PLACE. Conf. (2004).

[118] W.G. Tickner, Heat sealable resin blends, in: US Patent 4189519, 1980.

[119] S.E. Pirtle, L. Mergenhagen, J. Wooster, R. Cotton, The effects of resin selection and extrusion variables on peelable seal performance, in: TAPPI PLACE Conference CD, Paper PLA0468, 2004.

[120] S. Pasquali, Polybutene-1 for peeelable seals, in: J.F. Macnamera Jr. (Ed.), Film extrusion manual, 3rd (ed.), TAPPI Press, Atlanta, 2020, pp. 229–234.

[121] A. Yanidis, Blend of polybutylene and ionomer forming easy-open heatseal, in: US Patent 5547752 A, 1996.

[122] V. Brodie, III, Packaging films capable of being heat-sealed closed and thereafter peeled open, in: US Patent 5891500, 1999.

[123] L.K. Mergenhagen, T.J. Pope, C.A. Timper, Polyolefin plastomer-based peelable seal formulations, in: TAPPI PLC Conference Proceedings, 1996, pp. 63–71.

[124] G.L. Hoh, Blends of ionomer with propylene copolymer and articles, in: U.S. Patent 4539263A, 1985.

[125] G.L. Hoh, Blends of ionomer with propylene copolymer, in: U.S. Patent 4550141A, 1985.

[126] S.O. Cook, Flexible packaging films of high density polyethylene capable of forming easy openable seals, in: U.S. Patent 4188441, 1980.

[127] D. Falla, Peelable seal films with enhanced moisture barrier properties for flexible packaging applications, SPE ANTEC Conf. (2015).

[128] D.A. Busche, D.H. Bostian, Easy open package, in: US Patent 4944409 A, 1990.

[129] R.A. Bourque, et al., Multiple compartment pouch and beverage container with smooth curve frangible seal, in: U.S. Patent 7055683, 2006.

[130] R.A. Bourque, et al., Multiple compartment pouch and beverage container with frangible seal, in: U.S. Patent 7306095 B1, 2007.

[131] M. Nase, A. Zangel, B. Langer, H.J. Baumann, W. Grellmann, P. Poelt, Investigation of the peel behavior of polyethylene/polybutene-1 peel films using in situ peel tests with environmental scanning electron microscopy, Polymer 49 (2008) 5458–5466.

[132] M. Nase, B. Langer, W. Grellmann, Fracture mechanics on polyethylene/polybutene-1 peel films, Polym. Test. 27 (2008) 1017–1025.

[133] T.R. Mueller, Lidding applications for peelable seal roll applied coatings, TAPPI PLACE. Conf. (2004).

[134] V. Cushing, R. Cooper, Primerless heat seal coatings for film substrates, 2008 AIMCAL Fall Conf. (2008).

[135] J. Pascal, Peelable structures for PP, PS, PET and PVC. Understanding how multilayer construction influences seal and peel properties, TAPPI PLACE. Conf. (2008).

[136] DuPont, DuPont™ Appeel® lidding sealant resins. <http://www.dupont.com/content/dam/assets/products-and-services/packaging-materials-solutions/assets/L-14227_Appeel_brochure_dupont_09-03-2011_BD.pdf>, 2008 (accessed 26.07.14).

[137] D.D. Zhang, Peelable seals applications to dissimilar surfaces, TAPPI Polymers, Laminations & Coat. Conf. (1998) 341–350.

[138] T. Kendig, DuPont Appeel® lidding sealant resin, concepts for easy open packaging, Pack. Expo. (2010).

[139] D.J. Anzini, Creating the next generation of flexible recolosable packages, SPE FlexPackCon Conf./AIMCAL Conference (2014).

[140] P.B. Christoff, Reclosable bag, material, and method of and means for making the same, in: US Patent 4,617,683, 1986.

[141] S.J. Straus, E.J. Zuscik, W.B. Bower, A. Yanidis, Resealable packaging material, in: US Patent 5089320, 1992.

[142] L.S. Timm, T.E. Bublitz, Resealable packaging system, in: US Patent 5993962, 1996.

[143] R. Exner, O. Dagestad, Recloseable package, in: US Patent 8596867, 2013.

[144] Y.M. Trouilhet, J. Roulin, Multilayer films for recloseable packaging, in: US Patent 8617677, 2013.

[145] E.L. Haedt, P.J. Sheridan, J.J. Michaels, L.M. Nett, Peelable/resealable packaging film, in: U.S. Patent 7422782 B2, 2008.

[146] T.K. Cruz, Easy-open reclosable films having an interior frangible interface and articles made from them, in: U.S. Patent 8329276 B2, 2011.

[147] S.S. Ranganathan, B.L. Westmoreland, A.W. Moehlenbrock, A.J. Frost, V.A. Shaver, D.C. Karam, Easy-open, reclosable package, in: U.S. Patent 8541081 B1, 2013.

[148] Y. Trouilhet, Recloseable structures produced by co-extrusion coating PSA/ionomer, TAPPI Eur. PLACE. Conf. (2011).

[149] A.W. Moehlenbrock, S.S. Ranganathan, Easy open and reclosable package with discrete laminate with die-cut, in: U.S. Patent 8979370B2, 2015.

[150] A. Bardiot, N. Sajot, P. Gerard, C. Notteau, Hot-extrudable pressure-sensitive hot-melt adhesives and their use in multilayer films, in: US Patent 7622176, 2009.

[151] L. Reade, Sound method: ultrasonic sealing, Film. & Sheet Extrusion (2014)June 2014, pp. 23–27, viewed 02 August 2020 from. Available from: http://www.filmandsheet.com.

[152] S. Bach, N. Bunk, J.-P. Majschak, Effect of laminate structure and morphology on fracture mechanics and seam strength of sealed seams, Eur. TAPPI PLACE. Conf. (2013).

[153] J. Van Dun, S. Bensason, K. Thurling, Polyolefins for ultrasound sealing, Eur. TAPPI PLACE. Conf. (2013).

[154] K.S. Suresh, M. Roopa Rani, K. Prakasan, R. Rudramoorthy, Modeling of temperature distribution in ultrasonic welding of thermoplastics for various joint design, J. Mater. Process. Technol. 186 (2007) 138–146.

[155] M.N. Tolunay, P.R. Dawson, K.K. Wang, Heating and bonding mechanisms in ultrasonic welding of thermoplastics, Polym. Engr Sci. 23 (1983) 726–733.

[156] N. Tateishi, T.H. North, R.T. Woodhams, Ultrasonic welding using tie materials. Part I: analysis of process operation, Polym. Engr. Sci. 32 (1992) 600–611.

[157] S. Bach, K. Thuerling, J.-P. Majschak, Ultrasonic sealing of flexible packaging films - principle and characteristics of an alternative sealing method, Packaging Technol. & Sci. 25 (4) (2012) 233–248.

[158] D.L. Brebner, Troubleshooting Heat Seal Layers of Surlyn® Ionomer Resin. Precautionary Measures in Developing Coextruded Structures With Heat Seal Layers of Surlyn®, DuPont Technical Literature, n.d.
[159] R.T. Van Ness, Heat seal performance. The role of contaminates and processing variables, TAPPI J. 62 (1979) 79–82.
[160] J.F. Macnamera (Ed.), Film Extrusion Manual, third ed., TAPPI Press, Atlanta, GA, 2020.
[161] W. Durling (Ed.), Extrusion Coating Manual, 5th (ed.), TAPPI Press, Atlanta, Ga, 2017.
[162] M. Spalding, A. Chatterjee (Eds.), Handbook of Industrial Polyethylene and Technology, Wiley – Scrivener Publishing, Hoboken, NJ, 2017.
[163] J. Degroot, J. Arionus, Polyolefin plastomers, in: J.F. Macnamera Jr. (Ed.), Film extrusion manual, 3rd (ed.), TAPPI Press, Atlanta, 2020, pp. 215–222.
[164] B.A. Morris, J.D. Vansant, The influence of sealant modulus on the bending stiffness of multilayer films, TAPPI Polymers, Laminations Coat. Conf. (1997) 165.
[165] B.A. Morris, Optimizing flexible packaging with unique tools from DuPont, PackExpo (2011).

Further reading

DuPont, 2016 DuPont, Checklist From Ethylene Copolymer Training, 2016.

Chapter 8

Barrier of flexible packaging films

A critical function of packaging is to protect the product. This often requires the inclusion of special barrier layers or coatings within a multilayer structure to manage gas and liquid permeation, provide oil and grease resistance, prevent loss or impartation of organic molecules affecting taste and aroma, and block light. This chapter examines the importance of barrier packaging, especially with respect to food packaging, describes methods for measuring barrier performance, and provides typical permeation properties of commonly used barrier materials. The science of permeation is reviewed with an emphasis on factors that control permeation in flexible packaging. The chapter concludes with a critical assessment of four emerging technologies: oxygen scavenging, layer multiplier technology, barrier nanocomposites, and advanced coatings.

8.1 Why it is important

As described in Chapter 1, protecting the product is a fundamental function of a package. Packaging plays an important role in preserving the freshness of food by helping manage:

- Oxygen ingress;
- Moisture (ingress or egress);
- Carbon dioxide (for modified atmosphere packaging);
- Light;
- Loss of flavorants, nutrients and oils from the product to the package (known as scalping) or external environment;
- Loss of aromas from the product;
- Impartation of off-flavors, aromas, oils or deleterious compounds to the food from the package or external environment.

In pharmaceutical packaging, protecting active ingredients from moisture has been the primary barrier needed, although oxygen barrier is increasingly becoming important for biological medicines.

The importance of barriers to food freshness is covered in detail in the food chemistry literature [1–4]. Only a brief overview is given here along with a summary of some of the barrier needs. There are many factors that influence the shelf-life of food. Some of these are inherent to the food itself (intrinsic factors) and others relate to the environment (extrinsic factors) [1]. Table 8.1 provides a list of many of these factors. Packaging is designed to mitigate many of the extrinsic factors.

Interactions between the intrinsic and extrinsic factors increase the likelihood of adverse reactions that limit shelf life. These include chemical/biochemical, microbiological, and physical interactions.

One of the most important chemical interactions that affects food quality is oxidation. Foods high in fats, especially unsaturated fats, are susceptible to oxidation and rancidity. There are three chemical routes to oxidation of fatty acids [1]

1. Free radical oxidation in the presence of metal ion catalysts, heat or light;
2. Photo-oxidation where photosensitizers such as chlorophyll or myoglobin affect the energetic state of oxygen;
3. Enzymic oxidation catalyzed by lipoxygenase.

Once oxidation has been initiated by any of these routes, the reaction proceeds along the same lines: hydroperoxide intermediates are formed, which then react to form aldehydes, ketones and alcohols which give the off-flavor or odor associated with stale, rancid products.

The role of light, especially UV light found in fluorescent lighting in many retail stores, in the oxidation of foods has been well studied. Dairy products such as milk [5–13], cheese [14–17], yogurt [18] and sour cream [19], various oils [20–27], pasta [28], cookies [29], meat [30] and juices [31–33] are susceptible. These studies typically combine

The Science and Technology of Flexible Packaging. DOI: https://doi.org/10.1016/B978-0-323-85435-1.00013-2

TABLE 8.1 Factors affecting food quality.

Intrinsic factors	Extrinsic factors
• Water activity (water available) • pH, type of acid • Microbes present in the product • Availability of oxygen • Redox potential • Natural chemistry of the product • Added preservatives • Product formulation • Packaging interactions	• Time-temperature history during processing • Temperature, RH and light exposure during storage and distribution • Gas composition inside package • Consumer handling

Source: Adapted from H. Brown, J. Williams, Packaged product quality and shelf life, in: Coles, R., McDowell, D. and Kirwan, M. J. (eds.), Food Packaging Technology, Blackwell Publishing, CRC Press, Boca Raton, FL, 2003, pp. 65–95.

sensory panel and analytical testing to ascertain degradation mechanisms that result in off-flavor, color formation and loss of food quality. Riboflavin (vitamin B2) is particularly prone to photo-oxidation and often contributes to the degradation of other components [34–37]. Korman et al. [35] use it as a model compound to study the ability of different packaging structures to block UV light and slow degradation. Other investigations compare the effectiveness of different package forms using actual food products [8,10,11,14,18,24,25,29]. Generally, the more light the package blocks, the longer is the shelf life, as determined by sensory panel or color change. Lennersten and Lingnert [24] determine the specific wavelength of light that is most deleterious to color change of mayonnaise. Since different polymers absorb light at different wavelengths, such studies may lead to better packaging structures, although none of the materials used by Lennersten completely prevents color change.

Successful strategies for limiting oxidation include lowering the oxygen level to less than 1%, limiting light exposure and adding antioxidants. Packaging that provides oxygen barrier or scavenging, light barrier and enables modified atmosphere packaging (MAP) is used. Examples include:

- Using nitrogen flushed packaging for nuts to extend the time until the onset of rancidity.
- Providing light barrier for packaging of dairy products, oily food and juices, as described above.
- Vacuum packaging of fatty fish,
- MAP for fresh red meat to extend shelf life over simple plastic overwrap. MAP preserves the red color of meat longer. Vacuum packaging with a barrier film can be used for even greater shelf life but turns the meat purple. This low oxygen state is the freshest state of a complex oxidation process of meat known as the myoglobin cycle. Consumer perception that red meat is freshest, however, has been difficult to change.

Fresh fruits and vegetables are living and respiring. Respiration can lead to a depletion of reserves, metabolic collapse and disruption of tissue. This can be slowed by lowering the temperature and by modifying the atmosphere inside the package to extend shelf life. Optimal gas environments specific to each fruit or vegetable have been determined and can be found in the literature [3,38]. Ethylene scavengers are also sometimes used to slow the ripening process.

Microbial growth in food can consume nutrients, give off by-products such as gases or acids, and release enzymes. These may affect the texture, flavor, odor and appearance of the food. Some enzymes persist after the death of the microorganism and contribute to spoilage. Deleterious pathogens such as *Listeria monocytogenes* and *Clostridium botulinum* can cause food poisoning. Under ideal conditions bacteria populations double every 20 minutes and can reach $10^6/cm^2$ in 8 hours. The minimum number of microorganisms before food quality is compromised depends on the intrinsic and extrinsic factors. These need to be carefully controlled to limit or prevent the impact of microorganisms. Counts of $10^6-10^8/cm^2$ microorganisms per gm or cm^2 are often used as the upper limit, and 10^5 when dealing with pathogenic bacteria [2].

Packaging technology can be used to control or kill microbes. Examples include retort or in-package pasteurization to sterilize and MAP and vacuum packaging to exclude oxygen and/or introduce CO_2 which can retard the growth of some microbes. Packaging structures that enable these processes contribute to the shelf life of the product.

Considerable research has been conducted on antimicrobial packaging [38–46] but with few commercial adoptions. One challenge is that it generally requires the package to be in direct contact with the food to be effective, or that the active ingredient be volatile. Other challenges include the need for the active ingredient to be able to withstand

TABLE 8.2 Barrier needs by food type.

			Optimum barrier properties				
Food	**Typical time/temp.**	**Required gas composition**	O_2	CO_2	H_2O	**Light**	**Other**
Red meats	0°C–5°C, 6–14 d	High oxygen (70%–80%), high carbon dioxide (30%–20%)	High	High	High		
	0°C–5°C, 1–4 d	No controlled atmosphere	Low		High		
Other meats	0°C–5°C, 1 d-6 w	Low oxygen, high carbon dioxide	High	High	High		
Cured meat products	5°C, 4 w	No oxygen	High	High	High	High	
Cured fish products	5°C, 4 w	High carbon dioxide	High	High	High	High	
Fish, high fat	0°C–5°C, 1–7 d	Carbon dioxide (40%), oxygen (30%), nitrogen (30%)	High	High	Low		Odors
Fish, low fat	0°C–5°C, 1–7 d	Carbon dioxide (40%–60%), nitrogen (60%–40%)	High	High	Low		Odors
Eggs	2°C–12°C, 25 d-4 w						Aroma
Fluid milk	2°C–5°C, 2 d-8 m		High		High	High	Aroma
Fermented milk	2°C–5°C, 16–18 d		High	High	High	High	Aroma
Fresh cheese	<5°C, 1–8 w		High	High	High	High	
Semi soft and hard cheese	<5°C, 1–18 w		High	Low	High	Low or High	
Mold ripened cheese	<5°C, 1–8 w		High	High	High	High	
Butter, oil, dairy spreads, etc.	<5°C, 6 w or more		High		High	High	Odors, grease
Pasta (fresh)	2°C–5°C, 4 w	Low oxygen	High	High	High	High	
Fruits and Vegetables							
Root crops	0°C–25°C, from 1 w	No oxygen, no carbon dioxide	High	High	High	High	Odors
Most vegetables	0°C–25°C, from 1 w	Oxygen (1%–5%), no carbon dioxide	High	High	High	High	Odors
Most fruits and some vegetables	0°C–18°C, from 1 w	Oxygen (1%–5%), carbon dioxide (0%–5%)	High	High	High	High	Odors
Dry products							
Flour/grains	2°C – room temp., 3 y				High		
Powders, high fat	Room temp., 1 y	Low oxygen	High		High	High	
Powders, low fat	Room temp. >1 y	Low oxygen	High		High	High	
Breakfast cereals	Room temp, 1 y	Atmospheric/ low oxygen	High	Low	High		
Pastas (dried)	Room temp, 1–6 m				High		
Spices and herbs	Room temp, from 6 m		High		High	High	

(*Continued*)

TABLE 8.2 (Continued)

			Optimum barrier properties				
Food	**Typical time/temp.**	**Required gas composition**	O_2	CO_2	H_2O	**Light**	**Other**
Snack foods	Room temp., 1 y	Low oxygen	High		High	High	
Coffees/teas	5°C-Room temp., 1 y	Low oxygen	High	Low	High	High	
Breads	5°C-Room temp., 1 d-12 w	Low oxygen, high carbon dioxide	High	High	High		
Cakes	5°C-Room temp., 1 d-4 m	Atmospheric/ low oxygen, high carbon dioxide	High	High	High	High	
Crackers and cookies	Room temp, 4 m				High		
Chocolates	2°C-room temp, 6–12 m					High	
Beverages							
Water	Room temp, 1 y				High		Aroma
Juice	5°C-room temp, 5 d-1 y	Atmospheric, low oxygen	High		High	High	
Carbonated drinks (beer & soft drinks)	5°C-room temp, 6 m-1 y	High carbon dioxide, low oxygen	High	High	High	High	
Frozen foods							
High fat products	−18°C, 1 y	Low oxygen	High		High	High	
Low fat products	−18°C, 2 y	Low oxygen	High		High	High	
Others							
High fat products (dressing, sauce, etc.)	5°C-room temp., 1 y	Low oxygen	High		High	High	Aroma, grease
Ready meals	2°C–5°C, 3–21 d	Low oxygen, high carbon dioxide	High	High	High		

Source: Adapted from K. Petersen, P. V. Nielsen, G. Bertelsen, M. Lawther, M. B. Olsen, N. H. Nilsson, et al., Potential of biobased materials for food packaging, Trends Food Sci. Technol. 10 (1999) 52–68.

extrusion processing if it is incorporated into the package film (rather than as a coating or in a separate sachet insert), the migration rate of the active ingredient to the food (and matching this rate to the rate of microbial growth), possible interactions of the food product with the active ingredient that may change its antimicrobial efficacy, changes in physical or optical properties of the film, and regulatory hurdles. Beyond the use of silver zeolite masterbatches, very little commercial adoptions have taken place. (See Section 16.2.3 for more details on active packaging.)

Physical processes that limit shelf life include material damage during storage, transportation and distribution; insect infestation; and moisture effects. Good package design can limit each of these:

- Vibration, handling and compressive loads during stacking may cause packages to burst or open. The structural integrity of the package is important.
- Insect penetrators bore holes into the package; having a good aroma barrier can help prevent this. Insect invaders enter through holes already present, such as in poor seals. Sealant selection can help mitigate this.
- Moisture ingress can affect the texture of dry foods such as cereals or crackers. Moisture loss can affect food quality; examples are changes in the color of fresh red meat and dehydration and freezer burn in frozen foods. Moisture loss also can result in serious economic consequences for products sold on a weight basis. Proper choice of packaging materials helps control moisture.

TABLE 8.3 Degree of protection required by various foods assuming 1-year self-life at 25°C.

Food/beverage	Max O_2 gain (ppm)	Other gas protection needed?	Max water gain or loss	High oil resistance needed?	Good barrier to volatile organics needed?
Canned milk and flesh foods	1–5	No	3% loss	Yes	No
Baby foods	1–5	No	3% loss	Yes	Yes
Beers and wine	1–5	$<$20% CO_2 or SO_2 loss	3% loss	No	Yes
Instant coffee	1–5	No	2% gain	Yes	Yes
Canned soups, vegetables and sauces	1–5	No	3% loss	No	No
Canned fruits	5–15	No	3% loss	No	Yes
Nuts, snacks	5–15	No	5% gain	Yes	No
Dried foods	5–15	No	1% gain	No	No
Fruit juices and drinks	10–40	No	3% loss	No	Yes
Carbonated soft drinks	10–40	$<$ 20% CO_2 loss	3% loss	No	Yes
Oils and shortenings	50–200	No	10% gain	Yes	No
Salad dressings	50–200	No	10% gain	Yes	Yes
Jams, jellies, syrups, pickles, olives, vinegars	50–200	No	10% gain	Yes	No
Liquors	50–200	No	3% loss	No	Yes
Condiments	50–200	No	1% gain	No	Yes
Peanut butter	50–200	No	10% gain	Yes	No

Source: Adapted from G.L. Robertson, Shelf life of foods, in: Food Packaging. Principles and Practice, third ed., CRC Press, Boca Raton, FL, 2013, pp. 329–366.

The migration of molecules from the outside environment or from the package itself into the food can result in poor food quality. Similarly, the migration of components of the food into the package, known as scalping, can affect flavor or nutrient value. These molecules are typically larger than gas molecules and their migration in plastic packaging has not been well studied. Complex interactions may occur between the ingredients and the package, resulting in swelling or plasticization of the polymers that make up the package. This in turn may affect migration rates of the ingredients or other components in the food.

The barrier needs of several food types are summarized in Tables 8.2 and 8.3. These are based on data compiled by Petersen et al. [47] and Robertson [3]. They are presented merely as starting points and derive from work in the 1970s by Salame [48]. Changes in food formulation, preparation, and packaging technology influence shelf-life. The use of MAP, for example, has grown. Robertson [2,3] reviews accelerated shelf-life testing, modeling and other tools for helping determine barrier needs.

8.2 How to measure

8.2.1 Definitions

A permeant is a gas or liquid molecule that migrates through the package film. Permeation is the amount of permeant that is transported through a unit area of film in a given time. The permeation coefficient, P, is normalized to a constant thickness and a given temperature and pressure. Typically, the pressure and temperature are Standard Temperature and

TABLE 8.4 Conversion factors between some units used to report permeation of gases.

Unit	Multiply by this factor to get cm^3-mil/m^2-d-atm
cm^3-mil/m^2-d-atm	1
mL-mil/m^2-d-atm	1
cm^3-25μm/m^2-d-atm	1
cm^3-mil/100 in^2-d-atm	15.5
Barrier [10^{-10}mL-cm/cm^2-s-(cm Hg)]	2587
mL-μm/m^2-d-atm	0.0394
m^3-μm/m^2-d-atm	39370
in^3-mil/100 in^2-d-atm	254.2
mL-mm/m^2-d-atm	39.4
mL-mm/cm^2-d-atm	3.94E + 05
mL-mm/cm^2-s-atm	3.40E + 10
mL-mm/cm^2-d-Pa	3.99E + 10
mL-mm/m^2-d-kPa	3989
m^3-μm/m^2-d-kPa	3989173
mL-μm/m^2-d-kPa	3.99

Pressure, STP, which unfortunately varies depending on the standard being considered: for example, IUPAC uses 0°C, 100 kPa and NIST uses 20°C, 101.325 kPa.

For gases, P has the following units:

$$P = \frac{\text{(Quantity of permeant) (Film thickness)}}{\text{(Area) (Time) (Pressure drop across film)}} \tag{8.1}$$

The quantity of the permeant may be in volume, mass or molar units. There are many different units used in the scientific literature for P. For barrier packaging, the most common units for oxygen permeation are cm^3-mil/(100 in^2-day-atm) and cm^3-25-μm/(m^2-day-atm). Some conversion factors are given in Table 8.4. Note that the film thickness is in the numerator. A common mistake in converting units is to divide by the film thickness instead of multiplying. For example, suppose one measures the oxygen permeation of a 2-mil film to be 2 cm^3/(m^2-day-atm). To put this into standard units for P, the value needs to be normalized to a 1-mil film. Rather than dividing by 2, one must multiply by 2; $P = 4$ cm^3-mil/(m^2-day-atm). A way to remember this is to recognize that a 1-mil film should let more gas permeate through than a 2-mil film; hence the value for P should be higher than the measured results for the 2-mil film.

For oxygen permeation, P is typically called OPV or oxygen permeation value.

The transmission rate, TR, is defined as the amount of permeant that is transported through a film of unit area per time:

$$\text{TR} = \frac{\text{(Volume or mass)}}{\text{(Area) (Time)}} \tag{8.2}$$

For different gases, TR may take on a specific descriptor such as oxygen transmission rate (OTR). Other acronyms in use include O2TR and OGTR (oxygen gas transmission rate). The transmission rate is a property of a package and permeant under the specific conditions of the test (pressure, temperature, thickness, package structure, RH, etc.) These should be specified when reporting data. An equation for calculating TR for a multilayer film from individual layer permeation values and thicknesses is given in Section 8.4.

Another important parameter governing gas or liquid diffusion and permeability is the solubility parameter, S, which relates the concentration of a gas to its partial pressure through Henry's law:

$$c = Sp \tag{8.3}$$

where c = concentration of permeant, S = solubility coefficient, and p = partial pressure of the permeant.

The diffusion coefficient, D, relates the flux of the permeant, J, to the concentration gradient. This relationship is known as Fick's first law:

$$J = -D\frac{dc}{dx} \tag{8.4}$$

where J = flux of permeant (amount of permeant transporting through a unit area of film per time), D = diffusion coefficient, c = concentration of permeant (amount per unit volume), and x = dimension normal to the surface of the film (thickness direction).

Here dc/dx is the concentration gradient or the difference in concentration across the film of thickness X:

$$\frac{dc}{dx} = \frac{c_{\text{outside}} - c_{\text{inside}}}{X} \tag{8.5}$$

where c_{outside} = concentration of permeant on the outside of the film, c_{inside} = concentration of permeant on inside of the film, X = thickness of the film.

The value of D relates to how fast the permeant will diffuse through the film at a given concentration gradient. Higher values of D indicate faster diffusion.

Section 8.4 shows that P is the product of D and S ($P = DS$) and is a steady-state value. The importance of these parameters is explored in Section 8.4. This section describes how to measure P, the parameter most relevant for determining barrier performance in flexible packaging. Once P is known for each layer in a multilayer package, the transmission rate can be calculated. The transmission rate multiplied by time yields the amount of permeant that will be transported into the package over that time, which can be related to shelf-life performance.

8.2.2 Gas permeation

Several test standards have been developed for measuring gas permeation through films which are summarized in Table 8.5. Robertson [49] outlines four general methods for measuring gas permeability:

1. **Pressure increase:** As described in ASTM D1434, a flat film is clamped to a chamber with the test gas introduced on one side and the pressure on the other side measured with a mercury manometer. The pressure drop is related to the transmission rate.
2. **Volume increase:** Also described in ASTM D1434, the change in volume at constant pressure is measured. Although not used as much as other methods, rapid measurements can be made for high transmission rate films.
3. **Concentration increase:** ASTM D3985, F1307, and F1927 describe methods to measure the transmission rate by maintaining constant total pressure on both sides of a film clamped in a cell. On one side, partial pressure of the permeant gas is swept continuously across the film while an inert gas is maintained on the other side into which the permeant gas diffuses. The concentration of the permeant in the inert gas is measured by one of several techniques including chemical analysis, gas chromatography (GC), thermal conductivity or special electrodes. The relative humidity can be controlled. This method is probably the most commonly used method today given its speed and accuracy. Several companies sell test equipment including Ametek Mocon, Labthink International Inc. and OxySense.
4. **Detector film:** A detector film is sealed between two pieces of test film. The detector film is coated with a reagent that changes its absorption spectra as the gas is absorbed. The transmission rate is measured using a photospectrometer. Serna et al. [50] describe a method using a platinum-based sensor film inserted inside a sealed package. An external sensor measures the fluorescence decay that is related to OTR. The method is based on techniques for measuring headspace composition in MAP (ASTM F3136). The fluorescence decay method is shown to have good agreement with the concentration increase technique for low to moderate barrier films; high barrier packaging is not amenable to this technique. The technology is offered by Ametek Mocon.

OPV results are typically reported in units of cm^3-mil/(100 in^2-day-atm) or cm^3-mil/(m^2-day-atm). Here the atm in the denominator is the total pressure drop across the film. For food packaging, air is on the outside of the package.

TABLE 8.5 Summary of standard methods for measuring gas permeation [51].

Method	Title	What is measured?
Oxygen Permeation		
ASTM D3985	Method for determination of oxygen transmission rate	Measurement of O_2 transmission through films, sheets, laminates, etc.
ASTM F1307	Method for determination of oxygen transmission rate through dry packages using a coulometric sensor	Measurement of steady-state rate of transmission of oxygen gas into packages
ASTM F1927	Method for determination of O2GTR, permeability, and permeance at controlled relative humidity through barrier materials using ac coulometric detector	Determination of the rate of transmission of O_2 gas, at steady state, at a given temperature, and %RH level, through film, sheeting, laminates, etc.
JIS K 7126	Method for determining the gas transmission rate of any plastic material in the form of film, sheeting, laminate, etc.	Determination of the gas transmission rate of any plastic material in the form of film, sheeting, laminate, etc.
DIN-53380 Part 3	Testing of plastics—Determination of gas transmissions rate	Oxygen-specific carrier gas method for testing of plastic films and plastics moldings
ISO-14663−2	Determination of steady-state rate of transmission of oxygen	Determination of steady-state rate of transmission of oxygen gas through ethylene/vinyl alcohol copolymer in the form of film
Water Vapor		
ASTM F1249	Test method for determination of O2GTR, permeability, and permeance at controlled RH through barrier materials using a coulometric detector	Obtain reliable values for the water vapor transmission rate (WVTR) of plastic film and sheeting. The WVTR is an important property of packaging materials and can be directly related to shelf life and packaged product stability
ASTM D6701	Test method for determining WVTRs through nonwoven and plastic barriers (withdrawn)	Covers a procedure for determining the rate of water vapor transmission (WVT) ranging between 500 and 100,000 g/m^2/day through nonwoven and plastic barrier materials. The method is applicable to films, barriers consisting of single or multilayer synthetic or natural polymers, nonwoven fabric, and nonwoven fabrics coated with films up to 3 mm (0.1 in.) in thickness
ASTM E398	Test method for determination of O2GTR, permeability, and permeance at controlled RH through barrier materials using a coulometric detector	Covers a procedure for determining the rate of WVT ranging between 500 and 100,000 g/m^2/day through nonwoven and plastic barrier materials. The method is applicable to films, barriers consisting of single or multilayer synthetic or natural polymers, nonwoven fabric, and nonwoven fabrics coated with films up to 3 mm (0.1 in.) in thickness
JIS K-7129	Plastics—film and sheeting—determination of WVTR by instrumental method	Specifies an instrumental method for determining the WVTR of plastic film, plastic sheeting, and multilayer structures including plastics that have smooth surfaces without any embossed portions, using a humidity detection sensor method, infrared detection sensor method, and gas chromatographic sensor method
ISO-15106−2	Plastics—film and sheeting—determination of WVTR—Part2: infrared detection sensor method	Specifies an instrumental method for determining the WVTR of plastic film, plastic sheeting, and multilayer structures including plastics, using an infrared detection sensor. The method provides rapid measurement over a wide range of WVTRs

(Continued)

TABLE 8.5 (Continued)

Method	Title	What is measured?
TAPPI T-557	WVTR through plastic film and sheeting using a modulated infrared sensor	Covers a procedure for determining the rate of WVT through flexible barrier materials. The method is applicable to sheets and films up to 2.54 mm (0.1 in.) in thickness, consisting of single or multilayer synthetic or natural polymers and foils, including coated materials. It provides for the determination of (1) WVTR, (2) the permeance of the film to water vapor, and (3) for homogeneous materials, water vapor permeability coefficient
Carbon Dioxide		
ASTM F2476	Method for the determination of CO2TR through barrier materials using an infrared detector	Determination of the steady-state rate of transmission of carbon dioxide gas through plastics in the form of film, sheeting, laminates, coextrusions, or plastic-coated papers or fabrics. It provides for the determination of (1) CO2TR, (2) the permeation of the film to carbon dioxide gas (PCO2), and (3) carbon dioxide permeability coefficient (PvCO2) in the case of homogeneous materials.

Source: Reproduced with permission from M. Fereydoon, S. Ebnesajjad, Development of high-barrier film for food packaging, in: Ebnesajjad, S. (ed.), Plastic Films in Food Packaging, Elsevier, New York, 2013, pp. 71–92.

Since air only contains 21% oxygen, the OPV should be multiplied by 0.21 to obtain the oxygen transmission rate for the package. To avoid confusion, once this factor has been applied, the "atm" should be dropped from the units.

8.2.3 Moisture vapor transmission rate

ASTM E96 describes what is often called the cup method. A desiccant is placed in a standard cup and the flat film is sealed on top with wax. The container is placed in a controlled environment (typically 25°C, 75% RH or 38°C, 90% or 100% RH) and the weight gain is tracked over time. MVTR is equal to the slope of the weight vs. time curve divided by the area of the film. While simple and inexpensive, this technique is labor intensive, takes several days to complete and cannot handle films with very low transmission rates. The partial pressure inside the cup may also change as the desiccant absorbs water. This is particularly true of silica gel [49].

An alternative is to use the desiccant weight gain method on the actual package, which may provide results that correlate better to a packaging application. This method is described in ASTM D3079.

More rapid methods, such as ASTM F1249, have been developed using infrared detectors to measure changes in RH on the dry side of a film located in a test cell. The film is positioned in the cell so that it divides the cell into two chambers, top and bottom. Water is placed in the bottom chamber to create a 100% RH atmosphere. The top side of the cell is swept with dry air and the change in RH is measured with the IR detector. One disadvantage of this method is that the RH is limited to 100% unless special equipment is used that allows for control of RH.

8.2.4 Permeation of organic molecules

Relatively little data has been published in the scientific literature for the permeation of organic molecules through plastic films. Organic molecules such as favor or aroma molecules are large and complex. Their permeation often depends on their concentration and the presence and concentration of other permeants. Most gases are noninteracting with the polymer and the diffusion coefficient is independent of concentration. This allows relatively simple Fickian diffusion equations to be used to interpret and predict behavior. Experimental results with pure gases can be extrapolated to low concentrations typical of real packages. For organic molecules, diffusion is often non-Fickian. Interactions between the permeant and the polymer may swell the polymer, creating more free volume and faster diffusion. The permeant may also plasticize the polymer, allowing polymer chains to move and more readily open temporary holes for the transport of the permeant molecules. Even if the permeant does not cause swelling or plasticization, other organic molecules may be present that do, allowing the permeant to diffuse more rapidly.

Organic molecule permeation experiments are often done with pure components to facilitate detection of the permeant. This has several drawbacks. One is that the presence of co-permeants may open up free volume through swelling or plasticization effects that would not be accounted for in an experiment looking at pure components one at a time. Secondly, food quality frequently suffers from just a small loss or gain of a flavorant molecule. Extrapolating from a high concentration experiment to a low concentration is problematic with non-Fickian diffusion since the diffusion coefficient is a function of concentration. Finally, odor/taste issues are often very difficult to attribute to a specific molecule, and if one is implicated it often is at very low concentrations compared to other molecules that may be present. Headspace GC or other methods may reveal a plethora of candidate species that are present and the temptation is to pick a molecule that is present in abundance. Designing packaging film structures to minimize the permeation of that molecule may be counterproductive since the factors that influence its permeation rate may be totally different from the actual culprit molecule.

There is no universal method for measuring the permeation of organic molecules. Methods typically involve one of the following, often with pure compounds with the limitations described above [52]

- Measuring sorption into the polymer by following weight gain with time.
- Determining the amount of permeant released from the polymer once it has been saturated.
- Measuring the loss of permeant molecules from the aroma/flavor source.

Vaha-Nissi et al. [52] use a two-cell device with automatic sampling for GC analysis. They use a dilute blend of four aroma molecules to circumvent the concentration and co-permeant issues. Stevens [53] describes work at Mocon to measure aroma/flavor barrier. They create a three to five component aromo/flavor molecule "soup" using multidimensional GC. They test the permeation of the soup through the test material using an isostatic cell to measure individual and total permeation rates. They also couple human olefactory and analytical techniques to identify the key compounds to study. Tou et al. [54] describe a permeation cell with a mass spectrometer capable of parts per billion sensitivity to gaseous organic molecules which they use to measure the permeability of several flavorant molecules through two commercial films, a polyethylene film and a plasticized polyvinyl chloride film.

ASTM F119-82 describes a method for measuring the penetration rate of grease or oil through a film. The film is placed on top of ground or frosted glass. A weighted oil-soaked cloth is placed on top of the film. When the oil penetrates through the film, the glass becomes transparent as the oil wets the surface. The time for the glass to become transparent is recorded. Often these tests are run at elevated temperatures to accelerate the time for penetration. This test method is often referred to as the ground glass method.

Falla et al. [55] develop a grease penetration test modifying a method proposed by Wyser et al. [56,57]. The test film is placed on top of a fluorescent thin layer chromatography (TLC) plate. A stainless steel ring is place on top. The grease is placed inside and a metal weight/plug (2 kg) is placed on top. The assembly is put into an oven for accelerated testing. In their experiments, they used 40°C, 50°C, 60°C, and 70°C. After 48 hours the film and weight are removed, and the plate is photographed in a viewing box with UV light. The grease absorbs light at 254 nm and appears as dark regions on the plate in the photograph. Imaging software is used to convert the dark regions into gray scale, with 0 being no grease and 100 total coverage. Typical values were between 0 and 60.

Care should be taken to ensure the oil does not oxidize or otherwise change at the test temperature. Falla et al. used lard that does not require refrigeration. Also, the temperature should be chosen to either represent the end-use or be below any thermal transition (such as melting) that the polymer undergoes that would affect the correlation of the results with the actual use temperature. Falla ran many of their studies at 70°C, which was above the temperature at which some of the polymers they were testing started to melt. This may have affected their results (see Section 8.3.3).

8.3 Typical permeation values

8.3.1 Gas barrier

Typical oxygen permeation values for high barrier polymers are given in Table 8.6 and oxygen transmission rates for substrates in Table 8.7. Of the high oxygen barrier polymers, the most commonly used resins are ethylene vinyl alcohol (EVOH), polyvinylidene chloride (PVDC) and polyamide 6 (PA 6). These resins are described in Chapter 4. Here a few of the key attributes of these resins for barrier applications are highlighted.

TABLE 8.6 Typical oxygen permeation values and moisture vapor transmission rate values for commonly used polymers in flexible packaging.

Resin	OPV 23°C/0%RH cc-mil/m^2-d-atm	MVTR 38°C/90%RH g-mil/m^2-d	References
LDPE/LLDPE	6500–7800	12–19	[49,58]
HDPE	700–2300	1.5–12	[49,58]
EAA	3100–7750	15.5–25	[58]
EMAA	3100–7750	15.5–25	[58]
Ionomer	3100–7750	15.5–25	[58]
EPE	7750–12,400	14–19	[58]
EVA	9300–15,500	15.5–85	[58]
EMA	9300–15,500	15.5–139	[58]
ULDPE	9300–14,725	19–23	[58]
POP/POE	9300–31,000	20–31/31–46	[58]
COC	1000	1–2	[59]
PP	2300–6000	8–12	[49,58]
PVDC	1.2–14	0.6–3.2	[49,58]
PET	55–70	20–50	[49,58]
PA6	40–60	155–310	[49,58]
aPA (6I/6 T)	54	22	[60,61]
MXD6	2.5–8	50	[61,62]
EVOH (32 mol% ethylene)	0.3	60	[49,58,63]
EVOH (44 mol% ethylene)	1.2	23	[49,58,63]
Polyacrylonitrile	12	95	[61]
PLA	465	620.00	[58]
PS	5425	109–155	[58]
PB -1	5970	15.5–19	[58]

COC, Cyclic olefin copolymers; *EAA*, ethylene acrylic acid; *EMA*, ethylene methyl acrylate; *EMAA*, ethylene methacrylic acid; *EPE*, enhanced polyethylene; *EVA*, ethylene vinyl acetate; *EVOH*, ethylene vinyl alcohol; *HDPE*, high-density polyethylene; *LDPE/LLDPE*, low-density polyethylene/linear low-density polyethylene; *PA*, polyamide; *PB-1*, polybutene-1; *PET*, polyethylene terephthalate; *PLA*, polylactic acid; *POP/POE*, polyolefin plastomer/polyolefin elastomer, also known as mPE or metallocene plastomer; *PS*, polystyrene; *PVDC*, polyvinylidene chloride; *ULDPE*, ultralow-density polyethylene.

8.3.1.1 Ethylene vinyl alcohol

- EVOH comes in a variety of compositions, usually designated by the mol% ethylene in the polymer backbone, ranging from about 27 to 48 mol%. The remainder of the polymer chain contains vinyl alcohol groups which are responsible for its oxygen barrier performance. The lower the % ethylene, the lower the OPV. The limiting case of 0% ethylene is polyvinyl alcohol (PVOH) which has an excellent oxygen barrier but is not melt processable.
- The oxygen permeation of EVOH is sensitive to relative humidity. As the RH increases above 60%, the OPV increases substantially. The OPV of EVOH with lower mol% ethylene changes more rapidly as RH is increased. This is shown in Fig. 8.1.
- EVOH is a transparent, melt-processable resin. The processability improves with increasing mol% ethylene although the barrier performance decreases.
- EVOH is stiff, prone to flex cracking and difficult to thermoform. It is often sandwiched between layers of polyolefin or PA to improve performance.
- EVOH is used in coextruded barrier films for a wide variety of applications. Some example structures are given in Appendix B.

TABLE 8.7 Typical oxygen transmission rate and moisture vapor transmission rate values for barrier substrates.

Substrate	OTR 23 °C/0%RH cc/m^2-d-atm	MVTR 38° C/90%RH g/m^2-d	References
OPET	55–70	20	[49,63]
OPET, PVDC coated	3–8	8.5	[49,63]
met-OPET	0.65	0.5–1	[49,63]
PP cast	3000–4200	8–12	[49,63]
BOPP	1550–1700	5	[49,64]
met-BOPP	25	<1	[40,48]
PCTF - Aclar	5.4	0.24	[65]
Cellulose (cellophane), PVDC coated	8	6–9	[49,63]

Film thickness is 25 μm (1-mil).
BOPP, Biaxial orientated polypropylene; *met*, metalized; *OPET*, biaxial oriented polyester; *PCTFE*, polychlorotrifluoroethylene; PP, polypropylene; *PVDC*, polyvinylidene chloride.

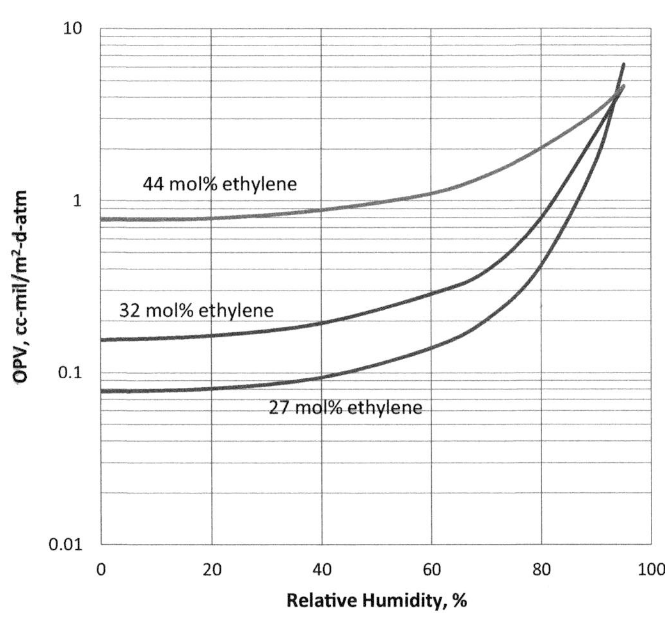

FIGURE 8.1 OPV of EVOH as a function of relative humidity. *Data from Kuraray, Eval EVOH properties, https://eval.kuraray.com/products-services/about-eval-evoh/properties/, n.d. (accesed 06.03.22). http://www.eval.eu/media/36916/ea%20-%20technical%20bulletin%20no%20110.pdf.*

8.3.1.2 Polyvinylidene chloride

- PVDC is used as a barrier coating or as a coextruded resin.
- PVDC is challenging to melt process since its decomposition temperature is close to its melting point. Temperature isolation and other equipment technologies have been developed to help process it.
- PVDC is one of the few barrier polymers that are not sensitive to RH. In addition to an excellent oxygen barrier, it has low MVTR.
- PVDC contains chlorine which which poses challenges in recycling operations.

8.3.1.3 *Polyamide*

- PA 6 and PA 6/6,6 are the predominant polyamides used in flexible packaging. PA 6/6,6 has improved flexibility, better clarity and gloss, improved dimensional stability, lower moisture sensitivity, and lower processing temperatures than PA6, but is more expensive. They have similar oxygen permeation performance.
- Longer chain polyamides, such as PA 12, absorb less water than PA 6 but have poorer oxygen barrier performance.
- Amorphous 6I/6T copolyamides are often blended with PA 6 to improve processibility, clarity and oxygen barrier performance.
- MXD6 polyamide has lower oxygen permeability than PA 6 but is not used very much in flexible packaging due to cost and poor flex crack performance.
- The OPV of PA 6, 6/6,6 and MXD6 increases with increasing RH; the OPV of 6I/6T amorphous PA decreases with increasing RH.

8.3.1.4 *Other resins and substrates*

- Liquid crystal polymers (LCPs) generally have low oxygen permeability but have not found much use in flexible packaging due to their high cost and poor processability. They have a very low molecular weight (MW) and viscosity and most have very high Tm, making them difficult to match with other polymers in coextrusion. In films, they exhibit highly anisotropic physical properties without the use of special rotating dies that have proven difficult to scale up to commercial production.
- Polyethylene terephthalate (PET) is only a moderate oxygen barrier. Orienting PET (to make OPET film) reduces permeation but this is still not sufficient for many high barrier applications. OPET is often coated with an organic (such as PVDC or PVOH) or inorganic (e.g., aluminum, AlO*x*, or SiO*x*) coating to achieve high barrier performance.
- Polyethylene naphthalate (PEN) is a polyester resin with about 3x the oxygen barrier of PET. The high cost and insecure supply of the monomer has limited its use.
- Polyethylene furanoate (PEF) is a newly available member of the polyester family derived from ethylene glycol and 2,5-furandicarboxylic acid (FDCA). FDCA can be sourced from biobased sugars and thus PEF is generating interest both for being derived from renewable carbon and its barrier performance. Reports show it has eleven times the oxygen barrier of PET [66,67], putting in the range of high barrier resins. Cost and availability are still issues. Early developments have been aimed at CO_2 barrier for carbonated beverage bottles.
- Aluminum foil has long been used in flexible packaging as a barrier layer. Gas permeability is essentially zero except for defects or cracks that may occur during manufacturing, handling, or flexing of the package during use.
- High barrier substrates are typically coated with a barrier material such as aluminum, SiO*x*, AlO*x*, or PVDC. The inorganic coatings are often susceptible to flex cracking and other abuse, resulting in pinholes and cracks that let gas permeate through. These films are typically laminated with other polymers to protect the barrier coating.

8.3.2 Moisture vapor barrier

Typical moisture vapor transmission rates (MVTRs) for various polymer films and substrates are provided in Tables 8.6 and 8.7. The high moisture barrier polymers are hydrophobic and typically have a high level of crystallinity. An exception is cyclic olefin copolymers (COCs) which are amorphous. Many of the high oxygen barrier polymers have a poor barrier to moisture vapor. A quick guide to moisture and oxygen barrier can be found in Fig. 8.2.

8.3.2.1 *Poly-chloro-tri-fluoro-ethylene*

Poly-chloro-tri-fluoro-ethylene (PCTFE) film has the highest moisture barrier of any clear thermoplastic film. It is sold by Honeywell under the tradename Aclar. It is used primarily in pharmaceutical blister package applications and in military ration packaging; its relatively high cost compared to functional alternatives such as polyolefins keeps it out of most other flexible packaging applications.

8.3.2.2 *Polyolefins*

Of the commonly used polymers in flexible packaging, polyolefins have the highest moisture barrier. Isotactic polypropylene (PP) homopolymer and high-density polyethylene (HDPE) have the best moisture barrier properties within the standard polyolefin family of resins due to their high crystallinity. Crystalline regions are generally thought to be

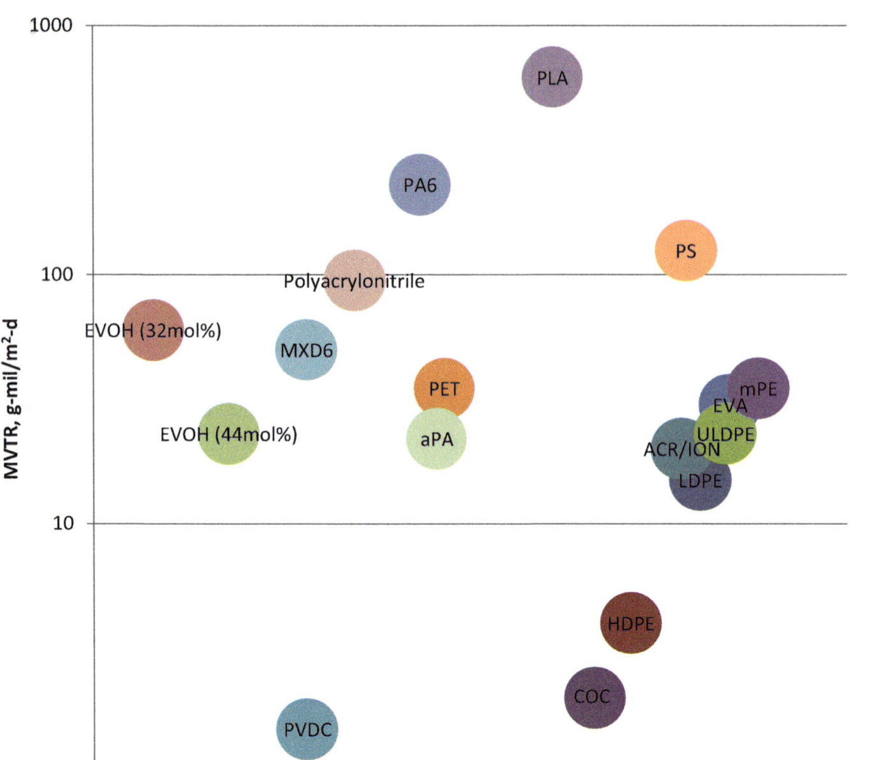

FIGURE 8.2 MVTR (38°C, 90% RH) and OPV (23°C, 0% RH) of common packaging resins. A key to resin acronyms is provided in Table 8.6.

impermeable to gases or vapors and their presence also restricts the mobility of nearby molecules. The higher the level of crystallinity in the polymer, therefore, the better the moisture barrier. Catalysis and additive advances have allowed the development of new grades of HDPE with lower MVTR [68].

COCs have excellent moisture barriers and are increasingly being used in pharmaceutical blister packaging.

8.3.2.3 Polyvinylidene choloride

Of the high oxygen barrier polymers, PVDC is the only one that also has an excellent moisture barrier. It is used as a barrier layer in coextruded films or as a coating onto PET or PP substrates. Other high oxygen barrier polymers, such as EVOH or PA, swell in the presence of water, opening up voids on a microscale that allow water to permeate. In PVDC, the halogen groups more strongly interact with themselves than water, making the permeation insensitive to water or relative humidity [69].

Metallized OPP and aluminum foil also make excellent moisture barriers and are used as components of laminations used in flexible packaging.

8.3.3 Grease and oil resistance

Most of the high oxygen barrier polymers such as EVOH, PA, PVDC, and even PET have excellent oil and grease resistance due to their high level of crystallinity and polarity. The oil and grease resistance of polyolefins vary with their crystallinity and polarity. Based on an old chemistry adage of "like dissolves like" this is not surprising since oils are simply low MW polyolefins and generally nonpolar. In the sections below the oil resistance of polyolefins is discussed based on work done at DuPont [70–72]. (The ethylene copolymer business of DuPont became part of Dow in 2019.)

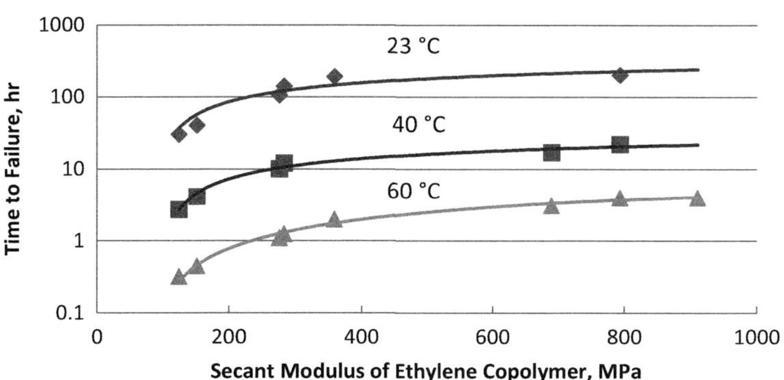

FIGURE 8.3 Penetration of mineral oil through polyolefin films using ASTM F-119 ground glass test. *Reproduced with permission from the Dow Chemical Company.*

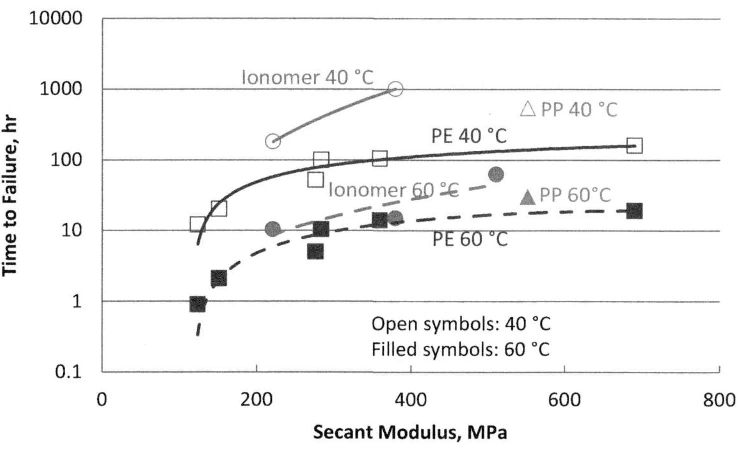

FIGURE 8.4 Penetration of peanut oil through polyolefin films using ASTM F-119 ground glass test. *Reproduced with permission from the Dow Chemical Company.*

8.3.3.1 Polyethylene

The oil resistance of PE is strongly affected by its crystallinity. In general, if the PE has lower crystallinity, the oil resistance is reduced. Crystallinity, in turn, is influenced by the density of the resin and film fabrication conditions (particularly quench conditions). For polyolefins and ethylene alpha-olefin copolymers like LLDPE the crystallinity increases with increasing density. For polar ethylene copolymers like EVA, the opposite is true due to the higher MW of the comonomer. One way to normalize the effect of crystallinity is to plot the oil resistance vs the secant modulus of the film[1]. Typical data for the penetration of mineral oil in PE are given in Fig. 8.3. Here the time to failure in a ground glass patch test (ASTM F-119—see Section 8.2.4) is plotted. A change in slope occurs at around 350 MPa (50 kpsi), equivalent to about 0.935 g/cc density for a polyethylene . The oil resistance levels off above this level of crystallinity. This suggests that an HDPE resin has about the same oil resistance as an MDPE resin. This work was done before the introduction of single-site catalyst technology. Recent work by Falla et al. [55] shows a bimodal metallocene-catalyzed HDPE has better grease resistance than a Ziegler–Natta HDPE. This grade of HDPE also has a superior moisture vapor barrier [68], indicating it has a very high level of crystallinity.

1. Modulus increases with crystallinity; VLDPE and EVA's typically have a modulus in the range of 34–140 MPa (5–20 kpsi), LDPE and LLDPE 140–410 MPa (20–60 kpsi), and HDPE 690–970 MPa (100–140 kpsi). Cast films typically have lower crystallinity and modulus than blown films of the same composition.

Similar results are found for peanut oil and are plotted in Fig. 8.4. Here grades of ionomer and a grade of PP are included for comparison.

Ethylene vinyl acetate (EVA) copolymer is expected to have comparable oil resistance to an PE with similar crystallinity. Again, secant modulus can be used as a common parameter. One of the data points in Figs. 8.3 and 8.4 is from a 5% VA copolymer (secant modulus equals 110 MPa or 16 kpsi); it fits on the same curves as the PE grades.

As the temperature increases, the oil resistance decreases. The greater the density of the PE, the less sensitive the oil resistance is to changes in temperature. This may reflect the differences in melting points. HDPE and LDPE have peak melting points (as measured by DSC) of 135°C and 110°C, respectively, but melt over a temperature range: some of the less perfect crystals melt at temperatures well below the peak melting point. In designing and interpreting accelerated testing at elevated temperatures care must be taken to consider if the polymer undergoes any thermal transitions at the test temperature that will affect the results. For example, at 60°C some of the crystals in LDPE will melt whereas as very few crystals of HDPE will have melted at this temperature. It should be noted that this observation is based on a study done before the introduction of metallocene PE, which has much narrower melting point ranges. Indeed, in the Falla et al. study, the metallocene HDPE is shown to have significantly better oil resistance at 70°C than the Ziegler–Natta HDPE; at 60°C both resins perform well. EAA and ionomer in their study do not perform as well as the HDPE, but the 70°C test temperature is not appropriate for these polymers given their lower melting points. Sufficient loss of crystallinity at this temperature likely will yield results that will not correlate with the actual performance at room temperature. Unless the end-use is exposed to such temperatures, the test temperature of an accelerated test should be chosen so that the polymers do not undergo any major changes that will affect the performance.

8.3.3.2 Acid copolymers

The oil resistance of acid copolymer resins (ACRs) increases with increasing acid level. As the acid level is increased, the crystallinity of an acid copolymer decreases. This contrasts with PE where decreasing crystallinity reduces oil resistance. This suggests that hydrogen bonding among the acid groups in the copolymer is important for oil resistance.

Fig. 8.5 shows the effect of acid level on the resistance to mineral oil penetration. Most grades of acid copolymer used in flexible packaging have an acid level in the range between 4 and 12 wt.% methacrylic acid (MAA) or between 3% and 9.5 wt% acrylic acid (AA). At 12 wt.% MAA, the oil resistance is substantially better than a standard LLDPE grade or metallocene plastomer, but not as good as HDPE (see Fig. 8.2). Data for EAA is not generated in this study, but EAA is expected to have a similar performance as EMAA at equivalent mole % acid. Because AA has a lower MW than MAA, the wt.% AA must be multiplied by 1.2 to convert from wt.% AA to wt.% MAA at the same mol%. Thus, a 10 wt.% AA EAA resin has about the same amount of acid on a molar basis as a 12 wt.% MAA EMAA resin.

8.3.3.3 Ionomers

Partially neutralizing an acid copolymer with a metal salt to make an ionomer typically improves its oil resistance. The greater the acid level and % neutralization, the higher the resistance. Sodium ionomers typically give higher resistance than equivalent zinc ionomers. In Dow's Surlyn ionomer product line, Surlyn 1707 (sodium, 0.9 MI) and 1605 (sodium,

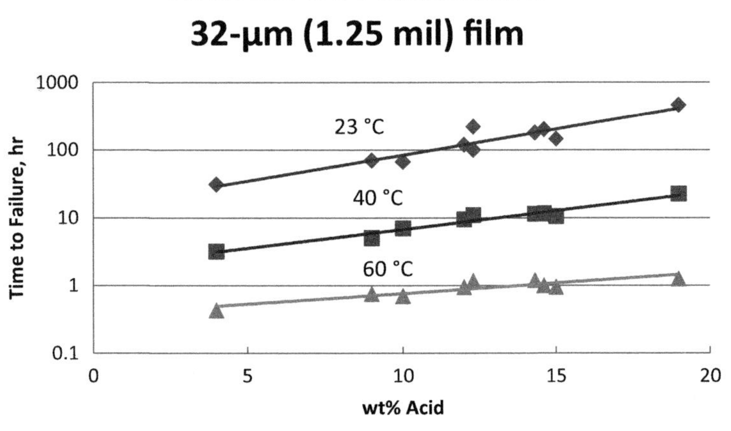

FIGURE 8.5 Mineral oil resistance of acid copolymers as a function of acid content (MAA) and temperature from ASTM F-119 ground glass test. *Reproduced with permission from the Dow Chemical Company.*

2.8 MI) have the highest oil resistance. Surlyn 1652 (zinc, 5 MI) has the lowest oil resistance but is still higher than standard acid copolymers and LDPE.

Ionomers typically have higher oil resistance than LDPE, LLDPE, and in many cases HDPE. See Fig. 8.4 for a comparison of peanut oil penetration. De Garavilla [71] compares the corn oil penetration resistance of several polyolefins. His results, shown in Fig. 8.6, show that ionomers have substantially higher oil resistance than other premium sealants such as VLDPE, mPE and EVA. The oil resistance of Surlyn ionomers is sensitive to temperature; like PE, the higher the temperature, the lower the oil resistance. The temperature dependence of the penetration of vegetable oils is illustrated in Fig. 8.7.

8.3.3.4 Polypropylene

One grade of PP is included in the DuPont study [70]. PP is found to have better oil resistance than HDPE and is comparable to the best ionomer grades. The sensitivity of the oil resistance of PP to temperature changes is greater than HDPE but about the same as ionomers. See Fig. 8.4.

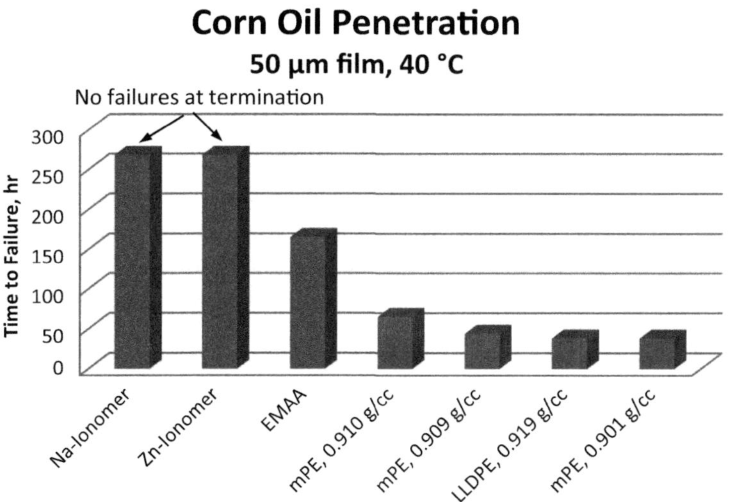

FIGURE 8.6 Oil resistance of several premium sealant resins. 50-µm film exposed to 40°C corn oil in ASTM F-119 ground glass test. Test terminated at 268 hours (before ionomer samples failed). *Data from J. de Garavilla, Ionomer, acid copolymer, and metallocene polyethylene resins: a comparative assessment of sealant performance, Tappi J. 78(6) (1995) 192. Reproduced with permission from the Dow Chemical Company.*

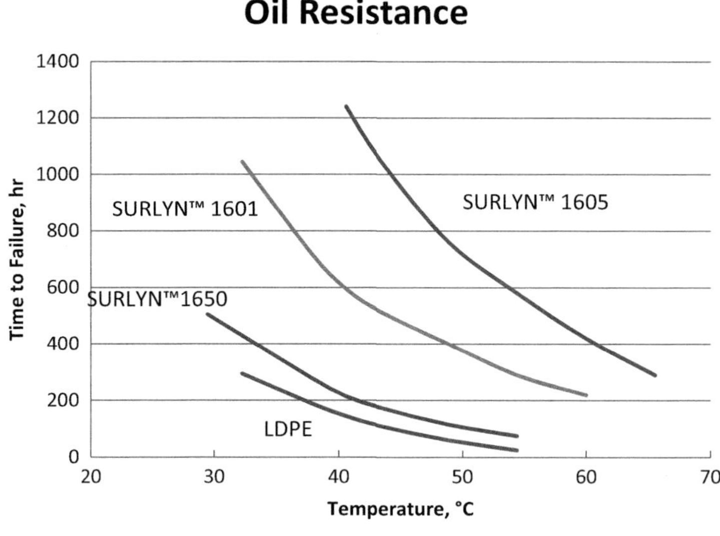

FIGURE 8.7 Oil resistance of several grades of Surlyn ionomer as a function of temperature. Samples from 25-µm extrusion coatings on paper. Tested in vegetable oil using ASTM F-119 ground glass test. *Composite data taken from DuPont, Barrier properties – oil resistance, Surlyn Update, 1(9) (1973). Reproduced with permission from the Dow Chemical Company.*

8.3.3.5 Film thickness

The DuPont study finds that the oil resistance of polyolefin films is nonlinear with respect to film thickness. A doubling of the thickness results in more than doubling of the oil resistance. (No explanation was given at the time. Subsequent studies have shown that the crystallinity at the surface of a blown film may be lower than the bulk. Thus, one explanation is that the thicker the film, the less impact the surface regions have on permeation. See Section 8.4.1.2.2 for further discussion.)

8.3.3.6 Oil type

Most of the study by DuPont was done with mineral and vegetable (primarily peanut) oils. For mineral oil, it was found that the MW of the oil can have a substantial impact on oil resistance. A low MW oil, such as No. 2 Fuel Oil (similar to kerosene), will permeate through the film much faster than standard mineral oil.

Vegetable oils are composed mainly of triglycerides and these can normally be expected to behave similarly regardless of the source. For example, it is found that soy oil and peanut oil give similar permeation rates. Other studies show peanut oil, corn oil, cotton seed oil and olive oil give similar results. However, vegetable oil can contain various levels of free fatty acid that can greatly alter the permeation rate and even the comparison between ionomers and PE. In one such case, the permeation rate of rice oil containing 18% free fatty acid is found to be considerably faster through an ionomer than through LDPE. (For comparison, the peanut oil used in the study has only 0.14% free fatty acid.) Consequently, the oil resistance of the ionomer to rice oil is lower than an LDPE with a density of 0.922 g/cc.

Falla et al. [55] cite the work of Wyser et al. [56,57] showing that chicken fat mixed with oleic acid is fast migrating. They use lard which is high in oleic acid and does not require refrigeration in their study.

8.3.4 Organic molecules

The migration of aroma and flavor molecules into or through packaging films is a challenge to measure due to interactions between the molecules and package, as is discussed in Section 8.2.4. What little data that exist in the literature should be viewed with caution; studies may be conducted at concentration ranges that do not reflect real packaging applications or in absence of other molecules that may affect the migration rates.

Many packaging engineers assume that the barrier to organic molecules correlates with the oxygen barrier. While high barrier polymers such as EVOH, PA and even PET often provide a good barrier to aroma and flavor molecules, this is not always the case.

D-limonene is one of the most studied molecules. Polymers that have shown to be good barriers include acrylonitrile, PET, amorphous PA, EVOH, and polylactic acid (PLA) [73,74].

The permeation of flavor and aroma molecules is discussed in more detail in Chapter 16 where product-package interactions in the end-use are considered.

8.3.5 Chemical resistance

Chemical resistance can be important for some packaging applications, such as in the packaging of chemicals, cleaning agents, reagents, and alcohol towelettes. Chemicals can affect the strength, flexibility, impact resistance, appearance, dimensions and weight of packaging films. For plastics, chemicals may attack the polymer chain through oxidation, react- to functional groups on the chain or cause depolymerization. Chemicals may swell or soften the polymer. Some chemicals may also cause stress cracking.

The chemistry of the polymer that makes up the package and the product are two prime factors affecting chemical resistance, but a number of other factors are important. In general, the chemical resistance will decrease with increasing temperature, a higher concentration of the chemical, higher pressure, longer exposure time and higher residual stress. Mixtures of chemicals may also affect the resistance in ways single chemicals do not. For example, one chemical may swell the polymer, allowing another to permeate and react with the polymer.

There are a number of references on the chemical resistance of polymers. A good general reference is the Chemical Resistance of Plastics and Elastomers [75]. A number of tables can be found online [76–80]. Most of the data used in these references are generated using thick plaques and may not represent what may occur in a thin film. Because of this and the many factors outlined above that affect chemical resistance, these references should only be used as a guide. Testing the actual package under conditions of its actual use and environment is always recommended.

8.3.6 Light barrier

The more opaque a film is, the better is its light barrier. Some transparent films, however, will preferentially absorb certain wavelengths of interest, blocking them from the product. For example, amorphous nylon (PA 6I/6T) absorbs light in the UV wavelength that is often used in supermarkets and is deleterious to a number of foods [81]. Intawiwat et al. [82] experiment with adding red, green and silver pigments and an optical brightener to PE film to block specific wavelengths harmful to dairy products (400–450 nm and 600–650 nm). The films are transparent and have varying degrees of success in blocking the targeted wavelengths.

Aluminum foil, some types of paper, and metallized films provide a good light barrier. Pigments and fillers are also used to create opacity. Rutile titanium dioxide, TiO_2, is a particularly effective additive for light barrier because of its high refractive index [83]. To create a good barrier to light, the film must bend the light wave. One way is to add an inorganic material that has a higher refractive index than the plastic film. Calcium carbonate, clay and barium carbonate have refractive indices too close to most plastics to scatter much light. Rutile, followed by anatase, TiO_2 is very effective. Particle size also affects scattering and opacity. The optimal particle diameter depends on the wavelength of light being scattered:

$$\boldsymbol{D} = \frac{2\lambda}{\pi(\varepsilon_1 - \varepsilon_2)} \tag{8.6}$$

where $\boldsymbol{D}$ = optimal particle size for a given wavelength, λ = wavelength of light being scattered, and $\varepsilon_1 - \varepsilon_2$ = refractive index difference between the polymer and particle.

For scattering of visible light, a range of particle sizes works best.

Opacity increases with increasing concentration of the pigment and film thickness [84]:

$$\text{Scattering} \propto \varepsilon X \tag{8.7}$$

where ε = refractive index, X = thickness of polymer film layer thickness, and c = concentration of pigment in polymer.

Increasing the concentration by two times is equivalent to doubling the film thickness. It is the total amount of pigment that controls scattering. Splitting the pigment layer into separate layers of the same total thickness separated by clear layers has no effect on the opacity.

8.4 Science of permeation

The permeation of gas or liquid molecules through flexible packaging films generally follows one of two modes:

1. Sorption of the permeant onto the film, diffusion through the film, followed by desorption on the other side. This is the mode for most polymer films.
2. Permeation or effusion through pinholes, cracks or other defects in the film, and perforations. This mode explains the behavior of aluminum foil, metallized films and glass (inorganic) coated films, as well as permeation through leakers in the seal area or from damage caused by puncture events.

8.4.1 Permeation through polymer films

The mechanism for permeation through a polymer film is illustrated in Fig. 8.8. The permeant is absorbed onto the surface, diffuses through the film and desorbs on the other side. The driving force for the transport of a permeant through a polymer film is the concentration gradient, or the difference between the concentrations on each side of the film divided by the film thickness, as given by Eq. (8.5).

Fick's First Law, given by Eq. (8.4), relates the diffusive flux to the concentration gradation through the proportionality constant D. D is known as the diffusion coefficient and has units of [length^2/time]. Larger values of D indicate greater diffusion. If D is independent of concentration then the diffusion is known as Fickian. If D changes with concentration it is known as non-Fickian. The diffusion of most inert gases through polymer films is Fickian. The diffusion of larger organic molecules, as well as moisture for some polymers, may be non-Fickian.

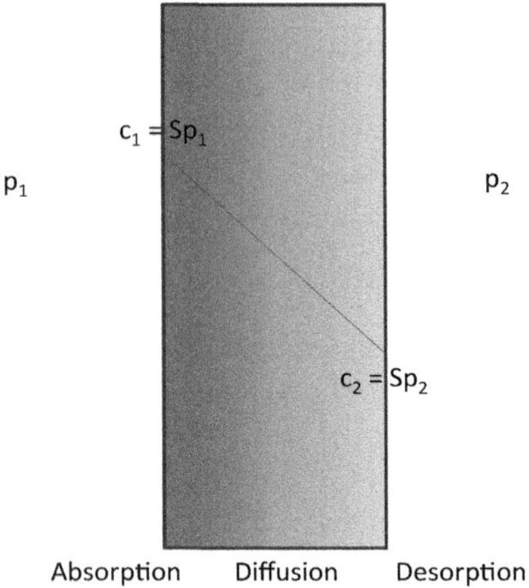

FIGURE 8.8 Mechanism for permeation through a polymer film.

Fick's Second Law relates the change in concentration with time to the derivative of the concentration gradient. For thin films it is assumed that the diffusion is only in the thickness dimension:

$$\frac{\partial c}{\partial t} = D\frac{\partial^2 c}{\partial x^2} \tag{8.8}$$

Here D is assumed to be independent of concentration (Fickian). Eq. (8.8) describes the unsteady-state diffusion of permeants through the film. It is particularly useful for describing the initial stages of diffusion before a steady state is reached.

When the diffusion reaches a steady state, the diffusion flux is constant and Eq. (8.4) can be integrated over the thickness to yield:

$$JX = -D(c_2 - c_1) \tag{8.9}$$

where c_1 and c_2 are the concentrations on each side of the film. Substituting the definition of J ($=Q/At$) into Eq. (8.9) gives the following relationship for Q (the amount of permeant):

$$Q = \frac{D(c_2 - c_1)At}{X} \tag{8.10}$$

Here A is the surface area of the film. For gases, it is convenient to use partial pressures rather than concentration. The concentration of a dilute gas is equal to the product of the partial pressure and the solubility coefficient of the permeant in the polymer (S). This relationship is known as Henry's Law and is introduced in Eq. (8.3).

Substituting Eq. (8.3) into (8.10):

$$Q = \frac{DS(p_2 - p_1)At}{X} \tag{8.11}$$

The product DS is known as the permeability coefficient, P:

$$P = DS. \tag{8.12}$$

Eq. (8.11) becomes:

$$Q = \frac{P(p_2 - p_1)At}{X} \tag{8.13}$$

Since permeability is a fundamental property of a polymer and is often used to compare and select polymers for barrier applications, reviewing the assumptions used in deriving P and their limitations is in order:

- **Diffusion is at a steady state:** For most gases, this turns out to be a good assumption. Ylvisaker [85] shows that a steady state is reached when $t > 0.5X^2/D$. Since most flexible packaging films are thin, a steady state is reached within hours. For example, Ylvisaker calculates the time to reach a steady state for moisture transport through an BOPP film to be 15 minutes. For large molecules, the diffusion is slower, particularly if there is any interaction between the permeant and polymer, and unsteady state calculations may need to be considered [49,85].

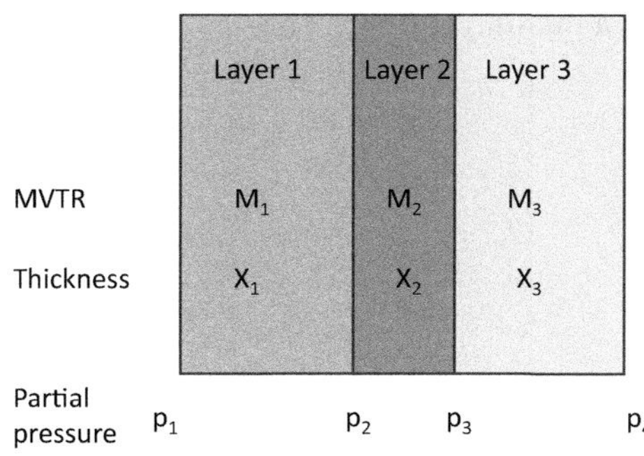

FIGURE 8.9 Permeation through three-layer film.

- **Diffusion is one-dimensional:** The thickness dimension of flexible packaging films is typically orders of magnitude smaller than either of the planar dimensions. This assumption is usually valid.
- **Diffusion is Fickian and the diffusion coefficient is independent of concentration:**. This is true for most gases such as oxygen, nitrogen and carbon dioxide. It fails to be true for water in some polymers, as well as some large organic molecules that interact with the polymer.

By definition, P is independent of thickness and is a material property rather than a package property. Conversely, the transmission rate (TR) is a package property that relates to the amount of permeant that is transported through a package film with a specific structure subject to a given environment (pressure, RH, etc.). For a monolayer film, TR is related to P is the following manner for gases:

$$\mathrm{TR} = \frac{P\Delta p}{X} \tag{8.14}$$

When the film thickness is halved, the transmission rate doubles.

8.4.1.1 Permeation through multilayer films

For multilayer films, a total permeation coefficient can be computed. Fig. 8.9 shows the key parameters for permeation through a 3-layer film. At a steady state, the amount of permeation, Q, through each layer is equal:

$$Q = Q_1 = Q_2 = Q_3 \tag{8.15}$$

Also, the area of each layer is the same:

$$A = A_1 = A_2 = A_3 \tag{8.16}$$

The pressure drop across the multilayer film is the sum of pressure drops across each layer:

$$p_1 - p_4 = (p_1 - p_2) + (p_2 - p_3) + (p_3 - p_4) \tag{8.17}$$

Rearranging Eq. (8.13):

$$\Delta p = \frac{QX}{tPA}. \tag{8.18}$$

Substituting into Eq. (8.17):

$$\frac{QX_T}{tP_TA} = \frac{QX_1}{tP_1A} + \frac{QX_2}{tP_2A} + \frac{QX_3}{tP_3A} \tag{8.19}$$

where X_i = thickness of layer i, X_T = total thickness of the multilayer film, P_i = permeability coefficient of layer i, and P_T = permeability of multilayer film.

Factoring out Q/tA:

$$\frac{X_T}{P_T} = \frac{X_1}{P_1} + \frac{X_2}{P_2} + \frac{X_3}{P_3} \tag{8.20}$$

Extending to n layers yields the general relationship for permeation through a multilayer film:

$$\frac{X_T}{P_T} = \sum_{i=1}^{n} \frac{X_i}{P_i}. \tag{8.21}$$

In terms of transmission rates:

$$\frac{1}{\mathrm{TR}_T} = \sum_{i=1}^{n} \frac{1}{\mathrm{TR}_i}. \tag{8.22}$$

A consequence of the form of Eq. (8.21) is that if one of the layers has a much lower P than the other layers, the other layers do not contribute much to the overall permeability of the film. For example, in a coextruded film with the structure (PE-tie-EVOH-tie-PE), the PE and tie layers do not contribute much to the oxygen barrier—it almost all comes from the EVOH layer. Functionally, however, the other layers may contribute to the barrier of the structure by reducing pinholes or flex cracks in the EVOH, or by helping to maintain layer continuity during thermoforming.

8.4.1.2 Factors that affect permeability through polymer films

The chemistry and structure of the polymer film, the chemistry, size and shape of the permeant molecule and the external environment affect the rate of permeation through a polymer film. Although P is the product of diffusion and solubility, P generally trends with D. The solubility coefficient varies by about two orders of magnitude for various polymers and permeants, whereas the diffusion coefficient may vary over six orders of magnitude [86].

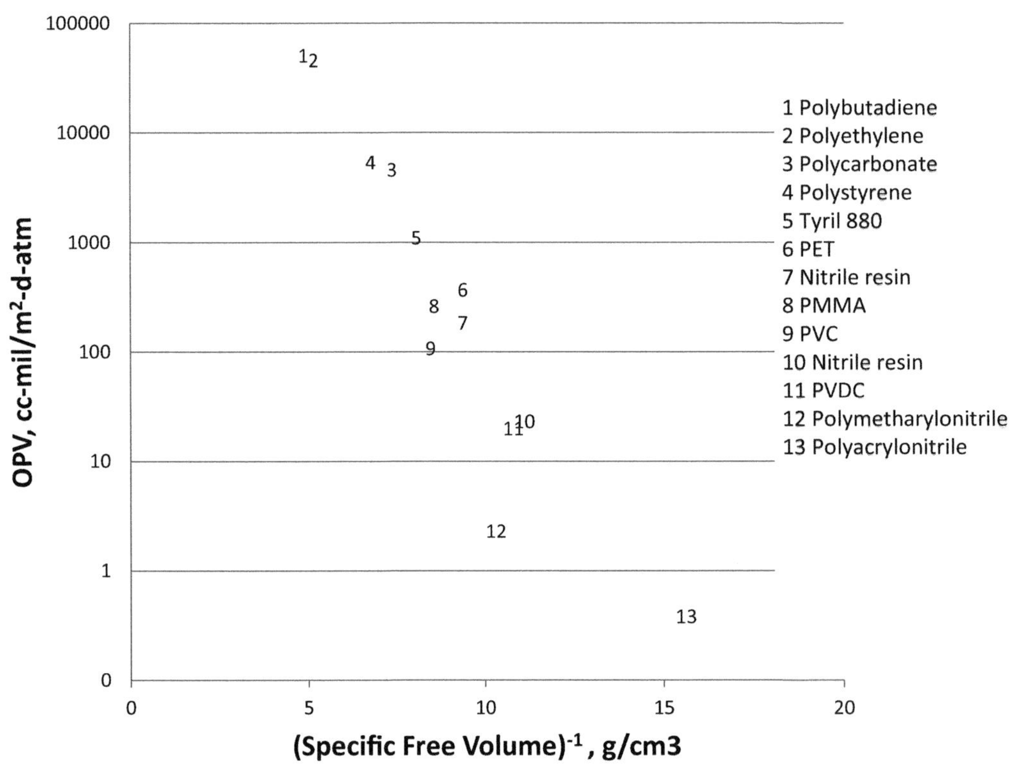

FIGURE 8.10 Influence of polymer-free volume on the permeability of oxygen. OPV, oxygen permeation value; PET, polyethylene terephthalate; PMMA, polymethyl methacrylate, PVC, polyvinyl chloride; PVDC, polyvinylidene chloride. *Based on the work of Lee WE. Selection of barrier materials from molecular structure. Polym. Eng. Sci. 20 (1980) 65–69. Reproduced with permission.*

TABLE 8.8 Effect of vinyl functional group on oxygen permeability of polymers.

Functional Group R in $(-CH_2-CHR-)_n$	OPV cc-mil/m^2-d-atm
$-OH$	1.6
$-CN$	0.62
$-Cl$	124
$-F$	233
$-COOCH_3$	259
$-CH_3$	2328
$-C_6H_5$	6468
$-H$	7502

Source: Reproduced with permission from G.L. Robertson, Optical, mechanical and barrier properties of thermoplastic polymers, in: Food Packaging. Principles and Practice, third ed., CRC Press, Boca Raton, FL, 2013, pp. 91–130.

8.4.1.2.1 Polymer factors

The diffusion of permeant molecules through polymer films can be thought of as a series of hops through temporary voids in the polymer microstructure. These voids are related to the free volume of the polymer, which is defined as the difference in molar volume at the system temperature and at absolute zero (0K). The voids are due to imperfections in molecular packing for polymers above their Tg and vibration and rotation of molecular segments for both rubbery and glassy polymers [69]. P generally increases with increasing free volume, as is shown in Fig. 8.10 based on the work of Lee [86].

Factors that decrease free volume (and gas permeation) include:

- Absence of bulky side groups on the polymer chain.
- Narrower MWD and less long chain branching.
- Higher cohesive energy density (CED). CED is introduced in Chapter 6. It is a quantitative measure of how strongly the polymer molecules interact with one another. Stronger interactions generally result in higher packing densities.
- More rigid polymer chains. Chain stiffness hinders molecular movement, reducing the creation of temporary voids for permeant molecules to hop into.
- Higher crystallinity. Crystals are generally impermeable and so by themselves they reduce permeation to the surrounding amorphous regions. Their orientation and morphology may create a tortuous path for the permeant molecule to pass through. Crystals can also inhibit the movement of nearby molecules in the amorphous regions of the polymer, affecting free volume.
- Higher polarity. Polarity generally increases chain stiffness and CED. An example of the effect of polymer functional group chemistry on oxygen permeation is given in Table 8.8. The side groups with the greater polarity provide the lowest permeation. Polarity also may increase the interaction with polar permeant molecules. Increased interaction slows diffusion.
- Higher Tg. The solubility of permeants in glassy polymers ($T <$ Tg) may be greater than in rubbery polymers ($T >$ Tg) due to the presence of unrelaxed voids that may form during quenching [69]. The lower chain mobility in the glassy region, however, reduces free volume and diffusion, and typically outweighs the solubility effect.
- Higher orientation. Orientation aligns the molecules and can increase packing density, which reduces free volume. Orientation may also induce crystallization and morphologies that are conducive to lower permeation. Orientation processes, such as tenter frame and double bubble processes, typically decrease the permeability but the effect varies with the polymer.
- Absence of plasticizers and lubricants. Plasticizers and lubricants increase the flexibility and molecular mobility of the polymer chains, increasing free volume. Plasticized versions of PVC, for example, may have OPV values that are 200 times greater than unplasticized PVC [49].
- Interaction between polymer chains. Beyond hydrogen bonding, ionic bonding, and other chemical interactions that influence CED and packing density, crosslinking reactions may take place. Crosslinking limits the mobility of

polymer chains, reducing free volume. Robertson provides an example where one crosslink out of 30 monomers leads to a 50% reduction in D [49].

8.4.1.2.2 Thickness

Permeability is normalized for the thickness of the film and so in theory is independent of thickness. This is true if the polymer itself is unchanged by thickness. The stress and cooling conditions in the film manufacturing process, however, may change as the thickness is altered. This may affect the orientation and crystalline morphology of the polymer. This in turn may affect the permeability [87,88]. Haley and Borke [88] provide an example for the MVTR of HDPE. They measure the MVTR of blown films ranging in thickness from 30 to 75 μm for 11 different grades of HDPE. The grades are manufactured with a variety of catalyst technologies and have a range of MW and MWD characteristics. The characteristics of a subset of resins from their study are shown in Table 8.9. These resins are selected to highlight the differences in their permeation behavior, which are plotted in Fig. 8.11. The MVTR of some of the resins are independent of thickness whereas others show an increase in MVTR with decreasing film thickness. The authors conclude that polydisperse HDPE grades with Mz greater than about 400,000–500,000 g/mol are more susceptible to morphological changes during film blowing and exhibit an increase in MVTR with decreasing thickness. (Mz is a version of MW that is weighted towards the high MW fraction of the polymer.) They correlate the change in MVTR with overall crystallinity.

Another factor that may influence the role of film thickness on permeability of LLDPE is comonomer distribution. Several studies have shown that blown film of LLDPE made by a Ziegler–Natta catalyst has an amorphous region at the surface [89,90]. This is believed to be due to surface segregation of the high-comonomer low MW fraction in these heterogenous polymers. Films made from homogenous LLDPE made with metallocene catalysts do not have an amor-

TABLE 8.9 Characteristics of HDPE grades used in Haley and Borke [88] study.

Sample	Type	Catalyst	Process	Density g/cc	MFI g/10 min	Mn g/mol	Mw g/mol	Mz g/mol
CC-1	Ethylene-butene	Chrome		0.949	0.3	12,000	140,000	568,000
SH-4	Homo-polymer	Vanadium/titanium	Solution	0.960	1.0	24,000	121,000	380,000
ZC-1	Ethylene-butene	Ziegler-Natta	Slurry cascade	0.949	0.1	10,000	296,000	1,574,000
ZH-2	Homo-polymer	Ziegler-Natta	Slurry cascade	0.957	1.3	22,000	123,000	407,000

MI, Melt index; *Mn*, number average molecular weight; *Mw*, weight average molecular weight; *Mz*, z-average molecular weight.

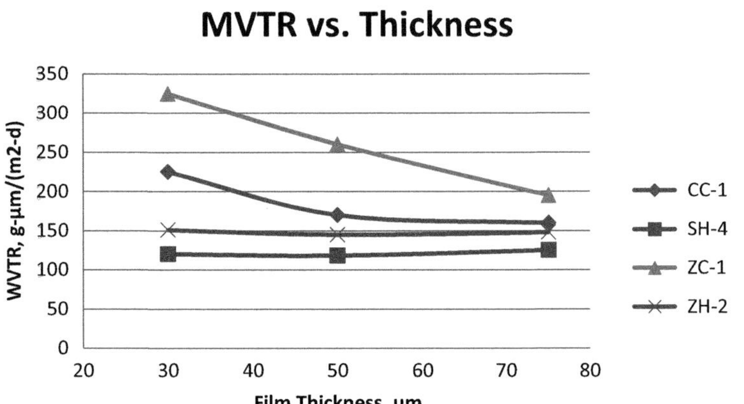

FIGURE 8.11 MVTR (normalized for thickness) as a function of blown film thickness. *Data from J.C. Haley, J.S. Borke, The interplay between polymer polydispersity and film gauge in HDPE barrier films: how polydispersity controls gauge dependence of film barrier properties, SPE ANTEC Conf., 2009, pp. 231–235.*

phous region at the surface. The consequences of the amorphous surface region on peel strength [91,92], adhesion [93–95], and slip agent migration [96] are described in Sections 7.5.3.2.2, 10.4.2.2.1, and 12.1.4.1, respectively. As the film gets thinner for heterogeneous PE, the amorphous region near the surface takes up a larger fraction of the total thickness. Since these amorphous regions have a higher permeability than the bulk due to their lower level of crystallinity, the measured permeability may be higher than what would be extrapolated from measurements on thicker films.

8.4.1.2.3 Permeant factors

The size, shape, chemistry, and, in some cases, the concentration of the permeant influences P. The relative diffusion, solubility and permeation factors for several gas molecules are given in Table 8.10.

TABLE 8.10 Effect of gas molecule on permeation, diffusion and solubility in polymer films.

Gas	P	D	S
N_2 (=1)	1	1	1
CO	1.2	1.1	1.1
CH_4	3.4	0.7	4.9
O_2	3.8	1.7	2.2
He	15	60	0.25
H_2	22.5	30	0.75
CO_2	24	1	24
H_2O	550	5	10

Nitrogen value is 1; other values are relative to nitrogen.
Source: From D.W. Van Krevelen, K. te Nijenhuis, Properties determining mass transfer in polymer systems, Properties of Polymers, Elsevier Science, Amsterdam, 2009, pp. 655–702.

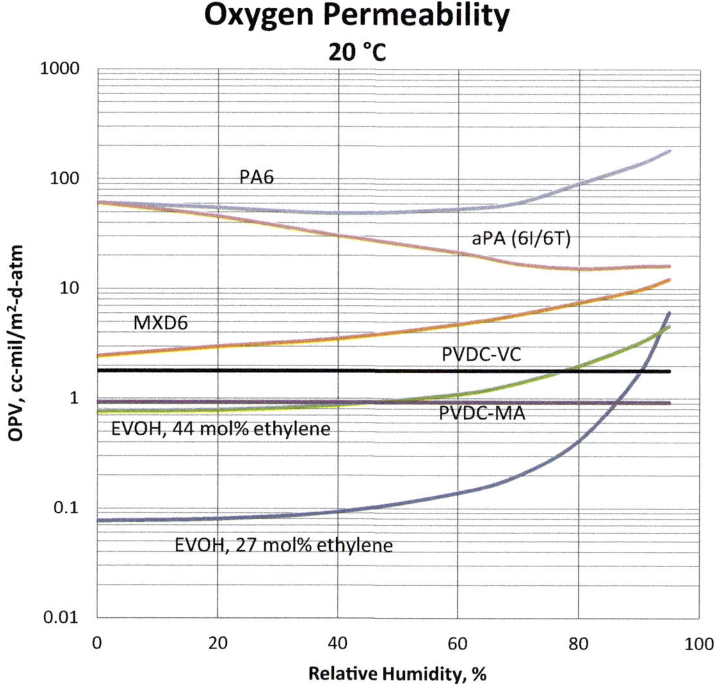

FIGURE 8.12 Effect of relative humidity on the oxygen permeability of some common barrier polymers.

Larger and bulkier molecules will diffuse more slowly, although they may have greater solubility and condense more readily onto the surface of the film. Cooksey [97] observes that molecules with eight or more carbons condense and are sorbed more easily than smaller molecules. Branched molecules are sorbed more easily than linear molecules.

The polarity of the permeant and the polymer affects the diffusion rate. The greater the affinity of the permeant to the polymer, the slower the diffusion. Cooksey [97] provides the example of carvone ($C_{10}H_{14}O$) and limonene ($C_{10}H_{16}$). Both have 10 carbons. Limonene is less polar and permeates faster through nonpolar films. The solubility parameter, introduced in Chapter 6, is a way to estimate polarity differences. The solubility parameter is related to the square root of the cohesive energy density and can be found tabulated in various references [98–100]. The solubility parameters for several polymers are listed in Table 6.2. In general, small differences in solubility parameters of the permeant and polymer indicate an affinity that may result in greater interactions. Auras et al. [74] provide a good grounding on the use of solubility parameters for flavor–polymer interactions as well as an example. They study the sorption of ethyl acetate and d-limonene by several polymers including PLA, polystyrene (PS), PET, PP, and PE. The difference in solubility parameters between the flavor molecule and polymer correlates fairly well with sorption/solubility (S) but not very well with permeability (P). They conclude that the solubility parameter method is useful for predicting S but does not account for specific interactions such as swelling, dissolution or immiscibility. This may account for the poor correlation with permeability.

Interaction between the permeant molecule and the polymer can lead to swelling or plasticization effects. This results in greater free volume and higher permeation, not only for the permeant molecule causing the effect but for other molecules that may be present. Such interactions are often more pronounced at higher permeant concentrations, leading to non-Fickian diffusion behavior. Moisture vapor permeation has this effect on some polar polymers which is discussed in more detail in Section 8.4.1.2.4.3.

8.4.1.2.4 Environmental effects

8.4.1.2.4.1 Temperature For gases, D, S, and P all tend to increase with increasing temperature, but for vapors the relationship is more complex. Here gas is defined as a substance like oxygen or nitrogen that is in a single gaseous state at room temperature. A vapor is a substance like water and many organic molecules that exists as a mixture of states (gas and liquid or solid) at room temperature.

P generally follows an Arrhenius relationship:

$$P = P_0 e^{-E_p/RT}. \tag{8.23}$$

where $P_0 =$ permeability coefficient at the reference temperature, $E_p =$ activation energy, R = ideal gas constant, and$T$ = temperature (K).

The activation energy is made up of contributions from D and S:

$$E_p = E_d + \Delta H_s \tag{8.24}$$

where $E_d =$ activation energy for diffusion and

$\Delta H_s =$ heat of sorption.

E_d is always positive; diffusion increases with increasing temperature. ΔH_s is generally positive for gases but negative for vapors: solubility increases with increasing temperature for gases and decreases for vapors. P may increase or decrease with temperature depending on the relative contributions of S and D. As a result, the order of performance of different barrier polymers may change with temperature [49].

The glass transition temperature impacts the temperature dependence of P. Cooksey [97] reports that for gas permeation, P increases by about 9%/°C at temperatures above the Tg of the polymer and about 5%/°C at temperatures below Tg.

8.4.1.2.4.2 Pressure Pressure has little impact on permeability unless the permeant has a strong interaction with the polymer (and then P increases with pressure). It is important to note that for gas permeation, it is the partial pressure of the permeant, not the total pressure, that is the driving force for diffusion. Robertson [49] reports on studies of high-pressure pasteurization (HPP), citing the work of Juliano et al. [101] and Guillardo et al. [102] These studies show that at the high pressures experienced in this process (300–600 MPa) gas transmission is more affected by distortion of the film that may result in leakers than by the effect of pressure on permeability.

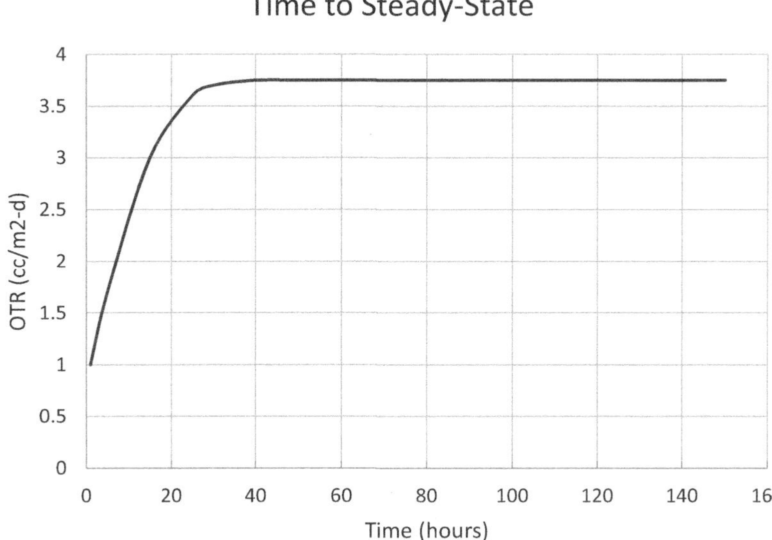

FIGURE 8.13 OTR (oxygen transmission rate) versus time for a (LLDPE-tie-PA6-EVOH-PA6-tie-LLDPE) film. Total thickness: 69 μm (2.7 mils). Layer thicknesses: (17/15/7/17/7/15/17). *From B.A. Morris, The role of polyolefins in "protecting" moisture sensitive barrier layers in packaging structures, in: TAPPI International Flexible Packaging & Extrusion Division Conference, 2018 with data from Y. Zheng, B. Walthers, S. Guerra, S. Bawiskar, Go beyond adhesion: the dual functionality of tie-layers, in: 2017 SPE FlexPackCon Conference, Tampa, FL, 2017. Reproduced with permission from the Dow Chemical Company.*

8.4.1.2.4.3 Relative humidity The permeability of some polymers is sensitive to water. Examples include PVOH, EVOH, PA6, PA 6,6/6 and acrylonitrile. In these polymers water acts as a plasticizer or lubricant, increasing the mobility of the polymer chains and increasing free volume. At higher relative humidity, the permeability increases, as is shown in Fig. 8.12.

The permeability of nonpolar polymers such as polyolefins is generally not sensitive to water. This is because the solubility of water in these polymers is too low to have a plasticizing effect. Highly polar polymers containing halogens (Cl or F atoms), such as PVDC and fluoropolymers, more strongly interact with each other than with water [69]. Consequently, their barrier properties are insensitive to RH.

One interesting exception in the polyamide family of resins is PA 6I/6T or amorphous nylon (sold as Selar PA by DuPont and Grivory by EMS). The oxygen permeation of this polyamide decreases with increasing RH. Studies show that this is due to the presence of both aliphatic and aromatic units along the polymer chain. Longer aliphatic chains increase the overall permeability, but are responsible for the drop in permeability with higher RH. The studies conclude that water molecules compete with gas molecules for pathways through the polymer, decreasing the permeability of the gas [69].

In flexible packaging, water-sensitive high-oxygen barrier polymers such as EVOH are typically sandwiched between other polymers, especially polyolefins. Many mistakenly believe that by placing the EVOH between good moisture barrier resins like PE that the EVOH is protected from water. This is only true during the first few minutes of the exposure until the transport of moisture reaches a steady state. For thin films, a steady state can be reached in a few hours [85,103–105]. An example is shown in Fig. 8.13 where a steady state is reached after 30 hours for a 69-μm film. Granger and Tigani [103] measure the OPV of EVOH monolayer and symmetric coextruded films of (LDPE-tie-EVOH-tie-LDPE) over a range of RH from 0 to 100%, where the RH is controlled to be the same on both sides of the film. The OPV results from the mono and coextruded structures fit on top of one another, as shown in Fig. 8.14, indicating that the LDPE is offering no protection from the moisture once a steady state is reached.

In a multilayer film, the relative humidity of the EVOH layer falls somewhere between the RH on the inside and outside of the package. The composition and thickness of layers surrounding the barrier layer never fully protect the barrier layer from seeing moisture but can influence the RH within this range. This can be important since the OPV of EVOH rapidly rises when the RH is above about 70% (see Fig. 8.1). Moving the barrier layer towards the side of the package that is exposed to lower humidity may be advantageous.

For a quick estimate of what RH to use in calculating the permeability for the barrier layer ($\%RH_B$), the change in RH throughout the package thickness can be considered to be linear and the RH of the barrier layer can be determined by its location:

$$\%RH_B = \%RH_{outside} - \frac{x_B}{X_T}(\%RH_{outside} - \%RH_{inside}) \tag{8.25}$$

where $x_B =$ distance from the outside film surface to the centerline of the barrier layer, $X_T =$ total thickness of the film.

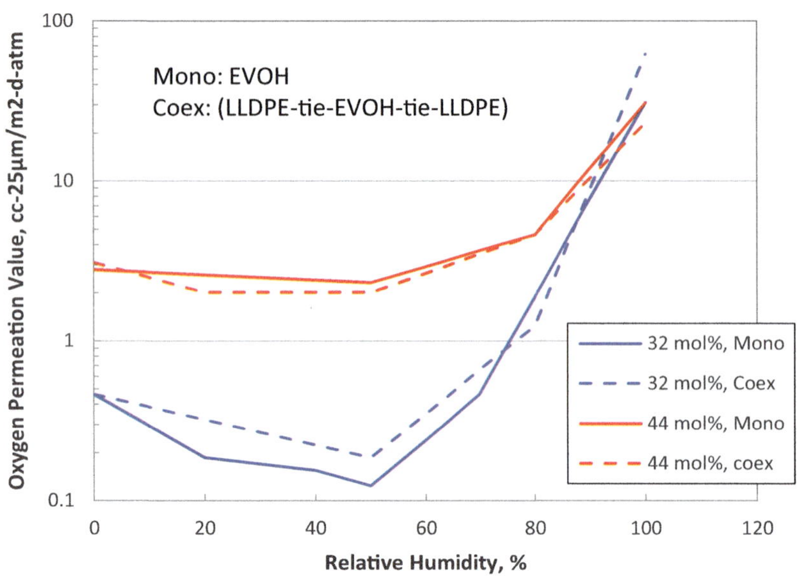

FIGURE 8.14 Effect of sandwiching EVOH between layers of PE on oxygen permeation value measured at 30°C. Coextruded structure is symmetric (LLDPE-tie-EVOH-tie-LLDPE). Data is normalized to 25 µm EVOH. Two types of EVOH are presented: 44 mol% ethylene and 32 mol% ethylene. *Data from V.M. Granger, M.A. Tigani, Permeation properties of moisture sensitive barriers materials under end use conditions, TAPPI J. 76 (1993) 98–102.*

For example, if the RH on the outside of the package is 70% and the RH on the inside is 30%, the barrier layer would see an average RH of 50% if it were placed in the middle of the film structure. As the barrier layer is moved closer to the inside of the film structure, the effective RH that it sees will become closer to that of the inside of the package. Snack foods and other dry foods such as crackers and cereal generally have a low RH inside the package. Liquids and moist food have a high RH inside the package (100% RH for water-based liquids). Barrier films for packaging such products will experience high RH in the barrier layer. Most refrigerated and freezer environments are at high RH (but at colder temperatures, which generally reduces gas permeation.)

Eq. (8.25) can be used for a quick estimate. A more accurate estimation needs to include the relative moisture transmission rates of each of the layers in the analysis. Consider a three-layer structure shown in Fig. 8.9 where the center layer is a moisture-sensitive oxygen barrier polymer such as EVOH and the outer layers are polyethylene: (PE-EVOH-PE). In reality, there will be tie layers but these are left out (considered part of the PE layers) to simplify the analysis. The tie layers are typically thin and made up of similar resins as the layers they are bonding to so this is a good assumption. By calculating the effective relative humidity of the barrier layer, one can select a representative permeability for that layer using data from the literature for P vs. RH, such as that found in Fig. 8.12. Normally the temperature is constant across the multilayer film and the partial pressure of water vapor is directly related to the relative humidity:

$$p = \frac{\%\mathrm{RH}}{100} p_s \tag{8.26}$$

where $p_s =$ saturated water vapor pressure at the given temperature.

In the following analysis the partial pressure, p, will be used with the understanding that the partial pressure can be replaced with relative humidity at any time using Eq. (8.26). The saturated water vapor pressure for air is tabulated in many standard engineering handbooks [106,107].

The partial pressure of moisture vapor in the center (barrier) layer is the average of the partial pressure at each interface (see Fig. 8.9):

$$p_c = \frac{p_2 + p_3}{2}. \tag{8.27}$$

The mass flow rate of moisture, Q/t, through an area of film A, is related to the MVTR of that layer. Since MVTR is generally measured at a particular pressure (based on the RH and temperature of the measurement) and film thickness, the MVTR is corrected to account for these differences:

$$\frac{Q}{t} = A\left(\frac{M\Delta p}{X}\right) \tag{8.28}$$

where M = MVTR, X = thickness, and Δp = partial pressure difference.

At equilibrium (steady-state)

$$\frac{Q_1}{t} = \frac{Q_2}{t} = \frac{Q_3}{t}. \tag{8.29}$$

Substituting Eq. (8.28) into Eq. (8.29):

$$\frac{M_1(p_1 - p_2)}{X_1} = \frac{M_2(p_2 - p_3)}{X_2} \tag{8.30}$$

$$\frac{M_2(p_2 - p_3)}{X_2} = \frac{M_3(p_3 - p_4)}{X_3}. \tag{8.31}$$

Solving Eqs. (8.27), (8.30), and (8.31) simultaneously yields

$$p_c = \frac{p_1\left[\frac{X_2}{M_2} + \frac{2X_3}{M_3}\right] + p_4\left[\frac{X_2}{M_2} + \frac{2X_1}{M_1}\right]}{2\left[\frac{X_1}{M_1} + \frac{X_2}{M_2} + \frac{X_3}{M_3}\right]}. \tag{8.32}$$

Eq. (8.26) can be used to convert Eq. (8.32) in terms of relative humidity:

$$\%\mathrm{RH}_c = \frac{\%\mathrm{RH}_1\left[\frac{X_2}{M_2} + \frac{2X_3}{M_3}\right] + \%\mathrm{RH}_4\left[\frac{X_2}{M_2} + \frac{2X_1}{M_1}\right]}{2\left[\frac{X_1}{M_1} + \frac{X_2}{M_2} + \frac{X_3}{M_3}\right]}. \tag{8.33}$$

The MVTR of oxygen barrier polymers like EVOH is much greater than that of polyolefin layers and the thickness of the barrier layer is usually small. Under these conditions, the term X_2/M_2 in Eq. (8.33) is small compared to the other terms and can be ignored. The equation simplifies to:

$$\%\mathrm{RH}_c = \frac{\%\mathrm{RH}_1\left[\frac{X_3}{M_3}\right] + \%\mathrm{RH}_4\left[\frac{X_1}{M_1}\right]}{\left[\frac{X_1}{M_1} + \frac{X_2}{M_2} + \frac{X_3}{M_3}\right]}. \tag{8.34}$$

Some example calculations are provided in Table 8.11 which reveal that when M_1 and M_3 are small, such as when the inside and outside layers are polyolefins, the RH essentially varies linearly across the thickness of the film and Eq. (8.25) is a good estimation of the RH of the barrier layer. This is illustrated in Fig. 8.15.

TABLE 8.11 Effect of layer composition and thickness on RH of barrier layer in a three-layer structure.

Example Structure	Outside RH1	Inside RH4	X1	M1	X2	M2	X3	M3	RHc
	%	%	μm	g/m²-d	μm	g/m²-d	μm	g/m²-d	%
(PE EVOH-PE)	50	100	45	18	10	70	45	18	75
(PE-EVOH-PE)	50	100	25	18	10	70	65	18	64.2
(PE-EVOH-PE)	50	100	65	18	10	70	25	18	85.8
(PE-EVOH-PE)	50	100	65	18	5	70	30	18	84.1
(PA-EVOH-PE)	50	100	45	200	10	70	45	18	55.2
(PA-EVOH-PE)	50	100	25	200	10	70	65	18	52.5
(PA-EVOH-PE)	90	30	25	200	10	70	65	18	87.0
(PE-EVOH-PLA)	50	100	45	18	10	70	45	600	97.3
(PE-EVOH-PLA)	50	100	25	18	10	70	65	600	94.5
(PE-EVOH-PLA)	50	100	65	18	10	70	25	600	98.5

Calculated from Eq. (8.33). RHc is the effective relative humidity in the center layer (barrier layer or layer 2).

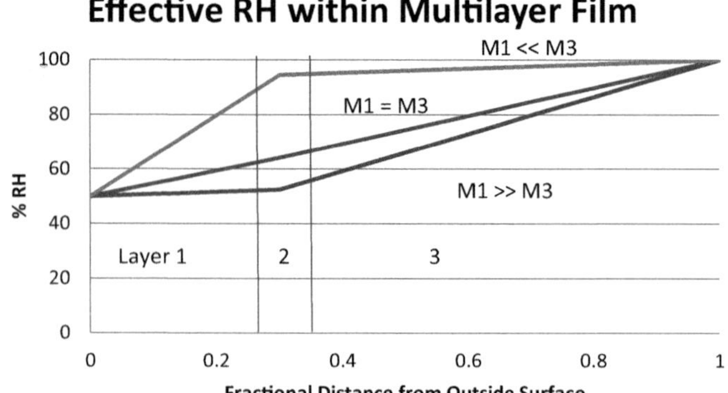

FIGURE 8.15 Effect of structure composition on RH of the barrier layer (layer 2). Assumes 100 μm film with three layers, with individual layer thicknesses 24, 10, and 65 μm, respectively. Inside RH is 50% and outside RH is 100%.

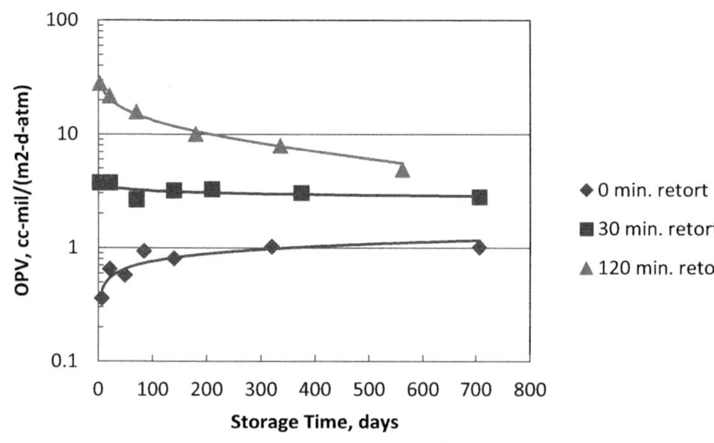

FIGURE 8.16 Effect of retort on oxygen permeability of EVOH. Structure: (PP-tie-EVOH-tie-PP) with a thickness (355/25/50/25/355 microns). Retort temperature: 121°C. Storage Conditions: 21°C, 100% RH outside, 75% RH inside. *Reproduced with permission from B.C. Tsai and B.J. Jenkins, Effect of retorting on the barrier properties of EVOH, J. Plastic Film. Sheeting 4 (1988) 63–71.*

When the moisture permeability of the outside layer (M_1) is large and the inside layer (M_3) is small, the structure acts as if the outside layer is not present. The RH of the barrier layer is nearly that of the outside of the package. This is typical of a (PA-EVOH-PE) barrier film. The reverse is true: if $M_3 >> M_1$ then the barrier layer RH is nearly that of the inside of the package. By optimizing the thickness and composition of the layers of the structure, the barrier layer permeation can be minimized. The analysis leads to the following general design guidelines for minimizing the RH and OTR of the barrier layer:

- Place the barrier layer nearest to the driest external environment.
- Place the low MVTR resin between the barrier and the wettest external environment.

The math gets rather tedious when extending Eq. (8.33) to more than three layers. When dealing with four or more layers, one can group the layers to the left of the barrier layer into a single layer using Eq. (8.21) as an approximation for layer 1 in Eq. (8.33). Similarly, the layers to the right of the barrier layer can be grouped into layer 3.

8.4.1.2.4.4 Retort sterilization Retort sterilization is a particularly demanding process for high barrier plastic packaging, combining elements of elevated temperature, pressure and relative humidity. Conditions vary with the specific application but a typical retort process involves exposing the packaged product to 121°C for 30 or more minutes in a pressurized steam environment. Under these conditions, EVOH is acutely prone to loss of barrier performance. Most structures using EVOH surround the barrier layer with layers of polyolefins such as PP or PE that at room temperature provide an effective moisture barrier. Temporary swings in ambient RH would not impact the EVOH performance since it typically takes on the order of an hour or more for the moisture permeation to reach a new steady state. At the elevated temperature found in retort, however, polyolefins readily transmit water to the EVOH. The EVOH becomes saturated and its oxygen permeation increases substantially.

Following retort, the EVOH layer eventually dries out to the equilibrium moisture content associated with the ambient RH and temperature [described by Eq. (8.33)]. This can take considerable time since once the package cools down, the polyolefin layers once again become good moisture barriers and impede the drying process. Even after drying, however, the permeability of the EVOH never fully recovers. This is shown in Fig. 8.16 for a (PP-tie-EVOH-tie-PP) plastic can. The OPV of cans exposed to retort (top and middle curves) never converge to the steady-state OPV of an untreated can (bottom curve) even after 700 days of storage. This phenomenon is known as retort "shock." Tsai and Jenkins [108] attribute this to a permanent change in the free volume that cannot be overcome under normal storage conditions.

Lopez-Rubi et al. [109] explore the effect of exposing monolayer EVOH film to elevated temperature and/or humidity. They use a variety of analytical techniques such as wide-angle X-ray scattering (WAXS), small-angle X-ray scattering (SAXS), Raman spectroscopy, Fourier transform infrared spectroscopy (FTIR), and differential scanning calorimetry (DSC). They find:

- Thermal exposure alone results in an increase in crystallinity, which should decrease oxygen permeability, although it is not measured in this study.
- Exposure to high humidity at room temperature essentially "melts" some of the imperfect crystals of the EVOH. These crystals do not return when the film dries out.
- Exposure to a combination of high temperature and humidity (120°C for 20 minutes to simulate retort) shows a significant change in the crystallinity and crystalline morphology. Visual inspection reveals damage from water vapor exiting the amorphous phase as well as partial melting of the crystalline phase. Retort appears to have resulted in a more heterogeneous crystalline structure with small-sized crystals. The crystalline structure itself is more defective, especially for grades of EVOH with low levels of ethylene.

In a follow-up study [110], multilayer films are tested. The structure is (PP-tie-EVOH-tie-PP) with thickness (90/10/10/10/90 microns). Grades of EVOH ranging from 26 to 48 mol% ethylene are used. The film samples are dried at 70°C for 1 week prior to conditioning. Conditioning states include no treatment, retort (120°C for 20 minutes in a steam autoclave), retort followed by drying in an oven at 70°C for 1 week, and thermal annealing (at low humidity). OTR is measured at 45°C and 0% RH, conditions chosen to speed up the drying process following retort. For the retorted samples, the testing is done continuously to generate curves similar to Fig. 8.16. The results show:

- A substantial increase in OTR immediately following retort was followed by a decline over 30 or so hours due to drying. The recovery time would likely be considerably longer if the testing had been done at room temperature.
- A permanent offset in OTR after retort for grades of EVOH with low ethylene content. This is similar to that found by Tsai and Jenkins for 32 mol% ethylene EVOH.
- For grades of EVOH with 44 mol% ethylene and higher, no permanent offset in OTR. The OTR returned to the value found for the untreated sample.
- Films that are dried at 70°C for a week after retort show a reduction in OTR compared with untreated samples. They attribute this to higher levels of crystallinity as shown by FTIR and scanning electron microscopy (SEM). They note that this is not a practical solution to retort shock.
- Annealing the films prior to retort (160°C for grades with less than 44 mol% ethylene and 140°C for grades with 44 or 48 mol% ethylene) for 20 minutes prevented some of the loss in crystallinity in the low ethylene samples.

A number of ideas have been proposed to reduce the impact of retort shock of EVOH, including:

- Using higher ethylene content grades of EVOH. Grades with 32 and 38 mol% ethylene are probably the most used grades in film applications. As shown by Lopez-Rubio et al. [110], switching to a 44 mol% grade essentially eliminates retort shock. The higher OPV of these grades, however, often make them unsuitable for many high barrier applications. These grades also will still exhibit a temporary increase in OPV immediately after retort.
- Pre-annealing the film at 140°C–160°C [110]. This would likely be challenging to do on a commercial scale.
- Post-drying at 70°C [110]. This is impractical for packaged foods.
- Adding a desiccant to the tie layers [111]. Farrell developed technology that was used in plastic retorted cups (PP-tie-EVOH-tie-PP) for many years. A desiccant was selected that absorbs water at the elevated temperatures of retort, just when the PP layers have their highest moisture transmission. The addition of the desiccant reduces the amount of water the EVOH layer sees during retort and significantly reduces oxygen ingress and retort shock. In a

thin film, this may be impractical since the desiccant may run out of absorption capacity during the retort recycle. The addition of desiccant also reduces optical clarity.

- Adding nanoparticles such as nanoclay (kaolinite) [112]. Laragon and Nunez show that oxygen and moisture permeation decrease with the addition of 4% clay and note that there is less damage after retort when the clay is added.
- Adding other polymers such as amorphous nylon, ionomer or polyolefin to the EVOH [113–115].

None of these approaches has proven to be highly successful. In most cases, the film structure must be designed with thicker EVOH layers to compensate for the change in performance during and after retort. Other barrier materials such as aluminum foil are often used for retort flexible packaging.

Saez [116] proposes replacing the outer PP layer with PA. This allows the EVOH to dry out faster following retort, lowering the permeability. He evaluates a series of 100 μm films with PA (30%) or EVOH (10%) or both. A low hydrolysis grade of PA is used. Results show a symmetric "high" barrier structure (PP-tie-EVOH-tie-PP) and an asymmetric PA structure (PA-tie-PP-PP-PP) have equivalent oxygen permeation following retort. An asymmetric high barrier structure (PA-EVOH-PA-tie-PP) has the lowest permeability following retort.

Kim et al. [117] find that blends of polyketone with 30–70 wt.% EVOH have oxygen permeation values lower than EVOH alone. They attribute this unexpected result to the combination of lamellae EVOH morphology and strong polar interactions between the polyketone and EVOH. Immediately following retort, the OTR of a blend of 30 wt.% EVOH with 70 wt.% polyketone is reported to only increase by 3x whereas the OTR of the EVOH control film increases by 74x. The blend also has physical properties that suggest it may be tougher and orient better than EVOH; the elongation at break for the blend is around 300% compared with around 20% for EVOH. Optical properties are not described; typically, immiscible polymer blends are not transparent. Nevertheless, such blends may find use in flexible packaging applications where mitigation of moisture sensitivity of EVOH is needed.

8.4.2 Permeation through defects, pinholes and perforations

High barrier flexible packaging structures containing aluminum foil, vacuum deposited metal coatings or "glass" oxide coatings (such as SiO*x* and AlO*x*) are impermeable except in regions where defects, cracks or pinholes may be present. The amount of permeation is governed by the number and size of the pinholes, as well as the permeation properties of the rest of the film structure. In MAP, holes or perforations are sometimes introduced to control the atmosphere inside the package, especially for fruits and vegetables that transpire during storage. Here the holes penetrate through the entire package structure and the amount of permeation is governed by the number and size of the holes and the diffusion coefficient of the permeant in air. The analysis for perforations can also be used to understand the effect of channel leakers in the seal area.

8.4.2.1 Permeation through pinholes in high barrier packaging

Defects resulting in pinholes in aluminum foil and metallized film can arise from the foil or metallization manufacturing process or from handling and use. Table 8.12 shows data provided by Murray [118] for the number of pinholes specified by an aluminum foil manufacturer. The number of defects increases with decreasing foil thickness. According to Murray "the number and size of the pinholes will depend on the filtration of inorganic materials from the molten

TABLE 8.12 Specifications for number of pinholes for various aluminum foils from a major supplier.

Foil gauge	Foil caliper, μm	Average number of pinholes	Maximum number of pinholes
28.5	7	423	1584
35	9	211	1056
50	13	85	528
75	18	21	106
100	25	0	0

Source: Reproduced with permission from L. Murray, The impact of foil pinholes and flex cracks on the moisture and oxygen barrier of flexible packaging, TAPPI PLACE. Conf., 2005.

TABLE 8.13 Oxygen transmission rates of some flexed and unflexed foils and films.

Substrate	Unflexed	Flexed
9-μm Aluminum foil 3-ply structure	0.47	50
12-μm met-OPET 2-ply structure	1.5	10.8

Units are cm^3/m^2-day at 23°C/50% RH.
Source: Based on data from J. Lush, D. Ferrari, An economic value model for converters comparing the total cost of metallized film vs. foil in flexible packages, TAPPI PLACE. Conf., 2006.

aluminum, the casting process and the rolling conditions such as lubricant type and lubrication function." Weis [119] and Yializis et al. [120] discuss how the process for metallizing OPP and OPET films can result in cracks and pinholes. The basics of the metallization process and some of the factors that influence quality are described in Section 2.5.

Lush and Ferrari [121] provide data, shown in Table 8.13, on the effect of flexing on the OTR of foil and metallized BOPP film. Flexing generally increases permeation through the creation of cracks. Weis [119] conducts several mechanical stress tests on metallized films and finds that stress has little impact on the oxygen barrier unless the film is elongated by more than 5% in the process. Typically in flexible packaging, foil and metallized films are laminated with polymer layers for protection during handling and use. The toughness and adhesion of the polymer layers to the foil or metallized film impact the flex crack resistance of the laminate. For example, acid copolymers generally provide better flex crack protection for foil than LDPE [122].

Besides flexing, the package may be subjected to other stresses during storage or the end-use that may affect pinholes and barrier performance. Weis found that metallized films stored for 3 years have greater permeation than when originally tested. He attributes this to elevated pinhole density caused by moisture and corrosion at key points on the surface.

Weis shows that the permeability of oxygen through metallized films is determined by the number of pinholes and is inversely proportional to the film thickness. Murray concludes that even with the pinholes, the permeation through the pinholes is smaller than through the edges of the film (heat seals) and that flexed packages with foil provided a better barrier than those without foil.

The gas permeability through barrier polymer films made by vacuum depositing inorganic coatings such as SiO*x* and AlO*x* is also controlled by defects [123–125]. Defects arise from contaminating particles such as antiblock agents, and scratches during handling of the films during subsequent converting processes and from mechanical loading (linear elongation of 2%–3% is enough to destroy the coating) [123]. Other factors that affect oxygen and moisture permeation through vacuum-deposited inorganic coatings include [123]

- **Thickness of the coating**. Industrially, coating thicknesses ranging from about 40 to 60 nm are used. At low coating thickness, typically below 10 nm, OTR increases rapidly with decreasing thickness. The surface of the film substrate is rough on a molecular level; thin coatings are not enough to smooth out the peaks and valleys and provide a complete coating. Above about 60 nm the coatings are more brittle and the defect rate controlling the permeability increases.
- **Thickness of the polymer film next to the coating**. Smaller defects in the coating reduce the relevant thickness of the film that influences permeation. See the discussion below.
- **Smoothness of the substrate film**. Surface roughness of the substrate film contributes to uneven coating. Pretreatment that smooths the polymer film surface is shown to decrease the number of defects and reduce permeability.
- **Barrier of the layer next to the inorganic coating**. High film barrier increases the overall barrier of the packaging structure. Coextrusion, precoating the film, or barrier adhesives can be used to increase the barrier performance of the layer adjacent to the coating layer. In one example applying an acrylate precoating reduces OTR of a BOPP film coated with an inorganic coating. The acrylic coating is polar and is more effective at reducing OTR than MVTR. In a second example, coextrusion of PET with polyethylene naphthalate (PEN), a higher barrier polymer, reduces MVTR of OPET coated films by an order of magnitude.
- **Design of the structure (film and coating)**. Moosheimer and Langowski [123] provide examples of inorganic coatings that are more effective at reducing OTR than MVTR. Since the film, precoating, adhesives and coatings all contribute to permeability and affect moisture and oxygen barrier in different ways there may be ways to optimize performance.

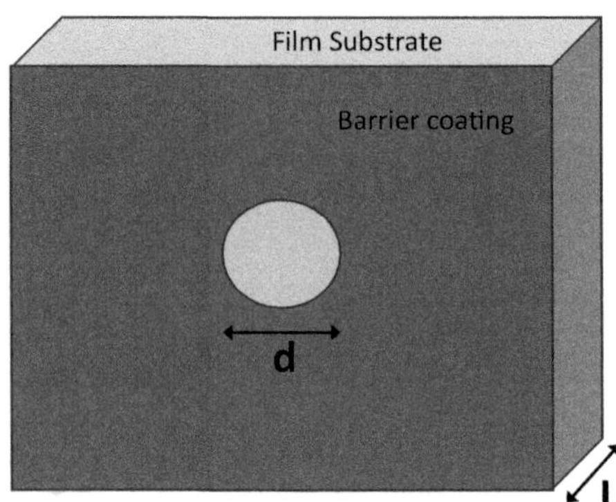

FIGURE 8.17 Structure of film with a hole in barrier coating considered in analysis by Rossi and Nulman [126].

- **Double-sided coating**. Czeremuszkin et al. [125] predict that double-sided coated films will have approximately two times the barrier of single-sided coated films of the same coating. This is borne out in experimental data unless the defects are very large. The coatings and films can be thought of as two coating/film combinations in series with each film ½ the thickness of the actual film substrate. Since thickness would have to be much thinner for the second coating to have a more than 1x effect, 2x is what is observed in commercial structures.

Rossi and Nulman [126] find that modeling the transmission of permeants through a defect in a barrier coating is complex. They consider a film structure as shown in Fig. 8.17 where the permeant flows through a hole of diameter d in a thin coating on top of a film of thickness *L*. Here the term "film" is used to denote the substrate that is coated by the barrier layer. The coating is assumed to be impermeable and the diffusion through the film is characterized by a diffusion coefficient *D*. They develop the equations for a circular hole but later generalize their arguments for any shape hole characterized by a single dimension. They find that the amount of permeant crossing the hole per unit time, *Q*, is given by two expressions, depending on the relative size of *L* with respect to *d*:

$$Q \propto Dd \text{ for } L \gg d \tag{8.35}$$

$$Q \propto D\frac{d^2}{L} \text{ for } L \ll d \tag{8.36}$$

where Q = amount of permeant passing through the hole in the barrier layer per unit time, d = diameter of hole or defect in the barrier layer, L = thickness of the film backing up the barrier layer, D = diffusion coefficient of permeant in the film.

The two-regime solution is a consequence of two different types of boundary conditions: zero flux at the surface where the barrier layer is intact, and a steady-state flux at the hole boundary. These results, along with numerical simulations by Rossi and Nulman and Hanika et al. [127] provide the following insights into the effect of substrate and defect on permeability in these systems:

- Permeation increases with an increasing number of defects and defect area.
- Defects are more important when the permeation of the permeant through the film is high.
- For large holes, permeation increases with hole area (d^2) and decreasing film thickness.
- For small holes, permeation is independent of film thickness and increases with the diameter of the hole.
- Many small defects lead to a higher permeability than a few large defects with the same area fraction of defects.
- Permeation increases with increasing spacing between defects, up to the point where the average spacing between defects is three times the diameter.

Hanika et al. develop a formula for predicting transmission through a film with a distribution of defects in the barrier layer based on numerical simulations:

$$\frac{Q_{\text{coated film}}}{Q_{\text{film}}} = \sum_{\text{all defects}} \left(\frac{\frac{A_D}{l^2}}{1 - \exp\left(-0.507\frac{\sqrt{A_D}}{L}\right) + 0.01\frac{A_D}{l^2}} \right) \tag{8.37}$$

where A_D = area of defect, l = distance between defects, and L = thickness of film.

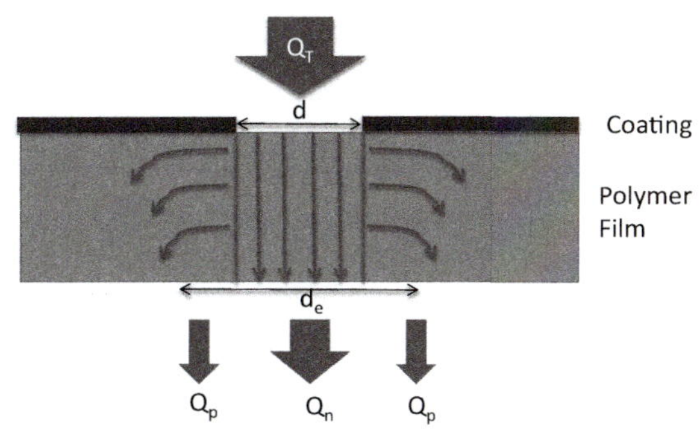

FIGURE 8.18 Concept of normal (Q_n) and lateral (Q_p) diffusion of permeant in a region near a defect in a barrier coating. The concept of diffusion in two dimensions leads to the effective defect diameter (d_e) being larger than the actual defect diameter (*d*). *Based on G. Czeremuszkin, M. Latrèche, A.S. da Silva Sobrinho, M.R. Wertheimer, A simple model of oxygen permeation through defects in transparent coatings, in: 42nd Annual Technical Conference Proceedings - Society of Vacuum Coaters, pp. 176–180.*

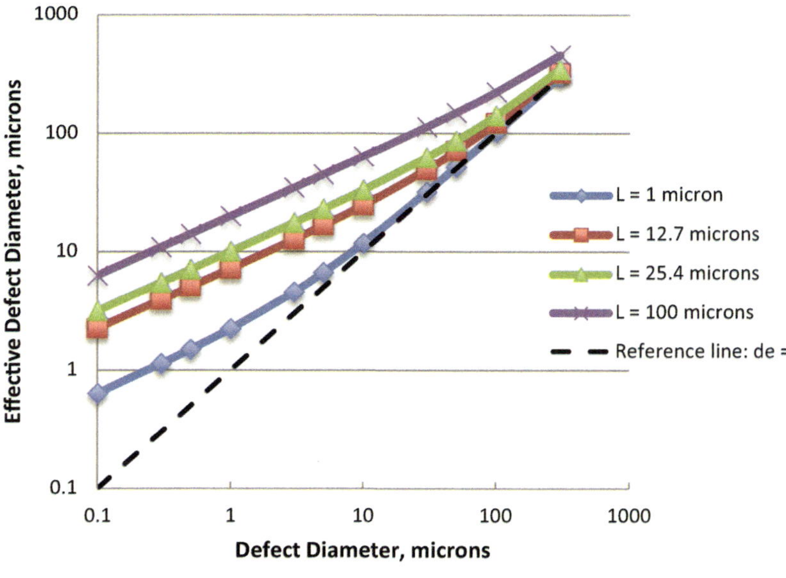

FIGURE 8.19 Effective defect diameter versus defect diameter based on Eq. (8.38) for several film thicknesses.

They determine the defect size distribution of a metallized BOPP film using optical microscopy, scanning electron microscopy and atomic force microscopy. Optical microscopy is found to be sufficient to measure the defect distribution. Their prediction of oxygen permeation using Eq. (8.37) agrees well with their experimental permeation results. Agreement is not as good for moisture permeation, which they attribute to a large number of domain boundaries on the metallized surface.

da Silva Sobrinho et al. [124] develop a method to visualize defects in ultrathin inorganic coatings that allows analysis of defect size and distribution across hundreds of square centimeters of film. The method is based on oxygen plasma etching and renders defects visible by an optical microscope. Atomic oxygen created by the plasma etches away the polymer below a defect. The inorganic coating itself is not susceptible to the etching process and stays intact. A staining technique is used to visualize the defect under a microscope. Measurements of defect density of SiOx and silica nitride coatings of PET film are found to correlate with OTR.

Czeremuszkin et al. [128] develop a simpler model that agrees with the numerical model of Rossi and Nulman. It accounts for the counter-intuitive observation that diffusion of the permeant through the region below a defect is greater than it would be if no barrier coating were present. In the case of no-barrier coating, diffusion is one-dimensional across the thickness of the film. With a barrier coating with a sparse distribution of defects the diffusion can be two dimensional in the region below the defect as depicted in Fig. 8.18. This increases the local amount of diffusion, although overall the amount of permeation is less than for the no-coating case since most of the area is coated.

Czeremuszkin et al. derive an expression for the effective defect diameter that has the same permeation rate as an uncoated film (one-dimensional diffusion). Their key simplifying assumption is that the diffusive driving force (concentration gradient) is the same in both the thickness and lateral dimensions near the defect edge. The result is Eq. 8.38:

$$d_e = \sqrt{4dL + d^2} \tag{8.38}$$

where $d_e =$ effective defect diameter, d = defect diameter, and L = thickness of film.

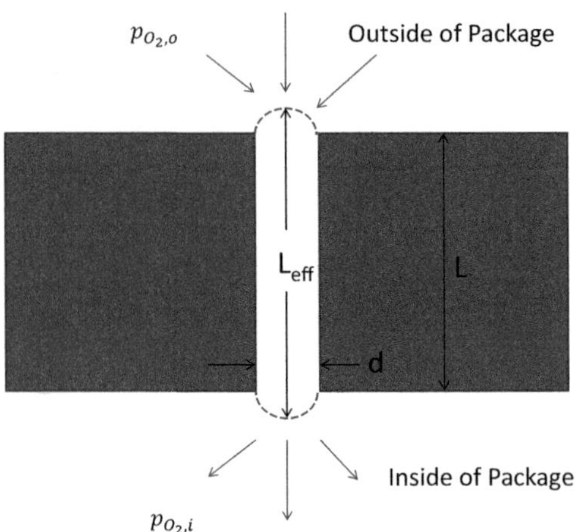

FIGURE 8.20 Schematic of the hole diffusion model. *Based on the analysis of D.R. Paul, R. Clarke, Modeling of modified atmosphere packaging based on designs with a membrane and perforations, J. Membr. Sci. 208 (2002) 269–283.*

The effective defect diameter based on Eq. (8.38) is plotted versus actual defect diameter for several film thicknesses in Fig. 8.19. The dashed curve represents the case where there is no barrier coating. Czeremuszkin et al. note that these results are independent of the diffusion coefficient or barrier properties of the film; the defect edge effect is simply due to geometric considerations.

Eq. (8.38) was derived for single defects. For the practical case where many pinholes or defects may be present, this equation can still be used if the defects are independent: the amount of total permeation is simply the sum of the permeation through each defect. This occurs when the separation distance between defects exceed d_e. If two small defects are separated by a distance less than $d_e/2$ they behave like a single larger defect. Thus, permeation through coated films with defects close to one another or agglomerated may be less than what would be expected based on their number. When the number density of defects is large enough so that the average spacing between defects is of the order $d_e/2$ then the coating may behave as if does not exist. Czeremuszkin et al. postulate that this may be why some nominally impermeable coatings sometimes show poor barrier performance despite large surface coverage. They provide the example of coatings with columnar growth morphology, which have significant porosity, as falling into this category

8.4.2.2 Permeation through perforations or channels

When the hole or channel extends through the whole package film, diffusion and convection of molecules through the hole must be considered. Solubility and Henry's law are no longer factors. A number of approaches have been used to model the mass transport through a hole in a film, including modified Fick's law, the Maxwell-Stephan equation, and Knudsen diffusion [129,130]. For the diffusion of oxygen the problem can be simplified by treating the gas phase as a pseudo-binary gas (O_2 and N_2) since the binary diffusion coefficients are similar [131,132]. Other gases such as CO_2 can be treated in turn. This avoids the complications of the more rigorous Maxwell-Stephan equation.

Following the approach of Paul and Clarke [132], the diffusion of gas from the outside to the inside of the package can be divided into three zones, as shown in Fig. 8.20. Diffusion to and from the hole is considered in a hemisphere extending from the hole. An effective length of the hole is derived to account for the diffusion in these regions:

$$L_{\text{eff}} = L + \frac{7}{6}d \tag{8.39}$$

where L = length of hole (thickness of film for the case of perforations, or length of channel in seal area) and d = diameter of the hole.

The correction factor 7/6 varies depending on the assumptions used in the entrance and exit of the hole. Del-Valle et al. [131], for example, find the factor to be ½.

For the case of oxygen, the molar transport through the hole is given by

$$Q = n_{\text{holes}}\left(\frac{\pi d^2}{4}\right)\frac{D_{O_2,\text{air}}\left(p_{O_2,o} - p_{O_2,\ i}\right)}{L_{\text{eff}}RT} \tag{8.40}$$

where Q = molar transport rate through the holes, n_{holes} = number of holes, d = diameter of hole, $D_{O_2,\text{air}}$ = diffusion coefficient of oxygen in air, $p_{O_2,o}$ = partial pressure of oxygen on outside of package, $p_{O_2,i}$ = partial pressure of oxygen on inside of package, R = gas constant, and T = temperature (absolute).

For other gases or gas mixtures, values are simply substituted for oxygen and air in Eq. (8.40). In deriving this equation, Paul and Clarke assume the pressure inside the package is nearly the same as on the outside so that the convective flow is small.

The diffusion coefficient used in Eq. (8.40) differs by several orders of magnitude from the diffusion through most polymer films. For example, $D = 0.18\ cm^2/s$ for oxygen in air at STP (298K, 1 atm). For diffusion of oxygen through LDPE, $D = 5 \times 10^{-5}\ cm^2/s$. Graham's law of effusion states that the rate of transport of a gas through a pinhole is inversely related to the square root of it molar mass. Therefore,

$$D \propto \frac{1}{\sqrt{\widetilde{M}}} \tag{8.39}$$

where $\widetilde{M}$ = molar mass of a gas molecule. Hence, the permeability of gasses through perforations has some selectivity based on the molar mass of the molecules. Lee and Renault [133], for example, find that longer pinholes with smaller diameters accelerate the formation of higher CO_2 and lower O_2 concentrations within a modified atmosphere package.

Eq. (8.40) shows that the amount of transport of gas through the hole increases with:

- Increasing number of holes.
- Increasing hole diameter.
- Decreasing hole length.
- Increasing diffusion coefficient (in the gas mixture).
- Increasing partial pressure difference between the inside and outside of the package.

Several investigators [130–143] have coupled models of gas transport through holes in MAP with models of product transpiration and the like. A general conclusion is that the number of holes and their dimensions along with the permeability of different films can be used to optimize MAP performance. Similarly, Chung et al. [144] model the permeation due to small leaks and conclude that pinholes and channel leakers have a significant impact on high barrier packaging and should not be ignored.

8.5 Emerging technologies

8.5.1 Oxygen scavenging

The concept of actively scavenging oxygen from a package dates back to the 1920s [145]. Combined with a passive barrier film structure (i.e., incorporating a barrier layer such as EVOH or metallized film), scavenging can significantly extend the shelf life of oxygen-sensitive products. Most applications have been inserts, such as sachets or labels, to flexible packaging or liners for closures of rigid packaging. Using the technology in the flexible film structure itself has had limited commercial success.

Rooney [145] provides a good review of oxygen-scavenging technology and the reader is referred to his work for details. Here a few of the technologies are highlighted [145–147]

- Powdered iron or other oxidation species such as sulfites. The oxygen is consumed by the oxidation of the iron. The reaction is fast and very effective with enough concentration of iron. It is mostly used in sachets, such as Mitsubishi Gas Chemical's Ageless sachets. Color and scavenging capacity issues have made it difficult to apply to a packaging film structure.
- Polydiene and cobalt. There have been numerous patent applications on this type of technology, mostly with polybutadiene, but limited commercial applications. One issue is the creation of aldehyde byproducts that can result in poor organoleptics.
- MXD6 nylon and cobalt. Cobalt cations form a complex with the benzylic groups of *meta*-xylylene. This complex traps oxygen and/or causes the oxygen to oxidize the benzylic groups. MXD6 provides some passive barrier as well.
- Cyclohexene and cobalt. Chevron Phillips developed technology based on polyethylene methacrylate with cyclohexenyl groups that reportedly give off few byproducts [148].
- Platinum or palatium based coatings. In the presence of hydrogen in MAP, these metals can scavenge oxygen.
- α-tocopherol and metal catalyst (such as iron, copper, manganese and cobalt) [149].
- Ascorbic acid-based systems, often enhanced with a copper or iron catalyst.
- Enzyme-based scavengers. Glucose oxidase with a catalyst is the most commonly considered [147,150].

A few iron-based (moisture activated) and UV or photo-activated systems have been commercialized in films or labels. Many of these have been discontinued, but scavengers and other active package concepts are an area of active research [147,151,152]. Some of the potential issues to consider when evaluating scavenging technology include:

- **Regulatory compliance of additives for food packaging**. Many of the technologies involve new chemistries or reactions. These chemistries must be in compliance with global food packaging regulations.
- **Capacity/shelf life of the scavenger**. The inability of a thin film to hold enough of the scavenger and still provide the other requirements of the package (strength, clarity, etc.) is one reason adding the scavenging technology to the film has only had limited commercial success. Sachets and rigid packaging have more capacity. One way to extend the shelf life of the scavenger is to combine the scavenger technology with a good passive barrier. This will reduce the amount of oxygen coming into the film from the outside that needs to be scavenged.
- **Initiation of the scavenging reaction**. Given the limited capacity of the film to hold the scavenger, ideally the scavenging reaction should be turned on only when it is needed. Some technologies are initiated by moisture. Others use UV light. Understanding the technology and matching it to the type of product being packaged and the facilities is an important consideration. For example, moisture initiation may not be a good approach for dry food.
- **Organoleptics**. Many of the technologies have failed due to the poor taste/odor of byproducts that are given off during the scavenging reaction. This should be tested with the product to be packaged.
- **Color**. Some of the technologies, such as iron, impart color to the plastic. Iron works well when hidden in a sachet insert, but not so well in a clear film.
- **Scavenging rate**. The rate of scavenging must match the speed of oxidation of the packaged article. If a food product spoils before the scavenger has even been initiated, the active barrier is wasted.

8.5.2 Layer multiplication

Layer multiplication involves a special coextrusion technology where layers are split and stacked on top of each other to double the number of layers. By repeating the process several times film structures with tens, hundreds or even thousands of layers can be made [153]. Sometimes referred to as micro- or nanolayer technology due to the thickness dimension of the layers, this technology has been evaluated for potential barrier applications. A schematic of the process is shown in Fig. 8.21. Much of the original development was done by Schrenk and co-workers at Dow Chemical [154–160] using a feedblock and a series of multipliers. In each multiplier, two or more layers are split down the middle and then stacked on top of each other, multiplying the number of layers by a factor of two. Adding additional multipliers increases the number of layers by a factor of 2^n where n is the number of multipliers.

Most of the early applications for layer multiplication technology were for optical films [161–167]. Various optical effects can be achieved through the proper selection of refractive indices of the polymers used in adjacent layers. Such films find use as optical filters and reflectors. Electrical capacitor films have also been fabricated.

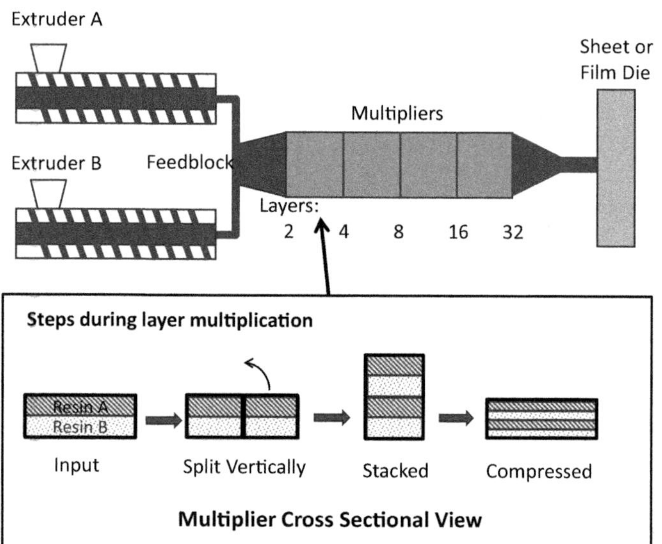

FIGURE 8.21 Schematic of layer multiplication technology.

When the original patents expired in the 2000s, several feedblock and flat die manufactures, including EDI/Nordsen and Cloeren [168], came out with versions of the technology, opening it up to potential use by a broader group of converters. Schirmer [169–172] developed a blown film process that involves flow through a series of thin "layer sequence repeater" plates to create up to about 30 thin layers; the technology is offered by Alpha Marathon. Dow Chemical [173] introduced annual die technology for a blown film that uses the feedblock/layer multiplier technique combined with a crosshead style die. The multilayer composite formed in the multipliers flows around the mandrel and combines on the other side to form the tube. To deal with this weld line, Dow designed the die to overlap the flow, doubling the number of layers in the region. Zumbrunnen [174] at Clemson University developed blown film technology based on chaotic advection to create microlayers in a controlled manner.

Layer multiplication has proven useful for studying the fundamentals of polymer flow, crystallinity and orientation in confined layers, polymer diffusion, and interfacial adhesion and reactivity. Dooley et al. [175–177] use the technique to show how the elastic behavior of polymers creates complex flow patterns in dies. Macosko's group at the University of Minnesota found the technique to be useful for its ability to amplify interfacial effects through the creation of hundreds of interfaces. They use it to study polymer interlayer adhesion and other interfacial phenomena [178–184]. Hiltner and Baer's group at Case Western Reserve University (CWRU) also used the technique to study interfaces, such as polymer diffusion and adhesion [185,186]. The focus of the CWRU group, however, is on the effect that confined micro and nanolayers have on film properties. They find that the crystalline structure of PE and PP when confined between layers of PS or other polymers in which they are immiscible, changes as the layers become thinner and approach the molecular dimension [187–190]. The amount of crystallinity decreases. For HDPE they find crystallinity decreases from 60% to 33% in one case and from 56% to 40% in another. A 3x increase in oxygen permeability is reported, which they attribute to the reduced crystallinity. A decrease in PET crystallinity and greater orientation of crystalline domains is also reported when confined to thin layers [191].

Schrenk and Alfrey [159] show that sandwiching a high modulus, low elongation material between layers of low modulus, high elongation material may result in a film that undergoes much larger ductile deformation before failing than what the stiffer material would do alone. The high elongation material acts to stop the propagation of cracks across the brittle layers, allowing the stress in the brittle materials to reach their ductile yield point. All the layers can be stretched to large elongations; the composite has both high modulus and high elongation. They give several examples that are explored in more detail in Chapter 9.

The CWRU group also reports an increase in ductility of brittle layers sandwiched between other polymers. For example, polycarbonate/styrene acrylonitrile (PC/SAN) and polycarbonate/poly methyl methacrylate (PC/PMMA) structures show improved toughness as the layer thickness is decreased. They attribute this to a change in deformation from crazing to shear yielding as the layers get thinner. Of critical importance is the interaction between the craze opening and the microlayer bands in the PC. Adhesion between the layers is also important [185,192,193].

Reports on the barrier properties of micro and nanolayer films have been mixed. Schrenk and Alfry [159] show that films with 125 alternating layers of PS and PE or PP have MVTR properties that can be predicted by Eq. (8.22). The MVTR decreases with an increasing amount of PE or PP in the structure and has a value similar to a structure with all the PP or PE in a single layer. At least for PE and PP, the thinness of the layers does not affect the permeability. As discussed above, CWRU found that the crystallinity of HDPE decreases with decreasing thickness, which generally increases MVTR.

Studies by Hiltner and Baer's group at CWRU [194–198] find that when polyethylene oxide (PEO) or polycaprolactone (PCL) is confined to nanolayers, the oxygen barrier improves by two orders of magnitude. In the original experiments, PEO was sandwiched between layers of ethylene acrylic acid copolymer or polystyrene. Layer multiplier technology is used to create films with 1000 or more layers, with the thickness of each layer in the nanometer realm. As the thickness dimension is reduced, the confinement changes the crystalline morphology from spherulites to truncated spherulites to lamellae oriented in the plane of the film. This results in a 100x improvement in a gas barrier. So far the technology does not seem to have been commercialized. One factor is that PEO is water-soluble and so the OPV is likely to be strongly influenced by relative humidity, although data by the CWRU group shows only a 2–3x increase in OPV at 85% RH compared with 0% RH. Another is that PEO can be difficult to process and matching of viscosities to prevent flow instabilities is critical when creating hundreds or even thousands of layers. (Flow instability in coextrusion is described in Chapter 13). Scaling up the technology may prove difficult. More importantly, even with a 100x reduction in OPV, the OPV of PEO and PCL is still not as low as commercial barrier polymers such as EVOH or PVDC.

Extending this to other polymers has proven to be challenging. The formation of the in-plane lamellae depends on the crystallization temperature and confining resin. The confining resin needs to be immiscible with the barrier layer.

If the quenching temperature is below a critical value for the polymer, the crystalline lamellae form on-edge (perpendicular to the plane of the film) rather than in-plane. The on-edge morphology increases gas permeation. For PEO and PCL the critical temperature is 25°C and 10°C, respectively [199–201]. For syndiotactic PP, the transition is at 90°C and the lamellae are almost randomly oriented [202]. Limited work with isotactic PP shows an even higher transition temperature [199]. For polyvinylidene fluoride (PVDF) the transition occurs at 115°C [203]. This suggests that for most polymers, if the in-plane lamellae form, they may do so at unrealistically high temperatures for commercial processes.

Lin et al. [204–206] demonstrate that in-plane lamellae morphology can also be formed in oriented multilayer films if the polymer is sufficiently confined to nanometer dimensions. They start with a PP-PEO (90/10 vol%) film with 16 microlayers of PEO. The PEO layers are biaxially stretched down to 38 nm in thickness using a laboratory film stretcher. The crystalline morphology changes from randomly oriented lamellae to in-plane lamellae. The OPV (23°C/0% RH) is reported to be 164 cc-mil/m^2-d-atm, about 7 times lower than their BOPP control. To put this into perspective, a metallized BOPP typically has an OPV of around 25 cc-mil/m^2-d-atm. The authors report the OPV of the PEO nanolayers to be 20 cc-mil/m^2-d-atm; EVOH has values 0.3–1.2 cc-mil/m2-d-atm. Concerns about moisture sensitivity (the OPV increases by a factor of two at 85% RH) led the investigators to also explore biaxially oriented PP-PCL films. Initial trails show that on-edge lamellae form, which the authors propose is due to nucleation of the PCL by the PP layers. They show that inserting PS between the PCL and PP as the confining layer: PP-[PS-PCL-PS]$_n$- PP does provide the desired in-plane lamellae morphology after stretching. An order of magnitude reduction in OPV without significant change in physical properties is obtained, but with two practical limitations. One is that the best results occurred after the films had been subjected to isothermal recystallization for 30–45 minutes at 33°C–41°C in a second oven after stretching. Secondly, the stretching process has a narrow operating window for creating continuous PCL and PS layers; discontinuity results in higher permeation.

Layer multiplier studies using conventional barrier polymers such as PA 6 or EVOH have not shown significant reductions in OPV when reducing the layer thickness to the micro or nanometer regime [207–212]. Oliver [207] found some reduction in OPV with thinner layers of EVOH compared with thicker layer controls in sheet but not in films. In many investigations the permeability is actually higher than conventional coextrusion of the same total amount of EVOH. Reduction in crystallinity and layer nonuniformities due to flow instabilities have been suggested as possible reasons. The nature of the tie resin has been shown in some studies to affect flow instability; development of tie resin technology to allow nanolayered films may be needed.

Two advantages have been reported for EVOH films. Both involve a reduction in brittleness of EVOH. EVOH is a very stiff polymer at room temperature, making it prone to flex cracking and difficult to draw during thermoforming and orientation processes. Several studies have found that by splitting the EVOH into tens or hundreds of layers, the flex crack resistance is improved. Medlock and Dolgovskij [210] compare a conventional 7-layer cast film structure with a 21-layer structure made with Cloeren's layer multiplier technology. Two 130μm conventional films are made as controls:

1. (58 μm LDPE – 4 μm tie – 7 μm EVOH – 4 μm tie – 58 μm LDPE).
2. (58 μm LDPE – 4 μm PA – 7 μm EVOH – 4 μm PA – 58 μm LDPE).

In structure 2, each LDPE layer is comprised of two separate layers, the one closest to the PA layer contains a tie resin concentrate. Twenty-one layer counterparts are made of each of these structures by splitting and multiplying the tie-EVOH-tie or PA-EVOH-PA core into 17 layers. Each layer is less than 1-μm thick. The sum of the thicknesses of each of the EVOH layers total that of the conventional controls so that the barrier properties can be directly compared. OPV measurements show that the 21-layer structures have essentially the same permeation as the 7-layer conventional structures, within experimental error. The films are subjected to Gelbo flex cycles and re-measured. The nanolayer film structures show superior flex crack resistance. After 500 flexes, the conventional films experience a substantial increase in OPV and pinholes. The tie-EVOH-tie (structure 1) nanolayer films show little change in OPV even at 1000 flexes. The PA-EVOH-PA nanolayer structure gives mixed results depending on the grade of EVOH used. The authors conclude that the lower modulus PE-based tie resin stops the propagation of cracks better than the stiffer PA.

Alipour et al. [211] also use a Cloeren process to make five and 19-layer films of (LDPE-tie-EVOH-tie-LDPE). Total thickness is 38 μm. Average thicknesses of the tie/EVOH layers are 3.2/8.7 μm and 0.7/1 μm, for the 5 and 19 layer films, respectively. Like Medlock and Dolgovskij, OPV is the same for both films. Tensile measurements show an increase in elongation at break and energy to break for the 19-layer structure. Increasing the number of layers is found to slow the rate of crystallinity of the tie resins, which may increase the elongation and energy to break.

Lee et al. [212] conduct a similar study using the Dow layer multiplier technology, making cast film structures with up to 35 layers and using statistical analysis to confirm the effect of the number of layers on the barrier performance after flexing. They make 5, 17, and 35-layer structures of (PE-[tie-EVOH-tie]-PE) where the core [tie-EVOH-tie] is split and recombined to achieve the desired number of layers. The PE and core are kept at 80% and 20%, respectively, of the total thickness of the structure. Within the core, the EVOH to tie ratio is varied from 30:70 to 70:30. The films are subjected to Gelbo flex cycles. In addition to OPV measurements, the number of pinholes is measured by using a dye test. The number of pinholes is found to decrease with an increasing number of layers (and decreasing layer thickness). They attribute the improved flex crack resistance of the nanolayer structures to (1) greater flexibility of the EVOH layer when coextruded into thinner layers and (2) a more tortuous path for the gas molecules to travel if only some of the EVOH layers are broken.

Investigators [208,209] have also reported that nanolayer barrier films are more ductile and easier to draw and orient. This is due to lower crystallinity and changes in orientation and crystalline morphology, as well as the synergistic effects when ductile materials are coextruded with brittle ones (see Chapter 9).

So far most of the applications for layer multiplication for packaging films have been taking advantage of the improved mechanical properties of the film, particularly for stretch films for pallet wraps and the like [209,213,214]. For example, Apeldoorn Flexible Packaging uses microlayer technology to make a stretch film that increases pallet stability at elevated temperatures, uses 30% less material and reduces product damage during transportation by 80% [213]. An early adopter of layer multiplier technology is Nike [215,216], which uses it to make flexible cushioning membranes for athletic shoes: structures containing microlayers of alternating gas barrier (EVOH) and elastomers (TPU) provide a superior barrier for the pressurized nitrogen as well as flexibility and strength. Up to 1000 microlayers are used with layer thicknesses as thin as 0.01 micron.

8.5.3 Nanocomposites and nanocoatings

Polymer nanocomposites are blends of polymer and nanosized inorganic particles. They have received considerable attention in recent years dating from work done at Toyota [217] in the 1980s with PA 6. Conventional inorganic fillers such as calcium carbonate, glass fibers or beads, or talc are micron-sized and are often added to polymers at levels of 20% and higher to impart stiffness or strength. The addition of the filler typically detracts from optical clarity and toughness. Nanoparticles have at least one dimension less than 100 nm. They are added at much lower levels, typically around 5% or less. At such levels and dimensions, they have little negative impact on optical clarity (because of the small size, light is not diffracted) and toughness, yet have been reported to impart improvements in stiffness, strength, flame retardancy, electrical conductivity and gas barrier.

Most of the early work on the gas barrier was with nylon-clay nanocomposites. The nylon is mostly PA6, with some work using MXD6 and amorphous PA. Other polymers include PET, PP, PE, and ionomers. The nanoparticle is most often montmorillonite (MMT) clay. Other particles include kaoline, metal oxides, sepiolite and layered silicates.

Dispersing the particles in the polymer matrix is critical for enhancing performance. Various surface treatments and coupling agents are used to enhance the dispersion of the nanoparticles [218–225]. As described in Chapter 6, MMT has a layered structure with the individual platelets bound to one another by attractive forces with a separation distance of about 3.5 Å. Manufacturers add surfactants that separate the platelets to about 10 Å. For the best properties, fully separating the platelets is required. This is called exfoliation (see Fig. 6.26). Partial separation of the layers into smaller domains is called intercalation. In terms of barrier performance, intercalation is better than no separation, and exfoliation gives the best results. Unfortunately, achieving a true exfoliated state is difficult. There are three methods that are typically used to disperse the particles. One is in situ polymerization of the polymer with the nanoclay [217,226]. The monomer can diffuse between the platelets and cause further separation as the polymer chain grows. This only works for certain polymers, such as PA6, that can be polymerized in the presence of the nanoclay. Other methods of dispersing the nanoparticle in the polymer include melt compounding and making aqueous or solvent dispersions. The latter process is used for coatings onto the surface of a film or other substrate.

Oxygen and carbon dioxide permeation have received most of the attention from investigators studying the barrier performance of nanocomposites. Most authors report decreases in gas permeation of about 67% or less (or 3x increase in barrier performance) for films made of nanocomposites [218,226–231]. Chaiko et al. [218–220] report a decrease in OPV of LDPE of 95% by optimally selecting the surfactant. They attribute this to an increase in crystallinity of the PE due to the nucleating effects of the clay (as well as the more traditional mechanism of the clay particles creating a more tortuous path for the permeant described below). Choi et al. [226] measure an 91% reduction in OPV for a PET-MMT nanocomposite formed in situ during polymerization. Layered deposition of nanoparticles from aqueous dispersions has been reported to yield improvements in the barrier of up to about 20x (95% reduction in OTR) [232,233].

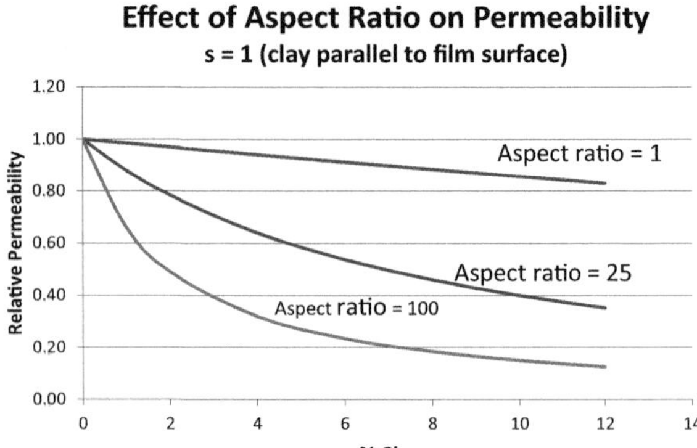

FIGURE 8.22 Model prediction of reduction in permeability as a function of nanoclay content and aspect ratio. *From Eq. (8.42).*

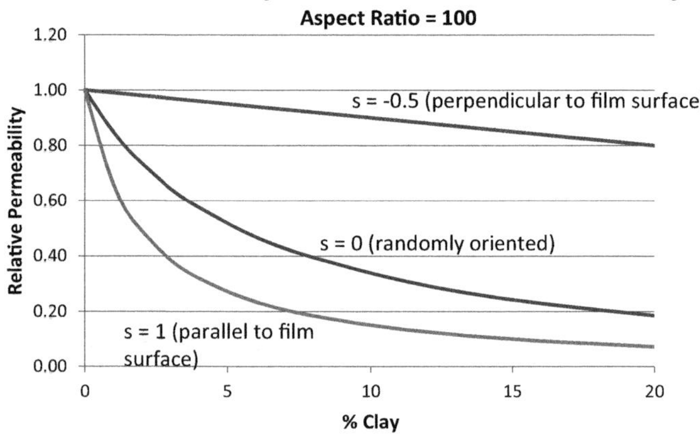

FIGURE 8.23 Model prediction of reduction in permeability as a function of nanoclay particle orientation and volume percent. *From Eq. (8.42).*

The reduction in permeability is generally thought to be due to the creation of a more tortuous path for the gas molecule through the film due to the alignment of the clay platelets parallel to the surface of the film. Many investigators [234–236] have derived models of the gas permeability of a nanocomposite as a function of the nanoparticle concentration, aspect ratio and orientation. An example is given in Eq. (8.42) [235]

$$P_{\text{nanocomposite}} = \frac{\phi_{\text{matrix}} P_{\text{matrix}}}{1 + 0.5\phi_{\text{clay}}\alpha_{\text{clay}}\frac{2}{3}\left(s + \frac{1}{2}\right)} \tag{8.42}$$

where P = permeability, ϕ = volume fraction, α_{clay} = aspect ratio of clay particle, s = orientation factor ($s = 1$ for particles parallel to surface, -0.5 for particles perpendicular to surface and 0 for randomly oriented particles).

Eq. (8.42) shows that permeability decreases with increasing aspect ratio, the volume fraction of the clay particles and orienting the particles parallel to the film surface, all consistent with a more tortuous path mechanism. These effects are highlighted in Figs. 8.22 and 8.23. Bhunia et al. [236] also find that increasing the length of the particle decreases the permeability.

Lan [227] uses Eq. (8.42) to compute the aspect ratio (defined here as the ratio of the dimension of the face of the platelet to its thickness) and volume concentration needed to achieve a 50% reduction in gas permeation assuming the platelets are aligned parallel to the film surface ($s = 1$). The results, given in Table 8.14, show that with a particle aspect

TABLE 8.14 Nanoparticle volume percent and aspect ratio relationship to provide a 50% reduction in gas diffusivity through a film.

Aspect ratio	Volume (%)
20	9
200	1
400	0.5

ratio of 200, only 1 vol% is needed to achieve a 50% reduction in permeability. The high aspect ratio of these particles is responsible for their barrier enhancing properties. The aspect ratio is determined by the natural dimensions of the clay and its dispersion in the polymer matrix. If there is no separation of the platelets then the particle is nearly spherical and the aspect ratio is approximately 1. Intercalation reduces the average thickness and increases the aspect ratio. Fully exfoliated clay exposes the nanometer thickness of the individual platelets and has the highest aspect ratio. Aspect ratios of 400 and higher are possible for exfoliated clay. Attrition of particles during processing may reduce the aspect ratio of the particles by shortening their length. The difficulty in achieving full exfoliation and minimizing attrition help explain why in practice is it difficult to achieve increases in barrier performance greater than 3x (67% reduction in permeation).

A 3x improvement in barrier may not be enough to bring a moderate barrier resin-like PA6 into the high-performance barrier category. High barrier polymers like EVOH and PVDC have O_2 permeability of 0.3 (dry) and 2.5 cc-mil/m^2-day-atm, respectively. OPET (30–80 cc-mil/m^2-day-atm) and PA6 (30–80 cc-mil/m^2-day-atm when dry) are at least an order of magnitude higher. Polyolefins range from about 3100 to 15,500 cc-mil/m^2-day-atm and so are a couple of orders of magnitude even higher. Therefore, a 3x reduction in OPV does not bring PA6 to the next level of performance needed to displace incumbent materials such as EVOH and PVDC for many applications.

Adding nanoparticles to more expensive barrier polymers such as EVOH or MXD6 has been reported [227,237]. By starting with higher-priced polymers, however, the improved barrier performance may not be cost-effective. Nanocomposites of these materials have also suffered from performance issues. For example, Aleperstein et al. [237] find that adding nano clay to EVOH reduces the mechanical properties of the EVOH. They attribute this to a reduction in the crystallinity of the EVOH.

Nevertheless, there are some applications where a 3x improvement could be significant. The one that gets the most attention is PET bottles where a somewhat small improvement in barrier may allow PET to capture new markets (beer, smaller bottles for carbonated beverages). Any solution, however, must be very cost-competitive as multiple technologies are pursuing this large market opportunity.

Another hurtle for the adoption of nanocomposites into mainstream barrier packaging applications is color. Even if the optical transparency is not negatively affected, many of the nano particles are natural clays and the like. They tend to impart a brownish tint to the film that may detract from the appearance of the package. Synthetic clays and nanoparticles with good color are available but typically at a higher cost.

Nanocomposites and coatings have also been developed for antimicrobial, antioxidant, oxygen scavenging and other active packaging applications. Antimicrobials include copper, copper oxide, titanium dioxide, zinc oxide, and silver [238–241]. Various biobased nanofibers, often derived from cellulose and starch, show promise for increasing the strength of polymers [241].

Nanocomposite barrier coatings have also been slow to be adopted commercially. PVOH-based nanocomposite coatings [233] are generally sensitive to moisture. Coatings also are susceptible to scratching and abrasion if they are on the outside of a package. Coated films, therefore, are typically laminated to other films so that the coating is protected in the interior of a package film structure.

Finally, nanomaterials are subject to increased scrutiny for potential health issues. Not enough research has yet been conducted to make definitive claims around health effects. One concern is if the nano particles become airborne. Their size makes it difficult for the human body's natural defenses to prevent them from getting into the lungs. When properly dispersed in polymers they should be protected from becoming airborne, although their fate in the environment is still uncertain. Another uncertainty is their migration from the package to food products and their effect on human health if ingested [242–244]. Regulatory bodies around the world are still working on how to deal with nanocomposites in food packaging. Nanocomposites have garnered considerable attention in the R&D community and these challenges are the subject of ongoing research.

8.5.4 Advanced coatings

In addition to nanocomposite coatings, a number of other coating technologies are in various stages of development or commercialization. Standard techniques to produce metallized, AlO*x* and SiO*x* coatings are described in Sections 2.5 and 4.3.2. Here two emerging technologies are briefly discussed: atomic layer deposition and "diamond-like carbon" coatings made by plasma-enhanced chemical vapor deposition. One of the driving forces is the desire by some end-users to eliminate metalized films so that metal detectors can be used on packaging lines to divert inadvertent metal from getting into the package. The presence of metal detectors at recycling facilities is also putting pressure on finding alternatives to metallized films.

8.5.4.1 Atomic layer deposition

Atomic layer deposition (ALD) uses sequential gas-phased chemical reactions to deposit a very thin coating, typically in the nanometer thickness realm, onto a substrate [245–250]. ALD is based on self-terminating reactions. The cyclical self-limiting reactions allow for controlled, very thin coatings to be made. Most of the work directed at reducing gas permeability involves depositing thin layers of aluminum dioxide coatings.

The main advantage of ALD over chemical vapor deposition and physical vapor deposition techniques is the ability to produce very thin coatings. Disadvantages include its slow speed. Most of the work to date has been with batch processes. Challenges remain to scale-up the process to high-speed continuous roll-to-roll processes necessary to produce quantities for packaging applications. It has found use primarily in the semi-conductor and energy conversion (solar cells) markets.

Hirvikorpi et al. [251,252] and Groner et al. [253] demonstrate that, at least on a lab-scale, ALD provides a reduction in gas permeation in a range suitable for packaging applications. In one study they coat AlOx onto various uncoated paper and polymer-coated paper. Oxygen transmission decreases by 50%–90% with coatings of 25–100 nm. In a second study, they compare ALD with three other barrier coating technologies: electron beam evaporation, magnetron sputtering and sol-gel. They coat PE and PLA substrates and find that ALD gives the best overall reduction in gas permeability. A later study [254] shows that a roll-to-roll process produces similar barrier performance as a batch process. They admit that this is just a feasibility study and that ALD is still a time-consuming and expensive process.

8.5.4.2 Diamond-like coatings

Plasma enhanced chemical vapor deposition (PECVD) can be used to produce a "diamond-like" carbon (DLC) coating. Such coatings are hard, scratch-resistant and have good barrier properties. The advantages of PECVD over chemical vapor deposition include lower temperatures (coatings can be made at room temperature so that the polymer substrate is not distorted), chemical stability, fewer toxic byproducts, quick processing time and high deposition rates [255,256].

Several companies have introduced technology to coat the inside of PET bottles with DLC, including Mitsubishi Heavy Machinery [257,258]. DLC has been used by Kirwin Brewery for PET beer bottles [259]. One of the limitations of using this technology for flexible films is that cracks form in the coating when the coated film is stretched by only a few percent [260]. Cracks increase gas permeability. Abbasa et al. [255] introduce silicon (Si) during the coating process to reduce internal stress and pinholing. They claim that this produces films with similar barrier performance as metallized films. Indeed, a competing technology was introduced by several firms, including Toyo Seikan Co., Ltd. [261], Toppan Printing Co., Ltd., and SIG Corpoplast GmbH Co. with Schott AG [262], which uses SiO*x* in a PECVD method to produce barrier coatings for PET bottles. Another drawback to DLC coatings is that they typically produce an off-color.

While PECVD is scaleable to high-speed coating suitable for packaging films, it still has its challenges. Both DLC and ALD are technologies to monitor to see if they develop into viable technology for barrier packaging films.

References

[1] H. Brown, J. Williams, Packaged product quality and shelf life, in: R. Coles, D. McDowell, M.J. Kirwan (Eds.), Food Packaging Technology, Blackwell Publishing, CRC Press, Boca Raton, FL, 2003, pp. 65–95.

[2] G.L. Robertson, Deteriorative reactions in food, Food Packaging. Principles and Practice, thirrd ed., CRC Press, Boca Raton, FL, 2013, pp. 293–328.

[3] G.L. Robertson, Shelf life of foods, Food Packaging. Principles and Practice, third ed., CRC Press, Boca Raton, FL, 2013, pp. 329–366.

[4] B. Blakistone, What suppliers need to know: a brief course in food chemistry, TAPPI Polymers, Laminations Coat. Conf. (1998) 153–161.

[5] J.B. Webster, S.E. Duncan, J.E. Marcy, S.F. O'Keefe, Effect of narrow wavelength bands of light on the production of volatile and aroma-active compounds in ultra-high temperature treated milk, Int. Dairy. J. 21 (2011) 305–311.

[6] J.B. Webster, S.E. Duncan, J.E. Marcy, S.F. O'Keefe, Controlling light oxidation flavor in milk by blocking riboflavin excitation wavelengths by interference, J. Food Sci. 74 (2009) S390–S398.

[7] L.J. Whited, B.H. Hammond, K.W. Chapman, K.J. Boor, Vitamin A degradation and light-oxidized flavor defects in milk, J. Dairy Sci. 85 (2002) 351–354.

[8] W. Cladman, S. Scheffer, N. Goodrich, M.W. Griffiths, Self-life of milk packaged in plastic containers with and without treatment to reduce light transmission, Int. Dairy J. 8 (1998) 629–636.

[9] A. Saffert, G. Pieper, J. Jetten, Effect of package light transmittance on vitamin content of milk. Part 2: UHT whole milk, Packaging Technol. Sci. 21 (2008) 47–55.

[10] T. Moyssiadi, A. Badeka, E. Kondyli, T. Vakirtzi, I. Savvaidis, M.G. Kontominas, Effect of light transmittance and oxygen permeability of various packaging materials on keeping quality of low fat pasteurized milk: chemical and sensorial aspects, Int. Dairy J. 14 (2004) 429–436.

[11] A. Karatapanis, A.V. Badeka, K.A. Riganakos, I.N. Savvaidis, M.G. Kontominas, Changes in flavour volatiles of whole pasteurized milk as affected by package material and storage time, Int. Dairy J. 16 (2006) 750–761.

[12] T.K. Dalsgaard, D. Otzen, J.H. Nielsen, L.B. Larsen, Changes in structure of milk proteins upon photo-oxidation, J. Agric. Food Chem. 55 (2007) 10968–10976.

[13] E.M. Becker, D.R. Cardoso, L.H. Skibsted, Deactivation of riboflavin triplet-excited state by phenolic antioxidants: mechanism behind protective effects in photooxidation of milk-based beverages, Eur. Food Res. Technol. 221 (2005) 382–386.

[14] G. Mortensen, J. Sorensen, H. Stapelfeldt, Effect of light and oxygen transmission characteristics of packaging materials on photo-oxidative quality changes in semi-hard Havarti cheeses, Packaging Technol. Sci. 15 (2002) 121–127.

[15] D. Deger, S.H. Ashoor, Light-induced changes in taste, appearance, odor, and riboflavin content of cheese, J. Dairy Sci. 70 (1987) 1371–1376.

[16] J.P. Wold, A. Veberg, F. Lundby, A.N. Nilsen, J. Moan, Influence of storage time and color of light on photooxidation in cheese: a study based on sensory analysis and fluorescence spectroscopy, Int. Dairy J. 16 (2006) 1218–1226.

[17] R. Alves, A. Van Dender, S. Jaime, I. Moreno, B. Pereira, Effect of light and packages on stability of spreadable processed cheese, Int. Dairy J. 17 (4) (2007) 365–373.

[18] C.S. Frederiksen, V.K. Haugaard, L. Poll, E.M. Becker, Light-induced quality changes in plain yoghurt packed in polylactate and polystyrene, Eur. Food Res. Technol. 217 (2003) 61–69.

[19] H. Larsen, S.B.G. Tellefsen, A.V. Dahl, Quality of sour cream packaged in cups with different light barrier properties measured by fluorescence spectroscopy and sensory analysis, J. Food Sci. 74 (2009) S345–S350.

[20] M. Thron, K. Eichner, G. Ziegleder, The influence of light of different wavelengths on chlorophyll-containing foods, Lebensm.-Wiss u.-Technol 34 (2001) 542–548.

[21] C.L. Hicks, C. Adbullah, Photoxidation of cinnamon oil, J. Food Sci. 52 (1987) 1041–1046.

[22] E. Psomiadou, M. Tsimidou, Stability of virgin olive oil. 2. Photo-oxidation studies, J. Agric. Food Chem. 50 (2002) 722–727.

[23] J.Y. Kim, D.S. Choi, M.Y. Jung, Antiphoto-oxidative activity of sesamol in methylene blue- and chlorophyll-sensitized photo-oxidation of oil, J. Agric. Food Chem. 51 (2003) 3460–3465.

[24] M. Lennersten, H. Lingnert, Influence of wavelength and packaging material on lipid oxidation and colour changes in low-fat mayonnaise, Lebensm.-Wiss. u.-Technol. 33 (2000) 253–260.

[25] A. Kanavouras, F.A. Coutelieris, Shelf-life predictions for packaged olive oil based simulations, Food Chem. 96 (2006) 48–55.

[26] A. Kanavouras, P. Hernandez-Munoz, F.A. Coutelieris, Shelf life predictions for packaged olive oil using flavor compounds as markers, Eur. Food Res. Technol. 219 (2004) 190–198.

[27] G. Ramirez, G. Hough, A. Contarini, Influence of temperature and light exposure on sensory shelf-life of a commercial sunflower oil, J. Food Qual. 24 (2001) 195–204.

[28] L. Piergiovanni, S. Limbo, The protective effect of film metallization against oxidative deterioration and discoloration of sensitive foods, Packaging Technol. Sci. 17 (3) (2004) 155–164.

[29] L.-X. Lu, F. Xu, Effect of light-barrier property of packaging film on the photo-oxidation and shelf life of cookies based on accelerated tests, Packaging Technol. Sci. 22 (2009) 107–113.

[30] H. Larsen, F. Westad, O. Sorheim, L.H. Nilsen, Determination of critical oxygen level in packages for cooked sliced ham to prevent color fading during illuminated retail display, J. Food Sci. 71 (2006) S407–S413.

[31] L. Manzocco, G. Kravina, S. Calligaris, M.C. Nicoli, Shelf life modeling of photosensitive food: the case of colored beverages, J. Agric. Food Chem. 56 (2008) 5158–5164.

[32] V.K. Haugaard, C.J. Weber, B. Danielsen, G. Bertelsen, Quality changes in orange juice packed in materials based on polylactate, Eur. Food Res. Technol. 214 (2002) 423–428.

[33] O. Solomon, U. Svanberg, A. Sahistrom, Effect of oxygen and fluorescent light on the quality of orange juice during storage at 8°C, Food Chem. 53 (1995) 361–368.

[34] J.M. King, D.B. Minn, Riboflavin photosensitized singlet oxygen oxidation of Vitamin D, J. Food Sci. 63 (1998) 31–34.
[35] N. Korman, N. Orbey, M. Ozilgen, A simple procedure for assessing the protective effect of transparent polymer films used for packaging foods containing light–sensitive nutrients, Polym. Int. 25 (1991) 163.
[36] I. Ahmad, Q. Fasihullah, A. Noor, I.A. Ansari, Q. Nawab Manzar Ali, Photolysis of riboflavin in aqueous solution: a kinetic study, Int. J. Pharmaceutics 280 (1–2) (2004) 199–208.
[37] J.D. Kanner, O. Fennema, Photooxidation of tryptophan in the presence of riboflavin, J. Agric. Food Chem. 35 (1987) 71–76.
[38] K. Cooksey, Effectiveness of antimicrobial food packaging materials, Food Addit. Contam. 22 (10) (2005) 980–987.
[39] J.H. Han, Antimicrobial packaging systems, in: S. Ebnesajjad (Ed.), Plastic Films in Food Packaging, Materials, Technology and Applications, Elsevier, New York, 2013, pp. 151–180.
[40] K. Cooksey, Utilization of antimicrobial packaging films for inhibition of selected microorganisms, ACS Symp. Ser. 753 (2000) 7–25.
[41] A. Balasubramanian, L.E. Rosenberg, K. Yam, M.L. Chikindas, Antimicrobial packaging: potential vs. reality - a review, J. Appl. Packaging Res. 3 (4) (2009) 193–221.
[42] S. Quintavalla, L. Vicini, Antimicrobial food packaging in meat industry, Meat Sci. 62 (3) (2002) 373–380.
[43] C. Realini, B. Marcos, Active and intelligent packaging systems for a modern society, Meat Sci. 98 (2014) 404–419.
[44] P. Suppakul, J. Miltz, K. Sonneveld, S.W. Bigger, Active packaging technologies with an emphasis on antimicrobial packaging and its applications, J. Food Sci. 68 (2003) 408–420.
[45] T. Huang, Y. Qian, J. Wei, C. Zhou, Polymeric antimicrobial food packaging and its applications, Polymers 11 (3) (2019) 560.
[46] A. Khaneghah, S. Hashemi, S. Limbo, Antimicrobial agents and packaging systems in antimicrobial active food packaging: an overview of approaches and interactions, Food Bioprod. Process. 111 (2018) 1–19.
[47] K. Petersen, P.V. Nielsen, G. Bertelsen, M. Lawther, M.B. Olsen, N.H. Nilsson, et al., Potential of biobased materials for food packaging, Trends Food Sci. Technol. 10 (1999) 52–68.
[48] M. Salame, The use of low permeability thermoplastics in food and beverage packaging, in: H.B. Hopfenberg (Ed.), Permeability of Plastic Films and Coatings to Gases Vapours and Liquids, Plenum Press, New York, 1974, p. 275.
[49] G.L. Robertson, Optical, mechanical and barrier properties of thermoplastic polymers, Food Packaging. Principles and Practice, third ed., CRC Press, Boca Raton, FL, 2013, pp. 91–130.
[50] A. Serna, I. López, A. Hernández, Method to measure oxygen permeance in sealed flexible packaging, SPE ANTEC Conf. (2018).
[51] M. Fereydoon, S. Ebnesajjad, Development of high-barrier film for food packaging, in: S. Ebnesajjad (Ed.), Plastic Films in Food Packaging, Elsevier, New York, 2013, pp. 71–92.
[52] M. Vaha-Nissi, T. Hjelt, A. Kokko, R. Kokkonen, A. Mikkelson, Tool for measuring diffusion of aromas through packaging materials, TAPPI PLACE. Conf. (2008).
[53] M. Stevens, Aroma/flavor barrier measurement, TAPPI PLACE. Conf. (2004).
[54] J.C. Tou, D.C. Rulf, P.T. DeLassus, Mass spectrometric system for the measurement of aroma/flavor permeation rates across polymer films, Anal. Chem. 62 (1990) 592–597.
[55] D. Falla, S.K. Goyal, B. Gillon, B. Quong, Multilayer polyethylene films having grease resistant properties, SPE ANTEC Conf. (2017).
[56] Y. Wyser, C. Pelletier, J. Lange, Nestlé Research Center, Novel Method for Testing the Grease Resistance of Plastic-Based Dry Pet Food Packaging, in: TAPPI PLACE Conference, Boston, MA, September 8–12, 2002.
[57] J. Lange, C. Pelletier, Y. Wyser, Novel method for testing the grease resistance for pet food packaging, Packaging Technol. Sci. 15 (2002) 65–74.
[58] T.I. Butler, B.A. Morris, PE-based multilayer film structures, in: S. Ebnesajjad (Ed.), Plastic Films in Food Packaging, Materials, Technology and Applications, Elsevier, New York, 2013, pp. 21–52.
[59] D.R. Constant, Cyclic olefin copolymer, a primer on basic structure, polymer attributes and barrier properties in films for flexible packaging, TAPPI PLACE. Conf. (2004).
[60] DuPont, 2005, DuPont™ Selar® PA3426 blends with nylon 6. http://www.dupont.com/content/dam/assets/products-and-services/packaging-materials-solutions/assets/selar_pa3426_nylon_blends.pdf (accesed 12.11.14).
[61] Kuraray, Eval EVOH properties, https://eval.kuraray.com/products-services/about-eval-evoh/properties/, n.d. (accesed 06.03.22).
[62] Mitsubishi Gas Chemical Company, Inc., ND, MX-Nylon, Nylon-MXD6. https://www.mgc.co.jp/eng/products/ac/nmxd6/ (accesed 08.08.20).
[63] DuPont Teijin Films, Oxygen and water vapour barrier properties of flexible packaging films. http://usa.dupontteijinfilms.com/informationcenter/downloads/Oxygen_And_Walter_Vapour_Barrier_Properties_of_Flex_Pack_Films.pdf, 2001 (accesed 08.08.20).
[64] Jindal Poly Films, Ltd., BOPP films. https://www.jindalpoly.com/products/bopp-films (accesed 08.08.20).
[65] Honeywell International, Inc., Honeywell Aclar films. https://www.packagingcomposites-honeywell.com/aclar/, 2011 (accesed 08.08.20).
[66] S.K. Burgess, O. Karvan, J.R. Johnson, R.M. Kriegel, W.J. Koros, Oxygen sorption and transport in amorphous poly(ethylene furanoate), Polymer 55 (18) (2014) 4748–4756.
[67] L. Sun, J. Wang, S. Mahmud, Y. Jiang, J. Zhu, X. Liu, New insight into the mechanism for the excellent gas properties of poly(ethylene 2,5-furandicarboxylate) (PEF): role of furan ring's polarity, Eur. Polym. J. 118 (2019) 642–650.
[68] N. Aubee, P. Lam, S. Marshall, A new family of sHDPE polymers for enhanced moisture barrier performance, J. Plastic Film. Sheeting 22 (4) (2006) 315–330.
[69] A. Singh, W.J. Koros, Permeation processes in barriers and membranes: differences and similarities, in: TAPPI Polymers, Laminations and Coatings Conference, 1998, pp. 361–371.

[70] R.J. Powell, DuPont internal report, 1968.
[71] J. de Garavilla, Ionomer, acid copolymer, and metallocene polyethylene resins: a comparative assessment of sealant performance, Tappi J. 78 (6) (1995) 192.
[72] DuPont, Barrier properties – oil resistance, Surlyn Update 1 (9) (1973).
[73] M.G. Sajilata, K. Savitha, R.S. Singhal, V.R. Kanetkar, Scalping of flavors in packaged foods, Comprehesive Rev. Food Sci. Food Saf. 6 (2007) 17–35.
[74] R. Auras, B. Harte, S. Selke, Sorption of ethyl acetate and d-limonene in poly(lactide) polymers, J. Sci. Food Agric. 86 (2006) 648–656.
[75] W. Woishnis (Ed.), Chemical Resistance of Plastics and Elastomers, fourth ed., William Andrew, St. Louis, 2007.
[76] CDF Corporation, Polyethylene chemical resistance chart, http://www.cdf1.com/technical%20bulletins/Polyethylene_Chemical_Resistance_Chart.pdf, 2004 (accesed 08.08.20).
[77] Lati High Performance Thermoplastics, Chemical resistance data. https://www.lati.com/en/downloads/literature/, 2002 (accesed 09.08.20).
[78] Plastics International, Unknown, Chemical resistance chart. http://www.plasticsintl.com/plastics_chemical_resistence_chart.html (accesed 09.08.20).
[79] Burkle GmbH, Chemical resistance. https://www.buerkle.de/en/downloads/chemical-resistance, 2014 (accesed 09.08.20).
[80] ThermoScientific, Unknown, Labware chemical resistance. http://www.thermoscientific.com/content/dam/tfs/LPG/LCD/LCD%20Documents/Product%20Manuals%20%26%20Specifications/Filtration%20Products/Filter%20Units%20and%20Bottletop%20Filters/Filter%20Units/D20480~.pdf (accesed 09.08.20).
[81] B.M. Graham, F. Luisa, M.A. Cezar, Transparent container, Patent Application WO2003022699 A1, 2003.
[82] N. Intawiwat, E. Myhre, H. Oeysaed, S.H. Jamtvedt, M.K. Pettersen, Packaging materials with tailor made light transmission properties for food protection, Polym. Eng. Sci. 52 (9) (2012) 2015–2024.
[83] P.M. Niedenzu, Choices of light attenuation in plastic film applications, TAPPI PLACE. Conf. (2008).
[84] P.M. Niedenzu, Review of titanium dioxide ability to scatter light, AIMCAL Conf. (2014). Myrtle Beach, SC.
[85] J. Ylvisaker, Mass transfer analysis of film barrier materials, TAPPI Polymers, Laminations Coat. Conf. (1998) 1181–1189.
[86] W.E. Lee, Selection of barrier materials from molecular structure, Polym. Eng. Sci. 20 (1980) 65–69.
[87] W.G. Todd, W.R. Podborny, J.V. Krohn, Maximize barrier performance of reduced-gauge HDPE films, SPE ANTEC Conf. (1999).
[88] J.C. Haley, J.S. Borke, The interplay between polymer polydispersity and film gauge in HDPE barrier films: how polydispersity controls gauge dependence of film barrier properties, SPE ANTEC Conf. (2009) 231–235.
[89] P. Brant, A. Karim, J.F. Douglas, F.S. Bates, Surface composition of amorphous and crystallizable polyethylene blends as measured by static SIMS, Macromolecules 29 (1996) 5628–5634.
[90] K.A. Chaffin, J.S. Knutsen, P. Brant, F.S. Bates, High-strength welds in metallocene polypropylene/polyethylene laminates, Science 288 (2000) 2187–2190.
[91] N.Z. Qureshi, E.V. Stepanov, G. Capaccio, A. Hiltner, E. Baer, Self-adhesion of polyethylene in the melt. I. Heterogeneous copolymers, Macromolecules 34 (2001) 1358–1364.
[92] N.Z. Qureshi, M. Rogunova, E.V. Stepanov, G. Capaccio, A. Hiltner, E. Baer, Self-adhesion of polyethylene in the melt. 2. Comparison of heterogeneous and homogeneous copolymers, Macromolecules 34 (2001) 3007–3017.
[93] L.M. Sherman, Metallocene PP & PE weld strongly to each other, Plast. Technol. (2001) 45.
[94] B. Poon, A. Chang, S.P. Chum, L. Tau, W.R. Volkenburgh, A. Hiltner, et al., Adhesion of polyethylene to polypropylene in multi-layer films, SPE ANTEC Conf. (2001) 1668–1672.
[95] B. Poon, S.P. Chum, A. Hiltner, E. Baer, Adhesion of polypropylene to metallocene/Ziegler-Natta polyethylene blends in microlayers, SPE ANTEC Conf. (2002) 1849–1853.
[96] S. Sakhalkar, D. Hirt, Surface segregation of erucamide in LLDPE films: thermodynamic analysis and experimental verification, SPE ANTEC Conf. (1998).
[97] K. Cooksey, Important factors for selecting food packaging materials based on permeability, Flex. Packaging Conf. (2004).
[98] D.W. Van Krevelen, K. te Nijenhuis, Cohesive properties and solubility, Properties of Polymers, Elsevier Science, Amsterdam, 2009, pp. 189–227.
[99] A.F.M. Barton, Handbook of Solubility Parameters and Other Cohesion Parameters, CRC Press LLC, New York, 1999.
[100] C.M. Hansen, Hansen Solubility Parameters: A User's Handbook, CRC Press LLC, New York, 1999.
[101] P. Juliano, T. Koutchma, Q. Sui, G.V. Barbosa-Cánovas, G. Sadler, Polymeric-based food packaging for high-pressure processing, Food Eng. Rev. 2 (2010) 274–297.
[102] V. Guillarda, M. Mauricio-Iglesiasa, N. Gontarda, Effect of novel food processing methods on packaging: structure, composition, and migration properties, Crit. Rev. Food Sci. Nutr. 50 (2010) 969–988.
[103] V.M. Granger, M.A. Tigani, Permeation properties of moisture sensitive barriers materials under end use conditions, TAPPI J. 76 (1993) 98–102.
[104] Y. Zheng, B. Walthers, S. Guerra, S. Bawiskar, Go beyond adhesion: the dual functionality of tie-layers, in: 2017 SPE FlexPackCon Conference, Tampa, FL, 2017.
[105] B.A. Morris, The role of polyolefins in "protecting" moisture sensitive barrier layers in packaging structures, in: TAPPI International Flexible Packaging & Extrusion Division Conference, 2018.

[106] J.P. Hecht, Psychrometry, evaporative cooling, and solids drying, in: D.W. Green, M.Z. Southard (Eds.), Perry's Chemical Engineering Handbook, ninth ed., McGraw Hill Education, 2018. section 12.

[107] A.M. Sadegh, W.M. Worek (Eds.), Marks' Standard Handbook for Mechanical Engineers, twelth ed., McGraw Hill, New York, 2017, pp. 134–135.

[108] B.C. Tsai, B.J. Jenkins, Effect of retorting on the barrier properties of EVOH, J. Plastic Film. Sheeting 4 (1988) 63–71.

[109] A. Lopez-Rubio, J.M. Lagaron, E. Gimenez, D. Cava, Morphological altercations induced by temperature and humidity in ethylene-vinyl alcohol copolymers, Macromolecules 36 (2003) 9467–9476.

[110] A. Lopez-Rubio, P. Hernandez-Munoz, E. Gimenez, T. Yamamoto, R. Gavara, J.M. Lagaron, Gas barrier changes and morphological alterations induced by retorting in ethylene vinyl alcohol-based food packaging structures, J. Appl. Polym. Sci. 96 (2005) 2192–2202.

[111] C.J. Farrell, B.C. Tsai, J.A. Wachtel, Drying agent in multi-layer polymeric structure, U. S. Patent 4,407,897, 1983.

[112] J.M. Lagaron, E. Nunez, Nanocomposites of moisture-sensitive and biopolymers with enhanced performance for flexible packaging applications, J. Plastic Film. Sheeting 28 (2011) 79–89.

[113] J.M. Largaron, E. Gimenez, B. Altava, V. Del-Valle, R. Gavara, Characterization of extruded ethylene-vinyl alcohol copolymer based barrier blends with interest in food packaging applications, Macromol. Symp. 198 (2003) 473–482.

[114] S.A. Lee, S.C. Kim, Laminar morphology development using hybrid EVOH-nylon blends, J. Appl. Polym. Sci. 67 (1998) 2001–2104.

[115] J.H. Yeo, C.H. Lee, C.-S. Park, K.-J. Lee, Rheological, morphological, mechanical, and barrier properties of PP/EVOH blends, Adv. Polym. Tech. 20 (2001) 191–201.

[116] S. Saez, Improve the performance of your flexible retort film, SPE ANTEC Conf. (2018).

[117] J. Kim, S. Oh, S.M. Cho, J. Jun, S. Kwak, Oxygen barrier properties of polyketone/EVOH blend films and their resistance to moisture, J. Appl. Polym. Sci. 137 (47) (2020) 49537.

[118] L. Murray, The impact of foil pinholes and flex cracks on the moisture and oxygen barrier of flexible packaging, TAPPI PLACE. Conf. (2005).

[119] J. Weis, Factors affecting barrier properties of metallized films, Verpack. Rundsch. 44 (1993) 23–29.

[120] A. Yializis, R.E. Ellwanger, J. Harvey, Loss of barrier in aluminum metalized polypropylene, Proc. Int. Conf. Vac. Web Coat., 10th (1996) 276–283.

[121] J. Lush, D. Ferrari, An economic value model for converters comparing the total cost of metallized film vs. foil in flexible packages, TAPPI PLACE. Conf. (2006).

[122] J.D. Vansant, Acid copolymer for extrusion coating, in: T. Bezigian (Ed.), Extrusion Coating Manual, fourth ed., TAPPI Press, Atlanta, 1999, pp. 235–239 (see also fifth ed.).

[123] U. Moosheimer, H.-C. Langowski, Permeation of oxygen and moisture through vacuum web coated films, in: 42nd Annual Technical Conference Proceedings - Society of Vacuum Coaters, 1999, pp. 408–414.

[124] A.S. da Silva Sobrinho, G. Czeremuszkin, M. Latrèche, M.R. Wertheimer, Detection and characterization of defects in transparent barrier coatings, in: 42nd Annual Technical Conference Proceedings - Society of Vacuum Coaters, 1999, pp. 316–319.

[125] G. Czeremuszkin, M. Latreche, M.R. Wertheimer, Permeation through defects in transparent barrier-coated plastic films, in: 44th Annual Technical Conference Proceedings - Society of Vacuum Coaters, 2001, pp. 178–183.

[126] G. Rossi, M. Nulman, Effect of local flaws in polymeric permeation reducing barriers, J. Appl. Phys. 74 (1993) 5471–5475.

[127] M. Hanika, H.-C. Langowski, U. Moosheimer, W. Peukert, Inorganic layers on polymeric films – influence of defects and morphology on barrier properties, Chem. Eng. Technol. 26 (2003) 605–614.

[128] G. Czeremuszkin, M. Latrèche, A.S. da Silva Sobrinho, M.R. Wertheimer, A simple model of oxygen permeation through defects in transparent coatings, in: 42nd Annual Technical Conference Proceedings - Society of Vacuum Coaters, pp 176–180.

[129] V. Del-Valle, E. Almenar, J.M. Lagarón, R. Catalá, R. Gavara, Permeation through porous polymeric films for modified atmosphere packaging, Food Addit. Contam. 20 (2003) 170–179.

[130] Z. Hussein, O.J. Caleb, U.L. Opara, Perforation-mediated modified atmosphere packaging of fresh and minimally processed produce—a review, Food Packaging Shelf Life 6 (2015) 7–20.

[131] V. Del-Valle, P. Hernandez-Munoz, R. Catala, R. Gavara, Optimization of equilibrium modified atmosphere packaging (EMAP) for minimally processed mandarin segments, J. Food Eng. 91 (2009) 474–481.

[132] D.R. Paul, R. Clarke, Modeling of modified atmosphere packaging based on designs with a membrane and perforations, J. Membr. Sci. 208 (2002) 269–283.

[133] D.S. Lee, P. Renault, Using pinholes as tools to obtain optimum modified atmospheres in packages of fresh produce, Packaging Technol. Sci. 11 (1998) 119–130.

[134] T. Hirata, Y. Makino, Y. Ishikawa, S. Katsuura, Y. Hasegawa, A theoretical model for designing modified atmosphere packaging with a perforation, Trans. ASAE 39 (1996) 1499–1504.

[135] T.J. Rennie, S. Tavoularis, Perforation-mediated modified atmosphere packaging. Part I. Development of a mathematical model, Postharvest Biol. Tech. 51 (2009) 1–9.

[136] D.A. Castellanos, D.R. Herrera, A.O. Herrera, Modelling water vapor transport, transpiration and weight loss in a perforated modified atmosphere packaging for feijoa fruits, Biosyst. Eng. 151 (2016) 218–230.

[137] J.C. Montanez, F.A.S. Rodriguez, P.V. Mahajan, J.M. Frias, Modelling the effect of gas composition on the gas exchange rate in perforation-mediated Modified Atmosphere Packaging, J. Food Eng. 96 (3) (2009) 348–355.

[138] J.C. Montanez, F.A. Rodríguez, P.V. Mahajan, J.M. Frías, Modelling the gas exchange rate in perforation-mediated modified atmosphere packaging: effect of the external air movement and tube dimensions, J. Food Eng. 97 (1) (2010) 79–86.

[139] D.S. Lee, J.S. Kang, P. Renault, Dynamics of internal atmosphere and humidity in perforated packages of peeled garlic cloves, Int. J. Food Sci. Technol. 35 (5) (2000) 455–464.

[140] S.K. Pandey, T.K. Goswami, An engineering approach to design perforated and non perforated modified atmospheric packaging unit for capsicum, J. Food Process. Technol. 3 (11) (2012).

[141] J. González-Buesa, A. Ferrer-Mairal, R. Oria, M.L. Salvador, Alternative method for determining O_2 and CO_2 transmission rates through microperforated films for modified atmosphere packs, Packaging Technol. Sci. 26 (7) (2013) 413–421.

[142] J. González, A. Ferrer, R. Oria, M.L. Salvador, Determination of O_2 and CO_2 transmission rates through microperforated films for modified atmosphere packaging of fresh fruits and vegetables, J. Food Eng. 86 (2) (2008) 194–201.

[143] G. Xanthopoulos, E.D. Koronaki, A.G. Boudouvis, Mass transport analysis in perforation-mediated modified atmosphere packaging of strawberries, J. Food Eng. 111 (2) (2012) 326–335.

[144] D. Chung, S. Papadakis, K.L. Yam, Simple models for evaluating effects of small leaks on the gas barrier properties of food packages, Packaging Tech. Sci. 16 (2003) 77–86.

[145] M.J. Rooney, Oxygen-scavenging packaging, in: S. Ebnesajjad (Ed.), Plastic Films in Food Packaging, Materials, Technology and Applications, Elsevier, 2013, pp. 139–149.

[146] J. Xu, J. Uradnesheck, B.A. Morris, DuPont Internal Report, 2009.

[147] A. Dey, S. Neogi, Oxygen scavengers for food packaging applications:a review, Trends Food Sci. Technol. 90 (2019) 26–34.

[148] T.Y. Ching, K. Katsumoto, S. Current, L. Theard, Ethylenic oxygen scavenging compositions and process for making same by esterification nor transesterification in a reactive extruder, PCT Application PCT/US1994/07854, 1995.

[149] Y. Byun, D. Darby, K. Cooksey, P. Dawson, S. Whiteside, Development of oxygen scavenging system containing a natural free radical scavenger and a transition metal, Food Chem. 124 (2) (2011) 615–619.

[150] S. Winestrand, K. Johansson, L. Järnström, L.J. Jönsson, Co-immobilization of oxalate oxidase and catalase in films for scavenging of oxygen or oxalic acid, Biochemical Eng. J. 72 (2013) 96–101.

[151] S. Yildirim, B. Röcker, M.K. Pettersen, J. Nilsen-Nygaard, Z. Ayhan, R. Rutkaite, Active packaging applications for food, Compr. Rev. Food Sci. Food Saf. 17 (2018) 165–199.

[152] C. Vilela, M. Kurek, Z. Hayouka, B. Röcker, S. Yildirim, M. Antunes, et al., A concise guide to active agents for active food packaging, Trends Food Sci. Technol. 80 (2018) 212–222.

[153] D. Langhe, M. Ponting, (eds.), Manufacturing and Novel Applications of Multilayer Polymer Films, William Andrew Publishing, pp. 46–116.

[154] W.J. Schrenk, D.S. Chisholm, K.J. Cleereman, T. Alfrey, Jr., Multilayered iridescent plastic articles, U.S. Patent 3576707, 1971.

[155] T. Alfrey, Jr., W.J. Schrenk, Highly reflective thermoplastic bodies for infrared, visible or ultraviolet light, U.S. Patent 3711176, 1973.

[156] W.J. Schrenk, Apparatus for multilayer coextrusion sheet or film, U. S. Patent 3884606, 1975.

[157] T. Alfrey, Jr., W.J. Schrenk, Multilayer coextrusion process for producing selective reflectivity, U. S. Patent 4094947, 1978.

[158] W.J. Schrenk, R.K. Shastri, H.C. Roehrs, R.E. Ayres, M.A. Barger, J.N. Bremmer, Method and apparatus for producing multilayer plastic articles, U.S. Patent 5628950, 1997.

[159] W.J. Schrenk, T. Alfrey Jr., "Some physical properties of multilayered films, Polym. Eng. Sci. 9 (1969) 393–399.

[160] J. Im, W.J. Schrenk, Coextruded multilayer sheet and film, J. Plastic Film. Sheeting 4 (1988) 104–115.

[161] R. Tangirala, E. Baer, A. Hiltner, C. Weder, Photopatternable reflective films produced by nanolayer extrusion, Adv. Funct. Mater. 14 (2004) 595–604.

[162] T. Kazmierczak, H. Song, A. Hiltner, E. Baer, Polymeric one-dimensional photonic crystals by continuous coextrusion, Macromol. Rapid Commun. 28 (2007) 2210–2216.

[163] C.-Y. Lai, E. Baer, The interdiffusion of polymeric gradient refractive index (GRIN) lens materials, SPE ANTEC Conf. (2011).

[164] M. Mackey, M. Wolak, J. Shirk, L. Flandin, A. Hiltner, E. Baer, Enhanced dielectric properties of micro and nanolayered films for capacitor applications, SPE ANTEC Conf. (2011).

[165] M. Ponting, A. Hiltner, E. Baer, Gradient multilayered films by forced assembly coextrusion, SPE ANTEC Conf. (2009).

[166] H.T. Tai, A. Hiltner, E. Baer, M. Wiggins, M. Sandrock, J. Shirk, Tunable reflective polymeric nanolayered films, SPE ANTEC Conf. (2005).

[167] M. Pointing, A. Hiltner, E. Baer, Gradient multilayer films by forced assembly, SPE ANTEC Conf. (2009).

[168] P.F. Cloeren, Interface control, U.S. Patent 6905324 B2, 2005.

[169] H.G. Schirmer, For extrusion of resin to form modular assembly of thin disks, U.S. Patent 5762971 A, 1998.

[170] H.G. Schirmer, Modular disk coextrusion die, U.S. Patent 6413595, 2002.

[171] H.G. Schirmer, R. Jester, G.D. Medlock, T. Schell, Nano-layer effects in blown barrier films, SPE Int. Polyolefins Conf. (2010).

[172] H.G. Schirmer, Layer sequence repeater module for a modular disk co-extrusion die and its usage, European Patent EP2639038A1, 2013.

[173] J. Dooley, J. Robacki, S. Jenkins, P. Lee, R. Wrisley, Producing microlayer blown film structures using layer multiplication and unique die technology, SPE ANTEC Conf. (2011).

[174] M.L. Zumbrunnen, D.A. Zumbrunnen, A new blown film die for controllably forming and extruding microlayers, polymer blends and composites, SPE ANTEC Conf. (2009).

[175] J. Dooley, K. Hughes, Coextrusion layer thickness variation – effect of polymer viscoelasticity on layer uniformity, SPE ANTEC Conf. (1995) 69–74.

[176] J. Dooley, K.S. Hyun, K. Hughes, An experimental study on the effect of polymer viscoelasticity on layer rearrangement in coextruded structures, Polym. Eng. Sci. 38 (1998) 1060–1071.

[177] J. Dooley, Viscous and elastic effects in polymer coextrusion, J. Plastic Film. Sheeting 19 (2003) 111–122.

[178] R. Zhao, C.W. Macosko, Polymer-polymer mutual diffusion via rheology of coextruded multilayers, AIChE J. 53 (2007) 978–985.

[179] L. Levitt, C.W. Macosko, T. Schweizer, J. Meissner, Extensional rheometry of polymer multilayers: a sensitive probe of interfaces, J. Rheology 41 (1997) 671–685.

[180] R. Zhao, C.W. Macosko, Slip at polymer-polymer interfaces: rheological measurements on coextruded multilayers, J. Rheology 46 (2002) 145–167.

[181] Z. Zhang, T.P. Lodge, C.W. Macosko, Interfacial slip reduces polymer-polymer adhesion during coextrusion, J. Rheology 50 (2006) 41–57.

[182] J.S. Schultz, T.P. Lodge, C.W. Macosko, The reactive formation of diblock copolymer at a polymer/polymer interface and its effect on interfacial structure, Mat. Res. Soc. Symp. (2000) 629. FF2.2.1-FF2.2.6.

[183] T. Saito, C.W. Macosko, Interfacial crosslinking and diffusion via extensional rheometry, Polym. Eng. Sci. 42 (2002) 1–9.

[184] C.W. Macosko, H.K. Jeon, T.R. Hoye, Reactions at polymer-polymer interfaces for blend c ompatibilization, Prog. Polym. Sci. 30 (2005) 939–947.

[185] A. Hiltner, S. Nazarenko, T. Ebeling, T. Schuman, E. Baer, Novel microlayered structures for adhesion and interdiffusion studies, Polymeric Mater. Sci. Eng. 75 (1996) 471–472.

[186] E. Baer, D. Jarus, A. Hiltner, Microlayer coextrusion technology, SPE ANTEC Conf. (1999).

[187] T.E. Bernal-Lara, R. Masirek, A. Hiltner, E. Baer, E. Piorkowska, A. Galeski, Morphological studies of multilayed HDPE/PS systems, J. Appl. Polym. Sci. 99 (2006) 597–612.

[188] T.E. Bernal-Lara, R.Y.F. Liu, A. Hiltner, E. Baer, Structure and thermal stability of polyethylene nanolayers, Polymer 46 (2005) 3043–3055.

[189] S.J. Pan, J. Im, M.J. Hill, A. Keller, A. Hiltner, E. Baer, Structure of ultrathin polyethylene layers in multilayer films, J. Polym. Sci. Part B: Polym. Phys. 28 (1990) 1105–1119.

[190] Y. Jin, M. Rogunova, A. Hiltner, E. Baer, R. Nowacki, A. Galeski, et al., Structure of polypropylene crystallized in confined nanolayers, J. Polym. Sci. Part B: Polym. Phys. 42 (2004) 3380–3396.

[191] F.J. Balta Calleja, F. Ania, I.P. Orench, E. Baer, A. Hiltner, T. Bernal, et al., Nanostructure development in multilayer polymer systems as revealed by X-ray scattering methods, Progr Colloid Polym. Sci. 130 (2005) 140–148.

[192] T.E. Bernal-Lara, A. Ranade, A. Hiltner, E. Baer, Nano-and microlayered polymers: structure and properties, in: G.H. Michler, F.J. Balta-Calleja (Eds.), Mechanical Properties of Polymers Based on Nanostructure and Morphology, Taylor & Francis, Baco Raton, 2005, pp. 629–681.

[193] J. Kerns, A. Hsieh, A. Hiltner, E. Baer, Mechanical behavior of polymer microlayers, Macromol. Symp. 147 (1999) 15–25.

[194] V.V. Pethe, Oxygen and carbon dioxide permeability of EAA/PEO blends and microlayers, Masters thesis, Case Western Reserve University, 2008.

[195] H.P. Wang, J.K. Keum, A. Hiltner, E. Baer, B. Freeman, A. Rozanski, et al., Confined crystallization of polyethylene oxide in nanolayer assemblies, Science 323 (2009) 757–760.

[196] E. Baer, A. Hiltner, Confined crystallization multilayer films, U.S. Patent Application 2010/0143709 A1, 2010.

[197] H. Wang, J.K. Keum, A. Hiltner, E. Baer, Confined crystallization of PEO in nanolayered films impacting structure and oxygen permeability, Macromolecules 42 (2009) 7055–7066.

[198] M. Pointing, J. Keum, B. Freeman, A. Hiltner, E. Baer, Substrate effects on the confined crystallization of polycaprolactone in coextruded nanolayered films, SPE ANTEC Conf. (2011).

[199] H.P. Wang, J.K. Keum, A. Hiltner, E. Baer, Crystallization kinetics of poly(ethylene oxide) in confined nanolayers, Macromolecules 43 (2010) 3359–3364.

[200] J.M. Carr, D.S. Langhe, M.T. Ponting, A. Hiltner, E. Baer, Confined crystallization in polymer nanolayered films: a review, J. Mat. Res. 27 (10) (2012) 1326–1350.

[201] H.P. Wang, J.K. Keum, A. Hiltner, E. Baer, Impact of nanoscale confinement on crystal orientation of poly(ethylene oxide), Macromol. Rapid Commun. 31 (2010) 356–361.

[202] D.S. Langhe, A. Hiltner, E. Baer, Melt crystallization of syndiotactic polypropylene in nanolayer confinement impacting structure, Polymer 52 (25) (2011) 5879–5889.

[203] M. Mackey, L. Flandin, A. Hiltner, E. Baer, Confined crystallization of PVDF and a PVDF-TFE copolymer in nanolayered films, J. Polym. Sci. B 49 (2011) 1750–1761.

[204] Y. Lin, Structure-property relationship of polyolefins used as packages and adhesives, Thesis, Case Western Reserve University. https://etd.ohiolink.edu/!etd.send_file?accession = case1290044976&disposition = inline, 2011 (accessed 17.08.20).

[205] Y. Lin, A. Hiltner, E. Baer, A new method for achieving nanoscale reinforcement of biaxially oriented polypropylene film, Polymer 51 (2010) 4218–4224.

[206] Y. Lin, A. Hiltner, E. Baer, Nanolayer enhancement of biaxially oriented polypropylene film for increased gas barrier, Polymer 51 (2010) 5807–5814.

[207] G. Oliver, Performance packaging utilizing layer multiplication, in: AMI Barrier Conference, Germany, 2009.

[208] G. Oliver, Micro-layer coextrusion methods and applications, TAPPI PLACE. Conf. (2012).

[209] O. Catherine, Sub-micron coextrusion technology and application examples, TAPPI PLACE. Conf. (2012).

[210] G. Medlock, Dolgovskij, Barrier performance of nanolayer EVOH film, TAPPI PLACE. Conf. (2012).

[211] N. Alipour, U.W. Gedde, M.S. Hedenqvist, S. Yu, S. Roth, K. Brüning, et al., Structure and properties of polyethylene-based and EVOH-based multilayered films with layer thicknesses of 150nm and greater, Eur. Polym. J. 64 (2015) 36–51.

[212] P.C. Lee, J. Dooley, J. Robacki, S. Jenkins, R. Wrisley, Flex barrier property enhancement in film structures using microlayer coextrusion technology, SPE ANTEC Conf. (2013).

[213] DuPont, 23rd DuPont packaging awards: sustainability and waste reduction drive winners, DuPont Packaging and Industrial Polymers. http://www2.dupont.com/Packaging_Resins/en_US/whats_new/23rd_packaging_awards_winners.html, 2011 (accessed 18.10.14).

[214] H. Schirmer, Nano-layers create thinner and wider tubular films, SPE FlexPackCon Conf. (2014).

[215] Plastics Technology, Microlayer films: New uses for hundreds of layers. http://www.ptonline.com/articles/microlayer-films-new-uses-for-hundreds-of-layers, 2006 (accessed 25.08.20).

[216] H.W. Bonk, D.J. Goldwasser, P.H. Mitchell, Flexible membranes, U.S. Patent 6082025 A, 2000.

[217] A. Okada, Y. Fukushima, M. Kawasumi, S. Inagaki, A. Usuki, S. Sugiyama, et al., Composite material and process for manufacture same, U.S. Patent 4,739,007, 1988.

[218] D.J. Chaiko, Nanocomposite packaging, in: Proceedings of the Electrochemical Society, 2003–2004, pp. 205.

[219] D.J. Chaiko, A.A. Leyva, Thermal transitions and barrier properties of olefinic nanocomposites, Chem. Mater. 17 (2005) 13–19.

[220] D.J. Chaiko, Thermally stable surfactants and compositions and methods of use thereof, US Patent 7420071B2, 2008.

[221] M.A. Osman, J.E.P. Rupp, U.W. Suter, Effect of non-ionic surfactants on the exfoliation and properties of polyethylene-layered silicate nanocomposites, Polymer 46 (2005) 8202–8209.

[222] M. Sirousazar, M. Yari, B.F. Achachlouei, J. Arsalani, Y. Mansoori, Polypropylene/montmorillonite nanocomposites for food packaging, e-Polymers 1 (2007) 305–313.

[223] S.M. Ali Dadfar, I. Alemzadeh, S.M. Reza Dadfar, M. Vosoughi, Studies on the oxygen barrier and mechanical properties of low density polyethylene/organoclay nanocomposite films in the presence of ethylene vinyl acetate copolymer as a new type of compatibilizer, Mater. Des. 32 (2011) 1806–1813.

[224] M. Shafiee, S.A. Ahmad Ramazani, M. Danaei, Investigation of the gas barrier properties of PP/clay nanocomposite films with EVA as a compatibilizer prepared by the melt intercalation method, Polym. Technol. Eng. 49 (2010) 991–995.

[225] H. Hosseinkhanli, A. Sharif, J. Aalaie, T. Khalkhali, S. Akhlaghi, Oxygen permeability and the mechanical and thermal properties of (low-density polyethylene)/poly (ethylene-co-vinyl acetate)/organoclay blown film nanocomposites, J. Vinyl Addit. Tech. 19 (2013) 132–139.

[226] W.J. Choi, H.-J. Kim, K.H. Yoon, O.H. Kwon, C.I. Hwang, Preparation and barrier property of poly(ethylene terephthlate)/clay nanocomposite using clay-supported catalyst, J. Appl. Polym. Sci. 100 (2006) 4875–4879.

[227] T. Lan, Nanocomposite materials for packaging applications, in: SPE ANTEC Conference. http://www.nanocor.com/tech_papers/Antec-Nanocor-Tie%20Lan-5-07.pdf, 2007 (accessed 18.10.12).

[228] A.Y. Goldman, Barrier liner material with nanocomposite for packaging applications, SPE ANTEC Conf. (2005) 1468–1471.

[229] S. Bonner, D. Sabandith, C. Swannack, W. Zhou, Nanocomposite film technology, SPE ANTEC Conf. (2001).

[230] P. Meneghetti, S. Qutubuddin, Synthesis, thermal properties and applications of polymer-clay nanocomposites, Thermochim. Acta 442 (2006) 74–77.

[231] R.K. Shah, R.K. Krishnaswamy, D.R. Paul, Blown films from LDPE-organoclay and ionomer-organoclay nanocomposites, SPE ANTEC Conf. (2006) 597–601.

[232] L.M. Sherman, Nanocomposites: less hype, more hard work on commercial viability, Plastics Technology. https://www.princeton.edu/~cml/assets/publicity/Nanocomposites200705.pdf, 2007 (accessed 04.11.14).

[233] H.S. Kravitz, F. Levitt, Barrier coatings for films and structures, US Patent 8080297 B2, 2011.

[234] L.E. Neilson, Models for the permeability of filled polymer systems, J. Macromol. Sci. (Chem.) A1 (5) (1967) 929–942.

[235] S. Abbott, N. Holmes, Nanocoatings: principles and practice, DesTech Publications. https://www.destechpub.com/product/nanocoatings-principles-and-practice/, 2013 (accessed 25.08.20).

[236] K. Bhunia, S. Dhawan, S.S. Sablani, Modeling the oxygen diffusion of nanocomposite-based food packaging films, J. Food Sci. 77 (2012) N29–N30.

[237] D. Aleperstein, et al., Experimental and computational investigations of EVOH/clay nanocomposites, J. Appl. Polym. Sci. 97 (2005) 2060–2066.

[238] M. Hoseinnejad, S.M. Jafari, I. Katouzian, Inorganic and metal nanoparticles and their antimicrobial activity in food packaging applications, Crit. Rev. Microbiol. 44 (2018) 161–181.

[239] H.M.C. Azeredo, C.G. Otoni, D.S. Correa, O.B.G. Assis, M.R. de Moura, L.H.C. Mattoso, Nanostructured antimicrobials in food packaging-recent advances, Biotechnol. J. 14 (2019) 1900068.

[240] T.V. Duncan, Applications of nanotechnology in food packaging and food safety: Barrier materials, antimicrobials and sensors, J. Colloid Interface Sci. 363 (2011) 1–24. Available from: https://doi.org/10.1016/j.jcis.2011.07.017.

[241] C. Vasile, Polymeric nanocomposites and nanocoatings for food packaging: a review, Materials 11 (2018) 1834.

[242] Z. Honarvar, Z. Hadian, M. Mashayekh, Nanocomposites in food packaging applications and their risk assessment for health, Electron. Physician 8 (6) (2016) 2531–2538. Available from: https://doi.org/10.19082/2531.

[243] J.Y. Huang, X. Li, W. Zhou, Safety assessment of nanocomposite for food packaging application, Trends Food Sci. Technol. 45 (2015) 187–199. Available from: https://doi.org/10.1016/j.tifs.2015.07.002.

[244] N. Bumbudsanpharoke, S. Ko, Nano-food packaging: an overview of market, migration research, and safety regulations, J. Food Sci. 80 (2015) R910–R923.

[245] R.W. Johnson, A. Hultqvist, S.F. Bent, A brief review of atomic layer deposition: from fundamentals to applications, Materialstoday 17 (2014) 236–246.

[246] H.C. Guo, E. Ye, Z. Li, M.-Y. Han, J.L. Xian, Recent progress of atomic layer deposition on polymeric materials, Mater. Sci. Eng., C: Mater. Biol. Appl., 70(Part 2) (2017) 1182–1191.

[247] T.O. Kaeaeriaeinen, P. Maydannik, D.C. Cameron, K. Lahtinen, P. Johansson, J. Kuusipalo, Atomic layer deposition on polymer based flexible packaging materials: growth characteristics and diffusion barrier properties, Thin Solid Films 519 (10) (2011) 3146–3154.

[248] V. Chawla, M. Ruoho, M. Weber, A.A. Chaaya, A.A. Taylor, C. Charmette, et al., Fracture mechanics and oxygen gas barrier properties of Al2O3/ZnO nanolaminates on PET deposited by atomic layer deposition, Nanomaterials 9 (1) (2019) 88/1–88/17.

[249] M. Vaha-Nissi, M. Pitkanen, E. Salo, J. Sievanen-Rahijarvi, M. Putkonen, A. Harlin, Atomic layer deposited thin barrier films for packaging, Cellulose Chem. Technol. 49 (7–8) (2015) 575–585.

[250] M. Vaha-Nissi, M. Pitkanen, E. Salo, E. Kentta, A. Tanskanen, T. Sajavaara, et al., Antibacterial and barrier properties of oriented polymer films with ZnO thin films applied with atomic layer deposition at low temperatures, Thin Solid Films 562 (2014) 331–337.

[251] T. Hirvikorpi, M. Vähä-Nissi, T. Mustonen, E. Iiskola, M. Karppinen, Atomic layer deposited aluminum oxide barrier coatings for packaging materials, Thin Solid Films 518 (2010) 2654–2658.

[252] T. Hirvikorpi, M. Vähä-Nissi, A. Harlin, M. Karppinen, Comparison of some coating techniques to fabricate barrier layers on packaging materials, Thin Solid Films 518 (2010) 5463–5466.

[253] M.D. Groner, S.M. George, R.S. McLean, P.F. Carcia, Gas diffusion barriers on polymers using Al_2O_3 atomic layer deposition, Appl. Phys. Lett. 88 (2006) 51907.

[254] T. Hirvikorpi, R. Laine, M. Vaha-Nissi, V. Kilpi, E. Salo, W.-M. Li, et al., Barrier properties of plastic films coated with an Al_2O_3 layer by roll-to-toll atomic layer deposition, Thin Solid Films 550 (2014) 164–169.

[255] G.A. Abbasa, P. Papakonstantinoua, T.I.T. Okpalugoa, J.A. McLaughlina, J. Filikb, E. Harkin-Jonesc, The improvement in gas barrier performance and optical transparency of DLC-coated polymer by silicon incorporation, Thin Solid Films 482 (2005) 201–206.

[256] G. Czeremuszkin, M. Latreche, M.R. Wertheimer, Plasma-enhanced chemical vapor deposition of transparent barrier coatings, in: 46th Annual Technical Conference Proceedings - Society of Vacuum Coaters, 2003, pp. 586–591.

[257] Mitsubishi Heavy Machinery Systems, Coating equipment. https://www.mhi-ms.com/products/foodpackaging/dlc/, n.d. (accessed 24.08.20).

[258] Mitsubishi Heavy Machinery Systems, Keep a drink delicious and fresh! A carbon coating that evolves PET bottled beverages. https://www.mhi-ms.com/research/sip/highlight/column_0001.html, n.d. (accessed 24.08.20).

[259] A. Shirakura, M. Nakaya, Y. Koga, H. Kodama, T. Hasebe, T. Suzuki, Diamond-like carbon films for PET bottles and medical applications, Thin Solid Films 494 (2006) 84–91.

[260] R. Asakawa, S. Nagashima, Y. Nakamura, T. Hasebe, T. Suzuki, A. Hotta, Combining polymers with diamond-like carbon (DLC) for highly functionalized materials, Surf. Coat. Technol. 206 (2011) 676–685.

[261] Toyo Seikan Co., Ltd., Barrier bottles. https://www.toyo-seikan.co.jp/e/technique/petbottle/barrierbottle/, n.d. (accessed 24.08.20).

[262] A.G. Schott, Barrier coatings. https://www.schott.com/magazine/english/info104/si104_09_pet.html, 2020 (accessed 24.08.20).

Chapter 9

Strength, stiffness and abuse resistance of multilayer flexible packaging films

The structural integrity of the package is critical for protecting the product from the packing operation through distribution and use. In multilayer flexible packaging, often one or more layers are specifically incorporated to bear the load when it comes to strength and integrity, but the other layers play a role as well. This chapter looks at various aspects of structural integrity such as tensile strength, flexural stiffness, tear, scratch, puncture, impact and flex fatigue resistance. The importance of these attributes is discussed, including the need for improved abuse resistance associated with the growth of business-to-consumer e-commerce distribution channels. Measurement techniques are described, followed by an exploration of how the design of the multilayer film structure impacts these properties.

9.1 Why it is important

9.1.1 Strength

The package needs to be able to hold the product and withstand any stress that occurs throughout the life of the packaged good. An important stress factor is the mass and density of the product being packaged. A film structure for packaging cement will require greater strength than one for packaging feathers. If the weight of the product per unit area of the package exceeds the tensile strength of the film then the package will likely fail. Even before the ultimate strength is exceeded, the film may yield and deform, which may be undesirable. Stiff, high yield strength layers such as oriented polyethylene terephthalate (OPET) are often incorporated into the package design to bear the load.

Other stresses may also be important. For example, some deli meats are cooked in their package. During cooking, elevated temperatures lower the tensile strength of the film. As the cooking proceeds, pressure may build up inside the package as water and other components volatilize, exerting stress on the package. This combination of higher stress and lower strength must be factored into the design of the film.

Forces experienced during shipping may cause a package to fail. An example is illustrated in Fig. 9.1 where a package filled at sea level is transported over a mountain pass. The lower ambient pressure at the higher elevation causes the package to bulge outward and puts stress on the film. If this stress exceeds the yield strength of the film, then the film will elongate; when the package returns to a lower elevation it may look distorted and esthetically unpleasing. Higher stress may result in failure. In this example, the mountain pass is at 3660 m (12,000 ft) and the pressure differential is 0.038 MPa. For the given dimensions of the packaging film, this results in stress of 388 MPa [1]. If the yield strength of the film is lower than this, it will yield.

The tensile strength of the film is not the only consideration. If the film is heat-sealed, the strength of the seal must be considered. The reader is referred to Chapter 7 for a discussion on heat seal and hot tack strength. Puncture resistance, flex cracking and other abuse factors are considered below.

9.1.2 Stiffness

The bending or flexural stiffness of a packaging film is important for several reasons. In some packages, it is a critical part of the package form. The standup pouch, for example, needs stiffness for the package to maintain its rigidity. The bending stiffness is also important for the handling of the film on automated packaging lines. If it is too stiff it may not form around shaping mandrels in form-fill-seal operations. If it is too flexible, it may not maintain its shape during filling and other operations.

The Science and Technology of Flexible Packaging. DOI: https://doi.org/10.1016/B978-0-323-85435-1.00012-0

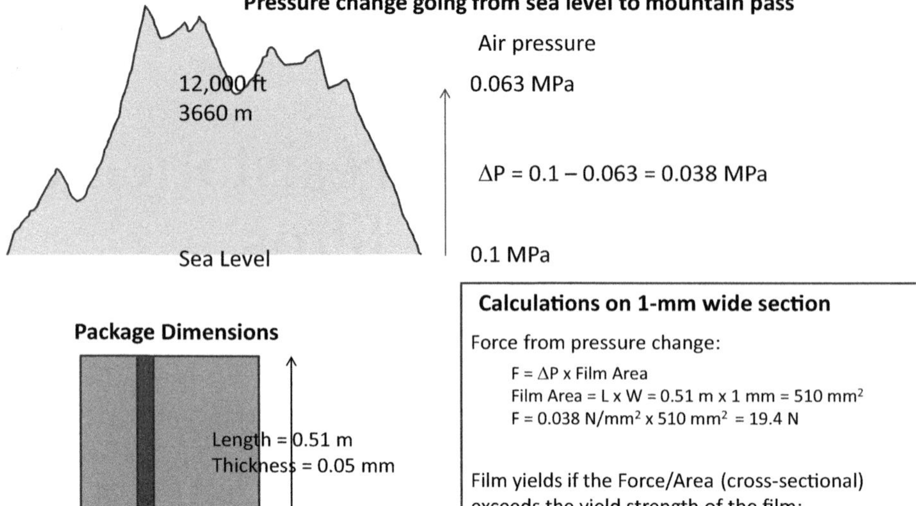

FIGURE 9.1 Example of shipping a flexible package over a mountain pass. The package will experience stress as the ambient pressure decreases with the elevation gain. If this stress exceeds the yield strength of the film, the film will elongate.

Bending stiffness is related to the "hand" or feel of the film and package. The consumer may associate a "flimsy" package with an inferior product. Downgauging (reducing the thickness) of a packaging film without maintaining the bending stiffness may be perceived by the consumer. McKenna et al. [2] and Bunker [3] found that consumers can detect a difference in film thickness of 15% in films of the same bending stiffness but that increasing stiffness while downgauging may allow further thickness reduction without consumer awareness. As is discussed in Section 9.4.2, the design of a multilayer film structure can have a substantial impact on its bending stiffness.

Bending stiffness should not be confused with tensile stiffness or elastic modulus. Tensile modulus (also known as the elastic or Young's modulus) is obtained from tensile tests and is a fundamental property of a material. When combining multiple materials in a multilayer film, however, the tensile modulus is simply the sum of the thickness times the modulus of each layer divided by the total thickness. Bending stiffness is a package property that is a complex function of the tensile modulus, thickness and location of the individual layers.

9.1.3 Abuse resistance

Abuse can come from the product within the package, or from external factors. Some examples include:

- Puncture of the package from bones in meat products or sharp edges of dry foods (e.g., noodles or crackers). This is generally a slow puncture process that occurs during shipping and distribution.
- Puncture of the package as the product is dropped into the package during a vertical filling operation. This often involves high-speed impact and scratching of the package by the product.
- Abrasion of the package from the product within. This often results from vibrations that occur during shipping and handling. Examples include coffee grinds and snack products.
- Abrasion of the package from outside the package. This may occur as packages rub against each other during shipping.
- Failure of a package during dropping. The package may be dropped during handling, for example as the filled package is placed into a secondary package for shipping. Dropping often involves high-speed impact.
- Tearing of the packaging film. Tear resistance can be an important aspect of many abuse events such as puncture and abrasion. Poor tear resistance can propagate cracks generated during these events, contributing to failure. In some cases, however, low tear strength is desirable; directional tear properties, where the film tears easily in a particular direction, may be used as an easy-open feature for a package.

- Flex fatigue. Flexing of the package during handling may cause cracking of brittle layers. This can reduce the shelf life of products in barrier packaging containing aluminum foil, metallized films or ethylene vinyl alcohol (EVOH). Incorrectly stacked packages on a pallet can cause wrinkles in the frontal surface area resulting in failures from vibration and sloshing during transportation.
- Environmental stress cracking. Exposure to certain chemicals, while the package is under stress, may cause cracks and failures in some materials.
- Distortion of the packaging film at elevated temperatures. Processes such as heat sealing can distort the film or cause the film to stick to the seal bar. Hot fill, retort sterilization, cook-in the package, and in-package pasteurization expose the package to elevated temperatures and pressures that may distort the packaging film.
- Cold temperature impact failure. Some food products are exposed to blast freezing after packaging, which can put stress on the package. Some polymers have good toughness at room temperature but become brittle at cold temperatures.

9.1.4 E-commerce

Prior to the COVID-19 pandemic, online shopping was growing at three times the rate of general retail and was projected to grow to $900 billion in sales in the United States by 2023 [4]. The pandemic has accelerated this transformation of the retail market. While household items, personal care products and beauty items are amenable to direct delivery, groceries have fostered four online business models [4]:

- Delivered basket, where the product is delivered directly to the consumer. The primary package is typically shipped in a secondary box and needs to survive cross-country transport.
- Click and collect, where the online order is picked up by the consumer at a physical store. Traditional primary packaging can typically be used since the consumers transport the product from the store to their home.
- Fresh delivery, where the order is delivered by the local store or third party within 24 hours. Traditional primary packaging can be used since distribution involves short local transportation.
- Meal kit, where a kit of preassembled ingredients is delivered to the consumer. This format often has significant secondary packaging, such as dry ice and air pillows to protect the contents.

E-commerce can involve up to three times the touchpoints as traditional physical retail (bricks and mortar). Distribution for a typical brick and mortar store may start at the consumer goods company with products safely packed in cases on pallets. These travel by truck to a distribution center (DC) and then to individual stores. The product may not be unloaded from the pallets until it safely arrives at the store. The consumer picks the product and transports it home.

An e-commerce supply chain may start with product in cases on pallets, but once it gets to the e-commerce DC it will be offloaded and put into inventory. When the online order is received the product will be packaged as a single item or in a multiitem box, perhaps with secondary protective packaging, and perhaps sent to a DC for air freight. After arriving at the destination city, it is offloaded at another DC. It may then travel by truck to a local DC, be sorted, and sent out for local delivery. With so many touchpoints, the need for abuse-resistant packaging increases accordingly.

Packaging designed for brick and mortar stores is often not optimal for e-commerce. Packaging for the retail shelf is often oversized, has expensive design esthetics, contains redundant features to prevent theft and is not sturdy enough to survive shipping [5]. An example is the bag-in-box commonly used for cereal. It is designed for retail stores. The box has a good shape factor for dense packing on pallets, is easily displayed on the retail shelf and has a large area for eye-catching graphics. In an e-commerce supply chain, these features may not be needed and the excess headspace in the package may cause the contents to jostle more during handling. By the time it gets to the consumer, the product may be crushed, resulting in a poor customer experience [6].

In the short term, brand owners are compensating by using more secondary protective packaging (air pillows, bubble wrap, etc.). This goes against public demand for less packaging waste and the use of more sustainable packaging materials. Increasing packaging density (the percent of secondary package volume occupied by products) is critical for reducing package waste, minimizing product damage and cost. About 24% of the average shipping parcel is air, with some e-commerce categories up to 40% [7]. Machine vision identification of package volumes and data-driven optimization algorithms are being used to increase packaging density. New biobased packaging materials have been introduced but are generally too expensive for widespread adoption. Excessive secondary packaging may also diminish the customer experience. The "unboxing experience" is the critical last impression the consumer has before using the product and often influences repurchase decisions. The customer experience is the new "shelf presence" and one of the key challenges for businesses to get right in order to make e-commerce successful [8].

The most effective way to reduce secondary packaging is to design better primary packaging suited for e-commerce. Breakage during transport will result in lost repeat business and with social media, may lead to critical reviews and tarnishing of the brand. Amcor provides examples of some common failure points for e-commerce packages [4]:

- Threaded caps on plastic bottles for liquid packaging loosen and product leaks; caps shatter or crack.
- Metal cans dent.
- Glass jars shatter.
- Trigger spays compress and eject liquid.
- Paper composites puncture and leak.
- Seals on stand-up pouches loosen and leak.

Some product categories and their traditional primary packaging formats are better suited for e-commerce than others. Bemis (now Amcor) [4,9] tested over 170 products using ISTA 3A standards (see Section 9.2.7) and found only 53 survived (69% failed). They rated 24 product categories on two criteria:

- Fit for delivery/e-commerce ready. This includes shipping economics (relative to the product's value), repackaging requirements (or use of secondary packaging) and shelf life.
- Degree of packaging change required. This includes survivorship (from the ISTA testing), fit with a subscription model (regular repeat purchase) and design upside.

They plot their scores for fit for delivery/e-commerce ready versus degree of packaging change required and find the product categories fall into three groups:

- Unleash commercial activity (high fit for delivered e-commerce, low degree of packaging change needed).
- Unlock packaging for growth (medium fit for delivered e-commerce, medium-high degree of packaging change needed).
- Fit for alternative distribution chain (e.g., fresh delivery or click and collect) (low fit for delivered e-commerce, high degree of packaging change needed)

The segmentation of various product categories into these growth scenarios are given in Table 9.1. Existing packaging formats can be used to grow products in the *Unleash commercial activity* category whereas packaging innovation is needed for products in *Unlock packaging for growth.* Products in *Fit for alternative distribution chain* require refrigeration or expensive secondary packaging that makes them more amenable to store pickup or delivery from a local brick and mortar store.

Redesigning the primary package for e-commerce often improves sustainability. Examples include:

- Reducing secondary packaging by making the primary package more survivable.
- Reducing product waste by improving product survival throughout the supply chain.
- Reducing package size, weight, etc.
- Lowering carbon footprint by switching from rigid packaging formats to flexible packaging.

The concepts of strength, stiffness and abuse resistance developed in this chapter are increasingly important for meeting the challenges of new distribution channels created by online buying.

9.2 How to measure

9.2.1 Mechanical properties

ASTM D882 [10] provides a standard method for measuring the tensile properties of the film. Strips are cut from the sample and placed in the gripping jaws of a tensile testing machine. The load is measured as the gripping jaws are pulled apart at a constant speed. At the same time, the distance or displacement between the jaws is measured. The load versus displacement data allows for the calculation and plotting of stress versus strain and the tensile properties as defined below. A common specimen width is 1-inch (25.4 microns) and the test speed (grip separation rate) may range from 0.5 to 20 inches/minute (12.7–508 mm/minute), with 1 and 10 inches/minute (2.54 and 254 mm/minute) the most common.

A typical stress versus strain curve for a ductile plastic is shown in Fig. 9.2. At low strain, the stress increases linearly. The deformation is elastic; the slope of the curve in this region is known as the elastic modulus and is related to

TABLE 9.1 Segmentation of product categories into e-commerce growth strategies.

E-commerce growth category	Strategy	Product categories
Unleash commercial activity	Typical package formats perform—drive growth with commercial development.	Towel/tissue Toothpaste Chips Litter Nuts & seeds Protein powder Diapers Seasonings Pet treats Razors
Unlock packaging for growth	Category growth is constrained by existing packaging formats. Packaging innovation is critical.	Soft drinks Condiments Pet food Dry foods Canned vegetables Shampoo Cereal Detergent Bleach & cleaner
Fit for alternative distribution chains	Refrigeration or expensive secondary packaging is needed to meet shelf life requirements. Better fit for fresh delivery and click & collect.	Fresh produce Chocolate Milk Meat & cheese

Source: Based on Amcor, Stand out and deliver. <https://www.amcor.com/insights/educational-resources/north-america-ecommerce-ebook>, 2020 (accessed 30.08.2020) analysis of fit for delivery/e-commerce ready and degree of packaging change required.

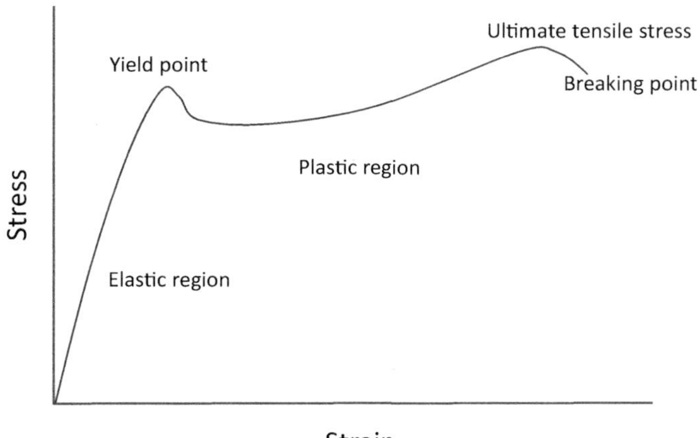

FIGURE 9.2 Characteristic stress versus strain curve for a plastic [11]. *Adapted from L. W. McKeen, Film Properties of Plastics and Elastomers, third ed., Elsevier, Amsterdam, 2012, pp. 27–34.*

the stiffness. The yield point represents the transition from elastic to plastic deformation. Below the yield point, the plastic will recover to its original shape once the strain is removed. Deformations above the yield point are not recoverable. The test specimen may neck-in at the yield point, giving a local maximum in the stress. Further increases in strain eventually result in the rupture of the specimen.

The stress versus strain curve differs from one polymer to the next but generally takes on one of the characteristic shapes shown in Fig. 9.3. Brittle plastics typically have a steep initial slope but fail at low strains and do not exhibit plastic deformation. Stiff and strong plastics have a steep slope, exhibit a yield point, have a high maximum or ultimate stress at break, but fail at a relatively low strain. Softer, tough polymers generally have a gentler slope but fail at much higher strains. Their maximum stress is lower but their high elongation before failure often translates into greater toughness. Representative curves for several polymers are shown in Fig. 9.4.

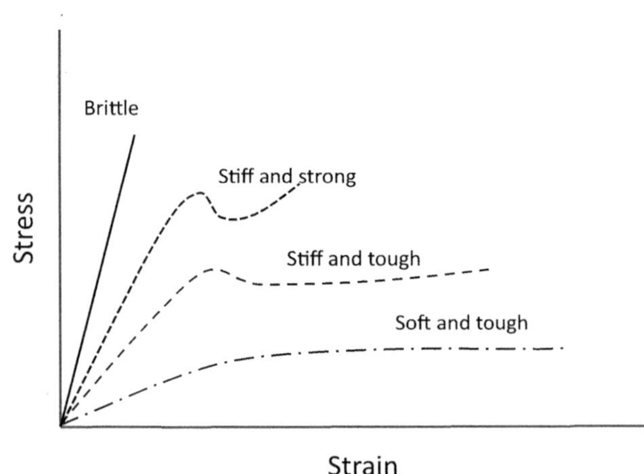

FIGURE 9.3 Characteristic stress versus strain curves for various polymer types. *Adapted from L. W. McKeen, Film Properties of Plastics and Elastomers, third ed., Elsevier, Amsterdam, 2012, pp. 27–34.*

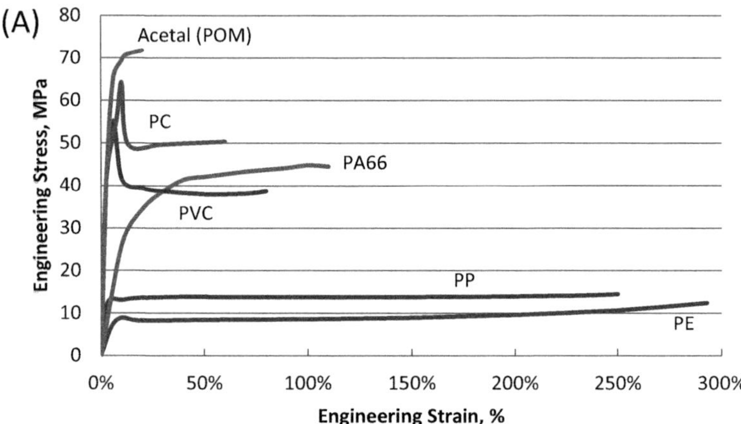

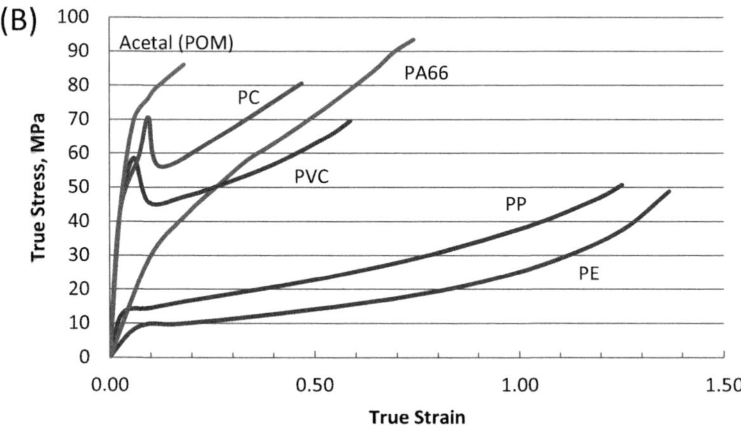

FIGURE 9.4 Representative stress versus strain curves for several polymers. (A) engineering stress and strain. (B) true stress and strain. *Data fromS. Bahadur, Strain hardening equation and the prediction of tensile strength of rolled polymers, Polym. Eng. Sci., 13 (1973) 266–272.*

The stress versus strain curve is used to calculate or define a number of mechanical properties of the film:

Strain or elongation: The change in length as the specimen is stretched between the moving grips of the tensile tester is known as the engineering or Cauchy strain or elongation:

$$\varepsilon = \frac{\Delta L}{L_0} \tag{9.1}$$

where ε = engineering strain, ΔL = change in length = $L - L_0$, L_0 = initial length of the specimen (or distance between grips), and L = length of specimen during the tensile test.

When ε is expressed as a percentage, it is known as the percentage elongation.

Strain ratio: Sometimes the tensile curve is plotted versus the strain ratio rather than the strain or elongation. The strain ratio is the ratio of the instantaneous length to the initial length:

$$\lambda = \frac{L}{L_0} \tag{9.2}$$

where λ = strain ratio, L = length, and L_0 = initial length.

True strain: The engineering strain is useful for small strains but as the strain becomes greater than about 5% the logarithmic strain, known as the true or Hencky strain, is a better representation of the strain. This takes into account the incremental stretching of the specimen.

$$\varepsilon_T = \ln(\lambda) = \ln(1 + \varepsilon) \tag{9.3}$$

where ε_T = true strain.

Engineering stress: The load measured by the tensile tester divided by the initial cross-sectional area (specimen width times thickness) is known as the engineering or nominal stress. The engineering stress is used to report most tensile properties in accordance with ASTM D882.

True stress: The true stress is the load divided by the instantaneous cross-sectional area. The true stress (σ_T) is related to the engineering stress (σ_E) in the following manner:

$$\sigma_T = \sigma_E\lambda = \sigma_E\exp(\varepsilon_T). \tag{9.4}$$

Yield stress or strength: As a viscoelastic material is stretched, initially the elongation is reversible; when the strain is released, it returns to its original length. At some critical strain, however, the material begins to deform plastically. Beyond this point, the material does not return to its original shape. This typically shows up as an inflection point on the stress versus strain curve. The engineering stress at this point is known as yield stress or yield strength.

Elongation at yield: This is the engineering strain or elongation where the material yields or begins to plastically deform.

Tensile strength: The maximum engineering stress is known as tensile stress or strength. This is also known as the ultimate tensile strength.

Tensile strength at break: The engineering stress at the point where the specimen breaks is known as the tensile strength or stress at break.

Elongation at break: The engineering strain at which the specimen breaks is known as the elongation at break. It is often expressed as a percentage.

Elastic modulus: The slope of a tangent drawn parallel to the initial linear portion (before the yield point) of the stress versus strain curve is known as the elastic, tensile or Young's modulus. The modulus is related to the stiffness of the material.

Secant modulus: For many films it is difficult to accurately determine the slope of the initial portion of the stress–strain curve. Rather than approximating a slope, a line is drawn from zero to the engineering stress at a specified strain (such as 1%). The slope of this line is the secant modulus at that strain, such as the 1% secant modulus.

Strain hardening parameter: The plastic deformation behavior of many ductile polymers can be represented by a two-parameter model [12]:

$$\sigma_T = \sigma_0\exp(m\varepsilon_T) \tag{9.5}$$

where σ_T = true stress, ε_T = true strain, σ_0 = constant, and m = strain hardening parameter.

The strain hardening parameter is the slope of a line drawn through the yield stress and the ultimate tensile strength.

9.2.2 Bending stiffness

The bending stiffness of a film is only partly related to the elastic or tensile modulus obtained from the tensile tests described in Section 9.2.1. For a monolayer film, the bending stiffness is proportional to the modulus times the cube of the thickness of the film:

$$E_b \propto EX^3 \tag{9.6}$$

where E_b = bending stiffness, E = elastic or tensile modulus, and X = thickness of film.

For a multilayer film, the relationship is more complex, as is described in Section 9.4.2. The geometry of the test frame also plays a role, so measuring the bending stiffness directly is preferable to relying on the tensile modulus of the film.

Lange et al. [13] describe several types of test methods for measuring the bending stiffness of films, some borrowed from the paper industry and some homegrown:

Bend Under Own Weight: In this type of test, the length of the film is extended over a ledge and the angle of the bend is measured. The extension is increased until a characteristic bending angle is achieved. The length of the extension is related to the bending stiffness. One such method is described in DIN 53 362 [14]. In their tests, Lange et al. find this method gives values lower than other methods which they attribute to large plastic deformation that underestimates the elastic stiffness. They also find that it does not work for curled films.

Cantilever Tests: These tests apply a load or weight to the free end of the film specimen that is anchored on the other end. The amount of deflection of the film at a given load is measured, which is related to the stiffness. Standard methods include ASTM D5342 [15], ASTM D747 [16] and ISO 2493 [17]. Cantilever tests usually work best for stiff films. Lange et al. report that this method gives reliable and accurate results for the films they tested.

3-Point Flex: These tests typically involve measuring the force to push a film through a slot. The slope of the force versus displacement is related to a bending modulus. Standard methods include ASTM D2923 [18] and BS 2782: Part 3, Method 332 A [19].

Many of these methods use homegrown equipment, such as the DuPont method described by Morris and Vansant [20,21]. Fig. 9.5 shows conceptually how the test is performed. In the DuPont method, a 10 cm × 10 cm film specimen is placed on a horizontal slotted flat plate, the dimensions of which are shown in Fig. 9.6. The slot has Teflon tape applied to both sides to minimize friction. The slotted plate is rigged on a tensile testing machine such that it rises toward a stationary beam as the crosshead moves at a fixed speed of 12.7 mm/minute (0.5 inches/minute). See Fig. 9.7 for a schematic of the rig. As the crosshead rises, the film contacts the beam and is forced into the slot. The force is recorded by the tensile tester. The force versus displacement curve is plotted. The slope of the force versus displacement curve is computed using the same method for computing the secant modulus (ASTM D882). This slope is called the Stiffness Factor, which is directly related to the bending modulus.

To obtain the most accurate results, Morris and Vansant take the following precautions:

- The stiffness factor is taken as the average of measurements of multiple specimens of the film with four orientations: film facing up and down in each of the machine (MD) and tranverse directions (TD).
- The slope of the force versus displacement curve is used rather than just the force.
- The specimen width is kept constant when comparing data; the film specimen width directly affects the force measurements.
- Some films (such as polyolefins) are dusted with talc to reduce the error from surface friction. Tack, high coefficient of friction (COF) or static laden films will give erroneously high readings because of the added force required to overcome these factors.
- A lead weight is used on top of the tensile tester to reduce vibration effects.
- Other factors that may alter the results are minimized, including variations in surface slip of the samples or sides of the instrument slot, improper machine alignment, incorrect full-scale load adjustment, incorrect sample size, wrinkled samples and excessive film curl.

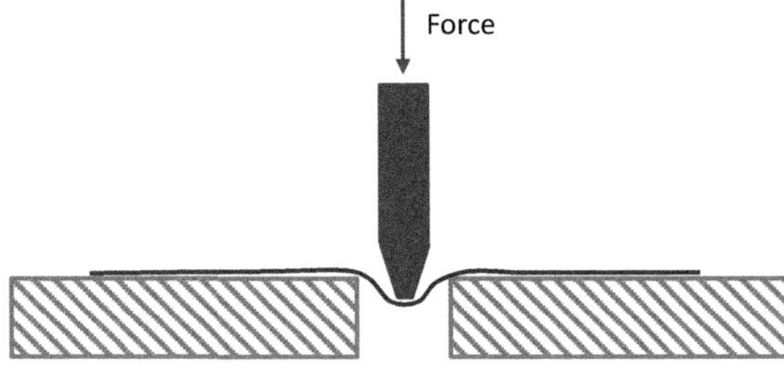

FIGURE 9.5 Schematic of 3-point flex bending test. *Adapted from B.A. Morris, J.D. Vansant, The influence of sealant modulus on the bending stiffness of multilayer films, TAPPI Polymers, Laminations Coat. Conf., 1997.*

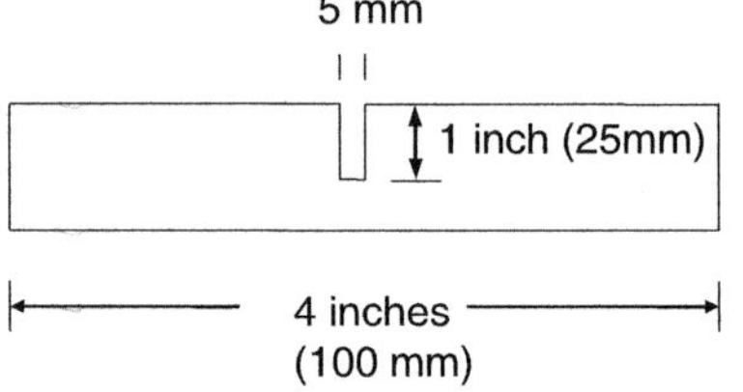

FIGURE 9.6 Side view of slotted plate used in the DuPont bending stiffness test method. *Reproduced with permission of The Dow Chemical Company.*

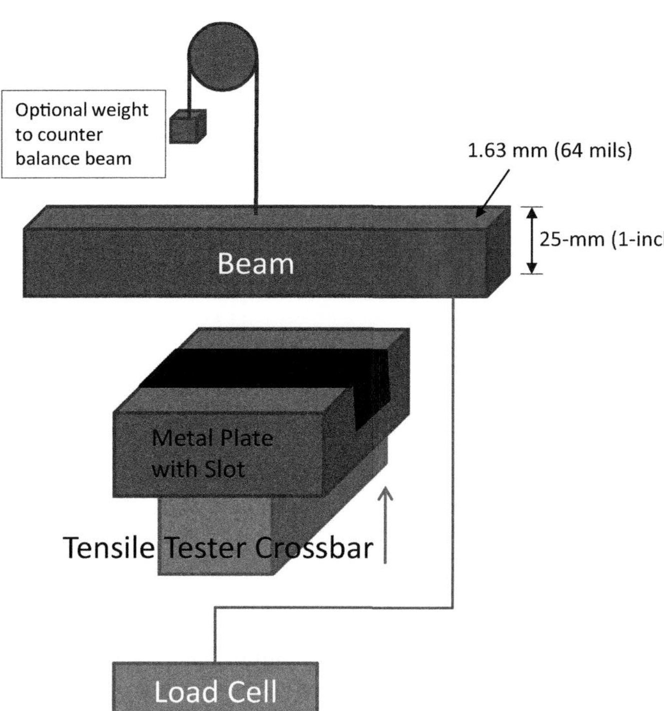

FIGURE 9.7 DuPont bending stiffness test. Slotted plate is attached to the crossbar of tensile tester. As the crossbar moves upward, the film is pushed against the beam. The load cell measures the force. The slope of the force versus displacement curve is the Stiffness Factor. *Reproduced with permission of The Dow Chemical Company.*

Morris and Vansant find that the DuPont method gives consistent results that agree well with their model predictions. Lange et al. evaluate a version of the 3-point flex test and find it works well but is sensitive to misalignment errors. For this reason, they prefer the cantilever test.

A common element of all the test methods is that they depend on the test geometry. To compare amongst the different tests, the results should be normalized against a standard film sample.

9.2.3 Tear strength

The Elmendorf tear resistance test is the most commonly used method for measuring the tear strength of packaging films. The method is described in detail in ASTM D1922 [22] and ISO 6383-2 [23]. The Elmendorf method is a high-speed notched tear test. The specimen is cut to a specified size and mounted to the test machine. A knife in the machine cuts a slit in the film where the tear will initiate. A weighted pendulum is released that swings through the specimen, propagating a tear. The machine measures the loss in kinetic energy of the pendulum as it tears through the specimen. This loss in energy is related to the tearing force. The tear resistance is reported in units of force divided by the thickness of the film. Typically, 10 specimens are tested and the results are averaged.

The trouser tear method is a slow-speed notched tear resistance test and is described in ASTM D1938 [24] and ISO 6338-1 [25]. A standard tensile tester is used with a crosshead speed of 200 or 250 mm/minute (8–10 inches/minute). The rectangular tear specimen is precut on one end in what resembles a pair of trousers (see Fig. 9.8). The two sides of the precut region are placed in the grips of the tensile tester and the force needed to propagate the tear is measured. The tear resistance is the force divided by the specimen thickness.

9.2.4 Puncture and impact resistance

Several tests are used in the packaging industry to measure puncture resistance. No one test represents all the types of failures that may be experienced by a package in the field. Indeed, some package puncture failures do not correlate with any lab test; packaging engineers have sometimes resorted to homegrown tests, such as drop tests with packages filled with screws, to try to simulate real-world issues. The standard laboratory tests can be categorized as slow puncture or high-speed impact/puncture tests.

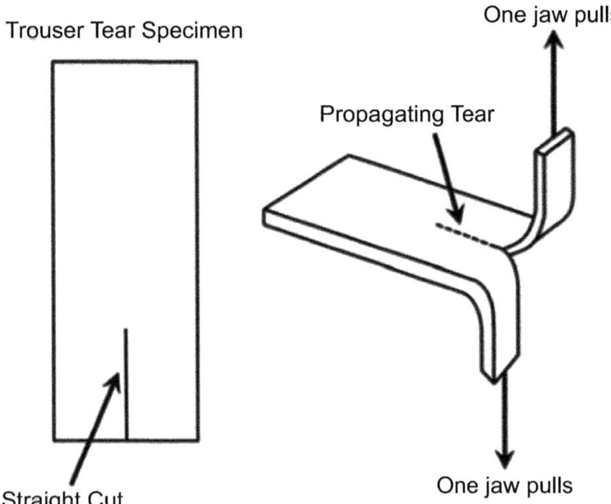

FIGURE 9.8 Trouser tear test sample configuration. *Reprinted with permission from L. W. McKeen, Film Properties of Plastics and Elastomers, third ed., Elsevier, Amsterdam, 2012, pp. 27–34.*

9.2.4.1 *Slow puncture*

The integrity of some packages is compromised by puncture from sharp objects in the package, such as bones or dry noodles. Slow puncture laboratory tests try to mimic the conditions that cause this type of failure. One such method is ASTM F1306 [26]. A specimen is cut from the sample and clamped into a frame mounted to a universal test machine (tensile tester). The mounted specimen is positioned perpendicular to and below a vertical probe that punctures the film during the test. The standard recommends the probe tip be hemispherical with a radius of 3.2 mm. The probe is forced into the specimen at a constant speed of 25.4 mm/minute (1 inch/minute). The test machine measures the peak force, displacement and energy to penetrate the film (area under the force vs displacement curve).

Other probe geometries and speeds are sometimes used. Lange et al. [27] review several methods and probe designs. Other, home-grown methods have been developed. One such method developed by DuPont [28] is the static puncture test, shown schematically in Fig. 9.9. Here the film specimen is mounted on a test rig and a vertical needle-like probe, 2 mm in diameter with a 45 degrees tip angle, is placed on top. A weight is placed on top of the probe to exert a known force. The time for the probe to penetrate the film is measured by recording when the probe hits a metal plate underneath the film and completes an electric circuit.

9.2.4.2 *High-speed puncture*

The falling dart impact test is a noninstrumented test that is simple to use but can be time-consuming and labor-intensive. The test is described in detail in ASTM D1709 [29] and ISO 7765-1 [30]. A film specimen is clamped horizontally into a test ring at the base of a drop tower. The dart is positioned above the specimen and dropped at a specified distance [method A calls for a 38-mm (1.5-inches) diameter dart dropped at 0.66 m (26 inches) and method B calls for a 51-mm diameter dart dropped at 1.5 m (60 inches)]. The result is pass or fail. If the dart penetrates through the film, the test is repeated with a fresh specimen with a higher weight dart. If the dart fails to penetrate the film, the test is repeated with a lower weight dart. The results from a series of determinations are used to calculate the weight at which 50% of the test specimens will fail, known as the Impact Failure Weight.

An instrumented version of the dart impact test, described in ASTM D4272 [31], measures the total energy of impact for rupturing the film. The velocity of a free-falling dart of a specified shape that has passed through the film specimen is measured by using a photoelectric speed trap. The kinetic energy is calculated and compared with the kinetic energy of the free-falling dart when no film is present. The loss in kinetic energy due to the impact and rupture of the film is used as an index of impact resistance. Special speed sensing sensors are needed for this test method.

ISO 7765-2 [32] covers a similar instrumented falling dart test but differs from ASTM D4272, so their results cannot be directly compared. The ISO method uses an instrumented impactor head with a load cell to measure the energy of impact directly. The ASTM method computes the kinetic energy from the velocity of the dart.

A pendulum impact test is described by ASTM D3420 [33]. A film is clamped vertically in a holder with a circular aperture. A pendulum swings a hemispherical probe (25.4 mm in diameter) through the film. The energy needed to burst through the film is measured by the loss in kinetic work from the expenditure of kinetic energy by the pendulum.

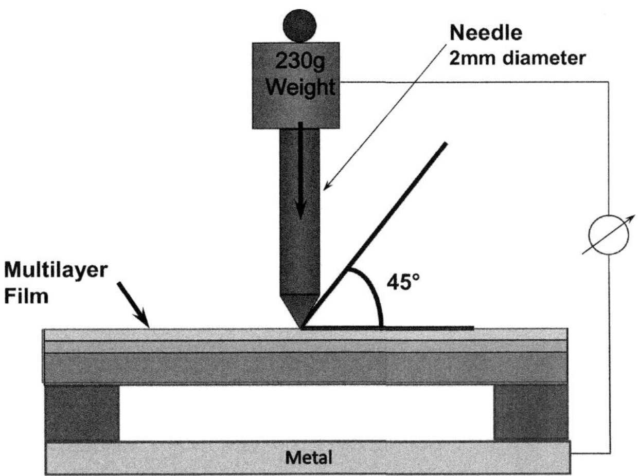

FIGURE 9.9 Schematic of DuPont static puncture test. *Reproduced with permission of The Dow Chemical Company.*

There are two methods that differ by the size of the diameter of the aperture: method A uses a 60 mm diameter sample size whereas method B uses 89 mm. Method B is commonly known as the Spencer impact test. The pendulum hits the film at a maximum velocity of 74 m/minute and maximum energy of 5 J. The impact strength is reported in units of energy (J). There is no equivalent ISO test.

ASTM makes a note in each of these test methods that the results are not necessarily equivalent, especially ASTM D1709 and ASTM D4272, which measure different aspects of the energy (initiation of failure for the former and initiation plus completion of failure for the latter). Also, ASTM cautions that impact measurements are not linear with film thickness. Results should be compared only for specimens that differ by no more than +/−15% in thickness.

A high-speed puncture test for plastic films is described in ASTM D7192 [34] and by McKeen [11]. The film specimen is clamped into a ring horizontally, exposing a circular area to the puncture probe. The probe consists of a 12.7 mm diameter rod with a hemispherical end of the same diameter positioned perpendicular to the circular area. A hydraulic actuator forces the probe into the specimen. Load cells and displacement sensors record the load versus displacement as the probe contacts and break through the specimen under what is essentially multiaxial deformation. The load versus displacement curve, such as that given in Fig. 9.10, provides information that can be used to calculate the puncture strength (load required to break the specimen divided by the cross-sectional area of the film) and the energy to puncture (related to the area under the curve). The test is often performed at several speeds, ranging from 2.5 to 250 m/minute, to determine the rate sensitivity of the film to impact.

Friction may impact dart impact results [35−38]. One example is from Mukherjee et al. [35]. They use a method based on ASTM D7192 where an instrumented probe of a hemispherical diameter of 12.7 mm falls under gravity to impact a clamped film. The drop height is adjusted to control the velocity at impact. Two types of probes are used, a steel probe (high friction) and one coated with polytetrafluoroethylene (PTFE) (low friction). Monolayer linear low-density polyethylene (LLDPE) films with thicknesses of 51 μm (2 mils) or 102 μm (4 mils) are tested. The rank order of the impact resistance of the films differed between the two probes with the low friction probe generally providing lower energy to failure. An analysis of the stress profile during impact shows that when the low-friction probe impacts the film, deformation starts from the tip of the probe. This contrasts with the high-friction probe where deformation begins at the edge. They conclude that the frictional properties of the probe have a significant effect on the impact measurement. Other investigators have spread talc on the film surface or probe to achieve frictionless contact [38].

The development of a new high speed, reverse impact laboratory test by Carajal et al. [39] that better simulates impact events during vertical-form-fill-seal filling operations is described in Section 9.4.3.2.

9.2.5 Scratch and abrasion resistance

Several scratch tests have been developed for the automotive industry to measure the scratch and mar resistance of coatings. These include ISO 1518 [40] and the 5-finger scratch test (Ford BN 108-13; General Motors GMN3943; Daimler-Chrysler LP-463DD-18-01 and PF-10938; and Nissan NEW M0159 Supplement U01-1) [41]. These methods apply a constant force to a scratch pin as it is traversed across the specimen. Different weights or forces are applied. The samples are examined for evidence of scratching. The force at which the scratch first becomes evident is recorded. These methods generally penetrate too deeply for most packaging films and typically have not been used for that purpose. Other scratch test devices include

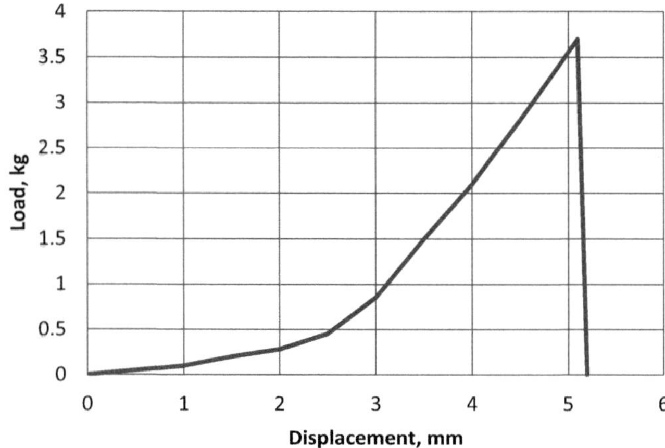

FIGURE 9.10 Typical output of high-speed puncture test: Load versus displacement curve.

pencil hardness, scratching machine, Taber tester, single-pass pendulum sclerometer, scratch apparatus, Revetest scratch tester, and needle test [42]. These also have not found widespread use in flexible packaging.

A method developed at Texas A&M University [42,43] incorporated into ASTM 7027 [44] has recently been applied to packaging films [45–47]. An instrumented scratch machine is used that consists of a stage and clamping system for holding the specimen, a load generator and a horizontal motion servo system. Load and position sensors, data acquisition and computer recording devices may also be used. The machine uses a stylus-type scratch indenter with a 1 mm diameter spherical tip. The specimen is usually a plaque or tensile bar at least 3 mm thick but thinner films can be tested with appropriate backing [45,48].

Two test modes are specified. In test mode A, the scratch is imparted on the specimen with a linearly increasing normal load from 2 to 50 N over a distance of 0.1 m at a speed of 0.1 m/second. This mode allows for the determination of the critical normal load for failure. The onset of failure is noted as the point where damage to the surface is first considered unacceptable. Preferably a quantitative method is used to determine the onset of failure. The depth of the scratch is one indication but measuring the width of the scratch has been found to be more accurate. This is done by using a flatbed scanner, or more accurately with a microscope. Generally, a criterion for the width of the scratch that represents unacceptable damage is determined in advance for the test. The test is run and the sample examined. The location of the point where the width of the scratch reaches the failure point is noted. The normal load for failure is then determined by calculating the load at that location knowing that the load varies linearly along the length of the scratch. The scratch deformation of many polymers recovers partially over time. It is recommended to measure the scratch width or depth after the scratched specimen has aged for 24 hours.

In test mode B, a constant load is applied over the same distance and speed as in mode A. This mode measures the constant load response and can be used to compute a scratch COF.

Different loads, speeds and stylus geometries may be used. Similar "nano" scratch tests that use stylus diameters of around 1 μm [48–50] have also been developed to evaluate the material response to scratch and mar. One of the advantages of the linear load technique is that it allows one to study the different damage mechanisms that may occur during a scratch event.

Blom [51] reviews abrasion testing for flexible packaging. In his study, Blom uses a Taber linear abrasion tester fitted with a flexible materials kit. A weighted probe with a 1.5 mm tip is rubbed across the surface for a cycle length of 25.4 mm (1 inch) at 30 cycles per minute. The cycles to failure are recorded. Different weights can be attached to the probe depending on the nature of the sample. This test method has been incorporated into ASTM F3300 [52].

A number of other abrasion tests are used in other industries, but most are not easily adapted to flexible packaging. A common test is a rotary abrasion test where a roughened surface of a wheel is rubbed across the surface. One problem with this method is that many of the softer polymers used in packaging tend to quickly "gum" up the roughened surface, which skews the results.

9.2.6 Flex fatigue

Flexing of packaging films can lead to the formation of pinholes. ASTM F392 [53] describes a method for subjecting film samples to a standard cycle of twisting motion. The test is generally referred to as the Gelbo flex test after one of the manufacturers of the test equipment. The method involves mounting the film specimen in the testing device and subjecting the specimen to a prescribed number of cycles. The specimen is removed and tested for pinholes.

9.2.7 Shipping tests for package durability

Several standard tests for simulating shipping and transportation of packages have been developed [51]. These include:

ASTM D4169-09, Standard Practice for Performance Testing of Shipping Containers and Systems.
ASTM D7386-12, Standard Practice for Performance Testing of Packages for Single Parcel Delivery Systems.
ISTA Method 3 A, Packaged-Products for Parcel Delivery System Shipments 70 kg (150 lb) or Less (standard, small, flat or elongated).
ISTA Method 3E, Unitized Loads of Same Product.
ISTA Method 4AB, Enhanced Simulation Performance Tests.
ISTA Method 6 series—test methods for specific companies such as Amazon, Sams Club, and Federal Express.

These methods require testing of actual packages and are typically done in the final stages of new product development. They are time consuming, expensive, iterative and do not yield results that are predictive (or suitable for modeling). As a result, they often lead to over engineered packages.

9.3 Typical values

Table 4.3 provides typical values of mechanical, tear and impact resistance properties of monolayer films. Many of these properties can be found on product datasheets provided by the material supplier or in compiled databases online, such as Campus [54], Prospector [55], Matweb [56], or Plastics Technology [57]. Examples with properties of multilayer flexible films are discussed in Section 9.4.

9.4 Engineering principles for multilayer films

A starting point for the prediction of the physical properties of multilayer films is the rule of mixtures. This rule simply states that the property of the multilayer film is the sum of properties of its individual layers, normalized for the relative thickness of each. For a two-layer film the relationship is

$$S = \varphi_1 S_1 + \varphi_2 S_2 \tag{9.7}$$

where S = property of the multilayer film, S_1 = property of layer 1, S_2 = property of layer 2, φ_1 = volume fraction of layer 1 (thickness of layer 1/total film thickness), and φ_2 = volume fraction of layer 2.

Generalizing for more than two layers:

$$S = \sum_{i=1}^{n} \varphi_i S_i \tag{9.8}$$

The rule of mixtures assumes that the properties of the composite are a linear combination of each of its components and that there are no interactions between the layers. In many packaging situations the rule of mixtures is sufficient. There are a number of deviations, however. Some of the reasons for these deviations include:

- Certain properties, such as bending stiffness, impact puncture resistance and abrasion resistance, are nonlinear with respect to thickness. Bending stiffness is taken up in detail in Section 9.4.2 and abrasion resistance in Section 9.4.3.4.
- Adhesion or changes in crystallization behavior at the interface between layers may provide reinforcing effects.
- A neighboring ductile layer may block the propagation of cracks initiated in a brittle layer.
- The load may be transferred to a stronger layer.
- Neighboring layers may confine the local deformation in a ductile layer, increasing the tendency of a crack to propagate. This explains why many multilayer films containing PE have lower tear strength than PE alone.
- The stress history imparted during film fabrication may differ between mono-layer films and coextruded films, resulting in differences in properties. This topic is taken up in more detail in Chapter 14.

There is no comprehensive theory to know when there will be a deviation from the rule of mixtures. Indeed, identical structures with layers made from different grades of the same generic resin may display different behavior. Table 9.2 provides an overview of the literature that is reviewed in the following sections. Many of the properties are found to follow the rule of mixtures in these studies (labeled L in the table), although many more deviate (labeled N).

TABLE 9.2 Summary of studies from the literature on multilayer film structures described in this chapter.

Reference	Structures	Yield strength	Tensile strength	Elongation at bk	Elastic modulus	Bending stiffness	Puncture	Impact (spencer or dart)	Scratch	Elmendorf tear	Flex failure	Abrasion
[58]	(LLDPE-LDPE-LLDPE)	L	L	L				N		N		
[59]	(LLDPE-LDPE-LLDPE)	L	N	L	L		N	L		N		
[60]	(LDPE-mPE$_A$-LDPE)		N		L							
[61]	(LDPE-mPE$_B$-LDPE)		L		N							
	(bLLDPE-mLLDPE-bLLDP-mLLDPE)							N		N		
	(oLLDPE-mLLDPE-oLLDP-mLLDPE)							L		L		
[62]	(LLDPE-LDPE-LLDPE)							L				
	(LLDPE-HDPE-LLDPE)							N				
[63]	(mLLDPE-LLDPE-mLLDPE)									N		
	(LLDPE-mLLDPE-LLDPE)									L		
[64]	(Ionomer-LDPE)		L					N		N	N	
	(PA66-ionomer)		L					N		N	N	
[65]	(EVA-EVOH-EVA)	L										
[66]	OPET (48 g)/adh/PE sealant OPET (48 gauge)/adh/Al (38 gauge)//PE sealant		N	N	L			N		N		

TABLE 9.2 (Continued)

Reference	Structures	Yield strength	Tensile strength	Elongation at bk	Elastic modulus	Bending stiffness	Puncture	Impact (spencer or dart)	Scratch	Elmendorf tear	Flex failure	Abrasion
[67]	BOPP/OPET									L		
	BOPP/cPP									L		
	BOPP/LLDPE									N		
[68]	(PA6-tie-PE)	L			L							
[69]	OPET/adh/Al/ adh/OPET	N	N	N								
	(LDPE-PS- LDPE)		N	N	L							
	(PP-PS-PP)		L	N	L							
[51]	OPET/EVA											N
	OPA or cPA/ EVA											N
[70]	OPET/Al/PE OPET/Al/PVC OPET/Al/PP OPET/Al/PA							N			N	N
[27]	Paper/Al/PE or PET/Paper/Al/PE						N					
	PET/Al/PE or OPP/Al/Ion						N					
[28]	(PA-tie-sealant)					N						
[39]	(HDPE-tie-PA- tie-sealant)							N				
[45]	OPET-met/adh/ PE								N			
	OPET-met/adh/ (PA-tie-PE)								N			
[46]	OPET-met/adh/ (PE sealant) OPET-met/extr/ (PE-sealant)								N			
[47]	PP, PA, PP/PA								N			

(*Continued*)

TABLE 9.2 (Continued)

Reference	Structures	Yield strength	Tensile strength	Elongation at bk	Elastic modulus	Bending stiffness	Puncture	Impact (spencer or dart)	Scratch	Elmendorf tear	Flex failure	Abrasion
[20]	OPP/sealant OPET/EVA/ sealant					N						
[27]	OPET-met/adh/ PE					N						
[2]	PE/PE/PE					N						
[21]	(LLDPE-tie-PA-tie-LLDPE)					N						
[71]	(HDPE-tie-PA-tie-sealant)					N						
[72]	(LLDPE-tie-EVOH-tie-LLDPE)										N	
[73]	(LLDPE-tie-EVOH-tie-LLDPE)										N	

Note: Properties labeled L follow the rule of mixtures (linear behavior) whereas properties labeled N do not (nonlinear behavior).

There is some bias in this list, however, since the goal of the ensuing sections is to highlight some of the factors that cause such deviation.

9.4.1 Mechanical properties

A number of investigators have studied the properties of multilayer polyethylene (PE) films [58–61,63,74–76], comparing them with monolayer films of the component resins, as well as blends with the same volume ratio. Several studies evaluate (A-B-A) or (A-B-A-B) coextruded blown film structures with the volume percent of B varying over the range of 0 to 100%. Different compositions of A and B are studied, such as LDPE and linear low-density polyethylene (LLDPE) [58], LDPE and metallocene polyethylene (mPE) [60,75], and Ziegler–Natta (ZN) LLDPE and mLLDPE [61]. Some of the mechanical properties of the coex structures follow the rule of mixtures and others show significant deviation, as summarized in Table 9.2. An example is shown in Figs. 9.11 and 9.12 taken from the work of Beagan et al. [75]. Here two different mPE grades are coextruded with LDPE skin layers. The tensile strength of the coextruded film structures with mPE resin A shows a substantial positive deviation from the rule of mixtures whereas those with mPE B show a slight negative deviation. The opposite is true for secant modulus (Fig. 9.12) where resin A shows a slight deviation from the rule of mixtures and resin B a substantial negative deviation. The crystallinity of the two mPE grades differs, as is indicated by the lower modulus of resin A. The two grades are not produced by the same manufacturer and so may have other differences as well.

Several investigators have suggested that co-crystallization at the interface of some polyethylenes is responsible for the deviations from the simple rule of mixtures. Differences in crystalline morphology have been reported. For example, Elkoun et al. [61] find that coextruded films with layers of ZN butene LLDPE (B) and a single-site LLDPE (S) deviate more from the rule of mixtures than one containing ZN octene-LLDPE (O) and S. They attribute the improved performance of the B-S coextrusion to the formation of transcrystalline regions at the interface between the dissimilar B and S (and less formation for the similar O and S).

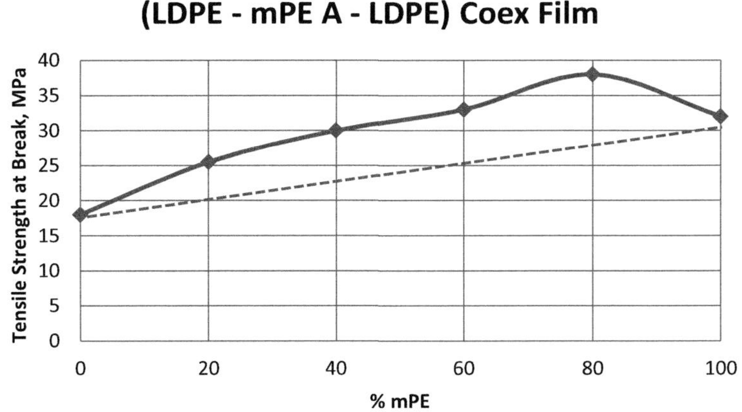

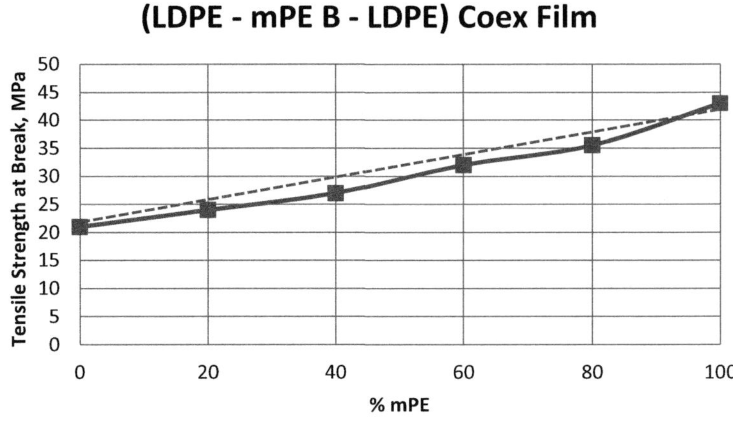

FIGURE 9.11 Tensile strength at break of coextruded films with different amounts of a core mPE layer. Top graph is for mPE A and bottom graph for mPE B. Dashed curves represent the values predicted by the rule of mixtures. *Adapted from C.M. Beagan, G.M. McNally, W.R. Murphy, The blending and coextrusion of metallocene catalyzed polyethylene in blown film applications, SPE ANTEC Conf., 1999.*

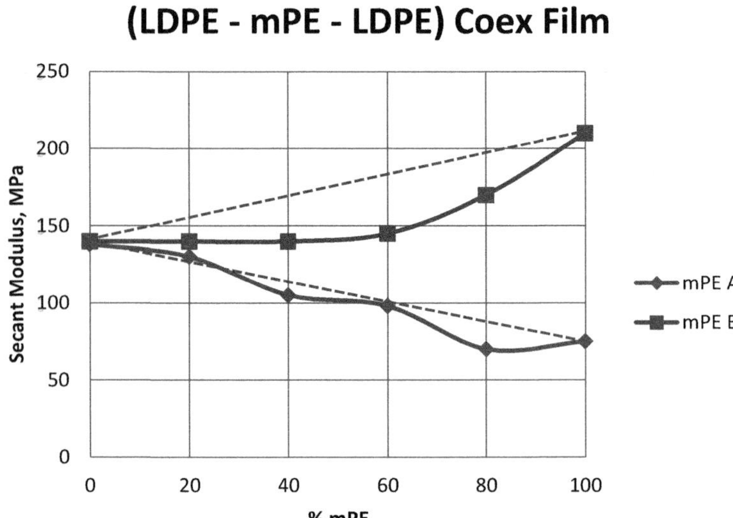

FIGURE 9.12 2% Secant modulus of coextruded films with different amounts of a core metallocene polyethylene layer. Dashed curves represent the values predicted by the rule of mixtures. *Adapted from C.M. Beagan, G.M. McNally, W.R. Murphy, The blending and coextrusion of metallocene catalyzed polyethylene in blown film applications, SPE ANTEC Conf., 1999.*

Zhang and Ajii [77,78] study PP-LLDPE blends and coextrusions. A homopolymer (hPP) and random copolymer (rPP) polypropylene are tested. The PE is an octene-LLDPE made by a ZN catalyst. A five-layer coextrusion blown film with the structure (PP-LLDPE-PP-LLDPE-PP) is compared with a blend of the same volume ratio of PP to PE (50:50). The film is made under conditions to impart high draw in the MD [high draw down ratio (DDR) and low frost line height]. The coextruded film generally has a higher percent elongation to break than the blend. Wide-angle X-ray scattering (WAXS) measurements and scanning electron microscopy (SEM) show differences in crystalline morphology and orientation. Two types of oriented lamellae crystalline structure may form at the PE-PP interface: an epitaxial structure where the PE crystallizes on the PP and a transcrystalline structure where the PP, with its higher crystallization temperature (Tc), induces an oriented crystalline structure on the polymer with the lower Tc (PE in this case). The homopolymer PP-LLDPE blend exhibits epitaxial structure whereas the random copolymer PP-LLDPE blend and coextrusion show transcrystalline behavior. The difference in crystalline morphology may influence adhesion at the interface.

Gaucher-Miri et al. [65] evaluate coextruded films of (EVA-EVOH-EVA). The films are 500-μm thick with the EVOH core ranging from 1% to 20% of the structure. The skin layers are a blend of EVA (9% VA) and VLDPE (10% of the blend). There is no mention of a tie layer. They conduct dynamic mechanical analysis (DMA) and tensile tests (mostly at 100 inches/minute [2500 mm/minute] or 10,000%/minute) at 55°C to 95°C.

Monolayer control films are first evaluated. The EVA deforms homogeneously in the tensile tests and exhibits strain hardening behavior. The EVOH control film exhibits neck-in at the yield point. The elongation at break increases with temperature, the opposite of what occurs with EVA, but breaks just after yielding at 55°C. The strain hardening parameter for EVOH is independent of temperature but decreases at high strain rates and increases with increasing relative humidity. Gaucher-Miri et al. manipulate the data to compare EVOH and EVA at the same true strain rate (to account for the necking of the EVOH). They find that EVOH has higher yield stress and a lower strain hardening parameter than EVA. By measuring the stress versus strain curve at strain rates ranging from 0.254 to 254 mm/minute they show that the EVOH transitions from homogeneous to heterogeneous (necking) deformation when the strain hardening parameter changes from being less than one to greater than one. This occurs at strain rates between 2.54 and 25.4 mm/minute.

The (EVA-EVOH-EVA) coextruded film deforms homogeneously (no necking) over the temperature range tested; the EVA layers control the deformation. Yield stress increases and the strain hardening parameter decreases with an increasing fraction of EVOH in the structure. The rule of mixtures works well and can be used to predict under what conditions the coextruded film will neck: when the strain hardening parameter drops below one, neck-in occurs.

Huang et al. [68] make (PA6-tie-PE) coextruded films with different layer ratios. They provide a good discussion of engineering versus true stress and strain and fitting or modeling the data with a two-parameter model (yield stress and strain hardening parameter). Yield stress, strain hardening parameter and tensile modulus are found to follow the rule of mixtures.

Schrenk and Alfie [69] describe the mutual mechanical reinforcement of multilayer films. When thin layers of high-modulus low-elongation material are sandwiched between thin, adhering layers of low-modulus, high-elongation material, the stress at break for the laminate is much higher than for either material alone. The high modulus material undergoes higher elongation in the laminate than in a monolayer film. The low modulus material acts to stop cracks from propagating laterally. These laminates exhibit high elongation and high modulus and have much higher energy to break than single materials.

In one study, they make OPET/adh/Al/adh/OPET laminates and vary the adhesive strength. Good adhesion produces the reinforcement effect, allowing the foil to reach the plastic yield region by inhibiting cracks that form transverse to the direction of pull. Poor adhesion results in a break-up of the foil layer into shorter segments during the tensile test.

They describe experiments with LDPE-polystyrene-LDPE films with three or 125 layers. Adhesion is poor between LDPE and polystyrene (PS), but as the thickness of the layers gets smaller separating the layers becomes more difficult. Despite the poor adhesion, the 3-layer films show mutual reinforcement when the tensile test is done in the MD. Here the PS layer is the high modulus, low elongation layer. In the TD, the 3-layer films show mutual destruction with failure of all the layers at low elongation.

The 125-layer films of LDPE and PS are biaxially stretched on a lab stretcher to 25-μm (1 mil) thickness. The ultimate tensile strength obeys the rule of mixtures in both directions with no mutual reinforcement. Elongation at break is low for all the compositions. The thin PE layers are too weak to stop cracks from propagating.

Films with 125-layers of alternating PP and PS are also made and biaxially stretched to 25-μm. Tensile strength increases linearly with increasing PP content. The elongation is 100% or more – much higher than for the PE-PS films—indicating mutual reinforcement is occurring.

They conclude:

- Mutual reinforcement depends on the adhesion between the layers—better adhesion makes reinforcement more likely.
- Mutual reinforcement depends on the orientation of the brittle layer.
- The high elongation layer must be strong enough to stop the propagation of cracks in the brittle layer in order for mutual reinforcement to occur.
- Tensile strength in some structures can be predicted from the volume fractions of the components using the rule of mixtures.

When no mutual reinforcement exists, Hu [70] explains how the tensile strength may be less than the sum of its components. Consider the two-layer structure shown in Fig. 9.13. As the laminate is stretched axially, stress develops in each layer that is the product of its modulus and the strain. In equilibrium, the stress on the laminate, σ_L, is simply the sum of the stress on each layer:

$$\sigma_L = \sigma_1 + \sigma_2 \tag{9.9}$$

When an additional load is added, the layer with the higher modulus carries more of the load until it reaches its ultimate tensile strain and breaks. When the first layer breaks, this puts more stress on the second layer, which then breaks when it reaches its ultimate tensile strength. If the tensile strength of the first layer (higher modulus) is greater than the second then the laminate will fail immediately after the first layer fails. If the tensile strength of the first layer is less than the second, the laminate will fail once the tensile strength of the second layer is reached: the tensile strength of the laminate is never greater than the sum of its components. Indeed, it will be closer to the tensile strength of the strongest layer.

Timm [66] studies the tensile properties of two laminate structures: OPET (13-μm)/adh/PE sealant and OPET (13-μm)/adh/Al (10-μm)//PE sealant. Six different sealant films are tested with varying stiffness and thickness (25–125 μm). He asks the question, are the properties the sum of its parts? The answer depends on the test method failure mode and dynamics. He provides a conceptual model as to why this is so:

- **Single-ply polymers:** Crystals bear most of the load in a tensile test, especially in the MD. Weaker amorphous regions are shielded from the load until cracks develop. The TD is less predictable—both amorphous and crystalline regions bear the load.
- **Laminate:** Adhesion is critical. The stiffer/stronger layer bears more of the load initially, in both MD and TD. The TD properties are more consistent.

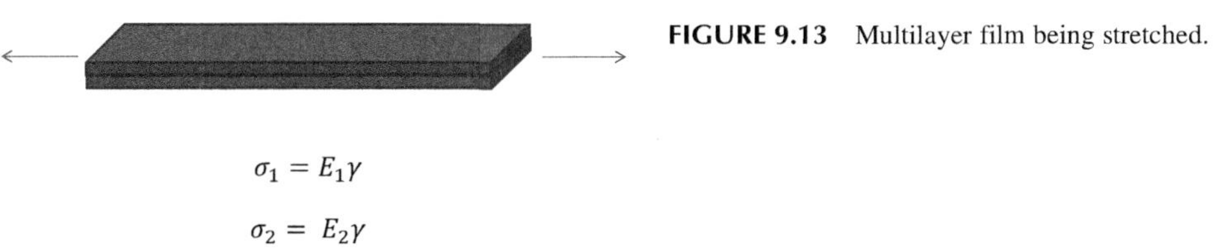

FIGURE 9.13 Multilayer film being stretched.

The results show that the tensile strength of the OPET film dominates until the sealant film thickness reaches about 100 μm (4 mils) where the sealant film starts to become as important as the OPET film. The elongation to break is dominated by the OPET film. The secant modulus, however, follows the rule of mixtures. Results of impact and tear tests from this study are discussed in Sections 9.4.3 and 9.4.4.

These examples show that the mechanical properties of multilayer films often follow the rule of mixtures and can be estimated by the linear sum of the properties of the component layers. There are many exceptions, however, that arise when mutual reinforcement or destruction occurs at the interface. Changes in crystallization, adhesion and the ability of the adjacent layers to stop the propagation of cracks are some of the mechanisms for these deviations. For mutual reinforcement to occur, the high elongation layers must have sufficient strength and adhesion to stop the propagation of cracks that develop in the brittle layers.

9.4.2 Bending stiffness

The bending stiffness of multilayer films is not a simple linear combination of the stiffness (or modulus) of the component layers. Instead, it is a complex function of the elastic modulus, location and thickness of each of the layers. A brief analysis is presented here, based on Morris and Vansant [20], followed by several examples of how it applies to flexible packaging.

To describe the bending stiffness of a thin film it is assumed that the film or sheet can be modeled as a beam, such as an I-beam. As a beam is bent, one side of the beam is in compression and the other is in tension, as shown in Fig. 9.14. The tensile or compressive stress is at a maximum at the outer surfaces, and zero at a point within the beam. This zero-stress point is called the neutral axis. A beam (film, sheet, etc.) is stiffer than another if it bends less under a given amount of stress.

For calculation purposes, the modulus in compression is assumed to be represented by the elastic modulus and equal to the modulus in tension. Compression moduli are difficult to measure and making this assumption makes the computation easier. The assumption is validated by the good agreement between the experimental work of Morris and Vansant and the model results.

The deflection, D, of a beam under an applied force, F (see Fig. 9.15), is described by Eq. (9.10):

$$D = \frac{F}{\sum_{i=1}^{n} E_i I_i} \left[\frac{L^3}{24} \right] \tag{9.10}$$

where $E_i =$ modulus of layer i, $I_i =$ contribution to the total moment of inertia from layer i, n = number of layers, and L = length of the beam.

Here the term in brackets is independent of the beam design (or film structure), but rather is a function of the way it is held and how the force is applied. Rearranging Eq. (9.10) and renaming the term in brackets H and its inverse H' yields

$$\frac{F}{D} = \left(\sum_{i=1}^{n} E_i I_i \right) H' \tag{9.11}$$

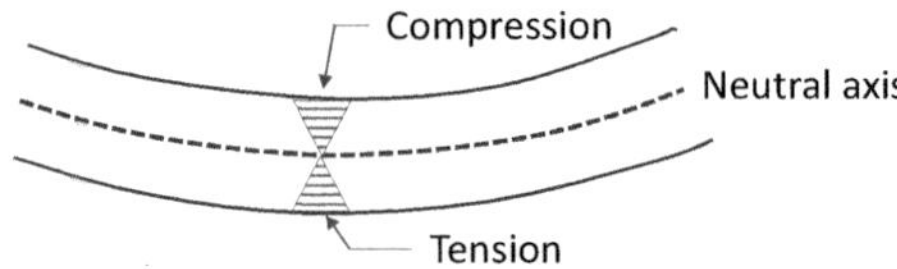

FIGURE 9.14 Beam under bending motion, rectangular cross-section of single material.

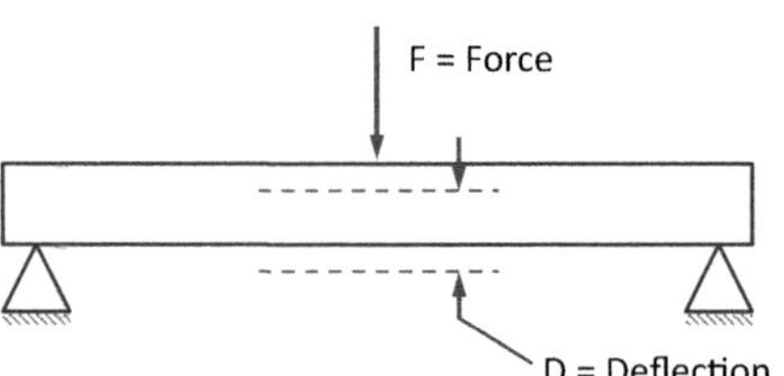

FIGURE 9.15 Force and deflection of a beam.

Eq. (9.11) is in the same form as that for a spring, where $(\Sigma E_i I_i)H'$ is the spring constant describing the relationship between applied force and deflection. $\Sigma E_i I_i$ represents a "stiffness factor" describing the composite stiffness of the beam, film, sheet, etc. H' describes how the force is applied. $\Sigma E_i I_i$ is calculated by first locating the neutral axis. The moment of inertia term for a given layer, I_i, is related to the cube of distance of that layer from the neutral axis. For a more detailed analysis, see Jones [79].

Morris and Vansant develop a computer model based on Eq. (9.11). The model input includes the elastic modulus, thickness and location of each layer. They validate the model by comparing the calculated stiffness factor to experimental measurements of bending stiffness using the DuPont 3-point bending method described in Section 9.2.2. Two types of structures are tested, a 3-layer coextrusion of (HDPE-tie-sealant) and a proprietary laminate. A graph comparing the test results with the model is shown in Fig. 9.16. The statistically determined trend line through the results has an R^2 value of 0.96, indicating an excellent fit. Lange et al. [13] develop a similar model and also report good agreement between model and experimental results.

There is an offset in the prediction (the slope of Fig. 9.16 is not equal to 1) because of the geometry of the test [H' in Eq. (9.11)]. The model, therefore, is most useful for computing changes in bending stiffness versus a control. The model is very helpful for understanding how material selection and structure design influence stiffness. An example is shown in Fig. 9.17. Here four 25-μm (1 mil) thick layers are combined to make a film. Two of the layers are made from a relatively stiff HDPE resin and two are made from a flexible polymer such as ethylene vinyl acetate (EVA) copolymer or mPE plastomer. If the layers are arranged so that two stiff layers make up the core, the stiffness factor is 1. Simply rearranging the layers so that the stiff layers make up the skin increases the bending stiffness by

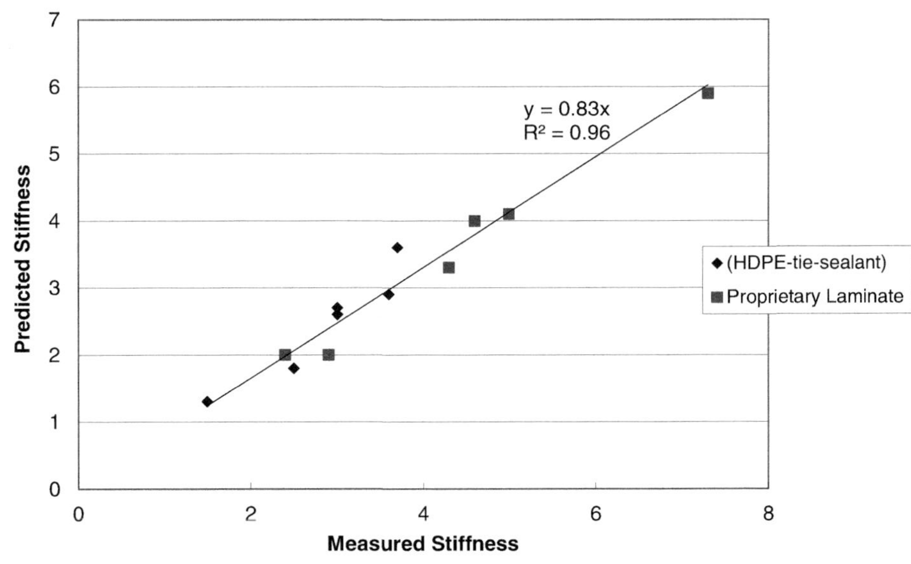

FIGURE 9.16 Comparison of model prediction and experimental results. *Adapted from B.A. Morris, J.D. Vansant, The influence of sealant modulus on the bending stiffness of multilayer films, TAPPI Polymers, Laminations Coat. Conf., 1997.*

Model insights: I-beam effect

Consider four 25-μm (1-mil) layers:

	Structure 1	Structure 2
	mPE	HDPE
	HDPE	mPE
	HDPE	mPE
	mPE	HDPE
Stiffness Factor	**1**	**4**

A factor of four difference in stiffness!

FIGURE 9.17 Example of the effect of layer arrangement on bending stiffness of a multilayer film. Assumptions: HDPE modulus = 690 MPa, mPE modulus = 69 MPa. *From B.A. Morris, J.D. Vansant, The influence of sealant modulus on the bending stiffness of multilayer films, TAPPI Polymers, Laminations Coat. Conf., 1997.Reproduced with permission from The Dow Chemical Company.*

a factor of 4! This is what Morris and Vansant call the I-beam effect. Separating stiff layers can be very effective for building in bending stiffness into a structure. Note that in a tensile test, these two structures would have the same elastic modulus. It is only in bending that the layer location makes such a difference.

In engineering terms, the further away from the neutral axis and stiffer the skin layer, the stiffer the structure. Many food packaging applications involve the combination of a stiff protective or barrier layer [such as nylon, OPET, biaxial oriented polypropylene (BOPP), aluminum foil, EVOH] and a sealant. In such structures, frequently the neutral axis resides in the stiff barrier or protective layer, which lessens its contribution to the overall bending stiffness of the structure. The sealant layer is always on the outside of the structure, often away from the location of the neutral axis. As a result, the modulus and location of the sealant layer can have a disproportionate effect on the overall bending stiffness. Understanding this effect can aid in anticipating potential problems with substituting relatively stiff sealant materials with ones with lower modulus. It also points to opportunities for downgauging and cost reduction without loss of stiffness by using relatively stiff sealants such as ionomers. See Figs. 4.9 and 7.56 for a comparison of the stiffness of various sealant resins.

The I-beam effect often comes into play when dealing with the lamination of films. An example is given in Table 9.3 where laminations made by adhesive and extrusion lamination are compared. The structure involves bonding BOPP to a sealant film. The base case (A) is for when the moderately stiff sealant film, such as a 25-μm (1 mil) LLDPE or ionomer film, is adhesively laminated to the BOPP. Extrusion laminating the two films, case B, substantially increases the bending stiffness, even though the LDPE adhesive has a lower modulus than the chemical adhesive. This is because the extruded adhesive is thicker and separates the two films further, enhancing the "I-beam" effect. This is confirmed in case C where the LDPE adhesive thickness is increased to 25-μm (1 mil). The BOPP must be increased from 18 to 38-μm (70−150 gauge) in the adhesive lamination (case D) to achieve the same increase in stiffness as extrusion laminating with 25-μm (1 mil) of LDPE. Lange et al. [13] obtain similar results with a PET/adh/PE-film lamination and conclude that the thickness of the adhesive layer overwhelms the effect of its modulus. The designer can use the model results to weigh the relative material and fabrication costs of the two processes to achieve the desired increase in stiffness.

The model can also be used to look at the effect of material substitution and downgauging on stiffness. Table 9.4 shows an example of a capping web for a meat package that has the structure OPET/EVA/ionomer with thicknesses 11.9/22.9/33 μm (0.47/0.9/1.3 mils). Such combinations of a high modulus film with a sealant are prevalent in the snack industry as well. Using typical values for the moduli of these materials, Morris and Vansant compute a stiffness factor (case 1 in Table 9.4). Case 2 shows the effect of substituting a lower modulus mPE sealant for the ionomer. The bending stiffness is reduced by 67%. Case 3 shows that the thickness of the mPE layer would have to be increased to 63.5-μm (2.5 mils) to achieve the same stiffness as case 1. This not only increases the cost but also may slow the transfer of heat to the interface during heat sealing, slowing packaging line speeds. Case 4 looks at substituting the ionomer with a higher modulus one. The sealant can be downgauged by 7.6-μm (0.3 mils) or 30% and still achieve the same stiffness as the base case (case 1). In case 5, the sealant is further downgauged by bulking up the EVA layer. This effectively moves the stiff sealant layer away from the neutral axis, resulting in a structure with potentially lower cost. As with any of these cases, the model only predicts what will happen to bending stiffness. Reducing sealant layer thickness may diminish the sealing properties and must be thoroughly tested under conditions simulating its actual use.

TABLE 9.3 Comparison of adhesive and extrusion laminated films.

Case	BOPP thickness microns (gauge)	Adhesive type	Adh modulus MPa (kpsi)	Adh thickness μm (mil)	Predicted stiffness factor
A	18 (70)	Adhesive	690 (100)	2.5 (0.1)	0.39
B	18 (70)	LDPE	138 (20)	12.7 (0.5)	0.70
C	18 (70)	LDPE	138 (20)	25.4 (1)	1.25
D	38 (150)	Adhesive	690 (100)	2.5 (0.1)	1.38

Structure is BOPP/adh/sealant film. Assumptions: BOPP modulus is 1590 MPa (230 kspi), sealant modulus is 275 MPa (40 kpsi) and sealant thickness is 25 μm (1 mil).
Source: Adapted from B.A. Morris, J.D. Vansant, The influence of sealant modulus on the bending stiffness of multilayer films, TAPPI Polymers, Laminations Coat. Conf., 1997.

TABLE 9.4 Model results, OPET//(EVA-sealant).

Case	Sealant	Sealant modulus MPa (kpsi)	Sealant thickness μm (mils)	EVA thickness μm (mils)	Total thickness μm (mils)	Predicted stiffness factor	Stiffness change (%)
1	IONOMER1	276 (40)	33 (1.3)	22.9 (0.9)	67.8 (2.67)	1.49	
2	mPE	69 (10)	33 (1.3)	22.9 (0.9)	67.8 (2.67)	0.49	−67
3	mPE	69 (10)	63.5 (2.5)	22.9 (0.9)	98.3 (3.87)	1.48	−1
4	IONOMER2	483 (70)	25.4 (1)	22.9 (0.9)	60.2 (2.37)	1.49	0
5	IONOMER2	483 (70)	10.2 (0.4)	45.7 (1.8)	67.8 (2.37)	1.48	−1

Assumptions: OPET modulus is 3450 MPa (500 kpsi), OPET thickness is 11.9 μm (0.47 mil) and EVA modulus is 69 MPa (10 kpsi).
Source: Adapted from B.A. Morris, J.D. Vansant, The influence of sealant modulus on the bending stiffness of multilayer films, TAPPI Polymers, Laminations Coat. Conf., 1997.

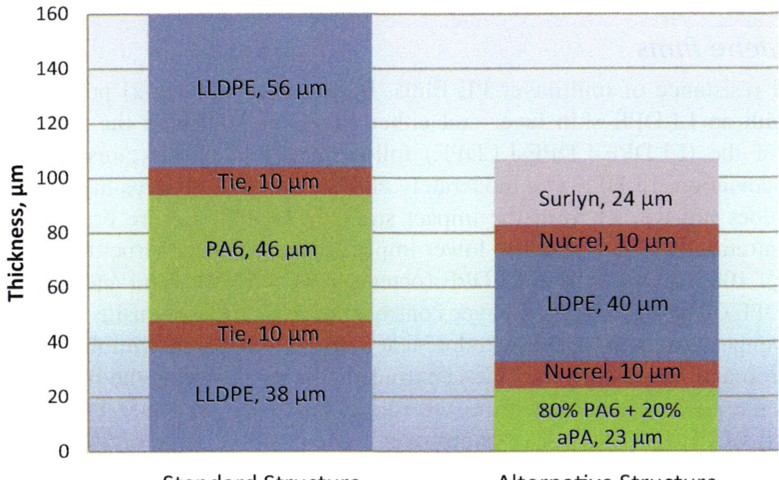

FIGURE 9.18 Redesign of thermoforming web for meat package using principles from bending stiffness model. *Reproduced with permission fromB.A. Morris, J. D. Vansant, K. Hausmann, More structural strength for thin films, Kunstoffe Int., May (2011) 42–46.*

McKenna et al. [2] vary the PE type and thickness of 3-layer PE films. Using a model as a guide, they design structures with lower thickness and a range of bending stiffness. A consumer panel perceives the stiffest films to be the thickest, independent of their actual thickness. The tests show that consumers can distinguish between films differing in bending stiffness by about 10%–15%. By judicious material selection and layer arrangement, film structures can be significantly downgauged while maintaining performance and consumer perception by matching the bending stiffness of the incumbent film.

These concepts are used by Morris, Vansant, and Hausmann [21] to redesign a package for meat products such as smoked ham. Fig. 9.18 shows the typical structure of a conventional forming web. The LLDPE sealant layer placed on the inside of the film is 56 μm thick, while a layer of only 24 μm of the significantly stiffer ionomer (Surlyn from Dow, shown on the right) provides the same mechanical properties. As part of this optimization they propose additional changes to the structure in order to reduce thickness and costs. This includes shifting the second stiff component, a polyamide 6 (PA6) barrier layer, to the outside of the structure so that the I-beam effect can be used to further raise the flexural stiffness. Additionally, the concept provides improvements to the thermoformability through blending amorphous polyamide (aPA) into the PA6 layer. Because this material increases the oxygen barrier this step also allows the PA thickness to be reduced. At the same time, the aPA adds to the stiffness of the PA layer. Lastly, the new concept calls for a core layer of low-cost LDPE, which has the effect that both of the stiff layers (Surlyn and PA) lie further from the neutral axis, which additionally optimizes the I-beam effect and overall helps to achieve the minimum thickness for the thermoforming film. Nucrel acid copolymer resin from Dow has proved to be an ideal tie layer for such structures.

In total, the new film concept is more than 30% thinner than the original structure at a slightly lower cost. The material costs are reported as being reduced by 4%, and if the structure is used in Europe, further savings can be obtained from a reduction in packaging levies. The switch to the ionomer sealant has additional advantages: an expected 25% higher film puncture resistance, and greater packaging integrity because the high hot tack provided by the ionomer sealant can reduce the danger of leakers, and the possibility of raising productivity due to the low temperature at which sealing begins.

9.4.3 Puncture, scratch, and abrasion resistance

The abuse resistance of multilayer films, whether measured by puncture, scratch or abrasion tests, often deviates from the rule of mixtures. Some of the deviations is due to nonlinear thickness effects. For example, if the abuse resistance increases with the square of the thickness, combining two layers of the same material and thickness will quadruple the abuse resistance whereas the rule of mixtures predicts a mere doubling. As will be illustrated in the examples below, other factors also play a role, such as the orientation of the film toward the indenter. Here the presence of a soft sealant layer or a hard structural layer on the surface can affect the deformation of the film as it is contacted by the probe, changing the puncture or scratch resistance. Other factors include interlayer adhesion, load transfer, changes in crystalline morphology and blunting of crack propagation by adjacent ductile layers.

9.4.3.1 Puncture resistance of polyethylene films

A number of studies have examined the impact resistance of multilayer PE films. Butler and Morris [62] provide an example of two three-layer coextruded films with an LLDPE skin layer and either LDPE or HDPE as the core. As shown in Fig. 9.19, the dart impact strength of the (LLDPE-LDPE-LLDPE) follows the rule of mixtures but the (LLDPE-HDPE-LLDPE) film shows a negative deviation. LLDPE is a moderately elastic resin that absorbs high levels of impact energy. LDPE, while not as elastic, does not detract from the impact strength. HDPE is more brittle than LLDPE and tends to form localized stress concentration sites that result in lower impact strength for the structure.

Siegman and Nir [58] compare blown films of 100% LDPE, 100% LLDPE (octene, ZN), a 50/50 blend and a coextrusion of (25% LLDPE−50% LDPE−25% LLDPE). The LLDPE monolayer control film has superior tensile strength, yield strength, % elongation at break, dart drop impact strength, and Elmendorf tear strength compared with the LDPE monolayer film. For the high-speed tests, dart drop and Elmendorf tear, the coextruded film outperforms the blend and approaches the values of LLDPE, suggesting some positive deviation from the rule of mixtures. Martinez et al. [59] also compare LLDPE-LDPE blends with (LLDPE-LDPE-LLDPE) coextrusions over a composition range of 0%−100% LLDPE. Dart impact strength follows the rule of mixtures fairly well, but puncture resistance of the coextruded films shows a positive deviation from the rule of mixtures at LLDPE concentrations above 50% and have considerably higher resistance than the blend. As described in Section 9.4.1, Siegman and Nir ascribe the deviations from the rule of mixture and differences between coextruded structures and blends to crystalline morphology changes. They show microscopy evidence of the formation of a row nucleated crystalline structure in the coextruded films at the LLDPE-LDPE

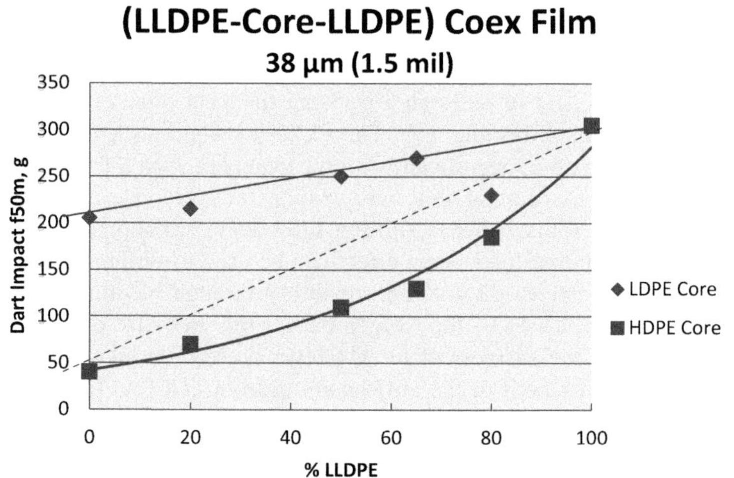

FIGURE 9.19 Dart impact of coextruded 3-layer film. Core layer is either LDPE (0.918 g/cc, 2.0 g/10 minute MI) or HDPE (0.965 g/cc, 0.7 MI). Outside layers are LLDPE (0.920 g/cc, 1 MI). Dashed curve is what would be predicted from the rule of mixtures for the HDPE-core structure. *Adapted from T.I. Butler, B.A. Morris, PE-based multilayer film structures, in: S. Ebnesajjad (ed.), Plastic Films in Food Packaging, Materials, Technology and Applications, Elsevier, New York, 2013, pp. 21−52.*

interface which they attribute to layer orientation during the film blowing process. Taylor and Baik [76] observe similar property advantages of coextrusion versus blends of a hexene ZN LLDPE and LDPE. Compositions of the blends and coextruded blown film contain 20% LDPE. The coextruded structure has considerably higher slow puncture and dart impact resistance than the blend.

It should be noted that the impact and puncture resistance of monolayer blown films of blends of LLDPE and LDPE generally do not follow the rule of mixtures. Patel et al. [80] find that dart impact resistance of blends of an octene LLDPE with LDPE decreases with increasing % LDPE faster than the blend rule. They attribute this to a change in orientation of the polymer chains as well as a reduction in the number of tie molecules when LDPE is added. Tie molecules are polymer chains that connect crystalline regions together [81] and contribute toward mechanical strength. Slow puncture resistance goes through a maximum at 10% LDPE, which they attribute to higher gloss (resulting in more friction with the probe), and decreases with increasing LDPE in an otherwise linear manner.

The puncture resistance of LLDPE depends on its molecular architecture. For example, Sadeghi and Ajji [82] measure the slow puncture resistance of four LLDPE cast films using a sharp probe. They use rheology, thermal analysis and x-ray scattering to discern the differences of the resins that nominally differ by catalyst type, density and melt flow index. A ZN LLDPE had higher puncture resistance than three metallocene LLDPEs of lower density. They hypothesize this is due to greater tie chain density and the presence of a high MW fraction in the ZN-LLDPE.

Elkoun et al. [61] evaluate blends and coextruded films of three different types of LLDPE: a butene ZN LLDPE (b), an octene ZN LLDPE (o) and an octene single-site catalyst LLDPE (s). Films with the construction (A-B-A-B-A) are made on a 5-layer coextrusion blown film line. Polymer s is blended or coextruded with b or o. The dart impact strength of the (b-s-b-s-b) coextrusion shows some negative deviation from the rule of mixtures whereas the (o-s-o-s-o) coextrusion follows the rule of mixtures fairly well. The deviation from the rule of mixtures for tear strength is more pronounced and is described further in Section 9.4.4.1. The b-s coextrusion has significantly better dart impact strength than the blend of the same composition whereas both the o-s coextrusion and blend follow the rule of mixtures. They attribute this to the formation of transcrystalline regions at the interface between the dissimilar b and s (and less formation of these crystalline regions for the similar o and s resins).

9.4.3.2 Puncture resistance of laminates

Combining polyolefin films with stiffer layers such as PA or PET can be less straightforward. Hausmann and Fleiger [28] evaluate three-layer blown films of (PA6-tie-sealant) with layer thicknesses (50/20/40 μm). The tie layer is an acid copolymer. Three sealants are tested, an LLDPE, an mPE plastomer and an ionomer. The ionomer is Surlyn 8140, an especially stiff grade with a modulus of around 480 MPa (70 kpsi). The films are tested according to DIN 53373 (similar to ASTM D1306 described in Section 9.2.4.1), using a 2-mm diameter probe with a 90 degrees angle and a test speed of 100 mm/minute. The sealant side of the film faces the probe. Surprisingly, the sealant selection made a significant contribution to the penetration resistance, as is shown in Fig. 9.20. The PA layer, known for its toughness, was expected to dominate. The mPE plastomer performs better than the LLDPE, but the stiff grade of ionomer performs the best. Similar results are found when the DuPont static puncture test is used, as is shown in Fig. 9.21. Here the time for failure for the coextruded film containing Surlyn 8140 as the sealant is five times that of the film with mPE as the sealant.

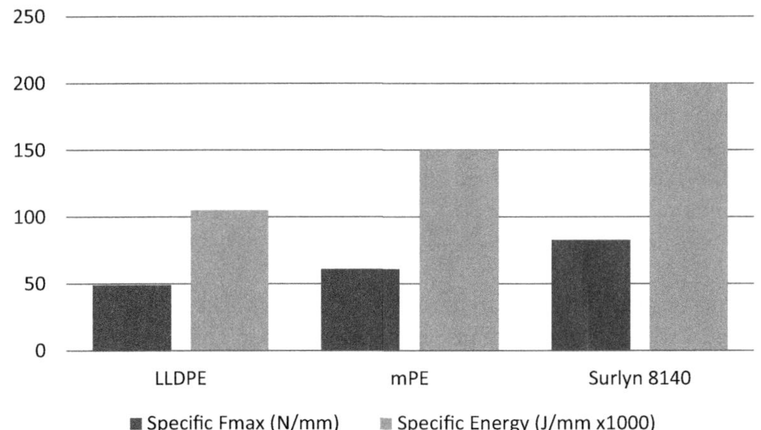

FIGURE 9.20 Effect of sealant type on penetration resistance of (PA6-tie-sealant) multilayer film. 2-mm probe diameter with 90 degrees tip angle, 100 mm/minute. *Reproduced with permission from The Dow Chemical Company.*

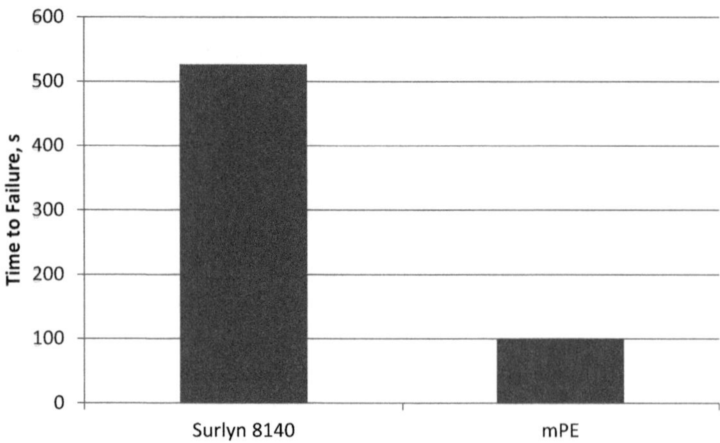

FIGURE 9.21 Effect of sealant type on time to failure in DuPont static puncture test of (Pa6-tie-sealant) multilayer film. Weight of probe is 230 g. Tested from sealant side. *Reproduced with permission from The Dow Chemical Company.*

The toughness of Surlyn ionomer is attributed to its stiffness and strain hardening behavior. In relative terms, the stiffness of Surlyn 8140 is 10–15 times that of the mPE plastomer. The soft plastomer exhibits little or no strain hardening (see Chapter 5). It has a high amount of creep elongation and fails quickly in a slow puncture test, which is essentially a measure of creep resistance. The stiff ionomer has a high degree of strain hardening due to its long chain branching and ionic interactions. This results in a low level of elongation and higher puncture energy.

Many in the industry believe that lower interlayer adhesion increases the puncture resistance of multilayer films by dissipating some of the energy through delamination. Darby et al. [83,84] find this is not true for slow puncture tests. They adhesively laminate films and measure puncture resistance as a function of curing time. Two structures are studied: 25-μm Al/adh/12-μm OPET (Al stands for aluminum foil) and 12-μm OPET/adh/50-μm LDPE film. Urethane-based adhesives are used. ASTM F1306-90, a slow speed test described in Section 9.2.4.1, is used to measure puncture resistance. They use two probes, one specified in the standard (needle with a 3-mm diameter tip) and the other hemispherical [27].

Peel strength measurements and puncture tests are conducted on samples as a function of adhesive cure time. In an initial experiment, they apply the adhesive by hand with a Mayer rod, and due to equipment limitations, test adhesion and puncture resistance on separate specimens exposed to the same cure time. A second experiment uses a pilot scale laminator. Testing for puncture and peel strength is done at the same time intervals on the same specimens. Both experiments give similar results. Control films and films with zero adhesion (using shampoo as the adhesive) are also tested.

The results show that with good bonding, the Al/adh/PET laminate behaves like a single material. Puncture resistance is similar to PET alone. With poor bond strength, there are two peaks on the force versus displacement curve, one associated with separation of the layers followed by puncture of the foil and the second to the puncture of the PET. The foil punctures at a lower load than the PET. The laminate fails at higher deformation but with about the same puncture force as the PET. The peel strength is found to increase with curing time, as expected. The puncture force, however, only slightly increases with cure time and the change is small enough that it may not be statistically significant. Similar results are found for the PET/adh/PE laminate. They conclude that at least in their tests, bond strength is not a factor for puncture resistance. No data is collected at high speed and so no conclusions can be drawn regarding high-speed puncture/impact tests.

Lange et al. [27] provide a good transition toward high-speed testing of laminate structures. They evaluate a number of different test methods, settling on a penetration probe method similar to ASTM 1306. They test six different probe geometries and test speeds ranging from 1 to 1000 mm/minute. The ranking of test structures remains the same over this speed range. Force to puncture, energy to puncture (area under force vs displacement curve), and puncture resistance factor (PRF) (force times work or energy to puncture) are compared with field experience for several structures. Energy to puncture and PRF give the best agreement with field experience. A set of test conditions are selected for the remaining tests based on these results.

They find that for paper laminates the penetration resistance is limited by the paper; the paper fails first (along with foil if it is in the structure) followed by the polymer. Failure of the laminate is defined by the first layer that fails. Polymer layers fail quickly afterwards due to stress concentration created by the failure of the paper layer. Puncture resistance is somewhat lower when tested from the polymer side of the paper laminates. Seyler [85] also reports slow puncture resistance depends on which side of an asymmetric multilayer film is facing the probe. Lange et al. explain this with beam theory, which says the stress will be uneven during penetration, with higher stress on the side furthest away from the probe. When the probe comes from the polymer side, the stress in the paper is higher and breaks sooner. Lange et al. find some benefit from putting the paper in the middle of the structure, which results in higher puncture resistance due to the whole structure failing, not just the paper. This suggests structure design plays a role.

Structures without paper (only polymer) show a one-step rupture mechanism in Lange's experiments. All the layers fail at the same time—any improvement in one layer directly translates into an improvement for the laminate. Puncture resistance of two to six times that of structures with paper are achieved. The direction of penetration still makes a difference for all-polymer structures. Puncture resistance is lower when the probe penetrates from the side with the stiffest layers (such as BOPP or OPET). This is consistent with scratch tests reported by Hare et al. [45] who provide an explanation involving differences in deformation behavior. See Section 9.4.3.3 for details.

Timm's [66] study of OPET/adh/PE film laminates, introduced in Section 9.4.1, includes measurements of dart drop. They find that the impact resistance is not a simple sum of the impact resistance of the component films. Increasing the thickness of the PE film increases the dart drop resistance. The composition of the PE film also gives some variation in the results.

Laboratory puncture tests such as dart impact, Spencer, and slow puncture often do not correlate well with actual field experience [39,85]. Empirical tests have been developed in an ad hoc fashion by package engineers to screen new film structures. For example, the package may be filled with screws or other sharp objects, dropped onto a hard surface and examined for pinholes. Such tests may provide rankings amongst film candidates but lack the rigor to compare structures amongst different labs or the quantitative output that can be used to validate predictive models. The lack of good lab tests has necessitated conducting expensive trial by error field testing on actual packaging lines, often cutting into production to test new structures.

A well-documented early attempt to develop meaningful lab tests and correlate the results with field tests was made by Hu [70] at the US Army Natick Laboratories. Hu describes the development of foil-based packaging structures for military rations. Many factors are important for foil laminate packaging such as good sealing. This study focuses on abuse resistance and specifically pinholes. Two types of pinholes are described—those that only puncture the foil layer and those that completely puncture through the whole package, which he calls complete-puncture pinholes. The latter is the most important for vacuum packaging (such as freeze-dried food) and the focus of the study.

Puncture, flexure and abrasion are the three modes of failure. A puncture can occur when sharp food such as a freeze-dried carrot pierces the package. Flexural issues arise in folds or in the seal area that can bend during transportation. Abrasion occurs when the package rubs against another during transportation. He notes that most lab tests do not simulate these modes of failure.

For puncture resistance Hu observes that previous work by Furno et al. [86] and others use a probe size (0.25-inches or 6-mm) that is much larger than the typical pinholes they observe in the actual packages. Hu settles on a slow puncture test, similar to ASTM 1306, with a needle-like probe with a 0.14–0.16 mm diameter and a test speed of 5 mm/minute. For flexure resistance, Hu uses an MIT Folding Endurance tester with a specially designed vacuum box. The film is flexed for a given number of cycles and then mounted on the vacuum box to test for pinholes (using a dye technique). Abrasion resistance is tested by cutting a square specimen from the film and folding it twice into a triangle. The tip of the triangle is subjected to an abrasive material on a rotating turntable. After a given number of rotations, the sample is removed and tested for pinholes.

Hu tests several film samples to select the best for field tests. In addition to the above-mentioned lab tests on the films and laminates, Hu tests filled packages. For puncture, the packages are vacuum packaged and tested for loss of vacuum after two weeks. Abrasion is tested on the filled packages with a modified belt abrasion machine. A good test of flexure on the filled packages is not found. Field testing is conducted on a US Army base with freeze-dried vacuum packaged food carried by soldiers while maneuvering an obstacle course.

An attempt is also made to correlate the package failure results to food texture and hardness. The maximum force (during compression), the force of rupture, ultimate strain and work of rupture measurements are made on the food, depending on the food product.

The results of the lab puncture testing of various laminates show that the puncture resistance increases with thickness. Hu observes different results depending on which side the puncture is initiated (inside—for heat seal side, or outside). He finds no definitive relationship for which side gives the highest puncture resistance and concludes that it may depend on the structure components and design. The puncture resistance of the laminates is less than the sum of the puncture resistance of its components. Hu explains this by relating it to how a laminate fails when stressed axially, as described in Section 9.4.1.

The lab flexure tests show that increasing film thickness decreases flex resistance. This is true of the laminates as well. The laminate flex resistance falls in between the flex resistance of its component films. Aluminum foil has poor flex resistance, as expected; adding an OPET layer increases it appreciatively.

Increasing thickness increases the abrasion resistance. The laminate abrasion resistance is much higher than the sum of the individual components.

Varying the test speed within the limited range of the lab tests does not change the rankings of the laminates. In general, the laminates perform well on one or two tests but not on all three. Compromises have to be made depending on which mechanisms are believed to be most important for the application.

Tests on the lab filled packages show:

- Higher hardness foods result in a greater number of package failures in the puncture and abrasion tests.
- Vacuum packaging creates more pinholes than air packaging.
- The rank correlation between the lab tests on the laminates and the filled packages is good, ranging from 0.715 to 0.99.

Hu subjects 10 states (two food types x five package structures) to the field test. The field failures correlate better with the product of puncture force and puncture energy (the PRF mentioned earlier when describing the study by Lange et al.) rather than either one alone. Hu suggests that the PRF may correlate better because some layers may have higher force than energy while other layers may have higher energy than force. Field failures correlate well with the instrumented PRF and the abrasion resistance lab tests but not well with flex resistance. This suggests that puncture and abrasion are the two main failure mechanisms for this application.

Carbajal et al. [39] develop a puncture test to simulate pinholes created during the filling operation of a snack food vertical-form-fill-sealing line. They recognize that food packaging has unique characteristics that are not captured in standard lab tests: the speed of the falling food, the diameter of punctured holes and possible deformation of the food product during the puncture event due to its low stiffness and strength. To create a test that better represents the conditions of a real filling operation, they conduct a preliminary study of the dynamics of single crackers (also known as a biscuit or wafer in some parts of the world) falling onto a suspended horizontal film. They use high-speed photography to examine the puncture event. A sequence of photos from the study is shown in Fig. 9.22 using a commercial square-shaped nonwoven cracker that is prone to creating punctures. They find the cracker does not significantly deform during the puncture event and pinholes form at a critical speed of 6.1 m/second. This last data is obtained by varying the speed of the cracker by adjusting the drop height and checking for pinholes by using a dye technique. This provides the basis for the test method, where a speed of 4.2 m/second (speed of the cracker after falling 1 m) is used.

The reverse impact test developed by Carbajal et al. is shown in Fig. 9.23, where a moving hydraulic frame with the mounted film contacts a stationary needle probe at a constant velocity (4.2 m/second). The probe is a hardened steel needle with a 1-mm diameter hemispherical tip, designed to simulate the size of the pinholes found in the cracker drop tests. The film is mounted horizontally on the hydraulic frame for ease of testing (tests conducted at different angles give the same ranking of results). The needle is attached to a force transducer that measures the force versus displacement during the impact event. The platform moves rather than the needle to reduce noise in the measurement. Noise is further reduced by computing the energy (work) versus displacement by integrating the area under the force curve.

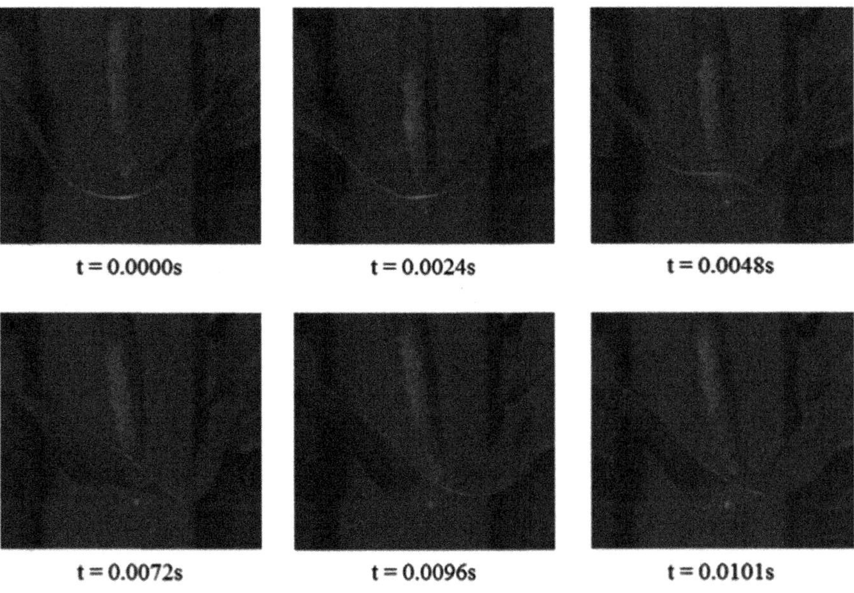

FIGURE 9.22 Photos from high-speed photography of cracker dropped onto a film. *From L.A. Carbajal, R. Jio, D.M. Hahm, B.A. Morris, R. R. Kendzierski, Impact puncture resistance of multilayer flexible food packages, SPE ANTEC Conf., 2015. Reproduced with permission from The Dow Chemical Company.*

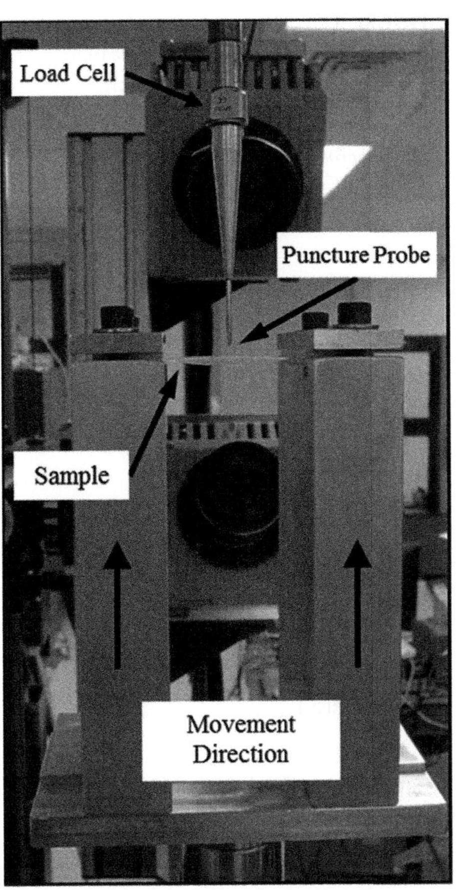

FIGURE 9.23 Reverse impact test. *From L.A. Carbajal, R. Jio, D.M. Hahm, B.A. Morris, R. R. Kendzierski, Impact puncture resistance of multilayer flexible food packages, SPE ANTEC Conf., 2015.Reproduced with permission from The Dow Chemical Company.*

The ultimate work of puncture is used to correlate with puncture resistance. The puncture event during the test is monitored with two high-speed video cameras (50,000 frames/second).

There are several features of this test that make it well suited to study the impact puncture of foods. One is the dimensions of the needle probe are similar to the small dimensions found in the cracker drop tests and much smaller than standard laboratory dart drop tests. Second, the speed of the test mimics that found in the product drop tests and is much higher than standard lab tests such as dart drop and Spencer impact. Finally, specific steps are taken to reduce the noise of the measurement to obtain reliable data.

The method is tested on a series of barrier snack films with the structure (HDPE-tie-PA6-tie-sealant). The PA layer thickness, total thickness, sealant (ionomer or EVA-ionomer blend) and adhesion of the tie layer to the outside PA layer are varied. The results show an increase in puncture resistance with increasing PA layer thickness (shown in Fig. 9.24). PA has excellent puncture resistance and these results are consistent with what is known from field studies on other film structures.

The results are also compared with a bullet drop test. This is similar to a screw drop test, an empirical test described earlier that is sometimes used to rank films; the projectiles simulate the falling food product. Rather than filling a pouch with bullets, Carjajal et al. drop individual bullets onto the film from different heights and measure the velocity at which pinholes form using a high-speed camera. Bullets are selected over screws due to their better flying stability, closeness to the weight of a cracker and similar tip diameter as measured pinholes from the earlier cracker impact tests. Excellent agreement between rankings of the films from the puncture test and the bullet drop test is found. The test method can be used to validate models to predict puncture resistance and study design features such as toughening individual layers, interlayer adhesion, splitting layers and sealant selection (see Section 18.2).

Drop tests are often conducted on filled packages. The performance of a packaging structure in these types of tests is influenced by many factors including:

- The film structure, as outlined above.

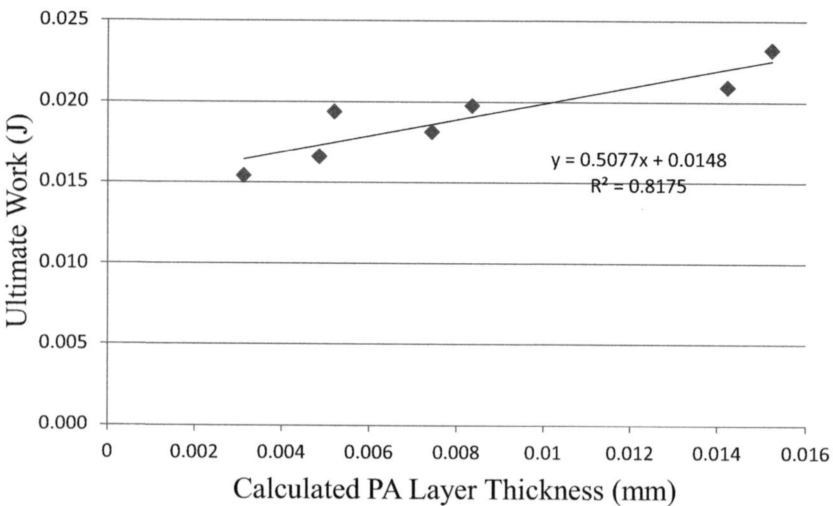

FIGURE 9.24 Effect of polyamide (nylon) layer thickness on high-speed puncture resistance. *From L.A. Carbajal, R. Jio, D.M. Hahm, B.A. Morris, R. R. Kendzierski, Impact puncture resistance of multilayer flexible food packages, SPE ANTEC Conf., 2015.Reproduced with permission from The Dow Chemical Company.*

- The height of the drop.
- The angle of the drop and where the impact occurs.
- The integrity of the seals.
- The weight of the product
- The nature of the product (liquid vs solid; sharpness and hardness of solid products, etc.)
- How full the package is (amount of headspace).

As is described in Chapter 18, finite element analysis is increasingly being applied to flexible packaging to better understand these factors and their effect on the stress imparted on the package at impact and to predict performance.

9.4.3.3 Scratch resistance

Scratch resistance tests have been proposed for improving the correlation between laboratory tests and field package failure studies. Many package failures originate from scratching from either the inside or outside of the package, for example, when a sharp food product is dropped into the bag on the filling machine, a bone or sharp product rubs against the inside of the package during shipping, or the outside of the package is subjected to sharp objects during handling or shipping. Hare et al. [45–47] use a scratch tester with a linearly increasing load to analyze damages in packaging films. As described in ASTM D7027 and Section 9.2.5, a probe is moved across the specimen at a constant velocity with a linearly increasing force and the resulting scratch is examined. In one study [45], two 75-μm (3 mils) laminates are tested: (1) OPET-met/adh/(PE-PE) and (2) OPET-met/adh/(PA-tie-PE). A probe with a 1-mm diameter steel tip is used with a testing speed of 1 mm/second over a force range of 0–40 N. The films are mounted onto a hard surface (PMMA). The tests are run until the probe penetrates through the entire sample.

Hare et al. report that structure 2 requires a higher scratch force before it punctures, agreeing with field trials. Structure 1 delaminates between the layers during the scratch test, contributing to the lower force for failure. Structure 2 maintains structural integrity throughout the test, distributing the stress more uniformly. Superior adhesion contributes to higher scratch resistance.

The scratch progresses through several modes of deformation as the load increases, from smooth deformation to slip-stick. Modes of deformation during scratching are discussed by several investigators [42,87,88]. Examination of the progression under a steadily progressive load can be useful in understanding damage mechanisms.

Scratching from the sealant side results in a higher force to penetrate than from the OPET side of the laminate. Hare et al. cite earlier work by Jiang et al. [89] to explain this. When a soft coating (or film in this case) lies on top of a hard surface the probe deforms the soft film but the deformation does not disturb the hard surface beneath. This is illustrated in Fig. 9.25. It is only after the soft surface is fully penetrated that the hard surface starts to see deformation and damage. When a hard coating lies on a soft surface the probe deforms the hard surface, which in turn deforms the soft film beneath, resulting in damage to both layers. Failure occurs at lower forces. Hare et al. show microscopy evidence that bears this out.

This is explored further in a follow-up study by Hamdi et al. [47]. Polypropylene (PP) film (65 μm), polyamide (PA) film (15 μm) and their laminate (PP/PA) is tested using the linear load scratch test (ASTM D7027). The critical load for the onset of

(A)

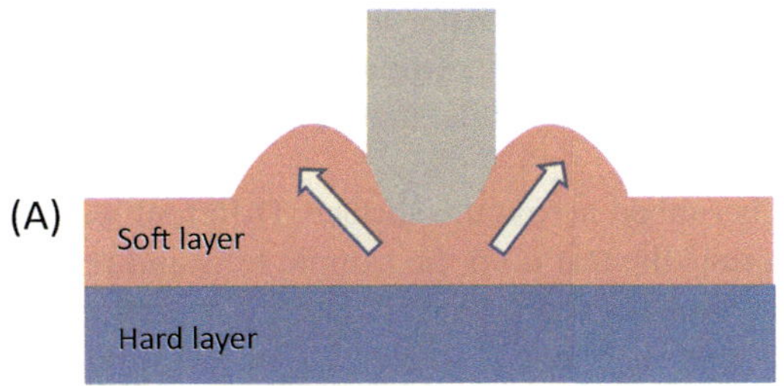

(B)

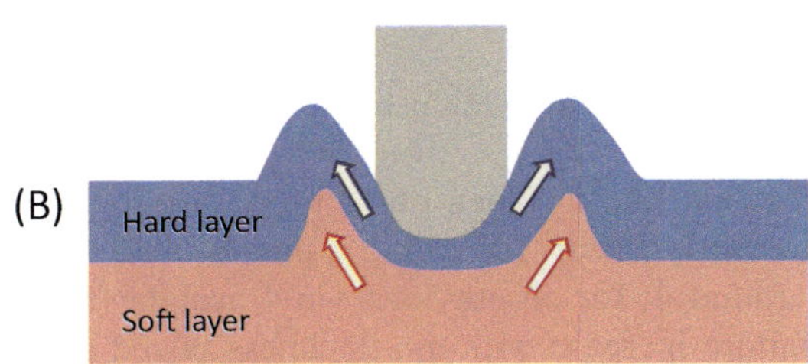

FIGURE 9.25 Illustration of deformation of (A) soft film on top of a hard film and (B) a hard film on top of a soft one. *Adapted fromB.A. Hare, A. Moyse, H.-J. Sue, Analysis of scratch-induced damages in multi-layer packaging film systems, J. Mater. Sci., 47 (2012) 1389–1398.*

slip-stick and tearing damage modes are in the following order: PA < PP < PA/PP (PA side up facing the probe) < PP/PA. The high scratch-resistance of the PP/PA laminate is attributed to its low surface friction and its stress/strain response. Finite element modeling (FEM) agrees with the experimental results and provides further insight. When PP is the top layer it absorbs the scratch energy more effectively and the stiff and strong PA bottom layer shields the interface with the aluminum backing from stress build-up and delamination. When the PA layer is on top, stress is concentrated at the backing interface resulting in delamination. The PP/PA laminate can better distribute the stress throughout the laminate.

In another study using the same test procedure, Hare et al. [46] test a laminate structure put together by two different methods, adhesive and extrusion lamination:

a. OPET-met/adh/(PE sealant film) (98 μm total).
b. OPET-met/LDPE (extrusion)/(PE-sealant film) (106 μm total).

Scratch tests are conducted on the OPET side of the structure. The results show that the extrusion laminated structure has higher scratch resistance than the adhesive laminated structure. Microscopy indicates that the adhesive laminated structure delaminates, concentrating the stress and causing failure at lower loadings. The extrusion laminated structure maintains its integrity with no localized delamination throughout the test, helping to distribute the stress over a wider area.

Factors that affect the linear load scratch test include:

- Hardness of the surface the film is mounted on during the scratch test. Fig. 9.25 illustrates the effect this has on the deformation of layers within the film.
- Probe shape. Like puncture tests, the size and shape of the probe influence the force of scratching.
- Film orientation (which layer is exposed to the probe). This is discussed above and illustrated in Fig. 9.25.
- Method for fixing the film to the surface (Hare et al. use vacuum).
- Interlayer adhesion. This is a factor in both of the Hare et al. studies. Some investigators use the scratch test as a way of measuring adhesion [90].
- Surface friction of the film or laminate.

The scratch test is most applicable to damage caused by scratch related events, such as puncture or tearing. It is probably best suited for abrasion and scratch damages such as sliding and vibration during transportation. It may not be well suited for investigating the mechanical integrity of seals where a peel test may be more appropriate.

9.4.3.4 Abrasion resistance

Not many studies of abrasion resistance of packaging films are reported in the literature. As is described in Section 9.4.3.2, Hu [70] uses an abrasion test method to evaluate foil-based laminates for the packaging of military rations. The lab method involves cutting a square specimen from the film and folding it twice into a triangle. An abrasive material attached to a rotating turntable is rubbed against the tip of the triangle. After a given number of rotations, the sample is removed and tested for pinholes. Hu finds that this and a puncture test correlate well to field testing of vacuum packaged freeze-dried food subjected to soldiers carrying the packages while maneuvering an obstacle course.

Blom [51] uses a linear Taber test described in Section 9.2.5. A probe is rubbed across the surface. Different weights can be attached to the probe depending on the nature of the sample. Blom runs the test until the sample fails and records the number of cycles to failure. The test results are variable and Blom recognizes more work is needed to understand the source of the variability (test method, material, etc.). Nevertheless, his work provides interesting observations and insights into the abrasion resistance of packaging films:

- Measurements in the TD of monolayer film often give higher values (cycles to failure) than in the MD. Biaxially oriented films have a fairly balanced abrasion resistance in TD and MD.
- Abrasion resistance is inversely related to the applied load, but it is not linear. The relationship is more exponential in nature. Some films are more sensitive to changes in applied load than others.
- Abrasion resistance increases with film thickness. The relationship is not linear. When plotted on a log-log plot, the cycles to failure versus thickness curves for monolayer films are fairly linear, indicating a power-law type of relationship. For OPET and PA films, the powerlaw exponent is around 4; for an ethylene vinyl acetate (EVA) film with 5% VA, it is around 3.7. This means that the cycles to failure increase with the thickness raised to the fourth power.
- Multilayer films have much higher abrasion resistance than the simple sum of the components, as is shown in Fig. 9.26. (Note that the vertical axis in this figure uses a logarithmic scale.) The nonlinear nature of the effect of thickness plays a role in this behavior, but it is more complex; applying the powerlaw model to the total thickness of the laminate overpredicts the cycles to failure. Blom says adhesion plays a role – stacked films without the adhesive have lower abrasion resistance, but no data is provided.
- The orientation of the multilayer film (sealant or nonsealant side facing the abrader) has an effect on some films. In general, the film fails in fewer cycles when the sealant side is facing the abrader. This differs from the puncture and scratch results of Lange et al. and Hare et al. described in Sections 9.4.3.2 and 9.4.3.3 where the nonsealant side facing the probe produced the lower results.

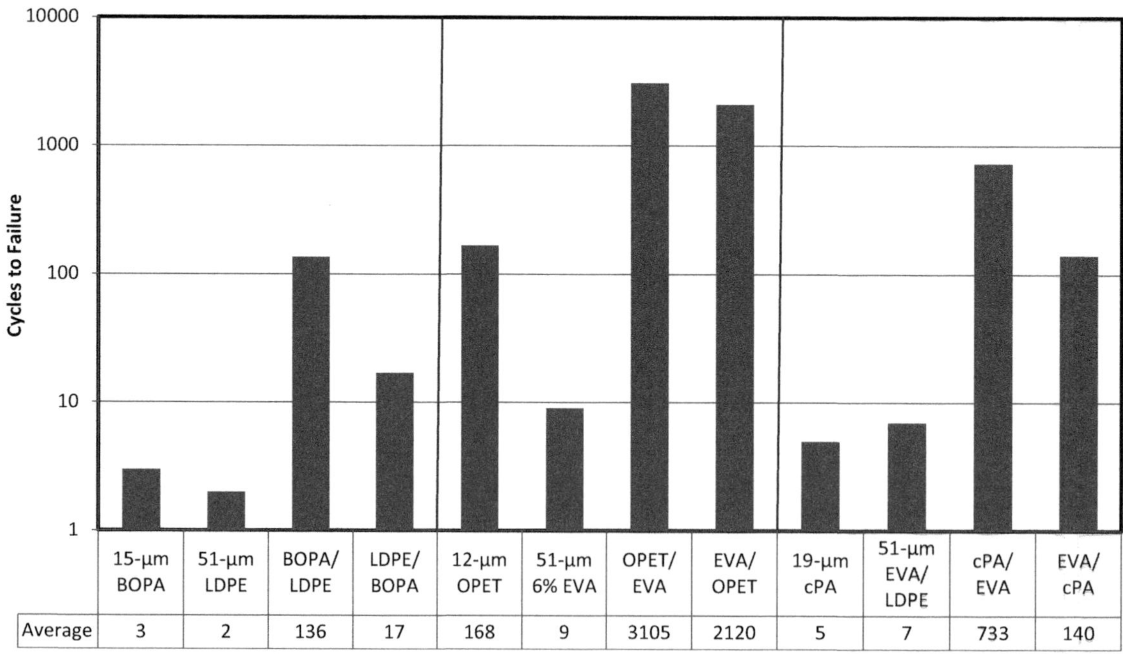

	15-µm BOPA	51-µm LDPE	BOPA/ LDPE	LDPE/ BOPA	12-µm OPET	51-µm 6% EVA	OPET/ EVA	EVA/ OPET	19-µm cPA	51-µm EVA/ LDPE	cPA/ EVA	EVA/ cPA
Average	3	2	136	17	168	9	3105	2120	5	7	733	140

FIGURE 9.26 Abrasion resistance of packaging films and their laminations. *Adapted from H. Blom, Flexible packaging durability studies, SPE FlexPackCon Conf., 2014.*

TABLE 9.5 Abrasion coefficient (Eq. 9.12) of some common packaging materials.

Material	Abrasion coefficient
OPET	39.3
OPA	21.3
Cast PA	14.4
Cast PP	7.6
EVA, 6% VA	4.1
EVA (5% VA)/LLDPE blend	2.9

Source: Data from H. Blom, Flexible packaging durability studies, SPE FlexPackCon Conf., 2014.

Blom proposes an abrasion coefficient as a material parameter:

$$A = \frac{\log(\text{cycles to failure}) \times \text{applied load}}{\text{thickness}} \tag{9.12}$$

Stiffer materials (PET, PA) have higher A than soft ones like EVA and PE. The COF may also play a role. Values for A reported by Blom are given in Table 9.5.

9.4.4 Tear resistance

The tear resistance of multilayer films made up of only stiff materials such as OPET or BOPP generally follows the rule of mixtures. Most flexible packaging film structures, however, contain a ductile polyethylene or ethylene copolymer sealant layer. These materials have some of the highest tear strength of films used in packaging due to plastic yielding as the tear propagates through the sample. Combining a ductile layer with a stiff layer typically results in tear strength that is considerably lower than a simple addition of the two layers. This section begins with a discussion of the tear resistance of polyethylene films and the special consideration of directional tear properties. The case of multilayer flexible film structures combining stiff and ductile layers is then addressed.

9.4.4.1 Polyethylene films

LLDPE has a low yield strength which results in a large zone of plastic deformation that precedes the propagation of a crack as the film is torn. This zone of deformation gives rise to high tear strength. The size of this zone of deformation can be predicted from fracture mechanics. Kinlock and Young [91] show that the radius of the zone of deformation in front of a crack tip is

$$r_y = \frac{1}{2\pi}\frac{G_a E}{\sigma_y^2} \tag{9.13}$$

where $r_y =$ radius of zone of deformation, $G_a =$ fracture energy or toughness, E = elastic or Young's modulus, and $\sigma_y =$ yield stress.

Values of r_y for some common polymers used in flexible packaging are computed in Table 9.6 from parameters compiled by Wu et al. [67]. The value of r_y for LLDPE is particularly large due to its high fracture toughness (for which Wu et al. cite several sources) and low yield strength. In their experiments, Wu et al. measure the width of the yield zone for an LLDPE film to be around 20 mm, somewhat lower than the calculated value in Table 9.6 but of the right order of magnitude. Chang et al. [89] also measure the yield zone width for several films made from LLDPE and blends of LLDPE and PP. These fall within a range of 10–22 mm for LLDPE and much lower for the blend. They find that the measured trouser tear resistance is directly proportional to the width of the yield zone.

Most PE films exhibit a certain degree of anisotropy in the tear strength. Generally, the different classes of PE behave as follows:

- LDPE: moderate tear resistance with some films with MD tear strength > TD tear strength and others the opposite, depending on the fabrication process—see Section 9.4.4.2.

TABLE 9.6 Fracture mechanics parameters for some common polymers used in flexible packaging.

Polymer	Tensile yield stress, σ_y	Young's modulus, E	Fracture energy, G_a	Theoretical radius of deformation zone, r_y
	MPa	GPa	kN/m	mm
LLDPE (high MW)	8	0.2	120	60
HDPE	30	1.0	6	1.1
PP	35	1.2	8	1.2
PC	65	2.8	1	0.1
PVC	50	3.5	4	0.9
PET	50	2.8	8	1.4
PA (66)	60	3.0	3	0.4
PMMA	75	3.4	0.4	0.04
PS	70	3.4	0.3	0.03

Source: Parameter values are taken from R.-Y. Wu, L.D. McCarthy, Z.H. Stachurski, Tearing resistance of multi-layer plastic films, Int. J. Fract., 68 (1994) 141–150 and computation of r_y from Eq. (9.13).

- LLDPE: high tear resistance with TD tear strength > MD tear strength.
- HDPE: high tear strength in TD, very low in the MD.

These differences in tear anisotropy can be explained by differences in molecular architecture, crystalline morphology and orientation. Key molecular parameters include the molecular weight distribution and amount of short and long-chain branching. The crystalline morphology relates to the ordered structures that form as the polymer chains organize into crystals. This is a function of the stress the polymer is under during quenching, as well as the time available for crystallization during the film fabrication process. Since most films are quenched while being uniaxially or biaxially stretched, the orientation of both the crystalline and amorphous regions may take place. Hence, the film fabrication process plays an important role in the ultimate shape of the crystals, the orientation of the polymer and the tear strength.

Given the strong dependence of the film fabrication process on the tear strength performance of PE it can be challenging to compare different resin families. Increasing the amount of MD orientation during filmmaking reduces MD tear strength and increases TD tear strength for linear PE; the opposite occurs for LDPE and ionomers. Biaxially oriented LLDPE has low tear strength in both MD and TD. Choi et al. [92] measure the intrinsic tear strength of PE by compression molding films without any preferred orientation. They find that the intrinsic tear strength correlates with the geometric mean of the MD and TD tear strength of the blown film. Using this as a general guide, the intrinsic tear resistance of various PE-types can be ranked in the following order: ZN LLDPE > mLLDPE > LDPE > HPDE. Within the LLDPE category, tear strength generally decreases with increasing density. The higher tear strength of ZN LLDPE over metallocene LLDPE is due to the presence of a high MW fraction in ZN LLDPE.

Many investigators have studied the crystalline morphology of polyethylene film using X-ray scattering, birefringence, electron microscopy and other techniques [93–107]. A complete review of the literature is beyond the scope of this book. Several structures have been proposed, including:

- **Low-stress Keller–Machin row nucleated structure** [94]: As the longest chains are extended in flow and cooled, stress-induced crystallization provides nucleation sites for the formation of lamella or plate-like structures of folded chains extending radially from the primary orientation direction, typically the MD. This is illustrated in Fig. 9.27. Because of the nearness of the nucleation sites, the lamellae extend radially or nearly perpendicular to the direction of orientation. The lamellae twist as they grow outward. They resemble twisted ribbons that extend perpendicular to the main chains. This is one of the most commonly reported structures for PE blown films.
- **High-stress Keller–Machin row nucleated structure:** This is similar to the low-stress structure, except that the lamellae do not grow as long and extend radially outward without twisting; the folded chains that make up the lamella structure remain parallel to the primary chain orientation. This structure has been reported primarily for high

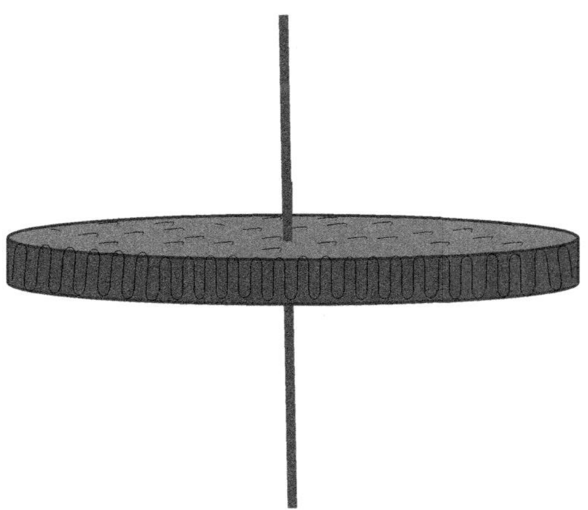

FIGURE 9.27 Illustration of plate-like lamellae structure of folded chains extending radially from a primary extended chain.

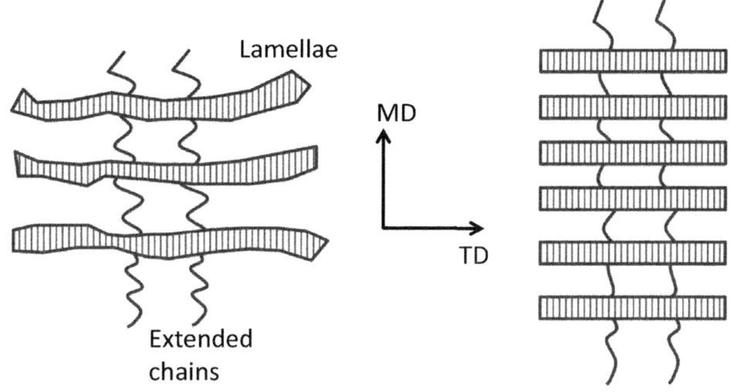

FIGURE 9.28 Comparison of low and high stress Keller-Machin row nucleated lamellae morphologies.

molecular weight HDPE films, although Lee et al. [108,109] have reported it for ionomers. The difference between the high and low-stress morphologies is illustrated in Fig. 9.28.

- **Spherulite-like superstructure:** Spherulites are crystalline structures that extend radially from a single point, forming spherical domains. Boundaries between spherulites can sometimes lead to weakness or brittleness. Zhang et al. [110,111] found spherulitic-like regimes with randomly oriented lamella in LLDPE films, while others have observed row nucleated morphologies. Subtle differences in molecular architecture and processing conditions can lead to different morphologies.

Three studies are highlighted here that reveal the role that crystalline morphology, orientation and molecular architecture have on tear strength anisotropy in PE films. The first is from Patel et al. [112]. They produce 25 μm (1 mil) monolayer blown films of eight grades of LLDPE, four made with a ZN catalyst and four analogs with a single-site catalyst of similar density and melt flow index (MFI). The films are made under conditions of low and high orientation in the MD using two different die configurations and sets of processing parameters as presented in Table 9.7. The films are characterized for crystalline and amorphous phase orientation using a variety of analytical techniques: optical birefringence (total orientation), wide angle x-ray scattering (WAXD) and Fourier transform infrared spectroscopy (FTIR) (crystalline phase orientation) and shrinkage tension (amorphous phase orientation). The results show:

- The a-axis crystalline orientation increases slightly with decreasing MFI (increasing MW) for the high orientation ZN-LLDPE films. mLLDPE-F has the least orientation which they attribute to it having the narrowest MWD and therefore the fastest relaxation time of the grades evaluated. No correlation is found for the low MD orientation films.
- All the films show orientation in the b-axis as expected. mLLDPE-F is the most isotropic.
- The low orientation films have shrink tension too low to measure. For the high orientation films, the shrink tension of ZN-LLDPE A-C increases substantially with decreasing MFI. mLLDPE-F has the least shrink tension, indicating it has the least amount of orientation.

TABLE 9.7 Experimental parameters for low and high machine direction orientation blown film states in study by Patel et al. [112].

	Low MD orientation	High MD orientation
Die diameter (cm)	15.2	15.2
Die gap (mm)	1	2.8
Output (kg/h)	52.3	85.5
Film thickness (microns)	25.4	25.4
Frost line height (cm)	71.1	71.1
BUR	3.5	2.0
DDR	8.7	41.8
Process time (s)	4.7	2.6

The Elmendorf tear results are shown in Table 9.8. In general, the ZN LLDPE samples have higher tear resistance than their corresponding mLLDPE. Increasing orientation in the MD decreases the MD tear strength and increases the TD tear strength: the ratio of MD/TD tear strength increases as the film is oriented less in the MD (and more in the TD). The ratio is never greater than 1 indicating the MD tear strength is always less than the TD tear strength. Dart impact strength is also tabulated in Table 9.8. The impact strength is higher for mLLDPE than ZN-LLDPE and decreases when the MD orientation is increased.

The second study is from Zhang et al. [110]. Their study includes one grade of LDPE, three grades of LLDPE (a ZN butene copolymer, a ZN octene copolymer, and metallocene octene copolymer) and one HDPE resin. Blown films are made from each of the resins keeping the output, blow-up ratio (BUR) and die gap constant while varying the DDR. DDR (also known as the take-up ratio) and BUR are defined by Eqs. (2.3) and (2.4). As is described in Chapter 2, the thickness ratio, BUR and DDR are inter-related (see Eq. 2.7), so that as they increase the DDR, the thickness decreases. The thickness of their films ranges from 25 to 100 μm. The frost line height is minimized. FTIR trichroism and SEM are used to determine the orientation and crystalline structure of the films. The tear strength is measured following ASTM D1922 (Elmendorf tear).

The HDPE film exhibits the low-stress Keller–Machin structure, but without twisting of the lamella. The LDPE film also shows a low-stress Keller–Machin lamella structure that becomes more twisted and intertwined as the DDR is increased. The LLDPE films exhibit spherulitic structures with relatively random lamellae arrangements. All three LLDPE types exhibited the same type of structure with evidence of regional orientation. Low DDR favors transcrystalline morphology (in the thickness dimension). The metallocene LLDPE shows the least amount of orientation.

In order for a lamella crystal structure to form, polymer chains must extend during flow to create nucleation sites (fibril crystals) that initiate the growth of the lamella outward. This is favored for polymers with high molecular weight (MW) fractions or long-chain branching, and long relaxation times. Zhang et al. measure the relaxation time for the polymers used in the study and find they have the following order: LLDPE < LDPE < HDPE. Both the HDPE and LDPE have long relaxation times. In addition, the HDPE has a high zero shear viscosity (high MW) and the LDPE has long chain branching, which contributes to chain entanglement and strain hardening. These set up conditions for the formation of oriented fibril nuclei followed by lamellae growth. The short relaxation time of the LLDPE allows the extended chains to relax prior to crystallization, making row nucleation or lamella formation less likely.

Zhang et al. compute the ratio of the MD to TD tear strength for each of the films and relate this to the crystalline structure. The HDPE shows very high tear resistance in the TD and extremely low tear strength in the MD. The LDPE film shows the reverse: high tear strength in the MD and low in the TD, with the MD:TD tear strength ratio increasing with increasing DDR. The LLDPE films are more balanced but with a predominance of tear strength in the TD. For LDPE, the twisted lamellae favor tear in the TD (between stacks of lamellae). For HDPE, the nontwisted lamellae result in weak columns of crystals and tear propagates in the MD between the stacks. The lack of row nucleated structure in LLDPE gives a fairly balanced tear resistance but with preference in the MD due to the local orientation of lamellae in the TD. This model for the tear resistance of PE films is illustrated in Fig. 9.29.

Zhang also measures the tensile strength of the films in the TD and MD. The tensile strength of the HDPE and LLDPE films is higher in the MD than in the TD which they attribute to the presence of the long fibril structure.

TABLE 9.8 Elmendorf tear and dart impact results from study by Patel et al. [112].

	Melt index	Density		MD	Elmendorf Tear, g/mil				Dart impact
Resin	g/10 min	g/cc	MWD	Orientation	MD	TD	MD/TD	Geometric Mean	g
ZN-LLDPE A	1	0.921	3.8	High	300	818	0.37	495	144
ZN-LLDPE A	1	0.921	3.8	Low	484	594	0.81	536	261
ZN-LLDPE B	0.83	0.917	3.8	High	380	810	0.47	555	267
ZN-LLDPE B	0.83	0.917	3.8	Low	505	599	0.84	550	513
ZN-LLDPE C	0.63	0.917	4.1	High	444	785	0.57	590	504
ZN-LLDPE C	0.63	0.917	4.1	Low	-	529			835
ZN-LLDPE D	1.1	0.94	3.6	High	31	627	0.05	139	60
ZN-LLDPE D	1.1	0.94	3.6	Low	40	131	0.31	72	65
mLLDPE E	0.84	0.921	3.2	High	229	716	0.32	405	312
mLLDPE E	0.84	0.921	3.2	Low	377	522	0.72	444	835
mLLDPE F	1.1	0.918	2.5	High	285	445	0.64	356	717
mLLDPE F	1.1	0.918	2.5	Low	312	360	0.87	335	835
mLLDPE G	0.94	0.917	3.8	High	388	656	0.59	505	276
mLLDPE G	0.94	0.917	3.8	Low	416	483	0.86	448	835
mLLDPE H	1.2	0.94	3.2	High	45	1014	0.04	214	60
mLLDPE H	1.2	0.94	3.2	Low	64	227	0.28	121	120

Lee et al. [113] also examine the crystalline morphology and orientation of different families of PE blown films and relate this to the tear strength. They use a grade of LDPE, two grades of LLDPE (a ZN octene copolymer and a metallocene hexene copolymer) and one grade of HDPE. These are blown into a film using a conventional process (short stalk). Thickness (50-μm) and the frost line height are held constant. The BUR is varied, which changes the DDR accordingly [see Eq. (2.7)]. The films are tested for Elmendorf tear strength (ASTM D1922) and characterized by WAXS and small angle x-ray scattering (SAXS).

WAXS and SAXS show that all the films have a Keller–Machin low-stress crystalline morphology: row nuclei are aligned in the MD and lamellae grow along the TD and twist as they grow. A Herman's orientation function, which is a measure of crystal orientation, is computed from the results which confirm that all the films have the same structure. This simplifies the interpretation of the tear strength. For the mPE grade, only isotropic patterns are found for the range of BUR investigated. The authors explain that the mPE is isotropic since it contains few very long chains: long chains contribute most to orientation.

As orientation is increased in the MD (BUR decreased), the tear strength of mPE and LLDPE remains isotropic, the LDPE films tear more readily in the TD (MD tear strength $>$ TD tear strength) and the HDPE films tear predominately in the MD (MD tear strength $<$ TD tear strength). The results are shown in Fig. 9.30. Here the isotropic baseline for MD/TD tear strength is around 0.6 rather than the expected 1. Lee explains that this is due to an artifact of the

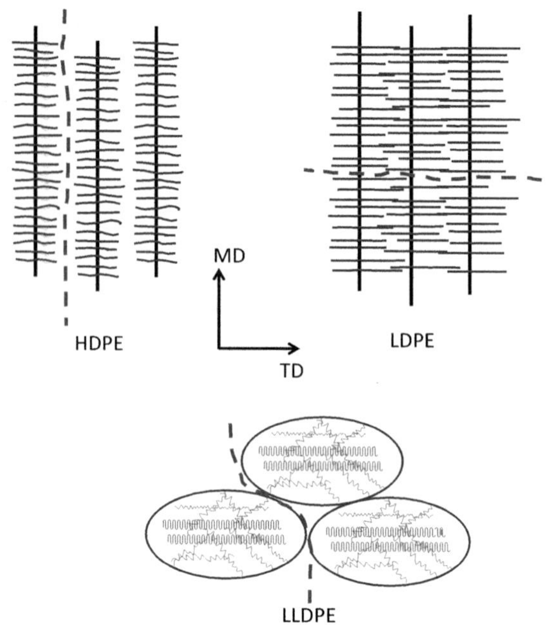

FIGURE 9.29 Model of tear resistance of high-density polyethylene (HDPE), low-density polyethylene (LDPE), and linear low-density polyethylene (LLDPE) blown film. HPDE has short lamellae perpendicular to the main orientation of the long nucleating chains in the machine direction (MD). Tear propagates along the interface between columns (dashed line). LDPE has long, intertwined lamellae perpendicular to the main orientation in MD. Tear propagates between lamellae. LLDPE in Zhang's experiments showed spherulitic structures with local orientation. Tear is fairly isotropic with some preference in the MD. *Adapted from X.M. Zhang, S. Elkoun, A. Ajji, M.A. Huneault, Oriented structure and anisotropy properties of polymer blown films: HDPE, LLDPE and LDPE, Polymer, 45 (2004) 217–229.*

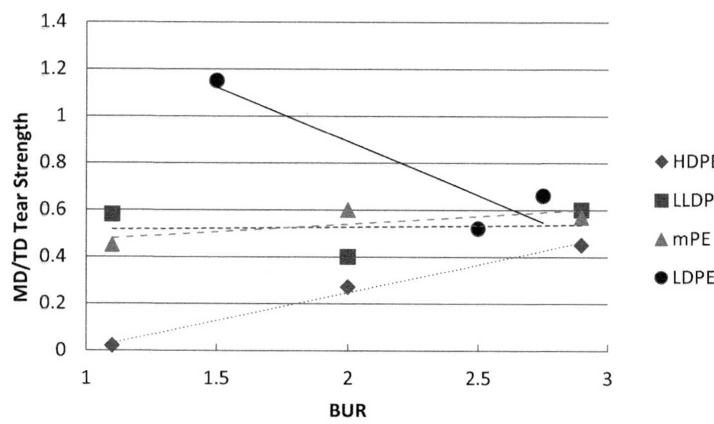

FIGURE 9.30 Results of blown film trial by Lee et al. [113] showing differences in machine direction/transverse direction tear strength ratio as a function of blow-up ratio (BUR). As BUR increases, orientation in the machine direction decreases. Results originally reported as a function of Herman's orientation factor but are replotted here in terms of BUR.

laboratory blown film line used, which produces nonuniform thickness in the hoop direction that favors MD tear. The trends, however, agree with those reported by Zhang et al. and others.

Lee argues that the tear results are consistent with the morphology and number of tie molecules. Tie molecules are polymer chains that connect crystalline lamellae together through the amorphous regions [81]. They bear the load, contributing toward the mechanical strength. This is a refinement of the model proposed by Zhang et al. For LDPE, tear prefers to propagate in the TD (between lamellae) since the density of tie molecules is low (Fig. 9.31). For LLDPE and HDPE, tear propagates between the lamellae columns (Figs. 9.29 and 9.31) or breaks up crystals (breaking van der Waals forces rather than the stronger covalent bonds) since the tie molecule density is much higher.

Lee also notes that the high tear strength of LLDPE often leads to other deformation that contributes to the tear strength. LDPE and HDPE exhibit much less plastic deformation, simplifying the analysis of their tear strength.

LLDPE is often melt blended with LDPE to enhance melt strength and processability. Zhang et al. [114] examine the effect of adding LDPE to an octene-LLDPE (ZN) on Elmendorf tear strength. Monolayer blown film (25 μm) over the full range of blend compositions is made while keeping processing conditions such as temperature, DDR, blow-up ratio, output, thickness, and frost line height constant. They use SEM and FTIR to study crystalline morphology. As previous work showed, their pure LDPE film has a row nucleated twisted lamellae structure. With few tie chains, MD tear strength is higher than TD tear strength. 100% LLDPE has a spherulitic-like superstructure with TD tear strength greater than MD tear strength. Adding 20% LDPE to LLDPE creates a row nucleated lamellae structure—the long

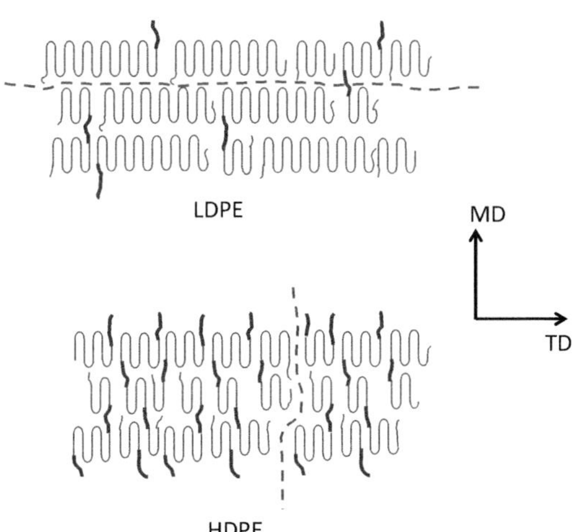

FIGURE 9.31 Model of tear resistance of low-density polyethylene (top) and high-density polyethylene (bottom). Dark lines represent tie chains and light lines crystalline lamella. Top illustration shows few tie molecules spanning the gap between lamella. Tear propagates between lamellae stacks in the transverse direction (dashed line). Bottom illustration shows many tie molecules linking the individual lamellae. Tear propagates between lamellae columns in the machine direction. *Adapted from L.-B.W. Lee, R.A. Register, D.M. Dean, Tear anisotropy in films blown from polyethylenes of different molecular architectures, J. Polym. Sci., Part. B: Polym. Phys., 43 (2005) 413–420.*

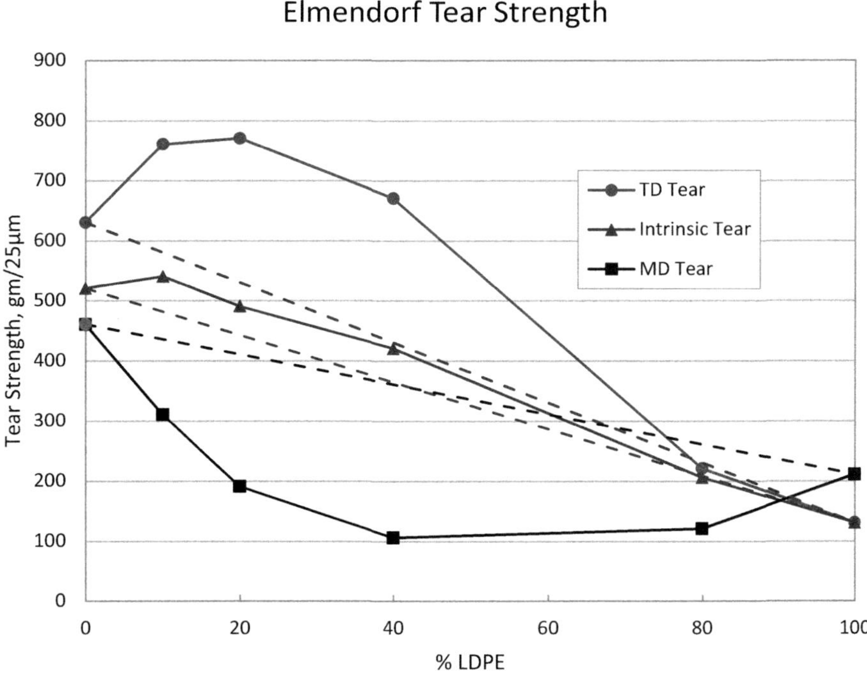

FIGURE 9.32 Machine and transverse direction Elmendorf tear strength of monolayer blown (51-μm thickness) and intrinsic tear strength of compression molded films of blends of ZN octene-LLDPE (0.920 g/cc, 0.95 g/10 min MFI (190°C/2160 g)) with LDPE (0.921 g/cc, 0.64 g/10 minute MFI (190°C/2160 g)). Dashed lines represent rule of mixtures. *Adapted from R.M. Patel, T.P. Karjala, N.R. Savargaonkar, P. Salibi, L. Liu, Fundamentals of structure-property relationships in blown films of LLDPE/LDPE blends, J. Plastic Film. Sheeting, 35 (2019) 401–421.*

change branching and longer relaxation time of the LDPE produces strain hardening behavior that nucleates crystal growth along the chain to produce the lamellae structure. The LLDPE chains, being more mobile (short relaxation time) grow radially outward in the TD. As more LDPE is added to create LDPE-rich blends, the lamellae become more twisted like LDPE.

TD tear strength is shown to have positive synergy at blend levels up to about 60% LDPE. MD tear has negative synergy in this same range. Tear anisotropy transitions from LLDPE like behavior (TD > MD) to LDPE-like behavior (MD > TD) at around 70% LDPE. The isotropic tear strength at this concentration, however, is low (but nearly on the simple rule of mixtures line). The change in tear strength is consistent with the changes in crystalline morphology and orientation.

Patel et al. [80] and Mezghani et al. [115] report similar trends in tear strength results for monolayer blown film of blends of LDPE with ZN octene-LLDPE and hexene-LLDPE, respectively. Patel et al. also measure the "intrinsic" tear resistance on compression molded films to eliminate the effects of film orientation during the filmmaking process. Their results are shown in Fig. 9.32. The intrinsic tear resistance is essentially linear indicating that the deviations from the rule of mixtures for MD and TD tear resistance of the blown films are related to orientation. The authors point out that it is well known that ZN LLDPE has a high-density high-MW fraction with a high concentration of tie chains whereas LDPE has much fewer tie chains. Adding LDPE to LLDPE reduces the number of tie chains which should reduce MD tear strength. Furthermore, the addition of LDPE will increase the amorphous phase orientation due to its long chain branching and strain hardening behavior. This also should act to reduce MD tear strength.

Several studies of multilayer polyethylene films reveal that tear strength is not additive. Siegman and Nir [58] and Martinez et al. [59] find that (LLDPE-LDPE-LLDPE) films deviate from the rule of mixtures. Elkoun et al. [61] find different combinations of LLDPE either give linear or nonlinear behavior as described in Section 9.4.3.1 for impact toughness. A coextrusion of butene-LLDPE (b-LLDPE) and metallocene-LLDPE (m-LLDPE) deviates from the rule of mixtures whereas a coextrusion of octene-LLDPE and m-LLDPE agree well with the rule of mixtures. The distinctions are attributed to differences in the crystalline structure.

Borse and Aubee [63] make three-layer structures of (A-B-A) with a butene-LLDPE and a metallocene LLDPE. When the b-LLDPE comprises the skin layers, the film follows the rule of mixtures. Placing the m-LLDPE on the outside and b-LLDPE on the core creates synergies above that of the rule of mixtures.

Guillotte [64] tests coextruded films of ionomer and LDPE and find the tear strength to be lower than either the PE or ionomer alone and do not change as the percent of ionomer is increased. He attributes this to less orientation in the ionomer layer during coextrusion since the LDPE freezes first and dictates the stretching of the polymer. The ionomer is molten while it is extended, reducing its orientation. This effect is explored in more detail in Chapter 14.

9.4.4.2 Directional tear technology

Some packages are designed to be torn open. While perforations and laser scoring are often used to facilitate easy tear performance, the choice of materials and the film fabrication process can sometimes be manipulated to provide a directional tear function. Directional tear refers to the ability of the film to tear in one direction, typically the machine or transverse film direction, no matter how the tear is initiated. For the tear to propagate readily in the TD, the MD tear strength needs to be substantially greater than the TD tear strength and the TD tear strength needs to be low. The reverse is true for directional tear performance in the MD.

Most applications requiring directional tear performance need the tear to propagate in the TD. This is due to the orientation of most packaging films on packaging equipment. For example, on a pouch machine, the opening is sealed in the TD. A notch may be incorporated into the package to help the consumer initiate the tear in the TD. As is described in Section 9.4.4.1, linear PE, such as LLDPE and HDPE, typically has much higher tear strength in the TD than the MD. This makes it difficult to use these materials for TD directional tear applications. A better starting point is LDPE, which often has weaker tear strength in the TD than MD. The TD to MD tear strength ratio of LDPE can be minimized through the selection of fabrication conditions. Studies dating back to the early 1960s [116] show that the TD to MD tear ratio decreases as the orientation in the MD is increased. This can be accomplished by increasing the DDR, decreasing BUR and minimizing the frost line height. This is shown in Fig. 9.30 where the MD/TD tear strength of LDPE is shown to increase with decreasing BUR and is further discussed in Section 14.1.1 and illustrated in Fig. 14.2 based on the work of Huck and Clegg [116].

The TD tear strength of LDPE can still be too high for effective directional tear due to plastic deformation, as described by Eq. (9.13). Meilhon and Wollak [117–119] propose the use of ionomers, either alone or in a coextruded blown film structure with LDPE. Typical film structures involve (A-B-A) coextrusion with ionomers as either the skin layers (for greater gloss) or as the core (for lower COF). The LDPE should be a high-pressure low-density resin, not an LLDPE or metallocene PE.

Ionomers are ethylene-acid copolymers whose acid groups have been partially neutralized with a metal salt such as zinc or sodium. They are similar to LDPE in that they are made by free-radical polymerization, which introduces long-chain branching along the polymer chains. Therefore, they have similar property trends. For example, ionomers have the same preference toward TD tear propagation as LDPE. Ionomers, however, typically have lower tear strength than LDPE, which makes them ideal for directional tear applications.

On the simplest level, the origin of the lower tear strength of ionomers likely arises from the glassy amorphous regions in ionomers. The acid–metal salt interactions tend to associate with nanometer-sized clusters or domains within the amorphous phase of the PE matrix. Ionomers, therefore, have crystalline and amorphous regions like PE, but also ionic domains, as illustrated in Fig. 4.8, [120]. In the solid-state, the molecular chains are restrained by the ionic interactions within the domains. The chains in the immediate vicinity of these domains are also somewhat restrained, creating a pseudo-glassy amorphous state. Furthermore, Longworth [120] proposes that the notch sensitivity of ionomers may be due to a stress concentration effect. The radius of the propagating crack is small and may be of the same order of magnitude as the distance between ionic clusters. The ionic phase is glassy and nondeformable. As the crack propagates, greater strain is put on the PE phase surrounding the ionic domains than when the ionic domains are not present. As a result, failure occurs at lower stress. This is consistent with a theory for tear strength put forward by Rohn [121], which assumes that tear resistance is related to the amount of molecular mobility available to respond to the formation of a tear or crack. The lower the mobility, the lower the tear strength.

Meilhon and Wollak find that ionomers are even more responsive to changes in blown film processing parameters than LDPE. Directional tear in the TD direction is favored by running the film line at high DDR, low BUR and low frost line height. Lee et al. [108,109] conduct fundamental studies to understand the origins of the directional tear. They study six polymers: high and low MFI ($\sim$1 and 10 g/10 minute) grades of LDPE, ethylene methacrylic acid (EMAA) copolymer (a precursor to ionomer), and ionomer. Blown films from each of the grades are made on a lab line, varying the BUR (1.5–3.5) and frost line height (8–50 cm), keeping the film thickness constant (50-μm). The take-up speed is adjusted to keep the film thickness constant so that as the BUR is increased, the DDR decreases [as per Eq. (2.7)]. Elmendorf tear strength in the MD and TD are measured and the film crystalline morphology characterized using SAXS and WAXS.

The LDPE film exhibits Keller–Machin low-stress orientation with lamellar stacking along the MD and twisting as they grow (see Fig. 9.28). The ionomers and EMAA have a structure corresponding to the Keller–Machin high-stress structure, which up to this point had only been reported in the literature for HDPE. EMAA and ionomers crystallize more slowly than LDPE and have a higher density of crystalline nuclei (since they freeze at a lower temperature where crystallization rapidly occurs). Lee et al. suggest that this results in the high-stress crystalline structure.

The low MFI ionomer shows a rotation of the crystalline lamellae when the film is made at high BUR. The authors say they are not aware of LDPE exhibiting this behavior. The rotation corresponds to a change in directional tear from TD to MD. Thus, ionomers under the right processing conditions can be made to have a directional tear in either the TD or MD directions.

Lee et al. compute Herman's orientation function (f), an indication of crystalline orientation, and find it correlates with preferred tear direction and the ratio of MD to TD tear strength (negative f gives preferred MD tear and positive f preferred TD tear for the ionomer and EMAA). The LDPE tear is either isotropic or in the TD direction. Lower MFI grades of all three resins have a steeper slope of f versus BUR indicating more orientation. There is not a perfect correlation between 1/MFI and MD/TD tear strength ratio since 1/MFI is a melt property whereas Herman's orientation function is a solid-state value.

Of the three processing variables, only BUR shows a significant effect on tear. Since take-up velocity (MD strain) is decreased as BUR (TD strain) is increased, the MD/TD strain ratio increases rapidly as BUR is decreased.

Based on the X-ray analysis, Lee proposes that the least resistance to tear propagation is between the lamellae—perpendicular to the nucleating lines and to the direction of laminar stacking, as is shown for LDPE in Fig. 9.29. Ionomers and LDPE have a similar molecular architecture with long-chain branching and a relatively low number of tie molecules. For the same melt viscosity, ionomers have a lower chain length (MW) since viscosity is built up by the ionic bonding between the metal salts and acid groups along the chain. Thus, ionomers are expected to have even fewer tie molecules than LDPE, which likely contributes to their lower tear strength. As the BUR is increased, the crystalline orientation of the ionomer and EMAA flips 90 degrees so that the extended polymer chains are in the TD and the parallel stacks of lamellae are in the MD, the new preferred direction of tear. Tear propagation is still between lamellae.

Sometimes, the directional tear is needed in the MD rather than TD. As shown in Section 9.4.4.1, linear PE such as HDPE and LLDPE tend to preferentially tear in the MD. But these polymers have their drawbacks. HDPE has poor heat seal performance and the overall tear strength of LLDPE may be too high for directional tear. As shown above, ionomers can be flipped to tear preferentially in the MD direction by running at high BUR. Alternatively, blends of ionomer and LDPE can be used to achieve MD tear [122]. These blends may be used as monolayer or in coextrusions with other layers that would not typically have linear tear properties on their own.

Translating directional tear from an ionomer or polyolefin film to a laminate can be challenging. In general, good adhesion is needed between the layers to transfer the tear propagation across the laminate. Loss of adhesion during the tear event can dissipate the tear force and cause the structure to become more resistant and less likely to tear in the preferred direction.

9.4.4.3 Multilayer films

PE films have some of the highest tear strengths found in flexible packaging, yet laminations of PE film to stiffer substrates such as OPET or BOPP typically have much lower tear strength [66,67]. Wu et al. [67] provide insight into this phenomenon. They begin by developing a theoretical model for the tear strength of multilayer films. They find that the tear strength of the composite is the sum of the tear strength of the individual layers; tear strength is predicted to follow the rule of mixtures. They conduct experiments with laminates of LLDPE film with biaxially oriented polypropylene (BOPP) using a two-sided adhesive tape (based on BOPP). The tear strength of the BOPP-LLDPE laminates is much lower than the LLDPE film alone. They also conduct tear strength measurements on BOPP-OPET and BOPP-cast polypropylene (cPP) laminates and find that these follow the simple rule of mixtures, as predicted by their model. This is illustrated in Fig. 9.33.

As noted in Section 9.4.4.1, the tear strength of PE, and LLDPE in particular, is enhanced by a large zone of deformation in advance of the crack tip. When the LLDPE film is laminated to the stiff BOPP, the BOPP constrains the deformation, limiting the zone of deformation. Wu conducts some rather elegant experiments confining the dimension of the zone of deformation by applying the double-sided tape to each side of the LLDPE-film test specimen. This leaves a well-defined width through which the tear can propagate. They measure the tear strength at several confined zone widths, the results of which are shown in Fig. 9.34. For large zone widths, the tear strength is the same as that of the unconfined specimen. As the zone width is constrained, the tear strength begins to decrease in a linear manner, as predicted by fracture mechanics. By extrapolating to zero width, they obtain a value for the "limiting tear resistance" or LTR. By using LTR in their model for the BOPP-LLDPE laminates, the predicted tear resistance of the laminate agrees well with the measured values.

This mechanism of the deformation of the PE layer being confined by stiffer layers in a laminate is confirmed in the work by Timm [66]. Timm measures the tear strength of an OPET/LLDPE laminate and finds that the tear strength is considerably less than the LLDPE alone. Increasing the thickness of the LLDPE sealant layer is more effective in increasing the tear strength of the laminate than changing the composition of the LLDPE sealant film to one with stronger tear strength. This is consistent with Wu's explanation since the thicker LLDPE film would allow a larger zone of deformation to form.

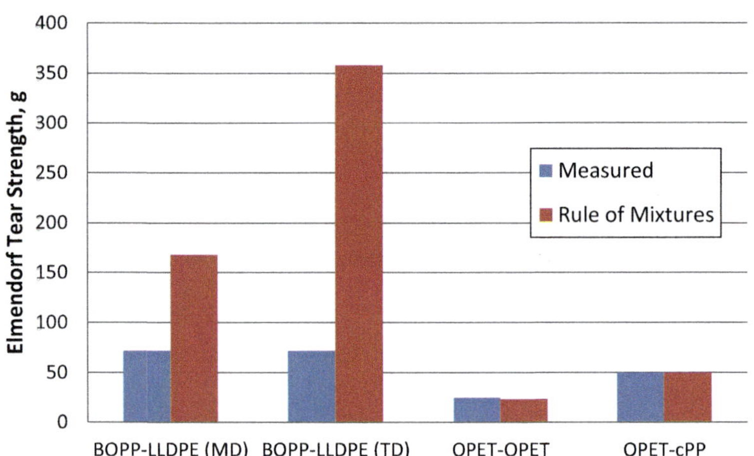

FIGURE 9.33 Elmendorf tear measurements of some laminates. *Data from R.-Y. Wu, L.D. McCarthy, Z.H. Stachurski, Tearing resistance of multi-layer plastic films, Intern. J. Fract., 68 (1994) 141–150.*

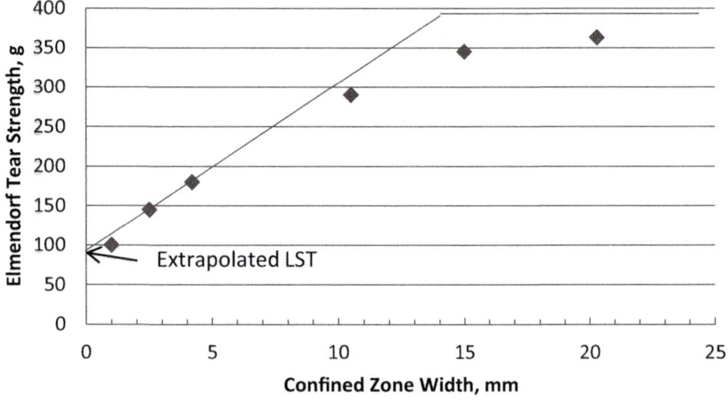

FIGURE 9.34 Confined tear resistance measurements on a LLDPE film. The edges on both sides of the test specimen are taped to confine the zone of deformation to a predetermined value. *Data from R.-Y. Wu, L.D. McCarthy, Z.H. Stachurski, Tearing resistance of multi-layer plastic films, Intern. J. Fract., 68 (1994) 141–150.*

9.4.5 Flex fatigue

Lab flex fatigue tests such as the Gelbo flex test and various vibration and shock tests are designed to provide insight into failures during handling and transportation. Donno [123,124] conducts actual transportation trials on tomato paste packaged in two laminate film structures, one using 14-μm EVOH and the other 12-μm OPET-AlOx as the barrier. The 108 oz (3.2 L) pouches are filled, placed in cases (six pouches per case), stacked on pallets (30 cases per pallet) and trucked for 4023 km (2500 miles). Following shipment there are no leakers; the seals remained intact. The highest degree of flex cracking is found near the corners of the pouches. The oxygen transmission rate (OTR) is measured on pouches near the corners before and after shipment. There is no statistically significant change in OTR from samples near the bottom of the pallet. The OTR of pouches taken near the top of the pallet is significantly higher after shipping, with the EVOH structure increasing by a factor of three and OPET-AlOx by a factor of 12.

Sundell et al. [125] conduct both road and laboratory simulated vibration tests. Six barrier bag-in-box packages from five suppliers with aluminum foil, met-OPET and EVOH as the barrier layers are evaluated. Three and 10-L sized pouches are filled with water for the lab tests and red wine for the 1000 km (621 miles) transportation trial. Laboratory two-dimensional vibration tests are conducted to simulate 800 km of travel. For the road test, two pallets of each package format are trucked 800 km on standard roads and an additional 200 km on city streets. The 3 L pouches are stacked four layers high and the 10 L pouches three layers high. The aluminum foil and EVOH barrier packages show little change in OTR after the lab vibration or road tests. There is no significant difference between the 3 L and 10 L formats. The packages with met-OPET, however, show a significant increase in OTR and evidence of stress cracking. Only one leaker is found, which is determined to be due to the abrasion of a welded seam from the box panel folded edge.

Donno et al. [124,126] look at the effect of headspace volume on the survivability of pouches filled with water (1 centipoise viscosity) or a more viscous 5% starch solution (39,000 centipoise) when subjected to laboratory vibration, compression and shock (drop) tests. Commercial four-side sealed retortable bags designed for packaging 1.5 kg of product are used. The laminate structure is OPET-AlOx/BOPA/CastPP. The filled pouches are retorted at 121°C (250°F) for 30 minutes in a lab retort autoclave. The drop test uses ASTM D5276 where pouches are dropped at increasing heights until failure. Drop is on the face of the pouch. The results of the compression tests show no significant effect of liquid viscosity on failure, refrigerated samples fail at a significantly lower force than samples stored at room temperature, and headspace volume has little effect on the maximum compressive force for failure (although the authors suggest there is a general trend toward higher force for failure with higher headspace volume). None of the samples fail the vibration test. In the drop test, increasing headspace volume increases the protection of the low viscosity product (water). Headspace has no significant effect on the drop height at failure for the more viscous starch solution.

In another study, Donno [124] subjects the same four-side sealed retort bags to transportation simulations under ISTA 3 standards for laboratory vibration and shock. Changes in OTR are used to assess performance. The bags are filled with water at two headspaces: 200 and 400 mL. The bags are thermally processed and packed five per container into regular slotted containers (RSC). Two lab simulations under the ISTA 3 protocol are conducted: 3A for small parcel and 3E for unit load delivery systems. An Oxysense dot is added to the package to nondestructively measure oxygen ingress via fluorescence. Results show a significant increase in oxygen ingress with time when the packages are subjected to the transportation simulation compared to a control. Higher headspace reduces the increase in OTR for the small parcel simulation. There is no effect of headspace for the unit load simulation. Donno concludes that higher headspace may be helpful for small parcels where there is more handling than for shipments on a pallet. Cracks are found in the barrier layer with microscopy. Donno postulates that the low viscosity fluid puts stress on the seals and allows the pouch to fold onto itself.

Doyon et al. [127] compare field transportation tests with laboratory Gelbo flex tests. The field transportation tests use five routes of different lengths (ranging from 414 to 809 km) with four loadings and unloadings per route. 4 L bag-in-box bags with met-OPET as the barrier layer are filled with wine. The bags are stored upside down to reduce pressure on the valves. The age of the package varies from three to eight months. Flex cracking is monitored by measuring OTR in four sections of the bags.

The lab test uses ASTM F392-93 (Gelbo test). OTR is measured after 0, 50, 100, 125, 200, and 500 cycles. The results are correlated with the field tests to estimate an equivalent number of cycles to obtain the same measured OTR of field samples.

The field trials reveal more mechanical damage near the edge of the package and the valve. OTR increases by a factor of 10–20 during the field trial. Routes with similar distances but older packages have considerably more damage and higher OTR; the distance of transport is statistically less significant than the age of packages. Doyon et al. conclude that loading/unloading causes more damage than the vibration/shock of transport. The Gelbo lab tests show OTR versus number of cycles follows a square root relationship, meaning OTR rises quickly at low cycles and then slows. This suggests that early logistical operations (filling, boxing, palletization and loading) are likely greater contributors to damage than vibration/shock during transport.

Sadeghi and Ajji [82] characterize the molecular architecture of three metallocene LLDPE and one ZN LLDPE and relate this to film properties. The mPE's range in density from 0.895 to 0.902 g/cc whereas the ZN LLDPE has a density of 0.934 g/cc. The mPE's differ in MW, MWD and the presence of sparse long chain branching. The ZN LLDPE has short-chain branching (SCB) mostly on short chains whereas the SCBs on the mPEs are more uniformly distributed on all the chains. Cast films are subjected to 2700 flex cycles and then examined for pinholes. The ZN LLDPE have six pinholes whereas the three mPE films have none. The authors propose the higher SCB on the longer chains is responsible for the better flex crack resistance of the mPE resins, citing Krishnaswamy et al. [128] and Lu et al. [129]. Krishnaswamy et al. make blends of two HDPE resins with different MW, one with more SCB on the high MW component and the other with more SCB on the low MW component. The bend with more SCB on long molecules has slower crystallization at high temperatures, higher ultimate mechanical properties and better slow crack resistance. They attribute this to more tie chains when the SCB is located on the high MW chains.

Lu et al. separate a hexene-LLDPE by MW into five fractions. The MW and SCB of each fraction are measured. The slow crack growth of each fraction is measured by sandwiching them between layers of PE. They find there is a critical MW above which resistance to slow crack growth develops. They conclude that a high MW fraction, or putting the SCB on longer chains, is needed to create enough tie chains to get good resistance to slow crack growth.

Flexing of barrier laminate structures containing aluminum foil or metallized film is a major cause of the formation of pinholes and barrier loss, as is described in Section 8.4.2. Armstrong and Chow [72] compare the OTR of an EVOH coextrusion with two aluminum foil structures, one with polyamide and the other with PE. The EVOH contains 32 mol % ethylene. Before flexing, the EVOH has a higher OTR than the foil structures. After flexing 50 cycles, its OTR increases by less than a factor of two. The OTR of the polyamide/foil structure increases by about a factor of 20; at 50 cycles its OTR is more than double that of the EVOH structure. The PE/Al structure fairs the worst with an increase by a factor of 50; its OTR is about a factor of six higher than the EVOH structure after 50 flexes. See Fig. 9.35. Metallized OPET and met-BOPP do not fair any better; both end up with significantly higher OTR after 50 flex cycles than an EVOH coextrusion with PE, with the met-BOPP showing the highest increase in OTR.

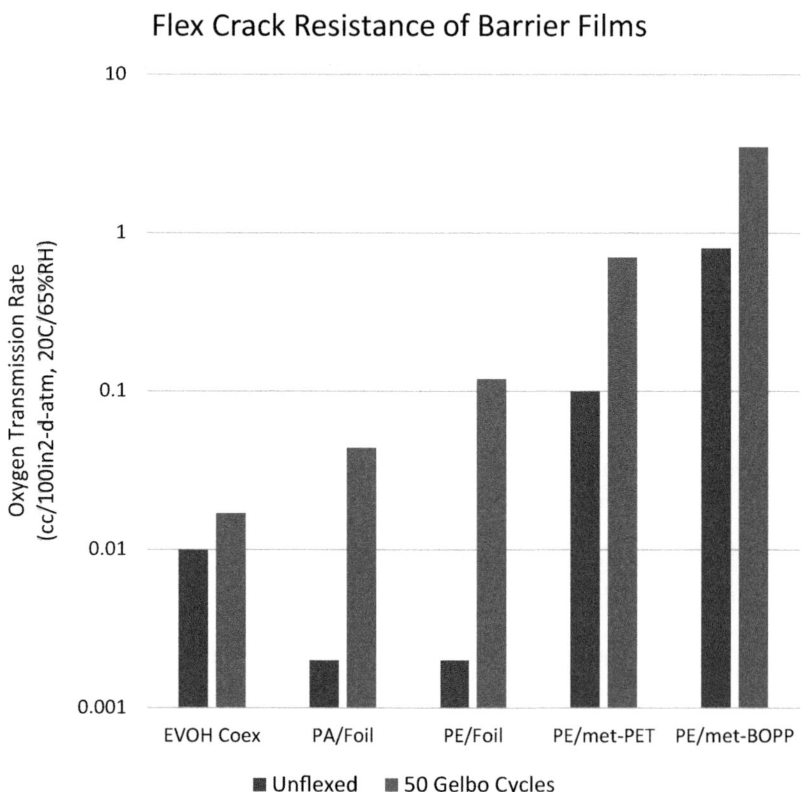

FIGURE 9.35 Oxygen transmission of barrier films before and after flexing (Gelbo flex test). EVOH coex: (PE-tie-EVOH-tie-PE) [65/12/5/12/65 μm], PA/foil: OPA/PE/Al/tie/PE [20/12/8/12/100 μm], PE/Foil: PE/tie/Al/tie/PE [65/12/8/12/38 μm], PE/met-OPET: PE/met-OPET/PE [12/15/32 μm], and PE/met-BOPP: PE/met-BOPP/PE [39/18/51 μm]. *Data from R. Armstrong, E. Chow, Improving performance of flexible barrier packaging for liquids through the use of modified ethylene vinyl alcohol (EVOH), in: SPE Polyolefins and FlexPack Conference, Houston, TX, February 2011.*

Several strategies have been used to increase flex crack resistance of barrier laminates. These include:

- Changing the adhesive. Ferrari [130] tests structures of PET/adh/met-PET/adh/PE using different adhesives: solventless (two-part polyester/polyether urethane), water-based (two-part acrylic urethane) and solvent (two-part curing polyester urethane). OTR is measured after 0, 20, and 270 Gelbo cycles. The structure with the solvent-based adhesive retains its barrier better after flexing than the structures with the other adhesives.
- Applying a coating. Ferrari also finds coating the met-OPET with an ultra-high barrier coating helps retain barrier performance after cycling. Cumberbatch and Reitz [131] show that coating PET with PVOH prior to metallization substantially improves OTR after flexing.
- Laminating or coextruding with protective polymers. Cumberbatch and Reitz [131] show that laminates of met-PET with PE have significantly better flex crack resistance than met-PET alone. Chow and Armstrong [72,73] show similar results for EVOH monolayer films and laminates with PE, as shown in Fig. 9.36. Ionomers and acid copolymers provide good flex crack resistance when coated onto foil. In one study [132], pinholes are measured in a paper/PE/Al/sealant laminate after 20 Gelbo cycles. PVDC, LDPE and ionomer sealants give >70, 9–20 and 0–2 pinholes, respectively. The toughness and good adhesion ionomers provide to the foil are thought to contribute to their good flex performance. EVOH is often sandwiched between layers of PA6 in coextruded film structures. This helps with flex resistance as well as aides in thermoforming. Chow and Armstrong [73] show that a "soft" PE-EVOH coextruded film provides better flex crack resistance than a "stiff" one. Here the structures are (LLDPE–LLDPE-tie–EVOH–LLDPE-tie–LLDPE) and (LLDPE + EVA–EVA-tie–EVOH–EVA-tie–LLDPE + EVA) at thickness (36/5/8/5/36 μm) for the stiff and soft films, respectively.
- Modifying the barrier polymer. Ionomers, talc, amorphous EVOH, elastomers (polyolefin or styrene-based) are some of the modifiers that have been added to EVOH to improve flex crack resistance [72,133–135]. Care should be taken with blends since optical, barrier and film quality (gels) may be affected.
- Reduce the thickness of the barrier layer (EVOH). Armstrong and Chow [72,73] show that reducing the thickness of EVOH increases flex crack resistance. This is shown in Fig. 9.36 for monolayer and coextruded film. This effect may be related to a reduction in the bending stiffness of the film. Taking this even further, Medlock and Dolgovskij [136] and Lee et al. [137] show that creating repeated microlayers (<1 μm) of EVOH using layer multiplication technology improves flex crack resistance (see Section 8.5.2).

9.4.6 Essential work of fracture

Tear strength is not an intrinsic material property as it varies with test geometry and other factors. Fracture toughness and stress intensity factors often used in the study of the fracture failure of bulk polymers fall short for thin ductile films used in flexible packaging. Recently the concept of essential work of fracture (EWF) has been applied to films in the academic literature as a way to measure fracture toughness [138–145].

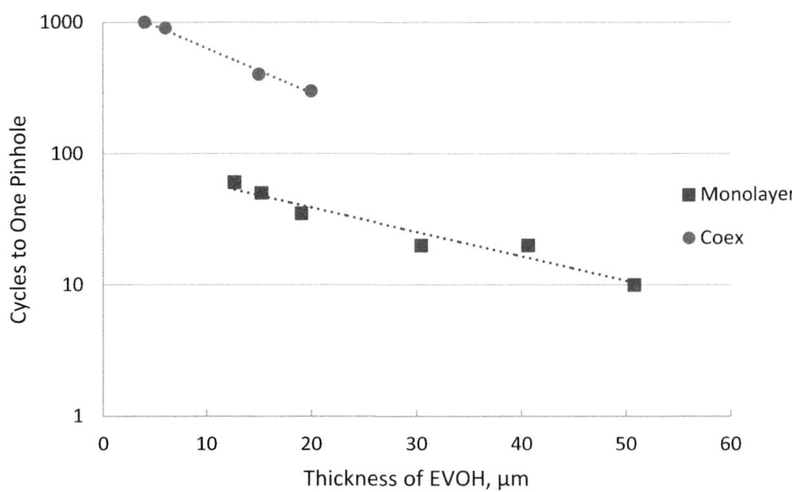

FIGURE 9.36 Flex crack resistance of ethylene vinyl alcohol (EVOH) monolayer and coextruded blown films as a function of thickness. The EVOH contains 32 mol % ethylene. The coextruded structure is (LLDPE-tie-EVOH-tie-LLDPE) with thickness [36/5/X/5/36 μm]. *Data from R. Armstrong, E. Chow, Improving performance of flexible barrier packaging for liquids through the use of modified ethylene vinyl alcohol (EVOH), in: SPE Polyolefins and FlexPack Conference, Houston, TX, February 2011 and E. Chow, R. Armstrong, Improving the performance of bulk packaging of liquids through the use of modified ethylene vinyl alcohol (EVOH), in: SPE Polyolefins and FlexPack Conference, Houston, Tx, February 2012.*

The concept of the work of fracture involves dividing the fracture into two zones, an inner region corresponding to the fracture process zone and an outer region corresponding to yielding (plastic zone) as illustrated in Fig. 9.37. The work of fracture (W_f) is partitioned into two components, the essential (W_e) and the nonessential (plastic) work of fracture (W_p). These two components correspond to the two fracture regions. W_e and W_p can be interpreted as the energy to initiate the crack and the energy for propagation, respectively. W_e is an intrinsic material property whereas W_p depends on the geometry.

The work of fracture test typically involves using a tensile tester to measure the force versus displacement of a double notched specimen, as shown in Fig. 9.37. The horizontal distance between the notches is the ligament length, ***l***. The work of fracture is given by Eq. (9.14):

$$W_f = W_e + W_p = w_e l X + \quad w_p \beta l^2 X \tag{9.14}$$

where $W_f =$ work of fracture, $W_e =$ essential work of fracture, $W_p =$ nonessential or plastic work of fracture, $w_e =$ specific essential work of fracture (per unit ligament length), ***l*** $=$ ligament length, X $=$ thickness of specimen in ligament region, $w_p =$ specific non-essentual work of fracture (per unit ligament length), and β $=$ plastic zone shape factor.

In terms of the specific work of fracture, W_f is divided by the cross-sectional area of the ligament, X***l***:

$$w_f = \frac{W_f}{Xl} = w_e + \beta w_p l \tag{9.15}$$

Specimens are double notched to different ligament lengths and the force versus displacement is measured. A typical outcome is shown in Fig. 9.38. The area under the force versus displacement curve is equal to w_f. The plot of w_f versus ***l*** yields a straight line with intercept w_e and slope βw_p.

The applicability of using the ESF analysis for thin films is subject to certain conditions described in detail in various references [146,147]. Three of these conditions are [143]:

- The ligament must be fully yielded prior to crack propagation.
- The load versus displacement curves must be self-similar, as is shown in Fig. 9.38.
- The ligament must be in pure plane stress.

Academia and industrial R&D groups [138–153] are starting to use this technique to relate the fundamental mechanics of crack initiation and propagation to the orientation and crystalline morphology of films. For example, Maspoch et al. [146] find that the specific EWF, w_e, is independent of film thickness but sensitive to orientation, whereas the reverse is true for the specific non-EWF. Rennert et al. [143] apply the technique to a series of LDPE blown films with thicknesses ranging from 15 to 400 μm made by varying the DDR. The fracture mechanics parameters obtained from the work of fracture testing are related to the crystalline morphology and orientation found from X-ray diffraction studies. Such studies should ultimately enhance the basic understanding of impact toughness, tear strength and peel strength in packaging films and lead to better models for predicting performance.

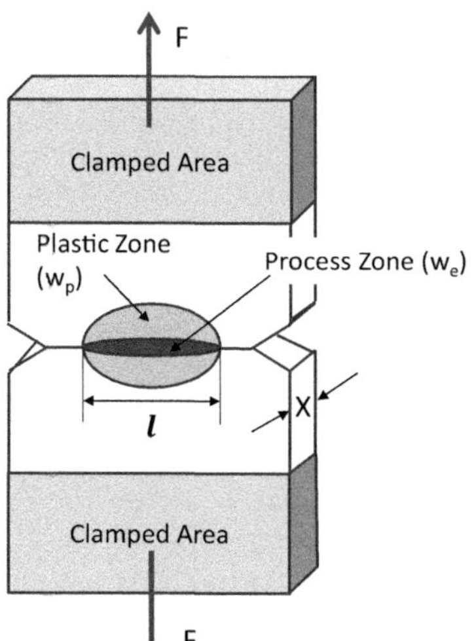

FIGURE 9.37 Illustration of double notched specimen during work of fraction test. *Adapted from M.L. Maspoch, V. Henault, D. Ferrer-Balas, J.I. Velasco, O.O. Santana, Essential work of fracture on PET films: influence of the thickness and the orientation, Polym. Test., 19 (2000) 559–568.*

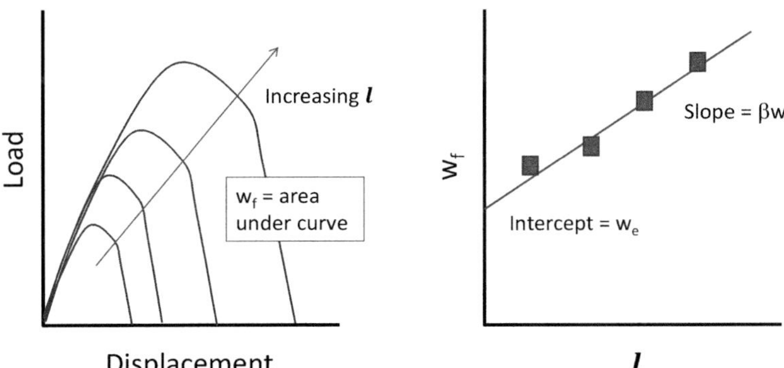

FIGURE 9.38 Example force versus displacement curves produced during work of fracture measurements. Area under the curve is the work of fracture, which when plotted versus the ligament length (right) yields a line with intercept w_e and slope βw_p.

9.4.7 Microlayer technology

Section 8.5.2 describes layer multiplication technology that allows for the creation of tens, hundreds or even thousands of layers in a coextruded film. In flat die extrusion, this is accomplished by systematically doubling the number of layers by splitting the flow stream and recombining the layers in a series of "multiplier" modules, as is shown in Fig. 8.19. For blown film, Schirmer [154,155] has developed technology using thin "layer sequence repeater" plates to subdivide the flow.

Typically microlayer or nanolayer films are produced with two or three layers repeatedly multiplied to create a core region with many thin layers. Skin layers that may be comprised of different resins encapsulate the core region. An example structure is: (A-[C-D]$_n$-B) where n is the number of multiplications. As the number of layers increases, their individual thicknesses decrease. Films with layer thickness in the submicron and even nanometer dimensions have been reported. As the layer dimension decreases, there may be changes in the physical properties brought on by changes in crystallinity, crystalline morphology, orientation and other factors.

As described in Section 9.4.1, Schrenk and Alfie [69] show that the mechanical properties of thin layers of stiff and ductile polymers may be mutually reinforced as the layer dimensions decrease. This occurs when there is good adhesion between the layers and the ductile material can stop cracks initiated in the brittle layer from propagating.

The research group of Hiltner and Baer [156–165] at Case Western Reserve University has published several papers exploring the effect of forced assembly of micro or nanolayers in coextruded film structures. This work is reviewed by Langhe and Ponting [166]. Some of their findings include:

- Improved toughness of polycarbonate/styrene acrylonitrile (PC-SAN) and polycarbonate/polymethyl methacrylate (PC-PMMA) microlayer structures, which they attribute to a change in deformation from crazing to shear yielding as the layer thickness is decreased. They note that differences in interlayer adhesion may explain some of the differences found between the two systems [156–158].
- Greater ductility of talc-filled PP. Microlayers of talc-filled PP are made by compression molding and by coextrusion (using the process illustrated in Fig. 8.19). The coextruded structures of alternating layers of talc-filled PP have better ductility than compression molded versions. They explain that the talc particles are more oriented in the coextrusion process, providing a greater cross-sectional area for the load-bearing PP. The ductility of alternating layers of unfilled PP and filled PP go through a maximum with an increasing number of layers [159,160].
- Reduction in crystallinity of HDPE sandwiched between layers of polystyrene (PS). As the HDPE layers become thinner (from 300 microns to 10 nm), % crystallinity decreases from 60% to 33%. The peak Tm decreases from 134°C to 130°C. The decrease in crystallinity is accompanied by a change in crystalline morphology. Nucleation occurs from the interface—bulk nucleation becomes insignificant as thickness decreases. Edge-on lamella form at low thickness as opposed to discoids (flattened spherulites due to the thickness dimension being less than the spherulite dimension). As a result of the changes in crystallinity and morphology, the gas transmission rate increases with an increasing number of layers [158,161–163].
- Change in the crystalline structure of PET. Films with 1056 layers of PET-PET or PET-PC were coextruded with layer thickness ranging from micro to nanometer dimensions. After extrusion the PET layers are amorphous. The samples are annealed at various temperatures to crystallize the PET. The nonconfined PET crystals in the PET-PET structures start to form at 117°C. Crystallization of PET confined between the layers of PC does not begin until 132°C. The confined PET exhibits lower % crystallinity and greater orientation in the crystalline domains [158,164].

- Change in crystalline morphology of PP. Films with alternating layers of PP and PS are coextruded with layer dimensions ranging from 650 to 10 nm. As the confined dimension of the PP layer is decreased, the crystalline morphology changes from spherulitic to flattened spherulitic (discoids). As the dimension approaches 10 nm, further distortions are found [165].

Section 8.5.2 describes studies that show layer multiplication technology can increase the ductility of PE-tie-EVOH coextruded films. Improvements in physical properties by utilizing microlayer technology have found their way into the commercial stretch and shrink films, as is described in Sections 8.5.2 and 11.2.4.5.

Layer multiplication technology is a good example of how the structure design and fabrication process influence film properties and may lead to deviations from the simple rule of mixtures. As the layer dimensions decrease and the layers are confined between other layers, changes in the crystalline morphology and orientation may arise that positively or negatively impact performance. The design of the structure (layer composition, layer placement and layer dimension) cannot be done in isolation of the process. Chapter 14 deals with processing effects in more detail.

References

[1] J.D. Vansant, Internal Communication, 2012.

[2] J. McKenna, G. Bunker, T. Fry, N. Savargaonkar, J. Wooster, Change the package, keep the feel, TAPPI PLACE. Conf. (2010).

[3] G. Bunker, Tips and techniques: how to downgauge film without losing the 'feel' of quality, Plast. Technol. (2010)viewed 30 Dec 2014 on. Available from: http://www.ptonline.com/articles/tips-and-techniques-how-to-downgauge-film-without-losing-the-feel-of-quality.

[4] Amcor, Stand out and deliver. <https://www.amcor.com/insights/educational-resources/north-america-ecommerce-ebook>, 2020 (accessed 30.08.2020).

[5] L. Macdonald, Unpacking the trends and challenges around eCommerce packaging in 2019 and beyond, Bridge Connect. (2019)viewed on 29 August 2020 at. Available from: https://www.ebridgeconnections.com/media/blog/january-2019/ecommerce-packaging-trends-2019.aspx.

[6] C. Smith, E-commerce packaging provides challenges/opportunities for CPG products, Parcel (2019)viewed on 29 August 2020 at. Available from: https://parcelindustry.com/article-5234-E-commerce-Packaging-Provides-Challenges_Opportunities-for-CPG-Products.html.

[7] DHL, Rethinking packaging. <https://www.dhl.com/global-en/home/press/press-archive/2019/rethinking-packaging-dhl-trend-report-discovers-how-e-commerce-era-drives-wave-of-sustainability-and-efficiency.html>, 2019.

[8] L.M. Pierce, 3 challenges of ecommerce packaging (from an insider), Packaging Dig. (2019)viewed on 29 August 2020 at. Available from: https://www.packagingdigest.com/trends-issues/3-challenges-ecommerce-packaging-insider.

[9] Bemis, Packaging for a new era of e-commerce, ebook, 2016.

[10] ASTM International, Standard test method for tensile properties of thin plastic sheeting, in: D882-12, ASTM International, West Conshohocken, PA, 2012.

[11] L.W. McKeen, Film Properties of Plastics and Elastomers, third ed., Elsevier, Amsterdam, 2012, pp. 27–34.

[12] S. Bahadur, Strain hardening equation and the prediction of tensile strength of rolled polymers, Polym. Eng. Sci. 13 (1973) 266–272.

[13] J. Lange, C. Pelletier, Y. Wyser, Modeling and measuring the bending stiffness of flexible packaging materials, PFFC Peer Reviewed Paper, 2002 March. Also:Y. Wyser, C. Pelletier, J. Lange, Predicting and determining the bending stiffness of thin films and laminates, Packaging Technology and Science, 14 (2001) 97–108.

[14] German Institute for Standardization, Determining the Flexural Rigidity of Plastic Film and Woven Textile Fabrics With or Without Plastic Coating by the Cantilever Method, German Institute for Standardization, 2003 (DIN 53 362).

[15] ASTM International, Standard Test Method for Resistance to Bending of Paper and Paperboard (Taber Type Tester in Basic Configuration), ASTM International, West Conshohocken, PA, 1997 (D5342-97).

[16] ASTM International, Standard Test Method for Apparent Bending Modulus of Plastics by Means of a Cantilever Beam, ASTM International, West Conshohocken, PA, 2010 (D747-10).

[17] International Organization for Standardization, Paper and Board - Determination of Bending Resistance, International Organization for Standardization, Geneva, Switzerland, 2010 (ISO 2493).

[18] ASTM International, Standard Test Method for Rigidity of Polyolefin Film and Sheeting, ASTM International, West Conshohocken, PA, 2013 (D2923-13).

[19] British Standards Institution, Method of Testing Plastics. Mechanical Properties. Stiffness of Plastics Film, BSI, London, England, 1976 (BS 2872, Part 3, Method 332A).

[20] B.A. Morris, J.D. Vansant, The influence of sealant modulus on the bending stiffness of multilayer films, TAPPI Polymers, Laminations Coat. Conf. (1997).

[21] B.A. Morris, J.D. Vansant, K. Hausmann, More structural strength for thin films, Kunstoffe Int. (2011) 42–46. May.

[22] ASTM International, Standard Test Method for Propagation Tear Resistance of Plastic Film and Thin Sheeting By Pendulum Method, ASTM International, West Conshohocken, PA, 2009 (D1922-09).

[23] International Organization for Standardization, Plastics – Film and Sheeting – Determination of Tear Resistance – Part 2: Elmendorf Method, International Organization for Standardization, Geneva, Switzerland, 1983 (ISO 6383-2).

[24] ASTM International, Standard Test Method for Tear-Propagation Resistance (Trouser Tear) of Plastic Film and Thin Sheeting By Single-Tear Method, ASTM International, West Conshohocken, PA, 2014 (D1938-14).

[25] International Organization for Standardization, Plastics – Film and Sheeting – Determination of Tear Resistance – Part 1: Trouser Tear Method, International Organization for Standardization, Geneva, Switzerland, 1983 (ISO 6383-1).

[26] ASTM International, Standard Test Method for Slow Rate Penetration Resistance of Flexible Barrier Films and Laminates, ASTM International, West Conshohocken, PA, 2010 (F1306-90(2008E1)).

[27] J. Lange, H. Mokdad, Y. Wyser, Understanding puncture resistance and perforation behavior of packaging laminates, J. Plastic Film. & Sheeting 18 (2002) 231–244.

[28] K. Hausmann, D. Fleiger, New Sealants and Peelable Film Concepts, in: Specialty Plastic Films '98, 14th Annual World Congress, Oct. 19–21, 1998, Dusseldorf, Germany, 1998.

[29] ASTM International, Standard Test Method for Impact Resistance of Plastic Film By The Free-Falling Dart Method, ASTM International, West Conshohocken, PA, 2009 (D1709-09).

[30] International Organization for Standardization, Plastics Film And Sheeting – Determination of Impact Resistance By The Free-Falling Dart Method, Part 1: Staircase Methods, International Organization for Standardization, Geneva, Switzerland, 1988 (ISO 7765-1).

[31] ASTM International, Standard Test Method for Total Energy Impact of Plastic Film By Dart Drop, ASTM International, West Conshohocken, PA, 2014 (D4272-14).

[32] International Organization for Standardization, Plastics Film and Sheeting – Determination of Impact Resistance by the Free-Falling Dart Method, Part 2: Instrumented Puncture Test, International Organization for Standardization, Geneva, Switzerland, 1994 (ISO 7765-2).

[33] ASTM International, Standard Test Method For Pendulum Impact Resistance of Plastic Film, ASTM International, West Conshohocken, PA, 2014 (D3420-14).

[34] ASTM International, Standard Test Method for High Speed Puncture Properties of Plastic Films Using Load and Displacement Sensors, ASTM International, West Conshohocken, PA, 2010 (D7192-10).

[35] B. Mukherjee, S. Biswas, P. Valavala, P. OConnell, P. Patton, R. Patel, Characterization of impact toughness of thin plastic films, in: 2020 SPE Virtual ANTEC Conference, 2020.

[36] A. Soever, M. Hopf, L. Frormann, Effects of lubrication in high speed falling dart testing of elastomer materials, Polym. Test. 30 (2011) 703–708.

[37] S. Goyal, V. Ker, T. Chuang, M. Gil, Measuring dart impact strength of polyethylene films, SPE ANTEC Conf. (2002) 2405–2509.

[38] A. Lustiger, J. Furmanaski, An anomaly in the drop dart testing of polyethylene film, SPE ANTEC Conf. (2019).

[39] L.A. Carbajal, R. Jio, D.M. Hahm, B.A. Morris, R.R. Kendzierski, Impact puncture resistance of multilayer flexible food packages, SPE ANTEC Conf. (2015).

[40] International Organization for Standardization, Paints and Varnishes – Determination of Scratch Resistance, Part 1: Constant-Loading Method, International Organization for Standardization, Geneva, Switzerland, 2011 (ISO 1518-1).

[41] Taber Machines Inc., Multi-finger scratch/mar tester. <http://www.testingmachines.com/pdf/10-45-multi-finger-scratch-tester.pdf>, n.d. (accessed 09.01.14).

[42] M. Wong, G.T. Lim, A. Moyse, J.N. Reddy, H.-J. Sue, A new test methodology for evaluating scratch resistance of polymers, Wear 256 (2004) 1214–1227.

[43] R.L. Browning, G.-T. Lim, A. Moyse, H.-J. Sue, H. Chen, J.D. Earls, Quantitative evaluation of scratch resistance of polymer coatings based on a standardized progressive load scratch test, Surf. & Coat. Technol. 201 (2006) 2970–2976.

[44] ASTM International, Standard Test Method for Evaluation of Scratch Resistance of Polymeric Coatings and Plastics Using An Instrumented Scratch Machine, ASTM International, West Conshohocken, PA, 2013 (D7027-13).

[45] B.A. Hare, A. Moyse, H.-J. Sue, Analysis of scratch-induced damages in multi-layer packaging film systems, J. Mater. Sci. 47 (2012) 1389–1398.

[46] B.A. Hare, H.-J. Sue, L.Y. Liang, P. Kinigakis, Scratch behavior of extrusion and adhesive laminated multilayer food packaging films, Polym. Eng. Sci. 54 (2014) 71–77.

[47] M. Hamdi, X. Zhang, H.-J. Sue, Fundamental understanding on scratch behavior of polymeric laminates, Wear 380–381 (2017) 203–216.

[48] L. Lin, G.S. Blackman, R.R. Matheson, A new approach to characterize scratch and mar resistance of automotive coatings, Prog. Org. Coat. 40 (2000) 85–91.

[49] L. Lin, G.S. Blackman, R.R. Matheson, Quantitative characterization of scratch and mar behavior of polymer coatings, Mater. Sci. Engr A317 (2001) 163–170.

[50] R. Consiglio, N.X. Randall, B. Bellaton, J. von Stebut, The nano-scratch tester (NST) as a new tool for assessing the strength of ultrathin hard coatings and the mar resistance of polymer films, Thin Solid. Films 332 (1998) 151–156.

[51] H. Blom, Flexible packaging durability studies, SPE FlexPackCon Conf. (2014).

[52] ASTM International, Standard Test Method for Abrasion Resistance of Flexible Packaging Films Using a Reciprocating Weighted Stylus, ASTM International, West Conshohocken, PA, 2018 (F3300-18).

[53] ASTM International, Standard Practice for Conditioning Flexible Barrier Materials For Flex Durability, ASTM International, West Conshohocken, PA, 2011 (F392/F392M-11).

[54] Campus website, <http://www.campusplastics.com/> (accessed 05.09.20).

[55] Prospector website, <http://www.ulprospector.com/en/na> (accessed 05.09.20).

[56] Matweb website, <http://matweb.com/> (accessed 30.08.2020).

[57] Plastics Technology website, <http://www.ptonline.com/> (accessed 05.09.20).

[58] A. Siegmann, Y. Nir, Structure-property relationships in blends of linear low- and conventional low-density polyethylene as blown films, Polym. Engg. Sci. 27 (1987) 1182–1186.

[59] F. Martinez, G.E. Mazuera, D.R. Parikh, Comparative analysis of thick gauge blended and coextruded polyethylene films, J. Plastic Film. & Sheeting 6 (1990) 44–62.

[60] C.M. Beagan, G.M. McNally, W.R. Murphy, The blending and coextrusion of metallocene catalyzed polyethylene in blown film applications, J. Plastic Film. & Sheeting 15 (1999) 329–340.

[61] S. Elkoun, M.A. Huneault, X.M. Zhang, K. McCormick, F. Puterbaugh, L. Kale, LLDPE blown films property enhancement through coextrusion, SPE ANTEC Conf. (2003) 1396–1400.

[62] T.I. Butler, B.A. Morris, PE-based multilayer film structures, in: S. Ebnesajjad (Ed.), Plastic Films In Food Packaging, Materials, Technology and Applications, Elsevier, New York, 2013, pp. 21–52.

[63] N. Borse, N. Aubee, Developing cost effective co-extruded film structures for flexible packaging, SPE ANTEC Conf. (2014) 1304–1308.

[64] J.E. Guillotte, Co-extrusion of blown film laminates, SPE J. 25 (1969) 20–24.

[65] V. Gaucher-Miri, G.K. Jones, R. Kaas, A. Hiltner, E. Baer, Plastic deformation of EVA, EVOH and their multilayers, J. Mater. Sci. 37 (2002) 2635–2644.

[66] D. Timm, How sealant film choice can affect a laminated structure's properties, TAPPI PLACE. Conf. (2005).

[67] R.-Y. Wu, L.D. McCarthy, Z.H. Stachurski, Tearing resistance of multi-layer plastic films, Intern. J. Fract. 68 (1994) 141–150.

[68] C.-H. Huang, J.-S. Wu, C.-C. Huang, Predicting the permeability and tensile properties of multilayer films from properties of individual component layers, Polym. J. 36 (2004) 386–393.

[69] W.J. Schrenk, T. Alfrey Jr., Some physical properties of multilayered films, Polym. Engr. Sci. 9 (1969) 393–399.

[70] K.H. Hu, Measurements on the resistance of flexible packaging materials to puncture, abrasion, and flexure and the relationship of these measurements to the performance of packages subjected to conditions causing pinhole formation, U.S. Army Natick Laboratories, Technical Report 73-32-GP, 1973.

[71] B.A. Morris, Application of modeling to speed flexible package development, J. Plastic Film & Sheeting 29 (2013) 249–270.

[72] R. Armstrong, E. Chow, Improving performance of flexible barrier packaging for liquids through the use of modified ethylene vinyl alcohol (EVOH), in: SPE Polyolefins and FlexPack Conference, Houston, TX, February 2011.

[73] E. Chow, R. Armstrong, Improving the performance of bulk packaging of liquids through the use of modified ethylene vinyl alcohol (EVOH), in: SPE Polyolefins and FlexPack Conference, Houston, Tx, February 2012.

[74] R. Baldizonne, G. Carianni, P. Mariani, A new Ziegler-Natta octene LLDPE for high performance blown film applications: benefits of coextruded vs. blended LDPE/LLDPE films, Eur. TAPPI PLACE. Conf. (2005).

[75] C.M. Beagan, G.M. McNally, W.R. Murphy, The blending and coextrusion of metallocene catalyzed polyethylene in blown film applications, SPE ANTEC Conf. (1999).

[76] J. Taylor, J.J. Baik, Benefits of coextruded LLDPE/LDPE film vs. blended LLDPE/LDPE film, J. Plastic Film. & Sheeting 16 (2000) 223–235.

[77] X. Zhang, A. Ajji, Polypropylene-polyethylene multilayer films, SPE ANTEC Conf. (2005) 2947–2951.

[78] X. Zhang, A. Ajji, Oriented structure of PP/LLDPE multilayer and blends films, Polymer 46 (2005) 3385–3393.

[79] R.M. Jones, Mechanics of Composite Materials, McGraw-Hill Book Co, New York, 1975.

[80] R.M. Patel, T.P. Karjala, N.R. Savargaonkar, P. Salibi, L. Liu, Fundamentals of structure-property relationships in blown films of LLDPE/LDPE blends, J. Plastic Film. Sheeting 35 (2019) 401–421.

[81] H.D. Keith, F.J. Padden Jr., R.G. Vadimsky, Intercrystalline links in polyethylene crystallized from the melt, J. Polym. Sci., Part. A-2 (Polym. Phys.) 4 (1966) 267–281.

[82] F. Sadeghi, A. Ajji, Application of single site catalyst metallocene polyethylenes in extruded films: effect of molecular structure on sealability, flexural cracking and mechanical properties, Can. J. Chem. Engr. 92 (2014) 1181–1188.

[83] D. Darby, K. Cooksey, R. Kaas, P. Gerard, Cheruvathur, Studying the relationship of adhesive bond strengths and puncture resistance of flexible packaging laminations, SPE FlexPackCon Conf. (2014) 37–43.

[84] R. Cheruvathur, The effect of bond strength of flexible laminates on puncture resistance, Masters Thesis, Clemson University, All Theses, 594. <https://tigerprints.clemson.edu/cgi/viewcontent.cgi?article = 1594&context = all_theses>, 2009 (accessed 06.09.20).

[85] R.J. Seyler, Slow rate penetration of packaging films, J. Plastic Film. Sheeting 6 (1990) 191–224.

[86] F.J. Furno, R.S. Webb, N.P. Cook, High speed puncture testing of thermoplastics, J. Appl. Polym. Sci. 8 (1964) 101–110.

[87] M. Wong, A. Moyse, F. Lee, H.-J. Sue, Study of surface damage of polypropylene under progressive load, J. Mater. Sci. 39 (2004) 3293–3308.

[88] V. Jardret, H. Zahouani, J.L. Loubet, T.G. Mathia, Understanding and quantification of elastic and plastic deformation during a scratch test, Wear 218 (1998) 8–14.

[89] H. Jiang, R. Browning, J. Whitcomb, M. Ito, M. Shimouse, T. Chang, et al., Mechanical modeling of scratch behavior of polymer coatings on hard and soft substrates, Tribol. Lett. 37 (2010) 159–167.

[90] T.-W. Seo, J.-I. Weon, Influence of weathering and substrate roughness on the interfacial adhesion of acrylic coating based on an increasing load scratch test, J. Mater. Sci. 47 (2009) 2234–2240.

[91] A.J. Kinloch, R.J. Young, Fracture Behavior of Polymers, Elsevier Applied Science Publishers, London, 1985.

[92] B.-H. Choi, A. deGroot, R. Patel, M. Demirors, K. Anderson, Investigation of tear properties of polyethylene blown and compression molded films, SPE ANTEC Conf. (2011).
[93] A.C. Chang, T. Inge, L. Tau, A. Hiltner, E. Baer, Tear strength of ductile polyolefin films, Polym. Eng. Sci. 42 (2002) 2202–2212.
[94] A. Keller, M.J. Machin, Oriented crystallization polymers, J. Macromol. Sci. (Phys.) B1 (1967) 41–91.
[95] W.F. Maddams, J.E. Preddy, X-ray diffraction orientation studies on blown film polyethylene films. I. Preliminary measurements, J. Appl. Polym. Sci. 22 (1978) 2721–2737.
[96] W.F. Maddams, J.E. Preddy, X-ray diffraction orientation studies on blown film polyethylene films. II. Measurements on films from a commercial blowing unit, J. Appl. Polym. Sci. 22 (1978) 2739–2749.
[97] W.F. Maddams, J.E. Preddy, X-ray diffraction orientation studies on blown film polyethylene films. III. High-stress crystallization orientation, J. Appl. Polym. Sci. 22 (1978) 2751–2759.
[98] P.H. Lindenmeyer, S. Lustig, Crystalline orientation in extruded polyethylene film, J. Appl. Polym. Sci. 9 (1965) 227–240.
[99] T.H. Kwak, C.D. Han, M.E. Vickers, Development of crystalline structure during tubular film blowing of low-density polyethylene, J. Polym. Sci. 35 (1988) 363–389.
[100] A. Haber, M.R. Kamal, Morphology and orientation in polyethylene tubular blown films, in: SPE ANTEC Conf., 1997, pp. 446–450.
[101] M. Gilbert, D.A. Hemsley, S.R. Patel, Effect of processing conditions on the orientation and properties of polyethylene blown film, Bristsh Polym. J. 19 (1987) 9–23.
[102] A. Gupta, D.M. Simpson, I.R. Harrison, Crystalline morphology in oriented HDPE blown films, Plast. Eng. (1993) 33–36. November 1993.
[103] S. Fatahi, A. Ajji, P.G. Lafleur, Structure-property correlations for PE blown films, SPE ANTEC Conf. (2005) 1766–1770.
[104] R.M. Patel, T.I. Butler, K.L. Walton, G.W. Knight, Investigation of processing-structure-properties relationships in polyethylene blown films, Polym. Engg. Sci. 34 (1994) 1506–1514.
[105] D. Godshall, G. Wilkes, R.K. Krishnaswamy, A.M. Sukhadia, Processing-structure-property investigation of blown HDPE films containing both machine and transverse direction oriented lamellar stacks, Polymer 44 (2003) 5397–5406.
[106] A.M. Sukhadia, The effects of molecular structure, rheology, morphology and orientation of polyethylene blown film properties, SPE ANTEC Conf. (1998).
[107] X. Zhang, V. Jean-Marie, A. Ajji, Processing-structure-property relationships in multilayer films, in: 59th Annual Technical Conference - Society of Plastics Engineers, Vol. 2, 2001, pp. 1456–1460.
[108] L.-B.W. Lee, R.A. Register, D.M. Dean, Origin of directional tear in blown films of ethylene copolymers, Polym. Mater. Sci. Eng. 83 (2000) 482–483.
[109] L.-B.W. Lee, R.A. Register, D.M. Dean, Origin of directional tear in blown films of ethylene/methacrylic acid copolymers and ionomers, J. Polym. Sci. Part. B: Polym. Phys. 43 (2005) 97–106.
[110] X.M. Zhang, S. Elkoun, A. Ajji, M.A. Huneault, Oriented structure and anisotropy properties of polymer blown films: HDPE, LLDPE and LDPE, Polymer 45 (2004) 217–229.
[111] X.M. Zhang, S. Elkoun, A. Ajji, M.A. Huneault, Effect of crystalline structure on tear resistance of LDPE and LLDPE blown films, in: 61st Annual Technical Conference - Society of Plastics Engineers, Vol. 2, 2003, pp. 1376–1380.
[112] R. Patel, T. Hermel-Davidock, R. Paradkar, B. Landes, L. Liu, M. Demirors, et al., Characterization of amorphous and crystalline orientation in polyethylene films, SPE ANTEC Conf. (2006) 612–616.
[113] L.-B.W. Lee, R.A. Register, D.M. Dean, Tear anisotropy in films blown from polyethylenes of different molecular architectures, J. Polym. Sci., Part. B: Polym. Phys. 43 (2005) 413–420.
[114] X.M. Zhang, A. Ajji, M.A. Huneault, Control of blown film tear strength anisotropy through blending of LDPE and LLDPE, in: 61st Annual Technical Conference - Society of Plastics Engineers, Vol. 2, 2003, pp. 1391–1395.
[115] K. Mezghani, S. Furquan, S.H. Tabatabaei, A. Ajji, Effect of blend ratio of h-LLDPE and LDPE on tear properties of blown films, Int. Polym. Process. 27 (3) (2012) 392–398.
[116] N.D. Huck, P.L. Clegg, The effect of extrusion variables on the fundamental properties of tubular polyethylene films, SPE ANTEC Conf. (1961) 18. -2.
[117] D. Meilhon, H.F. Wollak, Shrinkable thermoplastic packaging films, U.S. Patent 5,356,677, 1994.
[118] D. Meilhon, Ready-tear packaging, U.S. Patent 5,556,674, 1996.
[119] H.F. Wollak, B. Rioux, Thermoplastic film for packaging, U.S. Patent 6,572,979 B1, 2003.
[120] R. Longworth, The physical structure of Surlyn® ionomer resins, DuPont Intern. Rep. (1967).
[121] C.L. Rohn, Tear strength theory of semicrystalline polymer, J. Appl. Polym. Sci. 28 (1983) 3011.
[122] N.J. Otten, E.J. Presnell, Polymer film having linear tear properties, U.S. Patent 4,098,406, 1978.
[123] K. Dunno, Effects of transportation hazards on high barrier flexible packaging films, J. Appl. Packaging Res. 9 (2016). Article 1.
[124] K. Dunno, Effects of transportation hazards on package performance and food product shelf life, Clemson University PhD thesis, All Dissertations, Paper 1432. <https://tigerprints.clemson.edu/all_dissertations/1432/>, 2014 (accessed 04.09.20).
[125] H.A. Sundell, B. Holen, F. Nicolaysen, C. Hilton, O. Lokkeberg, Bag-in-box packaging for wine: Analysis of transport stress in barrier films, Packaging Technol. Sci. 5 (1992) 321–329.
[126] K. Dunno, W. Whiteside, R. Thomas, K. Cooksey, Effect of headspace volume of retort pouches on simulated transport hazards, Int. J. Adv. Packaging Technol. 3 (2015) 138–146.

[127] G. Doyon, A. Clement, S. Ribereau, G. Morin, Canadian bag-in-box wine under distribution channel abuse: material fatigue, flexing simulation and total closure/spout leakage investigation, Packaging Technol. Sci. 18 (2005) 97–106.

[128] R.K. Krishnaswamy, Q. Yang, L. Fernandez-Ballester, J.A. Kornfield, Effect of the distribution of short-chain branches on crystallization kinetics and mechanical properties of high-density polyethylene, Macromolecules 41 (2008) 1693–1704.

[129] X. Lu, N. Ishikawa, N. Brown, The critical molecular weight for resisting slow crack growth in a polyethylene, J. Polym. Sci. B Polym. Phys. 34 (1996) 1809–1813.

[130] D. Ferrari, Barrier packaging in the real world, Celplast Metallized Products. <http://www.celplast.com/packaging-insight/barrier-packaging-real-world-2/> (accessed 31.08.2020).

[131] G.M. Cumberbatch, R.R. Reitz, Polyvinyl alcohol coated polyester barrier films, in: TAPPI Polymers, Laminations & Coatings Conference, San Francisco, Vol. 2, 1998, pp. 913–926.

[132] DuPont, Properties of Surlyn ionomer resin for flexible packaging, DuPont product. Lit. (1978).

[133] A.I. Fetell, Nylon-containing ionomer improves the performance of ethylene vinyl alcohol (EVOH) copolymers in multilayer structures,", TAPPI Polymers, Laminations & Coat. Conf. (1997) 503–505.

[134] A.I. Fetell, Ethylene vinyl alcohol copolymer blends, WO 1997009380 A1, 1997.

[135] K.J. Shelat, N.K. Dutta, N.R. Choudhury, Interfacial interaction and morphology of EVOH and ionomer blends by scanning electron microscopy and its correlation to barrier characteristics, Langmuir 24 (2008) 5464–5473.

[136] G. Medlock, Dolgovskij, Barrier performance of nanolayer EVOH film, TAPPI PLACE. Conf. (2012).

[137] P.C. Lee, J. Dooley, J. Robacki, S. Jenkins, R. Wrisley, Flex barrier property enhancement in film structures using microlayer coextrusion technology, SPE ANTEC Conf. (2013).

[138] M. Nottez, S. Chaki, J. Soulestin, M.F. Lacrampe, P. Krawczak, Mechanical behavior and essential work of fracture of starch-based blown films, Int. Polym. Process. Soc. Conf. 30 (2014) S07–S597. pdf.

[139] J.S.S. Wong, D. Ferrer-Balas, R.K.Y. Li, Y.-W. Mai, M.L. Maspoch, H.-J. Sue, On tearing of ductile polymer films using the essential work of fracture (EWF) method, Acta Mater. 51 (2003) 4929. -2939.

[140] A.B. Martínez, A. Segovia, J. Gamez-Perez, M.L. Maspoch, Essential work of fracture analysis of the tearing of a ductile polymer film, Eng. Fract. Mech. 77 (2010) 2654–2661.

[141] T. Bárány, T. Czigány, J. Karger-Kocsis, Application of the essential work of fracture (EWF) concept for polymers, related blends and composites: a review, Prog. Polym. Sci. 35 (2010) 1257–1287.

[142] Y.-W. Mai, B. Cotterell, On the essential work of ductile fracture in polymers, Int. J. Fract. 32 (1986) 105–125.

[143] M. Rennert, M. Nase, R. Lach, K. Reincke, S. Arndt, R. Androsch, et al., Influence of low-density polyethylene blown film thickness on the mechanical properties and fracture toughness, J. Plastic Film. & Sheeting 29 (2013) 327–346.

[144] M.L. Maspoch, J. Gamez-Perez, A. Gordillo, M. Sanchez-Soto, J.I. Velasco, Characterization of injected EPBC plaques using essential work of fracture (EWF) method, Polymer 43 (2002) 4177–4183.

[145] D. Ferrer-Balas, M.L. Maspoch, A.B. Martinez, E. Ching, R.K.Y. Lie, Y.-W. Mai, Fracture behavior of polypropylene films at different temperatures: assessment of the EWF parameters, Polymer 42 (2001) 2665–2674.

[146] M.L. Maspoch, V. Henault, D. Ferrer-Balas, J.I. Velasco, O.O. Santana, Essential work of fracture on PET films: influence of the thickness and the orientation, Polym. Test. 19 (2000) 559–568.

[147] R. Lach, K. Schneider, R. Weidisch, et al., Application of the essential work of fracture concept to nanostructured polymer materials, Eur. Polym. J. 41 (2005) 383–392.

[148] C.-F. Lee, H.-J. Sue, D.M. Fiscus, Effect of processing parameters on essential work of fracture toughness of LLDPE blown films, Polym. Eng. & Sci. 55 (2015) 2403–2413.

[149] B.-H. Choi, M. Demirors, R.M. Patel, W.A. de Groot, K.W. Anderson, V. Juarez, Evaluation of the tear properties of polyethylene blown films using the essential work of fracture concept, Polymer, 51, 2732–2739.

[150] I. Kim, B.-H. Choi, Evaluation of fracture characteristics of polyethylene blown films using the essential work of fracture (EWF) with variable film orientations, in: 70th Annual Technical Conference of the Society of Plastics Engineers, Vol. 1, 2012, pp. 803–808.

[151] M.M. Hossain, C.-F. Lee, D.M. Fiscus, H.-J. Sue, Physical assessment of essential work of fracture parameters based on m-LLDPE blown films, Polymer 96 (2016) 104–111.

[152] P. Gupta, G.L. Wilkes, A.M. Sukhadia, R.K. Krishnaswamy, S. Wharry, T. Mansfield, Does the length of the short chain branch affect the mechanical properties of linear low density polyethylenes? An investigation based on films of copolymers of ethylene/1-butene, ethylene/1-hexene & ethylene/1-octene synthesized by a single site metallocene catalyst, PMSE Prepr. 91 (2004) 378–379.

[153] D. Fiscus, G. Gururajan, X. Chen, J. Schaefer, S. Yakovlev, K.R. Chen, Essential work of fracture of m-LLDPE blown films, in: Abstracts of Papers, 256th ACS National Meeting & Exposition, Boston, MA, United States, August 19–23, 2018, PMSE-689.

[154] H.G. Schirmer, For extrusion of resin to form modular assembly of thin disks, U.S. Patent 5762971 A, 1998.

[155] H.G. Schirmer, Modular disk coextrusion die, U.S. Patent 6413595, 2002.

[156] J. Kerns, A. Hsieh, A. Hiltner, E. Baer, Mechanical behavior of polymer microlayers, Macromol. Symp. 147 (1999) 15–25.

[157] T.E. Bernal-Lara, A. Ranade, A. Hiltner, E. Baer, Nano-and microlayered polymers: structure and properties, in: G.H. Michler, F.J. Balta-Calleja (Eds.), Mechanical Properties of Polymers Based on Nanostructure and Morphology, Taylor & Francis, Baco Raton, 2005, pp. 629–681.

[158] A. Hiltner, S. Nazarenko, T. Ebeling, T. Schuman, E. Baer, Novel microlayered structures for adhesion and interdiffusion studies, Polymeric Mater. Sci. Eng. 75 (1996) 471–472.

[159] C. Mueller, J. Kerns, T. Ebeling, S. Nazarenko, A. Hiltner, E. Baer, Microlayer coextrusion: processing and applications, Polym. Process. Eng. 97 (1997) 137.

[160] E. Baer, D. Jarus, A. Hiltner, Microlayer coextrusion technology, SPE ANTEC Conf. (1999).

[161] T.E. Bernal-Lara, R. Masirek, A. Hiltner, E. Baer, E. Piorkowska, A. Galeski, Morphological studies of multilayed HDPE/PS systems, J. Appl. Polym. Sci. 99 (2006) 597–612.

[162] T.E. Bernal-Lara, R.Y.F. Liu, A. Hiltner, E. Baer, Structure and thermal stability of polyethylene nanolayers, Polymer 46 (2005) 3043–3055.

[163] S.J. Pan, J. Im, M.J. Hill, A. Keller, A. Hiltner, E. Baer, Structure of ultrathin polyethylene layers in multilayer films, J. Polym. Sci., Part. B: Polym. Phys. 28 (1990) 1105–1119.

[164] F.J. Balta Calleja, F. Ania, I.P. Orench, E. Baer, A. Hiltner, T. Bernal, et al., Nanostructure development in multilayer polymer systems as revealed by X-ray scattering methods, Progr Colloid Polym. Sci. 130 (2005) 140–148.

[165] Y. Jin, M. Rogunova, A. Hiltner, E. Baer, R. Nowacki, A. Galeski, et al., Structure of polypropylene crystallized in confined nanolayers, J. Polym. Sci., Part. B: Polym. Phys. 42 (2004) 3380–3396.

[166] D. Langhe, M. Ponting, (Eds.), Manufacturing and Novel Applications of Multilayer Polymer Films, William Andrew Publishing, pp. 46–116.

Chapter 10

Adhesion in multilayer flexible packaging

Multilayer films require adhesion between the layers for the structure to perform as a whole. Many combinations of polymers and substrates used in multilayer flexible packaging do not adhere well to one another. Adhesive and tie resin technology is used to enhance the adhesion between such layers. This chapter provides an introduction to the adhesion of multilayer films—how adhesion is measured and the fundamental science and engineering of bonding at interfaces. The chapter concludes with a guide to tie layer technology for extrusion coating/lamination and coextrusion processes as well as a brief introduction to laminating adhesive technology.

10.1 Why adhesion is important

The concept of combining different materials into a common film structure rests on the premise that the layers will adhere to one another. Without sufficient adhesion, the layers will delaminate and not provide the function required. Tensile strength, stiffness and other physical properties of the film require good interlayer adhesion. Puncture, impact, tear and scratch resistance are influenced by the transfer of stress across the interface, which is influenced by adhesion. The question becomes, how much adhesion is sufficient? In some snack packaging structures, 50 g/25 mm is deemed sufficient. In some rigid packaging applications, 4000 g/25 mm may be specified. The requirement varies with the film structure and application. Even within a film structure, the adhesion requirement between some layers may be more demanding than others. For example, in a barrier film structure, the adhesion specification for layers on the outside of the barrier layer may be lower than for those on the inside; layer delamination between the barrier and the product may lead to loss of protection (e.g., oxygen or moisture ingress) whereas delamination of outer layers may only cause esthetic issues.

In some applications "inseparable" adhesion is not the best adhesion. Instead, controlling the adhesion within a certain range may be critical. For example, delamination peel-seal technology (see Section 7.7) relies on controlled adhesion between the sealant layer and an adjacent layer to provide an easy-open feature. In some cases, films may provide better resistance to tear or puncture if their layers delaminate during the event. In other cases, a film may be laminated onto a substrate to protect it during transportation but must be easily stripped off for retail use; adhesion in such cases is required to be low.

Adhesion depends on the fabrication method, the chemistry of the layers and the relative configuration of the layers to each other. If the multilayer film is fabricated from pre-made films, these films will require an adhesive or glue to bond them to one another. This is typically accomplished using an adhesive or extrusion lamination process. In adhesive lamination, a low viscosity adhesive is applied to one or more of the substrates and subsequently dried or cured. The adhesive may be solvent borne, waterborne or solventless. These are described in Section 10.5. In extrusion lamination, molten polymer bonds to the substrate whose strength increases upon quenching to the solid-state. This technology is explored in Section 10.4.1. An alternative to laminating pre-made films is to coextrude the layers in the melt. Specialized tie resins for bonding dissimilar layers in coextrusion are taken up in Section 10.4.2.

The fundamentals of adhesion, including thermodynamics, wetting, diffusion and chemical interaction are explored in Section 10.3. In adhesive lamination, wetting and web tension control are critical. When bonding a polymer melt to a solid substrate, as in extrusion coating and lamination, in addition to the chemistry of the polymer and substrate, the coating parameters such as temperature, line speed, air gap and drawdown are important. In coextrusion blown or cast film, temperature, contact time, and quench conditions are also important. These processing effects are explored briefly in Sections 10.4 and 10.5 and in greater detail in Chapter 15.

Many combinations of resins used to make coextrusion films for flexible packaging do not naturally bond to one another. This is illustrated in Table 10.1 where combinations that have poor adhesion are represented by a square. For these combinations, tie resins have been developed to provide adhesion; the tie resin is sandwiched between the two

The Science and Technology of Flexible Packaging. DOI: https://doi.org/10.1016/B978-0-323-85435-1.00011-9

TABLE 10.1 Adhesion of various polymers in coextrusion.

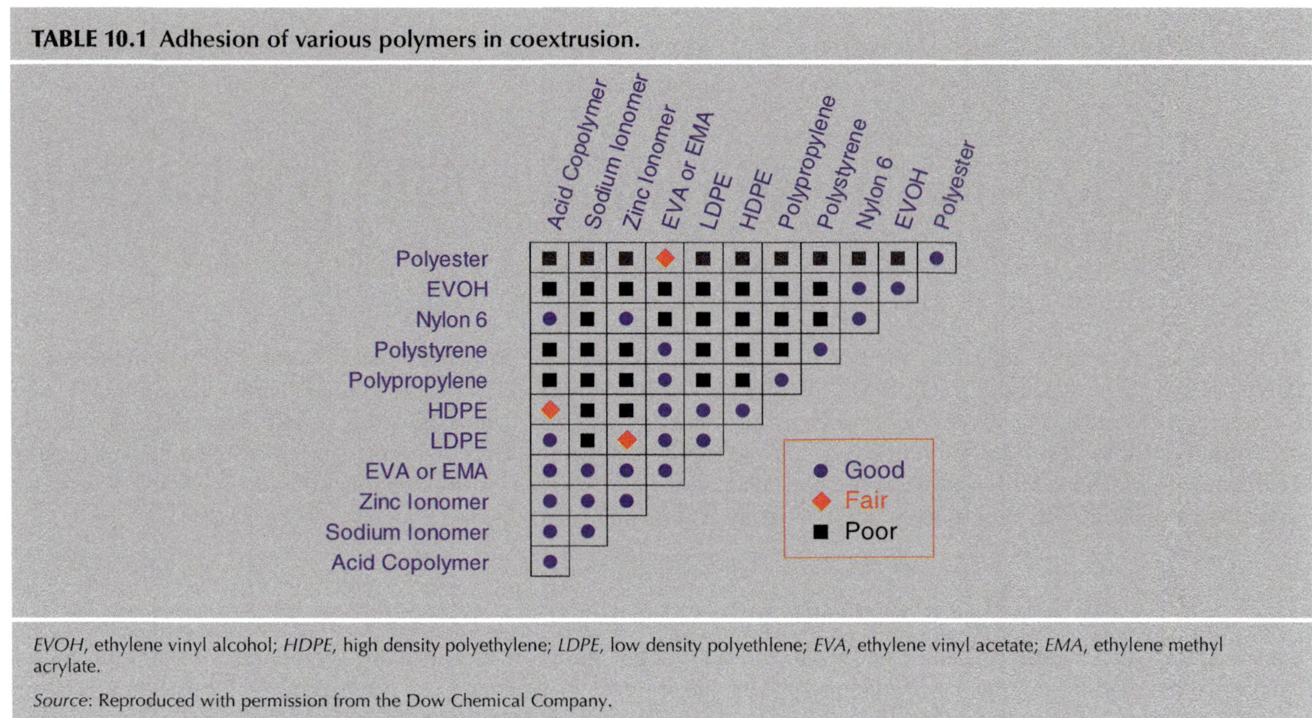

	Acid Copolymer	Sodium Ionomer	Zinc Ionomer	EVA or EMA	LDPE	HDPE	Polypropylene	Polystyrene	Nylon 6	EVOH	Polyester
Polyester	■	■	■	◆	■	■	■	■	■	■	●
EVOH	■	■	■	■	■	■	■	■	●	●	
Nylon 6	●	■	●	■	■	■	■	■	●		
Polystyrene	■	■	■	●	■	■	■	●			
Polypropylene	■	■	■	●	■	■	●				
HDPE	◆	■	■	●	●	●					
LDPE	●	■	◆	●	●						
EVA or EMA	●	●	●	●							
Zinc Ionomer	●	●	●								
Sodium Ionomer	●	●									
Acid Copolymer	●										

● Good
◆ Fair
■ Poor

EVOH, ethylene vinyl alcohol; *HDPE*, high density polyethylene; *LDPE*, low density polyethlene; *EVA*, ethylene vinyl acetate; *EMA*, ethylene methyl acrylate.

Source: Reproduced with permission from the Dow Chemical Company.

incompatible layers. Tie layer technology is described in Section 10.4. While the tie layer technology is critical, other factors including the structure design (what is being bonded to what, layer thicknesses), the fabrication process, and the end-use influence adhesion in coextrusion.

Polymer coatings onto solid substrates also depend on the chemistry. Oxidation of low-density polyethylene (LDPE) is often used to impart enough polarity to the PE to achieve adhesion to polar substrates such as paper and aluminum foil. This is not sufficient to obtain good bonds to oriented polyethylene terephthalate (OPET) or oriented polypropylene (OPP), as is highlighted in Table 10.2. Acrylates, vinyl acetate, acid, anhydride and other functionalities can be incorporated into the polymer, through copolymerization or grafting, to enhance adhesion to specific substrates. Even with such functionality, additional methods may be needed to achieve good adhesion. These include ozone treatment of the melt curtain (especially for LDPE); corona, ozone, flame, or plasma treatment of the substrate; and the use of chemical primers on the substrate.

10.2 How to measure adhesion

10.2.1 Peel test

The peel test is the most commonly used method for measuring adhesion in flexible packaging. The peel test comes in a variety of forms, varying in angle and apparatus. Mount [1] counts 23 different procedures listed in the ASTM index, plus another 40 under the term "adhesive bonding." Other organizations, such as the Association of Industrial Metallizers, Coaters and Laminators (AIMCALs) have additional procedures. Perhaps the most often used procedure is the "T-peel" test described in ASTM D1876 [2]. In this procedure, a rectangular specimen 25.4 mm by 305 mm is cut from the film. The layers are separated at the interface where the adhesion is to be measured. The separated ends become what is known as the peel arms. These peel arms are clamped in the jaws of a tensile tester (see Fig. 10.1) and pulled apart at a constant speed. The force to pull the peel arms apart is recorded. Peel strength is reported as the peel force divided by the specimen width, such as g/25 mm. Conversion factors for commonly used units are given in Table 10.3. The mode of failure is also recorded, along with any descriptors as to how the specimen failed. These may include terms such as a cohesive, adhesive, partial tear, etc. These descriptors can be important for determining why adhesion may have changed from one measurement or structure to another.

Delaminating the sample at the desired interface for measuring peel strength can be challenging. Patience is a virtue. Some techniques that have been employed include using tweezers or pliers and a magnifying glass, scoring the surface

TABLE 10.2 Adhesion of various polymers (extrudates) to solid packaging substrates in extrusion coating and lamination.

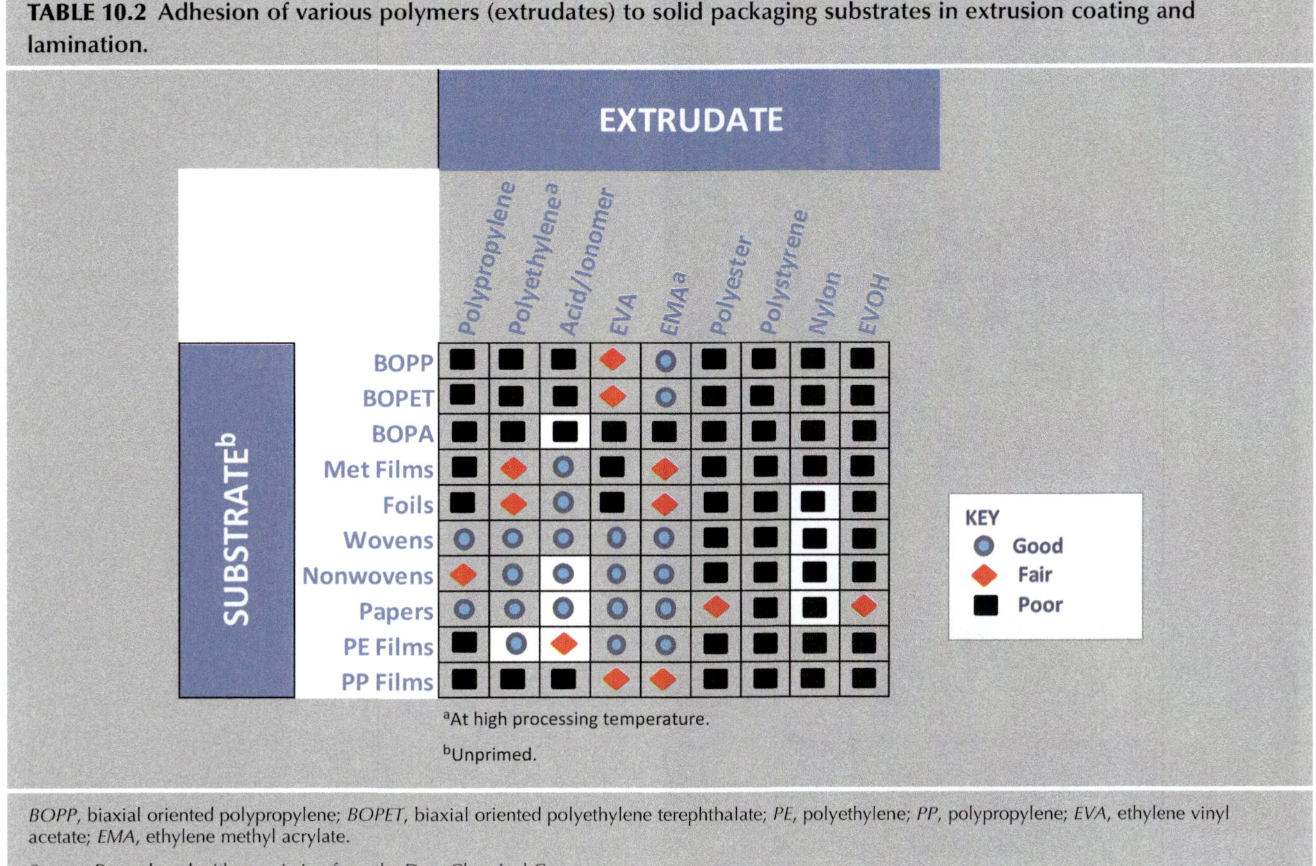

SUBSTRATE[b] / EXTRUDATE	Polypropylene	Polyethylene[a]	Acid/Ionomer	EVA	EMA[a]	Polyester	Polystyrene	Nylon	EVOH
BOPP	Poor	Poor	Poor	Fair	Good	Poor	Poor	Poor	Poor
BOPET	Poor	Poor	Poor	Fair	Good	Poor	Poor	Poor	Poor
BOPA	Poor	Poor	Poor	Poor	Poor	Poor	Poor	Poor	Poor
Met Films	Poor	Fair	Good	Poor	Fair	Poor	Poor	Poor	Poor
Foils	Poor	Fair	Good	Poor	Fair	Poor	Poor	Poor	Poor
Wovens	Good	Good	Good	Good	Good	Poor	Poor	Poor	Poor
Nonwovens	Fair	Good	Good	Good	Good	Poor	Poor	Poor	Poor
Papers	Good	Good	Good	Good	Good	Fair	Poor	Poor	Fair
PE Films	Poor	Good	Fair	Good	Good	Poor	Poor	Poor	Poor
PP Films	Poor	Poor	Poor	Fair	Fair	Poor	Poor	Poor	Poor

KEY: Good (circle); Fair (diamond); Poor (square)

[a]At high processing temperature.

[b]Unprimed.

BOPP, biaxial oriented polypropylene; *BOPET*, biaxial oriented polyethylene terephthalate; *PE*, polyethylene; *PP*, polypropylene; *EVA*, ethylene vinyl acetate; *EMA*, ethylene methyl acrylate.

Source: Reproduced with permission from the Dow Chemical Company.

of the film with a razor blade and carefully stretching the specimen [3], dipping the ends in solvent and heat sealing. One must be especially careful when using the solvent technique to ensure the solvent has not affected the adhesion or changed the physical properties of the peel arm by swelling or other means. Once separated, it is helpful to check to make sure separation has started at the target interface. Fourier transform infrared (FTIR) spectroscopy is one method to determine which layer has been exposed. For ethylene vinyl alcohol (EVOH) or polyamide (PA) layers, staining with a texture of iodine can be used as a quick test. Iodine turns these polymers brown.

The angle, peel speed, specimen width, and temperature all influence the peel strength result. ASTM D1876 calls for a peel speed of 254 mm/min (10 inches/min). Other commonly used test speeds include 50.8 mm/min (2-inches/min) and 25.4 mm/min (1-inch/min). Specimen width is 25.4 mm in the standard, but 15 and 10 mm are often used, especially in Europe and Japan. Standard conditions of 23°C and 50% RH are specified in the standard.

Converting results based on different specimen widths is straightforward; the peel strength varies linearly with width. For example, if a result is reported as 100 g/15 mm, the result for 25 mm width should be proportionally higher: 100 g/15 mm × 25/15 = 167 g/25 mm.

Results run at different test speeds and temperatures cannot be directly compared. The effect of these parameters on peel strength is examined in detail in Section 10.2.3. The peel angle is another important parameter. ASTM D1876 calls for placing the peel arms into the tensile tester jaws; the two peel arms form a 180 degrees angle relative to one another. The actual angle between the peel arms and the substrate is not controlled; it is allowed to establish whatever angle is dictated by the forces on the peel arms and substrates. The actual peel angle is important since it affects the amount of energy expended during the pull to bend and stretch the peel arms. The angle influences the concentration of stress as the peel front moves through the specimen. In scientific investigations of peel strength, the angle is typically controlled. This may be done simply by manually holding the peel tab during the pull so that it maintains a 90 degrees angle with the peel arms (a true "T"). A variation is to hold or mount the specimen so that the peel tab forms a 180 degrees angle

FIGURE 10.1 Photo of peel test. Picture by Vladimir Antonov. *Reproduced with permission from the Dow Chemical Company.*

TABLE 10.3 Unit conversion factors for peel strength.

	g/in	g/25 mm	g/15 mm	lb/in	N/25 mm	N/15 mm	J/m^2 (N/m)
g/in	1	1	0.6	0.0022	0.0098	0.0059	0.392
g/25 mm	1	1	0.6	0.0022	0.0098	0.0059	0.392
g/15 mm	1.67	1.67	1	0.0037	0.0163	0.0098	0.654
lb/in	454	454	272.4	1	4.44	2.67	178
N/25 mm	102	102	61.2	0.225	1	0.6	40
N/15 mm	170	170	102	0.375	1.67	1	66.7
J/m^2 (N/m)	2.55	2.55	1.53	0.0056	0.025	0.015	1

Note: Multiply by the factor to convert from the unit in the first column to the corresponding unit in the first row. For example, 1 g/25 mm = 0.0098 N/ 25 mm.

with one of the peel arms. This is known as a 180 degrees peel test. These configurations are illustrated in Fig. 10.2. Special sleds have been designed to explore other angles so that as the peel progresses, the angle remains constant. A variation on this is the German wheel, usually used in rigid structures, described in DIN 53357 [4]. Here the specimen is mounted to the circumference of a large wheel that freely rotates. The peel arm is clamped to the jaw of the tensile tester above the wheel. As the jaw pulls the peel arm from the substrate, the wheel rotates, keeping the angle at a constant 90 degrees.

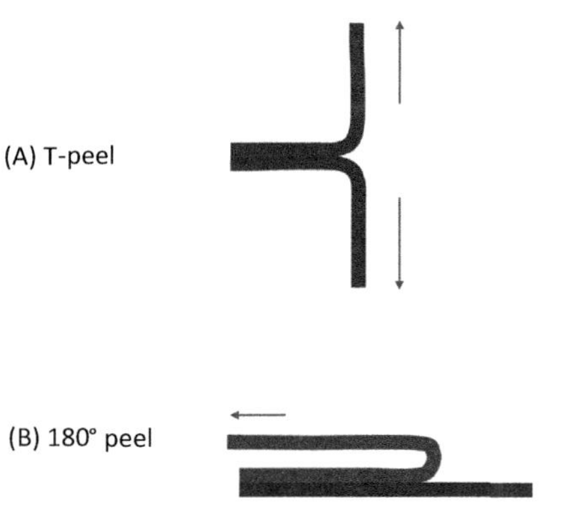

FIGURE 10.2 Peel angle configurations during peel tests. (A) T-peel configuration, (B) 180 degrees configuration.

10.2.2 Adhesion of coatings

Adhesion of thin coatings or ink on films is difficult to measure using a peel test. Various tape tests have been developed [5–7], where a pressure-sensitive tape is applied to the surface of the coating. When the tape is removed, the amount of area of the coating that is pulled from the substrate is recorded. A variation on this test involves scoring a grid pattern in the coating prior to applying the tape. When the tape is removed, the number of grids that have been removed can be counted for accurate measurement of the surface area removed.

The specific type of tape, and how it is applied and removed can affect the results of these types of tests. They generally are more useful as a go/no-go quality control test rather than a quantitative measurement of adhesion.

AIMCAL [1,8] has developed a laboratory test for measuring the adhesion of the metallized layer from the film substrate in metallized films. The test involves constructing a specified structure by heat sealing an ethylene acrylic acid (EAA) copolymer adhesive film to the metallized surface along with a backing layer. The adhesive film and backing layer are then pulled from the metallized film using a T-peel test. The test is designed to mimic the failure mode in a real packaging laminate, although the results are only good enough for quality control purposes. The actual adhesion in a real laminate will differ due to differences in lamination technique, structure, etc. Jesdinszki et al. [8] point out that when adhesion is good, the sealed EAA may elongate or tear during the peel test and provide poor reproducibility. They propose two new methods. One involves substituting an amorphous polyethylene terephthalate (PET) or PA for the EAA film. These have higher mechanical strength and elongate less during the peel test. In the second approach, a LDPE film is adhesively laminated to the metallized film in a laboratory lamination process. This has the advantage of creating a laminate structure similar to commercial structures. The metal adhesion is found to be similar to measurements on structures produced at a pilot scale.

The paper industry uses a burst test called a Perkins Southwest test to measure adhesion of polymer coatings to paper. This method is described in section 15.4.

10.2.3 Fracture mechanics analysis of peel test

Peel strength is a measure of the force required to separate the peel arms. On a fundamental level, it involves propagating a crack along the interface as new interfacial area is created. The engineering discipline of fracture mechanics, introduced in Section 7.5.6, can be used to analyze the peel test and gain insight into how the peel test parameters (imposed angle and rate) and viscoelastic properties of the peel arms affect the result [9,10]. Fig. 10.3 shows a schematic of the peel test where a force P is imposed at an angle θ, causing the peel arm to delaminate over a distance a. The width of the peel arm is b and thickness X. The external work imposed on the interface is equal to the peel force times the distance traveled as given by Eq. (10.1):

$$W_{ext} = Pd \tag{10.1}$$

where W_{ext} = external work, P = peel force, and d = travel distance of the peel force.

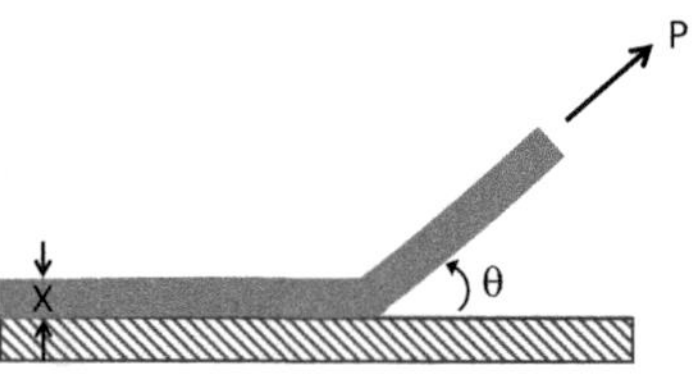

FIGURE 10.3 Peel test with peel force P, peel arm thickness X, peel arm width b, and peel angle θ.

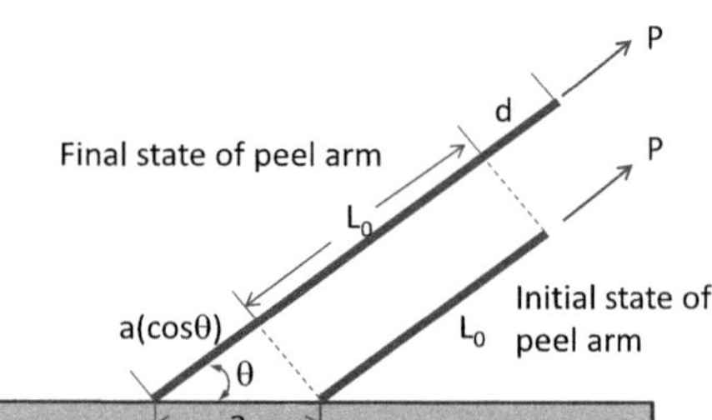

FIGURE 10.4 Geometry of peel test force–displacement.

After peeling, the length of the peel arm is simply the original length of the peel arm, L_O, plus the peeled distance a, and any extension of the peel arm during peeling:

$$L = L_O + a(1 + \varepsilon_a) \tag{10.2}$$

where L = length of peel arm after peeling, L_o = initial length of peel arm, ε_a = inelastic extension of peel arm ($= \Delta a/a$), and a = peeled distance.

Referring to Fig. 10.4, due to geometric considerations, the peel force travel distance is given by Eq. (10.3):

$$d = (1 + \varepsilon_a - \cos\theta)a \tag{10.3}$$

Substituting Eq. (10.3) into Eq. (10.1) yields

$$W_{ext} = P(1 + \varepsilon_a - \cos\theta)a \tag{10.4}$$

An energy balance shows that the external work is equal to the sum of the energy of creating a new interfacial area, the energy of extending the peel arm and the energy of bending the peel arm:

$$W_{ext} = U_a + U_e + U_b \tag{10.5}$$

$$U_a = \text{energy of fracture} = G_a ab \tag{10.6}$$

$$U_e = \text{energy of elongation} = G_e ab \tag{10.7}$$

$$U_b = \text{energy of bending} = G_b ab \tag{10.8}$$

where W_{ext} = external work G_a = fracture energy = energy to create new surfaces per unit area of peel propagation, G_e = energy dissipated during peel arm elongation per unit area of peel propagation, G_b = energy dissipated during bending of the peel arm per unit area of peel propagation, a = peel distance, and b = peel arm width.

Substituting Eqs. (10.4), (10.6) and (10.7) into Eq. (10.5) and rearranging gives the following general equation for peel strength:

$$\frac{P}{b} = \frac{G_a + G_e + G_b}{1 + \varepsilon_a - \cos\theta} \tag{10.9}$$

The fracture energy is the energy needed to create new interfacial area as the peel arms are pulled away from each other. This involves breaking bonds as well as overcoming local deformation during the creation of the new area at the peel front. The local deformation gives rise to viscoelastic energy dissipation, which can be significant for many polymers used in flexible packaging. The fracture energy is often written as the product of these two factors:

$$G_a = W_a(1 + \Phi(\mathrm{R}, \mathrm{T})). \tag{10.10}$$

where $W_a =$ work of adhesion, $\Phi =$ local viscoelastic energy dissipation, $R =$ rate of pull, and $T =$ temperature.

The work of adhesion encompasses wetting and bonding at the interface. This term is often smaller than Φ but since it has a multiplying effect, it is very important. The effect of peel test pull rate and temperature enter through their effect on local viscoelastic dissipation.

If the peel arms have infinite tensile stiffness (no extension or $\varepsilon_a = 0$) and zero bending stiffness (zero modulus), the material takes on the behavior of an "infinitely rigid string." This essentially ignores stretching and bending of the peel arm and Eq. (10.9) reduces to:

$$G_a^{\infty E} = \frac{P}{b}(1 - \cos\theta) \tag{10.11}$$

where $G_a^{\infty E} =$ fracture energy of an infinitely rigid string.

In this case, the measured peel strength (*P/b*) is directly related to the fracture energy. Often the elongation and bending of the peel arm is ignored in flexible packaging, but as is described below this is generally not a good assumption.

If the peel arm undergoes tensile deformation but the bending stiffness is still zero, Eq. (10.9) becomes:

$$G_a^{eb} = \frac{P}{b}(1 + \varepsilon_a - \cos\theta) - X\int_0^{\varepsilon_a} \sigma d\varepsilon \tag{10.12}$$

where $G_a^{eb} =$ fracture energy of an elastic peel arm with zero bending modulus, $X =$ thickness of peel arm, $\sigma =$ stress, and $\varepsilon =$ strain.

Here the integral term is simply the area under the stress–strain curve in a tensile test of the peel arm and represents the energy to elongate the peel arm.

The bending energy contribution to Eq. (10.9) can be significant if there is plastic yielding of the peel arm. Gent and Hamed [11] study the plastic yielding of OPET film during bending. They note that peel strength often is observed to go through a maximum at intermediate thickness. Gent and Hamed conclude that at low thickness, plastic yielding occurs, but the total energy that is dissipated is small. As thickness increases, more energy is dissipated, and peel strength increases with G_b. Eventually, the peel arm gets too stiff and less yielding occurs, which leads to a reduction in G_b and peel strength. At high thickness, no yielding occurs and G_b is small.

Kinloch et al. [10] use a beam theory approach to compute G_b. The peel arm goes through a loading and unloading cycle as the crack propagates and the peel arm is straightened out afterwards. Bending with large displacements can occur which they describe with beam theory. Both elastic and plastic deformation is considered. They derive a set of equations that can be easily solved iteratively using standard software tools. Their results are highlighted here but the reader is referred to the original paper for the appropriate equations.

Kinloch's model allows for the rotation of the "beam" representing the peel arm; the local angle at the peel front, θ_0, may differ from the imposed angle, θ. At the extremes, if $\theta_0 = 0$ then all the energy for delamination is transmitted during the bending of the peel arm. If $\theta_0 = \theta$ then the peel arm has zero bending modulus and acts as a string; all the energy is transmitted directly to the interfacial region. For most laminates $0 < \theta_0 < \theta$ and energy is transmitted both directly and through bending of the peel arm. Of particular interest is how the local bending angle is influenced by the stiffness and thickness of the peel arm since the local angle can affect the concentration of stress at the interface. For example, Pascal [12] shows that the coating thickness in extrusion coating affects the bending angle and peel strength.

As described by Eq. (10.9), the measured peel force, *P*, is not independent of the test geometry. It will vary with test angle, speed and the laminate structure. The fracture energy, G_a, is a better parameter for adhesion when all the factors are carefully considered. While computing G_a is often a useful exercise, the peel force measurement is sufficient in practice if its limitations are clearly understood. These limitations are explored in more detail below, using the experimental work of Kinloch et al. as an illustration.

10.2.3.1 Peel angle

Kinloch et al. [10] measure peel strength of four laminate structures characteristic of those found in flexible packaging (here Al = aluminum foil, PE = polyethylene, and OPET = biaxially oriented polyethylene terephthalate):

1. 6 μm Al//PE1. PE is extrusion coated onto the aluminum foil. The PE thickness ranges from 30 to 250 μm. For the peel measurement, the Al is bonded to a rigid support and the PE peeled away from it: Al is the substrate and PE the peel arm.
2. 6 μm Al//PE2. PE2 differs from PE1 in molecular weight (MW) distribution and orientation, resulting in different physical properties. Al is the substrate and PE the peel arm.
3. 12 μm OPET//(10 μm Tie1−20 μm PE1). This structure is made via coextrusion coating of the tie and PE onto the OPET film. During the peel test, the PE is mounted and becomes the substrate from which the OPET is peeled away (as the peel arm).
4. 12 μm OPET//(10 μm Tie2−20 μm PE1). This is the same as structure 3 except a different tie resin is used. Both tie resins are modified polyethylene. The PE and OPET are the substrate and the peel arm, respectively.

Kinloch et al. use a special sled to control the angle and test speed during the peel tests. The imposed angle is varied from 30 to 180 degrees. Peel speed is 5 mm/min. Tensile tests are conducted on the peel arms to obtain parameters for their model. Optical microscopy is used to measure the local peel angle, θ_0, during peeling. $G_a^{\infty E}$ is computed using Eq. (10.11) and G_a using their model, considering both extension and bending of the peel arm.

The local peel angle, θ_0, is plotted vs. the imposed peel angle, θ, in Fig. 10.5. There is good agreement between the measured values (shown in the figure) and the model calculation for θ_0 (not shown). For all four structures $0 < \theta_0 < \theta$ indicating the peel force is transferred to the interface directly and through bending as described earlier. The two structures with PE as the peel arm (1 and 2) give similar results, as do the two structures with OPET as the peel arm (3 and 4). The initial peel angle with PE as the peel arm is significantly higher at a given imposed angle than when OPET is the peel arm. This reflects the greater stiffness of the OPET. Structure 2 has somewhat higher θ_0 than structure 1. Structure 2 also has substantially higher G_a than structure 1 (see Table 10.4).

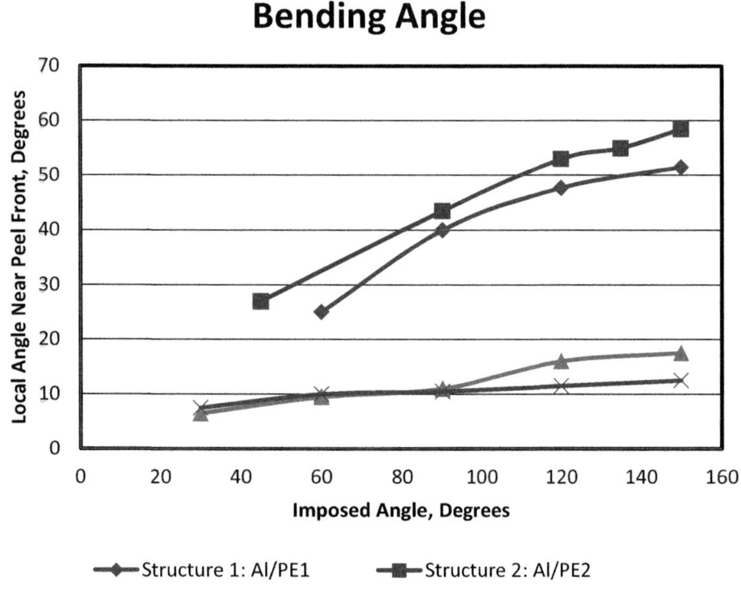

FIGURE 10.5 Effect of peel test angle (imposed angle) on the local angle near the peel front, θ_0. Peel speed is 5 mm/min [10]. *Data from A.J. Kinloch, C.C. Lau, J.G. Williams, The peeling of flexible laminates, Int. J. Fract. 66 (1994) 45–70.*

Fig. 10.6 shows peel strength (P/b), $G_a^{\infty E}$, and G_a as a function of imposed peel angle for structures 1 and 3. Structures 2 and 4 behaved similarly to structures 1 and 3, respectively, and are not shown. Peel strength decreases with increasing angle, as is expected from the (1- cosθ) dependence from geometric considerations. Yet the simple form of Eq. (10.11) is not sufficient to account for the angular effect, as is shown by the rise in $G_a^{\infty E}$ with increasing angle. As the angle increases, more elongation and bending occur, which must be taken into account. Kinloch's model provides values for G_a that are nearly independent of angle. While the measured peel strength varies with peel angle, the fracture energy is independent of angle if computed correctly. It is often challenging to compute the fracture energy from raw peel strength data. For the most part, the industry simply uses peel strength and therefore, care must be taken to ensure the angle is specified.

Values for the average value of G_a for the four structures are given in Table 10.4. G_a for structures 1 and 2 (PE peel arm) is considerably higher than that for structures 3 and 4 (PET peel arm). This reflects differences in bond strength, but also differences in viscoelastic properties of the peel arm. The flexible PE peel arm likely has greater local viscoelastic dissipation (Φ) than the OPET peel arm in structures 3 and 4. Comparing structures 1 and 2, G_a is considerably higher for structure 2. The bond strengths of PE1 and PE2 are likely similar since both grades rely on oxidation during coating to create polar species for adhering to the metal oxides on the foil. The substantial difference in G_a, therefore, likely reflects differences in viscoelastic dissipation.

10.2.3.2 Peel rate, temperature, and viscoelasticity

The effect of peel rate and temperature on peel strength is complicated due to several transitions the polymer may go through. For a detailed discussion, the reader is referred to Wu [9]. In general, as the temperature is increased at a constant peel rate, peel strength decreases and the failure mode may transition from cohesive to interfacial. As peel rate is

TABLE 10.4 G_a for the four structures [10].

Structure	Structure	Peel arm	G_a (J/m^2)
1	Al/PE1	PE	133 ± 6
2	Al/PE2	PE	228 ± 7
3	OPET/(Tie1-PE1)	OPET	46.7 ± 8.1
4	OPET/(Tie2-PE1)	OPET	24.1 ± 3.0

Source: Data from A.J. Kinloch, C.C. Lau, J.G. Williams, The peeling of flexible laminates, Int. J. Fract. 66 (1994) 45–70.

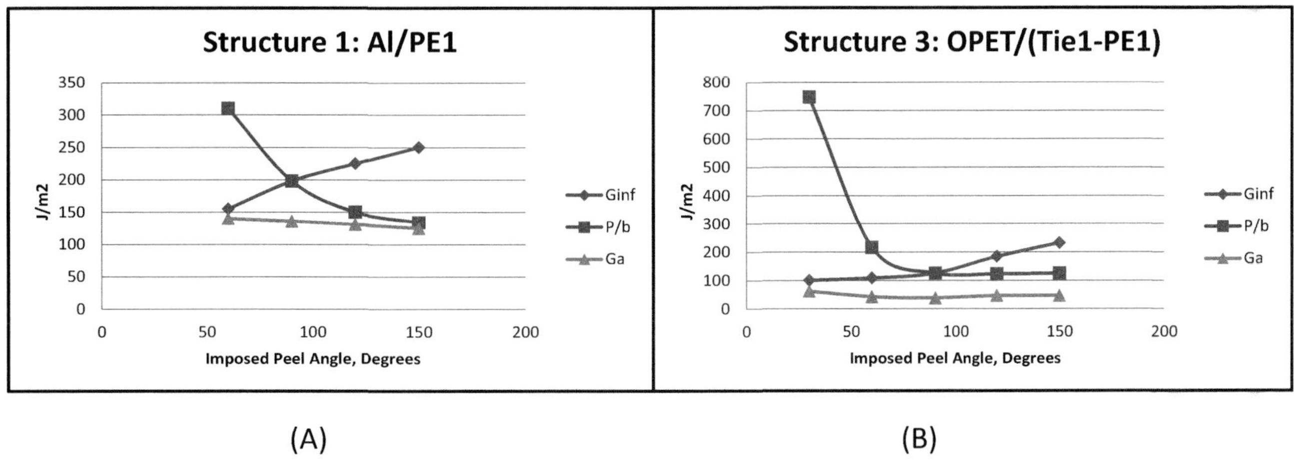

FIGURE 10.6 Effect of imposed peel angle and structure on measured peel strength (*P*/*b*), calculated fracture energy of an inelastic string ($G_a^{\infty E}$) and fracture energy (G_a) from Kinloch model. Peel speed is 5 mm/min [10]. (A) structure 1: Al/PE1, (B) structure 3: OPET/(Tie1-PE1). *Data from A. J. Kinloch, C.C. Lau, J.G. Williams, The peeling of flexible laminates, Int. J. Fract. 66 (1994) 45–70.*

increased at a constant temperature, peel strength typically increases but decreases sharply when the failure mode changes from cohesive to interfacial. These transitions generally correspond to viscoelastic transitions in the bulk adhesive and can be thought of as equivalent. Indeed, investigators have constructed master curves for G_a by using appropriate shift factors for the peel rate and temperature.

One way the peel rate affects the peel test measurement comes in through the local viscoelastic energy dissipation at the crack tip, Φ, as described in Eq. (10.10). Kinloch et al. [10] plot G_a for Structure 1 (Al/PE1, as described in Section 10.2.3.1) as a function of the crack speed, which is the test peel rate divided by $(1 + \varepsilon_a - \cos\theta)$. The PE thickness is 35 μm and data for several peel angles are plotted in Fig. 10.7. Included in Fig. 10.7 is data for the structure 51 μ Al//25 μ EMAA, where the EMAA (ethylene methacrylic acid copolymer) is Nucrel 0910, a 9 wt.% MAA grade from Dow. The structure and measurements are from a study by Dow (formerly DuPont). The peel strength measurements are conducted with a 180 degrees peel angle. From Eq. (10.11), $G_a^{\infty E}$ is simply $2(P/b)$ at this peel angle. $G_a^{\infty E}$ and G_a results are both plotted, where G_a is computed using the Kinloch model that accounts for both bending and stretching of the peel arm.

The results presented in Fig. 10.7 provide several insights into peel strength performance, including:

- The positive slope indicates that a higher peel rate increases the local viscoelastic energy dissipation at the crack tip [see Eq. (10.10)]. If the curve were extrapolated to zero speed, the result theoretically would be W_a. Kinloch et al. note that in previous work investigating styrene-butadiene (SB) rubber bonding to various polymer substrates, the curve extrapolated to values of around 0.05 J/m^2, which is the order of magnitude for secondary van der Waals forces. Whether these curves would actually extrapolate to such a low value is undetermined with the limited data provided, and the value of W_a is likely higher due to the presence of some chemical interaction between polar oxidative species on the PE formed during extrusion coating and the aluminum oxide on the surface of the foil. The slope, however, shows that the measured energy is at least an order of magnitude greater than the bond strength, underscoring the importance of the deformation zone at the peel front and the viscoelastic properties of the peel arm from which it arises.
- Comparing G_a for PE and EMAA gives an indication of the effect of bonding strength. Both curves have similar slopes and both peel arms have similar physical properties. Hence, the shift upward from PE to EMAA (bottom to middle curve in Fig. 10.7) reflects the stronger bonds formed by the acid groups in EMAA to the oxides on the aluminum surface.
- The difference between G_a and $G_a^{\infty E}$ (middle and top curves) for EMAA reflects the impact of bending and stretching the peel arm. Roughly half the measured peel strength for this structure is due to these factors.

The viscoelastic nature of the adhesive influences the peel strength. Consider the three adhesive materials illustrated in Fig. 10.8 taken from Wu [9]. Materials I, II, and III are rubbery, liquid-like and strain-hardening, respectively. The dashed horizontal lines represent two scenarios for the maximum interfacial stress. When the stress measured during peel testing exceeds the maximum interfacial stress of the structure, failure occurs. This happens when the stress–strain curve intersects the horizontal line. If the maximum interfacial stress is represented by the bottom line, then all three

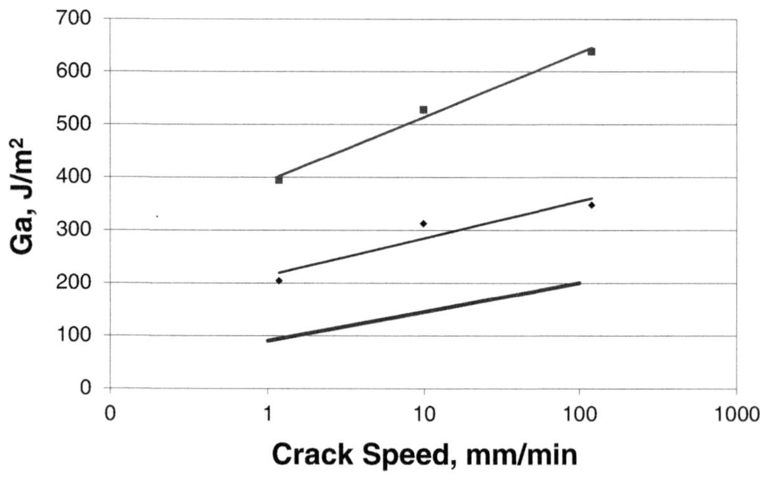

FIGURE 10.7 The effect of peel rate on peel strength. Bottom curve is G_a from measurements on the structure Al//35 μm PE1 [10]. The top and middle curves are data from Dow on the structure Al//25 μm EMAA; the top and middle curves are $G_a^{\infty E}$ and G_a, respectively. *Data from A. J. Kinloch, C.C. Lau, J.G. Williams, The peeling of flexible laminates, Int. J. Fract. 66 (1994) 45–70. The figure is reproduced with permission from the Dow Chemical Company.*

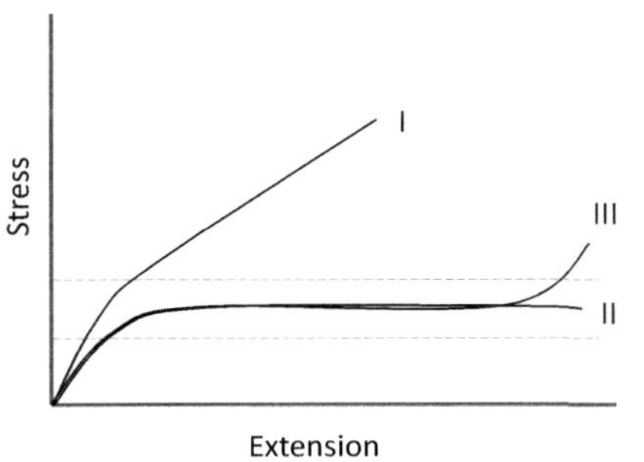

FIGURE 10.8 Illustration of stress–strain curves for rubbery (I), liquid-like (II), and strain-hardening (III) materials and their influence on peel strength. Dashed horizontal lines represent two scenarios for the maximum interfacial strength as described in the text [9]. *Adapted from S. Wu, Fracture of adhesive bond, in: Polymer Interface and Adhesion, Marcel Dekker, Inc., New York, 1982, pp. 475–569.*

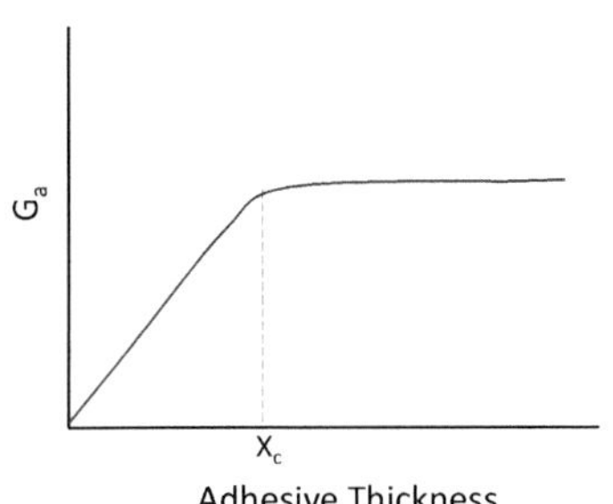

FIGURE 10.9 Effect of adhesive thickness on G_a.

materials fail adhesively (or interfacially) with about the same peel strength. If the maximum interfacial stress is represented by the top line, then material I fails first (adhesively). Materials II and III have a large deformation energy contribution that is given by the area under their stress–strain curve. Material II eventually fails cohesively and III adhesively. In this case, the measured peel strength of II and III will be considerably higher than material I.

The molecular structure of the adhesive influences the shape of the stress–strain curve, but so does the rate and temperature. Decreasing the temperature or increasing the rate may change the material from II to I, which changes the peel failure mode from cohesive to adhesive and decreases the peel strength.

In multilayer structures with adhesives, both the adhesive and the adjoining layer that make up the peel arm influence the peel strength. Gent and Hamed [11] show that the bending angle and thickness of OPET films affect the bending energy, which goes through an abrupt increase when the film yields. This bending force affects the measured peel strength, as is mentioned earlier. Gent and Hamed find, however, that when the adhesion of the OPET to the substrate is strong, the peel strength is much higher than the bending mechanics of the OPET layer would suggest. They conclude that deformation in the adhesive layer makes up the difference.

10.2.3.3 Thickness

Fracture mechanics predicts that the fracture energy, G_a, increases with increasing thickness up to a critical thickness, X_c, after which G_a is constant. This is illustrated in Fig. 10.9. The zone of deformation in front of the peel which gives rise to Φ extends across the entire thickness of the adhesive when the thickness is small. Indeed, the size of the zone of deformation is constrained by the adhesive thickness, as is illustrated in Fig. 10.10A. As the thickness is increased, the size of the zone of deformation increases along with Φ. When the thickness of the adhesive equals the critical thickness, the zone of deformation has reached its natural size. Further increases in the adhesive thickness result in no further increase in the size of the zone of deformation and G_a is constant (Fig. 10.10B).

The critical thickness of the adhesive is the radius of the plastic zone at the peel front (or crack tip in the language of fracture mechanics), r_y. From Kinloch and Young [14], a stress analysis for plane stress gives:

$$r_y = \frac{1}{2\pi}\left(\frac{K_I}{\sigma_y}\right)^2 \tag{10.13}$$

where r_y = radius of plastic zone at crack tip, K_I = stress intensity factor, and σ_y = yield strength.

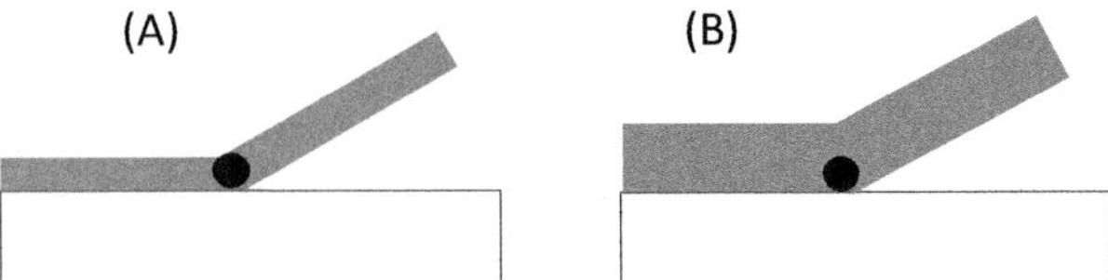

FIGURE 10.10 Cartoon of the influence of the thickness on the size of the zone of deformation in front of the advancing peel or crack. (A) thickness is less than natural size of zone of deformation, (B) thickness is greater than natural size of deformation [13]. *Reproduced with permission from B.A. Morris, Understanding why adhesion in extrusion coating decreases with diminishing coating thickness, J. Plast. Film Sheeting 24 (2008) 53–88.*

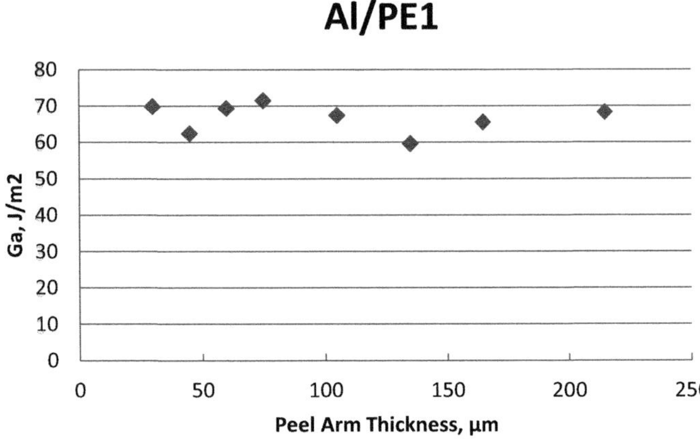

FIGURE 10.11 G_a as a function of PE1 thickness for the structure Al//PE1. Peel angle is 180 degrees and rate is 1 mm/min [10]. *Data from A.J. Kinloch, C.C. Lau, J.G. Williams, The peeling of flexible laminates, Int. J. Fract. 66 (1994) 45–70.*

The stress intensity factor comes from an analysis of the stress on the advancing peel front. It is defined as the maximum stress at the crack tip divided by the imposed stress and is typically > 1. It is used to account for the fact that failure generally occurs at loads considerably less than the idealized strength of the interface. The stress is thought to be amplified at the crack tip. When the stress analysis is equated with the energy analysis previously introduced, K_I is found to be directly related to G_a and E (modulus):

$$K_I^2 = EG_a. \tag{10.14}$$

Combining Eqs. (10.13) and (10.14) and letting $X_c = r_y$:

$$X_c = \frac{1}{2\pi}\frac{G_a E}{\sigma_y^2}. \tag{10.15}$$

Eq. (10.15) shows that the critical thickness increases with increasing modulus and decreasing yield strength. It also increases with increasing G_a, which is a circular argument: the size of the zone of deformation depends on the fracture energy, which depends on the amount of deformation. It makes sense if one considers that anchoring the peel arm to the substrate more strongly will allow for more deformation to occur as the peel arm is pulled away. The same argument can be made for bending and stretching the peel arm. If the adhesion is stronger, the peel arm can be pulled away at a steeper angle, resulting in more bending and possible yielding of the adhesive and adherend. Similarly, more stretching of the peel arm may occur when it is more firmly attached to the substrate.

Kinloch et al. [10] measure G_a for Al/PE1 (Structure 1 introduced in Section 10.3.2.1) with the peel arm (PE1) thickness ranging from 30 to 215 μm. The peel angle is 180 degrees and peel rate 1 mm/min. The results are plotted in Fig. 10.11. G_a is mostly constant. A calculation of X_c yields a value of 15–20 μm, suggesting that the thickness range explored by Kinloch is on the plateau region of the curve. Wu [9] reviews several studies that show an increase in fracture energy with an increasing thickness similar to that shown schematically in Fig. 10.8.

The effect of thickness is further explored in Sections 10.4.2.2.3 and 15.1.4.

10.3 Fundamentals

Adhesion is the state in which two dissimilar materials are in such intimate contact that mechanical force or work can be transferred across their interface. The interfacial bonding forces may arise from thermodynamics, mechanical interlocking, chemical bonding or electrostatic attraction. The mechanical strength of the adhered system is a function of interfacial attraction, the strength of the interface and the cohesive strength of the two bulk phases.

The adhesion at an interface predicted from theoretical considerations is generally much greater than what is measured. Fracture mechanics, introduced in Section 10.2.3, resolve this discrepancy by introducing the concepts of defects. The fracture energy involves the propagation of these defects or cracks and the generation of new interfacial areas. As described by Eq. (10.10), the fracture energy has contributions from the work of adhesion and viscoelastic deformation. The adhesion component arises from the reversible work of adhesion (thermodynamics), diffusion and chemical interaction. This section briefly explores each of these contributions to adhesion performance. The terms "adhesive" and "substrate" will be used in this section to describe bonding between layers. The adhesive may be a specially formulated material for bonding, such as a liquid adhesive used in lamination or a polymeric tie layer in coextrusion, or it may simply be one of the two layers at the interface. The substrate may be a solid substrate such as OPET film, paper or aluminum foil, or it may designate a molten polymer layer in a coextrusion, depending on the context of the discussion.

10.3.1 Thermodynamics of adhesion

10.3.1.1 Surface and interfacial tension

When two materials are joined, they form a region of finite thickness between them over which the composition and energy vary continuously from one bulk phase to the other. This is known as the interface. A tension is created at the surface of a material due to the energy difference between the material and the medium at the surface (such as air). When two materials are joined by an interface, an interfacial tension develops. The concept of interfacial tension or energy at a polymer—polymer interface is introduced in Section 6.4.2 and Fig. 6.15.

The surface energy of solids (also known as the surface tension, although the industry generally uses the term "surface energy" for solids like polymer films and "surface tension" for liquids) is not determined directly, but rather is typically measured by the response of a liquid or liquids on the surface, usually in the presence of air or a saturated vapor of the test liquid. Two techniques are commonly used. The practical test is the dyne test [16,17], which uses calibrated formulations of 2-ethoxyethanol and formamide. A test solution (often in the form of a marker pen) is applied onto the surface to estimate the surface energy. The second approach is to measure contact angles, which requires laboratory equipment. The contact angle is the angle a droplet forms on the surface of a solid as shown in Fig. 10.12. Young's equation relates the cosine of the contact angle to the interfacial tensions at the solid—liquid—vapor interface:

$$0 = \gamma_{SV} - \gamma_{SL} - \gamma_{LV}\cos\theta_c. \tag{10.16}$$

where γ_{SV} = interfacial tension between the solid and vapor phases, γ_{SL} = interfacial tension between the solid and liquid phases, γ_{LV} = interfacial tension between the liquid and vapor phases, and θ_c = contact angle.

In a method known as the Zisman plot, $\cos\theta_c$ is plotted vs. γ_{LV} for a series of homologous liquids. The value of γ_{LV} at a point extrapolated to $\cos\theta_c = 1$ is known as the critical surface tension or energy of the solid. Other methods involve measuring the contact angle of a polar (water) and nonpolar (methylene iodide) liquid on the substrate, from which the surface energy can be calculated by various averages. These methods tend to yield more accurate values according to Wu [18]. The critical surface energy from the Zisman plot is often smaller than the surface energy measured by other techniques. The surface energy and critical surface energy of several polymer substrates are given in Table 10.5.

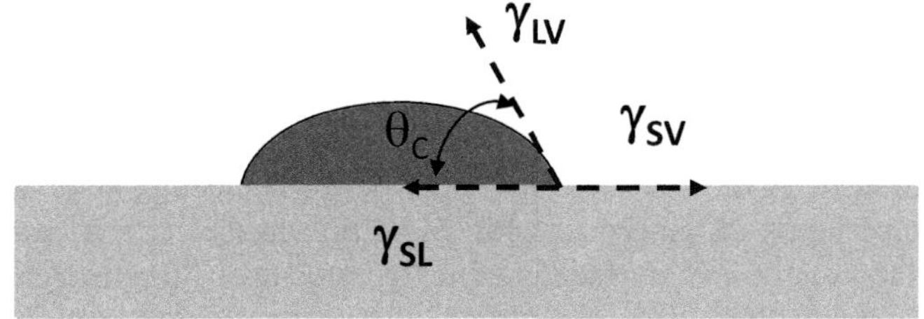

FIGURE 10.12 Contact angle and associative surface tensions.

TABLE 10.5 Surface energy of solid polymer films at 20°C [18].

Polymer	Surface energy (dynes/cm)	Critical surface energy (dynes/cm)
Polyethylene	35.7	31
Polypropylene	30.1	32
Polystyrene	40.7	33
Polyvinyl chloride	42.9	39
Polyethylene terephthlate	42.1	43
Polyamide 66	44.7	46

Source: Data from S. Wu, Surface tension and polarity of solid polymers, in: Polymer Interface and Adhesion, Marcel Dekker, Inc., New York, 1982, pp. 169–214.

TABLE 10.6 Surface tension of polymer melts [19].

Polymer	Temperature (°C)	Surface tension (dyne/cm)
Linear polyethylene	180	26.5
Branched polyethylene (LDPE)	180	24.6
Ethylene vinyl acetate (17.7 wt.% VA)	180	25.5
Ethylene vinyl acetate (26.6 wt.% VA)	180	25.4
Polypropylene	180	20.8
Polystyrene	180	29.2
PET	300	26.4
PA 66	300	28.3

Source: Data from S. Wu, Interfacial and surface tensions of polymer melts and liquids, in: Polymer Interface and Adhesion, Marcel Dekker, Inc., New York, 1982, pp. 67–132.

A pendant drop method is typically used to measure the surface tension of polymer melts. The surface tension of polymer melts decreases linearly with increasing temperature. A series of measurements at different temperatures can be used to extrapolate a value for the polymer solid. This method typically agrees well with the liquid homolog methods. Surface tensions of several polymer melts used in flexible packaging are given in Table 10.6.

The surface tension of a polymer decreases with decreasing density. Thermal transitions that change the density can affect the surface tension. During the glass transition, the surface tension undergoes a continuous change but the slope of the surface tension vs. temperature curve changes. For semicrystalline polymers, as the polymer is cooled from the melt crystallization increases the density, resulting in a discontinuous increase in the surface tension. Most polymer films have an amorphous region at the surface; as the polymer cools and solidifies, polymer chain segments not participating in forming crystals are pushed towards the surface. Thus, typical measurements of solid polymers reflect the surface energy for the amorphous phase (and have good agreement with extrapolations from the melt). Schonhorn [20] shows, however, that if the polymer is quenched against a nucleating surface, different degrees of crystallinity formed at the polymer surface change the surface energy. Table 10.7 provides data on polyethylene molded against different surfaces. The method involves placing 250 μm thick polyethylene film against a thin metal or salt surface, exposing the sample to 200°C for 30 minutes under nitrogen, cooling, and chemically dissolving away the metal or salt. The surface energy was calculated from contact angle measurements. The surface crystallinity of the PE was calculated based on values for amorphous and crystalline surface energies. PE molded against surfaces that provide no nucleating such as nitrogen and PET has a surface energy of 36.2 dynes/cm, consistent with amorphous PE. Molding surfaces comprising of various metals and salts promote crystallinity at the surface of the PE; surface energies greater than 60 dynes/cm were measured which correspond to surface crystallinity of greater than 90%. Wu [18] tabulates data for other polymers, such as PP and PA66, which have similar behavior when molded against nucleating surfaces.

TABLE 10.7 Effect of mold surface on surface energy and surface crystallinity of polyethylene [18,20].

Mold surface	Surface crystallinity (%)	Surface energy at 20°C (dynes/cm)
Nitrogen	0	36.2
Polytetrafluoroethylene	0	36.2
Polyethylene terephthlate	0	36.2
Copper	5.1	37.4
Nickel	53.3	51.3
Tin	60.1	53.8
Aluminum	63.2	54.9
Glass	63.2	54.9
Chromium	66.2	56.1
Mercury	86.4	64.8
Gold	96.3	69.6
Tantalum	78.5	61.4
NaCl	86.4	64.8
KBr	86.4	64.8
KCl	86.4	64.8
CaF_2	91.4	67.3

Source: Data from H. Schonhorn, Heterogeneous nucleation of polymer melts on high-energy surfaces. II. Effect of substrate on morphology and wettability, Macromolecules 1 (1968) 145–151 and adapted from S. Wu, Surface tension and polarity of solid polymers, in: Polymer Interface and Adhesion, Marcel Dekker, Inc., New York, 1982, pp. 169–214.

Note that Table 10.7 shows that chromium has a nucleating effect on PE. In practice, however, commercial films of PE made by casting onto a chromium-plated chill roll do not have the high surface energies found in the table. The rate of quenching in a commercial cast film process is very fast, which suppresses crystallization. Furthermore, Schonhorn makes the point that good wetting is required for surface nucleation to occur. Chrome plating is used to minimize wetting and chill roll sticking. These factors conspire to suppress the crystallinity at the surface.

Wu [19] describes methods for measuring interfacial tension and tabulates values for several pairs of polymers. An approximate relationship for the interfacial tension of polymer pairs [Eq. (10.7)] is derived in Chapter 6 based on the Good and Girifalco equation. This equation relates the interfacial tension to the difference in solubility parameters:

$$\gamma_{12} \cong 0.26(\delta_1 - \delta_2)^2 \tag{10.17}$$

where γ_{12} = interfacial tension, and δ_I = solubility parameter of material i.

The interfacial tension is strongly influenced by the polarity of the polymer, which is reflected in the solubility parameter. Values for solubility parameters for some common polymers used in flexible packaging are given in Table 6.2 and more extensive data is found in the literature [21,22].

10.3.1.2 Wetting and work of adhesion

The ability of the polymer or adhesive to spread across a substrate is known as wetting. In Section 7.5.2 the concept of the spreading coefficient is introduced:

$$\lambda_{12} = \gamma_2 - \gamma_1 - \gamma_{12} \tag{10.18}$$

where λ_{12} = spreading coefficient of substance 1 onto substance 2, γ_1 = surface tension (or energy) of substance 1 (adhesive), γ_2 = surface tension (or energy) of substance 2 (substrate), and γ_{12} = interfacial tension of substances 1 and 2.

For spontaneous wetting, $\lambda_{12} \geq 0$.

The thermodynamic work of adhesion, W_a, is the reversible work required to separate two surfaces in a direction perpendicular to their interface. W_a is related to the surface energies in the following manner:

$$W_a = \gamma_2 + \gamma_1 - \gamma_{12}. \quad (10.19)$$

There are two competing theoretical models of adhesion as it relates to wettability. In the interfacial defect model, loss of adhesion is thought to be due to stress concentration around defects, which may be considered to be micro unwetted voids. Improving wetting reduces defects. To maximize adhesion, the spreading coefficient is maximized. From Eq. (10.18), this can be accomplished by minimizing γ_1 and γ_{12}.

In the fracture energy model, the energy to propagate fracture (G_a) is proportional to W_a [see Eq. (10.10)]. Adhesion is maximized by maximizing W_a. This can be done by minimizing γ_{12} and maximizing γ_1.

The two models agree on minimizing the interfacial tension, γ_{12}, but disagree on the direction to move the adhesive surface tension, γ_1. Combining the two approaches, the following generalizations can be made for the conditions for optimal adhesion:

1. The first condition for optimum adhesion is to minimize the interfacial tension. This can be done by matching the polarity of the adhesive or coating to the substrate. By Eq. (10.17), matching solubility parameters is a way to do this.
2. To further optimize adhesion, the spreading coefficient can be driven to zero by matching surface tensions. Thermodynamically, the spreading coefficient does not need to be maximized; spontaneous wetting still occurs at $\lambda_{12} = 0$. Letting $\gamma_1 = \gamma_2$, W_a can be maximized while still maintaining $\lambda_{12} = 0$.

These models just consider the thermodynamics of adhesion. Irreversible contributions to the work of adhesion can be achieved through chemical bonding at the interface, which can overwhelm the thermodynamic contribution.

The viscosity is also important—the coating or adhesive must be able to flow to wet the surface. Schubert [23] observes that LDPE at 200°C takes about an hour to wet "wettable" aluminum foil in extrusion coating due to its high viscosity whereas ethanol will wet it immediately. Increasing temperature reduces the viscosity and helps with spreading. Increasing the contact time provides more time for spreading and adhesion. In extrusion coating, LDPE is often coated at >300°C to provide faster wetting.

10.3.1.3 Practical guidelines

A general guideline for good wetting is for the surface energy of the substrate to be about 10 dynes/cm greater than the coating. Often the surface is corona, plasma, flame or ozone-treated to clean the surface and raise the surface energy to promote good wetting, especially of inks. For example, Gilbertson [24] provides the following typical surface tensions for inks:

- Solvent-based inks: 27–30 dynes/cm.
- Water-based inks: 35–40 dynes/cm.
- UV ink monomers and oligomers: 30–36 dynes/cm.

Polymer substrates such as OPET, OPP and PA are often corona treated by the film manufacturer to promote adhesion. The treatment level may age down with storage time. A second treatment is often applied in-line prior to coating to get the best results. Too much treatment can cause problems if the treatment breaks down the molecules at the surface. Primers are sometimes used to prepare the surface for subsequent adhesion.

Aluminum foil is classified according to its wettability. When water forms a continuous layer across the width of the coil with no beading the foil is given a Class A rating. This corresponds to a surface energy of >72 dynes/cm and is the best rating for foil wettability. Class B foil wets with a solution of 90% water, 10% alcohol (47 dynes/cm) and Class C with a solution of 80% water, 20% alcohol (38.5 dynes/cm). The wettability may vary across the width of the coil (roll) due to the manufacturing process. Foil less than 60 μm thick is rolled to its final thickness by squeezing two layers together separated by oil. The coiled product is usually annealed to refine its crystalline grain structure, remove the oil and promote a homogenous oxide layer. This is called soft annealing. The oxide layer forms immediately after rolling but is refined in the annealing step. The oxide can be considered as having two layers, a bottom "barrier" layer next to the aluminum that provides a moisture barrier and corrosion protection, and a top layer that must bond to the coating. During annealing, the bottom layer becomes thicker and the top layer more compact as it loses water. Since annealing is done on the coil, the oxide layers are typically thicker near the edges of the coil than in the middle (along the width) since more oxygen can diffuse near the ends. Also, not all the oil is removed during annealing—about 10% remains. More oil is concentrated near the center—so wettability may be better near the edges than at the center of the web [25–27].

Corona treatment of foil can help with adhesion by helping to burn off the oil and raise the energy level, especially for unannealed foil [28]. Flame treatment is also often used to help clean the surface of the foil.

The smoothness, porosity, and chemical composition of paper and paperboard influence their ability to be wet and bonded. These are impacted by a number of factors in the papermaking process such as the softness of the wood pulp, the openness of the fiber structure, basis weight of the paper, calendering, and chemical treatments and coatings.

The quality of the surface of the substrate is important for good bonding. Solid substrates should be free of contaminants that may affect adhesion. Contaminants may arise from many sources, including:

- Oils left from the manufacturing process, such as in the rolling of aluminum foil.
- Various processing or slip agents that bloom to the surface with time. These low MW components sometimes interfere with adhesion.
- Break down of molecules from excessive treatment of the surface.
- Silicone sprays sometimes used to prevent chill roll sticking and other processing issues. These are particularly problematic for good adhesion and should be avoided.
- Reverse side transfer, where the coating surface picks up contaminants from the adjoining layer in the roll.

Films of polar ethylene copolymers such as ethylene vinyl acetate (EVA), ethylene methyl acrylate (EMA), EAA and ionomers typically have surface energies similar to nonpolar LDPE. One factor for this behavior is that these polymers are still mostly polyethylene in character. They are random copolymers with typically up to about 20 wt.% of the polar monomer. On a molar basis, less than 10% of the repeat units contain the polar monomer. Another factor is the polarity of the medium the film is exposed to. Studies [29–32] show that the surface energy of PE modified with polar groups increases when immersed in water, a polar medium. Schultz et al. [31] measure the contact angle of acrylic acid grafted polyethylene (PE-g-AA) films immersed in water or formaldehyde (which is less polar than water). The grafted PE has about 1 wt.% AA. The measurement uses a liquid droplet technique that allows the calculation of the polar and nonpolar contributions to the surface energy. For the PE control, the polar surface energy does not change when immersed in the liquid. The PE-g-AA surface energy does change, which Schultz attributes to the AA groups reorienting to the surface. At 20°C this takes several days (12 days to level off) but at 90°C it takes about an hour. The ratio of the polar surface energy of the PE-g-AA to the liquid is constant; the surface energy is lower when immersed in the less polar formaldhyde than it is in water.

Schultz et al. conduct adhesion experiments with the PE-g-AA by pressing films against aluminum foil. The PE-g-AA gives about eleven times the peel strength as the PE control. They attribute this to the AA groups reorienting to the interface with the polar aluminol on the surface of the foil. They show that a commercially made coating of PE-g-AA onto foil gives lower peel strength than their model system. They relate this to the short contact time for the commercial process—seconds vs. 15 minutes for their process. Putting the commercial laminate through their process raises the adhesion to the same levels found in their model system.

Cho et al. [32] measure contact angles of films pressed from EAA copolymers and their blends with LDPE in both air and water. From this, they compute the polar contribution to the surface energy. Their results are plotted in Fig. 10.13. The EAA ranges in AA content from 3 to 20 wt.%. The percent of EAA in the blends ranges from 5% to 83% so the % AA in the blend ranges from 1 to 7 wt.%. When measured in air, the polar contribution of the surface energy of all the EAA copolymers is zero, the same as for an LDPE control. For the water measurements, they immerse the films for one hour at 80°C in octane-saturated water prior to the measurement. The films are then placed in distilled water and, while immersed, the contact angle is measured with a droplet of octane. The polar contribution to surface energy increases with increasing acid level of the copolymer. When in contact with a polar medium, the acid groups reorient to the surface.

The EAA-LDPE blends show similar behavior. The polar surface energy measured in water increases with increasing wt.% AA in the blend until about 6–8 wt.% AA (in the blend), after which it plateaus to the pure EAA value. Peel strength is measured on laminates made by hot pressing the films to aluminum foil for 30 minutes at 170°C. Peel strength increases with increasing % AA in the blend, suggesting that the reorientation shown in the water immersion experiment simulates the reorientation that must occur when the film wets the polar aluminum surface.

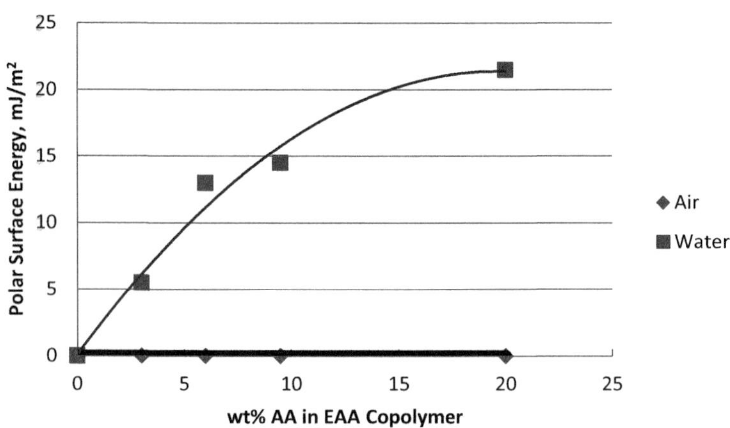

FIGURE 10.13 Effect of acrylic acid level on polar contribution to surface energy for EAA films tested in air or water [32]. *Reproduced with permission from C.K. Cho, K. Cho, C.E. Park, Relationship between surface rearrangement and adhesion strength of poly(ethylene-co-acrylic acid)/LDPE blends to aluminum, J. Adhes. Sci. Technol. 11 (1997) 433–445.*

The discussion so far has been about wetting a liquid adhesive or coating onto a solid substrate. In coextrusion, two molten polymers are combined. Wetting is typically not a limiting factor for adhesion in coextrusion. This is because the surface tensions of molten polymers are fairly similar (see Table 10.6) so the thermodynamics for spreading are favorable. Also, the flow geometry in coextrusion forces the melt streams into intimate contact.

The adhesion of many inks and coatings to films is driven by wetting and thermodynamics, but for interlayer adhesion in coextruded films, wetting is a necessary but typically not a sufficient condition for good adhesion. Without good wetting, other mechanisms such as diffusion and chemical interaction that build strength at the interface cannot take place.

10.3.2 Mechanical interlocking

Mechanical anchoring of the adhesive or coating to the substrate can be utilized to create a strong interface if the substrate has enough cavities and microscopic undercuttings. An example is bonding to paper and paperboard, where the flow into the pores and around the fibers creates a mechanical interlocking network.

Factors that favor mechanical interlocking include high surface roughness and porosity of the substrate, good wetting, and sufficient flow of the bonding layer. In flexible packaging, porous substrates such as paper, paperboard, and nonwovens may take advantage of this mechanism. Mechanical adhesion is widely utilized in other industries, such as chemical and mechanical etching of metals for adhesion.

10.3.3 Diffusion

Once close contact between the layers is obtained through wetting, diffusion of molecular chains across the interface may occur if the layers have enough compatibility and mobility. Diffusion builds strength at the interface by providing further intimate contact between chains, enabling short-range attractive forces (such as dispersive forces), entanglement and co-crystallization. The interface becomes more diffuse as penetration proceeds and ultimately the interface disappears. The mechanism of diffusion and buildup of strength at the interface for similar materials is the same as for auto-adhesion, explored in detail in Section 7.5.3.

In self or auto-adhesion, the two bonding layers are the same and so compatibility is not an issue. For dissimilar materials, the molecules must have enough chemical affinity for diffusion to take place. One way to quantify this is with solubility parameters: polymers with similar solubility parameters will be more compatible and likely to interdiffuse. Wu [33] derives the following relationship from a fracture energy balance:

$$\sigma_f = p\exp[-q(\delta_1 - \delta_2)] \tag{10.20}$$

where σ_f = fracture strength, p, q = constants, δ_i = solubility parameter of layer i.

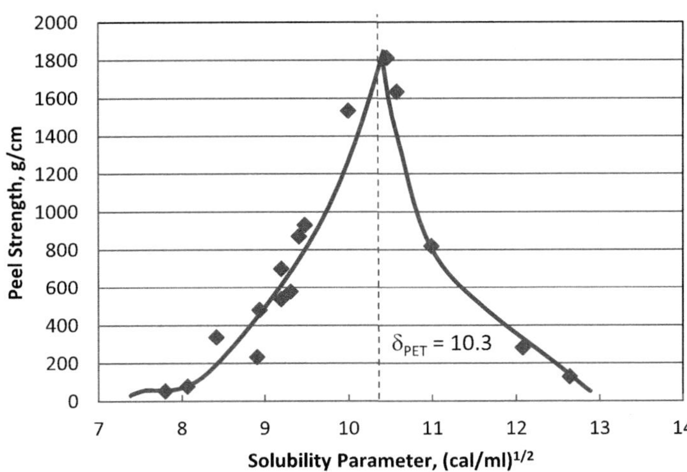

FIGURE 10.14 Peel strength vs. solubility parameter for various polymers bonded to polyethylene terephthlate. Peel strength was measured at 90 degrees angle at 5 cm/min. Bonding was in the melt and peel strength measured after cooling to room temperature [33]. *Adapted from S. Wu, Formation of adhesive bond, in: Polymer Interface and Adhesion, Marcel Dekker, Inc., New York, 1982, pp. 359–447.*

The validity of this relationship is shown in Fig. 10.14 where the peel strength of a series of polymers melt bonded to polyethylene terephthlate (PET) is plotted against their solubility parameter. The solubility parameter of PET is 10.3 $(cal/mL)^{1/2}$, where the maximum peel strength is observed.

Ward [34] uses a heat seal test to measure peel strength of 16 grades of polyethylene, ethylene copolymers and tie resins to two styrene copolymers (a styrene methyl acrylate copolymer and a styrene maleic anhydride (MAH) copolymer). He finds that the results correlate with the difference in solubility parameters, but that the mechanical properties are also important. EVA resins give the best performance.

Section 7.5.3 shows that diffusion and the resulting bond strength increase with increasing time and temperature and decreasing MW. Increasing time allows more molecular penetration, up to the point where the interface disappears in the case of auto-adhesion. Increasing temperature increases the thermal motion of the molecular chains, speeding up diffusion. Lower MW increases the amount of diffusion but reduces cohesive strength: the experimental data on the effect of MW is not definitive. In some cases, higher MW increases peel strength due to an increase in mechanical strength.

When chemical reactions occur at the interface, diffusion and strength buildup is constrained by the formation of bond sites. Lee and Wool [35–37] study a model polymer/substrate system: carboxylated polybutadiene bonding to an amine functionalized aluminum substrate. Both the carboxylate and amine densities are controlled. The adhesion of some of their structures ages up over a period of 41 days. Their explanation: initially the polymer has strong bonds to the solid surface but the density of chains near the surface is sparse compared to the bulk of the polymer. The cohesive strength between the chains attached to the surface and the nearby chains is weak since there is not much chain entanglement. With time, if the surface energetics is right, the near-surface region can fill in: nearby chains entangle, increasing the cohesive strength and hence the peel strength. The time scale for the near-surface diffusion of polymer chains is 100–10,000 times greater than the terminal relaxation time for bulk diffusion. Near the surface the polymer chains are constrained in their movement by the bond sites with the surface. Hence diffusion is considerably slower at the interface. In the model system, the terminal relaxation time is 2 seconds but the time scale for diffusion near the interface is many days.

They vary the amine concentration on the surface and find that the adhesion goes through a maximum at an intermediate level of surface functionality. The higher the amine level, the higher the surface energy. They find that the highest surface energy does not necessarily give the best adhesion. They propose this is due to competing mechanisms. As the surface functionality is increased, there is more bonding between the polymer and substrate: the polymer wets the surface better. This increases the adhesive strength at the interface. On the other hand, as the surface wetting is increased, the density of chains nearby is depleted, reducing the cohesive strength. If the surface energetics is right, over time these depleted regions can fill in and peel strength will increase, as described above. They argue, however, that for a high surface energy situation, there are so many polymer surface interactions that the molecular structure near the surface is permanently altered. The chains become ordered near the surface and cannot entangle as readily, resulting in lower cohesive strength. This results in the peel strength (at long times) being higher at an intermediate level of surface wettability.

Block copolymers may also provide adhesion at an interface by promoting diffusion between the layers. An example is olefin block copolymers (OBCs) developed by Dow Chemical. One version combines propylene blocks with ethylene blocks that allow adhesion between PE and PP [38–40]. Eagan et al. [41] show that multiblock copolymers of iPP/PE are effective compatibilizers for PE and PP. As is described in Section 10.4.2.2.1, however, getting the block copolymer to the interface in coextrusion may take much more time and be less effective than forming in situ block copolymers through interfacial reaction.

10.3.4 Chemical interaction

Bond strength can develop at the adhesive-substrate interface through specific chemical interactions. Chemical bonding as it relates to the bulk polymer is reviewed in Section 4.2.1.5. These same types of bonds can be used to promote adhesion at the interface between two dissimilar materials. The relative strength of these bonds is summarized in Table 10.8. The dispersive and polar bonds are collectively known as van der Waals forces. They are relatively weak and act over very short distances. They can be very numerous as atoms along the polymer chain can interact with other atoms of neighboring chains to collectively provide strength at an interface. Polymers that form a diffuse interface through molecular diffusion often rely upon these types of forces. Examples include:

- Heat sealing of polyethylene to itself.
- Bonding high-density polyethylene (HDPE) to a linear LDPE (LLDPE) or EVA sealant in coextrusion.
- Bonding of PE-based tie resins to PE in coextrusion.
- Bonding of EAA tie resins to LDPE in coextrusion coating.

If conditions do not favor diffusion, then specific chemical interactions must be promoted at the interface to achieve good bond strength. This may be accomplished by modifying the substrate, incorporating specific chemical functionality into the adhesive, or both. In general, the stronger and most durable bonds involve covalent, ionic or acid/base interactions, although van der Waals forces are also used. Some examples include:

- Heat sealing an EVA-based lidding film to a PET tray. Little intermolecular diffusion can occur since the Tg and Tm of PET is high. The polar vinyl acetate groups on the EVA bond with the polar groups on the PET through dipole–dipole and other polar van der Waals forces.
- Bonding of EVA to polyvinylidene chloride (PVDC) in coextrusion. This involves dipole–dipole interactions.
- Bonding of EMAA to PA6 in coextrusion. The acid groups in EMAA react with the amine end groups of the PA6 to form a covalent bond (imide or amide) or an ionic salt.
- Bonding MAH-containing tie resins to EVOH or PA in coextrusion. The anhydride reacts with the hydroxyl (OH) groups on the EVOH to form a covalent bond (ester) and to the amine end groups of PA to form an imide.
- Bonding EAA or EMAA tie resin to aluminum foil in coextrusion coating. The acid group in the acid copolymer reacts with the aluminum oxide to form an acid/base interaction.

TABLE 10.8 Chemical bond energies.

Type of bond	Range (kcal/mol)
Dispersive	0.02–10
Polar	
Dipole–dipole	1–5
Dipole-induced dipole	<0.5
Hydrogen bond	2–6
Chemical	
Covalent	15–150
Ionic	150–250
Donor–acceptor	
Bronsted acid/base	Up to 250
Lewis acid/base	Up to 20

- Oxidizing LDPE during extrusion coating to bond to aluminum foil. Oxidation forms polar species on the LDPE, such as acids, aldehydes and ketones, which interact with the aluminum oxide. The oxidation is typically accomplished through heat or ozone treatment.
- Priming OPET, OPP or PA film with a polyethylene imide (PEI) primer to promote adhesion to oxidized LDPE in extrusion coating. The imide group on the PEI provides an anchor to the film (which may be treated) and the oxidized PE (carboxyl groups formed during oxidation) forms amide linkages with the PEI.

In extrusion coating, adhesion of LDPE is generally improved by pretreating the substrate. When functional polymers are used, pretreatment may not be necessary. Leclere et al. [42] consider the effect pretreating paper has on bonding to MAH-functionalized polyethylene. The functionalized polyethylene is made via either copolymerization or chemical grafting and the amount of functionality is varied. Corona, ozone and flame treatment are studied. They find that low levels of anhydride functionality respond favorably to pretreatment, but higher levels do not.

Allen [43] and Bodine and Tanny [44] show that the adhesion of a tie layer (with anhydride functionality) to various substrates (PET, PP, paper, and aluminum) can be improved by first priming the substrate with PEI to form stronger chemical bonds at the interface. Alternatively, the adhesion of PE, PP, or EVA in monolayer extrusion coating can be improved by adding a tie layer concentrate to the resin and priming the substrate. Other performance properties may also be enhanced, such as the water resistance of the bond.

The reactants must be accessible to each other at the interface with enough time and temperature for chemical interaction to occur: the reacting functional groups must be able to find each other. This requires that they be present in sufficient number or concentration. They also must be mobile enough to reach the interface and have enough energy and time to react. Higher processing temperature favors greater chain mobility and faster reaction rates. Longer contact times at temperature allow for more diffusion of reactants to the interface and for more bonding to take place. Chain mobility and surface energetics also play a role. For example, MAH-functionalized tie resins bond very well to PA6 in the melt but poorly to solid PA films. The amine end groups on the PA6 are accessible to the anhydride in the melt but not in the solid-state. The amines may be buried beneath the surface of the PA in the solid-state. Steric hindrance and other factors may also affect accessibility. An example that is described more fully in Section 10.4 is the difference in the performance of anhydride-grafted polyethylene and anhydride copolymers. When the anhydride is copolymerized in the polymer chain, it is less accessible and reactive than when it dangles from the chain, as it does when it is grafted. The grafted polymer performs well with only a fraction of the anhydride concentration of the copolymer.

MW is another factor. If the functional group is on the end of the chain, reducing the MW increases the number of functional groups (by decreasing the average chain length and increasing the number of chains). Lowering MW also reduces the viscosity and increases the diffusion rate, which may help get the functional group to the interface. If the MW is too low, however, the strength of the interfacial region may be weakened.

Evidence of chemical interaction at the interface in the coextrusion of polymer melts is found in a number of studies. Botros et al. [45–49] examine the reaction of MAH-grafted polyolefins to EVOH through melt blends and coextrusion. FTIR spectroscopy studies on melt blends show a complete disappearance of the peaks associated with anhydride after 1 minute in an extruder, indicating the reaction is fast enough for industrial processes. The melt viscosity of the blends also increases suggesting interfacial reactions. Scanning electron microscopy (SEM) of peeled interfaces from coextruded films shows local regions with adhesive residue suggesting cohesive and adhesive failure at the interface. They conclude that the tie resin forms strong covalent bonds in local regions. They also find evidence of rupture of the EVOH layer in places, again suggesting strong bonds. This is confirmed with electron spectroscopy chemical analysis (ESCA) and FTIR. In a second study, the complex viscosity and dynamic modulus of blends of EVOH with polypropylene (PP) grafted with MAH and their components are measured. The complex viscosity and dynamic moduli (G' and G'') correlate with peel strength in coextrusion. Stronger interactions between the PP grafted with MAH (PP-g-MAH) and EVOH increase the viscosity and modulus and also peel strength. FTIR spectra of the blends show the creation of an ester peak when the anhydride reacts with EVOH. They find that the tie resin with the greatest adhesion in coextrusion also has the highest viscosity and modulus, and the largest ester peak in the blends.

Macosko's group at the University of Minnesota uses the rheology of multilayer structures to study interfacial interactions. Through compression molding or coextrusion layer multiplier technology (see Section 8.5.2), they create hundreds of layers that amplify the interfacial area, creating a sensitive system to study different types of interactions. Levitt et al. [50] use multilayer stacks and extensional viscosity measurements to study nonreactive pairs [polystyrene (PS)-PP, PS-amorphous polyamide (aPA)], graft reactions where the reaction just occurs at the end of the PA chain (aPA/PS-*co*-MAH) and crosslinking where multifunctional polymers react (PS-MAH and PS-oxazoline). They make films of the pure polymers and then compression mold multilayer stacks of the pairs with 20, 50, and 100 layers.

A Meissner type extensional rheometer (see Section 5.3.6) with elongations of >20 (Hencky strain of 3) at 0.1/second and 220°C is used to measure the force to stretch the laminates under controlled conditions. The sensitivity can be increased simply by increasing the number of layers. The measured force increase over the bulk average is related to the interfacial tension. For the un-grafted polymer pairs, the interfacial tension increases with the extension rate. The grafting interaction produces a substantially different rheology response, with a much larger increase in interfacial tension than for the nonreactive pairs. Crosslinking pairs show strain-hardening behavior with the apparent viscosity increasing by a factor of 200 over the homopolymer average.

Saito and Macosko [51] use coextrusion layer multiplier technology to create 256-layer sheets, which they pull in a uniaxial extensional rheometer. The sheet contains alternating layers of ethylene-*co*-glycidyl methacrylate (EGMA) copolymer and MAH-grafted polyethylene (PE-g-MAH). The EGMA contains 8 wt.% GMA (an epoxy group) and the PE-g-MAH contains 4 wt.% MAH. MAH opens up to form a diacid, which reacts readily with the epoxy group on the EGMA. The extension force increases substantially, from which the interfacial stress is extracted and correlated with the amount of interfacial crosslinking.

When reactions occur at the interface between glassy (amorphous) polymers, the adhesive fracture energy, G_c, is found to be a function of the chain length and areal bond density. At low chain length, below the entanglement MW, G_c is rather weak (as the chain can be pulled out) and increases linearly with increasing areal bond density. Near the entanglement MW, crazing can occur and G_c increases with the square of the bond density (with chain scission of one of the connector chains) [52]. Boucher et al. [53] extend this to semicrystalline polymers by studying the formation of diblocks of a functionalized polyolefin and PA6, which is commercially important for coextruded films for packaging. They use a PP grafted with succinic anhydride (PP-g-SA) with approximately one anhydride group per chain, which is further diluted with pure PP (5% grafted PP and 95% PP). The anhydride reacts with the amine end groups of PA6 to form an imide bond that creates the PP-g-PA diblock at the interface. 600-μm sheet of the functionalized PP and PA6 are compression molded together at temperatures between 180°C to 220°C for contact times up to 900 seconds. This temperature is above the melting point (Tm) of the PP and below the Tm of the PA6. They find if both polymers are molten the reaction is too fast to study the effect of areal bond density on adhesion. Adhesion is measured using the Asymmetric Dual Cantilever Beam (ADCB) method. This involves inserting a wedge (razor blade) between the layers and measuring the crack length that forms in front of it (after allowing the sample to equilibrate for 24 hours). The stiffness of the layers separated by the wedge acts as the driving force for crack propagation, which decreases as the crack propagates. The facture energy (critical fracture release rate, G_c) is calculated from the crack length, modulus and thickness of the layers, and thickness of the razor blade. X-ray photoelectron spectroscopy (XPS or ESCA) is used to measure the density of the PE-g-PA chains that are formed at the interface, after first dissolving and washing away the free PA chains. They conclude that the ultimate failure of PP-PA6 joints in this in situ diblock system is due to cleavage of the PP chains. They argue that the reaction rate between the anhydride and amine is very rapid and the bond density at the interface is determined by the diffusion rate of reactants to the interface. A log-log plot of G_c vs. areal bond density suggests a square relationship, like glassy polymers.

Subsequent investigators [54–58] using this technique find adhesion depends on:

- Annealing conditions (time and temperature for reaction and diffusion to take place).
- Amount of anhydride functionality in the PP phase. With too little anhydride functionality there are not enough functional groups in the bulk to diffuse to the interface. Above a threshold and the interface becomes saturated, blocking further anhydride groups from reaching the interface.
- MW of the functionalized PP. Higher MW bridges the gap between crystalline lamellae at the interface to transfer the stress far away from the interface. There is likely a limit as very high MW will slow diffusion.
- The molecular architecture of the grafted PP. This should be as similar as possible to the matrix resin to ensure fast diffusion and rapid reaction at the interface and cocrystallization upon cooling.
- An annealing temperature above the Tm of both the PP and PA is followed by rapid cooling to promote maximum epitaxy at the interface (orientation of the crystals). This is generally achieved in industrial coextrusion processes.

These studies are not truly representative of commercial extrusion lamination and coextrusion processes used in flexible packaging due to the long annealing times, slow cooling rates (due to the thick sheet), and for coextrusion film processes, temperatures being below the Tm of PA. For example, anhydride-functionalized tie resins generally do not bond well to PA film in extrusion coating/lamination, presumably due to the short contact times before solidification. There are also stretching flows in coextrusion processes that increase the contact area just prior to solidification. In subsequent experiments, Barraud et al. [59–63] study the same PP-g-PA interface as Boucher et al. under conditions that better mimic commercial processes. They use thin layers so that the interface can be heated to the set temperature within 20 seconds and cooled rapidly. The PP-g-SA/PA6 laminate is heated to 250°C, above the Tm of both substrates, held for a fixed amount of time and rapidly cooled. Areal bond density and G_c are measured using the same techniques as Boucher. Barraud et al. also develop an apparatus that allows the melt sandwich to be stretched at a speed from 1–200 mm/s over 15 cm. The results of these studies show:

- The areal bond density reaches saturation within about 30 s.
- The areal bond density influences the crystalline morphology near the interface.
- The molecular characteristics of both the PP and PA affect the saturated bond density.
- After stretching, the bond density is higher than expected based on just dilution (expansion of interfacial area). This additional reaction occurs due to diffusion and convection during orientation.

Du et al. [15] study bond creation and peel strength between HDPE-g-MAH and PA6 using layer multiplier or microlayer coextrusion. HDPE (0.948 g/cc, 0.8 MI) is blended with HDPE-g-MAH containing 0.1 wt.% MAH at levels up to 20% using a twin-screw extruder. The blended HDPE, designated HDPE*, is coextruded with PA6 with 2, 8, 16, 32, 64, and 128 layers, with the thickness of the HDPE layers twice the thickness of the PA6 layers. The total thickness of the sheet is 1 mm. Adhesion is measured by a peel test conducted at 23°C at a pull speed of 2 mm/min. They use the method of Boucher to measure the areal bond density by XPS after first dissolving away the unreacted PA6. Key observations from this work include:

- In 2-layer structures, peel strength and bond density increase with increasing PE-g-MAH in the HDPE* blend and levels off at around 10% PE-g-MAH. The authors suggest this is evidence of the reaction being diffusion-limited.
- The peel strength increases with the number of layers in the coextrusion, particularly between 16 and 32 layers. The XPS measurements also show an increase in anhydride-amine bond density at the interface with an increasing number of layers (see Fig. 10.15). Although the total amount of interfacial area increases in the system with increasing layers, the interfacial area of a specific HDPE*-PA6 interface is constant. They attribute the increase in bond density to increases in shear and elongation with increasing layers, which allows more reactants to reach the interface. This may also breakup the densely packed HDPE-PA6 copolymer at the interface allowing other HDPE-g-MAH chains to reach the interface and bond to the PA6.
- Measuring the peel strength at different layers in the 32-layer structure is challenging. The failure interface tends to jump between such thin layers. No significant differences in peel strength are observed. The shear rate, in theory, should vary along the height of the die flow channel, with higher shear rates at the channel wall and zero shear at the centerline. Based on the observations described in the previous bullet, this should result in higher bond density near the channel walls. The authors say in their experimental set-up, however, the middle layers see high shear rates due to the splitting and recombination process of the layer multiplication technology employed.
- SEM shows that in the structures with 8 and 16 layers, the interfaces between HDPE* and PA6 layers are distinct, but with 32 and higher layers, they became diffuse. An interphase forms at 32 layers and higher, consistent with the higher bond density. This corresponds to the large increase in peel strength between 16 and 32 layers.
- The SEMs of the fracture after peel testing show that the 16-layer structure behaves like the 2-layer structure in that it fails along a single interface with some of the PA pulling out and remaining with the HDPE* layer (with some crack propagation in the bulk of the PA). This differs from that observed by Boucher et al. where failure is within

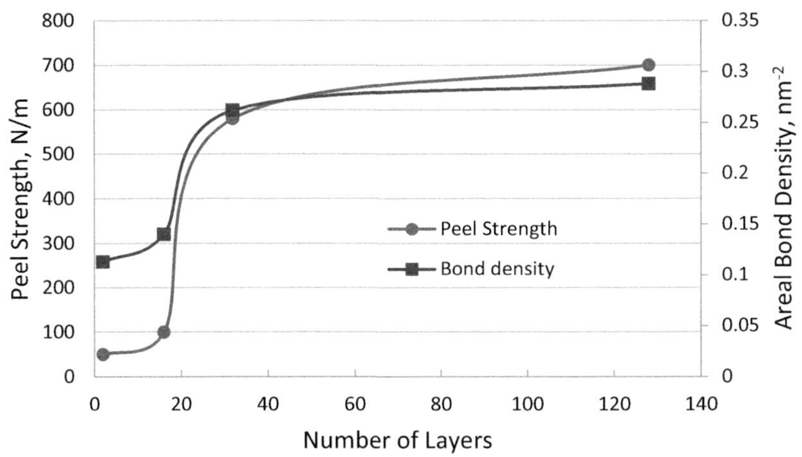

FIGURE 10.15 Peel strength and areal bond density in a coextruded sheet of HDPE-g-MAH and PA6 [15]. *Data from Q. Du, G. Jiang, J. Li, S. Guo, Adhesion and delamination failure mechanisms in alternating layered polyamide and polyethylene with compatibilizer, Polym. Eng. Sci. 50 (2010) 1111–1121.*

the polyolefin layer. In general, fracture tends to occur in the layer with lower yield stress. That would be the HDPE in this system, but because the PA layer is much thinner than the HDPE layer, it yields more readily in the PA layer. For 32 layers, there is jumping of the crack between layers, so extra energy is expended in breaking through the thin HDPE and PA6 layers, in addition to propagating along the interphase.

Song et al. [64] study the reaction of HDPE-g-MAH with PA6 using both a lab press (compression molding) and coextrusion. They coextrude bilayer sheet with different levels of anhydride (by blending down the HDPE-g-MAH with HDPE). Two dies are used, one that compresses the melt as it flows and another that does not. The bilayers have equal volume flow rate so that the interface is near the centerline of the flow channel where the shear rate and stress are zero. Laminations of similar compositions are also separately prepared with a laboratory press at different hold-up times. They use the ADCB method to measure adhesion and XPS to measure areal bond density after dissolving away the unreacted PA (per the Boucher procedure).

Their results show:

- G_c increases with increasing reaction time and concentration of PE-g-MAH in the blend with HDPE. G_c also increases with increasing processing temperature.
- The extent of the chemical reaction at the interface, as measured by the interfacial-grafted chain (PE-g-PA) concentration (areal bond density), Σ, increases with increasing reaction time and temperature.
- The fracture energy increases with increasing interfacial chain concentration, as is shown in Fig. 10.16. $G_c \propto \Sigma$ for small Σ and $G_c \propto \Sigma^2$ for larger Σ. They cite literature that G_c is proportional to Σ if simple chain scission or chain pull-out is the cause of the interfacial failure. If a small amount of plastic deformation becomes involved (as in glassy polymers), then G_c is proportional to Σ^2. They show that this is also true for semicrystalline polymers, like that observed by Boucher et al.
- Σ is a linear function of C_o (initial concentration of MAH) at the beginning of the reaction. Therefore, $G_c \propto C_o$ in the initial stages of the reaction.

The paper goes on to report on the acceleration of the reaction rate during coextrusion; the reaction rate is two orders of magnitude faster than that found in the compression molding experiments. This is consistent with the observations of Du et al. Botros [46] reports similar behavior for an anhydride-functionalized tie resin and EVOH. Song et al. [65] report acceleration of peel strength between polyurethane and an amine-modified polyethylene during coextrusion. Song et al. relate this to chain alignment during compression in the die. This phenomenon is discussed further in Chapter 15.

The studies described so far using the dissolution method directly show that chemical interaction at the interface correlates with enhanced peel strength or fracture energy. However, the dissolution technique is destructive, time-consuming and limited in the types of structures that can be studied. Recently, a nondestructive optical technique, sum frequency generation vibrational spectroscopy (SFG), has been shown to be sensitive to changes at reactive polymer interfaces, including the reaction of PE-g-MAH with EVOH and PA [66,67]. In SFG, a visible laser beam is overlapped with a tunable IR frequency laser beam at the interface to generate a signal. In theory, the SFG signal is only generated when the medium lacks inversion symmetry. Bulk polymer samples generally have such symmetry and so the technique is selective to solid surfaces and interfaces. The SFG signal at an interface is modulated by specific chemical groups

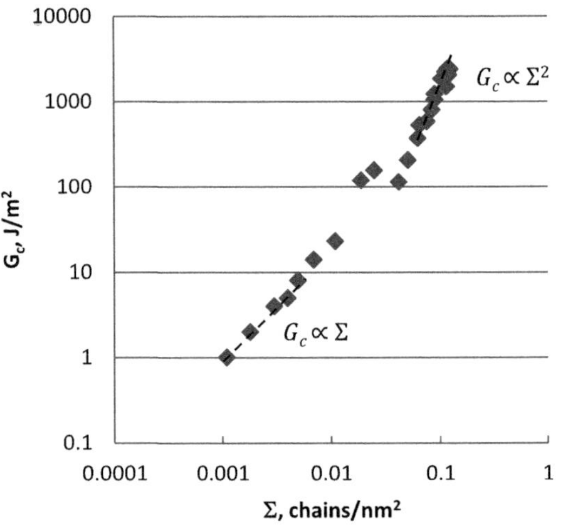

FIGURE 10.16 Fracture energy as a function of the surface concentration of reactive pairs (interfacial copolymer density, Σ) for thermal lamination or coextrusion of HDPE-g-MAh to PA6 [64]. *Adapted from J. Song, A.M. Baker, C.W. Macosko, R.H. Ewoldt, Reactive coupling between immiscible polymer chains: acceleration by compressive flow, AIChE J. 59 (2013) 3391–3402.*

and their orientation in a way that can be interrogated by this technique. The SFG spectra have the potential to follow reactions at film interfaces, although the early work has been on very controlled interfaces generated in the lab. These studies have been able to detect changes in the chemical signature and orientation at these reactive interfaces that correlate with changes in adhesion. While this work is still in the proof of concept stage, it may eventually be applied to real multilayer packaging films in the future. One limitation is that the structure must be optically transparent for the signals to penetrate to the interface.

10.3.5 Boundary layer

The strength of the interface is critical to good adhesion. If a weak layer forms near the interface, even if the bond strength is high, the peel strength may be poor due to cohesive failure within the "boundary" layer. Weak boundary layers can form for a number of reasons. Some examples include:

- **Low MW polymer:** If the adhesive or substrate contains a low MW component, this may create poor cohesive strength. Even if the low MW component contains functional groups that securely bond to the substrate, the lack of chain entanglement near the interface will cause the chains to be pulled away from the interface when stressed. One contributor to loss in MW is chain scission brought on by polymer degradation from too much heat exposure during processing, excessive oxidation (typically in extrusion coating), or overtreatment. The low MW may also be caused by insufficient curing of a primer or adhesive.
- **Surface contamination:** Oils, blooming agents, and various processing aids can interfere with adhesion and create a weak boundary layer. These may be byproducts of the substrate or polymer manufacturing process, such as oils from the rolling of aluminum foil. They may be slip, antiblock or chill roll release agents purposely added to modify the surface of the film or polymer. These act by migrating or "blooming" to the surface. Chapter 12 describes blooming agents in detail. Excessive amounts can form a "chalk" like appearance on the surface of film and cause adhesion problems, especially during printing. Various lubricating agents, such as stearates, can interfere with adhesion. Surfaces can pick up contamination from the reverse side of a multilayer film while wound in the roll. This is known as a reserve-side transfer. Contaminants may also come from outside the film, such as lubricating sprays used during lamination or coating to prevent chill roll sticking. Silicone sprays are especially problematic towards adhesion. Contaminants may act to block wetting, diffusion or chemical interaction directly, or by creating a weak boundary layer near the interface due to their low MW.

10.4 Tie resin technology

Obtaining good adhesion between polymer melts (coextrusion) or between a molten polymer and solid substrate (extrusion coating and lamination) can be challenging, depending on the specific substrates involved. As described in Section 10.1, often specialized tie resins or coating resins are needed to achieve the desired results. The chemistry of the tie resin is important, as is the chemistry of the substrate, the design of the structure, the converting process and the end-use environment [68–70]. Solid substrates may need to be cleaned, treated or even primed to be receptive to the coating. In coextrusion, layers need to have similar flow properties to avoid instabilities. This section discusses the chemistry of tie resins, factors that influence their performance and selection criteria. The effect of processing on interlayer adhesion is taken up in greater detail in Chapter 15.

10.4.1 Extrusion coating and lamination

In extrusion coating and lamination, the molten polymer bonds to a solid substrate by wetting, chemical interaction and in some cases mechanical interlocking. Further bond strength develops as the polymer solidifies upon cooling. This section discusses the critical properties of the substrate and polymer for effective bonding. The coating process can have a significant impact on the adhesion through its influence on chemistry and stress. These are taken up in detail in Chapter 15. How the structure is ultimately used—the temperature and chemical environment it is exposed to—also affects the adhesion. This is described in Chapter 16.

10.4.1.1 Substrate

Film, foil, and paper substrates can be thought of as three-dimensional structures with both surface and bulk properties. On a microscale, the surface is generally not molecularly homogenous. Polymer films have molecular chain length variations (MWD), chain-end and branching variation, additives, catalyst residues and degradation products. Most films are amorphous at the surface but changes in surface crystallinity can affect surface energy (see Section 10.3.1). Metal foils often have multiple morphologies, additives, process contaminants and oxides (see Section 10.3.1.1). Paper has fiber length variation, fiber thickness variation, fiber orientation, porosity, binders and additives. On a macro scale, the surface has both distinct surface energy and topography that influence adhesion.

The surface and bulk properties of the substrate are important for peel strength performance. Key aspects of the surface include:

- **Contact angle/surface energy or tension:** A general guideline used by practitioners (see Section 10.3.1.1) is for the substrate to have a surface energy of about 10 dynes/cm higher than the coating for good wetting. Since most coating resins are polyolefin-based with surface energies of around 25–30 dynes/cm, substrate surface energies of 40–45 dynes/cm are typically used. Often this requires treatment with corona, ozone, flame or plasma [24,28,71–79]. Films treated by the manufacturer may lose some of their surface energy during storage; in-line treatment is typically used to ensure the energy level is sufficient and to help clean the surface of the film.
- **Chemical functionality:** For a durable bond, the substrate needs a "hook" for the polymer to latch onto. The hook needs to match the chemical functionality of the polymer. Some substrates naturally have functional groups that can be used to attach to the polymer. An example is the oxides on the surface of aluminum, which can interact with the acid groups in oxidized polyethylene or acid copolymers. For other substrates, treatment or chemical primers [80,81] may be required to provide the functional groups for bonding. A summary table of general guidelines for primers presented by Bentley and Singer [80] is reproduced in Table 10.9.
- **Cleanliness:** Contamination at the surface can interfere with chemical bonding or create weak boundary layers. Sources of contamination include oils from the substrate manufacturing process, catalyst residues, lubricants, processing aids and other low MW species. Some additives, such as slip agents, migrate to the surface (see Chapter 12); the surface may change with time. Exposure to air, water, and organic contaminants in the environment may also affect adhesion. Olsson-Jacques et al. [82] show that a freshly exposed aluminum surface quickly absorbs over ten atomic layers of organic contaminants. The contamination is nonuniform and affects adhesion over time. Van den Brand, et al. [83] study the change in the surface of aluminum with time and its effect on bonding with carboxylic acid. A fresh aluminum surface is prepared by vacuum evaporation and aged in ambient air or in a clean dry environment. Aging in air results in the adsorption of water and organic species on the oxide surface of the aluminum that could not be removed by solvents. Adsorbed water causes hydroxylation and the growth of the oxide layer. It loses 60% of its capability to bond to carboxylic acid over the course of 20 hours of aging. Aging in a clean dry environment results in much less change in the oxide layer and only a 35% loss in bonding capability. One effective method for ensuring a clean surface is in-line treatment, such as flame, corona or plasma treatment. The high-energy treatment burns off most contamination.

TABLE 10.9 General guidelines on chemical primers [80].

	Substrate			Property		
Primer type	Paper	Metal	Plastic film	Heat resistance	Moisture resistance	Chemical resistance
Shellac	Poor	Excellent	Poor	Poor	Poor	Poor
Organic titanate	Good	Good	Good	Fair	Fair	Fair
Polyurethane	Very good	Excellent	Excellent	Excellent	Excellent	Excellent
Polyethylene imide	Very good	Good	Excellent	Excellent	Poor	Poor
Ethylene acrylic acid	Excellent	Excellent	Fair	Fair	Excellent	Good
Polyvinylidene chloride	Excellent	Fair	Excellent	Good	Very good	Fair

Source: Reproduced with permission from D. J. Bentley Jr., F.M. Singer, Chemical primers to enhance adhesion and other properties, in: T. Bezigian (Ed.), Extrusion Coating Manual, fourth ed., TAPPI Press, Atlanta, GA, 1999, pp. 185–200.

- **Pore structure and roughness:** The surface topology influences how much polymer penetration and mechanical interlocking can take place. Although not typically used in flexible packaging, chemical etching and physical abrasion can be used to create a roughened surface for bonding. In flexible packaging, porous substrates such as paper, paperboard and nonwovens take advantage of the ability of the polymer to flow into the pores and around the fibers. Many applications using paper or paperboard look for evidence of fiber tear when evaluating adhesion. This occurs when the polymer takes the fibers with it as the coating is peeled away from the substrate. This happens when there is sufficient mechanical interlocking. Some papers, however, have been designed with poor cohesive strength near the surface so that fiber tear can be more easily obtained. In this case, a weak boundary layer is purposely created, which may be appropriate for some applications, but not for others.

  Some of the important properties of the bulk include:

- **Dimensional stability:** The substrate needs to withstand the temperature during coating without shrinking or distorting. Dimensional changes can cause stress, which reduces peel strength. Trouilhet and Morris [84] give an example of a coextrusion coating of two layers of LDPE onto OPET: OPET/primer/(LDPE-LDPE). The initial structure has poor adhesion between the OPET and LDPE. The layer of LDPE that contacts the OPET is extruded at 310°C and the outer layer at 280°C to reduce taste and odor issues. The authors use a heat transfer model to calculate the temperature of the OPET layer during coating. They find that the OPET never melts but it does get above the temperature at which the OPET starts to shrink. By switching to an OPET film with a higher shrinkage temperature and better dimensional stability, the peel strength improved. Troulihet and Morris hypothesize that the shrinkage imparts stress on the interface, reducing the peel strength.

  A similar problem occurs when extrusion coating hot LDPE onto a metallized film such as met-OPP. If the supporting film shrinks when exposed to the hot polymer, the induced stress may crack or craze the thin metal layer. This is more of an issue with metallized PP or PE than for OPET.
- **Physical properties:** The thickness, stiffness (modulus) and yield strength of the substrate can affect the peel dynamics, as discussed in Section 10.2.3. Yielding can result in stretching of the peel arm during the peel test, contributing to higher peel strengths. The thickness and stiffness of the substrate may change the local angle at the peel front, affecting the peel energy.

10.4.1.2 Polymer

The characteristics of the coating polymer contribute to adhesion performance, including:

- **Type of chemical functionality:** Chemical functionality is often added to the polymer, the substrate or both to achieve the desired bond strength. LDPE is the most common coating resin used in extrusion coating and lamination. It is inexpensive, has good thermal stability and processes well at high line speeds. By itself, however, it lacks chemical functionality for bonding to polar substrates such as aluminum foil. To overcome this, LDPE is purposely oxidized during the process by running hot (above 300°C), increasing the time in the air gap, and in some cases applying ozone treatment [85,86] to the melt curtain. The oxidation creates polar species such as acids, aldehydes and ketones on the surface that help with bonding to polar substrates. These bonds can be variable since they rely on the coating process to form the oxidative species. Specific functionality can be built into the polymer chain through copolymerization or grafting reactions. Table 10.10 gives some examples of functional groups that bond to various substrates and commonly used ethylene copolymers that have these functional groups. Co- and terpolymers of ethylene with acid [87–94], anhydride [12,95] and acrylates [96–102] are the most commonly used specialty copolymers in extrusion coating for their wide range of adhesion performance. EVA [103,104] is sometimes used but requires low-temperature processing to avoid thermal degradation. Silane chemistry [105,106] has also been studied but has not gained widespread adoption for flexible packaging.
- **Amount of chemical functionality:** Generally, higher amounts of functionally give better peel strength. The amount of functionality on oxidized LDPE is fairly small. By some measurements [107], it is equivalent to about 1 wt.% acrylic acid. Commercial grades of EAA copolymers used in extrusion coating range in acid level from about 3–10 wt.%. High acid levels tend to have better product resistance, which can lead to longer shelf life for packaged goods. There are tradeoffs. For example, higher acid results in less compatibility with LDPE. EAA is often used as a tie layer to coextrusion coat LDPE onto foil: Al//(EAA-LDPE). High acid gives better foil adhesion in the presence of aggressive food products, but lower adhesion to the LDPE. Intermediate levels are often used as a compromise.
- **Flow characteristics:** The polymer must wet the substrate in a short amount of time between first contact with the substrate and solidification. This requires low viscosity. If the viscosity is too low, however, neck-in and other

TABLE 10.10 Examples of chemical functional groups that bond to various substrates and ethylene copolymers that contain these groups.

Substrate	Functionality	Example polymers
Al Foil	Acid, anhydride, acrylate	Oxidized PE, EAA, EMAA, Ionomer, E/MAH/MA, EMA
OPET	Acrylate, vinyl acetate	EMA, EVA
OPP	Acrylate, vinyl acetate	EMA, EnBA, EVA, VLDPE
Paper	Acid	Oxidized PE, EAA, EMAA, Ionomer
Inks	Acrylate, vinyl acetate, acid	EMA, EnBA, EVA, E/BA/MAA

Al, Aluminum; *EAA*, ethylene acrylic acid; *EMA*, ethylene methyl acrylate; *EMAA*, ethylene methacrylic acid; *EnBA*, ethylene n-butyl acrylate; *EVA*, ethylene vinyl acetate; *MAH*, maleic anhydride; *OPET*, oriented polyethylene terephthalate; *PE*, polyethylene; *VLDPE*, very low-density polyethylene.
Source: Reproduced with permission from the Dow Chemical Company.

processing problems may develop. Also, since viscosity is related to MW, low viscosity may cause the cohesive strength of the coating to be too weak for high peel strength performance. As described in Chapter 5, the melt flow index (MFI) is not the best measure of flow, but for reference, the typical MFI (190°C/2160 g) range for polyethylene extrusion coating grades is from 4 to 12 g/10 minute.

- **Thermal stability:** The polymer is often run well above its melting point to achieve good flow characteristics and to provide heat for chemical interaction at the interface. LDPE is also run hot to promote oxidation. The polymer must have thermal stability to be run at the desired temperature without significant degradation. Degradation can cause off-gassing, taste and odor issues in the final product, gel and blackspecs. Degradation can also cause the polymer chains to either crosslink (increasing viscosity) or scission (reduce MW and strength). The use of antioxidants can improve thermal stability but must be used with caution if oxidation is desired for adhesion. Gregory [108], for example, shows that adding 50 ppm of an antioxidant to LDPE reduces the peel strength to aluminum foil from 200–300 to 50 g/15 mm. One of the reasons LDPE is the mainstay of the extrusion coating industry is that it has excellent thermal stability without the use of antioxidants and good melt strength.
- **Additives:** Chill roll release, slip agents, antiblock, lubricants, processing aids, antioxdiants and other additives may affect adhesion. For example, Gregory [108] reports a 50% drop-off in adhesion of LDPE to foil when 3000 ppm of calcium stearate is added. Additives can affect adhesion in a number of ways. One is that they may interfere or react with the functional groups in the polymer. The effect of adding antioxidants to LDPE is one such example. Some additives are designed to bloom to the surface, such as chill roll release and slip agents. They may change the surface properties or block the wetting and bonding at the interface, although normal amounts generally do not cause problems. Low MW additives may preferentially bond to the substrate and create a weak boundary layer. It is advisable to test the effect of additives on adhesion to ensure they do not have an undesirable effect.

10.4.1.3 Examples from the literature

A few examples from the scientific literature are provided here to illustrate some of these factors. Ventresca et al. [109] study the effect of clay precoating treatments on the adhesion of LDPE to paperboard during extrusion coating. They find many factors affect adhesion, including:

- Type and level of binder in the clay coating (styrene butadiene (SB), styrene acrylate (SA) or a modified starch (MS)). For SB and SA, lower binder level reduced peel strength; for MS peel strength increased. At the highest LDPE oxidation level the MS system provided the highest peel strength.
- The split between clay and calcium carbonate. At low top-coat thickness, increasing the clay content reduced peel strength. The opposite was true for high top-coat thickness.
- Coat weight of precoats and top-coats: higher precoat or top-coat weight increased peel strength.
- Use of PEI primer or not. The primer provided higher peel strength when the oxidation of the LDPE was high.
- Time in air gap and melt temperature of the LDPE. High oxidation (long time in the air gap and high temperature) of the LDPE during coating gives the highest peel strength and overwhelms the other factors. At low levels of oxidation, the treatment chemistry plays a significant role in the peel strength results. At low levels of oxidation, there is also significant age-down of peel strength after 1 week.
- Paperboard properties. Peel strength increases with increasing surface roughness and porosity of the paperboard.

Ulren and Hjertbert [110] study the effect of functional group chemistry on the adhesion to aluminum. They prepare laminates by solution casting or hot-pressing films to the aluminum. They measure peel strength and correlate the results with FTIR measurements of the interaction at the interface. Three types of ethylene copolymers are tested: poly(ethylene-*co*-vinyltrimethoxysilane), EAA copolymer and ethylene butyl acrylate (EBA). Three levels of BA (0.7–3.7 mol%) and AA (1.2–3.7 mol%) are used. The strength of the interfacial interactions (as measured by FTIR and peel strength) correlates with the concentration and acidity/basicity of the group: AA is more efficient than BA. Silane functionality provides strong adhesion to the aluminum at much lower concentrations than the acid or ester groups. Silanols as well as Al–O–Si linkages are detected at the interface.

It should be noted that silanes are typically not used commercially in flexible packaging due to their thermal stability issues. Acid copolymers are used commercially and provide good adhesion to a number of polar substrates [87]. Schultz et al. [31] study the adhesion of PE-g-AA to aluminum foil and find that the acrylic acid interacts with the aluminol on the surface of the foil leading to the formation of carboxylates.

Several investigators report an increase in peel strength when Al/EAA or Al/EBA laminates are immersed in boiling water [31,111]. They attribute this to the hydration of the oxides, forming a layer of pseudoboehmite (an altered form of oxide) on the surface of the foil. The growth of this oxide layer occurs mostly in the interfacial cavities, increasing the contact area between the foil and polymer.

Schulz and Lavelle [112] describe the peel strength to aluminum foil of three acid/anhydride-functionalized polymers: HDPE-g-AA (1% AA) (bulk grafted); EVA terpolymer (12.7% VA, 0.39% MAH, 1.8% grafted AA); and a blend of PP + PP-g-MAH. The PP-graft concentrate has a low MW due to chain scission that occurs during grafting. Laminates of films of the polymers to Al foil are made by hot pressing for 15 minutes. They measure contact angles in water to different alkanes to get polar and nonpolar surface energies. An EVA control does not bond to the foil; the presence of the AA and MAH groups on the EVA provide adhesion. They find the acid groups reorient to the polar foil interface as is described in Section 10.3.1.3. The PP blend gives poor adhesion, underscoring the fact that chemical interaction is not sufficient for good peel strength. Here the low MW-grafted molecules create a weak cohesive boundary layer near the interface.

Pascal [12] provides an example of how the mechanical properties and thickness of the coating affect peel strength. He coextrusion coats LDPE onto 37-μm aluminum foil using 10-μm of a tie resin: 37 μ Al//(10 μ tie—LDPE). The thickness of the LDPE is varied from 0 to 100 μm. The peel strength to the foil is shown in Fig. 10.17. The peel strength increases with increasing LDPE thickness until the thickness reaches about 50 μm, after which it decreases sharply. He attributes this change to a change in the peel geometry during the peel test. In the zone below 50 μm, the local peel angle is more acute and plastic deformation dominates. Starting at 50 μm the angle becomes less acute and the peel strength is more dependent on the interfacial strength. The effect of coating thickness is explored with more depth in Section 15.1.4.

The effect of the bulk properties of the polymer can be seen when comparing the peel strength of LDPE, EBA and EAA tie resins to aluminum foil [113]. Laminates are made by first creating an OPET/EAA/Al laminate and subsequently extrusion laminating it to a PE film: 13 μm OPET/19 μm EAA/9 μm Al//18 μm Tie//50 μm LLDPE film. The lamination temperature is 321°C. Table 10.11 shows the results of peel strength measurements. A T-peel configuration is used with a pull rate of 300 mm/min. The samples are separated at the tie-Al interface. The order of peel strength performance is EBA > EAA > LDPE. The EAA is generally believed to provide the best adhesion to the foil. Indeed, high

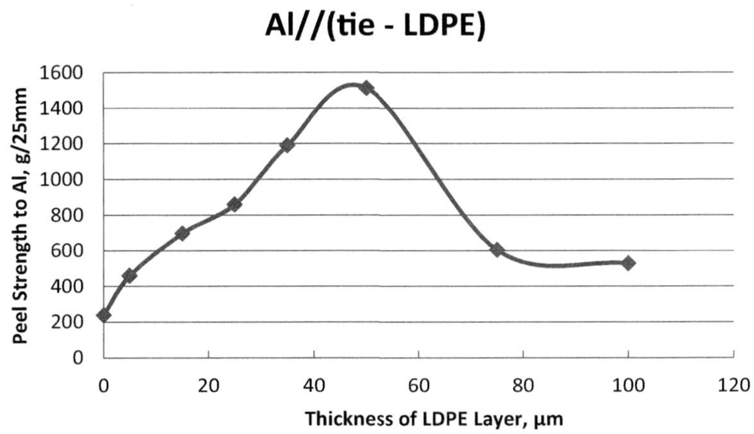

FIGURE 10.17 Peel strength as a function of low-density polyethylene thickness [12]. *Adapted from J. Pascal, Ultraversatile adhesives to broaden the possibilities of extrusion lamination, in: TAPPI PLACE. Conf., Cincinnati, OH, 2006.*

TABLE 10.11 Peel strength of LDPE, EAA and EBA tie layers to aluminum foil in extrusion lamination [113].

Polymer	Wt.% comonomer	Modulus (MPa)	Peel strength (g/25 mm)	Peel failure mode
LDPE	0	210	240	Peel
EAA	7	180	750	Peel
EBA	27	35	1470	Elongation/Peel

Source: Data from B.A. Morris, Y.M. Trouilhet, Lamination adhesion of foil to thermoplastic polymers, in: U.S. Patent 8,137,776 B2, 2004.

AA is needed to maintain adhesion to the foil in the presence of chemical attack from acid foods, as is discussed in Chapter 16. As described above, the AA forms a carboxylate with the oxides on the surface of the foil. LDPE relies on oxidation during the coating process to provide adhesion. At lower coating temperatures, the peel strength drops to zero. As mentioned earlier, the amount of acid functionality that forms during oxidation is equivalent to only about 1 wt.% AA. The peel strength of LDPE to aluminum foil is sensitive to processing conditions and typically has a lower value than the peel strength of EAA. EBA generally has poor adhesion to aluminum foil at lower coating temperatures. It too relies on oxidation. In addition, some of the BA groups can convert to AA groups, liberating ethyl alcohol, during extrusion. The amount of conversion is small and the level of AA is expected to be considerably less than the 7% AA on the EAA copolymer. So why does the EBA give higher peel strength performance than the EAA? The explanation lies in the failure mode. The EBA structure elongates as it peels from the aluminum. This creates substantially more energy for failure. The elongation relates back to the bulk properties of the polymer. The EBA is a much softer molecule, with a much lower tensile modulus, than the LDPE and EAA.

10.4.1.4 Choosing an extrusion lamination or coating adhesive resin

When choosing amongst lamination/coating polymers, several factors must be taken into account including cost, performance, mode of failure, processibility, regulatory status, and organoleptics. LDPE is the most widely used extrusion lamination resin due to its relatively low cost, ease of processing and ability to bond to many substrates, albeit in many cases with the help of oxidation of the melt curtain and priming of some substrates. Historically, MFI and density have been used to select LDPE grades for extrusion coating and lamination. Unlike linear PE, the density range offered by LDPE manufactured is confined to a fairly narrow range of about 0.915–0.930 g/cc, with higher density grades used for applications requiring a higher moisture barrier. With the density restricted, the MFI often becomes the driving factor for selecting a grade. Historically, most extrusion coating grades of LDPE had an MFI in the range of 4 to 12 g/10 minute, with the higher MFI grades used in higher speed processes where the melt curtain is drawn down more. Yet, as discussed in Chapter 5, MFI does not reflect subtle differences between suppliers or even grades offered by the same supplier. Differences in MW, MWD and long-chain branching affect the ability to draw the polymer down to thin coatings in high-speed processes and neck-in of the melt curtain. Often these are opposing attributes: grades that draw well typically have more neck-in. Clevenhag and Oveby [114] screen 17 grades of LDPE with MFI 7–9 g/10 minute and density 0.917–0.920 g/cc, long thought to be a uniform commodity for high-speed coating. Using pilot coating lines, they assess the resins for suitability in high-speed extrusion coating, evaluating drawability and neck-in among other attributes. They find that processibility does not correlate with MFI and develop their own laboratory screening protocols based on dynamic rheology measurements (G' and η_o) that better reflect subtle differences in MWD and long-chain branching.

Grades made on high-pressure autoclave manufacturing processes have historically been favored for extrusion coating over grades made on tubular reactor processes due to their higher level of long-chain branching. New advances in tubular reactor technology and chemistry have started to close the gap. With new technology come new paradigms. For example, Dow's advanced tubular reactor LDPE grades [115] process like traditional autoclave resins but have a lower MFI; the old guidelines do not apply to these resins. Like any choice of resin, it is helpful to consult with the manufacturer to select the best grade for a given application.

LDPE has its drawbacks for some applications. LDPE relies on oxidation from the extrusion process to achieve bonding to polar substrates like aluminum foil. Oxidization can be inconsistent at high line speeds and can result in off-flavor and odor issues. The bonds formed with the oxidized species are more susceptible to chemical attacks from demanding products. For demanding applications, acid copolymers, ionomers and other ethylene copolymers may be a consideration. Table 10.12 provides a selector guide for choosing amongst various ethylene copolymers for bonding to aluminum foil. Depending on the application, one polymer type may be favored over another.

TABLE 10.12 Relative ratings of some commonly used ethylene copolymers for extrusion coating/lamination to aluminum foil.

Attribute	LDPE	LLDPE	EAA, EMAA, ionomer	EMA, EBA, EEA
Low price	◇◇◇◇	◇◇◇◇	◇	◇◇
Green peel strength	◇◇	◇	◇◇◇	◇◇◇◇
Chemical resistance/shelf life	◇	◇	◇◇◇◇	◇
General processibility	◇◇◇◇	◇	◇◇◇	◇◇
Ability to run at low temperatures to reduce off-flavor/odor			◇◇◇◇	
Ability to run at high line speed and maintain adhesion	◇	◇	◇◇◇	◇◇
Regulatory status	◇◇◇◇	◇◇◇◇	◇◇◇◇	◇◇◇

Note: Structure is OPET/EAA/Al//Tie/LLDPE film. Lamination temperature is 321°C. A higher number of diamonds indicates better performance.
EAA, Ethylene acrylic acid copolymer; *EBA*, ethylene butyl acrylate copolymer; *EEA*, ethylene ethyl acrylate copolymer; *EMA*, ethylene methyl acrylate copolymer; *EMAA*, ethylene methacrylic acid copolymer; *LDPE*, low-density polyethylene; *LLDPE*, linear low-density polyethylene.

Table 10.2 provides additional information on the relative adhesion performance to various substrates of some commonly used polymers in extrusion coating and lamination. These are merely starting points. Actual results will depend on the processing conditions and overall structure design [116]. Processing considerations are taken up in detail in Chapter 15. Some substrates, such as PA film, require chemical primers for lasting bonds.

Cost is an important consideration when selecting the coating/lamination polymer. It is important to consider all aspects of the cost and not just the price of the resin. LDPE is usually the lowest price resin used in extrusion lamination/coating and for many applications it is the best choice. Since the adhesion of LDPE to polar substrates is dependent on oxidation, its performance is susceptible to variations in the extrusion coating process. A key parameter that is described in detail in Chapter 15 is the time in the air gap between the die exit and chill roll, which is indicative of the time available for oxidation to occur. Increasing time in the air gap generally increases adhesion. Increasing line speed decreases the time in the air gap, making it difficult to achieve good adhesion performance at high line speed. Oxidation of LDPE produces byproducts that contribute to odor and off-flavor issues for some food products. Hence, LDPE may be limited in yield and productivity, two key components of cost. One way to overcome these challenges is to use ozone treatment, which allows the LDPE to be run at lower temperatures and faster line speeds. Some of the organoleptic issues remain. Another approach is to modify the LDPE by blending it with a polar ethylene copolymer such as E/MAH/BA [15,63], EEA [70], or acid copolymer [117,118].

Morris [117,118] shows that compatibility between the modifier and the LDPE is important. He tests the adhesion to foil of a series of blends of acid copolymers with LDPE. He also makes blown films from blends of the same polymers and measures film haze. Film haze will decrease as the acid copolymer component of the blend is better dispersed (smaller domain size) which occurs as the compatibility between the components increases. The haze correlates with the inverse of the peel strength to foil. This suggests that a finer dispersion of the acid copolymer within the LDPE provides more acid functionality at the interface for bonding to the foil. Based on this finding, Morris develops a modifier designated AE (adhesion enhancer).

The performance of a 20% AE, 80% LDPE blend is shown in Fig. 10.18. Here 50-μm aluminum foil is coextrusion coated with 8-μm of the tie layer and 38-μm of LDPE. The performance of an LDPE control as the tie layer is poor, indicating the conditions were not favorable for oxidation. As is described in Chapter 15, the time in the air gap, while typical for commercial operations, is on the low side for good oxidative adhesion. The AE blend substantially improves the peel strength to around 600 g/25 mm. While not as good as a standard acid copolymer, which in this experiment gives a value of 1200 g/25 mm, this represents a significant gain in performance. To better understand how this boost in performance may play out in improved operational efficiency, Morris conducts a statistically designed experiment, varying the % AE in the blend (0%–20%), processing temperature (293°C–327°C), thickness of the coating (20–45 μm) and time in the air gap (60–100 ms). A quadratic response surface model is used to design and analyze the results. A statistical model is developed from the peel strength results.

All four variables are found to be statistically significant. Peel strength increases with increasing level of AE modifier in the blend, increasing temperature, increasing coating thickness, and increasing time in the air gap. The effect of

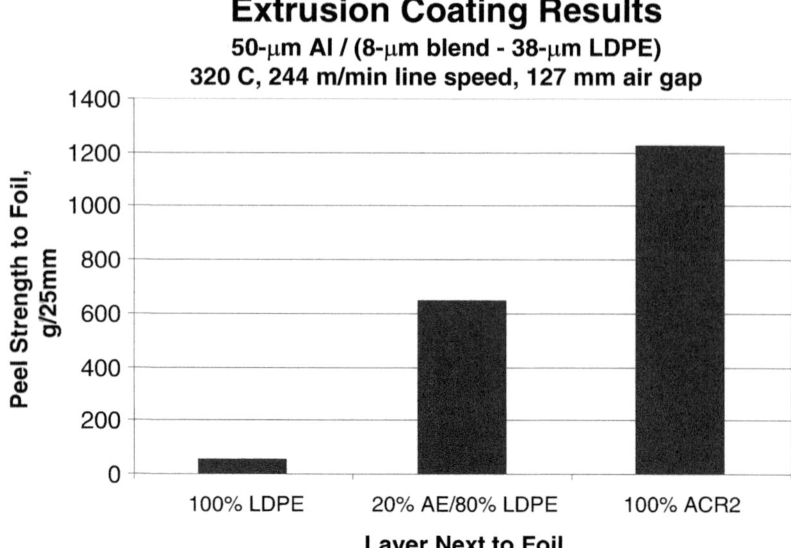

FIGURE 10.18 Performance of low-density polyethylene (LDPE), LDPE modified by blending 20% AE, and an acid copolymer (ACR2) in coextrusion coating. Peel strength is to the aluminum foil [117]. *Adapted from B.A. Morris, H.T. Thai, Modifying LDPE for improved adhesion to aluminum foil, in: TAPPI PLACE. Conf., 2004. Reproduced with permission from the Dow Chemical Company.*

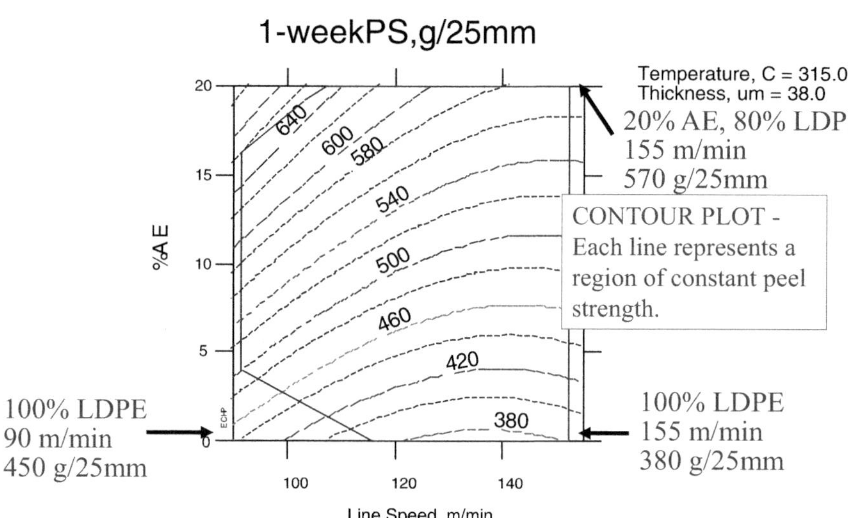

FIGURE 10.19 Contour plot of peel strength to aluminum foil showing the effect of varying % AE and line speed [117]. *Adapted from B. A. Morris, H.T. Thai, Modifying LDPE for improved adhesion to aluminum foil, in: TAPPI PLACE. Conf., 2004. Reproduced with permission from the Dow Chemical Company.*

temperature and TIAG is similar to that found with LDPE, suggesting an oxidation or stress mechanism [86]. The effect of thickness is discussed in detail in Section 15.1.4.

The benefits of adding AE to the LDPE can be illustrated using the output from the statistical model. The model predictions from the statistically designed experiment are mapped on contour plots, as shown in Figs. 10.19 and 10.20. Such a plot is similar to a topographical map where each line represents a line of constant elevation. Here each line represents a constant region of peel strength. Moving along the line is like walking along a ridge of a hill. Moving perpendicular to the line is like walking up or down a hill. The closer the lines are together, the steeper the hill.

The lower-left corner of Fig. 10.19 represents the case of running at 90 m/min with pure LDPE (0% AE). Here the coating temperature and thickness are set at the midpoint of the range investigated. Under these conditions, the peel strength is 450 g/25 mm. The effect of increasing line speed can be seen by moving horizontally across the plot. Increasing the line speed to 150 m/min reduces the peel strength to 380 g/10 minute (lower right corner). The effect of adding AE is found by moving vertically on the plot. At 20% AE and a line speed of 150 m/min (upper right corner) the peel strength is 580 g/25 mm. The productivity can be improved while maintaining the peel strength performance by adding AE.

Fig. 10.20 shows the effect of lowering the coating temperature, which may help reduce odor and off-flavors. Beginning in the lower right corner, the peel strength for 100% LDPE at 325°C is 450 g/25 mm. Moving along the

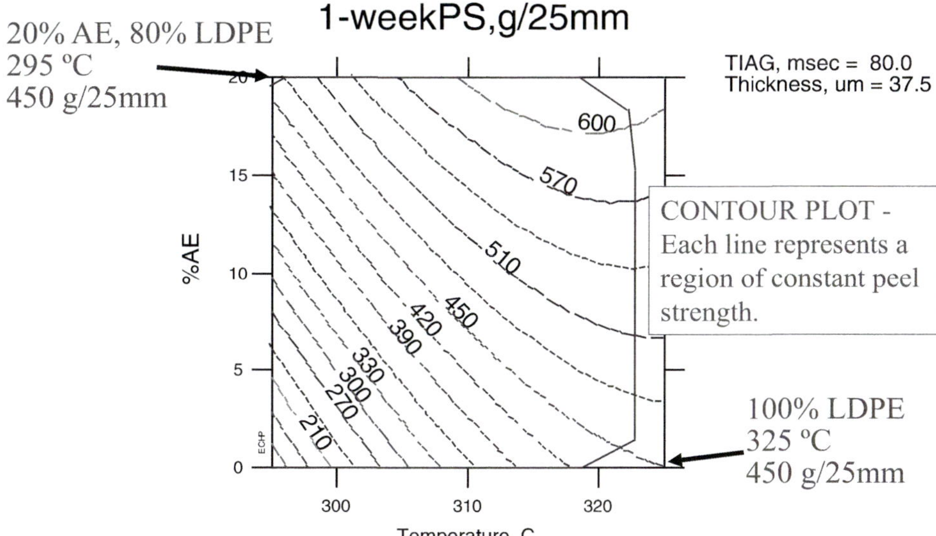

FIGURE 10.20 Contour plot of peel strength to aluminum foil showing effect of varying % AE and temperature [117]. *Adapted from B.A. Morris, H.T. Thai, Modifying LDPE for improved adhesion to aluminum foil, in: TAPPI PLACE. Conf., 2004. Reproduced with permission from the Dow Chemical Company.*

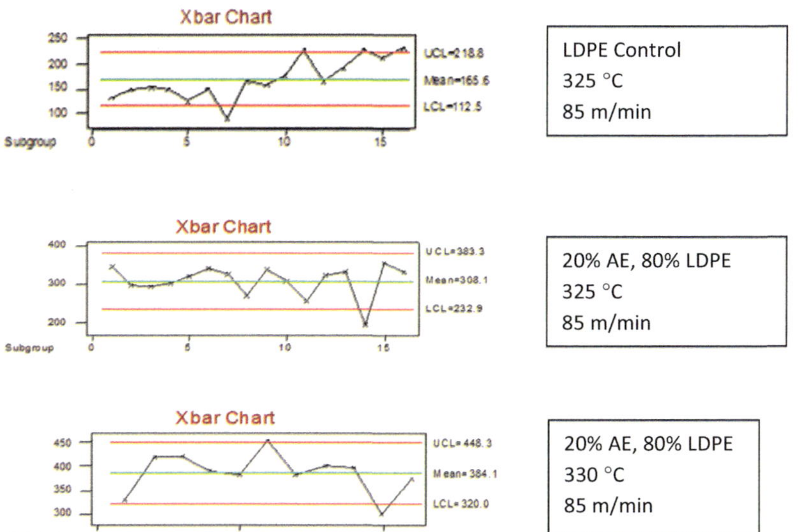

FIGURE 10.21 Control charts of peel strength performance at a converter run under three conditions [117]. *Adapted from B.A. Morris, H.T. Thai, Modifying LDPE for improved adhesion to aluminum foil, in: TAPPI PLACE. Conf., 2004. Reproduced with permission from the Dow Chemical Company.*

450 g/25 mm contour line to the upper left corner shows that adding 20% AE modifier will allow the temperature to be reduced to 293°C while maintaining peel strength. Lowering the coating temperature the same amount with 100% LDPE will reduce the peel strength to 150 g/25 mm (lower left corner).

Morris and Thai [117] go on to describe the benefits demonstrated during a trial at a converter. The converter produces a standard foil laminate for a dry food application. They have a quality specification of 250 g/25 mm for the peel strength to aluminum foil and were having trouble meeting it using LDPE. Control charts for three conditions are shown in Fig. 10.21. The top chart shows the preexisting situation using LDPE alone. None of the 15 production lots charted here meet the 250 g/25 mm specification. The yield is only 4% or there are 960,000 defects per million opportunities (DPMO).

The middle chart shows the case of adding 20% AE to the LDPE, keeping the line speed and temperature constant. The mean peel strength increases from 166 to 308 g/25 mm. Except for one data point (lot 14), all the results meet the 250 g/25 mm specification. The poor adhesion with lot 14 is found to be due to special causes—poor foil quality. Statistical analysis shows that by adding 20% AE, the yield increases from 4% to 86% (139,000 DPMO).

The lower chart shows the case for adding 20% AE as well as increasing the temperature by 5°C. Now the mean is 384 g/25 mm, yield 99.6%, and defects only 4,400 per million opportunities. The process reliability has been improved by four sigma.

This demonstrates that the use of a modifier can be a cost-effective approach for many applications. Situations, where product resistance or very high line speeds are needed, will still require specialty polymers used at full strength, such as acid copolymers and ionomers.

10.4.2 Coextrudable adhesives

10.4.2.1 Coextrudable tie resin technology

As described in Section 10.1 many combinations of polymers used in coextrusion do not bond well with each other. Resin suppliers have responded by developing tie resins that are sandwiched between the polymers to provide good adhesion. The general strategy is shown in Fig. 10.22. Frequently a polyolefin such as polyethylene is being bonded to a polar polymer such as PA or EVOH. The bonding mechanisms of diffusion and chemical interaction are primarily utilized. The tie resin is typically comprised of a matrix or base resin to which chemical functionality is added. The matrix resin provides bonding to one side of the structure via diffusion and the chemical functionality provides adhesion on the other side via chemical interaction [69].

Most commercial tie resins are based on polyolefins. This is to take advantage of the diffusion mechanism for bonding to polyolefin substrates. Polyolefins have no chemical functionality in which to interact and so diffusion is the primary way to achieve adhesion at the polyolefin-tie interface. Good compatibility is required. As is discussed in Section 10.3.3, matching polarities and solubility parameters is one way to ensure good compatibility and diffusion at the interface. In practice, if the substrate is polyethylene or an ethylene copolymer such as EVA or ionomer, the tie resin is based on polyethylene or an ethylene copolymer that is most compatible. Similarly, if the substrate is polypropylene then a polypropylene-based tie resin is typically chosen.

The polyolefin-based tie resin is modified so that it contains functional groups that can bond to polar substrates. A list of common coextrusion tie resin functional groups is given in Table 10.13. Some of these are the same as for solid substrates (Table 10.10). The main difference is that in coextrusion both the tie resin and the substrate are molten. Reactivity and accessibility of reactive species is higher in coextrusion. The functional group can be incorporated into the tie resin by direct copolymerization, chemical grafting, blending, or any combination of these methods. Acid copolymers are examples of direct copolymerization where acrylic or methacrylic acid is incorporated into the polyethylene backbone via free-radical polymerization. Grafting MAH onto a polyolefin chain is an example of the second approach. This is accomplished either in a reactive extrusion or solution process. A peroxide creates a free radical on the polymer chain to which the anhydride group attaches. The anhydride dangles from the polymer chain rather than being directly

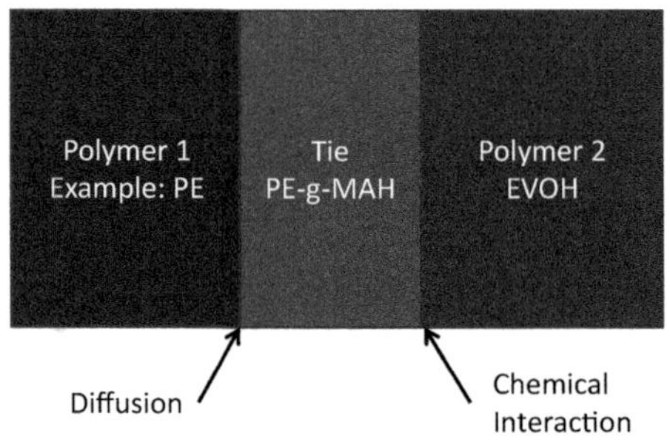

FIGURE 10.22 Design strategy for tie resins in coextrusion.

TABLE 10.13 Common functional groups used in coextrudable tie resins.

Functional group	Bonds to
Acid	PA, ionomers
Anhydride	PA, EVOH
Vinyl acetate	PVC, PVDC, PETg, PS, PP, ionomers
Acrylate (MA, EA, BA)	PET, PS, PP, Ionomers

BA, Butyl acrylate; *EA*, ethyl acrylate; *EVAOH*, ethylene vinyl alcohol; *MA*, methyl acrylate; *PA*, polyamide; *PET*, polyethylene terephthalate; *PETG*, polethylene terephthalate glycol; *PP*, polypropylene; *PS*, polystyrene; *PVC*, polyvinyl chloride; *PVDC*, polyvinylidene chloride.

incorporated into the backbone of the polymer. For this reason, grafted functional groups are typically more reactive than the same group directly polymerized into the chain. The blending approach is often employed by tie resin manufacturers or film converters who use tie resin concentrates. For economic reasons, it is less expensive to graft polymer and let it down in a base resin. Furthermore, most tie resins are formulated for specific adhesion attributes, in some cases using all three approaches. For example, a formulated anhydride-modified EVA-based tie resin contains a base resin with vinyl acetate groups incorporated by direct polymerization. The anhydride may be grafted onto a polyethylene or EVA polymer and then blended with the EVA and other ingredients to create the final formulation.

Tie resins can be further formulated with various modifiers such as rubber and tackifiers [119–123]. While these modifiers provide some level of compatibility or chemical interaction, in many cases they impact peel strength performance by increasing the viscoelastic dissipation, bending energy, or elongation of the peel arm. These factors are discussed in detail in Section 10.2.3.

10.4.2.2 Factors affecting peel strength in coextrusion

The chemistry of the tie resin and the substrates it bonds too are critical for adhesion performance, but the design of the entire multilayer film structure, how it is fabricated and the environment it is subjected to in its ultimate use is also important.

10.4.2.2.1 Tie resin matrix and adherend chemistry

Diffusion is the primary mechanism for adhesion of tie resins to nonpolar substrates such as polyethylene and polypropylene. Polyethylene and ethylene copolymers come in many varieties, as is described in Section 4.2.2. In general, polyethylene homopolymers (LDPE and HDPE) will bond well to each other and to linear alpha-olefin copolymers, commonly known as LLDPE. These are all nonpolar polymers. While metallocene-catalyzed polyethylene generally provides better adhesion than Ziegler–Natta catalyzed LLDPE, both will bond well to other types of polyethylene.

When it comes to polar ethylene copolymers, the type of copolymer and level of functionality affects adhesion. Most grades of EVA will bond well to polyethylene, as will EMA copolymers. Bonding to the more polar acid copolymers and ionomers can be more challenging. The adhesion of acid copolymers to LDPE decreases with increasing acid content. This is illustrated in Fig. 10.23 with data from a study by Hester et al. [124]. The peel strength of the laminates as made (green) and after storage for 1 week at 50% RH, 23°C decreases as the acid content increased from 4 to 9 wt. %. After exposure to acetic acid to simulate the packaging of acidic food products such as ketchup, salad dressings and orange juice the peel strength shows the opposite effect. The effect of the end-use environment is discussed in more detail in Section 10.4.2.2.5. Here it is sufficient to note that the selection of an acid copolymer for this application is often a compromise: high acid is favored for the best product protection and shelf-life performance, but low acid is favored to prevent delamination to the PE. Therefore, acid copolymers with approximately 7–9 wt.% acrylic or methacrylic acid are often used.

Ionomers are often more polar than acid copolymers. Their adhesion to polyethylene decreases with increasing acid content and neutralization. Some grades are more compatible with polyethylene than others, and can be used as tie resins to bond PE to PA. The type of polyethylene also affects the adhesion. HDPE has poor adhesion to many grades of ionomer; indeed, this property is used to an advantage for designing easy-open peel-seal packages that delaminate

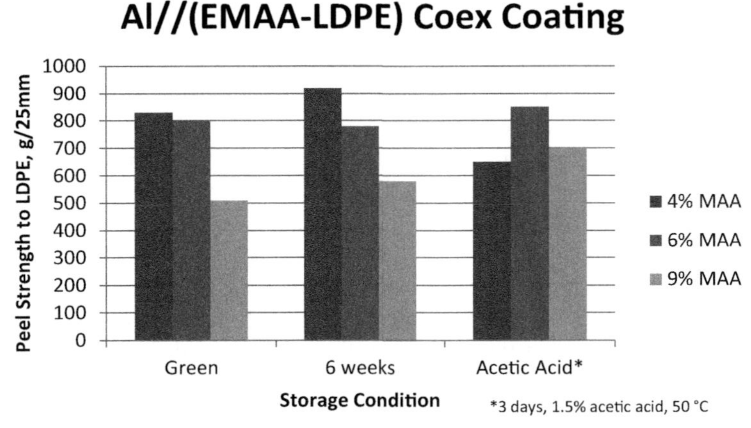

FIGURE 10.23 Peel strength performance of ethylene acid copolymers to polyethylene in coextrusion coating [124]. *Adapted from D.L. Hester, S.A. Sadik, Y. Trouilhet, Using acid copolymers as coextrusion tie layers, in: TAPPI Polymers, Laminations Coat. Conf., 1989, pp. 459–463.*

between the ionomer sealant and HDPE core (see Section 10.7.2). In the LLDPE family of resins, lower density generally improves adhesion to ionomer, and metallocene LLDPE provides better adhesion than Ziegler–Natta. Ziegler–Natta LLDPE, because of its broader comonomer distribution, typically has a fraction of chains that have little comonomer content. In effect, these chains act like islands of HDPE embedded in the LLDPE, negatively impacting the adhesion to ionomer.

Section 10.3.3 describes how solubility parameters can be used as a guide to compatibility and diffusion. The smaller the difference in solubility parameters, the more alike the polymers are and the more likely they will interdiffuse. One example where this has historically not been a good indicator is bonding polyethylene to polypropylene. Section 6.4.1 shows that the solubility parameters of these two nonpolar resins differ by less than 1 Hildebrand. Because there are no polar interactions between the molecules, this small difference in solubility parameters is not enough for miscibility. They should still be compatible, however. Yet in most coextrusion films, the adhesion between PE and PP is poor. The advancement of metallocene LLDPE has shed new light as to why. Research [125–128] shows that an amorphous layer forms at the surface of Ziegler–Natta LLDPE films. This is due to the broad MWD and comonomer distribution of these polymers: a fraction of the polymer contains relatively short chains with an abundance of comonomer. These low MW chains are excluded from the crystals and are forced towards the surface. The low MW amorphous region serves as an impediment to building strength at the interface. The longer PE chains most responsible for strength development must first diffuse through the weak amorphous region before entangling and co-crystallizing with the PP polymer chains. (See Section 7.5.3.2.2 for a discussion of this phenomenon and its effect on heat-sealing performance.) Chain entanglements formed in the amorphous region are weak and unanchored.

Metallocene-catalyzed LLDPE has been found to bond well to PP. The proposed explanation is that because of its uniform MWD and comonomer distribution, low MW amorphous regions do not form at the interface. Diffusing polymer chains become entangled and can participate in the PE or PP crystals. This creates an anchoring effect that builds strength at the interface.

EVA bonds well to PP in coextrusion. EVA is made via a free-radical polymerization process and generally has a very uniform comonomer distribution. This may help prevent the formation of the amorphous region at the interface. Also, EVA has a slightly higher solubility parameter than LDPE that is more of a match for PP.

One method to build strength between two incompatible polymers is to use a block copolymer. Block copolymers (see Section 4.2.1.6) have long sequences of a single repeat unit along the chain. They are most often found commercially in the styrene family, such as styrene-butadiene-styrene copolymers. These have long sequences of styrene and butadiene along the chain. If an appropriate copolymer can be synthesized such that one block is compatible with one polymer and the other block with another polymer, block copolymers can be used as tie resins. Cole and Macosko [129] find, however, that miscibility of the blocks with the polymers to be adhered to is not sufficient for good adhesion. They synthesize block copolymers for bonding PP or PE to PS. One of the copolymers does not participate in the polyolefin crystalline structure while the other one does. They find that co-crystallization gives better adhesion performance.

OBCs have shown good affinity between PE and PP in several studies. These are polymers with blocks of different olefinic compositions on the same polymer chain made by a unique catalyst technology involving chain shuttling [130]. Ayyer et al. [131] compare the adhesion performance of a polyethylene based OBC with two mVLDPE plastomers with density 0.855 and 0.876 g/cc, respectively. The OBC has soft segments with density 0.855 g/cc and an overall density of 0.876 g/cc. Coextruded sheet 2 mm thick with repeated microlayers of (HDPE-tie-PP) are tested using the three polyethylene grades as tie resins. The OBC provides higher peel strength than either of the two plastomers. Hu et al. [38] evaluate a PP-PE OBC consisting of 50% iPP and 50% LLDPE blocks as tie resins for bonding PE to PP in coextrusion cast films. The polymer outperforms several LLDPEs and polyethylene OBCs, giving good adhesion to both the PE and PP layers. Adding a MAH-grafted HDPE to the PP-PE OBC provides good adhesion of PE and PP to PA which is maintained at 85°C. Good adhesion before and after retort at 121°C is also obtained.

Using block copolymers as modifiers in tie resins may be less promising. Cole and Macosko [129] report that when blended with other polymers, already formed block copolymers take too long to get to the interface to be effective tie resins in coextrusion. A better strategy is to form block copolymers in situ at the interface via chemical reactions [129,132].

As described in Section 10.4.2.1, many tie resins designed to bond to EVOH or PA comprise a polyolefin matrix resin to which anhydride and other functionality are added. The matrix resin is selected primarily to bond to the polyolefin side of the film structure, as described above, but can also affect the peel strength to EVOH and PA. Fig. 10.24 shows an example of the effect of matrix resin for one tie resin formulation. LLDPE-based tie resins often outperform their LDPE and HDPE counterparts in blown film. The effect of PP matrix resin on peel strength is shown in

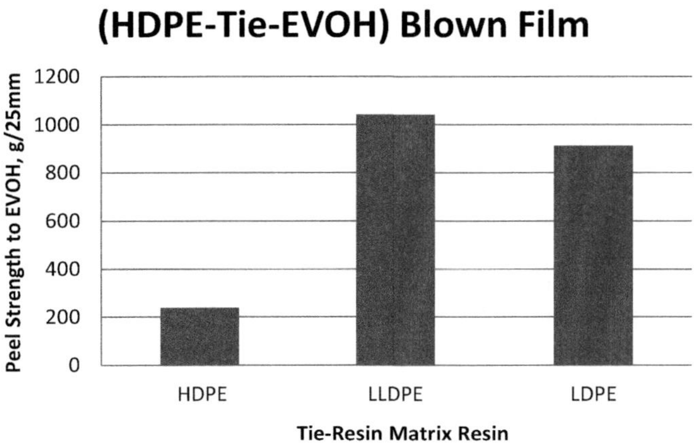

(HDPE-Tie-EVOH) Blown Film

Peel Strength to EVOH, g/25mm

1200
1000
800
600
400
200
0

HDPE LLDPE LDPE

Tie-Resin Matrix Resin

FIGURE 10.24 Peel strength to ethylene vinyl alcohol as a function of tie resin matrix resin type. All the tie resins are anhydride-modified polyethylene tie resins with the same general formulation except for the change of the matrix resin. Data is for blown film. *Reproduced with permission from the Dow Chemical Company.*

(PP-Tie-EVOH-Tie-PP) Blow Molded Bottles

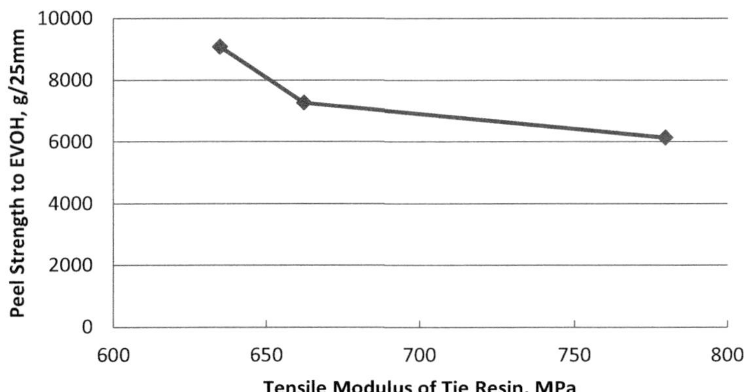

FIGURE 10.25 Effect of tie resin matrix resin modulus on peel strength to ethylene vinyl alcohol in blown molded bottles. Tie resins are anhydride-modified polypropylene resins with the same general formulation except the matrix resin [68]. *Adapted from B.A. Morris, Factors influencing adhesion in coextruded structures, TAPPI J. 75 (1992) 107–110. Reproduced with permission from the Dow Chemical Company.*

(PP-Tie-PA 6-Tie-PP) Cast Film

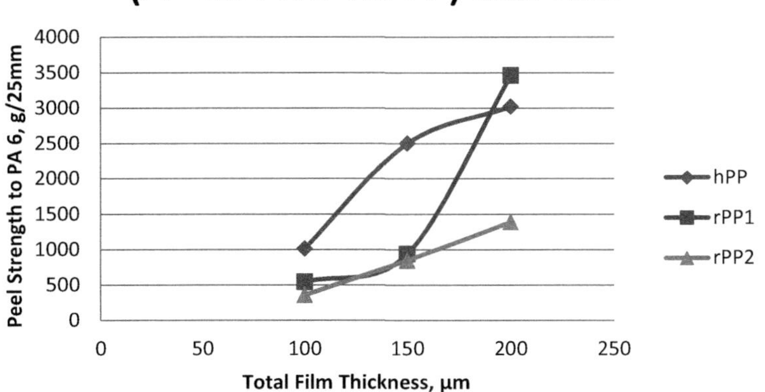

FIGURE 10.26 Effect of polypropylene (PP) tie resin matrix resin on peel strength to PA6 in cast sheet. hPP = homopolymer polypropylene; rPP = random PP-E copolymer polypropylene [133]. *Data from G.W. Kamykowski, Factors affecting adhesion of tie-layers between polypropylene and polyamides, J. Plast. Film Sheeting 16 (2000) 237–246.*

Figs. 10.25 and 10.26. Fig. 10.25 shows peel strength to EVOH decreasing with an increasing modulus of the base resin. Since homopolymer PP has a greater modulus than a random copolymer PP, this indicates a progression of higher peel strength with greater comonomer content. The opposite is shown in Fig. 10.26 where adhesion to PA for the tie resin with homopolymer PP as the matrix resin gives the best peel strength performance. All of these examples are anhydride-modified polyolefin formulations and within each set, the same grafted polyolefin is used as the source of the anhydride. The key takeaway is that the matrix resin upon which the tie resin is based upon can have a significant impact on the peel strength performance even though the actual bond strength comes from the anhydride. The performance may depend on the physical properties of the matrix polymer (such as modulus), the rheological properties (such as relaxation time [47] and strain-hardening behavior in extensional flow [134]), chemical compatibility between the grafted let-down resin and the matrix resin, and other formulation details. This is important for film converters who blend their own tie resins from tie resin concentrates. Fully formulated grades produced by a tie resin manufacturer have been optimized for matrix resin, graft source, and other additives. When the converter does their own blending, they take on the responsibility for understanding and testing the effect the matrix resin has on the performance in their structure.

The matrix resin can be modified through copolymerization or melt compounding. Rubber modification is often used to toughen polymers, such as the addition of an ethylene-propylene rubber (EPR) to toughen PP. Tougheners act by stopping cracks from propagating. Rubber modification is also sometimes used in tie layer formulations. Fig. 10.27 shows an example from Inoue et al. [122] where an EPR is added to an anhydride-modified PP tie resin formulation. The amount of anhydride is held constant. The peel strength to EVOH steadily increases with increasing EPR content until about 20 wt.% and then decreases. As described earlier, the patent literature is full of examples of different modifiers used in tie resins. The type of modifier and its level of addition affect the peel strength performance.

The chemistry of the adherend (what the tie layer is bonding to) is also important. The type of functionality on the adherend chain, whether the reactive group is located at the chain end or in the backbone, the concentration of the functional groups, and the ability for the functional group to get to the interface can all be important. Fig. 10.28 shows an example of the peel strength of a MAH-functionalized polyolefin to a blend of amorphous PA and PA6 in a coextruded blown film. Increasing the amount of PA6 in the nylon blend increases the peel strength. Presumably the end group chemistry is more favorable at higher levels of PA6 in the blend for greater reactivity and bond strength.

The physical properties of the adherend can also influence peel strength performance. An example is the evaluation of tie resin formulations for bonding to PA6 in a three-layer blown film. Initially films with the structure

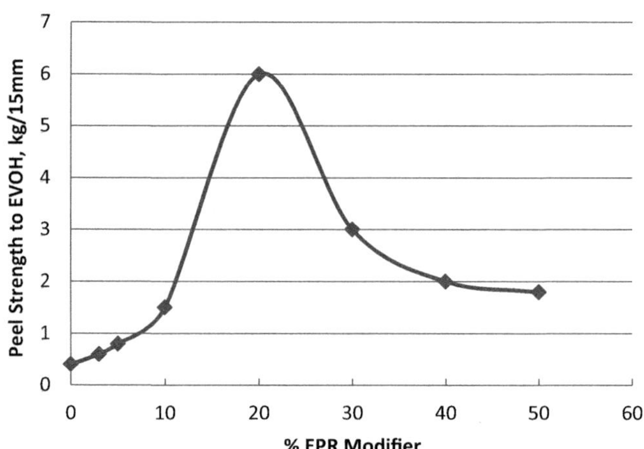

FIGURE 10.27 Effect of rubber modification of a polypropylene (PP)-based tie resin on peel strength to ethylene vinyl alcohol. The tie resin formulation contains PP, PP-g-MAH, and an ethylene-propylene rubber (EPR). The amount of PP-g-MAH is held constant at 10% and the amount of EPR varied. Formulation is melt mixed, pressed into a 0.5 mm sheet and thermally laminated to a 0.5 mm EVOH sheet at 180°C for 3 minutes [122]. *Data from T. Inoue, et al., Process for preparing laminated resin product, in: U.S. Patent 4,058,647, 1977.*

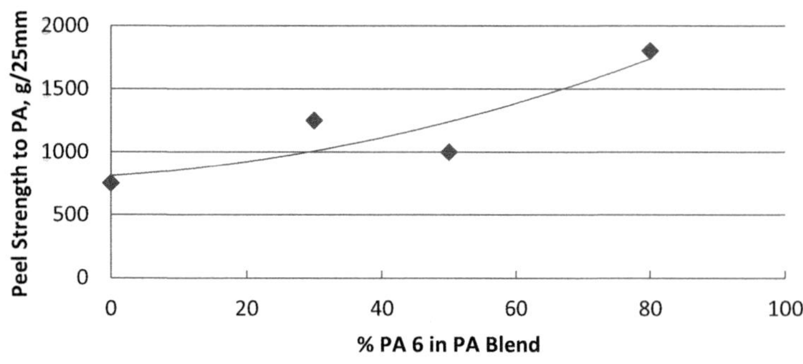

FIGURE 10.28 Effect of polyamide type on peel strength of an anhydride-modified ethylene copolymer tie resin [68]. *Adapted from B.A. Morris, Factors influencing adhesion in coextruded structures, TAPPI J. 75 (1992) 107–110. Reproduced with permission from the Dow Chemical Company.*

(PA6–tie–LDPE) were made and the peel strength between the tie and PA6 layers tested. The results showed no discrimination amongst the tie resin compositions. This is because the adhesion exceeded the yield strength of the LDPE; the LDPE peel arms elongated during the peel test. The peel test was measuring the tensile properties of the LDPE rather than the peel strength to PA. When the PE layer was switched to HDPE which has higher stiffness and yield strength, the peel arm no longer stretched during the test and differences in tie resin performance could be assessed.

Another example can be found in Fig. 10.29. Here four different tie resin compositions are compared for their peel strength to EVOH. A three-layer blown film structure is used: (EVOH–tie–sealant). In one structure an ionomer is the sealant and in the other an EVA. In all cases, the peel strength to the EVOH layer is reported. The ionomer-containing structure has lower peel strength, in general, than the EVA structure. This is because the ionomer is stiffer than the EVA. The EVA yields and elongates more during the peel test. The actual bond strength to the EVOH is the same, but the peel strength is higher with the EVA structure due to the energy to elongate the peel arm. Consequently, the choice of the sealant may affect the selection of the tie resin. In this set of data, Tie 1 gives the best performance for the ionomer sealant but Tie 2 provides the best peel strength when the EVA is the sealant.

10.4.2.2.2 Tie resin functionality

Choosing the appropriate chemical functionality to bond to the adherend is an important consideration in selecting tie resins. Table 10.13 provides a list of common functional groups and the polymers they bond to. Of particular interest in flexible packaging are tie resins that bond polyolefins to PA and EVOH. Historically acid copolymers and ionomers were used to bond to PA6 in the early days of coextrusion. Lightly neutralized zinc ionomers gave the best adhesion to nylon in internal studies at DuPont [135,136]. As shown in Fig. 10.30, weight percent acid in the 6%–12% range is needed to give good adhesion.

Anhydride-modified polyolefins have become the mainstay of tie resin technology for bonding to PA6 and EVOH [137–142]. The anhydride group reacts with the amine end groups of PA6 to form an imide [137] and with the

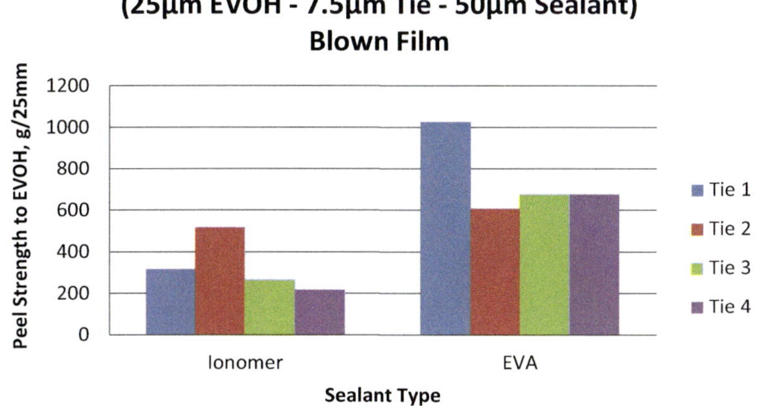

FIGURE 10.29 Effect of sealant type of peel strength of tie resins to ethylene vinyl alcohol. *Reproduced with permission the Dow Chemical Company.*

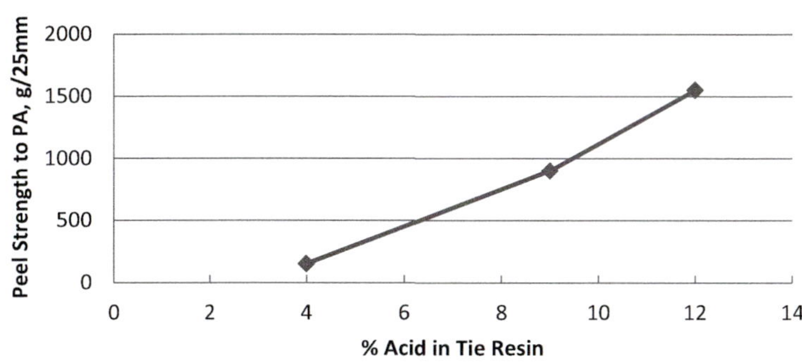

FIGURE 10.30 Effect of acid level on adhesion to PA6 in blown film. Tie layers are EMAA copolymers with wt.% MAA ranging from 4 to 12 wt.% [124]. *Data from D.L. Hester, S.A. Sadik, Y. Trouilhet, Using acid copolymers as coextrusion tie layers, in: TAPPI Polymers, Laminations Coat. Conf., 1989, pp. 459–463.*

hydroxyl group of EVOH to form an ester. These reactions are shown in Fig. 10.31. There is also some evidence for a reaction between MAH and the amide (NH) side groups in PA [143,144]. The amide group may hydrolyze to form NH_2 and a carboxylic acid. The NH_2 can then bond to the MAH. This 2-step process is thought to be much slower than MAH bonding the NH_2 end group. Macosko, et al. [132], for example, report that the reaction between MAH and the amine end groups is 99% complete in two minutes at 180°C.

Anhydride is typically grafted onto the polymer chain in a separate process after the polymer has been synthesized, rather than via copolymerization. Anhydride grafts tend to be more reactive than copolymers. For example, the patent literature teaches that anhydride levels in the range of 0.1−0.5 wt.% are used to bond to PA and EVOH when the anhydride is grafted onto a polyethylene backbone. This is an order of magnitude less than the amount of acid needed in an EMAA copolymer to bond to PA, as shown in Fig. 10.30. One supplier makes an ethylene-acrylate-MAH terpolymer with 3 wt.% MAH, but offers anhydride-grafted polymers for use as tie resins for PA or EVOH in coextrusion, presumably because they are more effective at one-10th the functionality. The difference is explained in Fig. 10.32 where the grafted anhydride group is less restricted than an anhydride group incorporated into the polymer chain via copolymerization. The greater mobility of the grafted anhydride group contributes to its enhanced reactivity.

Not all anhydride grafts are the same. Tie resin manufactures often have several technologies to choose from for optimizing peel strength for a given application. Fig. 10.33 shows an example of three different graft technologies evaluated in a cast film structure. Clearly, graft type 3 gives the best performance in this structure. In other applications, graft 1 or 2 may be better.

The amount of chemical functionality is an important parameter. Fig. 10.30 shows that increasing the acid content of an acid copolymer tie resin increases the peel strength to PA6. Fig. 10.34 shows a similar effect for anhydride. Here a PP-g-MAH concentrate is blended with PP as the tie resin. An increase in concentrate corresponds to an increase in anhydride content and greater peel strength. As described in Section 10.3.4, Song [64] directly measures the amount of anhydride-imide linkages in a coextrusion of HDPE-g-MAh and PA6 by dissolving away the free PA and measuring

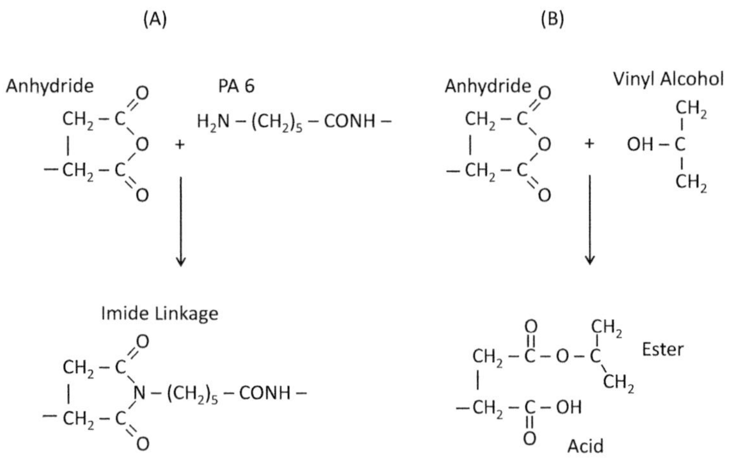

FIGURE 10.31 Chemical coupling of anhydride with (A) PA6 and (B) EVOH.

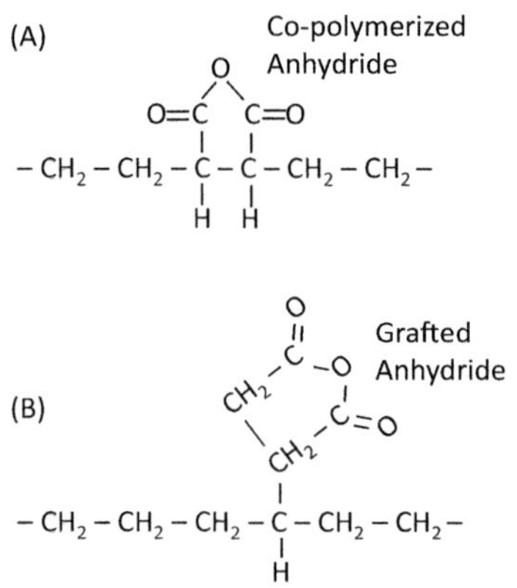

FIGURE 10.32 Chemical structure of ethylene-co-maleic anhydride copolymer (A) and maleic anhydride-grafted polyethylene (B).

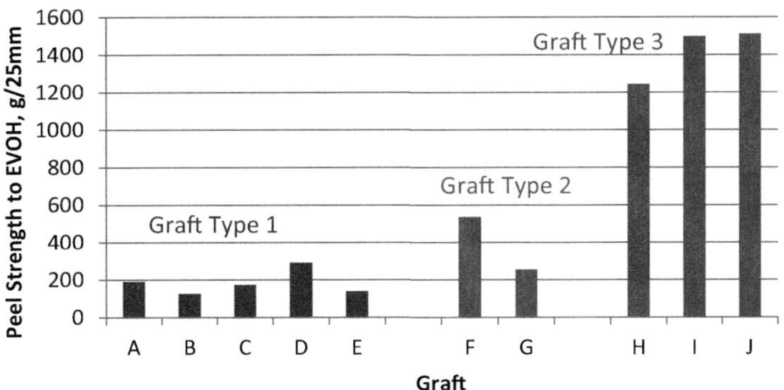

FIGURE 10.33 Influence of grafting technology of anhydride-functionalized tie resins. *Reproduced with permission from the Dow Chemical Company.*

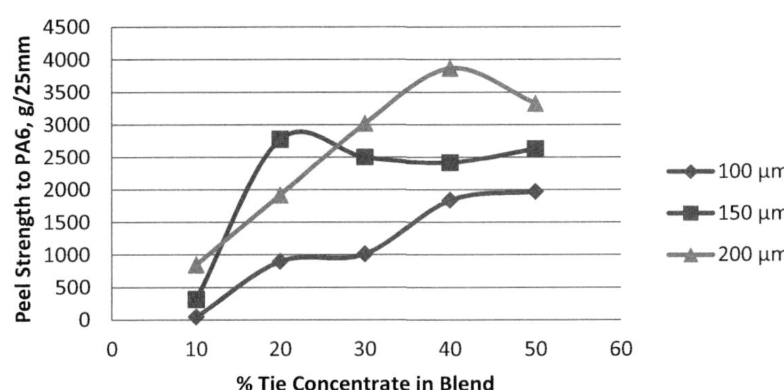

FIGURE 10.34 Effect of tie layer anhydride level on peel strength to PA. The tie layer is a blend of PP and a PP-g-MAH concentrate [133]. *Data from Data from G.W. Kamykowski, Factors affecting adhesion of tie-layers between polypropylene and polyamides, J. Plast. Film Sheeting 16 (2000) 237–246.*

the imide groups with XPS. The surface concentration of linkages at the interface is found to correlate with the initial concentration of anhydride.

Increasing chemical functionality has its limits. At some point diminishing returns or economics take their toll. Eventually, the bond strength may exceed the cohesive strength of the tie resin or adherend and further increases do not yield improved performance. Or the target peel strength is achieved and further increases just add cost to the tie resin. In some cases, the peel strength reaches a maximum and decreases with further increases in functionality. An example is PP-based tie resins. Grafting MAH onto PP generally reduces the MW of the PP due to the susceptibility of PP to chain scission. Adding too much low MW graft to a tie resin formulation can reduce its cohesive strength to the point where peel strength suffers.

10.4.2.2.3 Film structure

In addition to the chemistry of the tie resin and the layers it bonds to, the thickness and mechanical properties of the film structure can influence the peel strength. Figs. 10.26 and 10.35 show the effect of total structure thickness on peel strength. Here the structure is (PP–tie–barrier–tie–PP) with each tie layer being 4% of the structure [133]. The tie is an anhydride-modified PP and the barrier is PA6 or EVOH. The thickness is controlled by changing the line speed while keeping the output constant. The results show an increase in peel strength with increasing thickness. One explanation for this behavior comes from the fracture mechanics of the peel test (Section 10.2.3): as the peel arms become thicker and stiffer, the energy to bend the peel arms increases. A second explanation involves an increase in stress as the sheet is drawn down to thinner gauge. The effect of processing and stress are explored in detail in Chapter 15.

Fig. 10.36 shows the effect of varying the thickness of a (tie–ionomer) coating onto an OPET film. As the thickness increases, the peel strength to the OPET increases. Since the (tie–ionomer) is thin and flexible, the energy to bend the peel arm is likely not contributing much to the overall peel strength. The effect shown here is more likely a result of the mechanism described in Section 10.2.3.3 and Fig. 10.10 where increasing the thickness increases the zone of

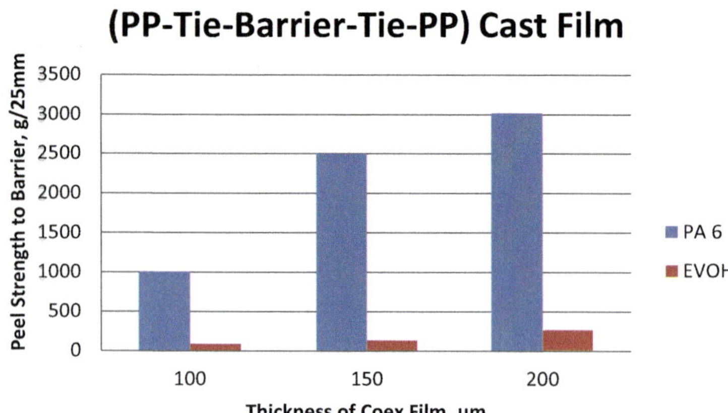

FIGURE 10.35 Effect of total film thickness on peel strength between a tie layer and barrier. Cast film of (PP–tie–barrier–tie–PP) with thicknesses of (38/4/16/4/38%) [133]. *Data from Data from G.W. Kamykowski, Factors affecting adhesion of tie-layers between polypropylene and polyamides, J. Plast. Film Sheeting 16 (2000) 237–246.*

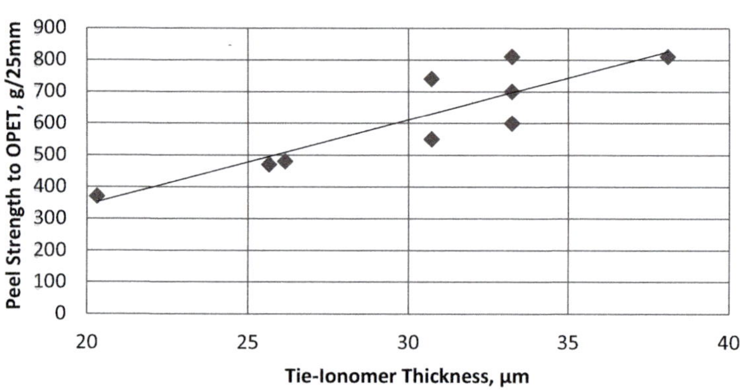

FIGURE 10.36 Effect of coextrusion coating thickness on peel strength to OPET film. *Reproduced with permission from the Dow Chemical Company.*

viscoelastic deformation at the peel front, increasing the fracture energy. Using typical values for the modulus and yield strength of ionomers, Eq. (10.15) predicts the critical thickness is around 150 μm. Since the (tie–ionomer) thickness is well below this value, the peel strength is on the rising portion of the peel strength vs. thickness curve (Fig. 10.9).

How thin is too thin for the tie layer thickness? This depends on the total structure, the strength of the chemical interaction at the interface, and the target peel strength needed for the application. Often it is limited by how well the converting equipment can control the thickness uniformity. About 2%–4% of the structure is frequently the lower limit, but some structures have stretched this limit.

Ayyar et al. [131] find that when the tie layer thickness becomes very thin the mode of failure changes and the peel strength decreases faster than is predicted by the change in viscoelastic deformation. They make coextruded tapes with a repeating structure of (PE–tie–HDPE) using layer multipliers to divide and stack the layers (see Section 8.5.2). For films with a tie layer thickness in the range of 2–14 μm, 65-layer structures are made. To study the regime of tie layer thickness from 0.1 to 2 μm, films with 257 layers are created. The results for an mVLDPE tie resin with a density of 0.855 g/cc are shown in Fig. 10.37. As the tie layer thickness decreases from 14 to 1 μm the fracture toughness (which is related to peel strength) decreases linearly, as expected from the fracture mechanics theory described in Section 10.2.3.3. The fracture toughness correlates with the measured and calculated zone of deformation length. Below 1 μm they find that the change in fracture toughness decreases faster than expected. SEM reveals that continuous plastic deformation in the thick tie layers gives way to fibril formation as the tie layer thickness drops below 1 μm.

The thickness and stiffness of other layers in the structure also affect peel strength. Some examples of the effect of stiffness of the tie resin are presented in Section 10.4.2.2.1. Fig. 10.38 shows an example where the increasing thickness of the adherend, in this case EVOH, increases the peel strength. Here the structure is a coextrusion coating of (LDPE–tie–EVOH–tie–LDPE) onto Tyvek, a spun-bond nonwoven PE fabric. The adhesion between the anhydride-modified polyolefin-based tie resin and EVOH is shown for two EVOH thicknesses. The stiffness of the peel arm may

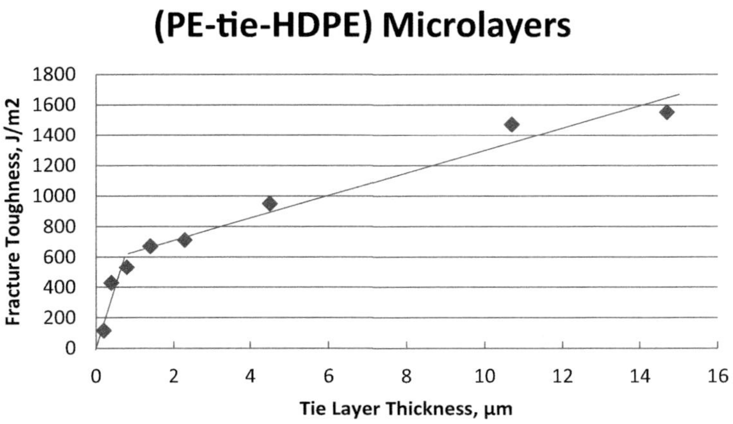

FIGURE 10.37 Effect of tie layer thickness on fracture toughness. Structure is repeated layers of (PP–tie–HPDE) where the tie is a metallocene plastomer with density 0.855 g/cc [131]. *Data from R.K. Ayyer, A.R. Kamdar, Y.J. Lin, P.S. Dias, B.C. Poon, A. Hiltner, et al., Effect of tie layer thickness on the adhesion of ethylene based copolymers to polyolefins, in: SPE ANTEC Conf., 2009, pp. 428–432.*

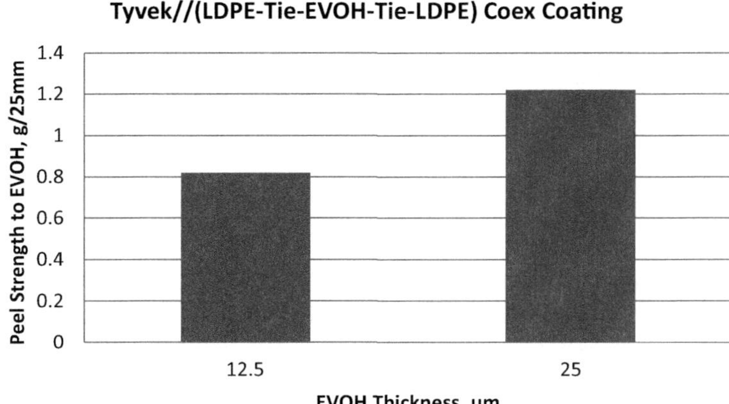

FIGURE 10.38 Effect of ethylene vinyl alcohol (EVOH) layer thickness on peel strength to EVOH in a coextrusion coating. *Reproduced with permission from the Dow Chemical Company.*

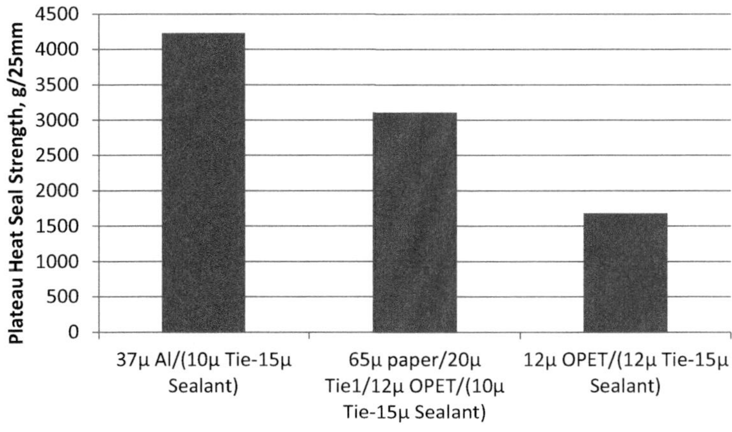

FIGURE 10.39 Effect of film structure on seal strength [145]. *Data from J. Pascal, Peelable structures for PP, PS, PET and PVC. Understanding how multilayer construction influences seal and peel properties, in: TAPPI PLACE. Conf., 2008.*

be one factor for the higher peel strength of the thicker EVOH layer. Another factor may be the slower cooling of the molten web when it contacts the chill roll during coating due to the thicker extrudate.

Fig. 10.17 in Section 10.4.1.3 gives an example from Pascal [12] showing the effect of LDPE thickness on the adhesion of a tie resin to aluminum foil. Pascal [145] provides another example of how the structure design can affect peel strength. Heat seal strength of three structures each with the same sealant and tie resin is shown in Fig. 10.39. The seal strength varies considerably amongst the three structures even though they have the same sealant. Differences in the adhesion of the tie resin to foil and PET film (which is likely primed, but this is not revealed in the paper) may account for some of the differences in performance. More than likely, however, the observed differences are due to variation in energy dissipation due to bending and yielding of the peel arms.

10.4.2.2.4 Processing

Contact time, temperature, interfacial area creation, and stress/orientation impact peel strength performance. These are determined by the converting process and specific operating conditions, which often can be optimized for adhesion. The effect of processing on adhesion is taken up in greater detail in Chapter 15. Here an overview of how the process affects adhesion and tie resin design is provided.

Anhydride-containing tie resins have a special temperature consideration. The anhydride group can convert to a diacid with the addition of water. This is an equilibrium reaction with the anhydride form favored at high temperatures typical of extrusion. During transportation and storage, the resin pellets absorb enough water to drive some of the anhydrides to the diacid form. The diacid form does not bond well to PA or EVOH. The anhydride can be reformed in the extruder. Tie resin suppliers generally recommend a starting temperature of 210°C or higher to ensure the reaction is complete. Once good adhesion performance is established, the temperature can be incrementally reduced to determine the best temperature for that particular process.

The performance of tie resins may differ from one process to another. An example is shown in Fig. 10.40 comparing the results of peel strength measurements on a series of cast and blown films. The structure is the same for each: (PP–tie–EVOH). Two tie resins are evaluated. The cast film gives higher peel strength for both tie resins than the blown film, which may be due to the greater thickness of this film. In the cast film, both tie resins give the same performance, but in the blown film tie A-2 substantially outperforms tie A-1. This underscores the importance of testing tie resins in the same process and structure as the commercial process.

In general, less orientation and longer contact times at higher temperatures favor greater adhesion. Many processing parameters influence these characteristics and can be used to help optimize adhesion:

- Adhesion is favored at higher temperatures, within the bounds of the thermal stability of the polymers. Higher temperature increases the diffusion rate of the polymer chains and enhances the rate of chemical reactions at the interface.
- Increasing contact time in the die after the layers have been combined and in the air gap between when the polymer leaves the die and is quenched also improves adhesion. For example, Morris [146] finds that increasing the time between the die exit and the frost line in blown film increases peel strength by up to ten times, as is shown in Fig. 10.41. Similar effects are found for coextrusion coating and cast film [147]. Compressive flow has been found to accelerate reaction rates at the interface [64,65,148]. These effects can be substantial and are described in detail in Chapter 15.
- Increasing line speeds reduces contact time, which can reduce peel strength.
- Increasing the quench rate or heat transfer rate can reduce the time the polymers remain in contact at elevated temperatures, reducing peel strength. Decreasing chill roll temperature in coextrusion coating, lamination or cast film or using inner bubble cooling in blown film increases the quench rate. An example of the effect of chill roll temperature is provided in Fig. 10.42. The peel strength to EVOH at two chill roll temperatures is shown, with the higher chill roll temperature providing greater peel strength.
- Orientation generally lowers peel strength by increasing the amount of interfacial area, which can reduce the density of binding sites at the interface as well as impose stress on the existing bonds. Increasing blow-up ratio or draw-down ratio in blown film or increasing line speed and draw-down ratio on flat film or coextrusion coating increases

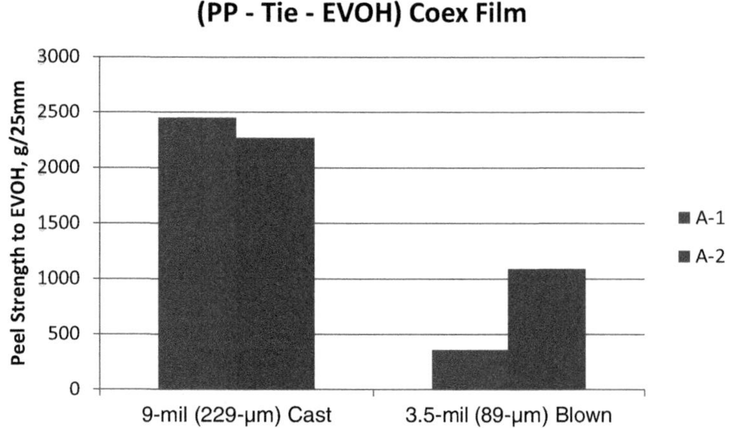

FIGURE 10.40 Effect of converting process on tie layer selection. *Reproduced with permission from the Dow Chemical Company.*

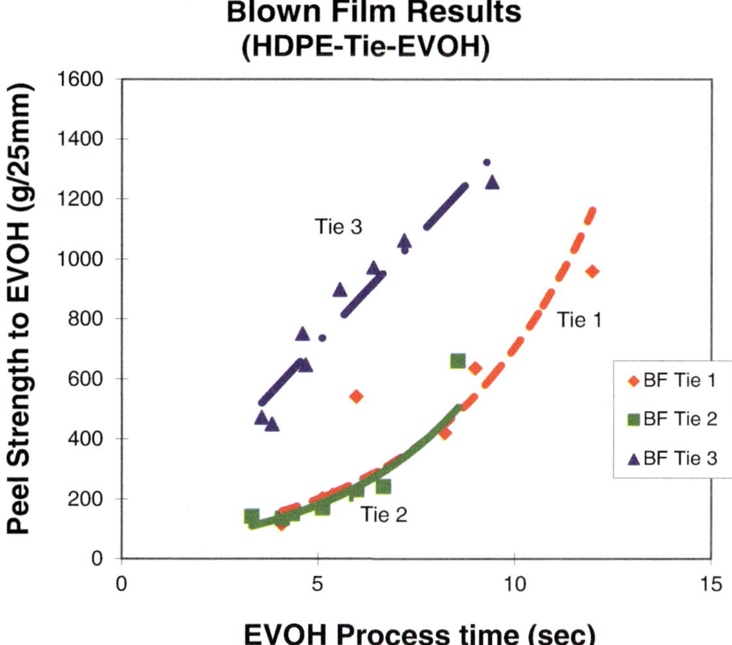

FIGURE 10.41 Effect of processing variables on peel strength to ethylene vinyl alcohol in blown film. Process time is the time the polymer spends in the bubble between the die exit and the frostline. It is defined by Eq (2.14) and is directly proportional to the contact time prior to crystallization in the bubble and inversely related to the stress [146]. *Reproduced from B.A. Morris, The effect of coextrusion blown film processing variables on peel strength performance, in: SPE ANTEC Conf., 1996, with permission from The Dow Chemical Company.*

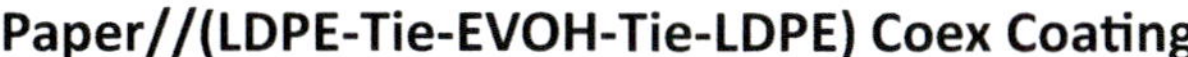

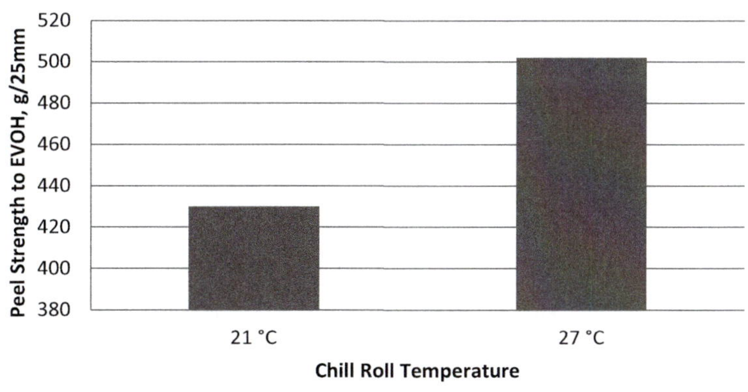

FIGURE 10.42 An example of the effect of chill roll temperature on peel strength to ethylene vinyl alcohol in coextrusion coating. The tie resin is an anhydride modified polyetylene. *Reproduced with permission from the Dow Chemical Company.*

orientation and may reduce peel strength. Peel strength measured in the direction of orientation is often less than in the transverse direction. Film orientation processes such as tenter frame, double-bubble, and machine direction orientation (MDO) often result in lower peel strength. An example of this effect for MDO is shown in Fig. 10.43 where peel strength of a tie layer to EVOH decreases with increasing draw ratio.

- The effect of thermoforming on peel strength is similar to orientation. As the film is stretched into its final shape, a new interfacial area is created and stress is imparted at the interface, negatively affecting peel strength. An example is shown in Fig. 10.44 for (PA6−tie−ionomer) blown film where thermoforming decreases the peel strength by a factor of three. Increasing the thermoforming temperature can help improve the peel strength, as is illustrated in Fig. 10.45 for a PP-based structure. As the forming temperature is increased, less stress is imparted on the interface and the molecules have more mobility to relax.

High orientation film processes such as tenter frame or double/triple bubble and MDO are especially challenging for adhesion. Paul et al. [150,151] show that the peel strength in coextruded films after orientation is maximized by using an orientation temperature that is greater than the Tg of the tie resin (for amorphous polymers) and near the Tm for semicrystalline ones. Annealing at elevated temperatures following the orientation also helps. They propose a heuristic model that helps explain these observations and points towards other ways to mitigate the drop-off in peel strength

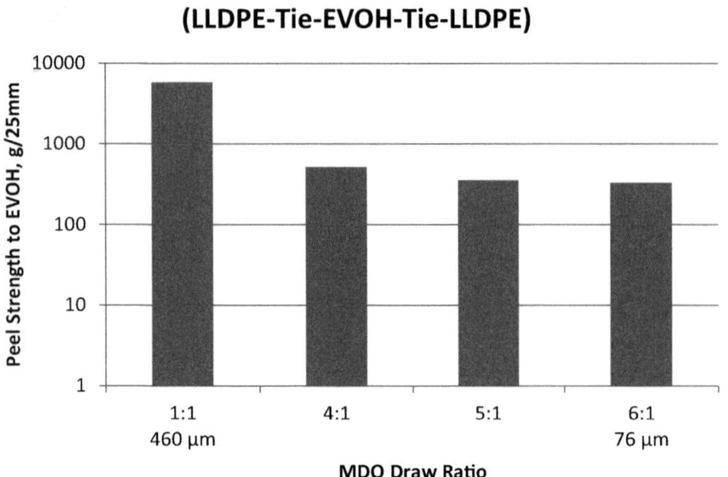

FIGURE 10.43 An example of the effect of machine direction orientation on peel strength. Plotted results are for peel strength of an anhydride-modified polyethylene tie resin to ethylene vinyl alcohol. Initial sheet thickness is 0.5 mm and the orientation temperature is 115°C. Note that peel strength is plotted on a log scale [149]. *Adapted from C.D. Lee, J.J. Strebel, J.D. Goudelock, Higher tie layer adhesion in machine direction oriented (MDO) barrier films, in: TAPPI PLACE. Conf., 2007.*

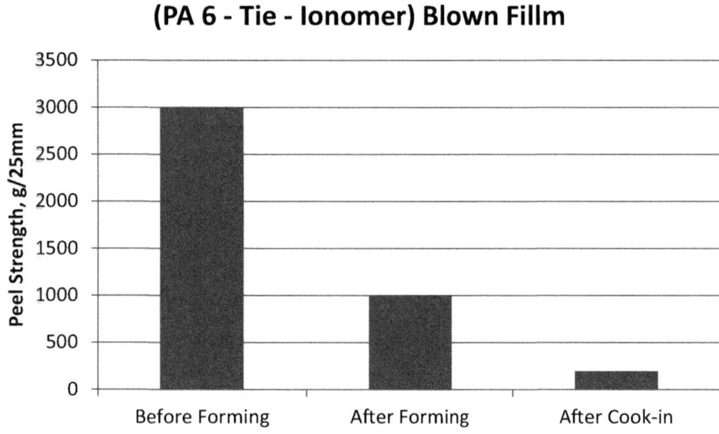

FIGURE 10.44 An example of the effect of orientation and end-use temperature on the adhesion of a tie resin to polyamide in a three-layer coextrusion. After thermoforming the film is subjected to cook-in-the-package conditions. *Reproduced with permission from the Dow Chemical Company.*

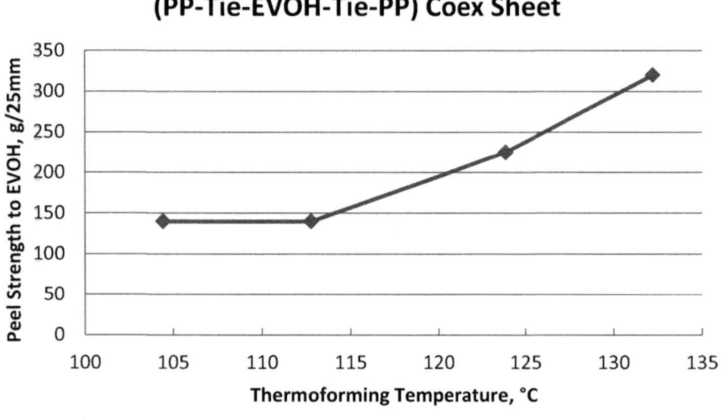

FIGURE 10.45 The effect of thermoforming temperature on the peel strength of an anhydride-modified PP-based tie resin to ethylene vinyl alcohol after forming [68]. *Adapted from B.A. Morris, Factors influencing adhesion in coextruded structures, TAPPI J. 75 (1992) 107–110. Reproduced with permission from the Dow Chemical Company.*

following orientation. As illustrated in Fig. 10.45A, when two immiscible polymers are joined together their chain segment densities vary across the interface. The thickness of the interface for random coil polymers is given by [152]

$$h \sim \sqrt{\frac{kT}{\chi}} \tag{10.21}$$

where h = thickness of interface, k = Boltzmann constant [1.38×10^{-23} J/K], T = temperature, K, and χ = interaction parameter.

The interaction parameter is introduced in Chapter 6 [Eq. (6.4)]. In general, adhesion is stronger with increasing entanglements and thickness of the interface.

Fig. 10.46B illustrates the case for homopolymer and random copolymers. Here there is little penetration of the immiscible polymers into the interphase region. During stretching, tension builds, chains become disentangled and there is less crossover. The thickness of the interface decreases. The formation of diblock copolymers at the interface through a chemical reaction (as in the bonding of tie resins containing MAH groups to PA or EVOH) has the effect of broadening the interface and increasing adhesion (Fig. 10.46C). Adhesion undergoes a more gradual decline during stretching as the area density of diblocks decreases (due to the formation of new interfacial area) and affine deformation brings the diblock junctions closer to the interface (reducing h). The model suggests minimizing tension will broaden the interface and add entanglements to improve adhesion. They outline several approaches to accomplish this:

- Annealing the film to relax tension in random copolymer segments
- Increasing the number of grafted chains (diblocks)
- Adding impact modifier to reduce strain in the semicrystalline network
- Increasing polarity of the tie layer to thicken the interface with polar polymers before stretching to create a more gradual decay with stretching.

They cite several studies where these approaches have proven to be successful [49,153,154]. Similarly, Uosaki [155] hypothesizes that lowering the modulus of the tie resin will help reduce stress during orientation to reduce breaking of bonds and diblocks at the interface. Working with a (PP–tie–EVOH) structure that is stretched 3 × 3 in a laboratory tenter frame stretcher (1.5 m/min, 100°C), he shows that lowering the modulus while maintaining crystallinity of the MAH modified PP-based tie resin increases the peel strength after orientation compared with conventional PP-based tie resins. He finds that this approach is more effective than increasing the MAH level.

Interfacial adhesion in highly oriented multilayer films is taken up again in Section 11.2.4.6.

All processing effects have limitations from an operation or performance perspective. For example, increasing extrusion temperature may cause gel and black specs, increasing chill roll temperature may result in chill roll sticking and blocking, increasing air gap in extrusion coating/lamination increases neck-in, decreasing line speed reduces productivity and output, and decreasing orientation may reduce the physical properties. By understanding the fundamentals of

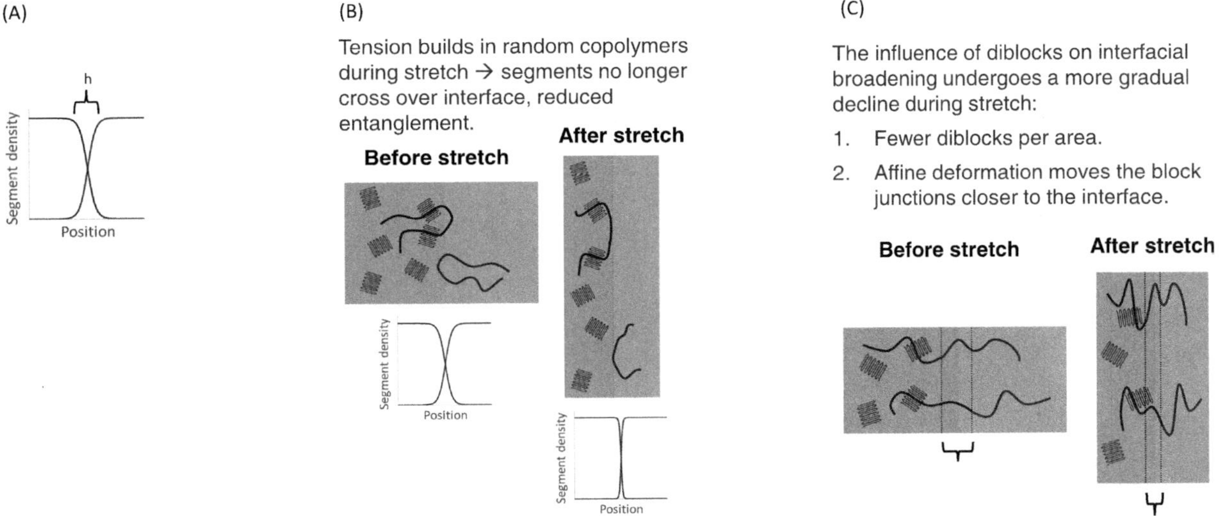

FIGURE 10.46 Model of effect of stretching on adhesion between two immiscible polymers. (A) typical chain density profile at interface, (B) interface with homopolymers or random copolymers, (C) interface with diblock copolymers formed by reaction [151]. *Reproduced from M. Paul, B. Morris, J. Weinhold, K. Hausmann, Tie layer adhesion chemistry for multilayer packaging, in: SPE ANTEC Conf., 2021, with permission from the Dow Chemical Company.*

why these parameters affect peel strength in the way that they do and the limits of the polymer involved, optimizing the process can often bring a boost in adhesion performance. Sometimes this cannot be done within the limits of the process and changes in material selection or film design are required to meet the specifications for film performance and interlayer adhesion. Fig. 10.47 shows an example of how this may play out. The graph plots the data from Fig. 10.41 in terms of output (keeping the frost line height constant). As the output of the converting line is increased, the peel strength decreases. By changing the tie resin from a standard grade to a high-performance grade, peel strength can be designed back into the film at high output. Similar plots can be produced for other processes such as coextrusion cast film [147].

10.4.2.2.5 End-use

How the package is ultimately used—the environment and conditions it is subjected to—affects adhesion performance. Stresses from temperature, humidity/moisture, and chemical attack (from inside or outside the package) can reduce peel strength performance. These factors should be considered when selecting a tie resin.

Both cold and hot temperatures can affect adhesion performance. Polymers undergo thermal transitions that can affect their physical properties such as modulus, which affect peel strength performance. The best tie layer at room temperature may not be the best choice at the temperature of use. An example is shown in Fig. 10.48 where the peel

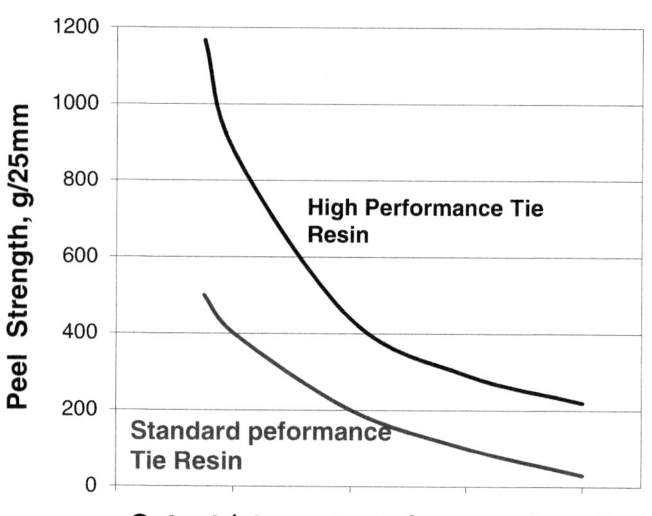

FIGURE 10.47 The effect of output on tie resin selection. Curves are replotted from data in Fig. 10.41. As output is increased, the peel strength decreases for a given tie resin. To maintain peel strength performance at higher output, sometimes a higher performance tie resin needs to be used. *Reproduced with permission from the Dow Chemical Company.*

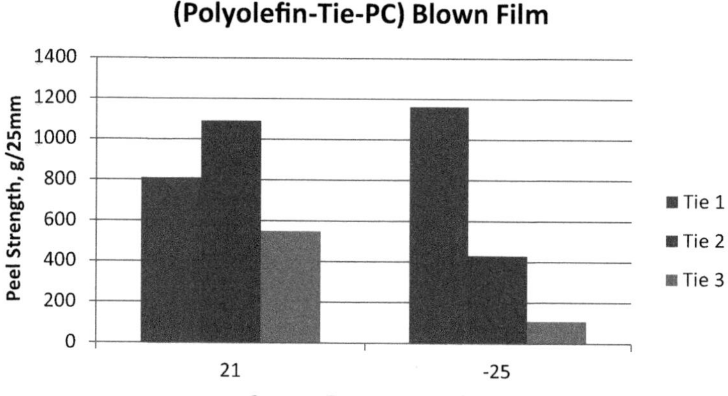

FIGURE 10.48 Effect of end-use temperature on tie resin selection. Peel strength of a coextruded blown film measured at room temperature after stored at 21°C or −25°C [68]. *Adapted from B.A. Morris, Factors influencing adhesion in coextruded structures, TAPPI J. 75 (1992) 107–110. Reproduced with permission from the Dow Chemical Company.*

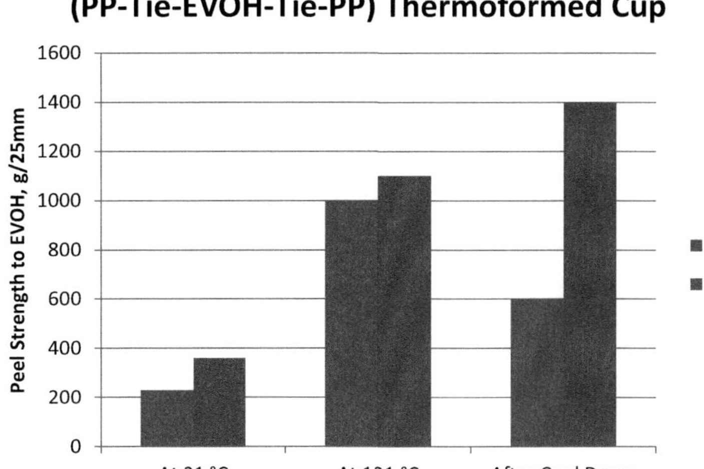

FIGURE 10.49 Effect of retort sterilization conditions on peel strength. *Reproduced with permission from the Dow Chemical Company.*

strength to polycarbonate of three tie resins is shown. At room temperature tie 2 has the best performance but at −25°C, tie 1 has the best performance.

Elevated temperature often decreases the peel strength. An example is shown in Fig. 10.44 for a (PA−tie−ionomer) coextruded blown film used in cook-in of deli meats. As discussed earlier, the peel strength to PA initially decreases due to thermoforming. The meat slurry is cooked at 80°C−90°C for several hours in the package. During this time the temperature and stresses can reduce the peel strength performance, as is shown by the third bar in the figure.

In some cases, peel strength can increase when exposed to elevated temperatures due to the relaxation of stresses. An example of this is shown in Fig. 10.49 for a (PP−tie−EVOH−tie−PP) thermoformed cup. Two anhydride-modified PP tie resins are evaluated under conditions simulating retort sterilization. In the initial cups, the peel strength of both adhesives is relatively low, presumably due to stresses imposed on the interface during solid-phase pressure forming. Both tie resins give about the same performance. Test strips cut from the cup are put on a tensile tester with an environmental chamber and heated up to 121°C to simulate retort. The peel strength measured at 121°C is shown in the middle set of bars. At this temperature, the tie resins are below their melting points, but the molecular chains become more mobile. The peel strength increases for both tie resins. This may be due to the relaxation of the stresses from thermoforming as well as the formation of new chemical bonds. The test strips are cooled back to room temperature and remeasured. The peel strength of tie 2 remains high but the peel strength of tie 1 drops. The performance of the two tie resins responds differently to the heat history imposed by the retort process.

Moisture can have an effect on bond strength. Section 10.4.1.3 describes the effect of boiling water on the adhesion of EAA where the form of the metal oxide on the surface of the foil changes. In coextruded structures, one or more of the layers may be plasticized or softened by moisture, which can affect peel strength through changes in the physical properties of the peel arms. Chemical interactions between some polymers may be negatively affected by water if the water outcompetes the functional groups on the polymer for the bond site.

Sometimes the bonds at the interface can undergo chemical attack from species in the product being packaged or from the outside. An example of the former is shown in Fig. 10.23. Here (LDPE−tie) is coextrusion coated onto aluminum foil. At room temperature, the dry peel strength to LDPE decreases with increasing acid level due to the decreasing compatibility between the acid copolymer and LDPE. But when the structures are subjected to acetic acid to simulate migration from acidic food products such as ketchup or salad dressing, the higher acid EMAA copolymers provide the best peel strength and hence self-life performance.

End-use considerations are taken up in greater detail in Chapter 16. The important takeaway is that the tie layer selection depends on the end-use requirements and conditions the package will be subjected. Care should be taken when selecting a tie resin to test it under the actual conditions of use.

10.4.2.3 Choosing a coextrudable tie resin

Table 10.14 outlines many of the factors that should be considered when selecting a tie resin. The first is the peel strength performance: will the tie resin provide the desired amount of adhesion to the two polymers? This depends on the tie resin formulation—the base resin and chemical functionality and any modifiers, as well as the film structure,

TABLE 10.14 Coextrudable tie resin selection criteria.

Selection criteria	Resin parameters
Peel strength	Adhesion (bonding, diffusion) Stress dissipative properties
Processing	Rheology Thermal stability Melt stability
End-use performance	Regulatory compliance MVTR Clarity Oil resistance Toughness Stiffness
Cost	Price Density Layer thickness Quality

Source: Reproduced with permission from the Dow Chemical Company.

process and end-use factors described in the previous sections. Section 10.4.2.1 For some common situations, such as bonding polyethylene to PA or EVOH, there are many choices. For other situations, there may be only a few choices to consider. Most tie resin manufactures provide selector guides and other product literature to help their customers find grades that bond to various substrates. This provides a starting point for fine-tuning the selection based on the overall needs of the packaging films. Often a discussion with the tie resin manufacturer's technical service group will help with this process.

The tie resin must be compatible with the process in which it will be used. To prevent flow instabilities during coextrusion, it helps to try to match the viscosity of the tie resin to the other layers in the structure. This is one reason tie resin manufacturers have multiple grades for bonding the same combination of resins. These grades may differ in their rheology to run on different converting equipment, such as a 1 or 2 g/10 minute melt index resin for blown film and a 4–8 g/10 minute grade for cast film or extrusion coating. A description of flow instabilities, strategies for minimizing flow instability, and limitations that these place on resin selection is provided in Chapter 13.

The tie resin must also have sufficient thermal stability for the converting process at the processing temperatures needed to run the substrate resins. For example, an EVA-based tie resin may run well in a blown film operation but may not have enough thermal stability for some extrusion coating or cast film operations running at temperatures above 230°C. EVA manufacturers generally recommend that EVA be run at temperatures below 240°C in coextrusion (235°C for mono extrusion). For high-temperature processes, EMA or PE-based tie resins may be more suitable due to their greater thermal stability. This sometimes is dictated by the substrate. For example, an EVA may be suitable for bonding to polyethylene terephthalate glycol (PETG) but not as suitable for use with PET homopolymer (Tm = 260°C), which must be run hotter.

The tie resin can contribute to melt strength of the coextrusion, which is important for maintaining bubble stability in blown film and preventing edge weave and neck-in in extrusion coating. Tie resins based on LDPE or polar ethylene copolymers generally have better melt strength than linear polymers like LLDPE.

The performance of the film in the end-use is the ultimate goal, and even though the tie resin is often a small component of the overall film thickness, it can contribute, or at least not take away, from this goal. Most tie resins take on properties like that of the resin they are based on. This can be used to help screen potential grades to use for a given application. Factors that may enter into the tie resin selection criteria include:

- Compliance with appropriate government regulations such as the US FDA, European Union, or other regional/country food packaging regulations.
- Moisture vapor transmission rate (MVTR). Some packages require low moisture permeation (high barrier) and the tie layer can contribute. In this case, a tie based on a medium or HDPE may be favored over one based on EVA or EMA.

- Clarity. The tie resin should not detract from the clarity of the package if transparency is an important criterion. This involves selecting tie resins that are based on polymers with high clarity and low haze. An EVA-based tie resin may be favored over an HPDE-based one. Rubber modification and other additive technologies may take away some of the tie resin's clarity. These should be kept in mind when selecting the tie resin grade.
- Oil resistance. For polyolefins, the greater the density the better the oil resistance. For situations where this is critical and the tie resin can be the difference, higher density PE-based tie resins or even ionomers may be used.
- Toughness. The tie resin may help contribute to the toughness of the film in concert with the other layers of the structure. Depending on the failure mode, this may involve selecting tie resins with very soft base resins such as EVA, EMA, or plastomer or very stiff base resins such as HDPE or PP.
- Stiffness. As is described in Chapter 9, the bending stiffness of a multilayer film is important for runnability on packaging lines and for consumer perception. The bending stiffness is highly influenced by layer tensile modulus, thickness, and location. Sometimes using a stiffer tie resin, such as one based on HDPE, can help increase the bending stiffness.

Finally, there is cost. The overall system cost should be considered, not just the price of the resin. The cost of using a higher-priced resin may be offset by a lower density, the ability to downgauge the thickness of the tie layer or increase line speed. Quality is also an important consideration. Fewer line breaks and less defects and product loss during film production can have significant payback. Inventory costs can also influence grade choice; a tie resin grade that can be used for multiple applications at the same plant may allow for reductions in the inventory of raw materials and cost savings.

To help reduce costs, tie resin manufacturers have introduced tie resin concentrates for some applications, such as bonding PE to PA or EVOH. The concentrates are typically anhydride-modified PE with a high level of MAH. They are blended by the converter into a polyolefin let-down resin on the film extruder at anywhere from 5% to 40% depending on application and concentrate. This cuts out a compounding step typically done by the tie resin manufacturer. The film converter essentially becomes the tie resin manufacturer and takes on the responsibility for formulating and an increased risk for quality. It is important to understand the special challenges when using tie resin concentrates. These include:

1. **The composition of the let-down resin (resin that is blended with the concentrate)**. As described in Section 10.4.2.2.1, the generic type (LDPE, LLDPE, HDPE, PP, etc.), as well as the specific grade (MWD, branch distribution, comonomer type, comonomer content, additives, etc.), of the polyolefin let-down resin can affect the peel strength performance. The matrix resin of a tie resin fully formulated by the tie resin manufacturer is often carefully chosen after many trials to ensure robust performance across a number of customers and applications. A converter who uses just what is on the plant floor as the let-down resin may have a suboptimum formulation. The physical, optical, chemical resistance, and other key properties of the tie resin are dictated by the matrix resin. These same properties need to be considered when choosing the let-down resin.
2. **Anhydride source**. There are several tie resin concentrate grades on the market. These may differ by the base resin used in the grafting process, the grafting process, the amount of anhydride, and other factors. The compatibility of the anhydride concentrate and the let-down resin may be important for adhesion. More demanding converting processes, such as cast film or extrusion coating, may require a more robust anhydride source. Most suppliers keep details of their concentrates confidential, so it can be difficult to compare concentrates from different suppliers prior to testing on the actual film line.
3. **Anhydride level**. The amount of anhydride needed depends on the substrate being bonded to. The tie resin should be formulated for robustness. Fig. 10.50 shows a typical anhydride vs. peel strength curve for the combination of concentrate and let-down resin. The shape of this curve may differ considerably from one concentrate and let-down resin to the next. As the anhydride level is increased, the peel strength generally increases until it reaches a plateau where further increases do not change the peel strength appreciably. The ideal target anhydride level is somewhere on the plateau far enough away from the onset of the plateau that typical quality fluctuations keep the peel strength within specifications. If the amount of concentrate used is too low and the anhydride level is on the rapidly rising section of the curve, the resulting peel strength can be highly variable. A good practice is to run a trial with several different levels of concentrate in the blend and test the peel strength performance to generate the curve. From this, the optimum blend concentration can be established based on quality and cost.
4. **Anhydride level consistency**. Consistency in peel strength performance is directly dependent on the consistency of the anhydride level of the final blend. Tie resin manufactures make pellets that can be directly tested for

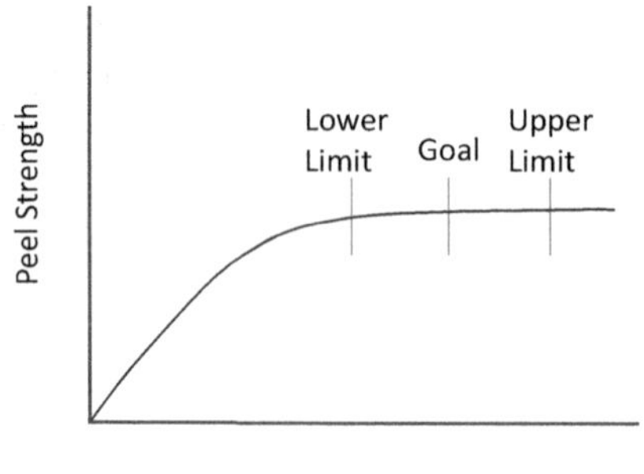

FIGURE 10.50 Theoretical effect of anhydride level on peel strength performance and ideal region to set specifications to avoid large variability in peel strength performance.

anhydride using FTIR and other analytical methods. They can use this data to ensure the process is under control and the product is within specifications. A converter making its own tie resin by blending with a concentrate does not isolate the tie resin as a pellet. Rather the tie resin ends up in a coextruded film structure where measuring the amount of anhydride is usually not possible. The converter must rely upon mass balance to control the process. The amount of variability will depend on the variability of the amount of anhydride in the concentrate and the error associated with the blending:

$$\%\text{Anhydride} = (B + e_b)(C + s_c) = B(C + s_c) + e_b(C + s_c) \tag{10.21}$$

where B = t arget % of concentrate in blend, e_b = error associated with blending, C = amount of anhydride in concentrate, and s_c = standard error of anhydride in concentrate.

The last term on the right-hand side of the equation, $e_b(C + s_c)$, represents the extra risk the converter takes on when blending rather than using a fully formulated tie resin. Modern loss-in-weight feeders and blenders help minimize e_b (see Chapter 6). The variability of the anhydride level in the concentrate, s_c, depends on the manufacturing capability of the concentrate supplier. This may vary by grade and among suppliers.

5. **Blending**. The quality of blending is important to ensure uniform distribution of the anhydride throughout the tie layer and to minimize local variation in anhydride functionality. Factors affecting the blend quality are described in Chapter 6. These include matching the shape, size, and density of the concentrate and let-down resin so that the pellets do not segregate prior to entering the extruder, using appropriate sized feeders and blenders to ensure good accuracy, minimizing the transfer length of pellet blends from the blender to the extruder, and using an extruder screw design sufficient for the task of melt blending. The percentage of the blend made up of concentrate can also be important. Very low blend percentages may hinder the ability of the blending system to adequately mix and distribute the anhydride uniformly. For example, a 1% blend represents just one pellet out of 100 pellets entering the feed hopper to the extruder. Small variations can create rather large deviations from the target composition, especially since single-screw extruders have little ability for back mixing. The concentrate supplier may have less concentrated grades for use at intermediate concentrations up in the 20%–40% range. Alternatively, sometimes fully formulated grades can be used as the concentrate.
6. **Processing**. All the processing issues described in Section 10.4.2.2.4 that affect peel strength also apply to concentrate formulations.
7. **Patents**. The tie resin intellectual property space is crowded. It is up to the converter to ensure they have the freedom to operate.
8. **Quality complaints**. Quality complaint handling becomes less straightforward when using concentrates. The tie resin supplier does not have control over what the converter does in their plant to properly blend the concentrate into the let-down resin. Often it is difficult to troubleshoot and find the root cause of adhesion issues when concentrates are used. This becomes a challenge for converters and tie resin suppliers alike.
9. **Safety**. Functionalized polymers may contain skin irritants and other health hazards. As the concentration increases, the potential for safety and health issues may increase. Care should be taken to read the safety data sheet and consult with the manufacture about proper handling.
10. **Testing**. Since there are no intermediate quality control tests on the tie resin itself to ensure consistency and quality, the converter may need to increase the frequency of quality control tests on the coextruded film.

A final note on choosing a tie resin is to test the candidates under conditions that simulate actual production and use. This includes running trials in the commercial structure on the production film line at production rates. Running a structure at low output to "get it right" may result in higher peel strength than when the process is ramped up to full production rates (see Fig. 10.47). It also means testing the film structure under conditions that simulate the end-use, such as thermoforming, or cold or hot temperatures. Lab tests can help screen candidates but ultimately the structure needs to be made and tested under actual conditions of manufacture and use.

10.5 Laminating adhesive technology

Adhesion lamination involves bonding together solid substrates such as paper, paperboard, aluminum foil, metallized films, and plastic films using a low viscosity adhesive. Low viscosity is needed to achieve good wetting of the substrate. This is achieved by dissolving or dispersing the adhesive polymer in a solvent or water or by using low MW monomers or prepolymers. After coating, the adhesive is dried or cured to high MW. There are three main classes of laminating adhesives:

1. **Solvent-borne adhesives**. These are typically based on urethane chemistry. They provide high bond strength to many substrates but are increasingly used only for the most demanding applications, such as retort packaging, due to environmental regulations and the cost of dealing with solvents.
2. **Waterborne adhesives**. These are typically acrylic, acrylate, or polyurethane-based. Initial formulations were not as high performing as solvent-borne adhesives and were only used in standard to mid-performance applications. Later generations have improved performance. These can be run on the same type of equipment as solvent-borne adhesives (typically gravure coating with drying stations).
3. **100% solids or "solventless" adhesives**. Typically based on urethane chemistry, these are applied as low viscosity monomers or prepolymers and subsequently cured, either by moisture, chemical reaction, or high energy [ultra-violet (UV) light or electron-beam radiation (e-beam)]. They can be one or two-part.

Extrudable laminating adhesives, where the adhesives are high MW thermoplastic resins, are covered in Section 10.4.1. The primary difference between extrusion lamination and adhesive lamination with 100% solids adhesives is the initial MW. Adhesives based on 100% solids chemistry have low MW and are typically applied at room or moderate temperature (50°C). Strength is achieved by curing the adhesive to high MW after lamination. In extrusion lamination, a high MW polymer is coated at a high temperature (typically >200°C) to achieve low viscosity. Strength is achieved by cooling the polymer below its freezing point to solidify the polymer.

Pressure-sensitive adhesives are often used for labels and hot melt adhesives for cartons. These will not be discussed in detail here.

As described in Chapter 2, there are two main processes for applying adhesives in lamination, dry and wet bonding. In the dry bond process, the adhesive is coated onto the first substrate and dried or partially cured (in the case of a pressure-sensitive adhesive). The second substrate is then laminated to the first with a heated nip roll, as illustrated in Fig. 2.19B. Solvent and waterborne adhesives are typically used in this process. Common coat weights and line speeds range from 2 to 3 g/m^2 (gsm) and 100 to 300 m/min, respectively [156]. Most formulations have 30%–50% solids for fast drying; solvent-borne formulations dry faster than waterborne adhesives and allow faster line speeds.

In the wet bond process (Fig. 2.19A), the adhesive is coated onto the first substrate, and drying or curing occurs after the second substrate is laminated and nipped to the first. For solvent or waterborne adhesives this requires that one of the substrates be porous and therefore this process is often used for paper or paperboard laminations. For UV cured adhesives, one of the substrates must be transparent to UV light. This is not a limitation for e-beam curable adhesives since e-beam radiation can penetrate opaque substrates. Some UV cured systems use a hybrid approach where the adhesive is partially cured with UV after coating onto the first substrate (dry bond process) and more completely cured following the nip (utilizing urethane or other reactive chemistry). This provides enhanced green strength during the lamination process.

Chemically cured adhesives may be applied as one-part or two-part. One-part formulations are designed to be applied as is. The curing chemistry is in a suspended state that is activated by heat (drying), water, or radiation. Two-part formulations are reactive and curing begins at the time they are mixed, which is typically just prior to application.

10.5.1 Chemistry

There are many chemistries used in adhesive formulations. Even within a chemical family, there are numerous combinations of monomers and prepolymers that may be used to achieve the desired performance attributes. For food packaging applications, these are limited by the need to comply with food packaging safety regulations. A general discussion on food packaging regulations is provided in Section 4.5. Regulations differ by country and will not be covered in detail in this general overview.

10.5.1.1 Urethanes

Urethane chemistry is one of the most versatile chemistries used in laminating adhesives [156–166]. They are used in solvent-borne, waterborne, and solventless adhesives. The basic chemistry involves the reaction of an isocyanate group (−N = C = O) with a hydroxyl group (−OH) to form a urethane linkage, as shown in Fig. 10.51. Polyurethane adhesives contain diisocyanate and a polyol (see Fig. 10.52). The chemistry and number of repeat units in the polyol determine the nature and flexibility of the adhesive. One common class of polyols is polyether. These have a repeat unit of (−C−O−C−) and when extended with isocyanate, the adhesive is called a polyether urethane. An example commonly used is polytetramethylene ether glycol. Polyethers with secondary hydroxyl rather than primary hydroxyl groups are typically used since they are less reactive. Mixtures of polyethers with different MW can be used.

The ether backbone imparts several useful properties to the adhesive, including [157]:

- Water white color;
- Hydrolysis resistance (holds up in wet environments);
- Microbe resistance—will not degrade in soil or with food contact;
- Good low-temperature flexibility;
- High flex and elongation—feels soft and rubbery.

The second class of polyols is polyesters. These are derived from the reaction of a diacid with an excess amount of di-alcohol to form hydroxyl-terminated polyesters. There are many possible monomers in the polyester family that may be used, such as aromatic (containing a ring structure) or aliphatic (straight carbon chain), linear or branched. This provides design flexibility. The repeat unit for polyesters is $-C-O-\overset{\overset{O}{\|}}{C}-$ and adhesives using this chemistry are called polyester urethanes. Polyester urethanes are more polar than polyether urethanes and generally have better adhesion to metals and other polar substrates. Some of the properties derived from the polyester backbone include [157]:

- Straw color;
- Chemical resistance—resists solvents and chemicals but is susceptible to attack by strong acids; resistant to UV (outdoor use); stable in the presence of water (hydrolytically stable);
- High-temperature resistance;
- Strong specific adhesion to polar substrates, such as metals;

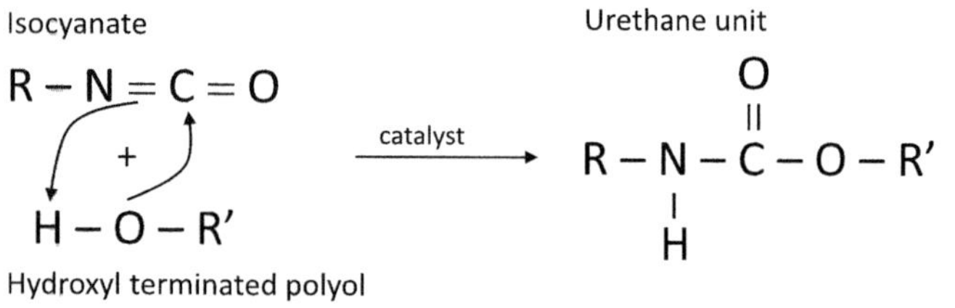

Isocyanate
R − N = C = O
+
H − O − R′
Hydroxyl terminated polyol
catalyst →
Urethane unit
$R-\underset{\underset{H}{|}}{N}-\overset{\overset{O}{\|}}{C}-O-R'$

FIGURE 10.51 Formation of a urethane unit from isocyanate and polyol. *R* and *R*′ represent chemical units.

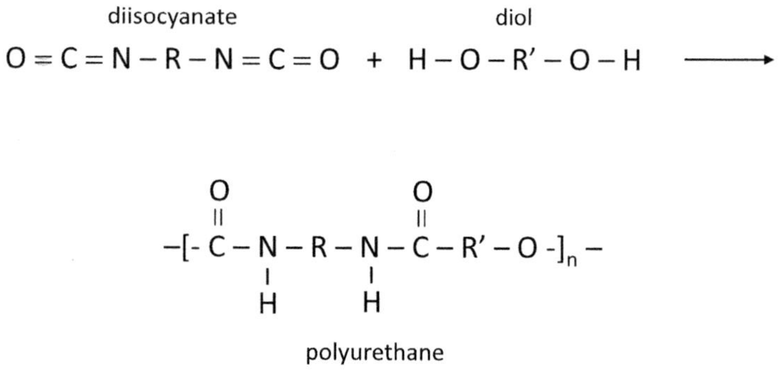

diisocyanate
diol
$O=C=N-R-N=C=O \ + \ H-O-R'-O-H \longrightarrow$

$-[-\overset{\overset{O}{\|}}{C}-\underset{\underset{H}{|}}{N}-R-\underset{\underset{H}{|}}{N}-\overset{\overset{O}{\|}}{C}-R'-O-]_n-$
polyurethane

FIGURE 10.52 Polyurethane chemistry.

- Poor flexibility/toughness at low temperatures (brittle);
- Crystallinity—the amount of crystallinity can be controlled to produce hard or soft formulations that vary in flexibility, chemical resistance, and adhesion.

Chain extenders may be added to modify the backbone. A common one is 1,4-butanediol. It can be used with polyether or polyester backbones as a chain extender or crosslinker and adds hard or soft segments depending on the technology. Trimethylolpropane, a liquid polyol with secondary hydroxyl groups, is another common chain extender. It helps with increasing tensile strength, hardness, elongation, modulus, and resilience [157].

There are many types of isocyanates for adhesive formulations but only a few comply with food packaging regulations. Isocyanates can be aliphatic or aromatic. The main aromatic isocyanates used are toluene diisocyanate (TDI) and diphenylmethane 4,4′ diisocyanate (MDI). Adhesives based on these isocyanates are known as TDI-based or MDI-based adhesives. Isophorone diisocyanate and hexamethylene diisocyanate are two common aliphatic isocyanates. Aromatic isocyanates are faster reacting by a factor of 2–10 but yellow in the presence of UV light. Aliphatic isocyanates are generally safer for use at elevated temperatures and, therefore, more likely to be approved for use in food packaging exposed to high temperatures such as in microwave cooking or retort. Aliphatic isocyanates are also more stable when exposed to UV light, but are generally more expensive than aromatic isocyanates.

Isocyanates react with water to make amines. Diamines further react with a diisocyanate to form a polyurea (see Fig. 10.53). Some amount of amine will always form, and the crosslinking reaction to create polyurea is the basis of one-part moisture curable adhesives. Aromatic isocyanates form primary aromatic amines (PAA), which have very low migration limits for meeting food packaging regulations. For food packaging applications, the converter must ensure a complete cure has occurred to eliminate PAAs.

The formation of the polyurethane from polyols and diisocyanates provides additional design variables for adhesives. These include the selection of the specific polyol, diisocyanate or prepolymer made from them [157]:

- The polyol may range from low to high MW and be used singular or in mixtures. The formation of a prepolymer with hydroxyl termination is also a possibility.
- The diisocyanate may range from low to medium MW, the latter being used in two-part adhesives.
- Prepolymers terminated with isocyanate are the basis of one-part moisture-cured adhesives. The isocyanate groups react with water to form polyurea (Fig. 10.53). This takes several days and is dependent on the amount of ambient moisture that diffuses into the adhesive.
- An acrylic polyol may be used as the base backbone or in mixtures with polyether or polyesters. These take on some of the characteristics of acrylic adhesives. These are called acrylic urethane, polyester–acrylic urethane, or polyether–acrylic urethane.
- Some polyurethane chemistries are terminated with an amine or epoxy group. Crosslinking occurs by reaction of these groups. These are still generally referred to as polyether, polyester, or acrylic urethanes.

10.5.1.2 Acrylics

Acrylic-based adhesives are derived from esters of acrylic acid. The most used are methyl, ethyl, butyl, 2-ethylhexyl acrylates, methyl methacrylate, and their blends. Acrylic esters are polymerized by heat in a solvent or in an emulsion to form long-chain polymers. These polymers are colorless; insoluble in aliphatic hydrocarbons; resistant to alkali, mineral oil, and water; resistant to degradation; and provide good bonding to many substrates. Acrylic adhesives are usually mid to high in MW and can be formulated to be hard or soft. For lamination, they are most often applied in the form of

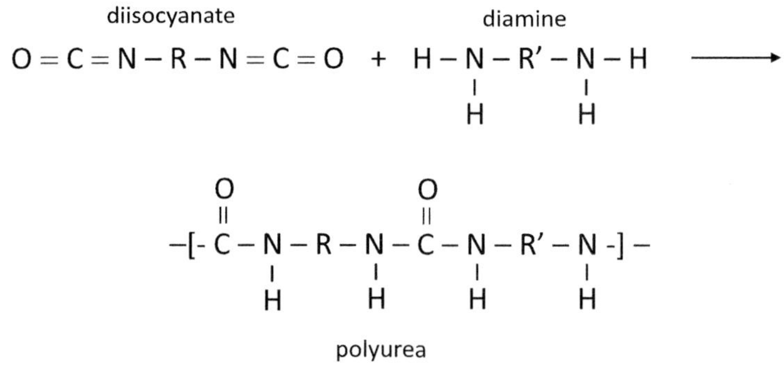

diisocyanate + diamine → polyurea

$$O=C=N-R-N=C=O \;+\; H-\underset{\substack{|\\H}}{N}-R'-\underset{\substack{|\\H}}{N}-H \longrightarrow -[-\overset{\substack{O\\\|}}{C}-\underset{\substack{|\\H}}{N}-R-\underset{\substack{|\\H}}{N}-\overset{\substack{O\\\|}}{C}-\underset{\substack{|\\H}}{N}-R'-\underset{\substack{|\\H}}{N}-]-$$

FIGURE 10.53 Polyurea formation from diisocyanate and diamine. Diamines may be created by reaction of polyols with water.

an emulsion in water. Acrylic water emulsions are typically formulated with tackifiers and other additives and have low viscosity and high solids compared with solvent-borne polyurethanes. One-part formulations can be with or with a crosslinker (often without). Without a crosslinker, they rely on the high MW of the polymer for tack, initial bond strength, and shear resistance [157,159,166].

For higher performance, waterborne isocyanates are used to form acrylic urethane polymers when crosslinked. Amine-epoxy termination is also used for crosslinking.

There are many acrylic chemistries but only a few comply with food packaging regulations.

10.5.1.3 Other chemistries

Adhesives based on natural materials are sometimes used in labels and cartons but are typically not used in lamination. Examples of these materials include [164]:

- Starch—used for cartons.
- Casein—derived from milk. Used for some labels.
- Animal glues—cartons. Mostly, have been replaced by starch or synthetic polymers.
- Natural rubber latexes—cold seals.

Polyvinyl acetate and polyvinyl alcohol are components of white glues for carton making. EVA and polyolefin elastomers are ingredients in hot melt adhesive formulations, usually in combination with tackifiers and wax or oil. These are used in cartons and typically not for film lamination.

PVDC is used for barrier coatings and adhesives. It provides both good moisture and oxygen barrier, as described in Chapter 8. Homopolymer PVDC is too brittle for use as a coating or adhesive. Comonomers are incorporated to give flexibility and adhesion. High MW PVDC is emulsified in water for adhesion applications. These emulsions typically have low viscosity and high solids. Solvent-borne PVDC formulations use low MW polymer and are seldom utilized in flexible packaging [157].

10.5.2 Types of adhesives

10.5.2.1 Solvent borne

Solvent-borne adhesives are the oldest adhesive technology but have seen declining use due to regulations on volatile organic compounds (VOCs) and increased cost for solvents and their capture and recovery. These adhesives, however, are still used in demanding applications such as retort, boil-in-bag, military, and medical packaging, and for bonding to digital inks [156,165]. They are often applied by gravure coating in dry bonding processes but can be used in wet bonding too. Since solvent evaporates faster than water, solvent adhesives can generally be run at faster line speeds. They also have better moisture resistance due to the lack of emulsifiers and surfactants [159]. Solvent-borne adhesives generally require less energy for drying than waterborne adhesives, but their overall environmental footprint (nonrenewable energy consumption and greenhouse gas emissions) is considerably higher due to the energy for solvent recovery [166].

Polyester urethane chemistry is often used, especially for polyester film applications. These also bond well to aluminum foil. Good thermal and chemical resistance is achieved with crosslinking with isocyanates. Polyurethane prepolymers terminated with isocyanate are used in one-part formulations where curing with moisture under ambient conditions takes 2–7 days [166]. Two-part polyurethane chemistries are also used that do not rely on moisture. Polyester urethane adhesives have good green strength, cure quickly and work well in fast processes. High solids formulations (50%) created for lower VOC generally have lower green strength [159]. Solvent-borne polyurethanes typically have better adhesion performance, temperature resistance, and water resistance than other adhesive technologies, hence their use in demanding applications.

10.5.2.2 Waterborne

Waterborne adhesives have increased in use due to the high cost of solvents, their high productivity and ease of handling, elimination of fire hazards, government regulation of VOCs, and minimal capital expense to switch from solvent-borne adhesives. Acrylic emulsions and polyurethane dispersions (PUDs) are the primary waterborne chemistries used for lamination. Early versions had poorer adhesion, water resistance, and heat resistance than solvent-borne adhesives. The introduction of crosslinking agents improved performance. Problems with foaming, short pot life (storage time in the coating process before the quality of the coating is compromised, usually because of viscosity rise), and clogging of anilox cells of the gravure coating process also had to be overcome. For common packaging film substrates such as

PET, PE, aluminum foil, and PA, waterborne adhesives have closed the gap in performance with solvent-borne adhesives [166].

The advantages and disadvantages of acrylic and PUD adhesives are given in Table 10.15. Acrylic emulsions are lower in cost, more widely used, and generally scavenge less slip additive from the sealant layer. Waterborne PU adhesives generally have higher performance. PUDs have good adhesion to a wide range of flexible packaging substrates which is enhanced when crosslinked. The MW can be varied to tune properties. Low MW PUD's have good tack and fast crystallization rates which benefit green strength.

When crosslinking agents are used, crosslinking of acrylic or PUD adhesives typically occurs at room temperature following drying. Green strength is generally high enough to slit the roll immediately after lamination even if curing is not complete. Despite the higher energy cost to dry waterborne adhesives, life cycle assessments show that the energy consumption of using water-based adhesives is about half that of solvent-borne adhesives [166].

Vinyl acetate ethylene (VAE) and vinyl acetate acrylic (VAA) copolymers are also used to make waterborne adhesives. The addition of ethylene to VAE lowers the glass transition temperature (Tg), which provides greater flexibility and better adhesion than polyvinyl acetate. VAA has a lower Tg than VAE and a better reaction with crosslinkers [159].

PVDC adhesives are predominantly water born and provide both moisture and oxygen barrier.

10.5.2.3 100% Solids

100% solids or solventless adhesives used in flexible packaging are typically based on polyurethane chemistry. Early versions developed in the 1960s were one-part adhesives that were cured by exposure to atmospheric moisture in the roll following the nip. They were limited by the substrates they bonded to and had challenges with bubbling, slow cure rate, and cloudiness. Two-part reactive formulations involving urethane and isocyanates were developed to overcome these deficiencies. This technology requires metering and mixing shortly before application since pot life is short (typically 20–30 minutes due to increases in viscosity). These adhesives eliminate the variability due to ambient moisture since they are no longer moisture-cured, but still suffer from high residual monomers and low initial bond strength. Improved versions using aliphatic isocyanates with higher viscosity require application temperatures of 50°C–70°C [159,160,162].

The adoption of 100% solids adhesives has driven developments in equipment design to overcome their limited green strength and pot life. Two-part adhesives have two characteristic curing time scales. The first is the practical time where green strength develops enough for the laminate to be handled in further converting steps (slitting, etc.). This typically is on the order of 24 hours. The second is the time scale for complete chemical curing to reduce residual monomer concentrations below migratory PAA regulations. This can take several days. Accurate metering and mixing technology are needed to precisely control the mixing ratio to ensure complete curing. Gravure coating typically used for solvent adhesives is not suitable for the higher viscosity and temperature control needed for 100% solids. The industry adopted the use of transfer roll coating, which uses alternating hard and soft rolls to spread the adhesive and has long been used in the paper coating industry, to deal with these issues [167]. Other coating techniques have also been applied (see Section 2.3.4). Evolution in web handling and tension control was also needed to deal with the poor green strength [167,168].

100% solids adhesives have shown both environmental and economic advantages over solvent-based adhesives. These include less air pollution, reduced liquid hazardous waste, lower equipment and installation costs, lower energy costs, smaller equipment footprint, reduced waste costs, and lower adhesive costs (including lower coat weight required for equal bond values and reduced freight costs), and higher productivity [167,169].

There has been some concern in the industry regarding using solventless adhesives with water-based inks. For the most part, however, experiments show they work [160].

10.5.2.4 Energy cured adhesives

Adhesives cured by high-energy sources, typically UV light or e-beam (EB) radiation, cure faster than conventional 100% solids adhesives. Most UV/EB curable adhesives are based on acrylic/acrylate chemistry with free-radical curing, although epoxies, polyurethanes, silicones, and vinyl esters can be used. Cure times are almost instantaneous. Aliphatic urethane acrylates are typically used for flexible packaging. A photoinitiator is added for UV crosslinking, along with oligomers and other additives. UV curable adhesives are generally not used for food packaging due to regulatory compliance hurtles for the photoinitiators and the need for at least one substrate to be transparent to UV light. E-beam curable adhesives have attracted more attention for flexible packaging [157,159,170–172].

TABLE 10.15 Advantages and disadvantages of adhesive technologies [157].

Form	Advantages	Disadvantages
Solvent-based polyurethanes	• Well known; lots of history • Grades available for all types of food packaging laminates • One-part—ease of handling • Two-part—enhanced heat and chemical resistance; some cases better adhesion • Adhesion to wide range of substrates • Newer chemistries allow higher % solids	• VOCs • May be expensive on dry weight basis • Health and safety concerns • Water may compromise cure • Rate of drying—limiting step • Time for cure • Two-part requires proper mixing ratio—chance of error
100% solids (solvent free) polyurethanes	• Little or no VOC • High application speed • Competitive cost on dry weight basis • Lower coat weight • Adhesion to wide range of substrates • Excellent clarity • Reduced waste of adhesive	• Requires dedicated laminator • Need transfer pump for one-part • Need meter-mix-disperse pump for two-part • Poor green adhesion; no shear resistance initially • Some require preheat to apply • Time for cure and bond to develop
Polyurethane dispersions (PUD)	• Medium to high solids at low viscosity • Little or no VOC • Similar to performance of mid-performance or general purpose solvent or 10% solids versions • Good bonds to wide range of substrates • Good to excellent clarity • Can be cost competitive • Reduced heat and safety concerns	• May generate foam • Typically not freeze/thaw stable • Requires careful mixing of ingredients—chance for error
Acrylic (water based)	• Higher % solids • Low viscosity—can be applied as is • Easy to dry, low film forming temperature • High transparency after drying • Some aroma barrier • One-part versions suitable for less demanding applications • Two-part versions provide enhanced performance • Economic on dry basis • Reduced scavenging of slip additives from the sealant layers	• One-part versions have low bond performance • One-part versions have low chemical and heat resistance • Two-part requires proper mixing ratio—chance of error
PVDC (water dispersion)	• High % solids • Low viscosity, apply as supplied • Easy to dry, low film forming temperature • No VOC • High transparency when dry • Very high shear and tunneling resistance • Bonds well to most plastic films • O_2 and moisture barrier in dry and humid environments • High aroma and chemical barrier • Economic on dry basis • 1-step	• Low pH—corrosive to soft metals • Requires higher nipping temperatures • Requires time to develop full barrier properties • Generally not used on foil or metallized films • Challenging to recycle due to chlorine content
Energy cured adhesives	• 100% solids, no volatiles • Instant or rapid cure • Reduced time in working inventory • Reduced waste • Fast application speeds • Reduced energy cost to apply (no drying)	• Need dedicated application and cure equipment • Cost of equipment and adhesive • Potential health and safety issues (skin sensitizer) • Food law compliance needs to be confirmed • Substrate transparency for UV cure • Effect of energy on polymer films

Source: Adapted from L. Jopko, Flexible packaging adhesives – the basics, TAPPI PLACE. Conf., 2004.

Cationic chemistry is also used, such as cationic curing of cycloaliphatic epoxy resins with onium salt photoinitiators [170]. It is slower to react than acrylate free-radical chemistry and curing is not instantaneous. Therefore, it can be used to UV cure prior to the nip to get around opaque substrate limitations. It is one-component chemistry with good pot life. Cationic chemistry has several drawbacks, however. It is inhibited by most printing inks containing alkaline substances and retarded by PA (nylon) films, there are limited raw materials, and byproducts are odorous. This technology is not used in food packaging due to these disadvantages [172].

There are four primary ways UV/EB adhesives are used [170]:

1. UV laminating adhesives are applied as a liquid to one substrate, nipped with the second, and cured with UV light through one of the substrates. At least one of the substrates must be transparent. They provide a fast, permanent cure that is resistant to heat and chemicals.
2. UV pressure-sensitive laminating adhesives (PSAs) are applied as a liquid onto the first substrate, cured to the pressure-sensitive adhesive state by UV light, and then nipped together with the second substrate. Since the cure is before the nip, they can be used with opaque substrates. The process requires a higher coat weight, has a slower cure rate (slower line speed), and remains tacky after curing. These PSAs provide stronger bonds to some substrates but generally have lower chemical and temperature resistance than UV laminating adhesives.
3. UV postcuring or dual-cure laminating adhesives are applied as a liquid onto the first substrate, partially cured with UV light to get good green tack, nipped with the second substrate, and aged to complete the cure. These are either based on cationic chemistry or incorporate polyurethane postcure chemistry. They may be used with opaque substrates but are sensitive to inks, substrates, and moisture. They often have a narrow processing window to balance UV exposure, time before the nip, and postcure time. Newer dual-cure systems use very low levels of isocyanate, are easy to handle as one-component compositions with no pot life issues, have good green strength, and completely cure within 24 hours.
4. EB curable laminating adhesives are applied as a liquid to the first substrate, nipped with the second substrate, and cured inline following the nip by EB. They are generally based on free-radical curing of acylates and contain no photoinitiators. EB can penetrate opaque substrates. The curing has a high conversion, and the cured adhesive has low odor, is permanent and tack-free, and has good chemical and heat resistance.

In flexible packaging, high-energy curing adhesives are used for labels. Compared with water or solvent-borne adhesives, UV/EB curable adhesives have:

- Low emissions (vs. solvent-borne);
- Smaller footprint (compact lamp vs. large dryer);
- Easy cleanup;
- Consistency in application;
- Fast processing speeds.

For laminations, two-part solventless laminating adhesives have several disadvantages that UV/EB curable adhesives can mitigate:

- 3–5 days to reach full cure—UV/EB cure is instantaneous;
- Need accurate mixing—UV/EB is one-part;
- Difficult to start and stop due to curing—UV/EB is easy cleanup;
- Requires heated equipment—UV/EV is cured at room temperature;
- Toxicity concerns over amines—UV/EV contain no amines.

Comparing UV and EB adhesive technology [172]:

- EB equipment is roughly 5 × the cost of UV curing equipment.
- EB has lower operating costs and maintenance costs than UV.
- EB is more versatile—can be used for opaque substrates.
- Accurate dosage is of critical importance for EB, less so for UV cure.

10.5.3 Applications

The type of adhesive and chemistry used for an application will depend on the specific application requirements, such as the substrates being bonded, performance criteria, end-use environment (e.g., hot fill, retort), regulatory requirements,

TABLE 10.16 Overview of adhesive chemistries grouped by performance level and application [161].

Performance level	Typical chemistries	Substrates/structures	Packaging applications
Standard	• Water-based: 1 and 2-part acrylate • Solventless: standard MDI-based 1 and 2-part	• OPP/cPP, OPP/met-OPP • OPP/PE, PET/PE • PA/PE, PA/(PE-EVOH-PE)	• Labels, wrap films, noodles • Snacks, cake, cookies • Meat, cheese vacuum packaging
Medium	• Solvent based: NCO terminated PU with OH coreactants; OH terminated polyester or PU adhesives with isocyanate coreagents • Solventless: Low monomer PU-based; MDI-based standard adhesives with special coreagents	• PA/PE, PA/(PE-EVOH-PE) • OPET/PE, OPET-met/PE, OPET/Al/PE	• Meat, cheese vacuum packaging • Coffee/powder vacuum packaging • Portion packaging, pouches, "warm" fill
High/Specialty	• Solvent based: Polyester or Polyester urethane adhesives with higher molecular weight and different, mostly aliphatic isocyanate coreactants • Solventless: aliphatic solventless; special low monomer adhesives	• OPET/Al/cPP, OPET/Al/OPET/cPP, OPET/Al/OPA/cPP, Al/cPP • OPET-SiOx/cPP or PET	• Fruit juice, stand-up pouches, detergents, refill, "hot fill" • Pouches, sterilization, boiling, pasteurization • Instant meals, retort sterilization, deep-drawable packaging, cold-formed, semirigid

Source: Adapted from H.-S. Schetschok, E.-M. Kupsch, Adhesive lamination vs. extrusion lamination, trends & technology update, in: TAPPI Eur. PLACE. Conf., 2011.

and the converter's equipment. Other factors that may weigh into the selection process include criteria for gas permeability, thermoforming capability, and chemical and heat resistance. Table 10.16 provides an overview of typical chemistries used for packaging applications segmented by performance needs. There are many variations and types of adhesives available on the market, with the performance of water-based and solventless adhesives advancing in recent years. Suppliers should be consulted for selecting the best adhesive for the application.

An example is the selection of high-performance adhesives for use in pouches and lidding films for retort packaging [158]. Retort is a sterilization process where the packaged product is subjected to elevated temperature and pressure. The protocols vary but a typical retort is 121°C for 30 minutes. Example structures include:

- Lidding: print/aluminum foil (Al)/adhesive/cast polypropylene (CPP), OPET/inks/adhesive/Al/adhesive/PA/adhesive/CPP.
- Trays: polyester (PET)/adhesive/Al/CPP.
- Foil containing pouches: OPET/adhesive/Al/adhesive/OPA/adhesive/CPP, OPET/inks/adhesive/OPA/adhesive/Al/adhesive/CPP.
- Nonfoil pouches: OPET/adhesive/barrier OPET; Barrier film/adhesive/OPA/adhesive/CPP.
- Nonfoil lidding: OPET/inks/adhesive/barrier film/adhesive/CPP; Barrier film/adhesive /CPP.

Regulatory compliance is an important criterion for selecting the right adhesives. The regulations vary by country/region but typically there are specific migration limits for residual monomers or their byproducts that must be met. In the US, the FDA has specific requirements for packages exposed to high temperatures, such as CFR177.1390 (for temperatures greater than 121°C) or CRP177.1390 (49°C $< T <$ 121°C). Structures, where there is no functional barrier between the adhesive and the food product, must meet these more stringent regulations. If the adhesive uses curing chemistry, care also must be taken to ensure the adhesive is mixed in the right ratio so that it fully cures since the regulations assume the cure is complete.

Solvent and solventless adhesives are typically used for retort applications. Solventless adhesives are constrained to about 2.4 gsm coat weight which may not be enough for demanding retort applications. Solvent adhesives can easily achieve 4.8 gsm and so are more often used for retort. Polyester or polyurethane monomers or prepolymers are typically used. Both are reacted with coagents such as isocyanate so the resulting chemistry is polyurethane.

10.5.4 Practical considerations

There are many aspects of the adhesive lamination process and adhesive chemistry that influence the final performance of the laminate. These include:

- **Substrate surface properties**. As previously described in Section 10.4.1.1 the cleanliness of the surface of the substrate is critical for good adhesion. It should be clear of oils and other contaminants that may affect wettability and adhesion. The surface energy of the substrate generally needs to be higher than the surface tension of the adhesive by about 10 dynes/cm. Surface treatments such as flame, corona, and plasma treatment are often used to clean the surface and raise the surface energy. Primers may also be used to promote adhesion to the substrate.
- **Adhesive interactions**. Monomers and other ingredients in the adhesive formulation may have unwanted interactions with residual solvents from inks, moisture, or other contaminants. This is particularly true for two-part isocyanate-urethane formulations. Isocyanates react with water to form amines and polyurea. Too much water can lead to poor performance. Solvents used should be urethane grade (<0.1% water and 0.5% alcohol) and the adhesives cured under ambient conditions (20%–80%RH). Some films can introduce water and degrade the performance of the adhesive. For example, PA (nylon) films must be dried prior to adhesive lamination. Some coatings on films can react with the adhesive. The isocyanate can react with alcohols leading to incomplete curing and lower temperature resistance. Sources of alcohol contamination may be from ink solvents (especially high boilers) or ink components. Isocyanate also rapidly reacts with amines, which can affect curing and performance. Amine contamination can come from some ink components and should be avoided. Other potential contaminants that may interact with the adhesive and deteriorate adhesion performance include slip, antistat, and antifog additives and barrier coatings. It is always advisable to check compatibility and talk to suppliers [158,173].
- **Viscosity**. Low viscosity is needed for good wetting of the substrate at high coating speeds. Some solventless adhesives are coated at elevated temperatures (~50°C) to reduce the viscosity of the coating. Reactive formulations have a limited pot life after mixing—about 20–45 minutes for solventless and 8–12 hours for solvent-borne adhesives [173].
- **Proper mixing**. For two-part reactive formulations mixing in the right ratio is critical. For urethane chemistries, excess hydroxyl will prevent proper curing and result in poorer product resistance. If excess isocyanate is added, the cure will be slower and will need moisture to finish. The resulting cure may be harder or brittle leading to package failure [158].
- **Coat weight**. Higher than needed coat weight results in poor economics, but if the coat weight is too low voids can form which in severe cases can cause delamination. Coat weight is a critical parameter measured by the converter—some converters have online measurements; others conduct offline measurements at startup.
- **Green strength**. Some adhesives provide good adhesion as soon as the substrates are combined (green adhesion) while others require several hours or even days for the adhesive strength to develop enough to conduct further converting steps. Dry bonded water and solvent-borne adhesives and partially cured pressure-sensitive adhesives tend to have good green strength. 100% solids adhesives often have poor green strength and take a day or more before they can be further processed. They also require good web handling and tension control. One problem associated with poor web handling and green strength is tunneling where local delamination occurs because the films have different extensibility and relaxation [159].
- **Cure time and temperature**. As mentioned previously, there are two cure times that are relevant for reactive adhesives. The first is the time to develop enough bond strength so that further converting processes such as slitting can take place. The second is the time to reach a full chemical cure to obtain the best properties. Some applications such as bottle labels may not need a complete cure, whereas food packaging adhesives generally need a complete cure to ensure residual monomers such as PAA are below regulatory migration limits. Complete cure for one-part moisture-cured urethane adhesives generally takes about 2–7 days [157]. Solventless two-part adhesives generally take about 2–5 days to cure, but retort adhesives using aliphatic isocyanates, which are slow to cure, take about 3–10 days [158,170]. Long times for curing contribute to storage costs. For example, an OPA/PE laminate with modern fast cure "smart" adhesive systems (two-part) still take a week to fully cure. Another example is a PA//PP retort structure using a 2-part solventless adhesive. The combined time to wait before slitting, cohesive strength to build before pouch making (heat sealing), time to ensure residual monomers are below migration limits and the time for additional crosslinking for heat resistance adds up to two-weeks from lamination to delivery. One of the features of high-energy cured adhesives (UV or e-beam) is the fast cure, which can cut the time to delivery down to two days [172]. Another aspect of chemical curing is the cure temperature. Retort applications may benefit from curing at a high temperature (50°C) to achieve complete cure and ultimate temperature resistance. A drop in warehouse temperatures in the winter months may affect the time needed for a complete cure.

10.5.5 New technology

New technology for solventless adhesive lamination was introduced by Dow and Nordmeccania in 2018. It combines equipment modifications with new fast cure chemistry. It uses a two-part adhesive (Dow's Symbiex adhesive) and two coating stations (Nordmeccania's Duplex SL One Shot laminator). Instead of mixing the two adhesives together prior to coating, each component is applied to a separate substrate. Mixing occurs as the substrates are nipped. The technology eliminates the need for a mixing head, the mixing ratio is accurately controlled by the coat weight and the adhesive formulations have been optimized for ease of application, good mixing in the nip, and fast cure. There is no "pot," so pot life is eliminated. Slitting can be done in 90 minutes and cleanup is easy [174,175].

References

[1] E.M. Mount, III, Adhesion Testing, November 2005, pp. 66–68. pffc-online.com.

[2] ASTM International, Standard Test Method For Peel Resistance of Adhesives (T-peel test), ASTM International, West Conshohocken, PA, 2008 (D1876-08).

[3] H. Zhou, Interlayer delamination and adhesion of coextruded films, J. Appl. Polym. Sci. 92 (2004) 3901–3909.

[4] Deutsches Institut fur Normung, Testing of Plastic Sheet; Adhesion Test, Berlin, Germany, 1982 (DIN 53357).

[5] ASTM International, Stand. Test. Methods Measuring Adhesive Tape test (2009) (D33590-09e2).

[6] ASTM International, Standard Test Methods For Assessing The Adhesion of Metallic and Inorganic Coatings By The Mechanized Tape Test, ASTM International, West Conshohocken, PA, 2010 (B905-00).

[7] ASTM International, Standard Practice For Evaluating Ink or Coating Adhesion to Flexible Packaging Materials Using Tape, ASTM International, West Conshohocken, PA, 2013 (F2252/F2252M-13e1).

[8] M. Jesdinszki, C. Struller, N. Rodler, D. Blondin, V. Cassio, E. Kucukpinar, et al., Evaluation of adhesion strength between thin aluminum layer and poly(ethylene terephthalate) substrate by peel tests—a practical approach for the packaging industry, J. Adhes. Sci. Technol. 26 (2012) 2357–2380.

[9] S. Wu, Fracture of adhesive bond, Polymer Interface and Adhesion, Marcel Dekker, Inc, New York, 1982, pp. 475–569.

[10] A.J. Kinloch, C.C. Lau, J.G. Williams, The peeling of flexible laminates, Int. J. Fract. 66 (1994) 45–70.

[11] A.N. Gent, G.R. Hamed, Peel mechanics for an elastic-plastic adherend, J. Appl. Polym. Sci. 21 (1977) 2817–2831.

[12] J. Pascal, Ultraversatile adhesives to broaden the possibilities of extrusion lamination, TAPPI PLACE. Conf. (2006). Cincinnati, OH.

[13] B.A. Morris, Understanding why adhesion in extrusion coating decreases with diminishing coating thickness, J. Plast. Film Sheeting 24 (2008) 53–88.

[14] A.J. Kinloch, R.J. Young, Fracture Behavior of Polymers, Elsevier Applied Science Publishers, London, 1985.

[15] Q. Du, G. Jiang, J. Li, S. Guo, Adhesion and delamination failure mechanisms in alternating layered polyamide and polyethylene with compatibilizer, Polym. Eng. Sci. 50 (2010) 1111–1121.

[16] ASTM International, Standard Test Method For Wetting Tension of Polyethylene And Polypropylene Films, ASTM International, West Conshohocken, PA, 2009 (D2578-09).

[17] Diversified Enterprises, Accu Dyne Test, Introduction and Overview, n.d. <http://www.accudynetest.com/cam_introduction.html>.

[18] S. Wu, Surface tension and polarity of solid polymers, Polymer Interface and Adhesion, Marcel Dekker, Inc, New York, 1982, pp. 169–214.

[19] S. Wu, Interfacial and surface tensions of polymer melts and liquids, Polymer Interface and Adhesion, Marcel Dekker, Inc, New York, 1982, pp. 67–132.

[20] H. Schonhorn, Heterogeneous nucleation of polymer melts on high-energy surfaces. II. Effect of substrate on morphology and wettability, Macromolecules 1 (1968) 145–151.

[21] M. Coleman, J. Graf, P. Painter, Specific Interactions and The Miscibility of Polymer Blends, Techonomic Publishing, Lancaster, PA, 1991.

[22] D.W. Van Krevelen, Properties of Polymers: Their Estimation and Correlation With Chemical Structure, Elsevier Scientific Publishing Co., Amsterdam, 1976.

[23] G. Schubert, Aluminum foil in quandary of high speed converting and adhesion performance, TAPPI Int. Flex. Packaging Extrusion Div. Conf. (2018).

[24] T. Gilbertson, Adhering flexo inks to flexible packaging substrates. The surface energy solution, SPE FlexPackCon Conf. (2014).

[25] G. Schubert, Adhesion of aluminum foil to coatings – stick with it, Eur. TAPPI PLACE. Conf. (2004).

[26] G. Schubert, Adhesion of coatings to aluminum foil, PFFC, September (2003) 148–150.

[27] G. Schubert, O. Plassmann, Adhesion to foil – more than just a one-sided story, TAPPI PLACE. Conf. (2008).

[28] G. Schubert, O. Plassmann, Shedding new light on corona-treated alu-foil, TAPPI PLACE. Conf. (2004).

[29] J.D. Andrade, S.M. Ma, R.N. King, D.E. Gregonis, Contact angles at the solid-water interface, J. Colloid Interface Sci. 72 (1979) 488–494.

[30] Y.C. Ko, B.D. Ratner, A.S. Hoffman, Characterization of hydrophilic-hydrophobic polymeric surfaces by contact angle measurements, J. Colloid Interface Sci. 82 (1981) 25–37.

[31] J. Schultz, A. Carre, C. Mazeau, Formation and rupture of grafted polyethylene/aluminum interfaces, Int. J. Adhes. Adhesives 4 (1984) 163–168.

[32] C.K. Cho, K. Cho, C.E. Park, Relationship between surface rearrangement and adhesion strength of poly(ethylene-co-acrylic acid)/LDPE blends to aluminum, J. Adhes. Sci. Technol. 11 (1997) 433–445.

[33] S. Wu, Formation of adhesive bond, olymer Interface and Adhesion, Marcel Dekker, Inc, New York, 1982, pp. 359–447.

[34] D.R. Ward, Thermodynamic parameters for predicting adhesion between functional polyethylenes to polystyrene copolymers, TAPPI PLACE. Conf. (2006).

[35] I. Lee, R.P. Wool, Optimum polymer solid interface design for adhesion strength, J. Adhes. 75 (2001) 299–324.

[36] I. Lee, R.P. Wool, Polymer adhesion vs substrate receptor group density, Macromolecules 33 (2000) 2680–2687.

[37] L. Gong, A.D. Friend, R.P. Wool, Polymer-solid interfaces: Influence of sticker groups on structure and strength, Macromolecules 31 (1998) 3706–3714.

[38] Y. Hu, J. Bonekamp, E. Garcia-Meitin, G. Marchand, K. Walton, R. Patel, Polypropylene based olefin block copolymers as tie layers for multilayer packaging, SPE ANTEC Conf. (2014).

[39] Y. Hu, G. Marchand, R. Patel, M. VanSumeren, E. Garcia-Meitin, S. Baker, et al., High performance PP/PE multilayer films enabled by PP based OBC, SPE ANTEC Conf. (2015).

[40] Y. Hu, G. Marchand, R. Patel, F. Olajide, Case studies of PP-based OBC for multilayer packaging, SPE ANTEC Conf. (2016).

[41] J.M. Eagan, J. Xu, R. Di Girolamo, C.M. Thurber, C.W. Macosko, A.M. LaPointe, et al., Combining polyethylene and polypropylene: enhanced performance with PE/iPP multiblock polymers, Science 355 (2017) 814–816.

[42] I.N. Leclere, B. Dinelli, J. Kuusipalo, Keys to good adhesion in coextrusion coating: Interactions between tie resin nature and pre-treatments, TAPPI Polymers, Laminations & Coat. Conf. (1997) 203–209.

[43] R. Allen, Using primers in combination with adhesive tie-layer resins or their blends to make structures with unique performance, TAPPI PLACE. Conf. (2005).

[44] J. Bodine, S.R. Tanny, Improved bonding performance of low temperature extrudate on primed substrates, SPE FlexPackCon Conf. (2006).

[45] M.G. Botros, Investigation of adhesion and failure mechanisms in tie-layer adhesives/EVOH systems, TAPPI Polym., Laminations & Coat. Conf. (1995) 331–347.

[46] M.G. Botros, Investigation of adhesion and failure mechanisms in tie-layer adhesive/EVOH systems, J. Plast. Film Sheeting 12 (1996) 195–211.

[47] M. Botros, C. Lee, D. Ward, T. Schloemer, New developments of Plexar® tie-layer adhesives in flexible and rigid applications, Eur. TAPPI PLACE. Conf. (2005).

[48] M. Botros, J. Merrick-Mack, Determining the most effective tie-layer in response to new material developments, Barrier Film. Conf. (2006).

[49] M. Botros, New-generation Plexar tie-layer adhesives selected for multilayer barrier applications2011 Eur. TAPPI PLACE. Conf. (2011).

[50] L. Levitt, C.W. Macosko, T. Schweizer, J. Meissner, Extensional rheology of polymer multilayers: A sensitive probe of interfaces, J. Rheol. 41 (1997) 671–685.

[51] T. Saito, C.W. Macosko, Interfacial crosslinking and diffusion via extensional rheometry, Polym. Eng. Sci. 42 (2002) 1–9.

[52] E.J. Kramer, L.J. Norton, C.-A. Dai, Y. Sha, C.-Y. Hui, Strengthening polymer interfaces, Faraday Discuss. 98 (1994) 31–46.

[53] E. Boucher, J.P. Folkers, H. Hervet, L. Leger, C. Creton, Effects of the formation of copolymer on the interfacial adhesion between semicrystalline polymers, Macromolecules 29 (1996) 774–782.

[54] E. Boucher, J.P. Folkers, C. Creton, H. Hervet, L. Léger, Enhanced adhesion between polypropylene and polyamide-6: role of interfacial nucleation of the β-crystalline form of polypropylene, Macromolecules 30 (1997) 2102–2109.

[55] C. Laurens, R. Ober, C. Creton, L. Léger, Role of the interfacial orientation in adhesion between semicrystalline polymers, Macromolecules 34 (9) (2001) 2932–2936.

[56] C. Laurens, C. Creton, L. Léger, Adhesion promotion mechanisms at isotactic polypropylene/polyamide 6 interfaces: role of the copolymer architecture, Macromolecules 37 (2004) 6814–6822.

[57] G. Jiang, H. Wu, S. Guo, Reinforcement of adhesion and development of morphology at polymer–polymer interface via reactive compatibilization: a review, Polym. Eng. Sci. 50 (2010) 2273–2286.

[58] K. Cho, F. Li, Reinforcement of amorphous and semicrystalline polymer interfaces via in-situ reactive compatibilization, Macromolecules 31 (1998) 7495–7505.

[59] T. Barraud, C. Creton, L. Leger, Formation of diblock copolymers at PP/PA6 co-extruded interfaces and their role in promoting adhesion, Proc. Annu. Meet. Adhes. Soc. 31 (2008) 132–134.

[60] T. Barraud, C. Creton, L. Leger, Characterizing the formation of diblock copolymers at PP/PA6 co-extruded interfaces and their role in promoting adhesion, Proc. Annu. Meet. Adhes. Soc. 32 (2009) 91–94.

[61] T. Barraud, F. Restagno, L. Léger, Characterizing the formation of diblock copolymers acting as adhesion promoters at PP/PA6 interfaces, in conditions relevant for co-extrusion, Proc. Annu. Meet. Adhes. Soc. 33 (2010) 356–358.

[62] T. Barraud, F. Restagno, S. Devisme, C. Creton, L. Léger, Formation of diblock copolymers at PP/PA6 interfaces and their role in local crystalline organization under fast heating and cooling conditions, Polymer 53 (22) (2012) 5138–5145.

[63] S. Devisme, Original test to characterize the stretching effect onto adhesion during film or sheet extrusion2011 Eur. TAPPI PLACE. Conf. (2011).

[64] J. Song, A.M. Baker, C.W. Macosko, R.H. Ewoldt, Reactive coupling between immiscible polymer chains: acceleration by compressive flow, AIChE J. 59 (2013) 3391–3402.

[65] J. Song, R.H. Ewoldt, W. Hu, H.C. Silvis, C.W. Macosko, Flow accelerates adhesion between functional polyethylene and polyurethane, AIChE J. 57 (2011) 3496–3506.

[66] M. Xiao, C. Mohler, C. Tucker, B. Walther, X. Lu, Z. Chen, Structures and adhesion properties at polyethylene/silica and polyethylene/nylon interfaces, Langmuir 34 (2018) 6194–6204.

[67] J.S. Andre, B. Li, X. Chen, R. Paradkar, B. Walther, C. Feng, et al., Monitoring interfacial reaction of maleic anhydride grafted polyolefins with ethylene vinyl alcohol copolymer in situ at buried solid/solid interface, submitted ACS Macro Lett. (2020).

[68] B.A. Morris, Factors influencing adhesion in coextruded structures, TAPPI J. 75 (1992) 107–110.

[69] B.A. Morris, I.-H. Lee, Tie polymers, in: J.F. Macnamara Jr. (Ed.), Film Extrusion Manual, 3rd (ed.), TAPPI Press, 2020, pp. 269–272.

[70] I.-H. Lee, B.A. Morris, Extrudable adhesive resins for coextrusion laminating and coating, in: W. Durling (Ed.), Extrusion Coating Manual, 5th (ed.), TAPPI Press, Atlanta, GA, 2017.

[71] T. Gilbertson, M. Plantier, Corona surface treatment, in: W. Durling (Ed.), Extrusion Coating Manual, 5th (ed.), TAPPI Press, Atlanta, GA, 2017.

[72] E. Laiho, Ylänen, Flame, corona, ozone – do we need all pretreatments in extrusion coating? in: T. Bezigian (Ed.), Extrusion Coating Manual, 4th (ed.), TAPPI Press, Atlanta, GA, 1999, pp. 89–98.

[73] J. DiGiacomo, Flame plasma surface treatment, in: W. Durling (Ed.), Extrusion Coating Manual, 5th (ed.), TAPPI Press, Atlanta, GA, 2017.

[74] R.A. Wolf, Advances in adhesion with CO_2-based atmospheric plasma surface modification, TAPPI PLACE. Conf. (2008).

[75] M.H. Hansen, M.F. Finlayson, M.J. Castille, J.D. Goins, The role of corona discharge treatment on improving polyethylene/aluminum – an acid/base perspective,", TAPPI Polymers, Laminations & Coat. Conf. (1992) 311.

[76] W. Eckert, Corona and flame treatment of polymer film, foil and paperboard, TAPPI PLACE. Conf. (2004).

[77] J. DiGiacomo, New developments in adhesion promotion using flame plasma surface treatment-a tutorial2020 SPE Virtual ANTEC Conf. (2020).

[78] S. Richards, The effects of surface treatment on heat seal and hot tack, TAPPI Int. Flex. Packaging Extrusion Div. Conf. (2018).

[79] D.A. Markgraf, Atmospheric plasma – the new functional treatment for extrusion coating and lamination processes, in: 9th TAPPI European PLACE Conference, Rome, Italy, May 12–14, 2003.

[80] D.J. Bentley Jr., F.M. Singer, Chemical primers to enhance adhesion and other properties, in: T. Bezigian (Ed.), Extrusion Coating Manual, 4th (ed.), TAPPI Press, Atlanta, GA, 1999, pp. 185–200.

[81] Mica Corp., Primer technology: converting with water. <http://www.tappi.org/content/06asiaplace/pdfs-english/converting.pdf>, n.d. (accessed 07.03.15).

[82] C.L. Olsson-Jacques, A.R. Wilson, A.N. Rider, D.R. Arnott, Effect of contaminant on the durability of epoxy adhesive bonds with Alclad 2024 aluminum alloy adherends, Surf. Interface Anal. 24 (1996) 569–577.

[83] J. van den Brand, S. Van Gils, P.C.J. Beentjes, H. Terryn, J.H.W. de Wit, Ageing of aluminium oxide surfaces and their subsequent reactivity towards bonding with organic functional groups, Appl. Surf. Sci. 235 (2004) 465–474.

[84] Y. Trouilhet, B.A. Morris, Prediction of temperature profiles across coating and substrate in the nip, TAPPI Polymers, Laminations Coat. Conf. (1999).

[85] P.B. Sherman, Ozonation of polymer melt for improved adhesion, in: T. Bezigian (Ed.), Extrusion coating manual, 4th (ed.), TAPPI Press, Atlanta, GA, 1999, pp. 75–88.

[86] B.A. Morris, N. Suzuki, The case against oxidation as a primary factor for bonding acid copolymers to foil, TAPPI Polymers, Laminations & Coat. Conf. (2000).

[87] W.H. Smarook, S. Bonotto, Adhesion of carboxyl-containing polyolefins, Polym. Engr. Sci. 8 (1968) 41–49.

[88] R.F. Bernacki, Adhesive polymers for extrusion coating and laminating, TAPPI Polymers, Laminations Coat. Conf. (1985) 311–318.

[89] B.A. Shah, The effect of interfacial chemical interactions in inter-layer adhesion of packaging structures, TAPPI Polymers, Laminations Coat. Conf. (1989) 789–792.

[90] G. Hoh, S.A. Sadik, J.A. Gates, "Aluminum foil adhesion of ethylene/methacrylic acid copolymer extrusion coatings, TAPPI Polymers, Laminations Coat. Conf. (1989) 361–366.

[91] M.F. Finlayson, B.A. Shah, The influence of surface acidity and basicity on adhesion of poly(ethylene-co-acrylic acid) to aluminum, J. Adhes. Sci. Techn. 5 (1990) 431–439.

[92] D.B. Charboneau, J.D. Domine, Acid terpolymers for extrusion coating and laminations, TAPPI Polymers, Laminations Coat. Conf. (1991) 281–288.

[93] J. Vansant, B. Morris, S. Marks, Acid copolymers for extrusion laminating and coating, in: W. Durling (Ed.), Extrusion Coating Manual, 5th (ed.), TAPPI Press, Atlanta, GA, 2017, pp. 235–239.

[94] J. Vansant, B. Morris, S. Marks, Ionomer resins for extrusion coating, in: W. Durling (Ed.), Extrusion Coating Manual, 5th (ed.), TAPPI Press, Atlanta, GA, 2017.

[95] L.L. Couturier, J. Schultz, Adhesion das les multicouches polymere-metal (Adhesion in polymer-metal multi-layers), Vide, les. Couches Minces – Suppl. 251 (1990) 87–89.

[96] G.L. Baker, R.F. Buesinger, The ethylene methyl acrylate copolymer, J. Plast. Film Sheeting 3 (1987) 112–117.

[97] B. Henn, It's time to get to know n-butyl acrylate copolymers, Plast. Technol. (1992) 71–74. June.

[98] D.S. Davis, EnBA: expanding the portfolio, SPE ANTEC Conf. (1995) 2155. -2119.

[99] E. Laiho, S. Kuusela, Specialties for extrusion coating—high pressure copolymers, TAPPI Polymers, Laminations Coat. Conf. (1995) 613–623.

[100] Y. Trouilhet, New acrylate based resins for extrusion coating, TAPPI Eur. PLACE. Conf. (2003).

[101] M.G. Baker, Ethylene acrylate copolymers for extrusion coating, in: W. Durling (Ed.), Extrusion Coating Manual, 5th (ed.), TAPPI Press, Atlanta, GA, 2017.

[102] B. Walther, D. Falla, D. Niemann, C. Zudercher, Adhesion properties and performance of ethylene-ethylacrylate copolymers, Eur. TAPPI PLACE. Conf. (2003).
[103] R.T. Van Ness, Ethylene vinyl acetate in extrusion coating, TAPPPI J. 59 (1976) 114–116.
[104] G.D. Cheney, S. Wever, Ethylene vinyl acetate resins for extrusion coating, in: W. Durling (Ed.), Extrusion Coating Manual, 5th (ed.), TAPPI Press, Atlanta, GA, 2017.
[105] T. Hjertberg, J.E. Lakso, Functional group efficiency in adhesion between polyethylene and aluminum, J. Appl. Polym. Sci. 37 (1989) 1287–1297.
[106] L. Ulren, T. Hjertberg, Adhesion between aluminum and copolymers of ethylene and vinyltrimethoxysilane, J. Appl. Polym. Sci. 37 (1989) 1269–1285.
[107] M.H. Hansen, M.F. Finlayson, M.H. Vaughn, Characterizing aluminum adhesion for low density polyethylene, TAPPI Polymers, Laminations & Coat. Conf. (1991) 349.
[108] B.H. Gregory, Extrusion Coating, A Process Manual, Trafford, Victoria, Canada, 2005, pp. 73–92.
[109] D. Ventresca, G. Welsch, K. Frey, The effect of paperboard coating treatments on polyethylene adhesion during extrusion coating2016 TAPPI PLACE. Conf. (2016).
[110] L. Ulren, T. Hjertberg, An FT-IR study on interfacial interactions in ethylene copolymers/aluminium laminates in relation to adhesion properties, J. Adhes. 31 (1990) 117–136.
[111] A. Stralin, T. Hjertberg, FTIR study on interfacial interactions between hydrated aluminium and polar groups in ethylene copolymers, Surf. Interface Anal. 20 (1993) 337–340.
[112] J. Schultz, L. Lavielle, Adhesion mechanisms of grafted polyolefins, Makromol. Chem., Macromol. Symp. 23 (1989) 343–353.
[113] B.A. Morris, Y.M. Trouilhet, Lamination adhesion of foil to thermoplastic polymers, in: U.S. Patent 8,137,776 B2, 2004.
[114] P.-A. Clevenhag, C. Oveby, Rheological indicators to predict the extrusion coating performance of LDPE, TAPPI PLACE. Conf. (2003).
[115] M. Biscoglio, Next generation of LDPE resins for extrusion coating applications, SPE ANTEC Conf. (2017).
[116] B.A. Morris, Adhesion in extrusion lamination and coating, in: W. Durling (Ed.), Extrusion Coating Manual, 5th (ed.), TAPPI Press, Atlanta, GA, 2017.
[117] B.A. Morris, H.T. Thai, Modifying LDPE for improved adhesion to aluminum foil, TAPPI PLACE. Conf. (2004).
[118] B.A. Morris, A blend of low density polyethylene and a copolymer of low acid-ethylene copolymer (such as acrylic acid or methacrylic acid); high speed coating, in: U.S. Patent 6,903,161 B2, 2005.
[119] I.-H. Lee, Ethylene-vinyl acetate copolymer blend, graft polymer, in: U.S. Patent 4,861,677 A, 1989.
[120] M. Hattori, et al., Adhesive resin composition, in: U.S. Patent 5,591,792, 1997.
[121] I.-H. Lee, Adhesive compositions based on blends of grafted polyethylenes and non-grafted polyethylenes and styrene container rubber, in: U.S. Patent 6,184,298 B1, 2001.
[122] T. Inoue, et al., Process for preparing laminated resin product, in: U.S. Patent 4,058,647, 1977.
[123] H. Matsumoto, H. Niimi, Grafted with an unsaturated carboxylic acid or derivative thereof, in: U.S. Patent 4,198,327 A, 1980.
[124] D.L. Hester, S.A. Sadik, Y. Trouilhet, Using acid copolymers as coextrusion tie layers, TAPPI Polymers, Laminations Coat. Conf. (1989) 459–463.
[125] K.A. Chaffin, J.S. Knutsen, P. Brant, F.S. Bates, High-strength welds in metallocene polypropylene/polyethylene laminates, Science 288 (2000) 2187–2190.
[126] L.M. Sherman, Metallocene PP & PE weld strongly to each other, Plast. Technol. (2001) 45. April.
[127] B. Poon, A. Chang, S.P. Chum, L. Tau, W.R. Volkenburgh, A. Hiltner, et al., Adhesion of polyethylene to polypropylene in multi-layer films, SPE ANTEC Conf. (2001) 1668–1672.
[128] B. Poon, S.P. Chum, A. Hiltner, E. Baer, Adhesion of polypropylene to metallocene/Ziegler-Natta polyethylene blends in microlayers, SPE ANTEC Conf. (2002) 1849–1853.
[129] P.J. Cole, Macosko, Polymer-polymer adhesion in melt-processed layered structures, J. Plast. Film Sheeting 16 (2000) 213–222.
[130] D.J. Arriola, E.M. Carnahan, P.D. Hustad, R.L. Kuhlman, T.T. Wenzel, Catalytic production of olefin block copolymers via chain shuttling polymerization, Science 2006 (312) (2006) 714–719.
[131] R.K. Ayyer, A.R. Kamdar, Y.J. Lin, P.S. Dias, B.C. Poon, A. Hiltner, et al., Effect of tie layer thickness on the adhesion of ethylene based copolymers to polyolefins, SPE ANTEC Conf. (2009) 428–432.
[132] C.W. Macosko, H.K. Jeon, T.R. Hoye, Reactions at polymer-polymer interfaces for blend compatibilization, Prog. Polym. Sci. 30 (2005) 939–947.
[133] G.W. Kamykowski, Factors affecting adhesion of tie-layers between polypropylene and polyamides, J. Plast. Film Sheeting 16 (2000) 237–246.
[134] C. Lee, W. Podborny, T. Schloemer, High-performance-tie-layer resins in flexible packaging applications: structure–performance relationships, SPE ANTEC Conf. (2009) 2332.
[135] R.J. Powell, DuPont Internal Report, 1971.
[136] M.S. Smith, DuPont Internal Report, 1979.
[137] C.R. Ashcraft, M.L. Kerr, Adhesion by molecular orientation, in: US Patent 6,650,721 A, 1987.
[138] J.W. Moriarty, Jr., Ethylene-maleic anhydride tie layer between polyamide and polyolefin layer, in: US Patent 4,668,571 A, 1987.

[139] D.A. Bivens, G.L. Duncan, Maleic-anhydride modified polymer lager coated with polymeric composition derived from vinylidene chloride, in: US Patent 4,064,315A, 1977.

[140] D.A. Krueger, T.W. Odorzynski, An adhesive of a blend of maleic anhydride propylene graft polymer and polypropylene, in: US Patent 4,617,240, 1986.

[141] D.A. Krueger, T.W. Odorzynski, Coextruded film of polypropylene, polypropylene blend, and nylon, in: US Patent 4,361,628A, 1982.

[142] E.J. Deyrup, Plastic composite barrier structures, in: U.S. Patent 4,973,625 A, 1990.

[143] P. Marechal, G. Coppens, R. Legras, J.-M. Dekoninck, Amine/anhydride reaction vs amide/anhydride reaction in polyamide/anhydride carriers, J. Polym. Sci.: Part. A: Polym. Chem. 33 (1995) 757–766.

[144] M. Van Duin, M. Aussems, R.J. Borggreve, Graft formation and chain scission in blends of polyamide-6 and -6.6 with maleic anhydride containing polymers, J. Polym. Sci. A Polym. Chem. 36 (1998) 179–188.

[145] J. Pascal, Peelable structures for PP, PS, PET and PVC. Understanding how multilayer construction influences seal and peel properties, TAPPI PLACE. Conf. (2008).

[146] B.A. Morris, The effect of coextrusion blown film processing variables on peel strength performance, SPE ANTEC Conf. (1996).

[147] B.A. Morris, Effect of process and material parameters on interlayer peel strength in coextrusion coating, film casting and film blowing, J. Plast. Film Sheeting 26 (2010) 343–376.

[148] C.W. Macosko, Flow accelerates interfacial reaction in coextrusion and compatibilization of polymer blends, SPE ANTEC Conf. (2015).

[149] C.D. Lee, J.J. Strebel, J.D. Goudelock, Higher tie layer adhesion in machine direction oriented (MDO) barrier films, TAPPI PLACE. Conf. (2007).

[150] M. Paul, B. Morris, J. Weinhold, K. Hausmann, S. Sengupta, Tie layer adhesion chemistry for multilayer packaging, SPE Int. Polyolefins Conf. (2020).

[151] M. Paul, B. Morris, J. Weinhold, K. Hausmann, Tie layer adhesion chemistry for multilayer packaging, SPE ANTEC Conf. (2021).

[152] E. Helfand, Y. Tagami, Theory of the Interface between Immiscible Polymers. II, J. Chem. Phys. 56 (1972) 3592–3601.

[153] H. Tanaka, H. Kawachi, M. Inaba, Y. Sawada, Adhesive polyethylene compositions and multi-layer laminated films using the same, in: US Patent 6210765B1, 2001.

[154] H. Tanaka, H. Kawachi, Adhesive polypropylene resin composition and multi-layer laminate body using the resin composition, in: US Patent 5695838, 1997.

[155] H. Uosaki, New developments in adhesive resins in oriented barrier//PP film applications, TAPPI PLACE. Conf. (2010).

[156] T.E. Rolando, Flexible Packaging – Adhesives, Coatings and Processes, Rappa Review Reports, iSmithers Rapra Publishing, 2000. <http://books.google.com/books?id = Wn493PrmhDEC&printsec = frontcover&source = gbs_ge_summary_r&cad = 0#v = onepage&q&f = false> (accessed 10.10.20).

[157] L. Jopko, Flexible packaging adhesives – the basics, TAPPI PLACE. Conf. (2004).

[158] T.R. Mueller, The role of adhesives in the retort pouch, TAPPI PLACE. Conf. (2006).

[159] E.M. Petrie, Laminating adhesives for flexible packaging, n.d. <https://citeseerx.ist.psu.edu/viewdoc/download?doi = 10.1.1.619.9118&rep = rep1&type = pdf> (accessed 10.10.20).

[160] R. Wolf, A technology decision – adhesive lamination or extrusion coating/lamination? TAPPI PLACE. Conf. (2010).

[161] H.-S. Schetschok, E.-M. Kupsch, Adhesive lamination vs. extrusion lamination, trends & technology update, TAPPI Eur. PLACE. Conf. (2011).

[162] H. Eichelmann, Laminating adhesives state-of-the-art and future technologies, Eur. TAPPI PLACE. Conf. (2005).

[163] G. Caimmi, Adhesive lamination, TAPPI PLACE. Conf. (2010).

[164] A. Emblem, M. Hardwidge, Adhesives for packaging, in: A. Emblem, H. Emblem (Eds.), Packaging Technology - Fundamentals, Materials and Processes, Elsevier, 2012, pp. 381–394.

[165] M. Donavan, The sticky side of packaging. Adhesives for flexible packaging, Industrial Print Magazine, <http://industrialprintmagazine.com/the-sticky-side-of-packaging/>, 2018 (accessed 10.10.20).

[166] T.R. Mueller, Medium performance acrylic waterborne laminating adhesives, TAPPI PLACE. Conf. (2007).

[167] W.R. Myer, Solventless laminating today, TAPPI PLACE. Conf. (2001).

[168] G. Caimmi, Laminators for 100% solids adhesives: technology and design, TAPPI PLACE. Conf. (2008).

[169] M. Leib, Enhanced economics of 100% solids adhesives, TAPPI PLACE. Conf. (2006).

[170] S.C. Lapin, UV/EB laminating adhesives for packaging applications, TAPPI PLACE. Conf. (2002).

[171] S.C. Lapin, Optimizing application properties of EB curable laminating adhesives for flexible packaging, TAPPI PLACE. Conf. (2005).

[172] A. Guilleux, Flexible packaging fast forward: radiation curable laminating adhesives, Eur. TAPPI Place. Conf. (2005).

[173] G.B. Kenion, Good manufacturing practices for converters, TAPPI PLACE. Conf. (2004).

[174] Packaging Europe, Dow's SYMBIEX lamination technology obtains global food approval for barrier applications. <https://packagingeurope.com/dow's-symbiex™-lamination-technology-obtains-global-food-app//>, 2018 (accessed 29.10.20).

[175] Dow, SYMBIEX solventless adhesives. <https://www.dow.com/en-us/brand/symbiex.html/>, n.d. (accessed 29.10.20).

Chapter 11

Thermoforming, orientation, and shrink properties of flexible packaging films

Many packaging films undergo thermoforming or orientation. During thermoforming, the film is stretched into a mold or cavity to form a package of the desired shape. Such stretching involves biaxial and planar orientation of the film typically below the melting point of one or more of the layers. In film orientation processes such as the tenter frame, double or triple bubble, and machine direction orientation (MDO) processes, the film is subjected to uniaxial, biaxial, or planar strain to enhance physical properties or impart shrink. Both processes are covered in this chapter, with a focus on the process and material parameters that affect performance. The chapter concludes with a description of oriented multilayer film technology.

11.1 Thermoforming

Thermoforming is an important packaging fabrication process used in both flexible and rigid packaging. In flexible packaging, forming is used to create cavities for the product to lie in. An example is thermoforming processes used in horizontal form fill seal machines. The film is heated to the forming temperature and forced into a cavity or mold with vacuum or the positive force of a plug or both. The temperature, material properties and process parameters (rate of the draw, draw ratio, etc.) influence the ability of the film to be successfully thermoformed. Most of the thermoforming literature involves sheet processes [1–5], with very little on films and even less on multilayer films. There is no single material or processing parameter that characterizes good and bad thermoformability. This section describes the important features of thermoforming and the factors that influence the ability of films to be thermoformed, borrowing from the sheet literature where appropriate.

11.1.1 Thermoforming part testing

Some of the properties of the thermoformed part that are critical to quality include:

- Uniformity of the layers after forming: Consistency of the draw, especially in the corners of the cavity is often a challenge. Thin areas can result in greater gas permeation and lower strength.
- Integrity of the layers after forming: This is particularly important for barrier layers. If the barrier layer breaks up or is stretched too thin during forming, the shelf life of the package may suffer. The strength of the materials and package also requires the integrity of the layers.
- Low residual stress: High stress frozen into the formed package can result in what is often referred to as "suck back" where the final shape of the cavity is not the same as the dimensions of the mold.

During the forming process, the forming force (vacuum or plug) and the surface temperature of the film or sheet are measured. The temperature is usually measured with an IR thermometer. The force and temperature are two key parameters to control for the consistent quality of the formed parts.

After forming, the thickness distribution in the formed part is typically measured on a representative set of samples. Several types of devices are used to measure the total thickness such as:

- Sonic techniques: for example, MagnaMike, Panametrics 25DL, ElectroPhysik Minitest 400,
- Magnetic flux: for example, ElectroPhysik Minitest 7400FH,
- Digital micrometer,
- Spiral micrometer,

The Science and Technology of Flexible Packaging. DOI: https://doi.org/10.1016/B978-0-323-85435-1.00002-8

- Optical microscope,
- Electron microscope.

The thickness is measured at several fixed locations on the part, such as near the lip, along a side wall, along the bottom, and in the corners. The results are typically reported as the standard deviation divided by the mean. The greatest variation is usually in the corners where the highest elongation has occurred.

For multilayer films, the layer distribution of the individual layers is also important to measure. This is typically done using an optical or electron microscope or atomic force microscope on cross-sectioned specimens. Depending on the structure, other techniques such as interferometry or IR spectroscopy may be able to distinguish the thickness of one or more of the critical layers, such as the barrier layer [6,7].

11.1.2 Material properties and testing

Several material properties have been shown to influence forming, including:

- Crystallinity. Lower crystallinity generally improves formability. The ideal material for forming is a rubbery elastic solid that deforms or flows uniformly with applied stress. Above their glass transition temperature (Tg), amorphous polymers exhibit this behavior often over a broad temperature range. Semicrystalline polymers typically have a much narrower thermoforming window that is dependent on the degree of crystallinity, the crystalline morphology and the thermal properties. Formability is generally improved with a lower degree of crystallinity and smaller crystal size. The use of nucleating agents, for example, may help improve formability by reducing the size of the crystals. The addition of a rubbery phase can improve formability. An example is impact copolymers of polypropylene (PP) where an ethylene-propylene rubbery phase is copolymerized into the PP homopolymer backbone. The resulting product has superior hot strength for greater sag resistance, improved formability, and better low-temperature toughness, but maintains most of its high-temperature performance since the homopolymer PP remains in the continuous phase [8].
- Hot strength. Higher strength at the forming temperature provides better sag resistance in the film or sheet. Higher molecular weight (MW) generally improves hot strength.
- Yield stress. Low yield strength helps give uniform orientation during forming. The temperature and speed of the process need to be adjusted to reach the yield point to achieve uniform thickness distribution [9].
- Recoverable strain (from creep tests, see Section 5.5.2). High recoverable strain makes it more difficult for the polymer to conform to mold shapes and increases residual stress (which affects physical properties) [10].
- Strain hardening (after yielding). Materials that strain harden exhibit higher stress as the strain is increased. (See Chapter 5 for a discussion on strain hardening in the melt.) Strain hardening helps control the uniformity of the thickness during forming through a "self-healing" effect. A region that locally experiences higher elongation than adjacent regions will experience higher stress. This acts to impede further stretching of that region, allowing the adjacent regions to catch up. High-strain hardening behavior is thus self-correcting. As described in Chapter 5, incorporating long-chain branching (LCB) and broadening the molecular weight distribution (MWD) are two ways to increase the strain hardening behavior of a polymer [11–14].
- Frictional properties. In plug assist thermoforming, the friction between the plug and film/sheet at the forming temperature influences the material distribution and quality of the final part. Friction is influenced by the polymer type, additives (see Section 12.1), temperature, and the composition of the plug [15–18].

Since thermoforming involves nonlinear biaxial deformation of the film at elevated temperatures it is generally difficult to measure polymer properties that simulate the thermoforming process. Several lab tests are used but each has its limitations:

- Hot tensile test. This involves measuring the tensile properties (stress versus strain) at the forming temperature using a tensile tester with an environmental chamber. This test is generally easy to perform and so probably is the most used test for formability. The deformation is uniaxial rather the biaxial, which may not adequately reflect thermoforming [19]. Also, the strain rate ($0.3-0.6\ s^{-1}$) is at the low end of the thermoforming process ($0.1-10\ s^{-1}$). For these reasons, using hot tensile tests to fit parameters for analytical predictive models is problematic. There are also practical measurement issues that affect the reproducibility, such as grip slip and stretching not confined to the neck-down region [20].
- Hot creep test. The creep test is described in detail in Section 5.5.2. The specimen is subjected to a load or stress, and the resulting strain is measured as a function of time. This is another commonly used test for formability. The

hot creep test allows higher shear rates than the hot tensile test but like the tensile test, the deformation is uniaxial. The creep experiment yields a characteristic retardation time, λ, which corresponds to the delay in response to the imposed stress. When the process time of the thermoforming process is less than 0.1λ, the material behaves like an elastic membrane, making it suitable for snap-back forming. When the process time is greater than 10λ, the material behaves like a viscous fluid, ideal for pressure forming [19].

- Dynamic mechanical (thermal) analysis (DMA or DMTA). This test is described in Section 5.3.5. The sample is subjected to oscillatory stress in the linear viscoelastic regime and the response is measured. The elastic (G') and loss (G'') moduli are two responses recorded by the test. The test can be either run at a constant temperature, varying the frequency, or constant frequency varying the temperature. DMA is very repeatable and is useful for obtaining model parameters, but as a direct indication of thermoformability it has several deficiencies. The test uses small deformations in the linear regime whereas thermoforming involves large, highly nonlinear deformations. DMA also does not duplicate the stretching and orientation found in thermoforming.
- Melt strength (Rheotens). This test is described in Section 5.3.6. It measures the melt tension on a filament extruded from a capillary rheometer at elevated temperatures drawn down between rotating rolls. This test is conducted in the melt, whereas thermoforming is done in the rubbery plateau region below the melting point. As such, the Rheotens test does not mimic the thermoforming process very well, although it can provide relative rankings for hot strength and sag resistance.
- Extensional viscosity. Several techniques used to measure this melt property are described in Section 5.3.6. Most are uniaxial measurements in the melt state, but some indication of strain hardening can be obtained with these measurements. To better simulate thermoforming and biaxial stretching, several investigators have devised a membrane inflation rheometer [21–24].
- Differential scanning calorimetry (DSC). The glass transition temperature, melt point, crystallization temperature, and heat of fusion are important attributes that can be measured with DSC. The fraction of molten resin at a given temperature can be computed by dividing the area under the melting peak up to that temperature by the total area under the peak (the heat of fusion).
- Optical or scanning microscopy. These techniques can be used to examine crystalline size and morphology.
- Instrumented lab-scale former. Dharia and Hylton [20,25] develop a small-scale thermoformability analyzer that mimics the forming process. The device heats, forms, and cools a single sheet under controlled conditions similar to a production thermoformer. Plug, vacuum, or vacuum with plug assist thermoforming modes can be used. The sheet temperature (monitored by an IR thermometer), soak time, plug speed, cooling time, maximum plug force, and maximum draw depth can all be controlled. The force to form is measured as a function of draw depth and time. Force versus depth plots provide useful information about the formability of the material. The thickness distribution in the final part can also be measured. The device can be used for materials development, process engineering, and quality control. The test is rapid, repeatable, and represents the thermoforming process. One limitation is the cost of the machine [26].
- Simulation coupled with lab forming. Hegemann et al. [15–17] determine material parameters for simulating plug-assisted thermoforming by conducting frictional measurements on a torsional rheometer and measuring force versus displacement during forming on a laboratory servo-hydraulic high-speed (10 m/s) testing machine heated to the forming temperature. Using commercial thermoforming simulation software, they reverse calculate the material parameters to obtain a fit between the model and experimental results.

11.1.2.1 Predictive criteria

Just as there is no single test for formability, there is no single criteria or test result that predicts performance. Many investigators have tried to correlate results from lab tests to lab or field-forming experience. Some of the criteria proposed by these investigators are highlighted here.

Wang and Boparai [27] and Ho et al. [28] use the difference in the area draw ratio (width times thickness after drawing divided by the initial width times thickness of the specimen, Eq. (11.4) below) in the machine and transverse direction of a hot tensile test as a guide for thermoformability. They define a dimensional thermoformability index (DTI) related to this difference:

$$\mathrm{DTI} = 100\sum_{i=1}^{n} \mathrm{Abs}(\tilde{y} - y_i)/n \tag{11.1}$$

where

$$y_i = \text{ADR}_i(TD) - \text{ADR}_i(MD) \tag{11.2}$$

$$\tilde{y} = \sum_{i=1}^{n} y_i/n \tag{11.3}$$

$$\text{ADR}_i = \frac{W_i X_i}{W_o X_o} \tag{11.4}$$

i = location of ith position on the tensile bar,
n = number of positions or specimens,
ADR = area draw ratio,
W_i = width of specimen at the ith position after drawing,
W_0 = initial width of specimen,
X_i = thickness of specimen at the ith position after drawing, and
X_0 = initial thickness of specimen.

In Eq. (11.1), Abs stands for absolute value. Multilayer films containing polyamide (PA) and polyethylene (PE) were tested for DTI using a hot tensile test at 95°C. Lower values of DTI (smaller difference between the TD and MD draw ratios) correlate with better formability based on experience and lab forming results. For example, a film containing a polyamide (nylon) copolymer has a lower DTI than one with PA6, as do films containing more PA at the expense of PE. This is consistent with industry experience and lab forming trials showing better formability with PA copolymer and greater PA content versus PE.

Ho et al. [28] use a hot tensile test to evaluate strain hardening. Using the slope of the true (Hencky) stress versus strain curve at two locations, they compute two moduli, one at low strain (0.3–0.5) and one at high strain (2). Multilayer films with lower low-strain modulus and higher high-strain modulus are expected to thermoform well. Low-strain modulus is associated with ease of filling the mold and a high high-strain modulus helps resist blowouts. This is borne out, at least qualitatively, in their experiments with multilayer films.

Kurzbeck and Munstedt [11,14] use transient elongational melt viscosity measurements to predict formability. Two grades of PP are compared, a standard linear PP and a PP containing LCB induced by e-beam irradiation. The branched PP exhibits strain hardening behavior in the transient elongational rheometer whereas the linear PP does not. A laboratory thermoforming test shows that the branched PP has more uniform thickness after forming than the linear PP.

Spence and Hylton [10] vary the applied stress and time in a hot creep test to simulate different parts of the forming process. A low stress (1916 Pa) and long time (10 seconds) test correlates with sag, and high stress (10 kPa) for a shorter time (5 seconds) is used to simulate part forming. Long retardation times measured in the first experiment correspond with long sag times on a commercial sheet line for polystyrene (PS), high-density polyethylene (HDPE), and PP sheet.

Several investigators develop criteria using DMA, albeit in different ways:

- In the frequency sweep mode, G'' is initially greater than G', indicating the viscosity dominates over elasticity at low frequencies (long times). G' increases faster than G'' with increasing frequency and the crossover point represents the point where elasticity starts to dominate. Elasticity is needed for sag resistance and viscous flow for part formation. Spence and Hylton [10] relate the crossover frequency (ω) to sag resistance. Materials with a crossover frequency near the sag frequency ($<0.1\ \text{s}^{-1}$) have good sag resistance. The sag frequency is simply the inverse of the heat soak time of the sheet (10 seconds in this example). The crossover frequency relationship is borne out in their experiments on a sheet line: PS and PP have good and poor sag resistance with crossover frequencies of 0.034 and $10\ \text{s}^{-1}$, respectively.
- The width of the viscoelastic plateau in the temperature sweep mode is related to the temperature window for forming [9]. The modulus is high at temperatures below the Tg or Tm (melting temperature) and decreases once it reaches the amorphous or melt region. Many investigators [28–30] have proposed that the temperature at which the elastic modulus (G') measured at a frequency of $1\ \text{s}^{-1}$ drops below 10 MPa is a good estimate for the minimum forming temperature. This is based on practical experience.
- DiRaddo and Meddad [31] compute a deformation sensitivity ratio ($=G''(\omega)/G'(\omega)$). The greater the ratio, the more viscous (and less elastic) is the material. DiRaddo and Meddad also compute a temperature sensitivity ratio ($=G(T_1)/G(T_2)$ at constant ω; $T_2 > T_1$); a higher ratio indicates greater sensitivity to process temperature. These ratios can be useful for understanding thermoforming issues.

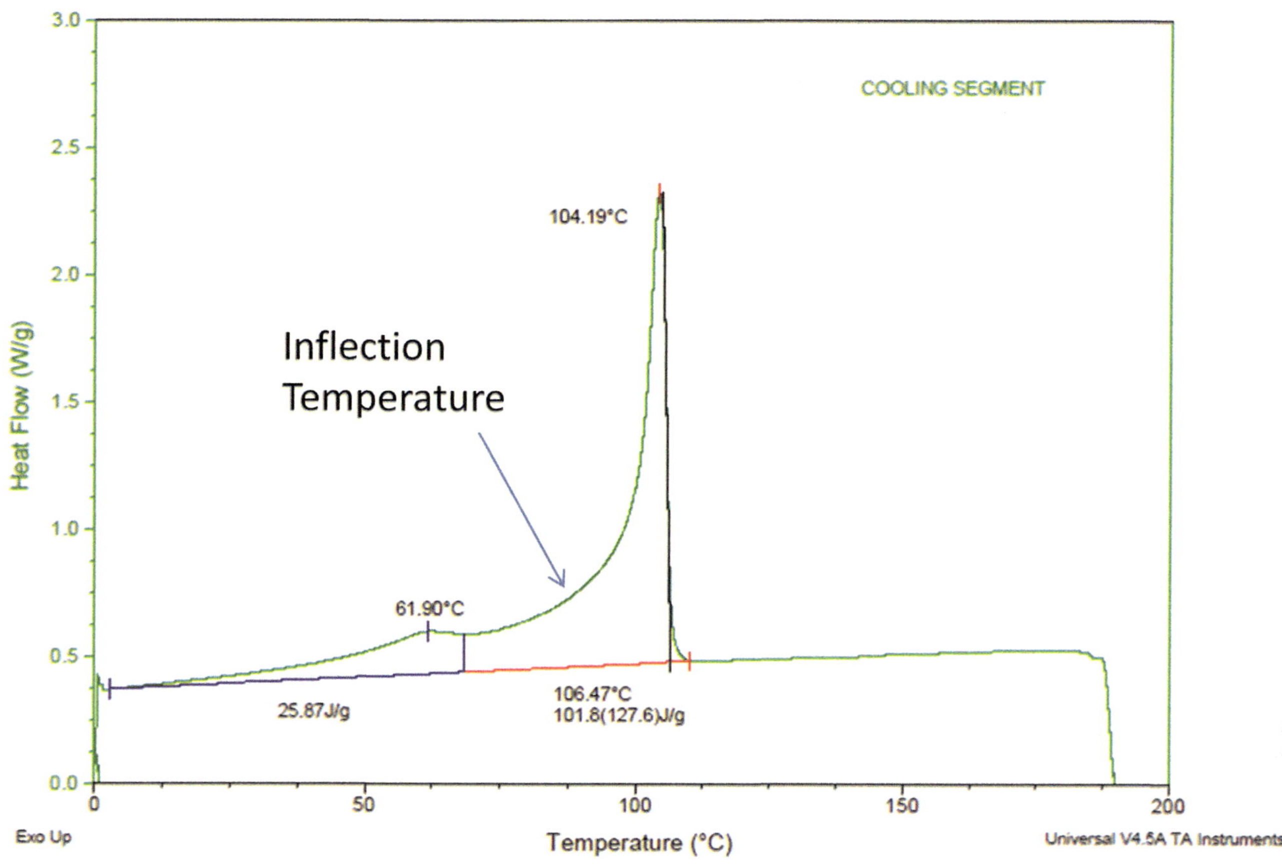

FIGURE 11.1 Cooling inflection temperature, as defined by Ward [32]. DSC cooling curve for LLDPE courtesy of Russ LaFon, DuPont.

Wang and Boparai [27] show that formability increases as the total crystallinity of the film structure decreases, as measured by the total heat of fusion using DSC. Ho et al. [28] find that formability of PA6-PE multilayer films improves as the fraction of the polymer that has melted at the forming temperature is decreased. One way they accomplish this is by incorporating more high-melting PA6 into the film structure. They propose that reducing the fraction of molten polymer at the forming temperature reduces blowouts.

Ward [32] correlates formability with the inflection point of the DSC crystallization peak. This is illustrated in Fig. 11.1. Monolayer films of various PE grades, ionomer, ethylene vinyl acetate (EVA), PA, and ethylene vinyl alcohol (EVOH) are thermoformed using a laboratory single cavity former. The films are evaluated for completeness of filling the cavity, presence of film breaks and uniformity, and gauge reduction at various points in the formed parts. The results show:

- Ionomer has the best forming characteristics with the most uniform gauge reduction and cell definition.
- Butene-LLDPE, PA, and LDPE do not form completely (at 105°C).
- Octene-LLDPE forms well but with slightly more thinning at the edges.
- EVOH and EVA form completely but cell definition and gauge thinning are not as good.
- Formability correlates with the temperature inflection point of the DSC crystallization peak. When the sheet temperature is varied, if the surface temperature is above the inflection temperature the film formed well. The approach is also extended to multilayer coextruded films.

11.1.3 Factors that affect thermoforming

Some of the key factors that affect thermoforming performance are discussed in the following sections.

11.1.3.1 Material parameters

These are discussed in Section 11.1.2.

11.1.3.2 Film structure

The composition, placement, and thickness of layers, as well as how the film is fabricated, may influence formability. As discussed in Section 11.1.3.3, the thermoforming temperature window of each of the materials in a multilayer structure must overlap. Careful selection of the layers is sometimes needed, especially for deep draw applications. Locating more formable layers next to difficult-to-form ones can help. For example, EVOH is often sandwiched between layers of PA6 to improve formability. The method used to fabricate the film, such as blown film, cast film, or lamination, can influence the formability. A rapid quench process such as cast film or water-quenched blown film reduces the crystallinity of films containing semicrystalline polymers and improves formability [33] (see Section 14.2). The film fabrication process also affects the balance of orientation and properties in the machine and transverse directions. Anisotropy can contribute to nonuniformities during forming.

Ward and Crowe [34] discuss the importance of retaining heat during forming. They thermoform barrier-forming webs of different thicknesses with the structure (PA–tie–tie–PA–EVOH–PA–tie–oLLDPE–oLLDPE) and find the thicker film has a broader forming temperature window and better retention of film gauge in the corners. They model the heat retention during forming and find higher heat retention correlates with better forming performance. They also propose structures with alternative PA or PE types with different heat capacities that may retain heat better.

11.1.3.3 Film temperature

The forming temperature and its uniformity across the film have a pronounced effect on thermoformability. The uniformity of the temperature depends on the heater type, heater profile, and heat-up time, which are all crucial variables for quality control. The target forming temperature will depend on the thermoforming temperature windows of each of the layers. For a material to deform, it must be soft enough, yet not too soft that sagging or other issues occur. Table 11.1 provides typical thermoforming temperature ranges for a variety of polymers. These are based on sheet melt forming processes. Some of the factors affecting this window include the intrinsic orientation of the sheet, hot strength, sag, thermal sensitivity and stability of the polymer, sheet geometry and thickness, uniformity of the heating, depth of draw, mold geometry, plug rate, and plug temperature [35]. The temperature must be hot enough to soften the polymer so that it can deform under vacuum or plug assist, yet not too hot that it suffers blowouts.

In packaging, solid-phase pressure forming (SPPF) is sometimes used. This process was developed for PP sheet by Shell in the 1980s [8,36]. It involves forming at a temperature just when the resin starts to soften. As shown in Fig. 11.2, this occurs at around 160°C for homopolymer PP. In this region, the complex viscosity starts to decrease. The forming window given in Table 11.1 reflects this since most PP sheet is solid-phase pressure formed to avoid

TABLE 11.1 Thermoforming temperature ranges for some common polymers.

Polymer	Thermoforming temperature (°C)	Maximum areal draw ratio
Polystyrene	150–190	8:1
High-impact polystyrene	160–205	8:1
ABS	150–205	10:1
Flexible for plasticized PVC	110–150	10:1
LDPE	125–175	6:1
HDPE	140–190	8:1
PP	140–165	6:1
PC	175–230	8:1
Amorphous PET	125–165	6:1

Source: Data from S.A. Ashter, Thermoforming of Single and Multilayer Laminates: Plastics Films Technologies, Testing, and Applications, Elsevier, Walthan, MA, 2014 and J.L. Throne, Thermoforming, Hanser Publishers, Munich, 1986.

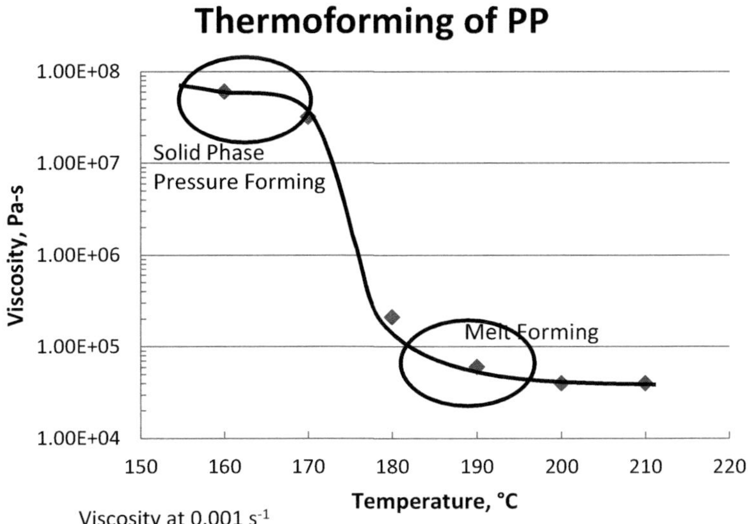

FIGURE 11.2 Complex viscosity at a frequency of 0.001 s^{-1} of polypropylene as a function of temperature from oscillatory rheology measurements highlighting thermoforming temperature regions. *Data from Shell Development Company, Polypropylene Technology Seminar, 1987.*

TABLE 11.2 Thermal properties and typical thermoforming temperatures for ethylene vinyl alcohol (EVOH) and common resins coextruded with EVOH.

Resin	Tg (°C)	Tc (°C)	Tm (°C)	Thermoforming temperature range (°C)	
				SPPF	Melt
EVOH	40–60	130–170	160–191	70–90	155–210
PS	100–120			110–135	135–210
PP	-10–0	120–145	165–175	140–160	170–210
PA 6	50	160–170	225–235	80–120	205 +
PET	80	160–180	250	80–120	

Source: Data from Kuraray Technical Bulletin No. 150, Thermoforming of EVAL® resin containing structures. <http://eval-americas.com/media/36418/tb_no_150.pdf>, 2000 (accessed 21.02.21); R. Chou, I.H. Lee, Behavior of ethylene vinyl alcohol (EVOH) amorphous nylon blends in solid-phase thermoforming, J. Plastic Film. Sheeting (13) (1997) 74–93; and E. Chow, High barrier forming films with EVOH, VII Int. TAPPI/CETEA Conf. Flex. Packaging, 2012.

excessive sag. Melt forming processes for PP operates at around 180°C–190°C. Chou and Lee [37] and Chow [38] provide typical forming ranges for EVOH and common polymers coextruded with EVOH. These are summarized in Table 11.2.

The forming temperature window for EVOH is particularly narrow since it is highly crystalline, crystallizes rapidly, and contains strong inter- and intramolecular interactions due to hydrogen bonding between –OH groups. The ease of forming and orienting EVOH improves as the mol% ethylene is increased, but at the expense of higher oxygen permeation. One way to visualize the challenge of forming multilayer structures containing EVOH is to look at the crystallization half time, as shown in Fig. 11.3. Here the half time represents the time for half the crystals to form at the given temperature. These results are for crystallization under quiescent conditions. In reality, crystallization is much faster due to strain-induced crystallization during forming: the stretching of the polymer chains speeds up the crystallization. Nevertheless, the data show how narrow the temperature window can be. The shaded region on the far left represents the forming window for PP in the SPPF process. Only the 44 and 38 mol% ethylene grades crystallize slowly enough to form in the same temperature window as PP. If the crystallization is too fast, as it is for the 32 and 27 mol% ethylene grades, local deformation occurs that leads to neck-in and breaks during forming.

Thermoforming EVOH-containing films often requires judicious selection of the EVOH grade and the other layers in the structure so they match up well with the forming window of EVOH. Common practices include using high

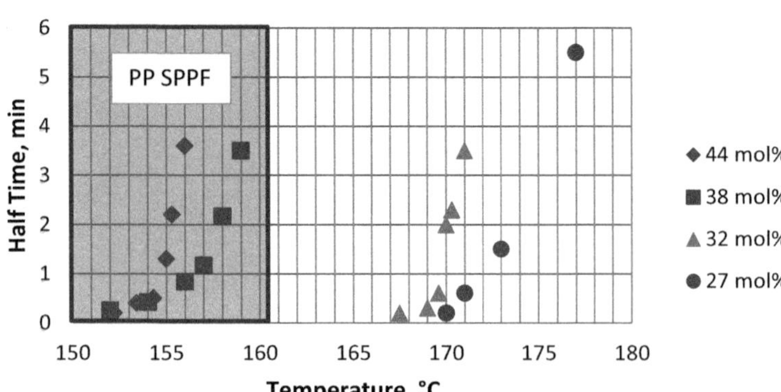

FIGURE 11.3 Crystallization half times for ethylene vinyl alcohol [39]. Shaded rectangle represents typical forming temperature range for PP in solid phase pressure forming (SPPF). *Data from Kuraray Technical Bulletin No. 150, Thermoforming of EVAL® resin containing structures. <http://eval-americas.com/media/36418/tb_no_150.pdf>, 2000 (accessed 21.02.21); R. Chou, I. H. Lee, Behavior of ethylene vinyl alcohol (EVOH) amorphous nylon blends in solid-phase thermoforming, J. Plastic Film. Sheeting 13 (1997) 74–93.*

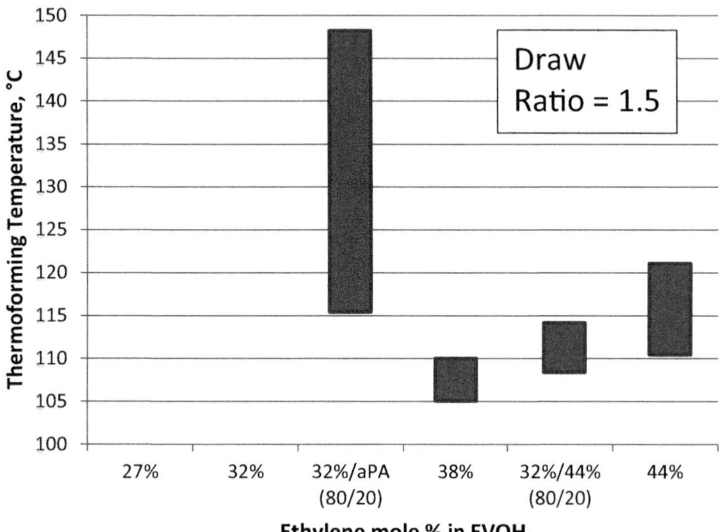

FIGURE 11.4 Temperature range over which a (PP–tie–EVOH–tie–PP) with layer thicknesses [660/50/100/50/660 μm] can be thermoformed without defects. *Reproduced with permission from R. Chou, I.H. Lee, Behavior of ethylene vinyl alcohol (EVOH) amorphous nylon blends in solid-phase thermoforming, J. Plastic Film. Sheeting 13 (1997) 74–93.*

ethylene grades (44 and 38 mol%) and limiting the draw to shallow depths. The high ethylene grades have wider thermoforming windows but also have a poorer gas barrier. EVOH suppliers have introduced special grades of EVOH with wider forming windows [38,40,41]. These are usually less crystalline and so the barrier performance may suffer. Film fabricators often blend grades of EVOH, such as blending a 32 mol% grade with a 38 mol% grade. This improves the forming window somewhat but also results in a reduction in barrier performance. Chou [37,42] and others [43–45] show that blending about 20% of an amorphous PA (aPA), such as Selar PA, a 6I/6 T copolymer from DuPont, with EVOH improves the forming window substantially without a loss in the barrier and allows the film to be drawn to much greater depths. One example is illustrated in Fig. 11.4 for a (PP–tie–EVOH–tie–PP) sheet with a draw ratio of 1.5. The sheet with unmodified 27 and 32 mol% ethylene grades of EVOH cannot be drawn without substantial neck-downs and breaks. Structures with 38, 32/44 blends, and 44 mol% ethylene can only be drawn over a 5 to 10°C window. The addition of 20% aPA to the 32 mol% ethylene grade of EVOH increases the forming window to 33°C. Fig. 11.5 shows the effect of the draw ratio. Blending 20% of the 44 mol% ethylene grade with the 27 mol% grade allows the structure to be thermoformed over a 10°C window for 0.6 and 1.5 × draw ratios, but is not sufficient for large draw ratios. The addition of aPA provides a much greater forming range for the shallow and moderate draw and allows the sheet to be formed, albeit over a narrow temperature range, for deep draw (3.3 ×).

Adding aPA to EVOH increases the haze. Fetel and others [46–48] show that adding a special PA-ionomer alloy (Surlyn AM7927 from Dow) improves the thermoforming window and drawability without a loss in clarity. The barrier performance, however, decreases by about the same as switching to a grade of EVOH with one level higher ethylene.

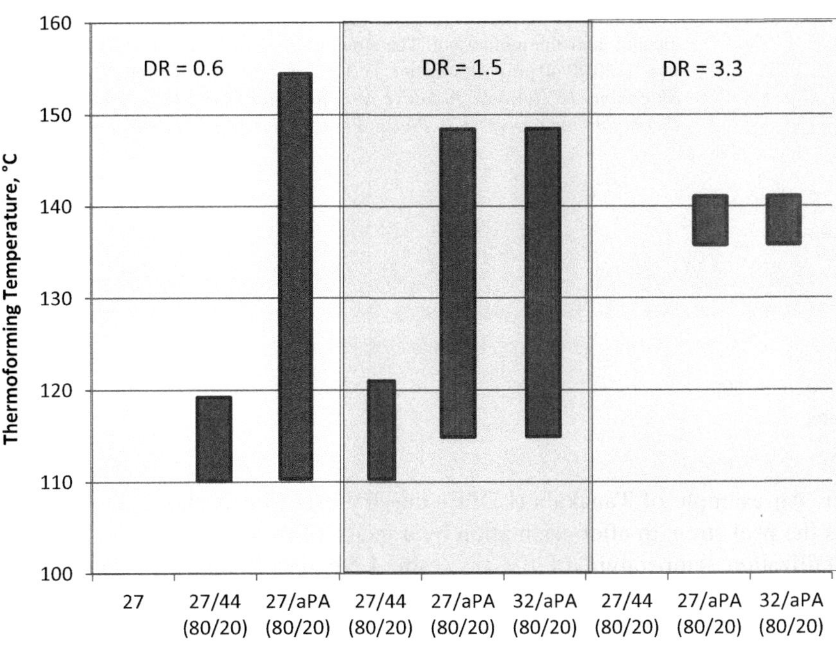

FIGURE 11.5 Temperature range over which a (PP–tie–EVOH–tie–PP) can be thermoformed without defects. For 1.5 and 3.3 × draw ratio, the layer thickness is the same as in Fig. 11.4. For 0.6 × draw, the layer thicknesses are [280/25/50/25/280 μm]. *Reproduced with permission from R. Chou, I.H. Lee, Behavior of ethylene vinyl alcohol (EVOH) amorphous nylon blends in solid-phase thermoforming, J. Plastic Film. Sheeting 13 (1997) 74–93.*

For example, adding the ionomer to a 32 mol% ethylene grade of EVOH provides barrier properties about the same as a 38 mol% ethylene, but with better formability.

Small differences in film temperature can affect thickness distribution in the final part. Temperature affects strain hardening and elongational viscosity, as well as the friction between the film and mold or plug. For complex shapes, nonuniform heating of the film may be optimal; increasing the temperature in regions where the polymer is stretched more may help with thickness distribution [49].

11.1.3.4 Processing parameters

For plug-assisted thermoforming, the plug shape/geometry (height, width, and depth), plug speed, and plug timing affect thermoforming [18,50–52]. For example, increasing the plug depth increases the draw ratio. The maximum draw ratio depends on the materials in the film structure, but in general, as the draw ratio is increased a critical draw ratio is reached above which breaks occur. The plug speed and timing may also affect breaks. In addition, the plug material is important: the material of construction affects how much the plug heats up during forming and its temperature, which in turn affects frictional properties and the film temperature.

For vacuum/pressure forming, the level and rate of change of the vacuum or pressure are significant factors.

In snap-back forming, the prestretch height, mold, and sheet temperature are important [53].

The mold temperature and its surface properties affect the friction between the film and tooling, which influences film thickness uniformity after forming [15–18].

11.1.3.5 Interlayer adhesion after forming

As described in Section 10.4.2.2.4, thermoforming tends to reduce interlayer adhesion. This is illustrated in Fig. 10.44 for a (PA6–tie–ionomer) coextruded film. Orientation during forming increases the interfacial area and decreases the bond density per unit area. Orientation also puts stress on the bonds that have formed. Improved adhesion can be obtained after forming and orientation by:

- Increasing the number of functional groups. For example, the amount of chemical functionality on a tie resin can be increased. Tanaka et al. [54] provide an example for an (LDPE–tie–EVOH) film, where the peel strength to EVOH after a 3 × 3 orientation at 80°C increases from about 10 to 30 g/15 mm by increasing the grafting ratio of the tie layer by a factor of ten.

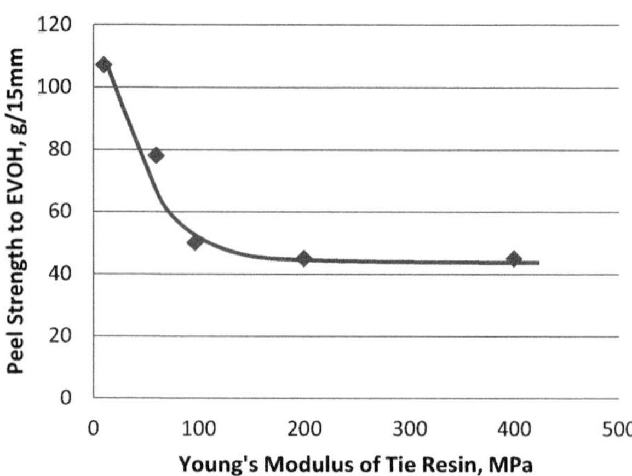

FIGURE 11.6 Effect of tie layer modulus on peel strength to ethylene vinyl alcohol after thermoforming. The structure is (LDPE−tie−EVOH) with thickness [160/40/40 μm]. Orientation is 3 × 3 at 80°C. *Data from H. Tanaka, H. Shigemoto, H. Hawachi, Adhesive resin for multi-layer films used for shrink or thermoforming processes, J. Plastic Film. & Sheeting 12 (1996) 278−286.*

- Decreasing the stiffness of the tie layer. An example of Tanaka's (LDPE−tie−EVOH) film is shown in Fig. 11.6, where decreasing the modulus increases the peel strength after orientation by a factor of two.
- Lowering the crystallization and crystallization temperature of the tie resin. Like decreasing the modulus, this decreases the stress during orientation. Switching from a PE-based tie resin to one based on VLDPE, EVA, or EMA may be advantageous. If the end-user can tolerate it, selecting a tie resin whose melting point is below the forming temperature often improves the peel strength after forming.
- Increasing the polarity of the tie layer to thicken the interface with the polar barrier layer prior to stretching [55,56].

Tie resin suppliers often have special grades that work better for thermoforming. Consulting with the supplier can be helpful.

11.1.4 Modeling

Several investigators [23,57−88] have developed numerical models to better understand the thermoforming process and predict performance. The deformation during thermoforming is large, nonlinear, rapid and complex, being primarily planar to equi-biaxial. Hyperelastic or viscoelastic constitutive models are typically used to simulate the rubbery polymer near its glass transition temperature (amorphous polymers) or melting point (semicrystalline polymers). Hyperelastic models include the Ogden and Mooney-Rivlin models. The Christensen and KBKZ models are often used for the viscoelastic models. Hyperelastic models generally work well for vacuum assist forming but not as well for plug-assist thermoforming, where the need to take into account the strain history makes the viscoelastic models more suitable [62,65,82].

Obtaining material parameters for the models is challenging [60,61,67,70,85,88]. The difficulties of testing under conditions similar to thermoforming have already been highlighted in Section 11.1.2. High temperatures, high strain rates, and complex deformations are problematic. Uniaxial tests are most often used, even though biaxial stresses found in thermoforming are often more than two times higher.

Most models use complex finite element analysis grids and algorithms. A membrane approximation is typically used, where the stresses in the thickness dimension are ignored. This typically agrees well with full three-dimensional analysis if the width/thickness ratio is >100 [57]. A draw back of the membrane approximation is the difficulty of correctly modeling the boundary conditions. Most of the studies reported in the literature further simplify the problem by assuming the temperature is constant (isothermal), although a few studies have coupled the thermal and flow analyses [58,63,64,73,77,79,80,83,85,87]. The model results show good agreement with experimental results for some examples, and not so good agreement for others. In general, the models are not truly predictive yet but do offer insight into material parameters and process conditions that influence thermoforming behavior and performance.

An example of how modeling can be used to screen new film structures is presented by Christopherson and Briere [81]. They use modeling to predict the puncture resistance of medical forming webs for a syringe packaging application. The acceptable failure rate during a simulated transit test is < 0.2%. About 50% of the failures typically come from abrasion of the film from the flanges of the syringe. Another 30% comes from puncture at the end. Typically films with a thickness of 65−150 μm are used to package a 10-mL syringe. After forming, the thickness in the corners is 15−35 μm. They compare two film structures with testing and modeling:

- (35-μm EVA(4.5%VA) - 80-μm Ionomer - 35-μm EVA)
- (30-μm PA - 5-μm tie - 65-μm LLDPE)

The films are formed with a 0.7-second cycle time at 14 packs per minute. The thickness at different locations on the formed package is measured using a MagnaMike accurate to +/− 1 μm. The formed package is cut at the flange and the force to puncture the end of the package at a 45-degree angle to the main axis of symmetry is measured using a 6.5-mm diameter metal probe at 508 mm/min (20 in/min).

To obtain the material parameters for the model, tensile tests are conducted on the films at 70°C, 80°C, and 90°C at uniaxial stretching rates of 10, 50, and 100 mm/min. The results are fit to a KBKZ model with Williams–Landel–Ferry (WLF) temperature dependence.

A finite element model is used to simulate forming and puncture. Postforming shrinkage (reversion) is not accounted for by the model (see Xu and Kazmer [78] and Song et al. [58] for discussions on warpage and shrinkage modeling). The shrinkage generally occurs at the top of the package where the film is thicker and has not yielded during forming. Since the puncture is measured in the corner away from the top flange area, the errors associated with this are expected to be small.

The model prediction of the thickness distribution and puncture agree well with the experimental results. The ionomer film thickness distribution is in better agreement with the model, suggesting less reversion. Both film structures give similar puncture results (and both are used commercially). The authors conclude that the model is a good tool to screen structures.

11.2 Orientation and shrink

Orientation involves stretching and aligning molecular chains during polymer processing. Polymer chains generally favor a random coil configuration as depicted in Fig. 11.7. When stressed, the chain may be stretched in the direction of the imposed strain. This aligns the molecules in the stretching direction. If the film is rapidly quenched, this orientation is frozen in place, which influences many film properties, as described in Section 11.2.1. If the stress has not been allowed to relax, the chains may exhibit a memory effect. When heated above the Tg or crystalline melting point, the chains may relax back toward their preferred random coil configuration, causing the film to shrink. This is the basis of shrink films and bags. If dimensional stability is desired, the oriented film is often annealed at an elevated temperature while under the imposed strain to allow the chains to relax while retaining their oriented configuration.

The terms stretching and orientation, while often used together, should be differentiated. Stretching is the amount of strain imposed by the polymer process. In blown film and double or triple bubble processes stretching is related to the draw-down ratio and blow-up ratio. In MDO, stretching is determined by the difference in roll speeds of the orientation rolls. In a tenter frame process, the transverse stretch is determined by the width ratio imposed by the chain or rail. The direction of the stretching is generally in either one (mono or uni) or two (biaxial) directions. These directions are typically designated the machine direction (MD) and transverse direction (TD), although the transverse orientation in a blown film or double bubble process is really in the hoop or circumferential direction.

Orientation refers to the alignment of the molecules, which affects the film's properties. The amount of orientation depends on the amount of imposed stretch, the rate of stretching, orientation temperature, annealing temperature (if the

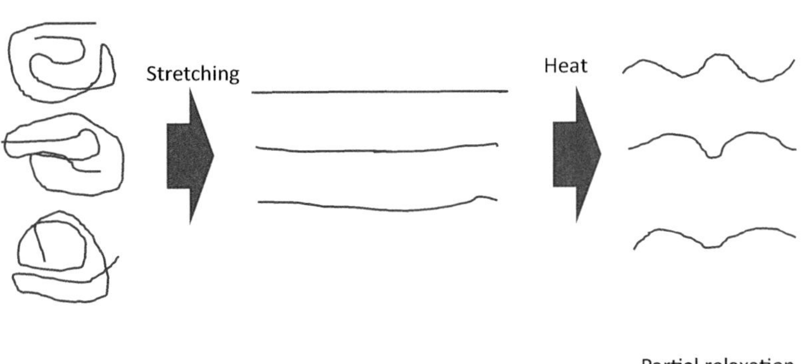

FIGURE 11.7 Stretching of polymer chains.

film is annealed after orientation), and the molecular chemistry and architecture of the polymer [chain stiffness, MW, MWD, amount of LCB, comonomer and comonomer distribution, hydrogen bonding, etc.]. Processes that stretch the polymer from the melt, such as cast film and blown film, generally provide less orientation than those that impose stretch from the solid or semisolid state, such as double or triple bubble, tenter frame, and MDO processes.

11.2.1 Why it is important

Orientation is important for many commercial package forms, especially those involving shrink:

- Shrink films, such as Cryovac and Clysar film, used in a variety of retail bundling and other applications.
- Barrier shrink bags, such as [EVA– polyvinylidene chloride (PVDC)–EVA] or PVDC free triple bubble films used for packaging raw meat.
- Retail skin packaging, an alternative to clamshell packaging. In carded skin packaging, polyvinyl chloride (PVC), PE or ionomer film is heated and draped over the product. Vacuum is applied through the card, forming the film over the product. The film shrinks as it solidifies, holding the product in place.
- Shrink labels.

Orientation is also used to improve the properties of nonshrink films. Examples include biaxial oriented polypropylene (BOPP), biaxial oriented polyester (BOPET), biaxially oriented polyamide (BOPA or BONY), and more recently Biaxially Oriented Polyethylene (BOPE). These are typically used in laminations as a printing surface and abuse resistance layer. BOPET, for example, is often reverse printed and laminated to a polyolefin sealant film. Such laminates are challenging to recycle which has generated renewed development of oriented PE films (MDO or tenter frame processes) to create all-PE structures (see Section 17.4.2). Oriented coextruded barrier films with EVOH are also under development.

These dimensionally stable films are oriented and then annealed to fix the orientation and prevent shrinkage. Some of the property enhancements include:

- An increase in optical transparency of semicrystalline polymers. Orientation often changes the crystalline structure, reducing the size or amount of large spherulites that refract light and cause haze. The overall amount of crystallinity may actually increase as strain helps induce crystallization. Because the size of the crystals is smaller and the crystalline structure undergoes a change into a more fibril structure, the transparency is improved.
- An increase in tensile strength and modulus as the molecules are more aligned. At low levels of orientation, properties such as strength and modulus correlate with both amorphous and crystal phase orientation. After the crystals become fully oriented, the amorphous phase may continue to orient, increasing the properties further [89].
- A decrease in oxygen and moisture permeation due to changes in crystalline morphology.

Orientation does not come for free. There are some tradeoffs, including:

- A decrease in elongation at break and slow puncture resistance, depending on the materials and process.
- A decrease in transverse direction tear strength, but this depends on the specific polymer and process.
- A decrease in interlayer adhesion of multilayer films due to a reduction in the number of bonds per unit area (due to the increase in interfacial area) and an increase in residual stress.

Another important class of films involves orientation during the packaging process. These include stretch films, pallet wraps and hoods, and cling films.

11.2.2 How to measure

11.2.2.1 Orientation

Several analytical techniques are used to measure the molecular orientation in films. These include:

- Birefringence [89–95]. Birefringence is related to changes in polarized light as it passes through a specimen. The difference in the refractive index of light polarized in parallel and perpendicular directions is an indication of the anisotropy of the material. This difference can be measured using polarized light optical microscopy. Birefringence correlates with orientation in both the amorphous and crystalline phase of polymer films.

- Dichroism or Fourier transform infrared spectroscopy (FTIR) [89,92–95]. Dichroism is a measure of absorption of electromagnetic waves by the specimen. In the most widely used method, polarized infrared light is used. The simplest representation of anisotropy is given by the dichroic ratio:

$$\mathscr{D} = \frac{\mathscr{A}_1}{\mathscr{A}_2} \tag{11.5}$$

where

$\mathscr{D}$ = dichroic ratio,
$\mathscr{A}_1$ = absorption of polarized light in the direction parallel to stretching of the polymer,
$\mathscr{A}_2$ = absorption of polarized light in the direction perpendicular to stretching.

Dichroism is a more powerful technique than birefringence since birefringence measures mean values over all components present. Specific frequencies of absorption bands using FTIR can be used in dichrosim that relate to different groups on the molecule, different segments, or even different polymers in polymer blends. Reviews of dichroism for oriented polymers are provided by Lefevre et al. [95], Read and Ward [96], and Jesse and Koenig [97].

- X-ray diffraction [89,92,95]. X-ray diffraction is a common technique for studying crystal structures and atomic spacing. In this technique, X-rays directed at the specimen are diffracted and detected by a monitor. The intensity and angle between the incident and diffracted rays provide information about the orientation of crystals. The wide-angle technique is useful for determining the orientation of the unit cell (generally noted as the a, b, and c axes) of the crystalline regions of a polymer. The results are often presented in the form of pole plots, which are stenographic projections of the intensities of the normals to the crystallographic planes [89].
- Shrinkage. Unrestrained linear thermal shrinkage of a film is related to the molecular orientation. See Section 11.2.2.2 for a description of this method.

Often more than one technique is applied due to challenges in measurement and interpreting results. For example, Ajji et al. [92,93] compare the orientation of PE and PP films using birefringence, FTIR (dichroism) and X-ray diffraction. They find that the FTIR measurements overestimate the *a*-axis orientation and in some cases can be wrong. They also question the FTIR results for amorphous phase orientation.

11.2.2.2 Shrink and shrink force

The unrestrained linear thermal shrinkage of a film is related to molecular orientation. Ajji and Zhang [90] find that shrinkage correlates with birefringence; both the amorphous and crystalline phase orientation contribute to shrinkage.

ASTM D2732 [98] describes a method for measuring shrinkage by immersing the film in a liquid bath. The liquid is chosen so that it does not plasticize the film. A polyethylene glycol-water mixture is often used. The liquid is controlled to a specified temperature +/− 0.5°C. A 100 mm × 100 mm film specimen is placed into a holder that allows free movement in the lateral directions but ensures the film is immersed in the bath. Details of the holder are given in the standard. The specimen is immersed in the liquid bath at a specified temperature for a given length of time. The film is removed and placed in a liquid bath at room temperature to quickly cool. The sample is removed and the linear dimensions in the machine and transverse directions measured.

The percent free shrinkage is given by Eq. (11.6):

$$\text{Unrestrained linear shrinkage, } \% = \left[\frac{L_0 - L_f}{L_0}\right] \times 100 \tag{11.6}$$

where

L_0 = initial length of side (100 mm)
L_f = length of the side after shrinking.

If the sample elongates, the % shrinkage will be negative in the direction of elongation.

Some investigations use an air or vacuum oven to expose the film to the specified temperature. Specimens are sometimes hung vertically (clipped to the upper edge) to ensure free shrinkage, or placed horizontally on a surface roughened with sand, cornmeal, or other granular media to help ensure the film has the freedom to shrink laterally. In general, an oven takes longer to heat the film and the temperature may not be as uniform as a liquid bath. Curl also can be a problem with oven heating of specimens.

The shrink force is defined by ASTM D2838 as the force per original unit width in a specified direction developed by a film under restraint while attempting to shrink at a given temperature. The shrink tension is the shrink force per unit of original cross-sectional area. The orientation release stress is the maximum shrink tension developed in a specified direction throughout its range of shrink temperatures while totally restrained from shrinking [99].

ASTM D2838 describes a test method that involves clamping a 1-inch (25.4 mm) wide film specimen in a shrink tension holder, one arm of which contains a strain gauge. The holder is immersed in a hot liquid bath and the force exerted by the film is measured by the strain gauge. Two methods are described. In Procedure A, the film is totally restrained from shrinking. In Procedure B, the film is shrunk by a predetermined amount before being restrained.

Shrink tension affects the appearance and performance of a film or package. It may also be used as a measure of orientation and is influenced by the film fabrication conditions. Procedure A is useful for determining the amount and direction of orientation, the orientation release stress and the maximum stress the film can exert at a given temperature. Procedure B better represents a packaging application since the film is seldom totally restrained. The procedure can be used to estimate the force that will be exerted on the packaged item and to predict the appearance of the package after shrinkage.

The orientation release stress is determined by carrying out shrink tension measurements using Procedure A at several temperatures and plotting the results. The maximum tension is the orientation release stress. Generating the full shrink tension versus temperature curve is typically only carried out to characterize a film made under an established set of processing conditions. For quality control and product development purposes, running the test at one or two temperatures is often sufficient.

ASTM D2838 works well for measuring the shrink force of high shrink films, such as those made on a double bubble process. Many applications require low shrink forces, such as when the product being packaged is deformable. Jin et al. [100] describe blown films of blends of LLDPE and LDPE that are used for this purpose. The LLDPE provides strength and LDPE shrinkage (due to LCB). The shrink force tester described in ASTM D2838 is not sensitive enough to measure the shrink force of low shrink films or the shrink in the transverse direction. Jin et al. develop a low shrink tension test that uses a solids dynamic mechanical analyzer (TA Instruments RSA III). The DMA device is used in the strain-controlled dynamic temperature ramp mode. The strain and tension are set so that the clamp distance remains constant as the temperature is ramped from room temperature to the melting temperature of the film. The measured force is related to the shrink tension.

They find that unlike a double bubble film with shows a shrinking force from the outset (as the sample is heating up from room temperature), a low shrink force film first undergoes an expansion. The expansion is due primarily to thermal expansion but is complex. They find they cannot compensate for this by preloading the film, so they end up subtracting it out as a baseline to calculate the shrink force near the melting point. They provide results for a series of LLDPE-LDPE monolayer films, which show that increasing the LDPE content increases the shrink force and reduces the temperature at which the maximum shrink occurs.

11.2.3 Factors that affect orientation and shrink

11.2.3.1 Polymer characteristics

MW, MWD, and LCB influence orientation and shrinkage behavior. In general, linear polymers draw down and orient more easily but generate less shrinkage than polymers with LCB. As mentioned in Section 11.2.2.2, LDPE is often blended with LLDPE to increase shrinkage. Broader MWD and higher MW (for increased chain entanglement) also tend to provide a wider processing window and increase shrinkage. Typical PP grades suitable for biaxial orientation have melt flow rates (MFRs) ranging from 1 to 5 g/10 min (230°C/2160 g) [101]. This is a lower MFR, or higher MW, than grades generally used in injection molding or fiber spinning. The polydispersivity index, a measure of MWD (see Section 4.2.1.2), of standard Ziegler Natta PP used in orientation is typically in the range of 8–12. DeMeuse [101] finds that narrower MWD grades made via single-site catalysis provide better mechanical properties upon orientation, but the stretching temperature window is very narrow.

One consequence of increased chain entanglement is strain hardening behavior, described in Chapter 5 and mentioned in Section 11.1 as an important parameter for thermoforming. Strain hardening is prevalent in polyethylene with LCB but can also result from strain-induced crystallization during the orientation process. Strain hardening increases the stress generated during stretching, which may result in more residual stress after quenching. The relaxation of residual stress results in shrinkage once the film is reheated.

Ajji et al. [102] evaluate the orientation of five LLDPE resins of different molecular architecture. The primary difference in MWD shows up in the high MW fraction in the z-average MW (Mz) measurement. The resins have similar density (~0.920 g/cc) but vary in crystallinity, which reflects differences in comonomer distribution. This is substantiated by temperature rising elution fractionation (TREF) measurements where some resins showed a significantly higher fraction at high elution temperature, which indicates a higher fraction of homopolymer. The authors propose that grades with greater fractions at lower TREF elution temperature have more chains of a quasicrystalline nature where the polymer is not quite solid or melt during orientation.

The LLDPE is biaxially oriented into the film from a cast sheet on a lab tenter frame stretcher and evaluated with two commercial LLDPE films made by the double bubble process. The authors measure orientation, shrink and shrink force after orientation. There is considerable scatter in their data but they conclude that a randomly distributed, quasi-crystalline high MW chain provides a higher shrink force, higher orientation, and greater shrinkage upon recovery than a LLDPE with lower MW and more distinct and separate amorphous (high comonomer) and homopolymer contents.

The chemistry of the polymer also plays a role. Hydrogen bonding results in greater shrinkage and shrink force. Polymers with a high level of hydrogen bonding, such as PVC, polyvinylidene chloride (PVDC), PA6, acid copolymers, and ionomers, generally provide a high level of shrink. As an example, consider the family of acid copolymers and ionomers. Higher acid levels provide greater hydrogen bonding. Ionomers, which are acid copolymers whose acid groups have been partially neutralized with a metal salt, contain hydrogen and ionic bonding. Increasing the acid level increases the shrinkage, as shown in Fig. 11.8. Here the shrinkage of two grades of ionomer with differing acid levels is plotted versus draw ratio. Increasing neutralization of the ionomer also increases shrinkage, as shown in Fig. 11.9. Both increasing acid and neutralization increase strain hardening behavior, as is described in Chapter 5.

The degree of crystallinity, rate of crystallization, and crystalline morphology influence drawability and orientation. In two-stage orientation processes, such as the tenter frame and double bubble processes, the sheet or tube is typically rapidly quenched in the first stage. For slow crystallizing polymers such as PET, the rapid quench results in an amorphous sheet or tube. For more rapidly crystallizing polymers such as PE, PP, and PA6, the rapid quench gives rise to smaller spherulites. Nonuniform cooling can result in regions of differing crystallinity or crystalline morphology, which will affect the uniformity of the film after stretching [103].

DeMeuse [101] reviews the literature for stretching of PP, showing that the lower density of the sheet prior to orientation makes stretching easier. Lower density corresponds to lower crystallinity.

EVOH is particularly difficult to orient due to its high level of crystallinity, which confines the mobility of its polymer chains. It crystallizes very rapidly, as described in Section 11.1.3.3 and Fig. 11.3, and undergoes further crystallization during stretching. Drawing of the crystal regions leads to local deformation and tearing. Sequential (MD and TD) orientation is particularly problematic since it crystallizes too fast during the first orientation step. Simultaneous orientation has been more successful [104]. As described in Section 11.1.3.3, special grades of EVOH, as well as various modifiers, have been developed that help improve the drawability of EVOH.

PET and PA6 are slower to crystallize and have been sequentially oriented commercially, starting with a rapidly quenched sheet. Ajji and Zhang [105] find that strain-induced crystallization of a PET film does not occur until the draw ratio is greater than 2×2 on a lab tenter frame stretcher. PA 6,66 is often blended with PA 6 to reduce crystallinity to improve the drawability [106].

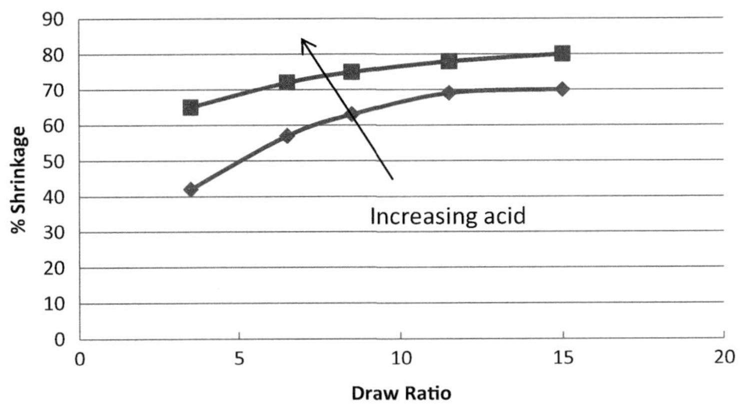

FIGURE 11.8 Effect of acid level and draw ratio on shrinkage of ionomer strands drawn in the melt. Shrinkage is measured at 95°C, which is above the melting point of the ionomers. *Reproduced with permission from The Dow Chemical Company.*

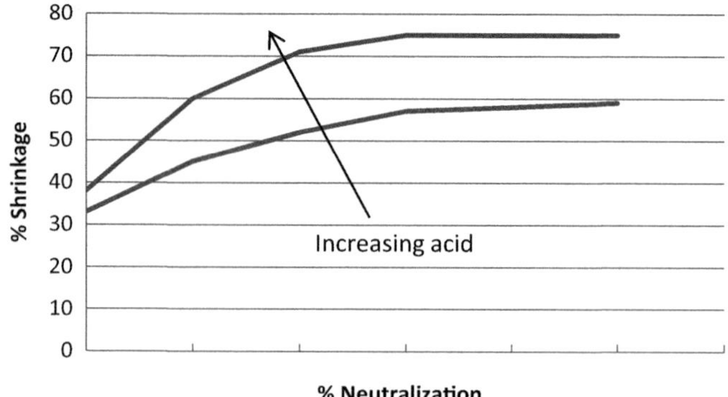

FIGURE 11.9 Effect of neutralization of acid groups on shrinkage of ionomer strands drawn in the melt. Shrinkage is measured at 95°C, which is above the melting point of the ionomers. *Reproduced with permission from The Dow Chemical Company.*

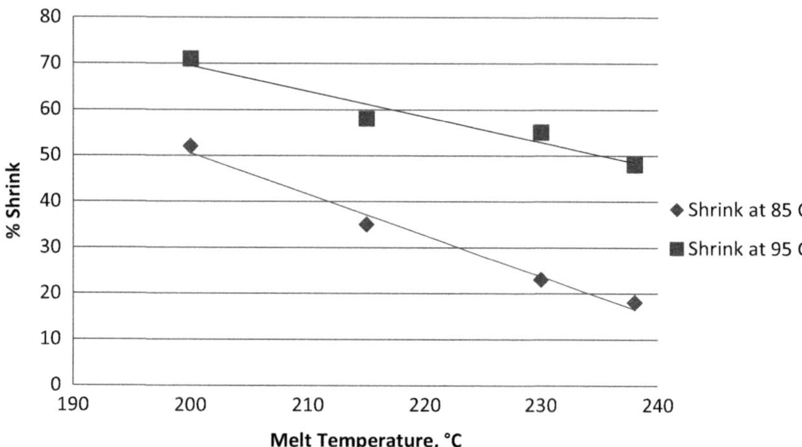

FIGURE 11.10 Shrinkage of ionomer blown film as a function of melt temperature during the film making process and shrinkage temperature. DDR = 7. *Reproduced with permission from The Dow Chemical Company.*

11.2.3.2 Processing

A number of processing factors influence the orientation and shrinkage of polymer films, including the stretch temperature, stretch ratio, stretch rate, annealing temperature (and time), and whether the stretching occurs from the melt (cast and blown film) or the solid (tenter frame, double/triple bubble, MDO), and whether the stretching occurs sequentially or simultaneously in the MD and TD.

11.2.3.2.1 Stretching temperature

The stretching temperature should be above the Tg for amorphous polymers and near the Tm for semicrystalline ones. For multilayer films, the temperature needs to be above the Tg of the layer with the highest Tg and below the melting temperature of the highest melting point layer; the stretching temperature is determined by the layer with the highest stretching temperature even if it is not the thickest layer [107,108]. A few examples from the patent literature put some further bounds on this:

- Nisamoto and Yamazaki [109] teach that for a multilayer barrier shrink bag with PVDC as the core and polyolefin skins, the film can be stretched at 110°C provided the intermediate layers have melting points from 70°C to 100°C. The stretch window is about 15°C. Furthermore, for good stretchability, the laminate must have at the stretch temperature tensile strength of 5–25 kg/cm^2 and greater than 500% elongation.
- Ciocca and Forloni [110] show that the orientation temperature for a (PE–tie–EVOH–tie–PE) film is between 110°C and 125°C. The PE layers are a blend of three different density polyethylene grades.
- Lustig and Vicik [111] teach that the stretching temperature must be below the melting temperatures of the outer layers and preferably between about 10°C–20°C below the lowest melting point of the outer layers. Structures containing (PA–tie–LLDPE) layers have stretching temperatures ranging from 100°C to 110°C.

- Forloni [112] reports that the stretching temperature for a multilayer barrier film is 120°C–135°C. An example is given for the structure (LLDPE–tie–EVOH–tie–LLDPE) on a sequential tenter frame. The MD stretching is conducted at 110°C and the TD at 125°C.
- Tanaka and Kawachi [113] teach the use of stretching temperatures between 70°C and 130°C for PP compositions.

Generally, lowering the stretching temperature increases the stress and orientation. This translates to greater shrinkage. An example of an ionomer blown film is given in Fig. 11.10.

11.2.3.2.2 Stretch ratio

The amount of stretch is typically characterized by a stretch ratio. In blown film, the stretch in the machine direction is often calculated by the ratio of the haul-off speed to the velocity of the polymer as it exits the die, known as the take-off ratio or draw-down ratio. In the transverse direction, the blow-up ratio is used, which is the ratio of the final bubble diameter to the die diameter. Formulas used to calculate each of these are provided in Section 2.2.1.3.

In cast film, the orientation is primarily in the machine direction as characterized by the draw-down ratio (Section 2.2.2.3) defined as the ratio of the line speed to the velocity of the polymer as it exits the die. Similarly, for MDO processes, the amount of stretching can be represented by a single number, either the ratio of the linear speeds of the rolls in the stretching chamber [Eq. (11.7)] or the ratio of the initial thickness of the film or sheet to the final thickness. For biaxial stretching, the stretching ratio is typically given in both directions, such as 2×2, meaning the stretch is 2:1 in the MD and 2:1 in the TD.

In general, increasing the stretching or draw ratio will increase the orientation and shrinkage. This is demonstrated for ionomer blown film in Fig. 11.8. As the stretching increases, the amorphous and crystalline regions (if they are present) orient. At high stretch ratios, the crystals become fully oriented, but the amorphous phase may continue to be oriented [89]. The increasing stretch ratio also impacts mechanical and optical properties. For MDO of HDPE, Parent et al. [114] and Tabatabaei et al. [115] find that the modulus first decreases and then increases with increasing stretch ratio. They attribute this to a transition from spherulitic/lamellae to fibrillar crystalline morphology. Haze and MD tear resistance decrease but puncture resistance increases with increasing stretch ratio. Bobovitch et al. [116–118] vary the MDO stretch ratio while keeping the TDO stretch ratio constant while making LLDPE films in a double bubble process. MD tensile strength increases while TD tensile strength decreases with increasing draw ratio. MD tear resistance decreases while TD tear remains mainly constant. They find these mechanical property trends correlate with the *c*-axis orientation of the lamellar crystalline structure as determined by x-ray diffraction and microscopy studies.

11.2.3.2.3 Stretch rate

Increasing the stretch rate (or rate of strain or elongation) increases the stress imposed on the molecules, resulting in higher orientation. For blown and cast film, the process time introduced in Chapter 2 is a good surrogate for the rate of stretching in the MD: the rate of stretching is inversely proportional to the process time. Here the process time is defined as the time the polymer spends between the die exit and when it begins to crystallize or solidify. It is in this region of melt orientation processes where the stretching occurs, so it makes sense that the shorter the process time, the faster the rate of stretching. The importance of process time is described in more detail in Chapters 14 and 15.

For an MDO process, the stretch ratio is large and the Hencky or true rate of strain is the appropriate characteristic parameter describing the stretch rate. [See Chapter 9 for a discussion of true versus engineering (also called technical stretch) strain.] Elongation is imparted by operating a driven nip roll faster than the trailing nip roll (see Fig. 11.11)

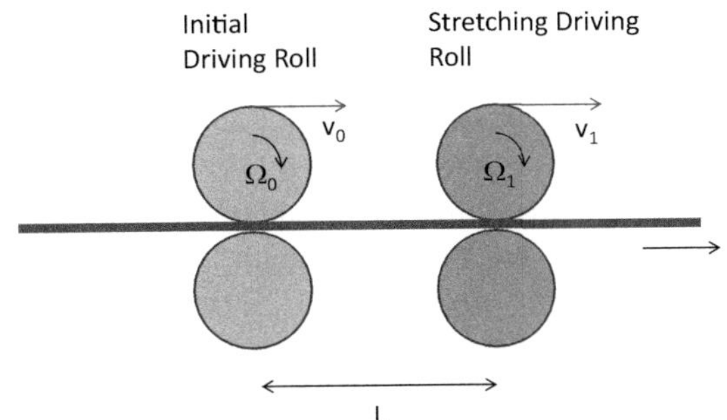

FIGURE 11.11 Schematic of roll-driven MDO process. The differential speed and distance between the two sets of rolls determine the draw ratio and draw rate as given in Eqs. (11.7) and (11.8).

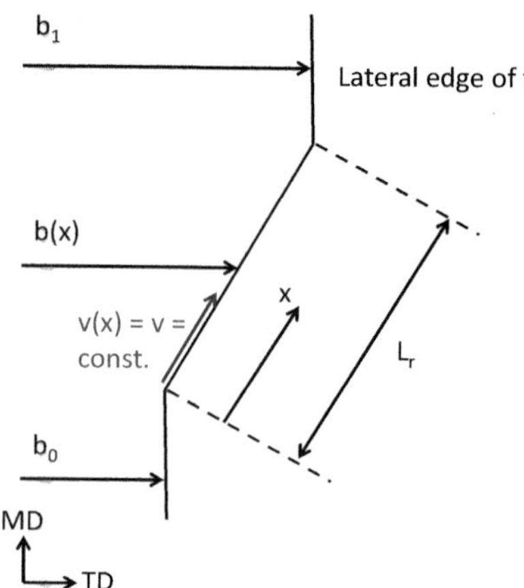

FIGURE 11.12 Parameters in Eqs. (11.9) and (11.10) for orientation in the transverse direction (TD) of the tenter frame film orientation process.

The velocity of the film is assumed to increase linearly between the rolls and the Hencky stretch rate is directly proportional to the difference in roll velocities and inversely related to distance between the stretching rolls:

$$\text{MD Stretch Ratio} = \frac{v_1}{v_0} = \frac{D_1\Omega_1}{D_0\Omega_0} \tag{11.7}$$

$$\text{Hencky MD Stretch Rate} = \frac{v_1 - v_0}{L} = \frac{\pi D_1\Omega_1 - \pi D_0\Omega_0}{L} \tag{11.8}$$

where

v_1 = linear velocity of driving roll imparting the stretch,
v_0 = linear velocity of the initial driving roll,
D_1 = diameter of the driving roll imparting the stretch,
D_0 = diameter of the initial driving roll,
Ω_1 = rotational speed (e.g., rpm) of driving roll imparting stretch,
Ω_0 = rotation speed of initial driving roll, and
L = distance between rolls.

For a tenter frame process, the stretch rate is determined by Eq. (11.8) in the MD. For the TD, the clip or rail velocity (v) is constant and the change in width (b) is assumed to change linearly along the rail length L_r (see Fig. 11.12). The stretch ratio and Hencky stretch rate are given by Eqs. (11.9) and (11.10).

$$\text{TD Strech Ratio} = \frac{b_1}{b_0} \tag{11.9}$$

$$\text{Hencky TD Strech Rate} = \frac{v}{L_r}\left(\frac{b_1}{b_0} - 1\right) \tag{11.10}$$

where

b_0 = initial film width,
b_1 = final film width,
v = rail velocity, and
L_r = rail length.

11.2.3.2.4 Annealing temperature

Annealing is an important part of the film stretching process. Most films require dimensional stability. BOPP and BOPET films used for lamination and other nonshrink applications are heat set or annealed following orientation. This involves exposing the film to an elevated temperature for a given amount of time to relax stress [103]. For polyolefin films, the annealing temperature is typically between Tg and Tm [115,119,120]. Annealing is usually done while the film is constrained to preserve the orientation, although some amount of shrinkage may be allowed [107,120]. In addition to reducing shrinkage, annealing reduces shrink tension [121].

Annealing may be done on hot rolls, or in the case of the double bubble process in a third bubble where the temperature and tension can be carefully controlled. For example, Hausmann et al. [107] describe a triple bubble orientation process for making dimensionally stable multilayer oriented film with a low-temperature sealing layer, such as the structure (PET−tie−PP−tie−ionomer). The polymers are coextruded into a tube and quenched in a 15°C water bath (first bubble). The tube is reheated in a water bath at 92°C for 5 seconds and inflated (second bubble) with a draw ratio of 3.5 in the TD and 2.6 in the MD. The bubble is collapsed and then re-inflated (third bubble) for annealing. Here it is exposed to 300°C air and a stretch ratio of 0.72. They provide general guidelines for the stretching and annealing temperature:

- The stretching temperature (second bubble) should be above the Tg of the polymer layer with the highest Tg and less than the melting point of the polymer with the highest melting point.
- The annealing temperature (third bubble) should be above the stretching temperature in the second bubble and generally needs to be above the Tg of the polymer layer with the highest Tg and less than the melting point of the polymer with the highest melting point.

Annealing is also often used during the production of shrink films. Annealing relaxes the stress introduced by the stretching process up to the annealing temperature. In the end-use, the temperature of the shrink tunnel must exceed the annealing temperature for shrinkage to occur.

Tabatabaei et al. [115] find that the modulus of MDO HDPE film increases with increasing annealing temperature. They attribute this to an increase in crystallization. Barham et al. [120] mention that when oriented polyolefin films are allowed to shrink during the annealing process, the stiffness decreases, just the opposite of Tabatabaei. The annealing temperature can be one of many variables that are tricky to get right. Schut [122] cites this as one reason why the use of MDO has not taken off as fast as some have predicted. She cites an example of PA−EVOH films where MDO increases barrier performance, but if it is not done right the film becomes brittle.

11.2.3.2.5 Melt versus solid-state stretching

Conventional blown and cast film processes impose some amount of orientation, with blown film typically providing a more balanced orientation in the MD and TD than cast film (mostly MD). In these processes, the stretching occurs primarily in the melt, although in blown film most of the stretching occurs at temperatures near the point where the polymer begins to freeze. This is explored in more detail in Chapter 14. While the amount and rate of stretching can be quite high, the stress that develops is limited because the polymer is fluid. As described in Chapter 5, polymers are viscoelastic. The amount of stress that develops is characterized by the modulus times the stretch ratio (elastic component) and viscosity times the rate of stretching (viscous component). The modulus and viscosity undergo a rapid decrease as the polymer transitions from the "solid" to liquid state: near the Tg for amorphous polymers and the Tm for semicrystalline polymers. Hence, substantially greater stress can be generated if the stretching is done while the polymer is in the "solid" state, as in the double/triple bubble, MDO and tenter frame processes. Higher stress results in greater polymer orientation and property enhancement in these processes compared to blown and cast film processes.

11.2.3.2.6 Sequential versus simultaneous orientation

For biaxial orientation processes, the order in which the sheet or tube is stretched in the MD and TD may have advantages. In the double bubble and blow film processes, the MD and TD stretching occurs simultaneously. Early tenter frame processes were sequential, but newer technologies have been developed that simultaneously stretch the sheet in both directions [103,123,124]. In sequential orientation, the MDO is typically done first, followed by TDO. The orientation after the first orientation must be low enough so that the film can be stretched in the second direction. Often the film will need to be heated to a higher temperature for the second stretch. BOPP is typically made sequentially; high throughput and good quality can be achieved [123]. Some polymers, however, cannot be stretched sequentially. For example, EVOH crystallizes too fast during the first orientation to be oriented in the second direction. Simultaneous orientation is the only way to make biaxially oriented films with these polymers [124].

The simultaneous stretch tenter frame process has a higher capital cost than double/triple bubble processes but can have higher throughput. The tenter frame process has greater temperature control than the double bubble process that is partially overcome by the triple bubble process, which may allow for increased orientation [103]. The tenter frame process also has flexibility for profiling the stretch—linear, exponential, logarithmic, etc.

Briel [123] provides some examples of how changing the stretch protocol can influence properties. In one experiment, polyolefins are stretched in a two-stage process: simultaneous MD and TD followed by MDO at a cooler temperature. The cooler second stage stretching provides greater shrinkage. McLeod [125] varies the stretching protocol for a sequential biaxial PP orientation process. In one experiment the area draw ratio is kept constant at 6 × 6 and the TDO draw speed is varied. Friel finds that increasing the TD draw speed increases the amount of shrinkage without changing the mechanical properties, which remain isotropic. In the second set of experiments, the draw ratios are varied. Anisotropic mechanical properties (as well as lower permeability) are achieved when the draw ratios are not equal (such as 5 × 8) but the shrinkage is the same as the 6 × 6 film.

11.2.3.2.7 Irradiation

Shrink films are often irradiated with electron beams, either before or after orientation. Irradiation prior to orientation helps provide stability during orientation, allowing for greater draw ratios. Irradiation after stretching helps retain the orientation [126]. Irradiation is also used to improve mechanical properties (such as resistance to piercing and abrasion), increase shrinkage and obtain higher seal strength [127]. Biddle [128] provides a history of two early approaches to shrink films:

- Cryovac used irradiation before blowing the film. The irradiation crosslinks the PE so that it is less fluid (higher viscosity) at the melting temperature. Tensile strength, shrink tension, and shrink percent are improved over nonirradiated film.
- DuPont used a different approach for Clysar. HDPE is typically not used for shrink film because it crystallizes too fast and so cannot be stretched very much. Blending in LDPE slows the rate of crystallization allowing more stretch, yet the crystals that do form act like crosslinks to enhanced the properties.

Zhang et al. [126] find that varying the radiation sequencing (before or after orientation) changes the shrinkage and shrink force characteristics. They evaluate commercial shrink films provided by a major supplier that have different e-beam sequences: NonIrr (no irradiation), PreIO (prior to orientation), and PostIO (after orientation). The three-layer films are made from blends of LLDPE and LDPE, are 0.6 mils (15 μm) thick and biaxially oriented at 400%/second at draw ratio of 5 × 5. Zhang et al. measure shrink force and shrinkage as a function of temperature. By plotting the force versus shrinkage at each temperature they create what they call "fictitious" stress–strain curves. By fictitious, they mean that the data come from two different experiments and so are not true stress–strain curves. Nevertheless, they conclude the approach is promising for a better understanding of how the manufacturing conditions affect shrinkage, but a more systematic study is needed on monolayer films.

11.2.4 Oriented multilayer films

11.2.4.1 Substrate films and laminations

Multilayer films used in flexible packaging often contain one or more oriented polymer film layers, such as oriented polypropylene (OPP), oriented polyester (OPET), and oriented polyamide (OPA) films. These are typically used as the structural layer in multilayer laminates. (Note that "O" is used as a prefix here to signify oriented film, either mono or biaxial. Later "BO" is used to specify biaxial oriented film.) Here the OPP, OPET, or OPA is first formed and subsequently, additional layers (such as sealing or barrier layers) are added by adhesive or extrusion lamination. Oriented films may also be coated for sealing or to improve barrier performance. Examples include PVDC-coated OPET for improved barrier and EVA-coated OPET for sealing. Metallization can also be thought of as a form of coating. Details on lamination and coating can be found in Chapter 10.

Commercial OPP or OPET substrate films are often made up of several layers, all of which have been oriented. For example, OPP may have a core layer made from homopolymer PP and an outer layer made from a propylene-ethylene random copolymer for sealing, or an outer layer made from a modified PP for improving the adhesion to metal during metallization. Further examples and a more detailed description of OPP technology can be found in Breil [129]. These films are made via coextrusion followed by orientation via a tenter frame, double/tripple bubble or MDO process. Since the polymers used in these films are similar to each other, the techniques and challenges of orientation are similar to that of monolayer films.

Interest in oriented PE films is increasing with the desire to produce mono-material packaging options for enhanced recyclability. Oriented PE would replace OPET, OPP, or OPA in laminated structures and provide many of the same benefits such as heat stability, toughness, transparency, and printability. The substrate film is often printed with the print surface turned inward during lamination to protect the print from abuse (reverse printing). MDO-PE and BOPE are stronger, more transparent, and have a higher gloss than PE blown film but have poorer tear resistance. Property

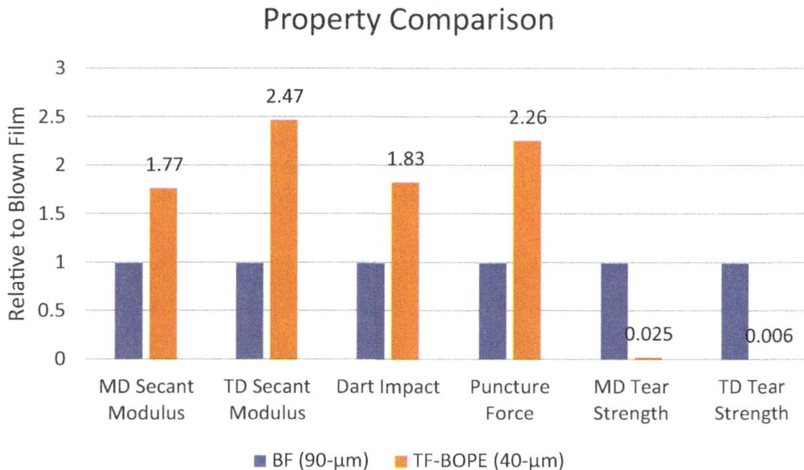

FIGURE 11.13 Properties of tenter frame biaxially oriented LLDPE (40-μm BOPE versus 90-μm blown film of same density resin). Blown film = 1 on scale. BOPE stretch ratio is 5 × 9 (MD × TD). *Data from Y. Lin, J. Xu, J. Pan, X.-B. Yun, M. Demirors, Biaxially oriented polyethylene (BOPE) films fabricated via tenter frame process and applications thereof, 2018 SPE ANTEC Conf., 2018.*

comparisons of a biaxially oriented LLDPE film made on a laboratory tenter frame process and a blown film of the same density are shown in Fig. 11.13. Lin et al. [130] propose that during orientation in the semisolid state (as in the MDO, double bubble and tenter frame processes), large crystals are broken up into smaller ones that do not contribute to tear resistance. They are able to increase tear resistance somewhat by changing the molecular architecture of the LLDPE so that more crystals form at temperatures below the orientation temperature. In addition to low tear strength, oriented PE will distort at lower temperatures than oriented PET, PP or PA due to its lower melt point. This poses challenges to downstream processes such as zipper incorporation and heat sealing. Changes in zipper design and switching to low seal initiation temperature sealants such as plastomers may be needed, as described in Section 17.4.2.

BOPET, BOPA, BOPP, and BOPE each have their relative strengths and weaknesses as a substrate for laminations. Substitution can be achieved through careful design of the packaging structure. Lin et al. [131] describe the substitution of BOPA with BOPE in a 3-ply structure for detergent packaging. The incumbent structure is 12-μm BOPET/12-μm BOPA/70–130-μm PE. The new structure is 12-μm BOPET/25-μm BOPE/60–120-μm PE. The BOPE is higher in thickness to match the stiffness of the BOPA; the PE sealant film thickness is reduced accordingly to achieve the same total thickness. 1-L liquid detergent pouches are made and subjected to quality testing. The new structure passed all four drop and compression tests.

Conventional PE is difficult to orient on a sequential tenter frame process [131]. Special grades have been developed with improved stretchability. For example, Lin et al. [131] describe an LLDPE grade that can be stretched 5 × in the MDO and 9 × in the TDO, making it suitable for sequential tenter frame lines designed for PP. Such developments, however, so far have been mostly limited to low to medium density ranges. Dow recently introduced a new HDPE (0.967 g/cc density) grade for the tenter frame process under their Elite brand of high-performance PEs [132]. Higher density PE provides greater stiffness that more closely matches BOPP and BOPET films and better temperature resistance during heat sealing.

MDO processes are friendlier to higher density PE grades. Fard et al. [133] describe five-layer MDO films with the structure: (LLDPE–HDPE–LLDPE–HDPE–[LLDPE + HDPE]). Here the LLDPE grades are special high-performance grades with tailored molecular architectures for processability and toughness (Enable and Exceed XP from ExxonMobil). Characterization studies show:

- The performance-LLDPE grade with LCB provides higher orientation than HDPE.
- Film stiffness increases with increasing stretch ratio and decreasing stretching temperature.
- Haze decreases with increasing stretch ratio and by placing the performance-LLDPE resin on the surface.
- Design of the structure (placement and composition of the layers) yields properties of MDO-PE comparable with BOPET and BOPA (see Fig. 11.14).

Lawrence et al. [134] study the effect of MDO on multilayer sealant films. Three-layer blown films are prepared with the general structure (A–B–C) with a fixed layer ratio of (20/60/20%). Layer A is the sealant layer, a 0.916 g/cc density mLLDPE. Layer C is the backing layer, an LLDPE (0.920 g/cc). The B layer is an LLDPE whose density is varied over a range of 0.917–0.936 g/cc. The blown films are made at their maximum thickness for the film line, ranging from 339 to 460 μm (13–18 mil) depending on the resins used. The MDO draw ratio is increased until the samples fail,

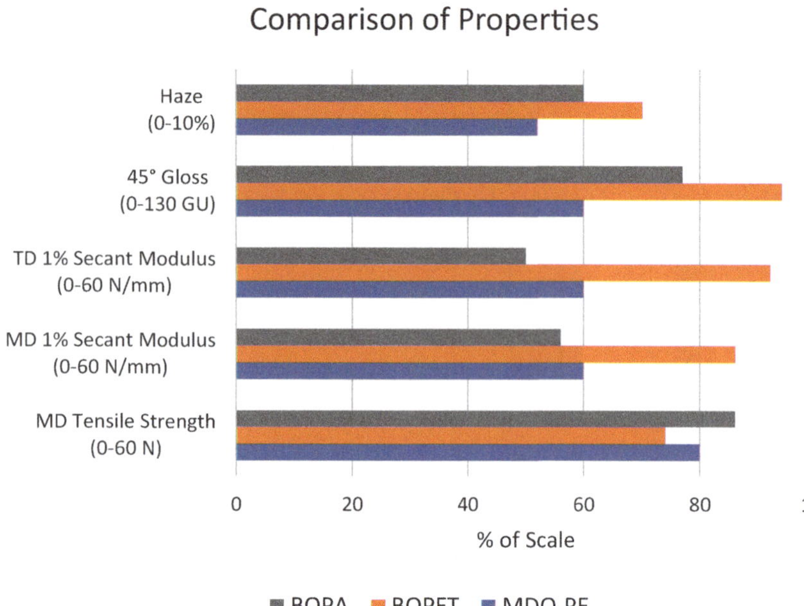

FIGURE 11.14 Comparison of machine direction orientation linear low-density polyethylene film with biaxial oriented polyester and biaxilly oriented polyamide. *Data from A. S. Fard, H. Ibrahim, J. Schaefer, V. Boudara, J. Throckmorton, C. Rysermans, Machine direction orientation of performance polyethylene films: application in recyclable pouches, SPE Int. Polyolefins Conf., 2021.*

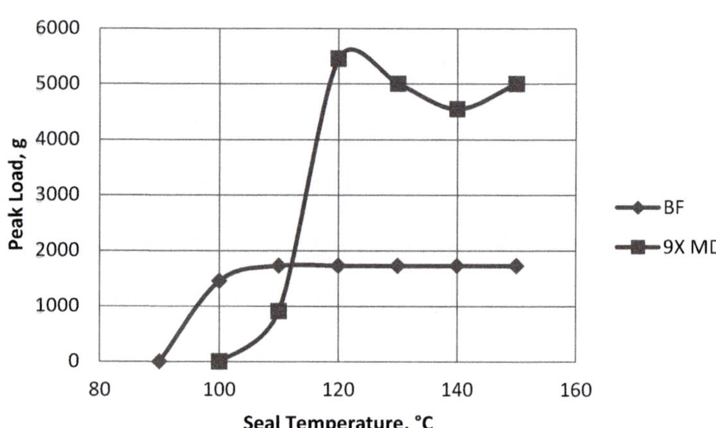

FIGURE 11.15 Heat seal strength for an LLDPE 3-layer film with thickness 40 μm (1.55–1.60 mil). The "BF" film is fabricated on a conventional blown film process. The MDO film has a draw ratio of 9. *Adapted from C. Lawrence, K.T. Forziati Jr., J. Garnett IV, Machine direction orientation and its effect on multilayer sealant film, SPE FlexPackCon Conf., 2013.*

with samples collected for property testing at 76 μm (3 mil) and the maximum draw ratio. Conventional blown films of the same thickness are made as controls.

The results show:

- Normalized dart drop impact increases with increasing orientation. This runs counter to conventional wisdom, but previous work has shown that the film design impacts this. Increasing the density of the core layer reduces the effect of orientation on the dart drop strength.
- Secant modulus substantially increases with increasing orientation. They cite literature that attributes this to changes in crystalline morphology from a laminar to a fibrillar structure.
- MD tear is substantially lower in the oriented samples than in the standard blown film counterparts. The magnitude of the difference is not as great at the highest draw ratios – some of the tear strength is recovered with further orientation. They attribute this to neck-in changing the crystalline morphology.
- Film haze decreases with increasing orientation.
- Ultimate heat seal strength is substantially increased with orientation. The oriented films do not elongate as much as the blown films during the heat seal measurement.
- The heat seal initiation temperature is increased with orientation. They attribute this to the crystalline structure becoming more perfect during orientation, increasing the melting temperature. These effects are illustrated in Fig. 11.15.

TABLE 11.3 Advantages and disadvantages of blown film MDO-PE versus tenter frame BOPE.

Advantages of MDO (assumes higher density PE resins used)	Disadvantages of MDO
Higher stiffness	Slightly worse optical performance
Slightly higher heat resistance	Lower toughness
Lower capital expense for new line	Lower output of line
Asset can also produce blown film	Higher edge trim waste
	Higher gauge variation

Source: Adapted from J. Wang, Oriented PE film solutions and their role in sustainable and recyclable flexible packaging, in: AMI Polyethylene Films Conference, 2020.

- OPV and MVTR decrease with increasing orientation. The effect is more pronounced for oxygen and for the lower density core structures.

Gerrits et al. [135] describe multilayer LLDPE film sequentially biaxially oriented on a laboratory tenter frame stretcher. The precursor cast film contains five layers with 0.912 g/cc LLDPE plastomer as one of the outside layers for heat sealing. The other layers comprise a 0.921 g/cc LLDPE resin, with slip and antiblock added to the other outside layer. The precursor thickness is 1200-μm. The film is first stretched 5x in the MD followed by 9x in the TD for a final thickness of 26-μm. Like Lin et al. [131], they find that compared with blown film the BOPE has increased impact resistance, puncture resistance, and tensile strength and reduced haze, tear strength, and elongation at break. DSC shows essentially no change in melt point. X-ray and microscopy reveal that the core layers have a fibrillar (shish-ka-bob) morphology with the fibral orientation preferentially in the TD. Unlike Lawrence et al. [134], they find the seal initiation temperature to be similar to blown films. Analysis shows the sealant layer to have a random crystalline morphology similar to that found when the polymer is quenched from the melt. They attribute this to the orientation temperature of the TDO process being higher than the Tm of the sealant; the sealant layer is melted out during the TDO process and is unoriented.

Both tenter frame BOPE and MDO-PE will likely have a role in the future of sustainable package design. A comparison of the two processes is provided in Table 11.3. This comparison assumes that MDO can use PE resins with higher density (0.95 g/cc and above) than the BOPE process (0.95 g/cc and below). As described above, resin suppliers are working on new resin technology that may challenge this assumption in the future.

11.2.4.2 Multilayer barrier shrink film and bags

Multilayer barrier shrink bags are used for a variety of meat and cheese applications. PVDC was one of the earliest polymers used for barrier shrink bags. One of the first applications was to protect poultry from freezer burn, taking advantage of its ability to shrink at 100°C. The poultry was placed in a blown film bag, the bag evacuated and sealed, and then immersed in boiling water. The bag shrank around the product for protection [128]. PVDC is still the predominant barrier resin used for barrier shrink bags for subprime meats and many other applications [136]. Typically it is coextruded with EVA and oriented in a double/triple bubble process. Some common structures include (EVA–PVDC–EVA) and (PE–EVA–PVDC–EVA–PE), with about 10% of the structure PVDC [106,137]. PVDC provides excellent barrier to oxygen, moisture, aroma, and flavors and shrinks up to 40%–50% at 85°C [106,136].

PVDC has several limitations due to its poor thermal stability. It is prone to gel and black spec formation during high-temperature extrusion, evolves corrosive byproducts that require special materials of construction and good ventilation and is known to discolor when exposed to irradiation. Hence special care must be taken to process PVDC into films [138]. Since it contains chlorine, PVDC also is problematic for recycling technologies. Other barrier structures have been developed using PA and EVOH. EVOH alone does not process well on a double/triple bubble process. In general, grades with higher ethylene content process easier, but at the expense of barrier performance. EVOH suppliers have made progress in developing special grades with improved stretching characteristics by lowering the level of crystallinity [40,136,138]. One approach has been to manipulate the head-to-head adjacency of hydroxyl groups along the polymer chain [137]. Polymer blending has also been used to improve the stretchability of EVOH, such as those described in Section 11.1.3.1 for improving the formability of EVOH. Sandwiching EVOH between layers of PA also helps stretchability. Most commercial non-PVDC barrier shrink bags use PA or a combination of PA−EVOH−PA. Copolymers of PA6, such as PA6/66, generally stretch better than PA6 due to their lower crystallinity [139]. The addition of PA6I/6T amorphous polyamide to PA6 is also used to improve stretchability [111,127,140] Structures include (PE−tie−PA−tie−PE) and (PE−tie−PA−EVOH−PA−tie−PE) [99]. In general, the non-PVDC barrier bags do not shrink as much as the PVDC bags, tend to be stiffer, and as a result hold a small share of the overall barrier shrink bag market. Alternative structures continue to be developed, such as the incorporation of an ionomer layer to improve shrinkage and reduce stiffness [138].

PA is used in barrier shrink bags for sausage casings and some cheese packaging. Some sample structures include (PA−tie−PE), (PA−tie−PA), (PA−tie−PA−tie−PE) or (PA−tie−PA−EVOH−PA−tie−PE−Ionomer) [106]. The PA is often a blend of PA6 and PA 6,66 for improved stretching. The PA provides a moderate oxygen barrier, good aroma and flavor barrier, excellent temperature resistance, and good abuse resistance.

11.2.4.3 Multilayer nonshrink barrier films

In recent years many investigators have looked at MDO, double/triple bubble, and tenter frame processes to orient coextruded barrier films for improved performance with the objective to reduce cost. An inline orientation process following the production of the coextruded sheet or tube has two inherent advantages over standard multilayer laminates. The first is the elimination of the lamination step by producing the structure in a single step. Rather than first producing an oriented substrate film, such as OPET, and subsequently adhesive or extrusion laminating it to a sealant film, these two processes can be consolidated into a single processing step. Second, the improved strength, barrier, and other properties of the oriented film hold the prospect of decreasing the thickness of the film, reducing material usage while still meeting performance specifications.

Several studies show improved barrier and mechanical strength performance of coextruded barrier films following MDO. Tabatabaei et al. [115] find that the orientation of HDPE generally improves its moisture barrier. Breese and Beaucage [141] evaluate MDO of HMW-HDPE films and find reductions in moisture vapor transmission rate (MVTR) and oxygen permeation value (OPV) correlates with increasing draw ratio and orientation as measured by birefringence. Froio et al. [142] present results from oriented (LDPE−tie−EVOH−tie−HDPE) films that show a 33% reduction in OPV, a 600% increase in MD tensile strength, and a 75% increase in modulus.

Hatfield and Breese [143] orient (LLDPE−LLDPE−tie−EVOH−tie−LLDPE−LLDPE) blown film using a Collins MDO unit. The blown film has a 2:1 BUR. The EVOH is a standard 38 mol% ethylene grade and the LLDPE is a hexene copolymer. The total blown film thickness is 200 μm (8 mils) with individual layer thicknesses of [17/25/4/8/4/25/17%]. During MDO, the film is preheated, drawn, and annealed, all at 100°C. Draw ratio is varied from 1 to 6.5. The results of oxygen permeation measurements show a reduction in OPV (measured at 50% relative humidity [RH]) of 40% when the draw ratio is increased from 1 to 6.5. The OPV also becomes less sensitive to relative humidity, normally a limitation with EVOH. As described in Chapter 8, the OPV of EVOH increases substantially as the RH is increased from 50% to 100%. Hatfield and Breese find that the reduction in OPV with MDO orientation is greater at higher humidity. In the RH range of 0% to 50%, the OPV decreases by 40% with orientation. At 90% RH, the reduction is 67%. The OPV of the oriented EVOH is about the same as PVDC.

Hatfield and Breese also measure the optical properties of the MDO film. The haze decreases to 10%, a 35% reduction. The gloss increases from 50% to 80% with orientation.

The impact resistance and tear strength are often negatively impacted by MDO, in addition to an increase in seal initiation temperature. These drawbacks can frequently be lessened through changes in the design of the structure and the processing conditions. For example, changing the sealant layer to a lower crystalline copolymer may help retain the seal initiation temperature.

Breese [144,145] provides several examples of how MDO can reduce cost:

- Source reduction. The increase in modulus and strength allows for downgauging (reducing thickness) without sacrificing package integrity. An example of a heavy-duty shipping sack is provided. A conventional 127 μm (5 mil) hexene-LLDPE blown film provides the right combination of properties to meet specifications. A specially formulated MDO film meets the requirements with 10% lower thickness.
- Material replacement. An example is the replacement of a 48 gauge (12.5 μm) OPET film with a 60 gauge (15 μm) polyolefin MDO film. Despite the higher gauge, the yield of the polyolefin film is higher because of its lower density. The MDO also brings advantages in MD break strength, anisotropic tear, heat sealability, and resistance to moisture. Disadvantages of the polyolefin film include poorer thermal dimensional stability, higher haze, higher oxygen permeability, and lower TD properties.

- Elimination of processing steps. Replacing a laminate with a coextrusion can reduce processing steps. Breese provides an example of replacing a high barrier laminate with a coextruded MDO film but does not provide details of the structures.

Schut [122] observes that MDO of multilayer films has not been widely adopted despite the potential advantages. The process remains tricky. Beyond monolayer stretch film it is difficult to get right. Several companies in the early 2000s invested heavily only to abandon their lines. Lines for specific applications, such as breathable hygienic films for diaper liners, self-adhesive labels, synthetic paper, PET strapping, PVC food wrap, and fibrous HDPE ribbon for weaving sacks were adopted by the 1990s. These lines were not very flexible, and lab lines were too small for the development of wide products. Newer lines allow more flexibility for product development on production lines, but the number of variables is huge: number and temperature of preheat rolls, type of roll heating, size of the gap in draw rolls, number of times the film is drawn, temperature, dwell time for annealing, annealing temperature, temperature and number of cooling rolls. Annealing conditions are particularly important to achieve dimensional stability. The sheer number of variables may be inhibiting the use of the process.

Blown film equipment suppliers offer in-line MDO options. Resin suppliers have also become more interested in MDO, which bodes well for the future. Some properties can be achieved that are not possible with biaxial orientation. For example, EVOH can be oriented in one direction but breaks up when sequentially oriented biaxially. HDPE provides a good deadfold when machine direction oriented, but not when biaxial oriented. Bimodal grades of HMW-HDPE have been developed that provide better drawability. Resins filled with minerals such as calcium carbonate impart breathability (through-hole formation) after stretching: after orienting in the MD, the film is rapidly quenched and then MD-oriented a second time to open up the holes behind the mineral particles.

Dimensionally stable biaxial oriented multilayer barrier films can be made with the double/triple bubble or tenter frame processes with the proper annealing or heat setting following the orientation. Hausmann et al. [107] provides some of the details for annealing, which is covered in Section 11.2.3.2.4. The value proposition for these films includes light-weighting (by replacing thicker laminates with thinner films due to their superior physical and barrier properties), elimination of metallization that interferes with recycling and use of metal detectors, and reducing subsequent lamination and converting steps. Alexy [146] and Egar [147] propose several applications for oriented coextruded barrier films made on a double bubble process with subsequent annealing in a third bubble. The simultaneous stretching in this process allows for the use of low ethylene (high barrier) grades of EVOH that will not orient sequentially. Example applications include:

- Lidding film. A conventional film has the structure OPET//(Polyolefin (PO)–tie–PA–EVOH–PA–tie–PO) [12 μm//40–65 μm]. Important features include puncture resistance and barrier. A proposed replacement structure is a biaxial oriented coextruded film with the structure (PET–tie–PO–tie–PA–EVOH–PA–tie-PO) with a total thickness of only 20–25 μm.
- Metallized balloon film. The conventional structure is BOPA//adh/PE/met [6–12/10/1 μm]. The proposed replacement structure is a triple bubble film with the structure (PA–tie-PO–tie–PA–EVOH–PA–tie–PO), 18–20 μm.
- Coffee package. The conventional structure is OPET-met/PE [12 /1/70–110 μm] with total thickness ranging from 80 to 120 μm. The proposed replacement is a 40–50 μm biaxial oriented film with the structure (PET–tie–PO–tie–PA–EVOH–PA–tie–PO).

The PET coex structure in the coffee package example is also proposed for standup pouches. Similar structures are reported by Hausmann et al. [107] and Shiffmann [148,149].

Breil [124] discusses oriented multilayer barrier films made on a simultaneous linear stretcher (tenter frame). One example of a PP film has the structure (cPP–hPP–tie–EVOH–tie–hPP–cPP) with thicknesses [0.7/8.5/0.4/1.3/0.4/8.5/0.7 μm] after orientation. The oxygen transmission rate (OTR) is reported to be 2.08 cm^3/m^2-d-bar, which is in line with metallized films and barrier laminates. The film has low cost, is transparent, and provides enhanced shelf life.

11.2.4.4 Shrink films and labels

Shrink films are used for overwraps and a variety of retail and food packaging applications. PVDC was one of the first materials used in shrink films. Other polymers include LDPE, irradiated LDPE, PE copolymer, PP, PET, and PVC [128]. Performance comparison of these materials is provided in Table 11.4. Up until the early 2000s, LDPE was preferred over LLDPE for shrink films because LLDPE did not shrink enough in the transverse direction. LLDPE resin suppliers adjusted the MW and MWD distribution to introduce grades for the application [114]. Currently, a variety of

TABLE 11.4 Comparison of shrink film types.

Film type	Advantages	Disadvantages
LDPE	Strong heat seals Low-temperature shrink Medium shrink force Low cost	Narrow shrink temperature range Low stiffness Poorer optical properties Sealing wire contamination
Irradiated LDPE	Higher tensile strength, shrink tension and % shrinkage than LDPE. Retains shrink tension for longer times than LDPE.	May have lower tear strength
LLDPE	High seal strength	Lower shrinkage than LLDPE Higher shrinkage temperature
PP	Good optical appearance High stiffness High shrink force Good durability	High shrink temperature High sealing temperature Brittle seals High shrink force – not suitable for fragile products
PVC	Lowest shrink temperature range Wide shrink temperature range Excellent optical appearance Controlled stiffness (by plasticizer content) Lowest shrink force for wrapping fragile products	Weakest heat seals Least durable after plasticizer lost Toxic and corrosive gas emissions from heat seal process Plasticizer health issue perceptions Durability issues at low temperatures Low shrink force hinders use as a multiple-unit bundling film Low film slip causes machine wraps
Multilayer Coextrusion	Excellent balance of properties, as determined by the layer construction	No significant performance disadvantages if the structure is designed well – the strengths of one ply compensates for the deficiencies of another

Source: Adapted from G. Biddle, Films, shrink, in Bakker, M., (ed.), The Wiley Encyclopedia of Packaging Technology, John Wiley & Sons, Inc., New York, 1986, pp. 335–338.

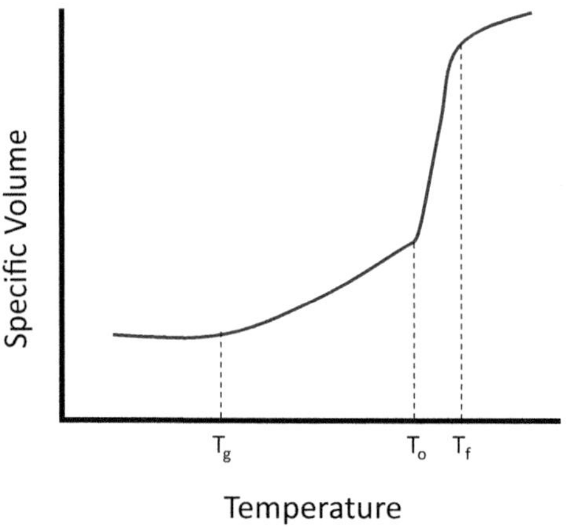

FIGURE 11.16 Specific volume change with temperature for a semicrystalline polymer. *Adapted from B.E. Obi, M. Leano, Correlating the melting of semicrystalline polymers to the shrink wrapping process in shrink-film packaging applications, SPE ANTEC Conf., 2015.*

PE types are used in shrink films, including LLDPE, LDPE, blends of LLDPE and LDPE, coextrusions of LLDPE and LDPE, blends of HDPE and LDPE and blends of HDPE, LDPE, and LLDPE [100,126,128,150–152].

Shrink films are typically fabricated using blown film, double bubble, or tenter frame processes. The performance depends on the processing conditions, as described in Section 11.2.3. Obi and Leano [153] describe the fundamental features of shrink films. In a shrink film packaging process, the heat and time in the shrink tunnel provide the thermal exposure to relax the orientation stress in the film, resulting in shrinkage. For PE films, the typical oven temperature is from 180°C to 215°C, and time in the oven 10 seconds. When the film heats up, the oriented polymer chains partially

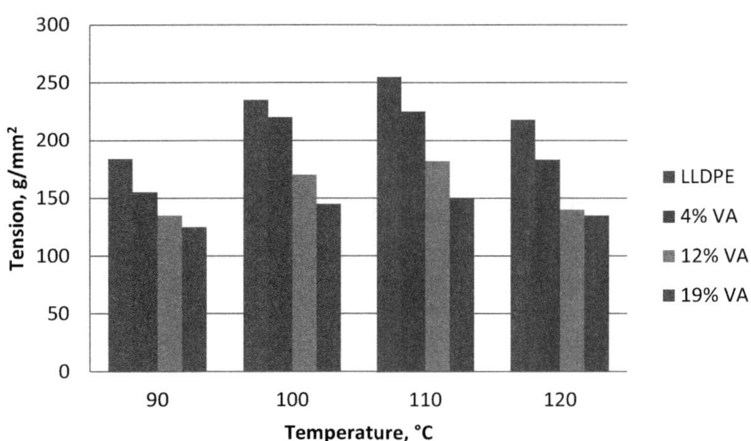

FIGURE 11.17 Shrink tension of (LLDPE-EVA-LLDPE-EVA-LLDPE) films as a function of vinyl acetate (VA) content and temperature. *Data from A.L. Bobovitch, A. Sagron, Y. Unigovski, A. Jarashneli, E. M. Gutman, Stress relaxation in low-shrink-force polyolefin films, Poly. Engr. Sci. 44 (2004) 1716–1720.*

revert back toward their random coil configuration. As a result, the film shrinks against the product. As the film cools, the polymer contracts due to crystallization and normal thermal contraction (as described by the thermal expansion coefficient). This is shown in Fig. 11.16 where the specific volume versus temperature is plotted for a semicrystalline polymer like PE or PET. The specific volume is the inverse of density and increases as the temperature increases; as more thermal energy is provided, the molecules move more, expanding the volume they occupy. The specific volume rises rapidly near the melting point of the polymer as the crystals lose their ordered structure and the chains expand. Conversely, as the film cools near the freezing point, the contraction gives rise to the shrink force. If the tunnel temperature is too low, not enough shrinkage occurs and the contraction force may be too low. If the tunnel temperature is too high, the shrinkage will occur, but the shrink force will decline.

Semicrystalline polymers retain much of their stiffness or modulus at temperatures above their Tg until the melting point is reached. This is because the crystals act as reinforcements. Once the crystals melt, the modulus drops precipitously. The polymer melts over a temperature range with the smaller crystals melting first. The larger crystals have the greatest reinforcing effect. In general, the heat tunnel is set below the peak melting point of the polymer so that all the crystals do not melt. If the temperature is too high, the polymer will end up as a viscous puddle.

DSC measurements can be used to calculate the cumulative percent melting at a given temperature. Obi and Leano show for a particular LLDPE-based shrink film, the shrink tension rapidly rises as the oven temperature increases above 109°C. DSC shows that 55% of the crystals have melted at this temperature. The tension peaks at 115°C where 69% of the crystals are melted. By 120°C, 86% have melted, leaving 14% of the crystals (the largest) to reinforce the film. Obi and Leano conclude that the polymer should have a broad melting temperature range to work well as a shrink film.

The shrink tension is a key parameter. The shrink tension generally decreases with decreasing crystallinity. For instance, films made from EVA and ionomers have lower shrink tension than LDPE or LLDPE and may be used in coextrusions or blends to produce shrink films for packaging fragile products. The work of Bobovitch et al. [154] illustrates this. Five-layer films with the structure (LLDPE-EVA-LLDPE-EVA-LLDPE) and thickness [10/20/40/20/10%] are made on a double bubble process. The stretch ratio is 5:1 in both MD and TD. Three different grades of EVA are used with 4, 12, and 19 wt.% VA. Fig. 11.17 shows the results of shrink tension measurements. As the VA content in the EVA layers is increased, the shrink tension decreases. Increasing VA content decreases the crystallinity of the EVA. Ionomers are also known for their low shrink force, allowing them to be used in skin packaging of fragile materials.

Shrink labels and sleeves have the ability to shrink and conform to the shape of the container. Key attributes include the ability to maintain shape and shrink under the application of heat, shrink force, shrink geometry, and printability. Commonly used polymers in shrink labels include PVC, polyolefins, polyethylene terephthlate glycol (PETG), styrene polymers, and cyclic olefin copolymers (COC) [155]. PVC is low in cost but because of its high density, it does not float in recycling separation systems. Halogens are also an issue for reprocessing. PETG has a high growth rate but is the most expensive, has shrink properties that are very sensitive to temperature, and does not float. Oriented polystyrene is primarily used in Japan. It is less stiff than PVC or PETG and due to temperature sensitivities, logistics can be challenging. OPP is not generally used because of its low shrinkage and stiffness [156].

Recycling has been a driving force toward polyolefin-based shrink labels. With their density less than water, polyolefin labels float when ground and put into a water floatation separation system. Enhanced polyolefin systems are described by Jester [156]. They are typically multilayer, achieve medium shrink (around 50%), and are recycle friendly.

Cyclic olefin copolymers (COC) may be blended or used as a separate layer. One structure is a (A-B-A) coextrusion with A being COC-rich and B a polyethylene-rich core. The COC adds stiffness and printability, and because it is amorphous, helps with orientation.

11.2.4.5 Stretch and cling films

Stretch films are extendable, elastic-plastic films that when stretched around a product, generate residual tension that contracts the film to the contour of the product. Stretch films and hoods have increasingly replaced tapes, shrink films, and metal banding for palletization and bundling. Some advantages of stretch films in these applications include [157]

- Lower handling costs;
- Labor savings;
- Transportation savings;
- Protection;
- Inventory control (easier identification of the product due to the clarity of the film).

A stretch film is stretched and oriented by the user at room temperature. The cold draw increases strength and if the stretch occurs in the elastic range of the polymer, the molecules attempt to recoil back to their original configuration once the strain is released [158]. This contrasts with shrink films where the film is stretched by the film producer at an elevated temperature with the residual stress frozen in place. This stress is relieved when the film is exposed to an elevated temperature by the user, resulting in shrink.

Cling films have historically been made from LDPE, EVA, and PVC. Metallocene LLDPE is starting to be used due to its higher mechanical properties (tear, impact, tensile strength). PE requires additives, such as polyisobutylene (PIB) or polyolefin elastomers (POE), for cling whereas EVA is naturally tacky. The tackiness of EVA increases with increasing VA content but the VA content has limits. In blown film, high VA (above about 18 wt.%) causes blocking; in cast film it causes chill roll sticking, which can lead to line tension variability and differential cooling and mechanical performance [158]. Metallocene LLDPE provides greater strength and dart impact resistance over conventional LLDPE, allowing for downgauging [157].

PIB is a sticky liquid that is difficult to handle. Polyolefin elastomers have been found to be a cleaner and more precise way to achieve cling—up to 30% is added to surface layers. ExxonMobil promotes its propylene-based elastomer Vistamaxx as an even more efficient additive for cling with loadings as low as 10% needed for some applications [159].

Both cast and blown film are used to make stretch films, with about 65%–70% in cast film as of 2004 [160]. Cast film is used mostly for machine wrap whereas blown film for hand wrap and some specialty high puncture resistant applications. Machine wrap has been growing faster than hand wrap as end-users automate their operations. Some of the properties that are important for stretch films include strength, puncture resistance, tear resistance, stretch percentage at break, load retention, haze, gloss, and cling [157]. The cast film process provides high clarity [158].

Stretch and cling film have moved from mono to multilayer. Blends and coextrusions are increasingly being used to optimize and customize the performance of stretch and cling films for a variety of applications and to control costs. For example, coextruded films can be made with cling on one side and slip on the other or with different colors [157]. A typical three-layer coextrusion has the layer ratio 1:3:1 with the core an elastic layer (EVA or thermoplastic olefin) and skins a polyolefin such as mLLDPE (with slip and antiblock) [161]. While blown film stretch film has kept to more simple structures, cast film has gone to more and more layers, including nanolayer technology discussed below. The early 2000s saw a movement to seven-layer and by the mid-2010s 9 and 11-layer cast lines became the norm with nanolayer lines growing [159]. This has allowed downgauging and through optimization of layer design, cost reduction. For example, increasing the number of layers in cast film provides better puncture resistance; LLDPE becomes less splitty. Introducing PP-based layers increases tear resistance. Optimization of the materials and fabrication process allows the use of less expensive materials while increasing performance in this very cost-competitive and increasingly demanding market. Stretch films in recent years have been downgauged from 25–35-μm to 10–15-μm while stretch percent has increased from 200% to 300% and machine wrapping speeds have almost doubled [159].

A few insights on structure–property relationships can be found in the literature. Shelley and Kokel [161] evaluate three-layer blown films for stretch hood applications. The core layer is made from blends of a reactor-made PP-based thermoplastic olefin (TPO) with EVA (6.6% VA) or LDPE. The ratio of TPO to EVA or LDPE ranges from 100:0 to 0:100. The film is blown at 3:1 BUR with a thickness 100 or 80 μm. They use a residual stress test (ASTM D5748) instead of stress relaxation to better simulate the strains imposed during pallet wrapping. The test involves imposing an initial extension of 100%, reducing the strain to 75%, and measuring the change in stress after 180 seconds. They use an elastic recovery test as an indication of snapback, important once the stretch hood is released (so it can hold in the product). This test involves measuring the stress after imposing an initial strain of 100% for 10 seconds and releasing it to zero. Increasing the TPO content, and switching from LDPE to EVA in the blend, increases the elastic recovery. Elmendorf tear resistance and dart impact resistance also increase with an increasing amount of TPO in the core layer blend. The authors postulate that these properties may correlate with the amount of amorphous phase in the core.

Small et al. [162] study the effect of PIB concentration, extrusion conditions, storage temperature, and aging on mechanical properties, crystallinity, and surface tack. PIB migrates to the surface after extrusion to provide cling. This migration is similar to slip additive migration and is affected by many of the same factors: compatibility between the additive and polymer, film morphology, extrusion variables, and film conditioning. Further discussion of slip migration is provided in Chapter 12. Small et al. make blown and cast films of a 70/30 blend of LLDPE and LDPE containing 0%–4% PIB. For the blown film, three different blow-up ratios (1.5, 2, and 2.5) are used. Results show that at these low levels of PIB there is little effect on the mechanical properties of the film. Tack increases with PIB level as expected. Tack also increases with storage time (up to about 20 days) and storage temperature.

Small et al. find that the cast film does not have greater tack than the blown film. Previous investigators propose that the small PIB molecules migrate through the amorphous regions. Since cast film has more amorphous regions than blown film, Small expected the tack to be greater for the cast film. They suggest that co-crystallization between the LLDPE and LDPE in the cast film may explain their observation. Evidence for co-crystallization shows up in DSC measurements. Other factors, not considered by Small, may also be involved, such as loss of additive due to temperature (cast film typically is produced at higher temperatures than blown film), differences in polymer surface roughness, and crystallinity effects (cast film typically has a lower level of crystallinity than blown film, which affects the solubility of the additive in the polymer.) These factors are discussed in Chapter 12.

In a second study, Small et al. [158] compare (EVA–mLLDPE–EVA) [2.5/20/2.5 μm] cast films with monolayer mLLDPE films. The EVA ranges from 6 to 18 wt.% VA. From 0%–20% of a PIB masterbatch (52% PIB in LLDPE) is added to the EVA layers. The VA content and PIB level have only a small effect on most of the tensile properties of the film, but the modulus is found to decrease with increasing VA in the EVA layer. Higher PIB levels increase elongation at break and decrease modulus. Increasing VA level also increases elongation at break. The (EVA–mLLDPE–EVA) films give better tack and comparable mechanical properties to mLLDPE monolayer films. By adding PIB just in the surface layer there is a potential for cost reduction using the coextruded structure. Tack development is also faster. A trade-off is the lower modulus of the coextruded film.

In a third study, Small et al. [163] consider the effect of varying the LLDPE type (comonomer, density and melt flow rate (MFR)) on mechanical and tack properties of monolayer cast films. The PIB level is fixed at 8% addition of the masterbatch. Measurements include DSC for crystallinity, FTIR for crystalline orientation, trouser tear, film tack and tensile properties. They find that the crystalline orientation decreases with decreasing MFR. As orientation decreases, modulus, percent crystallinity and tear resistance decrease. Octene-LLDPE has higher tear resistance than hexene-LLDPE. Tack is influenced both by crystallinity and orientation: decreasing crystallinity increases tack due to more migration of the PIB to the surface and higher orientation results in less tack. In some cases, orientation dominates crystallinity.

Microlayer technology, described in Sections 8.5.2 and 9.4.7, has been adopted for stretch films for pallet wrapping and other applications. AFP, an early adopter, commercialized a 27-layer structure in 2009 with a split sandwich structure of 11 microlayers on each side of a core and intermediate skin layers. They had reportedly considered buying a 9 or 11-layer line but found that the microlayer technology gave stronger films. The two microlayer sections were each 10% of the structure, the LLDPE core 40%, skin layers 10%, and intermediate layers 10%. Each microlayer is about 1% of the overall structure. A 70-gauge (17.8 μm) film stretched 250% has layers only 0.057 μm thick, putting the dimensions into the nanometer range (57 nm). For this market, these films are often called nanolayer films. Putting the same material through all the extruders and multiplication steps creates a plywood effect, strengthening the film. AFP found using the process for an all-LLDPE film gave the same strength as a multi-material film [9000 psi (62 Mpa)] and higher than for a typical stretch wrap film [6000 psi (41.4 MPa)]. The big gains with the multi-material film are in holding strength and tear resistance [164].

Going to multilayers enhances stretchability, tear resistance, and puncture resistance, allowing the use of less expensive materials such as butene-LLDPE. Equipment supplier Nordson EDI reports 34% greater stretch with a 48 microlayer LLDPE stretch film over conventional films. Productivity can also be enhanced as gels are less likely to cause line breaks due to the other layers encapsulating them [165]. Nanolayer films allow almost double the wrapping speed by the end-user [159].

11.2.4.6 Adhesion

Stretching and orientation of multilayer films can substantially reduce interlayer adhesion. For example, up to 90% of the preorientation bond strength can be lost during MDO of barrier films with conventional maleic anhydride-based tie resins. There are several factors that have been implicated or postulated for this loss of adhesion, including:

- The creation of a new interfacial area during stretching, reducing the number of chemical bond linkages (created prior to orientation) per unit area.
- The thinning of the structure, resulting in lower tensile strength and stiffness.
- Strain hardening of some tie resins during stretching, reducing their elasticity and ability to absorb and dissipate interfacial stress.
- Strain hardening of the structure during stretching due to bonds formed between the tie and barrier resin
- Increasing crystallinity, contributing to less elasticity of the tie and barrier.
- A change in crystalline morphology during stretching, influencing mechanical properties that affect peel strength.
- Frozen-in stress created during orientation and rapid quenching.
- Differential stress between the polyolefin and barrier during stretching, concentrating the stress at the interface.

Other factors may favor adhesion. Adhesion of miscible polymers is generally not negatively impacted by orientation. For example, Gerrits et al. [135] find in their biaxial oriented film study evidence for epitaxial crystalline growth into the stacked lamellar morphology of the LLDPE core layer at the interface with an mPE plastomer sealant layer. They postulate that this co-crystalline morphology increases the adhesion between the layers. Song et al. [166,167] find that converging flow accelerates chemical reaction at the interface, which they postulate is due to an increase in collision rate of reactants during stretching. This is discussed in Section 15.2.4. For the most part, however, the peel strength is substantially reduced after orientation. An example of loss of adhesion in (LLDPE−tie−EVOH−tie−LLDPE) film during orientation in an MDO process is given in Fig. 10.44. This is taken from Lee et al. [168] who found evidence for a number of the factors listed above in their study. Many of these factors can be mitigated through the design of the tie resin [55,56]. Close consultation with the tie resin manufacturer is important when dealing with oriented structures. Further discussion of the factors that influence adhesion in orientation is provided in Section 10.4.2.2.4.

References

[1] G. Gruenwald, Thermoforming. A Plastics Processing Guide, 2nd (ed.), CRC Press, 1998.

[2] J.L. Throne (Ed.), Understanding Thermoforming, Hanser Pub Inc, 2008.

[3] P. Schwazmann, Thermoforming: A Practical Guide, 2nd (ed.), Hanser Publications., 2018.

[4] J. Throne, Thermoforming, in: M. Kutz (Ed.), Applied Plastics Engineering Handbook, 2nd (ed.), William Andrew Publishing, 2017, pp. 345−375.

[5] J. Florian, Practical thermoforming, Principles and Applications, 2nd (ed.), CRC Press, 2019.

[6] G. Joseph, C. McGowan, Near infrared sensor technology & the modern approach, TAPPI PLACE. Conf. (2004).

[7] T. Blalock, S. Heveron-Smith, Multi-layer film measurement using dual-light interferometry, TAPPI PLACE. Conf. (2004). Indianapolis, IN.

[8] C. Hilton, Thermoforming polypropylene: an overview, SPE Polyolefins Conf. (2002) 219−233.

[9] N. MaCauley, E. Harkin-Jones, W. Murphy, Dynamic mechanical thermal analysis: a method for assessing the thermoformability of plastic sheeting, IChemERes. Event-Eur Conf. Young Res. Chem. Eng. 2 (1995) 677−6799. 1st.

[10] T. Spence, D. Hylton, Rheological studies of commercial thermoforming materials, SPE ANTEC Conf. (1988) 681−689.

[11] H. Muenstedt, S. Kurzbeck, J. Stange, Importance of elongational properties of polymer melts for film blowing and thermoforming, Polym. Eng. Sci. 46 (9) (2006) 1190−1195.

[12] A.D. Gotsis, B.L.F. Zeevenhoven, A.H. Hogt, The effect of long chain branching on the processability of polypropylene in thermoforming, Polym. Eng. Sci. 44 (5) (2004) 973−982.

[13] M. Yamaguchi, K. Suzuki, Enhanced strain hardening in elongational viscosity for HDPE/crosslinked HDPE blend. II. Processability of thermoforming, J. Appl. Polym. Sci. 86 (1) (2002) 79−83.

[14] S. Kurzbeck, H. Munstedt, Elongational properties of polymer melts and correlations with their processing behavior in blown film and thermoforming, in: Progress and Trends in Rheology V, Proceedings of the European Rheology Conference, 5h Potoroz, Slovenia, September 6−11, 1998, pp. 402−403.

[15] B. Hegemann, D. Liebing, P. Eyerer, Development of a measurement technique for tailored material characterization and validation of thermoforming simulation, SPE ANTEC Conf. (2004) 909−913.

[16] B. Hegemann, P. Eyerer, N. Tessier, T. Bush, K. Kouba, Polymer-polymeric friction at temperatures and rates simulating the thermoforming process, SPE ANTEC Conf. (2003).

[17] B. Hegemann, P. Eyerer, N. Tessier, T. Bush, Various plug assist materials and their effect on the thermoforming characteristics of polymeric sheet, Annu. Technical Conf. - Soc. Plast. Eng. (2002) 707−711. 60th(Vol. 1).

[18] R.A. Morales, M.V. Candal, O.O. Santana, A. Gordillo, R. Salazar, Effect of the thermoforming process variables on the sheet friction coefficient, Mater. & Des. 53 (2014) 1097–1103.
[19] J.L. Throne, Thermoforming, Hanser, Publishers, Munich, 1986.
[20] A. Dharia, D. Hylton, Novel method for rapid determination of thermoformability, SPE ANTEC Conf. (2005) 1214–1219.
[21] D.D. Joye, G.W. Pohlein, C.D. Denn, A bubble inflation technique the measurement of viscoelastic properties in equal biaxial extensional flow, Trans. Soc. Rheol. 16 (1972) 421.
[22] D.D. Joye, G.W. Pohlein, C.D. Denn, A bubble inflation technique the measurement of viscoelastic properties in equal biaxial extensional flow II, Trans. Soc. Rheol. 17 (1973) 287.
[23] A. Derdouri, R. Connolly, E. Verron, B. Peseux, Material constants identification for thermoforming simulation, SPE ANTEC Conf. (1998) 672–675.
[24] L.R. Schmidt, J.F. Carley, Biaxial stretching of heat-softened plastic sheets: Experiments and results, Polym. Engr. Sci. 15 (1975) 51–62.
[25] A. Dharia, Applications of thermoformability analyzer, SPE ANTEC Conf. (2006) 2605–2609.
[26] J.A. Grande, A simple, repeatable, realistic test of thermoformability, Plast. Technol. Mag. (2005) 55–57. October.
[27] X.C. Wang, M. Boparai, Identification of thermoform-abilty indicators for multilayer films, SPE ANTEC Conf. (2010) 2055–2059.
[28] K. Ho, D. Ward, K. Kuklisin, A. Murphy, Multi-layer blown films for thermoformed food packaging applications, SPE ANTEC Conf. (2011).
[29] T.C. Yu, Thermoforming behavior of olefinic instrument panel skins, SPE ANTEC Conf. (1999).
[30] D.G. Baird, D.I. Collias, Polymer Processing: Principles and Design, Butterworth-Heinemann, Boston, 1995, pp. 286–291.
[31] R. DiRaddo, A. Meddad, Sensitivity of operating conditions and material properties for thermoforming process, Plastics, Rubber Compos. 29 (2000) 163–167.
[32] D. Ward, The importance of thermal transfer and heat retention in designing thermoformable coextruded film structures, TAPPI PLACE. Conf. (2010).
[33] K. Xiao, I.-H. Lee, R. Armstrong, Comparison of water-quench vs air-quench blown film processes – Part II: thermoformability, SPE ANTEC Conf. (2012).
[34] D. Ward, J. Crowe, Heat transfer modelling in multilayer films used for flexible packaging, SPE ANTEC Conf. (2018).
[35] S.A. Ashter, Thermoforming of Single and Multilayer Laminates: Plastics Films Technologies, Testing, and Applications, Elsevier, Walthan, MA, 2014.
[36] Shell Development Company, Polypropylene Technology Seminar, 1987.
[37] R. Chou, I.H. Lee, Behavior of ethylene vinyl alcohol (EVOH) amorphous nylon blends in solid-phase thermoforming, J. Plastic Film. Sheeting 13 (1997) 74–93.
[38] E. Chow, High barrier forming films with EVOH, VII Int. TAPPI/CETEA Conf. Flex. Packaging (2012).
[39] Kuraray Technical Bulletin. No. 150, Thermoforming of EVAL® resin containing structures. <http://eval-americas.com/media/36418/tb_no_150.pdf>, 2000 (accessed 21.02.21).
[40] I. Matsui, H. Onishi, T. Yamamoto, Study on orientable EVOH, TAPPI PLACE. Conf. (2003).
[41] Eval Americas, Eval™ SP – oriented for future, Kuraray. <http://www.docstoc.com/docs/79126600/EVAL-SP-Oriented-for-the-future>, n.d. (accessed 31.05.15).
[42] R.T. Chou, G.I. Deak, Blends of ethylene vinyl alcohol copolymer and amorphous polyamide, and multilayer containers made, U.S. Patent 4,990,562, 1991.
[43] W.A. Oberle, Process for thermoforming EVOH barrier sheets and products relating hereto, EP0699125 B1, 1998.
[44] E. Giménez, J.M. Lagarón, L. Cabedo, R. Garava, J.J. Saura, Study of thermoformability of ethylene-vinyl alcohol copolymer based barrier blends of interest in food packaging applications, J. Appl. Polym. Sci. 96 (2004) 3851–3855.
[45] E. Giménez, J.M. Lagarón, L.M. Maspoch, L. Cabedo, J.J. Saura, Uniaxial tensile behavior and thermoforming characteristics of high barrier EVOH-based blends of interest in food packaging, Polym. Engr. Sci. 44 (2004) 598–608.
[46] A.I. Fetell, Nylon-containing ionomer improves the performance of ethylene vinyl alcohol (EVOH) copolymers in multilayer structures, TAPPI Polymers, Laminations & Coat. Conf. (1997) 503–505.
[47] A.I. Fetell, Ethylene vinyl alcohol copolymer blends, WO 1997009380 A1, 1997.
[48] K.J. Shelat, N.K. Dutta, N.R. Choudhury, Interfacial interaction and morphology of EVOH and ionomer blends by scanning electron microscopy and its correlation to barrier characteristics, Langmuir 24 (2008) 5464–5473.
[49] J.K. Lee, T.L. Virkler, C.E. Scott, Influence of initial sheet temperature on ABS thermoforming, Polym. Engr Sci. 41 (2001) 1830–1844.
[50] G.W. Harron, E.M.A. Harkin-Jones, P.J. Martin, Influence of thermoforming parameters on final part properties, SPE ANTEC Conf. (2000).
[51] A. Aroujalian, M.O. Ngadi, J.-P. Emond, Effect of processing parameters on compression resistance of a plug-assist vacuum thermoformed container, Adv. Polym. Technol. 16 (2) (1997) 129–134.
[52] P. Collins, E.M.A. Harkin-Jones, P.J. Martin, The role of tool/sheet contact in plug-assisted thermoforming, Int. Polym. Process. 17 (4) (2002) 361–369.
[53] C. Bordonaro, T. Virkler, P. Galante, B. Pineo, C. Scott, Optimization of processing conditions in thermoforming, SPE ANTEC Conf. (1988) 696–700. 56th.
[54] H. Tanaka, H. Shigemoto, H. Hawachi, Adhesive resin for multi-layer films used for shrink or thermoforming processes, J. Plastic Film. & Sheeting 12 (1996) 278–286.

[55] M. Paul, B. Morris, J. Weinhold, K. Hausmann, S. Sengupta, Tie layer adhesion chemistry for multilayer packaging, SPE Int. Polyolefins Conf. (2020).

[56] M. Paul, B. Morris, J. Weinhold, K. Hausmann, Tie layer adhesion chemistry for multilayer packaging, SPE ANTEC Conf. (2021).

[57] G. Nam, K.-H. Hah, J. Lee, Three-dimensional simulation of thermoforming process and its comparison with experiments, Polym. Eng. Sci. 40 (2000) 2232–2240.

[58] Y. Song, K. Zhang, Z. Wang, F. Diao, Y. Yan, R. Zhang, Coupled thermomechanical analysis for plastic thermoforming, Poly Eng. Sci. 40 (2000) 1736–1746.

[59] D. Laroche, F. Erchiqui, Experimental and theoretical study of the thermoformability of industrial polymers, J. Plastic Film. & Sheeting 15 (1999) 287–296.

[60] P. Novotny, P. Saha, K. Kouba, Fitting of K-BKZ model parameters for the simulation of thermoforming, Intern. Poly Process. 14 (1999) 291–295.

[61] P. Novotny, K. Kouba, P. Saha, The role of material in the simulation of the thermoforming process, in: Progress and Trends in Rheology V, Proceedings of the European Rheology Conference, 5 h Potoroz, Slovenia, September 6–11, 1998, pp. 367–368.

[62] P. Novotny, P. Saha, Optimization of thermoforming, SPE ANTEC Conf. (1999) 841–843.

[63] R. DiRaddo, D. Laroche, A. Bendada, T. Ots, Optimization of thermoforming with process modeling, SPE ANTEC Conf. (1999) 844–849.

[64] R. DiRaddo, P. Girard, S. Chang, Model-based control of material distribution in thermoformed parts, SPE ANTEC Conf. (2002).

[65] J. Lappin, E. Harkin-Jones, P. Martin, Finite element modeling of the plug-assisted thermoforming process, SPE ANTEC Conf. (1999) 826–830.

[66] G. Nam, H. Rhee, J. Lee, Finite element analysis of the effect of processing conditions on thermoforming, SPE ANTEC Conf. (1998) 690. -69.

[67] D. Laroche, F. Erchiqui, Experimental and theoretical study of the thermoformability of industrial polymers, SPE ANTEC Conf. (1998) 676–680.

[68] B. Debbaut, O. Homerin, 3-D numerical simulation of thermoforming: prediction of thickness and extensions, SPE ANTEC Conf. (1997) 720–725.

[69] B. Koziey, J. Procher, J. Tan, J. Vlachopoulos, New results in finite element analysis of thermoforming, SPE ANTEC Conf. (1997) 714–719.

[70] N. Martin, E. Harkin-Jones, P. Martin, Characterizing the biaxial properties of materials used in thermoforming and blow molding, SPE ANTEC Conf. (2002) 732–736.

[71] G. Nam, K. Ahn, J. Lee, Numerical and experimental studies of 3-dimensional thermoforming process, SPE ANTEC Conf. (1999) 836–840.

[72] R. Christopherson, B. Debbaut, Y. Rubin, Simulation of pharmaceutical blister pack thermoforming using a non-isothermal integral model, SPE ANTEC Conf. (2000) 773–777.

[73] C.-H. Wang, H. Nied, Solution of inverse thermoforming problems using finite element simulation, SPE ANTEC Conf. (2000) 768–772.

[74] M. Rachik, J. Roelandt, A unified approach for thermoforming numerical simulation, SPE ANTEC Conf. (1999) 831–835.

[75] K.Y. Tshai, E.M.A. Harkin-Jones, P.J. Martin, G. Menary, A unified approach of modeling polypropylene for thin-gauge solid-phase thermoforming applications, SPE ANTEC Conf. (2004) 919–923.

[76] B. Van Mieghem, F. Desplentere, A. Van Bael, J. Ivens, Improvements in thermoforming simulation by use of 3D digital image correlation, eXPRESS Polym. Lett. 9 (2015) 119–128.

[77] J.L. Throne, Parametric study of extrusion and thermoforming conditions for CPET, J. Plastic Film. & Sheeting 5 (1989) 222–240.

[78] H. Xu, D.O. Kazmer, Thermoforming shrinkage prediction, Polym. Engr. Sci. 41 (2001) 1553–1563.

[79] A. Yousefi, A. Bendada, R. Diraddo, Improved modeling for the reheat phase in thermoforming through an uncertainty treatment of the key parameters, Polym. Engr. Sci. 42 (2002) 1115–1129.

[80] C. O'Conner, P. Martin, G. Menary, J. Sweeney, P. Caton-Rose, P. Spencer, Development of a constitutive model of polypropylene for thermoforming, AIP Conf. Proc. 1353 (part 1) (2011) 874–879.

[81] R. Christopherson, M. Briere, Simulation of the puncture resistance of a thermoformed syringe pack, SPE ANTEC Conf. (2001).

[82] B.L. Koziey, M.O. Ghafur, J. Vlachopoulos, F.A. Mirza, Computer simulation of thermoforming, in: D. Bhattacharyya (Ed.), Composite Sheet Forming, Elsevier Science, 1997, pp. 75–89.

[83] A. Makradi, S. Belouettar, S. Ahzi, S. Puissant, Thermoforming process of amorphous polymeric sheets: modeling and finite element simulations, J. Appl. Polym. Sci. 106 (3) (2007) 1718–1724.

[84] M.K. Warby, J.R. Whiteman, W.-G. Jiang, P. Warwick, T. Wright, Finite element simulation of thermoforming processes for polymer sheets, Math. Comput. Simul. 61 (2003) 209–218.

[85] C.P.J. O'Connor, P.J. Martin, J. Sweeney, G. Menary, P. Caton-Rose, P.E. Spencer, Simulation of the plug-assisted thermoforming of polypropylene using a large strain thermally coupled constitutive model, J. Mater. Process. Technol. 213 (2013) 1588–1600.

[86] H. Hosseini, B.B. Vasilivich, A. Mehrabani-Zeinabad, Rheological modeling of plug-assist thermoforming, J. Appl. Polym. Sci. 101 (2006) 4148–4152.

[87] D. Ahmad, P. Wang, N. Hamila, P. Boisse, Numerical analysis of non-isothermal viscoelastic materials for thermoforming of polymer films, Key Eng. Mater. 554–557 (2013) 1692–1698.

[88] B. Debbaut, F. Fradet, H. Metwally, Validation of a new material model for thermoforming and blow molding simulation, SPE ANTEC Conf. (2015) 2673–2677.

[89] J.L. White, M. Cakmak, Orientation, Encyclopedia of Polymer Science and Technology, John Wiley & Sons, 2010.

[90] A. Ajji, X. Zhang, Correlations between orientation and some properties of polymer films and sheets, J. Plastic Film. & Sheeting 18 (2002) 105–116.

[91] F. Chambon, S. Srinivas, Orientation analysis of cast and blown polyethylene films, J. Plastic Film. & Sheeting 15 (1999) 297–307.

[92] A. Ajji, S. Elkoun, X. Zhang, Biaxial orientation characterization in PE and PP using WAXD X-ray pole figures, FTIR Spectroscopy, and birefringence, J. Plastic Film. & Sheeting 21 (2005) 115–126.

[93] A. Ajji, X. Zhang, S. Elkoun, Biaxial orientation in HDPE films: comparison of infrared spectroscopy, X-ray pole figures and birefringence techniques, Polymer 46 (11) (2005) 3838–3846.

[94] X. Zhang, A. Ajji, V. Jean-Marie, Processing-structure-properties relationship of multilayer films. I. Structure characterization, Polymer 42 (2001) 8179–8195.

[95] T. Lefevre, C. Pellerin, M. Pezolet, Characterization of molecular orientation, Compr. Anal. Chem. 53 (2008) 295–335.

[96] B.E. Read, I.M. Ward, Structure and Properties of Oriented Polymers, Wiley-Interscience, New York, 1975.

[97] B. Jasse, J.L. Koenig, Orientation measurements in polymers using vibrational spectroscopy, J. Macromol. Sci. Rev. Macromol. Chem. C17 (1979) 61–135.

[98] ASTM International, Standard Test Method for Unrestrained Linear Thermal Shrinkage of Plastic Film and Sheeting, ASTM International, West Conshohocken, PA, 2014 (D2732-14).

[99] ASTM International, Standard Test Method For Shrink Tension and Orientation Release Stress Of Plastic Film And Sheeting, ASTM International, West Conshohocken, PA, 2009 (D2838-09).

[100] Y. Jin, T. Hermel-Davidock, T. Karjala, M. Demirors, J. Wang, E. Leyva, et al., Shrink force measurement of low shrink force films, SPE ANTEC Conf. (2008) 1364–1368.

[101] M.T. DeMeuse, Biaxial Stretching of Film, Principles and Applications, Woodhead Publishing, Oxford, 2011, pp. 37–40.

[102] A. Ajji, J. Auger, J. Huang, L. Kale, Biaxial stretching and structure of various LLDPE resins, Polym. Eng. Sci. 44 (2) (2004) 252–260.

[103] T.P. Hanschen, Films, orientation, Encyclopedia of Polymer Science and Technology, John Wiley and Sons, Inc, 2002, pp. 559–577.

[104] J. Breil, Speciality oriented films: new markets, new applications, new developments, Specialty Plastic Films (2004) 2004.

[105] A. Ajji, X. Zhang, Biaxial orientation of multilayer LDPE/PET films: structure and properties, SPE ANTEC Conf. (2004) 2112–2116.

[106] J. Strobie, Producing co-extruded high barrier heat shrinkable packaging films, TAPPI PLACE. Conf. (2002).

[107] K. Hausmann, Y.M. Trouilhet, J. Schiffman, Low temperature sealing films, U.S. Patent Application 2013/0164516 A1, 2013.

[108] E.M. Mount, III, Uniformly stretching polymer sheets and multilayer films, Converting Quarterly "Substrate Secrets". <http://www.convertingquarterly.com/blogs/substrate-secrets/id/2271/uniformly-stretching-polymer-sheets-and-multilayer-films.aspx>, 2011 (accessed 07.06.15).

[109] Y. Nishamoto, K. Yamazaki, Heat-shrinkable laminate film, US Patent 4424243, 1984.

[110] P. Ciocca, R. Forloni, Multilayer oxygen barrier packaging film, US patent 6,221,470 B1, 2001.

[111] S. Lustig, S.J. Vicik, Oriented multilayer film and process for making same, US Patent 5077109, 1991.

[112] R. Forloni, Heat-shrinkable multilayer thermoplastic film, US Patent 6299984 B1, 2001.

[113] H. Tanaka, H. Kawachi, Adhesive polypropylene composition and multi-layer laminate layer body using the resin composition, US patent 5695838 A, 1997.

[114] L. Parent, P. Cigana, A. Ajji, P.J. Carreau, Effect of machine direction orientation conditions on structure and properties of HDPE films, Annu. Technical Conf. - Soc. Plast. Eng. (2009) 2342–2344.

[115] S.H. Tabatabaei, L. Parent, P. Cigana, A. Ajji, P.J. Carreau, Effect of machine direction orientation conditions on properties of HDPE films, J. Plastic Film. & Sheeting 25 (2009) 235–249.

[116] A.L. Bobovitch, S. Elkoun, A. Ajji, Orientation, structure and properties of double-bubble oriented LLDPE films, J. Plastic Film. & Sheeting 22 (2) (2006) 133–143.

[117] A.L. Bobovitch, S. Elkoun, A. Ajji, Orientation, structure and properties of double-bubble oriented LLDPE films, Annu. Technical Conf. - Soc. Plast. Eng. (2005) 1757–1761.

[118] A.L. Bobovitch, R. Tkach, A. Ajji, S. Elkoun, Y. Nir, Y. Unigovski, et al., Mechanical properties, stress-relaxation, and orientation of double bubble biaxially oriented polyethylene films, J. Appl. Polym. Sci. 100 (5) (2006) 3545–3553.

[119] P.C. Gillette, Biaxially oriented polyolefin film is post-treated by annealing to improve dimensional stability in limited temperature range for various time periods, U.S. Patent US 5152946 A, 1992.

[120] P.J. Barham, J.A. Odell, F.M. Willmouth, Heat-treating polyolefin films, UK Patent Application, GB 2055855 A, 1981.

[121] T. Pakula, M. Trznadel, Thermally stimulated shrink forces in oriented polymers: 1. Temperature dependence, Polymer 26 (1985) 1011–1018.

[122] J.H. Schut, MDO films: lots of promise, big challenges, Plast. Technol. (2005)Feb 2005, viewed 21 Feb 2021 at. Available from: http://www.ptonline.com/articles/mdo-films-lots-of-promise-big-challenges.

[123] J. Breil, New developments for biaxially stretched polyolefin film with linear motor driven simultaneous stretching technology, SPE ANTEC Conf. (2002).

[124] J. Breil, Specialty oriented films: new markets, new applications, new developments, Specialty Plastic Films (2004) 2004.

[125] M.A. McLeod, Drawing forces and film properties in semi-sequentially stretched polypropylene, SPE ANTEC Conf. (2005) 1752–1756.

[126] W. Zhang, H. Yang, R. Moffitt, J. Tung, Characterization of crosslinked, heat-shrinkable packaging films, SPE ANTEC Conf. (2007) 537–542.

[127] M. Zanella, A. Gini, P. Prandi, A multilayer heat-shrinkable plastic film with barrier characteristics, WO 2001003922 A1, 2001.

[128] G. Biddle, Films, shrink, in: M. Bakker (Ed.), The Wiley Encyclopedia of Packaging Technology, John Wiley & Sons, Inc, New York, 1986, pp. 335–338.

[129] J. Breil, Multilayer oriented films, in: J.R. Wagner (Ed.), Multilayer Flexible Packaging, Technology And Applications For The Food, Personal Care And Over-The-Counter Pharmaceutical Industries, Elsevier, Oxford, 2010, pp. 231–237.

[130] Y. Lin, M. Demirors, J. Pan, X.-B. Yun, Structure-property relationship of biaxially oriented polyethylene (BOPE) films made via double bubble film fabrication process, SPE ANTEC Conf. (2014).

[131] Y. Lin, J. Xu, J. Pan, X.-B. Yun, M. Demirors, Biaxially oriented polyethylene (BOPE) films fabricated via tenter frame process and applications thereof2018 SPE ANTEC Conf. (2018).

[132] K. Hausmann, Rethinking flexible packaging in a circular world using PE films, Sustainability Packaging Europe Conf. (2020).

[133] A.S. Fard, H. Ibrahim, J. Schaefer, V. Boudara, J. Throckmorton, C. Rysermans, Machine direction orientation of performance polyethylene films: application in recyclable pouches, SPE Int. Polyolefins Conf. (2021).

[134] C. Lawrence, K.T. Forziati Jr., J. Garnett IV, Machine direction orientation and its effect on multilayer sealant film, SPE FlexPackCon Conf. (2013).

[135] G. Gerrits, R. Kleppinger, R. Deblieck, Morphology of biaxially oriented polyethylene films with sealing performance, SPE Int. Polyolefins Conf. (2021).

[136] G. Medlock, Enhanced EVOH for improved orientability and shrink, TAPPI PLACE. Conf. (2010).

[137] K.D. Lind, J. Eckstein, R.J. Blemberg, R.L. Blemberg, G.H. Walbrun, Heat shrinkable barrier films, U.S. Patent US 6074715 A, 2000.

[138] I.-H. Lee, H.E. Schenck, Heat shrinkable multilayer film or tube that exhibits shrink stability after orientation, US Patent 8202590 B2, 2012.

[139] Ube, Ube Nylon Extrusion Application, Ube Industries, Ltd., 2005viewed 26 Jul 2015 from. Available from: https://www.ube.es/archivos/pdfs/extrusion.pdf.

[140] T. Ueyama, T. Itoh, Heat-shrinkable multilayer film, European Patent EP 1190847 A1, 2002.

[141] R. Breese, G. Beaucage, Effects of machine direction orientation (MDO) on the moisture and oxygen barrier properties of HMW-PE films, TAPPI PLACE. Conf. (2005).

[142] D. Froio, S. Schirmer, J.A. Ratto, M. Burke, G. Pigeon, The effect of machine direction orientation on multilayer high barrier films, SPE ANTEC Conf. (2012).

[143] E. Hatfield, D.R. Breese, Step change improvements in EVOH barrier using MDO, Equistar (2006)viewed on 07 Jun 2015 at. Available from: http://www.lyondellbasell.com/techlit/techlit/Tech%20Topics/Equistar%20Industry%20Papers/EVOH_Barrier_Using_MDO.pdf.

[144] R. Breese, Economic benefits of utilizing MDO films in flexible packaging, TAPPI PLACE. Conf. (2007).

[145] R. Breese, Benefits of machine direction-oriented (MDO) films in flexible-packaging applications, Converting Q. (2011)2011, quarter 2, pp. 32–36. Viewed on 07 June 2015 at. Available from: http://www.convertingquarterly.com/magazine/back-issues.aspx.

[146] F. Alexy, Replacing aluminum in high barrier packaging with Triple Bubble® blown film extrusion, Multilayer Packaging Films 2012 Conf. (2012).

[147] A. Edgar, All encompassing extrusion technology for producing a wide spectrum of simulataneously bioriented films, 2018 SPE ANTEC Conf. (2018).

[148] J. Schiffmann, Multi-layered two-dimensional or tubular food casing or food film, U.S. Patent Application US 2013/0044980 A1, 2014.

[149] J. Shiffmann, Multilayer sheet- or tube-type food casing or food film, U.S. Patent Application US 2014/0044902 A1, 2014.

[150] D.Y. Chiu, G.E. Ealer, F.H. Moy, J.O. Bühler-Vidal, Unipol II LLDPE – gas phase LLDPE for the shrink market, J. Plastic Film. & Sheeting 15 (1999) 153–178.

[151] A. Torres, N. Colls, F. Méndez, Properties predictor for HDPE/LDPE/LLDPE blends for shrink film applications, J. Plastic Film. & Sheeting 22 (2006) 29–37. Also 2005, SPE ANTEC Conference, pp. 2986–2989.

[152] M.M. Werlang, Effect of polymer processing conditions on shrinkability of LDPE and LLDPE films, and on optical properties of LLDPE films, SPE ANTEC Conf. (2003) 3271–3274.

[153] B.E. Obi, M. Leano, Correlating the melting of semi-crystalline polymers to the shrink wrapping process in shrink-film packaging applications, SPE ANTEC Conf. (2015).

[154] A.L. Bobovitch, A. Sagron, Y. Unigovski, A. Jarashneli, E.M. Gutman, Stress relaxation in low-shrink-force polyolefin films, Poly. Engr. Sci. 44 (2004) 1716–1720.

[155] C.-W. Chen, H.T. Pham, A.J. Poslinski, Understanding of shrink films through viscoelastic behavior of films, SPE ANTEC Conf. (2013).

[156] R. Jester, COC enhanced polyolefin films for shrink label sleeves and labels, SPE Polyolefins Flex. Packaging Conf. (2011).

[157] M.W. Wainer, Stretch film properties – effects of equipment and process variables, J. Plastic Film. & Sheeting 18 (2002) 279–286.

[158] C.M. Small, G.M. McNally, W.R. Murphy, G. Garrett, The effect of vinyl acetate concentration and polyisobutylene concentration on the properties of metallocene polyethylene/ethyl vinyl acetate co-extruded film for stretch and cling film applications, SPE ANTEC Conf. (2003) 3167–3171.

[159] T. Iaccino, What you need to know to make world-class stretch film, Plast. Technol. (2015)viewed online on 14 Feb 2021 at. Available from: https://www.ptonline.com/articles/heres-what-you-need-to-know-to-make-world-class-stretch-film.

[160] J.H. Schut, Stretching film's limits, Plast. Technol. (2004).

[161] C. Shelley, N. Kokel, Elastic recovery behavior of three layer film structures comprised of TPO and various ethylene copolymers, SPE ANTEC Conf. (2009) 2775–2780.

[162] C.M. Small, G.M. McNally, A. Marks, W.R. Murphy, The effect of extrusion processing conditions and polyisobutyene concentration on the properties of polyethylene for stretch and cling film applications, SPE ANTEC Conf. (2002).

[163] C.M. Small, G.M. McNally, W.R. Murphy, G. Garrett, The effect of polymer properties on the mechanical behavior and morphological characteristics of cast polyethylene film for stretch and cling film applications, SPE ANTEC Conf. (2003) 3266–3270.

[164] M. Naitove, First microlayerd commodity films, Plast. Technol. (2009)viewed 12 Jun 2015 at. Available from: http://www.ptonline.com/articles/firstmicrolayeredcommodityfilms.

[165] AIMCAL, MicroLayer cast film process yields stretch wrap with properties and potential to save on wrapping, AIMCAL. <http://www.aimcal.org/membernews/articles/id/910/microlayercastfilmprocessyieldsstretchwrapwithpropertiesandpotentialtosaveonwrapping.aspx>, 2008 (accessed 12.06.15).

[166] J. Song, R.H. Ewoldt, W. Hu, H.C. Silvis, C.W. Macosko, Flow accelerates adhesion between functional polyethylene and polyurethane, AIChE J. 57 (2011) 3496–3506.

[167] J. Song, A.M. Baker, C.W. Macosko, R.H. Ewoldt, Reactive coupling between immiscible polymer chains: acceleration by compressive flow, AIChE J. 59 (2013) 3391–3402.

[168] C.D. Lee, J.J. Strebel, J.D. Goudelock, Higher tie layer adhesion in machine direction oriented (MDO) barrier films, TAPPI PLACE. Conf. (2007).

Chapter 12

Frictional and optical properties of flexible packaging films

The slipperiness or tackiness of a packaging film impacts its ability to be handled on high-speed packaging lines. The incorporation of low melting sealant layers, which are naturally tacky, into multilayer flexible packaging films brings challenges for controlling the frictional and blocking properties. As in single layer films, migrating and nonmigrating additives are typically used to control the surface properties of multilayer films. For migrating additives, special consideration must be taken for how the presence of adjacent layers influences migration and performance. While frictional properties are the subject of this chapter, the principles are similar for other migrating additives such as antifog and antistat additives.

Similarly, the optical properties of multilayer films can be more than the simple sum of the properties of the constituent layers. The surface and interfacial properties must be taken into account, whether they are positive or negative.

This chapter explores the frictional and optical properties of films and how multilayer constructions may differ from single-layer films.

12.1 Frictional properties

In many flexible packaging applications, the film must easily slide over film and metal surfaces. This ease of sliding is generally characterized by the coefficient of friction (COF), which is the ratio of the force required to slide a film over a surface to the total force perpendicular to that surface. The COF is typically measured by sliding a weighted sled over the film, as described in Section 12.1.2. A COF of around 0.2 is often desired for running on high-speed packaging lines. Many common polymers used in packaging have inherently high COF values, including polyethylene (PE). The problem can be even more severe with low-melting polymers, such as ultra low-density polyethylene (ULDPE) and metallocene plastomers, polyolefin elastomers (POEs), ethylene vinyl acetate (EVA) copolymers and ionomers. It is a common practice in the industry to add fatty acid amides to polyolefins to reduce and control the COF. These low MW additives bloom to the surface to form a lubricating layer. The amount and rate at which these additives migrate to the surface affect the performance of films. If too little of the additive ends up on the surface, the COF may be too high. If too much comes to the surface, adhesion, appearance and heat seal performance may suffer.

Nonmigrating additives have also been developed to help control the COF of films but due to the higher addition levels these additives typically require, they have not been cost-effective for many packaging applications. With the ability to make multilayer structures, placing these additives in a thin surface layer make this technology more attractive.

Films that have been wound into a roll must be able to be unwound. Some polymers, particularly soft or tacky ones, tend to bind to each other in the roll. The inability to unwind the film is known as film blocking. Migrating and nonmigrating additives are used to help prevent the blocking of films. Blocking can also occur in the nip of blown film lines as the bubble is collapsed if the inside layer is tacky. In all-PE structures, for example, low melting sealant layers are used which when placed on the inside of the bubble are prone to blocking. Similar additive strategies are used to mitigate this challenge.

12.1.1 Why frictional properties are important

The frictional properties of films affect their ability to be handled in packaging operations. Two related but independent properties are important: COF and antiblocking behavior. The first relates to how easily the film slides past other films or metal surfaces. The second relates to how easy it is to unwind a roll. Often polymers that have high COF also tend

The Science and Technology of Flexible Packaging. DOI: https://doi.org/10.1016/B978-0-323-85435-1.00021-1

to block, but this is not always the case. The types of additives used to control COF are not usually the same as those utilized for preventing blocking, and the two may be synergistic or antagonistic.

The COF of films are often categorized into ranges:

1. High slip: 0.1–0.2;
2. Medium slip: 0.21–0.4;
3. Low slip: 0.41–0.7;
4. No slip: >0.7.

Some applications require low slip, others medium to high slip. For example, films run on vertical form fill seal lines require a low COF on the sealant side and a medium to high COF on the outside. The film needs to slide easily along the metal forming mandrel on one hand, but requires some film-to-film friction to help with the feeding of the film [1].

At the extreme, a roll that cannot be unwound cannot be used. But even limited blocking can distort the film as it is removed from the roll if the film is stretched. Blocking forces may also result in increased static discharge during unwinding.

12.1.2 How it is measured

12.1.2.1 Coefficient of friction

The frictional characteristics of a film may be measured on a specimen sliding over itself, usually called film-to-film, or to another substrate. Typically film-to-film and film-to-metal properties are measured. ASTM D1894 [2] describes a method for determining the COF of plastic films by measuring the force to pull a weighted sled across the film. A characteristic of the sliding force is that often the initial or starting force is higher than the ongoing force needed to move the sled. The static COF is defined as the force required to slide the film over the surface divided by the force applied normal to the surface (as determined by the weight of the sled), the moment the motion starts. The kinetic or dynamic COF is that ratio once that motion is in progress.

The test comprises of two surface specimens, one mounted to the sled and one to the plane so that during the test the specimen mounted to the sled slides across the specimen mounted to the plane. The method describes several apparatuses, one of which is depicted in Fig. 12.1. Here the apparatus is mounted to a tensile tester so that when the jaws separate the sled moves across the plane at a constant speed. The force transducer of the tensile tester measures the sliding force. The normal force is simply the weight of the sled. The COF is calculated using Eqs. (12.1) and (12.2):

$$\mu_s = \frac{F_s}{B} \tag{12.1}$$

$$\mu_k = \frac{F_k}{B} \tag{12.2}$$

where

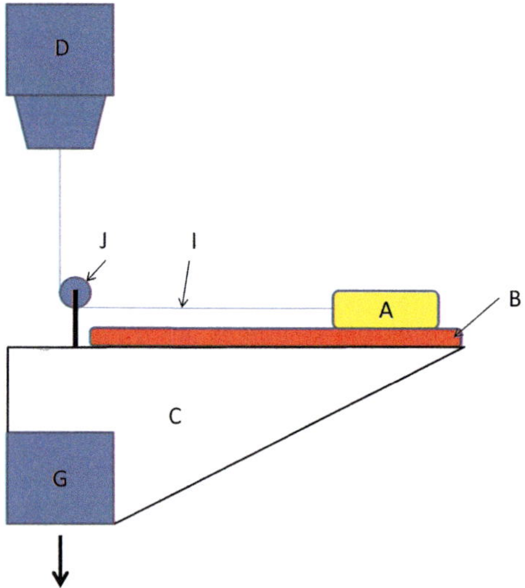

FIGURE 12.1 Schematic of COF measurement apparatus. The sled (A) moves across the plane (B) as the crosshead of the tensile tester (G) moves downward. The stationary force gauge (D) measures the sliding force. The sled is attached to the gauge with a nylon monofilament (I) that loops around a low friction pulley (J). The plane is attached to a supporting base (C). *Adapted from ASTM International, 2014, "Standard Test Method for Static and Kinetic Coefficients of Friction of Plastic Film and Sheeting," (D1894-14), ASTM International, West Conshohocken, PA.*

μ_s = static coefficient of friction,
F_s = initial force reading, g,
B = weight of sled, g,
μ_k = kinetic (or dynamic) coefficient of friction, and
F_k = average force reading during uniform sliding of the surfaces, g.

The test is usually carried out at room temperature (RT), but can be conducted at elevated temperature by heating the plane.

As described in Section 12.1.4 many plastic films use migrating additives to control the surface frictional properties. The performance of these additives is sensitive to temperature and aging effects. When making measurements on films containing these types of additives, care must be taken to control the temperature of the test and storage history of the specimen.

Maltby [3] shows that the sled weight can affect the COF results for tacky polymers. The ASTM standard D1984 calls for a sled weight of 200 g whereas BS 824, a British standard, uses a 700 g weight. Neither may represent the load in an actual application. Maltby presents results for a series of films with different levels of a slip additive using the two test standards. The high sled weight gives on average a COF that is about 40% higher than the lower sled weight, but the differences are fairly small and both methods provide the same trends. More interesting are the results of tests comparing the COF versus sled weight for a "slippery" film and a "tacky" film. The COF of the slippery film is unaffected by the sled weight, as shown in Fig. 12.2. The tacky film, while showing low COF under the standard weight, shows a rapid increase in COF with increasing weight. The important lesson here is that the COF measurement will not always mimic the real application.

12.1.2.2 Blocking

ASTM D3354 [4] describes a method for measuring the blocking force between two films. The method uses two metal flat plates with rigid adapters mounted in parallel to a universal testing machine (tensile tester). A specimen of two unblocked films is cut so that the films extend 40 mm from the front and back sides of the plates. The specimen is placed between the two plates. The free ends are carefully separated and taped to the top and bottom plates with double-sided tape. The configuration is illustrated in Fig. 12.3. The plates are first pressed together so the films are in contact. The plates are moved away from each other by the motion of the tensile tester at a constant speed of 5 mm/minute until the films are completely separated or the distance between the plates reaches 19 mm. The force of separation is recorded. A similar method is described for a constant rate of load device.

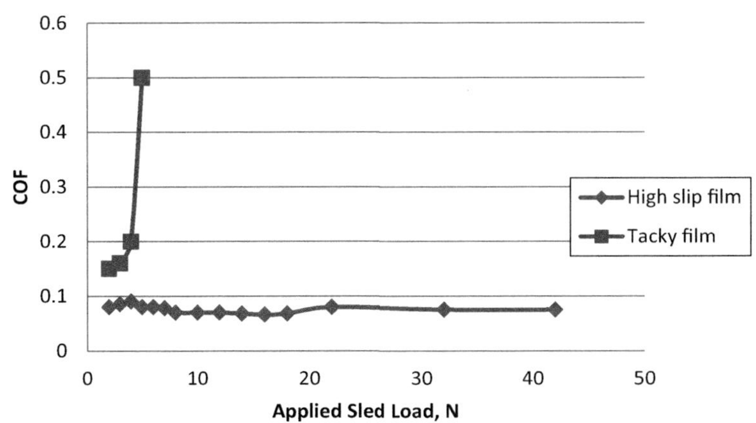

FIGURE 12.2 Effect of sled weight on coefficient of friction measurement. *Reproduced with permission from Maltby, A., 1998, "The effects of polyolefin formulation and processing variables on slip agent performance (I)," TAPPI Polymers, Laminations & Coat. Conf., pp. 1241–1250.*

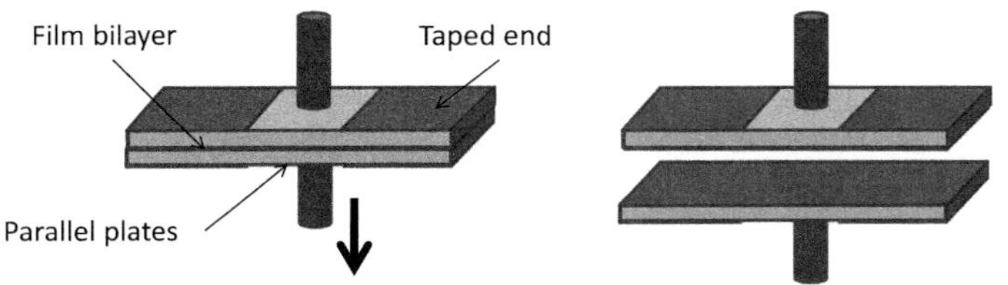

FIGURE 12.3 Blocking force measurement.

12.1.3 Additive technology

12.1.3.1 Why friction arises and how additives work

The surface of any material is never truly smooth—there are always asperities or protrusions from the surface present that are larger than the size of individual molecules. Collectively, these asperities give rise to surface roughness. When two surfaces are brought together, they touch only on the asperities; the real area of contact is governed by the deformation properties of the asperities (see Fig. 12.4). This area is influenced by the load or normal force (W) exerted on the surfaces and the yield pressure (p) or hardness of the surface:

$$A_r = \frac{W}{p} \tag{12.3}$$

where

A_r = real area of contact,
W = applied load, and
p = yield pressure of the material.

As the load is applied, a high unit pressure is exerted over the asperities, causing adhesion or junction formation at the asperity interfaces. These junctions give rise to the resistance to sliding or friction and are sheared during the sliding action. If these adhesive junctions have a shear strength s, the force to shear the junctions is given by

$$F = A_r s \tag{12.4}$$

where

F = sliding force
A_r = real area of contact
s = shear strength of junctions.

Combining Eqs. (12.3) and (12.4),

$$F = Ws/p. \tag{12.5}$$

Since the COF is defined as the sliding force divided by the normal force or load,

$$\mu \equiv F/W, \tag{12.6}$$

it follows that

$$\mu = \frac{s}{p}. \tag{12.7}$$

The COF is equal to the shear strength of the interface divided by the interfacial hardness, two material parameters. This turns out to be sufficient to describe the behavior of metals; for polymer films the load and surface roughness also play a role, as described below.

Eq. (12.7) suggests several strategies for reducing COF as illustrated in Fig. 12.5 [5]. One is to interpose a low shear strength lubricating layer between the asperities [decreasing the numerator of Eq. (12.7)]. This is one of the reasons fatty acid amides are effective as slip additives: their low surface energy reduces the adhesion between the asperities, effectively lowering the shear strength.

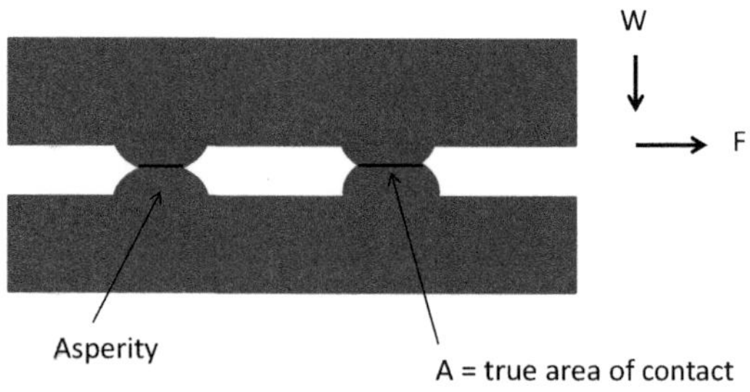

FIGURE 12.4 Illustration of true area of contact between surfaces. The real contact area is determined by the deformation behavior of asperities that make actual contact.

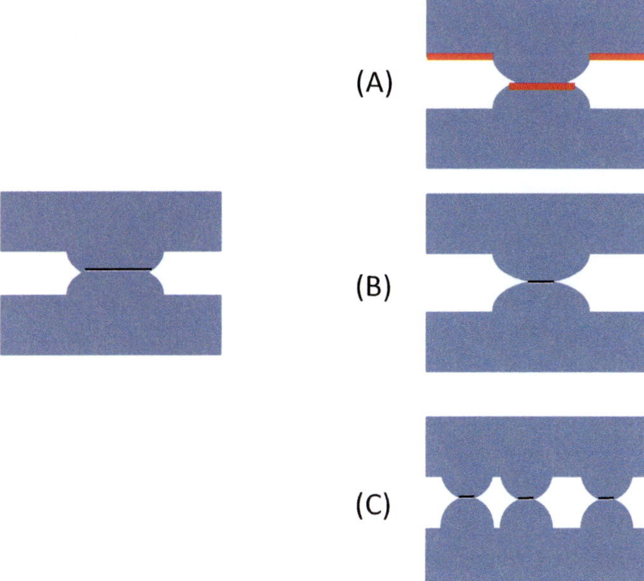

FIGURE 12.5 Illustration of strategies to reduce COF: (A) Introduce a lubricant at the interface (*red*) with weak shear strength or adhesion, (B) increase the hardness of the asperities, thereby reducing the real contact area, and (C) increase the number of asperities (surface roughness).

Another strategy is to increase the hardness of the surface [increasing the denominator of Eq. (12.7)]. Lubricants that can crystallize, such as the longer chain fatty acid amides, have been found to be more effective than liquid lubricants for reducing the COF of polymer films; the crystalline structure helps increase the hardness of the surface [5]. Even partial coverage of the surface can reduce COF [5–7]. Inorganic additives, such as talc or diatomaceous earth, as well as thermoset silicone powders, employ this strategy. These nonmigrating additives also increase the number of asperities and surface roughness, which helps reduce COF and blocking.

This last effect arises from the viscoelastic nature of polymers. Owens [8] finds that the effect of load and surface roughness on the frictional behavior of polymer substrates can be related to a single material parameter, the deformation index. This parameter describes the manner in which the real area of contact varies with the load and roughness. He proposes the following relationships:

$$\mu = k_1 W^{\left(\frac{2}{m}\right)-1} \tag{12.8}$$

$$\mu = k_2 N^{\left(\frac{2}{m}\right)-1} \tag{12.9}$$

where

W = load,
N = number of asperities (related to surface roughness),
m = deformation index, and
k_1, k_2 = constants.

The value for m varies from 2 for plastic materials to 3 for purely elastic ones. Metals deform plastically and therefore the COF of metal against metal is independent of load and surface roughness [the exponents in Eqs. (12.8) and (12.9) reduce to 0 when $m = 2$]. Rubbers have been experimentally found to have a deformation index of 3, conforming to the theory. For cellophane coated with a copolymer of vinylidene chloride-acrylonitrile (the K coating developed in the 1920s to make cellophane moisture resistant), Owens measures the value of m to be 2.35, consistent with a viscoelastic material. From Eq. (12.9), the COF the K-film is proportional to $1/N^{0.65}$. Other polymer films should have values of m that fall between 2 and 3, which in practice means that as the surface roughness increases, the COF will decrease.

12.1.3.2 Migrating additives

The most commonly used slip agents are fatty acid amides with chain lengths ranging from about 16 to 24 carbon atoms. Some of the more commercially important varieties are listed in Table 12.1. There are three general classes of amides used as slip agents: primary amides, secondary amides and *N,N*-ethylene bisamide structures. Table 12.1 shows the chemical structures of these classes of amides. The primary amides are further categorized as saturated and

TABLE 12.1 Examples of commercially important fatty acid amides and their classifications.

Type	Chemical structure	Examples
Primary, saturated	R–CO–NH_2 where R is –C–C–C–...	Stearamide (C18:0) Behenamide (C22:0)
Primary, unsaturated	R–CO–NH_2 where R is ... –C=C–C–...	Oleamide (C18:1) Erucamide (C22:1)
Secondary	R′–CO–NH-R	Oleyl palmitamide (C18:1 C16:0) Stearyl erucamide (C18:0 C22:1)
N,N-ethylene bisamide	R′–CO–NH–CH_2–NH–CO–R	Ethylene-bis-stearamide (C18:0) Ethylene-bis-oleamide (C18:1)

Note: R and R′ represent carbon chains of a general length. The nomenclature CXX:Y used in the examples indicates XX carbons in the molecule and Y double bonds.

unsaturated amides. Saturated amides have no carbon double bonds. Unsaturated amides have one or more carbon double bonds, which are more reactive or thermally unstable than single bonds. The commercially important unsaturated amides all have only one double bond; more than one results in too much instability. Erucamide and oleamide are unsaturated primary amides that are commonly used as slip agents for polyolefins, but are more volatile and less thermally stable than some of the other amides. Stearamide and behenamide are examples of primary saturated amides. They typically do not provide very good slip performance but are used as release and antiblock agents.

Secondary amides include oleyl palmitamide and stearyl erucamide. These have better thermal stability than the primary amides but are slower to migrate and do not provide as low a COF. Ethylene-*bis*-stearamide and ethylene-*bis*-oleamide are examples of *N,N*- ethylene bisamides. They too are more thermally stable than primary amides. The selection of the amide will depend on the polarity of the polymer (which affects migration rate), the processing temperature (which determines the need for thermal stability) and the ultimate COF needed (the amide crystalline structure affects the ultimate COF that can be achieved once it reaches the surface).

Fatty acid amides are derived from natural sources, typically vegetable oils or tallow (animal fat). The oils contain a mixture of acids of different chain lengths and saturation. These must be purified, typically using fractional distillation, and converted to amides to produce commercially viable slip agents. For example, erucamide is derived from rapeseed oil. The raw oil contains about 50% erucic acid. After purification, the fraction of erucic acid or its derivative erucamide ranges from about 80% to 94% [9]. Higher levels of purity are not economically viable. Walp [9] finds that the COF of LLDPE with two grades of erucamide, one with 80% purity and the other with 93%, are the same. Maltby [10] reports similar results. Nevertheless, it is important to understand that the composition may vary with the manufacture and source of the plant or animal.

Fatty acid amide slip agents are typically added into the polymer at 200 to 2000 ppm, either by the resin manufacturer or by the converter using a masterbatch. They are soluble in the polymer melt but not in the solid crystals. As the polymer crystallizes, the polymer crystals push the slip molecules away from the bulk toward the surface. The affinity or solubility of the slip agent with the polymer influences how fast the additive will migrate to the surface. Fatty acid amides are more soluble in polar polymers (such as EVA) and polymers with low levels of crystallinity (such as plastomers); they take longer to migrate to the surface in these polymers than standard LDPE. Some slip additives that migrate quickly in nonpolar polyethylene are too soluble in polar polymers and take too long to migrate. For these polymers, less polar amides, such as secondary amides and *bis*-stearamides, are often used instead of primary amides. Once on the surface, the slip molecules organize into bilayers and, with sufficient quantity, crystals. The crystalline structure of the amide has a substantial impact on COF. Some have favorable structures, such as erucamide, leading to lower COF. Others like mono saturated primary amides are poor slip agents but help reduce blocking.

Fatty acid amides are the most economical slip additives. They provide low-cost control of the frictional properties of the film when used optimally. They do have several drawbacks, however, including:

1. The COF takes time to develop due to the migrating action of the additive.
2. Too much amide on the surface can lead to adhesion issues for subsequent printing or lamination steps.
3. Too much amide can affect film appearance (described as chalking) and sealing properties.
4. Slip can be lost to adjacent layers in coextrusion, making it difficult to predict COF performance.

5. Slip agents can transfer to the adjacent surface in the wound roll, causing adhesion or COF issues on the opposite side of the film.

Despite these issues, fatty acid amides are the most widely used slip additives in flexible packaging. Factors affecting their performance are described in greater detail in Section 12.1.4.

Other migrating additives include antifog agents (e.g., ethoxylated alkylphenols, polyethoxylated oleic acid esters, and sorbitan esters of fatty acids), antistats (e.g., glycerol mono stearate) and cling additives (e.g., polyisobutylene).

12.1.3.3 Nonmigrating additives (slip and antiblock)

Inorganic additives such as talc, diatomaceous earth and silica (natural and synthetic) are often used as antiblock agents. They are typically added at levels ranging from 1000 to 10,000 ppm (0.1−1 wt.%). They reduce blocking by providing a rough surface to the film, which reduces the contact area when films are brought together. The size, shape and hardness of the additive and how well it is dispersed in the polymer influence the antiblocking performance. Larger particles may be more efficient at preventing blocking but generally increase the film haze. Irregular-shaped particles may be more effective at providing a bumpy surface than more regular-shaped particles. These inorganic additives typically do not reduce the COF of a film on their own but may allow a migrating slip agent to be more effective; in some cases, the combination of a slip agent and antiblock may provide lower COF at a lower concentration than the slip agent alone. Other factors influencing the selection of antiblocking additives include:

1. Particle hardness which affects machine wear.
2. Film clarity needs. Film clarity is a function of antiblock loading, film gauge, and the difference in refractive indices (RIs) of the polymer and particle. If the RI's are closely matched, the light will not bend as it moves from polymer to particle, reducing film haze and improving transparency. Higher loadings and film thickness act to increase haze simply because there are more particles for the light to encounter as it travels through the film.
3. Iron content. Iron may degrade some organic additives and cause color formation.

Talc is one of the most cost-effective inorganic antiblock additives and has the greatest usage for packaging films. Diatomaceous earth is perhaps the most effective at low concentrations and is used where optical clarity is important [11].

Glass spheres significantly reduce COF in polyolefin films [12]. They are most effectively used in thin skin layers that are less than one micron thick, such as in oriented polypropylene film. They can reduce the COF to 0.35−0.45.

Thermoset silicone powders are also used as nonmigrating slip additives. They are utilized like glass spheres but are more effective, reducing the COF to 0.25−0.35. These are considered the premier nonmigratory slip additives. The COF can be further reduced by adding migratory fatty acid amide slip additives.

Thermoplastic silicones are also used as slip additives, particularly in skin layers in coextrusion. Ryan et al. [13] describe masterbatches of ultra-high molecular weight siloxane in LLDPE which produce COF values in LLDPE film of around 0.25 when added at a 2% loading (4% of a 50% masterbatch). Sarargaonkar and McKinley [14] describe similar results with a nonmigratory masterbatch. They claim no effect on heat sealability of a 50-μm LDPE cast film when 10% or 20% of the masterbatch (containing 10% or 20% slip additive) is added. Hays [15] describes silicone masterbatches (25% silicone) added at 3% to LLDPE or LDPE film that provide film-to-film and film-to-steel COF values of 0.3 and 0.2, respectively. These nonmigratory masterbatches are said to limit reverse-side transfer to printing surfaces in the roll and are not affected by heat during storage like migratory fatty acid amide slip additives.

Constant [12] finds that cyclic olefin copolymers (COCs) impart medium slip performance. Adding 10%−20% of COC into LDPE reduces the COF from 0.6 to around 0.3. Increasing the COC concentration lowers COF but increases the haze, as does increasing the T_g of the COC.

The advantages of nonmigratory slip additives include:

1. The COF does not change with time. Migratory additives require time to bloom to the surface after fabrication.
2. COF does not change with storage temperature. The performance of migratory additives changes with temperature (see Section 12.1.4.6).
3. The additive is thermally stable, nonvolatile and can be used in high-temperature processes such as cast film or extrusion coating.
4. The problems adhesion, printing and sealing problems associated with low MW silicone lubricant oils are avoided with high MW silicone additives. The reserse-side transfer within a roll to the printing surface is limited. When using migratory slip additives, their effect on wetting and printing must be considered [16].

5. Cost can be mitigated by using the additive in only the outside layers in a multilayer film.

Despite these advantages, most of the industry uses migratory additives today.

12.1.4 Migratory slip additive performance factors

COF performance depends on the nature of the polymer surface, the amount of additive that migrates to the surface and the properties of the resulting surface (particularly the properties of the amide itself).

12.1.4.1 Polymer surface

The roughness of the polymer surface is shown to be important from a fundamental perspective in Section 12.1.3 [see Eq. (12.9)]: a smooth surface provides greater real contact area between the surfaces, resulting in higher friction. This has been observed in practice:

1. In general, films with higher gloss may require a higher slip concentration or a longer time to achieve the desired COF value [17].
2. In some cases using migrating slip agents in combination with inorganic antiblock additives such as diatomaceous earth which increase surface roughness can improve COF performance, either allowing less slip to be used or achieving faster COF development [17–19].
3. Using a matted chill roll in extrusion coating provides lower COF than a one with a mirror finish; the surface texture with the matted finish contributes to surface roughness [20,21].

Some of the improved performance may be due to the direct effect of increased surface roughness on reducing the real contact area. Slip additive migration may be another factor. There is a thermodynamic driving force for the slip molecule to come to the surface to minimize free energy (the slip molecule lowers the surface energy of the film). This driving force is eliminated if the film surfaces are so tightly bound in the roll that the interface disappears. Roughening the surface helps preserve this interface, facilitating the migration of the slip additive in the wound roll after the film has been produced.

Polymer surface crystallinity is another factor. Sakhalkar and Hirt [22] use electron microscopy to show that the surface region of a Ziegler–Natta (ZN) LLDPE blown film is amorphous. The fatty acid amide is more soluble in this amorphous region than in the bulk, contributing to the accumulation of the slip additive at the surface. The amount and morphology of polymer crystals at the surface is influenced by the polymer chemistry and molecular structure, as well as how the film is made. Sections 7.5.3.2.2 and 10.4.2.2.1 discuss how the molecular architecture of LLDPE affects surface crystallinity. Blown films of metallocene LLDPE, with their uniform comonomer distribution, do not have the surface amorphous region found in ZN LLDPE blown film. The ZN catalyst technology produces polymer chains with a broad MWD and comonomer distribution with the shorter chains having an abundance of comonomer. The comonomer creates short-chain branches that do not participate in the crystalline structure. As crystals form during quenching, these short chains are pushed toward the surface, forming the amorphous surface region. mLLDPE films typically require higher levels of a slip additive than ZN-LLDPE films. The lack of the amorphous surface region to help accumulate the slip additive at the surface is one factor contributing to this.

The film fabrication process can also contribute to differences in surface and bulk crystallinity. Kitchel [23] observes that slip additives in a multilayer LLDPE blown film has a tendency to migrate toward the outside of the bubble; the COF is lower on the outside surface than the inside surface if the slip additive is added to all layers uniformly. No explanation is given, but disparate cooling conditions on each side of the bubble may be contributing to crystallinity differences. In cast film, the polymer melt is quenched against a cold roll. Section 10.3.1 describes how contact with surfaces of different surface energy during quenching can induce changes in surface crystallinity. This suggests a cast film made in contact with a chrome chill roll may have a different surface crystallinity than a blown film whose contact surface is air. However, the reduction in overall crystallinity due to the much faster quench rate in cast film may overwhelm this effect.

12.1.4.2 Migration of slip additives

The thermodynamic driving force for additive migration and accumulation at the surface is threefold [22]. The first is the solubility of the slip additive in the polymer; if the additive is not soluble, it is forced to the surface. It is widely accepted that amides are not soluble in the crystalline regions of polymers. Thus, as the polymer melt crystallizes, the slip additive is forced toward the surface. Sharma and Beard [6] report that after solvent stripping of an amide slip additive from the surface of a PE blown film, no additional slip migration occurs, suggesting that solubility of the amide in the bulk of the PE is low. Work by Hirt and his students at Clemson University, however, suggests that for LLDPE, the solubility limit is high enough in the amorphous regions that solubility is not a significant driving force [24–29].

The solubility of the additive is a function of its chain length, functional groups and saturation, and the chemistry of the polymer. The classes of fatty acid amides are described in Section 12.1.3.1. The primary amides tend to be more polar than the secondary amides and ethylene bisamides and so are less soluble in nonpolar polymers such as polyolefins. Oleamide and erucamide, two primary amides, are the two most commonly used slip additives in polyolefins. But even amongst the polyolefins, there are differences in the migration of these amides. In general, LLDPE requires higher slip levels and more time to reach the ultimate COF than does LDPE. Polypropylene (PP) takes even higher levels of slip to achieve the same COF [3,30–33]. Metallocene plastomers require very high levels of slip additive [34]. At such high levels, heat-seal properties may be affected [35].

For polar ethylene copolymers such as EVA, ethylene methyl acrylate copolymers (EMA), acid copolymers and ionomers, the solubility of oleamide and erucamide is higher, reducing the driving force for migration to the surface. The migration rate in these polymers can be several times slower than for LDPE. For example, erucamide in an EVA containing 3.8% VA migrates three to five times slower than in a polyethylene of the same crystallinity. The interaction of erucamide with acid copolymers and ionomers is too great for effective performance and alternative additives are typically used. Higher levels of slip additive are also required. Even so, the development of low COF in ionomer films can take several days vs hours for LDPE. Resin suppliers can offer advice on the type of slip agent and amount to use for a given resin.

The second driving force for surface accumulation of the slip additive is a reduction of surface energy of the system favoring the formation of slip at the surface. This occurs as the amide forms a polar network structure on the surface.

The third driving force is related to morphological differences between the bulk and surface of the polymer. As described in Section 12.1.4.1, ZN LLDPE blown film has been found to have an amorphous region near the surface. Sakhalkar and Hirt [22] calculate that the greater solubility of the slip additive in the amorphous surface region than in the bulk is a significant driving force for slip accumulation at the surface.

The rate of migration of the slip additive is of particular concern for processors and is a function of the thermodynamic driving force and the diffusion coefficient. Several investigators have found the diffusion of slip molecules in polymers to be Fickian or nearly Fickian, meaning the diffusion coefficient is independent of concentration. This simplifies the analysis considerably. The diffusion coefficient relates the diffusive flux to the concentration gradient, in a relationship known as Fick's first law. This was introduced in Chapter 8, where the diffusive flux is the amount of the permeant transporting through a unit area of film per time:

$$J = -D\frac{dc}{dx} \tag{12.10}$$

where

J = diffusion flux = Q/At,
Q = amount of permeant,
A = area normal to flow, in this case the surface area of film,
t = time,
D = diffusion coefficient, and
dc/dx = concentration gradient in thickness dimension (x).

The larger the value of D, the faster the rate of diffusion.

The diffusion coefficient is a function of the molecular weight of the slip additive, the molecular shape of the additive and the percent crystallinity of the polymer. Higher MW of the additive decreases the diffusion coefficient. Oleamide (C18:1),[1] for example, is known to migrate faster to the surface of LDPE than erucamide (C22:1) [32,36]. Foldes [37] finds that additives with linear structures diffuse more slowly than spherical molecules, but the rate of change with temperature is more pronounced. Some authors state that the higher the percent crystallinity of the polymer, the slower the diffusion since crystals block the path of the slip molecule [38]. Others have found, however, that higher crystallinity (brought about by slower quench rates) increases the diffusion rate [29,39]. Klein and Briscoe [29] propose a theory that explains this behavior in terms of entanglements in the disordered amorphous regions between the

1. The nomenclature CXX:Y refers to XX carbon atoms and Y double bonds. Oleamide has 18 carbon atoms, with one C = C double bond, or level of unsaturation.

polymer crystalline lamellae. Tie molecules and fixed entanglements between adjacent lamellae are less abundant in the higher crystalline polymer, resulting in more slip mobility. This mobility overwhelms the effect of more crystals (fewer diffusion paths) which results in higher diffusion rates. Once again, the polymer molecular architecture (MW, MWD, long-chain branching, short-chain branching, short-chain branching distribution, % crystallinity and crystalline morphology) is seen to influence properties.

Several authors relate the diffusion rate to free volume theory [40–42]. This theory, introduced in Section 8.4.1.2.1, describes temporary "voids" that form in the polymer that allow the slip molecules to pass through. These voids or "free volume" are due to imperfections in the packing of polymer molecules and rotational and vibrational motions. Free volume is influenced by molecular architecture (MWD, long chain branching), presence of bulky side groups, chain rigidity, polarity, crystallinity and orientation. External factors may also influence the free volume. Muire and Hirt [43] show that externally applied pressure (as in a tightly wound roll) reduces the diffusion rate. They attribute this to a reduction in free volume for diffusion. Schumann et al. [44] also show that less slip additive makes it to the surface when the roll tension is higher in a wound roll. Many observers attribute this to a disappearance of the interface, reducing the driving force for accumulating at a surface that no longer exists. It is likely both effects contribute to the decrease in the effectiveness of slip agents in tightly wound rolls.

There are two general methods for measuring the diffusion coefficient of slip additives in polymers. One involves measuring the mass absorption of a film or plaque exposed to an additive at the surface, the so-called "diffusion-in" experiment. A variation of this technique is to measure the diffusion through a stack of films. Several authors describe work with a variety of additives using these techniques [27,29,39–41,45,46]. Gagliardi et al. [28] describe the difficulties with these techniques: they require measurement of long-time equilibrium additive concentrations, they introduce possible handling errors and, specifically for the film stack technique, they introduce a number of polymer–polymer interfaces.

The second technique is the "diffusion-out" experiment, usually done with FTIR microspectroscopy on the cross-section of a film. With this technique, the polymer is mixed with the slip additive and then made into a film, much like in commercial film applications. The slip additive distribution is measured as a function of time. A number of authors describe work using this technique [25,26,28,42,47–51]. It can be challenging, however, to associate a specific IR peak to the slip molecule [49].

In a variation on the diffusion-out experiment, Rawls et al. [47] use a surface washing technique to measure the surface concentration of erucamide on LLDPE films, followed by a bulk extraction to measure the remaining erucamide in the bulk. The technique involves washing the surface for 5–20 seconds in ethyl ether. The washings are analyzed by gas chromatography to determine the amount of erucamide on the surface. The surface concentration is plotted versus wash time and extrapolated back to zero. This provides a value for the surface concentration that has not been affected by the solvent.

The surface concentration of slip additive can be directly related to the COF. Fig. 12.6 provides an example from the work of Ramirez et al. [52]. The concentration of erucamide on the surface of an LLDPE film is measured and plotted against the COF. The COF is found to monotonically decrease until a concentration of around 0.5 μg/cm^2, after which further accumulation of slip at the surface has no effect.

The results of diffusion experiments typically show a partitioning of the slip additive between the bulk and surface. With time, the concentrations reach equilibrium. Table 12.2 shows data from Rawls et al. [47] for the portioning of erucamide in an LLDPE film after 168 hours (7 days) using the surface washing technique. Around 25%–50% of the erucamide ends up on the surface.

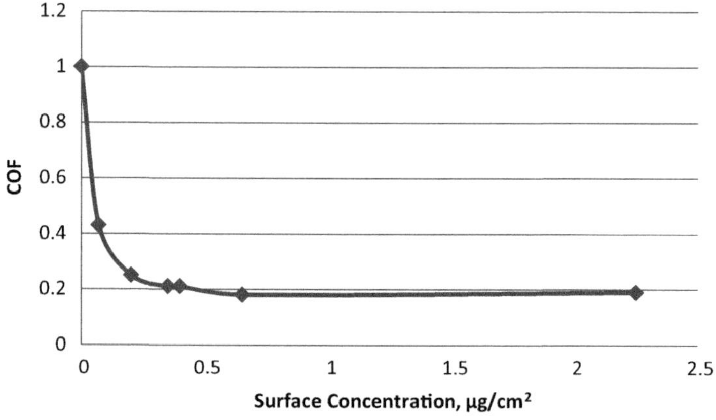

FIGURE 12.6 Correlation of kinetic COF with surface concentration of erucamide on a LLDPE film. *Adapted from Ramirez, M. X., Walters, K. B., Hirt, D. E., 2005, "Relationship between erucamide surface concentration and coefficient of friction of LLDPE film," J. Vinyl & Addit. Technol., pp. 9–12.*

TABLE 12.2 Fraction of erucamide in the bulk and on the surfaces of LLDPE film.

Film thickness (μm)	Wt.% added	Bulk concentration (μm/cm²)	Surface concentration[a], (μm/cm²)	% of total on surfaces (both sides)
27	0.6	3.7	3.8	51
27	1.2	8.3	6.7	45
27	1.8	15.5	6.5	30
27	2.5	22.5	9.5	30
50	1.8	32	10	24

Note: Bulk and surface concentrations are reported as the mass divided by the total surface area (both major faces, ignoring the edges).
[a]*Both surfaces.*
Source: Data from Rawls, A., Hirt, D. E., Havens, M. R., Roberts, W. P., 2002, "Evaluation of surface concentration of erucamide in LLDPE films," J. Vinyl & Addit. Technol., 8, pp. 130–138.

The diffusion and accumulation of slip additives as a function of time are described by the one-dimensional diffusion equation (assuming Fickian diffusion):

$$\frac{\partial C}{\partial t} = D\frac{\partial^2 C}{\partial x^2} \tag{12.11}$$

where
C = concentration of additive in the bulk,
t = time,
D = diffusion coefficient, and
x = spatial dimension normal to the surface of the film.

Eq. (12.11) is subject to the following initial and boundary conditions. See Fig. 12.7 for a diagram of the variables.

Initial condition:
Initially, the concentration of the slip additive is uniform throughout the bulk of the film and zero at the surface:

$$C = C_0 \text{ at t} \leq 0,\ 0 < x < X \tag{12.12}$$

$$\Gamma = 0 \text{ at } x = 0, X,\ t < 0 \tag{12.13}$$

where

C_0 = initial bulk concentration (mass/volume),
Γ = surface concentration (mass/surface area), and
X = film thickness.

Boundary conditions:
At the surfaces, the change in surface concentration is balanced by the diffusive flux. This results in an accumulation of slip additive at the surface [53].

$$\frac{\partial \Gamma}{\partial t} = D\frac{\partial C}{\partial x} \text{ at } x = 0 \tag{12.14}$$

$$\frac{\partial \Gamma}{\partial t} = -D\frac{\partial C}{\partial x} \text{ at } x = \text{X}. \tag{12.15}$$

To solve the equations, the surface concentration must be related to the bulk concentration. The simplest way is to assume a linear relationship:

$$\Gamma = MC = \frac{m}{S_g \rho_s} C \tag{12.16}$$

where

ρ_s = solid density (mass/volume) of polymer,

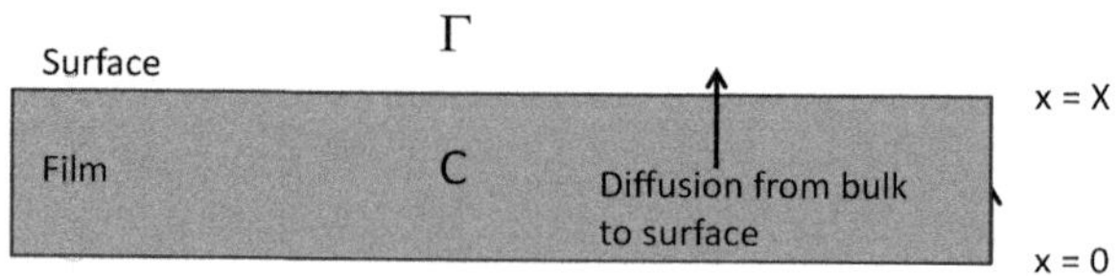

FIGURE 12.7 Diffusion in a film.

S_g = surface area/unit mass of polymer, and
m = partition coefficient of additive between the surface and bulk.

For a smooth surface,

$$S_g \rho_s = \frac{1}{X} \tag{12.17}$$

so that

$$\Gamma = mXC. \tag{12.18}$$

For a rough surface

$$S_g \rho_s = \frac{a}{X}, \text{ where } a > 1. \tag{12.19}$$

In Eq. (12.19), the parameter "a" is a roughness factor. Note that Eq. (12.17) and (12.19) assume diffusion to a single side of the film. If both sides are condidered, X should be replaced by 0.5X.

These equations can be solved numerically for the concentration at the surface and in the bulk as a function of time and location within the film. The input parameters are:

1. C_0 (g/cm^3): initial bulk concentration of slip additive,
2. X (μm): film thickness,
3. D (cm^2/second): diffusion coefficient,
4. m: partition coefficient,
5. a: roughness coefficient.

Joshi and Hirt [53] solve these equations using a finite difference method and get reasonable agreement between the model predictions and experimental results for the partitioning of the slip between the bulk and surface. Building upon Joshi and Hirt, the DuPont (now Dow) finite-difference model developed for heat transfer (see Section 12.7.6.1) was repurposed to handle diffusion; the basic equations are the same and only the boundary conditions must be modified. Fig. 12.8 shows the effect of varying the diffusion coefficient: higher D at constant m results in the surface reaching its equilibrium surface concentration faster. Fig. 12.9 shows the effect of increasing m at constant D. The equilibrium surface concentration increases. Fig. 12.10 illustrates the effect of increasing the film thickness at constant D and m. The equilibrium surface concentration is proportional to the film thickness.

The utility of the model is that it allows one to predict the level of slip accumulating at the surface as a function of time, film thickness and initial slip concentration. Presumably, one then could correlate slip surface concentration with COF, using empirical relationships such as the example in Fig. 12.6. While the model provides insight into slip migration and COF development, it has limitations for predicting actual COF in packaging films. Some of the challenges that still need to be resolved include:

1. Measurements of D and m are tedious and influenced by a number of factors, such as storage temperature, polymer chemistry, and film fabrication process.
2. The correlation of COF with slip additive surface concentration will differ with each polymer and slip additive.
3. The effect of surface roughness has not been explicitly considered. Subtle changes in process conditions may change the surface roughness and completely change the COF versus surface concentration relationship.
4. The effects of coextrusion have not been worked out. For example, what is the proper boundary condition for interfaces between dissimilar materials? Do slip molecules accumulate near the interface?

12.1.4.3 *Amide structure*

The structure of the amide on the surface determines the ultimate COF. Allen et al. [54] use electron microscopy to directly image the formation of regions of slip additive accumulation on the surface of LDPE with time and correlate

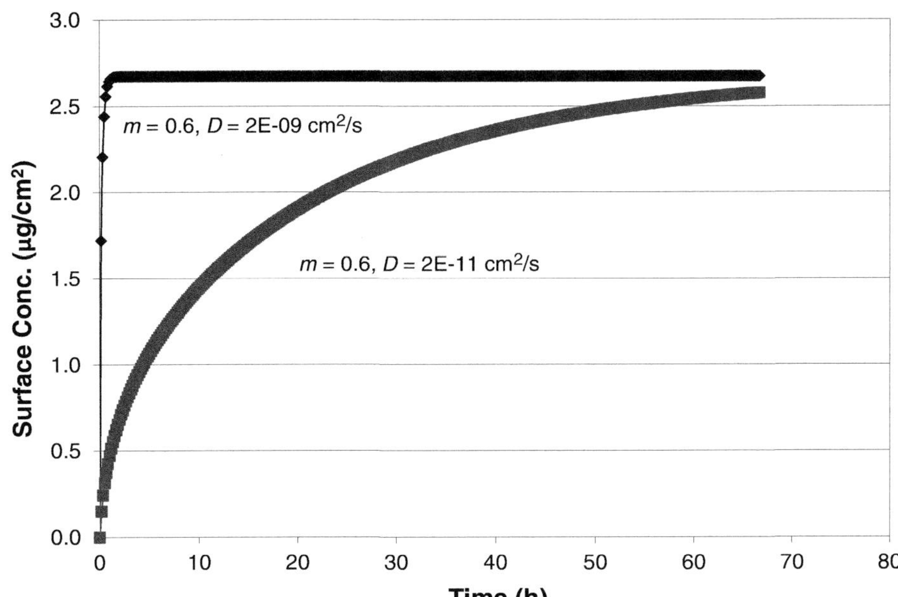

FIGURE 12.8 Model calculation of the migration of slip molecules to the surface of a 50 μm film. The partition coefficient (m) and initial additive concentration (C_0) are held constant. *Reproduced with permission from The Dow Chemical Company.*

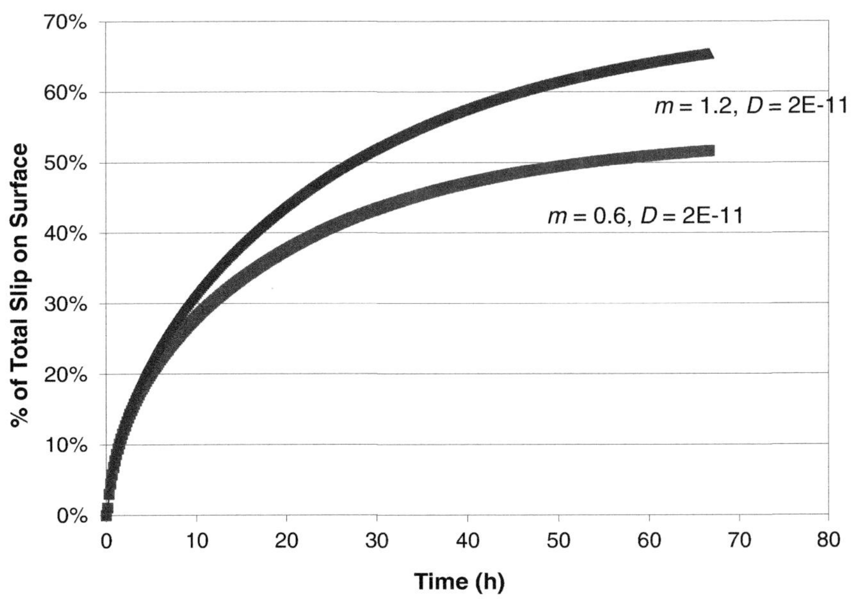

FIGURE 12.9 Model calculation of the migration of slip molecules to the surface of a 50 μm film. The diffusion coefficient (D) and initial additive concentration (C_0) are held constant. *Reproduced with permission from The Dow Chemical Company.*

the observations with previous measurements of COF. Initially, very small discrete concentrations of slip additive appear on the surface. With time, the number and size of the regions increase. The regions eventually coalesce and form a more or less continuous lubricating layer. The concentration at which coalescence occurs corresponds to the onset of low COF. Owens [5] observes the crystalline structure of fatty acid amides that migrate to the surface of PVDC coated cellophane. Ramirez et al. [52,55] use atomic force microscopy (AFM) to probe the surface of LLDPE films after erucamide has bloomed to the surface. They find the surface may not be completely covered even at high concentrations of additive. The AFM study shows evidence of crystals and discreet patches on the surface. Poison et al. [56] report similar behavior for PE coextruded films.

Chen et al. [57] study the concentration and morphology of erucamide on the surface of monolayer PE blown film using X-ray photoelectron spectroscopy (XPS), AFM and time-of-flight secondary ion mass spectroscopy (ToF SIMS). The XPS spectra of the surface of the films are nearly identical to pure solvent cast erucamide. They conclude that at equilibrium the film is fully covered. They follow the rate of diffusion of the erucamide to the surface using contact

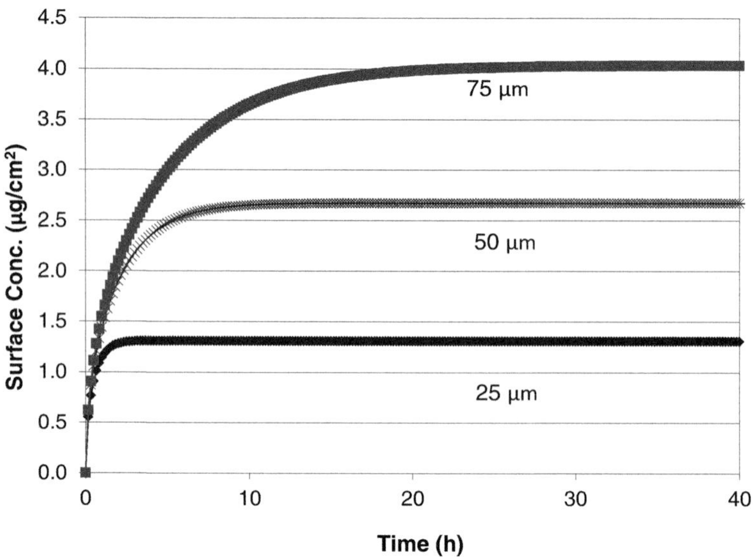

FIGURE 12.10 Model calculation of the migration of slip molecules to the surface of a film as a function of film thickness. $m = 0.6$ and $D = 2 \times 10^{-10}$ cm^2. *Reproduced with permission from The Dow Chemical Company.*

angle measurements and XPS. Both show near-complete coverage within about 50 minutes. The initial diffusion rate is Fickian, although there is an initial delay due to the surface washing technique employed. The migration rate depends on the PE grade, but no details are given. AFM shows erucamide crystal growth at the surface starting at about 20 minutes with complete coverage by about 40 minutes, corresponding to the XPS and contact angle results. Initiation of crystal formation occurs for an LDPE sample when 80% of the film is theoretically covered by a single layer of erucamide. Crystals then grow vertically and reach equilibrium in 150 minutes. The ToF SIMS results indicate a fairly even distribution of erucamide with local unevenness likely due to patchy crystalline growth.

In a later paper, Chen et al. [7] examine LLDPE film with erucamide as the slip additive stored at RT and at 60°C. Initially, the samples aged at RT have complete coverage of erucamide with crystal heights of around 10 nm. After greater than 1 year of RT aging, large crystals with heights of around 100 nm have formed, with bare PE in between; islands of erucamide crystals on top of the bare PE have formed.

Sharma and Beard [6] use XPS to characterize the surface structure of LDPE blown films containing stearamide or oleamide. The oleamide is found to have a complete structure or coverage, whereas the stearamide only partially covers the surface. This is consistent with the performance of the two additives: for oleamide, the complete layer corresponds to good surface lubrication and slip properties. For stearamide, the partial layer that forms is sufficient to reduce large-scale surface interactions and, hence, reduce blocking. Ramirez et al. [55] use AFM to study erucamide and behenamide on the surface LLDPE films. A film containing 1000 ppm erucamide shows incomplete coverage with a COF of 0.2 whereas a film with a similar amount of behenamide exhibits surface saturation but considerably higher COF. AFM shows distinct differences in crystalline morphology with the erucamide having a plate-like structure. Maltby [3] uses molecular modeling to show why some fatty amides are better than others for slip performance. The very regular crystalline structure of behenamide may inhibit its slip performance. Partial disruption of the crystalline structure, as occurs with mono unsaturated erucamide, may be responsible for lower COF performance.

Since the structure of the amide dictates the COF value once enough slip has migrated to the surface, it can be very challenging to obtain a medium value of COF with a low COF amide like erucamide. If mid-range values of COF are needed for an application, the best approach may be to switch to another type of fatty acid amide that provides such performance.

12.1.4.4 Film fabrication

Fatty acid amides are susceptible to degradation by oxidation, especially the unsaturated types. The high temperatures of some converting processes, such as extrusion coating and cast film, may affect the performance of these additives. Fatty acid amides generally melt in the range of 80°C–110°C and can become volatile at polymer processing temperatures. This can result in the loss of additive.

As described in Section 12.1.4.1, the film fabrication process can affect the roughness of the surface. A matted chill roll will typically provide lower COF than a polished chill roll. Conditions favoring high gloss in blown film may negatively affect COF performance.

The rate of quenching affects the amount of crystallinity. Fast quench processes such as cast film and water-quench blown film may have lower levels of crystallinity than the same structure made on a conventional air-quench blown film process. Higher levels of crystallinity affect the solubility of the slip additive. Similarly, orientation during film making may induce crystallization or change the crystalline morphology (see Chapter 11), which can influence slip additive solubility and diffusion.

The tension on the roll during wind-up affects slip migration. Muire and Hirt [43] attribute this to a reduction in polymer-free volume, which hinders slip migration. Schumann et al. [44] attribute this to a loss of a well-defined interface between film layers in the roll, reducing the driving force for surface accumulation.

12.1.4.5 Film thickness and additive level

For a monolayer film, decreasing the film thickness while keeping the initial additive concentration constant decreases the reservoir of additive that can migrate to the surface. At equilibrium or long times, the amount of slip that accumulates on the surface increases in proportion to the film thickness. For example, if the film thickness is halved, the weight percent of slip that needs to be added to achieve the same COF is doubled. This result arises from a simple mass balance (see Fig. 12.11):

Initial Mass = (Mass on Surface + Mass in Bulk)$_{\text{at equilibrium}}$

$$C_0 V = 2\Gamma_\infty A + C_\infty V \tag{12.20}$$

where

C_0 = initial additive concentration (mass/volume),
V = volume ($=XA$),
X = film thickness,
A = film surface area (one side; total surface area is 2A),
Γ_∞ = surface concentration of additive (mass/surface area) at equilibrium, and
C_∞ = additive concentration in bulk at equilibrium (mass/volume).

Eq. 12.18 shows the relationship between Γ and C for a single surface of area A. Since both sides of the film are considered here, the surface area to volume ratio ($S_g\rho$) is 2/X and the expression for Γ_∞ is reduced by 1/2 (see Fig. 12.11). Substituting this expression into Eq. (12.20):

$$C_0 XA = mC_\infty XA + C_\infty XA \tag{12.21}$$

By rearranging Eq. (12.21), the relationship between the initial and equilibrium concentration in the bulk is found to be simply related to the partition coefficient:

$$C_\infty = C_0/(1+m). \tag{12.22}$$

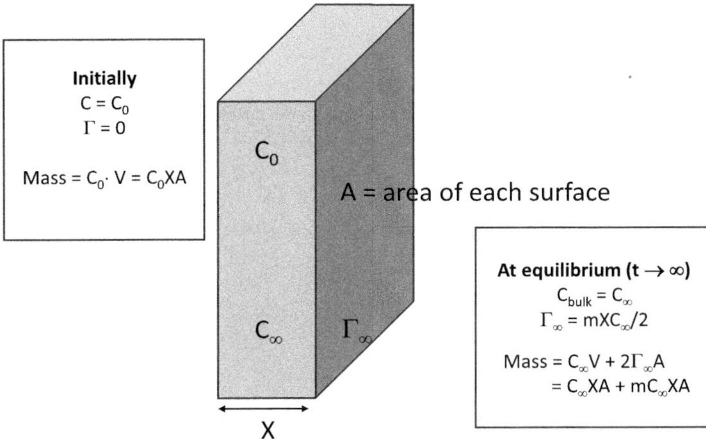

FIGURE 12.11 Mass balance for slip additive redistribution between bulk and surface over time. Conservation of mass is assumed. The analysis is for monolayer film with equal accumulation at both surfaces with total surface area 2A.

The equilibrium surface concentration is found by substituting Eq. (12.22) into Eq. (12.18):

$$\Gamma_\infty = \left[\frac{m}{2(1+m)}\right] XC_0. \tag{12.23}$$

If the partition coefficient, m, is assumed to be constant for a given film type and additive, then

$$\Gamma_\infty \propto XC_0. \tag{12.24}$$

Eq. (12.24) shows that the surface concentration increases with increasing thickness and initial additive concentration. Since the COF is determined by the surface concentration of the additive, two films with the same surface concentration (Γ_∞) should have the same COF. For equal surface concentration, Eq. (12.24) becomes:

$$X_1 C_{0,1} = X_2 C_{0,2}. \tag{12.25}$$

Eq. (12.25) can be rearranged to calculate the amount of additive ($C_{0,1}$) that should be put into a film of a new thickness (X_2) to achieve the same COF as a reference film of a known thickness (X_1) and additive level ($C_{0,1}$):

$$C_{0,2} = C_{0,1}(X_1/X_2). \tag{12.26}$$

The utility of Eq. (12.26) is illustrated in Fig. 12.12. Here the COF is plotted for a series of films of different thicknesses made from three LDPE resins. The resins have the same density (0.922 g/cc) and three levels of slip additive (C_0). The top set of data (290 ppm) was fitted to a curve and used as a reference. Once the reference curve is known (film 1), the COF of any other film (film 2) with a different concentration of additive can be calculated. Eq. (12.26) is rearranged to compute an "effective" thickness, X_1:

$$X_1 = X_2(C_{0,2}/C_{0,1}) \tag{12.27}$$

X_1 represents the thickness on the reference curve that has the same COF as the point on the film 2 curve (X_2, $C_{0,2}$). The COF for film 2 is read off the reference curve at thickness X_1. The two dashed curves for the 625 and 928 ppm data points are computed in this way. The agreement with the actual measured COF is excellent.

Several investigators [47,58] have reported experimental data for the effect of film or coating thickness on COF that agree with Eq. 12.26. The relationship, however, does not always hold true in practice. For example, Edwards and Foster [59] analyze COF measurements from extrusion coating of LDPE onto kraft paper and find the COF was inversely proportional to the thickness and the square of the initial slip concentration: $C_{0,2} = C_{0,1}(X_1/X_2)^{0.5}$.

As described in Section 12.1.4.2, once the surface concentration of additive reaches a critical value, the COF reaches its ultimate value and further increases in surface slip concentration have no effect on the COF. Increasing the initial additive concentration, however, may help speed up the time for the film to reach this ultimate value. Adding too much slip additive, though, can cause problems, which are discussed in Section 12.1.4.9.

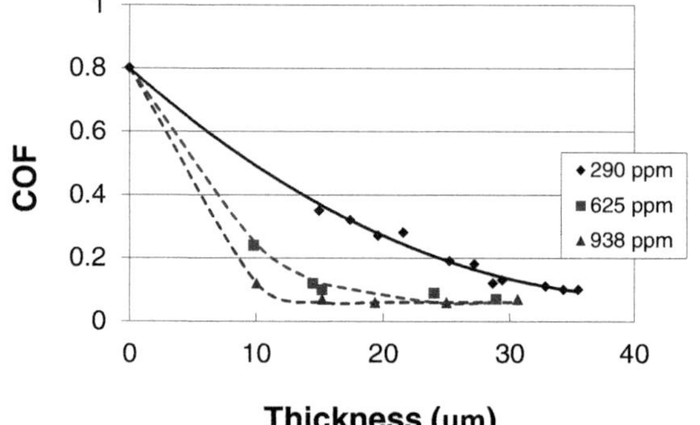

FIGURE 12.12 Effect of film thickness and slip additive level on COF of LDPE blown film with three levels of slip additive. Sold curve is a curve fit of the data points. Dashed curves were computed using Eq. (12.27). *Data courtesy of The Dow Chemical Company.*

12.1.4.6 Storage temperature

Many have observed that film stored at elevated temperatures, such as might be encountered in a warehouse during the summer months, has higher COF. One polymer supplier noted that they get more complaints on COF in the summer than at any other time of the year. Several explanations for this have been investigated:

1. A widely expressed rationale for this behavior is that the slip additive becomes more soluble in the film at elevated temperatures. As a result, some of the slip additive migrates from the surface back into the film, increasing the COF. Soutar [19] describes an incident where the COF is recovered by placing the film in cold storage for 10 days. Shuler et al. [60] investigate the fate of erucamide on the surface of trilayer LLDPE films with a plastomer skin layer during storage. Storing the films at 55°C for several days results in a significant reduction in the erucamide surface concentration as measured by both FTIR and surface washing techniques. After investigating several of the possibilities mentioned here, they conclude that migration back into the PE is the most reasonable explanation. Interestingly, the COF in these experiments actually decreases, which Shuler et al. hypothesize is due to a change in the crystalline morphology of the erucamide.
2. Rawls et al. [47] suggest slip additives may evaporate when films are stored at elevated temperatures. They claim LLDPE film stored at 60°C shows significant loss of erucamide due to sublimation or evaporation, even though the melting point is around 80°C. About 10% is lost over a 7-day period whereas substantially none is lost at RT. Shuler et al. [60], however, conclude that evaporation loss of samples stored at 55°C is insignificant. Chen et al. [7] draw a similar conclusion that evaporative loss of erucamide in their LLDPE films at 60°C storage conditions is negligible.
3. Chemical changes in the slip additive on the surface or in the bulk are investigated by Shuler et al. They conclude that a small portion of the erucamide on the surface of their films degraded and outline the potential chemical pathways. Chen et al. [57] find evidence in XPS measurements of some chemical change in erucamide at the surface of PE films aged at 60°C and 200°C for 15 hours and 15 minutes, respectively. The effect is more pronounced for the samples exposed to 200°C. In a later paper, Chen et al. [7], downplay the decomposition mechanism and say the previous study shows the decomposition of erucamide at 60°C is negligible. Sharma et al. [61] also find no significant evidence of oxidation at storage temperatures up to 70°C in their study of erucamide in PE film.
4. Chen et al. [7] investigate the concentration and morphology of erucamide on the surface of LLDPE stored at RT for up to a year, and at 60°C in a nitrogen environment for 72 hours. XPS is used to measure the surface concentration. The morphology is measured with AFM. Initially, the samples aged at RT have complete coverage of erucamide with crystal heights of around 10 nm. After greater than 1 year of RT aging, large islands of crystals with heights of around 100 nm have formed on top of the bare PE. After annealing at 60°C for 72 hours, the large crystals have melted. The erucamide does not appear to recrystallize on the PE surface (even after remaining at RT for 2 weeks). XPS shows a substantial increase in erucamide on the surface following the annealing. They postulate that the annealing unlocks the migration of erucamide in the bulk of the PE that had been trapped by the PE crystals.
5. Sharma et al. [61] use attenuated total reflectance infrared spectroscopy (ATR-IR), XPS and AFM to investigate the mechanisms behind the increase in COF of a PE film containing erucamide upon exposure to 45°C, 60°C, and 75°C for up to 2 months. The COF of films stored at 60°C and 75°C increases substantially whereas those stored at 45°C less so. This corresponds to changes in erucamide surface concentration: the films exposed to 60°C and 70°C show a significant reduction but not those stored at 45°C. Also, when cooled to RT, the erucamide surface morphology for the films stored at 60°C and 75°C changes from stacked crystals to bilayers. The morphology remains as stacked crystals for films stored at 45°C after cooling. They also note that the crystallinity of the polyethylene increases at the storage conditions but are not sure if this is related to the change in COF.
6. Dulal et al. [62] find the ease of opening (torque) HDPE screw-caps is a function of storage temperature, storage time and erucamide concentration. Morphology of the erucamide at the surface (AFM measurements) is found to change with storage time, with flat placoid-like structures at 38°C and raised compositions at 50°C with some areas of the surface uncovered. They propose that these loosely adhered structures are responsible for lubricity. Surface concentration was measured using GC-FID.

Based on these investigations, the change in COF during storage at elevated temperatures is complex and still not completely understood. For the practitioner, the important thing to note is that the COF may change during storage.

12.1.4.7 Effect of other additives

Slip additives are often not the only additive used in the film. Other additives are described in Chapter 4 and include antioxidants, antiblocks, antifog agents, chill roll releases and antistat agents. Some of these, such as antifog, chill roll release and antistat additives, may bloom to the surface much like slip additives. The slip additive must compete with the other additives for surface coverage. Reinecke and Navarro [63], for example, study the migration of erucamide and co-additives in a three-layer film with the structure (PE—PE—PE), where the additives are all added to the third layer. The co-additives are ethylene-oxide oligomers. The hydrophilic nature of the co-additive impedes the migration of erucamide to the surface. The MW and number of polar groups of the co-additive are important factors.

Inorganic antiblock additives are often used in films with marginal blocking performance. Soutar [17] and Zahedi et al. [18] cite several studies showing synergy between adding silica and amide slip agents; adding silica not only reduces blocking but allows less slip additive to be added to achieve the target COF. Radosta and Riley [64] show similar results for treated talc. Many suppliers of resins requiring low COF provide grades with both slip and antiblock additives.

Some types of additives, such as calcium carbonate, have been found to absorb slip agents, increasing COF [36].

High levels of silica and other inorganic additives can result in high haze. This can be alleviated by confining the inorganic to a thin skin layer. Inorganic additives have also been reported to react with some slip agents, resulting in degradation byproducts with color or odor/taste issues [65].

Llop et al. [66] examine ways to control the migration of erucamide in PE films by adding coagents or plasma surface treatments. ATR-IR measurements are used to measure the amount of erucamide and coagents at the surface. They find a block copolymer of polyethylene and polyethylene oxide is the most efficient coagent to reduce the migration of the erucamide to the surface. Corona treatment is also found to reduce the erucamide concentration at the treated side of the film. They use polarity to explain the results: the polar coagent helps increase the solubility of the erucamide in the polyethylene and the treatment creates a polar surface that diminishes the driving force for diffusion.

Hoang et al. [67] look at interactive effects among phenolic and phosphite antioxidants, calcium stearate antacid, a silica antiblock agent and erucamide slip additive on the process stability, color formation and long-term stability of a metallocene LLDPE blown film. Erucamide is found to have a positive effect on process stability that the authors hypothesize is due to a lubrication effect. The experiments also show a yellowing effect from the interaction between the silica and erucamide.

12.1.4.8 Coextrusion and multilayer structures

COF development in multilayer films containing migrating slip additives can be complex. The additives can diffuse into adjacent layers, or be transferred to the back side of films while wound in the roll. This leaves less of the additive for accumulating onto the desired low COF surface. The amount of adjacent layer migration is influenced by the solubility of the additive in the layers; fatty acid amide slip additives are more soluble in polar polymers such as tie layers and ionomers than PE. Further complicating matters is the effect of temperature on amide solubility and diffusion. In general, slip additives placed in one layer are not likely to stay in that layer. This needs to be accounted for when deciding how much additive to use and in which layers to place it.

Sankhe and Hirt [68] measure the erucamide concentration across a 50-μm bilayer film of LLDPE. The structure is (25-μm LLDPE—25-μm LLDPE); only one layer contains erucamide (at 1% loading). The erucamide concentration profile across the thickness of the film is measured using synchrotron-based FTIR micro spectroscopy in order to get sufficient spacial resolution for the films. Previous work [69] measuring erucamide profiles had been restricted to thick films (600 μm) due to the limited resolution of conventional FTIR spectroscopy. In the previous work, significant concentration gradients were found when the slip additive was added to only one layer compared with adding equal amounts of slip to all layers. In the thin-film work, they find that the concentration is nearly uniform across the entire thickness of the bilayer film, even 1 day after fabrication. They postulate that the slip migrates into the adjacent layer during coextrusion; the characteristic diffusion time increases with the square of the film thickness so it is not surprising that the concentration evens out quickly in thin films. The surface concentration of erucamide is measured using ATR-IR. Both sides of the bilayer film see an almost equal rise in surface concentration with storage time (up to 3 days) that coincides with a reduction in the bulk concentration. A follow-up study [70] replaces one of the LLDPE layers with a plastomer. Again, the erucamide concentration equilibrates across the thickness of the film even if all the erucamide is added to just one layer. ATR-IR reveals a greater concentration of erucamide on the plastomer surface than on the LLDPE surface.

Kitchel [23] tests a series of three-layer LDPE blown films with 500 ppm erucamide added to different layers. The total film thickness is 50 μm with layer percentages of (20/60/20). The COF is measured on the front and back sides of the film with the goal to maximize the difference. The greatest differential in COF is obtained when all the slip is added to the outside skin layer (with respect to the blown film bubble). In a second experiment, the overall film thickness varied from 25 to 150 μm. Increasing the thickness helps increase the COF differential.

Glover [58] extrusion coats LDPE containing a fatty acid amide slip additive onto two different substrates, kraft paper and aluminum foil. The COF of the coating is higher for the paper coatings, which he attributes to the porous paper absorbing some of the slip additives, leaving less to migrate to the surface of the LDPE.

Bryant [71] measures the COF as a function of time on a series of three-layer blown films containing different layer combinations of LLDPE and a sodium ionomer. Each layer is 25 μm thick for a total film thickness of 75 μm. A fatty acid amide is placed in one or more of the layers to study the effect of migration between the adjacent polar (ionomer) and nonpolar (LLDPE) layers. Higher levels of slip additive are needed to achieve low COF when at least one of the layers is an ionomer, reflecting the greater solubility of the amide in the ionomer. Interestingly, lower COF is obtained when the slip additive is added to the ionomer layer than when only added to an LLDPE layer. The presence of a polar layer also affects the response of the COF with increasing storage temperature (from 23°C to 38°C). The COF changes with temperature are more pronounced with the nonpolar LLDPE film than with those containing ionomer. Bryant attributes this to differences in solubility.

Soutar [19] provides a number of anecdotes regarding migrating slip additive in multilayer films. Tie resins in particular are shown to have an effect on COF, presumably due to their polar nature. He hypothesizes that the chemical functionality of the tie resin attracts the amide molecules. In one example, the layers are peeled apart and no accumulation of amide is found at the interface between the tie and PE layers. This suggests the slip additive resides in the bulk of the tie layer.

Soutar also discusses issues with the backside transfer of amides or other additives from the front surface of the film to the adjacent back surface of the film wound in a roll. This has led to contamination issues involving poor printing or heat sealing.

Poisson et al. [56] examine three-layer barrier blown films with the structure (PA 6,66–tie–PE). The tie is a maleic anhydride-modified PE. To this base case structure, several variations are considered by adding 50% by weight EVA (VA = 9 wt.%) to the tie layer, PE sealant or both. The layer thicknesses are (30/10/50 μm). Erucamide or oelamide is added to the sealant layer at concentrations ranging from 0–2500 ppm. AFM and FTIR are used to measure the slip concentration profile across the thickness of the film and the surface concentration. IR could not be used to measure the slip level in the PA layer since the IR peaks of the PA and amide overlap. The amount of slip additive in the PA layer is estimated by subtracting the measured amounts on the surface, in the sealant and in the tie layer from the total amount added.

Erucamide gives the best COF performance of the two slip additives tested. The oleamide shows some surprising COF results, such as increasing initially with time, which may be due to the polarity of the EVA. The COF with erucamide reaches the 0.2 goal quickly, even at 500–ppm concentration. Heat seal performance is not affected until high levels of erucamide are added; at 2500–ppm the heat seal initiation temperature increases by 30°C.

In Poisson's experiments with erucamide most of the slip ends up near the surface after the samples are aged for 1000 hours. COF correlates with the fraction of erucamide on the surface. The presence of EVA increases the COF at all levels of erucamide, but the difference is not substantial and is overcome by adding a small amount of additional erucamide. With oleamide not as much slip ends up near the surface, and surface concentration does not correlate with COF.

Poisson et al. present data on how much slip additive ends up in each layer. When EVA is added to the tie or PE layer, this substantially affects the concentration of slip additive in each of the layers and at the surface. A significant amount of slip additive is found in the polar tie layer and PA layers, although the amount reported for the PA layer seems unrealistic. It is likely this is due to the measurement technique. The FTIR may lack the resolution to accurately measure the concentration in the thin PE and tie layers. This error in turn is compounded by the subtraction method used to arrive at the PA layer concentration. Nevertheless, the tie layer measurements do confirm what others have surmised about the tie layer absorbing slip agents.

As the complexity of coextruded film structures increases, a better understanding of slip additive migration between layers and to the surface will be needed.

The effect of lamination adhesives should also not be ignored. The curing process and the chemistry can influence the COF performance of the laminated structure. Polyether based adhesives typically have a more negative impact than polyester-based adhesives.

12.1.4.9 Problems with too much slip additive

Too much fatty acid amide slip accumulating at the surface can have consequences, such as:

1. The film may take on a chalky or hazy appearance.

2. The film may become difficult to adhere to or print.
3. The slip additive may interfere with corona treatment. Treatment often attracts more slip additive to the surface (due to its polarity), causing COF to decrease, but contributing to the loss of treatment and adhesion issues.
4. Odor issues from degradation byproducts may increase due to the higher concentration.
5. In some cases, high slip additive leads to poor heat seal performance [35,48,56]. This is particularly true of grades of PE where high levels of slip additive are needed, such as for metallocene plastomers [35]. For example, Wooster and Simmons [35] find seal initiation temperatures increases by as much as 11°C for a metallocene plastomer blown film. More polar polymers, such as ionomers, have been found to tolerate high levels of slip without negative effects on heat seal [17,71].

In some polymers, especially polar ones, the slip additive can take a long time to come to the surface. It is tempting to add higher levels of additive; this may reduce the time to reach the desired COF but may also bring on problems over time as more slip additive comes to the surface.

12.1.4.10 Tips and classic references

Some practical tips for dealing with COF issues include:

1. Work with your resin supplier to find the best additive and level for your application. Many resin suppliers have grades with slip and antiblock additives already incorporated, or sell or can recommend masterbatches for this purpose.
2. Trying to control the COF to a medium level with a low-COF migrating type additive can be very challenging. The slope of the COF versus slip additive concentration curve is very steep at low concentrations, as illustrated in Fig. 12.6. It is difficult to control the amount of slip on the surface in this region. Once a sufficient amount of the slip additive has reached the surface, the COF is determined primarily by the crystalline structure of the amide. Consider switching to another amide that provides the intermediate COF at equilibrium.
3. Use only as much slip agent as needed—too much can lead to adhesion and printing problems.
4. Combining an inorganic antiblock and a fatty acid amide type slip agent may provide the targeted COF at lower loadings. Some combinations may cause color or odor issues, however.
5. Extruding at high temperatures can volatilize or degrade fatty acid amide slip additives. Select additives with the appropriate thermal stability when using high-temperature processes like extrusion coating.
6. Increasing the film surface roughness can improve the effectiveness of migrating slip additives. Reducing the film gloss or the smoothness of the chill roll may help reduce COF.
7. Storing film at elevated temperatures, such as in a warehouse, can cause the COF to increase. Consider controlling the temperature swings the film may experience during storage and transport, particularly during the summer months.
8. Fatty acid amid slip additives can migrate between layers in coextruded films, or be transferred from the backside of adjacent layers in a wound roll. Consider adding slip additives to more than one layer if migration is an issue.
9. Consider using nonmigrating slip additives if changes in COF due to coextrusion or storage, or adhesion performance are issues with migrating additives.
10. Polar polymers such as EVA or ionomers require higher levels of fatty acid amide slip additives and the COF takes longer to develop than for PE.
11. When selecting a masterbatch, make sure the carrier resin is compatible with the polymer it is being blended into. For example, if the polymer is LLDPE, use a carrier that is PE or EMA based. If the polymer is an ionomer, use a carrier that is an acid copolymer based. Poor compatibility can lead to heat seal problems and increase haze.

Slip additive concentrations are typically reported in the units of ppm (parts of slip additive per million parts of polymer). To convert to wt%, use the conversion 1000 ppm = 0.1 wt.% (10,000 ppm = 1 wt.%). Most of the time the converter will be adding slip to the polymer using a masterbatch. To calculate the wt.% masterbatch to add, Eq. (12.28) can be used:

$$\text{wt\% MB} = \frac{\text{Goal Slip Concentration [ppm]}}{100 \times \text{wt\% Slip Additive in MB}} \tag{12.28}$$

For example, if the goal is to add 1000 ppm of slip additive into LDPE, and if the masterbatch contains 10% slip additive, then the amount of masterbatch to add is 1000/(100 × 10) = 1%.

Fatty acid amide technology for lowering the COF of polyolefin and other plastic films has been around for over 50 years. There are a number of classic references that provide more detailed examples and practical observations of performance, many of which have been cited in this chapter. These include:

1. Owens [5,8,72]: In a series of papers, Owens describes work done at DuPont on the use and understanding of fatty acid amide slip agents in PVDC coated cellophane. Many of the observations and analysis are relevant to any plastic film.
2. Tabor et al. [54,73]: Reviews early work on slip additives and describes performance of oleamide and stearamide in polyethylene.
3. Thompson [36]: Provides a good overview of COF control for PE films.
4. Itzkoff and Woodell [21]: Describes the effect of extrusion coating processing parameters on COF.
5. Edwards and Foster [59]: Provides some simple calculations to predict COF performance in extrusion coating.
6. Glover [58]: Good overview of slip additive technology in extrusion coating of LDPE.
7. Soutar [19]: Provides examples of the complexity of COF in multilayer films and the effect of storage temperature.
8. Breuer et al. [30,31]: Good introduction to fatty acid amides for polyolefin films.
9. Maltby, et al. [3,10,32,34,74]: Good reading on slip agents for polyolefin films.
10. Hirt et al. [22,24–28,39,43,47,49,52,53,60,68,69]: Sound fundamental work on migration of erucamide in LLDPE film.
11. Dow Chemical studies on erucamide migration, morphology and aging [7,57,61].

12.2 Optical properties

The ability to see through a film and the sparkle or gloss reflected from a film's surface are part of the set of properties collectively known as optical properties. This chapter deals primarily with high clarity films, which are important in many packaging applications where the product is meant to be seen. Some applications call for a light barrier, which is usually achieved by incorporating an opaque layer (such as aluminum foil) into the structure, metallization of a surface of the film or incorporating pigments. Light barrier technology is reviewed in Chapter 8. Likewise, printing, glossy coatings and pigmented films are not covered in this chapter. Optical films such as polarizers, reflectors, diffusers, and light filters generally fall outside the realm of packaging; however, one of the techniques used to make optical filters is briefly described at the end of this chapter as an example of where technology from other markets may bring new ideas to packaging.

12.2.1 Definitions and measurement

The scattering of light as it passes through an essentially transparent film results in a loss of contrast (related to haze) and definition of details (clarity). Haze refers to the scattering of light at wide angles (2.5–90 degrees) whereas clarity is the scattering at narrow angles (0–0.1 degree), as shown in Fig. 12.13. Clarity is distance-dependent and is sometimes known as "see-through clarity," referring to the ability of a person to see far objects through a hand-held sample of the film. ASTM D1746 [75] provides a method for measuring the clarity of a plastic film. The sensitivity of the measurement decreases as the angle between the incident light beam and detector decreases. To provide accurate results, the method calls for using a device that controls this angle to within 0.1 degrees. The results are reported as a percent transmittance on a scale of 0–100.

The haze of a film is often observed as "smokiness" in the film. ASTM D1003 [76] defines haze as the percent of transmitted light that is scattered so that it deviates from its intended direction by more than a specified angle (2.5 degrees in the standard). This method is for low-haze films. If the haze is greater than 30%, ASTM E2387 or D1044

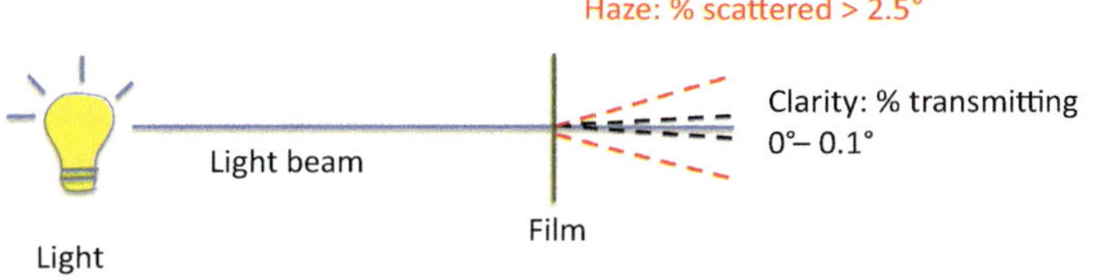

FIGURE 12.13 Concept of clarity and haze of films.

may be more appropriate. An equivalent method is ISO 14782 [77]. The ASTM method has two procedures, one using a haze meter and the other a spectrophotometer. Results are reported as percent transmittance that has been scattered.

Haze can originate in the bulk of the film or from surface roughness. To help separate the two effects, some investigators mask the surface of the film by applying oil, such as mineral oil or silicone oil, which has a refractive index (RI) similar to the plastic film. This results in a measurement of just the internal haze. The surface haze is then calculated by subtracting the internal haze from the total haze (measured on the specimen with no oil on the surface).

Gloss is a measure of surface reflectance or the shininess of the surface when observed at a specified angle. It is strongly influenced by surface roughness. A glossmeter (see ASTM D2457 [78]) measures the specular reflectance of white, unpolarized light at a specific angle from the incident beam. Several angles are used for gloss measurements, with the most common for films being

1. 20 degrees: high gloss surface,
2. 45 degrees: medium gloss surface.
3. 60 degrees: medium gloss surface.
4. 85 degrees: low gloss surface.

For transparent packaging films, 20 degrees gloss is the most often used. Results are reported as the specular reflectance, in percent. Because transparent films have two reflecting surfaces, gloss values can exceed 100%.

12.2.2 Haze and transparency

In general haze and transparency are inversely related—an increase in haze reduces transparency. Most studies have focused on the haze of film, especially on the haze of polyolefin films.

Haze in a polymer film arises from interior or bulk effects (internal haze) and surface roughness (surface haze). Crystallinity, gel, voids, phase separation of blends and various additives or fillers increase internal haze by providing sites for light to refract and scatter. In general, any disparity greater than the wavelength of light (400–700 nm) and a RI different from the polymer will refract light and contribute to internal haze. Surface haze is caused by surface roughness. Crystallinity, certain additives and flow instabilities are known to contribute to surface roughness and haze. The film fabrication process may affect both internal and surface haze.

12.2.2.1 Monolayer polyethylene film

Studies have conclusively shown that the haze of monolayer PE blown and cast films are is dominated by surface haze. Huck and Clegg [79] propose that the surface roughness of PE film is due to two mechanisms: extrusion haze due to flow instabilities such as sharkskin and crystallization haze from strain-induced crystallization near the surface. Many investigators since Huck and Clegg have confirmed their hypothesis and refined the general knowledge of the origins of surface haze. Perron and Lederman [80] and Stehling et al. [81] show that the extrusion haze is related to the elasticity of LDPE, which can be related to the MWD, especially the presence of a high molecular weight component. Other studies have shown surface haze to be dominated by elasticity in LDPE but crystallinity in HDPE [82]. Ashizawa et al. [83] show that haze correlates directly with the surface roughness of a series of HDPE, LLDPE, and LDPE blown films, with the HDPE films having the highest roughness and haze. Patel et al. [84] later confirm this relationship with more sophisticated AFM measurements.

Johnson et al. [85–88] find that elasticity also influences the crystalline morphology near the surface of LLDPE films. Comparing an mLLDPE that has high haze with a ZN-LLDPE that has lower haze, they find that the crystalline morphologies differ. The mLLDPE has what they call a spherulitic superstructure, which results in a high level of surface roughness. The ZN-LLDPE has a fibrillar or lamellar morphology with lower surface roughness. They propose that the narrower MWD of the mLLDPE results in short relaxation times, allowing for the superstructure morphology to develop. The broader MWD of the ZN-LLDPE, and in particular the high MW fraction, increases the elasticity and relaxation time, impeding the development of the superstructure morphology. Using a series of polymers, they correlate surface haze with recoverable shear strain from stress relaxation experiments (see Chapter 5), which is a measure of elasticity. When this parameter is small, the relaxation time is short and the spherultic superstructure morphology forms. At intermediate values, the spherulitic structure gives way to row nucleated lamellar structures and haze decreases. At even higher values, extrusion haze takes over and surface roughness and haze increase. Both low and high melt elasticity resins can have high haze but due to very different reasons.

Andreassen et al. [82] evaluate the blown film haze of two metallocene LLDPE (mLLDPE) and two ZN LLDPE resins of different MWD and elasticity. The order of elasticity is narrow MWD mLLDPE < broad MWD mLLDPE < narrow MWD ZN-LLDPE < broad MWD ZN-LLDPE. The haze of the two mLLDPE and the narrow MWD ZN-LLDPE films are about the same and substantially less than the haze of the broad MWD ZN-LLDPE film. The haze correlates with the measured surface roughness of the films.

Bafna et al. [89] show that the orientation of crystals at the surface contributes to differences in roughness and haze in HDPE blown film. They evaluate grades of single-site catalyzed (narrow MWD) and ZN catalyzed (broad MWD) HDPE with similar degrees of crystallinity. Blown film of the ZN HDPE has 60% haze versus 16% for the single-site grade. X-ray scattering studies show the two films have similar lamellar crystalline structures but differences in crystalline orientation that correlate with extensive surface roughness in the ZN film not seen in the single-site catalyzed HDPE film.

Other studies described in Sections 7.5.3.2.2, 10.4.2.2.1, and 12.1.4.1 show that blown films of Z-N LLDPE have an amorphous region near the surface due to their broad comonomer and MWD distribution. The shorter chains with an abundance of comonomer and short-chain branches are more concentrated near the surface, creating an amorphous region not found in metallocene LLDPE blown films. This region influences heat seal, adhesion and slip additive migration. Presumably, it also influences surface roughness and haze, although it is not clear how to reconcile this with the studies showing surface crystallinity dominates haze in these films.

Several approaches have been explored to reduce the haze of PE films. LDPE generally has a lower haze than ZN-LLDPE blown film, likely due to its higher level of elasticity. EVA copolymers have lower levels of haze than LDPE due to their lower level of crystallinity. Blending a small amount of LDPE into some types of LLDPE has lowered haze, presumably by increasing the elasticity of the LLDPE [90–93]. Similarly, Cooke and Tikuisis [93] show that adding a small amount of a high MW HDPE to LLDPE reduces haze. Blending LLDPE with other grades of LLDPE or rigid polymers, manipulating the molecular architecture by adding a small amount of long chain branching, and incorporating various additives such as sorbitol has been found to reduce haze in some situations [82,94–99].

Polymer processing also affects haze. When crystallinity is the dominant mechanism for surface roughness (e.g., LLDPE and HDPE), reducing overall crystallinity is important and increasing the cooling rate by reducing the frost line height reduces haze [79,83,100,101]. Ashizawa et al. [83] find that increasing draw-down ratio (DDR) or blow-up ratio (BUR) decreases internal haze but increases surface haze somewhat. Cast film and water quench blown films [102] typically have lower haze than air-quenched blown film due to the faster quench rate of these processes, leading to lower overall crystallinity. The haze of PE cast films is still dominated by surface roughness effects [85,103]. For example, the surface of the chill roll can have a substantial effect on surface roughness and haze; a smooth surface produces a much clearer film than a matte finish, although the matte finish will have fewer problems with sticking.

When elasticity or extrusion haze dominates, lowering output or decreasing the rate of cooling helps. Ionomers, for example, have a lower level of crystallinity than LDPE but higher elasticity. The haze of ionomer films is almost always dominated by elastic flow irregularities rather than crystallinity. Increasing the melt temperature and the frost line height generally helps lower the haze and increase the gloss of ionomers, just the opposite of what is typically done for PE. The slower cooling rate allows more time for the surface irregularities caused by elasticity to relax and smooth out.

12.2.2.2 Coextrusion

Since surface haze is the primary cause of haze in PE film, burying the high-haze resin in a multilayer coextrusion is often an effective strategy for reducing the haze of a film. The same surface roughness that develops on the free or outer surface of the film may not develop at an interior interface of a coextruded film. One explanation for this is that the level of stress and strain may be different. The outside of the film experiences a high level of stress in the die channel since stress is at its maximum at the die wall. At the die exit, the polymer on the outside must accelerate from essentially no flow to the average draw speed of the molten web. This may give rise to strain-induced crystallization. The amount of acceleration of the polymer at an interface closer to the interior of the film is not great and is, therefore, less susceptible to strain-induced crystallization. The crystalline structure or morphology that forms at a polymer-polymer interface may also differ from that at a polymer–air interface due to differences in free energy and physical constraints on the chain movement.

Several investigators have shown that coextruded structures with the same overall material composition but with different layers on the surface may have lower haze [103–107]. An example from Elkoun et al. [104] is shown in Fig. 12.14. Five-layer blown film with different layer combinations of three LLDPE resins is tested. One resin is a

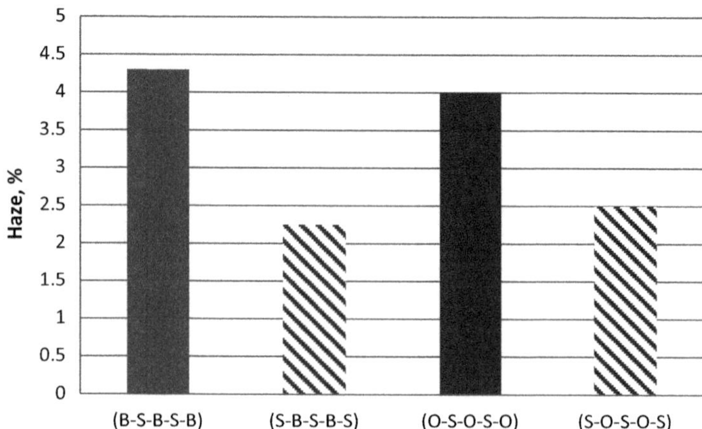

FIGURE 12.14 Effect of film structure on coextruded blown film haze. B = bLLDPE, O = oLLDPE, S = mLLDPE. Each film structure contains 50% S split equally among two or three layers. *Adapted from Elkoun, S., Huneault, M. A., Zhang, X. M., McCormick, K., Puterbaugh, F., Kale, L., 2003, "LLDPE blown films property enhancement through coextrusion," SPE ANTEC Conf., pp. 1396–1400.*

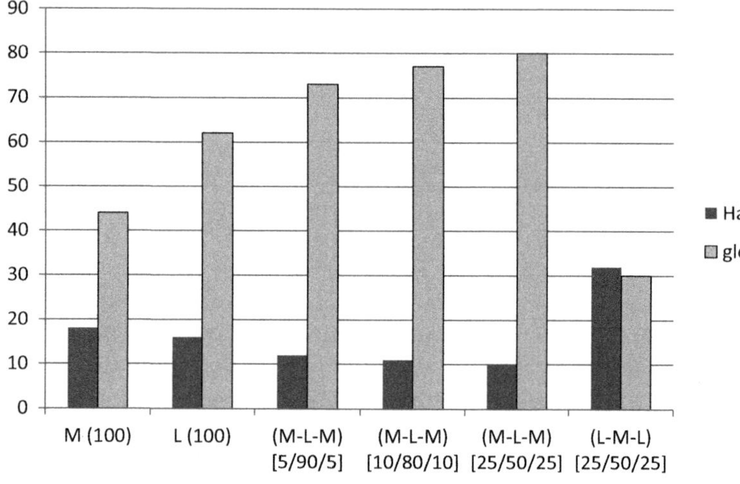

FIGURE 12.15 Example of synergistic and antagonistic effects of coextrusion on optical properties of a blown film. M = single site catalyzed LLDPE, L = Ziegler–Natta butene-LLDPE. *Data from Borse, N. and Aubee, N., 2014, "Developing cost effective co-extruded film structures for flexible packaging," SPE ANTEC Conf., pp. 1304–1308.*

conventional ZN butene-LLDPE (0.918 g/cc, 1 g/10 minute MFI), labeled B. Another is an advanced ZN octene-LLDPE (0.920 g/cc, 1 g/10 minute MFI), labeled O. The final grade is a metallocene LLDPE (0.917 g/cc, 1 g/10 minute MFI), labeled S. The typical haze for these three grades in a monolayer 25-μm blown film is 3.5%, 8.5%, and 2.5% for B, O, and S, respectively. The five-layer films contain alternating layers of S with either O or B, with the amount of S comprising 50% of the total structure. As shown in Fig. 12.14, when the lower haze S grade is on the outside of the film, the haze is lower than when the higher haze O or B grade is on the outside. The haze of coextruded films with the S skin layers also is similar to the typical value for a 100% S film.

Coextrusion does not always produce predictable results based simply on the haze of the outside layers. Fig. 12.15 shows results of haze and gloss measurements on three-layer blown film structures from Borse and Aubee [105]. They evaluate combinations of a ZN butene-LLDPE [0.921 g/cc, 0.8 g/10 minute MFI], labeled L, and a single site catalyzed LLDPE [0.916 g/cc, 0.65 g/10 minute MFI], labeled M. The first two sets of bars represent the control monolayer films. The next three sets of bars show data for a coextrusion with the M resin on the outside, with different layer ratios. The haze is significantly lower and gloss higher than the two controls; having M on the outside has a positive synergistic effect on the optical properties. The final set of bars shows the case where the L resin is on the outside. Here the haze is higher and the gloss is lower than either of the controls.

Another example of synergic and antagonistic effects is shown in Fig. 12.16. Zhang and Ajji [106] evaluate five-layer blown film with different combinations of PP and LLDPE. The LLDPE is a ZN octene-LLDPE. The results for two PP grades are shown in the figure, a clarified random copolymer and a homopolymer. The bars on the far left and right represent the results for monolayer controls. The bars in the middle represent different combinations of layers of PP (P) and LLDPE (L). In each case, 50% of the total composition is PP and 50% LLDPE. As expected, the coPP control film has significantly lower haze than the hPP control. This is due to the lower crystallinity of the coPP as well as

(A)

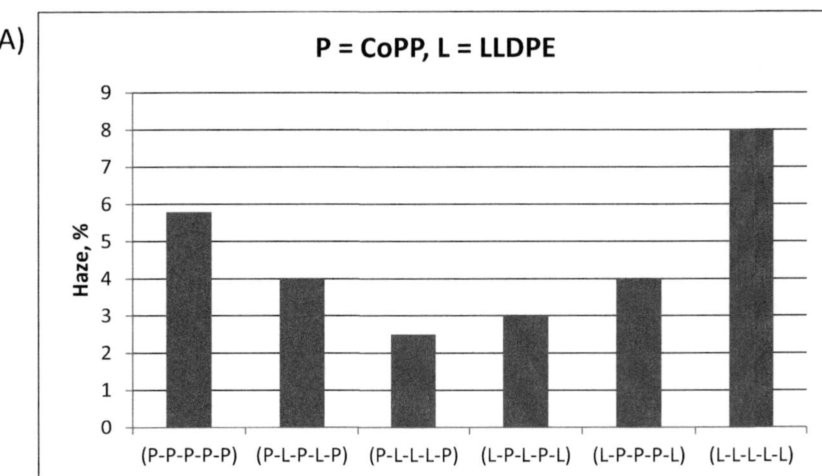

(B)

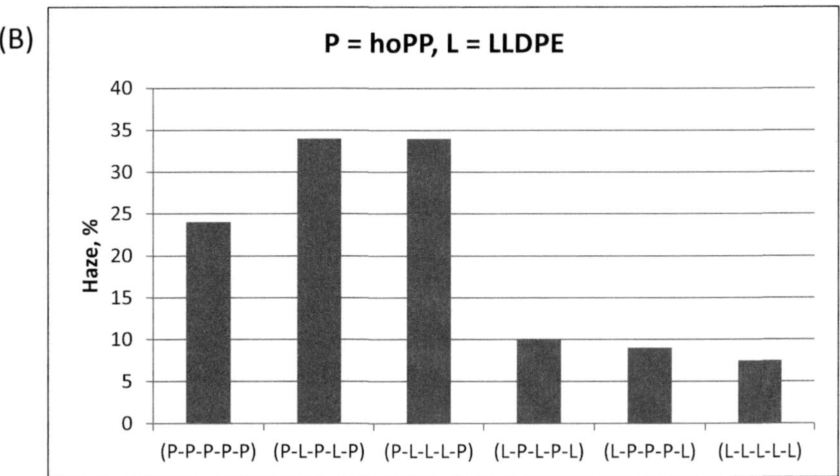

FIGURE 12.16 Effect of coextrusion blown film layer composition and structure on haze. P = random copolymer PP (Graph A) or homopolymer PP (Graph B). L = octene LLDPE. Structures with both PP and LLDPE layers contain a total of 50% PP and 50% LLDPE distributed amongst the individual layers. *Data from Zhang, X. and Ajji, A., 2005, Polypropylene-polyethylene multilayer films, SPE ANTEC Conf., pp. 2947–2951.*

the addition of the clarifier additive. The coPP also has a somewhat lower haze than the LLDPE control. All the coPP-LLDPE multilayer structures (graph A) have lower haze than either of its component control films. In contrast, graph B shows that the multilayer structures with hPP on the surface have higher haze than either the hPP or LLDPE control, whereas the structures with LLDPE on the surface have haze similar to the LLDPE control.

12.2.2.3 Other factors

Many other factors influence the haze and clarity of blown or cast film, including:

1. *Tie resin selection for coextruded films.* Every layer in a coextrusion can detract from the optical clarity of the structure, even a thin tie resin. For many common barrier films involving bonding PE to PA or EVOH there are many different tie resins to choose from. These are functionalized polymers typically based on different types of PE (HDPE, LLDPE or LDPE) or ethylene copolymers (EVA or EMA). See Chapter 10 for details. If high clarity and low haze are critical attributes, then selecting a tie resin based on its clarity and haze should be a consideration. Tie resins generally have similar clarity and haze as the polymer on which they are based. For high transparency, an EVA based tie resin may be a better choice than an HDPE based one, as long as it meets all the other requirements of the structure (such as barrier or temperature resistance). The tie resin may also contribute to a type of flow instability (see Section 13.4.5) believed to be related to chemical interactions at the interface, resulting in a local increase in molecular weight. This type of instability often affects the clarity of the film, especially the see-through clarity. Adjusting the amount of chemical bonding groups in the tie resin may help mitigate this type of instability.
2. Antiblock and other additive selection. As described in Section 12.1.3, antiblock additives are typically inorganic particles such as silica (diatomaceous earth) or talc. Their size, shape and concentration can affect haze and

transparency [108]. They are generally in the size range where they diffract light, so optimizing compositions for minimal usage can be important for minimizing haze. Coextrusion may allow them to be used in a thin skin layer, decreasing their effect on haze.

3. *Nucleating or clarifying agents*. Nucleating agents seed the crystallization process so that when the polymer cools from the melt, more crystals form (nucleate). With many more crystals growing, individual crystals cannot become as large. If the crystal size is limited to below the wavelength of light (400–700 nm) then haze is reduced. The most successful nucleating and clarifying agents have been for PP.
4. *Orientation*. Orientation can induce crystallization. In some films, such as biaxial oriented PET, the crystalline morphology is such that the film, although highly crystalline, is very transparent.
5. *Polymer blends*. Chapter 6 describes blending technology for flexible packaging. Most polymer-polymer blends are immiscible so that one of the polymers forms a separate phase or domain within the other. If these domains are larger than the wavelength of light and their RI does not match that of the continuous polymer phase, they will contribute to haze. The domain size is a complex function of viscosity ratio, surface energy and mixing. It is very challenging to obtain domain sizes that are small enough to maintain transparency. Therefore, it is important to consider the compatibility of polymers being blended into film layers, such as masterbatch carrier resins. Even polymers of the same product class, such as LLDPE and LDPE may be immiscible depending on their composition.
6. *Refractive index*. If two materials have RIs that are the same, their blends will be transparent to light even if they are immiscible. Unfortunately, it is rare that the RIs of two immiscible polymers, or a polymer and an inorganic additive, match sufficiently to avoid haze issues.

12.2.3 Gloss

Gloss is highly dependent on surface roughness; smoother surfaces typically have higher gloss. Factors for minimizing surface haze in polyolefins are discussed in Section 12.2.2. These same factors generally will help maximize gloss, such as:

1. Reducing surface imperfections due to flow instability (such as sharkskin and melt fracture) and elastic effects. The rheology of the polymer contributes to this behavior. For LDPE, selecting a grade with a narrower MWD or less long chain branching may help. Processing is also a factor: changes in melt temperature, frost line-height and output may have an impact. As discussed in Section 12.2.2, different polymers may respond in opposite ways, so understanding the dominant mechanism for low gloss will help guide the approach.
2. Reducing overall crystallinity of the polymer. Surface crystallinity is another contributor to surface roughness and low gloss. Reducing the overall crystallinity of the polymer may help. Ionomers, for example, have low levels of crystallinity and very high gloss. Faster quench rates also generally help reduce the crystallinity of polymers, which may help increase gloss. Cast film generally has a higher gloss than blown film. When casting film, the surface of the chill roll can impact gloss—a polished mirror surface will generally provide a higher gloss than a matte one.
3. Minimizing other surface imperfections, such as a gel.
4. Optimizing the use of inorganic antiblock additives, which by design roughen the surface.

12.2.4 Specialty microlayer films

Microlayer film technology is described in Sections 8.5.2, 9.4.7, and 11.2.4.5. It involves combining two or more layers and repeatedly spitting and recombining them to produce film structures with tens to hundreds of layers. The concept is illustrated in Fig. 8.21. Some of the earliest commercial adoptions of this technology were for films for optical filters. By selecting polymers with the right combination of RIs or other properties, specialty films can be made that selectively reflect certain wavelengths of light or have specific dielectric properties for use as optical filters, components for LCD displays, capacitors and other electronic devices [109–117].

One intriguing application for this technology that may have use for flexible packaging is described by Im and Schrenk [109]. They make a film with 400 alternating layers of PP and PC which blocks 80% of UV light while retaining good transparency of visible light. A two-layer control film of the same total thickness and PP and PC content does not block the UV light. They develop the theoretical basis for this effect based on the difference in RIs of layers, layer thickness and number of layers.

References

[1] T.A. Clark, J.R. Wagner Jr., Film properties for good performance on vertical-form-fill-seal packaging machines, J. Plastic Film Sheet 18 (2002) 145–156.

[2] ASTM International, Standard Test Method for Static and Kinetic Coefficients of Friction of Plastic Film and Sheeting, ASTM International, West Conshohocken, PA, 2014 (D1894-14).

[3] Maltby, A., The effects of polyolefin formulation and processing variables on slip agent performance (I), TAPPI Polymers, Laminations & Coatings Conference Proceedings, 1998, pp. 1241–1250.

[4] ASTM International, Standard Test Method for Blocking Load of Plastic Film by the Parallel Plate Method, ASTM International, West Conshohocken, PA, 2015 (D3354-15).

[5] D.K. Owens, Friction of polymer films. I. Lubrication, J. Appl. Polym. Sci. 8 (1964) 1465–1476.

[6] Sharma, A. and Beard, B., Mode of action of surface active slip and antiblocking additives, SPE ANTEC Conference, 1997, p. 3019.

[7] Chen, J., Pham, H., Savargaonkar, N., Ge, S., Aging dependent slip agent surface morphology of LLDPE films, SPE ANTEC Conference, 2011.

[8] D.K. Owens, Friction of polymer films. II. Effects of deformation properties, J. Appl. Polym. Sci. 8 (1964) 1477–1482.

[9] L. Walp, H. Tomlinson, Effect of erucic acid percentage on slip properties of erucamide, J. Plastic Film Sheeting 20 (2004) 275–287.

[10] Maltby, A., Effect of amide composition on performance as a plastics additive, TAPPI PLACE. Conference, 2002.

[11] D. Rosato, Advanced slip & antiblock friction control in packaging film, SpecialChem (2008)viewed 21 Mar 2021 at. Available from: http://polymer-additives.specialchem.com/tech-library/article/advanced-slip-antiblock-friction-control-in-packaging-film.

[12] Constant, D. R., Cyclic-olefinic copolymers as non-migrating, polymeric slip additives in LDPE cast film, SPE ANTEC Conference, 2002.

[13] K.J. Ryan, K.E. Lupton, P.G. Pape, V.B. John, Ultra high molecular weight functional siloxane additives in polymers: effect on processing and properties, J. Vinyl Addit. Technol. 6 (2000) 7–19.

[14] S. Savargaonkar, B. McKinley, Slip agents: extended performance range for polyolefins, Plast. Technol. (2006)viewed on 21 Mar 2021 at. Available from: http://www.ptonline.com/articles/slip-agents-extended-performance-range-for-polyolefin-films.

[15] Hays, T., Next generation slip additives for low CoF PE films, AMI Mulitlayer Packaging Films Conference, 2017.

[16] R.R. Tamm, Effect of film additives on printing, in: TAPPI Polymers, Laminations & Coatings Conference, San Francisco, August 30–September 3, 1998, pp. 1067–1071.

[17] DuPont, Processing and properties of Surlyn® ionomer resins by blown and cast film processes, STL from TSL, 70-4 (Revised Oct 1979), 1979.

[18] A.R. Zahedi, A. Ranji, S. Asiaban, Optimizing COF, blocking force, and printability of low density polyethylene, J. Plastic Film Sheeting 22 (2006) 163–176.

[19] A.M. Soutar, The fugitive slip, TAPPI Polymers, Laminations Coat. Conf. (1994).

[20] Edwards, R., Foster, B., Coefficient of friction in polyethylene extrusion coatings: An analysis, TAPPI Polymers, Laminations & Coatings Conference Proceedings, 1993.

[21] Itzkoff, M. L., Woodell, G. L., Effect of production parameters on COF of extrusion coated substrates, TAPPI Paper Synthetics Conference Proceedings, 1983, pp. 111–116.

[22] Sakhalkar, S., Hirt, D., Surface segregation of erucamide in LLDPE films: thermodynamic analysis and experimental verification, SPE ANTEC Conference, 1998.

[23] Kitchel, F., Differential C.O.F. in coextruded blown films, TAPPI Polymers, Coating Laminations Conference, 1988.

[24] Sankhe, S., Hirt, D., Roberts W., Havens, M., Evaluation of additive concentration profiles in multilayer films," SPE ANTEC Conference, 1999.

[25] Rawls, A., Joshi, N., Sakhalkar, S., Hirt, D., Evaluation of surface concentration of additives in LLDPE films, SPE ANTEC Conference, 1997, pp. 3029.

[26] N. Joshi, D. Hirt, Evaluating bulk-to-surface partitioning of erucamide in LLDPE films using FT-IR microspectroscopy, Appl. Spectrosc. 53 (1999) 11 (Erratum, 1999, 53, pp. 261A).

[27] N. Joshi, D. Hirt, Studying additive diffusion in polymer films using FTIR microspectroscopy, SPE ANTEC Conf. (1998).

[28] C. Gagliardi, L. Murie, D. Hirt, Measuring the diffusivity of additives in polymer films using in situ FTIR-ATR spectroscopy, SPE ANTEC Conf. (1999).

[29] J. Klein, B. Briscoe, Diffusion of long molecules through solid polyethylene. II. Measurements and results, J. Polym. Sci. 15 (1977) 2065–2074.

[30] T. Breuer, N. Martin, H. Tomlinson, E. Leonard, Effect of oleochemical additive lubricants on surface properties of polyolefin films, SPE ANTEC Conf. (1983) 896.

[31] T. Breuer, N. Martin, H. Tomlinson, S. Gloyer, H. Crawford, Fatty amide slip and antiblock additives in polypropylene and linear low density polyethylene, SPE ANTEC Conf. (1984) 141.

[32] A. Maltby, R. Marquis, "Slip additives for film extrusion, J. Plastic Film Sheeting 14 (1988) 111.

[33] P. Patel, N. Savargaonkar, COF & blocking in mLLDPE film: 'why don't my recipes work? TAPPI Polymers, Laminations Coat. Conf. (1999) 1071.

[34] K. Coupland, A. Maltby, Modification of the surface properties of polyethylene plastomer films by the use of additives, J. Plastic Film. Sheeting 13 (April 1997) (1997) 142.

[35] J. Wooster, B. Simmons, Optimizing COF, seal performance, and optical properties of polyolefin plastomers, TAPPI Polymers, Laminations Coat. Conf. (1995). and 1996 SPE ANTEC Conference.

[36] K.I. Thompson, Coefficient of friction testing and factors that affect the frictional behavior of polyethylene films, TAPPI J. 71 (1998) 157–161. September.

[37] E. Foldes, Mobility of additives in ethylene polymers, Polym. Bull. 34 (1995) 93–99.

[38] K. Coupland, A. Malty, D. Parker, AddCon Asia, Paper 15, 1997.

[39] N. Joshi, S. Sakhalkar, D. Hirt, Mass transfer of slip and antifogging agents in LLDPE films, SPE ANTEC Conf. (1996) 2406.

[40] S. Hu, D. Lin-Vien, R. French, Probing the concentration profiles of additives in polymers in IR microspectroscopy: the diffusion of Cyasorb UV531 in polypropylene, Appl. Spectrosc. 46 (1992) 225.

[41] I. Quijada-Garrido, J. Barrales-Rienda, G. Frutos, Diffusion of erucamide (13-cis-docosenamide) in isotactic polypropylene, Macromolecules 29 (1996) 7164–7176.

[42] E. Foldes, Transport of small molecules in polyolefins. II. Diffusion and solubility of Irganox 1076 in ethylene polymers, J. Appl. Poly. Sci. 48 (1993) 1905–1913.

[43] L. Muire, D. Hirt, Effect of applied external pressure on the surface concentration of erucamide in LLDPE films, SPE ANTEC Conf. (1999).

[44] B. Schumann, J. Wooster, S. Murray, The effect of film winding tension and melt temperature on COF and other properties of PE blown film, TAPPI Polymers, Laminations Coat. Conf. (1998) 633.

[45] P. Passiniemi, General theory for determination of diffusion coefficients of solvents and gases in polymers, Polymer 36 (1995) 341.

[46] G.T. Fieldson, T.A. Barbari, The use of FTIR-ATR spectroscopy to characterize penetrant diffusion in polymers, Polymer 34 (1993) 1146.

[47] A. Rawls, D.E. Hirt, M.R. Havens, W.P. Roberts, Evaluation of surface concentration of erucamide in LLDPE films, J. Vinyl Addit. Technol. 8 (2002) 130–138.

[48] E. Foldes, A. Szigeti-Erdei, Migration of additives in polymers, SPE ANTEC Conf. (1997) 3024.

[49] M. Ramirez, D.E. Hirt, B. Roberts, M. Havens, N. Miranda, COF of LLDPE films as a function of erucamide surface concentration, SPE ANTEC Conf. (2000).

[50] R.J. Silverwood, F. Sadeghi, A. Ajji, Study of kinetic of slip agent migration in polyethylene sheets, SPE ANTEC Conf. (2012).

[51] S.-S. Chang, Migration of low molecular weight components from polyolefins, Macromol. Symp. – Intern. Union. Pure Appl. Chem. 28 (1982) 734–735.

[52] M.X. Ramirez, K.B. Walters, D.E. Hirt, Relationship between erucamide surface concentration and coefficient of friction of LLDPE film, J. Vinyl Addit. Technol. (2005) 9–12.

[53] N. Joshi, D. Hirt, Theoretical analysis of additive distribution in polymer films, SPE ANTEC Conf. 2802 (1994) 2798.

[54] D. Allen, B.J. Briscoe, D. Tabor, Lubrication of polymers by oleamide and stearamide. 2, Wear 25 (1973) 393–397.

[55] M.X. Ramirez, D.E. Hirt, L.L. Wright, AFM characterization of surface segregated erucamide and behenamide in linear low density polyethylene film, Nano Lett. 2 (1) (2002) 9–12.

[56] C. Poisson, V. Hervais, M.F. Lacrampe, P. Krawczak, T. Falher, C. Gondard, et al., Optimization of polyethylene/binder/polyamide extrusion blow-molded films. III. Slippability improvement with fatty acid amides, J. Appl. Polym. Sci. 115 (2010) 2332–2345.

[57] J. Chen, J. Li, T. Hu, B. Walther, Fundamental study of erucamide used as a slip agent, J. Vac. Sci. Technol. A Vacuum Surfaces Films 25 (4) (2007) 886–892.

[58] J.H. Glover, Slip migration in extrusion coating of LDPE, TAPPI Polymers, Lamination Coat. Conf. (1987) 231–239.

[59] R. Edwards, B. Foster, Coefficient of friction in polyethylene extrusion coatings: an analysis, TAPPI Polymers, Laminations Coat. Conf. (1993).

[60] C.A. Shuler, A.V. Janorkar, D.E. Hirt, Fate of erucamide in polyolefin films at elevated temperature, Polym. Engr. Sci. 44 (2004) 2247–2253.

[61] R. Sharma, M.L. Pacholski, K.B. Laughlin, J. Ngunjiri, J. Sparks, V. Kalihari, et al., Impact of elevated temperature on surface properties of erucamide-containing polyethylene films, SPE Int. Polyolefins Conf. (2017).

[62] N. Dulal, R. Shanks, D. Chalmers, B. Adhikari, H. Gill, Migration and performance of erucamide slip additive in high-density polyethylene bottle caps, J. Appl. Polym. Sci. 135 (43) (2018) 46822. app.

[63] H. Reinecke, R. Navarro, How to control migration of slip agents in polyethylene films, Plast. Res. Online, Soc. Plast. Eng. (2011).

[64] J.A. Radosta, W.D. Riley, Treated Talc as an effective anti-block for LDPE blown film, J. Plast. Film Shtg 7 (3) (1991) 247–258.

[65] C.W. Peloso, M.J. O'Connor, S.W. Bigger, J. Scheirs, "Characterizing the degradation of the polymer slip additive erucamide in the presence of inorganic antiblock agent, Polym. Degrad. Stab. 62 (1998) 285–290.

[66] C. Llop, A. Manrique, R. Navarro, C. Mijangos, H. Reinecke, Control of the migration behavior of slip agents in polyolefin-based films, Polym. Engr. Sci. 51 (2011) 1763–1769.

[67] E.M. Hoang, C.M. Liauw, N.S. Allen, E. Fontan, P. Lafuente, Effect of additive interactions on the thermo-oxidative stabilization of a film grade metallocene LLDPE, J. Vinyl Addit. Technol. 10 (3) (2004) 149–156.

[68] S.Y. Sankhe, D.E. Hirt, Studying additive migration in LLDPE and POP films using synchrotron-based FTIR microspectroscopy, SPE ANTEC Conf. (2002).

[69] S.Y. Sankhe, D.E. Hirt, Evaluation of additive concentration profiles in multilayer films, SPE ANTEC Conf. (1999).

[70] S.Y. Sankhe, D.E. Hirt, Using synchrotron-based FT-IR microspectroscopy to study erucamide migration in 50-μm-thick bilayer linear low-density polyethylene and polyolefin plastomer films, Appl. Spectrosc. 57 (1) (2003) 37–43.

[71] K. Bryant, Understanding coefficient of friction in coextruded films, TAPPI Polymers, Laminations Coat. Conf. (1998) 969.

[72] D.K. Owens, Friction of polymer films. III. Migration of lubricants, J. Appl. Polym. Sci. 14 (1970) 185–192.

[73] B.J. Briscoe, V. Mustafaev, D. Tabor, Lubrication of polythene by oleamide and stearamide, Wear 19 (1972) 399–414.

[74] A. Maltby, M. Read, Effect of temperature on the slip and blocking of polyolefins containing various fatty acid amides, TAPPI Polymers, Laminations Coat. Conf. (1999) 1079.

[75] ASTM International, Standard Test Method for Transparency of Plastic Sheeting, ASTM International, West Conshohocken, PA, 2015 (D1746-15).

[76] ASTM International, Standard Test Method for Haze and Luminous Transmittance of Transparent Plastics, ASTM International, West Conshohocken, PA, 2013 (D1003-13).

[77] International Organization for Standardization, Plastics − Determination of Haze for Transparent Materials, International Organization for Standardization, Geneva, Switzerland, 1999 (ISO 14782).

[78] ASTM International, Standard Test Method for Specular Gloss of Plastic Films and Solid Plastics, ASTM International, West Conshohocken, PA, 2013 (D2457-13).

[79] N.D. Huck, P.L. Clegg, The effect of extrusion variables on the fundamental properties of tubular polyethylene film, SPE ANTEC Conf. 18-2 (1961) 1−14.

[80] P.J. Perron, P.B. Lederman, Effect of molecular weight distribution on polyethylene film properties, Polym. Eng. Sci. 12 (5) (1972) 340−345.

[81] F.C. Stehling, C.S. Speed, L. Westerman, Causes of haze of low-density polyethylene blown films, Macromolecules 14 (1981) 698−708.

[82] E. Andreassen, A. Larsen, K. Nord-Varhaug, M. Skar, H. Oysaed, Haze of polyethylene films - effects of material parameters and clarifying agents, Polym. Eng. Sci. 42 (5) (2002) 1082−1097.

[83] H. Ashizawa, J.E. Spruiell, J.L. White, An investigation of optical clarity and crystalline orientation in polyethylene tubular film, Polym. Engr. Sci. 24 (1984) 1035−1042.

[84] R. Patel, V. Ratta, P. Saavedra, J. Ji, Surface haze and surface morphology of blown film compositions, J. Plastic Film Sheeting 21 (2005) 217−231.

[85] M.B. Johnson, G.L. Wilkes, A.M. Sukhadia, D.C. Rohlfing, Optical properties of blown and cast polyethylene films: surface vs bulk structural considerations, J. Appl. Polym. Sci. 77 (2000) 2845−2864.

[86] G.L. Wilkes, M.B. Johnson, A.M. Sukhadia, D.C. Rohlfing, Optical haze properties of polyethylene blown films: Part 1 − surface vs bulk structural considerations, SPE ANTEC Conf. (2001).

[87] A.M. Sukhadia, D.C. Rohlfing, M.B. Johnson, G.L. Wilkes, Optical haze properties of polyethylene blown films: Part 2 − the origins of various surface roughness mechanisms, SPE ANTEC Conf. (2001).

[88] A.M. Sukhadia, D.C. Rohlfing, M.B. Johnson, G.L. Wilkes, A comprehensive investigation of the origins of surface roughness and haze in polyethylene blown films, J. Appl. Polym. Sci. 85 (11) (2002) 2396−2411.

[89] A. Bafna, G. Beaucage, F. Mirabella, G. Skillas, S. Sukumaran, Optical properties and orientation in polyethylene blown films, J. Polym. Sci. B Polym. Phys. 39 (23) (2001) 2923−2936.

[90] C.T. Lue, T.H. Kwalk, D. Li, C.R. Davey, D.Y. Chiu, Haze improvement with addition of HP-LDPE or HDPE. Part I − blown film property comparison, SPE ANTEC Conf. (2004) 2140−2143.

[91] H.Y. Chen, J. Li, R.L. Sammler, C.T. Lue, R. Kolb, T.H. Kwalk, Haze improvement with addition of HP-LDPE or HDPE. Part II − Mechansitic understanding, SPE ANTEC Conf. (2004) 2117−2121.

[92] K. Jordens, The influence of small amounts of LDPE on the morphology and resulting haze of LLDPE blown films, SPE ANTEC Conf. (2002).

[93] D.L. Cooke, T. Tikuisis, Addition of branched molecules and high molecular weight molecules to improve optical properties of LLDPE film, J. Plastic Film Sheeting 5 (1989) 290−307.

[94] L. Liu, T. Karjala, S. Ge, M. Demirors, Relationship between surface haze and crystalline morphology of polyethylene blown films, SPE ANTEC Conf. (2012).

[95] J. den Doelder, T. Karjala, M. Demirors, The relationship between tubular low-density polyethylene (LDPE) blown-film optics and molecular structure, SPE ANTEC Conf. (2009) 2806−2810.

[96] M.J. Cran, S.W. Bigger, The effect of metallocene-catalyzed polyethylene on the physicomechanical properties of film blends with conventional linear low-density polyethylene, J. Plastic Flim Sheeting 22 (2006) 121−132.

[97] H. Mavridis, Blending of high-performance LLDPEs with butene-LLDPEs and LDPEs, J. Plastic Film Sheeting 17 (2001) 253−265.

[98] W. Chinsirikul, A. Fuongfuchat, N. Kerddonfag, S. Phiboonkulsumrit, C. Winotapun, P. Khumsang, Preventing film blocking and optimizing haze of metallocene PE blown films through blending with rigid polymers, Intern. Polym. Process. XXIII (2004) 152−160.

[99] T.I. Butler, S.Y. Lai, R. Patel, A blown film comparison between conventional LLDPE and POP, J. Plastic Film Sheeting 10 (1994) 248−261.

[100] M.M. Werlang, Effect of polymer processing conditions on shrinkability of LDPE and LLDPE films, and on optical properties of LLDPE films, SPE ANTEC Conf. (2003) 3271−3274.

[101] P.F. Smith, I. Chun, G. Liu, D. Dimitrievich, J. Rasburn, G.J. Vancso, Studies of optical haze and surface morphology of blown polyethylene films using atomic force microscopy, Polym. Eng. Sci. 36 (16) (1996) 2129−2134.

[102] K. Xiao, I.-H. Lee, R. Armstrong, Comparison of water-quench vs air-quench blown film processes − part I: flat film properties, SPE ANTEC Conf. 70 (2012) 972−977.

[103] P. Scarfato, Di Maio, L. Garofalo, E. Incarnato, L., Three-layered coextruded cast films based on conventional and metallocene poly(ethylene/ α-olefin) copolymers, J. Plastic Film Sheeting 30 (2014) 284−299.

[104] S. Elkoun, M.A. Huneault, X.M. Zhang, K. McCormick, F. Puterbaugh, L. Kale, LLDPE blown films property enhancement through coextrusion, SPE ANTEC Conf. (2003) 1396−1400.

[105] N. Borse, N. Aubee, Developing cost effective co-extruded film structures for flexible packaging, SPE ANTEC Conf. (2014) 1304−1308.

[106] X. Zhang, A. Ajji, Polypropylene-polyethylene multilayer films, SPE ANTEC Conf. (2005) 2947–2951.

[107] R. Baldizonne, G. Carianni, P. Mariani, A new Ziegler-Natta octene LLDPE for high performance blown film applications: benefits of coextruded vs. blended LDPE/LLDPE films, Eur. TAPPI PLACE. Conf. (2005).

[108] D.L. Davidson, D. Swartz, Unique antiblocks for high clarity polyethylene film, SPE ANTEC Conf. (2004) 1111–1116.

[109] J. Im, W.J. Schrenk, Coextruded microlayer sheet and film, J. Plastic Film Sheeting 4 (1988) 104–115.

[110] A. Hiltner, S. Nazarenko, T. Ebeling, T. Schuman, E. Baer, Novel microlayered structures for adhesion and interdiffusion studies, Polym. Mater. Sci. Engg 75 (1996) 471.

[111] R. Tangirala, E. Baer, A. Hiltner, C. Weder, Photopatternable reflective films produced by nanolayer extrusion, Adv. Funct. Mater. 14 (2004) 595–604.

[112] T. Kazmierczak, H. Song, A. Hiltner, E. Baer, Polymeric one-dimensional photonic crystals by continuous coextrusion, Macromol. Rapid Commun. 28 (2007) 2210–2216.

[113] C.-Y. Lai, E. Baer, The interdiffusion of polymeric gradient refractive index (GRIN) lens materials, SPE ANTEC Conf. (2011).

[114] M. Mackey, M. Wolak, J. Shirk, L. Flandin, A. Hiltner, E. Baer, Enhanced dielectric properties of micro and nanolayered films for capacitor applications, SPE ANTEC Conf. (2011).

[115] M. Ponting, A. Hiltner, E. Baer, Gradient multilayered films by forced assembly coextrusion, SPE ANTEC Conf. (2009).

[116] H.T. Tai, A. Hiltner, E. Baer, M. Wiggins, M. Sandrock, J. Shirk, Tunable reflective polymeric nanolayered films, SPE ANTEC Conf. (2005).

[117] D. Langhe, M. Ponting (Eds.), Manuf. Nov. Appl. Multilayer Polym. Films, William Andrew Publishing, pp. 46–116.

Part V

Effect of the converting process on film properties

Part IV focuses primarily on how the resin or substrate properties affect film properties. Such properties cannot be isolated from the process used to make the film. The converting process has profound effects on orientation, crystallization, interlayer adhesion, and other aspects of product performance. While process–structure–property relationships have been well studied in the literature for monolayer films, coextruded films have received much less attention. This section describes in detail the effects the converting process has on film properties with a focus on coextrusion blown film, cast film, and extrusion coating and lamination.

The properties of polymer films are dependent upon the process by which they are made. A polymer has inherent characteristics determined by its chemical makeup: length of its molecules, presence of bulky side groups, nature of bonds that form between molecules, etc. These characteristics, in response to the stress and temperature history imposed by the process, create macroscopic features that determine the film properties. These features include ordering of the molecules into crystals, the distribution or morphology of these crystals, the strength or density of tie molecules that bind the crystalline regions to each other, and orientation of molecular chains.

The stress imposed by the process is a function of the strain, strain rate, and the temperature at which the strain is imposed. A process that stretches the polymer near the melting temperature imposes more stress than a process that imposes the same strain at higher temperature where the polymer can flow. The time scale of the process where the stress is applied must be compared with the fundamental time scales of the polymer. These include:

- relaxation time of the polymer versus process time (Deborah number)
- crystallization time versus process time (quench rate)
- adhesion versus process time (kinetics of reaction and diffusion vs new area generation).

The fundamental flow characteristics must also be considered, especially for coextrusion of dissimilar materials. Flow instabilities may arise depending on the flow geometry, imposed stress, and characteristics of the polymers. Such instabilities directly affect what materials can be successfully coextruded and may limit the choice of materials or the design of the coextruded structure.

This section is organized into three chapters. Chapter 13 describes how the process affects film quality and properties. Thermal instability gives rise to gel, flow instability produces dimensional variability and appearance issues, and differences in shrinkage and crystallinity produce curl. These issues are scrutinized, as they are vital to making functional film. The stress–strain history and quench rate influence film properties through their effect on orientation and crystallinity. These factors are illustrated in Chapter 14 by examining features of the blown film process:

- differences in stress history between monolayer and coextrusion processes and their effect on properties
- differences between air and water quench blown film
- the development of elongated morphology in polymer blends.

Chapter 15 provides a comprehensive look at how the process affects adhesion to substrates in extrusion coating/lamination and interlayer adhesion in coextrusion film and coating processes.

Chapter 13

Effect of processing on quality of flexible packaging films

The film fabrication or coating process affects film quality in many ways. Quality issues can arise when the process is not under control: surges in output and variability in processing parameters such as temperature or draw rates, to name a few, can affect product quality. Several references deal with process troubleshooting that the reader is referred to for methods to ensure good extrusion, film fabrication, and coating operations [1–11]. This chapter focuses on some of the unique processing issues that are encountered in extrusion, and more specifically coextrusion: thermal instability and gel formation, draw-down instabilities, die drool or die lip buildup, flow maldistribution, flow instability issues when polymers of different rheological characteristics are brought together, and curl of multilayer film comprised of layers that shrink by different amounts during quenching. The fundamental causes of these issues are described along with possible ways to mitigate them through process troubleshooting and optimizing material selection in the package design.

13.1 Thermal stability

Processing polymers above their thermal stability temperature limit may lead to oxidation and degradation that can create gel, black specs, and organoleptic issues. Butler [12] and Spalding et al. [11,13,14] provide good overviews on troubleshooting gel issues. In addition to long holdup times and other general contributing factors for gel formation (see below), thermal stability problems in coextrusion are sometimes encountered when combining polymers with substantially different melting points. For example, coextruding an ethylene vinyl acetate (EVA) based tie resin with polyethylene terephthalate (PET) is problematic since the PET must be processed at greater than 280°C to melt the polymer. Suppliers of EVA recommend not processing EVA above about 230°C due to its limited thermal stability. The fundamental stability issue can be addressed by either using an ethylene methyl acrylate (EMA) copolymer-based tie resin, which has greater stability, or substituting the PET with a lower melting copolymer such as polyethylene terephthalate glycol (PETG).

Determining the source of gel can be a challenge in coextrusion, particularly if it is large enough to distort the layers around it or affects more than one layer. Microscopy can be used to try to identify the layer. FTIR is useful for identifying the composition of the gel, although this can be difficult if the gel particle is grossly decomposed. Hot stage microscopy can help determine if the gel particle is crosslinked or simply unmelt.

Sources of gel and defects include:

- Unmelted particles of the same composition as the layer. The extrusion screw design and processing conditions may need to be optimized if this is the case.
- Foreign contamination, such as fibers, insects, and other particles.
- Foreign contamination from other resin particles/pellets. This is sometimes called polymer cross-contamination. This may arise from contamination at the resin manufacturer or the converter and typically occurs when pellets of one polymer inadvertently mix with another. Pellets of the same product family but significantly different in molecular weight (MW) may have difficulty melting completely and cause this type of gel. Pellets from resins of differing chemistry may not melt completely if they have a higher melting point, thermally degrade if they have poor thermal stability, or in extreme cases react with the main pellets to cause crosslinking. Valves on pellet blenders, hoppers and transfer devices are notorious as locations for hang-ups where polymer from several runs prior may be present and slough off into the main pellet conveying stream.

The Science and Technology of Flexible Packaging. DOI: https://doi.org/10.1016/B978-0-323-85435-1.00005-3

- Fines, angle hair and longs. These are typically generated during transport of the polymer pellets through transfer pipes. Because they have different shapes, they may process differently resulting in unmelt, gel or black specs.
- Degraded polymer. This can occur anywhere along the manufacturing process, from resin manufacturer to converter. Degradation is generally due to long hold-up times at high temperatures. Oxidation is often involved [15]. Antioxidants are often added by the resin manufacturer to help stabilize the polymer during extrusion. An introduction to thermo-oxidative degradation and stabilization is provided in several general references [16–18] and summarized in Section 4.2.16.1.

The gel can be present in the incoming resin as received from the resin supplier or be created during the converting operation. Some questions to answer to determine the source of the gel include:

- Does the gel occur randomly or is it steady? A steady stream of gel particles coming out of the die may indicate a hang-up somewhere in the extruder or die.
- Is the gel distributed uniformly throughout the melt curtain or bubble or in a single location? Uniform or random location of the gel may indicate the gel was in the resin prior to extrusion. If the gel is in a single location, however, it is more likely to have originated in the extruder or die.
- Does the amount of gel increase when the polymer throughput is increased? If the gel is in the incoming resin, increasing throughput should increase the amount of gel that comes through.
- Does the gel stop when a different lot is introduced into the extruder? To definitively determine if there is a bad lot, reintroduce the trouble lot and see if the gel returns.

Preventing gel formation in the extruder involves streamlining flows. Spalding et al. [11,13,14] provide several examples of poorly designed extruder screw sections that prolong hold-up, including poor solids conveying that leaves downstream sections partially filled, restricted flow at the entrance to barrier flighted screws, small flight radii that lead to recirculating flows, fluted mixing elements with channels that are too deep, and transfer lines with diameters too large for enough wall shear stress to keep the polymer at the wall renewed. Other regions that are challenging to purge and keep clear of stagnated polymer flow include screw tips with hourglass adaptors, breaker plates, valved adaptors, static mixers, adapter pipes with 90° elbows, internal flow surfaces of coextrusion feedblocks, nested spiral and stacked-plate blown film die entry ports and keyhole-style T-slot cast film dies [19,20].

In addition to streamlining flow, minimizing temperature and using antioxidant additives, inerting the feed hopper with nitrogen helps reduce gel formation [13,21]. Investigators have also found the addition of polymer process aids (PPAs) helps mitigate gel; studies suggest that the PPA 1) migrates to the wall of the channel to help prevent holdup of oxidized polymer at the wall and 2) helps avoid unmelt by stabilizing the break-up of the solid bed in the melting section of the screw [22–24].

Resin transitions can also contribute to gel and defects. Thoroughly cleaning hoppers and transfer lines when switching resins and good purging practices are critical for preventing gel. Functional polymers such as acid copolymers, ionomers and tie resins often have an affinity to stick to metal surfaces, making purging more challenging. They also tend to react with polymers that may have hung up from previous runs, pushing them out in a shower of gel. Because of this, they may be blamed for a past problem.

The following steps help ensure efficient transitions [19,20]:

- Empty the hopper prior to introducing the new resin, but do not allow the extruder to run empty while running at production speed. The polymer carries away the heat from the process. If the polymer feed is reduced while the extruder is still operating at full speed, the heat will be absorbed by less polymer mass, causing the temperature of the remaining polymer to spike higher. This can cause degradation. In severe cases, off-gassing from degradation may cause pressure to build up inside the extruder, which can be dangerous.
- Introduce the new resin [or an inert purge resin such as polyethylene (PE) or polypropylene (PP)].
- If the new resin is run at higher temperature settings than the old one, allow the new polymer to be completely introduced before raising the temperature settings. This helps the prior resin from seeing too much temperature. (If the new resin has a higher melting point than the operating temperature of the old resin, this step will have to be modified. Slowing down the line before raising the temperature is recommended in this case.)
- Likewise, if the new resin is run at a colder temperature than the old resin, lower the temperature first before introducing the new resin.
- Vary the speed of the extruder in random intervals for 1–5 minutes each. The variable speed is important for dislodging hung-up resin. By changing the screw speed, the shear rate is varied and with it the viscosity. A higher viscosity helps push out a lower viscosity resin. Different polymers respond differently to changes in shear rate; by

Worksheet for Disco Purge Procedure

MAXIMUM EXTRUDER RPM = ____________ RPM

1st min:	30% of MAX =	0.30 x ________	RPM = ________	RPM
2nd min:	90% of MAX =	0.90 x ________	RPM = ________	RPM
3rd min:	50% of MAX =	0.50 x ________	RPM = ________	RPM
4th min:	15% of MAX =	0.15 x ________	RPM = ________	RPM
5th min:	70% of MAX =	0.70 x ________	RPM = ________	RPM
Minutes 6-10:	15 to 20% of MAX =			
		0.15 x ________	RPM = ________	RPM
	to	0.20 x ________	RPM = ________	RPM

Minutes 11 to 15: Repeat the cycling steps of the first 5 minutes.

FIGURE 13.1 Worksheet for "disco" purge procedure. *Reproduced with permission from The Dow Chemical Company.*

varying the screw speed the chances of hitting upon a favorable shear rate for dislodging the incumbent resin is increased. DuPont named this the disco purge procedure (after the 1980s dance craze). An illustration of the disco purge method is given in Fig. 13.1. This procedure has been found to be more efficient than running the purge resin at a constant screw speed [25].

- Repeat the procedure if necessary. More than one iteration of the disco purge procedure may be required, especially for more complex dies such as those found in blown film.
- Purge compounds can be effective for cleaning the extruder. These compounds often contain foaming agents, inorganic scrubbing additives such as silica, and reactive compounds to help remove degraded polymer deposits. Carefully follow the instructions from the supplier. Purging these compounds out can be difficult and may take more resin than normal purging. The use of purge compounds, however, is warranted when the polymer has degraded in the extruder or die and when a particularly thorough cleanup is needed.

Efficient purging and resin transitions are related to the changeover time required between resins on film lines. Thurber et al. [26] study the changeover time for a laboratory blow film line, transitioning from LDPE to LLDPE. They use an optical tracer technique to monitor the concentration of a blue dye added to the resin just prior to its introduction to the extruder. They find that throughput is the primary factor influencing the changeover time; increasing throughput decreases changeover time. Staging formulations with increasing viscosity shortens the changeover time. They also find that changeover volume (amount of resin needed for the changeover: throughput × changeover time) is independent of changeover time. This means that throughput can be increased to speed up the transition without costing additional resin volume. They did not use a varying throughput (disco) transition protocol, which may have sped up the transition further.

Degradation of some polymers can create byproducts that are detrimental to equipment or human health, or cause taste and odor issues in the end-use application. PVDC, for example, can degrade into corrosive chlorine compounds. Coextrusion lines running PVDC are typically designed to isolate the temperature that the PVDC layer sees to help prevent degradation. Thermal degradation of polymers may evolve fumes containing carbon monoxide, acrolein and other byproducts. Proper ventilation is required to ensure worker safety [27]. Oxidative byproducts such as aldehydes and ketones may affect the taste/odor of the package.

13.2 Die lip buildup (die drool)

Die lip buildup, also known as die drool, die drip and plate-out, is the unwanted deposit and accumulation of material on the die face during extrusion. The material can slough off onto the film or coating resulting in product defects or line breaks. Increased machine downtime to clean the die of accumulated drool hurts productivity. Die drool is a problem with monolayer and coextrusion processes and has been reviewed by Gander and Giacomin [28] and Musil and Zatloukal [29,30]. Many factors influence die lip buildup; Zatloukal and Musil [30] provide an extensive list shown in Table 13.1.

TABLE 13.1 Factors contributing to die lip buildup.

Category	Factors
Die design	Abrupt corners at die lips "Dead corners" inside extrusion die (thermal degradation) High-surface energy die wall materials High surface roughness Short die land length
Polymer material	Low molecular weight fractions Multimodal polymers (low molecular weight tail) High elasticity of polymer Linear architecture of polymer High filler content High concentration/incompatible additives or pigments High concentration of volatiles (e.g., moisture) Dissimilar material properties in blends High content of stearates, ester lubricants, extending oils or waxes Chemical reactions between polymer and metal barrel/screw/die (PVC or PVDC resins)
Processing conditions	High processing temperature (thermal degradation) Drawdown Intensive cooling of die exit
Other factors	Die swell Shark skin Wall slip Vortices or secondary flows (thermal degradation) Slip-stick flow Pressure fluctuations in screw Excessive screw/barrel wear

Source: Adapted from M. Zatloukal, J. Musil, Die drool phenomenon in polymer extrusion, SPE ANTEC Conf. (2014) 1048–1054.

The root causes of die lip buildup are not completely understood and are likely due to multiple mechanisms, some specific to particular polymer systems. Musil and Zatloukal categorize die drool into internal and external origins. External die drool is related to the die exit region only and is more easily rectified; flaring of the exit angle generally reduces this type of drool [31,32]. Internal die drool is related to the die exit as well as flow channels upstream of the die exit, such as the extruder screw, barrel, transition pipes, and die.

Flow-induced fractionation of the polymer is one of the proposed mechanisms. As polymers flow, low MW components tend to migrate toward the high shear rate region near the die wall. Musil and Zatloukal [33,34] show that for high-density polyethylene (HDPE) significant fractionation only occurs very near the die wall and that fractionation and die drool increase with increasing linear chain length. Increasing the polymer chain length is associated with increasing melt elasticity and shear viscosity. Long-chain branching (LCB) is found to reduce die drool. These polymer characteristics also influence slip-stick flow behavior, which they conclude is associated with die drool for this polymer. Slip-stick is a phenomenon where the flow changes from zero velocity at the die wall to a finite velocity (known as the slip velocity) and back. Musil and Zatloukal also find that thermal degradation of HDPE inside the processing equipment upstream of the die exit can increase melt elasticity and die lip buildup [35].

Hogan et al. [36] investigate whether die swell is related to die drool. Die swell is the expansion of the polymer as it exits the die; it increases with increasing elasticity of the polymer but is present even in nonelastic (Newtonian) materials. In a series of experiments with capillary dies they find that as the length to diameter (L/D) ratio is increased the die swell decreases (as expected) but die drool increases. Their experiments with different shear rates and residence times provide further evidence that shear-induced migration of low MW species toward the die wall contributes to die lip buildup.

Chai et al. [37] extrude five LLDPE resins for one hour over a range of throughputs and measure the amount of die lip buildup. They find die deposit increases with polymer die swell up to a point; at high throughput, swell continues to increase but die deposit levels out. Die swell is thought to influence die drool since it brings the polymer melt closer to the edge of the die. Chai et al. postulate that at high flow rates, the chains become elongated and the angle of contact with the die surface is not as favorable for deposit. Die lip buildup in their experiments increases with increasing viscosity. They propose that the low MW fraction migrates toward the surface of the die. Short chains relax more quickly than longer chains resulting in die deposits. Higher viscosity results in more shear stress and heat that drives more of the low MW fraction toward the surface. Like Musil and Zatloukal, the LDPE they include in their study does not exhibit die lip buildup and blending LDPE into their LLDPE resins decreases the amount of die drool proportionately. They also find that die buildup and melt fracture (also called extrudate distortion, see Section 13.4.1) do not occur at the same time, also observed by other investigators. At the very high flow rates needed for melt fracture, the die buildup is suppressed; high buildup occurs at intermediate flow rates.

Giacomin et al. [38–41] propose die drool is the result of cohesive fracture of the fluid creating an internal slip surface near the wall of the channel. The velocity of this drool layer is low – it creeps along and accumulates on the die surface at the exit of the channel. It may also degrade, causing further drool products. Heating the slip layer increases its flow and reduces die drool, as does increase the cohesive fracture energy of the fluid.

Holtzen and Musiano [42] study the effect of pigments on die lip buildup. The addition of pigments such as TiO_2 is known to exacerbate die drool. They conclude that plate-out can originate at flow disturbances upstream of the die lips, typically in a region of abrupt change, such as the entrance region of the die. Viscoelastic polymers tend to form vortices or dead zones behind flow disturbances. Holtzen and Musiano's experimental studies show that the flow dead zones or vortices that form behind burrs or even degraded polymer can give rise to plate-out. They postulate that polymer trapped in the vortices can degrade over time into more polar species (through oxidation) that adhere to the die walls and eventually come out as die drool. Their experiments show streaks of degraded polymer working their way up the die channel in accordance with this theory.

The tendency for vortices to form depends on the channel geometry and the viscoelastic nature of the polymer. They use the Deborah number (De) to characterize the viscoelastic nature of the flow. The Deborah number is the ratio of the characteristic relaxation time of the polymer to the characteristic time of the process:

$$\mathrm{De} = \frac{\lambda}{t_p} \tag{13.1}$$

where De = Deborah number, λ = polymer relaxation time, and t_p = process time.

Increasing polymer throughput (decreasing t_p) increases De. Increasing the polymer MW and amount of LCB (which increase λ) also increases De. Vortices will form more readily in flows with high De. By varying the throughput in their experiments, they find that increasing De increases the amount of die lip buildup.

Their work suggests two reasons why pigment particles tend to be implicated in die lip buidup problems. The first is that the particles bind to some of the polymer molecules, reducing the polymer's mobility. This increases the relaxation time of the polymer and increases De. The ability of the pigment particle to bind to the polymer depends on the surface coating on the pigment. Holtzen and Musiano's study finds that surface coatings on TiO_2 have a significant impact on die drool. Secondly, some pigments reduce the thermal stability of the polymer; polymer caught up in the vortex is more likely to degrade and cause die lip buildup.

Another implication of Holtzen and Musiano's work is that the entrance geometry of the die channel is important: a more streamlined flow will reduce the tendency for dead zones to form. Elbows and sudden expansion of flow channels should be avoided. Elimination of burrs and other protrusions in the flow channel is also essential. Paying attention to the materials of construction, specifically, the surface finish may help reduce die lip buildup. Lowering processing temperatures to help prevent polymer degradation may also help minimize die lip buildup.

The die exit region has garnered considerable attention for both the origin of die drool and the means to mitigate it. As the polymer emerges from the die, the flow at the die wall transitions from no-flow to free-surface flow. This is illustrated in Fig. 13.2. This sudden acceleration has been implicated in causing die lip buildup. Gander and Giacomin [28] show that this boundary between the solid wall, liquid polymer and air is a high-stress region; they suggest high stress may fractionate or rupture the polymer, releasing low MW components. Zatloukal and Chaloupkova [43,44] use finite element modeling to show that this acceleration creates a negative pressure that extends inside the die. They propose that the negative pressure may suck low MW species toward the wall and surfaces of the die at the exit.

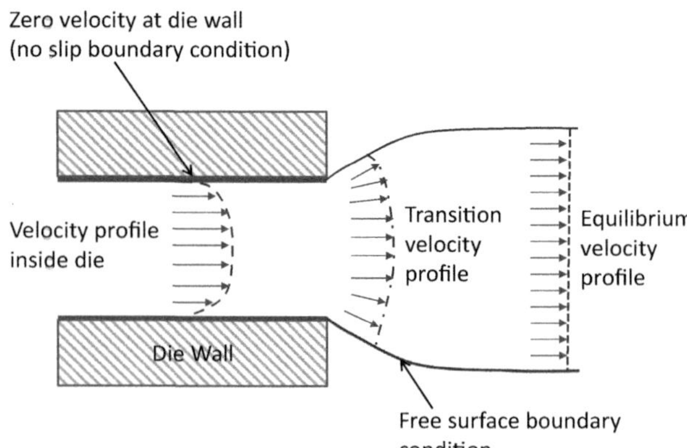

FIGURE 13.2 Illustration of polymer flow rearrangement at the die exit.

Flaring or chamfering the die exit region has been found to be an effective means to reduce die drool [31,32]. Gander and Giacomin explain this in terms of stress. In a converging flow channel, the stress will overshoot, exacerbating the die drool (causing more fractionation). In a flared or diverging channel, the stress will undershoot, reducing the fractionation. Zatloukal and Musil [30] show that chamfering reduces the negative pressure at the die exit, with a 45-degree angle having the best results.

The addition of polymer processing aids, particularly fluoroelastomers, has been shown to reduce die lip buildup [28,37,45]. Modifying the die surface with a low surface energy coating (such as a fluoropolymer) has also proven successful [46]. These techniques reduce the adhesion between the polymer and channel wall. This relaxes the no-slip condition at the die wall; slip flow along the wall reduces the shear rate at the wall and the driving force for fractionation. Furthermore, the presence of slip flow reduces the amount of flow rearrangement and acceleration at the die exit. This reduces the stress and the negative pressure, further decreasing the driving force for migration of low MW species to the die surfaces. Reduced stress also reduces melt swell (and stress-related degradation that may lead to additional swell), making the contact angle between the polymer and die less favorable for deposit.

Moisture entrained with the polymer, or generated during extrusion, can contribute to die lip buildup. Most polyolefins contain negligible moisture due to their hydrophobic nature, but various mineral fillers, pigments or other additives may absorb moisture that can be volatilized during extrusion. Polar polymers, such as acid copolymers and ionomers pick up moisture if not handled properly prior to extrusion. Acid copolymers can also generate moisture during extrusion when adjacent acid groups react to form anhydride groups; higher acid grades are more prone to moisture evolution due to the greater probability that acid groups will come into contact. As the polymer exits the die, the pressure drops and allows the water vapor (and other volatiles) to escape. As it does, the polymer may rupture or small bits of the polymer may be entrained and impinge upon the external die surface. Avoiding moisture issues can be as easy as ensuring open boxes of hydrophilic resins are kept closed during operations, or in the case of acid copolymers, using the proper screw design, processing at optimal conditions and selecting a grade that is less prone to problems.

Many other factors have been implicated in die lip buildup, as delineated in Table 13.1. More than one may be involved in a specific problem and a systematic approach to resolving issues is recommended. The reader is referred to the review articles mentioned at the beginning of this section for more ideas on how to address the problem.

13.3 Moisture related issues

Condensation polymers such as polyester and polyamide are susceptible to chain scission if moisture is present during extrusion. This reduces the MW and can make the polymer brittle. Also, lower MW reduces the melt viscosity, which can lead to flow instability issues during coextrusion (see Section 13.4). Most polyester resins need to be dried prior to extrusion. The resin supplier can provide guidance on proper drying conditions. Polyamide resins are packaged dry in moisture-resistant packaging; they generally can be used without drying if they are processed immediately after opening the bag or box and care is taken to keep moisture out during processing.

As described in Section 13.2, moisture absorbed by the polymer prior to extrusion can contribute to die lip buildup. Moisture can also show up as a bubble pattern in the film, known as a lace curtain. In addition to polayamide, ethylene vinyl alcohol and ionomers are prone to moisture absorption. The water molecules are soluble in the molten polymer under pressure in the die. When the pressure is released at the die exit, the water expands, forming bubbles. Generally, there is a threshold temperature and water concentration at which this occurs; as the processing temperature is increased, the maximum amount of water than can be entrained with the polymer without generating bubbles decreases. As a result, resin moisture tends to present more problems in high temperature processes such as extrusion coating.

13.4 Flow maldistribution and instability issues

The fabrication of coextruded films and coatings involves complex flows at high elongation and shear rates. These give rise to various flow issues that affect product quality. The specific issues described below have their origins in the viscoelastic nature of polymers or in the hydrodynamics of the process. However, perhaps the largest cause of quality issues in film and coating is extruder instability (such as surging and temperature nonuniformity) and this should not be overlooked. Good melt quality from the extruder is essential for downstream converting operations. This is true for coextrusion or monolayer extrusion. The pressure variation should be less than 1% and the temperature uniform to ensure good film converting. This can generally be achieved through proper screw design and processing conditions. The reader is referred to several useful books that cover this topic [1–6,11].

13.4.1 Sharkskin and extrudate distortion

This family of defects has a variety of names and causes, some of which are still unknown or debated in the scientific community. They have been studied for well over 50 years and are subject to periodic reviews, including those by Koopmans et al. [47] and Vergnes [48]. These defects have dramatically different characteristics for linear and branched polymers. One way to characterize these defects is to consider the behavior of the extrudate emanating from a die as the flow rate is increased. As depicted in Fig. 13.3 (and pictured in Fig. 5.10), the extrudate of linear polymers such as HDPE and linear low-density polyethylene (LLDPE) at a low flow rate is smooth. As the flow rate is increased, the surface starts to become rough. This is typically seen as a loss of gloss and transparency. With higher flow, the surface defect takes on a regular pattern of ridges on the surface perpendicular to the flow. This surface texture is commonly called sharkskin.

Sharkskin originates at the die exit. It is associated with the abrupt transition from the confined flow in the die to the free surface flow at the die exit described in Section 13.2 and illustrated in Fig. 13.2. There are two explanations still being debated in the literature. One involves slip-stick behavior at the die wall. The other involves rupture of the skin of the extrudate due to the elongational stresses arising from the acceleration in velocity of the polymer as it changes from parabolic flow to plug flow at the die exit.

With further increases in flow rate, the pressure becomes unstable, with regular oscillations. The flow curve, depicted in Fig. 13.3A, is discontinuous: the flow switches between regions of smooth flow and sharkskin. This is known as oscillatory or slip-stick flow. The latter describes what has been experimentally observed as the mechanism for the behavior. As the pressure oscillates, the flow at the wall jumps from the traditional no-slip condition to a finite slip velocity. The cause of this seems to be an entanglement–disentanglement transition of polymer chains adsorbed onto the wall [48].

At higher flow rates, the flow may become stable and smooth before eventually, at still higher flow rates, it becomes chaotic. The chaotic flow is known as extrudate distortion or gross melt fracture. This severe distortion in many polymer systems appears to originate in the die entrance region.

Branched polymers like LDPE, as well as polystyrene (PS) and some PP grades, do not exhibit sharkskin or slip-stick flow. Instead, they transition to helical flow, where the extrudate has a regular screw-like appearance. As shown

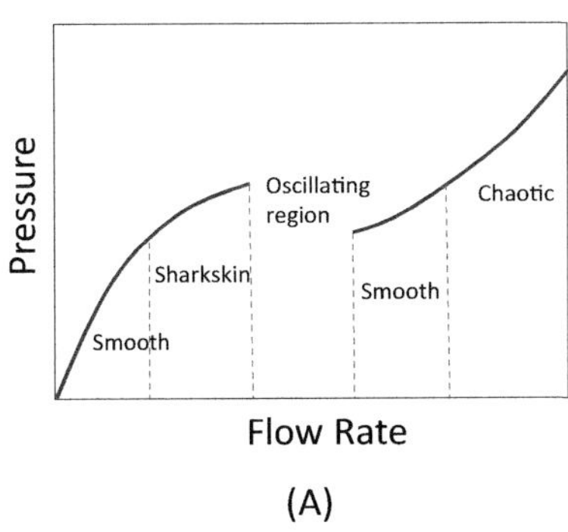

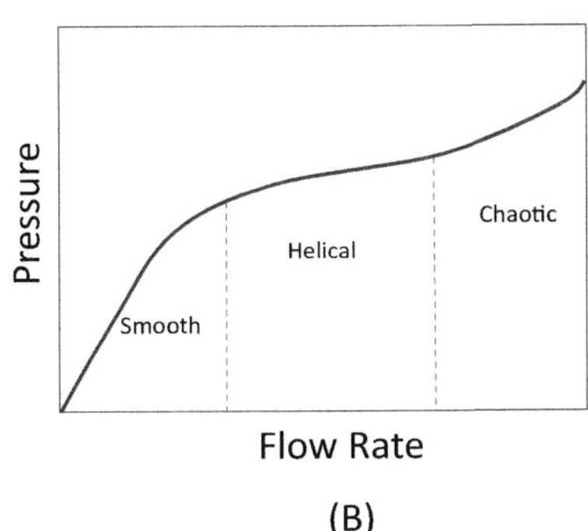

FIGURE 13.3 Extrudate distortion flow curves for linear (A) and branched (B) polymers. *Adapted from B. Vergnes, Extrusion defects and flow instabilities of molten polymers, Int. Polym. Process. (2015) 3–28.*

in Fig. 13.3B, at a higher flow rate, the flow becomes chaotic, exhibiting behavior similar to the linear polymers at high flow rates.

The phenomenon that is responsible for these defects may be specific to each polymer. There are a number of factors that have been found to influence these defects, including [49]:

- Chemical nature of the polymer.
- Polymer MW and molecular weight distribution (MWD).
- Polymer branching.
- Die entrance geometry.
- L/D ratio of capillary or die.
- Material of construction of the capillary or die.
- Temperature.
- Flow rate.

As stated previously, LCB prevents the formation of sharkskin. Indeed, when LLDPE was first introduced in the 1970s, it could only be run on blown film equipment designed to run LDPE at severely reduced rates due to sharkskin and bubble instabilities. Modifications to the equipment, such as utilizing wider die gaps, were required to run LLDPE. Monchai et al. [50] show that when LDPE is subjected to extensive shearing in an internal mixing device in what is known as shear modification, it exhibits sharkskin behavior like linear PE (and the flow rate for the onset of severe melt fracture is greatly increased). They propose that this behavior is due to the alignment of the chains during shear modification of LDPE, which reduces its elongational viscosity (and strain hardening behavior).

Burghelea et al. [51] compare the surface instabilities of an LDPE and LLDPE resin and relate the differences to their extensional rheological behavior. Extensional viscosity measurements show the LLDPE exhibits no strain hardening behavior and steady-state is reached quickly. The LDPE shows extensive strain hardening and does not reach steady state in the time scale of the measurement. The terminal stress at break for the LDPE is 4.6 times higher than that for the LLDPE; at a given strain, the LDPE can sustain much higher tensile stress than LLDPE. They measure the flow fields for the two resins in the die exit region using laser Doppler velocimetry and find they are qualitatively similar—flow accelerates near the channel wall at the exit and slows down near the center as depicted in Fig. 13.2. From the velocity measurements, they compute the axial flow gradients (which are related to the stress) near the wall and find they are quite different. A tensile stress boundary layer develops near the wall with a stress imbalance between this boundary layer and the bulk flow. The magnitude of the imbalance is higher for the unstable LLDPE flow (the stress in the boundary layer is about $2\times$ higher than in the bulk) and exceeds the melt strength of the LLDPE, giving rise to sharkskin. For the stable LDPE case, the imbalance is less (the boundary layer is only about 30% higher than the bulk) and does not exceed the melt strength of the LDPE. Shear stresses are found to have little effect since they are significantly smaller than the tensile stresses.

Several methods used to eliminate or postpone the onset of sharkskin include:

- Reducing the output. This is not a desirable fix but can be a short-term solution while other methods are evaluated.
- Increasing the die gap. This can cause some cooling issues, which have been resolved by equipment design improvements over the years. LLDPE is typically run commercially today with a wider die gap than LDPE. It also may lead to some property changes.
- Heating the die lips. This reduces the viscosity and stress on the wall.
- Increasing the MWD or decreasing the MW of the linear polymer. This can result in loss of film strength and other property changes.
- Utilizing coextrusion to add lower viscosity skin layers. This lowers the stress at the wall to reduce sharkskin while maintaining a higher MW core layer to retain the film physical properties.
- Lowering the surface energy of the metal surface. Brass and some other metals have been found in the lab to suppress sharkskin but Vergnes [48] comments that this is not done commercially. Coating with fluoropolymer or other techniques to enhance slip at the wall has also been proposed.
- Using PPAs. These are described in Section 13.2. Typically made from fluoroelastomers, these are added in small amounts to the polymer. After an induction time of several minutes, the PPA coats the die walls. The PPA is continuously added to the polymer to replenish the coating that is sloughed off during flow. The generally accepted mechanism is that the coating aids in the slip of the polymer flow along the die wall (known as slip flow). This helps reduce the magnitude of flow acceleration at the die exit. Ramamurthy [52] proposes, however, that the PPA works by increasing the adhesion at the die wall.

The use of PPAs is also the most practical method to postpone slip-stick or oscillatory flow to higher throughput. Other methods have been tried in the lab but are not practical for commercial operations.

For both linear and branched polymers the helical instability and chaotic or gross melt fracture have been shown to originate in the entrance region of the die. The main method for reducing the onset of these instabilities is to reduce the entrance angle. Vergnes [48] reports that the use of a smooth convergent entrance whose angle varies continuously between the reservoir and the die land increases the throughput (at the onset of gross melt fracture) by a factor of six over an abrupt 90 degrees entry. For a long time, it was thought that the disruption of eddies or secondary flow at an abrupt entrance was responsible for these instabilities; but it has been shown that such eddies do not exist for linear polymers and are not needed for gross melt fracture to occur. Extensional stresses may be involved; Vergnes finds evidence for a critical extensional stress above which distortion develops.

Another method that has been successful in reducing extrudate distortion is increasing the temperature. The lower viscosity reduces the stress. A longer die land can also sometimes help.

13.4.2 Drawdown: neck-in, edge bead, edge weave, draw resonance and edge tear

As the melt curtain in drawn down to its final thickness in high-speed cast film and extrusion coating processes, several flow phenomena may occur that affect quality and productivity. These include neck-in (a reduction in the width as the melt curtain traverses the air gap between the die exit and chill roll), edge bead (thickening at the edge of the melt curtain), edge weave (fluttering or instabilities at the edges), draw resonance (a flow instability that manifests as thickening or width oscillations) and edge tear (tearing of the melt curtain).

13.4.2.1 Neck-in and edge bead

As the molten polymer is drawn from the die exit to the chill roll in extrusion coating and cast film, the width of the web decreases. This is known as neck-in and is illustrated in Fig. 13.4. Neck-in affects productivity since the processer does not have the ability to utilize the full width of the die. It also contributes to a thickening of the web near the outside edge of the web. This is known as edge bead and is sometimes referred to as a dog bone profile. This too affects productivity since the edges that fall out of specification must be slit and removed (edge trim). Examples of neck-in and edge beads are shown in Figs. 13.5 and 13.6 for an LLDPE and LDPE resin, respectively.

Neck-in and edge bead can be reduced by the following actions:

- Increasing the melt strength or elasticity of the polymer. Polyolefins with LCB (LDPE, EVA, EMA, EAA, EMAA, and ionomers) exhibit less neck-in than linear polymers such as LLDPE and HDPE. Broader MWD and higher MW (lower MFI) may also help.
- Decreasing the melt temperature. This increases the melt strength.
- For linear polymers, decrease the draw-down ratio (DDR) [the ratio of the line speed to the velocity at the die exit and can be estimated by the ratio of the die gap to the final film or coating thickness—see Eq. (2.16)]. Increasing the coating or film thickness (at constant die gap) or decreasing the die gap (at constant coating or film thickness) decreases the draw ratio. A die gap that is too narrow may cause sharkskin and melt fracture, however (see Section 13.4.1).

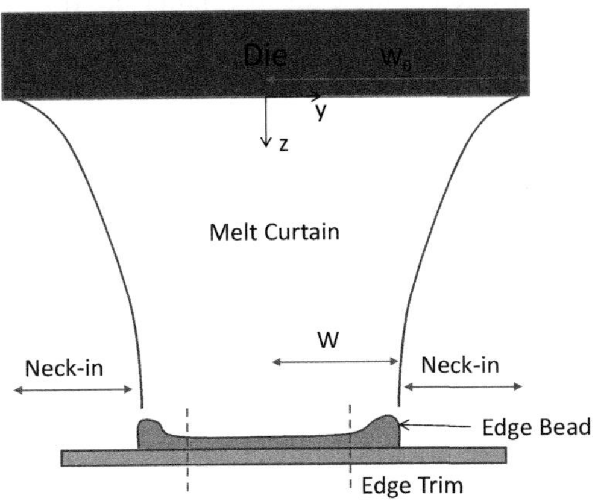

FIGURE 13.4 Definition of neck-in, edge bead and coordinate variables in extrusion coating and cast film.

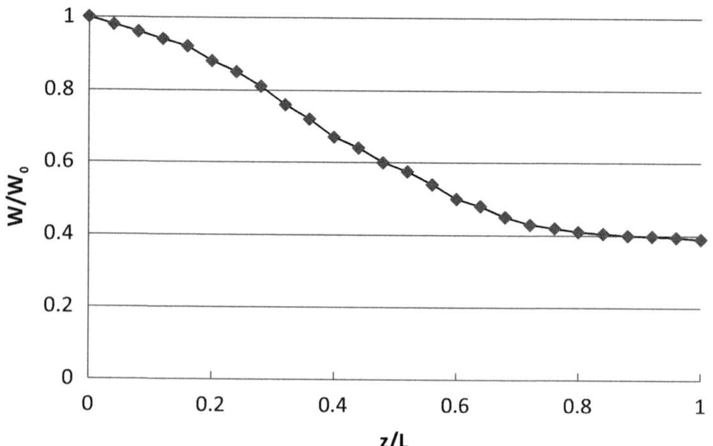

FIGURE 13.5 Change in width (2W) of an LLDPE extrusion cast film as a function of distance from the die exit. $2W_0$ = die width. L = distance from die exit to chill roll (air gap), z = distance from die exit [53]. *Experimental data from D. Silagy, Y. Demay, J.F. Agassant, Stationary and stability analysis of the film casting process, J. Non-Newtonian Fluid Mech., 79 (1998) 563–583.*

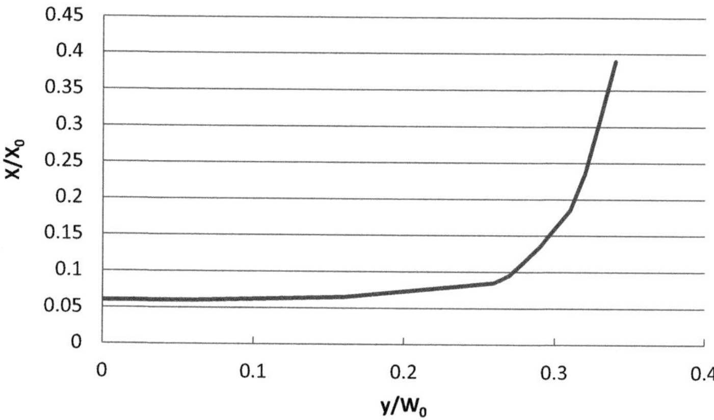

FIGURE 13.6 Thickness profile across a cast film of low-density polyethylene (LDPE) illustrating edge bead. Data is from the final film ($z = L$ at the chill roll). W_0 = half die width, y = distance from centerline along the width of the web ($y/W_0 < 1$ due to neck-in), X = thickness, X_0 = thickness of die gap. Draw ratio = 15. *Experimental data from D. Silagy, Y. Demay, J.F. Agassant, Stationary and stability analysis of the film casting process, J. Non-Newtonian Fluid Mech., 79 (1998) 563–583.*

- Decreasing the air gap or increasing the line speed (at a constant DDR, which occurs when the die gap and final film or coating thickness are held constant). This reduces the characteristic process time, which for this process is the time in the air gap and increases the Deborah number. Higher De is an indication of higher elasticity and melt strength.
- For highly elastic polymers, reducing the stress in the die exit region by modifying the die design to allow for more relaxation. Zatloukal and Barborik [54–56] show that greater stress at the die exit increases neck-in for polymers with long relaxation times (which is typical of polymers with LCB).

Modern extrusion coating and cast film dies utilize edge bead reduction technology to minimize the edge bead [57,58]. This technology involves thinning the flow at the edges to compensate for thickening during drawdown. Several versions exist, each with their pros and cons. In one version, rods are inserted at the ends to contour the flow at the edges (see Section 2.3.1.1). These can be adjusted for different polymers and operating conditions. Other designs use flexible die lips. Leakage flow can be an issue and should be considered when selecting the design and operating these types of dies.

Neck-in and edge bead arise from an unbalance of forces at the edge of the melt curtain due to the free surface [59]. This is illustrated in Fig. 13.7. As the melt curtain is stretched in the machine direction, it contracts in the transverse and thickness dimensions; the drawdown in the air gap gives rise to a contraction force. Near the centerline, the contraction force is opposed by the melt tension of the web. Adjacent molecules along the width help to resist the contraction in the width dimension. This resistance is related to the elasticity or melt strength of the polymer. Debroth and Erwin [60–62] propose that the extension of the melt curtain near the centerline behaves like a planar extension, where the width is fixed, and the contraction occurs only in the thickness dimension:

$$X_L = X_0 \frac{v_0}{v_L} = X_0/\mathrm{DDR} \tag{13.2}$$

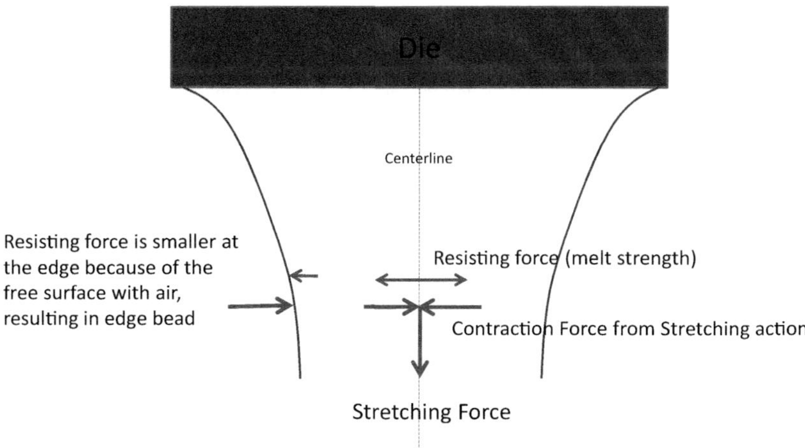

FIGURE 13.7 Forces on melt curtain leading to edge bead.

where X_L = thickness of melt curtain near centerline at the chill roll, X_0 = thickness of melt curtain at die exit (die gap), v_0 = velocity of the melt curtain at die exit, v_L = velocity of film or coated substrate at chill roll (or line speed), and DDR = drawdown ratio (the ratio of the line speed to the velocity at the die exit, v_f/v_0).

At the edges, the resisting force is unbalanced. On one side is a polymer that can help transfer the tension across the web, but on the other side, the contraction is only opposed by the air pressure. As a result, the polymer beads up in response to the contracting force. At the edge, the extension resembles tensile extension:

$$d_L = X_0\sqrt{v_0/v_L} \tag{13.3}$$

where d_L = thickness of the edge bead.

Combining Eqs. (13.2) and (13.3) yields an equation for the bead ratio (ratio of the edge thickness to the centerline thickness):

$$\frac{d_L}{X_L} = \sqrt{\frac{v_L}{v_0}} = \sqrt{\text{DDR}}. \tag{13.4}$$

Debroth and Erwin measure the bead ratio for an LDPE and find it agrees well with Eq. (13.4). Canning and Co [63] study LDPE and LLDPE resins and find similar agreement with Eq. (13.4). Other investigators have shown that the amount of neck-in correlates with the ratio of the planar to uniaxial extensional viscosity [64–66].

Many investigators have found the presence of LCB reduces neck-in [66–71]. Seay and Baird [69] study several metallocene LLDPE (mLLDPE) resins with different levels of sparse LCB and an LDPE control (with high levels of LCB). They find that for the linear mLLDPE, higher levels of sparse LCB reduce neck-in at a low DDR and is more effective than broader MWD. At high DDR, all the LLDPE resins converge to the same amount of neck-in. The LDPE control shows reduced neck-in over the entire range of DDR investigated. Modeling shows that incorporating LCB on the shorter chains of the mLLDPE is more effective at reducing neck-in at high DDR.

Numerical simulation of the cast film process provides further insights into the origins of neck-in and edge bead. Barborik and Zatloukal [66] provide a comprehensive review of the literature and cast film modeling is discussed in more depth in Chapter 18. These models show that the elasticity and strain hardening behavior of the polymer and processing parameters influence neck-in and edge bead. An example from the work of Debbaut et al. [72] is shown in Fig. 13.8 where the width or neck-in ratio as a function of DDR for a Newtonian and a Giesekus fluid is computed.

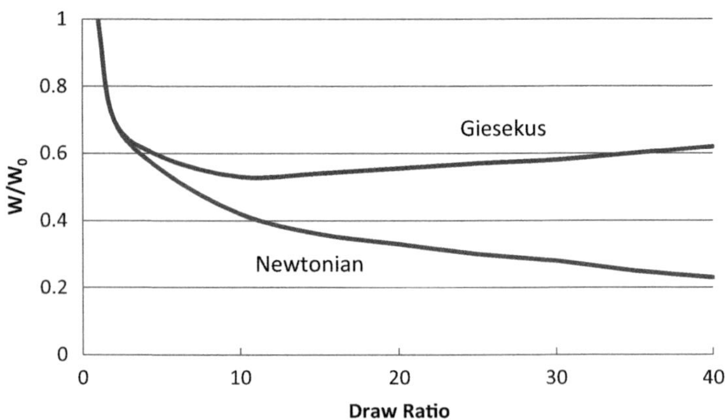

FIGURE 13.8 Effect of draw down ratio and polymer viscoelasticity on neck-in. The Newtonian fluid is purely viscous whereas the Giesekus fluid has viscous and elastic behavior. *From numerical simulations of B. Debbaut, J.M. Marchal, M.J. Crochet, Viscoelastic effects in film casting, Z. angew Math. Phys., 46 (Special Issue: Theoretical, experimental, and numerical contributions to the mechanics of fluids and solids) (1995), S679–S698.*

A Newtonian fluid is purely viscous whereas a Giesekus fluid includes elasticity (which manifests as viscosity thinning in shear and strain hardening in extension) that generally gives rise to melt strength. The viscoelastic fluid has less neck-in. The experimental work of Canning and Co [63] with an LDPE and two LLDPE resins validates this. The LDPE exhibits extensional hardening behavior and its neck-in decreases with increasing DDR, in agreement with the Giesekus fluid of Debbaut's model. The neck-in of the LLDPE resins increases with increasing DDR, agreeing with Debbaut et al.'s analysis of a Maxwell fluid, which exhibits shear thinning but not extensional hardening. Similar findings are reported by other investigators [54–56,66,68–75].

Fig. 13.9 shows the simulation for constant DDR, but different Deborah numbers (De) (Eq. 13.1). De = 0 corresponds to a Newtonian fluid and increasing values of De represent greater elasticity or melt strength. As De increases, the neck-in decreases (neck-in ratio increases), as indicated by the long lines along the y/W_0 axis. Also, as De increases, the shape of the thickness profile changes; the thickness remains steady over a wider swath (until the very edge of the melt curtain). This is important for product quality and productivity, as less of the edge will need to be trimmed. Note that the actual thickness ratio (X/X_0) at the edge of the web is about the same for the different Deborah numbers.

Figs. 13.8 and 13.9 also show how processing affects neck-in. DDR is directly related to the ratio of the die gap to the coat weight or film thickness (Eq. 2.16). If the die gap is kept constant, increasing the coating or film thickness will decrease DDR. Fig. 13.8 shows that this will change the amount of neck-in depending on the elasticity of the polymer. Likewise, the Deborah number in Fig. 13.9 is proportional to the line speed and inversely proportional to the air gap. At constant DDR and coating or film thickness, Fig. 13.9 indicates that increasing the line speed or decreasing the air gap will reduce neck-in. This is shown more explicitly in Table 13.2, which shows some unpublished data by Foederer and Morris [76] collected while conducting a temperature in the air gap study. LDPE is extrusion coated onto paper at a melt temperature of 324°C. The die width and gap are kept constant at 91 cm (36 inches) and 760 μm (30 mils), respectively. The coat weight (final thickness) is varied from 6.4 to 25.4 μm (0.25–1 mil), the air gap from 17.8 to 28 cm (7–14 inches) and line speed from 152 to 457 m/minute (500–1500 fpm). The neck-in is recorded as the total reduction in width from both edges at the chill roll. For example, if the final coating width was 81 cm, the neck-in was recorded as 10 cm (91–81 cm).

The results are plotted in Fig. 13.10 as a function of line speed. The plot of main effects from statistical analysis (Fig. 13.11) shows that increasing air gap and reducing coat weight substantially increase the neck-in. The effect of line speed is less pronounced in this study. The results are consistent with the numerical analysis of Debbaut et al.

Neck-in and edge bead have been found to be influenced by DDR, MW, MWD, relaxation time, the ratio of second to first normal stress difference at die exit, LCB, strain hardening in uniaxial extension, planar to uniaxial extensional viscosity ratio, take-up length (air gap), the velocity at the die exit and temperature [66]. The numerical simulations described above provide insight but are challenging for the practitioner to utilize especially with the non-Newtonian constitutive equations that often require sophisticated rheological measurements. Zatloukal and Barborik [56,77] reduce numerical simulations into simpler analytical equations involving normal stress differences, relaxation time, planar to axial extensional viscosity ratio and uniaxial extensional strain hardening. Since these material properties are not readily available, even these equations are not practical for most industrial processors. In industry, the Rheotens melt tension measurement (see Section 5.3.6) is often used as a measure of melt strength. Polymers with higher melt tension generally have lower neck-in. Polymers with higher MW, broader MWD or LCBs typically have higher melt strength.

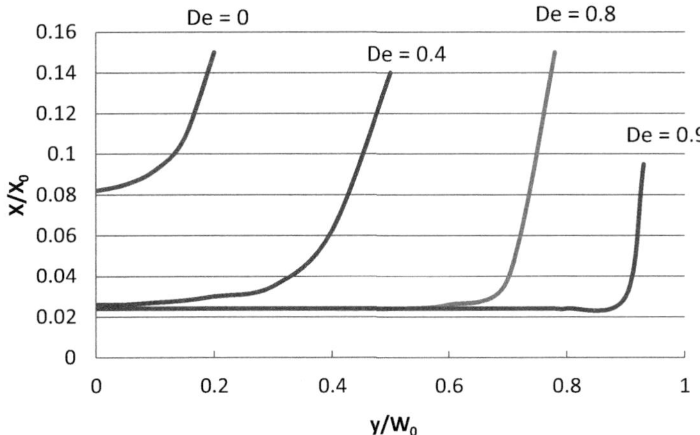

FIGURE 13.9 Effect of Deborah number (De) on neck-in and edge bead. Here y is the distance in the width dimension from centerline. Thickness ratio is taken at the chill roll (z = L). Draw down ratio is constant at 40. *From numerical simulations of B. Debbaut, J.M. Marchal, M.J. Crochet, Viscoelastic effects in film casting, Z. angew Math. Phys., 46 (Special Issue: Theoretical, experimental, and numerical contributions to the mechanics of fluids and solids) (1995), S679–S698.*

TABLE 13.2 Neck-in data from an extrusion coating trial.

Line speed (m/min)	Coat weight (μm)	Air gap (cm)	Neck-in (cm)
152	6.4	18	11
152	6.4	28	16
152	12.7	18	9
152	12.7	28	14
213	12.7	28	12
213	12.7	18	8
152	25.4	18	8
152	25.4	28	12
305	25.4	28	11
305	25.4	18	7
457	25.4	18	7
457	25.4	28	11

Notes: LDPE was coated at 324°C using a Cloeren edge bead reduction die (57 mm plug and 32 mm flag settings). The LDPE had a density of 0.923 g/cc and melt flow index (190°C/2160 g) of 4.5 g/10 minutes. Deckled die width was 91 cm and die gap 760 μm.
Source: Data courtesy of B.M. Foederer, B.A. Morris, Use of IR technology to model temperature loss in the air gap, TAPPI PLACE. Conf. (2014).

Ionomers, with their ionic interactions, have exceptionally high melt strength for polyolefins. Increasing the melt temperature reduces melt strength and increases neck-in. Chai et al. [78] and Clavenhag and Oveby [79,80] propose using dynamic linear viscoelastic measurements (G' and G''). As described in Chapter 5, the elastic modulus (G') is very sensitive to changes in MWD and LCB at low values of loss modulus (G''). Chai et al. discuss how the crossover point where $G' = G''$ is related to MW and MWD but is not measurable for some PE grades due to the onset of melt fracture. They propose using the value of G' at the frequency where $G'' = 500$ Pa (shorthand written as $G'(G'' = 500$ Pa)). They find neck-in and maximum draw-down (DDR where line breaks occur) correlate well with $G'(G'' = 500$ Pa) as well as with melt tension from Rheotens measurements. Clavenhag and Oveby evaluate several commercial grades of LDPE with similar melt flow index and density but differ widely in processibility (such as the amount of neck-in). They find a good correlation between neck-in from pilot-scale extrusion coating trials with lab measurements of zero shear viscosity (found from curve fitting dynamic viscosity measurements) and $G'(G'' = 500$ Pa). These rheological measurements that are more closely associated with elasticity and melt strength provide a better tool to predict neck-in performance from standard laboratory measurements.

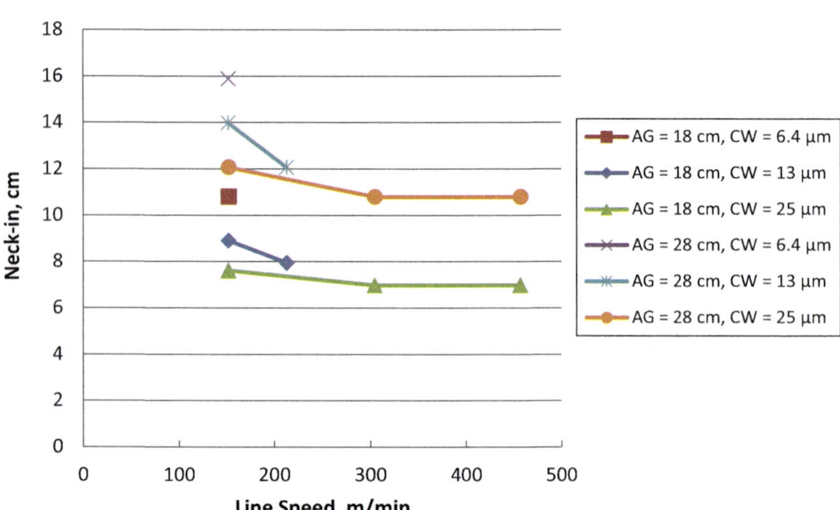

FIGURE 13.10 Scatterplot of neck-in experimental results from Table 13.2. *AG*, air gap, *CW*, coat weigth (thickness). *Data courtesy of B.M. Foederer, B.A. Morris, USe of IR technology to model temperature loss in the air gap, TAPPI PLACE. Conf. (2014).*

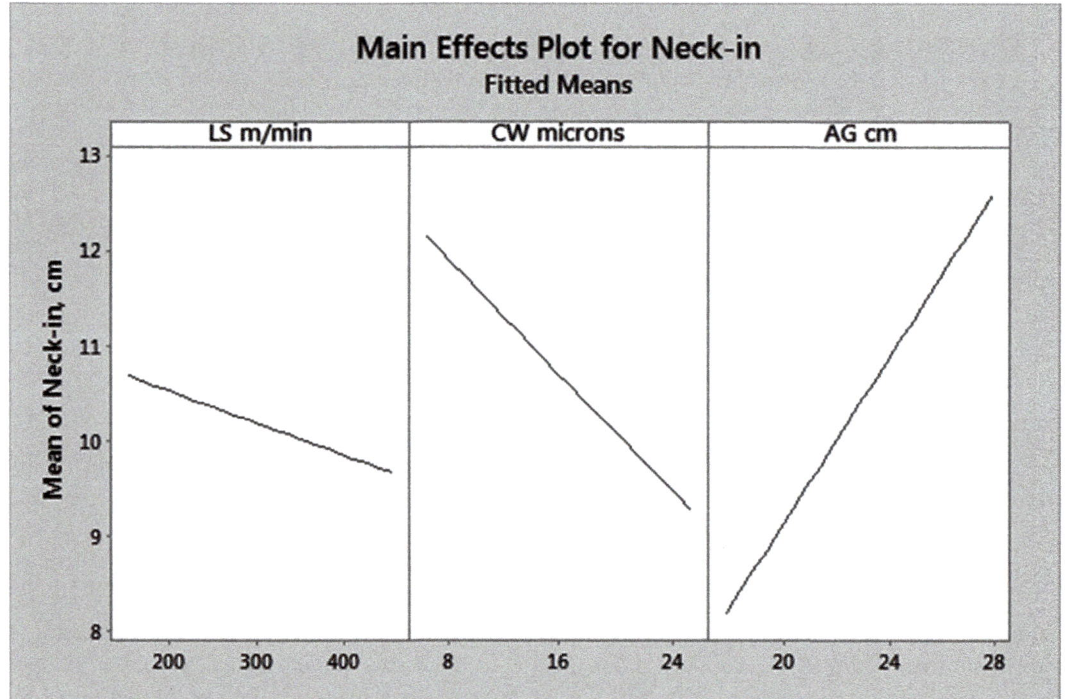

FIGURE 13.11 Main effects plot from statistical analysis of neck-in data from Table 13.2. *LS*, line speed, *CW*, coat weigth (thickness), *AG*, air gap.

13.4.2.2 Edge weave, draw resonance and edge tear instabilities

At high DDRs, edge weave or draw resonance instabilities may develop in the molten web. These instabilities may have different origins and characteristics depending on the process and polymer. Examples include:

- Weaving of the coating. If the edge of the melt curtain moves beyond the substrate, it may stick to the nip roll and cause line breaks and wrap-ups.

- Periodic variation of the polymer output when the DDR exceeds a critical threshold. This is known as draw resonance and is a common problem in high-speed fiber spinning and film/coating operations. For films and coatings, the thickness, width or both may periodically vary. It is a hydrodynamic phenomenon and even occurs for Newtonian fluids. The critical DDR for Newtonian fluids is around 20 and can be higher for non-Newtonian fluids [81].

Many investigators have applied linear perturbation methods to the equations describing flow during film casting. These methods allow for the calculation of the draw ratio where draw resonance first occurs, called the critical draw ratio (DDR_c). These studies have found that draw resonance stability increases with increasing polymer elasticity, and more specifically hardening in tension [82–85]. Shear and extension thinning may decrease stability by reducing the critical draw ratio. Die dimensions and air gaps also play a role [84]. Faster cooling or higher heat transfer coefficients may improve stability [86,87].

These theoretical studies as well as experimental work show that the viscoelastic effects are complex. Burghelea et al. [88] investigate the stability of film casting of PP using real-time measurements of the width and velocity profile across the melt curtain using particle tracing techniques. They vary DDR and find three regimes:

- Small DDR ($DDR < DDR_c$)—stable,
- Large DDR ($DDR > DDR_c$)—periodic oscillation in width and velocity,
- Very Large DDR ($DDR >> DDR_c$)—chaotic fluctuations in width and velocity (nonperiodic). This regime is not well studied but is responsible for edge breaks.

DDR_c decreases with increasing Wissenberg number ($Wi = 2\lambda v_0/L$ where λ = relaxation time, v_0 = die exit velocity and L = air gap) indicating elasticity is a destabilizing factor. This agrees with other investigators for shear-thinning and extension thinning fluids. Elasticity has been found in other studies, however, to be a stabilizing factor for extension hardening fluids. Indeed, draw resonance is generally severe for linear polymers like LLDPE and HDPE which have extensional thinning behavior. Branched LDPE is extensional hardening with higher elasticty and melt strength; its stability is considerably improved.

Strategies for minimizing this type of instability include:

- Ensuring the extruder is producing a good quality melt. Pressure fluctuations should be minimized and the temperature uniform across the die width. Thermocouples in the extruder or transfer pipes can be unreliable. IR thermometry can be used to get an accurate reading of the temperature profile in the melt curtain. Hot spots at the edges can result in loss of melt strength, leading to edge weave issues [89].
- Reducing temperature. Lower temperature increases melt strength. In some cases, the excessive temperature may cause crosslinking or other changes in the polymer that contribute to instabilities.
- Decreasing the die gap. This reduces DDR, although as with any of these tactics, care should be taken to ensure the change does not cause other issues such as excessive pressure, high stress or sharkskin.
- Increasing the film or coating thickness. This also reduces DDR.
- Decreasing the take-up speed, but this reduces throughput.
- Switching to a higher melt strength polymer. Higher MW (lower MFI), LCB, polar interactions (as in ionomers and acid copolymers) and broader MWD may help.
- Coextruding with a high melt strength resin such as LDPE or ionomer. Bain and Co [90] show that coextruding thin extensional hardening resin (LDPE) skin layers with a linear (LLDPE) core improve the stability of the process.
- Blending with a higher melt strength polymer such as LDPE. Frey et al. [91] show data for edge weave of linear PE and blends of linear PE with LDPE. The addition of LDPE reduces the edge weave considerably.
- Managing the airflow in the air gap. This is mostly for extrusion coating where the air gap is often long. As the melt curtain is accelerated through the air gap it entrains air. The air is forced down onto the chill roll and has nowhere to go but up or out. This can cause air circulation patterns that cool the web unevenly.
- Encapsulating the core resin with a high melt strength polymer. New die technology allows a material to be fed to the edge of the die via a satellite extruder. This allows the encapsulation material to help stabilize the core resin. Frey et al. [91] report mixed results when applying this to linear PE.

Another type of instability is edge or curtain tear. Tears can initiate at the edge of the melt curtain as it is drawn down. This may cause line breaks and loss of productivity. Many of the remedies for this instability are the opposite of those for edge weave and draw resonance:

- Change the temperature. Lower viscosity may be needed to help draw the polymer down. Some polymers, however, may crosslink at high temperatures. Increasing their temperature may worsen the problem, as crosslinking tends to

increase the likelihood of tearing. So as a general guideline, try raising the temperature and if that does not help, try lowering the temperature.
- Reduce the DDR (this is the same for both types of instability). This can be achieved by decreasing the die gap or increasing the coating or film thickness (by lowering the line speed at constant output).
- Lower the melt strength. Narrower MWD, lower MW (higher MI) and less LCB may help.
- Use an IR thermometer to check the temperature distribution across the curtain width. Air circulation patterns created as air is entrained by the melt curtain as it is drawn down in the air gap may suck in cooler air at the edges. This may cause the edges to be cooler than the rest of the web. If this is the case, try heating up the die near the edges or better managing the airflow.

In extrusion coating, a draw-down test is used to characterize the drawability of a polymer. This consists of lining out the coating line at a standard line speed and coating thickness and gradually ramping up the line speed while keeping the output constant. The maximum speed prior to the onset of either line breaks or edge weave is recorded. Higher values of speed indicate better drawability, meaning the polymer can be drawn down to thinner coatings at high line speeds. Drawability is often a balancing act between edge weave and edge tear issues; look for a sweet spot where both are acceptable.

Stary et al. [92] look at failure and rupture of an LDPE melt when drawn in uniaxial extension (using a Munstedt type extensional rheometer). They vary the rate of extension and temperature to achieve Weissenberg numbers from 0.6 to 1113. The work shows three regimes: (1) low *Wi* where failure is by neck-in and is ductile, (2) intermediate *Wi* where competition between necking and strain hardening occurs (failure occurs at secondary necks), and (3) high *Wi* where strain hardening dominates and changes the rupture behavior.

13.4.3 Bubble instability in blown film

As the output is increased on a blown film line, the bubble can become unstable, causing gauge variation, and ultimately line breaks and downtime. Butler [93] reviews the bubble stability literature and proposes six types of instability:

1. Periodic oscillations of bubble diameter;
2. Helical motion;
3. Bubble tears at die;
4. Bubble flutter;
5. Bubble sits down on air ring;
6. Variation in frost line height.

These are illustrated in Fig. 13.12. The first is particularly prevalent when processing linear polymers such as LLDPE.

A number of investigators have studied the effect of processing conditions on bubble stability. Factors that improve stability include: decreasing the melt temperature [95,96]; increasing the blow up ratio (BUR) [94,97]; changing the

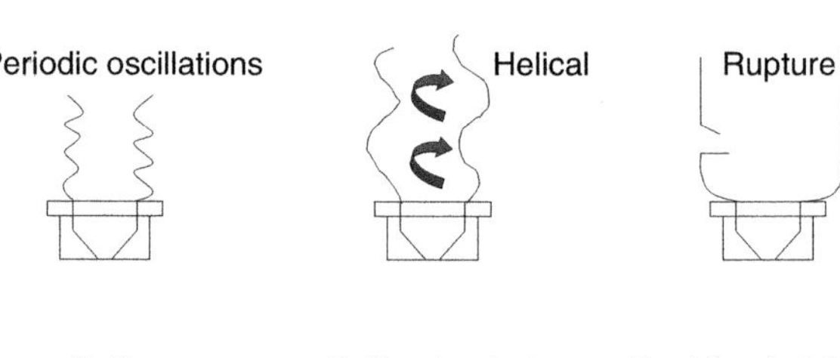

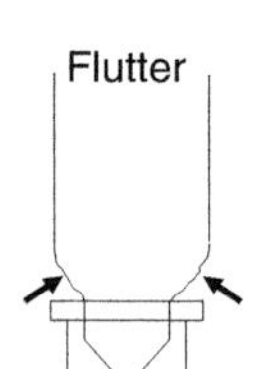

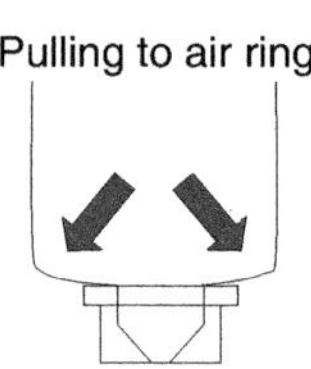

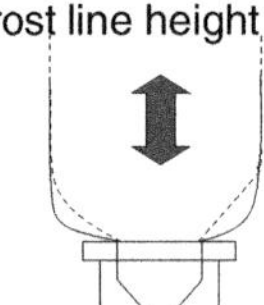

FIGURE 13.12 Types of bubble instability, as categorized by Butler [94]. *Adapted from P. Tas, L. Nguyen, S. Marshall, M. Kashanian, Melt strength, maximum output, and physical properties of SURPASS TM1 sLLDPE–LDPE blends in blown film, J. Plastic Film. Sheeting, 21 (2005) 265–279.*

frost line-height [93]; improving equipment design or operation [93]; and modifying the rheology of the resin [98–100]. Yoon and Park [101] conduct a linear stability analysis and find bubble stability is limited by both the BUR and thickness reduction ratio (ratio of final film thickness to die gap). The analysis also suggests there is an optimum frost line height. Their model agrees with observations of an LLDPE blown film line. Some of these approaches are limited by practical or physical considerations. For example, lowering the melt temperature of the polymer may be limited by the design of the extruder screw [93] and has been shown to negatively impact film properties [102].

The rheology and structure of the polymer are considered by many investigators. Han and Park [95] find that the presence of LCB improves stability. Broader MWD and increasing comonomer length [99] (butene < hexene < octene) have been shown to increase bubble stability of LLDPE. Field et al. [103] show explicitly that bubble stability improves with increasing melt strength of the PE, as measured by the Rheotens method (see Chapter 5). Ghijels et al. [104] also relate melt strength to bubble stability showing that melt strength increases with decreasing MFI for a given family of polyethylenes. At a fixed MFI, they find melt strength to be in the following order: autoclave LDPE > tubular LDPE > LLDPE. Munstedt et al. [105] correlate bubble stability with strain hardening in uniaxial extensional viscosity measurements, which is related to melt strength. Fang et al. [106] show that elasticity and elongational viscosity are important. In their studies, the order of decreasing bubble stability is LDPE > HDPE > mPE > LLDPE. Kolarik et al. [107] conduct experiments and develop a numerical model which shows melt strength and strain hardening in extensional flow are important for bubble stability; there are optimum values of melt strength and strain hardening that can provide the widest operating window and minimum film thickness. At a given strain hardening parameter, the model predicts increasing melt strength increases the operating window for stability.

A common technique used in the blown film industry to improve bubble stability and increase output is to blend LDPE into LLDPE [100,102,103,108–110]. Typically 10%–50% LDPE is used for sealant webs for food and specialty packaging applications and >60% LDPE for optical and shrink properties in collation shrink films and masking films. Several investigators have reported a synergistic effect with blends of 40%–80% LDPE having higher melt strength than either LDPE or LLDPE alone. Compositions with 20%–30% LDPE generally match the melt strength of LDPE alone. Patel et al. [108] and Ghijsels et al. [109] explain this effect by invoking a model proposed by Bersted [111,112], which says that viscosity at low shear rates increases with increasing LCB due to an increase in chain entanglements. At very high levels of LCB (such as that found in pure LDPE), however, the molecule becomes more compact. This leads to a reduction in coil size (radius of gyration) and fewer entanglements. Adding a small amount of LLDPE expands the coil size and increases the entanglements. As a result, the blends have higher melt strength.

Adding LDPE to LLDPE increases the melt strength of the LLDPE but can negatively impact film properties. One study by Lue [113] on 50-μm (2 mils) blown film shows reductions in tensile strength, dart impact strength and MD Elmendorf tear resistance of 19%, 62%, and 87%, respectively, when 25% of a 0.6 MFI LDPE resin is added to a hexene Ziegler–Natta (ZN) LLDPE (1.0 MI). This is shown in Fig. 13.13. Patel et al. [108] report a reduction in MD tear strength with increasing LDPE level that is the mirror imagine of the increase in melt strength (see Section 9.4.4.1 and

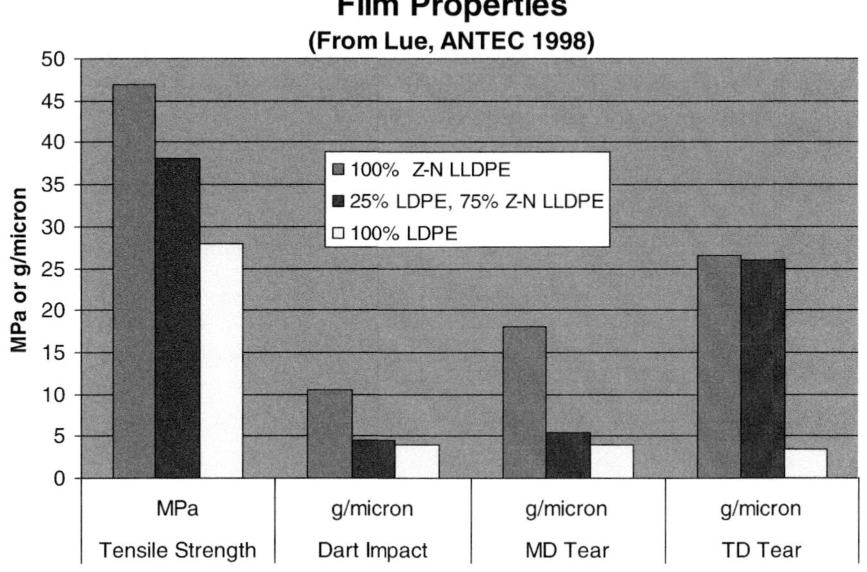

FIGURE 13.13 Effect of adding 25% LDPE to a hexene Ziegler–Natta LLDPE on blown film properties. *Data from C.T. Lue, Easy processing metallocene polyethylene, SPE ANTEC Conf., 56(2) (1998) 1816–1819.*

Fig. 9.32). They also report a reduction in dart impact strength; they attribute both effects to an increase in chain orientation that coincides with a reduction in tie chains between crystalline lamellae as the amount of LCB is increased in the blend. Advances in LDPE technology have been introduced that help mitigate some of the property losses of the blends. For example, Karjala et al. [114,115] describe advanced LDPE grades in Dow's Agility product line that allow less LDPE to be used in the blend to increase film output, which results in improved film physical properties.

Additives that enhance the melt strength of LLDPE have been developed that offer the processer a different set of trade-offs [116,117]. Blended at 1–4 wt.%, these can be thought of as chain extenders that help create long-chain-like structures during processing. An additive described by Morris [116] allows gauge reduction of up to 20% due to reductions in gauge variation while maintaining most LLDPE film properties, including dart impact, puncture resistance, and tensile properties. There is a moderate decrease in MD tear strength and an increase in haze. Both are believed to be related to an increase in molecular relaxation time due to the chain extension. This results in an increase in MD orientation, causing a reduction in MD tear strength. The increase in haze is a consequence of surface roughness due to unrelaxed stress. Williamson et al. [117] describe a similar additive technology that allows downgauging by 20% by reducing the amount of LDPE in the blend. The additive boosts melt strength, MD shrink tension and shrink force while maintaining puncture resistance and optical properties. Such additives are not suitable for every situation; they only improve output when production is limited by melt strength considerations such as bubble stability. They require more extruder power than LLDPE–LDPE blends and should not be used when extruder torque is limiting output.

Resin producers continue to improve the performance of LLDPE. Advances in reactor design and catalyst technology have helped push the processing envelope of LLDPE. Reactors in series that produce bimodal MWD and metallocene LLDPE with sparse LCB are two examples. Improvements in blown film equipment have also helped boost the output of LLDPE blown film lines. These include better air ring and cooling designs that help keep the bubble stable.

Coextrusion of LLDPE with LDPE or other resins with superior melt strength is another avenue for reducing bubble instability and enhancing output without as much trade-off in film properties.

13.4.4 Layer rearrangement in coextrusion

Distortion or rearrangement of layers in coextrusion can occur due to the viscous and elastic properties of the polymer and the geometry of the flow channel. This can lead to poor thickness control, and loss of barrier and other properties.

13.4.4.1 Shear viscosity effects

When two polymers of different shear viscosity flow down a channel, the lower viscosity resin will preferentially flow toward the region of the highest shear rate. Han [118] attributes this behavior to the principle of minimum viscous energy dissipation. The result is that if the flow channel is sufficiently long and the viscosity difference sufficiently large, the lower viscosity polymer will encapsulate the higher viscosity polymer. This is because the shear rate is highest near the channel wall. This encapsulation effect gives rise to layer maldistribution in coextrusion. This is illustrated in Fig. 13.14 where layer A tends to encapsulate layer B as they travel down the channel.

This is often more of a problem in flat die extrusion where the polymers travel together for long distances after being combined in a feedblock. In blown film, the polymers are usually combined closer to the die exit. Also, there are no edges in the annual flow in blown film, which helps reduce the distortion effects.

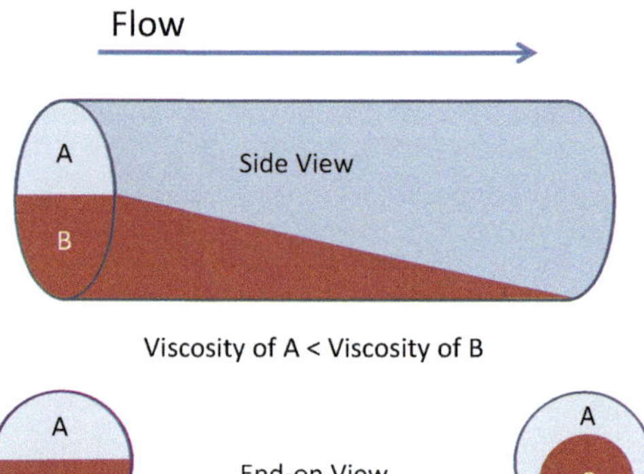

FIGURE 13.14 Illustration of encapsulation effect in steady-state flow. Layer A has a lower viscosity than layer B.

Some methods used to minimize this behavior include:

- Selecting polymers that match in viscosity.
- Minimizing the travel distance after the polymers are combined. Multimanifold dies, for example, are less prone to this since the polymers are combined just prior to the die exit.
- Contouring the shape of the feed channel to compensate for the flow rearrangement. This is illustrated in Figs. 13.15 and 13.16 for a three-layer (A–B–A) structure. Some feedblock designs use inserts to contour the channel [120,121]. Other feedblocks have rotating pins that contour the channel. It may take some trial and error to come up with an acceptable contour that compensates for the maldistribution.

13.4.4.2 Elasticity effects

Khan and Han [122,123] estimate the effect of differences in polymer elasticity on interface displacement and layer rearrangement. Their calculations show that the more elastic fluid will tend to encapsulate the less elastic one. They conclude, however, that viscosity plays the dominant role in determining the shape of the interface in steady flow [118].

In a series of elegant experiments and numerical simulations, Dooley and coworkers [124–131] show that polymer elasticity can lead to substantial layer rearrangement in coextrusion, particularly near the edges of a flat die. They eliminate the viscosity effects by conducting coextrusion experiments with the same polymer. The layers are pigmented to see the flow patterns. (The pigment is shown to not affect the flow.) They use several different feedblock configurations to visualize the flow down a channel. In the simplest, two layers of different color with a layer ratio of 20/80 are coextruded down a channel of different geometries. After establishing a steady flow, they stop the extrusion, freeze the

Viscosity of A < B

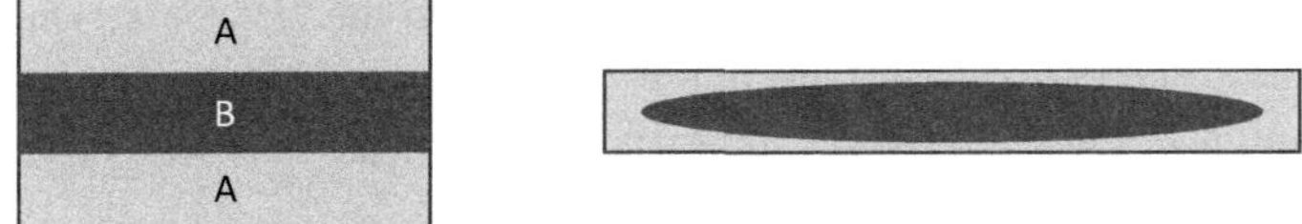

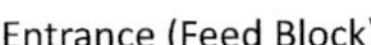

Film Layer Distribution

FIGURE 13.15 Layer redistribution in a three-layer coextrusion for the case where the skin layer (A) has a lower viscosity than the core layer (B). This results in A encapsulating B (upper diagram). If the feed channel is shaped with a "bow tie" contour, the layer distribution in the final film may be uniform (bottom diagram) [119]. *Adapted from T.I. Butler, Effects of flow instability in coextruded films, TAPPI J., 75(9) (1992) 205–211.*

Viscosity of A > B

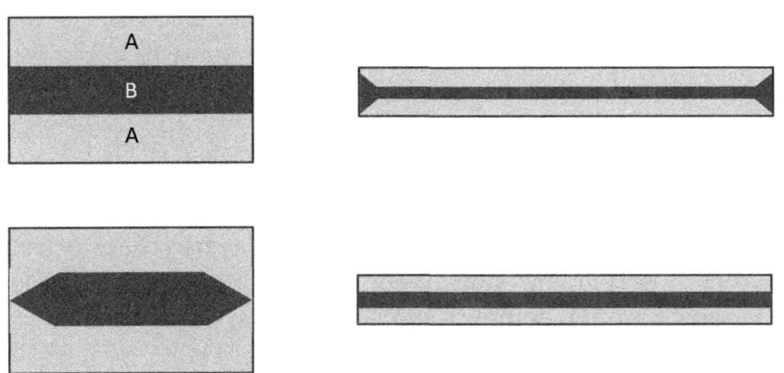

Entrance (Feed Block)

Film Layer Distribution

FIGURE 13.16 Layer redistribution in a three-layer coextrusion for the case where the skin layer (A) has a higher viscosity than the core layer (B). This results in a dog bone or bow tie distribution in the final film. If the feed is contoured as in the bottom diagram, the final film distribution may be uniform. *Adapted from T.I. Butler, Effects of flow instability in coextruded films, TAPPI J., 75(9) (1992) 205–211.*

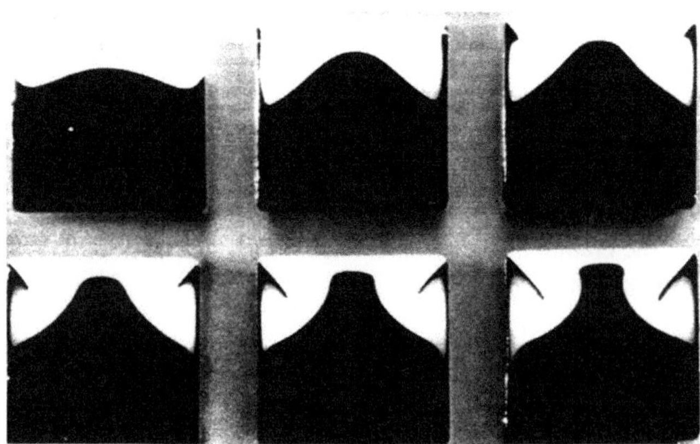

FIGURE 13.17 Example of layer rearrangement due to secondary flows of elastic polymers. Cross-section of two-layer coextruded polystyrene (colored white and black) structure as it progresses through a square channel. Progression is from upper left to lower right. *From with permission J. Dooley, K.S. Hyun, K. Hughes, An experimental study on the effect of polymer viscoelasticity on layer rearrangement in coextruded structures, Polym. Engr. Sci., 38 (1998) 1060–1071.*

polymer in place, and open the channel to see the extrudate at different locations along the channel. In other experiments, they use special feedblocks to coextrude 165 layers or 49 strands to aid in the visualization of the flow.

Several different geometries are studied, including square, rectangular, teardrop, circular and a rectangular manifold (to simulate what may happen in a die). Three polymers are used: polycarbonate (PC), PS and LDPE. Rheology measurements show that the PS and LDPE are more elastic than PC.

Their results, some of which are illustrated in Fig. 13.17, show:

- PS and LDPE exhibit substantial layer distortion while flowing down noncircular channels. PC shows little or no distortion.
- The special feed block configurations with 165 layers or 49 strands show secondary flow in the corners of the square and rectangular channels that cause the distortion. This is confirmed by numerical simulation. The secondary flows or vortices are due to the second normal stress difference (a measure of elasticity). Numerical simulations show that the amount of secondary flow correlates with the ratio of the second normal stress difference to the shear viscosity [132–134]. (Borzacchiello et al. [134] point out that these secondary flows are due to normal stress differences and not elastic effects associated with polymer chain relaxation.)
- The geometry of the channel influences the amount of distortion:
  - Circular channels do not show distortion. The radial symmetry prevents the formation of secondary flows.
  - Square channels have the greatest amount of distortion.
  - Rectangular channels have less distortion and only in the corners.
  - Teardrop-shaped channels have six regions of secondary flows vs. eight for rectangular and square.
  - A rectangular manifold channel shows distortion in the corners, especially near the end of the channel. This geometry is similar to coat hanger style dies and suggests that more layer rearrangement may take place near the edges of the die than in the center. This is what is often seen in practice.
  - The entrance geometry (vertical or back) and manifold design of a coathanger die affect the flow patterns [130].

The size of the secondary flows is several orders of magnitude smaller than the bulk flow in the channel, except near the walls or corners. In the coextrusion of two polymers of different viscosity and elasticity, the viscous encapsulation velocity dominates near the entrance of the channel. The encapsulation velocity decreases as the polymers flow down the channel and the layers reach their equilibrium configuration. The elastic secondary velocity is steady throughout the channel so that far down the channel the viscous and elastic flows become the same order of magnitude [134,135].

These results, as well as others [136,137], show that the viscosity and elasticity of the polymer must be taken into account during coextrusion. The selection of the polymers as well as the channel and die design are important.

13.4.5 Interfacial instabilities

In coextrusion, defects have been observed that are associated with the instability of flow between layers. These defects have been described in a number of ways, including chevrons, wood-grain, orange peel, waves, ground-glass appearance and loss of see-through clarity. Their origins may differ and are often complex and poorly understood.

Interfacial instabilities have been categorized into four types:

1. **Zig-zag**. This is characterized by a high frequency, low amplitude wave or zig-zag. It may start out as an increase in surface roughness and haze and in its more severe form resemble wood grain patterns. Experiments have shown

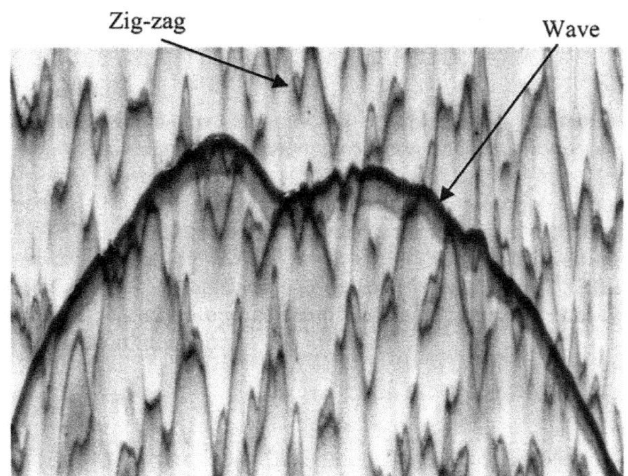

FIGURE 13.18 Photo of zig-zag and wave-type instabilities in a coextruded cast film [139]. *Reproduced with permission from M. Zatloukal, M. Martyn, J. Vlcek, P. Saha, Viscoelastic coextrusion flow modeling by advanced constitutive equations: the influence of the die design on interfacial instabilities, Eur. TAPPI PLACE. Conf. (2005).*

that this type of instability originates in the die land near the die exit [138]. Fig. 13.18 shows a photograph of a film with zig-zag instability.

2. **Wave**. This is characterized by a low frequency, large amplitude wave that often extends across the width of the film or sheet (see Fig. 13.18). Several investigations show that this instability originates in the region where the polymer flow streams merge [140–143].
3. **Scattering**. This instability is found in microlayer coextrusion involving greater than 100 layers with thicknesses less than 10 μm. Distinct layers are required. The mechanism is unknown but Ramanathan et al. [141] speculate that a normal stress mismatch drives secondary flows near the die wall. Ramanathan et al. [144] show that adding a protective layer between the die wall and the microlayers can eliminate this instability. Zhu et al. [145] relate breakup of layers to dewetting described by a simple model balancing viscous and capillary forces. The bursting speed is found to be related to the viscosity ratio and the interfacial tension. They note several unanswered questions from their analysis.
4. **Interfacial adhesion**. This type of instability typically shows up as a loss of see-through clarity or as a "ground glass" appearance as the observer looks through the film. It occurs in barrier packaging films where a reactive tie resin bonds to a barrier resin such as polyamide (PA) or ethylene vinyl alcohol (EVOH), although most reported studies find it is more prevalent with EVOH [146–148]. A proposed mechanism is a local increase in MW at the interface between the reacting polymers which creates viscosity mismatches and local stresses that may cause instabilities. For example, Levitt et al. [149] amplify the effect of interfacial interactions by measuring the extensional stress on specimens containing up to 100 layers of pairs of polymers. Combinations of pairs that form crosslinks at the interface have stress values that are two orders of magnitude higher than nonreactive pairs. Bondon et al. [146] study both PA6 and EVOH coextruded film structures containing maleic anhydride grafted polypropylene (PP-g-MAH) tie resins. The PA6–tie interface is sharp and smooth whereas the EVOH–tie interface is thicker (15 vs 60 nm) and rough, corresponding to the see-through clarity of the films. They attribute the differences to the type of reaction at the interface. The PA6–tie interface is consistent with the formation of di-block copolymer (reaction of the anhydride with chain ends of the polyamide) whereas the EVOH–tie interface is indicative of a graft reaction along the EVOH chain. This type of instability can often be controlled by reducing the amount of reaction at the interface by decreasing the process temperature and contact time or the chemical functionality in the tie resin [146,147]. There is a limit to how effective this can be without adversely affecting adhesion. Special tie resin formulations have been developed to address this issue with mixed success [148].

Zig-zag and wave instabilities are the most studied types and the subject of the rest of this section.

13.4.5.1 Fundamental studies

Interfacial instability has been studied for well over 40 years. Schrenk et al. [140] and Han and Shetty [150] find evidence for a critical interfacial shear stress above which the onset of zig-zag instability occurs. Their value for the critical stress is about 5×10^4 dynes/cm. Later investigators find that this value varies with polymer type, MWD and other factors [151]. There does not seem to be a universal value for the critical stress. Indeed, Wilson [152] says that critical stress is not supported by theory and may even be misleading. Nevertheless, this concept is widely accepted, perhaps because it is easy to determine and known methods can be used to reduce stress. Zatloukal et al. [153,154] observe that modeling and experiments show that the onset of zig-zag instability in the die land occurs at a total stress value similar in magnitude to the critical stress for the onset of wave instability in the merge region (discussed below). They suggest that this may be an indication that both types of instability arise from the same source.

Zatloukal and de Witte [155–158] study the effect of adding a PPA to PE on zig-zag instabilities. Fluoropolymer PPAs are thought to migrate to the channel wall and provide slip due to incompatibility with the polymer. Boron nitride and nanoclay PPAs do not migrate and provide internal "slip" within the bulk flow. They make three-layer blown film with the structure (mLLDPE–HDPE–mLLDPE) and add migratory PPA to different layers. The onset of zig-zag instability increases to a higher flow rate (stability is enhanced) when the PPA is added to the skin layers but the stability is reduced when it is added to all three layers or to just the HDPE core. Their analysis of the experimental results coupled with modeling suggests when PPA is only added to the skin layers, it coats the walls and reduces the stress state of the skin layer, reducing flow instabilities. If PPA is added to all three layers or just the core layer, PPA in the middle layer migrates to the interface with the skin layer and acts as a stress generator when the flow converges in the die.

Perturbation methods are used to study flow instabilities in the engineering discipline of fluid mechanics. A periodic wave is introduced, either experimentally or to model equations. If the wave grows as it travels through the medium, in this case as the polymer flows through the transfer pipes and die, the system is unstable. Yih [159] conducted one of the earliest studies of this type using Newtonian fluids. He found that the viscosity ratio, thickness ratio and initial wave frequency are important. Others have looked at viscoelastic polymers and find that the elasticity ratio and the flow channel geometry are also important [152,160–167].

Many investigations have focused on the region where the polymers merge. Changes in the channel geometry influence stability [141,143,168–173]. The acceleration of one flow stream into the other during the merge is hypothesized as being responsible for the wave instability, but simple velocity ratios have proven insufficient to predict the onset of instability [168–173]. Elongational viscosity and elasticity are thought to be involved [173,174]. Xiao [168,169] evaluates three different types of PE in blown film of (LDPE-tie-PE) and finds the more elastic PE to be more prone to instability. Ramanathan et al. [141] find cast sheet/film structures with a large extensional viscosity ratio or a skin layer with a large extensional viscosity to be more prone to instability. Perdikoulias and Tzoganakis [170,171] find LDPE coextruded with itself can generate instabilities; grades with broader MWD are more prone to instability. LDPE grades with broader MWD generally have more LCB, greater elasticity and higher elongational viscosity due to strain-hardening behavior. The stability of the broad MWD grades seems to be independent of the channel geometry and processing conditions, whereas narrower MWD grades exhibit instability that is associated with interfacial stress (which is related to channel geometry and processing conditions).

Zatloukal et al. [139,175–178] conduct numerical simulations of flow in the merge area to develop predictors of instability. The results are compared against experimental data of Perdikoulias and Tzoganakis [170,171] and Martyn et al. [142,143,179]. They propose criteria that go by several names in successive papers, depending on which stress components they include. In an early version, it is known as the total normal stress difference (TNSD), which is essentially the normal stress difference (elasticity) averaged over the flow stream for each layer. Later papers also propose adding the tangential stress to the criteria (total stress difference). TNSD is positive when the major layer is stretched more than the minor layer and negative when the minor layer is stretched more. They correlate the sign and shape of this parameter with the onset of instability. If the sign goes negative or the shape is discontinuous, the flow is likely to be unstable. They conclude that skin layers that exhibit strain-hardening behavior in extension are more prone to instability.

The measurement of elongational viscosity and elasticity of polymer melts is challenging, as described in Chapter 5, especially under high extension rate conditions that simulate real polymer processing. Numerical methods using constitutive equations that can handle the elasticity characteristics of the polymer typically require material data that is not easily obtained, extensive knowledge of the flow channel geometry that is usually not available except to the equipment manufacturer and knowledge of numerical techniques. While these techniques have advanced the understanding of flow instabilities, they often are too difficult to use as everyday tools for troubleshooting or selecting specific resins or structures.

13.4.5.2 Methods to improve stability

The theory discussed above suggests several methods for improving stability. One is to try to match the velocities of the polymer streams as they are combined in the feedblock or die. For example, some feedblocks have vanes that either automatically adjust the flow or can be manually adjusted. Other feedblocks have dividing plates to help distribute the velocity. Fixed geometry dies, such as blown film dies, generally have little leeway to adjust the velocity of the streams beyond changing the output of the extruder. Problems associated with mismatched velocities often occur as the converter tries to make structures that are beyond the design optimum of the die. In other words, they try to put too much or too little through one of the layers. As the number of layers has increased in modern coextrusion systems, this has become less of an issue. Thick layers can be subdivided into several layers, each with its own extruder and flow channel, so that the process runs closer to its optimum. This allows the flow in each layer to better match with the other layers.

Another method to improve stability is to reduce interfacial stress [119,140,150]. As shown in Fig. 13.19, stress varies linearly across the height of the channel. The stress is highest at the channel wall and goes to zero somewhere in the channel; for monolayer flow, the stress is zero down the centerline of the channel. Fig. 13.19 outlines several ways to reduce the interfacial stress, including:

- **Increasing the temperature**. Higher temperature reduces the viscosity of the polymersand reduces stress.
- **Reducing the viscosity of the skin layers**. This may be accomplished by raising the temperature or switching to a lower viscosity grade.
- **Moving the interface away from the surface (thicker skin layer)**. Since the stress is highest at the channel wall, moving the interface away from the wall reduces stress.
- **Increasing the channel height**. Stress is related to the pressure drop across the channel length. Opening the cannel up reduces the pressure drop and the stress.
- **Reducing the output**. This is always an unfavorable action but may help to temporarily fix a problem until anther solutions can be found.

Matching viscosities is another approach that has been widely used with practical success. It is supported by perturbation theories. While the fundamental studies suggest matching elongational viscosity may help, data are hard to find. Elongational viscosity data is generally not supplied by resin manufacturers and is difficult to measure. In some cases, it may be enough to consider the general family of polymers the resin belongs to. LDPE, for example, is known to strain harden during extension whereas LLDPE does not; therefore, LDPE is likely to have higher elongational viscosity than LLDPE at the same melt flow index.

At low extension rates, elongational viscosity is equal to three times the shear viscosity; matching shear viscosity may at least provide directional trends toward matching elongational viscosity. Shear viscosity is much easier to measure, and data are generally readily available. A key question is what viscosity to use. A rather crude method is to match the melt flow rates (MFRs) of the polymers. This is not often very accurate since the viscosity changes with temperature and shear rate. A better method is to use the viscosity for the temperature and shear rate at the interface. The shear rate varies across the flow channel; the first step is to determine to location of the interface. Although the location of the interface in the final film is known, this location is not the same as that in the flow channel. This is due to the external constraints or boundary conditions on the flow. In the channel, the flow is bound by the walls; flow takes on a parabolic shape with zero velocity at the walls. At the channel exit, the boundary condition changes to a free surface. The individual layer flows change in response, shifting the location of the interfaces.

Another complication is that the shear rate is discontinuous at the interface. This can be seen in Fig. 13.20, which shows the velocity profile of a two-layer flow. The velocity is continuous at the interface (the velocity of layer Aequals

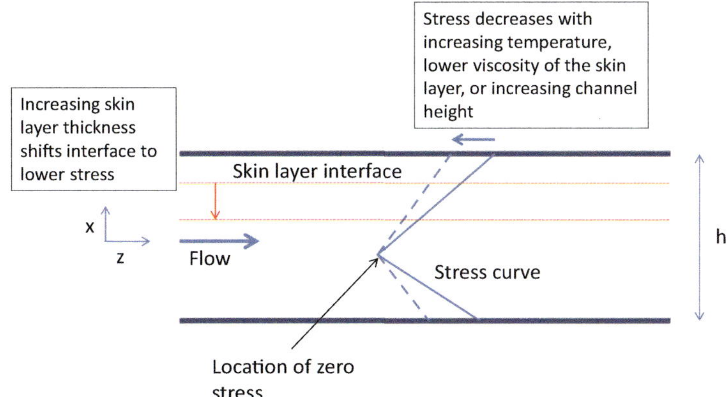

FIGURE 13.19 Ways to reduce interfacial stress in coextrusion. Depiction is of the stress profile in a rectangular flow channel of height h.

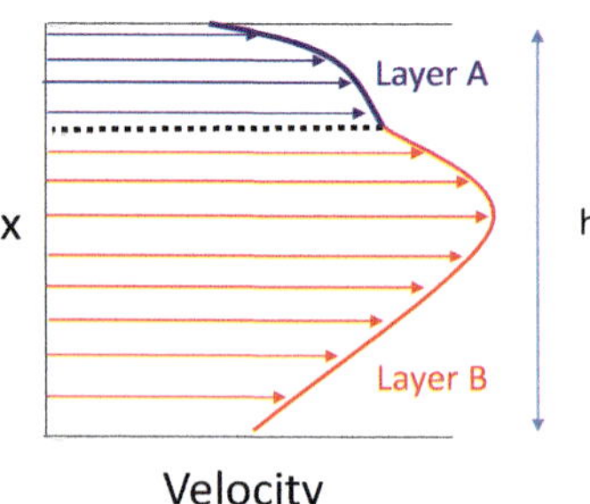

FIGURE 13.20 Velocity profile for a two-layer coextrusion in a rectangular channel.

the velocity of layer B). This is a standard assumption in fluid mechanics called no-slip. The slope of the velocity curve is equal to the shear rate at a given point along the flow curve. At the interface, the two slopes are not equal.

An even better approach is to determine the shear stress at the interface. As mentioned earlier, the stress is continuous at the interface and varies linearly across the channel height. If the location of the interface can be determined, and the stress profile is known, the stress at the interface can be computed. It then becomes a simple matter of picking off the viscosity of each of the layers from their respective rheology curves, plotted as viscosity vs. shear stress (illustrated in Fig. 6.22 or Fig. 13.21). Shear stress is equal to the viscosity times the shear rate as shown in Eq. (13.6).

The location of the interfaces and the stress profile can be computed by solving the equations of motion for multilayer flow down a rectangular channel. Several references give the solution, including Han [118] and Han and Shetty [180] for isothermal power-law fluids, Puissant et al. [181] for nonisothermal power-law fluids and Nordberg and Winter [182], Mavridis and Shroff [174] and Pearson [183] for general nonisothermal viscoelastic fluids. The equations are nonlinear and require numerical methods to solve. The equations are more manageable if the flow is assumed to be Newtonian ($n = 1$ in the power-law model) and isothermal. While not truly representing the flow of polymers, such an approach still provides reasonable estimates that have proven successful in minimizing stability problems [184]. A seminumerical approach can be applied that involves starting at one channel wall and numerically marching across the height of the channel in small increments. At each step the flow rate for the first layer can be computed from the model equation and compared with the real flow rate obtained from the layer thickness in the film (times the line speed). When they match, the interface is found. This is repeated for each interface.

To illustrate the usefulness of such a tool, consider a five-layer coextruded film with the structure (Ionomer–tie–PA6–tie–HDPE). The throughput, dimensions of the channel, layer thicknesses and temperatures are presented in Table 13.3. The viscosity vs. shear stress curves for the four resins at their respective temperatures are shown in Fig. 13.21. Without some bounds on the shear stress in the channel, it is difficult to determine how well matched the viscosities are. A simple method can be used to estimate the stress profile in the channel by first calculating the shear rate at the wall for Newtonian flow in a rectangular channel:

$$\dot{\gamma}_w = \frac{6Q}{wh^2} \tag{13.5}$$

where $\dot{\gamma}_w =$ shear rate at the wall, $Q =$ volumetric throughput (mass throughput divided by the average density of the melt), $w =$ width of channel, and $h =$ height of channel.

The shear stress at each wall is determined from the rheology data for each of the skin layers, in this case the ionomer (wall 1) and HDPE (wall 2). This can be determined by replotting the rheology data in Fig. 13.21 as viscosity vs. shear rate by recognizing

$$\tau = \eta\dot{\gamma} \tag{13.6}$$

where $\tau =$ shear stress and $\eta =$ viscosity.

The results are shown in Fig. 13.22. The last piece of information that is needed is the location of the plane of zero shear stress, ξ. Since the stress profile is linear, ξ can be estimated by simple proportion:

$$\frac{\xi}{1-\xi} = \frac{\tau_{w1}}{\tau_{w2}} \tag{13.7}$$

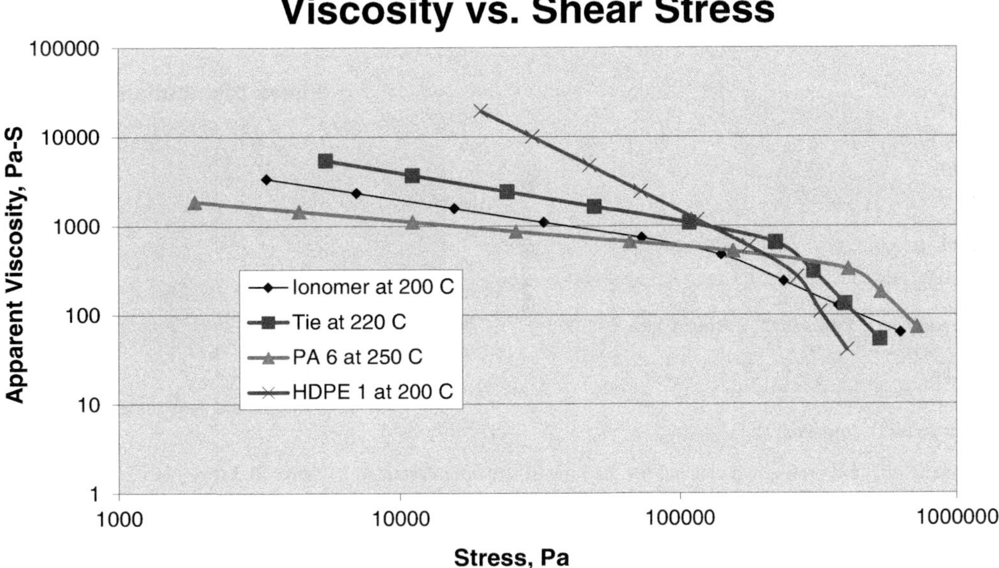

FIGURE 13.21 Viscosity versus shear stress curves for the four resins and temperatures in Table 13.3 (Scenario 1). Data is from capillary rheometry measurements, extrapolated to the given temperature using a double power-law model with Arrhenius temperature dependence. *Reproduced with permission of The Dow Chemical Company.*

TABLE 13.3 Input parameters for a film structure (Scenario 1).

Total throughput, kg/h	260		
Width of channel, mm	1100		
Height of channel, mm	1.80		
		Thickness	**Temperature**
Layer	**Resin**	**μm**	**°C**
1	Ionomer	20	200
2	Tie	7.5	220
3	PA 6	12	250
4	Tie	7.5	220
5	HDPE 1	50	200
Total thickness		97	

where ξ = dimensionless distance from channel wall 1 (distance divided by the channel height), τ_{w1} = shear stress at wall 1, and τ_{w2} = shear stress at wall 2.

The estimated stress profile is given in Fig. 13.22. The stress ranges from 0 to 140,000 Pa. The curves in Fig. 13.21 show that over this range the ionomer and tie resin viscosities are well-matched, but the tie–PA and tie–HDPE combinations are not. To get more quantitative, the location of each interface needs to be estimated. This is done using the model described above. The results of the model calculation are shown in Table 13.4. A general guideline based on practical experience is to keep the viscosity ratio between 3:1 and 1:3. The results show that the first tie–PA interface has a viscosity ratio of 3.7 and has the potential to be a source of flow instability. The second PA–tie interface is also

Estimate of Stress in a Flow Channel

Total Throughput, kg/hr:	260
Average Resin Density, g/cc:	0.7
Width of Channel, mm:	1100
Height of Channel, mm:	1.8

Estimated Wall Shear Rate, 1/s: 174

		Temperature	Wall
Wall	**Grade**	**Degrees C**	**Stress, kPa**
1	Ionomer	200	106
2	HDPE 1	200	143

Location of Zero Shear Stress (% from Wall 1): 43%

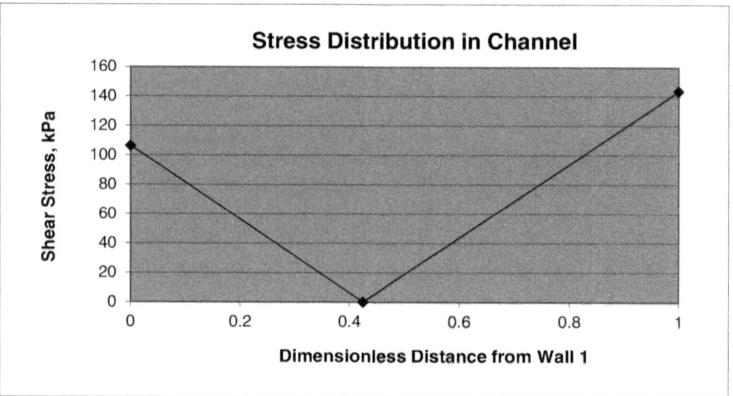

FIGURE 13.22 Estimate of stress profile in a channel for the flow of structure described in Table 13.3 (Scenario 1). *Reproduced with permission of The Dow Chemical Company.*

TABLE 13.4 Results of a multilayer flow model for the input conditions found in Table 13.3 (Scenario 1).

		Location of zero shear stress			0.402		
Interface	Resin	Distance from wall	Shear stress (kPa)	Shear rate (1/s)	Viscosity (Pa-s)	Viscosity ratio	Comment
Wall 1	Ionomer	0.000	84.8	124	682		
1	Ionomer	0.304	20.7	15	1357	0.53	
	Tie			8	2554		
2	Tie	0.372	6.4	1	4792	3.67	Mismatch
	PA 6			5	1305		
3	PA 6	0.480	16.4	17	994	0.34	
	Tie			6	2894		
4	Tie	0.550	31.2	15	2054	0.22	Mismatch
	HDPE 1			3	9239		
Wall 2	HDPE 1	1.000	126.1	124	1014		

Source: Reproduced with permission of The Dow Chemical Company.

very close to the bounds of the stability region and it too should be considered. Finally, the tie–HDPE interface also shows considerable viscosity mismatch. The viscosity ratios are plotted in Fig. 13.23.

There are several ways to address viscosity mismatches. These include:

- Adjusting the temperature of one or more layers. This changes the viscosity of the polymer. There are limitations to this approach. One is that resins may have thermal stability limits above which gel formation or other problems may occur (see Section 13.1). Another is that the temperature must be well above the melting point or glass transition temperature of the polymer to achieve flow. These set limits on what temperatures can be used. Perhaps a greater limitation, however, is the heat transfer that occurs between layers as they travel down the channel. This tends to equilibrate the temperature across the layers and, therefore, maintaining large differences in temperature between layers may be impractical. Some models, such as the one developed by Mavridis and Shroff, incorporate heat transfer into the calculation. For those models that do not have heat transfer implicitly built into the equations, it can be useful to run the simulation at several temperatures to understand how sensitive the solution is to temperature effects.

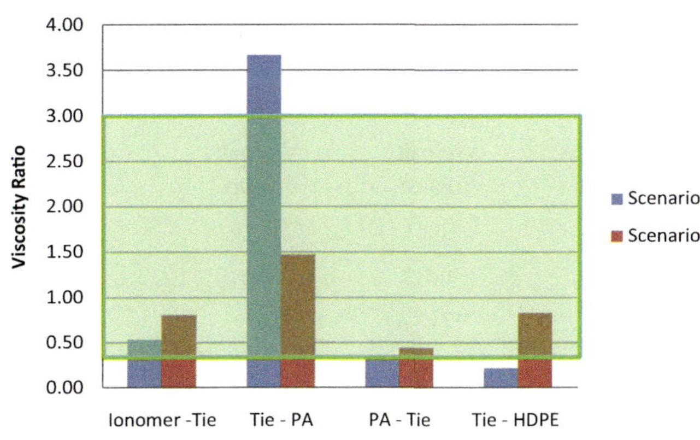

FIGURE 13.23 Viscosity ratios at each interface from scenarios 1 and 2. Shaded (green) region represents a region of stability where the viscosity ratio falls between 1:3 and 3:1.

TABLE 13.5 Input parameters for a film structure (Scenario 2).

Total throughput, kg/h	260		
Width of channel, mm	1100		
Height of channel, mm	1.80		
		Thickness	Temperature
Layer	Resin	μm	°C
1	Ionomer	20	200
2	Tie	7.5	240
3	PA 6	12	245
4	Tie	7.5	225
5	HDPE 2	50	250
Total thickness		97	

- Switching one or more of the polymers to a grade with higher or lower MW (or melt viscosity).
- Increasing or decreasing the channel height (simulating adjusting the die gap, for example). This changes the shear rate and stress profile in the channel, which may affect the viscosity ratios since different polymers have different shear thinning characteristics.
- Increasing or decreasing the throughput. This also changes the shear rate and stress in the channel.
- Changing the thickness of one or more of the layers. Increasing the thickness of a skin layer, for example, moves the interface further from the channel wall, which decreases the shear stress. Since the polymers may have different responses to changes in shear stress, moving the interface may change the viscosity ratio.

The model is useful for quickly evaluating the approaches. For this example, the first two approaches are used to bring the viscosities into better alignment. Changing the temperatures alone is not enough to eliminate the viscosity mismatches, particularly between the tie and HDPE. Replacing the HDPE with a lower MW grade (higher MFI) along with adjusting the temperatures of the layers results in a good match at each interface. This scenario is shown in Table 13.5 and the results in Table 13.6. As shown in the bar graph in Fig. 13.23 the viscosity ratios in scenario 2 all fall within the desired range. This is also seen in the viscosity vs. shear stress curves for these resins at the new temperature profiles (Fig. 13.24).

More sophisticated numerical models can be employed if details of the flow channel are known. For example, Vlcek et al. [185] use a 2D finite element model to analyze wave flow instabilities for three-layer feedblock coextrusion of (PETG-LDPE based tie-LDPE) [20/10/70%]. Experiments over a 4-month period failed to resolve the problem whereas 6-hours of modeling did. The model revealed that the PETG is higher in viscosity than the LDPE; switching to another

TABLE 13.6 Results of a multilayer flow model for the input conditions found in Table 13.5 (Scenario 2).

		Location of zero shear stress			0.547		
Interface	**Resin**	**Distance from wall**	**Shear stress (kPa)**	**Shear rate (1/s)**	**Viscosity (Pa-s)**	**Viscosity ratio**	**Comment**
Wall 1	Ionomer	0.000	84.8	124	682.1		
1	Ionomer	0.289	40.0	41	981.2	0.81	
	Tie			33	1213.4		
2	Tie	0.348	30.9	23	1358.1	1.47	
	PA 6			33	923.8		
3	PA 6	0.436	17.2	16	1097.2	0.44	
	Tie			7	2494.1		
4	Tie	0.489	9.0	3	3469.4	0.83	
	HDPE 2			2	4176.5		
Wall 2	HDPE 2	1.000	70.3	124	565.0		

Source: Reproduced with permission of The Dow Chemical Company.

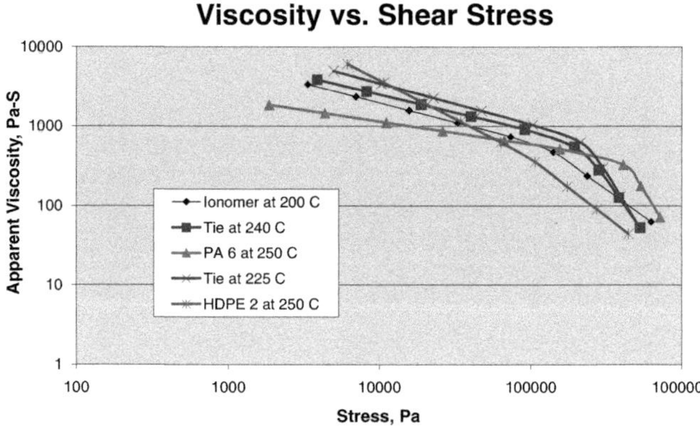

FIGURE 13.24 Viscosity versus shear stress curves for the five layers in Table 13.5 (Scenario 2). Data is from capillary rheometry measurements, extrapolated to the given temperature using a double power-law model with Arrhenius temperature dependence. *Reproduced with permission of The Dow Chemical Company.*

grade of LDPE fails to solve the problem since its shear-thinning behavior does not match the PETG. The model pointed toward changing the LDPE to LLDPE, which is less shear thinning, which fixed the issue.

Minimizing stress and matching viscosity are complementary approaches often used together to solve interfacial instability issues. Using one alone may not be sufficient, although both often result in similar tactics. The same model that is used to compute viscosity ratios also yields information about the interfacial stress. A few tactics do differ between the two approaches:

- The stress approach suggests decreasing the viscosity of the skin layer, even if it results in a mismatch in viscosity with the layer adjacent to it. In practice, this has proven effective in many situations.
- The stress approach suggests increasing the temperature across the board to reduce interfacial stress. This may be impractical in situations where melt strength (bubble stability) or thermal stability issues are a concern.

13.5 Curl

Curl is a term used to describe the tendency of a packaging film to roll up and not lay flat. Curl is undesirable because of the handling problems it creates on high-speed packaging lines. In extreme cases, curl may also detract from the appearance of the final package. Curl is particularly problematic in coextruded blown films when the layers are not symmetrically balanced. Films containing barrier resins such as EVOH and polyamide are often difficult to make flat.

13.5.1 Causes

In coextrusion blown film, the primary cause of curl is the stress created when the layers shrink by different amounts as the film is quenched, or afterwards during postquench (or secondary) crystallization. The shrinkage of a polymer during quenching from the melt depends on its freezing point, crystallinity and coefficient of thermal expansion, as well as the quench rate of the process. For semicrystalline polymers, the volume undergoes a rapid reduction at the freezing point due to the formation of crystals. This is illustrated in pressure–volume–temperature (PVT) data for HDPE in Fig. 13.25. Here the specific volume (inverse of density) is plotted versus temperature as the polymer is slowly cooled from the melt [186–188]. This decrease in volume is designated ΔV_c (change in volume due to crystallization). The volume also changes due to thermal expansion and contraction, which is indicated by the positive slope leading up to the crystallization temperature.

Fig. 13.26 shows PVT data for three commonly used polymers in packaging films, HDPE, EVA (18 wt.% VA) and EVOH (38 mol% ethylene). Their different freezing temperatures and ΔV_c clearly show up in this type of plot. Such data is useful for qualitatively understanding why some films curl more than others. For example, many dry food packaging films (e.g., pouches for cereal and crackers) have the structure (HDPE–EVA), where the HDPE serves as a moisture barrier and the EVA gives low-temperature sealability. These films typically have a very small amount of curl toward the EVA layer. While the HDPE undergoes a significant volume change as it crystallizes, the EVA is still molten when this occurs. It can flow to relieve stresses caused by the shrinking HDPE layer. When the EVA itself freezes, its volume change is much smaller, giving less of a tendency for curl. Furthermore, the EVA's modulus is considerably smaller than that of HDPE; the stiffness of the HDPE helps prevent curl.

For barrier applications, a layer of EVOH (and tie layers) can be added to a typical dry foods packaging structure: (HDPE–tie–EVOH–tie–sealant). This structure has a strong tendency to curl toward the HDPE layer. The PVT data

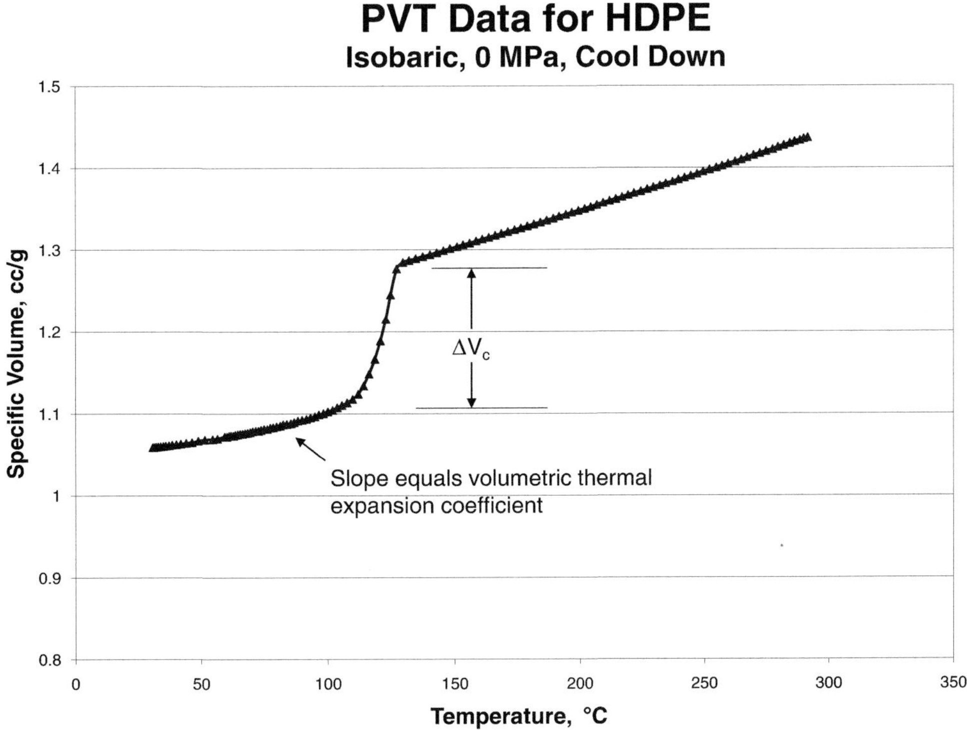

FIGURE 13.25 Pressure–vapor–temperature data for high-density polyethylene. *Reproduced with permission from B.A. Morris, Reducing curl in multilayer blown film: experimental results, model development, and application to a cereal liner film, J. Plastic Film. Sheeting, 19 (2003) 31–54.*

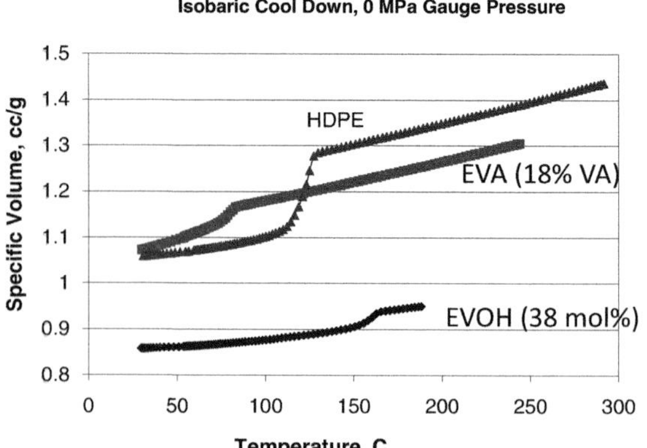

FIGURE 13.26 Pressure–vapor–temperature data for ethylene vinyl alcohol, high-density polyethylene, and ethylene vinyl acetate. *Reproduced with permission from B.A. Morris, Reducing curl in multilayer blown film: experimental results, model development, and application to a cereal liner film, J. Plastic Film. Sheeting, 19 (2003) 31–54.*

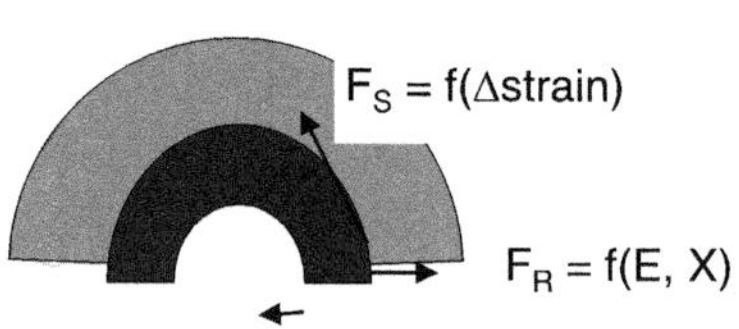

FIGURE 13.27 Forces acting on a two-layer film due to differential shrinkage. *Reproduced with permission from B.A. Morris, Reducing curl in multilayer blown film: experimental results, model development, and application to a cereal liner film, J. Plastic Film. Sheeting, 19 (2003) 31–54.*

show why. The EVOH freezes first so that as the HDPE freezes, its shrinkage gives rise to stresses that cannot be relieved by melt flow of the EVOH layer. Also, the moduli of EVOH and HDPE differ only by a factor of about two. The combination of high-differential shrinkage and comparable moduli results in curl.

The PVT curves show what happens to the polymers as they are quenched when primary crystallization occurs. After quenching, some polymers will undergo further crystallization called secondary crystallization. The crystals that form are often less perfect than the primary crystals, filling in some of the space between spherulites, for example. They still can produce shrinkage that affects curl. Polyamide (PA), in particular, is known to undergo secondary crystallization when exposed to moisture. Moisturization is one technique used to control the curl of PA containing films, as described below [189–191].

The actual mechanism for curl is residual stress caused by differential shrinkage, and in some cases orientation from the film blowing process. Several investigators in the fields of coatings and composites [192–194] have developed models for relating curl to residual stress between two thin films. The models rely on classical beam and plate theory of continuum mechanics. As depicted in Fig. 13.27, the residual stress is countered by a resisting force related to the stiffness and thickness of the layers in the structure. When the residual stress is strong enough to overcome the resisting force, the film bends or curls. A force and momentum balance is used to calculate these forces. In the appendix to this chapter, the approach of Shen and Suresh [192] is used to develop a model that relates the curl of films to the modulus, thickness, location and shrinkage of each layer.

Figs. 13.28 and 13.29 show the results of the model for two-layer films. A convenient way to characterize the curl of films is the curvature, which is the reciprocal of the radius of curvature (see Fig. 13.30). The radius of curvature is the radius of a circle that circumscribes the curling film; its reciprocal is related to the severity of the curl. A flat film has a curvature of zero and films that curl tightly have large values of curvature. Fig. 13.28 plots the curvature vs. the ratio of the stiffness (modulus) of the two layers. Here β represents the difference in shrinkage between the polymers (the ratio of the expansion coefficients). As β approaches 1, the differential shrinkage goes to zero. As β approaches 0, the differential shrinkage approaches infinity. Fig. 13.28 shows the model results for

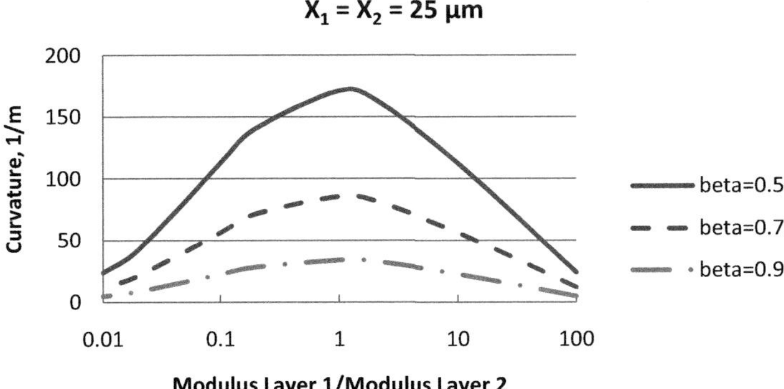

FIGURE 13.28 Model results for two-layer films showing effect of differential shrinkage (β) and modulus on curl. X = thickness. *Reproduced with permission from B.A. Morris, Reducing curl in multilayer blown film: experimental results, model development, and application to a cereal liner film, J. Plastic Film. Sheeting, 19 (2003) 31–54.*

Curvature of 2-Layer Films
beta = 0.75
Curvature, 1/m
200
150
100
50
0
0.01
0.1
1
10
100
Modulus Layer 1/Modulus Layer 2
X1/X2=0.5
X1/X2=1
X1/X2=2
X1/X2=10

FIGURE 13.29 Model results for two-layer film showing the effect of thickness and modulus on curl. X = thickness. *Reproduced with permission from B.A. Morris, Reducing curl in multilayer blown film: experimental results, model development, and application to a cereal liner film, J. Plastic Film. Sheeting, 19 (2003) 31–54.*

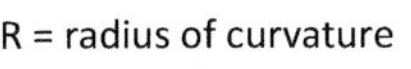

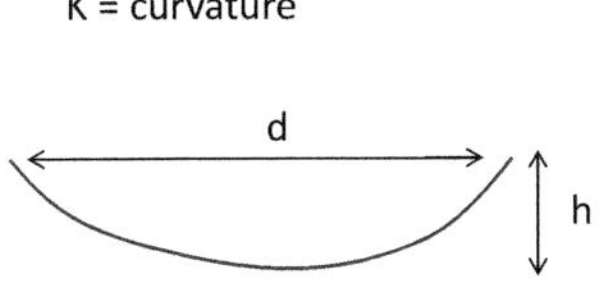

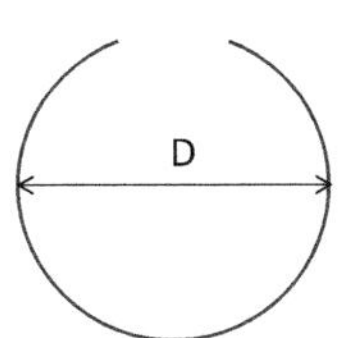

For nearly flat film:

$$K = \frac{1}{R} = \frac{1}{\left(\frac{h}{2} + \frac{d^2}{8h}\right)}$$

For highly curled film:

$$K = \frac{1}{R} = \frac{2}{D}$$

FIGURE 13.30 Measuring curl.

the case where the two layers are equal in thickness. The greatest curl occurs when the stiffness of the two layers is equal. As the stiffness of one layer dominates the other, the curl decreases. This agrees with practical experience: consider the painting of a metal part, such as an automotive door. As the paint dries, it shrinks relative to the door, but the door does not curl because it is so much stiffer than the paint. For the paint, the residual stress manifests itself in other ways, such as poor adhesion or pitting.

Fig. 13.28 also shows that as the differential shrinkage increases, the curl increases, as expected. In Fig. 13.29, the effect of varying the thickness of the film layers is demonstrated. The stiffness and thickness of the layers interact to varying degrees to resist the residual stress.

13.5.2 Strategies for reducing curl

With the residual stress mechanism as a foundation, several strategies are outlined for minimizing curl in film coextrusion:

1. Reduce the crystallinity of the polymers in the film structure, particularly the layer the film is curling toward.

Reducing the crystallinity of a polymer decreases ΔV_c and modulus, both of which can impact curl. Typically, it is not the first layer to freeze that the film curls toward, even if that layer undergoes a large change in volume during crystallization. Its shrinkage occurs when the other layers are still molten and can flow in response. Later in the quenching process, other layers begin to crystallize, imparting stresses against the already frozen first layer. It is these stresses that give rise to curl.

Morris [186,187] presents data for two-layer coextruded blown films of various layer ratios prepared on a laboratory blown film line. Table 13.7 shows the relative amount of curl for each combination of materials. Structures containing PA or EVOH curl tightly, whereas those containing only low crystalline polymers curl loosely. Interestingly, although the EVOH–HDPE film curls severely, when the two layers (which have poor adhesion to each other) are separated, the resulting films are flat. This validates the hypothesis that residual stress at the interface is involved.

In a second study, Morris conducts a statistically designed experiment looking at five factors on the curl of two-layer films with the structure (EVOH–polyolefin). The variables are:

- Polyolefin (PO) crystallinity (high: HDPE or low: EVA);
- BUR of the blown film process (1.6–2.9);
- Ratio of thickness of PO to EVOH (0.5:1–2:1);
- Frost line height of blown film process (10–25 cm);
- Modification of EVOH by adding Surlyn AM7927 at 0%–20%. AM7927 is an additive manufactured by Dow that improves the thermoformability of EVOH and is expected to reduce the crystallinity and modulus of EVOH.

The parameters that are kept constant include the extrusion processing temperatures, final total film thickness (113 μm), haul-off speed, and EVOH grade (38 mol% ethylene). Because the haul-off speed and film thickness are constant, as the BUR is increased the DDR decreases [from 8.6 to 4.8—see Eq. (2.7)].

The analysis of the results is summarized in Table 13.8. In all cases, the curl is toward the polyolefin layer. The crystallinity of the polyolefin layer has the most statistically significant impact followed by the BUR and the thickness ratio. Frost line height is only significant for MD curl (not TD curl). EVOH modification has only a small impact on curl. Overall, it is found that the curl can be reduced (in order of greatest to least impact) by decreasing the crystallinity of the polyolefin layer (HDPE to EVA), increasing frost line height (MD only), reducing the BUR and decreasing the polyolefin/EVOH thickness ratio.

In a third study, Morris looks at the effect of changing the crystallinity of the tie resin in the three-layer structure: (38-μm HDPE–13-μm tie–25-μm EVOH). The HDPE has a density of 0.960 g/cc and MFI (190°C/2160 g) of 0.75 g/10 minutes. The EVOH is the 38-mol % ethylene grade used in the previous studies. The tie resins are all anhydride modified ethylene copolymers. The base resins and corresponding freezing points are shown in Table 13.9 along with the curl results. To measure TD curl, 25-mm wide strips are cut in the MD (and vice versa for measuring MD curl). From these strips, 5-mm strips are cut and the radius of curvature measured. The curvature is reported in Table 13.9.

TABLE 13.7 Curl of two-layer blown film.

Structure	Curls toward	Severity
(HDPE–EVA)	EVA	Low
(HDPE–Ionomer)	Ionomer	Low
(Ionomer–EVA)	EVA	Flat
(HDPE–EVOH)	HDPE	Very High
(HDPE-PA)	HDPE	Very High

Source: Reproduced with permission from B.A. Morris, Reducing curl in multilayer blown film: experimental results, model development, and application to a cereal liner film, J. Plastic Film. Sheeting, 19 (2003) 31–54.

TABLE 13.8 Results of statistically designed experiment evaluating curl of two-layer blown film.

Factors that reduce curl	Effectiveness
↓ Crystallinity of PO (HDPE ⇒ EVA)	* * *
↓ BUR (↑ DDR)	* *
↓ PO/EVOH thickness ratio	*
↑ Frost line height (MD curl)	***
↑ EVOH modifier (Surlyn AM7927)	Not Significant

Note: Asterisks represent statistical significance, which a higher number of asterisks meaning greater significance.
Source: Reproduced with permission from B.A. Morris, Reducing curl in multilayer blown film: experimental results, model development, and application to a cereal liner film, J. Plastic Film. Sheeting, 19 *(2003)* 31–54.

TABLE 13.9 Effect of changing tie resin on curl of coextruded blown film.

Tie resin	Type	DSC Tc (°C)	K (m^{-1})	
			Mean	St. dev.
Tie1	EVA	58	123	9
Tie2	EVA	67	86	7
Tie3	LDPE	95	147	15
Tie4	LLDPE	112	152	55

Note: Film structure is (HDPE–tie–EVOH).
Source: Data from B.A. Morris, Reducing curl in multilayer blown film: experimental results, model development, and application to a cereal liner film, J. Plastic Film. Sheeting, 19 (2003) 31–54.

A tie resin with a low melting point may be fluid enough as the HDPE crystallizes to relieve the stress created by the differential shrinkage of the HDPE against the already solid EVOH. The data in Table 13.9, however, show that the tie resin melt point has only a small effect on the curl (all the values are indicative of a very tight curl). Furthermore, as the films aged (see Fig. 13.31), the curl becomes more severe and all the films end up with about the same curl.

Aging is found to have a significant impact on curl. Samples of two-layer films containing PA or EVOH are aged in different moisture environments (0% RH desiccator at room temperature, 50% RH at room temperature, and 90% RH, 40°C). In general, the curl becomes more severe with time, but less so in high humidity environments. The modulus of the layers is found to increase with time, but not as much as the curl. This suggests that at high humidity, relaxation of stresses at the interface may be involved rather than a secondary crystallization or plasticization (from water) effect. In the case of PA, the presence of moisture can cause secondary crystallization [190–192], which causes further shrinkage with time. In either case, aging is just one of the many factors that make understanding curl difficult.

2. Match the shrinkage of the layers so that they cancel out.

This is a more theoretical than practical strategy since it is often difficult to achieve, but it provides a direction for modifying a structure to reduce curl. It explains why symmetric structures have little or no curl. For example, the structure (LLDPE–tie–EVOH–tie–LLDPE) where the layers on each side of the EVOH are the same should have no curl. As the film is quenched, the EVOH freezes first, followed by the LLDPE and tie resins on each side. Both sides shrink, but by the same amount so that the stress balances out. Increasing the thickness of one of the LLDPE layers or increasing its density (which increases crystallinity and modulus) will cause the film to curl in that direction.

When creating unbalanced structures, trying to add some of the more crystalline polymer on the other side of the first layer to freeze may help. With the increased number of layers often available on modern coextrusion blown film lines, this is becoming more practical. In the past, with only three or five layers, this was a challenge. Consider a five-

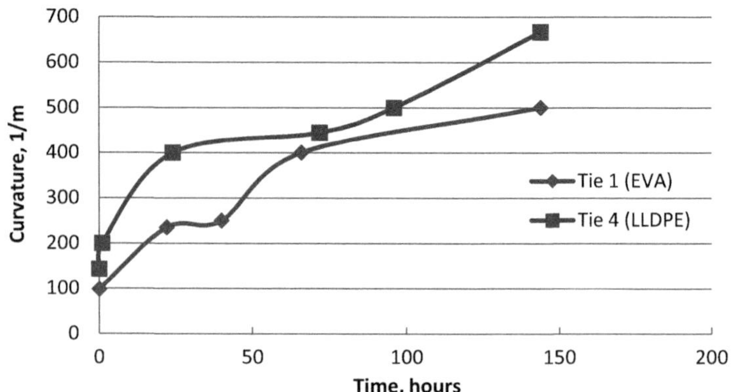

FIGURE 13.31 Effect of aging on curl of three-layer films from Table 13.9. *Data from B.A. Morris, Reducing curl in multi-layer blown film: experimental results, model development, and application to a cereal liner film, J. Plastic Film. Sheeting, 19 (2003) 31–54.*

layer barrier structure: (HDPE–tie–EVOH–tie–EVA) with thicknesses (35/3/5/3/20 μm). This structure tends to curl toward the HDPE layer. Reducing the thickness of the HDPE layer would reduce the curl, but the thickness of this layer is needed to achieve moisture barrier specifications. With only five layers, many have tried increasing the crystallinity of the tie layer on the sealant side of the structure, perhaps switching from an EVA-based tie resin to an LLDPE or even HDPE-based tie resin. This has had limited success due to the thinness of this layer. With a nine-layer line, one has the flexibility to build in other layers on the sealant side to try to balance out the shrinkage and curl.

Another way to balance the shrinkage involves polyamide coextruded films. Moisturizing the PA can cause secondary crystallization and shrinkage that may balance curl. This is discussed in more detail below.

3. Match the freezing points or crystallization rates of the layers so that shrinkage occurs at the same time.

In many coextruded films, the layers have different freezing points. The layer with the highest freezing point crystallizes first while the other layers are still molten and can flow in response. This layer provides a solid surface for the subsequent layers to shrink against, causing curl. Therefore, the order in which freezing takes place can affect the amount of curl. If the layers shrink at the same time, most of the shrinkage will cancel out.

For polyolefins, the freezing point varies with the class of polymer. The order from highest to lowest is: polypropylene homopolymer (PPh) > polypropylene random copolymer (PPr) > HDPE > LLDPE > LDPE. Specialty copolymers, such as metallocene plastomer, EVA and ionomer have lower freezing points than LDPE; their freezing points decrease with increasing comonomer content.

The freezing point of EVOH decreases with increasing ethylene content. For polyesters and polyamides (PAs), copolymers often have lower freezing points and crystallinity than the homopolymer. For example, PETG has a much lower melting and freezing point than PET.

Goetz [190,191] shows that the crystallization rate of polyamides varies greatly with PA type and whether nucleating agents are added. Nucleating agents such as talc provide sites for crystals to form, generally increasing the temperature at which crystallization begins and the number of crystals that form (by several orders of magnitude). While the total amount of crystallinity ends up being about the same as un-nucleated PA, the size of the crystals is substantially reduced. Nucleated polyamide blown film tends to have substantially less haze than un-nucleated blown PA.

PA 6 crystallizes faster than PA 6/66 copolymers, and nucleated PA 6 crystallizes faster yet. A PA grade that crystallizes slowly may overlap with the crystallization of PE in (PA–tie–PE) structures. For this reason, nucleated PA 6 has higher curl than un-nucleated PA 6 in such structures. PA 6/66 has even less curl, and blends of PA 6/66 with amorphous PA (6I/6T, such as DuPont's Selar PA), crystallize in the same range as PE. Their crystallinity often cancels out the shrinkage of the PE and produce flat film. This is illustrated in Fig. 13.32.

4. Increase the quench rate.

Increasing the quench rate during film fabrication suppresses the total amount of crystallinity, reducing overall shrinkage. It also lowers the temperature at which crystallinity occurs – through what is known as supercooling. This has the effect of clustering the freezing points of the layers, allowing some of the shrinkage to overlap and cancel out.

Cast films of the same composition typically have less curl than blown films due to the higher quench rate. Likewise, water quench blown film curls less than air quench blown film (see Chapter 14). Crist et al. [189] show that a cast film of PA 6 is almost amorphous, whereas a blown film of un-nucleated PA 6 has large spherulite crystals.

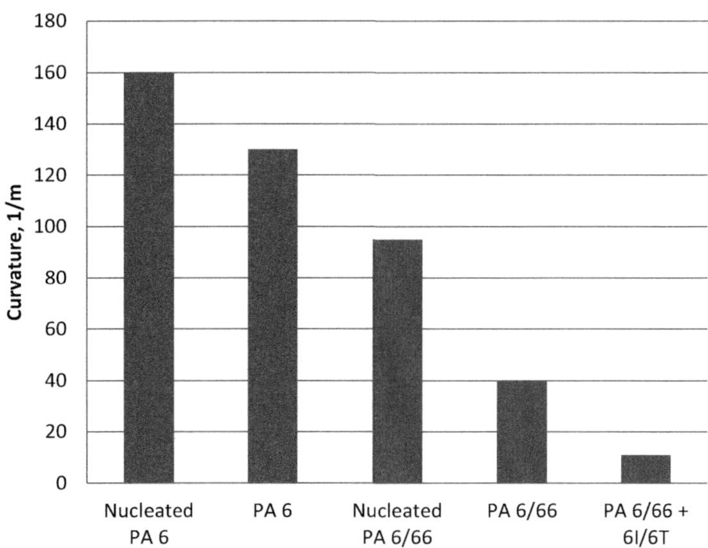

FIGURE 13.32 Curl of (PA−tie−LDPE) blown film (air quenched), total thickness 100 μm. *Data from W. Goetz, Curl in multilayer PA/PE film, Eur. TAPPI PLACE. Conf. (2001).*

Compared with blown films, cast films of PA 6 are nearly transparent, have a lower density, tensile strength and stiffness and better formability.

5. Enhance postquench crystallization.

Most of the crystallization of semicrystalline polymers occurs as the polymer solidifies from the melt during the film making process. Some polymers will continue to crystallize even at room temperature, although most do so at a very slow rate. Postquench crystallization, known as secondary crystallization, generally produces crystals that are less perfect and smaller than the primary crystals formed during quenching. This reordering of the molecules increases the density and shrinks the film. Some of the changes that occur include increased stiffness, dimensional change, and changes in curl. The aging effects shown in Fig. 13.31 may be related to postquench crystallization, among other things.

Postquench crystallization is more likely to occur in films that experienced rapid quench rates, which suppresses the primary crystallization. A polymer has a certain preferred degree of crystallinity based on the ordering of the molecules into the lowest energy configuration. A rapid quench freezes the molecules in place before they have a chance to reach this preferred state. The driving force is still present to form crystals, yet the energy hurdle for reconfiguration of the polymer chains can be large particularly if the ambient temperature is below the glass transition temperature (Tg) of the polymer. Molecular motion at room temperature may further be slowed by the presence of crystals that confine chain movement.

Polyamide is a special case. PA 6 and its copolymers absorb water, which plasticizes this highly polar polymer, disrupting the hydrogen bonds that bind the molecular chains together (the covalent bonds between atoms within the chain are not affected). Depending on the amount of moisture absorbed, the Tg decreases from about 55°C to −10°C. As the Tg drops below the ambient temperature, molecular motion increases, allowing secondary crystallization to occur. The secondary crystallization does not significantly increase the density but does cause 1%−2% shrinkage over time. This time can be substantially accelerated by heating the polymer while it is exposed to moisture. Often this is done by dipping the film into a heated water bath after quenching.

In a coextruded structure that is curling away from the PA, secondary crystallinity can help balance the shrinkage and reduce curl. If the film is already balanced, secondary crystallinity may induce curl toward the PA layer. Goetz [190,191] provides examples of three-layer (PA−tie−PE) blown films exposed in-line to a postquench water bath. The water is controlled to different temperatures and the residence time in the bath is 5 seconds. Excess water is removed from the surface of the films following the moisturization process and the films are conditioned at 23°C and 50% RH for 24 hours prior to measuring curl.

The results are shown in Table 13.10. With no water bath exposure, the results are similar to Fig. 13.32; changing the PA grade to one with a higher natural level of primary crystallinity or rate of crystallinity increases the curl toward the PE layer. Exposure to room temperature water is enough to switch the curl of the film containing PA 6/66, the least crystalline PA, toward the PA. Those films with PA types with greater curl toward the PE layer require higher temperature water (more secondary crystallization) to reduce the curl. The curl of the film samples containing PA 6 could not

TABLE 13.10 Effect of moisturization on curl of (PA–tie–LDPE) blown films.

Waterbath	PA 6/66	PA 6/66 nucleated	PA 6 + PA 6/66 (50:50)	PA 6	PA 6 nucleated
None	40	80	100	180	200
25°C	−100	0			
40°C			0		
50°C				50	100
70°C				20	50

Notes: Total film thickness is 100 μm. Results are reported as Curvature in units of m^{-1}. Positive and negative values indicate curl toward LDPE and PA layers, respectively.
Source: Data from W. Goetz, Curl in multilayer PA/PE film, Eur. TAPPI PLACE. Conf. (2001).

be completely eliminated even with exposure to boiling water. Control of the PA type, through a selection of copolymers or blends, and moisturization conditions allows for the production of curl-free films.

6. Balance the modulus, thickness, location, and shrinkage of the layers.

This strategy is often the most practical approach. Materials selection is usually constrained by cost and functionality. For example, reducing the density of the HDPE layer may reduce curl, but at the cost of reducing the moisture barrier. Likewise, the option of significantly altering the blown-film process to achieve a faster quench may be cost-prohibitive for a given application. Also, changing quench conditions can have an impact on film properties (Chapter 14) and adhesion (Chapter 15). Thus, the converter is often left with trying to balance and rearrange layers to reduce curl using a trial and error approach that can both be costly and have no assurance of success. Clearly, a predictive model of the effect of thickness, layer arrangement, modulus and shrinkage on curl would be of benefit to this process.

13.5.3 Modeling

One approach is to use statistically designed experiments to develop empirical models. A statistical model for the two-layer film developed by Morris was described earlier (Table 13.8). Kucejko and Pratt [195] use a DOE (design of experiments) approach to reduce curl in a five-layer film with the structure (EVA–HDPE–LDPE–LDPE–HDPE). The nominal layer thicknesses are (10/20/50/20%) where the LDPE core (50%) represents the combination of the third and fourth layers. An eight-state screening design is run varying:

- EVA melt processing temperature;
- Outside HDPE layer melt processing temperature;
- Total thickness;
- % HDPE (outside layer);
- % EVA (inside layer);
- Frost line height of the blown film process.

The HDPE and EVA layer thickness are varied at the expense of the LDPE core. The results show that reducing the % HDPE has the largest effect on reducing curl. Increasing output and decreasing thickness also help reduce curl. Kucejko and Pratt use their statistical model as a guide for making a film for a customer that has no curl.

Statistical-based models are very useful for a specific set of variables and provide predictions inside the experimental design space. They are also helpful when many variables may be significant, and the underlying mechanisms are not well understood. A more broadly useful tool is one based on first principles that can be applied to different structures.

Such a model has been developed by Morris [186–188] using classic beam theory. The is derived for two layers in the appendix to this chapter, but it has extended to any number of layers. Only the curl that develops during the quenching of the film is considered (no secondary crystallization). The model inputs include the thickness, location in the multilayer structure and tensile or secant modulus of each of the layers. These inputs are either well known or can be easily obtained. A more challenging parameter to obtain is the thermal expansion coefficient, α, for each of the layers. Morris first assumed that α is the sum of the change in volume due to crystallization (from the PVT data) and the coefficient of linear thermal expansion in the solid state (obtained from thermal mechanical analysis data):

$$\alpha_i = \alpha_T + \frac{\Delta Vc}{\Delta T} \tag{13.8}$$

where α_i = coefficient of thermal expansion for layer i, α_T = coefficient of linear thermal expansion (solid-state), ΔVc = volumetric change during crystallization, ΔT = temperature range for the analysis.

The final input parameter is the temperature range, ΔT, over which the polymer cools during quenching. The calculation is assumed to begin after the first layer freezes since any thermal contraction that occurs before that point is compensated by flow of the polymers. ΔV_c of the first layer to freeze is ignored and ΔT is taken as the difference between the crystallization temperature of the second layer to crystallize and room temperature.

This approach predicts film curvature that is substantially greater than what is measured, although it does correctly predict the direction of the curl. There are several reasons why this simple model does not yield accurate quantitative results:

- Quench rate. The PVT data is generated at a cooling rate of 2°C/minute, whereas commercial blown-film lines quench the polymer at around 500–1500°C/minute. A fast quench has the effect of reducing the temperature at which crystallization begins and the extent of crystallization due to undercooling of the melt. Spruiell and coworkers [196–198] developed a fast-quench device for measuring crystallization kinetics with cooling rates up to 5000°C/min and published work on HDPE, LDPE, and PP. At fast quench rates, they find that crystallization becomes heat-transfer limited. The exothermic heat of crystallization is balanced by the limited heat convection away from the film surface, giving rise to a temperature plateau several investigators have found in blown film (see Fig. 14.3, for example) [199,200]. Most of the crystallization occurs in the plateau region and the crystals do not impinge until near the end of the plateau, suggesting much of the stress due to shrinkage during crystallization (ΔV_c) may relax. This would account for some of the overprediction of curl in the model. Spruiell et al. find that increasing the quench rate reduces the plateau temperature (which they equate to the freezing point), reduces the time for nucleation to occur, decreases the width of the plateau, increases the rate of crystallization and reduces overall crystallinity.
- Strain-induced crystallization. As the blown film crystallizes, it is also being stretched and oriented. Strain is believed to accelerate the onset of crystallization by aligning the molecules and forming crystals along the chain. These crystals then nucleate the formation of chain-folded lamellae or "rows" of crystals. Most of the literature on strain-induced crystallization involves high-speed fiber spinning [201] where orientation is about an order of magnitude greater than for blown film. Lagasse and Maxwell [202] show, however, that even small amounts of strain/stress can substantially increase nucleation rates. Doufas and McHugh [203] develop a model of strain-induced crystallization in blown film. Faster quench rates act to delay crystallization, whereas strain/stress acts to speed it up. More work is needed to understand the crystallization kinetics when both occur simultaneously, as in the blown-film process.
- Relaxation of strain (creep). Since the crystals do not impinge until most of the crystallization has taken place, some relaxation of strain imposed by shrinkage during crystallization is likely to happen. The amount of relaxation during crystallization and after is difficult to predict and may have a substantial influence on curl.
- Anisotropic shrinkage. It is assumed that the shrinkage is isotropic, yet the experiments described earlier show that factors such as BUR and DDR influence curl. This suggests that shrinkage and curl may be anisotropic.
- Secondary crystallization. As described above, some polymers continue to crystallize over time after they are quenched. Shrinkage due to secondary crystallization may balance out some of the curl. This is ignored in the model.
- Excessive deformation. The model is based on equations developed for small deflections of beams. Some structures, particularly ones involving EVOH or nylon, curl very tightly. The model may no longer apply, particularly if the materials yield.

Morris works around these limitations by introducing a relaxation factor to Eq. (13.8):

$$\alpha_i = \alpha_T + \frac{\Delta Vc}{\Delta T}\Re \tag{13.9}$$

where $\Re$ = relaxation factor.

The relaxation factor is determined experimentally from two-layer film curl data; the model is run backwards to compute α from the curl results. By making measurements on different combinations of resins, a database of relaxation factors is built. Once $\Re$ is known for each resin, the model can be extended to any combination of resins and number of layers. To test the model, the curl of the (HDPE−tie−EVOH) structures described earlier is computed. The layer thicknesses are 38, 13, and 25 μm, respectively. The measured curvature is around 140/m, independent of tie layer, and ages up to 500/m. Using an EVA to mimic the tie resin, the model predicts a curvature of 220/m, which falls within the range of the experimental results. Furthermore, changing the parameters of the model to reflect different tie resins (such as LLDPE) does not appreciably change the results. This also agrees well with the experimental results.

To illustrate the utility of the model, Morris analyzes a barrier cereal liner film with the structure (HDPE−tie−EVOH−tie−sealant). The HDPE, typically 60% or more of the structure, provides a moisture barrier. The EVOH, around 7% of the structure, provides oxygen and aroma barrier. The sealant is typically an EVA or ionomer formulated for ease to open (peel-seal). The tie layers comprise about 8% of the total structure. This combination of layers is often very difficult to make without significant curl. Qualitatively, this is because the EVOH freezes first and cannot flow in response to the substantial shrinkage of the HDPE layer as it freezes.

A sensitivity analysis is conducted using the model to predict which parameters most impact the curl for this structure. The results are given in Table 13.11 and Fig. 13.33. To simplify the analysis, the tie layers are ignored. The parameters are the thickness, modulus and relative shrinkage (α) of each of the layers. The range for each parameter is chosen to represent practical limits for that layer. The total thickness is kept constant. The analysis suggests that curl can be reduced in this structure by making the HDPE layer thicker, stiffer and shrink less and making the EVOH layer thinner, less stiff and shrink more. The HDPE layer has the greatest impact on the curl and the sealant layer has the least.

To validate the model predictions, eight films are made on a five-layer blown-film line and their curl measured. The blown-film line is equipped with five-extruders (63-mm diameter for the inside and outside layers, 50-mm diameter for the middle layers) feeding a 203-mm diameter coextrusion die equipped with a dual-lip air ring. The blow-up ratio is 2.0, DDR 14, haul-off speed 19.8 m/minute, frost line-height 56 cm and die gap 1.88 mm. Total film thickness is 69 μm. The structure is (HDPE−tie1−EVOH−tie2−Sealant). Final extruder barrel temperature and die temperature set points are 227°C.

Variables include:

- Sealant type (Table 13.12)—Three sealants are used, an EVA (18 wt.% VA) and two ionomers commonly used in cereal liner and cracker packaging structures.
- EVOH type (Table 13.13)—A standard grade of EVOH (38 mol% ethylene) is evaluated along with two modifiers, an amorphous nylon and an ionomer, originally designed to enhance the thermoformability of EVOH. Adding amorphous nylon to PA 6 is known to reduce curl but not much was known about its effect on curl of EVOH structures. The modifiers are added at 20% to the EVOH.

TABLE 13.11 Results of parametric study of factors affecting curl of (HDPE−EVOH−sealant) film. K is the curvature as defined in Fig. 13.30.

Layer	Attribute	Attribute range	K (m^{-1})
Sealant	Modulus, MPa	14−345	175−172
	α, m/m-°C	0.00002−0.0002	189−164
	Thickness (HDPE floats), μm	7.5−25	119−200
EVOH	Modulus, MPa	690−3450	88−217
	α, m/m-°C	0.00002−0.0001	208−123
	Thickness (HDPE floats), μm	2.5−7.5	133−263
HDPE	Modulus, MPa	690−1730	227−147
	α, m/m-°C	0.00005−0.0004	−106−417
	Thickness, μm	71−88.5	125−213

Note: Base case structure: (40 μm HDPE−4 μm EVOH−19 μm Sealant). Base case $K = 175$/m. Positive values indicate curl is toward the HDPE layer and negative layers toward the sealant.
Source: Adapted B.A. Morris, Reducing curl in multilayer blown film: experimental results, model development, and application to a cereal liner film, J. Plastic Film. Sheeting, 19 (2003) 31−54.

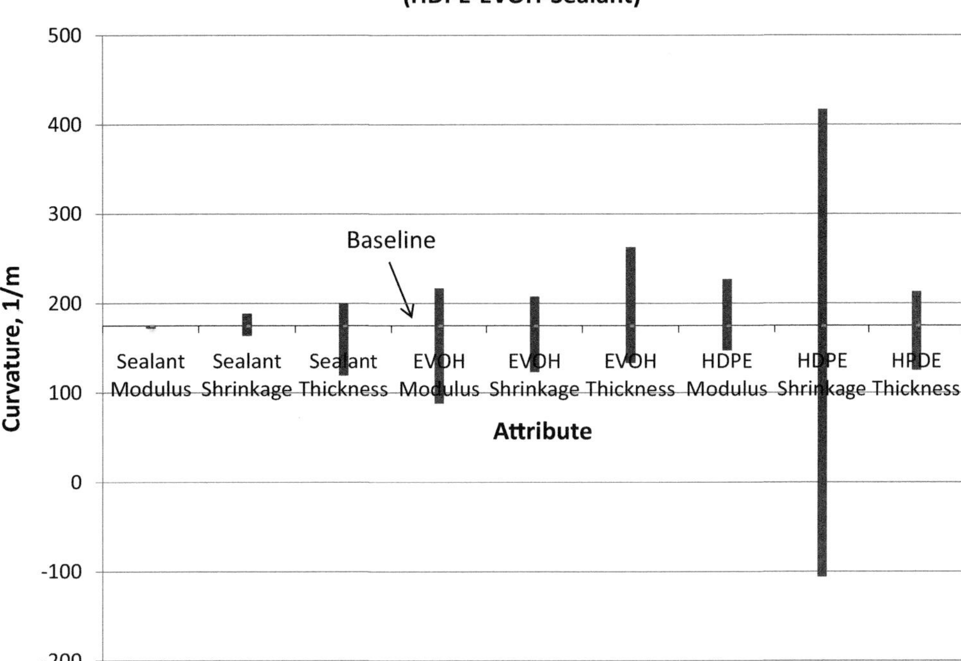

FIGURE 13.33 Sensitivity analysis of (HDPE–EVOH–sealant) film. *Adapted from B.A. Morris, Reducing curl in multilayer blown film: experimental results, model development, and application to a cereal liner film, J. Plastic Film. Sheeting, 19 (2003) 31–54.*

TABLE 13.12 Sealants used in blown film trial.

Code	Resin	Modulus (MPa)	Description
EVA	Elvax 3165	41	18%VA EVA
ION1	Surlyn 1650	275	Zn-Ionomer
ION2	Surlyn 1855	150	Zn-Ionomer
TIE2	Bynel 3930	83	EVA-Based Tie

Source: Reproduced with permission from B.A. Morris, Reducing curl in multilayer blown film: experimental results, model development, and application to a cereal liner film, J. Plastic Film. Sheeting, 19 (2003) 31–54.

TABLE 13.13 Ethylene vinyl alcohol used in blown film trial.

Code	Resin	Modulus (MPa)	Description	Manufacturer
EVOH1	Eval LCH101A	2270	38-mol % ethylene	Kuraray
EVOH2	80% Eval LCH101A, 20% Surlyn AM7927	1565	Blend	Kuraray, Dow
EVOH3	80% Eval LCH101A, 20% Selar PA	2130	Blend	Kuraray, DuPont

Source: Reproduced with permission from B.A. Morris, Reducing curl in multilayer blown film: experimental results, model development, and application to a cereal liner film, J. Plastic Film. Sheeting, 19 (2003) 31–54.

- HDPE type (Table 13.14)—A standard grade and a talc-filled grade of HDPE are evaluated. The mineral filler reduces the shrinkage and increases the stiffness of the HDPE layer, providing a convenient way to test the insights from the model sensitivity analysis.

TABLE 13.14 High-density polyethylene used in blown film trial.

Code	Resin	Modulus (MPa)	Description	Manufacturer
HDPE	Sclair 19A	1140	0.96 g/cc	Nova Chemical
Talc-HDPE	Zemid 641	1380	30% talc	DuPont Canada
TIE1	Bynel 4003	830	HDPE-Based Tie	Dow

Source: Reproduced with permission from B.A. Morris, Reducing curl in multilayer blown film: experimental results, model development, and application to a cereal liner film, J. Plastic Film. Sheeting, 19 (2003) 31–54.

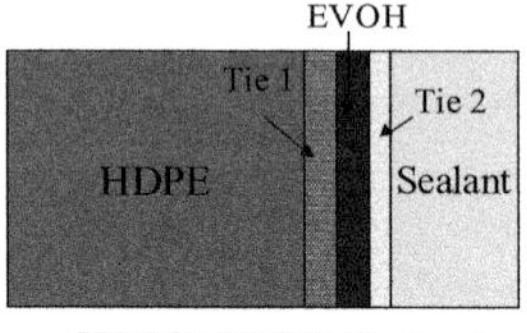

FIGURE 13.34 Structures in curl experiment. *Reproduced with permission from B.A. Morris, Reducing curl in multilayer blown film: experimental results, model development, and application to a cereal liner film, J. Plastic Film. Sheeting, 19 (2003) 31–54.*

- HDPE thickness (63%–84%)—Because of limitations on the HDPE extruder, this is accomplished by changing the HDPE-based Tie1 thickness from 4% to 25%. The standard and thick-HDPE structures are illustrated in Fig. 13.34.
- EVOH thickness (7%–4%).
- Sealant thickness (30%–12%)—To keep the total thickness constant, the sealant thickness is reduced as the HDPE thickness is increased. Tie2 (4%) is considered part of the sealant layer for the analysis.

The design of the experiment is described in Table 13.15 along with the curvature results. The results are also plotted in Figs. 13.35–13.37. The results show:

- In the standard cereal liner structure, the curl is very tight toward the HDPE layer. The curl is in the MD. No value is recorded for the TD since the curl is too strong in the MD to get a meaningful measurement in the TD.
- Changing the sealant does not impact the curl very much, as predicted by the model. ION2 gave the least amount of curl of the sealants tested. Its performance may be enhanced by postquench crystallization that is not accounted for by the model.
- Modifying the EVOH resin does not significantly reduce the curl in this structure. In other structures, however, this approach may be effective.
- Increasing the thickness of the HDPE layer significantly reduces curl, as predicted by the model.
- Using a mineral-filled HDPE significantly reduces curl. As shown by the model, this is due primarily to less shrinkage, although an increase in stiffness is also beneficial. The direction of curl changes from MD to TD.
- A combination of these factors yields essentially flat film.
- Overall, there is excellent agreement between the experimental results and the model predictions.

The films containing the talc-HDPE had poor tear-resistance and barrier properties. A calcium carbonate-filled HDPE may be a better choice for practical application. Subsequent measurements on such a film showed excellent tear resistance and MVTR as good as a standard grade of HDPE.

These evaluations show that the model makes accurate predictions despite simplifying assumptions that include:

- Ignoring postquench crystallization.

TABLE 13.15 Design of curl experiment and curvature results.

Name	Sealant	EVOH	HDPE	HDPE Thickness	Curvature (m^{-1}) Green MD	Green TD	1-week MD	1-week TD	Model
Standard	EVA	EVOH1	HDPE	Standard	250	0	250	0	188
Sealant 2	ION1	EVOH1	HDPE	Standard	238	0	250	0	200
Sealant 3	ION2	EVOH1	HDPE	Standard	182	0	192	0	192
EVOH2	EVA	EVOH2	HDPE	Standard	238	0	250	0	156
EVOH3	EVA	EVOH3	HDPE	Standard	333	0	250	0	182
Thick PE	EVA	EVOH1	HDPE	Thick	147	0	98	0	91
Talc PE	EVA	EVOH1	Talc-HDPE	Thick	0	105	0	48	59
Combined	ION2	EVOH2	Talc-HDPE	Thick	0	50	0	25	45

Source: Adapted from B.A. Morris, Reducing curl in multilayer blown film: experimental results, model development, and application to a cereal liner film, J. Plastic Film. Sheeting, 19 (2003) 31–54.

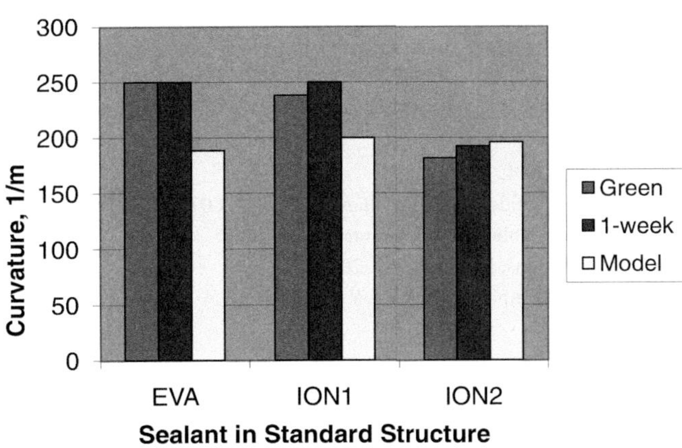

FIGURE 13.35 Curl results from blown film trial: effect of sealant type. *Data from B.A. Morris, Reducing curl in multilayer blown film: experimental results, model development, and application to a cereal liner film, J. Plastic Film. Sheeting, 19 (2003) 31–54.*

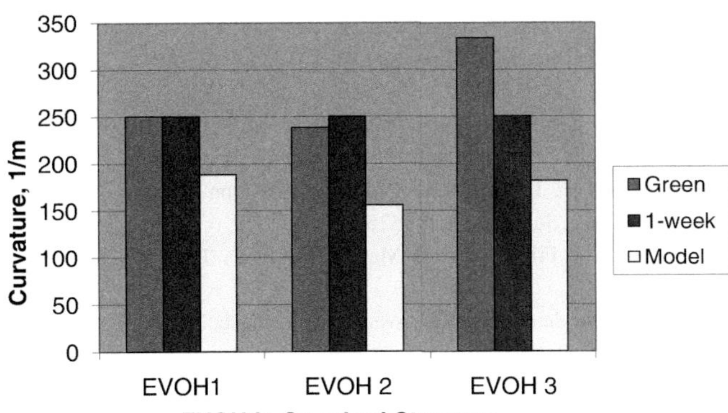

FIGURE 13.36 Curl results from blown film trial: effect of ethylene vinyl alcohol type. *Data from B.A. Morris, Reducing curl in multilayer blown film: experimental results, model development, and application to a cereal liner film, J. Plastic Film. Sheeting, 19 (2003) 31–54.*

- Using a single relaxation "correction" factor to account for a number of physical attributes that are difficult to determine, including the effect of quench rate and stress on crystallization kinetics and the amount of stress relaxation.
- Assuming that shrinkage of adjacent layers occurring before a layer freezes-off does not contribute to the differential strain.

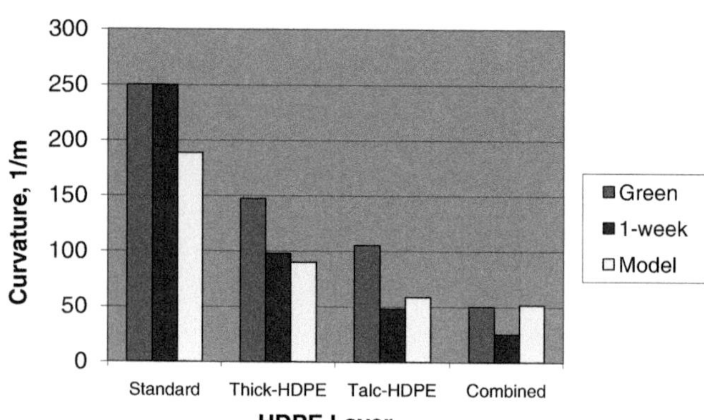

FIGURE 13.37 Curl results from blown film trial: effect of high-deinsity polyethylene type and thickness. *Data from B.A. Morris, Reducing curl in multilayer blown film: experimental results, model development, and application to a cereal liner film, J. Plastic Film. Sheeting, 19 (2003) 31–54.*

- Ignoring processing effects although processing conditions such as BUR and DDR are found in experiments to affect curl.
- Assuming that shrinkage is isotropic.
- Ignoring plastic yielding and other nonlinear effects due to large deformations in the continuum mechanics description of curl.

Despite these limitations, the model is a useful tool for designing structures that minimize curl.

References

[1] J. Vlachopoulos, J.R. Wagner (Eds.), The SPE Guide On Extrusion Technology and Troubleshooting, The Society of Plastics Engineers, Brookfield, CT, 2001.

[2] C. Rauwendaal, Polymer Extrusion, fifth ed., Hanser Publications, Cincinnati, OH, 2014.

[3] H.F. Giles Jr., E.M. Mount III, J.R. Wagner Jr. (Eds.), Extrusion: definitive Process. guide Handb, William Andrews, 2007, pp. 453–463.

[4] C. Chung, Extrusion of Polymers: Theory and Practice, second ed., Hanser Gardner Publications, Cincinnati, OH, 2000.

[5] Z. Tadmor, C. Gogos, Principles of Polymer Processing, second ed., John Wiley & Sons, Hoboken, NJ, 2006.

[6] Z. Tadmor, I. Klein, Engineering Principles of Plasticating Extrusion, Van Nostrand Reinhold Co, New York, 1970.

[7] W. Durling ed., Extrusion Coating Manual, fifth ed., TAPPI Press, Atlanta, GA.

[8] P.G. Lafleur, B. Vergnes, Polymer Extrusion, John Wiley & Sons, Hoboken, NH, 2014.

[9] E.M. del Pilar Noriega, C. Rauwendaal, Troubleshooting the Extrusion Process: A Systematic Approach to Solving Plastic Extrusion Problems, third ed., Hanser Publications, Cincinnati, OH, 2019.

[10] J.F. Macnamra Jr. (Ed.), Film Extrusion Manual, third ed., TAPPI Press, Atlanta, 2020.

[11] G.A. Campbell, M.A. Spalding, Troubleshooting & Analysis of Single-Screw Extrusion, Hanser Gardner Publications, New York, 2010, pp. 484–490.

[12] T.I. Butler, Gel troubleshooting, in: T.I. Butler (Ed.), Film Extrusion Manual, TAPPI Press, Atlanta, 2005, pp. 275–286.

[13] M.A. Spalding, E. Garcia-Meitin, S.L. Kodjie, Gel troubleshooting, in: J.F. Macnamra Jr. (Ed.), Film extrusion manual, third ed., TAPPI Press, Atlanta, 2020.

[14] M.A. Spalding, E. Garcia-Meitin, S.L. Kodjie, G.A. Campbell, Troubleshooting and mitigating gels in polyolefin film products, SPE ANTEC Conf. (2013).

[15] T. Andersson, B. Stålbom, B. Wesslé, Degradation of polyethylene during extrusion. II. Degradation of low-density polyethylene, linear low density polyethylene, and high-density polyethylene in film extrusion, J. App. Polym. Sci. 91 (2004) 1525.

[16] T. Finnegan, R. King III, Additives for film products, in: J.F. Macnamra Jr. (Ed.), Film Extrusion Manual, third ed., TAPPI Press, Atlanta, 2020.

[17] K. Keck-Antoine, E. Lievens, J. Bayer, J. Mara, D.-S. Jung, S.-L. Jung, Additives to design and improve the performance of multilayer flexible packaging, in: J. Wagner (Ed.), Multilayer Flexible Packaging, second ed., Elsevier, Oxford, 2016, pp. 53–76.

[18] H. Zweifel, R. Maier, M. Schiller, Plastics Additives Handbook, sixth ed., Hanser Plublications, Cincinnati, 2009.

[19] S.B. Marks, B.A. Morris, Extrudable polymers: purging and resin transactions, in: J.F. Macnamra Jr. (Ed.), Film Extrusion Manual, third ed., TAPPI Press, Atlanta, 2020.

[20] S.B. Marks, B.A. Morris, J.D. Vansant, Extrudable polymers: purging and resin transactions, in: W. Durling (Ed.), Extrusion Coating Manual, fifth ed., TAPPI Press, Atlanta, Ga, 2020.

[21] M.A. Spalding, A.C. Neubauer, Mitigating resin degradation via nitrogen inerting for single-screw extruders, SPE ANTEC (2021).

[22] S.S. Woods, S.E. Amos, The use of polymer processing aids to reduce gel formation in polyolefin plastomer extrusion, in: Polymers, Laminations & Coatings Conference, vol 2, San Francisco, August 30–September 3, 1998, pp. 675–685.

[23] S.S. Woods, S.E. Amos, The use of polymer processing aids to reduce gel formation in polyolefin plastomer extrusion, Annu. Technical Conf. - Soc. Plast. Engineers, 57th 1 (1999) 28−33. Vol.
[24] K.R. Slusarz, J.P. Christiano, S.E. Amos, An experimental investigation of the effect of polymer processing additives (PPA) on the melting and gel formation mechanisms in a single screw extruder, SPE ANTEC Conf. (2000).
[25] J.D. Vansant, Theory and practice of extruder systems purging, Proceedings, TAPPI PLACE. Conf. (2007).
[26] C. Thurber, M. Bishop, E. Marchbanks, K. Olson, H. Kim, J. Wang, Changeover time for a lab-scale blown film line, SPE ANTEC Conf. (2018).
[27] DuPont, Proper use of local exhaust ventilation during processing of plastics, At proper use of local exhaust ventilation during processing of plastics (dupont.com), 2011 (accessed 03.04.21).
[28] J.D. Gander, A.J. Giacomin, Review of die lip buildup in plastics extrusion, Polym. Engr. Sci. 37 (1997) 1113−1126.
[29] J. Musil, M. Zatloukal, Historical review of die drool phenomenon in plastics extrusion, Polym. Rev. 54 (2014) 16−34.
[30] M. Zatloukal, J. Musil, Die drool phenomenon in polymer extrusion, SPE ANTEC Conf. (2014) 1048−1054.
[31] F. Ding, L. Zhao, A.J. Giacomin, J.D. Gander, Flaring dies to suppress die drool, Polym. Eng. Sci. 40 (10) (2000) 2113−2123.
[32] J.A. Rakestraw, M.G. Waggoner, Polymer extrusion die and use thereof, U.S. Patent 5,458,836 A, 1995.
[33] J. Musil, M. Zatloukal, Experimental investigation of flow induced molecular weight fractionation during extrusion of HDPE polymer melts, Chem. Engr. Sci. 66 (2011) 4814−4823.
[34] J. Musil, M. Zatloukal, Experimental investigation of flow induced molecular weight fractionation phenomenon for two linear HDPE polymer melts having same M_n and M_w but different M_z and M_{z+1} average molecular weights, Chem. Engr. Sci. 81 (2012) 146−156.
[35] J. Musil, M. Zatloukal, Investigation of die drool phenomenon for HDPE polymer melt, Chem. Engr. Sci. 65 (2010) 6128−6133.
[36] T.A. Hogan, P. Walla, B.C. Dems, Investigation of the relationships between die build up and die swell, Polym. Engr. Sci. 49 (2009) 333−343.
[37] C.K. Chai, G. Adams, J. Frame, Polyethylene die deposit - measurement, formation mechanism and routes to reduction, Annu. Technical Conf. - Soc. Plast. Engineers, 59th 1 (2001) 401−406. Vol.
[38] A.M. Schmalzer, A.J. Giacomin, Die drool theory, J. Polym. Eng. 33 (1) (2013) 1−18.
[39] P.H. Gilbert, A.J. Giacomin, Slip heating in die drool, Can. J. Chem. Eng. 93 (3) (2015) 580−589.
[40] P.H. Gilbert, A.J. Giacomin, Slip heating in die drool with viscous dissipation, Int. Polym. Process. 30 (1) (2015) 141−146.
[41] G.S. Hoy, A.J. Giacomin, P.H. Gilbert, Die drool and polymer degradation, Polym. Technol. Eng. 55 (3) (2016) 242−258.
[42] D.A. Holtzen, J.A. Musiano, Die lip plate-out: a proposed mechanism, SPE Regional Technical Conf. (St. Louis) (1996).
[43] M. Zatloukal, K. Chaloupkova, Effect of die design and wall slip on die drool phenomenon for metallocene based LLDPE: theoretical and experimental investigation, SPE ANTEC Conf. (2007).
[44] K. Chaloupkova, M. Zatloukal, Analysis of die drool phenomenon for metallocene based linear polyolefins, SPE ANTEC Conf. (2006).
[45] C.-M. Chan, Viscosity and the formation of die drool at the polymer-metal interfaces, Int. Polym. Process. 10 (3) (1995) 200−203.
[46] S.J. Kurtz, S.R. Szaniszio, Reduction of die drool in filled resins and product improvement, U.S. Patent 5,008,056, 1991.
[47] R. Koopmans, J. Den Doelder, J. Molenaar, Polymer Melt Fracture, CRC Press, Boca Raton, 2011.
[48] B. Vergnes, Extrusion defects and flow instabilities of molten polymers, Intern. Polym. Process. (2015) 3−28.
[49] J.M. Dealy, K.F. Wissbrun, Melt Rheology and Its Role in Plastics Processing, Theory and Applications, Chapman & Hall, 1995, pp. 336−340.
[50] S. Monchai, N. Mieda, V. Anh Doan, S. Nobukawa, M. Yamaguchi, Effect of shear history on flow instability at capillary extrusion for long-chain branched polyethylene, J. Appl. Polym. Sci. 124 (2011) 429−435.
[51] T.I. Burghelea, H.J. Griess, H. Munstedt, Comparative investigations of surface instabilities ("sharkskin") of a linear and a long-chain branched polyethylene, J. Non-Newtonian Fluid Mech. 165 (2010) 1093−1104.
[52] A.V. Ramamurthy, Wall slip in viscous fluids and influence of materials of construction, J. Rheology 30 (1986) 337.
[53] D. Silagy, Y. Demay, J.F. Agassant, Stationary and stability analysis of the film casting process, J. Non-Newtonian Fluid Mech. 79 (1998) 563−583.
[54] M. Zatloukal, T. Barborik, Effect of extensional viscosity, elasticity and die exit stress state on neck-in phenomenon during extrusion film casting: theoretical study, SPE ANTEC Conf. (2016).
[55] M. Zatloukal, T. Barborik, Effect of die exit stress state, Deborah number and extensional rheology on neck-in phenomenon, SPE ANTEC Conf. (2018).
[56] T. Barborik, M. Zatloukal, Effect of die exit stress state, Deborah number, uniaxial and planar extensional rheology on the neck-in phenomenon in polymeric flat film production, J. Non-Newtonian Fluid Mech. 255 (2018) 39−56.
[57] T. Bezigian, A survey of current bead reduction die technology, TAPPI Polymers, Laminations & Coat. Conf. (1997) 55−67.
[58] T. Bezigian, A survey of current bead reduction die technology, in: T. Bezigian (Ed.), Extrusion Coating Manual, fourth ed., TAPPI Press, Atlanta, 1999, pp. 55−64.
[59] J.M. Dealy, K.F. Wissbrun, Melt Rheology and Its Role in Plastics Processing, Theory and Applications, Chapman & Hall, 1995, p. 553.
[60] T. Dobroth, E. Erwin, Causes of edge beads in cast films, SPE ANTEC Conf. (1985) 89−92.
[61] T. Dobroth, E. Erwin, Causes of edge beads in cast films, Polym. Engr. Sci. 26 (1986) 462−467.
[62] T. Dobroth, E. Erwin, Elimination of edge beads in cast polymer films, SPE ANTEC Conf. (1986) 893−900.
[63] K. Canning, A. Co, Edge effects in film casting of molten polymers, J. Plastic Film. & Sheeting 16 (2000) 188−203.
[64] N. Toft, M. Rigdahl, Extrusion coating with metallocene-catalysed polyethylenes, Int. Polym. Process. 17 (2002) 244−253.
[65] S. Shiromoto, Y. Masutani, M. Tsutsubuchi, Y. Togawa, T. Kajiwara, A neck-in model in extrusion lamination process, Polym. Eng. & Sci. 50 (1) (2010) 22−31.

[66] T. Barborik, M. Zatloukal, Steady-state modeling of extrusion cast film process, neck-in phenomenon, and related experimental research: a review, Phys. Fluids 32 (2020). published online on June 5, 2020.
[67] J. Kuusipalo, M. Vaha-Nissi, A. Savolainen, Study of rheology and adhesion in extrusion coating of different polyethylenes, in: Polymers, Laminations & Coatings Conference, Vol. 2, San Francisco, August 30—September 3, 1998, pp. 769—777.
[68] K. Chikhalikar, S. Banik, L.B. Azad, K. Jadhav, S. Mahajan, Z. Ahmad, et al., Extrusion film casting of long chain branched polypropylene, Polym. Engr Sci. 55 (2015) 1977—1987.
[69] C.W. Seay, D.G. Baird, Sparse long-chain branching's effect on the film casting behavior of PE, Int. Polym. Process. 24 (1) (2009) 41.
[70] H. Pol, S. Banik, L.B. Azad, S. Thete, P. Doshi, A. Lele, Nonisothermal analysis of extrusion film casting process using molecular constitutive equations, Rheol. Acta. 53 (2014) 85—101.
[71] H. Pol, S.S. Thete, P. Doshi, A.K. Lele, Necking in extrusion film casting: the role of molecular architecture, J. Rheology 57 (2013) 559—583.
[72] B. Debbaut, J.M. Marchal, M.J. Crochet, Viscoelastic effects in film casting, Z. Angew Math Phys, 46 (Special Issue: Theoretical, experimental, and numerical contributions to the mechanics of fluids and solids) (1995), S679—S698.
[73] S.S. Thete, P. Doshi, H.V. Pol, New insights into the use of multi-mode phenomenological constitutive equations to model extrusion film casting process, J. Plastic Film. & Sheeting 33 (1) (2017) 35—71.
[74] C. Sollogoub, Y. Demay, J.F. Agassant, Non-isothermal viscoelastic numerical model of the cast-film process, J. Non-Newtonian Fluid Mech. 138 (2-3) (2006) 76—86.
[75] H.V. Pol, S.S. Thete, Necking in extrusion film casting: numerical predictions of the Maxwell model and comparison with experiments, J. Macromol. Science, Part. B 55 (2016) 984—1006.
[76] B.M. Foederer, B.A. Morris, Use of IR technology to model temperature loss in the air gap, TAPPI PLACE. Conf. (2014).
[77] M. Zatloukal, T. Barborik, Understanding of neck-in phenomenon in production of flat polymeric films, SPE ANTEC Conf. (2020).
[78] C.K. Chai, P. Hollacher, O. Plassmann, M. Tronchon, Rheological study of melt elasticity on extrusion coating with polyethylenes, in: 9th TAPPI European PLACE Conference, Rome, Italy, May 12—14, 2003, pp. 45—83.
[79] P. Clevenhag, C. Oveby, Rheological indicators to predict the extrusion coating performance of LDPE, TAPPI PLACE. Conf. (2003).
[80] C. Overby, Rheological indicators of LDPE to predict processing performance in extrusion coating, Ann. Trans. Nordic Rheological Soc. 13 (2005) 187—190.
[81] J.M. Dealy, K.F. Wissbrun, Melt rheology and its role in plastics processing, Theory and applications, Chapman & Hall, 1995, p. 554.
[82] D. Silagy, Y. Dewy, J.-F. Agassant, Study of the stability of the film casting process, Polym. Engr Sci. 36 (21) (1996) 2614—2625.
[83] D. Rokadea, S. Chougaleb, P. Patila, T. Bhattacharjeec, D. Gawandea, H. Pola, et al., Controlling draw resonance in the extrusion film casting process using polymer nanocomposites: an experimental study and numerical linear stability analysis, J. Plast. Film. Sheeting (2020). published online Dec 2020.
[84] N.R. Anturkar, A. Co, Draw resonance in film casting of viscoelastic fluids: a linear stability analysis, J. Non-Newtonian Fluid Mech. 28 (3) (1988) 287—307.
[85] H. Ishihara, Analysis of draw resonance instability in the film casting process, in: T. Kanai, G.A. Campbell (Eds.), Film Processing, 1999, pp. 210—224.
[86] S. Smith, D. Stolle, Numerical simulation of film casting using an updated Lagrangian finite element algorithm, Polym. Eng. Sci. 43 (5) (2003) 1105—1122.
[87] P.J. Lucchesi, E.H. Roberts, S.J. Kurtz, Reducing draw resonance in LLDPE film resins, Plast. Eng. 41 (5) (1985) 87—90.
[88] T.I. Burghelea, J.J. Griess, H. Muenstedt, An in situ investigation of the draw resonance phenomenon in film casting of a polypropylene melt, J. Non-Newtonian Fluid Mech. 173-174 (2012) 87—96.
[89] A. Christie, Extrusion coaters: stop that edge weave,", Plast. Technol. (2004)viewed 18 Apr 2021 at. Available from: http://www.ptonline.com/articles/extrusion-coaters-stop-that-edge-weave.
[90] B. Bian, A. Co, Tension in multilayer film casting of polymer melts, SPE ANTEC Conf. (1998) 113.
[91] K.R. Frey, G.D. Jerdee, C.D. Cleaver, The use of encapsulation dies for processing linear polyolefin resins in extrusion coating, SPE ANTEC Conf. (2001).
[92] Z. Stary, M. Papp, T. Burghelea, Deformation regimes, failure and rupture of a low density polyethylene (LDPE) melt undergoing uniaxial extension, J. Non-Newtonian Fluid Mech. 219 (2015) 35—49.
[93] T.I. Butler, Blown film bubble instability induced by fabrication conditions, SPE ANTEC Conf. 58 (1) (2000) 156—164.
[94] P. Tas, L. Nguyen, S. Marshall, M. Kashanian, Melt strength, maximum output, and physical properties of SURPASS TM1 sLLDPE—LDPE blends in blown film, J. Plastic Film. Sheeting 21 (2005) 265—279.
[95] C.D. Han, J.Y. Park, Studies on blown film extrusion. III. Bubble instability, J. Appl. Polym. Sci. 19 (1975) 3291.
[96] C.D. Han, R. Shetty, Flow instability in tubular film blowing. I. Experimental study, IEC Fundam. 16 (1977) 49.
[97] P. Tas, L. Nguyen, S. Marshall, M. Kashanian, Bubble stability and maximum output of SURPASS sLLDPE resin in blown film, TAPPI PLACE. Conf. (2004) 320—334.
[98] T. Kanae, J.L. White, Kinematics, dynamics and stability of the tubular film extrusion of various polyethylenes, Polym. Eng. Sci. 24 (1984) 1185.
[99] D.M. Kalyon, D.-W. Yu, F.H. Moy, Rheology and processing of linear low density polyethylene resins as affected by alpha-olefin comonomers, Polym. Eng. Sci. 28 (1988) 1542.
[100] T.J. Obijeski, K.R. Puritt, Improving the output and bubble stability of thick gauge blown film, SPE ANTEC Conf. (1992) 150.

[101] K.-S. Yoon, C.-W. Park, Stability of a blown film extrusion process, Intern. Polym. Process. XIV (1999) 342–349.

[102] J. Taylor, J.J. Baik, Benefits of coextruded LLDPE/LDPE film versus blended LLDPE/LDPE film, SPE ANTEC Conf. 58 (1) (2000) 209–215.

[103] G.J. Field, P. Micic, S. Bhattacharya, Melt strength and film bubble instability of LLDPE/LDPE blends, Polym. Int. 48 (1999) 461–466.

[104] A. Ghijels, J.J.S. Ente, J. Raadsen, Melt strength behavior of PE and its relation to bubble stability in film blowing, Intern. Polym. Process. V. (1990) 284–286.

[105] H. Münstedt, T. Steffl, A. Malmberg, Correlation between rheological behaviour in uniaxial elongation and film blowing properties of various polyethylenes, Rheol. Acta 45 (2005) 14–22.

[106] Y. Fang, P.J. Carreau, P.G. Lafleur, Rheological effects of polyethylenes in film blowing, Polym, Engr. Sci. 43 (7) (2003) 1391–1406.

[107] R. Kolarik, M. Zatloukal, M. Martyn, The effect of polyolefin extensional rheology on non-isothermal film blowing process stability, Int. J. Heat. Mass. Transf. 56 (1-2) (2013) 694–708.

[108] R.M. Patel, T.P. Karjala, N.R. Savargaonkar, P. Salibi, L. Liu, Fundamentals of structure-property relationships in blown films of LLDPE/LDPE blends, J. Plastic Film. Sheeting 35 (4) (2019) 401–421.

[109] A. Ghijsels, J.J.S.M. Ente, J. Raadsen, Melt strength behavior of polyethylene blends, Int. Polym. Process. 7 (1992) 44–50.

[110] N. Savargaonkar, R. Patel, T. Karjala, P. Salibi, L. Liu, Fundamentals of abuse performance of LLDPE/LDPE blends in blown film applications, SPE ANTEC Conf. (2014) 880–884.

[111] B.H. Bersted, J.D. Slee, C.A. Richter, Prediction of rheological behavior of branched polyethylene from molecular structure, J. Appl. Polym. Sci. 26 (1981) 1001–1014.

[112] B.H. Bersted, On the effects of very low levels of long chain branching on rheological behavior in polyethylene, J. Appl. Polym. Sci. 30 (1985) 3751–3765.

[113] C.T. Lue, Easy processing metallocene polyethylene, SPE ANTEC Conf. 56 (2) (1998) 1816–1819.

[114] T. Karjala, L. Kardos, A. Shah, Agility™ performance LDPE as a blend component in high throughput and high bubble stability blown film applications, SPE ANTEC Conf. (2016) 978–982.

[115] T. Karjala, L. Kardos, Y. Lin, Agility™ performance LDPE in blends elevates film mechanicals and extrusion output to the next level, SPE ANTEC Conf. (2017) 841–845.

[116] B.A. Morris, Modifier for increasing output and reducing gauge variation in LLDPE blown film, SPE Polyolefins Conf. (2008).

[117] A. Williamson, J. Gomes, M. Oleveira, T. Karjala, T. Abel, B. Watson, et al., Performance enhancement of polyethylene: Dowlex™ GM AX01, SPE Int. Polyolefins Conf. (2021).

[118] C.D. Han, Multiphase Flow in Polymer Processing, Acedemic Press, New York, 1981, pp. 354–413.

[119] T.I. Butler, Effects of flow instability in coextruded films, TAPPI J. 75 (9) (1992) 205–211.

[120] J. Dooley, B. Stout, An experimental study of the factors affecting the layer thickness uniformity of coextruded structures, SPE ANTEC Conf. (1991) 62–65.

[121] J. Powers, J. Dooley, Coextrusion layer thickness distribution improvements via equipment design, SPE ANTEC Conf. (1997) 328–333.

[122] A.A. Khan, C.D. Han, On the interface deformation in the stratified two-phase flow of viscoelastic fluids, Trans. Soc. Rheology 20 (1976) 595–621.

[123] A.A. Khan, C.D. Han, A study on the interfacial instability in the stratified flow of two viscoelastic fluids through a rectangular duct, Trans. Soc. Rheology 21 (1977) 101–131.

[124] J. Dooley, Dietsche, Numerical simulation of viscoelastic polymer flow - effects of secondary flows on multilayer coextrusion, Plast. Eng. 52 (April) (1996) 37–39. 4.

[125] J. Dooley, R. Ramanathan, Coextrusion layer thickness variation – effects of geometry on layer rearrangement, SPE ANTEC Conf. (1994) 89–93.

[126] J. Dooley, B.T. Hitlon, Layer rearrangement in coextrusion, Plast. Eng., 50 (Feb) 25–27.

[127] J. Dooley, K. Hughes, Thickness variation in coextruded layers – effect of polymer viscoelasticity on layer uniformity, TAPPI J. 79 (4) (1996) 235–241.

[128] J. Dooley, K. Hughes, Coextrusion layer thickness variation – effect of polymer viscoelasticity on layer uniformity, SPE ANTEC Conf. (1995) 69–74.

[129] J. Dooley, K.S. Hyun, K. Hughes, An experimental study on the effect of polymer viscoelasticity on layer rearrangement in coextruded structures, Polym. Engr. Sci. 38 (1998) 1060–1071.

[130] J. Dooley, K. Hughes, Analyzing the flow through dies containing different channel geometries, SPE ANTEC Conf. (1996) 236–240.

[131] P.D. Anderson, J. Dooley, H.E.H. Meijer, Viscoelastic effects in multilayer polymer extrusion, Appl. Rheology 16 (4) (2006) 198–205.

[132] L. Dietsche, J. Dooley, Numerical simulation of viscoelastic polymer flow – effect of secondary flows on multilayer coextrusion, SPE ANTEC Conf. (1995) 188–193.

[133] B. Debbaut, T. Avalosse, J. Dooley, K. Hughes, On the development of secondary motions in straight channels induced by the second normal stress difference: experiments and simulations, J. Non-Newtonian Fluid Mech. 69 (1997) 255–271.

[134] D. Borzacchiello, E. Leriche, B. Blottiere, J. Guillet, On the mechanism of viscoelastic encapsulation of fluid layers in polymer coextrusion, J. Rheology 58 (2) (2014) 493–512.

[135] B. Debbaut, J. Dooley, Secondary motions in straight and tapered channels: experiments and three-dimensional finite element simulation with a multimode differential viscoelastic model, J. Rheology 43 (1999) 1525–1545.

[136] J.L. Zryd, J. Dooley, Relative effects of viscous and elastic forces on layer thickness uniformity in coextrusion, SPE ANTEC Conf. (1998) 217.
[137] R. Shanker, R. Ramanathan, W.J. Schrenk, Comparison of experimental velocity measurements in the manifold of an extrusion die with numerical simulation, SPE ANTEC Conf. (1994) 75–79.
[138] K. Matsunage, T. Kajiwara, K. Funatsu, Numerical simulation of multi-layer flow for polymer melts – a study of the effect of viscoelasticity on interfacial shape of polymers within dies, Polym. Engr. Sci. 38 (1998) 1099–1111.
[139] M. Zatloukal, M. Martyn, J. Vlcek, P. Saha, Viscoelastic coextrusion flow modeling by advanced constitutive equations: the influence of the die design on interfacial instabilities, Eur. TAPPI PLACE. Conf. (2005).
[140] W.J. Schrenk, N.L. Bradley, T. Alfrey Jr., H. Maack, Interfacial flow instability in multilayer coextrusion, Polym. Engr. Sci. 18 (1978) 620–623.
[141] R. Ramanathan, R. Shanker, T. Rehg, S. Jons, D.L. Headley, W.J. Schrenk, 'Wave' pattern instability in multi-layer coextrusion – an experimental investigation, SPE-ANTEC Tech. Pap. 54 (1996) 224–228.
[142] M.T. Martyn, T. Gough, R. Spares, P.D. Coates, Visualization of melt interface in a co-extrusion geometry, SPE ANTEC Conf. (2001).
[143] M.T. Martyn, P.D. Coates, M. Zatloukal, J. Perdikoulias, Visualisation and analysis of LDPE melt flows in a coextrusion geometry, in: TAPPI PLACE Conference, Boston, MA, 1084, 2002. Also SPE ANTEC Conference, 2002.
[144] R. Ramanathan, W.J. Schrenk, J.A. Wheatley, Coextrusion of multilayer articles using protective boundary layers and apparatus therefor, U.S. Patent 5,269,995 A, 1993.
[145] Y. Zhu, A. Bironeau, F. Restagno, C. Sollogoub, G. Miquelard-Garnier, Kinetics of thin polymer film rupture: model experiments for a better understanding of layer breakups in the multilayer coextrusion process, Polymer 90 (2016) 156–164.
[146] A. Bondon, K. Lamnawar, A. Maazouz, Experimental investigation of a new type of interfacial instability in a reactive coextrusion process, Polym. Eng. & Sci. 55 (11) (2015) 2542–2552.
[147] S. Vuong, L. Leger, F. Restagno, Controlling interfacial instabilities in PP/EVOH coextruded multilayer films through the surface density of interfacial copolymers, Polym. Eng. & Sci. 60 (7) (2020) 1420–1429.
[148] M. Botros, C. Lee, D. Ward, T. Schloemer, New developments of Plexar® tie-layer adhesives in flexible and rigid applications, Eur. TAPPI PLACE. Conf. (2005).
[149] L. Levitt, C.W. Macosko, T. Schweizer, J. Meissner, Extensional rheometry of polymer multilayers: a sensitive probe of interfaces, J. Rheology 41 (1997) 671–685.
[150] C.D. Han, R. Shetty, Studies on multilayer film coextrusion II. Interfacial instability in flat film coextrusion, Polym. Engr. Sci. 18 (1978) 180–186.
[151] W. Kopytko, M. Zatloukal, J. Vlcek, Coextrusion of LDPEs in the cast film process (zig-zag interfacial instabilities), in: 62nd Annual Technical Conference - Society of Plastics Engineers, Vol. 1, 2004, pp. 372–375.
[152] G.M. Wilson, Interfacial instability in multilayer extrusion processes, SPE ANTEC Conf. (1993).
[153] M. Zatloukal, W. Kopytko, J. Vlcek, P.S. Ha, Investigation of zig-zag type of instabilities in coextrusion, SPE ANTEC Conf. (2005).
[154] M. Zatloukal, W. Kopytko, P. Saha, M. Martyn, P.D. Coates, Theoretical and experimental investigation of interfacial instability phenomena occurring during viscoelastic coextrusion, Plastics, Rubber Compos. 34 (9) (2005) 403–409.
[155] M. Zatloukal, J. de Witte, The effect of process aids on interfacial instabilities in coextrusion flows: theoretical and experimental investigation, Annu. Technical Conf. - Soc. Plast. Engineers, 62nd 1 (2004) 381–385. Vol.
[156] M. Zatloukal, J. de Witte, Effect of process aides on interfacial instabilities in coextrusion, SPE ANTEC Conf. (2005).
[157] M. Zatloukal, J. De Witte, Influence of process aids on zigzag type of interfacial instabilities in multilayer flows: theoretical and experimental investigation, Plastics, Rubber Compos. 34 (4) (2006) 149–154.
[158] M. Zatloukal, J. de Witte, C. Lavalle, P. Saha, Investigation of PPA interactions with polymer melts in single layer extrusion and coextrusion flows, Plastics, Rubber Compos. 36 (6) (2007) 248–253.
[159] C.S. Yih, Instability due to viscosity stratification, J. Fluid Mech. 27 (1967) 337–352.
[160] G.M. Wilson, B. Khomami, An experimental investigation of interfacial instabilities in multilayer flow of viscoelastic fluids, Part I. Incompatible polymer systems, J. Non-Newtonian Fluid Mech. 45 (1992) 355–384.
[161] G.M. Wilson, B. Khomami, An experimental investigation of interfacial instabilities in multilayer flow of viscoelastic fluids, Part II. Elastic and nonlinear effects in incompatible polymer systems, J. Rheology 37 (1993) 315–339.
[162] G.M. Wilson, B. Khomami, An experimental investigation of interfacial instabilities in multilayer flow of viscoelastic fluids, Part III. Compatible polymer systems, J. Rheology 37 (1993) 341–354.
[163] E.J. Hinch, O.J. Harris, J.M. Rallison, The instability mechanism for two elastic liquids being co-extruded, J. Non-Newtonian Fluid Mech. 43 (1992) 311–324.
[164] R. Valette, P. Laure, Y. Demay, J.-F. Agassant, Experimental investigation of the development of interfacial instabilities in two layer coextrusion dies, Intern. Polym. Process. XIX (2) (2004) 118–128.
[165] R. Valette, P. Laure, Y. Demay, A. Fortin, Convective instabilities in the coextrusion process, Intern. Polym. Process. XVI (2) (2001) 192–197.
[166] P.K. Ray, J.C. Hauge, D.T. Papageorgiou, Nonlinear interfacial instability in two-fluid viscoelastic Couette flow, J. Non-Newtonian Fluid Mech. 251 (2018) 17–27.
[167] H.K. Ganpule, B. Khomami, An investigation of interfacial instabilities in the superposed channel flow of viscoelastic fluids, J. Non-Newtonian Fluid Mech. 81 (1) (1999) 27–69.

[168] K. Xiao, Eliminating interfacial instability at PE/tie interface using a multi-layer stacked coextrusion die, SPE ANTEC Conf. (2004) 362–366.

[169] K. Xiao, M. Zatloukal, Multilayer die design and film structures, in: T. Kanai, G.A. Campbell (Eds.), Film Processing Advances, Hanser Publishers, 2014, pp. 67–110.

[170] J. Perdikoulias, C. Tzoganakis, Interfacial instability phenomena in blown film coextrusion of polyethylene resins, SPE ANTEC Conf. 53 (1995) 176–180.

[171] J. Perdikoulias, C. Tzoganakis, Interfacial instabilities during coextrusion of LDPEs, SPE ANTEC Conf. (1997) 351–355.

[172] M.T. Martyn, P.D. Coates, M. Zatloukal, Visualisation and analysis of polyethylene coextrusion melt flow, AIP Conf. Proc. 1152 (2009) 96–109.

[173] M.T. Martyn, P.D. Coates, M. Zatloukal, Influence of coextrusion die channel height on interfacial instability of low density polyethylene melt flow, Plastics, Rubber Compos. 43 (1) (2014) 25–31.

[174] H. Mavridis, R.N. Shroff, Multilayer extrusion: experiments and computer simulation, Polym. Engr. Sci. 34 (1994) 559–569.

[175] J. Perdikoulias, M. Zatloukal, C. Tzoganakis, Predicting the onset of interfacial instabilities in coextrusion flows, TAPPI Polymers, Laminations & Coat. Conf. (1999) 1003–1010.

[176] M. Zatloukal, C. Tzoganakis, J. Vlcek, P. Saha, Numerical simulation of polymer coextrusion flows, Intern. Polym. Process. XVI (2) (2001) 198–207.

[177] M. Zatloukal, J. Vlcek, P. Saha, C. Tzoganakis, Viscoelastic stress calculation in multi-layer coextrusion dies: die design and extensional viscosity effects on the onset of 'wave' interfacial instabilities, Polym. Engr. Sci. 42 (2002) 1520–1533.

[178] M. Zatloukal, W. Kopytko, A. Lengalova, J. Vlcek, Theoretical and experimental analysis of interfacial instabilities in coextrusion flows, Polym. Engr. Sci. 98 (2005) 153–162.

[179] M.T. Martyn, R. Spares, P.D. Coates, M. Zatloukal, Imaging and analysis of wave type interfacial instability in the coextrusion of low-density polyethylene melts, J. Non-Newtonian Fluid Mech. 156 (2009) 150–164.

[180] C.D. Han, R. Shetty, Studies on multilayer film coextrusion I. The rheology of flat film coextrusion, Polym. Engr. Sci. 16 (1976) 697–705.

[181] S. Puissant, B. Vergnes, Y. Demay, J.F. Agassant, A general non-isothermal model for one-dimensional multilayer coextrusion flows, Polym. Engr. Sci. 32 (1992) 213–220.

[182] M.E. Nordberg III, H.H. Winter, A simple model of nonisothermal coextrusion, Polym. Eng. Sci. 30 (7) (1990) 408–415.

[183] S.D. Pearson, Coextrusion flow analysis, TAPPI J. 77 (6) (1994) 193–197.

[184] B.A. Morris, Application of modeling to speed flexible package development, SPE ANTEC Conf. (2012).

[185] J. Vlcek, W. Kopytko, M. Zatloukal, B. Toure, The effect of material selection in feed-block coextrusion of a three layer film, Annu. Technical Conf. - Soc. Plast. Engineers, 63rd (2005) 128–131.

[186] B.A. Morris, Reducing curl in multilayer blown film: Experimental results, model development, and application to a cereal liner film, J. Plastic Film. & Sheeting 19 (2003) 31–54.

[187] B.A. Morris, Reducing curl in multilayer blown film. Part I: experimental results, model development and strategies, TAPPI Polymers, Laminations Coat. Conf. (2001).

[188] B.A. Morris, Reducing curl in multilayer blown film. Part II: application of predictive modeling to a barrier cereal liner film, TAPPI Polymers, Laminations Coat. Conf. (2001).

[189] M.D. Crist, K. Do, E. Noon, Curl in nylon multi-layer film, TAPPI Polymers, Laminations & Coat. Conf. (2000) 289–303.

[190] W. Goetz, Curl in multilayer PA/PE film, Eur. TAPPI PLACE. Conf. (2001).

[191] W. Goetz, Curl in multilayer PA/PE film, TAPPI PLACE. Conf. (2002).

[192] Y. Shen, S. Suresh, Elastoplastic deformation of multilayered materials during thermal cycling, J. Mater. Res. 10 (1995) 1200–1215.

[193] M. Benabdi, A. Roche, Mechanical properties of thin and thick coatings applied to various substrates. Part I. An elastic analysis of residual stresses within coating materials, J. Adhes. Sci. Technol. 11 (1997) 281–299.

[194] E. Corcoran, Determining stresses in organic coatings using plate-beam deflection, J. Paint. Technol. 41 (1969) 635–640.

[195] R.M. Kucejko, J.B. Pratt, Curl in five layer coextruded film, Film. Extrusion TAPPI Short. Course Notes (1991) 97.

[196] Z. Ding, J. Spruiell, Interpretation of the nonisothermal crystallization kinetics of polypropylene using a power law nucleation rate function, J. Polym. Sc.i B: Polym. Phys. 35 (1997) 1077–1093.

[197] Z. Ding, J. Spruiell, An experimental method for studying nonisothermal crystallization of polymers at very high cooling rates, J. Polym. Sci. B: Polym. Phys. 34 (1996) 2783–2804.

[198] P. Supaphol, J. Spruiell, Nonisothermal bulk crystallization studies of high-density polyethylene using light depolarizing microscopy, J. Polym. Sci. B: Polym. Phys. 36 (1998) 681–692.

[199] T.I. Butler, R. Patel, S.Y. Lai, J. Spuria, Blown-film frost line-freeze line interactions, TAPPI Polymers, Laminations Coat. Conf. (1993) 13.

[200] B.A. Morris, The effect of coextrusion on bubble kinematics, temperature distribution and property development in the blown-film process, SPE ANTEC Conf. (1998).

[201] A. Ziabicki, H. Kawai (Eds.), High-speed fiber spinning, Wiley, New York, 1985.

[202] R. Lagassa, B. Maxwell, An experimental study of the kinetics of polymer crystallization during shear flow, Polym. Engr. Sci. 16 (1976) 189.

[203] A. Doufas, A. McHugh, Modeling flow-induced crystallization in film blowing, SPE ANTEC Conf. (2000).

Chapter 14

Effect of the blown film process on film properties

The properties of polymer films cannot be separated from the process by which they are made. The stress and strain history imposed on the polymer as it is melted, extruded through the die, drawn to its final shape and quenched affects the orientation, crystallinity and morphology of the film. These in turn impact the mechanical, optical, barrier and other film properties. Three examples involving blown film are presented that illustrate the importance of processing on flexible packaging films:

- The effect of stress–strain history on film mechanical properties. In particular, the differences in strain history between monolayer and coextruded films and how this translates into property changes are highlighted.
- The impact of quench rate on properties. Key differences between air quench vs water quench blown film and their influence on the properties of barrier coextruded film are discussed.
- The influence of orientation on blend morphology.

14.1 Stress–strain history in blown film

In the air-quench blown film process, the polymer exits the die in the form of a tube. Air is inserted inside the tube to inflate the tube outwards into a bubble as it is drawn upwards via a nipped set of rolls at the top of the film tower. A schematic of the process is shown in Fig. 14.1. Cooling air is blown across the bubble to quench the hot polymer. The shape of the bubble is determined by the inflation pressure, line speed, cooling rate and the rheology of the polymer. The bubble shape influences the strain history, crystallinity, orientation, frozen-in stress and other key attributes of the film and its properties.

The frost line is the point where the bubble no longer changes shape: both the diameter and velocity reach their final value at the frost line. It is often associated with crystallization, since for many films a hazy band can be seen across the bubble near the frost line. Studies with online X-ray scattering and Raman spectroscopy, however, show that less than 5% of the crystallinity occurs before the frost line [1–6]. Nevertheless, the region between the die exit and frost

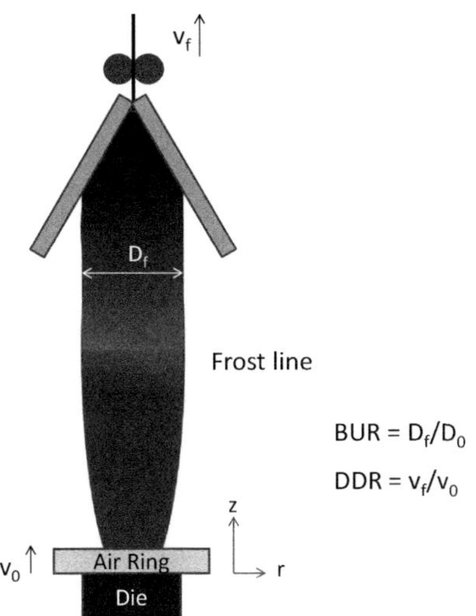

FIGURE 14.1 Schematic of blown film process.

The Science and Technology of Flexible Packaging. DOI: https://doi.org/10.1016/B978-0-323-85435-1.00014-4

line has a profound effect on the stress history and property development of the film. It is only near the frost line where the bubble becomes stiff enough to sustain the stretching needed for the film to reach its final bubble diameter and velocity. High strain rates occur in this region that give rise to stress that is frozen in place as the film crystallizes. This stress orients the molecular chains, which in turn influences the extent of crystallization and its morphology.

Experimental studies by Zhang et al. [4,5] have captured the development of the crystalline network structure during film blowing using a special in-line X-ray scattering apparatus. A laboratory film line is constructed inside a synchrotron radiation source for the measurements. They describe four zones of development from the die exit to positions above the frost line for a blend of 50% octene-linear low-density polyethylene (o-LLDPE), 25% butene-LLDPE and 25% low-density polyethylene (LDPE). In zone 1, a deformable precursor crystalline network structure emerges from the entangled chains during cooling and elongation. Further stretching leads to an acceleration of crystallization and the formation of a nondeformable scaffold at the end of zone 2. This is the frost line where the bubble shape stops changing—the modulus is sufficient to sustain the forces of bubble expansion. Above the frost line in zone 3 the scaffold fills in. The crystallization rate is still high even though it is nondeformable. Crystallization rate begins to tail off at the end of this zone leading into zone 4 where there is a further decrease in crystallization rate. The formation of the deformable crystal-based network is found to be mainly determined by the molecular architecture rather than processing parameters [drawdown ratio (DDR) and blow-up ratio (BUR)], but process parameters do influence the critical value of crystallinity for the creation of the nondeformable scaffold at the frost line.

14.1.1 Monolayer blown film

The importance of the stress–strain history during processing of blown film has been well known since the early days of film fabrication. Huck and Clegg [7] published one of the earliest comprehensive studies of the effect of output, BUR and frost line height (FLH) on the mechanical and optical properties of LDPE film. Statistical analysis was used to identify key trends; an example of how the tear properties change with processing conditions is shown in Fig. 14.2. In addition to mechanical properties, an extensive investigation of how processing parameters affect optical properties is described (see Chapter 12). The authors relate the mechanical properties to changes in the shape of the bubble and orientation measurements using birefringence. Numerous other investigators have studied the effect of processing variables on blown film crystallinity and property development, including online measurements of orientation and crystallinity [5–27]. For example, Shirdodkar et al. [15] find that dart impact and machine direction (MD) Elmendorf tear strength of LLDPE films increases with increasing process time, defined as [see Eq. (14.3)], and correlate this with orientation. The specific response of a polymer to processing conditions will depend on its molecular characteristics, such as molecular weight (MW), molecular weight distribution (MWD), branching and crystallinity, as well as how these characteristics affect the rheology of the polymer [6,8,12,24,28–36].

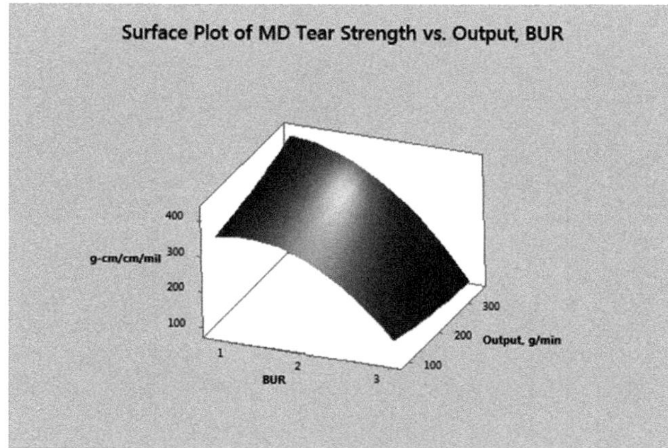

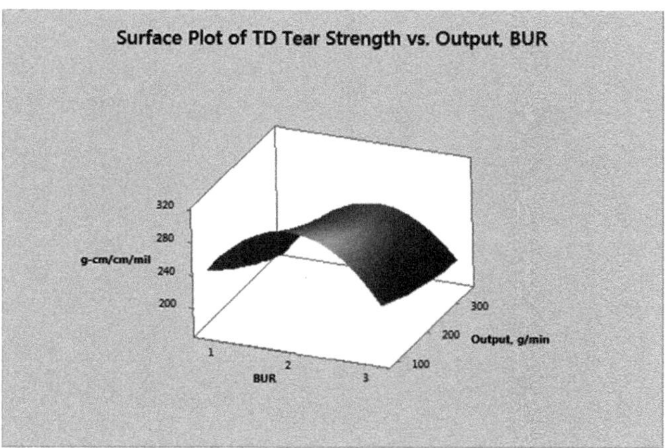

FIGURE 14.2 Effect of output and blow-up ratio on machine direction (left) and transverse direction (right) Elmendorf tear strength of low density polyethylene blown film. Frost line height is 14 inches (35.6 cm). *Data from N.D. Huck, P.L. Clegg, The effect of extrusion variables on the fundamental properties of tubular polythene film, SPE ANTEC Conf. 18 (2) (1961) 1–14.*

Scale-up guidelines have been proposed based on the analysis of stress, heat transfer and crystallization during film blowing [15,37–40]. Dimensionless groups are used to estimate process parameters that will duplicate the properties of a film produced on one line to that on another line, such as a small pilot-scale line. One such set of rules derived by Kanai et al. [37] is summarized in Table 14.1. Here the BUR, take-up ratio or DDR, and FLH/die diameter (D_O) are held constant. Studies [30,34,35] show that when using these rules, the deformation history of the two processes is the same. Shirodkar et al. [15] and Butler et al. [38,39] propose holding process time constant as a scale-up rule. As shown in Table 14.1, this condition is a natural outcome of the Kanai rules. Butler et al. propose scaling the output with the die diameter (see Section 2.2.1.4 and Table 2.3), which avoids changing the thickness of the film when the die gap is held constant.

Numerous studies have examined the effect of strain rate, stress and temperature in the region between the die exit and nip roll on the microstructure of the blown film: orientation, crystallinity and crystalline morphology. These microstructural features are then related to macro properties such as mechanical properties and permeability. Most of the studies have been on monolayer polyolefin film.

The temperature profile along the length of the bubble can be measured using IR thermometry. A typical measurement is shown in Fig. 14.3. The temperature steadily decreases until a plateau is reached. The onset temperature for the plateau is often called the crystallization line-height (CLH). Other definitions are shown in Fig. 14.3. The end of the plateau is called the freezing line height (FZH) in the early literature and was believed to be associated with the end of crystallization. Later, online measurements using small-angle X-ray scattering [1,2,4,5] or Raman spectroscopy [20–22] showed that the polymer continues to crystallize well beyond the end of the plateau, as shown in Fig. 14.3. The presence of the temperature plateau is best explained by Spruiell and coworkers [42–44], who duplicated the effect in very rapid quench laboratory experiments (5000°C/min): the crystallization becomes heat transfer limited at rapid cooling rates. The exothermic heat of crystallization is balanced by the limited heat convection away from the film surface, giving rise to a temperature plateau.

The velocity of the polymer in the machine and transverse directions has been measured by a number of investigators [45–48]. A simple technique is to mark the bubble near the die exit and record the position of the mark using a video camera. By playing back the video one frame at a time, and knowing the number of frames per second, the position of the marker can be determined as a function of time. The velocity is calculated as the change in position with

TABLE 14.1 Scale-up rules for blown film based on Kanai et al. [37] analysis.

Parameter	Line 1	Line 2
Scaling parameters (held constant)		
Melt temperature	Constant	Constant
BUR (D_f/D_O)	Constant	Constant
DDR (v_f/v_O)	Constant	Constant
FLH/D_O	Constant	Constant
Process parameters		
Die diameter (D_O)	$D_{O,1}$	$D_{O,2}$
Film diameter (D_f)	BUR*$D_{O,1}$	BUR*$D_{O,2}$
Die gap (X_O)	$X_{O,1}$	$X_{O,2}$
Film thickness (X_f)	$X_{f,1}$	$X_{f,1}*X_{O,2}/X_{O,1}$
Frost line height (FLH)	FLH_1	$FLH_1*D_{O,2}/D_{O,1}$
Haul-off speed (v_f)	$v_{f,1}$	$v_{f,1}*D_{O,2}/D_{O,1}$
Process time (t_f)	$FLH_1*\ln(DDR)/(v_{f,1}-v_{O,1})$	$t_{f,1}$ (constant)
Output (M)	$\pi D_O*BUR*X_f*v_f$	$M_1*(D_{O,2}/D_{O,1})^2*(X_{O,2}/X_{O,1})$

Note: The symbol * indicates multiplication.

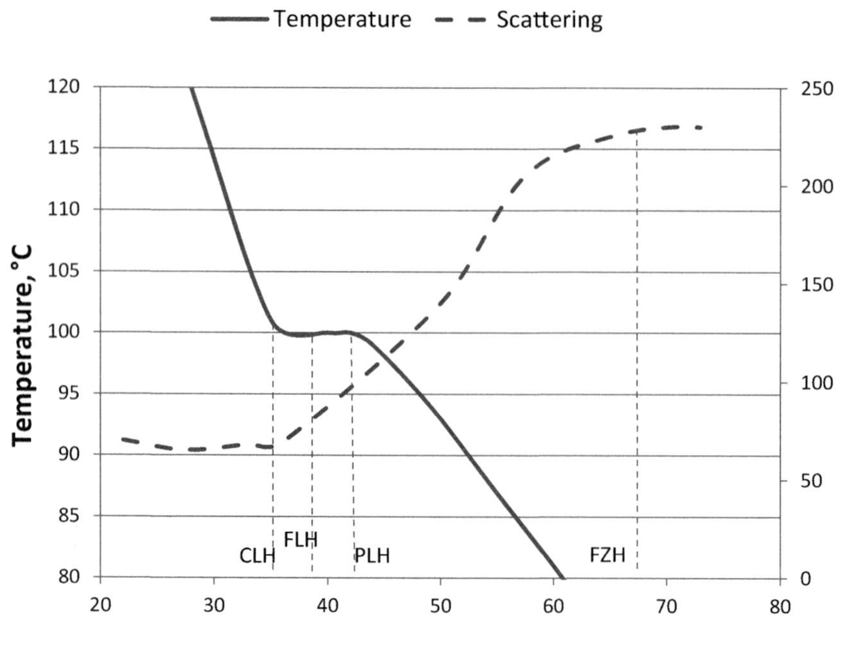

FIGURE 14.3 Crystallization and temperature profiles in blown film for a linear low-density polyethylene resin. Temperature was measured by IR and crystallinity by online small angle X-ray scattering. The data is smoothed. Scattering intensity is related to % crystallinity. *CLH*, Crystallization line height, *FLH*, frost line height, *PLH*, plateau line height, *FZH*, freezing line height. *Data and definitions are from M.D. Bullwinkel, G.A. Campbell, D.H. Rasussen, C.J. Brancewitz, Structure development during film blowing, SPE ANTEC Conf. (2001).*

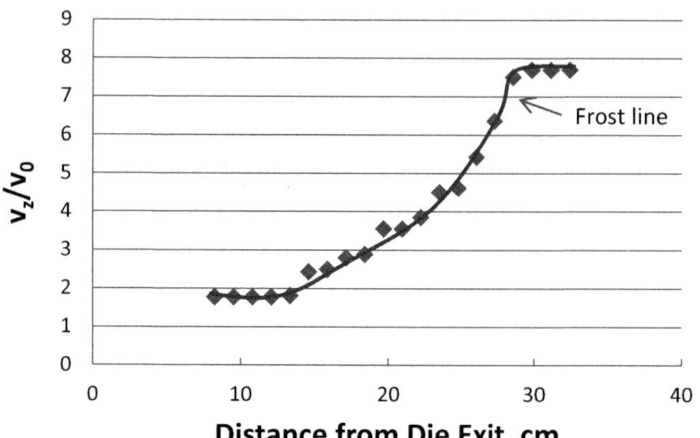

FIGURE 14.4 Machine direction velocity profile for a monolayer 50-μm high-density polyethylene blown film. Die diameter = 50 mm, frost line height = 27.5 mm, process time = 6.4 s. *Data from B.A. Morris, The effect of coextrusion on bubble kinematics, temperature distribution and property development in the blown film process, J. Plastic Film. Sheeting, 15(1) (1999) 25–36.*

time. Similarly, the velocity in the transverse direction can be determined by measuring the change in the diameter with time and position along with the bubble. An example MD velocity profile for a monolayer high-density polyethylene (HDPE) film is shown in Fig. 14.4.

The rate of strain or deformation can be computed from the velocity profiles:

$$\dot{\varepsilon}_1 = \frac{1}{v}\frac{dv}{dt} \tag{14.1}$$

$$\dot{\varepsilon}_2 = \frac{1}{v}\frac{dr}{dt} = \frac{v_z}{r}\frac{dr}{dz} \tag{14.2}$$

where $\dot{\varepsilon}_1$ = rate of strain in the z-direction (machine), $\dot{\varepsilon}_2$ = rate of strain in the r-direction (cross or transverse), v = velocity, t = time, z = machine direction, and r = radial direction.

The velocity data is fit to a polynomial and differentiated to find the rate of strain. An example of the rate of strain in the MD for a monolayer HDPE blown film is shown in Fig. 14.5. The maximum rate of strain occurs near the frost line.

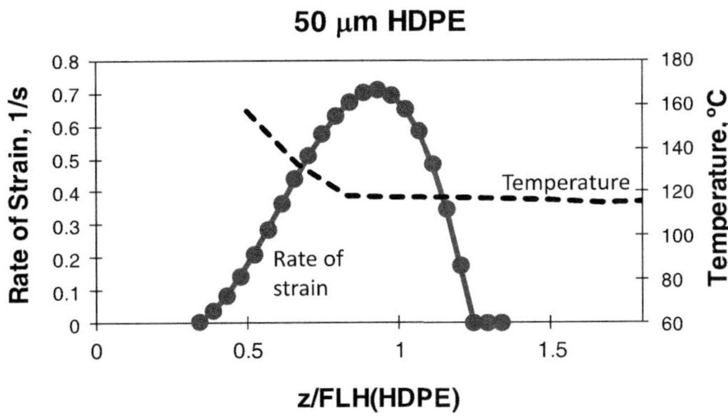

FIGURE 14.5 Machine direction rate of strain from velocity measurements on a monolayer high-density polyethylene blown film line. Temperature profile is superimposed. X-axis is the vertical distance from the die exit (*z*) divided by the frost line height. *Data from B.A. Morris, The effect of coextrusion on bubble kinematics, temperature distribution and property development in the blown film process, J. Plastic Film. Sheeting, 15(1) (1999) 25–36.*

The velocity and rate of strain data show that the change in bubble shape and velocity ceases near the frost line. The stress is related to the rate of strain, which reaches a maximum at the frost line and is subsequently frozen in. Many investigators have tried to correlate film properties with the deformation near the frost line with limited success [32,45,47,50–61]. Online birefringence measurements show some orientation occurs after the plateau region after the rate of strain has fallen to zero [60]. The actual stress is a complex function of the total strain (DDR and BUR), rate of strain, temperature, crystallization and orientation. Several investigators have tried to model the blown film process to determine the stress using mathematical or numerical simulation [3,28,29,41,53,54,59,62–71]. These models, for the most part, have resorted to simplifications to deal with the complications resulting from the rapid change in temperature, the melt viscoelasticity, thermal and flow-induced crystallization and the rheology of the semi-crystalline phase. For example, as the melt begins to crystallize, the relationship between stress and deformation changes. The crystallization rate itself is complicated by the rapid cooling rate and strain imposed by the process. The rapid cooling rate, on the order of 1000°C–5000°C/minute, suppresses crystallization. Most measurements of polymer crystallinity are conducted using a differential scanning calorimeter (DSC) where the cooling rate is typically 10°C–20°C/minute and can only go as high as about 100°C/minute. Special laboratory set-ups are needed to determine crystallization rates at cooling rates typical of the blown film process [42–44]. Ding and Spruiell [43] find that increasing the quench rate reduces the plateau temperature (which they equate to the freezing point), reduces the time for nucleation to occur, decreases the width of the plateau, increases the rate of crystallization and reduces overall crystallinity.

Complicating things further is flow or strain-induced crystallization. As the molecules line up in flow, the rate of crystallization may increase, sometimes by orders of magnitude [72–74]. This phenomenon is known to occur in high-speed fiber spinning [75], but also has been found to be important in blown film. Measurements of the rate of crystallization under quiescent conditions (no-flow) can considerably underestimate the amount of crystallinity which develops.

The molecular architecture affects the contribution of strain-induced crystallization. Zang et al. [5] and Zhao et al. [8] use a special in-line X-ray scattering apparatus to study the crystallization of polyethylene near the frost line. They find significant differences between a metallocene (mLLDPE) and an LDPE. mLLDPE is a linear polymer with little or no long chain branches whereas LDPE has substantial long chain branching (LCB). Characteristic relaxation times for mLLDPE are significantly shorter than LDPE. Zhao et al. find the mLLDPE crystallization is dominated by supercooling of the melt (temperature-induced crystallization) whereas the LDPE has both a temperature and strain-induced crystallization component. They attribute this to the difference in relaxation time behavior.

Doufas and McHugh [3,76] develop a model that tries to deal with all of these complications. Their model incorporates viscoelasticity of the melt and solid, nonisothermal dynamics as well as strain and thermal-induced crystallization. The model correctly predicts how processing conditions affect crystallinity and orientation in the film, and the position of the frost line, bubble shape and velocity profiles. The model also computes the locked-in stress at the frost line, which Doufas [76] shows correlates well with measured mechanical properties.

The Doufas and McHugh and other models still have their limitations, including:

- It is challenging to obtain the material parameters needed as inputs to the model. Many are difficult to measure. Indeed, Doufas uses data from the actual film line he was modeling to fit the parameters of his model. From this, he can correctly predict the effect of changing processing conditions. He also proposed that one could run lab-scale trials to obtain the model parameters for a given polymer, from which the model could predict the properties of films made on other lines. The models, however, are still a long way from being truly predictive from simple laboratory measurements on new polymers.

- The models do not predict the development of crystalline morphology (spherulites, lamella, etc.), which is known to be affected by polymer processing and to influence polymer properties [4–6,12,16,18,19,33,34,36,66,77–79]. Doufas [76] points out, however that morphological features can be inferred by the stress levels that are predicted by the model.
- Models for general coextrusion have not been developed. Coextrusion adds an additional level of complexity with interfacial interactions and changes in the bubble kinematics, as described in the next section.

14.1.2 Coextrusion blown film

Almost all the fundamental studies relating process to structure and properties have been conducted on monolayer film. Morris [49,80] was one of the first to study the deformation history of coextrusion blown film. His results show that the coextrusion process can significantly alter the deformation history compared with monolayer film, which in turn can change some properties. This work is discussed in detail below. It has been subsequently validated with more sophisticated online measurements by Giriprasath and Ogale [81,82] and illustrates some of the fundamental differences between coextrusion and monolayer extrusion.

Films are produced on a small-scale coextrusion blown film line equipped with a 3-layer 50-mm diameter Brampton Slimline die and dual air ring. The die is fed by two 25-mm and one 38-mm diameter extruders. Two structures are made, a monolayer 50-μm HDPE film and a 2-layer (22.5 μm EVOH–45 μm HDPE) film [EVOH = ethylene vinyl alcohol]. In the coextrusion, the HDPE layer is on the outside of the bubble. The HDPE is Sclair 19 A, a 0.75 MFI (190°C/2160 g), 0.960 g/cc density grade supplied by Nova Chemical. The EVOH is EVAL LCF101, a 1.5 MFI, 32 mol% ethylene grade supplied by Kuraray.

The melt temperature (230°C), BUR (2:1), DDR (8:1 for monolayer, 6:1 for coex) and final thickness are kept constant in each series of films. The process time (t_f), as defined by Eq. (14.3) (derived in Section 2.2.1.3), is varied by changing FLH and haul-off speed (v_f):

$$t_f = \frac{\text{FLH}}{v_f - v_o} ln\left(\frac{v_f}{v_o}\right) \tag{14.3}$$

where t_f = process time, FLH = frost line-height, v_f = haul-off speed, and v_0 = velocity of polymer at die exit.

The output of the extruders is changed with haul-off speed to keep the final film thickness constant.

There are two distinct visual "frost lines" observed when making the coextruded films, one for the HDPE layer and one for the EVOH layer. The results reported here are based on the FLH for the HDPE layer.

The temperature of the bubble as a function of vertical distance from the die exit (z-direction) is measured using an IR thermometer (3.43 μm frequency). Velocity and bubble shape profiles are measured using a video camera and tracer technique described above [45–48]. Velocity profiles were fit to a fourth-order polynomial and differentiated to find the rate of strain profiles.

Fig. 14.6 shows the temperature profiles for a series of 2-layer (EVOH-HDPE) coextruded films and one HDPE mono film. The curves have been normalized by dividing the vertical distance (z) by the FLH of the HDPE. The curves collapse onto a master curve, indicating the initial cooling rate is solely a function of the imposed quenching conditions. Note that state 7, a mono-layer HDPE film, has the same initial cooling rate as the coextruded films. Since the HDPE layer is on the outside of the coextrusion and the IR frequency of 3.43 μm does not penetrate very deeply into the layer [83], the temperature measurement is not influenced by the temperature of the EVOH.

The width of the crystallization plateau differs from state to state. Ghaneh-Fard et al. [60] find the width of the plateau to be a function of FLH and flow rate. Butler et al. [38] attribute this to inefficiencies in cooling as the frost line is moved further from the air ring. This relationship is refined in Fig. 14.7, where an excellent fit is found when the width of the plateau region in units of time [(CLH-PLH)/v_f] is plotted against process time.

Results of the rate of strain calculations in the MD for a series of HDPE monolayer films are given in Fig. 14.8. The maximum rate of strain occurs near the CLH, as found by previous investigators: in mono-layer extrusion, the maximum rate of change in bubble shape and velocity occurs when the polymer has enough melt strength, which takes place near the crystallization plateau (this coincides with the formation of a deformable precursor crystalline network structure or zone 1 of the study by Zhang et al. [4,5] described earlier). The magnitude of the maximum rate of strain is a function of the process time, as shown in Fig. 14.9 for the MD. With increasing

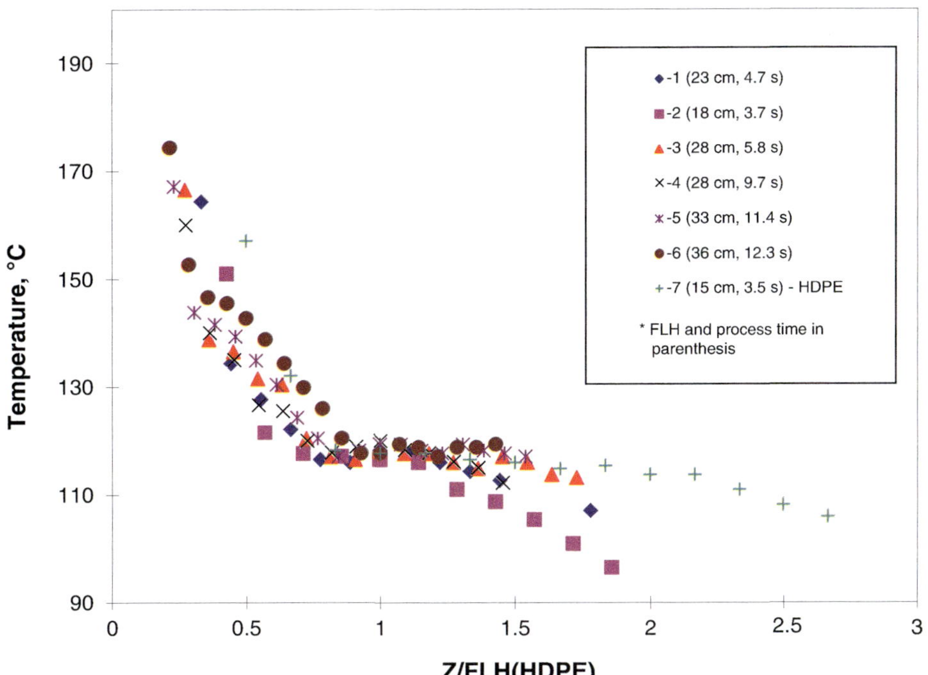

FIGURE 14.6 Normalized bubble temperature profile for (EVOH-HDPE) coextrusion or HDPE monolayer blown films. *Reproduced with permission from B.A. Morris, The effect of coextrusion on bubble kinematics, temperature distribution and property development in the blown film process, J. Plastic Film. Sheeting, 15(1) (1999) 25–36.*

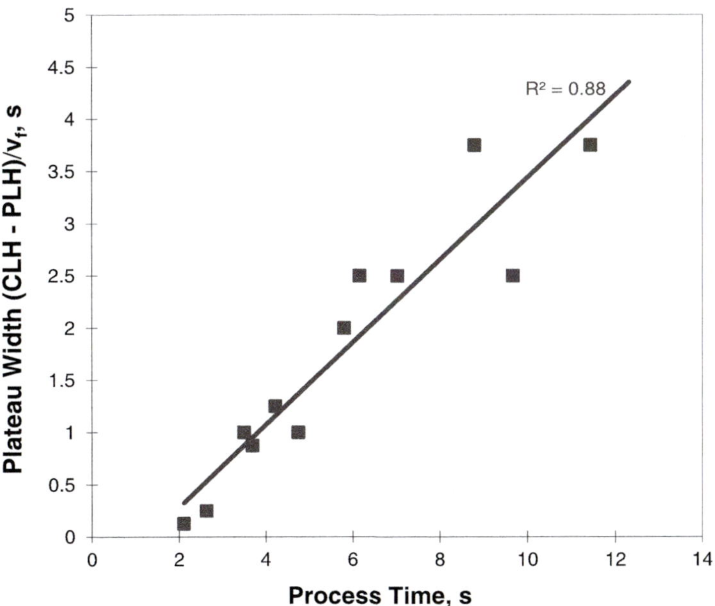

FIGURE 14.7 Width of temperature plateau (in units of time) plotted against process time (t_f). *Reproduced with permission from B.A. Morris, The effect of coextrusion on bubble kinematics, temperature distribution and property development in the blown film process, J. Plastic Film. Sheeting, 15(1) (1999) 25–36.*

process time, the maximum rate of strain (and hence stress) decreases. The TD rate of strain has the same features and is not shown here.

In coextrusion, the location of the maximum rate of strain is controlled by the material that freezes first. This is shown in Fig. 14.10 for the (EVOH-HDPE) coextrusion where the maximum rate of strain of the HDPE is shifted from near the CLH of the HDPE (z/FLH_{HDPE} of around 0.9) to the CLH of the EVOH (z/FLH_{HDPE} of around 0.4). EVOH has a significantly higher crystallization temperature than HDPE (155 vs 115°C by DSC). After the EVOH begins to freeze off, there is no significant change in the bubble shape or velocity of any of the layers. This can influence polymer properties in two ways. First, the maximum rate of strain of the HDPE layer at a given process time may be somewhat higher in this coextrusion, since the amount of stretching occurs over a shorter time span. This is shown in Fig. 14.9, but the effect is not substantial.

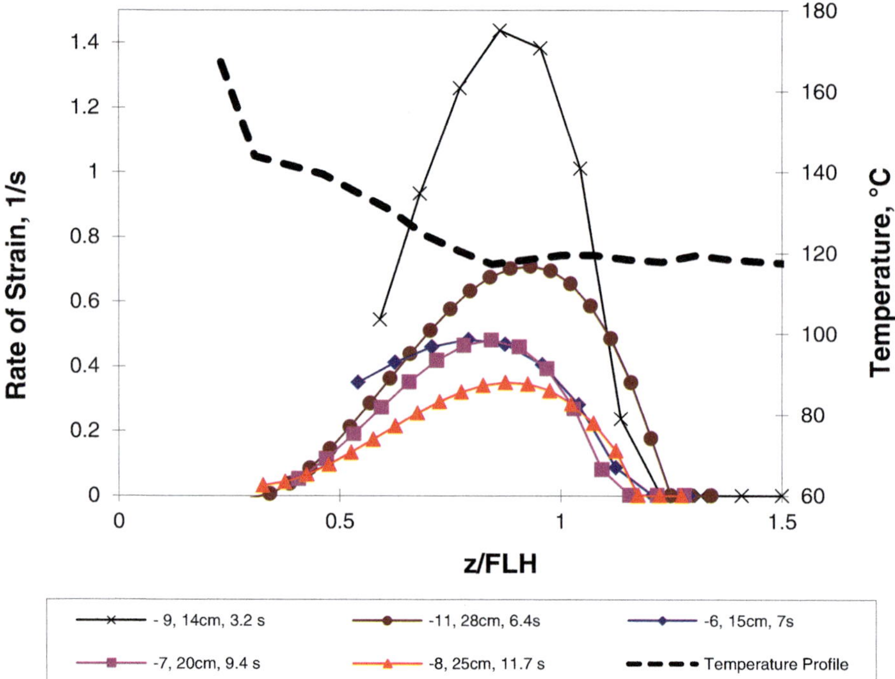

FIGURE 14.8 Machine direction rate of strain for 50-μm high-density polyethylene monolayer films at different frost line heights and process times. Curves are designated by experimental state number, frost line height and process time (see Tables 14.2 and 14.3). *Reproduced with permission from B.A. Morris, The effect of coextrusion on bubble kinematics, temperature distribution and property development in the blown film process, J. Plastic Film. Sheeting, 15(1) (1999) 25–36.*

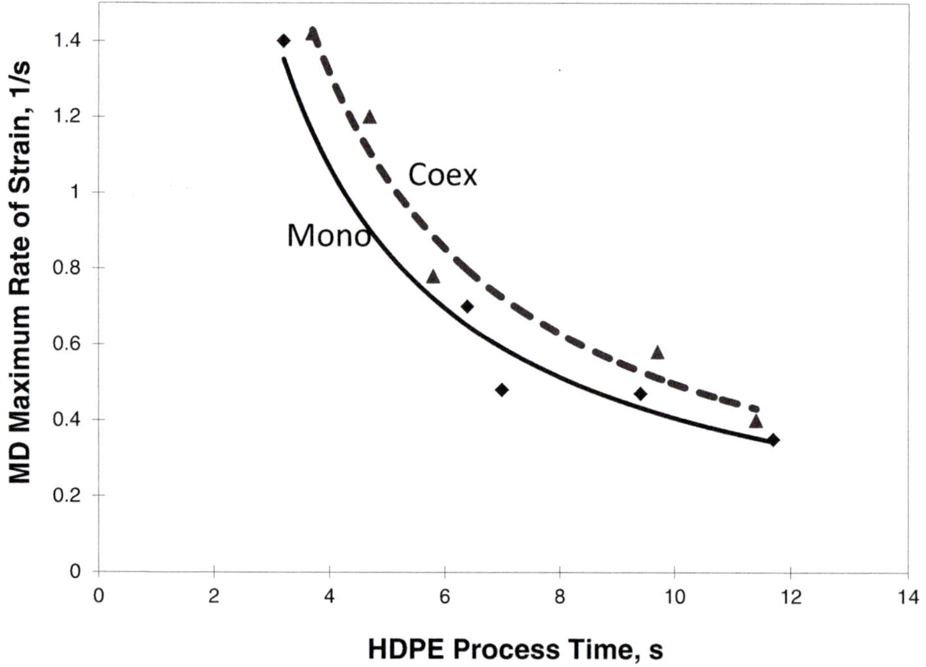

FIGURE 14.9 The effect of process time on machine direction maximum rate of strain. *Reproduced with permission from B.A. Morris, The effect of coextrusion on bubble kinematics, temperature distribution and property development in the blown film process, J. Plastic Film. Sheeting, 15(1) (1999) 25–36.*

Secondly, and perhaps more importantly, the location of the maximum rate of strain of the HDPE is now shifted to a region where the temperature of the bubble is significantly higher. The HDPE is still fully molten and can flow and relax in response to the strain. The imposed stress is also lower due to the lower viscosity at the higher temperature: stress is equal to the elongational viscosity times the rate of strain.

Because of these effects, the film properties of a material within a coextruded structure may differ from a mono extruded film. This is illustrated in Tables 14.2 and 14.3 and Figs. 14.11–14.13 where several properties of approximately 50-μm mono- and co-extruded HDPE film are compared. The coextrusion is with EVOH where the EVOH is

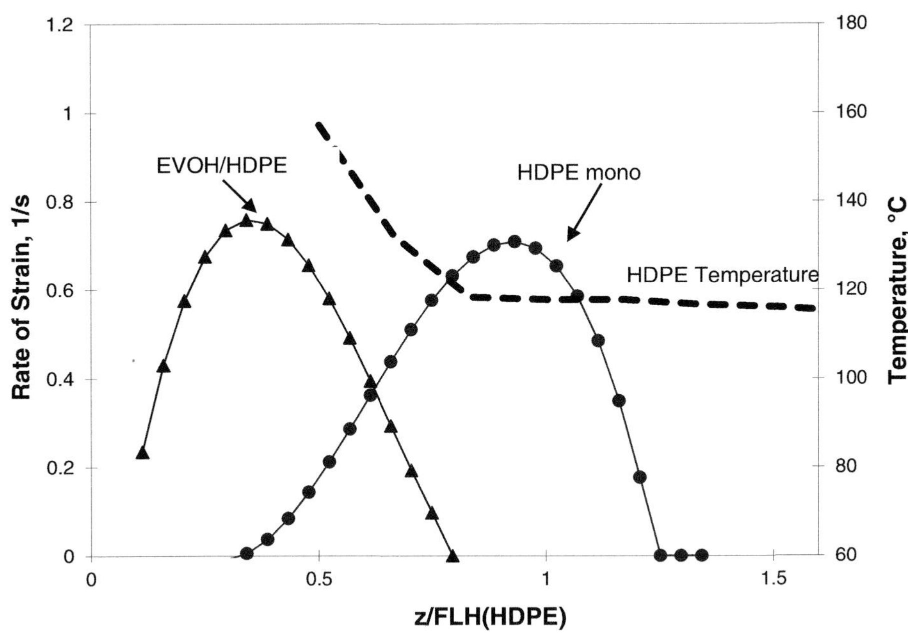

FIGURE 14.10 Effect of coextrusion on rate of strain. *Reproduced with permission from B.A. Morris, The effect of coextrusion on bubble kinematics, temperature distribution and property development in the blown film process, J. Plastic Film. Sheeting, 15(1) (1999) 25–36.*

TABLE 14.2 Properties of 50-μm monolayer high-density polyethylene films.

State No.	Haul-off speed (mm/s)	FLH (cm)	Process time (s)	MD Yld Str. Mean (Std Dev) (MPa)	MD Elong @ Bk Mean (Std Dev) (%)	TD Elmend. Tear Mean (Std Dev) (g/25μm)
6	0.35	15	7.0	30.6 (0.6)	948 (100)	62.1 (5.8)
7	0.35	20	9.4			
8	0.35	25	11.7	30.9 (0.9)	1038 (106)	45.3 (4.4)
9	0.71	14	3.2	29.7 (0.6)	687 (300)	210 (22.3)
10	0.71	11	2.6	30.7 (0.8)	779 (135)	297 (54.9)
11	0.71	28	6.4	30.8 (1.3)	991 (181)	62.7 (5.1)

Source: Reproduced with permission from B.A. Morris, The effect of coextrusion on bubble kinematics, temperature distribution and property development in the blown film process, J. Plastic Film. Sheeting, 15(1) (1999) 25–36.

TABLE 14.3 Properties of 45-μm high-density polyethylene (HDPE) film made by coextrusion with ethylene vinyl alcohol (EVOH). HDPE is separated from the EVOH prior to testing.

State No.	Haul-off speed (mm/s)	FLH cm	Process time (s)	MD Yld Str. Mean (Std Dev) (MPa)	MD Elong @ Bk Mean (Std Dev) (%)	TD Elmend. Tear Mean (Std Dev) (g/25μm)
1	0.71	23	4.7	28.8 (1.5)	914 (67)	51.4 (6.3)
2	0.71	18	3.7	29.6 (0.4)	847 (40)	63.2 (9.5)
3	0.71	28	5.8	29.5 (1.0)	927 (58)	52.1 (7.2)
4	0.42	28	9.7	29.7 (1.2)	897 (88)	41.1 (3.3)
5	0.42	33	11.4	28.9 (2.6)	884 (40)	36.7 (4.3)
6	0.42	36	12.3	29.8 (0.7)	1000 (53)	37.9 (8.3)

Source: Reproduced with permission from B.A. Morris, The effect of coextrusion on bubble kinematics, temperature distribution and property development in the blown film process, J. Plastic Film. Sheeting, 15(1) (1999) 25–36.

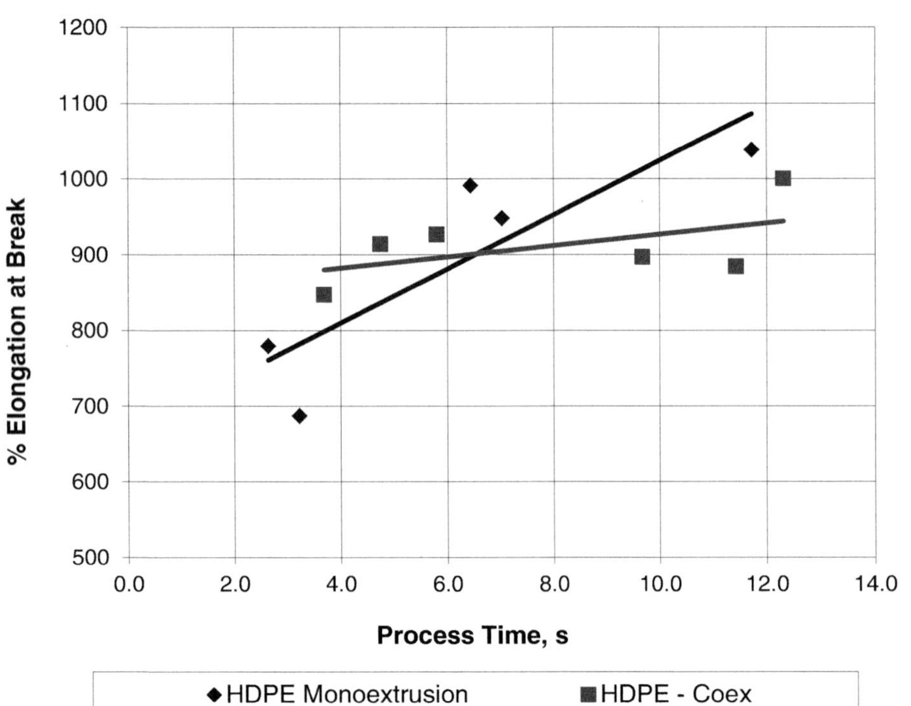

FIGURE 14.11 Machine direction elongation at break for high-density polyethylene films made by mono and coextrusion processes. *Reproduced with permission from B.A. Morris, The effect of coextrusion on bubble kinematics, temperature distribution and property development in the blown film process, J. Plastic Film. Sheeting, 15(1) (1999) 25–36.*

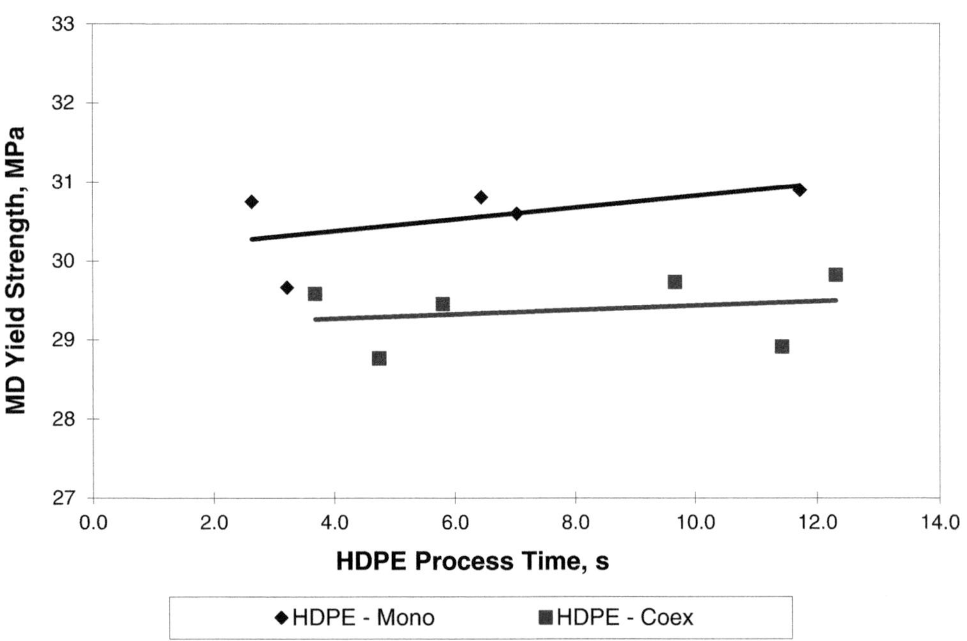

FIGURE 14.12 Machine direction yield strength for high-density polyethylene films made by mono and coextrusion processes. *Reproduced with permission from B.A. Morris, The effect of coextrusion on bubble kinematics, temperature distribution and property development in the blown film process, J. Plastic Film. Sheeting, 15(1) (1999) 25–36.*

removed from the HDPE before testing (EVOH and HDPE do not bond well to one another and the EVOH is stripped off without changing the HDPE). Fig. 14.11 shows that the elongation at break of mono film is more sensitive to changes in process time than coextruded film. Fig. 14.12 displays lower yield strength in coextruded film. Fig. 14.13 shows a significant difference in TD tear properties as a function of process time.

The observed differences in physical properties may reflect fundamental differences in the amount of orientation that develops in these films. Shirodkar et al. [15] show that increasing the process time in monolayer extrusion significantly reduces orientation. This is because the stress on the polymer is highest near the frost line where small changes

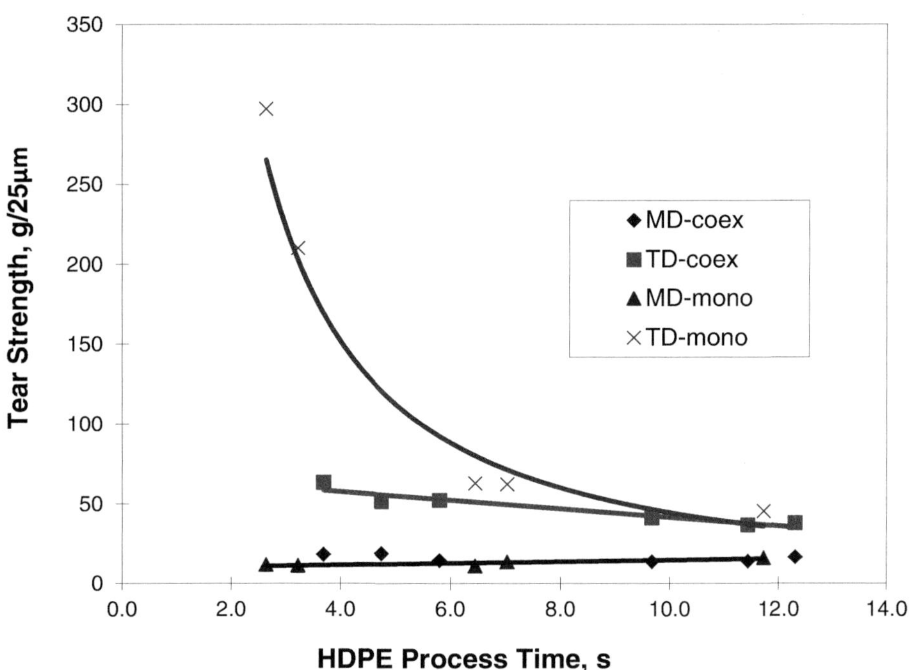

FIGURE 14.13 Elmendorf tear strength for high-density polyethylene films made by mono and coextrusion processes. *Reproduced with permission from B.A. Morris, The effect of coextrusion on bubble kinematics, temperature distribution and property development in the blown film process, J. Plastic Film. Sheeting, 15 (1) (1999) 25–36.*

in the rate of strain and time for relaxation make a significant difference in stress. In the (EVOH-HDPE) coextrusion, the stress (and orientation) in the HDPE layer is not as sensitive to changes in process time since the maximum rate of strain no longer is near the frost line. The stress on the HDPE is lower and changes in process time do not influence physical properties as much. These effects may have an impact on other properties as well, such as barrier performance and interlayer adhesion.

Increasing process time (keeping BUR and DDR constant) decreases the amount of stress imparted during orientation by reducing the rate of strain (see Fig. 14.9). The effect of processing parameters on the film properties of a polymer can be significantly changed when coextruding the polymer with a second polymer, even if the amount of the second polymer is comparatively small, provided the second polymer has a higher freezing point. In such a case, coextrusion changes the magnitude of the maximum rate of strain and shifts its location to regions of higher temperature. The results suggest that using monolayer film properties to predict coextrusion film properties may not be valid: when designing new film structures, the simple addition of individual layer film properties, such as physical and barrier properties, may be inadvisable. The material properties and processing parameters interact in complex ways to influence the bubble kinematics and temperature distribution in coextrusion. Such considerations must be considered to fully predict film performance.

14.2 Air versus water quench blown film

The quench rate has a substantial impact on the crystallinity of a plastic film, which influences physical, optical, and barrier properties as well as drawability and thermoformability. The conventional blown film process uses air as the cooling medium. An air ring distributes the primary cooling air around the outside of the bubble. Air may also be circulated inside the bubble to further enhance the cooling rate. While the quench rate in conventional air quench blown film lines is quite high, on the order of 1000°C/minute or more, it is still slow compared with cast film cooling where the melt is contacted with a cold roll. While quenching with air remains the predominant cooling technique for blown film lines, several equipment suppliers also offer water quench. In a water quench line, the tube is extruded downward from the die. The tube is blown out and through an aperture with flowing water, called a water ring, which cools the bubble from the outside. Because water is a more effective heat transfer medium than air, the quench rate approaches that of the cast film process.

The advantages of water quench over air quench blown film include lower crystallinity, higher clarity, enhanced thermoformability and reduced curl. Water quench blown film lines provide a better balance of orientation than cast film lines and the opportunity for greater product diversity. Their versatility versus air quench lines is more limited since the water ring is sized for a fixed BUR; to change the final film width or lay flat, the water ring must be swapped out. Water quench blown film lines also typically run at a lower output than cast film lines [84–87].

Some of the crystallinity issues with air quench blown film can be overcome with resin selection. For example, changing from polyamide 6 (PA 6) to polyamide copolymers such as PA 6/66 or adding an amorphous PA helps improve thermoformability and reduce curl. Water quench provides the opportunity to reduce the use of these more expensive resins.

Xiao, Armstrong and Lee [84,85] compare the optical, barrier and thermoforming properties of identical barrier films made on commercial-scale water and air quenched blown film lines. Their work provides an excellent example of how the film fabrication process affects properties, but also how secondary processes, in this case thermoforming, may also change performance. They made 9-layer barrier structures on two nearly identical Brampton Engineering blown film lines. The characteristics of these lines are summarized in Table 14.4.

Details of the structures and resins used in the study are provided in Tables 14.5 and 14.6. Three types of structures are made. Structure A is a symmetric structure with LLDPE on the outside and PA-EVOH-PA as the core. Structure B is nonsymmetric with PA on one of the outside layers and mPE on the other, but with the same PA-EVOH-PA core. Sandwiching EVOH between layers of PA is often done to improve the formability of the film. Structure BT is a variation on structure B made with the layers reversed on the film line so that the PA layer is on the inside. The final structure, C, is symmetric with LLDPE as the outside layers but only EVOH as the barrier core. Structure C is included to evaluate the performance of a modified EVOH grade designed for improved thermoformability (EVOH3). The die diameter and output of the two lines are identical and all the films are 100-μm thick. No information is given about the individual layer thicknesses.

The films are tested for their optical properties (haze), moisture vapor transmission rate (MVTR) and oxygen transmission rate (OTR) prior to thermoforming. These results are designated as the "flat film" results. The films are thermoformed on two types of thermoforming equipment, a ZED batch former and a Multivac R5200 Solid Phase Pressure former. The barrier properties of the thermoformed packages are tested, and an assessment of the thermoforming processing window is made from these trials.

The flat film haze results are graphed in Fig. 14.14. The haze of the air quenched films is significantly higher than for the water quenched films. This was expected and is believed to be due to higher crystallinity in the air quenched

TABLE 14.4 Characteristics of the water quench and air quench blown film lines used in the experiments by Xiao et al. [84,85].

	Air quench line	Water quench line
Extruders/Layers	9	9
Extruder diameter (number of extruders)	63.5 mm (2); 50 mm (7)	63.5 mm (9)
Cooling	Inner Bubble Cooling (IBC) and Air Ring	Water Ring on outside of bubble
Die Diameter	450 mm	450 mm
Haul-off	Oscillating	Oscillating

TABLE 14.5 Blown film structures made and tested in quench study by Xiao et al. [85].

Structure	1	2	3	4	5	6	7	8	9
A	LLDPE	LLDPE	Tie	PA 6	EVOH1	PA 6	Tie	LLDPE	LLDPE
A2	LLDPE	LLDPE	Tie	PA 6	EVOH2	PA 6	Tie	LLDPE	LLDPE
B	mPE	Tie	PA 6	EVOH1	PA 6	Tie	LLDPE	Tie	PA 6
BT	PA 6	Tie	LLDPE	Tie	PA 6	EVOH1	PA 6	Tie	mPE
C	LLDPE	LLDPE	LLDPE	Tie	EVOH1	Tie	LLDPE	LLDPE	LLDPE
C2	LLDPE	LLDPE	LLDPE	Tie	EVOH3	Tie	LLDPE	LLDPE	LLDPE

Note: Layer 1 is on the inside of the bubble and layer 9 on the outside.

TABLE 14.6 Description of resins used in Xiao et al. [85] blown film evaluation.

Material	Characteristics
LLDPE	Octene, 0.918 g/cc density, 1 g/10 min MI
mPE	0.902 g/cc density, 1 g/10 min MI
Tie	Maleic anhydride modified PE
PA	PA 6
EVOH 1	38 mol% ethylene
EVOH 2	32 mol% ethylene
EVOH 3	Modified 32 mol% ethylene

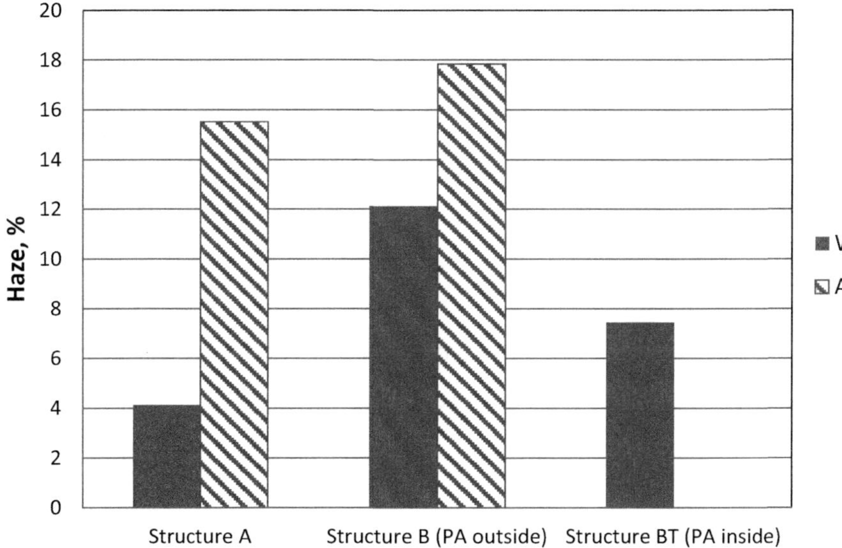

FIGURE 14.14 Results of haze measurements on flat films (prior to thermoforming) made via water quench (WQ) and air quench (AQ) blown film processes. *Adapted from K. Xiao, R. Armstrong, I.-H. Lee, Comparison of water-quench vs air-quench blown film processes – part I: flat film properties, SPE ANTEC Conf., (2012) 972–977.*

films. Indeed, optical birefringence and wide-angle X-ray scattering (WAXS) measurements confirm that the air quenched samples have significant crystallinity while the water quenched films have very little.

One surprising result is the lower haze of the water quenched nonsymmetrical structure when the PA layer is on the inside of the bubble (structure BT vs B). The haze drops from 12% to 7% when the PA layer is moved from the outside to the inside. The authors hypothesize that when the PA is on the outside of the bubble, it acts as an insulating layer. The latent heat of crystallization of the PA layer slows the transfer of heat away from the interior layers, allowing them slightly more time to crystallize. When the PA is on the inside of the bubbles, the remaining layers are more directly exposed to the cooling water, increasing the effective rate of cooling. The overall level of crystallinity is still very low, as the WAXS data indicate. This explanation is consistent with the permeation results described below. The authors do not explore other possible mechanisms for the change in haze, such as possible differences in surface roughness, or plasticization of the outside PA layer by water, allowing more crystallization in that layer.

The permeation results for the flat films are provided in Figs. 14.15 and 14.16. Both the MVTR and OTR give similar trends. The water quenched film has 30%–50% higher moisture or oxygen permeability than its air-quenched counterpart. The permeability of a semicrystalline polymer is known to decrease with increasing crystallinity; the crystals act as impenetrable barriers to the gas molecules. Therefore, this result is not surprising. Once again, the position of the PA layer in the nonsymmetric film (B vs BT) has an impact. The water quenched structure B has lower permeability than BT. Like the haze results, this suggests that having PA on the outside results in higher crystallinity. Note that the permeability of the water quenched B structure is still greater than the air quenched B structure.

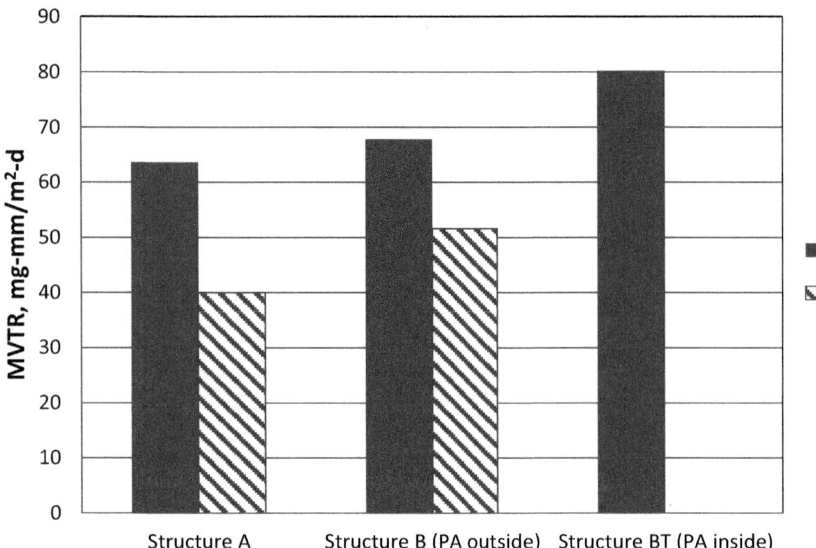

FIGURE 14.15 Moisture vapor transmission rate results comparing air quenched and water quenched blown films. Results are for flat films, prior to thermoforming. *Adapted from K. Xiao, R. Armstrong, I.-H. Lee, Comparison of water-quench vs air-quench blown film processes – part I: flat film properties, SPE ANTEC Conf., (2012) 972–977.*

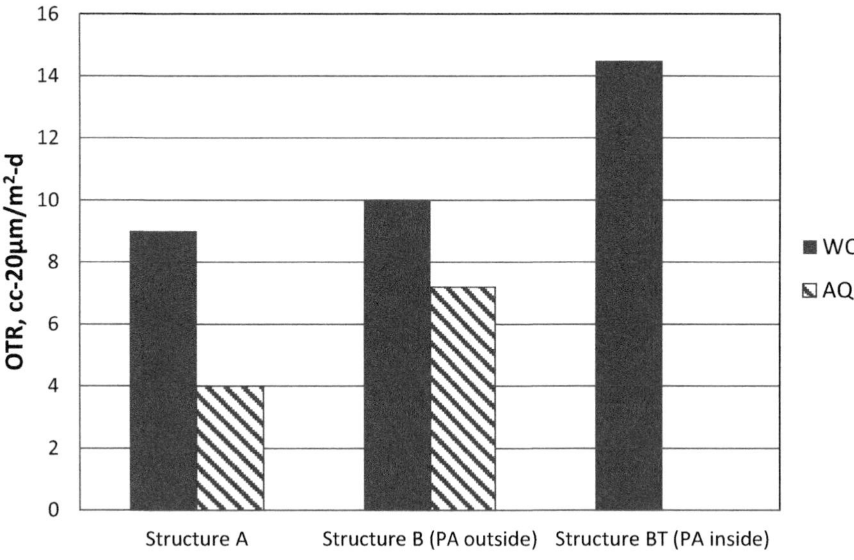

FIGURE 14.16 Oxygen transmission rate results comparing air quenched and water quenched blown films. Results are for flat films, prior to thermoforming. Test conditions are 85% RH at 30°C. *Adapted from K. Xiao, R. Armstrong, I.-H. Lee, Comparison of water-quench vs air-quench blown film processes – part I: flat film properties, SPE ANTEC Conf., (2012) 972–977.*

The films are thermoformed on the ZED former at a draw depth of 4-cm with a corner radius of 0.6-cm. The OTR is measured by purging the inside of the formed cavity with nitrogen, sealing off the opening, controlling the outside atmosphere of the package to 100% RH and inside to 65% RH, and measuring the change in concentration of O_2 inside the package with time. The results for structures A and A2 are plotted in Fig. 14.17. After thermoforming, packages made from the water quenched films have lower or equal OTR than their air quenched counterparts. This is the opposite trend seen for the flat films. The water quenched films, with their lower crystallinity, thermoform better: the appearance is superior, and the gauge distribution is more uniform with less thinning in the corners. The more uniform gauge distribution is likely the reason for the improved barrier performance of the water quenched films.

The thermoformability of the films is evaluated on the Multivac former. In this continuous roll-fed process, the 325-mm wide film is heated between top and bottom plates prior to forming. For the trial, both plates are set at the same temperature. Either a two or four pocket mold is used. Most of the trial is with the 4-pocket, deep draw mold with dimensions 135-mm × 125-mm × 85-mm. The heating and forming time are held constant at 1 and 2 seconds, respectively, and the temperature is varied to determine the forming window.

Three criteria are used to assess the formability:

- Depth and quality of the pocket or cavity draw

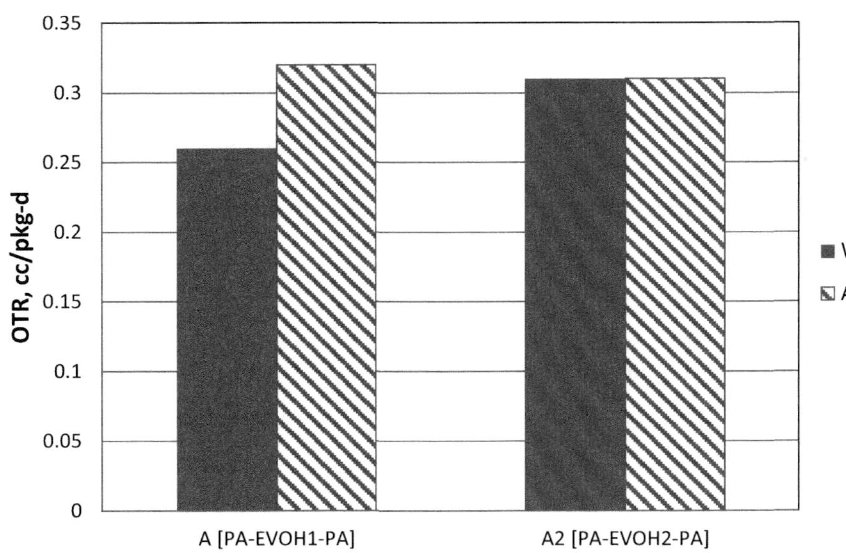

FIGURE 14.17 Oxygen transmission rate results on thermoformed packages. *Adapted from K. Xiao, I.-H. Lee, R. Armstrong, Comparison of water-quench vs air-quench blown film processes – part II: thermoformability, SPE ANTEC Conf., (2012) 978–983.*

TABLE 14.7 Summary of thermoformability assessment of air quenched C and A films.

Temperature (°C)	Film C EVOH1 AQ	Film C2 EVOH3 AQ	Film A (PA-EVOH1-PA) AQ
90			XX
95			
100	XX	X	XX
105			
110	X	X	XX
115			
120	X	X	X
125			
130	X	X	Ø
135	X	X	
140	X		Ø
145			
150	X		

Key: XX—100% could not form or were pocket blow outs; X—≥ 50% of pockets could not form, formed minimally or with pocket blow outs; Ø—100% formed, but with some variability in pocket appearance; O—good forming with consistent pockets.
Source: Adapted from K. Xiao, I.-H. Lee, R. Armstrong, Comparison of water-quench vs air-quench blown film processes – Part II: thermoformability, SPE ANTEC Conf., (2012) 978–983.

- Number of cavities formed
- Degree of cavity breaks.

For cavities not fully formed, a gauge distribution across the pocket cross-section is measured.

The C films, where the EVOH is not sandwiched between layers of PA 6, did not form well. The results for the air quenched versions are summarized in Table 14.7. About 50% or more of the cavities do not form or formed with breaks or blow-outs. Gauge distribution of the cavities that did form show noticeably thinner EVOH in the corners. Even the modified grade of EVOH (EVOH3) does not perform well enough for the film structure, mold design and process conditions considered in this trial. Comparing this with Film A, the authors conclude that encapsulating the EVOH with PA is critical for successfully thermoforming these structures.

Tables 14.8 and 14.9 show that the water quenched versions of films A and B thermoform better than their air quenched counterparts. They form more evenly at lower forming temperatures and form over a wider temperature window. Gauge distribution is more uniform for the WQ films, especially in the corners. One reason is that the forming temperature can be lower, which results in thicker corners.

This study illustrates how the film and packaging fabrication processes can have a significant impact on final package properties.

TABLE 14.8 Summary of thermoformability assessment of A films.

Temperature (°C)	Film A (PA-EVOH1-PA) AQ	Film A (PA-EVOH1-PA) WQ
90	XX	XX
95		X
100	XX	O
105		
110	XX	O
115		
120	X	O
125		
130	Ø	Ø
135		
140	Ø	

Source: Adapted from K. Xiao, I.-H. Lee, R. Armstrong, Comparison of water-quench vs air-quench blown film processes – Part II: thermoformability, SPE ANTEC Conf., (2012) 978–983.

TABLE 14.9 Summary of thermoformability assessment of B films.

Temperature (°C)	Film B AQ	Film B WQ
90	XX	X
95	XX	X
100	X	X - Ø
105	X	X - Ø
110	Ø	O
115		
120	Ø	O
125	Ø - O	
130	Ø - O	O
135		
140		

Source: Adapted from K. Xiao, I.-H. Lee, R. Armstrong, Comparison of water-quench vs air-quench blown film processes – Part II: thermoformability, SPE ANTEC Conf., (2012) 978–983.

14.3 Development of blend morphology in blown film

Chapter 6 describes why most polymer-polymer blends are immiscible, forming separate phases with the minor component imbedded in the major component. Blend properties are influenced by the size, shape, and distribution of the phases, which collectively is known as the blend morphology. Polymer processing impacts the blend morphology through impartation of shear and elongation. Several morphologies have been described for immiscible blends in blown film, including spheres and elongated particles, such as fibrils and ribbons. Forming elongated morphologies is a two-step process:

1. Sphere-like dispersed domains are formed during extrusion. Shear flows are the predominant type of flow in the extruder and die. These flows are inherently weaker than elongational flows that dominate once the polymer exits the die and are less likely to form elongated domains. The dispersed domain size is a function of:
   - Dispersed phase concentration;
   - The ratio of the viscosity of the dispersed phase to the continuous phase (henceforth referred to simply as the viscosity ratio);
   - Interfacial tension;
   - Continuous phase viscosity;
   - Shear stress.
2. Stretching and orienting the spherical domains in the elongational flow fields at the extruder exit (blowing and drawing) forms elongated structures. The morphology is influenced by:
   - Initial domain size;
   - Polymer elasticity;
   - Minor component percentage;
   - Draw ratio and other film blowing parameters.

Fig. 14.18 shows an example of an elongated morphology in a blown film. Here polybutene-1 (PB-1) has been added to a polyolefin to create a peelable sealant. Pirtle et al. [88] suggest that too much PB phase orientation in a peel-seal blend can produce unwanted "stringy seals." Morphology control, therefore, is important for packaging applications.

Eq. (6.8) is helpful for understanding how the material characteristics and processing parameters affect the morphology in shear flow. It is repeated here:

$$Ca = \frac{\eta_c \dot{\gamma} D_d}{2\Gamma_{12}} \tag{14.4}$$

where Ca = capillary number, η_c = viscosity of the continuous phase, $\dot{\gamma}$ = shear rate, D_d = diameter of droplet (discontinuous phase), and Γ_{12} = interfacial tension between the two polymers.

The capillary number is the ratio of the drag force exerted by the process to breakup the minor phase into smaller droplets to the interfacial force holding it together. As is described in Chapter 6, there is a critical capillary number for droplet breakup that is a function of the viscosity ratio of the polymers. Fig. 14.19 shows that in shear flow, as the viscosity ratio increases, $Ca_{critical}$ decreases until the viscosity ratio reaches a value of around one. Since the droplet size (D_d) is proportional to Ca, the droplet size reaches a minimum at a viscosity ratio of around one. Eq. (14.4) also shows that the domain size decreases with decreasing interfacial tension and increasing shear rate at constant Ca (and constant viscosity ratio).

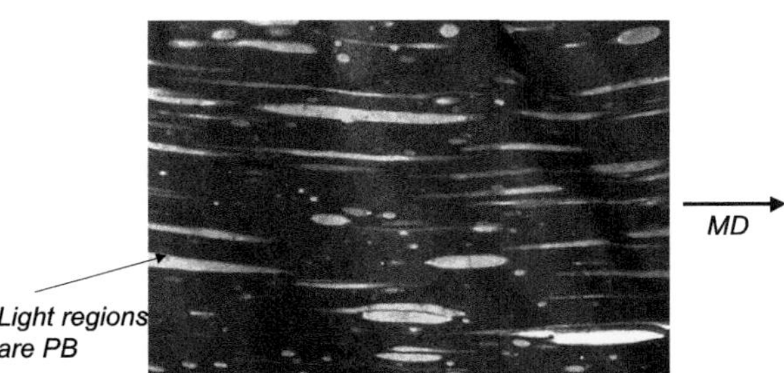

FIGURE 14.18 Elongated second phase (PB-1 in a polyolefin matrix) found in blown film. *Reproduced with permission of The Dow Chemical Company.*

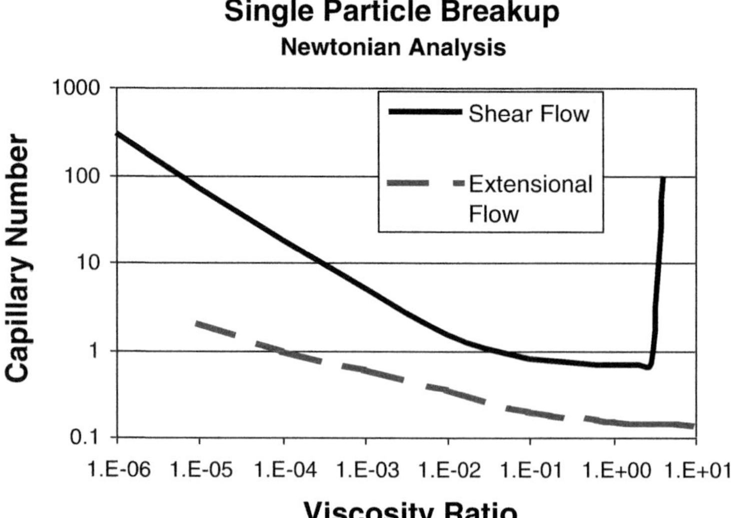

FIGURE 14.19 Critical capillary number for the break-up of Newtonian droplet in shear and elongational flow. See discussion in Section 6.4.2. *Data taken from H.P. Grace, Dispersion in high viscosity immiscible fluid systems and application of static mixers as dispersion devices in such systems, Chem. Eng. Commun. 14 (1982) 225–277.*

Some of the key material characteristics and processing factors influencing morphology development in blown film are now considered.

14.3.1 Material characteristics

14.3.1.1 Viscosity ratio

A viscosity ratio near one typically gives the smallest dispersed phase domains in the shear flow during extrusion. The smaller the domain size, the harder it is to deform and elongate the particle. Studies by David et al. [89–91] of blends of PE with styrene block copolymers show that the dispersed phase viscosity needs to be less than the continuous phase viscosity to obtain elongated morphologies. For viscosity ratios greater than one, the dispersed phase cannot be deformed enough to create the elongated morphologies.

14.3.1.2 Interfacial tension

As described by Eq. (14.4), for a given capillary number, the lower the interfacial tension, the smaller the dispersed phase domain diameter. Chapter 6 describes ways to reduce interfacial tension, such as promoting specific interactions by matching solubility parameters or adding compatibilizers. Smaller domain size tends to make elongated morphologies (such as laminar structures) more difficult to obtain [92–95].

14.3.1.3 Minor phase concentration in blend

Increasing the dispersed phase volume fraction (ϕ_d) generally increases the size of the dispersed domains, resulting in more fibril or laminar structures. The larger domain sizes may be due to the coalescence of smaller domains, which becomes more significant as the concentration increases. Lopez-Barron et al. [95] study blends of PA6 and LDPE where LDPE is the continuous phase. The domain size of the PA6 increases with increasing concentration for uncompatibilized blends but is constant for a compatibilized system. The compatibilizer reduces interfacial tension and acts to reduce coalescence. David et al. [89] find co-continuous fibrils form as ϕ_d is increased. This is determined experimentally by extracting films with a solvent that selectively dissolves the minor component. As the volume fraction is increased, more minor phase polymer is extracted, suggesting co-continuous morphologies are forming. The onset of co-continuous fibril morphology occurs at a lower volume fraction than predicted by Eq. (6.9), which David et al. [89,91] attribute to elongational effects.

14.3.1.4 Polymer elasticity (non-Newtonian behavior)

Getlichermann and David [91] find that a viscosity ratio less than one is not sufficient to create elongated morphologies. If the dispersed phase is Newtonian in behavior, elongated morphologies are not observed after blowing and drawing.

A polymer's non-Newtonian (elastic) behavior helps to stabilize the "threads" that form during elongation, allowing more elongation without the threads breaking up into small particles.

They suggest that dispersed phase tension thickening is one attribute that should help stabilize the thread and help create elongated morphologies. Tension thickening or strain hardening refers to elongational viscosity behavior. In general, linear polymers, such as LLDPE, have transient elongational viscosity that decreases with time (tension thinning), whereas the elongational viscosity of polymers with long-chain branching (such as LDPE) tends to increase with time (tension thickening). Thus, polymers with long-chain branching are more susceptible to the formation of elongated morphologies during film blowing.

14.3.2 Processing factors

14.3.2.1 Extruder speed

As extruder speed is increased, the shear rate increases, which should reduce the dispersed phase domain size. As discussed previously, smaller domains formed in the extruder and die are more difficult to elongate after they exit the die. Lee and Kim [93,94], however, find just the opposite for a LDPE–EVOH blend. They attribute this to the shorter extruder residence time at a higher extruder screw speed. The EVOH, which is the dispersed phase and melts at temperatures 50°C higher than the LDPE, has less time to melt fully and be dispersed. This gives rise to a larger domain size at the extruder exit and a greater tendency toward elongated laminar morphology.

14.3.2.2 Extruder temperature

The extruder and die temperatures affect the morphology through their impact on the viscosity of both components. As the temperature is increased, η_c decreases. As described by Eq. (14.4), this results in less stress being applied to breaking up the droplets, giving larger domain sizes. Also, the polymers' viscosity-temperature dependence may differ, altering the viscosity ratio.

14.3.2.3 Shear stress in extruder, adaptor and die

The flow in an extruder, adapter pipes and die, known as pressure-driven flow, creates a shear stress distribution across the flow channel. The highest shear stresses occur at the wall. The wall also experiences the longest residence times since the flow rate at the wall is low. This may create a distribution of morphologies across the channel. This is found to be true by Lee and Kim [93] for LDPE–EVOH blends and it has also been documented for blends involving PB-1 and LDPE [96].

Lopez-Barron et al. [95] observe a core-skin effect with uncompatibilized blends of PA6 and LDPE, with the skin having small elongated domains and the core larger particles. They attribute this to a discontinuity at the die wall at the die exit. The polymer near the die wall in the die flow channel has zero or nearly zero velocity (no-slip boundary condition) and accelerates almost instantaneous to the free surface velocity at the die exit, as illustrated in Fig. 14.20. This creates an intensive amount of stretching that may lead to the observations. Their compatibilized blends show no such effect, likely because the domain size is much smaller and more resistant to elongation.

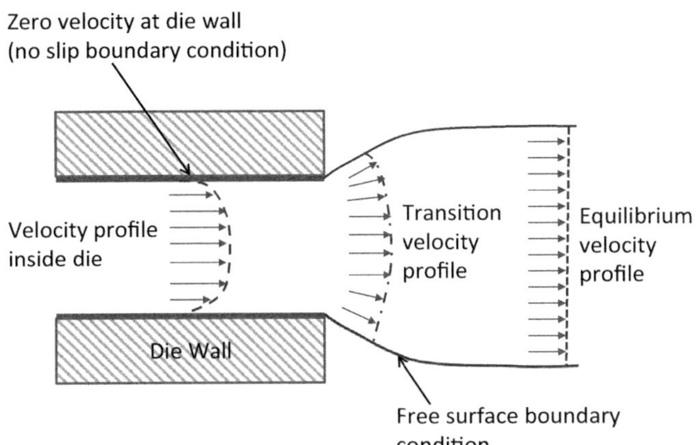

FIGURE 14.20 Illustration of transition from confined channel flow to free surface flow at the die exit. Polymer near the channel wall accelerates to the free surface velocity, resulting in a stretching action.

14.3.2.4 Screw design

For laminar or ribbon-like morphologies, the literature teaches not to over-mix the blend [97]. A careful screw design may be needed to accomplish this.

14.3.2.5 Draw ratio

The minor component is typically spherical in shape as it exits the extruder [89–95,98]. It is the elongational flow fields imposed by the drawing and blowing process that give rise to the highly elongated structures. The degree of elongation depends on the process. David et al. [89] find the convergence/divergence in a capillary die geometry only gives rise to elongated ellipsoids. A flat die produces fibrils in the MD and a blown film die gives fibrils and ribbons with two-dimensional orientation in the machine and transverse direction (TD) as well as co-continuous fibrils.

The draw in the MD is characterized by the DDR or take-up ratio (TUR), which is defined as the haul-off speed divided by the polymer melt velocity as it exits the die (see Section 2.2.1.3). The BUR characterizes the draw in the TD or hoop dimension. BUR is defined as the final bubble diameter divided by the die diameter. The DDR is inversely proportional to BUR times the final film thickness divided by the die gap [Eq. (2.7)]:

$$\mathrm{DDR} = \frac{v_f}{v_0} \approx \frac{X_o}{(\mathrm{BUR})X_f}. \tag{14.5}$$

where DDR = draw down ratio or take-up ratio, v_f = haul-off speed, v_0 = velocity of polymer at die exit, BUR = blow up ratio = D_f/D_0 (bubble diameter/die diameter), X_o = die gap, and X_f = final film thickness.

DDR decreases with increasing BUR and film thickness and decreasing die gap. [In Eq. (14.5), the density change between solid and melt is ignored.]

David et al. [89] find that the fibril diameter is not a straightforward function of draw ratio. Without any drawing, the particles in their study are spherical with a diameter of 0.5 μm. At a DDR of 5 and 20, the fibril diameters are 0.1 and 0.2 μm, respectively.

Robledo-Ortiz et al. [98] show that the stretching force increases with increasing DDR and decreasing FLH. The elongation of the minor phase domain (PA6 in their case) increases in the MD with increasing stretching force. The elongation in the hoop and radial (thickness) dimensions are balanced. Kim and Chun [94] show that balancing the orientation in the MD and TD provides lamellar morphologies in LDPE-EVOH (10 wt.%) blends that impart better barrier performance.

14.3.2.6 Frost line height and process time

Robledo-Ortiz et al. show that stretching force increases strongly with decreasing FLH. The elongation in the MD of PA6 in their PA6-LDPE blends increases accordingly. The FLH is a component of the process time, as shown in Eq. (14.3).

The effect of process time on the blend morphology has not been fully addressed in the literature. It encompasses critical aspects of the film blowing process, including the cooling and crystallization time. More important for this discussion is that the process time is inversely proportional to the elongation rate, as shown in Fig. 14.10. Stress is related to the draw ratio and to the rate of elongation. Therefore, process time is inversely related to the stress imposed by the process; increasing process time should reduce the elongation of the minor phase.

References

[1] M.D. Bullwinkel, G.A. Campbell, D.H. Rasussen, C.J. Brancewitz, Structure development during film blowing, SPE ANTEC Conf. (2001).

[2] M.D. Bullwinkel, G.A. Campbell, D.H. Rasmussen, J. Krexa, C.J. Brancewitz, Crystallization studies of LLDPE during tubular blown film processing, Int. Polym. Process. 16 (2001) 39–47.

[3] A.K. Doufas, A.J. McHugh, Simulation of film blowing including flow-induced crystallization, J. Rheology 45 (2001) 1085–1104.

[4] Q. Zhang, L. Li, F. Su, Y. Ji, S. Ali, H. Zhao, et al., From molecular entanglement network to crystal-cross-linked network and crystal scaffold during film blowing of polyethylene: an in situ synchrotron radiation small- and wide-angle x ray scattering study, Macromolecules 51 (2018) 4350–4362.

[5] Q. Zhang, W. Chen, H. Zhao, Y. Ji, L. Meng, D. Wang, et al., In-situ tracking polymer crystallization during film blowing by synchrotron radiation X-ray scattering: the critical role of network, Polymer 198 (2020) 122492.

[6] E.M. Troisi, M. van Drongelen, H.J.M. Caelers, G. Portale, G.W.M. Peters, Structure evolution during film blowing: an experimental study using in-situ small angle X-ray scattering, Eur. Polym. J. 74 (2016) 190–208.

[7] N.D. Huck, P.L. Clegg, The effect of extrusion variables on the fundamental properties of tubular polythene film, SPE ANTEC Conf. 18-2 (1961) 1–14.

[8] H. Zhao, Q. Zhang, L. Li, W. Chen, D. Wang, L. Meng, et al., Synergistic and competitive effects of temperature and flow on crystallization of polyethylene during film blowing, ACS Appl. Polym. Mater. 1 (6) (2019) 1590–1603.

[9] G. Gururajan, A.A. Ogale, Real-time Raman spectroscopy during LDPE/PP bilayer blown-film extrusion, Plastics, Rubber Compos. 34 (5/6) (2005) 271–275.

[10] J. Wang, C. Thurber, D. Monaenkova, P. Agarwal, E. Marchbanks, K. Olson, et al., Preliminary study of birefringence distribution in blown film, SPE ANTEC Conf. (2018).

[11] J.J.C. Picot, Molecular orientation in film extrusion of high-density polyethylene, Polym. Engr. Sci. 24 (1984) 415–420.

[12] A. Haber, M.R. Kamal, Morphology and orientation in polyethylene tubular blown films, SPE ANTEC Conf. (1987) 446.

[13] T.I. Butler, R. Patel, Blown film bubble forming and quenching effects on film properties, J. Plastic Film. & Sheeting 9 (1993) 181. Also 1993, SPE ANTEC Conference, pp. 51–57.

[14] R.M. Patel, T.I. Butler, Effect of fabrication conditions and molecular orientation on polyethylene blown film properties, TAPPI Polymers, Laminations Coat. Conf. (1993) 303.

[15] P.P. Shirodkar, V. Firdaus, H. Fruitwala, Scale-up of LLDPE blown film extrusion, TAPPI Polymers, Laminations Coat. Conf. (1994) 405–413. Also, Plastics Engineering, Dec 1994, pp. 27.

[16] D.M. Simpson, I.R. Harrison, A study of the effects of processing parameters on the morphologies and tensile modulus of HDPE blown films: application of composite theories on a molecular level to characterize tensile modulus, J. Plastic Film. Sheeting 10 (Oct) (1994) 302–325.

[17] R.M. Patel, T.I. Butler, K.L. Walton, G.W. Knight, Investigation of processing-structure-properties relationships in polyethylene blown films, Polym. Engr. Sci. 34 (1994) 1506–1514.

[18] M. van Gurp, B.J. Kip, J.P.C. van Heel, S. de Boer, On the development of orientation in LDPE blown films, J. Plastic Film. Sheeting 10 (1994) 156–176.

[19] X.M. Zhang, S. Elkoun, A. Ajji, M.A. Huneault, Oriented structure and anisotropy properties of polymer blown films: HDPE, LLDPE and LDPE, Polymer 45 (2004) 217–229.

[20] S. Cherukupalli, R. Paradkar, A.A. Ogale, A.J. McHugh, L. Henrichsen, Online measurements and heat transfer issues in blown film extrusion, SPE ANTEC Conf. (2002).

[21] S.S. Cherkupalli, A.A. Ogale, Online measurements of crystallinity using Raman spectroscopy during blown film extrusion, Polym. Engr. Sci. 44 (2004) 1484–1490.

[22] S.S. Cherkupalli, S.E. Gottlieb, A.A. Ogale, Real-time Raman spectroscopic measurement of crystallization kinetics and its effect on the morphology and properties of polyolefin blown films, J. Appl. Polym. Sci. 98 (2005) 1740–1747.

[23] S. Fatahi, A. Ajji, P.G. Lafleur, Structure-property correlations for PE blown films, SPE ANTEC Conf. (2005) 1766–1770.

[24] S. Fatahi, A. Ajji, P.G. Lafleur, Correlation between structural parameters and property of PE blown films, J. Plastic Film. Sheeting 21 (4) (2005) 281–305.

[25] N. Legros, A. Ghaneh-Fard, K.C. Cole, A. Ajji, M.M. Dumoulin, Tensile properties and orientation evolution with processing conditions in polyethylene blown films, SPE ANTEC Conf. (1998).

[26] L.K. Henrichsen, A.J. McHugh, Analysis of film blowing with flow-enhanced crystallization, Intern. Polym. Process. XXII (2007) 179–189.

[27] A. Ghaneh-Fard, Effects of film blowing conditions on molecular orientation and mechanical properties of polyethylene films, J. Plastic Film. Sheeting 15 (3) (1999) 194–218.

[28] C.D. Han, J.Y. Kwack, Rheology-processing-property relationships in tubular blown film extrusion. I. High-pressure low-density polyethylene, J. Appl. Polym. Sci. 28 (1983) 3399–3418.

[29] J.Y. Kwack, C.D. Han, Rheology-processing-property relationships in tubular blown film extrusion. II. Low-pressure low-density polyethylene, J. Appl. Polym. Sci. 28 (1983) 3419–3433.

[30] A.M. Sukhadia, Trade-offs in blown film processing-structure-property behavior of LLDPE type resins from chromium, metallocene and Ziegler-Natta catalysts, SPE ANTEC Conf. (1994) 202.

[31] M.A. Kennedy, A.J. Peacock, M.D. Faila, J.C. Lucas, L. Mandelkern, Tensile properties of crystalline polymers: random coplymers of ethylene, Macromolecules 28 (1995) 1407.

[32] G.A. Campbell, A.K. Babel, Physical properties and processing conditions correlations of the LDPE/LLDPE tubular blown films, Macromol. Symp. 101 (1996) 199.

[33] T.-H. Yu, G.L. Wilkes, Orientation determination and morphology study of high density polyethylene (HDPE) extruded tubular films: effect of processing variables and molecular weight distribution, Polymer 37 (1996) 4647–4687.

[34] T.-H. Yu, G.L. Wilkes, Influence of molecular weight distribution on the melt extrusion of high density polyethylene (HDPE): effects of melt relaxation behavior on morphology and orientation in HDPE extruded tubular films, J. Rheology 40 (1996) 1079–1093.

[35] R.K. Krishmaswamy, A.M. Sukhadia, The relative influences of process and resin time scales on the MD tear strength of polyethylene blown films, SPE ANTEC Conf. (1999).

[36] D. Godshall, G. Wilkes, R.K. Krishnaswamy, A.M. Sukhadia, Processing-structure-property investigation of blown HDPE films containing both machine and transverse direction oriented lamellar stacks, Polymer 44 (2003) 5397–5406.

[37] T. Kanai, M. Kimura, Y. Asano, Studies on scale-up of tubular film extrusion, SPE ANTEC Conf. (1986) 912.

[38] T.I. Butler, S.Y. Lai, R. Patel, Scale-up factors effecting crystallization in polyolefin blown film, TAPPI Polymers, Laminations Coat. Conf. (1994) 289.
[39] T.I. Butler, Using dimensionless scale-up variables to predict LLDPE blown film properties, TAPPI Polymers, Laminations Coat. Conf. (1995) 581–601.
[40] T.H. Kwack, Processing variables and the scaling parameters in blown film extrusion, SPE ANTEC Conf. (1999).
[41] T.I. Butler, Predicting blown film residual stress levels influence on properties, TAPPI PLACE. Conf. (2006).
[42] Z. Ding, J. Spruiell, Interpretation of the nonisothermal crystallization kinetics of polypropylene using a power law nucleation rate function, J. Polym. Sci. B: Polym. Phys. 35 (1997) 1077–1093.
[43] Z. Ding, J. Spruiell, An experimental method for studying nonisothermal crystallization of polymers at very high cooling rates, J. Polym. Sci. B: Polym. Phys. 34 (1996) 2783–2804.
[44] P. Supaphol, J. Spruiell, Nonisothermal bulk crystallization studies of high-density polyethylene using light depolarizing microscopy, J. Polym. Sci. B: Polym. Phys. 36 (1998) 681–692.
[45] D.M. Simpson, I.R. Harrison, The use of deformation rates in the scale-up of polyethylene blown film, J. Plastic Film. Sheeting 8 (1992) 192–206.
[46] T.A. Huang, G.A. Campbell, Deformation history of LLDPE/LDPE blends on blown film equipment, Adv. Polym. Technol. 5 (1985) 181.
[47] R. Farber, J. Dealy, Strain history of the melt in film blowing, Polym. Engr. Sci. 14 (1974) 435.
[48] T.I. Butler, R. Patel, S.Y. Lai, J. Spuria, Blown film frost line-freeze line interactions, TAPPI Polymers, Laminations Coat. Conf. (1993) 13.
[49] B.A. Morris, The effect of coextrusion on bubble kinematics, temperature distribution and property development in the blown film process, J. Plastic Film. Sheeting 15 (1) (1999) 25–36.
[50] D.M. Simpson, I.R. Harrison, The use of deformation rates in the scale-up of polyethylene blown film, SPE ANTEC Conf. (1991) 203–205.
[51] C.D. Han, J.Y. Park, Studies on blown film extrusion. II. Analysis of the deformation and heat transfer process, J. Appl. Polym. Sci. 19 (1975) 3277.
[52] C.D. Han, J.Y. Park, Studies on blown film extrusion. I. Experimental determination of elongational viscosity, J. Appl. Polym. Sci. (1975) 19.
[53] T. Kanai, J.L. White, Kinematics, dynamics and stability of the tubular film extrusion of various polyethylenes, Polym. Engr. Sci. 25 (1984) 1185–1201.
[54] T. Kanai, J.L. White, Dynamics, heat transfer and structure development in tubular film extrusion of polymer melts: a mathematical model and predictions, J. Polym. Engr 5 (1985) 135–157.
[55] K.S. Yoon, C.W. Park, A study on two-layer blown film coextrusion, SPE ANTEC Conf. (1991) 2315.
[56] A.K. Babel, G.A. Campbell, Correlating the plastic strain with the properties of the low density polyethylene blown film, J. Plastic Film. Sheeting 9 (1993) 246.
[57] A.K. Babel, G.A. Campbell, Physical properties and processing conditions correlations of the LDPE/LLDPE tubular blown films, TAPPI Polymers, Laminations Coat. Conf. (1994) 415–428.
[58] A.K. Babel, G.A. Campbell, A model linking process variables to the strength of blown films produced from LDPE and LLDPE, TAPPI J. 78 (5) (1995) 199–204.
[59] G.A. Campbell, B. Cao, A.K. Babel, Kinematics, dynamics and physical properties of blown film, in: T. Kanai, G.A. Campbell (Eds.), Film Processing, 1999, pp. 113–140.
[60] A. Ghaneh-Fard, P.J. Carreau, P.G. Lafleur, Study of kinematics and dynamics of film blowing of a linear low density polyethylene, SPE ANTEC Conf. (1996) 111–115.
[61] P.J. Carreau, A. Ghaneh-Fard, P.G. Lafleur, Rheological effects in film blowing, SPE ANTEC Conf. (1998).
[62] J.R.A. Pearson, C.J.S. Petrie, The flow of a tubular film. Part 1. Formal mathematical representation, J. Fluid Mech. 40 (1970) 1–19.
[63] J.R.A. Pearson, C.J.S. Petrie, The flow of a tubular film. Part 2. Interpretation of the model and discussion of solutions, J. Fluid Mech. 42 (1970) 609–625.
[64] C.J.S. Petrie, A comparison of theoretical predictions with published experimental measurements on the blown film process, AIChE J. 21 (1975) 275–282.
[65] G.A. Campbell, B. Cao, Blowing film modeling – die to frostline, TAPPI Polymers, Laminations Coat. Conf. (1986) 247–251.
[66] T.H. Kwack, C.D. Han, M.E. Vickers, Development of crystalline structure during tubular film blowing of low-density polyethylene, J. Appl. Polym. Sci. 35 (1988) 363.
[67] B.K. Ashok, G.A. Campbell, Two-phase simulation of tubular film blowing of crystalline polymers, Int. Polym. Process. 7 (1992) 240–247.
[68] J.C. Pirkle Jr., R.D. Braatz, A thin-shell two-phase microstructural model for blown film extrusion, J. Rheology 54 (2010) 471–505.
[69] R. Kolarik, M. Zatloukal, Modeling of nonisothermal film blowing process for non-Newtonian fluids by using variational principles, J. Appl. Polym. Sci. 122 (2011) 2807–2820.
[70] E.W. Kuijk, P.P. Tas, P. Neuteboom, A rheological model for prediction of polyethylene film properties, SPE ANTEC Conf. (1998) 30–34.
[71] S. Muke, H. Connell, I. Sbarski, S.N. Bhattacharya, Numerical modelling and experimental verification of blown film processing, J. Non-Newtonian Fluid Mech. 116 (1) (2003) 113–138.
[72] S. Vleeshouwers, H.E.H. Meijer, A rheological study of shear induced crystallization, Rheol. Acta 35 (1996) 391–399.
[73] R. Lagassa, B. Maxwell, An experimental study of the kinetics of polymer crystallization during shear flow, Polym. Engr. Sci. 16 (1976) 189.
[74] M. Sentmanat, O. Delgadillo-Velazquez, S.G. Hatzikiriakos, Crystallization of an ethylene-based butene plastomer: the effect of uniaxial extension, Rheol. Acta. 49 (2010) 931–939.

[75] A. Ziabicki, H. Kawai (Eds.), High-Speed Fiber Spinning, Wiley, New York, 1985.
[76] A.K. Doufas, A microstructural flow-induced crystallization model for film blowing: validation with experimental data, Rheol. Acta 53 (2014) 269–293.
[77] A. Keller, M.J. Machin, Oriented crystallization in polymers, J. Macromol. Sci. (Phys.) B1 (1967) 41–91.
[78] M. Gilbert, D.A. Hemsley, S.R. Patel, Effect of processing conditions on the orientation and properties of polyethylene blown film, Br. Polym. J. 19 (1987) 9–23.
[79] A. Gupta, D.M. Simpson, I.R. Harrison, Crystalline morphology in oriented HDPE blown films, Plast. Eng. Novemb. (1993) 33–36.
[80] B.A. Morris, The effect of coextrusion on bubble kinematics, temperature distribution and property development in the blown film process, SPE ANTEC Conf. (1998) 118–122. Also TAPPI Polymers, Laminations and Coatings Conference, pp. 831–837.
[81] G. Giriprasath, A.A. Ogale, Processing-structure relationship in bilayer polyolefin blown films, SPE ANTEC Conf. (2006) 830–834.
[82] G. Giriprasath, S. Gotlieb, A.A. Ogale, Process-structure-property relationships in bicomponent blow film extrusion using Roman spectroscopy, SPE ANTEC Conf. (2005) 158–162.
[83] B. Cao, P. Sweeney, G.A. Campbell, Simultaneous surface and bulk temperature measurement of polyethylene during film blowing, J. Plastic Film. Sheeting 6 (1990) 117.
[84] K. Xiao, R. Armstrong, I.-H. Lee, Comparison of water-quench vs air-quench blown film processes – part I: flat film properties, SPE ANTEC Conf. (2012) 972–977.
[85] K. Xiao, I.-H. Lee, R. Armstrong, Comparison of water-quench vs air-quench blown film processes – Part II: Thermoformability, SPE ANTEC Conf. (2012) 978–983.
[86] L. Ederleh, T. Bergmann, I. Putsch, New packaging solutions thanks to rapid cooling at extrusion process, Kunststoffe 104 (5) (2014) 26–31.
[87] M. Andrews, A comparison of extrusion processes and their effect on barrier film properties, blown, cast, and water quench extrusion technology, AMI Mulitlayer Packaging Films Conf. (2017).
[88] S. Pirtle, L. Mergenhagen, J. Wooster, R. Cotton, The effects of resin selection and extrusion variables on peelable seal performance, TAPPI PLACE. Conf. (2004).
[89] C. David, M. Trojan, R. Jacobs, M. Piens, Morphology of polyethylene blends: I. From spheres to fibrils and extended co-continuous phases in blends of polyethylene with styrene-isoprene-styrene triblock copolymers, Polymer 32 (3) (1991) 510.
[90] C. David, M. Getlichermann, M. Trojan, R. Jacobs, Morphology of polyethylene blends. II: Formation of a range of morphologies in blends of polyethylene with increasing content of styrene-isoprene-styrene triblock copolymers during extrusion-blowing, Poly. Eng. Sci. 32 (1) (1992) 6.
[91] M. Getlichermann, C. David, Morphology of polyethylene blends: III. Blend films of polyethylene with styrene-butadiene or styrene-isoprene copolymers obtained by extrusion-blowing, Polymer 35 (12) (1994) 2542.
[92] R. Ronzalez-Nunez, B. Favis, R. Carreau, Factors influencing the formation of elongated morphologies in immiscible polymer blends during melt processing, Poly. Eng. Sci. 33 (13) (1993) 851.
[93] S.-Y. Lee, S.-C. Kim, Morphology and oxygen barrier properties of LLDPE/EVOH blends, Intern. Polym. Processing, XI (1996) 238.
[94] S.W. Kim, Y.H. Chun, Barrier property by controlled laminar morphology of LLDPE/EVOH blends, Korean J. Chem. Eng. 16 (4) (1999) 511–517.
[95] C.R. Lopez-Barron, J.R. Robledo-Ortiz, D. Rodrigue, R. Gonzalez-Nunez, Film processability, morphology, and properties of polyamide-6/low density polyethylene blends, J. Plastic Film. Sheeting 23 (2007) 149–169.
[96] Basell Product Literature, Polybutene-1 for Use in Seal-Peel Applications, Customer Presentation Aid, 2001.
[97] P.M. Subramanian, Permeability barriers by controlled morphology of polymer blends, Poly. Eng. Sci. 25 (1985) 483.
[98] J.R. Robledo-Ortiz, D. Ramiriz-Arreola, R. Gonzalez-Nunez, D. Rodrigue, Effect of freeze-line position and stretching force on the morphology of LDPE-PA6 blown films, J. Plastic Film. Sheeting 22 (2006) 287–314.

Chapter 15

Effect of processing on interlayer adhesion

Adhesion is influenced by many factors, including the physical and chemical nature of the substrates and polymers being bonded, the process by which the structure is put together, and the environment to which the finished film is exposed in the end-use. These elements are discussed in detail in Chapter 10. This chapter highlights two areas where the process strongly affects adhesion: (1) adhesion of the polymer to substrate in extrusion coating and lamination and (2) interlayer adhesion in coextrusion cast film, blown film and extrusion coating/lamination.

15.1 Adhesion to substrates in extrusion coating

A number of elements influence the adhesion of polymers to solid substrates in extrusion coating and lamination, including:

- The composition and quality of the substrate and polymer.
- Extrusion and flow of the polymer through the die.
- Oxidation and drawing in the air gap.
- Wetting and cooling in the nip.
- Adhesion testing methodology.
- Environmental conditions during testing and in the end-use.

The effect of composition and quality of the polymer and substrate is covered in Chapter 10. The substrate must be free of surface contamination and have acceptable surface energy for wetting of the coating. Often flame, corona or plasma treatment is used to clean and raise the surface energy of the substrate.

Fig. 15.1 outlines the processing factors that affect adhesion in extrusion coating and lamination. The critical regions of the process for adhesion are:

- The extruder and die where the polymer is melted and shaped into the melt curtain. Temperature and flow uniformity of the polymer as it exits the die are key for ensuring good gauge distribution and wetting later in the process.
- The air gap where stresses are imparted, new interfacial area created and cooling begins as the polymer is drawn to its final thickness. For some polymers, oxidation and other chemical changes occur that promote adhesion. Ozone treatement of the melt curtain may be used to add oxidative species for improved adhesion to polar substrates such as aluminum foil, metallized films and paper.
- The nip and chill roll where the polymer is contacted with the substrate and rapidly quenched.

15.1.1 Extruder and die

The quality of the melt is important for good adhesion. Factors that ensure good melt quality include uniform temperature and thickness across the width of the melt curtain, constant output (no surging), a lack of flow instabilities and low gel or defects. Ensuring good melt quality starts with the extruder. A screw design that balances melting, mixing and conveying without overheating the polymer is critical. An introduction to screw design is found in Section 2.1 where the reader is directed to several general references on extrusion and screw design for details. Barrel temperatures, screw speed and back pressure also contribute to melt quality. The polymer at the extruder exit should be fully melted, uniform in temperature and free of substantial pressure fluctuations.

The Science and Technology of Flexible Packaging. DOI: https://doi.org/10.1016/B978-0-323-85435-1.00003-X

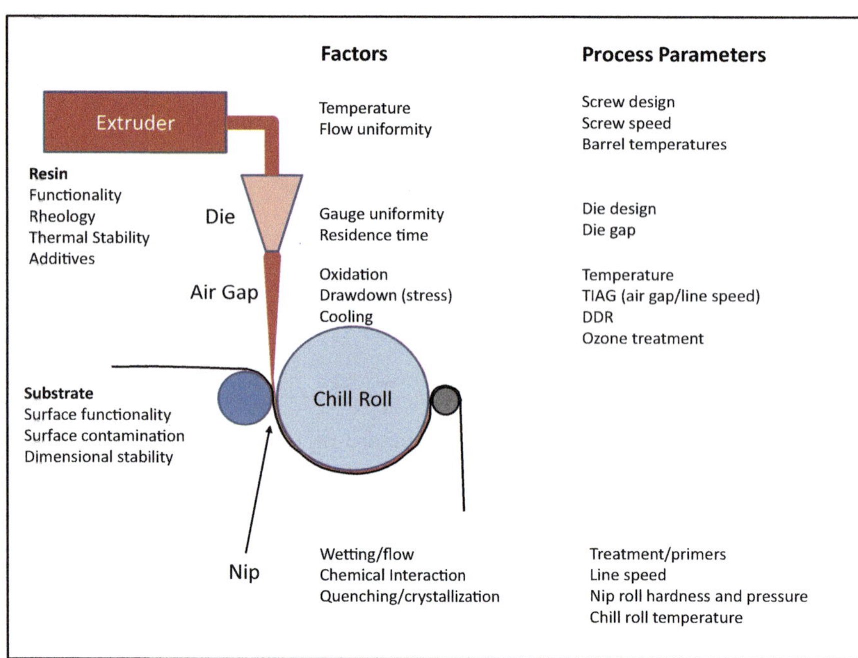

FIGURE 15.1 Factors of the extrusion coating process that influence adhesion. *TIAG*, time in the air gap; *DDR*, draw down ratio.

The die must distribute the polymer flow so that the gauge and temperature are uniform at the die exit. Distributing the flow evenly across the width of wide dies can be a challenge. Various designs have been utilized, many using a manifold to help control the pressure drop across the die. Dies also typically have a preland region at the die exit that allows the polymer to relax prior to exiting the die. This can help relieve stress that may be detrimental to peel strength performance.

Streamlining the flow in the extruder, transfer pipes and die helps prevent hang-ups and long residence time that can lead to excessive polymer degradation. Degradation can cause gel and other defects that detract from the quality of the coating. Degradation also changes the viscosity of the resin, which may affect its ability to wet the surface of the substrate.

Some amount of thermo-oxidative degradation may be desired to promote the formation of oxidative species for adhesion. Oxidation of low-density polyethylene (LDPE) forms polar functional groups such as acids, aldehydes and ketones that help with bonding to polar substrates. The oxidation process generally begins in the extruder and die, but most oxidation takes place in the air gap [1–3]. The oxidation of LDPE is explored in detail in Section 15.1.2.1.

15.1.2 Air gap

The air gap is the region between the die exit and the chill roll where the polymer is drawn down to its final thickness. Drawing imparts orientation and stress on the polymer, which can influence the adhesion to the substrate. The polymer may also undergo oxidation in the air gap, which for nonpolar polymers such as LDPE is critical for adhesion to polar substrates, such as aluminum foil, metallized films, chemically primed films and paper. The amount of oxidation depends on the residence time and temperature.

The residence time in the air gap is historically known as the time in the air gap (TIAG) [4]. The simplest and most widely used expression for TIAG is given by Eq. (15.1):

$$\mathrm{TIAG} = \frac{L}{v_f} \tag{15.1}$$

where L = length of the air gap (distance between die exit and nip) and v_f = final velocity or line speed (determined by the speed at the nip).

In English units, this can be written:

$$\mathrm{TIAG(ms)} = \frac{L[\text{inches}]}{v_f\left[\frac{ft}{\min}\right]} \times 5000 \tag{15.2}$$

In metric units:

$$\text{TIAG(ms)} = \frac{L[\text{mm}]}{v_f\left[\frac{m}{min}\right]} \times 60. \quad (15.3)$$

While easy to calculate, TIAG fails to describe the actual physics of flow. As illustrated in Fig. 15.2, TIAG assumes that the velocity of the melt curtain at the die exit instantaneously accelerates to the final velocity. Other expressions have been proposed, as described in Table 15.1. One method uses the average velocity (at the die exit and the nip). Another assumes the velocity changes linearly from the die exit to the nip. Chapter 2 derives an expression for the residence time (which shall be called the process time in this chapter) assuming a Newtonian velocity profile [8]:

$$t_f = \frac{L}{v_f}\left[\frac{\text{DDR} - 1}{\ln(\text{DDR})}\right] \quad (15.4)$$

where t_f = process time (time between die exit and nip) and DDR = draw down ratio = v_f/v_0 (v_0 = velocity at die exit).

Eq. (15.4) is a good compromise between ease of use and assuming a velocity profile that matches the actual physics of the flow in the air gap. As shown in Fig. 2.17, a Newtonian velocity profile agrees fairly well with the velocity profile for LDPE as measured by Canning et al. [9]. More exact models require the use of non-Newtonian constitutive equations that are more difficult to compute using a hand calculator or simple spreadsheet program, and often require material parameters that are challenging to measure. Example calculations for process time are provided in Table 15.2.

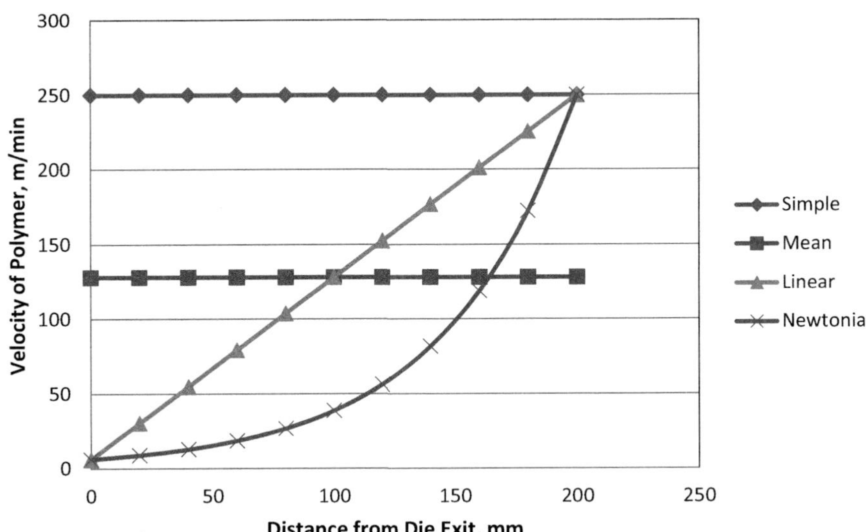

FIGURE 15.2 Velocity profiles in the air gap from various models for process time. Line speed = 250 m/min, die gap = 0.5 mm, air gap = 200 mm, coating thickness = 12 μm. Curves correspond to process time models shown in Table 15.1. The simple model assumes the velocity at the die exit is the same as the line speed. The mean model uses the average of the line speed and velocity at the die exit. The linear model assumes the velocity increases linearly as a function of the distance from the die exit and the Newtonian model assumes that the flow behaves like a Newtonian fluid ($v(z) = \text{DDR}^{z/L}$ where z = distance from die exit).

TABLE 15.1 Calculating time in the air gap or process time.

Model	Equation	Reference
Simple (TIAG)	$t_f^{simple} = TIAG = \frac{L}{v_f}$	Antonov and Soutar [4]
Mean	$t_f^{mean} = \frac{L}{0.5(v_0 + v_f)} = \frac{L}{0.5v_f}\left(\frac{DDR}{DDR+1}\right)$	Morris [5]
Linear	$t_f^{linear} = \frac{L}{v_f}\left(\frac{DDR}{DDR-1}\right)\ln(DDR)$	Morris [6,7]
Newtonian	$t_f^{Newtonian} = \frac{L}{v_f}\left(\frac{DDR-1}{\ln(DDR)}\right)$	Morris [7,8]

L = air gap; v_f = line speed; v_0 = velocity at die exit; DDR = draw down ratio = v_f/v_0.

TABLE 15.2 Sample process time calculations.

	Extrusion coating time in the air gap		
Input parameters	**Unit**	**Value**	**Symbol**
Resin density (solid)	g/cc	0.92	ρ_f
Resin density (melt)	g/cc	0.73	ρ_o
Die width	mm	1000	w_o
Coating width	mm	900	w_f
Line speed	m/min	400	v_f
Die gap	μm	1000	X_o
Air gap	mm	400	L
Coating thickness	μm	12	X_f
Calculated values			**Equation**
Velocity at die exit	m/min	5.4	$v_0 = v_f/\mathrm{DDR}$
DDR		73	$\mathrm{DDR} = \frac{X_f w_f \rho_f}{X_0 w_0 \rho_0}$
TIAG (simple model)	ms	60	$\mathrm{TIAG} = \frac{L}{v_f}$
t_f (mean velocity)	ms	118	$t_f^{\mathrm{mean}} = \frac{L}{0.5(v_0 + v_f)}$
t_f (linear model)	ms	261	$t_f^{\mathrm{linear}} = \frac{L}{v_f}\left(\frac{\mathrm{DDR}}{\mathrm{DDR}-1}\right)\ln(\mathrm{DDR})$
t_f (Newtonian model)	ms	1012	$t_f^{\mathrm{Newtonian}} = \frac{L}{v_f}\left(\frac{\mathrm{DDR}-1}{\ln(\mathrm{DDR})}\right)$

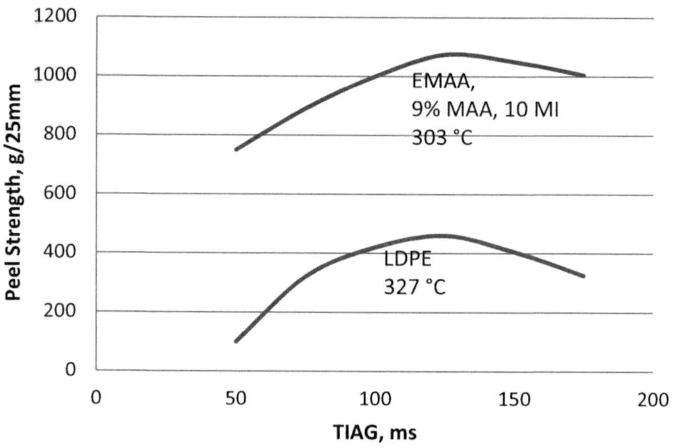

FIGURE 15.3 Peel strength to aluminum foil as a function of time in the air gap. *From V. Antonov, A. Soutar, Foil adhesion with copolymers: time in the air gap, in: TAPPI 1991 Polymers, Laminations & Coatings Conference Proceedings, TAPPI PRESS, Atlanta, 1991, p. 553. Reproduced with permission from The Dow Chemical Company.*

Many investigators have found that adhesion in extrusion coating increases with the time in the air gap. Antonov and Soutar [4] were among the first to summarize the data and show this for a wide variety of polymers and substrates. One set of data from their paper is reproduced in Fig. 15.3 where peel strength to aluminum foil increases with TIAG up to about 120 ms. A number of mechanisms may be responsible for this behavior, including increased time for oxidation of the melt and a reduction in stress. Countering these is an increase in cooling of the melt curtain. These are explored in the following sections.

15.1.2.1 Oxidation

LDPE is the largest volume resin used in extrusion coating and lamination, yet it does not bond well by itself to polar substrates such as paper, aluminum foil, metalized film and chemically primed films. It is well recognized that LDPE undergoes oxidation during the extrusion coating process to create functional groups that favor adhesion. The mechanism for thermo-oxidation of LDPE is the formation of alkyl radicals (R*) through the splitting of covalent bonds, followed by reaction with oxygen to form hydroperoxides (ROOH). This occurs in a chain reaction: [1,2,10–18]

$$RH \rightarrow R* \rightarrow RO_2* \rightarrow ROOH \rightarrow \text{Oxygenated products.}$$

Here R represents the polymer chain. The products of oxidation include carboxylic acids, aldehydes, ketones and alcohols.

The evidence that LDPE undergoes oxidation during the coating process and that the oxidation products are responsible for its adhesion to polar substrates is straightforward and incontrovertible:

- LDPE only adheres to polar substrates if the melt temperature is above about 285°C. This is shown by many investigators to hold for bonding LDPE to paper or foil. An example is presented in Fig. 15.4. In general, temperatures above about 310°C are used to process LDPE for extrusion coating applications.
- The onset of thermo-oxidative degradation of LDPE occurs at around 285°C and the rate of oxidation increases exponentially as the temperature is further increased. Several investigators [20–25] have linked the formation of oxidative species with processing temperature.
  - Laillo and Ylanen [20] correlate melt temperature with a carbonyl index (CI) that reflects the number of C = O groups that have formed. CI is obtained from Fourier transform infrared spectroscopy (FTIR) measurements by calculating the ratio of the peak height at 1720 cm^{-1} to a reference C-H peak at 2660 cm^{-1} [2]. As the melt temperature increases above 280°C, the CI increases substantially, as shown in Fig. 15.5.
  - Ristey and Shroff [25] show similar results for two grades of LDPE. The grade with the lower density undergoes greater oxidation. This is explained by the chemistry: the lower density polymer has more short-chain (methyl) branches and the branch points are more susceptible to oxidative reaction.
  - Andersson et al. [2] show similar results, which are reproduced in Fig. 15.6.

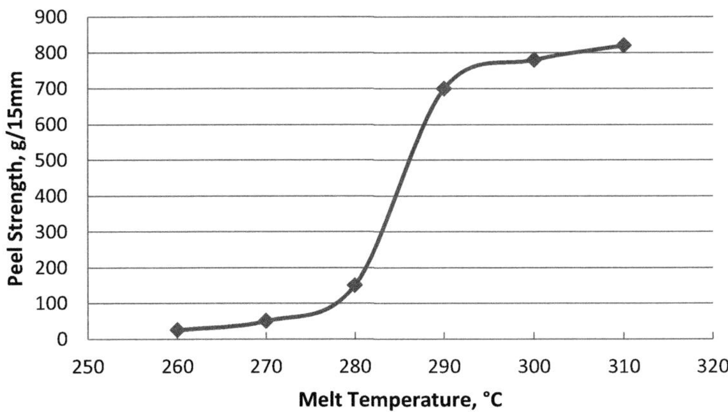

FIGURE 15.4 Peel strength of LDPE to aluminum foil as a function of melt temperature. Line speed: 100 m/min. 20-μm coating. *Reproduced with permission from P.B. Sherman, Ozonation of polymer melt for improved adhesion, in: Bezigian, (ed.), Extrusion Coating Manual, fourth ed., TAPPI Press, Atlanta, Ga, 1999, pp. 75–88.*

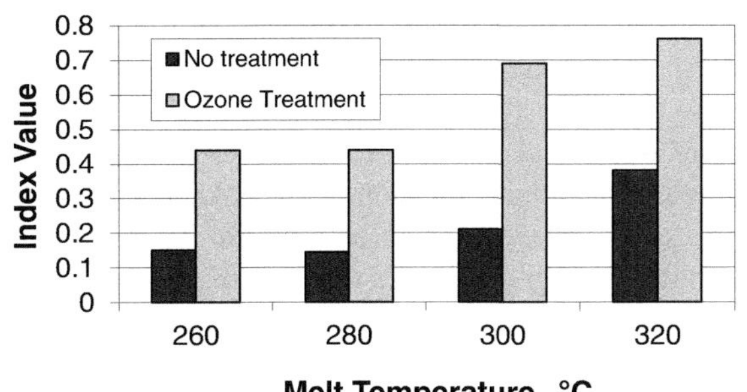

FIGURE 15.5 Oxidation of LDPE, as measured by carbonyl index, as a function of melt temperature and ozone treatment. *Reproduced with permission from E. Laiho, T. Ylanen, Flame, corona, ozone – do we need all pretreatments in extrusion coating, in: TAPPI Polymers, Laminations & Coat. Conference, 1997, p. 93.*

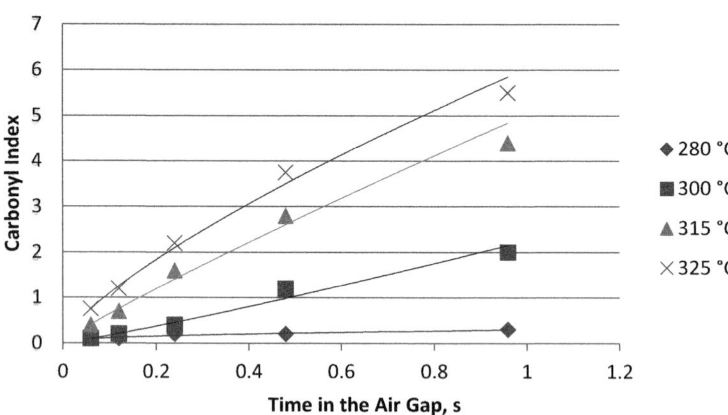

FIGURE 15.6 Oxidation of LDPE as a function of time in the air gap and temperature. Carbonyl index is a measure of oxidation obtained from Fourier transfrom infrared measurements. *Data from T. Andersson, B. Stalbom, B. Wesslen, Degradation of polyethylene during extrusion. II. Degradation of low-density polyethylene, linear low-density polyethylene, and high density polyethylene in film extrusion, J. Appl. Polym. Sci., 91 (2004) 1525–1537.*

   - Briggs et al. [21] use Electron Spectroscopy for Chemical Analysis (ESCA), also known as X-ray photoelectron spectroscopy (XPS), a more sensitive surface analysis technique than FTIR. They find that the increase in adhesion of PE to aluminum with increasing processing temperature correlates with an increase in oxygen and nitrogen species at the surface of the PE.
   - Some oxidative species are more effective for bonding to polar substrates than others. Hansen et al. [22] look specifically at the potential for these species to form acid-base interactions (the aluminum oxides on the surface of foil are basic) using a technique called flow microcalorimetry. They show that the acid functionality of LDPE fibers increases with increasing fabrication temperature (288°C–315°C).
- Decreasing the line speed or increasing the air gap increases the adhesion of LDPE to various substrates.
   - Croilius et al. [26] relate this to the web exposure time or the time the polymer spends in the air gap.
   - Antonov and Soutar [4] later define the air gap divided by the line speed as the Time In the Air Gap (TIAG) as given by Eq. (15.1). Referring back to the typical plot shown in Fig. 15.3, the peel strength normally goes through a maximum at about 100 ms. The region below 100 ms is of most interest to converters since it represents the region of higher productivity and output. In this region, increasing TIAG typically increases the adhesion to polar substrates. Sherman [19] suggests that this is due to an initiation effect; a certain time or temperature is required to initiate the oxidation reaction. At low TIAG, there is insufficient time for oxidation to begin.
   - Andersson et al. [2] relate TIAG directly to the CI by casting film into a water bath at different heights above the water while maintaining a constant line speed and output. The air gap ranges from 10 to 80 mm, corresponding to a range in TIAG of 0.12 to 1.92 seconds (v_f = 5 m/minute). The plots of CI versus TIAG are shown in Fig. 15.6. The results show a monotonous increase in surface oxidation with increasing air gap (or equivalently TIAG).
   - Ristey and Shroff [25] coat LDPE onto paper varying melt temperature, coat weight and air gap while maintaining constant line speed (900 ft/minute or 234 m/minute). Increasing the air gap increases the CI and adhesion to paper. They find a certain minimum CI is needed to obtain good paper adhesion.
   - Hertzberg et al. [24] measure peel strength of LDPE to aluminum as a function of CI and find that the peel strength increases with increasing oxidation but plateaus at high values of CI. This is similar to a peel strength versus TIAG plot. Hertzberg et al. attribute the plateau to a balancing of forces acting on the adhesion. The increased oxidation increases the polar functionality but at high levels causes chain scission. They measure the molecular weight distribution (MWD) of the oxidized LDPE and find a significant increase in the low MW fraction with increasing oxidation. Poor cohesive strength as a result of chain scission may account for a loss in adhesion. This explanation is consistent with other investigations. Ristey and Shroff [25] measure the change in MWD and melt flow index (MFI) of their coated LDPE. MFI is inversely related to MW. Plots of MFI versus melt temperature are shown to have a U shape with a minimum of around 330°C. At lower temperatures, the polymer degrades primarily by crosslinking while at higher temperatures chain scission occurs. Andersson et al. [2]. report similar results using an in-line rheometer during extrusion: chain extension or crosslinking dominates at a lower temperature (<325°C) whereas chain scission dominates at higher temperatures. Linear low-density polyethylene (LLDPE) and high-density polyethylene (HDPE) show chain scission behavior over the whole temperature range.
   - Foster [27] recasts Ristey and Shroff's data in terms of TIAG and shows that CI increases with TIAG.

- The temperature at which the adhesion of LDPE to polar substrates begins its uprise also corresponds to an increase in the generation of off-gases. Andersson et al. [10–12] show that the smoke at the die exit comprises oxidative byproducts, including volatile alcohols, acids, aldehydes and ketones, the same functionality found in the LDPE films. The concentration of the byproducts increases with increasing processing temperature and correlates with off-flavor of water samples subjected to the films, as determined by a human taste panel.
- Many investigators and processors have found treating the LDPE melt curtain with ozone promotes adhesion to polar substrates at lower extrusion temperatures, as shown in Fig. 15.5, or higher line speeds (shorter TIAG). This is attributed to enhanced surface oxidation [19,28–31].
- Adding antioxidants to LDPE can inhibit adhesion [21,32]. One example is provided in Fig. 15.7 based on a study by Zambetti et al. [32] using vitamin E as the antioxidant.

One of the drawbacks of LDPE is that its adhesion is dependent on the process: if sufficient oxidation does not occur the adhesion will suffer. Excessive temperature causes off-flavor issues due to oxidative byproducts and also loss of adhesion due to chain scission. There is a relatively narrow temperature range for processing LDPE for optimum adhesion, typically between about 310°C and 330°C.

Several approaches are used to mitigate these issues. One is to ozone treat the melt curtain [15,24,25,28]. For example, Hammond [28] shows that adhesion of LDPE to a chemically primed film can be achieved at substantially lower processing temperature or higher line speeds by using ozone treatment.

Another approach is to use functionalized polymers, such as acid copolymers and ionomers. These contain acid functionality incorporated into the backbone of the polyethylene chain during polymerization. There is no need for oxidation for these polymers to bond to polar substrates. Antonov and Soutar [4], however, show that these polymers exhibit an increase in peel strength to aluminum foil and paper with increasing TIAG, just like LDPE but starting at a much higher level of bond strength. This is illustrated in Fig. 15.3. Based on this evidence, many in the industry have assumed that oxidation is also involved in the adhesion of acid copolymers to polar substrates. Morris and Suzuki [29] examine this phenomenon and conclude that oxidation is not the underlying mechanism for these polymers. Their arguments and experimental results are summarized here:

- Oxidation does occur in acid copolymers during extrusion coating. In some unpublished work at DuPont in the 1980s, Golike [33] shows that oxidative species form during extrusion coating of acid copolymers at elevated temperatures (above 280°C). The amount of oxidative species formed (as measured by headspace GC) increases with increasing melt temperature and correlates with changes in organoleptics. He concludes that the mechanism for the oxidation of acid copolymers is similar to that of LDPE.

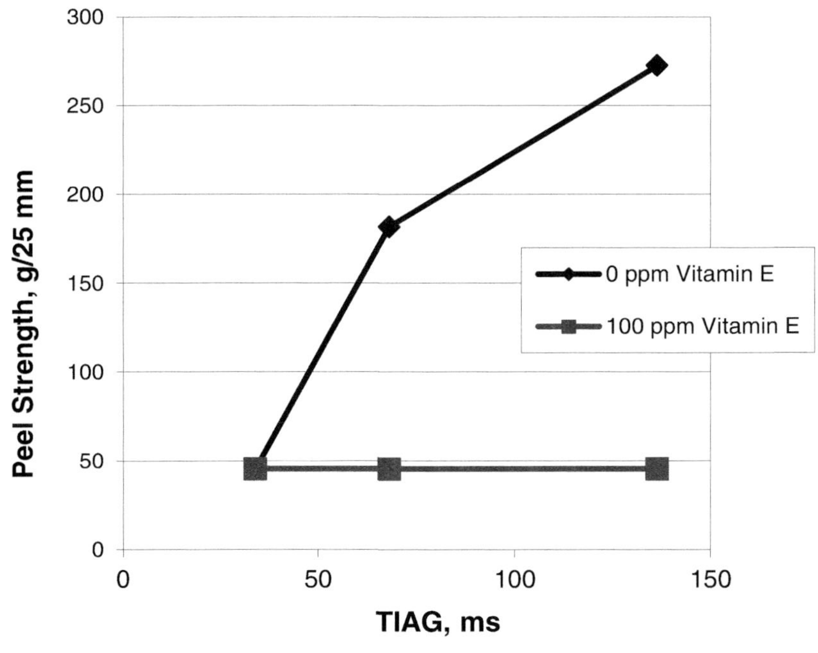

FIGURE 15.7 Effect of adding an antioxidant (vitamin E) to low-density polyethylene (LDPE) on adhesion of LDPE extrusion coated onto aluminum foil. *Data is taken from P. Zambetti, S. Baker, D. Kelley, Alpha tocopherol as an antioxidant for extrusion-coating polymers, TAPPI J., 78 (1995) 167. Graph is from W.J. Ristey, R.N. Shroff, The degradation and surface oxidation of polyethylene during the extrusion coating process, TAPPI Pap. Synth. Proc., 1998 and reproduced with permission from The Dow Chemical Company.*

- The effect of melt temperature on the adhesion of ethylene acid copolymer resins (ACRs) is complex. Low-acid ACRs (typically 7% and lower) show an increase in peel strength to foil with increasing temperature, similar to the trend found for LDPE. Higher acid polymers tend to be less sensitive to a temperature above a certain minimum needed for wetting and spreading of the polymer onto the substrate. Antonov and Soutar [4] report adhesion to aluminum foil of an ethylene methacrylic acid (EMAA) copolymer with 12 wt.% MAA and a melt flow index (MFI) of 7 g/10 minute (190°C/2160 g) does not correlate with melt temperature over a range of 243°C–304°C (470–580°F). This and other studies suggest that above about 280°C, the adhesion of high-acid ACRs does not substantially change with increasing temperature. Yet, organoleptic testing shows that more oxidative species form in acid copolymers when they are processed at these higher temperatures.
- One possible explanation comes from Hansen et al.'s flow microcalorimetry work [22]. They show that oxidation adds less than about 1 wt.% acid functionality to LDPE. While this is significant for LDPE, this may be too little to make a difference for high-acid copolymers that already have 9–12 wt.% acid functionality. Hester et al. [34] find that the effect of acid level on the adhesion of acid copolymers to foil is nonlinear (see Fig. 15.8); below 9 wt.% MAA adding additional acid functionality has a pronounced effect on adhesion. Above 9% acid, the effect is minimal.
- Morris [6] studies the effect of wetting and stress on the peel strength of acid copolymers to foil. This work is reviewed in Section 15.1.2.2. Wetting of the surface is quantified by measuring the contact angle (the angle between a liquid drop and the solid surface) of water on the coating surface. Higher contact angle indicates less wettability. The contact angle is related to the surface energy and should change when oxidative species are formed at the surface. He finds, however, that measurements of the contact angle of water on the coating surface do not correlate with peel strength. The results suggest that at the TIAG and temperatures used in the study, oxidation is insignificant. Yet, large differences in peel strength are measured. Morris' analysis suggests that the increase in peel strength with increasing TIAG is associated with a reduction in stress. Specifically, TIAG is inversely related to the rate of elongation during coating. Increasing the rate of elongation (decreasing TIAG) increases the stress imparted on the polymer. Residual stress can negatively impact adhesion [35].
- Morris and Suzuki [29] conduct four experiments to better understand the mechanism for the rise in peel strength with increasing TIAG for acid copolymers:
  - In the first, two acid copolymers and a LDPE control are treated with ozone as they are coated onto aluminum foil. The structure is kraft paper (50 μm)/LDPE (20 μm)/Al foil (9 μm)/resin (20 μm). The resin is melted in a 65 mm diameter extruder and fed through an internal deckled die with a die gap of 0.8 mm and a width of 500 mm. The air gap is 120 mm and line speed 80 m/minute for a TIAG of 90 ms. Four variables are tested: the resin type (two ACRs and one LDPE), the resin melt temperature, and the ozone concentration and flow rate. The resins are described in Table 15.3. Peel strength is measured using a T-peel test at a pull rate of 300 mm/minute. Samples are measured at 1, 7, and 14 days after coating. The results of the 1-day peel tests are shown in Table 15.4 and Fig. 15.9. (7- and 14-day results show similar trends.) The results indicate that ozone treatment has no positive effect on the adhesion of acid copolymers to foil. Indeed, at the lower processing temperatures, the peel strength is lower with ozone than without. This may be due to a cooling effect resulting from the flow of ozone onto the melt curtain. This contrasts with the LDPE case where the ozone treatment substantially increases the peel strength performance. The results suggest that increasing oxidation with ozone treatment does

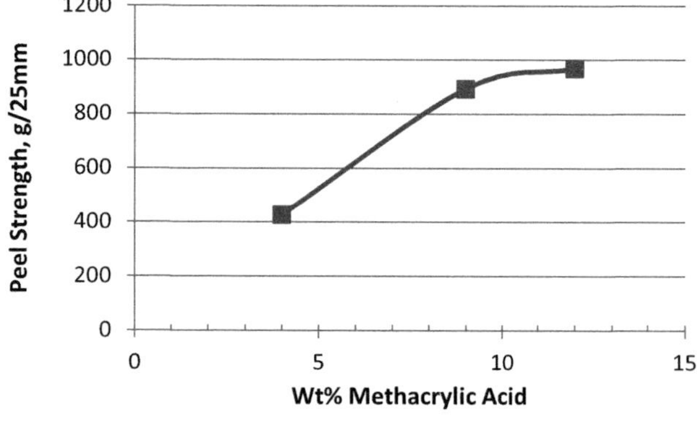

FIGURE 15.8 Effect of acid level on adhesion of acid copolymer (EMAA) to aluminum foil in extrusion coating. Structure is 25 μm EMAA/50 μm Al foil. Peel strength measured 1 week after coating. *Data from D. Hester, S. Sadik, Y. Trouilhet, USing acid copolymers as coextrusion tie layers, in: TAPPI Polymers, Laminations & Coat. Conference, 1989, p. 459.*

TABLE 15.3 Description of resins used in Morris and Suzuki study of the effect of oxidation on peel strength of acid copolymers to aluminum foil.

Code name	Supplier	Grade name	% Acid (MAA)	MFI (g/10 min)
N0908	Mitsui-DuPont (now Mitsui-Dow)	Nucrel 0908 C	9	8
N0407	Mitsui-DuPont (now Mitsui-Dow)	Nucrel AN4214C	4	7
N0910	DuPont (now Dow)	Nucrel 0910	9	10
N0910 + A/O	DuPont (now Dow)	Nucrel 0910 with 1000 ppm antioxidant	9	10
N1214	DuPont (now Dow)	Nucrel 1214	12	14

Source: From B.A. Morris, N. Suzuki, The case against oxidation as a primary factor for bonding acid copolymers to foil, in: TAPPI Polymers, Laminations & Coat. Conference, 2000. Reproduced with permission from The Dow Chemical Company.

TABLE 15.4 Effect of ozone treatment on peel strength to aluminum foil.

			Peel strength, g/25 mm		
			At extrusion temperature		
Resin	**Ozone conc. g/m^3**	**Ozone flow m^3/h**	**250°C**	**270°C**	**290°C**
N0908	0	0	550	570	650
	15	0.5	520	600	630
	15	1.0	450	570	600
	15	2.0	330	550	600
	25	0.5	470	580	620
	25	1.0	320	580	600
			Peel strength, g/25 mm		
			At extrusion temperature		
Resin	**Ozone conc. g/m^3**	**Ozone flow m^3/h**	**270°C**	**285°C**	**300°C**
N0407	0	0	500	520	530
	15	1.0	470	500	530
	25	1.0	430	480	500

			Peel strength, g/25 mm
Resin	**Ozone conc. g/m^3**	**Ozone flow m^3/h**	**290°C extr. temp.**
LDPE	0	0	30
	15	1.0	250
	15	2.0	230
	25	1.0	250
	25	2.0	230
	35	1.0	250
	35	2.0	250

Notes: Measurements were made 1 day after extrusion coating. Structure is kraft paper/LDPE/Al/Resin. Resins are described in Table 15.3.
Source: From B.A. Morris, N. Suzuki, The case against oxidation as a primary factor for bonding acid copolymers to foil, in: TAPPI Polymers, Laminations & Coat. Conference, 2000. Reproduced with permission from The Dow Chemical Company.

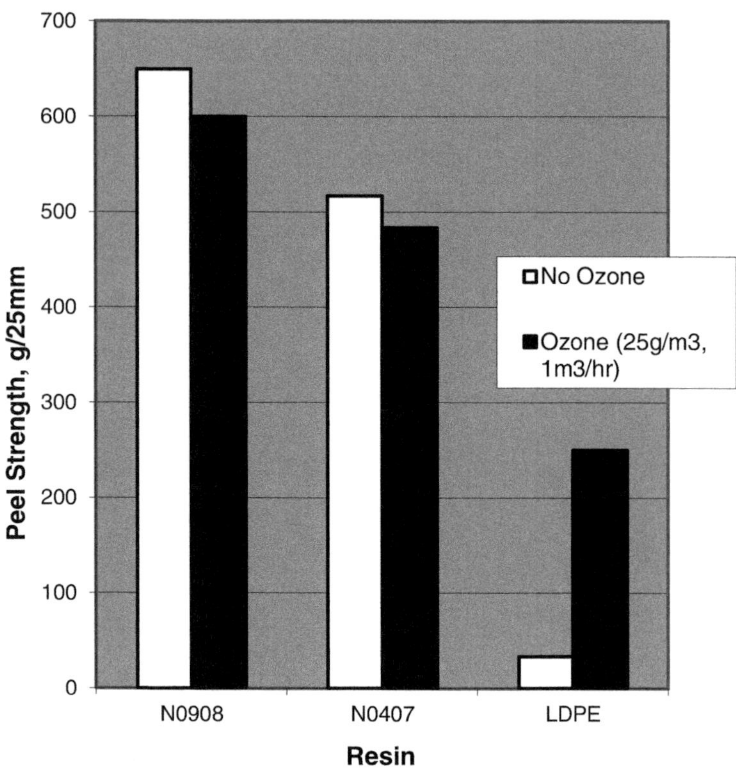

FIGURE 15.9 Effect of ozone treatment of melt curtain during extrusion coating on peel strength to aluminum foil for two acid copolymer resins and low-density polyethylene (LDPE). Structure is kraft paper/LDPE/Al/Resin. Resins are described in Table 15.3. *From B.A. Morris, N. Suzuki, The case against oxidation as a primary factor for bonding acid copolymers to foil, in: TAPPI Polymers, Laminations & Coat. Conference, 2000. Reproduced with permission from The Dow Chemical Company.*

not enhance the adhesion of acid copolymers to foil. The results also show that even with ozone treatment, the adhesion of LDPE to foil is substantially less than the acid copolymer.

- In the second experiment, 1000 ppm of a hindered phenol antioxidant is added to an acid copolymer (N0910, described in Table 15.3). The N0910 is melt compounded with a concentrate made from the same base resin using a single-screw compounding extruder at 190°C. For a control, the N0910 is put through the same extruder without the concentrate so that both resins have the same heat history. Both versions are extrusion coated onto 50 μm aluminum foil at a coating thickness of 25 μm. The melt temperature is set at 300°C and ranges from 290°C to 308°C (measured by both an IR thermometer and thermocouple). The die gap is 0.63 mm. The air gap (127 to 254 mm) and line speed (61 to 244 m/minute) are varied over a range so that the TIAG varies from 31 to 250 ms. Peel strength results are shown in Table 15.5 and Fig. 15.10. The results show that adding an antioxidant to the acid copolymer has no substantial effect on peel strength performance. These contrast with studies from the literature for LDPE (see Fig. 15.7). Electron Spectroscopy for Chemical Analysis (ESCA) is used to measure the amount of surface oxygen on the coatings. This technique measures the elemental composition near the surface of a material by bombarding the surface with high-energy electrons. The sampling depth is determined by the detection angle: 30 and 90 degrees correspond to approximately 50 and 100 Å, respectively. The ESCA results are shown in Fig. 15.11. The unmodified N0910 shows an increase in surface oxidation whereas the addition of antioxidant effectively suppresses oxidation. Both, however, give the same trend of increasing peel strength with increasing TIAG.
- The third experiment uses a high MFI grade (N1214) to coat foil at 230°C, well below the onset oxidation temperature. The same structure and set-up are used as for the antioxidant experiment. The N0910 run at 300°C is used for comparison. The peel strength results are shown in Fig. 15.12. Both the 1214 (230°C) and 0910 (300°C) show the characteristic increase in peel strength to foil with increasing TIAG. ESCA results are shown in Fig. 15.13. As expected, the N1214 shows no increase in surface oxygen concentration as a function of TIAG. The results show that greater oxidation is not required for the peel strength of acid copolymers to increase with increasing TIAG.
- The fourth experiment examines the effect of orientation on peel strength. Peel tests are conducted in the machine direction (MD) and transverse direction (TD) and compared with the mechanical properties of the coatings (% elongation at break). Returning to the same structure as in experiment 1, N0908 is coated onto the foil

TABLE 15.5 Results of antioxidant and low temperature extrusion coating experiments.

Line speed	Air gap	TIAG	Green peel, g/25 mm		1-week Peel, g/25 mm		At% O by ESCA	
m/min	mm	ms	Mean	Std	Mean	Std	30°	90°
50 μm foil/N0910 (no antioxidant) coated at 300°C								
244	127	31	507	73	668	60	1.4	0.7
122	127	63	718	31	890	19	2.1	1.2
61	127	125	624	31	768	20	2.3	1.5
244	254	63	777	61	951	74	3.3	2.2
122	254	125	886	87	1130	107	2.6	1.8
61	254	250	776	33	893	78	2.2	1.3
50 μm foil/N0910 + A/O coated at 300°C								
244	127	31	444	65	502	58	1.9	1.2
122	127	63	449	82	470	91	2.3	1.3
61	127	125	724	50	910	33	2.3	1.3
244	254	63	748	34	801	33	2	1.2
122	254	125	812	74	950	69	2	1.1
61	254	250	931	86	1149	87	*	*
50 μm foil/N1214 coated at 230°C								
244	127	31	562	56	525	70	2.2	1.4
122	127	63	823	43	799	69	2.6	1.3
61	127	125	681	53	681	43	1.7	0.8
244	254	63	768	47	787	50	1.9	0.9
122	254	125	879	29	857	52	1.9	0.8
61	254	250	814	65	784	64	1.8	0.8

*Value not reported due to sample contamination.
Source: From B.A. Morris, N. Suzuki, The case against oxidation as a primary factor for bonding acid copolymers to foil, in: TAPPI Polymers, Laminations & Coat. Conference, 2000. Reproduced with permission from The Dow Chemical Company.

side of the structure Paper/LDPE/Al. Coating thickness is 20 μm. The die gap is 0.8 mm. The melt temperature is 290°C. The air gap is 110 mm and line speed is varied from 80 to 200 m/minute (resulting in TIAG ranging from 83 to 33 ms). To obtain samples of the coating for elongation measurements, 12 μm oriented polyethylene terephthalate film (OPET) slip sheets are inserted between the foil and coating during part of the run. The coating is then peeled away from the OPET for testing. The results of peel strength to foil measurements in the machine and transverse direction are shown in Fig. 15.14. In the MD, the peel strength increases with increasing TIAG, consistent with previous findings. In the TD, however, the peel strength is unchanged with increasing TIAG. If oxidation were the predominant mechanism, the peel strength should increase in both directions. The elongation at break of the coatings in the MD increases with TIAG whereas it is relatively flat in the TD, as shown in Fig. 15.15. This corresponds with the peel strength results. Two explanations are proposed. One involves stress. As TIAG increases, the stress decreases (see Section 15.1.2.2). Lower stress results in less orientation of the polymer, resulting in higher elongation prior to breakage. Lower stress is also known to increase peel strength. The second explanation is related. The peel strength measurement captures the energy required to pull the specimen apart. This energy includes not only the bond strength between the coating and substrate but also the energy needed to bend and elongate the peel arm. A fracture energy analysis shows that in these structures the energy to

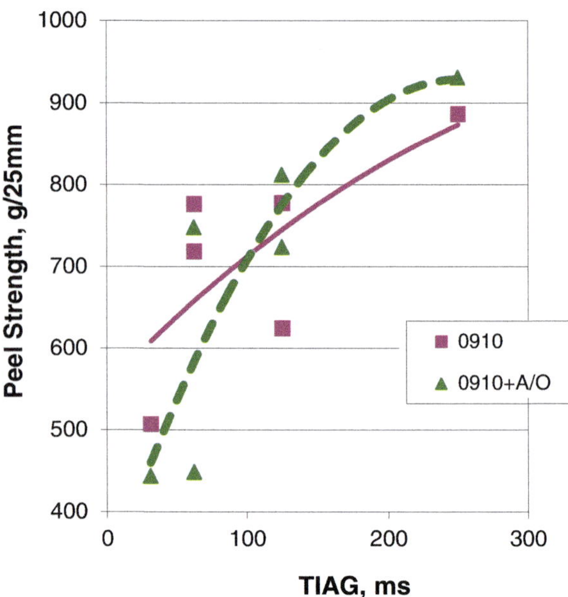

FIGURE 15.10 Effect of antioxidant on peel strength of an acid copolymer resin to aluminum foil. Results are for green peel strength. *From B.A. Morris, N. Suzuki, The case against oxidation as a primary factor for bonding acid copolymers to foil, in: TAPPI Polymers, Laminations & Coat. Conference, 2000. Reproduced with permission from The Dow Chemical Company.*

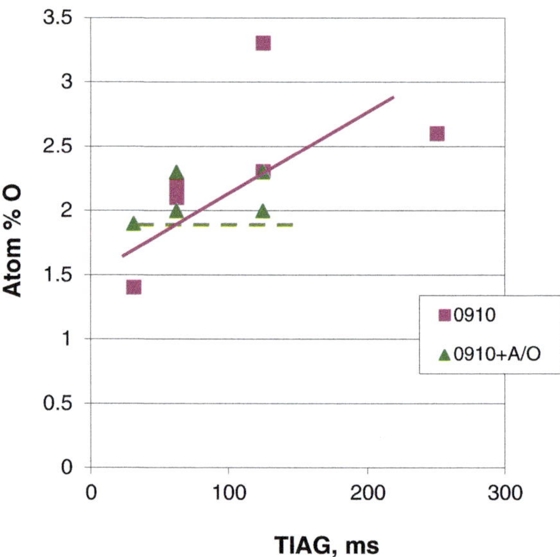

FIGURE 15.11 Effect of antioxidant on surface concentration of oxygen (ESCA results) after extrusion coating. *From B.A. Morris, N. Suzuki, The case against oxidation as a primary factor for bonding acid copolymers to foil, in: TAPPI Polymers, Laminations & Coat. Conference, 2000. Reproduced with permission from The Dow Chemical Company.*

elongate the peel arm plays a substantial role [30]. Decreasing the % elongation of the peel arm reduces the peel strength. (See Section 10.2.3 for a more detailed discussion on the fracture mechanics of peel strength measurements.)

The practical implication of this study is that acid copolymers may provide more consistent peel strength performance to polar substrates than LDPE. They can be coated at lower temperatures, reducing organoleptic concerns, and at faster line speeds while still maintaining acceptable adhesion performance. As discussed in Chapter 16, they also have better product resistance than LDPE, extending the package shelf life of difficult products such as ketchup and salad dressings.

Acid copolymers and ionomers have their drawbacks, the most important being their higher price than LDPE. Another approach to improving adhesion is to blend a functionalized ethylene copolymer with LDPE to enhance the adhesion to LDPE, making it more robust to processing variations. In Section 10.4.1.4, the development of an acid functionalized ethylene copolymer modifier (AE) for blending with LDPE is described [36]. The additive allows for higher line speeds and lower processing temperatures than LDPE alone, and helps improve operational reliability (see Fig. 10.20).

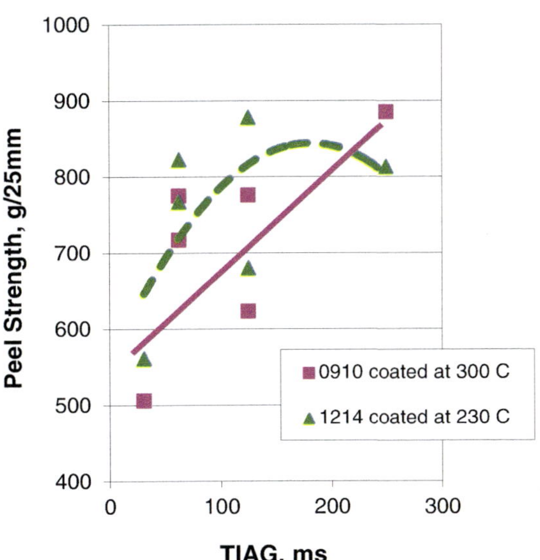

FIGURE 15.12 Effect of processing temperature on peel strength of acid copolymer resins to aluminum foil. *From B.A. Morris, N. Suzuki, The case against oxidation as a primary factor for bonding acid copolymers to foil, in: TAPPI Polymers, Laminations & Coat. Conference, 2000. Reproduced with permission from The Dow Chemical Company.*

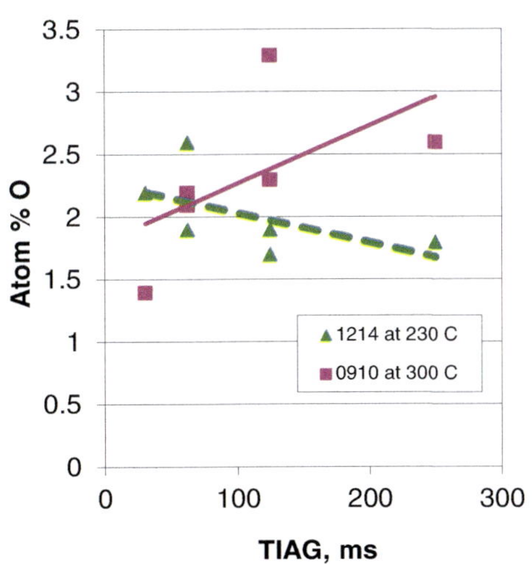

FIGURE 15.13 Effect of processing temperature on surface concentration of oxygen (ESCA results) after extrusion coating. *From B.A. Morris, N. Suzuki, The case against oxidation as a primary factor for bonding acid copolymers to foil, in: TAPPI Polymers, Laminations & Coat. Conference, 2000. Reproduced with permission from The Dow Chemical Company.*

The modifier approach is not for every situation. In particular, it does not provide the strong bond strength and resistance to aggressive products that the pure acid copolymer provides. But it is a good illustration of how understanding the manner in which adhesion performance is impacted by the process can lead to alternatives with better performance.

15.1.2.2 Stress and orientation

The polymer is subjected to stress and orientation as it is extruded through the die and drawn down to its final coating thickness in the air gap. The stress arises from shear forces in the extruder and die, from stretching or extensional forces as the melt curtain is drawn down and from shrinkage forces during crystallization. Morris [6] hypothesizes that these stresses may influence peel strength by impacting tensile properties of the coating and residual stress at the substrate-polymer interface. Stress is not one of the control parameters on an extrusion coating line that can be easily varied to directly test this hypothesis. Relationships between stress and common processing parameters such as air gap, die gap, melt temperature and line speed are first required. Morris develops these relationships and conducts experiments changing processing parameters in a systematic way to alter the stress imposed on the polymer. Peel strength to aluminum foil in these experiments is found to vary by a factor of three, strongly suggesting that stress plays a role in peel strength performance. Stress developed during both extrusion and drawing is found to be important. This work is summarized in this section.

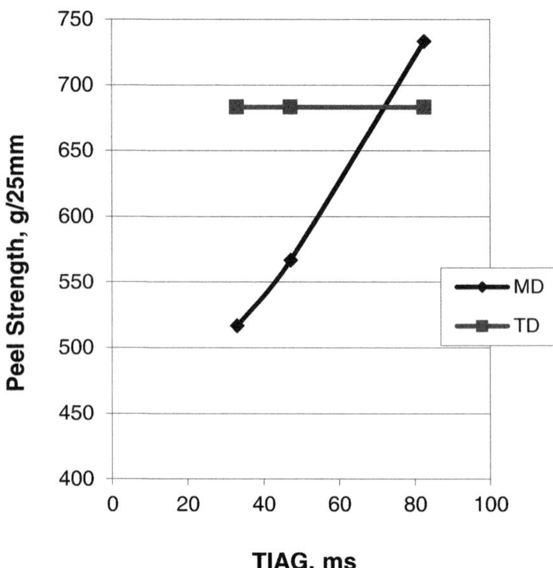

FIGURE 15.14 Effect of orientation on peel strength of N0908 acid copolymer to aluminum foil. Structure is kraft paper/LDPE/Al/Resin. *From B.A. Morris, N. Suzuki, The case against oxidation as a primary factor for bonding acid copolymers to foil, in: TAPPI Polymers, Laminations & Coat. Conference, 2000. Reproduced with permission from The Dow Chemical Company.*

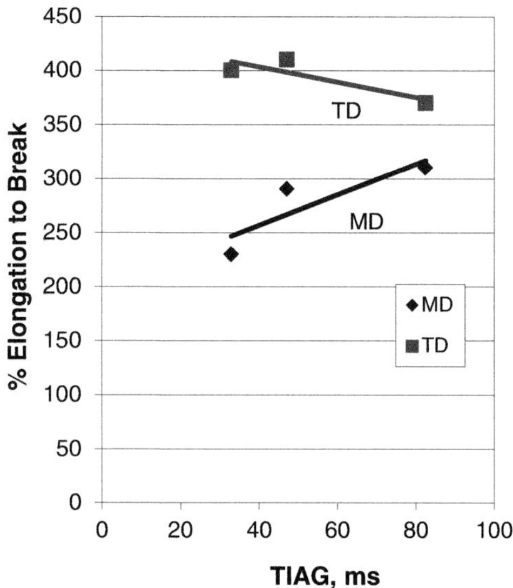

FIGURE 15.15 Effect of TIAG on % elongation at break of coatings from Fig. 15.14. *From B.A. Morris, N. Suzuki, The case against oxidation as a primary factor for bonding acid copolymers to foil, in: TAPPI Polymers, Laminations & Coat. Conference, 2000. Reproduced with permission from The Dow Chemical Company.*

As described in Section 10.2.3, peel strength is a measure of the work or energy needed to pull the specimen apart. It is the sum of the energy necessary for creating new surface area at the front of a crack traveling along the interface (this energy is known as the fracture energy) and the amount of energy dissipated by elongating and bending the peel arm. The fracture energy includes contributions from various adhesive forces such as thermodynamic and chemical interactions, as well as local deformation at the tip of the crack. The resistance to deformation is a function of the viscoelastic and tensile properties of the material. Changes in molecular orientation and crystalline morphology brought about by stress may affect these properties. For example, in polyethylene films, the crystalline regions organize into lamellae or flat plates. High stress in the molten polymer has been shown to give a more ordered and twisted lamellae structure [37]. Such changes influence the bulk tensile properties of the polymer [38].

Another factor influencing peel strength is residual or frozen-in stress at the interface. This internal stress acts to reduce the fracture strength of the bond [35]. Internal stress may result when the oriented chains created during extrusion and drawing do not have time to relax back to their natural random coiled configuration due to the rapid rate of cooling at the chill roll. The ability of the stress to relax can be characterized by comparing the time scale of the process to the characteristic relaxation time of the polymer. The ratio of these two-time scales is a dimensionless number known as the Deborah number (introduced in Chapter 5):

$$\text{De} = \frac{\lambda}{t_f} \tag{15.5}$$

where De = Deborah number, λ = relaxation time of the polymer, and t_f = characteristic time of the process.

Methods for measuring the polymer relaxation time are given in Section 5.5.1; the polymer relaxation time is a function of the polymer chemistry and molecular architecture (MW, MWD, branching, etc.) and temperature. For example, linear PE (e.g., LLDPE) generally has a shorter relaxation time than branched PE (LDPE). The characteristic process time for the purposes of this discussion is the time in the air gap described above (Table 15.1). When the Deborah number is small (De $<<$ 1), the polymer has sufficient time to relax during the time scale of the process ($t_f >> \lambda$) and little stress is frozen-in. When De is large (De $>>$ 1), more of the internal stress will be frozen in place.

Stress may also arise from shrinkage of the polymer as it crystallizes as is quenched on the chill roll. Increases in stress and orientation during the coating process may result in enhanced crystallinity (through flow or strain-induced crystallization) and greater shrinkage (see Section 13.5.3). For the purposes of this analysis, it is assumed that the stress due to shrinkage is either constant (independent of processing conditions) or is proportional to the net stress imparted during extrusion, drawing and relaxation.

The net stress, therefore, is the sum of the stress imposed during extrusion (up to the die exit) and drawing (in the air gap) minus any stress that has relaxed. In polymer processing both the strain and rate of strain imposed by the process give rise to stress. An engineering analysis of the extrusion coating process is given in the appendix to this chapter where equations are developed relating processing variables to stress. These relationships are summarized below for the case of when the resin and coating thickness are held constant. Here the symbol "~" can be read as "goes as" or "scales as."

Stress imparted in the extruder and the die (σ_{ext})

$$\sigma_{\text{ext}} \sim m(T)\frac{v_f^n}{X_0^{2n}} \tag{15.6}$$

Stress imparted during drawing in the air gap (σ_d)

$$\sigma_d \sim T,\ \dot{\varepsilon},\ \varepsilon_{\text{tot}} \tag{15.7}$$

$$e_{\text{tot}} \sim \ln(\text{DDR}) \sim \ln(X_o) \tag{15.8}$$

$$\dot{\varepsilon} = \frac{v_f}{L}\left(\frac{\text{DDR}-1}{\text{DDR}}\right). \tag{15.9}$$

Stress relaxation

$$\text{Relaxation} \sim \lambda(T), t_f \tag{15.10}$$

$$t_f = \frac{L}{v_f}\left(\frac{\text{DDR}}{\text{DDR}-1}\right)\ln(\text{DDR}) \sim \frac{L}{v_f}\ln(X_o) \tag{15.11}$$

where σ_{ext} is the stress imposed by the extrusion process (pressure driven flow in the die), σ_d is the stress imposed by drawing in the air gap, ε_{tot} is the total elongational strain at end of air gap, $\dot{\varepsilon}$ is the elongational rate of strain (in air gap), λ is the relaxation time of the polymer, T is the temperature of molten polymer, v_f is the line speed (velocity at the chill roll), X_0 is die gap, m, n are the power-law viscosity model consistency index and exponent [Eq. (15A.3)], and DDR is the draw-down ratio.

Astute observers will notice that a linear velocity assumption is used to compute Eq. (15.11) for the process time (see Table 15.1).

To examine the effect of stress on peel strength performance, 25 μm (1 mil) of an ethylene acrylic acid (EAA) copolymer (9 wt% AA with a MFI of 10 g/10 minutes) is extrusion coated onto 50-μm (2 mils) aluminum foil under various processing conditions. A pilot-scale line with a 107-cm wide T-slot die, externally deckled to 71 cm and fed by a 114-cm diameter extruder is used for the coating. The air gap (13–34 cm), line speed (60–240 m/minute) and die gap (0.38–1.01 mm) are varied, keeping coating thickness, deckled die width and chill roll temperature constant. Goal melt temperature is 282°C (540°F); temperatures of the molten web measured with an IR thermometer range from 282°C to 289°C at the die exit and from 276°C to 282°C near the chill roll.

Peel strength of the acid copolymer to foil is measured by cutting 25-mm wide strips in the machine direction from the center of the web. These are separated between the acid copolymer and the foil and pulled in a tensile tester in a T-peel configuration at 305 mm/minute. Polar contact angle (using distilled water), tensile properties (ASTM 882), and wide-angle x-ray scattering (WAXS) measurements are conducted on selected samples.

The experimental design and results are given in Table 15.6. Overall, peel strength varies by a factor of three, suggesting that stress is important to peel strength performance. Because of the interrelationships between stress imposed during extrusion (σ_{ext}) and draw-down in the air gap (σ_d, characterized by elongational rate of strain and total strain), it is difficult to isolate specific effects from the data. A statistical design of experiment (DOE) software program is used to fit the data to a quadratic model. Contour plots of constant peel strength are given in Fig. 15.16. The statistical analysis shows the following trends:

- In general, peel strength decreases with increasing elongational rate of strain ($\dot{\varepsilon}$). The effect of rate of strain is stronger at higher elongational strain (ε_{tot}).
- At constant rate of strain and σ_{ext}, peel strength decreases with increasing elongational strain.
- The effect of σ_{ext} is more pronounced at low total elongational strain (left plot). This suggests the elongational stress during drawdown dominates the shear stress during flow in the die. The elongational rate of strain [Eq. (15.9)] can be increased by increasing line speed (increasing σ_{ext}) or reducing air gap (constant σ_{ext}). The former approach (increasing σ_{ext}) results in lower peel strength than the latter when the total elongational strain is low.

Polar contact angles (using water) are measured on the outside surface of the acid copolymer for selected states to test whether oxidation may play a role in the peel strength results. As shown in Table 15.7, the polar contact angle does not correlate well with either time in the air gap (which represents the time for oxidation) or peel strength. The poor correlation is not what would be expected if oxidation were a key mechanism for adhesion in these experiments.

During experiment 5 (states 5.1–5.7 in Table 15.6), oriented polyester (OPET) film slip sheets were inserted between the foil and the coating so that samples of the coating could be isolated for testing. Table 15.8 shows the tensile properties of these samples. The DDR is 33.7 (die gap = 1 mm) and melt temperature is around 282°C. The bulk tensile properties of the coating show little correlation with the peel strength results and are discussed further below.

Wide-angle X-ray scattering (WAXS) is used to assess relative molecular orientation differences between two of the samples, states 5.3 and 5.7. As shown in Table 15.8, these samples differ widely in $\dot{\varepsilon}$ and σ_{ext} as well as in tensile properties. Also, they give the largest difference in peel strength within the 5-series. The WAXS results, however, show no differences in orientation. The orientation in the plane of the film is random. There is a slight orientation of the molecules in the thickness of the film (parallel to the machine direction) in both samples. The total amount of crystallinity of the samples also appears to be similar.

The peel strength results are consistent with the stress hypothesis. This can be seen when the results are cast in terms of the variables that influence stress ($\dot{\varepsilon}$, ε_{tot}, and σ_{ext}). Fig. 15.17 shows all of the peel strength data plotted versus rate of strain, $\dot{\varepsilon}$. As $\dot{\varepsilon}$ increases, the peel strength decreases. Since $\dot{\varepsilon}$ is directly related to the stress imposed during drawing, these results show that increasing stress results in lower peel strength. There is considerable scatter in the data, however, suggesting other parameters are also involved. Much of this scatter is likely related to the effects of varying total strain and σ_{ext}.

Extensional or elongational flow that occurs in the drawing process is known as a strong flow compared to shear flow in the die [39]. The effect of elongational flow, characterized by total elongational strain (ε_{tot}) and elongational rate of strain ($\dot{\varepsilon}$), should dominate σ_{ext} when ε_{tot} is high. At low ε_{tot}, the stress during drawing diminishes and σ_{ext} becomes important. The statistical model results support this. Thus, it is likely that stress imposed in both the die and the air gap plays a role in peel strength performance.

A model by Kinloch et al. [40] that computes the contributions of bending and elongating the peel arms to the peel strength measurement (see Section 10.2.3 for details) is used to analyze the effect of the differences found in the tensile property results shown in Table 15.8. The resultant changes in work required to bend and elongate the peel arm are not enough to explain the peel strength results. The WAXS data also show no differences in the orientation of the coatings. This suggests that bulk changes in tensile properties may not explain these peel strength results. Local changes in viscoelastic properties or changes in residual stress, however, may play a key role.

Relating this back to the discussion at the beginning of this section on the increase of peel strength with increasing TIAG (Fig. 15.3), increasing TIAG reduces stress by decreasing the rate of strain ($\dot{\varepsilon}$), which is found to be significant in the experimental work of Morris. Other factors that influence stress are also important for peel strength performance.

TABLE 15.6 Results from extrusion coating trial evaluating factors influencing stress on peel strength.

											Peel strength, gm/25mm			
							σ_{draw}		Web temperature, °C		Green		1-week	
Expt	Die Gap mm	Line speed m/min	Air Gap cm	L/V_f ms	Process Time ms	σ_{ext} Relative*	ε_{tot} ln(DDR)	ε' 1/s	at die	at chill roll	Mean	Std	Mean	Std
1.1	0.64	61	13	125	400	0.66	3.0	7.3	286	279	499	345	672	309
1.2	0.64	122	25	125	400	1.00	3.0	7.3	286	278	704	32	754	182
1.4	0.64	244	25	63	200	1.52	3.0	14.5	293	287	595	163	804	168
2.1	0.64	122	13	63	200	1.00	3.0	14.5	286	284	390	118	536	154
2.2	0.64	122	18	88	280	1.00	3.0	10.4	286	283	649	114	858	109
2.3	0.64	122	25	125	400	1.00	3.0	7.3	286	279	785	36	863	59
2.4	0.64	122	34	169	540	1.00	3.0	5.4	287	276	1035	104	1180	104
4.1	0.38	122	13	63	172	1.85	2.5	13.6	288	287	404	68	754	50
4.2	0.38	122	18	88	241	1.85	2.5	9.7	288	283	699	36	885	59
4.3	0.38	122	25	125	344	1.85	2.5	6.8	287		649	14	817	45
4.4	0.38	122	34	169	465	1.85	2.5	5.0	288	278	663	45	822	23
4.5	0.38	61	13	125	344	1.22	2.5	6.8	290		558	127	844	95
4.6	0.38	171	34	121	332	2.26	2.5	7.0	283	280	504	218	731	209
4.7	0.38	244	25	63	172	2.80	2.5	13.6	282		250	27	472	132
5.1	1.02	122	13	63	227	0.57	3.5	15.1	286	283	554	209	817	54
5.2	1.02	122	25	125	453	0.57	3.5	7.5	287	282	436	145	749	68
5.3	1.02	122	32	156	566	0.57	3.5	6.0	287	280	840	182	958	114
5.4	1.02	122	36	175	634	0.57	3.5	5.4	289	280	817	127	1017	177
5.5	1.02	61	13	125	453	0.38	3.5	7.5	287	285	722	241	908	95
5.6	1.02	183	34	113	408	0.73	3.5	8.4	288	281	658	241	849	132
5.7	1.02	244	25	63	227	0.86	3.5	15.1	289	282	322	73	622	82

*Extruion stress computed using Eq. 15.6 with $n = 0.6$ relative to expt 1.2 (122 m/min line speed, 0.64 mm die gap).
Source: From B.A. Morris, The influence of stress on peel strength of acid copolymers to foil, in: TAPPI Polymers, Laminations and Coatings Conference, 1998. Also B.A. Morris, J. Reinforced Plastics and Composites, 21 (2002) 1243–1255. Reproduced with permission from The Dow Chemical Company.

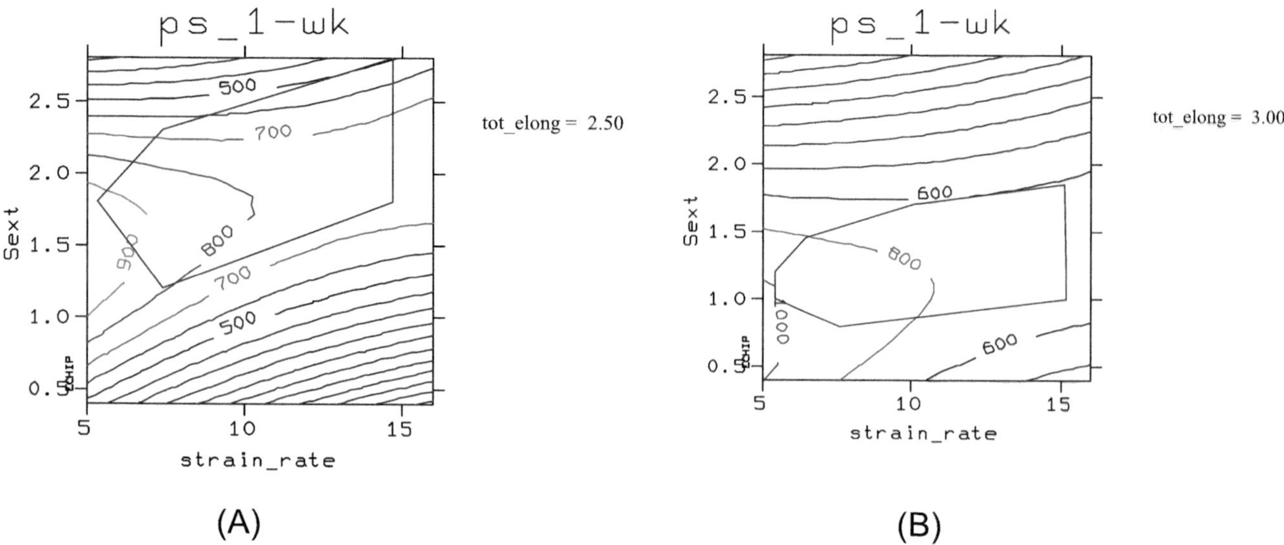

FIGURE 15.16 Contour plots of peel strength. Vertical and horizontal axes are extrusion stress (σ_{ext}) and rate of elongational strain ($\dot{\varepsilon}$), respectively. Lines represent regions of constant peel strength (measured 1-week after coating). The total elongational strain (ε_{tot}) is held constant at 2.5 (graph A) and 3.0 (graph B). *From B.A. Morris, The influence of stress on peel strength of acid copolymers to foil, in: TAPPI Polymers, Laminations and Coatings Conference, 1998. Also B.A. Morris, J. Reinforced Plastics and Composites, 21 (2002) 1243–1255. Reproduced with permission from The Dow Chemical Company.*

TABLE 15.7 Contact angle and peel strength results for seelcted experimental states.

Sample	TIAG (ms)	T_o (°C)	Peel strength - 1-week (gm/25 mm)	H_2O contact angle (degrees)
2.1	63	286	540	103.4
2.4	169	286	1180	112.4
5.1	63	286	820	108.5
5.4	175	289	1020	109.4
5.7	63	289	620	109.4

Source: From From B.A. Morris, The influence of stress on peel strength of acid copolymers to foil, in: TAPPI Polymers, Laminations and Coatings Conference, 1998. Also B.A. Morris, J. Reinforced Plastics and Composites, 21 (2002) 1243–1255. Reproduced with permission from The Dow Chemical Company.

15.1.2.3 Cooling

The melt curtain is cooled by air convection and radiant heat transfer as it transverses the air gap. The amount of cooling increases with increasing time in the air gap. Cooling decreases the amount of oxidation, increases the viscosity and slows chemical reaction at the melt-substrate interface. These all act to reduce adhesion, by decreasing wettability or chemical interaction. Too much cooling, therefore, may be detrimental to the formation of good bond strength.

The amount of cooling of the melt curtain can be insignificant unless the coating thickness is small. This can be shown by modeling and experimental measurements. A heat balance on a differential element in the melt curtain illustrated in Fig. 15.18, yields the following equation:

$$MC_pdT = -2hw(T - T_{air})dz \tag{15.12}$$

where M = mass flow rate, C_p = heat capacity, T = temperature of the web, h = heat transfer coefficient, w = width of the molten curtain, T_{air} = temperature of ambient air, and z = distance from die exit.

Solving for T as a function of z by integrating the resulting differential equation yields

$$\frac{T - T_{\text{air}}}{T_0 - T_{\text{air}}} = \exp\left(-\frac{2hz}{\rho_f X_f C_p v_f}\right) \tag{15.13}$$

TABLE 15.8 Tensile properties of 25-μm EAA coating (separated from foil by slip sheets), DDR = 33.7, T_0 = 281°C −287°C.

Sample	$\dot{\varepsilon}$ (1/s)	Relative σ_{ext}	1-week peel strength (gm/in)	Yield strength (MPa)	Tensile strength @ break (MPa)	Elongation at break (%)	Secant modulus (MPa)
5.1	15.1	0.57	820	9.8	18.0	359	140
5.2	7.5	0.57	750	9.5	22.4	436	129
5.3	6.0	0.57	960	9.4	23.4	460	126
5.5	7.5	0.38	910	9.2	20.5	433	141
5.6	8.4	0.73	850	9.2	22.9	446	128
5.7	15.1	0.86	620	8.3	20.0	392	130

Source: From From B.A. Morris, The influence of stress on peel strength of acid copolymers to foil, in: TAPPI Polymers, Laminations and Coatings Conference, 1998. Also B.A. Morris, J. Reinforced Plastics and Composites, 21 (2002) 1243−1255. Reproduced with permission from The Dow Chemical Company.

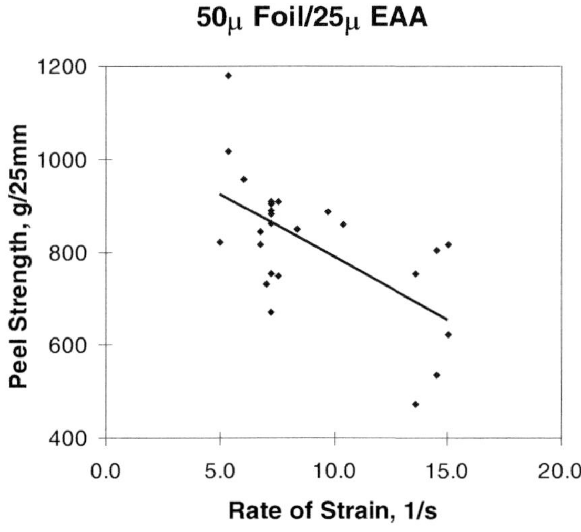

FIGURE 15.17 Effect of elongation rate of strain on peel strength to foil. *From B.A. Morris, The influence of stress on peel strength of acid copolymers to foil, in: TAPPI Polymers, Laminations and Coatings Conference, 1998. Also B.A. Morris, J. Reinforced Plastics and Composites, 21 (2002) 1243−1255. Reproduced with permission from The Dow Chemical Company.*

where v_f = line speed, X_f = coating thickness, ρ_f = density of coating (solid polymer), and T_0 = temperature of polymer at the die exit.

Eq. (15.13) simplifies the following equation when considering just the temperature at the nip, $T(z = L)$:

$$\frac{T - T_{air}}{T_0 - T_{air}} = \exp\left(- B\frac{\text{TIAG}}{X_f}\right) \text{ at } z = \text{L}, \tag{15.14}$$

where $B = \frac{2h}{\rho_f C_p}$.

Eq. (15.14) shows that the temperature at the nip decreases with increasing TIAG and decreasing coating thickness. This is illustrated in Fig. 15.19 assuming $B = 0.0073$ μm/ms, a typical value for LDPE.

The value of the heat transfer coefficient, h, is needed to use Eqs. (15.13) or (15.14). The heat transfer coefficient should have contributions from both convective and radiant heat transfer. Early investigators used correlations for h found in the general heat transfer literature that suggest the convective contribution ranges from about 5−30 W/m^2-K (1−5 Btu/hourr-ft^2-°F), and the radiant contribution is about 13 W/m^2-K (2.4 Btu/hourr-ft^2-°F) [41]. In the late 1960s, Van Ness [42] used the newly introduced IR thermometer to measure the molten web temperature in the air gap for the first time. As shown in Fig. 15.20, he found that the models considerably overestimated the temperature decrease in the air gap and hence the value of h.

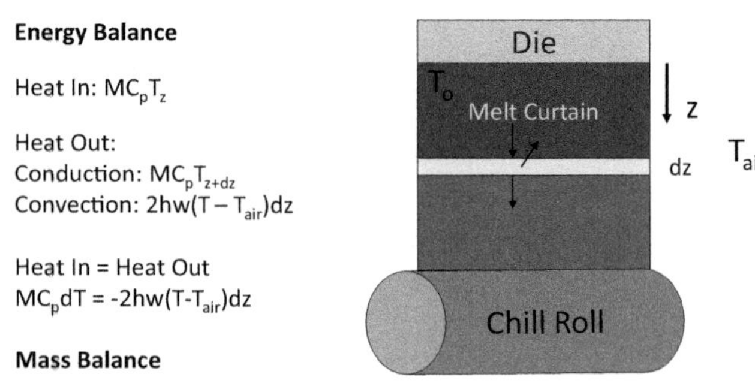

FIGURE 15.18 Heat balance on a differential element of the melt curtain in extrusion coating.

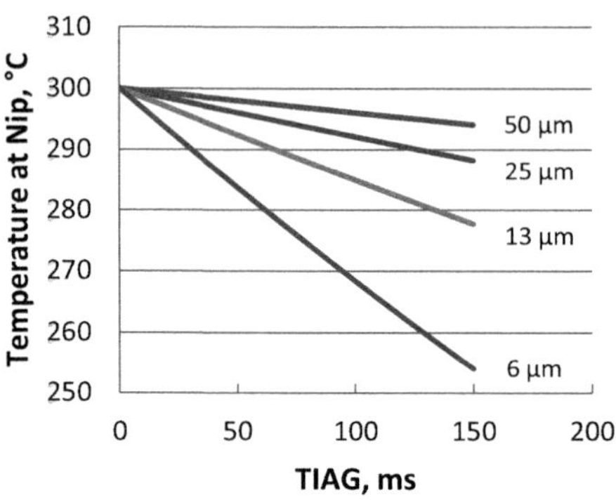

FIGURE 15.19 Temperature change in the air gap as a function of TIAG and coating thickness [from Eq. (15.14)]. Assumes the temperature at the die exit is 300°C and the surrounding air temperature is 25°C. *From Morris BA. Adhesion in extrusion coating: time in the air gap revisited. In: TAPPI PLACE conference; 2016.Reproduced with permission from The Dow Chemical Company.*

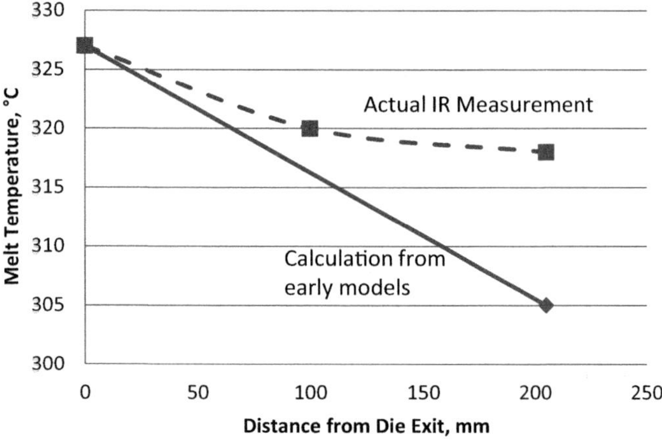

FIGURE 15.20 Comparison of temperature in the air gap from early models (solid line) with IR thermometer readings (dashed line). *From R.T. Van Ness, Infrared thermometry – a useful tool for the coating industry, TAPPI, 51(12) (1968) 39A-46A. Reproduced with permission from The Dow Chemical Company.*

Morris [8] finds that if h is set at 5.7 W/m^2-K (1 Btu/hour-ft^2-°F), the model agrees well with measured values of the temperature near the chill roll at several line speeds and coat weights, as well as with Van Ness's original measurements. This is shown in Fig. 15.21. In a follow-up study using an IR camera, Foederer and Morris [43] measure the temperature profile in the air gap on two different extrusion coating pilot lines over a wide range of processing conditions (150–450 m/minute, 6.35–25.4 μm coating thickness, 179–279 mm air gap, 8–33 draw ratio). The structure is paper/LDPE. A typical temperature scan is illustrated in Fig. 15.22. Their experimental results agree with the model using a value of h = 9.9 W/m^2-K (1.6 Btu/hour-ft^2-°F), as shown in Fig. 15.23. This agrees well with the results of Morris [8].

The low value of h is puzzling. Devisme et al. [44]. developed a model of the extrusion coating process that includes cooling in the air gap. They include both convective and radiant contributions, resulting in a value of h = 31 W/m^2-K. Their model agrees well with their experimental data for coating of PP onto aluminum foil: (PPr–tie)/Al. Their coating

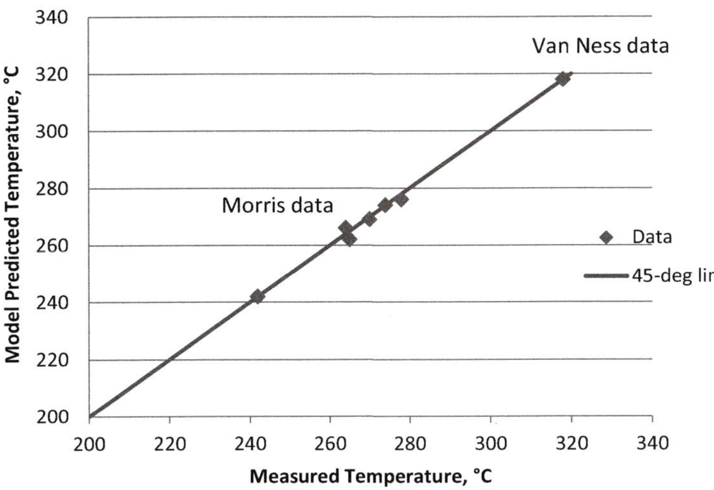

FIGURE 15.21 Comparison of experimental data from Morris [8] and Van Ness [40] with model results, assuming a convective heat transfer coefficient of 5.7 W/m^2-K (1 Btu/h-ft^2-°F). *Reproduced with permission from B.A. Morris, Understanding why adhesion in extrusion coating decreases with diminishing coating thickness, J. Plastic Film. & Sheeting, 24 (2008) 53–88.*

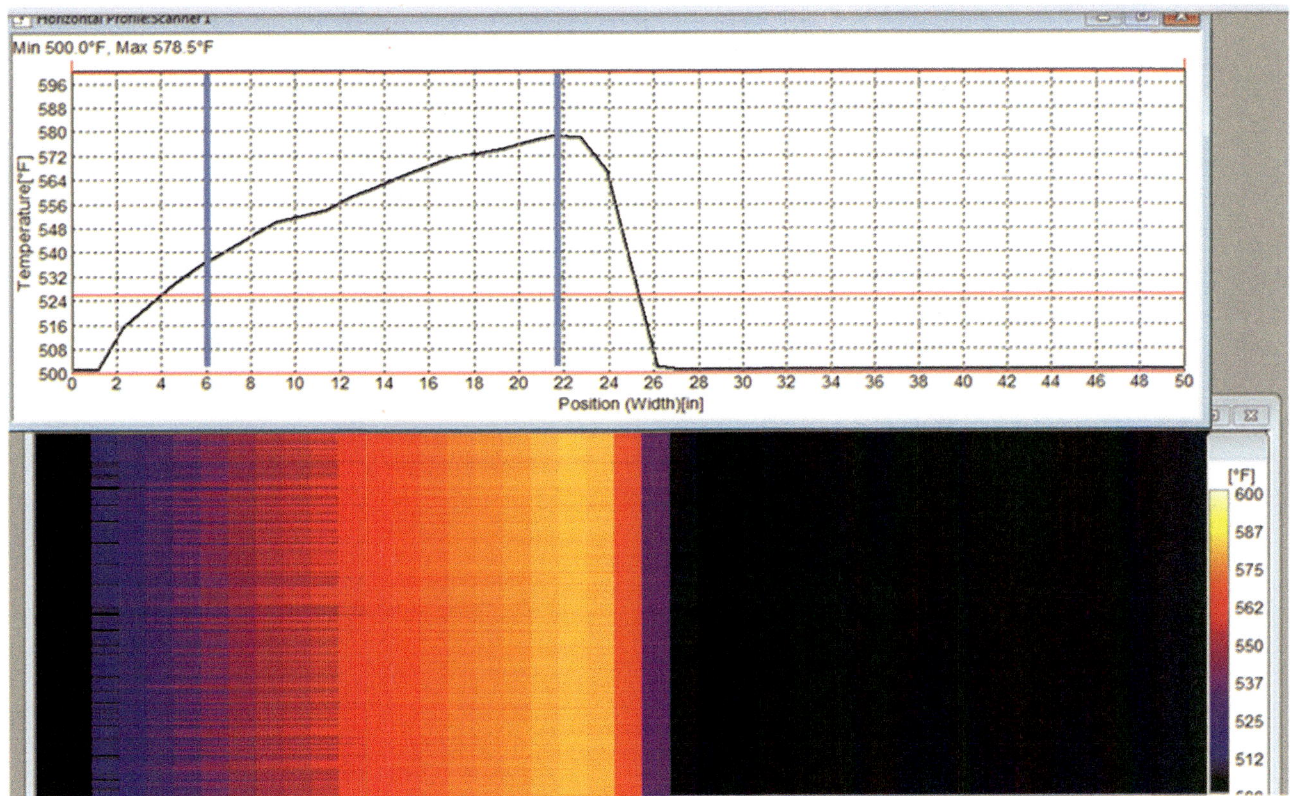

FIGURE 15.22 Example of a scan of the air gap with an IR thermometer. Conditions were 152.4 m/min line speed, 7.6 micron coating thickness and 178 mm air gap. *From B.M. Foederer, B.A. Morris, Use of IR technology to model temperature loss in the air gap, in: TAPPI PLACE. Conference, 2014. Reproduced with permission from The Dow Chemical Company.*

thickness is 80 μm, considerably thicker than most commercial coating operations and what was used in the work of Van Ness, Morris and Foederer and Morris. They also use a much slower line speed: 0.08–0.416 m/minute. The DDR is only 4. Other parameters of the experiment are: 60 mm air gap, 250 mm die width and 0.3 mm die gap. Under slow operating conditions, therefore, the heat transfer coefficient is in better alignment with conventional heat transfer theory. Perhaps at the higher speeds typical of commercial operations, other factors come into play not accounted for by the model. Nevertheless, the model predicts the correct trends in the data and can be used to make quantitative predictions provided the value for h is fit under conditions that are similar to the actual operation.

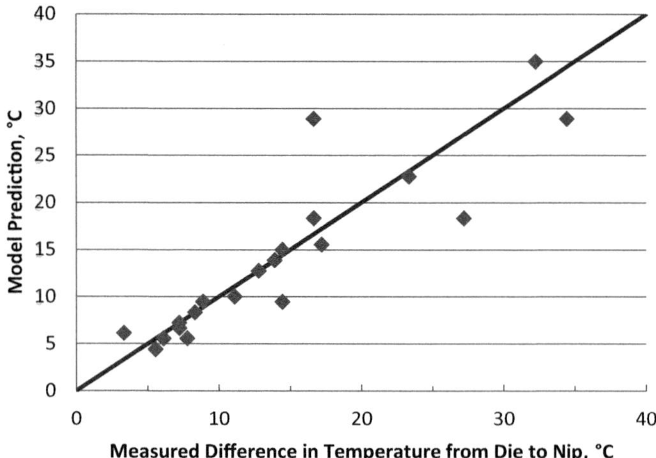

FIGURE 15.23 Comparison of model ($h = 9.9$ W/m^2-K) and experimental results from Foederer and Morris study. Data is for the difference in temperature at the die exit and at the nip (predicted [y-axis] and measured [x-axis]). The line represents an exact fit. *From B.M. Foederer, B.A. Morris, Use of IR technology to model temperature loss in the air gap, in: TAPPI PLACE. Conference, 2014. Reproduced with permission from The Dow Chemical Company.*

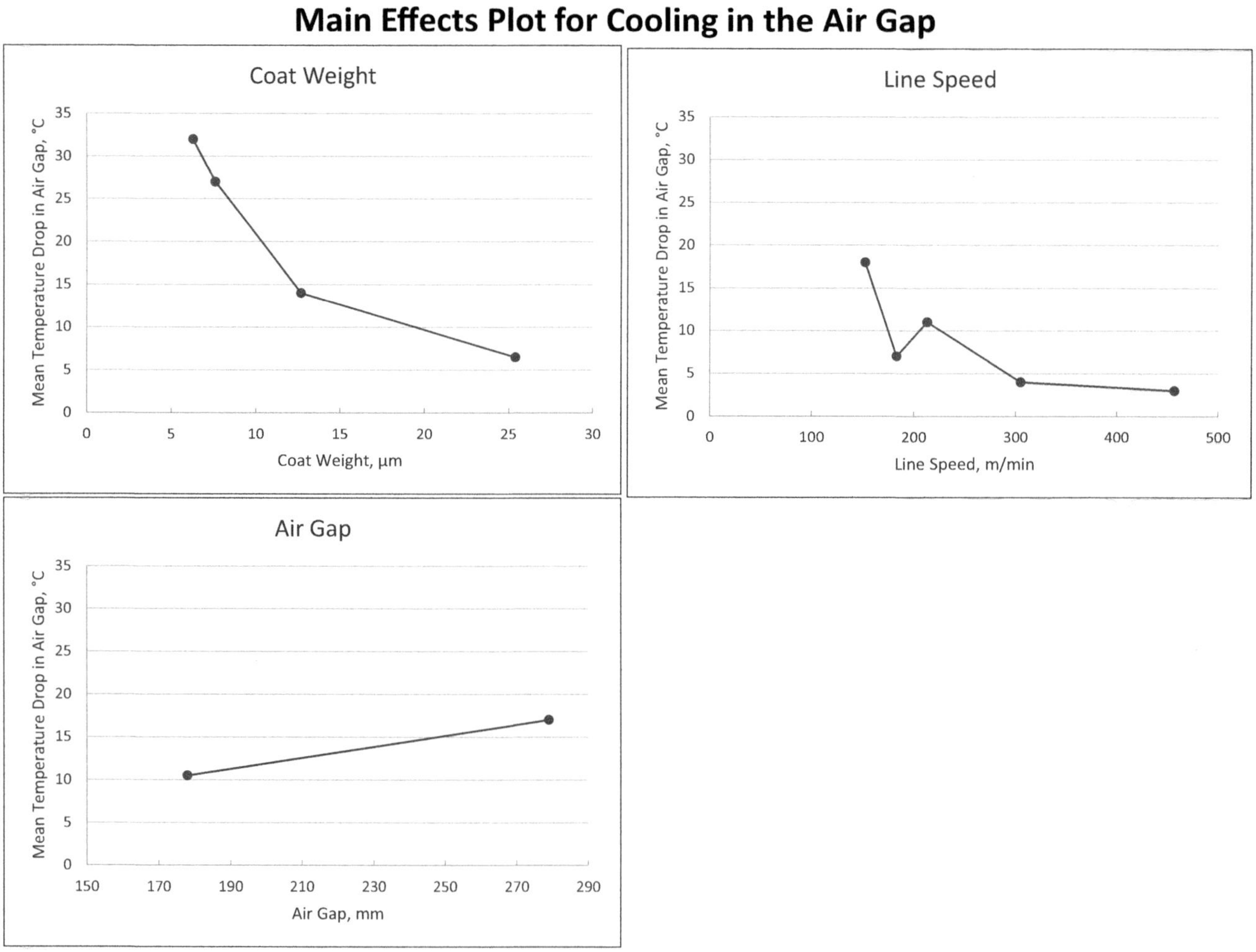

FIGURE 15.24 Main effects plot for cooling in the air gap (change in temperature between die exit and chill roll). *From B.M. Foederer, B.A. Morris, Use of IR technology to model temperature loss in the air gap, in: TAPPI PLACE. Conference, 2014. Reproduced with permission from The Dow Chemical Company.*

Foederer and Morris statistically analyze their data using an analysis of variance (ANOVA) method. The main effects plot, given in Fig. 15.24, shows that the cooling in the air gap decreases with increasing coating thickness and line speed and decreasing air gap. (In this graph, as well as in Fig. 15.23, the difference between the temperature at the die exit (T_0) and the temperature at the nip (T_L) is plotted. A greater value of this change in temperature indicates more cooling.) The air gap and line speed effects can be combined into the single parameter TIAG ($=L/v_f$). This is done in Fig. 15.25, which shows that for a given coating thickness, as the time in the air gap is increased, the amount of cooling (ΔT) increases. This is in agreement with Eq. (15.14).

Figs. 15.24 and 15.25 show that of the variables tested, the coating thickness has the greatest impact on cooling in the air gap. This is an indication of the importance of the mass flow rate, and is illustrated in Fig. 15.26 by plotting the temperature drop in the air gap versus the line speed for a 25-μm coating. Since the die gap and coat weight are kept constant, as the line speed is increased the mass flow rate increases (by increasing the extruder rpm). As the mass flow rate is increased, the amount of cooling decreases. This also agrees with the model.

These results show that at typical coat weights found in extrusion coating and lamination, the temperature does not change substantially in the air gap. This is particularly true for high output lines since the temperature drop in the air gap is related to the mass throughput. At very low coat weights, the temperature drop begins to become significant and may be a factor in adhesion. The simple model developed here can be used to predict under what process conditions this may be a concern.

Both the model and the IR measurements give an average temperature across the thickness of the melt curtain at a given distance from the die exit (although IR measurement is more influenced by the temperature near the surface). This provides information regarding the flowability of the polymer as it contacts the substrate in the nip, but the surface temperature is also important if chemical reactions are taking place. An analysis shows that temperature is essentially

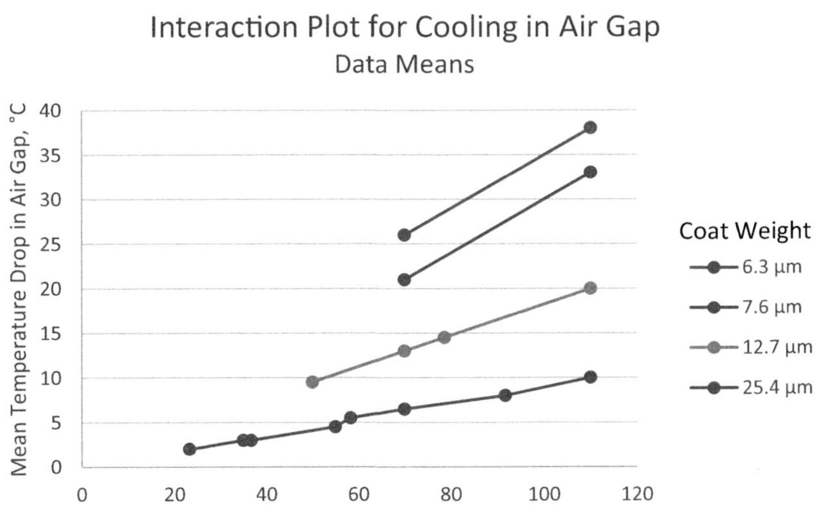

FIGURE 15.25 Interaction plot of mean temperature difference (between die exit and at nip) in the air gap versus time in the air gap (TIAG). *From B.M. Foederer, B.A. Morris, Use of IR technology to model temperature loss in the air gap, in: TAPPI PLACE. Conference, 2014. Reproduced with permission from The Dow Chemical Company.*

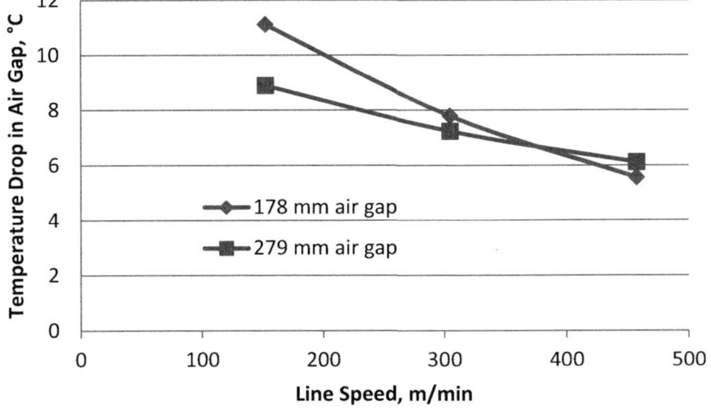

FIGURE 15.26 Effect of line speed on temperature change in the air gap. Data for 25-μm coating. Since coat weight is constant, line speed is directly proportional to extruder output. *From B.M. Foederer, B.A. Morris, Use of IR technology to model temperature loss in the air gap, in: TAPPI PLACE. Conference, 2014. Reproduced with permission from The Dow Chemical Company.*

uniform across the thickness of the curtain. The Biot (Bi) number is a dimensionless number that is the ratio of convection of heat from the melt curtain to the conduction of heat within the curtain:

$$\mathrm{Bi} = h\ell/k \tag{15.15}$$

where Bi = Biot number, h = heat transfer coefficient (9.9 W/ m^2-K), ℓ = characteristic length scale (the thickness of the melt curtain), and k = thermal conductivity (0.26 W/m-K for LDPE).

When Bi << 1, the thermal conduction within the film is faster than the thermal convection at the surface and the temperature equilibrates across the thickness. For a web thickness of 0.75 mm (30 mils) typical of the die gap, Bi = 0.03. Therefore, even at the die exit, the temperature is constant across the thickness of the web.

A uniform temperature profile across the width of the melt curtain is important for ensuring uniform adhesion across the breadth of the coated substrate. Airflow in the air gap can be complex, resulting in nonuniform convective cooling of the melt curtain. The air near the center of the curtain is entrained with the moving curtain and is forced down towards the chill roll. Once it gets there it has to turn, flowing upwards, typically near the edges. Also, there may be ventilation ducts near the die that affect the flow patterns. Both neck-in and the practice of thinning out the edges at the die exit to counter neck-in can result in the temperature at the edges of the curtain being different from the centerline temperature. The use of an IR thermometer or camera can be helpful for characterizing the temperature across the width of the melt curtain [43]. Countermeasures such as heating the edges of the die may help achieve more uniform temperature profiles.

Airflow patterns near the nip may push the curtain onto the chill roll prior to the substrate. Adjusting the position where the melt curtain is pinned to the substrate (known as the horizontal lay-on) so that the melt contacts the substrate first is favorable for adhesion. This allows the polymer more time to wet the substrate before being quenched.

15.1.2.4 General guidelines

The conditions in the air gap have an impact on adhesion performance. In general, adhesion can be enhanced by the following factors:

- Increasing TIAG by increasing the air gap or reducing the line speed. This increases the amount of oxidation (critical for LDPE adhesion) and reduces stress. Antonov and Soutar [4] suggest running with a TIAG between 50 and 100 ms. Table 15.9 provides values of the air gap at different line speeds that fit within this range. Of the two factors that make up TIAG, the air gap is the better parameter to adjust from an operational point of view. Too long of an air gap, however, can lead to issues with excessive neck-in or cooling in the gap.
- Increasing the melt temperature of the polymer at the die exit. This increases oxidation, reduces stress, improves wetting and increases the energy available for chemical interaction. If the temperature is too high, however, thermo-oxidative degradation may cause gel and black specs (crosslinking), embrittlement (chain scission) and taste and odor issues.
- Treating nonpolar polymers such as LDPE with ozone (to adhere to polar substrates like aluminum foil, metallized films, primed films, and paper).
- Adjusting the horizontal lay-on so that the polymer contacts the substrate prior to the chill roll.
- Reducing the die gap. This reduces the DDR and the stress generated in the air gap. Too narrow of a die gap, however, can lead to excessive stress in the die land.
- Stabilizing the airflow pattern or increasing the temperature of the die near the edges to ensure the melt curtain temperature is uniform across its width.

TABLE 15.9 Air gap ranges to obtain an optimum TIAG between 50 and 100 ms.

Line speed		TIAG = 100 ms		TIAG = 50 ms	
m/min	ft/min	mm	inch	mm	inch
100	328	167	6.6	84	3.3
200	656	333	13.1	167	7.5
300	984	500	19.7	250	9.8
400	1312	667	26.2	333	13.1
500	1640	833	32.8	416	16.4

15.1.3 Nip

The nip is the region between the chill roll and rubber nip roll. The speed of these rolls determines the line speed, which in turn influences the time and amount of drawdown in the air gap described in the previous section. The polymer melt contacts the substrate just prior to the nip and is rapidly solidified by the cold roll. The brief time in the nip is critical for adhesion; contact, flow, wetting, and chemical interaction must all take place as the polymer freezes off.

Wetting is covered in Chapter 10. In general, the substrate needs to have higher surface energy than the molten polymer and be free of contamination. Often corona, ozone, plasma or flame treatment is applied to the substrate prior to the nip to help clean and prepare the surface for wetting. Chemical primers can be applied to the substrate to promote adhesion. These must be properly dried or cured prior to the nip.

The flow in the nip region is very complex. The elongational flow in the air gap as the melt curtain is drawn down in thickness is followed by shear flow from the squeezing action of the nip roll. For porous substrates such as paper, it may be favorable for the polymer to flow into the pores to provide mechanical interlocking. Even for nonporous substrates, the surface roughness may require the polymer to flow and wet small-scale crevices. This occurs as the polymer is rapidly changing temperature. The stress on the polymer has little time to relax in the short time before the polymer freezes. The stress can induce orientation and crystallinity, further complicating the rheology and wetting in the nip region.

Devisme et al. [44] develop a model of the temperature in the nip region, incorporating assumptions around barriers to heat transfer between the polymer and chill roll and strain-induced crystallization. Their approach uses several extrapolations and assumptions, especially around strain-induced crystallization kinetics, but provides good agreement with their experimental results. As described earlier, their experiments are run under slower conditions than those typically used in commercial extrusion coating operations. Nevertheless, the model provides insight into the effect high cooling rates and stretching has on crystallization in extrusion coating. It also suggests that the heat transfer coefficient between the polymer and chill roll may need to be adjusted for surface roughness.

Trouilhet and Morris [45] develop a model for calculating the temperature as a function of time in the nip. While it does not incorporate many of the complexities included by Devisme et al., it provides useful insight into a number of adhesion and processing issues. The model calculates the temperature profile in a multilayer film as a function of time. Because the dimensions along the surface of the film (machine and transverse directions) are large compared to the thickness, a one-dimensional model is sufficient. For simplicity, the one-dimensional unsteady-state heat conduction equation with constant coefficients is used:

$$\frac{\partial T}{\partial t} = \alpha \frac{\partial^2 T}{\partial x^2} \tag{15.16}$$

where T = temperature, t = time, x = distance in the thickness dimension, α = thermal diffusivity = $\rho k/C_p$, ρ = density, k = thermal conductivity, and C_p = heat capacity.

The thermal diffusivity is assumed to be constant, although the model allows for variation as a function of temperature. Such data is often hard to find and so the variation is ignored as a first approximation. Curvature is also neglected since the coated layers and substrate(s) are thin compared to the diameter of the rolls.

Initial and boundary conditions are needed to solve Eq. (15.16).

Initial Condition

At $t = 0$, the melt curtain is assumed to be at T_0 and the substrate at ambient temperature.

Boundary Conditions

1. At the external surfaces, x = 0 and X (X is the thickness of the total multilayer structure).
   Three types of boundary conditions are allowed for in the model:
   - Type 1: Temperature is specified at the surface.

$$\text{At } x = 0 \text{ or } X,\ T = T_1. \tag{15.17}$$

   - Type 2: The surface is insulated from heat loss or gain. In mathematical terms, this means the change in temperature is zero.

$$\text{At } x = 0 \text{ or } X,\ \frac{dT}{dx} = 0. \tag{15.18}$$

- Type 3: The surface of the film is being heated or cooled by convection with an outside medium such as air or water. Here it is convenient to introduce a heat-transfer coefficient, h. The boundary conditions can be written as follows:

$$\text{At}x = 0, \; -k\frac{dT}{dx} = h(T - T_{\text{inf}}) \tag{15.19}$$

$$\text{At } x = X, \quad k\frac{dT}{dx} = h\left(T - T_{inf}\right). \tag{15.20}$$

Here T_{inf} is the temperature of the cooling or heating medium far away from the surface.

Note that the type 3 boundary condition is the most general of the three. When $h = 0$, the second boundary condition is obtained. Letting h approach infinity yields the first boundary condition. Typical values for h are given in many engineering texts. Radiant heat transfer may also be modeled by using an effective heat-transfer coefficient.

2. For multiple layers, the heat flux at each interface is equal. For example, at the interface between layers 1 and 2:

$$k_1\frac{dT}{dx}\bigg|_1 = k_2\frac{dT}{dx}\bigg|_2. \tag{15.21}$$

A finite difference scheme is used to solve the above equations. This method is chosen over analytical solutions (such as those found in Carslaw and Jaeger [46]) because of the geometry imposed by multiple layers. Also, the finite difference method lends itself well to more complex situations, such as geometries beyond films with parallel sides, incorporating the temperature dependence of the thermal parameters and considering the effect of phase changes. There are two general methods for solving finite-difference problems, explicit and implicit. The particular scheme used here is an implicit algorithm. Implicit algorithms are generally more difficult to program than explicit algorithms, but they do not have the convergence problems of explicit schemes. For the most part, the methodology given by Carnahan, Luther and Wilkes [47] is followed.

As polymers transition from the melt to the solid-state, they release energy, known as the enthalpy of crystallization. This can affect the change of temperature within the coating or film. Such effects are generally small for polymers because most polymers are not very crystalline, especially when they are rapidly quenched. Thus, this model will capture most of the features of temperature change, even if the materials are undergoing phase transformations. In some cases, however, a more general model incorporating enthalpy effects is used. The methodology of Shamsundar and Rooz [48] is followed by using a nonlinear Gauss-Seidel iteration scheme to solve the enthalpy equations.

Several examples where the model provides useful insight are highlighted below.

15.1.3.1 Time in nip and solidification

One of the early adoptions of extrusion coating was for coating LDPE onto paper and paperboard. A typical structure is 50-μm paper/25-μm LDPE. Fig. 15.27 shows the results of modeling the temperature development across the thickness of this structure as a function of time in the nip. The temperature the melt curtain is assumed to be initially uniform at 330°C. The chill roll temperature is 20°C. At time $t = 0$, the melt first comes in contact with the paper and the chill-roll and heat exchange by conduction begins. For boundary conditions, the chill-roll side of the LDPE is assumed to be 20°C (type 1 boundary condition). A type 3 boundary condition could be used here to account for the imperfect contact between the chill roll and polymer, but the type 1 boundary condition is found to be sufficient. For the paper surface against the rubber nip-roll, a type 2 boundary condition (insulated) is used. This is illustrated in Fig. 15.28.

An LDPE with a density of 918 kg/m^3 crystallizes (solidifies) at 91°C (and melts at 106°C) under quiescent conditions. While the crystallization kinetics will differ due to the stress imposed by the process and rapid cooling rate, this crystallization temperature can be used for general discussion. Fig. 15.27 shows that the entire LDPE layer reaches 91°C and solidifies in about 3 ms. This solidification time, t_s, is a useful parameter for characterizing the time available for wetting and adhesion but must be put into perspective by comparing it with the residence time in the nip, t_n.

The time in the nip is given by

$$t_n = \frac{d}{v_f} \tag{15.22}$$

where $t_n =$ time in the nip, $d =$ deformation of the rubber nip roll, and $v_f =$ line speed.

The deformation is a function of the force applied to the nip-roll, the hardness and thickness of the rubber, and the diameter of the rubber roll and chill roll.

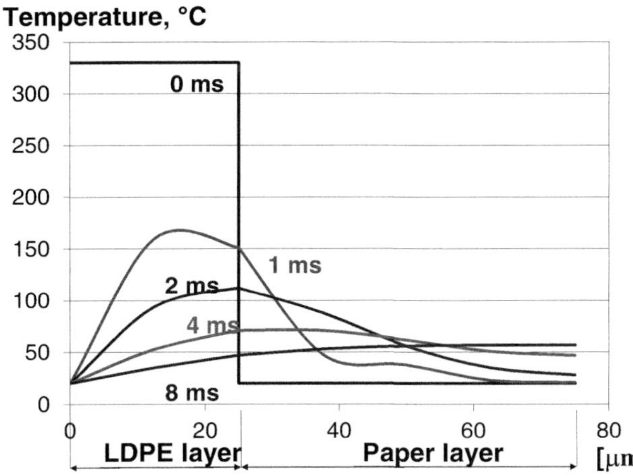

FIGURE 15.27 Model predictions of temperature profiles in the nip at different snapshots in time. Structure is 25-μm LDPE initially at 330°C coated onto 50-μm paper initially at 20°C. X-axis is the distance in the thickness dimension from the chill roll. *From Y. Trouilhet, Y. Morris, Prediction of temperature profiles across coating and substrate in the nip, in: TAPPI Polymers, Laminations Coat. Conference, 1999. Reproduced with permission from The Dow Chemical Company.*

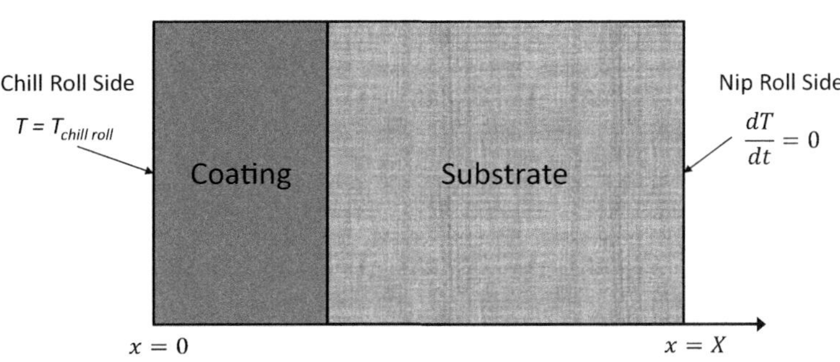

FIGURE 15.28 Boundary conditions for modeling the nip region of extrusion coating. *From Y. Trouilhet, Y. Morris, Prediction of temperature profiles across coating and substrate in the nip, in: TAPPI Polymers, Laminations Coat. Conference, 1999. Reproduced with permission from The Dow Chemical Company.*

For example, if the deformation is 2 cm and the line speed is 240 m/minute, then the time in the nip is 5 ms. Under these typical processing conditions, the LDPE layer solidifies completely in the nip since the time to solidify (3 ms) is smaller than the time in the nip (5 ms).

The melt solidifies under high pressure in the nip. The pressure in the nip rises to a maximum in the middle of the nip as shown in Fig. 15.29. The maximum pressure in the nip is given by

$$P_m = 2\frac{f}{d} \tag{15.23}$$

where $P_m =$ maximum pressure in the nip and $f =$ force applied to the nip roll per width of substrate.

For typical values of $d = 2$ cm and $f = 30{,}000$ N/m, the maximum pressure in the nip reaches 3 MPa.

The line tension also exerts pressure on the coated layer around the chill-roll, but this is typically negligible compared with the nip pressure. This can be seen from a force balance in the vertical direction:

$$p\mathrm{Dw} = 2\mathrm{T} \tag{15.24}$$

where p [N/m^2] = pressure on the coating, D [m] = diameter of chill-roll, w [m] = width of substrate, and T [N] = tension on the substrate.

For a typical line with 2000 N of tension, 0.7 m chill-roll diameter and 1.5 m width of substrate, the pressure applied on the coating once it leaves the nip is only 0.004 MPa, which is about 1000 times less than the maximum pressure in the nip.

Trouilhet and Morris propose the following condition for good adhesion:

$$t_n/2 < t_s < t_n. \tag{15.25}$$

When this condition is met the melt solidifies under high pressure. When $t_s < t_n/2$, the melt does not experience the full pressure: wetting and flow are not optimal. For solidification times longer than the residence time in the nip, the pressure is released too early; bubble formation, chill roll sticking and other problems may arise.

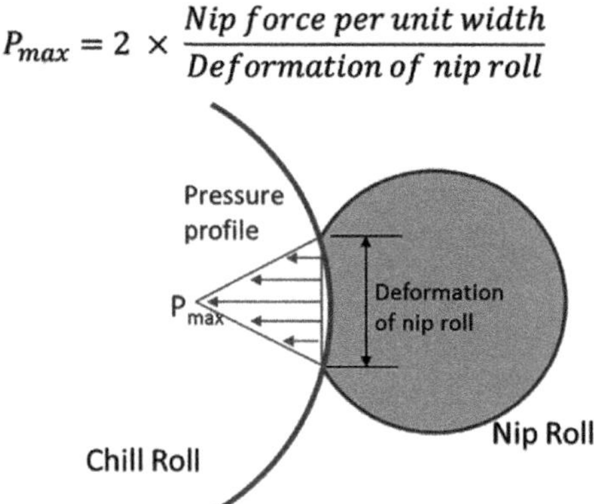

FIGURE 15.29 Pressure distribution in nip region. *From Y. Trouilhet, Y. Morris, Prediction of temperature profiles across coating and substrate in the nip, in: TAPPI Polymers, Laminations Coat. Conference, 1999. Reproduced with permission from The Dow Chemical Company.*

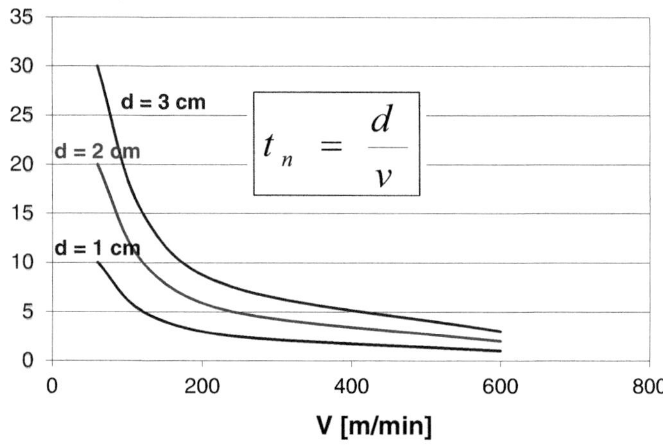

FIGURE 15.30 Time in the nip (t_n) as a function of line speed and nip roll deformation. *From Y. Trouilhet, Y. Morris, Prediction of temperature profiles across coating and substrate in the nip, in: TAPPI Polymers, Laminations Coat. Conference, 1999. Reproduced with permission from The Dow Chemical Company.*

Various process parameters affect time in the nip and solidification time. Fig. 15.30 shows how t_n varies with line speed and deformation of the rubber nip roll [as described by Eq. (15.22)]. Time in the nip varies from 1 to 30 ms when the line speed and deformation are varied from 600 m/minute and 1 cm to 60 m/minute and 3 cm.

The solidification time (t_s) is calculated using the temperature in the nip model. Fig. 15.31 shows the calculated values as a function of LDPE thickness and polymer melt temperature (at first contact with the substrate). Typical melt temperatures range from a maximum of about 240°C for EVAs to about 330°C for PEs that need to be oxidized. At 25-μm coating thickness, solidification time varies from 1.5 to 3 ms. The solidification time at a given initial melt temperature increases approximately with the square of the coating thickness, ranging from 0.75 ms at 12 μm to 12 ms at 50 μm.

Solidification time is also shorter if the polymer solidifies at a higher temperature (higher freezing point) than LDPE. This is the case for polypropylene (PP), polyamide (PA) and polyethylene terephthlate (PET) and is one of the reasons these polymers are more difficult to extrusion coat than LDPE.

For short solidification times, using a harder rubber surface on the nip roll may reduce time in the nip (by reducing the deformation) and improve adhesion to paper by ensuring the maximum nip pressure is exerted prior to solidification [as expressed by Eq. (15.25)].

15.1.3.2 Other examples

Trouilhet and Morris provide other examples of how a simple temperature model for the nip region can provide insight into extrusion coating problems. These examples were collected from actual problems in the industry.

- A two-layer coextrusion coating onto a primed oriented polyester (OPET) substrate was exhibiting poor adhesion. The two layers are both LDPE, with the layer adjacent to the OPET extruded at 310°C to oxidize and bond to the primed surface of the OPET and the outside layer run at 280°C for improved organoleptics. For a line speed of 120 m/minute and a nip deformation of 2 cm, time in the nip is calculated to be 10 ms. The temperature profiles

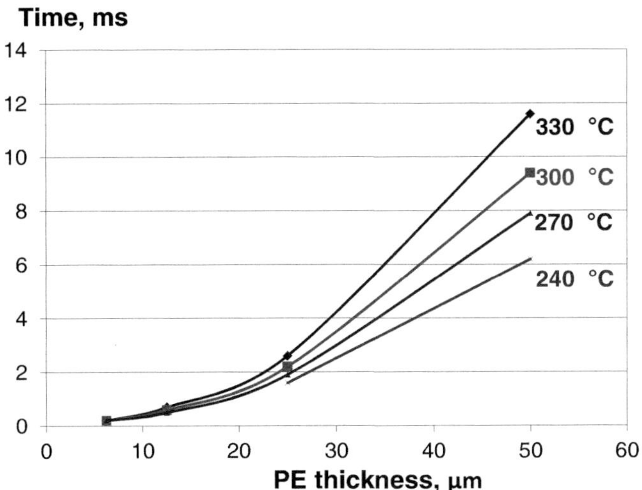

FIGURE 15.31 Solidification time (t_s) of a LDPE coating onto 50-µm paper at various melt temperatures. The solidification temperature is assumed to be 91°C. *From Y. Trouilhet, Y. Morris, Prediction of temperature profiles across coating and substrate in the nip, in: TAPPI Polymers, Laminations Coat. Conference, 1999. Reproduced with permission from The Dow Chemical Company.*

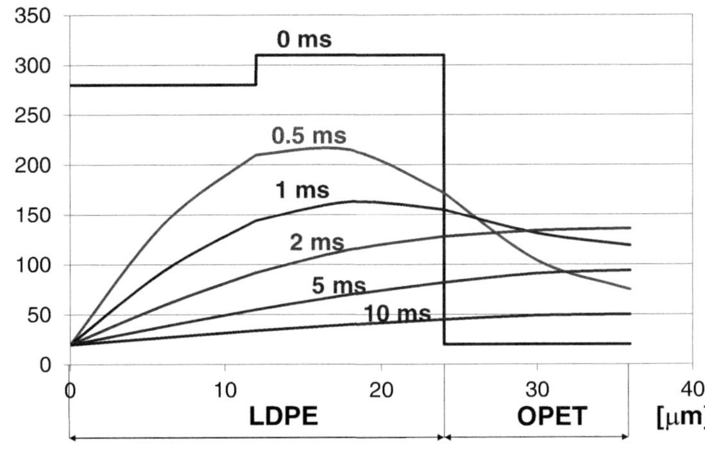

FIGURE 15.32 Model predictions of temperature profiles in the nip for coextrusion coating of LDPE onto OPET substrate. *From Y. Trouilhet, Y. Morris, Prediction of temperature profiles across coating and substrate in the nip, in: TAPPI Polymers, Laminations Coat. Conference, 1999. Reproduced with permission from The Dow Chemical Company.*

predicted by the model for 1, 2, 5 (middle of nip where the nip pressure is at its maximum) and 10 ms (end of nip) are given in Fig. 15.32. Part of the heat from the melt is transferred to the chill roll (set at 20°C) but part of the heat transfers to the OPET film shortly after contact.

The temperature of the two surfaces of the OPET film, one in contact with the LDPE and the other in contact with the rubber nip roll, is plotted as a function of time in Fig. 15.33. The temperature of the LDPE/OPET interface is at first higher than that at the OPET/rubber interface; heat flows from the hot LDPE into the OPET film which increases in temperature. The OPET film never melts, even at the PE/OPET interface as it is first contacted with the hot polymer. The OPET film reaches temperatures up to 136°C, however, which causes shrinkage of the film. It is hypothesized based on the model insight that the shrinkage was causing stresses at the interface and poor adhesion. The film was replaced with an OPET with higher dimensional stability at these temperatures and the adhesion improved.

- A barrier multilayer film with the structure (PE-tie-EVOH-tie-PE) [20/5/10/5/20 µm] is extrusion laminated with LDPE melt to an LDPE film (110 µm) in contact with the chill roll: the final structure is Barrier-Film/LDPE/PE-film. The laminate exhibits curl when the extrusion layer is thick (50 µm) but not when it was thin (25 µm). Curling in a multilayer structure occurs when the layers undergo different volume changes, such as when at least one layer changes density while the others do not. A layer of a film can increase in density if it undergoes postquench crystallization (the formation of secondary crystals after quenching). Postquench crystallization may occur if the polymer experiences temperatures above its glass-transition temperature (Tg). The Tg of an EVOH with 32 mole% ethylene content is 69°C. The model prediction for the average temperature the EVOH experiences during the lamination

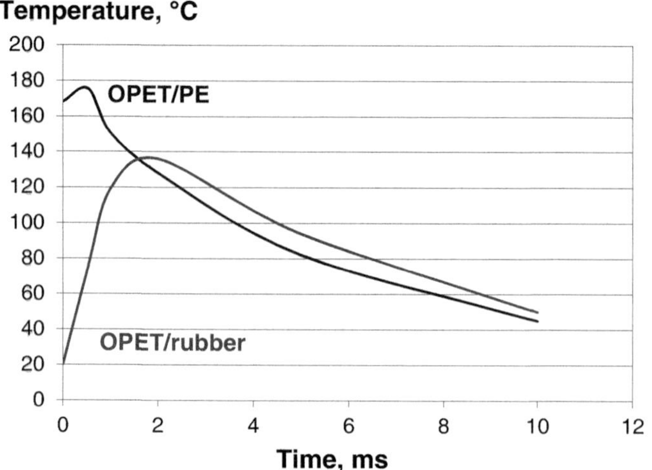

FIGURE 15.33 Model prediction of temperature of OPET film at the LDPE and rubber interfaces as a function of time in the nip. Contact of the molten LDPE with the OPET and chill roll occurs at t = 0. *From Y. Trouilhet, Y. Morris, Prediction of temperature profiles across coating and substrate in the nip, in: TAPPI Polymers, Laminations Coat. Conference, 1999. Reproduced with permission from The Dow Chemical Company.*

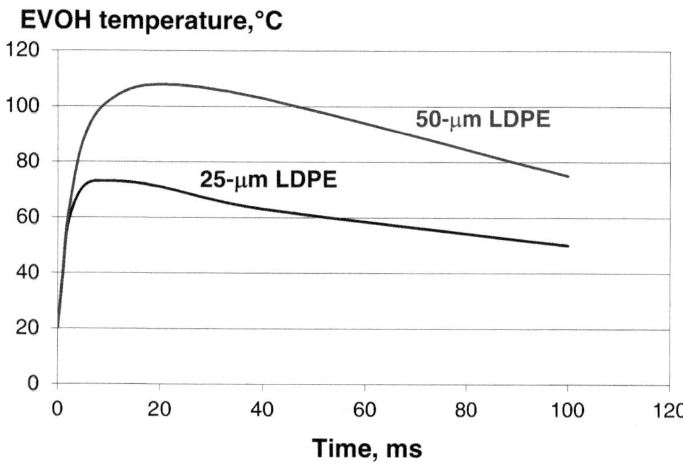

FIGURE 15.34 Model prediction of the temperature of the EVOH layer during lamination. Structure is 110 μm LDPE-film/LDPE/(20μm-LDPE-5μm-tie-10μm-EVOH-5μm-tie-20μm-LDPE). *From Y. Trouilhet, Y. Morris, Prediction of temperature profiles across coating and substrate in the nip, in: TAPPI Polymers, Laminations Coat. Conference, 1999. Reproduced with permission from The Dow Chemical Company.*

process is plotted as a function of time for the two thicknesses of LDPE in Fig. 15.34. The thinner extrusion layer keeps the EVOH temperature below Tg whereas the thick extrusion layer transfers enough heat to the EVOH layer to bring it to temperatures well above Tg for prolonged times. This explains the difference in curl behavior of the two laminates.

- Bubbles may form and get trapped in the laminating layer during extrusion lamination. An example is shown in Fig. 15.35. Among other things, bubbles affect the transparency and appearance of the laminate. Bubbles can arise from air or moisture entrapped in the resin that is allowed to expand as the pressure is released at the nip exit. This can occur if the polymer is still molten at the end of the nip ($t_s > t_n$). An example is a meat packaging film with the structure: PA6-film/LDPE/Ionomer-film. Two versions of different thickness are made by the converter:
  - thin laminate: 25/25/25 (total 75 μm) and
  - thick laminate: 100/100/100 (total 300 μm).

  Bubbles form in the thick laminate but not in the thinner laminate. The thick laminate is produced at 60 m/minute line speed and the thin laminate at 240 m/minute. For a 2-cm nip deformation, t_n is 5 and 20 ms for the thin and thick laminates, respectively.

  Temperature profiles in the laminate leaving the nip are given in Figs. 15.36 and 15.37 for the thinner and thicker laminates, respectively. Despite the longer residence time in the nip (20 ms instead of 5 ms), the PE temperature of the thicker laminate at the exit of the nip is higher. This is due to the thicker ionomer film between the melt and the chill-roll acting to insulate the cooling effect of the roll, as well as the higher mass of hot LDPE. The model shows that the LDPE is still molten at the end of the nip for the thicker laminate but not for the thinner one, consistent with the bubble behavior.

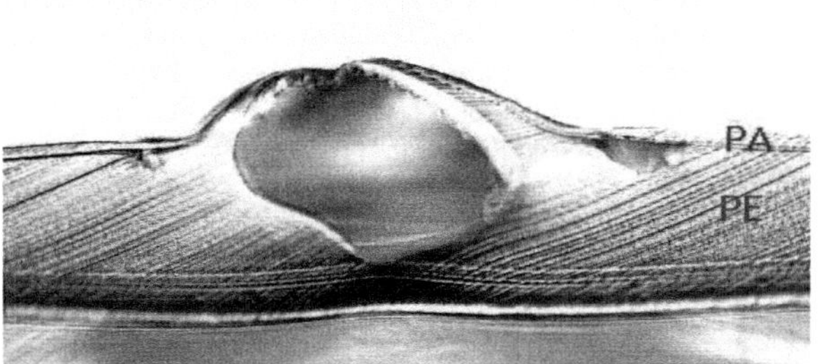

FIGURE 15.35 Photomicrograph of bubble in extrusion lamination layer. *From Y. Trouilhet, Y. Morris, Prediction of temperature profiles across coating and substrate in the nip, in: TAPPI Polymers, Laminations Coat. Conference, 1999. Reproduced with permission from The Dow Chemical Company.*

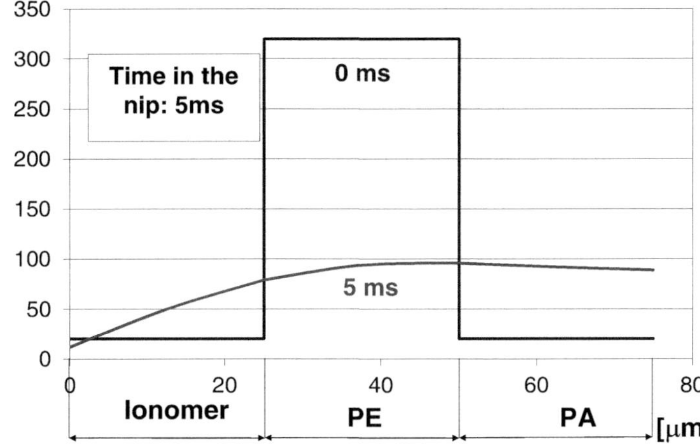

FIGURE 15.36 Model prediction of the temperature profile at the beginning and end of the nip region for the thin laminate structure: 25μm-PA/25μm-LDPE/25μm-ION. *From Y. Trouilhet, Y. Morris, Prediction of temperature profiles across coating and substrate in the nip, in: TAPPI Polymers, Laminations Coat. Conference, 1999. Reproduced with permission from The Dow Chemical Company.*

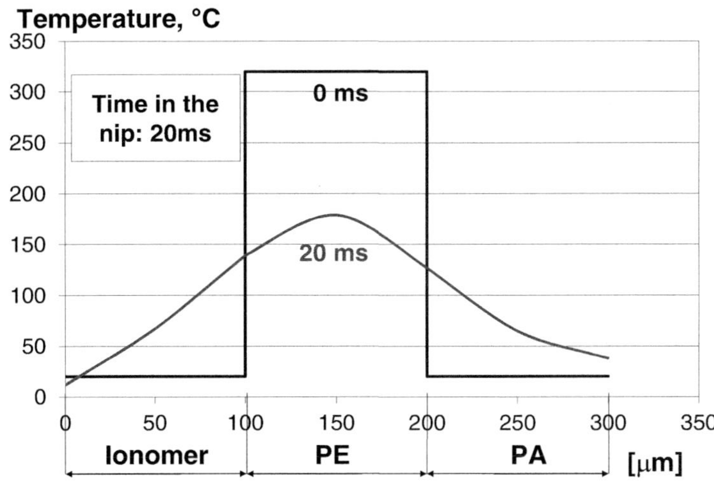

FIGURE 15.37 Model prediction of the temperature profile at the beginning and end of the nip region for the thick laminate structure: 100μm-PA/100μm-LDPE/100μm-ION. *From Y. Trouilhet, Y. Morris, Prediction of temperature profiles across coating and substrate in the nip, in: TAPPI Polymers, Laminations Coat. Conference, 1999. Reproduced with permission from The Dow Chemical Company.*

15.1.4 Case study: effect of coating thickness on peel strength

As the coating thickness is reduced, the adhesive strength generally decreases. This was known from the early days of extrusion coating: Goslin and Sweeney's [49] and Guillotte and McLaughlin's [41] extensive work on the coating of paper with LDPE in the early 1960s is highlighted in Figs. 15.38 and 15.39. Here the adhesion is reported as %

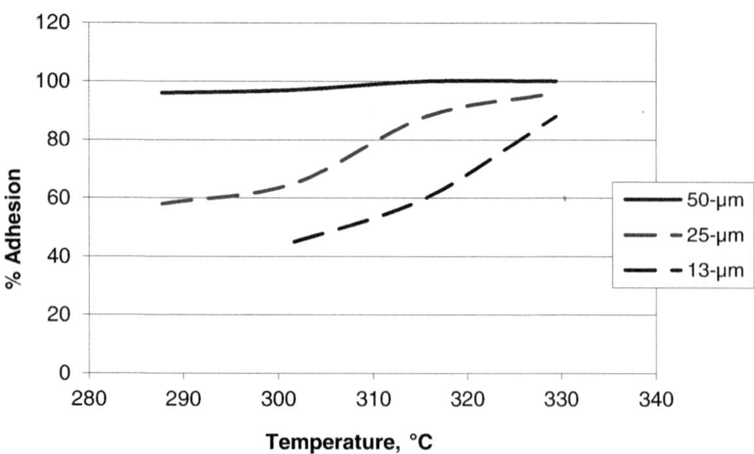

FIGURE 15.38 Adhesion of LDPE to kraft paper. Conditions: 68 kg/h, line speeds of 33.5, 65.5, and 134 m/min for 50, 25, and 13 μm coatings, respectively. This corresponds to TIAG's of 125, 64, and 31 ms. *Data from J.P. Goslin, H.F. Sweeney, The interaction of processing variables, base materials, and resins in polyethylene extrusion coating, TAPPI, 43(5) (1960) 434–447. Reproduced with permission from B.A. Morris, Understanding why adhesion in extrusion coating decreases with diminishing coating thickness, J. Plastic Film. & Sheeting, 24 (2008) 53–88.*

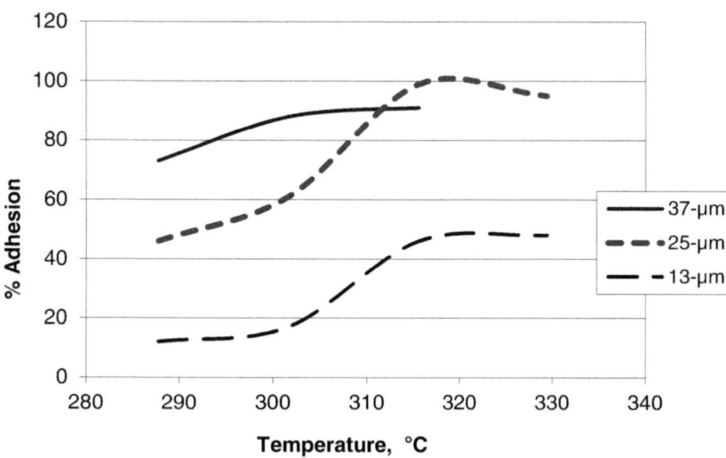

FIGURE 15.39 Adhesion of LDPE to kraft paper. 114-mm air gap, 91.4 m/min line speed, 75 ms TIAG. *Data from J.E. Guillotte, T.F. McLaughlin, Jr., Calculating the adhesion of polyethylene coatings to kraft paper, TAPPI, 45(3) (1962) 200–208. Reproduced with permission from B.A. Morris, Understanding why adhesion in extrusion coating decreases with diminishing coating thickness, J. Plastic Film. & Sheeting, 24 (2008) 53–88.*

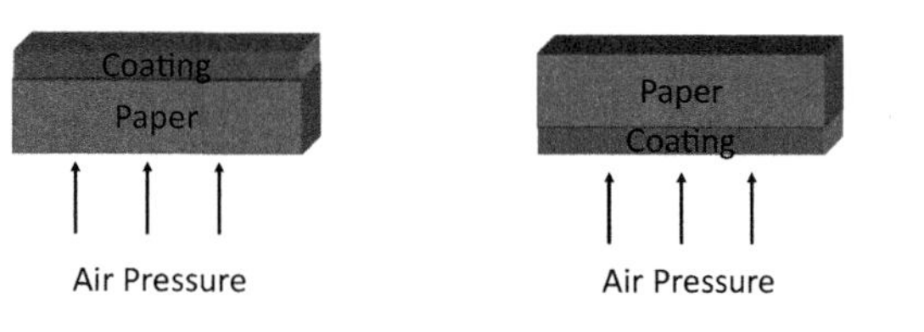

FIGURE 15.40 Illustration of the Perkins Southwick test method for measuring adhesion of LDPE to paper. Two measurements are made. In the first, air pressure is used to measure the pressure to separate the coating from the paper (coating side away from the air source). In the second, the burst pressure of the coating is measured (coating side facing the air source). The ratio is reported as % adhesion.

adhesion (adhesive force divided by the burst strength of the coating, as is illustrated in Fig. 15.40) so that the reduced physical strength of thinner coatings is normalized out. For a given coating temperature, lower coating thickness gives lower % adhesion. Similar results were found for the coating of nonporous substrates such as aluminum foil. Antonov and Soutar's [4] review in 1991 shows several examples of this, one of which is reproduced in Fig. 15.41.

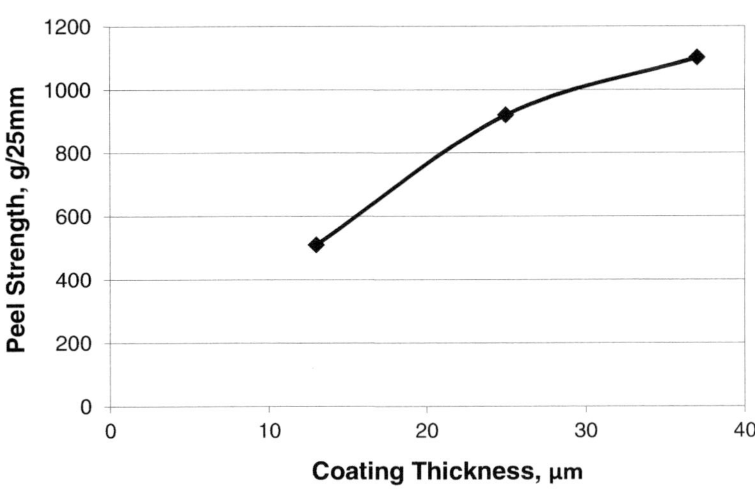

FIGURE 15.41 Peel strength of an acid copolymer to aluminum foil. Melt temperature = 303°C. TIAG = 80 ms. *Data from V. Antonov, A. Soutar, Foil adhesion with copolymers: time in the air gap, in: TAPPI 1991 Polymers, Laminations & Coatings Conference Proceedings, TAPPI PRESS, Atlanta, 1991, p. 553. Reproduced with permission from B.A. Morris, Understanding why adhesion in extrusion coating decreases with diminishing coating thickness, J. Plastic Film. & Sheeting, 24 (2008) 53–88.*

The mechanism for the reduction in adhesion with decreasing coating thickness may involve a number of factors including:

- Shorter time for oxidation in the air gap.
- Faster cooling in the air gap.
- Greater stress imposed during drawing in the air gap.
- More rapid cooling in the nip.
- Changes in the peel test measurement.

Following the work of Morris [8,50–52], experiments and modeling are used to determine which of these factors are most important. The case study is divided into three parts:

- Section 15.1.4.1 explores the reduction in the adhesion of LDPE to paper and other porous substrates. Several hypotheses are tested, including a reduction in oxidation time, faster cooling in the air gap, and more rapid quenching in the nip. A model of molten polymer penetration into the substrate shows that the greatest effect is cooling in the nip; thinner coatings have less time to flow into the substrate interstices once the chill roll contact is made. The model results agree well with experimental adhesion data from the literature.
- Section 15.1.4.2 examines the peel strength to aluminum foil and other nonporous substrates. Several hypotheses are proposed for why peel strength decreases in these structures, including a reduction in the air gap time, faster air gap cooling, more rapid nip quenching, and stress imposed during drawing. Modeling and experimental results show that cooling in the nip and stress have the greatest impact.
- Section 15.1.4.3 analyzes the peel test to understand why the peel strength of better adhering adhesives are more sensitive to changes in coating thickness. The analysis shows that changes in the critical dimension of the deformation region at the peel front may be responsible.

The case study illustrates how an understanding of process-property fundamentals leads to specific recommendations for improvement.

15.1.4.1 Porous substrates

For porous substrates, such as paper, paperboard, and nonwovens, it is generally thought that adhesion is created by molten polymer penetration into the pores and around the fibers. This sets up conditions for good physical or mechanical bonds, although chemical bonds may also play a role. Early investigators proposed that web cooling in the air gap might be responsible for reduced adhesion with thinner coatings [35,43]. Thicker coatings retain more of their heat in the air gap so that they are more fluid when they contact the substrate. Cooling in the nip region may also affect penetration time and velocity.

Three potential factors are explored here:

- Time in the air gap (TIAG),
- Cooling in the air gap,
- Cooling in the nip.

For each factor, a hypothesis is proposed and tested with experimental data, modeling results, and logic.

15.1.4.1.1 Time in the air gap

Hypothesis: A significant portion of the decrease in adhesion performance is due to a reduction in TIAG.

Antonov and Soutar [4] show that adhesion decreases with decreasing TIAG for a variety of resins and substrates, including paperboard. For LDPE coatings, this is believed to be due to less time for oxidation as is described in Section 15.1.2.1.

TIAG, as defined by Eq. (15.1), is simply the air gap (L) divided by the line speed (v_f): $TIAG = L/v_f$. Often a processor may reduce the coating weight by increasing the line speed, keeping everything else constant. TIAG in this case will decrease with decreasing thickness. This is how Goslin and Sweeney conduct their experiment. As they reduce the thickness from 50 to 13 μm, the TIAG is reduced from 125 to 31 ms. This is in the range where Antonov and Soutar find that adhesion typically decreases with decreasing TIAG.

Increasing the line speed, however, is not the only way to reduce coating thickness. Other investigators, such as Guillotte and McLaughlin, have independently varied thickness, TIAG, and temperature. The curves plotted in Fig. 15.38 represent results at constant TIAG. These show that without changing TIAG, the adhesion still decreases with decreasing thickness. This is further reinforced by conducting a statistical analysis of variance (ANOVA) on the Guillotte and McLaughlin data. The main effects are plotted in Fig. 15.42. On average, thickness has a much greater effect on % adhesion than either TIAG or coating temperature. Based on this analysis, the hypothesis is rejected:

- TIAG may have some effect on the reduction in adhesion with decreasing thickness but is not sufficient to explain it.
- The fact that adhesion increases with increasing TIAG (despite an increase in cooling, as shown in the next section) suggests chemical bonding due to oxidation may be contributing to the overall adhesion of PE to paper.

Eq. (15.4) is a more rigorous equation for the process time in the air gap that better reflects the physics of the flow. TIAG does not account for changes in the melt curtain DDR as thickness is altered. Once again there are two ways to reduce the coating thickness. One is to keep the throughput constant and increase the line speed. This increases DDR and the change in process time [Eq. (15.4)] is less than predicted by TIAG (L/v_f). The second method is to adjust the air gap or throughput to keep TIAG constant (as in the Guillotte and McLaughlin data). In this scenario, the actual process time in the air gap [Eq. (15.4)] increases with decreasing coating weight. In either case, this does not alter the conclusions.

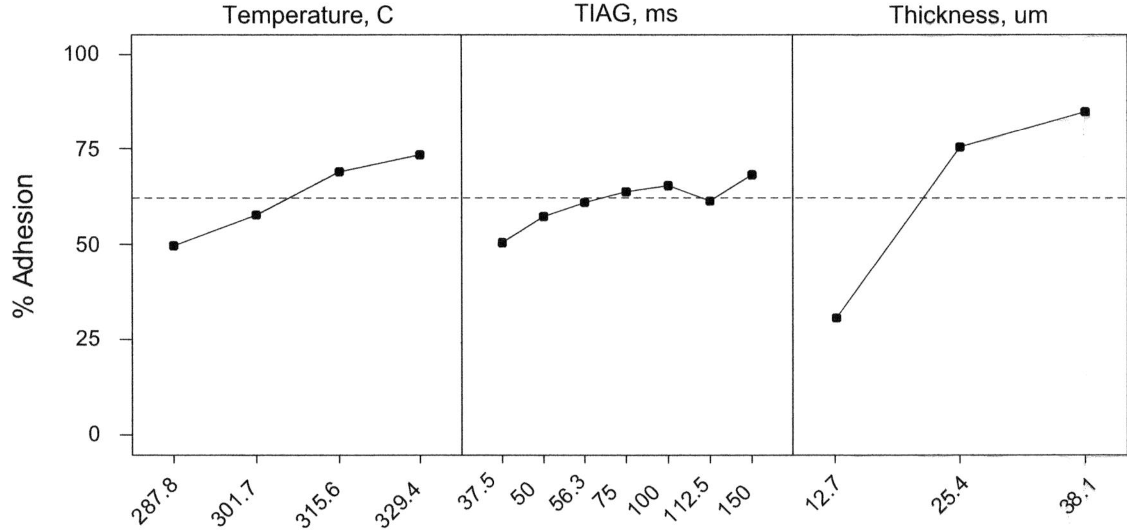

FIGURE 15.42 Main effects plot from ANOVA analysis of Guillotte and McLaughlin [41]. *Data Reproduced with permission from B.A. Morris, Understanding why adhesion in extrusion coating decreases with diminishing coating thickness, J. Plastic Film. & Sheeting, 24 (2008) 53–88.*

15.1.4.1.2 Cooling in the air gap

Hypothesis: Thinner coatings cool down more in the air gap, resulting in higher viscosity, less penetration of the coating into the porous substrate and therefore less adhesion.

The model developed in Section 15.1.2.3 is used to predict the temperature decrease in the air gap under the conditions used by Goslin and Sweeney (Fig. 15.38). As shown in Fig. 15.43, for a die exit temperature of 329°C, decreasing the thickness from 50 to 13 μm reduces the temperature at contact with the paper from 325°C to 312°C.

The relative penetration velocity of the polymer melt into the paper at first contact is computed by assuming that the flow is fully developed and an average pore can be represented as a cylinder. For pressure-driven flow of a power-law fluid down a cylindrical channel (see Table 5.4): [53]

$$v_{avg} = \frac{n}{3n+1}\left[\frac{R^{n+1}}{2m}\left(\frac{\Delta P}{l}\right)\right]^{1/n} \tag{15.26}$$

where $v_{avg} =$ average velocity in the pore, n = power law exponent [Eq. (15.27)], m = power law constant [Eq. (15.27)], R = radius of cylinder representing the average pore, and $\Delta P/l =$ pressure gradient.

Here a power law model is used to represent the viscosity:

$$\eta = m\dot{\gamma}^{n-1} \tag{15.27}$$

where η = shear viscosity and $\dot{\gamma} =$ shear rate.

An Arrhenius relationship is used to relate m to changes in temperature:

$$m = A\exp\left[a\left(\tfrac{1}{T} - \tfrac{1}{T_r}\right)\right] \tag{15.28}$$

where A = constant, a = temperature exponent (related to the activation energy), T = temperature, and $T_r =$ reference temperature.

Combining Eqs. (15.26)–(15.28), using typical values for LDPE ($a = 0.02$, $n = 0.4$), and assuming constant geometry (R) and pressure gradient in the nip ($\Delta P/l$), the penetration velocity of the molten polymer at first contact with the substrate in the nip is

$$V_p = \overline{B}\exp\left[\left(\frac{a}{n}\right)T_s\right] = \overline{B}\exp(0.05T_s) \tag{15.29}$$

where $V_p =$ penetration velocity, $\overline{B}$ = constant, and $T_s =$ temperature of the melt curtain at first contact with the substrate.

Fig. 15.44 shows the relative penetration velocity at first contact as a function of coating thickness and temperature (at the die exit). The results have been normalized to the penetration velocity of the 50-μm coating at 288°C, which gives enough penetration for 100% adhesion (see Fig. 15.38). If penetration is controlling adhesion, as the hypothesis implies, and if initial penetration velocity is the factor controlling penetration, then as soon as the relative penetration

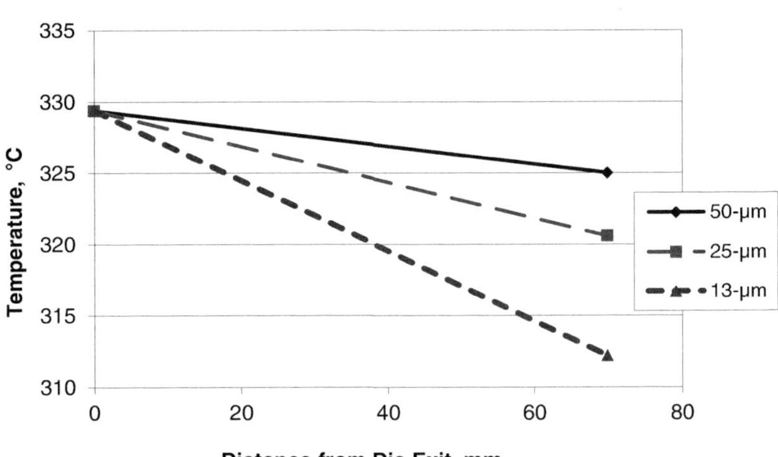

FIGURE 15.43 Calculation of temperature in the air gap using equation 15.13 based on processing parameters from Goslin and Sweeney [49]. The temperature at the die exit is 329°C. *Reproduced with permission from B.A. Morris, Understanding why adhesion in extrusion coating decreases with diminishing coating thickness, J. Plastic Film. & Sheeting, 24 (2008) 53–88.*

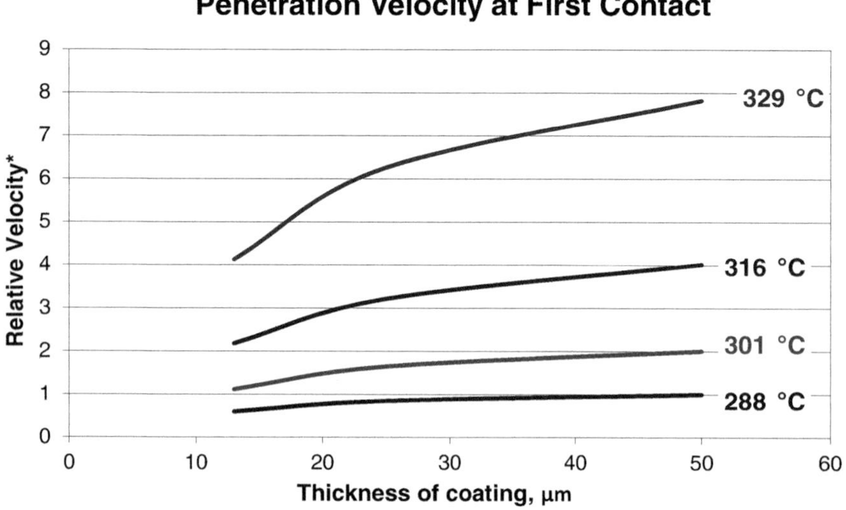

FIGURE 15.44 Initial penetration velocity from Eq. (15.29) based on cooling in the air gap. Conditions simulate data from Goslin and Sweeney [49]. *Reproduced with permission from B.A. Morris, Understanding why adhesion in extrusion coating decreases with diminishing coating thickness, J. Plastic Film. & Sheeting, 24 (2008) 53–88.*

velocity exceeds 1, the adhesion should be 100%. The plot for 13-μm coating thickness shows that the penetration velocity should be sufficient for 100% adhesion at temperatures above 301°C.

Comparing this with Fig. 15.38, however, shows that even at 329°C (where the penetration velocity is 4 times than assumed necessary for 100% adhesion), the measured adhesion is still less than 100%. Thus, the hypothesis is rejected. While cooling in the air gap changes the penetration in the right direction, there is not a sufficient decrease in temperature and increase in viscosity to explain the adhesion results, at least by itself[1].

A second hypothesis related to cooling in the air gap involves oxidation: *Thinner coatings cool down more in the air gap, resulting in less oxidation of LDPE and poorer adhesion.* The TIAG analysis (Section 15.1.4.1.1) suggests that chemical bonding may play a role in adhesion of LDPE to paper. Yet it is likely that polymer penetration into the pores of the paper and mechanical interlocking is a more important adhesion mechanism. This is suggested by the 50-μm coating data in Fig. 15.38, which shows 100% adhesion at 290°C, near the minimum temperature required for oxidation (see Figs. 15.4–15.6). Furthermore, the calculation in Fig. 15.43 shows that the 13-μm coating only cools to 312°C when the die exit temperature is 329°C, well above the 290°C threshold for 100% adhesion suggested by the 50-μm data if oxidation were dominant. Thus, this hypothesis is rejected: while oxidation and chemical bonding may enhance the adhesion of LDPE to paper, it is not sufficient.

15.1.4.1.3 Cooling in the nip

Hypothesis: Faster cooling in the nip for thin coatings provides less time for penetration and for adhesion to develop.

Section 15.1.3 describes how cooling in the nip is rapid and the time available for penetration before the polymer solidifies is short. The solidification time (t_s) decreases with approximately the square of the coating thickness.

Total penetration is the product of the velocity and time: $P = V_p t$. In the air gap analysis, the time element is ignored. Time is an integral part of the current analysis. The time in the nip for the polymer to cool to its crystallization temperature and start to freeze is defined at t_s (see Section 15.1.3). Well before the crystallization temperature, however, the viscosity becomes so high that flow essentially stops. In order to compute the relative penetration as a function of time, the temperature of the polymer as a function of time in the nip is first calculated using the unsteady state heat conduction model developed in Section 15.1.3. This model shows that the temperature at the PE-paper interface decreases approximately exponentially with time in the nip.

1. A sensitivity analysis on Eq. (15.29) shows that increasing n or decreasing $\overline{B}$ reduces the change in velocity with temperature. While the slope of the curves may change, the conclusions are unaltered. The model results still show that there is enough penetration velocity of the 13-μm coating at around 300°C to achieve 100% adhesion, which is inconsistent with the adhesion measurements by Goslin and Sweeney.

The penetration velocity at a given time is computed by replacing T_s in Eq. (15.29) with the temperature from the heat conduction model at each time step. The total penetration is found by integrating the velocity over time:

$$P = \int_0^{t_s} V_p(t)dt. \tag{15.30}$$

A typical penetration curve is given in Fig. 15.45. Most of the penetration occurs in the first 20% of t_s.

The relative penetration of coatings under the conditions run by Goslin and Sweeney and Guillotte and McLaughlin is computed. For the first part of the analysis, cooling in the air gap is ignored and the temperature at first contact with the substrate (T_s) is assumed to be equal to the die exit temperature. Later the effects of cooling in the nip and in the air gap are combined.

Fig. 15.46 shows the results of the calculation for the Guillotte and McLaughlin data. It is assumed that there is a minimum penetration threshold for 100% adhesion. Below the threshold, the adhesion increases linearly with penetration distance. Above the threshold, no further improvement in adhesion can be obtained (adhesion is greater than the coating burst strength, which corresponds to 100% adhesion). For the Guillotte and McLaughlin data, this threshold appears to be the penetration achieved for the 25-μm coating at 316°C. This penetration is set to 1 and all the other penetration distances are computed relative to this. Relative penetration values greater than 1 should have 100% adhesion; values less than 1 should have relative adhesion that linearly corresponds to the relative penetration.

The results show that

- Increasing the coating temperature by 15°C doubles the penetration distance.

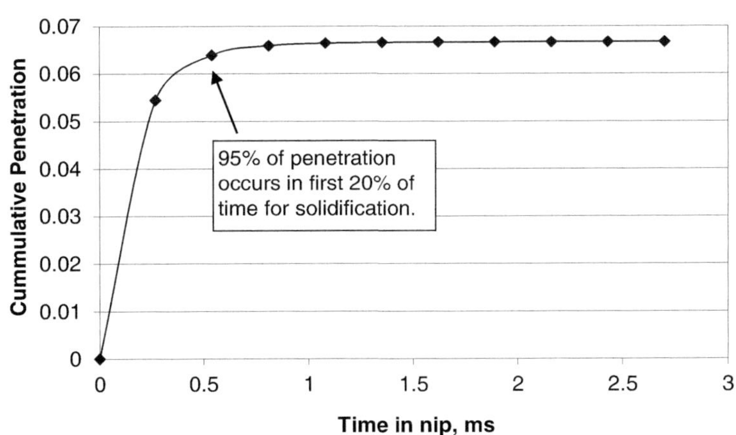

FIGURE 15.45 Penetration as a function of time in the nip for 25-μm coating onto paper. Initial temperature is 325°C. $t_s = 2.7$ ms. *Reproduced with permission from B.A. Morris, Understanding why adhesion in extrusion coating decreases with diminishing coating thickness, J. Plastic Film. & Sheeting, 24 (2008) 53–88.*

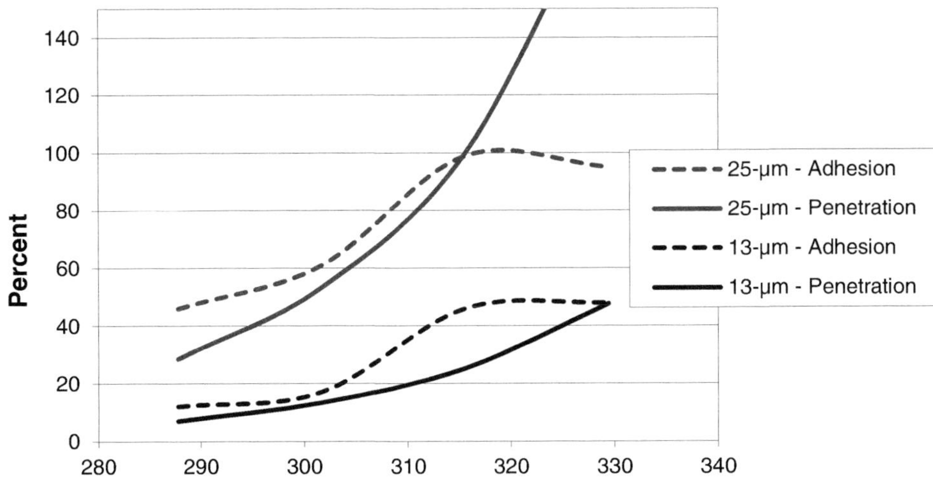

FIGURE 15.46 Comparison of penetration calculation with adhesion data of Guillotte and McLaughlin [41]. Cooling in the air gap is ignored. 100% penetration is set at 316°C and 25 μm, the onset of 100% adhesion in the Guillotte and McLaughlin data. *Reproduced with permission from B.A. Morris, Understanding why adhesion in extrusion coating decreases with diminishing coating thickness, J. Plastic Film. & Sheeting, 24 (2008) 53–88.*

- Penetration increases with roughly the square of the coating thickness (in direct relation to t_s).
- Doubling the thickness is equivalent to increasing the coating temperature by about 30°C.

The penetration calculation tracks well with the adhesion data, but the adhesion is somewhat higher than predicted solely by penetration. One reason for this is that chemical bonding (from oxidation of the melt curtain) improves the adhesion and is not accounted for by the penetration model. Several simplifying assumptions used to develop the model may be invalid and play a role. These include:

- Adhesion is a linear function of penetration.
- The flow is fully developed (even though the flow is transient and takes place in a few milliseconds before the polymer freezes).
- Entrance effects are negligible and can be ignored.
- The pressure gradient, $\Delta P/l$, is constant.

Finally, the effect of cooling in the air gap is combined with cooling in the nip. The output from the air gap cooling model is used as a starting point for the penetration calculation. The results are shown in Fig. 15.47. The trends are the same as for the cooling in the nip penetration model with the actual values of the penetration distance somewhat smaller since the starting temperatures are lower[2].

15.1.4.1.4 Conclusions and practical implications

The results of the analysis are consistent with the hypothesis that cooling in the nip plays a substantial role in the change in adhesion with thickness. The conclusions for adhesion to porous substrates are

- Faster quench times in the nip lead to less penetration of the molten polymer and less mechanical adhesion for thin coatings. This effect appears to be more important than cooling in the air gap.
- Penetration models predict the right trends in adhesion performance, although they tend to under predict the amount of penetration/adhesion. Penetration approximately increases with the square of the coating thickness and doubles with every 10°C −15°C increase in temperature.
- Chemical bonding may also be important, as suggested by the TIAG analysis and the penetration model results.

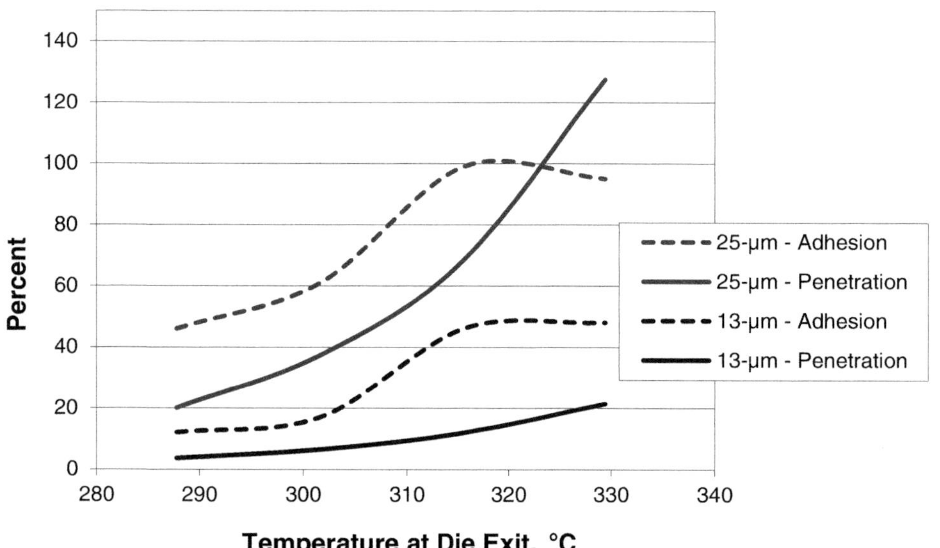

FIGURE 15.47 Comparison of penetration calculations with adhesion data of Guillotte and McLaughlin [41]. Cooling in the air gap is included. 100% penetration is set at 316°C and 25-μm, the onset of 100% adhesion in the Guillotte and McLaughlin data. *Reproduced with permission from B.A. Morris, Understanding why adhesion in extrusion coating decreases with diminishing coating thickness, J. Plastic Film. & Sheeting, 24 (2008) 53–88.*

2. An order of magnitude calculation of the actual penetration at which 100% adhesion occurs gives values of around 5 μm [assumes $R = 5$ μm, $\Delta P = 0.4$ MPa, $l = 125$ μm in Eq. (15.26)].

The study suggests several strategies for improving the thin coating adhesion to porous substrates. These include:

- Increasing the coating temperature. This increases the polymer fluidity and the available penetration time before flow ceases. Most LDPE is already coated near its temperature limit, however. Higher temperatures may lead to excessive gel formation, chain scission (particularly for PP, but under severe conditions LDPE can unzip), smoking, and generation of oxidative species that contribute to off-flavors and odors.
- Increasing the coating resin MFI. This allows more flow and penetration at a given temperature. High MFI has its drawbacks, however. These include higher neck-in and poorer heat seal and hot tack strength.
- Adding chemical functionality to the coating or substrate. Materials such as acid copolymers and ethylene acrylate copolymers have chemical functionality that may help improve the adhesion strength to the substrate. Treatments such as corona, ozone or plasma also increase functionality.
- Processing with a longer air gap. This promotes LDPE oxidation that adds to the polymer–substrate chemical adhesion. Drawbacks to this approach are increased air gap cooling and increased neck-in.
- Heating the substrate. Flame treating paperboard has been used to help promote adhesion during coating with PET.
- Increasing the nip pressure to increase the penetration driving force. This may also be accomplished by increasing the nip roll hardness. Both approaches, however, may decrease the time in the nip available for penetration and should be balanced with solidification time (see Section 15.1.3.1).
- Decreasing the polymer freezing point. One reason that LDPE is a better coating resin than PP or PET is that it has a lower freezing point that allows more time in the nip before the polymer freezes and penetration ceases. Ethylene copolymers such as ethylene vinyl acetate (EVA), ethylene methyl acrylate (EMA) and ethylene (meth)acrylic acid (EMAA or EAA), have lower freezing points than LDPE.
- Increasing the chill roll temperature. Usually, there are practical limits to doing this due to chill roll sticking.

15.1.4.2 Nonporous substrates

There are several possible mechanisms related to the coating process that may explain the observed decrease in adhesion with decreasing thickness to nonporous substrates such as aluminum foil. These include:

- Less time in the air gap (reduces oxidation and increases stress).
- Greater cooling in the air gap (increases viscosity and reduces wetting and chemical interaction).
- Faster cooling in the nip (reduces time for wetting and chemical interaction).
- Greater stress and orientation during drawing (increases residual stress).

As was done for porous substrates, a hypothesis is proposed for each of these potential mechanisms and tested using a combination of modeling and experimental results. One set of data that is used to test several of the hypotheses comes from a study involving blending an additive into LDPE to enhance its adhesion to aluminum foil [36]. This work is described in Section 10.4.1.4. A statistically designed experiment (DOE) was conducted varying:

- The coating temperature at the die exit.
- Coating thickness.
- TIAG.
- Percent of the adhesion-enhancing (AE) additive in the blend.

Peel strength to aluminum foil was the dependent variable. The results were fit to a statistical model, which is used in the analysis below and designated the AE DOE. An example of a plot of peel strength versus coating thickness from the model is shown in Fig. 15.48. The results were surprising in that the modified LDPE is more sensitive to changes in coating thickness than LDPE alone. This was the motivation for the study.

15.1.4.2.1 Time in the air gap

Hypothesis: Peel strength decreases with decreasing coating thickness due to a reduction in TIAG.

Section 15.1.2 describes how increasing time in the air gap increases the peel strength of LDPE and acid copolymers to aluminum foil. For LDPE coatings, this is believed to be due to greater oxidation time; for acid copolymers a reduction in stress may be involved.

There are several studies that show peel strength decreases with decreasing thickness even at constant TIAG. An example is reproduced from Antonov and Soutar [4] in Fig. 15.49. Here the effect of thickness is much more

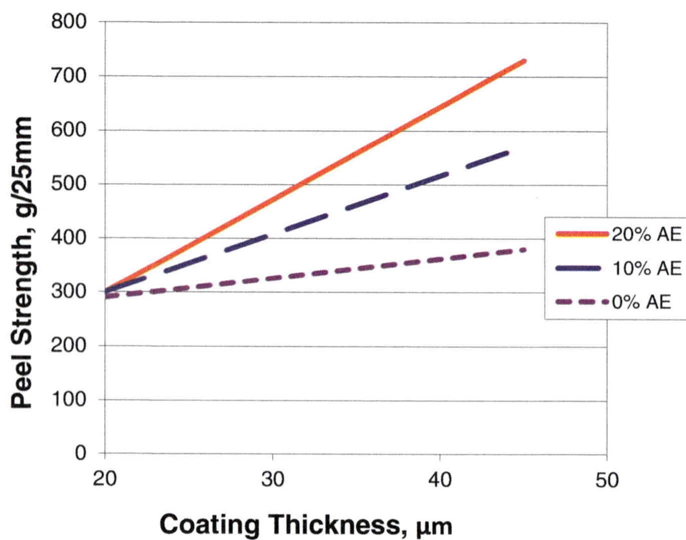

FIGURE 15.48 Results from DOE varying coating thickness, coating temperature, TIAG and % AE modifier in blend with LDPE for extrusion coatings onto 50 μm aluminum foil. Here temperature (310°C) and TIAG (80 ms) are kept constant and peel strength of the coating to aluminum foil as a function of thickness and %AE is plotted based on a statistical model of the experimental results. *Data from Morris [36]. Reproduced with permission from B.A. Morris, Understanding why adhesion in extrusion coating decreases with diminishing coating thickness, J. Plastic Film. & Sheeting, 24 (2008) 53–88.*

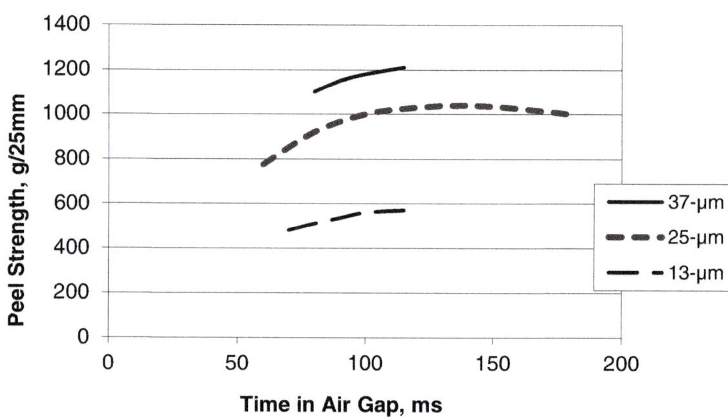

FIGURE 15.49 Adhesion of an acid copolymer to aluminum foil for three coating thicknesses. Melt temperature = 303°C. *Data from V. Antonov, A. Soutar, Foil adhesion with copolymers: time in the air gap, in: TAPPI 1991 Polymers, Laminations & Coatings Conference Proceedings, TAPPI PRESS, Atlanta, 1991, p. 553. Reproduced with permission from B.A. Morris, Understanding why adhesion in extrusion coating decreases with diminishing coating thickness, J. Plastic Film. & Sheeting, 24 (2008) 53–88.*

pronounced than the effect of TIAG. The same is true for the AE DOE, shown in Fig. 15.48. In this experiment, the statistical analysis shows that thickness has a greater impact than TIAG. This hypothesis, therefore, is rejected.

As discussed in Section 15.1.2, there are more accurate ways to compute the actual time in the air gap than TIAG; these methods, such as Eq. (15.4), take into account the drawdown of the polymer. The results of such calculations, however, show that at constant TIAG and die gap, the actual time in the air gap increases with decreasing thickness, making the hypothesis even more difficult to support.

15.1.4.2.2 Cooling in the air gap

Hypothesis: Thinner coatings cool down more in the air gap, resulting in less wetting (due to higher viscosity) and chemical reaction when they come into contact with the substrate.

Sections 15.1.2.3 and 15.1.4.1.2 show that thinner coatings do cool down more in the air gap than thicker ones, but the actual amount of cooling in the air gap is relatively small. Nevertheless, this hypothesis can be tested directly by applying the web temperature model [Eq. (15.13)] to the AE DOE results. Fig. 15.50 shows the web temperature model results for two coatings at 45 and 20 μm, each with a 325°C die exit temperature. The thicker web cools to 321°C as it travels through the 152 mm air gap to the chill roll. The thinner web cools to 315°C.

Fig. 15.51 shows statistical DOE model predictions for the peel strength of the thick and thin coatings. These are 800 and 300 g/25 mm, respectively. To find out whether cooling in the air gap contributes to this decrease in adhesion, the web temperature model is used to determine what the die exit temperature needs to be for the thick coating to cool to the same temperature as the thin coating. This is represented by the lower dashed curve in Fig. 15.50. A die exit

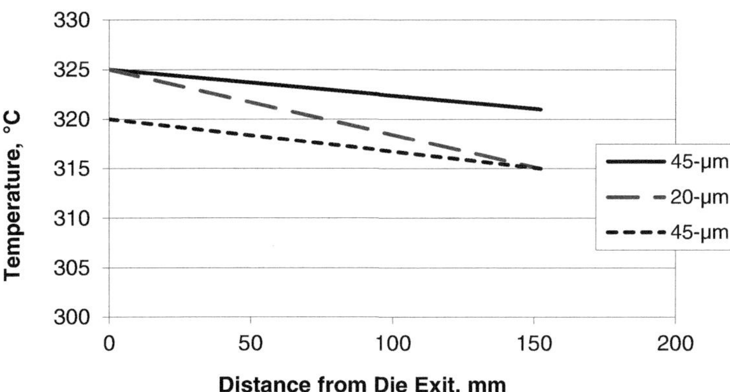

FIGURE 15.50 Web temperature model [Eq. (15.13)] applied to conditions of AE DOE (20% AE + 80% LDPE, 152 mm air gap, 91 m/min line speed, 100 ms TIAG). $H = 5.7$ W/m^2-K. *Reproduced with permission from B.A. Morris, Understanding why adhesion in extrusion coating decreases with diminishing coating thickness, J. Plastic Film. & Sheeting, 24 (2008) 53–88.*

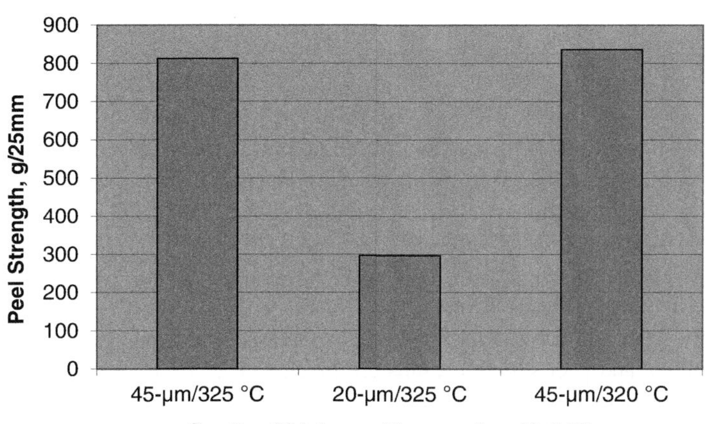

FIGURE 15.51 AE DOE model predictions for peel strength using coating thickness, die exit temperatures and processing parameters from Fig. 15.49. *Reproduced with permission from B.A. Morris, Understanding why adhesion in extrusion coating decreases with diminishing coating thickness, J. Plastic Film. & Sheeting, 24 (2008) 53–88.*

temperature of 320°C for the thick coating will duplicate the final temperature in the air gap of the thin coating. Plugging this value into the DOE model reveals no change in the peel strength, as illustrated by the final bar in Fig. 15.51. The hypothesis is rejected.

15.1.4.2.3 Cooling in the nip

Hypothesis: Thin coatings cool faster in the nip, providing less time for wetting and chemical reaction and, hence, less adhesion to nonporous substrates.

Fig. 15.52 shows the results of the unsteady state heat conduction model developed in Section 15.1.3 applied to an extrusion coating onto a foil laminate with the structure: [nip roll side] 13-μm OPET/19-μm LDPE/9-μm Al/Coating [chill roll side]. The coating is initially at 315°C and the foil laminate at 23°C. For boundary conditions, the outside coated surface is assumed to be in direct contact with the chill roll at 10°C and the outside foil laminate surface (OPET film) is in contact with a rubber nip roll (insulated). The results in Fig. 15.52 show the temperature at the foil surface as a function of time for 30 and 18-μm coating thicknesses. The foil surface temperature quickly rises from room temperature after the hot polymer contacts it. It reaches a maximum and slowly cools back down.

A key parameter proposed by Trouilhet and Morris [45] and described in Section 15.1.3 is the solidification time, t_s. This represents the time until the polymer reaches its freezing temperature (91°C for LDPE). Once the polymer begins to crystallize, the polymer chain movement is severely constrained, curtailing further flow, wetting, and rearrangement needed for chemical bonding. From Fig. 15.52, t_s for the 30 and 18-μm coatings are approximately 3 and 0.8 ms, respectively. This confirms that the thicker coatings have considerably more time for adhesion to develop before the onset of crystallization.

One way to try to determine the effect of the cooling rate in the nip on adhesion is to try to slow down the quench rate of the thin coating. Fig. 15.52 indicates how an experiment can be designed to do this. The bottom curve represents the temperature profile when an 18-μm coating is applied. The cooling of the thin coating can be slowed by inserting a

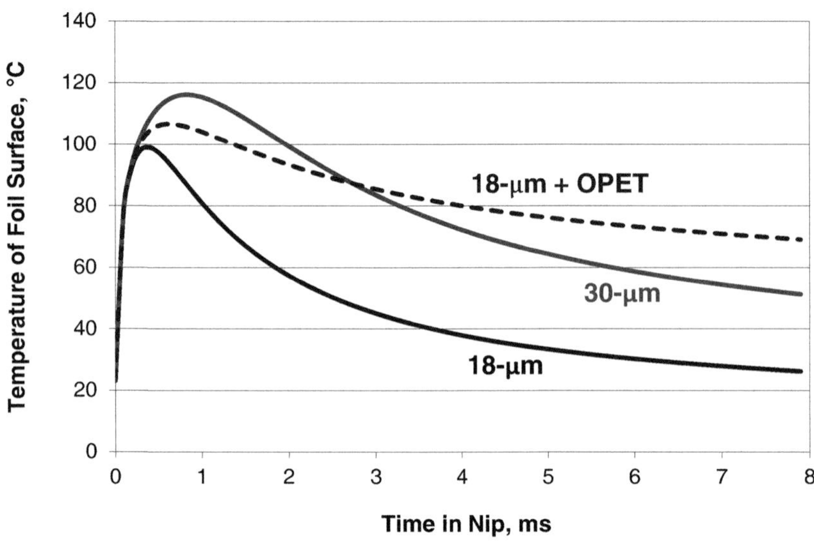

FIGURE 15.52 Heat transfer model calculations for the temperature of the foil surface during the coating of a foil laminate vs time in the nip. Dashed curve represents a way to slow the cooling of the 18-µm coating by inserting an OPET film between the coating and chill roll. Structure: 13-µm OPET/19-µm PE/9-µm Al/Coating; Coating temperature: 315°C; chill roll temperature: 10°C. *Reproduced with permission from B.A. Morris, Understanding why adhesion in extrusion coating decreases with diminishing coating thickness, J. Plastic Film. & Sheeting, 24 (2008) 53–88.*

film between the coating and chill roll. By selecting an appropriate film that has no adhesion to the coating, the film can be later removed for adhesion measurements. Fig. 15.52 shows the effect of using a 37-µm OPET film for this purpose in what essentially becomes an extrusion lamination. The structure is:

Nip Roll Side [13-µm OPET/19-µm LDPE/9-µm Al/ 18-µm coating /37-µm OPET] Chill Roll Side.

The presence of the OPET film insulates the interface between the coating and foil, extending t_s from 0.8 to about 3 ms. This is the same as the 30-µm coating without the insulating OPET layer.

The three structures in Fig. 15.52 were made on a pilot-scale extrusion lamination line at DuPont. The coating resin is a blend of 20% AE modifier and 80% LDPE. Coating conditions and peel strength results are given in Fig. 15.53. As expected, the 18-µm coating has significantly less peel strength than the 30-µm coating. The peel strength of the 18-µm coating produced with the insulating layer of OPET (removed for the peel strength measurement) falls between the two. The results show that nip cooling is important for peel strength performance. The fact that the insulated 18-µm coating peel strength is not as high as the 30-µm coating, despite similar t_s values, suggests a couple of mitigating factors. One is that t_s may not be the proper scaling parameter for peel strength. A closer look at Fig. 15.52 shows that the insulated temperature curve does not match the 30-µm coating curve peak temperature. The higher 30-µm peak coating temperature may play a role. Second, the results suggest that other factors, besides nip cooling, may influence the effect thickness has on peel strength performance.

The hypothesis cannot be rejected. Cooling in the nip is important but is not likely the sole driving force behind the loss in adhesion at low thickness.

A study by Divisme et al. [54] reaches similar conclusions. These investigators look at the adhesion of maleic anhydride grafted polypropylene (PP-g-MAH) to aluminum foil in a laboratory extrusion coating process. The structure is (rPP – PP-g-MAH)/Al where rPP is a random PP copolymer. Thicknesses are 60/20/37-µm. Their set-up differs from commercial extrusion coating processes used for flexible packaging in three ways: the process is slow (the line speed is only about 30 m/minute or 90 feet/minute whereas commercial lines typically run at 100–500 m/minute), the DDR is only 4 (versus 20 or greater), and a series of three chill rolls are used (most commercial lines use only one). The three chill rolls are important for slowing the cooling of the coating since the maleic anhydride must first hydrolyze to maleic acid to bond to the aluminum oxides on the surface of the foil. Maleic anhydride and maleic acid are in equilibrium; the acid forms at low temperatures when water is present. The elevated temperatures during extrusion favor the anhydride form. In commercial operations using maleic anhydride tie resins for bonding to aluminum foil, the structure often must be annealed following coating to allow the acid to form to achieve good adhesion.

Divisme et al. keep the air gap (6 mm) and DDR (4) constant. In some cases, they add an 80-µm layer of polystyrene (PS) to act as an insulator to reduce the rate of cooling, much like Morris did using OPET. They vary the melt temperature (232°C –295°C measured 2 cm after the die exit), chill roll and nip roll temperatures and line speed (0.083–0.667 m/s). In addition to peel strength they measure the bond density using ESCA at the aluminum foil-coating interface by first dissolving away the aluminum with acid or the PP with xylene. Microscopy and WAXS are used to characterize the percent crystallinity, crystalline morphology and orientation. Tensile properties of the coating are also measured.

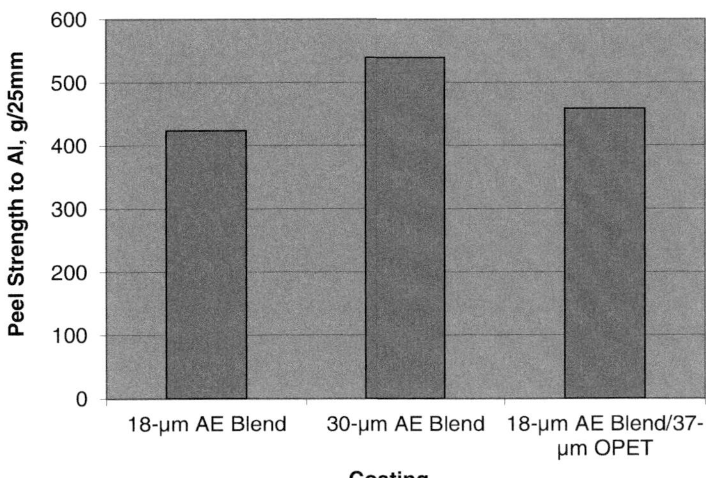

FIGURE 15.53 Peel strength results from coating trial duplicating structures from Fig. 15.52. Structure is OPET/PE/9-μm Al/coating. Coating conditions: 122 m/min, 127 mm air gap, 75 ms TIAG, 332°C coating temperature, 10°C chill roll temperature, 60 psi nip pressure. *Reproduced with permission from B.A. Morris, Understanding why adhesion in extrusion coating decreases with diminishing coating thickness, J. Plastic Film. & Sheeting, 24 (2008) 53–88.*

They find no direct correlation between peel strength and the processing parameters, which speaks to the complexity of these systems. They do find that peel strength generally increases with bond density. Some states with the same bond density have significantly different peel strength, which they attribute to differences in stress and orientation and its effect on mechanical properties (see next section). They define a parameter, ψ, to characterize the temperature history of the experimental states:

$$\psi = \int_0^{t_0} T(t)dt \tag{15.31}$$

where T = temperature at the foil-polymer interface at time t, t = elapsed time since the die exit, and t_0 = time at which the polymer starts to crystallize.

The temperature at the interface is determined by heat transfer modeling. Both bond density and peel strength increase with ψ. This is basically a way of saying that adhesion in this system is dominated by the amount of bonds that form at the interface, which in turn is determined by time at elevated temperatures. Once the polymer starts to crystallize the mobility decreases and reaction essentially stops. Increasing melt temperature, chill roll temperatures, nip roll temperature and adding the polystyrene insulating layer to slow cooling result in higher peel strength. A particularly influential factor in their experiments is the contact time on the chill rolls prior to the onset of crystallization, essentially t_s in Morris's analysis. They have unusually long t_s in their experiments due to the way they set their three chill roll temperatures but their results nevertheless show that decreasing the coating thickness reduces t_s and therefore the bond density and peel strength.

15.1.4.2.4 Stress

Hypothesis: Reducing the coating thickness puts more stress on the polymer, decreasing the peel strength.

Here the analysis from Section 15.1.2.2 and the Appendix is used to test the hypothesis. The stress imparted in the extruder and die is only weakly dependent upon thickness. Stress imparted during draw down in the air gap is related to three factors: the elongational strain rate, the total elongational strain and the relaxation time. The strain rate is inversely proportional to TIAG, which has already been considered in Section 15.1.4.2.1. The total strain, ε_{tot}, is equal to the logarithm of the DDR as shown in Eqs. (15.8) and (15A.12). DDR is approximately equal to the die gap divided by the coating thickness [see Eq. (15A.11)]. Since the die gap is often kept constant while changing thickness, the DDR, and, hence, stress, increases with decreasing coating thickness.

One possible representation for the characteristic process relaxation time is the air gap process time (t_f) as defined by Eq. (15.4). If TIAG (L/v_f) is held constant, then $t_f \sim$ DDR/ln(DDR). Since DDR increases with decreasing coating thickness (at constant die gap), by this time scale, the time for relaxation increases when the thickness is decreased. The earlier analysis showed that TIAG does not affect adhesion as strongly as thickness does. This suggests that not much relaxation takes place in the air gap while the polymer is being drawn down. The time available for relaxation after the polymer contacts the chill roll and drawdown has ceased may be a more relevant parameter for the process relaxation

TABLE 15.10 An analysis of stress, relaxation time and adhesion under the conditions of the AE DOE.

Parameter	45-μm coating	20-μm coating	% Change
Peel strength, g/25 mm	732	296	-60%
DDR	11.2	25.4	
ε_{tot} [ln(DDR)]	2.42	3.23	33%
TIAG, ms	80	80	
t_f, ms	338	603	78%
t_s, ms	5.9	0.7	-88%

Source: Reproduced with permission from B.A. Morris, Understanding why adhesion in extrusion coating decreases with diminishing coating thickness, J. Plastic Film. & Sheeting, 24 (2008) 53–88.

time. This time is simply the solidification time (t_s) introduced earlier – the time between when the polymer first contacts the chill roll and the onset of crystallization

Table 15.10 presents an analysis of the changes in stress, relaxation time, and adhesion under conditions that simulate the AE DOE. A 60% adhesion reduction occurs when changing the thickness from 45 to 20 μm, which is associated with an increase in stress (33% increase in ε_{tot}) and a reduction in time available for relaxation of the stress in the nip (88% decrease in t_s). Both associations support the hypothesis that stress plays a role in reducing peel strength with reductions in thickness.

Divisme et al. [54] measure the modulus and yield stress of their coatings (see Section 15.1.4.2.3), which they find correlate with the percent crystallinity and crystalline morphology. These are influenced by processing conditions related to stress (line speed, chill roll and nip roll temperatures, rate of elongation). Runs with the same bond density but substantially different peel strength are qualitatively explained by the differences in mechanical properties.

15.1.4.2.5 Conclusions and practical implications

Based on this analysis, the hypotheses that cooling in the nip and stress play a significant role in the change in adhesion with thickness are not rejected. Neither mechanism can explain all the observations. In particular, understanding the increased sensitivity of higher-performing adhesives, such as the AE-LDPE blends, to changes in coating thickness is still elusive and is taken up in Section 15.1.4.3.

The work suggests several strategies for improving the thin coating adhesion to nonporous substrates. These include:

- Reduce the cooling rate in the nip
  - Increase the coating temperature. Most coating resins, however, are already processed near their temperature limit. Higher temperatures may lead to other problems, as discussed in Section 15.1.4.1.4 and Chapter 13.
  - Lower the polymer freezing temperature. Copolymers such as acid copolymers, ethylene methyl acrylate (EMA), ethylene butyl acrylate (EBA), and EVA have lower freezing points than LDPE, as do some grades of metallocene polyethylene.
  - Heat the substrate via flame treatment or other means to increase the time before the polymer freezes.
  - Increase the coating resin MFI to allow more flow and wetting before the polymer freezes. Higher MFI will increase neck-in, however.
  - Increase the chill roll temperature. Usually, there are practical limits to doing this due to chill roll sticking.
- Reduce stress
  - Decrease the die gap to reduce DDR and stress during drawing. Unfortunately, many modern edge bead reduction dies do not allow this.
  - Run with a longer air gap. This reduces stress and for nonpolar polymers such as LDPE promotes oxidation. The primary drawback is increased neck-in.
- Add chemical functionality to the coating or substrate. Resins such as acid copolymers and ethylene acrylate copolymers have chemical functionality that may help improve the adhesion strength to the substrate. Treatments such as corona, ozone, or plasma also increase functionality. Priming the substrate is another approach.

15.1.4.3 Analysis of peel test

The analysis so far has failed to discern the mechanism behind the observation in Fig. 15.48 from the AE DOE [36] that the sensitivity of an adhesive's peel strength performance to changes in coating thickness increases as the bond strength increases. In this section, additional experimental results are shown that suggest that these findings are general: for an acid copolymer, which has strong bonds to foil, the peel strength is even more sensitive to changes in thickness than LDPE or the modified LDPE. An analysis of the peel strength measurement provides an explanation.

15.1.4.3.1 Experiment

Three resins of increasing acid functionality are extrusion coated onto a foil laminate to confirm the earlier work showing that the peel strength to foil of better performing adhesives is more sensitive to changes in coating thickness. The three resins are LDPE, a blend of 20% AE modifier [36] and 80% LDPE, and an ethylene acrylic acid (EAA) copolymer (9 wt.% AA, 10 g/10 min MFI). The coating conditions are:

- Die: Cloeren edge bead reduction die,
- Substrate: 13-μm OPET/19-μm tie/9-μm Al,
- Coating thickness: 20–50 μm (coated onto the foil side of the substrate),
- Coating temperature: 330°C for LDPE and AE-LDPE blend; 288°C for EAA,
- Die gap: 0.51 mm,
- Line speed: 122 m/minute,
- Air gap: 127 mm,
- Chill roll temperature: 10°C, and
- Nip pressure: 0.4 Mpa.

The peel strength between the coating and aluminum foil is measured in the machine direction. The results are plotted vs coating thickness in Fig. 15.54 and show:

- The peel strength of the coating to aluminum foil is in the following order: LDPE < AE + LDPE < EAA.
- The peel strength is more sensitive to changes in coating thickness when there is a stronger bond to the foil. This is indicated by the steeper slope for the EAA and AE + LDPE curves compared with the LDPE curve.

15.1.4.3.2 Peel test analysis

The peel test is a measure of the energy to pull the coating away from the substrate. The application of fracture mechanics to the peel test is described in Section 10.2.3. The peel force has contributions from elongating and bending the peel

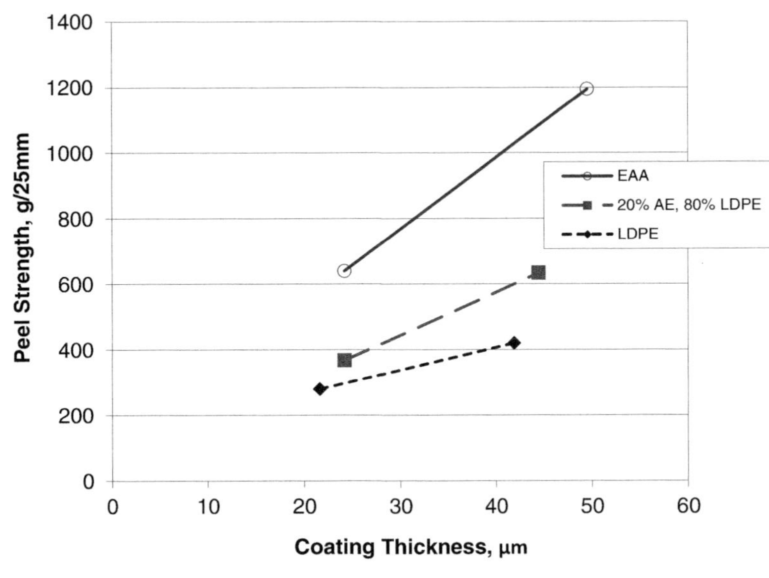

FIGURE 15.54 Peel strength to aluminum foil results from coating trial. Structure: 13-μm OPET/19-μm tie/9-μm Al/coating *Reproduced with permission from B.A. Morris, Understanding why adhesion in extrusion coating decreases with diminishing coating thickness, J. Plastic Film. & Sheeting, 24 (2008) 53–88.*

arms as well as the energy to create new interfacial area (known as the fracture energy). A mathematical expression is derived in Section 10.2.3 and reproduced here as Eq. (15.32) for convenience:

$$\frac{P}{b} = \frac{G_a + G_e + G_b}{1 + \varepsilon_a - cos\theta} \tag{15.32}$$

where P = peel force, b = width of peel arm, G_a = fracture energy = energy to create new surfaces, G_e = energy dissipated during peel arm elongation, G_b = energy dissipated during bending of the peel arm, ε_a = inelastic extension of peel arm, and θ = peel angle.

The coating thickness impacts each element of the peel force:

- Elongation of peel arm (G_e): G_e is related to the peel arm thickness (X) times the area under the stress–strain curve:

$$G_e = X\int_0^{\varepsilon_a} \sigma d\varepsilon. \tag{15.33}$$

  Here σ is the stress. The peel arm tensile strain, ε_a, is a nonlinear function of the thickness, which results in G_e undergoing a maximum at intermediate values of thickness. At low thickness, the peel arm may yield and ε_a may be large. This makes the integral term large but since X is small, G_e is small. As thickness increases, G_e increases. As the thickness is further increased, at some point the peel arm stiffness prevents yielding and ε_a becomes small. The integral term becomes small and even though X is large there is a reduction in G_e.
- Bending of peel arm (G_b): G_b arises from peel arm bending and can be significant if plastic yielding occurs. Gent and Hamed [55] study the plastic yielding of OPET film during bending. They note that peel strength often has been observed to go through a maximum at intermediate thickness. Gent and Hamed conclude that at low thickness, plastic yielding occurs, and the total energy that is dissipated is small. As thickness increases, more energy is dissipated and peel strength increases with G_b. Eventually, the peel arm gets too stiff and less yielding occurs, which leads to a reduction in G_b and peel strength. At high thickness, no yielding occurs and G_b is small.

  Gent and Hamed also propose that the adhesion strength at the interface can affect G_b. Low adhesion is less likely to lead to substrate yielding. High adhesion can lead to large energy dissipation within the substrate in the highly deformed region near the detachment. This can lead to an increase in peel strength in two ways:
  - Directly, through an additional force needed to propagate the bend in the peel arm, and
  - Indirectly, by causing larger deformation with the higher peel forces, resulting in greater energy loses in the peel arm.

  A similar argument can be made for G_e.

  Pascal [56] shows that the thickness affects peel strength through changes in the bending angle during peel strength measurements of extrusion-coated structures. 37-μm aluminum foil is coated with a coextrusion coating of (10-μm tie resin–LDPE) where the LDPE layer is varied from 15 to 100 μm. Peel strength to the foil is measured using an imposed T-peel configuration but with the peel tab (unpeeled portion of the laminate) unrestrained so that the angle of the pull changes with the stiffness of the coating. The peel strength goes through a maximum as the coating thickness is increased, as shown in Fig. 15.55. Near the maximum, the initial peel angle, θ_o, goes through a transition from a higher (obtuse) angle to a lower angle (90 degrees) that correlates with the decrease in peel strength. Analysis suggests that the angle influences the deformation mode during peeling. Due to shear deformation at the higher angle, the plasticity of the adhesive layer plays a greater role for the thin coatings characteristic of extrusion coating. At the lower angles of the thick coatings (characteristic of extrusion lamination), forces are mostly in the opening direction and the adhesive bond strength is more important. An adhesive that works well for extrusion coating may not necessarily work well for extrusion lamination.
- Fracture energy (G_a): As described in Section 10.2.3, G_a can be decomposed into the product of two factors, the work of adhesion and the local energy dissipation at the peel front:

$$G_a = W_a(1 + \Phi(R, T)) \tag{15.34}$$

  where W_a = work of adhesion or bonding, Φ = local viscoelastic energy dissipation, R = rate of pull, and T = temperature.

  W_a encompasses the thermodynamic and chemical adhesion that occurs at the interface. This term is typically quite small compared to the value of G_a but can be important because of its multiplying effect. In the experimental

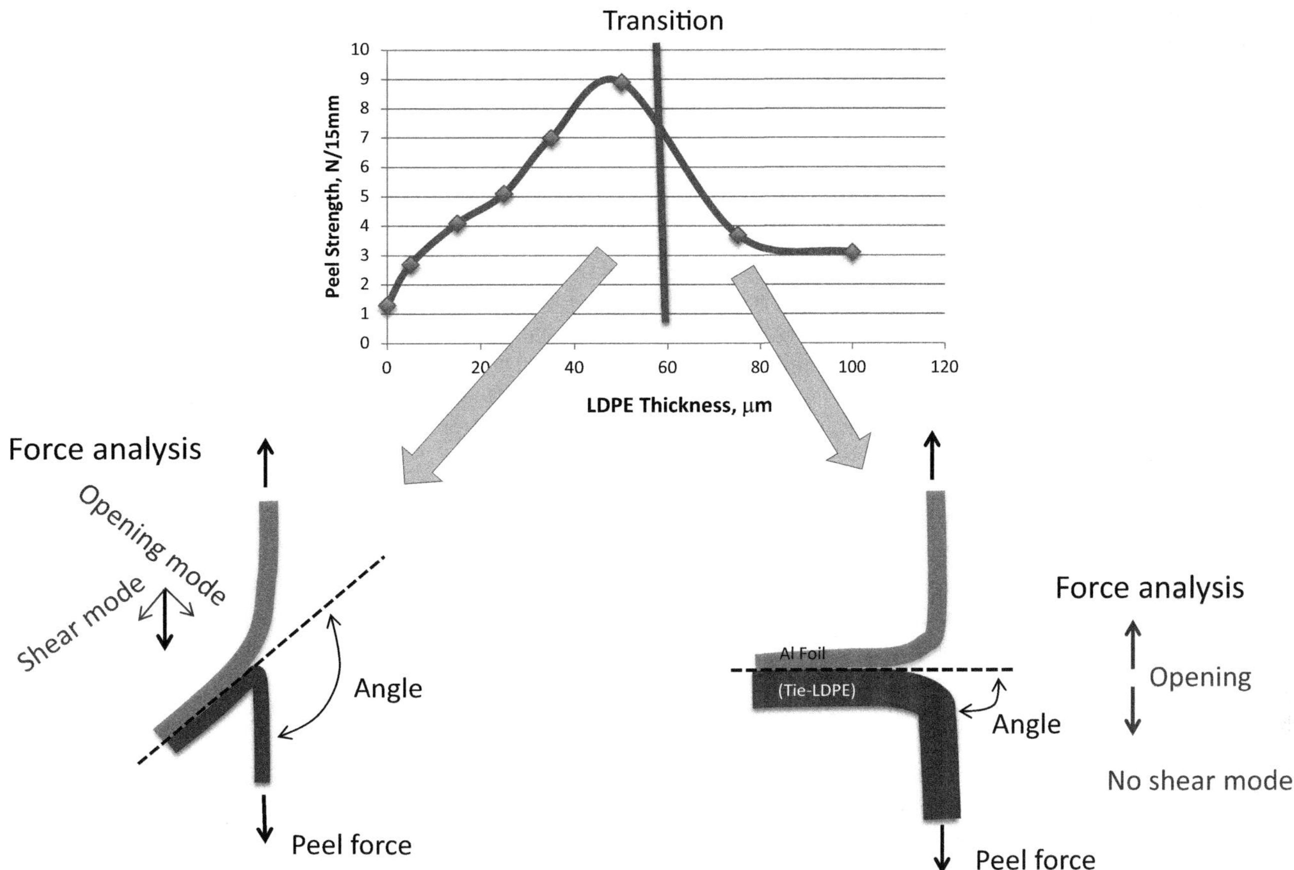

FIGURE 15.55 Effect of coating thickness on peel strength. Structure is 37-μm Al/(10-μm tie resin − LDPE) where the thickness of LDPE is varied from 0 to 100 μm. Peel angle goes through a transition near the maximum peel strength that influences the amount of shear force during the peel test. *Adapted from J. Pascal, Ultraversatile adhesives to broaden the possibilities of extrusion lamination, in: TAPPI PLACE. Conference, 2006.*

results, the AE−LDPE blend and the EAA copolymer have higher chemical functionality than LDPE. Their W_a is expected to be greater than that for LDPE.

The second term in Eq. (15.34), Φ, is the local energy that is dissipated as the detachment front, generally referred to as a crack, moves through. As described in Section 10.2.2.3, Φ arises from a zone of deformation created in front of the crack. At low thickness, the size of this zone is constrained by the layer thickness. As the thickness is increased, Φ increases as the deformation zone is allowed to expand. Φ and G_a increase with increasing thickness until this deformation zone reaches its natural size and is no longer constrained. This is the critical thickness, X_c. Further increases in thickness do not expand the deformation zone and G_a plateaus at a constant value. This is illustrated in Figs. 10.9 and 10.10.

The critical value of the thickness is derived in Section 10.2.2.3 and is related to the peel arm tensile properties and G_a:

$$X_c = \frac{1}{2\pi}\frac{G_a E}{\sigma_y^2} \tag{15.35}$$

where X_c = critical value of thickness where G_a transitions from a linear change to a plateau, E = tensile or Young's modulus, and σ_y = yield strength.

The dependence of X_c on G_a makes sense in lieu of Gent and Hamed's work described above. Higher G_a (through, e.g., an increase in W_a) implies that the deformation zone is larger, which in turn should increase X_c.

15.1.4.3.3 Application of peel strength model to experimental eata

The analysis of the peel test mechanics shows that thickness significantly affects the measured peel force. Moreover, all three components of the peel energy equation, G_e, G_b, and G_a, are a function of the adhesive strength, which may explain the behavior in Figs. 15.48 and 15.54. To apply these concepts to the foil coating structures, a model developed by Kinloch, Lau, and Williams [40] is used. The model provides analytical expressions for G_e, G_b, and G_a that are solved iteratively using a computer for conditions that simulate the AE DOE.

Model inputs include the imposed peel angle and the peel arm tensile properties (Young's modulus, plastic yield strain (e_y), and strain hardening parameter (α)). The strain hardening parameter is the slope of a line drawn through the yield stress and the ultimate tensile stress (see Section 9.2.1). Since Kinloch et al. studied LDPE-foil laminates, their tensile data are used in the analysis. It is assumed adding AE to LDPE does not significantly change the tensile properties. A "T-peel" configuration is used in this study to measure peel strength, a geometry that is not explicitly included in the model. The imposed peel angle, θ, is set to 90 degrees as an approximation.

Fig. 15.56 shows the results of the model for LDPE. The peel strength values (converted to the units of energy per area and designated G_{tot}) as a function of thickness are taken from the AE DOE results (bottom curve in Fig. 15.48). The results show G_e is small, reflecting experimental observations that the peel arms did not elongate beyond the yield point. G_a is large but not very dependent on thickness. G_b has the greatest dependence on thickness.

Fig. 15.57 shows the results for the 20% AE, and 80% LDPE blend (top curve from Fig. 15.48). Again G_e is small and will be ignored for the rest of this discussion. Here both G_a and G_b increase substantially with thickness. This may account for the blend's steeper peel strength curve in Fig. 15.48.

Comparing Figs. 15.56 and 15.57 shows that G_a for the blend more strongly changes with thickness than for LDPE. Using Eq. (15.35), X_c is estimated to be around 10–20 μm for LDPE and 50–80 μm for the AE blend. Since the coating thickness is in the range of 20–50 μm, this suggests that the LDPE coatings are near the plateau region, whereas the AE-blend coatings are still on the slope. This is illustrated in Fig. 15.58. Increasing the polymer-to-substrate bonding by adding chemical functionality increases X_c. Over the thickness range of interest, this change in critical thickness helps explain the peel strength sensitivity to thickness. X_c for the EAA coating is estimated to be 80–120 μm, which is consistent with its steeper slope in Fig. 15.54.

As noted earlier, G_b increases with increasing thickness for both LDPE and the AE-LDPE blend (see Figs. 15.56 and 15.57). The AE-blend slope, however, is considerably larger. Following the argument of Gent and Hamed, it appears that adding AE to LDPE increases the energy dissipated during peel arm bending. This is achieved through a greater anchoring effect. Comparing the initial peel angles supports this. As shown in Fig. 15.59, model calculations indicate that the peel angle remains constant for the AE blend coating as the thickness is increased. For the LDPE coating, the initial peel angle decreases as thickness is increased. For a given thickness, the peel arm of the AE-blend coating is pulled back at a higher angle than for LDPE, expending more energy during the peel strength measurement. In a study by Pascal [56] described earlier, the peel strength increased with the increasing thickness of the coating and then abruptly decreased at around 70 μm (see Fig. 15.55). The stiffness of the peel arm at a higher coating thickness

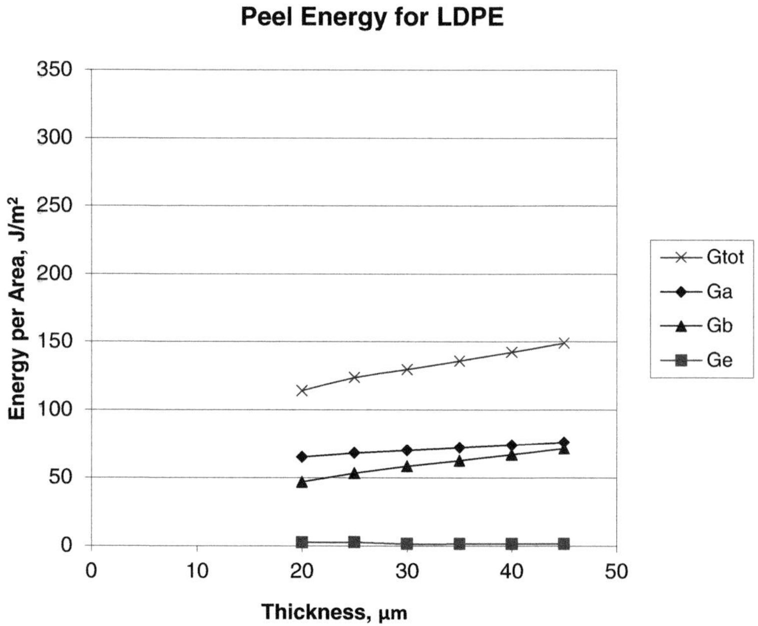

FIGURE 15.56 Results of energy analysis for LDPE coatings. Peel strength and thickness data taken from AE DOE model (Fig. 15.48) with 80 ms TIAG and 310°C coating temperature. Peel Energy model parameters: $E = 150$ MPa, $\alpha = 0.087$, $e_y = 6.2\%$, $\theta = 90$ degrees. *Reproduced with permission from B.A. Morris, Understanding why adhesion in extrusion coating decreases with diminishing coating thickness, J. Plastic Film. & Sheeting, 24 (2008) 53–88.*

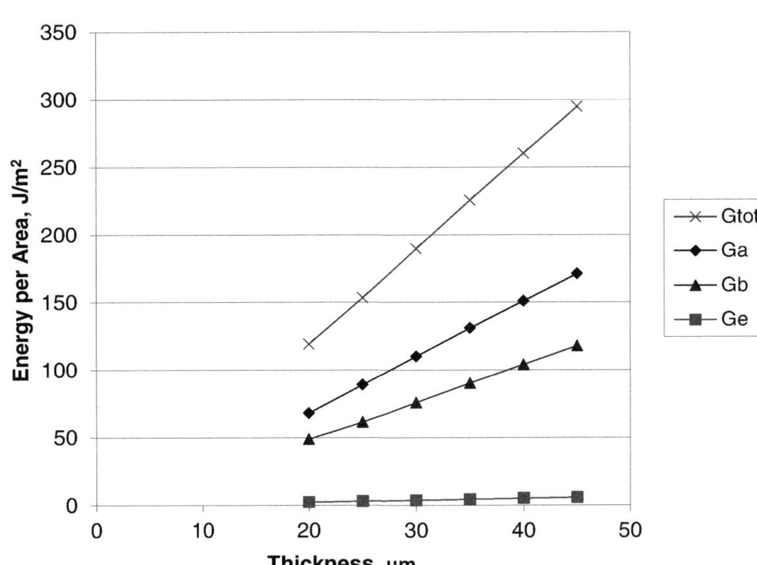

FIGURE 15.57 Results of energy analysis for 20% AE, 80% LDPE coatings. Peel strength and thickness data taken from AE DOE model (Fig. 15.48) with 80 ms TIAG and 310°C coating temperature. Peel Energy model parameters: $E = 150$ MPa, $\alpha = 0.087$, $e_y = 6.2\%$, $\theta = 90$ degrees. *Reproduced with permission from B.A. Morris, Understanding why adhesion in extrusion coating decreases with diminishing coating thickness, J. Plastic Film. & Sheeting, 24 (2008) 53–88.*

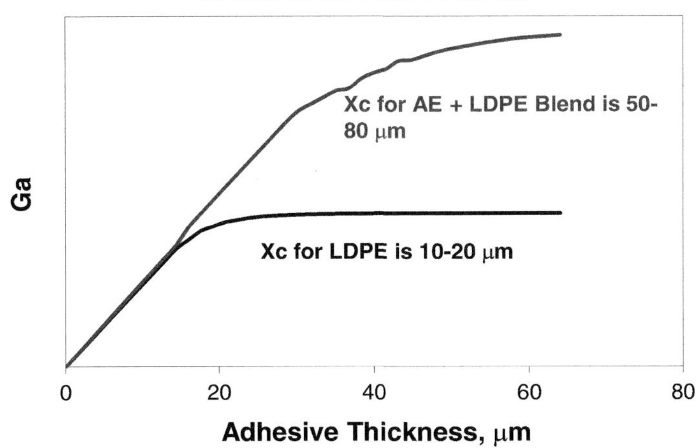

FIGURE 15.58 Scenario where G_a curves for LDPE and AE-LDPE blends diverge because of differing values of X_c. *Reproduced with permission from B.A. Morris, Understanding why adhesion in extrusion coating decreases with diminishing coating thickness, J. Plastic Film. & Sheeting, 24 (2008) 53–88.*

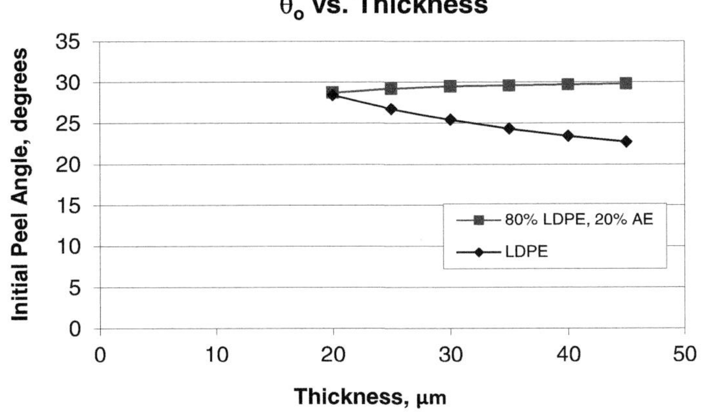

FIGURE 15.59 Comparison of initial peel angle (θ_0) for LDPE and AE-LDPE blend. Peel strength and thickness data taken from AE DOE model (Fig. 15.48) with 80 ms TIAG and 310°C coating temperature. Peel Energy model parameters: $E = 150$ MPa, $\alpha = 0.087$, $e_y = 6.2\%$, $\theta = 90$ degrees. *Reproduced with permission from B.A. Morris, Understanding why adhesion in extrusion coating decreases with diminishing coating thickness, J. Plastic Film. & Sheeting, 24 (2008) 53–88.*

significantly reduced the initial angle during the peel strength measurement and changed the deformation mode. In the AE-LDPE coating experiments, this transition was not reached.

15.1.4.3.4 Conclusions and practical implications

The peel strength test mechanics help explain why thin coatings have low peel strength performance and higher bond strength leads to greater sensitivity of peel strength to changes in thickness. The energy dissipated during peel arm bending decreases with decreasing thickness. Moreover, both the fracture and bending energy are strongly influenced by adhesion. Higher adhesion increases the initial peel angle, resulting in greater bending energy. In addition, the natural deformation zone size is enlarged, shifting the critical thickness (X_c) and onset of the fracture energy plateau region to higher thickness. Within the coating range thickness explored in the experiments, this has the effect of putting the LDPE on the plateau region and the AE-blend and acid copolymer on the sloping region. The greater bond strength provided by the chemical functionality in the AE blend and acid copolymer gives rise to a steeper climb in peel strength over a greater thickness range than for LDPE.

This work shows that increasing bond strength between the polymer coating and substrate does not always result in higher peel strength. The peel strength depends on many factors beyond the bond strength, including:

- The local energy dissipation (Φ) at the peel front.
- The local peel angle.
- Whether the polymer yields or elongates.

These factors depend on stiffness, thickness, viscoelastic response, and the yield stress of the polymer. The physical properties, in turn, do not affect each peel strength component in the same way. For example, reducing stiffness may reduce the peel arm bending energy, but increase the fracture energy. The situation becomes even more complicated for extrusion laminations where the second substrate properties may become a factor.

Despite these complexities, the findings suggest several general strategies for increasing the peel strength at low coating thickness. For LDPE, increasing the bond strength by increasing oxidation (increasing temperature or using ozone treatment) or adding modifiers may help. But as is shown with the AE modifier work, this may be most effective for thick coatings (or for laminations with stiff substrates). Without changing the physical properties, the bond strength may not be increased enough to significantly improve the peel strength of thin coatings. Acid copolymers have even stronger bond strength to aluminum foil than modified polyethylene. They are less crystalline than LDPE and, hence, have lower modulus and yield stress. Thus, their peel strength may be higher than LDPE at a given thickness, even though they may be more sensitive to changes in thickness. Further physical property changes can be obtained by incorporating soft comonomers such as acrylates or acetates into the ethylene backbone. EMA, EVA, or acid terpolymers (ethylene-acid-acrylate) are examples. Primers may be needed for some polymers to achieve good bond strength to the substrate.

In general, the polar ethylene copolymers, such as acid copolymers and EMA, have the right combination of stiffness, yield stress, and chemical functionality to improve peel strength over LDPE at low coating weights. One word of caution is that increasing the peel strength does not necessarily imply improved performance in a specific application; like any laboratory test, peel strength must represent the mode of failure in the application for this to be true.

15.2 Interlayer adhesion in coextrusion

The mechanisms for adhesion in coextrusion are described in Chapter 10. Sufficient time and temperature for interchain diffusion and chemical interaction at the interface are required for good adhesion. Complicating matters is the new interfacial area created as the melt is drawn to its final thickness just prior to quenching: as the interfacial area increases, the density of interfacial bonds is diluted. Reaction during draw down may compensate for this dilution effect. Experimental and modeling results suggest compressive and elongational flows during extrusion and draw down can accelerate new bond formation through convection and diffusion of reactive species to the interface. Orientation and stress imparted during draw down, however, may negatively impact peel strength. This section explores how the time, temperature, flow and stress history affect interlayer peel strength performance in coextrusion coating, cast film and blown film processes.

15.2.1 Initial process time studies

The region between the die exit and the start of solidification (frost line for blown film and contact with the chill roll for cast film and extrusion coating) is a critical region for interlayer adhesion. The layers are drawn down to their final thickness with a commiserate increase in interfacial area. The polymer chains are subjected to orientation and stress and the polymer begins to cool. Following up on work by Shirodkar et al. [57], Butler [58] and others showing process time in blown film correlates with a number of mechanical properties of film in monolayer extrusion, Morris [59] conducted an experiment showing process time also affects interlayer adhesion in coextrusion blown film. The equation for process time is provided in Chapters 2 and 13 and reproduced here:

$$t_f = \frac{h_f}{v_f - v_o}\ln\left(\frac{v_f}{v_o}\right) \tag{15.36}$$

where t_f = process time between die exist and frost line, h_f = frost line height, v_f = line speed, and v_0 = velocity of polymer at die exit.

In terms of more commonly used parameters, the blown film process time can be written as

$$t_f = \left[\frac{h_f}{v_f\left(1 - \frac{X_f\rho_f \mathrm{BUR}}{X_o\rho_o}\right)}\right] ln\left(\frac{X_o\rho_o}{X_f\rho_f \mathrm{BUR}}\right). \tag{15.37}$$

where BUR = blow-up ratio (ratio of final bubble diameter to die diameter), X_0 = die gap, X_f = final film thickness, ρ_0 = density of polymer at die exit (melt density), and ρ_f = density of polymer in final film (solid density).

A three-layer blown film line is used to produce films with the structure (HDPE-tie-EVOH) with layer thicknesses (37.5/12.5/20 μm). Melt temperature, blow-up ratio (BUR), die gap and film thickness are kept constant. The process time is varied by changing the frost line height (by changing the air quench speed) and the line speed (take-up speed). When the line speed is altered, the extruder output is adjusted to keep the final film thickness constant. Since BUR and die gap are held constant, the take-up ratio, TUR (or DDR as it is sometimes called), remains constant [see Eq. (2.70)]. Increasing line speed in this manner keeps the logarithm term in Eq. (15.36) constant (since TUR = v_f/v_0), but since the denominator of Eq. (15.36) contains the difference in velocities ($v_f - v_0$), the process time decreases. (If the extruder output were not adjusted, then the BUR would need to be decreased or die gap increased to keep the film thickness constant.)

The HDPE has a density of 960 kg/m^3, a MFI of 0.75 g/10 minute (190°C/2160 g), melt point (Tm) of 137°C and freezing point (Tc) of 115°C. The EVOH is a 32mole% ethylene grade with an MFR (210°C/2160 g) of 3.0 g/10 minute, Tm of 185°C and Tc of 157°C. Three of the tie layers are listed in Table 15.11. These are anhydride-modified LLDPE resins that are often used to bond PE to EVOH in blown film (see Section 10.4). The anhydride functionality reacts with the OH groups of the EVOH to form an ester. ADH1 and ADH2 are identical except for the amount of anhydride functionality. ADH2 has three times the amount of ADH1. ADH3 has an intermediate level of anhydride but is rubber modified, which gives it a lower modulus than the other two grades.

Peel strength between the tie resin and EVOH is measured using a T-peel configuration, pulled at 305 mm/minute after one-week of aging at 50% RH, 23°C. Plotted vs process time in Fig. 15.60, the results show a tenfold increase in peel strength with increasing process time over the range explored. The results underscore the importance of processing on adhesion in coextrusion.

Morris finds similar behavior for a variety of combinations in coextrusion blown film, including the adhesion of anhydride functionalized EVA-based tie resins to EVOH [59], PP-based tie resins to EVOH [60] and PE-based tie resins to PA 6 [61]. These are all combinations of immiscible polymers involving a chemical reaction at the interface. An increase in peel strength also occurs in nonreacting pairs where diffusion is the mechanism for adhesion. For example, bonding PE to ionomer is found to show this characteristic behavior.

TABLE 15.11 Tie resins used in coextrusion blown film trial.

Grade	Type	MFI (190°C/2160 g) g/10 min	Melt Pt °C	Freeze Pt °C	Modulus MPa	Anhydride %
ADH1	LLDPE	1.1	125	109	455	Standard
ADH2	LLDPE	1.1	129	115	434	3X standard
ADH3	LLDPE	3.1	130	112	207	2X standard

Source: From Morris [59]. Reproduced with permission from The Dow Chemical Company.

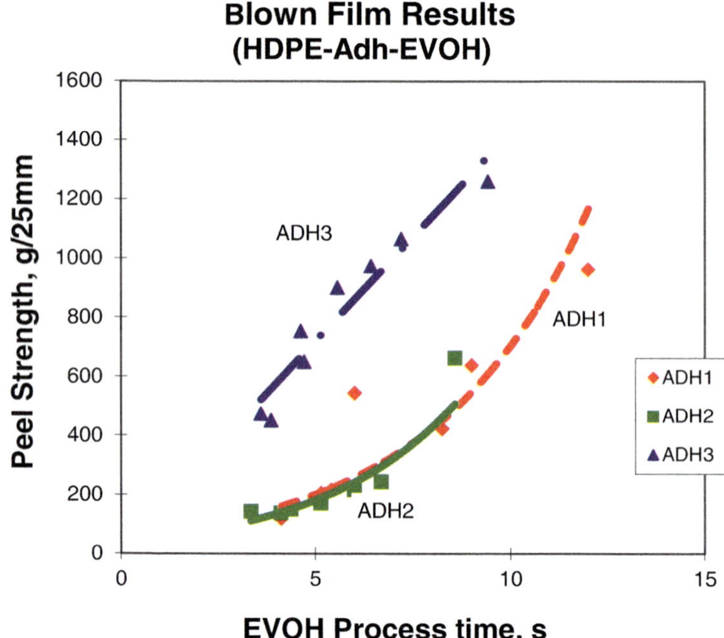

FIGURE 15.60 Peel strength to EVOH versus process time for blown film. The three-layer film has three frost lines, one for each layer. Here the process time is computed based on the EVOH frost line height, but the trends are the same for the process time computed on any of the frost line heights. From Morris [59]. *Reproduced with permission from The Dow Chemical Company.*

Morris [7,61] later extends the study to coextrusion coating and cast film. Here the characteristic process time is the time between the die exit and chill roll, given by Eq. (15.4). In one study, analog structures to that of the blown film study are made on a pilot-scale coextrusion coating line. The structure is kraft paper//19-μm LDPE//(5-μm tie–6.3-μm EVOH–5-μm tie–25-μm LDPE). The kraft paper is coated with LDPE in one pass and coextrusion coated with the EVOH structure in a second pass. Corona treatment is applied to the LDPE coated paper to ensure good adhesion between the tie layer and LDPE, since adhesion between the EVOH and tie layer is studied. The melt temperatures for the EVOH, tie and LDPE layers are 232°C, 260°C, and 260°C, respectively. The air gap (13–25 cm) and line speed (61–183 m/minute) are varied to change process time. Coating thickness, air gap (0.5 mm) and DDR (around 18) are kept constant. The tie resins, described in Table 15.12, are similar to those used in the blown film study, except for a somewhat higher MFI for processing in extrusion coating. The LDPE is a standard extrusion coating grade and the EVOH a 44 mol% ethylene grade.

The peel strength is measured between the outside tie resin and EVOH layers. The pull speed is 305 mm/minute in a T-peel configuration. The results are shown in Fig. 15.61. Like the blown film results, the peel strength increases with increasing process time. Morris [61] also finds the peel strength of LDPE to EAA copolymer increases with increasing process time in coextrusion coating.

While the peel strength increases with an increasing process time for both processes and all six tie resins, there are differences that merit closer examination. In the blown film study, the peel strength curves for ADH1 and ADH2 fall on top of one another, even though ADH2 has three times as much chemical functionality. This suggests that the reaction kinetics between the EVOH and anhydride groups on the tie resins is faster than the process time. If the chemical reaction were the limiting factor, the peel strength curve would be expected to be shifted upwards for ADH2, especially at short process times. Fig. 10.50 illustrates that tie resins are often designed with an excess amount of functionality to avoid variability in the application; the peel strength reaches a plateau as the functionality is increased. This will depend on the process as discussed below, but at least for these blown film experiments ADH1 and ADH2 appear to be on the plateau.

A stress mechanism is consistent with the results. As described in Chapter 14, increasing the process time in blown film decreases the stress during blowing. The process time is inversely related to the rate of strain. This can influence peel strength in a several ways. First, the reduced orientation resulting from the lower stress at a long process time may change the physical properties of the film. As described in Sections 10.2.3 and 15.1.4.3, the bending and elongation of the peel arm during the peel test affects the peel energy. Chapter 14 describes how the coextrusion process affects such properties as elongation at break and yield strength. More directly, stress imparted during the orientation of the film can be frozen in during quenching if the polymer has not had time to relax. Wu [29] describes how residual stress generally

TABLE 15.12 Tie resins used in coextrusion coating experiment.

Grade	Type	Description
Tie A	LLDPE	Standard anhydride level
Tie B	LLDPE	3X standard anhydride level
Tie C	LDPE	2X standard anhydride, rubber modified

Source: Data from B.A. Morris, Effect of time, temperature and draw down on interlayer peel strength during coextrusion coating, film casting and film blowing, in: TAPPI PLACE. Conference, 2008.

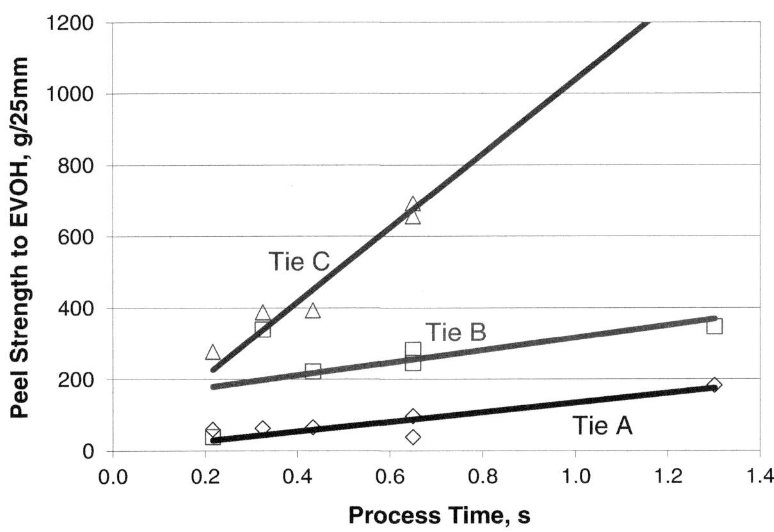

FIGURE 15.61 Peel strength to EVOH versus process time for coextrusion coating. Tie resins are described in Table 15.12. *From B.A. Morris, Effect of time, temperature and draw down on interlayer peel strength during coextrusion coating, film casting and film blowing, in: TAPPI PLACE. Conference, 2008. Reproduced with permission from The Dow Chemical Company.*

reduces peel strength. Finally, subtle changes in the viscoelastic properties can affect the fracture energy—the energy required to create new interfacial area as the peel is made.

The work of Zhang et al. [62] Provides an example of the effect of residual stress on peel strength. The interlayer adhesion between various layers of a 20-layer coextruded sheet of repeating layers of PS/PMMA is measured. These are two immiscible polymers with poor adhesion. The shear stress during extrusion is higher near the channel wall. Peel strength measurements on the quenched sheets show that the interfacial adhesion near the outside of the sheet, where the residual stress is highest, is about one-third lower than near the center.

Patham et al. [63] analyze the role of interfacial stress on adhesion between compatible polymers in coextrusion. The adhesion between compatible polymers is determined by chain interdiffusion across the interface, which in turn is influenced by the initial configuration of chain ends before contact. Chains initially extended parallel to the interface will have less interdiffusion than those perpendicular to the interface. They make a three-layer cast film with the structure (cap layer−PC−[PC + ABS]) using a multimanifold die and use a finite element model to compute the stress at the interface where the layers first come together. The model uses the nonlinear Phan-Thien-Tanner viscoelastic constitutive equation that includes normal stress in shear and strain hardening in extension. The results show more chain extension parallel to the interface in the PC-ABS layer relative to the PC layer, which is more pronounced depending on the grade of ABS used. The grade with the least amount of chain extension has the highest peel strength. The higher amount of chain stress at the interface corresponds to lower peel strength.

The behavior of ADH3 in Fig. 15.60 further supports the stress hypothesis. The peel strength curve for this tie resin is parallel to the curves for ADH1 and ADH2 (same slope) but is shifted upwards to higher levels of peel strength. This can be explained by examining the equation for fracture energy. As described by Eq. (15.34), the fracture energy is often written as the product of two factors, W_a and Φ. Here W_a represents the bonding at the interface (from either chemical interaction or diffusion) and Φ represents the local energy dissipated during the deformation of the polymer as the peel front propagates through. Since Φ is a function of the viscoelastic properties of the polymer, a rubber modified tie resin is likely to have a greater value of Φ than a nonrubber modified one.

The anhydride level in ADH3 is intermediate to that of ADH1 and ADH2. Since ADH1 and ADH2 fall on the same peel strength versus process time curve, this suggests that W_a is constant. The effect of rubber modification is to increase G_a through an increase in Φ. The curve for ADH3 is simply shifted upwards with the same slope.

The coextrusion coating results have different behavior. Comparing Tie A and B, increasing the anhydride level results in a shift upwards to higher values of peel strength but does not change the slope of the curves. The rubber modified Tie C has a greater slope than the curves for Tie A and B. Both of these observations suggest the formation of adhesive bonds may be limited by the time scales involved in coextrusion coating (and, since the process is similar, for cast film). The first result, increasing the anhydride level from Tie A to Tie B, directly supports this: if the extent of reaction is rate limited, increasing the reactant concentration should result in more bonds forming. This shifts the curve for Tie B upwards. This differs from the blown film results and is a consequence of the shorter process times for the coating process. If the reaction is not complete, increasing the process time should allow for more reaction (or diffusion in the case of nonreacting interfaces) to take place.

The second observation can be shown to support the reaction hypothesis by referring to Eq. (15.34). Unlike the blown film case, W_a increases with increasing process time due to more time for reaction. The product of an increasing W_a with a higher value of Φ due to rubber modification should result in a higher slope. This is exactly what is seen for Tie C in Fig. 15.61.

15.2.2 Time, temperature and area creation

The differences highlighted by Figs. 15.60 and 15.61 are the result of different time scales and temperature profiles for the blown and coating/cast film processes, and how these relate to new interfacial area creation. Here cast film and coextrusion coating are lumped together since they have similar time scales.

There are several time scales to consider. One is the contact time prior to the die exit. While significant adhesion occurs in the die (and feedblock if one is used), this time scale will be ignored here (but is taken up in Section 15.2.4) since substantial new interfacial area is created after the die exit as the film/coating is drawn down to its final thickness. This dilutes the initial bond density. Two postdie time scales are considered—the time between the die exit and when the polymer first reaches its crystallization or solidification temperature (process time) and the time it takes crystallization to be completed once it has begun (crystallization time). The former has already been shown to be important. The latter may be important because up to the point where the polymer fully crystallizes there may be some mobility of the molecules that can affect adhesion. The process time for blown film is about an order of magnitude greater than for cast film and coextrusion coating: 1–10 seconds versus 0.1–2 seconds. This can be seen directly by comparing Figs. 15.60 and 15.61.

The crystallization time for blown film is related to the temperature plateau near the frost line, shown in Fig. 14.3. As a polymer crystallizes, it liberates heat (the latent heat of crystallization). In the blown film process, the cooling of the melt overshoots, resulting in a plateau in the temperature as the heat transfer from the surface cannot keep up with the heat generated by the crystallizing polymer (see Chapter 14). The width of this plateau is found to increase with increasing process time, as shown in Fig. 14.7. While investigators have found that the polymer continues to crystallize beyond the plateau region, the mobility of the polymer chains is likely hindered enough to use the plateau time as the crystallization time. The study by Morris [64] described in Chapter 14 shows that the crystallization time is on the order of 2–5 seconds. This will vary depending on the inherent crystallinity of the polymer and whether inner bubble cooling is used.

The crystallization time is difficult to measure for cast film and extrusion coating. Instead, the melt curtain temperature and nip heat conduction models developed in Sections 15.1.2 and 15.1.3 are used. Once the melt curtain contacts the chill roll, heat is rapidly removed. The results of the model simulating the conditions of the coextrusion coating trial are given in Fig. 15.62. The temperature drops only a few degrees in the air gap and very rapidly cools once contact with the chill roll is made. A close-up of the model calculations after contact with the chill roll is shown in Fig. 15.63. There is only a brief temperature plateau as the latent heat of crystallization is removed. The crystallization time is on the order of 5 ms, three orders of magnitude less than for blown film.

The increase in the interfacial area during drawing can dilute the density of bonds that have formed per unit area. The relationship between when the new area generation occurs, temperature (needed to form new bonds), and time available for new bond generation is important. The increase in the interfacial area is related to the elongation of the

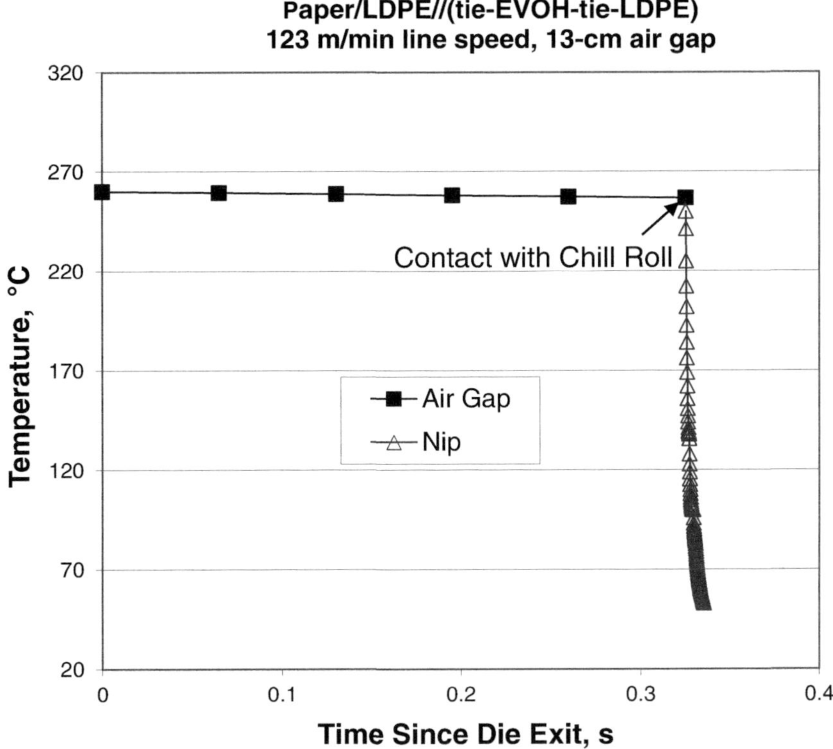

FIGURE 15.62 Calculation of temperature profile in the air gap and nip for the coextrusion coating experiment described in Fig. 15.61. *From B.A. Morris, Effect of time, temperature and draw down on interlayer peel strength during coextrusion coating, film casting and film blowing, in: TAPPI PLACE. Conference, 2008. Reproduced with permission from The Dow Chemical Company.*

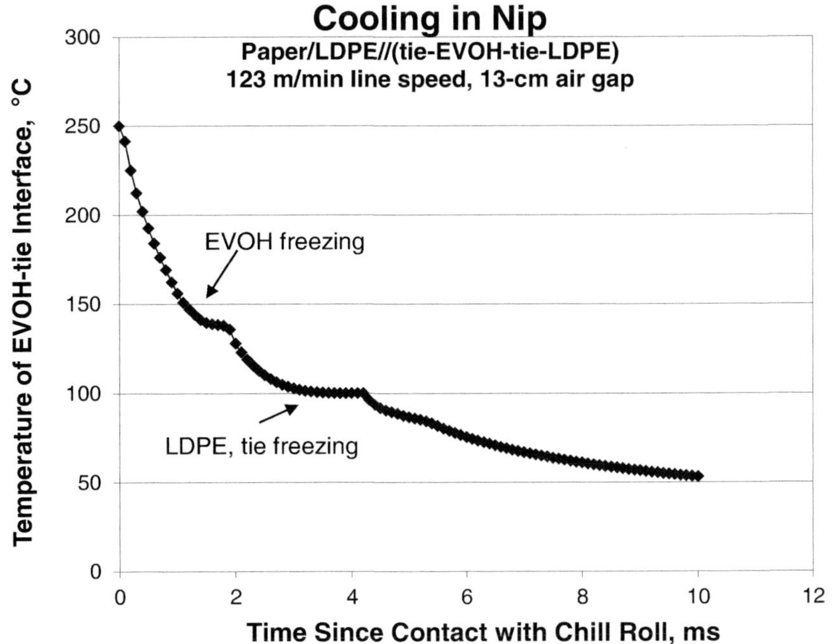

FIGURE 15.63 Close-up of temperature profile in the nip from Fig. 15.62 computed from heat conduction model. *From B.A. Morris, Effect of time, temperature and draw down on interlayer peel strength during coextrusion coating, film casting and film blowing, in: TAPPI PLACE. Conference, 2008. Reproduced with permission from The Dow Chemical Company.*

polymer during processing. As the polymer is stretched, its thickness decreases. To conserve mass, the area increases proportionally:

$$\frac{A(z)}{A_0} = \frac{X_0}{X(z)} \frac{w_0}{w(z)} \frac{\rho_0}{\rho_f} \tag{15.38}$$

where $A(z)$ = interfacial area as a function of distance from the die exit, A_o = initial interfacial area at die exit, $X(z)$ = thickness of the melt curtain as a function of distance from the die exit, w_0 = initial width at die exit for flat dies and initial circumference for tubular dies, and $w(z)$ = width of the melt as a function of distance from the die exit.

For the rest of the analysis the density ratio is ignored. For polyolefins, the density ratio is typically 0.7–0.8.

For extrusion coating and cast film, a simple mass balance shows that the ratio of the area at the nip to the area at the die exit is equal to the DDR; for blown film the area ratio (at frost line to die exit) is equal to the product of the BUR and take-up ratio or DDR:

$$A_f/A_o = \text{DDR} \quad \text{for extrusion coating and cast film.} \tag{15.39}$$

$$A_f/A_o = (\text{DDR})(\text{BUR}) \text{ for blown film.} \tag{15.40}$$

Increasing the DDR directly increases the amount of the new interfacial area created after the die exit.

Fig. 15.64 shows the calculated change in the interfacial area in the air gap for the extrusion coating trial. Superimposed on the graph is the temperature profile from Fig. 15.62. The interfacial area is created while the polymer is still very hot, which is important for bond formation and interlayer diffusion. The creation of new interfacial area, however, exponentially increases with distance from the die exit (z) so that a large increase in area occurs just before the polymer contacts the chill roll ($z/L = 1$). From Fig. 15.62, once contact with the chill roll is made, quenching occurs quickly, crystallization time is short, and there is little time for new bond formation.

Chapter 14 describes a study by Morris [64] that uses an IR thermometer and video techniques to measure the temperature, velocity and bubble diameter profiles along the bubble in the blown film process. Data for a coextrusion of (HDPE-EVOH) from this study is shown in Fig. 15.65, along with the calculation for the area ratio, which is the product of the bubble diameter and MD velocity profiles. The area ratio exponentially increases with distance from the die exit, similar to what is found for cast film/extrusion coating. The increase in interfacial area ceases at around $z/L = 1$, where L is the frost line height for the EVOH layer (which freezes first). The temperatures are much lower at this point than what was found in the extrusion coating example. The time available for bonding, however, is much longer. The HDPE layer does not start to freeze until about $z/L = 1.5$ or about 3 seconds after the interfacial area stops increasing. If a tie layer were present, it would not start freezing until even later since it would typically have an even lower freezing point. This is because, in coextrusion blown film, the bubble shape and velocity profiles are determined by the layer that freezes first (see Section 14.1.2). This can have an impact on the physical properties of the film. It may also have an effect on adhesion, shifting the point where the interfacial area ceases to grow to a region of higher temperature and, hence, greater mobility and reactivity of the molecules, and allowing more time for adhesion to take place before crystallization begins. Since many commercially important coextruded blown films involve a barrier polymer with a high

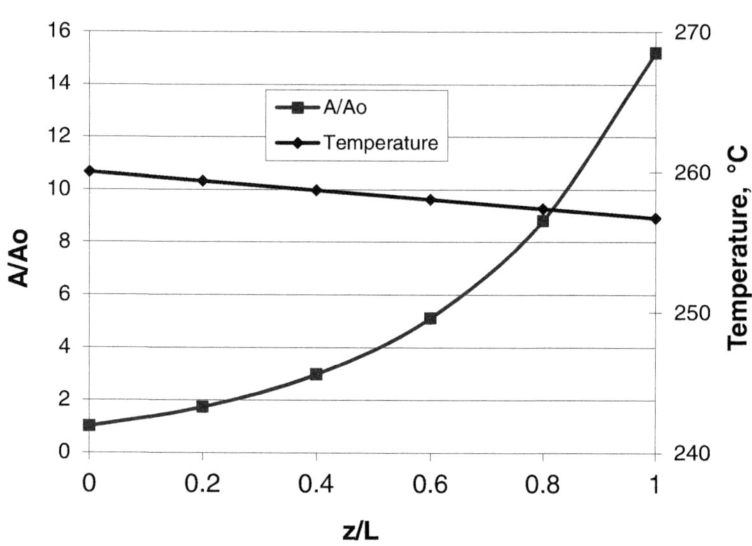

FIGURE 15.64 New interfacial area creation and temperature in the air gap in coextrusion cast film or coextrusion coating. The quantity z/L is the dimensionless distance from the die exit (L = air gap distance). *From B.A. Morris, Effect of time, temperature and draw down on interlayer peel strength during coextrusion coating, film casting and film blowing, in: TAPPI PLACE. Conference, 2008. Reproduced with permission from The Dow Chemical Company.*

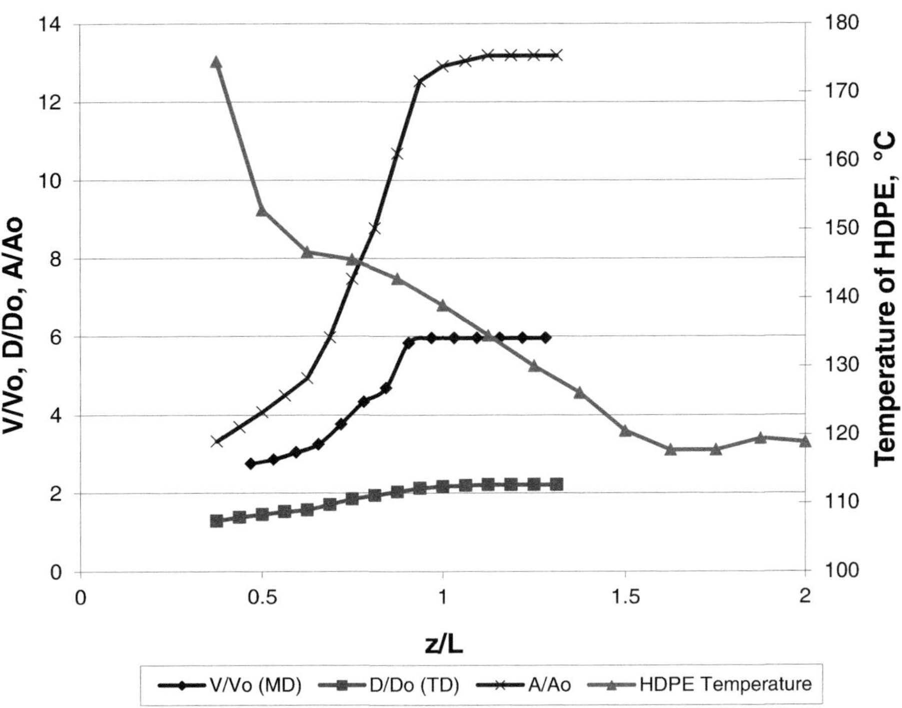

FIGURE 15.65 Temperature, bubble diameter, velocity in the machine direction and new interfacial area creation in blown film as a function of distance from the die exit. Based on data from Morris [64]. z = distance from die exit and L = EVOH frost line height. *From B.A. Morris, Effect of time, temperature and draw down on interlayer peel strength during coextrusion coating, film casting and film blowing, in: TAPPI PLACE. Conference, 2008. Reproduced with permission from The Dow Chemical Company.*

freezing point (e.g., PA or EVOH) and a polyolefin-based tie resin with a much lower freezing point, this effect may be fairly widespread.

To summarize, cast film/extrusion coating is characterized by high temperatures and an exponential increase in interfacial area in the air gap followed by very rapid freezing of the polymer layers. There may be little time for bond formation in the newly created interfacial regions. Blown film also shows a rapid increase in the interfacial area near the frost line of the layer that freezes first. The time available for bonding before crystallization is complete is several orders of magnitude longer than for cast film/extrusion coating. These differences in time, temperature, and area creation during drawing may help explain the differences in the behavior of the peel strength versus process time curves. The development of adhesion in these processes is complex, and other factors besides the reaction kinetics and diffusion also likely play a role, including the development and relaxation of stress and orientation during drawing and quenching and the effect these may have on crystallinity and crystalline morphology.

The work has several practical implications for converters of coextrusion cast film or extrusion coating. One is that tie resin formulations for cast film or extrusion coating may require higher levels of functionality than those used in blown film. The work shows that premium tie resins with rubber and other modifiers may be particularly beneficial for these processes.

Song et al. [65] provide a formula for determining the effective contact time between layers during coextrusion taking into account the bulk motion (convection) and interfacial area creation:

$$t_{\mathrm{eff}} = \frac{1}{Q} e^{\int_0^z f(z)dz} \left\{ \int_0^z e^{-\int_0^z f(z)dz} X(z) w(z) dz \right\} \tag{15.41}$$

where t_{eff} = effective reaction time or contact time, z = machine direction, $X(z)$ = thickness, $w(z)$ = width, and $f(z)$ = areal dilation.

The areal dilation is given by

$$f(z) = \frac{1}{X(z)} \frac{dX(z)}{dz} \tag{15.42}$$

Eq. (15.41) is derived from a conservation expression with the simplifying assumption of plug flow.

Song et al. illustrate the approach by considering the reaction time in a sheet line, calculating the contributions from the feedblock to the die entrance, the spreading in the die body, flow in the die land and finally the drawdown from the die exit to the chill roll. Laboratory results for a coextrusion of polyurethane with an amine-modified polyethylene are presented. The urethane reacts with the amine groups to form a block copolymer at the interface. The peel strength is found to correlate with the effective reaction time, as is shown in Fig. 15.66. As is described in Chapter 10, Song et al. also measure the number of reacted groups at the interface and find the density of bonds per area increases with the effective reaction time.

A significant portion of the effective contact time in Song's experiment occurs in the draw down region after the die exit. They find the peel strength decreases with increasing DDR, as shown in Fig. 15.67. In these experiments, the volumetric flow rate, Q, is held constant and DDR is increased by increasing the line speed, v_f. Although Song's effective contact time is not exactly the same as TIAG or t_f [Eq. (15.4)], the relationship between TIAG and DDR can be used to understand the trend in Fig. 15.67. Since $\mathrm{DDR} = v_f/v_0$ and $Q = X_0 w_0 v_0$, Eq. (15.1) can be re-written in terms of DDR:

$$\mathrm{TIAG} = \left[\frac{L X_0 w_0}{Q}\right] \frac{1}{\mathrm{DDR}} \tag{15.43}$$

where TIAG = time in air gap = L/v_f, L = air gap, X_0 = die gap, w_0 = die width, Q = volumetric flow rate, and DDR = draw down ratio.

Under the conditions of Song's experiment, the term in brackets is constant. The peel strength is inversely related to DDR, which in turn is inversely related to TIAG. This suggests that peel strength increases with time in the air gap. Replotting the data in terms of TIAG and t_f in Fig. 15.68 confirms this. This is the same relationship found by Morris [7].

Kiparissoff-Bondil et al. [66] make coextrusion cast sheet/films with the structure (PA6−[PP + PP-g-MAH]−PA6) at several thicknesses and layer ratios. Bond density at the PA6-tie interface is measured by dissolving away the PA6

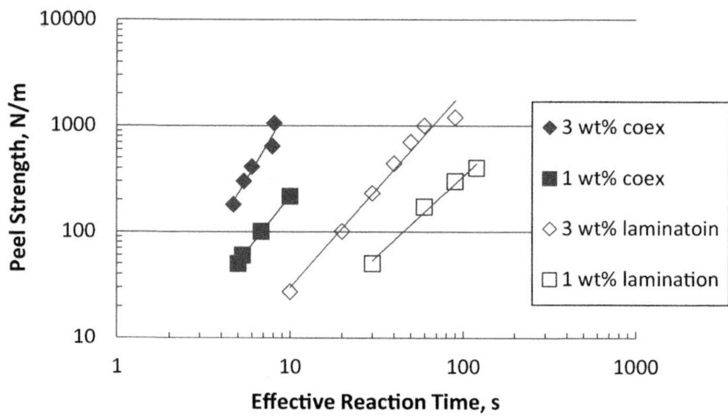

FIGURE 15.66 Peel strength versus effective reaction time for bonding of polyurethane to an amine modified polyethylene via coextrusion and thermal lamination. Weight percent refers to the amount of amine functionality on the polyethylene. *Reproduced with permission from P.P. Shirodkar, V. Firdaus, H. Fruitwala, Scale-up of LLDPE blown film extrusion, Plast. Eng. (1994) 27.*

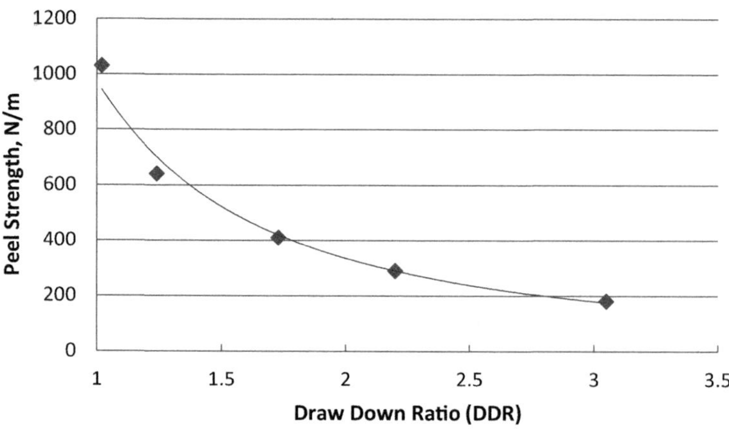

FIGURE 15.67 Effect of draw down ratio (DDR) on peel strength performance for a coextrusion of polyurethane and amine modified PE. *Data from J. Song, R.H. Ewoldt, W. Hu, H.C. Silvis, C.W. Macosko, "Flow accelerates adhesion between functional polyethylene and polyurethane, AIChE J., 57 (2011) 3496–3506.*

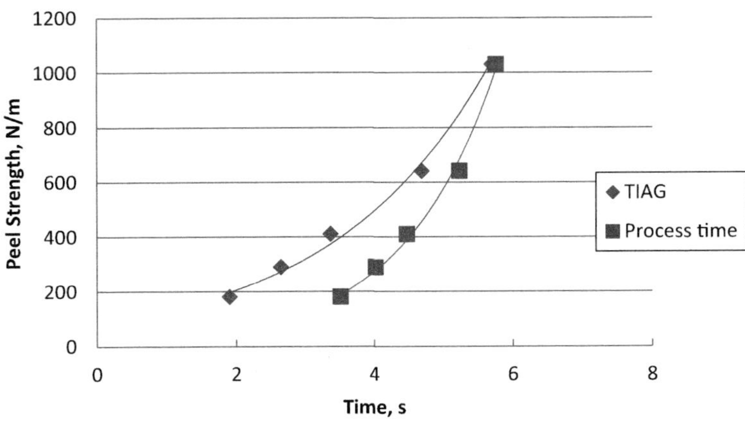

FIGURE 15.68 Data from Fig. 15.67 replotted in terms of TIAG and process time [Eq. (15.4)]. *Data from J. Song, R.H. Ewoldt, W. Hu, H.C. Silvis, C.W. Macosko, "Flow accelerates adhesion between functional polyethylene and polyurethane, AIChE J., 57 (2011) 3496–3506.*

and using ESCA to measure the density of reacted groups. At long contact times, the bond density saturates at a finite level due to steric hindrance and entropy effects. Their measurements of bond density for their coextruded structures are considerably less than the saturation density, which they attribute to the dilution effect due to stretching (draw down). They model the coextrusion process and determine the surface area ratio as a function of distance from the point where the layers meet. They couple this with diffusion of PP-g-MAH to the interface (but neglect convection). Their model results qualitatively agree with their experimental results but under predict their bond density measurements. They suggest including convection may provide better agreement.

Barraud et al. [67–69] from the same research group include convection in their model. For their experimental study they make (PA6–[PP + PP-g-MAH]–PA6) sheet structures using a vacuum lamination process. The specimens are annealed at 250°C for 1 hour to ensure bond density saturation. To simulate draw down in the film fabrication process they built an apparatus to quickly heat, stretch and cool the specimens. The process involves immersing the sheet in a hot oil bath at 250°C for 30 seconds, quickly removing the specimen from the oil bath (0.01 second), stretching (at various rates and times) and cooling with air jets (cooled with liquid nitrogen). The cooling rate simulates contact with a chill roll. Two types of stretching experiments are conducted: (1) constant stretch velocity but varying stretch distance and (2) constant stretch distance at varying stretch velocity. At very fast stretch velocity the stretching is essentially isothermal, followed by decreasing temperature with cooling. Cooling is faster with samples stretched by a greater distance since the resulting films are thinner.

They define an "open time" as the time before the sample is cooled to the crystallization temperature of the PP. This is the time available for reaction. They propose that the reaction stops once the PP starts to freeze. Diffusion and convection of MAH to the interface can take place prior to that. Their results show that bond density decreases with stretching distance but not as much as is predicted from a simple dilution effect (through the increase in interfacial area). They conclude that some reaction takes place during stretching. The case where the stretching distance is held constant and the stretch velocity is increased is similar to the process time studies of Morris described earlier. At high velocity (short open time) the bond density is 3x lower than the saturated bond density—there is not enough open time for reaction to compensate for the dilution of bond density due to new interfacial area creation. At low velocity (long open time) the reaction compensates for the dilution due to area creation. This is in qualitative agreement with the process time studies of Morris (Section 15.2.1) and the time scale analysis of the cast and blown film processes above.

The model developed by Baurrad et al. includes diffusion and convection of maleic anhydride to the interface. It accounts for finite reaction kinetics once the PP-g-MAH has reached the interface and slowing down of the reaction kinetics as the concentration at the interface approaches saturation (steric hindrance effect). The results are consistent with the time scale analysis described above:

- Bond density initially decreases with time from the dilution of the initial bond density due to stretching. Bond density then increases as bonds are formed from the reaction as the PP-g-MAH reaches the interface. The minimum bond density decreases with increasing stretch velocity (decreasing process time). This corresponds to the cast film/extrusion coating results of Morris.
- When the stretch time is much longer than the reaction time there is little reduction in bond density—the dilution effect from new area creation is compensated by new reaction. This corresponds to the blown film results of Morris.

The model results also show that increasing the diffusion coefficient of the PP-g-MAH (simulating lower molecular weight) changes the curves—there is less of an initial drop in bond density during stretching and a faster recovery.

15.2.3 Process time master curve

The peel strength versus process time curves for blown film and coextrusion coating in Figs. 15.60 and 15.61 have different time scales and slopes. Other results reported by Morris [59,60] indicate that different tie resins have different slopes. This is illustrated in another set of experiments by Morris [7,70,71]. Two sets of films are made, one by blowing and the other by casting.

The blown film study is conducted using the same equipment and procedures as in the original work [52]. A three-layer blown film line equipped with a 76 mm (3 inch) Brampton Slimline (stackable type) die with a 50 mm (2 inch) opening and a Uni-Flow dual air lip air ring is used. Three 25-mm (1-inch) diameter extruders feed the die. No inner-bubble cooling is used. Melt temperature, die gap (1 mm), final film thickness and blow-up ratio (2:1) are kept constant. Process time is varied by changing frost line height (through adjusting quench air velocity and temperature) and haul-off speed.

The test structure is (HDPE−tie−EVOH). The layer thicknesses are 37.5/12.5/20 μm (1.5/0.5/0.8 mils), respectively. Final barrel temperatures are 230°C for all three extruders, as is the die temperature. The HDPE has a density of 960 kg/m^3, a MFI of 0.75 g/10 minute (190°C/2160 g), a melt point (Tm) of 137°C and freezing point (Tc) of 115°C. The EVOH is a 32mole% ethylene grade with an MFR (210°C/2160 g) of 3.0 g/10 min, Tm of 185°C and Tc of 157°C. A description of the tie resins is given in Table 15.13. All are anhydride-modified polyethylene or ethylene copolymers.

The cast film samples are made on a coextrusion coating line. The 50-μm (2 mils) structure is (LDPE−tie−EVOH−tie−LDPE) with nominal layer distribution of (13/4/6/4/23 μm). The film is coextrusion coated onto 48-gauge (12-μm) OPET with the thinner LDPE side on the OPET surface. [The OPET is stripped away from the film before testing.] The tie and LDPE extruders are set at 260°C (500 °F) and the EVOH extruder at 232°C (450 °F). The Cloeren edge bead reduction die is set at 252°C (485 °F). The air gap is varied from 13 to 25 cm (5-10 inches) and line speed from 122 to 244 m/minute (400-800 feet/minute). The die gap is 0.69 mm (27 mils). The LDPE has a density of 0.918 g/cc and MFI (190°C/2160 g) of 7 g/10 minute. The EVOH contains 44 mol% ethylene and has a MFI (190°C/2160 g) of 5.5 g/10 minute. The tie resins are described in Table 15.13.

For the blown films, peel strength is measured in the machine direction between the EVOH and tie layer. One-inch strips are pulled in a T-peel configuration at a crosshead speed 305 mm/minute. Results reported are an average of five determinations.

The cast films are too thin to conduct standard peel tests. The LDPE is too stretchy and breaks before the pull tabs can be established. Instead, the cast films are separated from the OPET and heat-sealed lengthwise (perpendicular to the machine direction) at the thin LDPE-LDPE interface at 132°C (270 °F), 0.5 s dwell and 0.27 MPa (40 psi). Values of peel strength represent an average of at least three 12.7 mm (0.5 inch) wide strips cut from the sealed area and tested at a pull rate of 150 mm/minute (6-inches per minute), corrected to reflect 25-mm (1 inch) wide strips. Testing is carefully observed to note when the LDPE seal layers break through to the tie-EVOH interface and seal strength is recorded from that point forward. Iodine staining is used to confirm the failure mode. The long seal length is necessary in order to allow time for the seal to break through to the weakest interface.

The results from the two trials are plotted in Fig. 15.69. The curves have distinct characteristics:

- As in previous studies, the peel strength to EVOH increases with increasing process time.

TABLE 15.13 Tie resins used in coextrusion blown and cast film study.

Resin	Base resin	Process	MFI* (g/10 min)	Young's modulus (MPa/kpsi)
Tie 1	EVA (9% VA)	BF	1	132/19
Tie 2	EVA (25% VA)	BF	2	34/5
Tie 5	LLDPE	BF	1	455/66
Tie 6	mLLDPE	BF	1	414/60
Tie 7	LLDPE (rubber modified)	BF, CF	3	338/49
Tie 8	LLDPE	CF	4	428/62

*2160 g/190°C.
Source: Reproduced with permission from B.A. Morris, Effect of process and material parameters on interlayer peel strength in coextrusion coating, film casting and film blowing, J. Plastic Film. & Sheeting, 26 (2010) 343−376.

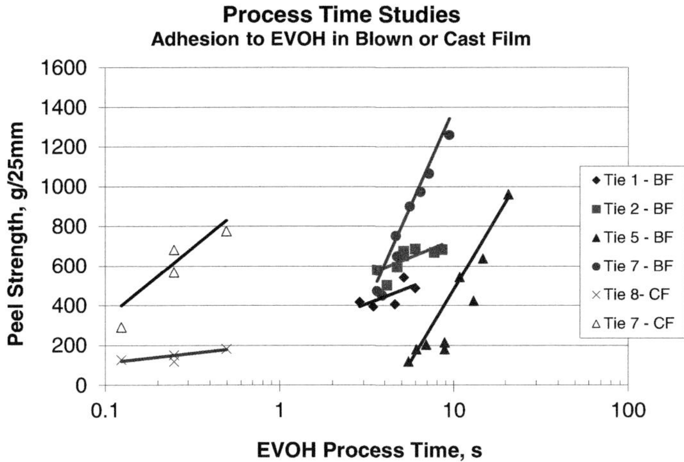

FIGURE 15.69 Peel strength to EVOH results from blown and cast film coextrusion trial. (BF = blown film; CF = cast film). *Reproduced with permission from B.A. Morris, Effect of process and material parameters on interlayer peel strength in coextrusion coating, film casting and film blowing, J. Plastic Film. & Sheeting, 26 (2010) 343–376.*

- The process time for blown film is longer than for cast film.
- In the blown film results, tie resins based on linear polymer (LLDPE) give a steeper slope than those based on LDPE or EVA, which contain long-chain branching (LCB).
- Lower modulus tie resins of the same chemistry have curves that are shifted upwards to higher values of peel strength (compare Tie 5 with 7 and Tie 1 with 2).
- Tie resins of similar composition exhibit a steeper slope in blown film than in cast film (compare Tie 5 with 8 and Tie 7 with itself in the two processes).

At first glance, each tie resin and process has its own peel strength versus process time curve, making general use of this information difficult. The trends in the data, however, suggest a way to transform the data into a useful master curve using concepts from a stress analysis.

The process time in blown and cast film is inversely proportional to the rate of strain (see Sections 14.1.2 and 15.1.2.1). Increasing the process time decreases stress by decreasing the rate of strain. The blown film results suggest that the tie resins based on linear polyethylene are more sensitive to changes in stress than the branched polymers since the slopes of the linear polyethylene tie resins curves are steeper. The influence of LCB and broader MWD on the response of the peel strength to process time is likely related to differences in elongational viscosity and relaxation behavior. Elongational viscosity is known to increase with increasing LCB due to entanglement effects. Likewise, the ability of a polymer to relax is hindered by LCB. An estimate of average relaxation time can be obtained from the inverse of the frequency where the storage modulus (G') and the loss modulus (G'') are equal in dynamic mechanical measurements (DMA). Results of DMA measurements at 190°C for three of the tie resins are given in Table 15.14. The EVA-based tie resin has about an order of magnitude larger relaxation time than the two LLDPE-based ones. The flatness of the peel strength versus process time curve for the EVA-based tie resin may be due to the inability of the EVA to respond to greater times for relaxation during the process.

The Deborah number (De) is introduced in Eq. (15.5) as the relaxation time of the polymer divided by the process time. The De number relates to the ability of the polymer to react to stresses during the time scale of the process. Large De (>1) signifies the polymer behaves like an elastic solid whereas small De (<1) indicates the polymer acts like a viscous fluid. Polymer relaxation time decreases with increasing temperature and generally follows an Arrhenius relationship:

$$t_{\text{relaxation}}(T) = a_T(T)t_{\text{relaxation}}(T_{\text{ref}}) \tag{15.44}$$

where $a_T(T) = e^{E_a/RT} - e^{E_a/RT_{ref}}$ T = characteristic temperature of the process [K], T_{ref} = reference temperature [K], 190°C = 463 K, R = ideal gas constant, and E_a = activation energy.

The characteristic temperature for the blown film process is taken to be 125°C. This is based on the analysis in Chapter 14 showing that the maximum rate of elongation occurs near the frost line of the layer that freezes first in the coextrusion. For this structure, the EVOH freezes first and has a freezing temperature of around 125°C where the tie resin is still molten. The characteristic temperature for the cast film structures is taken to be 260°C. This is based on the model calculations presented in Fig. 15.62. Using Eq. (15.44) with an activation energy of 30 kJ/mol, relaxation times for each of the tie resins at the characteristic temperature for the cast and blown film processes are computed and recorded in Table 15.15.

TABLE 15.14 Relaxation time data for three tie resins from Fig. 15.67.

Tie	Base resin	Frequency @ $G' = G''$, s^{-1}	Estimated t_{relax} (190°C), s
Tie 1	EVA	7.5	0.130
Tie 5	LLDPE	100	0.009
Tie 6	mLLDPE	140	0.007

Source: Reproduced with permission from B.A. Morris, Effect of process and material parameters on interlayer peel strength in coextrusion coating, film casting and film blowing, J. Plastic Film. & Sheeting, 26 (2010) 343–376.

TABLE 15.15 Characteristic relaxation time for tie resins at the characteristic temperature of the blown or cast film process.

Tie	Base resin	Process	t_{relax}, s	Vertical shift factor
Tie 1	EVA	BF	0.464	1.25
Tie 2	EVA	BF	0.464	0.67
Tie 5	LLDPE	BF	0.032	25
Tie 6	LLDPE	BF	0.025	20
Tie 7	LLDPE	BF	0.018	5
Tie 7	LLDPE	CF	0.0018	5
Tie 8	LLDPE	CF	0.0032	25

Notes: Data is shifted to 125°C for blown film and 260°C for cast film using Eq. (15.44). *BF* = blown film; *CF* = cast film.
Source: Reproduced with permission from B.A. Morris, Effect of process and material parameters on interlayer peel strength in coextrusion coating, film casting and film blowing, J. Plastic Film. & Sheeting, 26 (2010) 343–376.

Morris [7,71] finds that all the curves from Fig. 15.69 collapse onto a single master curve when the x and y-axes are transformed, as shown in Fig. 15.70. For the x-axis, the process time is divided by the relaxation time of the tie resin (at the characteristic temperature of the process). This should be recognized as the inverse of the Deborah number. Dividing the process time by the relaxation time of the tie resin normalizes the response of the tie resin to changes in stress. This accounts for the different slopes of the linear and branched tie resins.

The vertical shift factors for the y-axis were chosen to obtain a good fit. Fig. 15.71 shows that these vertical shift factors correlate well with Young's modulus of the tie resin. This suggests that the y-axis is related to the peel strength times the modulus. Other material properties such as yield strength may also correlate with the vertical shift factors but no attempt was made to establish such a relationship.

The origin of the vertical shift factors is likely from the local viscoelastic energy dissipation of the polymer. A more explicit equation is derived by equating the energy balance from Eq. (15.34) with a force balance (stress analysis) on the peel test. The result, derived in Section 10.2.3, is

$$K_I^2 = G_a E \tag{15.45}$$

where K_I^2 = stress intensity factor, G_a = fracture energy, and E = Young's modulus.

K_I is equal to the maximum stress at the crack tip (or peel front) divided by the imposed stress. Fracture occurs when the maximum local stress equals the local cohesive or adhesive strength. Typically failure at the interface occurs at imposed loads that are considerably less than the idealized strength of the interface. The stress intensity factor characterizes the amplification of the stresses at the crack tip that leads to failure. Thus the stress intensity factor takes into account the adhesive strength at the interface and the material parameters of the tie resin.

There are a number of limitations to the analysis. The relaxation times are estimates based on linear viscoelastic measurements when elongation flow dominates in these processes. The thickness of the tie and other layers likely play

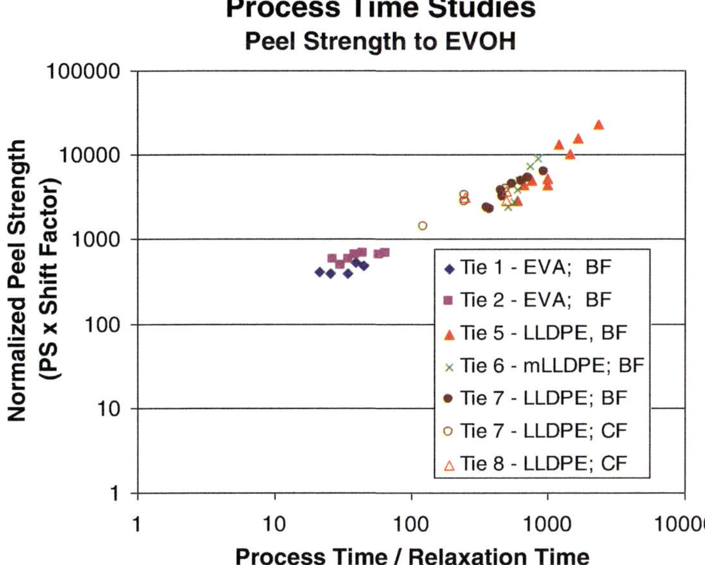

FIGURE 15.70 Transformed data into master peel strength versus process time curve. *Reproduced with permission from B.A. Morris, Effect of process and material parameters on interlayer peel strength in coextrusion coating, film casting and film blowing, J. Plastic Film. & Sheeting, 26 (2010) 343–376.*

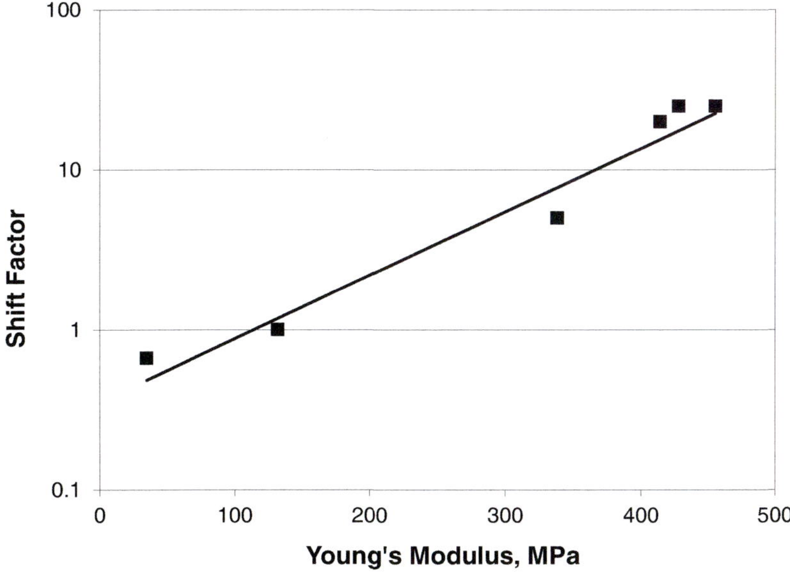

FIGURE 15.71 Vertical shift factors for Fig. 15.70 plotted vs Young's modulus. Modulus was measured from tensile bars. *Reproduced with permission from B. A. Morris, Effect of process and material parameters on interlayer peel strength in coextrusion coating, film casting and film blowing, J. Plastic Film. & Sheeting, 26 (2010) 343–376.*

a role. The tie resins themselves are often complex multicomponent alloys that defy the simple characterizations given here. The chemical reaction rate effects found to be important in cast film (see Section 15.2.1) are ignored, although Eqs. (15.44) and (15.45) may implicitly take these into account.

Nevertheless, Fig. 15.70 represents a powerful tool for better understanding of the effect of process and material parameters on adhesion in coextrusion. For the first time, a correlation for peel strength performance in similar films made on two very different processes has been obtained. The approach takes into account the differences in the stress-strain history (through t_f and $t_{\text{relaxation}}$), temperature (through the shift in $t_{\text{relaxation}}$), and time (through t_f) of the process as well as differences in molecular architecture (LCB, MWD, crystallinty and modulus) of the tie resin (through E and $t_{\text{relaxation}}$). The work suggests that stress may be as important as the kinetics of chemical reaction and diffusion at the interface for determining the process time response of the cast film process. The work shows that increasing output (inversely related to process time) reduces peel strength, but that this effect can be mitigated to some extent by the appropriate choice of the tie resin (modulus and relaxation time). The work also points to how tie resins might be optimized for a given process and set of processing parameters.

15.2.4 Flow accelerated reaction rate

Song et al. [65] find that flow accelerates the chemical interaction at the interface in coextrusion. The formation of block copolymers through a reaction at the interface between polyurethane and amine-modified polyethylene is studied in both coextrusion and thermal lamination. The coextrusion results are discussed in Section 15.2.2, where the peel strength is shown to correlate with effective reaction time. The thermal lamination experiments are conducted using a laboratory press at different hold-up times, followed by quenching in ice water. The lamination is conducted so that the layers are not squeezed down in thickness during the process. The results for the polyurethane-amine modified PE are shown in Fig. 15.66 for both the coextrusion and thermal lamination. The time to achieve a given peel strength value is two orders of magnitudes shorter for the coextrusion. Song et al. directly relate peel strength to the formation of block copolymers at the interface using X-ray photoelectron spectroscopy (XPS, also known as ESCA), as described in Section 10.3.4.

In a follow-up experiment, Song et al. [72] coextrude PA 6 with maleic anhydride grafted HDPE (HDPE-g-MAH) using a coextrusion sheet line with two die configurations. In one die, the flow is compressive—as the polymer is spread to its final width, it is also compressed in the thickness dimension. This is typical of a standard cast film die. The second die is designed to be noncompressive. Thermal laminations of the same materials are also produced using a laboratory press.

The PA 6/HDPE-g-MAH layer combination is typical of many flexible packaging structures. The anhydride reacts with the amine end groups on the PA 6 to form an imide. Like the polyurethane-amine modified PE combination, XPS shows that the creation of an in situ block copolymer at the interface through the formation of the imide group correlates with the development of peel strength at the interface. The peel strength is found to develop 100 times faster in the compressive die than the noncompressive die, suggesting the formation of the block copolymer is accelerated by the compressive flow. The reaction rate in the compressive die is about 80 times that found in the thermal lamination experiment. The reaction rate of the noncompressive die is about the same as the lamination. Scanning electron microscopy (SEM) shows similar behavior in the morphology for the compressive coextrusion and longtime lamination (high peel strength). Adhesion measurements, XPS, and SEM all show evidence of acceleration of adhesion in coextrusion with the compressive die. Furthermore, experiments on the lab press where the thickness of the laminate is reduced during lamination (compression) have accelerated reaction rates comparable to the coextrusion with the compressive die.

Song et al. discuss possible causes for the accelerated reaction rate during coextrusion:

- **Shear stress**. The interface between the two layers is deliberately placed near the centerline of the flow channel where the shear stress is zero and there is no mixing. This eliminates shear stress as a factor. The DDR is small in their experiments, suggesting that elongational stress is also negligible.
- **Pressure**. This is not expected to be a significant effect for liquid–liquid reactions over the pressure range experienced in their experiments.
- **Crystallinity at interface**. Song et al. measure crystallinity and finds no significant difference in samples made with the different dies. Also, the stress-strain curves for coextrusion and lamination samples are nearly identical.
- **Diffusive flux**. Compressive flow reduces the diffusion length scale. A dimensionless group analysis (Damkohler number, the ratio of reaction rate to diffusion rate) shows, however, that this change is not substantial enough to change the regime from reaction limited to diffusion limited. They conclude that this is not the cause.
- **Flow-induced orientation**. This is known to induce crystallization kinetics as well as break and promote reactions of groups in the polymer backbone.

Song et al. expand upon the last point by computing the Wissenberg number (Wi), which is equal to the shear rate times the relaxation time of the polymer. The orientation of the chains increases with increasing Wi and total strain [ln (DDR)]. The DDR ranged from 1.07 to 3.08 in their experiment and therefore total strain does not vary over a wide range. The Wi number, however, is substantially higher in the compressive die (83) than the noncompressive die (2.7), indicating compressive flow in the die significantly increases chain orientation.

Song et al. speculate that the enhanced reaction rate during orientation is the result of functional polymers being driven to the interface and stretched, increasing the collision probability of the functional groups. Song et al. cite examples from the literature that indicates the compressive/elongational flow in coextrusion or polymer compounding accelerates the reaction rate at the interface. Macosko et al. [73] show the reaction rate between components of a heterogeneous blend during mixing is 1000-fold greater than under quiescent conditions, and still around 300-times greater after accounting for the increase in interfacial area. Bostros [74] reports similar behavior when mixing an anhydride functionalized tie resin and EVOH. The diffusion rate near an interface is an order of magnitude slower than in the bulk [75]. Flow seems to compensate for this by forcing the reactive species to penetrate into the interfacial region, increasing the concentration of reactants at the interface.

Appendix: Relating stress to process control variables in extrusion coating/lamination and cast film processes

Nomenclature

DDR	draw down ratio $= v_f/v_o \sim X_o/X_f$
G	modulus
X_o	die gap
X_f	final coating thickness
L	air gap
m	powerlaw viscosity preexponent
n	powerlaw viscosity exponent
Q	volumetric flow rate
T	temperature of molten web
T_o	temperature of web at the die exit
TIAG	time in the air gap [Eq. (15.1)]
t	time
t_f	characteristic process time between the die exit and nip (air gap time)
V	velocity of an element of polymer
v_f	line speed
v_o	initial speed of polymer as it exits the die
w	width of die or melt curtain (neck-in is ignored)
z	position (distance from die exit)
ε_{tot}	total strain or elongation of the molten web
$\dot{\varepsilon}$	rate of strain or elongation
$\dot{\varepsilon}_{avg}$	average rate of strain (in elongation)
$\dot{\gamma}$	shear rate
$\dot{\gamma}_{app}$	apparent shear rate
η	shear viscosity
η_e	elongational or extensional viscosity
σ_{ext}	stress during extrusion
σ_d	stress during drawing of the molten web

Analysis

The stress on the molten polymer in the extrusion coating and cast film processes is the sum of the stress imposed during extrusion, σ_{ext}, and drawing, σ_d, minus the amount of stress that has relaxed. To represent stress during extrusion, consider pressure-driven flow through two flat plates (as in the land region of the die, where the shear rate is typically the highest). The stress is the product of the shear viscosity and shear rate:

$$\sigma_{ext} = \eta\dot{\gamma} \tag{15A.1}$$

For a non-Newtonian fluid, a power-law model can be used to relate viscosity to shear rate:

$$\eta = m\dot{\gamma}^{n-1} \tag{15A.2}$$

In Table 5.4, an expression for the shear rate at the wall is derived which has the form:

$$\dot{\gamma}_w = \frac{4n+2}{n}\frac{Q}{wX_0^2} \tag{15A.3}$$

Recognizing that $Q = wX_f v_f$ and combining Eqs. (15A.1), (15A.2), and (15A.3), the following relationship for stress imparted in the die land is obtained:

$$\sigma_{ext} = \left(\frac{4n+2}{n}\right)^n \frac{m v_f^n X_f^n}{X_0^{2n}} \tag{15A.4}$$

Note that m is a function of temperature and polymer molecular weight.

The stress imparted during drawing, σ_d, is more difficult to characterize. For a liquid, the stress is proportional to the rate of strain, but for a solid, the stress is related to the strain:

$$\sigma_d = \eta_e \dot{\varepsilon} \quad \text{for a liquid, and} \tag{15A.5}$$

$$\sigma_d = G\varepsilon_{\text{tot}} \quad \text{for an elastic solid.} \tag{15A.6}$$

Since polymer melts are viscoelastic with characteristics of both liquids and solids, it is likely that the stress is related to both $\dot{\varepsilon}$ and ε_{tot}. Both modulus and viscosity are functions of temperature. When dealing with a single material, Eq. (15A.5) and (15A.6) can be combined into the following expression:

$$\sigma_d = f(T, \varepsilon_{\text{tot}}, \dot{\varepsilon}) \tag{15A.7}$$

Relationships for $\dot{\varepsilon}$ and ε_{tot} in terms of extrusion coating and cast film variables are needed.

The velocity of an element of polymer within the molten web is equal to its change in position with time:

$$V = \frac{dz}{dt}. \tag{15A.8}$$

The instantaneous rate of strain is equal to the change in velocity with respect to distance:

$$\dot{\varepsilon} = \frac{dV}{dz} = \frac{1}{V}\frac{dV}{dt}. \tag{15A.9}$$

The total strain is found by integrating the instantaneous rate of strain over time:

$$\varepsilon_{\text{tot}} = \int_0^{t_f} \dot{\varepsilon}\, dt = \int_0^{t_f} \frac{1}{V}\frac{dV}{dt}\, dt = \ln\left(\frac{v_f}{v_0}\right). \tag{15A.10}$$

The ratio of the final velocity to initial velocity is known as the DDR and is related to the ratio of the die gap to final coating thickness:

$$\text{DDR} = \frac{v_f}{v_o} \cong \frac{X_o}{X_f} \tag{15A.11}$$

Here the relationship between DDR and the film gauge is determined by mass balance between the die exit and the chill roll, neglecting density differences and neck-in. Combining Eqs. (15A.10) and (15A.11), the following expression for the total strain is obtained:

$$\varepsilon_{\text{tot}} = \ln(\text{DDR}). \tag{15A.12}$$

Eq. (15A.9) gives the instantaneous rate of strain. For the purpose of this analysis, the average rate of strain will suffice. The average rate of strain is equal to the total change in velocity divided by the change in distance. For the region between the die exit and the chill roll this becomes

$$\dot{\varepsilon}_{\text{avg}} = \frac{\Delta V}{\Delta z} = \frac{v_f - v_0}{L}. \tag{15A.13}$$

Eq. (15A.11) can be used to put (15A.13) into terms of DDR:

$$\dot{\varepsilon}_{\text{avg}} = \frac{v_f}{L}\left(\frac{\text{DDR} - 1}{\text{DDR}}\right). \tag{15A.14}$$

The amount of stress relaxation is a function of the characteristic relaxation time of the polymer (a material parameter influenced by MW, MWD, branching, etc.), processing temperature and the time of the process. A characteristic time for the process can be defined as the total strain divided by the average rate of strain:

$$t_f = \frac{\varepsilon_{\text{tot}}}{\dot{\varepsilon}_{\text{avg}}} \tag{15A.15}$$

Substituting (15A.12) and (15A.14) into (15A.15) gives the following relationship for process time:

$$t_f = \frac{L}{v_f}\left(\frac{\text{DDR}}{\text{DDR} - 1}\right)\ln(\text{DDR}). \tag{15A.16}$$

Note that this is the form of the linear velocity model in Table 15.1. Derivation of the Newtonian fluid model for process time can be found in Chapter 2 [Eq. (2.23)].

References

[1] T. Andersson, B. Wesslen, "Degradation of LDPE, LLDPE and HDPE in film extrusion, TAPPI PLACE. Conf. (2004).
[2] T. Andersson, B. Stalbom, B. Wesslen, Degradation of polyethylene during extrusion. II. Degradation of low-density polyethylene, linear low-density polyethylene, and high density polyethylene in film extrusion, J. Appl. Polym. Sci. 91 (2004) 1525–1537.
[3] B.H. Gregory, Extrusion Coating, A Process Manual, Trafford, Victoria, Canada, 2005, pp. 73–92.
[4] V. Antonov, A. Soutar, Foil adhesion with copolymers: time in the air gap, TAPPI 1991 Polymers, Laminations & Coatings Conference Proceedings, TAPPI PRESS, Atlanta, 1991, p. 553.
[5] B.A. Morris, Adhesion, Extrusion Coating Manual, 5th edition, TAPPI Press, 2016.
[6] B.A. Morris, "The influence of stress on peel strength of acid copolymers to foil, in: TAPPI Polymers, Laminations and Coatings Conference, 1998. Also B.A. Morris, J. Reinforced Plastics and Composites, 21 (2002) 1243–1255.
[7] B.A. Morris, Effect of process and material parameters on interlayer peel strength in coextrusion coating, film casting and film blowing, J. Plastic Film. & Sheeting 26 (2010) 343–376.
[8] B.A. Morris, "Understanding why adhesion in extrusion coating decreases with diminishing coating thickness, J. Plastic Film. & Sheeting 24 (2008) 53–88.
[9] K. Canning, B. Bian, A. Co, Film casting of a low density polyethylene melt, SPE ANTEC Conf. (1999) 386–390.
[10] T. Andersson, B. Wesslen, J. Sandstrom, "Degradation of low density polyethylene during extrusion. I. Volatile compounds in smoke from extruded films, J. Appl. Polym. Sci. 86 (2002) 1580–1586.
[11] T. Andersson, T. Nielsen, B. Wesslen, Degradation of low density polyethylene during extrusion. III. Volatile compounds in extruded films creating off-flavor, J. Appl. Polym. Sci. 95 (2005) 847–858.
[12] T. Andersson, M.H.J. Holmgren, T. Nielsen, B. Wesslen, "Degradation of low density polyethylene during extrusion. IV. Off-flavor compounds in extruded films of stabilized LDPE, J. Appl. Polym. Sci. 95 (2005) 583–595.
[13] T. Andersson, B. Wesslen, "Degradation of low density polyethylene during extrusion. V. Effects of oxygen content in air gap, J. Appl. Polym. Sci. 96 (2005) 1767–1775.
[14] F. Gugumus, Physico-chemical aspects of polyethylene processing in an open mixer 6. Discussion of hydroperoxide formation and decomposition, Polym. Degrad. Stab. 68 (2000) 337–352.
[15] F.M. Rugg, J.J. Smith, R.C. Bacon, Infrared spectrophotometric studies on polyethylene. II. Oxidation, J. Polym. Sci. 13 (1954) 535–547.
[16] K. Barabas, M. Iring, T. Kelen, F. Tudos, Study of the thermal oxidation of polyolefins. V. Volatile products in the thermal oxidation of polyethylene, J. Polym. Sci., Polym. Symposia 57 (1977) 65–71.
[17] M. Iring, S. Laszlo-Hedvig, K. Barabas, T. Kelen, F. Tudos, Study of the thermal oxidation of polyolefins. IX. Some differences in the oxidation of polyethylene and polypropylene, Eur. Polym. J. 14 (6) (1978) 439–442.
[18] D.J. Carlsson, G. Bazan, S. Chmela, D.M. Wiles, K.E. Russell, Oxidation of solid polyethylene films: effects of backbone branching, Polym. Degrad. Stab. 19 (3) (1987) 195–206.
[19] P.B. Sherman, Ozonation of polymer melt for improved adhesion, in: Bezigian (Ed.), Extrusion Coating Manual, 4th (ed.), TAPPI Press, Atlanta, Ga, 1999, pp. 75–88.
[20] E. Laiho, T. Ylanen, "Flame, corona, ozone – do we need all pretreatments in extrusion coating, TAPPI Polymers, Laminations & Coat. Conf. (1997) 93.
[21] D. Briggs, D. Brewis, M. Lonieczko, X-ray photoelectron spectroscopy studies of polyethylene-aluminum laminates, Eur. Polym. J. 14 (1978) 1.
[22] M. Hansen, M. Finlayson, M. Vaughn, "Characterizing aluminum adhesion for low-density polyethylene, TAPPI Polymers, Laminations & Coat. Conf. (1991) 349.
[23] D. Brewis, D. Briggs, Adhesion to polyethylene and polypropylene, Polymer 22 (1981) 7.
[24] T. Hjertberg, B. Sultan, E. Sorvik, The effect of corona discharge treatment of ethylene copolymers on their adhesion to aluminum, J. Appl. Polym. Sci. 37 (1989) 1183.
[25] W.J. Ristey, R.N. Shroff, The degradation and surface oxidation of polyethylene during the extrusion coating process, TAPPI Pap. Synth. Proc. (1978).
[26] V. Croilius, W. Ebeling, R. Parsons, The effect of processing variables on the adhesion strength of polyethylene-coated aluminum foil, TAPPI J. 45 (1962) 352.
[27] B.W. Foster, Effect of extrusion coating parameters on activation of a primed film, TAPPI PLACE. Conf. (2002).
[28] D.R. Hammond, Improving oxidation of the extrusion melt at higher line speeds and lower melt temperature, Eur. TAPPI PLACE. Conf. (2013).
[29] B.A. Morris, N. Suzuki, The case against oxidation as a primary factor for bonding acid copolymers to foil, TAPPI Polymers, Laminations & Coat. Conf. (2000).
[30] J. Kuusipalo, M. Vaha-Nissi, A. Savolainen, Study of rheology and adhesion in extrusion coating of different polyethylenes,in: TAPPI Polymers, Laminations & Coatings Conference, San Francisco, August 30-September 3, 1998, Vol. 2, 1998, pp. 769–777.

[31] S. Richards, The effects of surface treatment on heat seal and hot tack, TAPPI Int. Flex. Packaging Extrusion Div. Conf. (2018).
[32] P. Zambetti, S. Baker, D. Kelley, Alpha tocopherol as an antioxidant for extrusion-coating polymers, TAPPI J. 78 (1995) 167.
[33] R. Golike, 1987, DuPont internal report.
[34] D. Hester, S. Sadik, Y. Trouilhet, Using acid copolymers as coextrusion tie layers, TAPPI Polymers, Laminations & Coat. Conf. (1989) 459.
[35] S. Wu, Polymer Interface and Adhesion, Marcel Dekker, Inc, New York, 1982, p. 465.
[36] B.A. Morris, H.T. Thai, Modifying LDPE for improved adhesion to aluminum foil, TAPPI PLACE. Conf. (2004).
[37] A. Keller, M.J. Machin, Oriented crystallization in polymers, J. Macromol. Sci. (Phys.) B1 (1) (1967) 41–91.
[38] T.H. Kwack, C.D. Han, M.E. Vickers, Development of crystalline structure during tubular film blowing of low-density polyethylene, J. Appl. Poly Sci. 35 (1988) 363–389.
[39] J.M. Dealy, K.F. Wissbrun, Melt Rheology and its Role in Plastics Processing, Theory and Applications, Van Nostrand Reinhold, New York, 1990.
[40] A. Kinloch, C.C. Lau, J.G. Williams, The peeling of flexible laminates, J. Fract. 66 (1994) 45–70.
[41] J.E. Guillotte, T.F. McLaughlin Jr., Calculating the adhesion of polyethylene coatings to kraft paper, TAPPI 45 (3) (1962) 200–208.
[42] R.T. Van Ness, Infrared thermometry – a useful tool for the coating industry, TAPPI 51 (12) (1968) 39A–46A.
[43] B.M. Foederer, B.A. Morris, Use of IR technology to model temperature loss in the air gap, TAPPI PLACE. Conf. (2014).
[44] S. Devisme, J.-M. Haudin, J.-F. Agassant, D. Rauline, F. Chopinez, Numerical simulation of extrusion coating, Intern. Polym. Process. XXII (2007) 90–104.
[45] Y. Trouilhet, Morris, Prediction of temperature profiles across coating and substrate in the nip, TAPPI Polymers, Laminations Coat. Conf. (1999).
[46] H.S. Carslaw, J.C. Jaeger, Conduction of Heat in Solids, Oxford University Press, 1959.
[47] B. Carnahan, H.A. Luther, J.O. Wilkes, Applied Numerical Methods, Wiley, New York, 1969.
[48] N. Shamsundar, E. Rooz, Numerical methods for moving boundary problems, in: W.J. Minkowycz, et al. (Eds.), Handbook of Numerical Heat Transfer, John Wiley & Sons, New York, 1988.
[49] J.P. Goslin, H.F. Sweeney, The interaction of processing variables, base materials, and resins in polyethylene extrusion coating, TAPPI 43 (5) (1960) 434–447.
[50] B.A. Morris, Understanding why adhesion in extrusion coating decreases with diminishing coating thickness, Part I: penetration of porous substrates, SPE ANTEC Conf. TAPPI PLACE. Conf. (2005).
[51] B.A. Morris, Understanding why adhesion in extrusion coating decreases with diminishing coating thickness, Part II: non-porous substrates, SPE ANTEC Conf. TAPPI PLACE. Conf. (2006).
[52] B.A. Morris, Understanding why adhesion in extrusion coating decreases with diminishing coating thickness, Part III: analysis of peel test, SPE ANTEC Conf. TAPPI PLACE. Conf. (2007).
[53] J. Vlachopoulos, Melt flow through channels, Introduction to Polymer Processing, 2nd (Ed.), McMaster University, 1994.
[54] S. Devisme, J.-M. Haudin, J.-F. Agassant, R. Combarieu, D. Rauline, F. Chopinez, Adhesion in polypropylene/aluminum laminates made by extrusion coating, J. Appl. Polym. Sci. 112 (2009) 2609–2624.
[55] A.N. Gent, G.R. Hamed, Peel mechanics for an elastic-plastic adherend, J. Appl. Polym. Sci. 21 (1977) 2817–2831.
[56] J. Pascal, Ultraversatile adhesives to broaden the possibilities of extrusion lamination, TAPPI PLACE. Conf. (2006).
[57] P.P. Shirodkar, V. Firdaus, H. Fruitwala, Scale-up of LLDPE blown film extrusion (Dec) Plast. Eng. (1994) 27.
[58] T.I. Butler, Using dimensionless scale-up variables to predict LLDPE blown film properties, TAPPI Polymers, Laminations Coat. Conf. (1995) 581.
[59] B.A. Morris, The effect of processing variables on peel strength in coextruded blown film, SPE ANTEC Conf. (1996).
[60] B.A. Morris, Peel strength issues in the blown film coextrusion process, TAPPI Polymers, Laminations Coat. Conf. (1996).
[61] B.A. Morris, Effect of time, temperature and draw down on interlayer peel strength during coextrusion coating, film casting and film blowing, TAPPI PLACE. Conf. (2008).
[62] J. Zhang, T.P. Lodge, C.W. Macosko, Interfacial slip reduces polymer-polymer adhesion during coextrusion, J. Rheol. 50 (2006) 41–57.
[63] B. Patham, N. Agashe, R. Davda, H. Asthana, Inter-layer adhesion in co-extruded structures: role of melt interfacial viscoelastic stresses during flow through the die, in: 65th Annual Technical Conference - Society of Plastics Engineers, 2007, pp. 2433–2437.
[64] B.A. Morris, The effect of coextrusion on bubble kinematics, temperature distribution and property development in the blown film process, J. Plastic Film. & Sheeting 15 (1) (1999) 25–36.
[65] J. Song, R.H. Ewoldt, W. Hu, H.C. Silvis, C.W. Macosko, Flow accelerates adhesion between functional polyethylene and polyurethane, AIChE J. 57 (2011) 3496–3506.
[66] H. Kiparissoff-Bondil, S. Devisme, D. Rauline, F. Chopinez, F. Restagno, L. Leger, Evidences for flow-assisted interfacial reaction in coextruded PA6/PP/PA6 films, Polym. Eng. & Sci. 59 (S1) (2019) E44–E50.
[67] T. Barraud, S. Devisme, H. Hervet, D. Brunello, V. Klein, C. Poulard, et al., Convection and diffusion assisted reactive coupling at incompatible semi-crystalline polymer interfaces, J. Phys. Mater. 3 (3) (2020) 035001.
[68] S. Devisme, Original test to characterize the stretching effect onto adhesion during film or sheet extrusion2011 Eur. TAPPI PLACE. Conf. (2011).
[69] T. Barraud, F. Restagno, L. Léger, Characterizing the formation of diblock copolymers acting as adhesion promoters at PP/PA6 interfaces, in conditions relevant for co-extrusion, in: 33rd Proceedings of the Annual Meeting of the Adhesion Society, 2010, pp. 356–358.

[70] B.A. Morris, Correlation of interlayer peel strength to post die processing times in coextrusion blown and cast film, SPE Coextrusion Topcon (2008). Cincinnati, OH, October 21–22, 2008.

[71] B.A. Morris, The effect of post die process time on adhesion in coextrusion blown and cast film, TAPPI PLACE. Conf. (2010).

[72] J. Song, A.M. Baker, C.W. Macosko, R.H. Ewoldt, Reactive coupling between immiscible polymer chains: acceleration by compressive flow, AIChE J. 59 (2013) 3391–3402.

[73] C.W. Macosko, H.K. Jeon, T.R. Hoye, "Reactions at polymer-polymer interfaces for blend compatibilization, Prog. Polym. Sci. 30 (2005) 939–947.

[74] M.G. Botros, Investigation of adhesion and failure mechanisms in tie-layer adhesive/EVOH systems, J. Plastic Film. & Sheeting 12 (1996) 195–211.

[75] X. Zheng, M.H. Rafailovich, J. Sokolov, et al., Long-range effects on polymer diffusion induced by a bounding interface, Phys. Rev. Lett. 79 (2) (1997) 241–244.

Part VI

End use considerations

Chapter 16

End-use factors influencing the design of flexible packaging

A flexible package must hold up under its conditions of use. This chapter explores how package interactions with its physical environment (temperature, pressure, humidity, and radiation) and with its contents may change the physical, optical or barrier properties of the package in ways that may influence performance. While most interactions are unwanted, active packaging purposely seeks to extend the shelf life of products through specific interactions. Some film properties change with time during storage or use through phyiscal aging phenomena related to molecular rearrangements. These end-use factors, as well as cost, must be taken into account when designing films for flexible packaging. The cost involves not just the materials and film converting costs, but the total cost of using the package, including packaging productivity, package failures during packing and the end-use, and product shelf life.

16.1 Environmental effects on package performance

16.1.1 Temperature

Films may be exposed to conditions in the end-use not accounted for by room temperature tests. These may involve high or low temperatures. Some examples of high-temperature exposure include

- Retort sterilization, where the filled package is subjected to temperatures of 121°C (250°F) or higher for 30–60 minutes in a steam autoclave. This process was originally developed for food packaged in metal cans.
- In-package pasteurization where the temperature may exceed 90°C. Many prepared meat products are pasteurized in the package following filling as extra insurance against dangerous microbes such as listeria and E. coli.
- Hot fill of the product during packaging. In a classic hot fill and hold process, the food is heated to 90°C–95°C (194°F–203°F) for 15–30 seconds to sterilize the product prior to packaging. The product is cooled to 82°C–85°C (180°F–185°F) before it is filled into the package. The filled package is sealed and held for 2–3 minutes to sterilize the inside of the package before being quickly cooled to preserve the quality and taste of the product. Many shelf-stable high-acid liquid food products are packaged in this way, including sauces, salsa, fruit drinks and teas.
- Cook-in where the product is literally cooked in the package. Processed deli meats are prepared in this way. Various meat cuts are blended together with binders, fillers, extender ingredients and water and molded into the package. The meat-water slurry is cooked for several hours at 80°C–100°C (176°F–212°F), depending on the product. This is a rather demanding application not just because of the elevated temperatures; the package must withstand stress induced by the cooking process and prevent the migration of oily components of the food through the package. Furthermore, marinade and other liquids may be injected into the meat during cooking. The vacuum inside the package may pull out some of the liquids that are injected or naturally formed during cooking, giving an unsightly liquid "purge" in the package. Good adhesion of the film material to the meat protein and the ability to tightly surround the meat during cooking help reduce purge. Functionalized polyethylene (PE), such as acid copolymers, ionomers and maleic anhydride-modified linear low polyethylene (LLDPE) are often used as the sealant layer.
- Packaging that is designed to be microwaved by the consumer. These packages often contain susceptors that help generate and concentrate the heat during microwaving. Temperatures can get very hot. Some packages are designed with lidding or other components that give way when high temperatures are reached. These are used for self-venting packages to allow steam and other gases to escape during cooking so that dangerous levels are not reached inside the package.

The Science and Technology of Flexible Packaging. DOI: https://doi.org/10.1016/B978-0-323-85435-1.00015-6

- Dual ovenable packaging. Here the product is designed to be cooked in a microwave or conventional convective oven. Crystalline polyethylene terephthlate (CPET) trays are often used for this type of package for their temperature resistance.

Examples of low-temperature environments a package may experience include cold temperature storage and transport of frozen foods in the distribution chain, as well as storage by the consumer in a home freezer. Consumer freezers are typically kept at −18°C (46°F) or lower. Industrial blast freezers may subject the package to evaporative temperatures of −30°C to −40°C (−22°F to −40°F), although the core temperature will typically be −7°C to −18°C (−19°F to 0°F).

Polymers undergo fundamental thermal transitions that affect their physical properties. For amorphous polymers such as polystyrene (PS), amorphous polyethylene terephthlate (PET) and amorphous polylactic acid (PLA), the glass transition temperature (Tg) is an important benchmark. The Tg is the transition between the glassy and rubbery phases of the polymer. Below Tg the polymer chains are constrained and do not flow. As the temperatures increase through the glass transition temperature, the stiffness and other key physical properties of the polymer significantly decrease. Standard grades of PLA, for example, have a Tg of around 55°C (131°F). Since temperature may exceed 55°C during standard shipping, packages made from amorphous PLA must be shipped in refrigerated trucks. This limits their use for many applications. Crystalline versions of PLA films/trays can be made with higher temperature resistance, but the crystallinity reduces the transparency of the package. The Tg for various amorphous polymers can be found in Table 16.1.

TABLE 16.1 Typical glass transition temperatures and melting points of resins used in flexible packaging. Based on differential scanning calorimeter measurements. There is no data for Tm of amorphous polymers since these polymers have no melting transtion. No data for Tg of semicrystalline polymers just means the property was not measured.

Polymer	Tg (°C)	Tm (°C)
Amorphous polymers		
APET	67	
PS	100	
PLA	55–60	
PA 6I/6 T	132	
PVC	80	
Semicrystalline polymers		
EVA		75–100
Ionomer		80–100
Acid copolymer		85–105
LDPE	−85 to −130 (depends on source)	110
HDPE	−70 to −125 (depends on source)	130–135
LLDPE – Ziegler-Natta	−78	120–125
LLDPE – metallocene		55–120
PP homopolymer	−18	160–165
PP random copolymer		135–155
PA 6	57 (dry), <0 wet	220
PA 6/66		195
EVOH, 44 mol% ethylene	53 (dry)	165
EVOH, 32 mol% ethylene	57 (dry)	183
PET (crystalline) or OPET	70–80	254–267

Source: Reproduced with permission from The Dow Chemical Company.

For semicrystalline polymers such as PE, polypropylene (PP), polyamide (PA), and CPET, the melting point (Tm) is the critical temperature. The melting of semicrystalline polymers typically occurs over a range of temperatures; the onset of the decrease in physical properties with increasing temperature may occur at a temperature below the peak melting point. This is illustrated in Fig. 16.1 where the storage modulus of a homopolymer PP from dynamic mechanical analysis (DMA) is plotted versus temperature. The Tm of PP homopolymer is around 161°C, but the modulus drops well below this.

The melting points of commonly used polymers in flexible packaging are given in Table 16.1. In multilayer film structures, the sealing layer is often selected to have a lower Tm than the heat resistant or structural layer. Depending on the nature of the temperature exposure (whether the film is under stress), the package may maintain its integrity even if Tm for one of the layers is exceeded, as long as the melt strength or hot tack of that layer is sufficient.

For oriented films or laminates containing oriented films, exposure to temperatures exceeding the orientation or annealing temperature may cause the film to shrink as oriented polymer chains relax back toward their random coiled state. This is desirable for shrink films and shrink bags but may cause distortion, delamination and other challenges for package formats requiring dimensional stability.

In addition to thermal transitions that may affect the mechanical properties of the polymer and render it unfavorable for a high-temperature application, there may be regulatory hurdles. Specific regulatory requirements may be needed for applications subjected to an elevated temperature such as microwave or convection oven cooking or retort. Migratory species in polymer films may diffuse into the food product faster at elevated temperatures, depending on the polymers. The regulations have specific testing protocols for high-temperature use. See Chapter 4 for more information on regulatory requirements.

Elevated temperatures in the end-use increase oxygen and moisture permeation and the migration of small molecules to or from the package (scalping and impartation, described in Section 16.2). These effects need to be planned for when designing the packaging structure, or the shelf life may be compromised.

For low-temperature environments, embrittlement of the polymer may become an issue. When the temperature drops below the Tg of the polymer, toughness often substantially decreases. Polymers with low Tg may be favored, or be used as a modifier. For example, PP has a Tg of around −18°C, whereas the Tg of PE is around −80°C or lower. PE is tougher than PP at freezer temperatures. The PP may be modified with a rubbery phase to improve its cold temperature toughness, either through copolymerization (impact copolymers) or by blending in a toughener such as ethylene propylene diene rubber (EPDM), ethylene–propylene rubber (EPR) or metallocene polyethylene (mPE) plastomer/elastomer. Similarly, PET (Tg = 80°C) is often modified with ethylene methyl acrylate (EMA) copolymer (Tg ~ −40°C) to prevent CPET trays from shattering if they are dropped by the consumer as they remove them from the freezer.

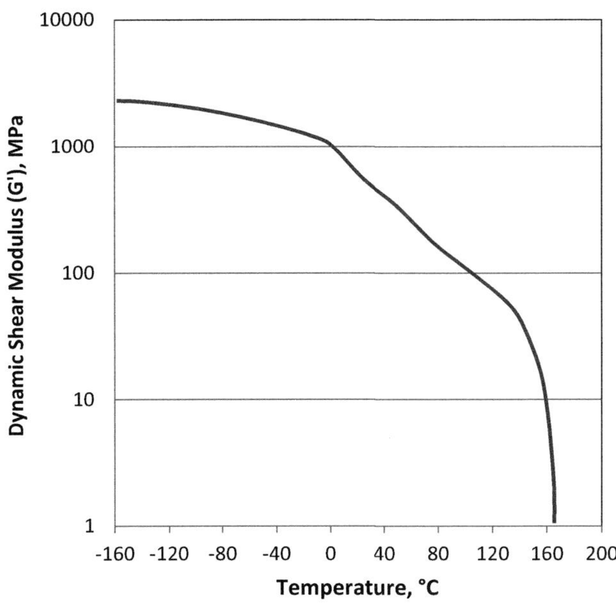

FIGURE 16.1 Storage modulus (*G*′) as a function of temperature for semicrystalline PP homopolymer [1]. *Data from T. Osswald, Mechanical behavior of polymers, Underst. Polym. Processing: Process. Gov. Equations Hanser (2011).*

16.1.2 Humidity

Moisture acts as a plasticizer for some polymers, affecting their gas permeation and mechanical properties. The effect of relative humidity on the oxygen permeation of ethylene vinyl alcohol (EVOH), PA and other polymers is described in detail in Section 8.4.1.2.4.3. Moisture pick-up in the end-use increases the permeation of gas molecules through these films, as is illustrated in Fig. 8.1 for EVOH. Barrier performance can be compensated by

- Selecting barrier polymers that are less sensitive to relative humidity.
- Increasing barrier layer thickness.
- Moving the barrier layer toward the surface exposed to the lower humidity environment, since the effective relative humidity varies within the thickness of the package films. Design equations to estimate the relative humidity the barrier layer sees in its end-use are provided in Section 8.4.1.2.4.3.

The combination of high relative humidity and temperature can be particularly problematic for the barrier performance of water sensitive barrier polymers such as EVOH. During retort sterilization, the packaged product is exposed to 121°C in a saturated steam environment for 30–60 minutes. A number of factors conspire against maintaining barrier performance under these conditions:

- Permeation increases with increasing temperature.
- Permeation increases with increasing relative humidity. Polyolefin layers that may otherwise provide some moisture protection of the oxygen barrier layer over short times (until equilibrium is established) are no help; their moisture barrier becomes almost nonexistent at retort temperatures.
- The heat and water change the crystalline structure of EVOH so that following retort, the oxygen barrier performance does not return to its original value [2–5]. This is known as retort shock and is described in detail in Section 8.4.1.2.4.4. Not only does extra oxygen enter during the retort process itself, but because the EVOH does not fully recover, the additional oxygen that is allowed into the package following retort (after the package returns to room temperature) must be considered when designing the package to meet shelf life requirements.

Alternative barrier materials and packaging structures may offer improved performance over EVOH for retort of flexible packaging. Some examples of oxygen permeation of several structures before and after retort are shown in Fig. 16.2. Other problems may occur under these demanding conditions, however. Schubert [7] describes aluminum foil based laminates used for retort stand-up pouches with the structure PET/adh/Al/adh/OPA/adh/PP. The adhesive is typically a 2-component polyurethane-polyester. The adhesion often improves during retort—moisture passes through the polymer layers and helps with the bonding of silanol groups in the adhesive to oxides on the aluminum foil. Bubbles can form, however, where the adhesive did not cover the aluminum foil (due to "dry spots" caused by the adhesive being applied to the printing ink). The moisture and oxygen in the bubble causes oxide growth on the foil surface. The spots can expand and evolve hydrogen.

Alternative sterilization processes such as high-pressure sterilization (Section 16.1.3) and irradiation (Section 16.1.4) are being developed that may avoid retort shock.

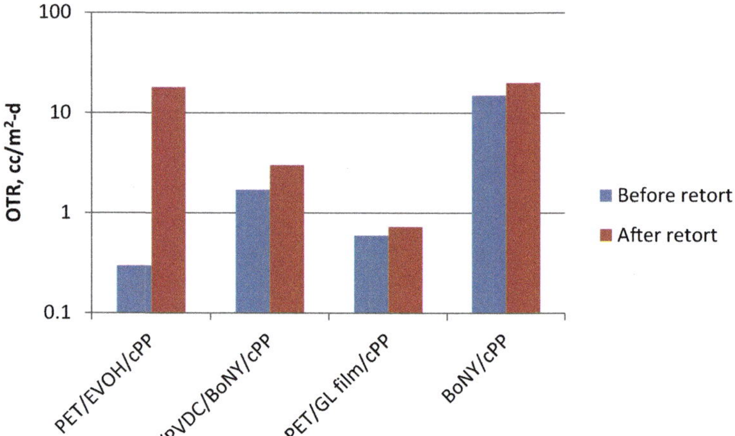

FIGURE 16.2 Oxygen transmission rate (OTR) of multilayer films before and after retort (125°C, 20 minutes). OTR is measured at 25°C, 70% RH. cPP = cast polypropylene, GL film is silica deposited laminated film from Toppan Printing Co., Ltd., and BONY is oriented polyamide film [5,6]. *Reproduced with permission from K.K. Mokwena, J. Tang, Ethylene vinyl alcohol: a review of barrier properties for packaging shelf stable foods, Crit. Rev. Food Sci. Nutr. 52 (2012) 640–650. Data from R.J. Hernandez, J.R. Giacin, Factors affecting permeation, sorption and migration processes in package-product systems, in: I.A. Taub, R.P. Singh (eds.), Food Storage Stability, CRC Press, Boca Raton, FL, 1998, pp. 269–330.*

Since moisture can act as a plasticizer for some polymers, such as PA 6, the stiffness or modulus of films containing these polymers may decrease over time if the package is subjected to a high relative humidity environment. The loss in stiffness is usually associated with a gain in toughness. Indeed, for some applications, films containing PA 6 are purposely subjected to a high moisture environment to provide toughness. Section 13.5 describes how moisturization can also be used to promote secondary crystallization in PA 6, which induces changes in dimension and curl.

16.1.3 Pressure

Pressure variations due to changes in altitude during shipping can put stress on package seals, as is described in Chapter 7. A "ballooning" effect occurs when a packaged sealed at sea level is transported to high elevations as the surrounding air pressure is diminished, allowing the air molecules inside the package to expand. End-use seal strength specifications should be set sufficiently high to prevent delamination during such changes in altitude.

In general, the differences in pressure that a typical package is subjected to in its use have little effect on the mechanical and barrier properties of flexible packaging films. A new technology, however, for in-package pasteurization or sterilization of food products is gaining favor called high-pressure processing (HPP) where the packaged product is subjected to extreme pressures of 300–800 MPa (43.5–116 kpsi). The high pressure (HP) inactivates microbes by disrupting the cell wall and the cell's ability to repair the damage. Time at pressure depends on the acidity and water activity of the food product, as well as the type of microbe to be deactivated. Gram-positive microbes (e.g., listeria) are more difficult to deactivate than Gram-negative ones. Spores are particularly resistant to HP [8–14].

In high-pressure pasteurization, pressure alone is used. The packaged food is adiabatically pressurized at room temperature for 1–15 minutes in a pressurized chamber. This inactivates vegetative microorganisms and is used for chilled or high-acid foods. Since adiabatic pressurization increases temperature (on the order of 2–4°C/100 MPa) the final temperature is in the range of 30–50°C (86–122°F). The use of HP pasteurization is growing for extending the shelf life of refrigerated foods such as processed meats, salsa and various fruits and vegetables. HPP has advantage over thermal pasteurization processes in that it preserves the natural flavor, texture and nutrition of the food.

To inactivate enzymes and spores, such as for low-acid shelf stable foods, temperature is combined with HP in what is called high-pressure sterilization. The packaged product is preheated to 60–90°C (140–194°F), depending on the product, prior to pressurization. Final temperature can be as high as 130°C (266°F) due to adiabatic pressurization. Since the thermal exposure is much shorter than thermal sterilization (retort) the quality of the food product is better preserved.

The best packages for high-pressure processing turn out to be flexible or semirigid packages with little or no headspace. The package deflects up to 15% during the high-pressure treatment. As long as the package has a conformable lid or sidewall and is sealed well, it generally holds up. HPP, therefore, seems ideally suited for flexible packaging. Shelf life extensions of two times or more have been reported, depending on the product. As the shelf life is extended by the prevention of microbial growth, spoilage by oxidation becomes more of an issue. The substitution of oxygen barrier film structures for nonbarrier ones may be needed.

The package must withstand the high-pressure process (flexing, etc.) without substantial change in mechanical, optical, barrier and migration potential properties. Investigators have found that most properties of individual plastics are not significantly affected by the high-pressure treatment. Mensitieri et al. [15], for example, show that most properties are within 10%–15% of their original value. Multilayer flexible packaging structures, however, may be susceptible to delamination that reduces barrier performance. Studies are mixed but in general show that delamination is more prevalent when heat is combined with HP (sterilization) and for laminates with metallized films, vacuum-deposited inorganic coatings (AlO_x or SiO_x) or in some cases aluminum foil (as opposed to all plastic coextruded films or laminations) [15–20].

Several mechanisms have been proposed for the delamination. Bull et al. [16] propose that desorption of gases absorbed during pressurization may be involved. This coincides with observations that packages with minimal headspace fare better than those with substantial headspace. Fraldi et al. [19] consider several hypotheses including differences in thermal expansion coefficients, thermal failure of the adhesives, changes in Tg and differences in the mechanical properties of the layers (i.e., stiffness). They discount the first three using experiments and sound arguments. They go on to explore the final hypothesis through experiments with two-layer laminates made from combinations of PP, PA, and PET films. They also develop analytical and numerical models. In their experiments, they subject the laminates to pressure treatments of 200, 500, and 700 MPa at room temperature (10 minutes) or 90°C (5 minutes). Only the PP/PET laminates that have been subjected to high temperature (over all the pressures) are delaminated. At both 25°C and 100°C, PP and PA have much closer moduli than they do to PET. The modeling work shows that differences in moduli increase the shear stresses in the region near the seal area which can result in delamination. They conclude that the hypothesis that delamination is caused by a mismatch in moduli of the film layers is supported by the experimental and modeling results.

The loss in barrier performance seems to be limited to delamination effects; for example, Lopez-Rubio et al. [21] show that the barrier performance of EVOH remains intact following HP pasteurization or sterilization treatments. They subject EVOH-containing packaging structures, (PP–tie–EVOH–tie–PP) or (PP–EVOH–PP), to various HP and thermal treatments: 400 and 800 MPa, 40°C and 75°C, 5 and 10 minutes. They also subject the films to a thermal sterilization (retort) treatment as a control: 120°C for 20 minutes. They measure the oxygen transmission rate (OTR) of the films before and after the treatments. As described in Section 8.4.1.2.4.4 and 16.1.2, EVOH undergoes a permanent increase in OTR following retort. Lopez-Rubio et al.'s results reflect this: the OTR of a 26 mol% ethylene EVOH grade increases by over 500 times immediately after retort and, even after allowing the sample to dry out for 150 hours following the retort treatment, the OTR is three times greater than prior to retort. Differential scanning calorimeter (DSC), Fourier transform infrared spectroscopy (FTIR) and X-ray scattering measurements show that the crystallinity of the EVOH decreases during the retort treatment. The samples subjected to HPP, however, show a slight decrease in OTR and an increase in crystallinity. HP sterilization seems to be a way to avoid the retort shock issues associated with using EVOH in retort and may open up the possibility of new structures for consideration. Today, most flexible packaging for retort uses aluminum foil laminates.

Investigations of the migration of flavor molecules during high-pressure treatments have been mixed and are reviewed by Marangoni Junior et al. [20] and Caner et al. [22]. Most studies show little or no change in the migration potential of packaging structures. Those with changes generally show a decrease in migration believed to be due to an increase in crystallization of one or more of the polymer layers.

HP can increase the melting point of semicrystalline polymers like PE and PP [15,23,24]. Seeger et al. [23] find the melting point of high-density polyethylene (HDPE), low-density polyethylene (LDPE), PP and ethylene vinyl acetate (EVA) copolymers increases linearly with increasing pressure up to 330 MPa (the limit of their experiments). The melting point is found to increase by 11°C–17°C for every 100 MPa increase in pressure. Sansone [24] reports an increase in the melting point of an LLDPE resin from 110°C to 140°C when the pressure is increased from atmospheric to 200 MPa. Mensitieri et al. [15] point out that the HP process may extend the temperature range over which a polymer may be used. In traditional thermal sterilization (retort), the melting point of LLDPE is too low. If an HP sterilization process is used instead, then PE may be a material to consider.

The use of HP sterilization has several sustainability benefits. By inactivating microbes and spores it extends the shelf life of the food product and helps prevent food waste (and it does this in a way that minimizes the impact on flavor, texture and quality). It also may allow the use of alternative packaging structures that are easier to recycle, such as coextruded films of EVOH and PP or even PE, where foil laminates are used today (as in retort flexible packaging).

16.1.4 Irradiation

Electron beam and gamma-ray (from radioactive isotopes cobalt-60 and cesium-137) radiation are used for medical device sterilization, food irradiation and sterilization and sanitation of packaging materials. At the typical dosage required to reduce microbial contamination, no heat is generated, which is beneficial for retaining the nutritional and sensory value of food. Most commercial operations prefer to sterilize the device or food product after packaging to avoid re-contaminating the product during packaging. The package, therefore, needs to withstand the effects of radiation.

Radiation creates free radicals that can cause crosslinking or chain scission of polymers, which affects mechanical properties such as tensile strength, elongation, and impact resistance. Chemical changes may also occur, such as the creation of peroxides through a reaction with oxygen. The peroxides may undergo further reaction over time, sometimes in the presence of moisture. The effects of radiation may not be immediately apparent and require extensive testing when considering new materials. Chemical interactions with the polymer, or various additives that may be in the polymer such as antioxidants, may create low molecular weight (MW) species that alter the odor, taste or safety of the product. For these reasons, government agencies regulate the use of materials that are used in irradiated packaging. In the US, irradiation of food packaging is regulated by the US FDA under Title 21 of the Code of Federal Regulations (CFR), Part 179.25 (General provisions for food irradiation). Packaging materials must comply with 21 CFR 179.45 (packaging materials for use during the irradiation of prepackaged foods). Unfortunately, many modern packaging materials and structures are not listed in CFR 179.45. New packaging structures/materials not already listed, or the subject of a Food Contact Notice, require a premarket safety assessment. This includes chemistry, toxicology and environmental components.

For selecting packaging materials exposed to radiation sterilization of medical devices, some general rules can be considered [25]:

- Materials designed for medical devices are generally durable to radiation.
- Higher MW and narrower molecular weight distribution (MWD) grades generally have greater radiation resistance.
- Aromatic polymers are more resistant than aliphatic ones.
- Amorphous polymers are more resistant than semicrystalline polymers.
- Higher levels of antioxidant improve radiation resistance. (Caution, for food packaging some antioxidants may create byproducts that are not on the regulatory lists.)
- Lower density polymers are more resistant than higher density ones.
- For semicrystalline polymers, lower levels of crystallinity improve radiation resistance. An example is presented in Fig. 16.3 showing the elongation at break for HDPE, LDPE and an ionomer before and after exposure. The order of higher to a lower level of crystallinity is HDPE > LDPE > ionomer, which is the same order as the effect of radiation on elongation.
- Small pendent side groups can improve radiation resistance. For example, the alpha-olefin side chains in LLDPE lower its density and crystallinity and improve radiation resistance over HDPE.
- Materials with low oxygen permeation are more radiation-resistant.
- A mnemonic is to avoid materials that are APT to fail [acetal, PP (unstabilized) or Teflon (PTFE)].
- Materials must be compliant with appropriate regulations.

The effect of radiation depends on its dosage and the exposure environment (oxygen, moisture, temperature). There are two main units of radiation dosage, Gy and Rad. 1 Gy is equal to 1 J of energy absorbed per 1 kg of matter. 1 Rad is equal to 100 ergs absorbed per gram of matter. Therefore 10 kGy = 1 MRad. Most medical devices are sterilized with 25 kGy or less. End-users of medical device packaging typically specify that the material must withstand 25 or 50 kGy, the latter in case of two passes through the sterilization process. For food irradiation, the maximum dosage is typically 10 kGy, except for some spices that are exposed to 30 kGy. A list is shown in Table 16.2. At these relatively low doses, most flexible packaging structures retain their mechanical properties and heat seal performance [29–32]. Hoh and Cumberbatch [27], for example, show that the mechanical properties of ionomers are not impacted by radiation exposure up to 50 kGy.

Goulas et al. [33] study the effect of radiation on multilayer flexible packaging. Five structures are tested for changes in mechanical properties, gas permeation and migration into aqueous simulants. The structures are:

1. (PP–tie–EVOH–tie–70/30 LDPE/LLDPE), 45 μm;
2. (LDPE–tie–EVOH–tie–LDPE), 55 μm;
3. (Ionomer–tie–EVOH–tie–LDPE), 75 μm;
4. (PA–tie–LDPE), 89 μm;
5. (LDPE–tie–PA–tie–ionomer), 135 μm.

The films are subjected to 5, 10, and 30 kGy. No significant difference in mechanical properties is found between the irradiated films at 5 and 10 kGy and nonirradiated controls. At 30 kGy there is a reduction in tensile strength for

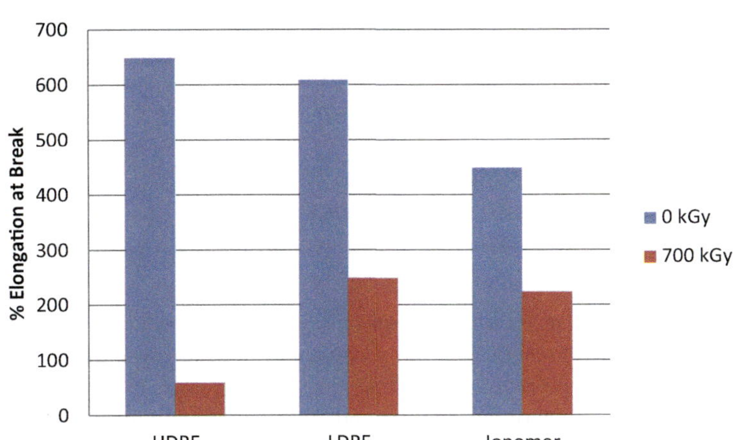

FIGURE 16.3 Effect of radiation on elongation at break of ethylene polymers. Polymers are subjected to 700 kGy of radiation [26,27]. *Data from B.J. Lyons, F. Weir, The effect of radiation on the mechanical properties of polymers, in: E. Dole (ed.), The Radiation Chemistry of Macromolecules, Vol II, Academic Press, 1973 and figure from G. Hoh, G.M. Cumberbatch, Stability of Surlyn® films to radiation, TAPPI Polymers, Laminations Coat. Conf. (1990) 251–292. Reproduced with permission from The Dow Chemical Company.*

TABLE 16.2 Foods and dosage levels permitted to be irradiated by the US FDA (21 CFR 179.26) [28].

Food product	Purpose	Maximum dosage (kGy)
Pork carcasses	Control of *Trichinella spiralis*	1 (0.3 minimum)
Fresh foods	Growth and maturation inhibition	1
Food	Disinfestation of arthropod pests	1
Enzyme preparations	Microbial disinfection	10
Dry or dehydrated spices/seasonings	Microbial disinfection	30
Fresh or frozen uncooked poultry	Control food-borne pathogens	3
Frozen, packaged meats for NASA space program	Sterilization	44 minimum
Refrigerated uncooked meat products	Control food-borne pathogens and extend shelf life	4.5
Frozen uncooked meat products	Control food-borne pathogens and extend shelf life	7
Fresh shell eggs	Control *Salmonella*	3
Seeds for sprouting	Control of microbial pathogens	8
Fresh or frozen molluscan shellfish	Control of Vibrio bacteria and other food-borne microorganisms	5.5
Fresh iceberg lettuce and spinach	Control food-borne pathogens and extend shelf life	4

Source: Data from U.S. Code of Federal Regulations, 21 CFR 179.26. <http://www.accessdata.fda.gov/scripts/cdrh/cfdocs/cfcfr/cfrsearch.cfm?fr = 179.26> (accessed 14.06.21).

films 2, 4, and 5 (10%–38%) and a decrease in percent elongation at break for films 2 and 4 (42%–48%). Young's modulus is unchanged. Overall, the authors conclude that the mechanical properties do not change very much compared with the controls.

No significant differences are found in the gas permeability [MVTR (23°C, 100%RH), OTR (23°C, 60%RH), CO_2 (23°C, 60%RH)] between the irradiated films and nonirradiated controls.

Migration studies are conducted with three simulants: distilled water, 3% aqueous acetic acid, and iso-octane. The films are cut into strips and immersed into the simulant in a beaker (two-side contact, a worst-case scenario). Storage conditions are 10 days at 40°C for water and acetic acid and 2 days at 20°C for iso-octane.

Results of the migration into aqueous simulants (distilled water and 3% acetic acid) show:

- No significant difference with nonirradiated controls for distilled water at all doses.
- For 3% acetic acid, no difference at 5 and 10 kGy. At 30 kGy, the two films containing ionomers show a 12%–22% decrease in overall migration.
- Overall migration of all the films in distilled water was well below the 10 mg/dm^2 limit for EU regulations for food packaging. The same is true for acetic acid except for the two ionomers films, which were just above this limit. They attribute this to the affinity ionomers have to acetic acid.

Results of the iso-octane migration studies show:

- At 5 and 10 kGy no significant difference between irradiated and nonirradiated films.
- At 30 kGy some differences emerge but all the samples are well below the EU migration limit.
- Migration is generally higher in iso-octane than in aqueous simulants. They attribute this to the swelling of the polymer.

Komolprasert [29] reviews the literature on the effect of radiation on the creation of regulated byproducts in different packaging materials. Most byproducts are below the regulated threshold for daily consumption, but some associated with the break-down of additives in the polymers are not listed in the regulations. One class of additives, hindered phenol antioxidants, has been found to degrade into species that affect organoleptics and may have regulatory hurdles. These must be considered when selecting materials and designing packaging structures.

Demertzis et al. [34] measure the solvent extractable byproducts in six packaging materials before and after exposure to 44 kGy: PE, PP, PET, PA, PS, and PVC. The byproducts are measured and identified with gas chromatography/mass spectroscopy (GC/MS). PE and PP show an increase in extractables due to oxidative degradation of the polymer, oligomers and additives. PET, PA, and PS show little change in extractables compared with the nonirradiated controls. The PVC used in the study was not resistant to radiation—it released HCl and a large amount of volatile substances could be extracted. Other studies have found PVC films to be stable during irradiation [32].

Barakat et al. [35] study the effect of e-beam sterilization on cyclic olefin copolymers (COC), which are used in pharmaceutical packaging. They detected the formation of oxidation byproducts and crosslinking but little change in thermal properties. Analysis of extracted solutions of the COC following irradiation showed a reduction in extracted antioxidant but an increase in the generation of degradation products that may be important for pharmaceutical applications. An increase in surface polarity of the COC following radio-sterilization was also observed.

Pasbrig [36] discusses the influence of radiation sterilization on flexible packaging materials for pharmaceutical and medical packaging. Most studies have been conducted on the packaging film but most applications involve sterilizing the packaged product. Some changes in the film such as COF and heat sealability are not important once the product has been packaged. Odors created during sterilization can be an issue; the use of modified atmosphere packaging can help. A decline in the physical properties of the film can affect package integrity during handling and shipping. Stress fractures within the inner layer of the package can lead to product migration. Typically changes to optical properties are barely visible and not a concern.

16.2 Packaging–product interactions

The product may interact with the package to affect the performance of the package, or the quality of the product in a positive or negative way. Three areas of food packaging are explored here:

1. Loss of package integrity from product component migration into the package and causing interlayer delamination or stress cracking
2. Organoleptic issues resulting from the scalping of flavor components from the food into the package or migration of components from the package into the food
3. Shelf life extensions from active packaging.

16.2.1 Loss of package integrity

A primary concern with product-packaging interactions is product loss over time, which can result in loss of vitamins and nutrients, color changes, and altered taste and aroma. These are taken up in more detail in Section 16.2.2. The migration of molecules from the product, such as aroma or flavor molecules, into the package may also directly affect the physical properties of the package, or in the case of multilayer packaging cause delamination between layers. The types of effects are illustrated in Fig. 16.4. Food components such as fatty acids may migrate to the barrier interface and outcompete for reaction sites, causing loss of adhesion and delamination. This may lead to loss of barrier and

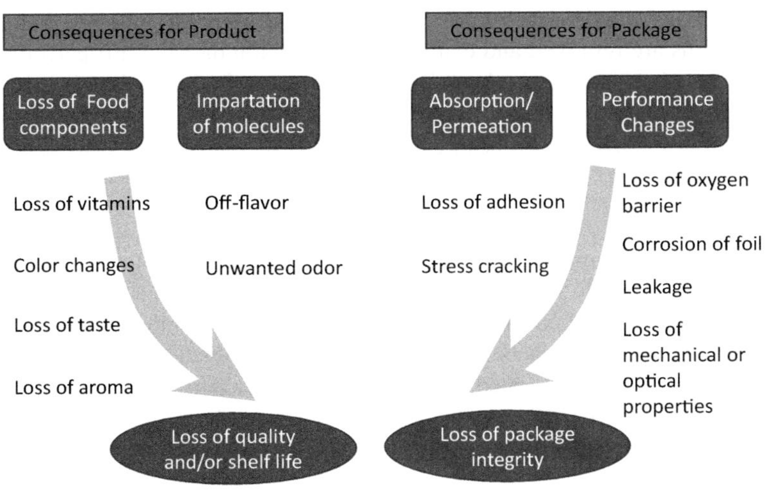

FIGURE 16.4 Negative consequences of product–package interactions. *Adapted from G. Pieper, Interaction phenomenon in multilayer flexible packaging, in: P. Ackermann, M. Jagerstad, T. Ohlsson (eds.), Foods and Packaging Materials – Chemical Interactions, Vol. 162, Special Publication – Royal Society of Chemistry, 1995, pp. 95–100.*

impact product quality. Environmental stress cracking (ESC) may also lead to a loss of barrier, corrosion of foil layers and leakage.

As described in Chapter 8, the migration of organic molecules in plastic films is complex and not well studied. Migration depends on the molecule first being absorbed by the film, which often involves swelling and other mechanisms in concert with other ingredients in the product. The chemistry of both the polymer and the organic molecules plays a role. This is followed by diffusion of the molecule into the film. How these molecules may change the physical properties of the package and interlayer adhesion is even less studied. Several examples from the literature are described here to illustrate the effects and challenges.

16.2.1.1 Barrier and physical properties

The absorption of flavor and other molecules into the packaging film may affect its mechanical, barrier and optical properties. Some molecules may act as plasticizers that reduce the stiffness of the package but increase its toughness. For example, water acts as a plasticizer for PA 6. Water may also directly impact the gas permeability of the film—the effect of relative humidity on the oxygen permeation of polar polymers like EVOH and PA is described in Chapter 8. Three examples of how the migration of organic molecules from the product into the package affects barrier properties in other ways are given here.

Van Willige et al. [38] study the influence of absorption of flavor molecules on the oxygen permeation of LDPE, PP, PC, and PET packaging films. The work is done with mono-material films to simplify the analysis. The films are immersed in a cocktail of flavor compounds containing limonene, hexyl acetate, nonanone and decanal at 40°C. The PE and PP samples are exposed for up to 8 hours; the PC and OPET films are exposed for up to 21 days. Specimens are removed, dried and tested for the amount of each compound absorbed. The OTR is measured on a parallel set of samples exposed under the same conditions.

The OTR of the LDPE and PP films increase with increasing absorption of the flavor compounds. The authors attribute this to swelling. The OTR of PC decreases slightly which they attribute to the filling in of micro voids. No effect of absorption on OTR of PET is found – they suggest a more sensitive permeation cell is needed to study the effect on PET.

Johansson and Leufven [39] study the effect of absorbed vegetable oil on the OTR of amorphous PET (APET), HDPE, and PP films. The films are exposed to rapeseed oil for 40 days. Periodically specimens are removed from the treatment, cleaned and the OTR measured. The OTR of the APET was unaffected by storage in oil. The OTR of the HDPE and PP films increase with storage time. The effect is more pronounced for HDPE than for the PP films. The amount of oil that is absorbed is also measured; the order of oil absorption was HDPE >> APET > PP.

Korbonits and Morris [40] evaluate the abrasion resistance of paperboard composite canisters used to package a potato snack product. The canister has a layer of aluminum foil that provides oxygen, moisture and light barrier. The structure is: PE/print/paperboard/LDPE/Al/sealant. Here the sealant not only provides a tamper-evident, easy opening hermetic seal but also must protect the aluminum foil from damage during distribution. The sharp edges of the snack product act like little saws that can easily abrade away the sealant, cut through the foil and cause a loss of barrier performance. Korbonits and Morris evaluate several sealant films by making 50-μm (2-mil) monolayer cast films and carefully inserting them into the canister against the inside wall (after the product was carefully removed), reinserting the product, and sealing the canister. The filled canisters are then subjected to a controlled shaking action using a shaker table to simulate movement during distribution. The films are removed and examined for abrasion damage. PE performs poorly in this test and ionomers, which have good oil resistance, are the best. The investigators find that the oil in the product swells the PE and makes it more susceptible to abrasion damage. The combination of excellent oil and abrasion resistance of the ionomers gives it the advantage. Indeed, ionomers have been used in the application for over 30 years.

16.2.1.2 Peel strength and delamination

Interactions between the product and package that lead to delamination of multilayer packaging structures can cause product loss and poor aesthetics. Often it is challenging to identify the specific molecule involved. Ortez and Tena [41] investigate the delamination failure of a cosmetics product packaged in the structure: OPET/adh/Al/adh/LLDPE film. The adhesive is a two-part solvent-based polyurethane adhesive. Sachets from the structure are filled with the product and stored at 40°C for 10 days. The sachets are then manually separated and tested to see what is migrating. They develop methods using headspace sold-phase microextraction GC-MS to detect several compounds in the inner layers and to analyze the cosmetic product. From these, phenoxy ethanol, β-linalool, menthol and p-propenyl anisole are suspected of causing delamination.

Schroeder et al. [42] study the effect of flavorant sorption on adhesion in two adhesive laminated packaging structures:

1. 15-μm PVDC ctg/46-μm OPET/ink/adh/51-μm EVA (6% VA). The adhesive is a one-component polyurethane-based on 4,4′-diphenyl methane diisocyanate (MDI).
2. same as 1) except the sealant layer is LLDPE instead of EVA.

Two food products are tested, a natural strength lemon juice and a branded hot sauce. Around 3 wt.% citric acid and acetic acid solutions are used to simulate lemon juice and hot sauce without organic components like d-limonene, respectively. The laminates are immersed in the food or simulant in a test cell such that only the food contact side is exposed. The samples are stored at 45°C for 30 days and the peel strength periodically measured.

Their results show

- Both laminates delaminate when exposed to the food products, but not to the simulants. This suggests low MW flavor components, such as d-limonene in the lemon juice, are responsible for the loss of adhesion.
- The locus of failure is between the ink and adhesive layers for structure 2) (LLDPE) and the loss of cohesion within the adhesive layer for structure 1) (EVA). This is determined by electron spectroscopy for chemical analysis (ESCA), also known as X-ray photoelectron spectroscopy (XPS), and scanning electron microscopy (SEM).
- Peel strength reduction is a poor predictor of delamination; significant peel strength reduction occurs in some samples that do not delaminate.
- The greatest peel strength reduction occurs when the laminates are exposed to the food. Some reduction occurs just from heating or with the simulant.
- d-limonene is the likely cause of delamination for lemon juice but the hot sauce is too complicated to implicate a specific molecule.

Olafsson et al. [43–47] investigate the effect of organic acids on the adhesion of LDPE to aluminum foil. Laminates with the structure LDPE/paperboard/LDPE/Al/LDPE are prepared by extrusion coating/lamination by Tetra Pak, a major manufacturer of barrier laminates for liquid packaging. Various acids are stored at room temperature in pouches made from the laminates. Periodically the laminates are tested for peel strength, examined for evidence of delamination and analyzed with FTIR and ESCA. They find:

- The LDPE delaminates from the foil within 3 days when exposed to 3 wt.% acetic acid. Over the course of the next 3 days, the peel strength recovers to about half its original value. FTIR reveals the formation of aluminum acetate at the interface that leads to the partial recovery of the peel strength (by forming hydrogen bonds with the polymer).
- Laminates exposed to propionic acid show complete delamination after 8 days and do not recover. Propionic acid has a longer chain length (extra methyl group) than acetic acid. FTIR shows more propionic acid migrates than acetic acid, which the authors attribute to its greater solubility in LDPE (due to its longer chain length).
- Citric acid, lactic acid and water have no effect on the peel strength. Compared with acetic acid, citric acid has an extra hydroxyl group and lactic acid has an extra carboxyl group, both of which should impede their migration in LDPE due to an increase in polarity. Higher polarity decreases solubility in the nonpolar LDPE.
- A series of fatty acids of different chain length (C10–C18) and saturation (0–3 carbon double bonds) are tested. The amount of fatty acid sorbed by the inner LDPE layer increases with increasing chain length for saturated fatty acids. The sorption decreases with an increasing degree of unsaturation. Most of the sorption occurs within the first two days of exposure.
- Pouches containing mono unsaturated fatty acids are found to quickly delaminate between the LDPE and foil. Pouches containing C18:2 (18 carbon atoms, two double bonds) and C18:3 showed a substantial drop in peel strength but recover to 50%–75% of the control over time. A second study with saturated fatty acids shows no delamination but a drop-off in peel strength by about 40%.
- Separate permeation measurements show that the permeation increases with increasing chain length for saturated fatty acids. The permeation decreases with increasing unsaturation.

In discussing the results, Olafsson et al. conclude

- The chemical structure of the acid affects sorption, migration and the ability of the acid to affect adhesion. The very polar organic acids are not expected to diffuse very quickly through nonpolar LDPE; Olafsson suggests they may diffuse as dimers. Acetic and propionic acids are linear and so diffuse faster than citric or lactic acid.
- The partial recovery of some of the bond strengths is due to the formation of metal salts at the interface.

- Fat absorption from fatty foods swells PE, causing an increase in migration [48–50].
- Free fatty acid concentration in foods is generally initially low but can increase with time due to hydrolysis. The concentration in oils and fats vary widely: 0.02%–0.05% in refined oils such as rapeseed oil, 0.5%–1% in virgin olive oil, 0.002% in fresh milk (4% fat), and 0.16% in butter.
- Adhesive forces between LDPE and Al are believed to be mainly due to acid-base interactions, including hydrogen bonds, and van der Waals forces between the aluminum oxides on the foil and oxidized functional groups on the LDPE.
- Delamination is believed to be due to the fatty acid penetrating through the LDPE and accumulating at the foil interface, creating a weak boundary layer of acid-base interactions between the low MW fatty acid and oxide surface of the foil.
- The sorption (and migration) behavior of the different fatty acids can be explained by differences in molecular structure:
  - Longer chain length gives rise to more van der Waals interactions and greater solubility in LDPE.
  - Greater unsaturation makes the molecule more rigid, which results in slower diffusion.
- The different adhesion results from the two fatty acid studies can be explained as being due to differences in the solutions (the first contains water, and the second just ethanol as the solvent). Water in the first study may be responsible for the delamination—it strongly forms hydrogen bonds that may interfere with the adhesion, whereas the bonds with ethanol are not as strong.

To improve product resistance, ethylene acrylic or methacrylic acid copolymers (EAA or EMAA) or ionomers are often used as tie resins between LDPE and foil. Product resistance generally increases with increasing acid level and thickness of the acid copolymer layer [51–53]. Bernacki [51] extrusion coats 25-μm of sealant onto the foil side of a paper-Al laminate to produce the structure: 50 lb kraft paper/12-μm LDPE/0.00035 g Al/25-μm sealant. The sealant is either monolayer or a coextrusion of LDPE and EAA. Pouches are made from laminates and filled with products. The products tested included ketchup, orange juice, a branded seasoned snack dip, blue cheese salad dressing, mustard, vegetable oil, wintergreen tobacco, shampoo, a branded topical pain relief lotion, lemon juice and iodine. Peel strength between the coating and foil is measured after storage for 3 weeks at 49°C and after 6 weeks at 49°C and 23°C. The results show:

- Monolayer EAA (9 wt.%) exhibits the best product resistance (least amount of reduction in peel strength).
- The peel strength in pouches containing the branded pain relief lotion and seasoned snack dip decreases when stored at 49°C, but increases in the pouches containing the other products, suggesting the other products are not very aggressive.
- EAA and ionomers offer improved product resistance over LDPE.
- The polymers with higher acid levels provide better product resistance.
- In the coextrusion structures, a 9 wt.% AA EAA grade offers protection when used as a tie layer comprising 50% and 20% of the sealant coating whereas a 3% AA grade does not offer protection at 20%.
- None of the pouches holds up to iodine. The branded pain relief lotion is also very aggressive. These are consistent with Antonov and Soutar [52] who list a number of products that should not be packaged in unprimed foil/acid copolymer laminates:
  - Anything with pH below 2.5–3. Certain condiments containing vinegar; for example, ketchup in US is suitable but differs around the world. Some chili sauces are not suitable.
  - Methylsalicylate. Low concentrations for flavoring may be fine but products for sore muscles (such as the branded pain relief lotion) are not.
  - Curry and some pepper products.
  - Providone – iodine.
  - Turpentine.
  - Chloroform.
  - High concentrations of flavoring oils, such as peppermint.

Accelerated shelf-life testing at elevated temperatures can be misleading and should be confirmed with long-term storage tests. An elevated temperature may anneal the sample and increase the bond strength if the product is benign. Elevated temperatures may also cause product interactions to take place that would not be at room temperature.

Pieper [37] shows shelf life data with acetic acid stored in commercial LDPE/paperboard/LDPE/Al/(ACR-LDPE) laminate structures. Here ACR stands for acid copolymer resin, which includes EAA, EMAA and ionomers. Fig. 16.5

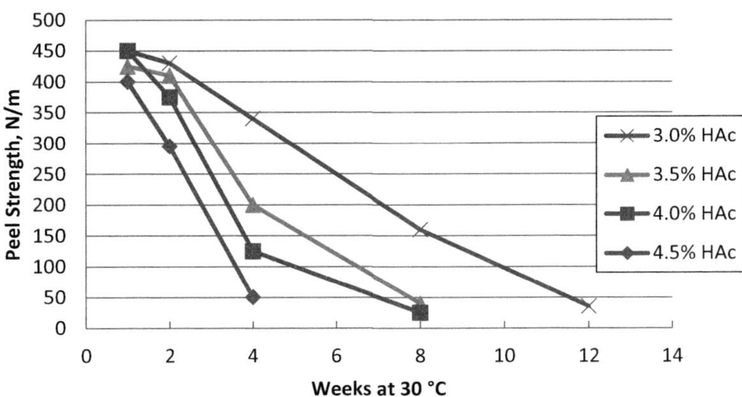

FIGURE 16.5 Effect of acetic acid concentration on peel strength between an ACR or ionomer tie resin and aluminum foil in a barrier laminate structure: LDPE/paperboard/LDPE/Al/(ACR-LDPE). *Reproduced with permission from G. Pieper, Interaction phenomenon in multilayer flexible packaging, in: P. Ackermann, M. Jagerstad, T. Ohlsson (eds.), Foods and Packaging Materials – Chemical Interactions, Vol. 162, Special Publication – Royal Society of Chemistry, 1995, pp. 95–100.*

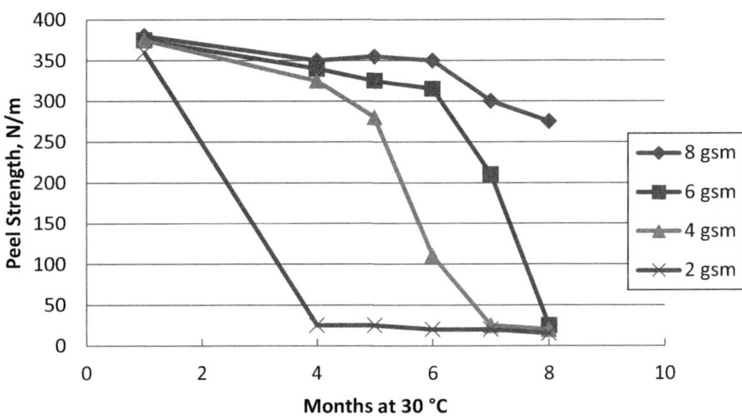

FIGURE 16.6 Effect of tie layer thickness on peel strength durability in the presence of orange juice stored at 30°C. Peel strength is measured between the ACR tie layer and aluminum foil barrier layer. Structure: LDPE/paperboard/LDPE/Al/(ACR-LDPE). *Reproduced with permission from G. Pieper, Interaction phenomenon in multilayer flexible packaging, in: P. Ackermann, M. Jagerstad, T. Ohlsson (eds.), Foods and Packaging Materials – Chemical Interactions, Vol. 162, Special Publication – Royal Society of Chemistry, 1995, pp. 95–100.*

shows the effect of acetic acid concentration on shelf life, as measured by peel strength between the ACR and foil. Increasing the acetic acid concentration in the product reduces the shelf life. One strategy to extend shelf life is to increase the acid content of the tie layer, as described by Bernacki. This must be balanced by the loss in adhesion between the ACR and LDPE. Most commercial applications use acid copolymers with acid levels in the range of 6–9 wt.%. Hester et al. [53] show that while increasing the acid level in the ACR reduces the initial adhesion to LDPE, higher acid helps reduce the loss in adhesion when exposed to acetic acid.

Another strategy for increasing shelf life is to increase the thickness of the ACR layer. Fig. 16.6 shows the effect of increasing the ACR thickness from 2 to 8 gsm (~2–8 μm) from the work of Pieper. Here the peel strength between the ACR and foil is reported for orange juice stored for up to 8 months at 30°C.

Pieper also considers inserting other polymers between the tie and foil to protect the adhesion. Permeability of acetic acid in orange juice is measured for PA, PET, EVOH, and (LDPE-EAA). The PET has the lowest permeability followed closely by the LDPE-EAA and then the PA. Surprisingly, EVOH performs poorly, showing that the acetic acid permeability does not track with oxygen permeation.

Retort sterilization is especially challenging, even for foil-based packaging. Fat and fatty acids can quickly migrate through polyoelfins at elevated temperatures due to their affinity. Long-chain fatty acids do not react with aluminum at room temperature, but they do above 100°C. Aluminum soaps or carboxylates form which may cause delamination. The formation of soaps is also partly responsible for the adhesion of acid copolymers to foil. When low MW fatty acids are present, they may out-compete the acid copolymer for bonding with the oxides at the foil surface, resulting in a weak boundary layer and delamination [7].

Schubert [7] simulates the retort of an aluminum foil laminate in the presence of various food simulants. Triglycerides are found to migrate into the coating. These are present with phospholipids (natural emulsifiers) and smaller fatty acids. The migration is detected by attenuated total reflectance infrared spectroscopy (ATR-IR) at penetration depths of up to 3-μm. SEM images show layer destruction, fat-filled voids and defects in the coating after retort. When the simulants contain fatty acids, delamination occurs and defects similar to what is found in commercial packages are found. At the delaminating interface, additional aluminum soaps are found. Schubert concludes that delamination failure is the result of progressive replacement of the polymer—oxide bonds with fatty acid—aluminum soap bonds.

Technology to purposely separate layers for recycling of waste multilayer laminates also provides insight. Uegdueler et al. [54] conduct a fundamental study on the use of carboxylic acids for this purpose. They test five adhesively laminated structures containing various adhesives (solvent or solvent-free polyurethane) and printing inks (PVC, PU, or nitrocellulose-based):

- OPET/ink/adh/Al/adh/cPP (PVC-based ink);
- OPET/ink/adh/Al/adh/cPP (PU-based ink);
- OPET/ink/adh/PE (white);
- OPET/ink/adh/PE (transparent);
- Lacquer/chem primed OPET/ink/adh/PE.

The laminate structures are exposed to carboxylic acid under different conditions to dissolve the PU adhesives and separate the layers. The first part of the study evaluates different acids. The rate of delamination is determined by the diffusion rate through the films, which is found to decrease with a longer chain length of the acid. Formic acid is found to have the highest diffusion rate through monolayer PET or cPP film. The rest of the study is conducted using formic acid. They find:

- Increasing the exposure temperature from 50°C to 75°C at least doubles the amount of diffusion in all five films. PP is found to be substantially more resistant to diffusion than PE, which they attribute to the high crystallinity of the PP. They also postulate that lower Tg of the PE may contribute to greater chain mobility and higher diffusion rates.
- Lowering the concentration of the acid from 100 to 75 vol% significantly decreases the diffusion rate, but the effect of temperature is more pronounced.
- The delamination rate is also influenced by the dissolution rate of the adhesive in the acid. They measure this rate for different concentrations of formic acid and temperature for the two adhesives. The dissolution rate for the solvent-free adhesive is considerably less than for the solvent-borne PU adhesive. The dissolution rate is higher at higher temperatures and acid concentration as expected.
- A kinetic model that takes into account diffusion through the films and dissolution of the adhesive shows that diffusion through the film (frontal) is more important than through the sides.

16.2.1.3 Environmental stress cracking

ESC is a term used to describe the cracking or crazing of polymers at modest applied stress (well below the tensile strength of the polymer) when exposed to specific organic liquids. In flexible packaging, ESC can lead to package failure, especially in regions of high stress. In addition, cracks that form on the inside of the package may lead to greater migration of molecules from the product into the package. It is especially a problem for amorphous glassy polymers (i.e., PC, PS, PMMA) but is still a concern for semicrystalline polymers like PE. For example, Pieper [37] finds that fats, oils and formulated tomato products (with salts and acids) are effective stress cracking agents. When packaging these products special caution is advised.

Several mechanisms may contribute to ESC [55]:

- The organic liquid may lower the surface energy of the polymer (compared with contact with air), which in turn may decrease the energy needed for craze or crack formation. Early investigators favored this mechanism but it is now thought to be a minor contributor.
- Dissolution in marginal solvents can swell the polymer and add internal stresses that lead to cracking/crazing.
- Many investigators have related ESC to differences in solubility parameters between the polymer and the organic liquid [56,57]. The concept of solubility parameters is introduced in Section 6.4.1. In general, the closer the match, the more likely cracking will occur; better compatibility with the polymer increases the sorption of the challenge molecule into the polymer. The relationship is not definitive; the size and shape (cyclic vs aliphatic) of the challenge molecule also affect the rate of absorption.
- The challenge fluid may plasticize the polymer in the region of the tip of a chance nick or flaw at the surface of the polymer, leading to the magnification of the stress (through the concept of the stress intensity factor introduced in Section 10.2.3) that causes the crack to propagate [58]. For PE, investigators propose a similar mechanism: stress-induced swelling and then plasticization of the amorphous regions, which increases the stress intensity factor.

Polymers with smaller molar volume (inverse of density) generally are less susceptible to stress cracking. For example, increasing the density of PE generally increases stress crack resistance (ESCR): HDPE has better ESCR than LDPE. Slowing the rate of cooling or annealing HDPE during fabrication, which increases crystallinity, generally improves ESCR. PP typically has higher ESCR than HDPE. Adding an impact modifier (rubbery phase) to the polymer can also improve ESCR by reducing crack propagation. Larger rubber particles (2–6 microns) are generally better than small particles.

In addition to density, the crystalline morphology and tie molecule concentration are important for ESCR of PE. The crystals in PE form plate-like structures called lamellae in films and coatings. Tie molecules hold the lamellae together. Pieper [37] explains ESC in packaging films using the model of Lustiger and Markham [59], which attributes the cracking of polymers under long periods of low-stress levels to untangling of tie molecules between crystalline lamellae. LDPE can be an effective contact layer if a grade is selected with good environmental stress crack resistance. In general, this means a high concentration of tie molecules. High MW grades are preferred for this reason.

Mannheim et al. [60] conduct transportation vibration simulation studies on orange juice aseptically packaged in three structures:

- PE/adh/met-PET/adh/PE;
- (PE–tie-EVOH–tie–PE);
- PET-met/adh/met-PET/adh/PE.

The third structure has the lowest initial oxygen permeation value (OPV), but the orange juice packaged and stored at 25°C deteriorates the fastest in this package. Flex cracking during the vibration test increases oxygen permeability from 1.6 to 16.8 cc/d/bag. The combined effects of oxygen permeability and product interactions reduce the shelf life of citrus products in this package compared with glass packaging. Transportation simulation is important for understanding how well the package will survive abuse—pinholes, flex cracking, puncture or seal failure may occur, any one of which can deteriorate shelf life. Section 9.4.5 gives more examples of transportation studies and the effect of package design on flex fatigue.

16.2.2 Organoleptics

Taste and odor changes to the food product can arise from the migration of low MW species from the package (or from the outside environment through the package) into the product or by migration of flavor or aroma components from the product into the package. The former is known as flavor impartation and the latter flavor scalping. Material selection, film structure design and film fabrication processing can impact both of these.

16.2.2.1 Impartation

Low MW species may diffuse through the flexible film and desorb into the product. In some cases, active components such as antioxidants or antimicrobials are intended to migrate to the surface and interact with the product. This is discussed in Section 16.2.3. Here, unwanted migration that may negatively impact sensory perception or cause regulatory concerns is briefly described.

The unwanted migratory species may arise from the substrate or resin, from the environment, or be created during processing [61]. Examples include:

- Residual monomer, catalyst, low MW oligomers, or processing aids (such as slip agents, plasticizers and lubricants) in the polymer resin. Sometimes the resin itself is clean but a masterbatch containing pigment or processing aid contains additives not accounted for.
- Mineral oils from recycled paperboard used in a secondary package, for example in bag-in-box packaging.
- Uncured components of adhesives.
- Solvent and other components of inks.
- Oils on the surface of aluminum foil.
- Various additives from paper or paperboard.
- Oxidative byproducts from extrusion of the polymer.
- Reactive byproducts from irradiation. (See Section 16.1.4.)
- Off-odor causing compounds from recycled plastic incorporated into the structure.

There also may be migratory species that are regulated from a food or product safety perspective. Some components have been banned for use in certain applications, such as bisphenol A (BPA) containing polymers for baby bottles. Often various simulants are used to conduct migratory studies to satisfy regulatory agencies such as the US FDA and the European Union Commission (EU 10/2011). These are described in Chapter 4.

Communication along the supply chain is important for ensuring the package does not create issues from impartation at the end-user. It is important to note that materials with food regulatory compliance may still cause organoleptic issues. Often these components are byproducts or residuals from the manufacturing process that are not controlled by the resin manufacturer. Other components may come from unforeseen sources, such as additive manufacturers, adhesives and printing components. In multilayer films and laminates, components from the outside layer may transfer to the inside product-contact layer while the film is wound in the roll. This in turn may provide a source of impartation to the product during packaging. Incorporation of recycle content, especially postconsumer recycle, into the film structure may be a source of uncontrolled migratory species (see Chapter 17) [62–64]. Finally, the film converting process itself can generate unwanted migratory species.

The amount of impartation depends on many factors, including:

- The concentration of migratory species in the package structure (or the environment if the origin of the species is outside the package).
- The concentration of the migratory species in the product. Ultimately, it is the concentration difference that drives diffusion from the package to the product.
- The solubility of the migrating species in the packaging material. If the species is strongly bonded to the packaging material it is less likely to migrate out. This depends on the chemistry of the migratory species and the material.
- The size and shape of the species. Larger molecules generally diffuse more slowly than smaller ones. Linear molecules may diffuse faster than ones with branches or bulky side groups. Longer chain aliphatic molecules may have greater solubility in PE.
- The presence of a barrier layer in a multilayer package structure between the layer containing the species and product (functional barrier in the language of regulators).
- The chemistry of the product. Acidic or fatty foods may have a different affinity for the migratory species than water. This affects the solubility and desorption of the species from the package to the food itself. Most migrants have a hydrophobic character and so more readily desorb into fatty foods. Some components of the product may plasticize or swell the polymer, allowing more migration to take place.
- Temperature. Higher temperature generally favors faster diffusion rates. Hot fill, retort, in-package sterilization or even storage in a nonrefrigerated warehouse in the summer may contribute to greater migration. Refrigeration will slow migration.
- Relative humidity. Some polymers are plasticized by water, allowing migrating species to diffusion faster through them.

The oxidative degradation of PE during cast film and extrusion coating and its effect on organoleptics has been well studied in a series of papers by Andersson et al. [65–70] and others [71–73]. The thermo-oxidative degradation of PE results in the formation of alkyl radicals (R*) through cleavage of covalent bonds, followed by a reaction with oxygen to form hydroperoxides (ROOH) in a chain reaction: RH → R* → RO2* → ROOH → Oxygenated products. The oxidation may start in the extruder but mostly occurs in the air gap between the die exit and chill roll (see Section 15.1.2.1). Some of the key findings include:

- Oxidation creates degradation byproducts consisting of alcohols, aldehydes, ketones and carboxylic acids in the $C_4 - C_{12}$ range that can be found in the resulting films/coatings. (These same functional groups are also bound to the longer polymer chains and are responsible for the increase in adhesion of LDPE to polar substrates, as described in Section 15.1.2.1.)
- Smoke collected at the die exit contains volatile aldehydes, ketones and carboxylic acids. More byproducts are found with thicker coatings/films.
- The amount and type of degradation byproducts depend on the exposure time in the air gap and temperature. At short times the byproducts are mainly branched alkanes; at longer times acids, alcohols, ketones and aldehydes form.
- The extrusion degradation byproducts correlate with off-flavor in water packaged in the films, as determined by a sensory panel. The presence of aldehydes and ketones has the clearest effect on off-flavor, with C_7–C_9 ketones and C_5–C_8 aldehydes having the greatest impact.

- The amount of oxidation, as measured by the carbonyl index (an infrared spectroscopy method described in Section 15.1.2.1 measuring the amount of carbonyl groups that form) correlates with the degree of off-flavor determined by the sensory panel. The carbonyl index increases with increasing temperature, the oxygen concentration in the air gap and time in the air gap.
- Adding scavengers can help reduce the off-taste. At low processing temperatures, reactive additives, such as maleic anhydride grafted polyolefins, can help control the release of off-flavor components. Zeolites are also effective, even at higher extrusion temperatures.
- Antioxidants added to the PE are effective for preventing off-flavor, but they must be selected with processing temperature in mind. The thermal instability of the antioxidant can jeopardize the intended effect. A mixture of antioxidants is found to be most effective.

Secondary packaging may impart off-flavor. Whitfield et al. [74] investigate the cause of musty-flavored dried fruit packaged in a PE liner-film inside a paperboard box. Using GC-MS, they identify the source of the odor as 2,4,6-trichloroanisole and 2,3,4,6-tetrachloroanisole and trace their origin to the paperboard boxes manufactured from waste paper pulp. Heydanek et al. [75] trace undesirable flavor impartation in cereal to the multicolored graphics on inserts and coupons inside the package. Classic packaging overwraps are found to be an insufficient barrier to the volatiles isolated from these inserts. A PVDC coated PET film is found to be a good barrier.

16.2.2.2 Scalping

Scalping involves the migration of aroma or flavor components from the product into the package. The resulting loss of components affects the sensory quality of the product. Selecting materials and designing packaging structures that minimize scalping may help extend the shelf life of the product.

The interplay between the package and product can be very complex and is the subject of several review articles [76–84]. The barrier of packaging structures to organic molecules such as flavor and aroma molecules is introduced in Chapter 8, along with a general discussion of oil and chemical resistance. Migration involves sorption and partition of the migrant molecule between the product and package and diffusion. Partition relates to the relative concentration of the species in the package and the product at equilibrium and is related to solubility. Diffusion is the time-dependent movement of the species between the two mediums. Partitioning is influenced by temperature, relative humidity, the chemistry of the food product (e.g., pH, fat content and chemical structure), molecular size and structure of the migrant, and the Tg, crystallinity and polarity of the polymer [78,79].

This complexity is illustrated in the examples from the scientific literature provided in Table 16.3. Many factors are involved that influence the loss of flavor and aroma components from the product to the package, including:

- External factors, such as time, temperature and relative humidity during storage.
- Composition of layers in the package structure, especially the polarity of the food contact layer and its surface area and thickness. Higher crystallinity in semicrystalline polymers tends to reduce sorption. For amorphous polymers, the glass transition temperature (Tg) is important: rubbery amorphous polymers above their Tg generally have greater sorption than glassy polymers below their Tg.
- Chemistry (polarity), size, structure and concentration of the migrating molecule.
- Chemistry of the food medium (pH, presence of fat or oil, etc.).
- Temperature of the food medium.
- Presence of other compounds that may swell the polymer, such as oils.

16.2.3 Active packaging

Active packaging is a term used to describe packaging that purposely interacts with the product to extend shelf life or enhance the product quality. The interaction provides active rather than passive protection. Passive and active packaging are often used together, such as the combination of an oxygen barrier layer (passive) with an oxygen scavenger (active). Whereas scalping and impartation involve unwanted migration of molecules, active packaging usually involves the migration, sequestering and emission of molecules to produce a desired outcome.

Some examples of active packaging include:

- Oxygen scavenging – see Section 8.5.1. Sachets (packaging inserts) for oxygen scavenging are one of the largest applications of active packaging.
- Odor scavenging to reduce undesirable aromas from food that has not yet spoiled.

TABLE 16.3 Summary of several product-package interaction studies from the scientific literature.

References	Goal	Experimental details	Key findings
[37,85]	Measure absorption of 19 aroma compounds in orange juice by 3 packaging structures	Packaging structures tested 1. LDPE/board/LDPE/AP/EVOH/AP/LDPE (4/9 gsm for last two layers) [Note: gsm = grams per square meter] 2. LDPE/board/LDPE/Al/AP/PET/AP/LDPE (0.5/26 gsm) 3. LDPE/board/LDPE/Al/AP/LDPE (8/34 gsm) AP = adhesive polymer. Samples are immersed in orange juice stored in glass jars for 24 weeks at 4°C	• Package structure 3 with most amount of LDPE absorbed the most d-limonene and aroma compounds • Some compounds show little absoption, others up to 50% • No significant difference found by sensory panel
[86]	Study absorption of aqueous aroma compounds by packaging	Film strips of LDPE and two grades of ionomers are stored in glass jars containing aqueous aroma compounds, including limonene, citral, ethyl butyrate and linolool. Essential oils are also tested	• Sensory panel finds significant differences between the control and test samples for all essential oils investigated but not for benzaldehyde and citral • Limonene has the greatest affinity for the LDPE and ionomers • Orange oil sorbs considerably less than limonene, which is attributed to the chemistry • Conclusion: compound characteristics are more important than the polymer for the materials studied
[60,87]	Compare package materials for absorption of flavor components in orange juice	1st study: orange juice stored at 25°C for 12 weeks in PE/board/Al/PE cartons and compared with glass	• Greater retention of ascorbic acid and less browning of orange juice occurs in the glass containers • Larger amounts of d-limonene are found in the carton • A taste panel is able to differentiate between orange juice packaged in glass and the cartons after 2.5 months of storage
[60,87]	Compare effects of LDPE processing on flavor absorption	2nd study: Two LDPE films are immersed in model solutions. The two films represent differences in processing for cartons and liners: 1. High process temperature, corona treated (carton production) 2. Lower process temperature, noncorona treated (liner film)	• The PE film accelerates ascorbic acid degradation with the treated film having the greatest effect • The liner film shows no adverse effect on browning or ascorbic acid degradation, but a loss in d-limonene • The taste panel cannot discern differences with a control
[88]	Compare absorption of flavor components in several packaging materials	Packaging materials containing LDPE, PET, PVDC and EVOH are stored in orange juice for 28 days at 25°C. Absorption of d-limonene, α-pinene, ethyl butyrate, and octanal measured	• The absorption of ethyl butyrate and octanal is the same for all four structures • Addition of a layer of EVOH or PET reduces the absorption of d-limonene and α-pinene by 50 and 20%, respectively
[89]	Compare absorption of flavor components in several packaging materials	Strips of LDPE, PET and PC are stored in orange juice and model simulants for 29 days at 20°C in the dark	• In a model simulate with 7 flavorants: o Absorption is much greater in LDPE than PET or PC

(Continued)

TABLE 16.3 (Continued)

References	Goal	Experimental details	Key findings
			o In LDPE, the absorption of valencene is much greater than for decanal, hexyl actate, octanal and nananone o In PC and PET, valencene is the lowest absorbed and decanal the highest o The polarities of the flavorant molecule and polymer have a considerable impact on absorption • In orange juice o Limonene is absorbed readily by the LDPE, while myrcene, valencene, pinene and decanal absorbs much less o In PC and PET, only small quantities of limonene, myrcene and decanal are absorbed • Loss of flavor compounds from the simulant or orange juice does not affect taste perception by a sensory panel from any of the three packaging materials
[90]	Compare absorption of flavor components in several packaging materials	Fresh orange juice stored at 4°C for 14 days in HDPE, PS and polylactic acid (PLA) packages	• Loss of ascorbic acid is greater in HDPE and PS than for PLA • Color changes less in PS and PLA than for HDPE • No detectable loss of d-limonene in PS and PLA while a large loss is found for HDPE
[91]	Study food interactions with PP film	PP film immersed in an acidic model solution with 10 aroma compounds	• Sorption protects the most reactive terpenes from degradation and minimizes formation of compounds that could behave as off-flavors (e.g., α-terpineol)
[92]	Study aroma absorption in packaging films	LLDPE, ionomer and PET films	• Sorption of aroma compounds is selective • d-liminone has 2.5 times higher absorption in LLDPE than does ethyl caproate • Selectivity may account for perception differences in sensory panel evaluations • In general, scalping of aroma compounds was found to be in the order of coPET < ionomers < LDPE
[93]	Study the effect of sorbed oil on aroma loss in packaging	Measures absorption of aroma compounds in the presence of oil in ULDPE, ionomers and co-PET packaging materials	• The oil apparently swells the polymer, increasing the solubility but not the diffusivity (the holes generated by swelling are filled by the long oil molecules) • The permeability of nonpolar compounds is considerably increased but not polar compounds

(Continued)

TABLE 16.3 (Continued)

References	Goal	Experimental details	Key findings
			• This has consequences for packaging of fatty foods – fat sorption may affect the rate of sorption of nonpolar aroma compounds
[94]	Study the effect of food formulation (fat) on aroma loss in packaging	Measured sorption of 6 food aroma compounds from 4 fatty food simulants (95% ethanol, sunflower oil, n-heptane, and iso-octane) in the same three packaging materials: ULDPE, ionomers, Co-PET	• Aroma scalping is highly dependent on the food simulant • Polar aroma components are more sorbed into polymers in the presence of a nonpolar fatty food simulant, and vice versa • In general, co-PET has the lowest partition coefficient for nonpolar components while ULDPE shows the lowest partition for polar aromas
[95]	Study the effect of PE molecular architecture on permeability of five organic compounds found in ham aroma.	Solubility, diffusion coefficient and permeability of 5 aroma compounds measured for 3 PE films: ULDPE and 2 mPE plastomers. The ULDPE has higher Mw, broader molecular weight distribution (MWD), higher level of crystallinity and lower oxygen permeability than the 2 mPE resins	• Solubility of permeant molecules is found to decrease with higher polarity • Diffusion coefficient increases with reduced molecular size • Overall, the 2 mPE films show higher permeability to the molecules than the ULDPE
[96]	Test scalping of PE grades	14 LLDPE sealant resins are tested for scalping of limonene, decanal and 2 (E)-nonenal. Sunflower oil and aqueous simulants spiked with the flavorants are used	• With sunflower oil as the simulant no significant scalping is observed • Using aqueous simulants, the degree of scalping is inversely related to the water solubility of the compound • LLDPE with the lowest density scalps the most but the difference between the resins is not considered substantial enough to expect sensorial differentiation between the sealant resins in the density range of 0.902–0.919 g/ml
[97]	Study interaction of permeants on polymer properties and absorption		• The permeant can change the mechanical properties of the polymer, increasing chain mobility and diffusion • One example is the effect of RH on PVOH. The polar water molecule interacts with the polar bonds in the PVOH, swelling the structure and retarding interaction with polar permeants
[98]	Evaluate the effect of film coatings and printing on absorption of d-limonene.	Two commercial OPP are used: 1. Flame treat/PP/Pololefin sealant 2. Acrylic ctg/OPP/PVDC ctg (sealant) Two inks: nitrocellulose-based and polyamide (PA) based. Printing is done on the flame treated side of film 1. For film 2, nitrocellulose ink is printed on acrylic side and PA-based ink on PVDC side (separate samples). Unprinted controls are also tested. Only the nonprinted side of the films are exposed to the limonene	• The printing and ink type is found to have only a small impact on permeation • The largest effect is the coating type with the acrylic coating providing a better barrier to limonene than PVDC

(Continued)

TABLE 16.3 (Continued)

References	Goal	Experimental details	Key findings
[99]	Measure partition (sorption) of flavor compounds in packaging materials	Three probe compounds (d-limonene, neral and geranial) are tested in LDPE, EVOH and co-PET	• The co-PET has significantly less sorption than the other two films. EVOH has lower sorption than LDPE, but the co-PET shows almost no sorption of d-limonene • Sorption occurs within a day in EVOH and LDPE but only slowly over 24 days in co-PET
[100]	Study the influence of polymer crystallinity on the interaction of PLA with the aroma compound ethyl acetate	PLA film and ethyl acetate used	• An increase in crystallinity results in a decrease in ethyl acetate sorption • The aroma compound plasticizes the PLA: increasing the concentration decreases the Tg from around 60°C to well below room temperature
[101]	Determine the effect of RH on absorption of flavor compounds	Sorption of α-pinene and ethyl butyrate in EVOH and LDPE films	• Sorption in EVOH increases with increasing RH • Even in the worst case (100% RH), however, EVOH outperforms LDPE as a barrier to organic vapors by at least 500,000-fold
[102]	Study the effect of RH on permeation of flavor compounds into packaging materials	Sorption, diffusion and permeability of alcohols and aldehydes in LLDPE, HDPE and EVOH films at different RH measured	• Higher humidity decreases the diffusion of alcohols and increases that of aldehydes • High humidity has a greater effect on LLDPE than HDPE (due to differences in crystallinity) and the most pronounced effect on EVOH (due to its polarity)
[103]	Study the sorption of esters in PP packaging materials	PP film and esters	• The size and shape of the permeant molecule influences the sorption and diffusion rate
[104,105]	Study interaction of food components on scalping	Natural food products and simulants with various oil and flavor components; LLDPE packaging film	• Complicated interactions of food ingredients may suppress migration of flavor molecules into packaging materials • A small amount of oil in emulsions such as milk are found to dissolve nonpolar flavor molecules such as limonene, decanal and linolool so that they are unavailable to be absorbed by a LLDPE packaging film
[106]	Determine the effect of storage time and temperature on sorption of flavor compounds in packaging materials	10-component model solution used. Packaging materials included LLDPE, OPP, PET, PC and polyethylene napthalate (PEN)	• The absorption rate and amount increases with increasing temperature (4 to 40°C in the study) • Depending on storage temperature, polyolefin (LLDPE and OPP) absorption is 3 to 2400 times that of polyesters (PET, PC and PEN)

- CO_2 scavenging or emission. One application is for scavenging CO_2 from fresh roasted or ground coffee in hermetically sealed packages to prevent pressure build and bursting [62]. Another example is the use of CO_2 emitters to extend the shelf life of modified atmosphere packaging.
- Moisture scavenging, typically in the form of desiccant sachets.

- Modified atmosphere packaging (see Section 3.1.4). This is the largest use of active packaging, although it does not involve an active interaction between the package and product.
- Ethylene scavenging/emitting for plant growth regulation.
- Antimicrobial activity.
- Anti-oxidation. An example is adding vitamin E to the package film with the intent of it migrating into the food product to act as a preservative.

A number of articles provide good reviews of the technology [107–117]. Most successes have been at the academic level, or with the use of sachet inserts. Commercial applications in packaging films have small and narrow markets in flexible packaging [118].

There are a number of challenges, both technical and financial. Section 8.5.1 describes the technology and challenges for oxygen scavenging. Two additional examples are highlighted here, odor scavenging and antimicrobials. The first has been more commercially successful than the second.

16.2.3.1 Odor scavenging

Ebner [112] provides a good grounding on the issues with odor scavenging that is summarized here. Many food products develop "confinement" odors – unpleasant odors that the consumer experiences when the package is first opened. These may arise from a number of sources, including:

- Sulfurous compounds from amino acid or carbohydrate breakdown.
- Amines.
- Aldehydes and ketones from lipid oxidation or carbohydrate degradation.
- Fatty acids.

The basic causes are oxidation, enzymatic reactions and bacteria growth, all of which can occur during product storage. Despite the odors, the product may be perfectly safe.

Many food types may benefit from reducing confinement odor, including poultry, bone-in meats, seafood, smoked and processed meats, baked goods, coffee, nuts, baking mixes and snack foods. To design an optimal system the odor profile needs to be known. Each situation may need a different solution.

An odor scavenger system must meet a number of critical to quality (CTQ) attributes or basic requirements, including [112]:

- Removes odors quickly.
- Does not remove desirable odors.
- Does not mask spoilage.
- Is readily incorporated into a package structure, label, sachet, etc. This usually means being extrudable or coatable.
- Is colorless.
- Complies with food packaging regulations.
- Is invisible to the consumer.

The chemistry of removing odor molecules suggests three possible approaches:

- Adsorbing or binding the odor molecule onto the surface of a scavenging agent. Examples of scavenging agents include activated charcoal and silica gel.
- Absorbing the odor molecule into the pores of a solid. Examples include zeolites, molecular sieves and cyclodextrins.
- Reacting with the odor molecule on the surface of the film to tie it up (chemisorption). An example is the reaction of hydrogen sulfide with iron or copper to form a metal sulfide.

The scavenging agent or solution may be extruded into the packaging film, coated onto the surface of the film, or contained in an absorbent pad or sachet. If it is incorporated into the package during extrusion it must be able to withstand the temperature of extrusion without significant degradation or volatilization. Ebner provides a list of typical odor scavengers and gives an example of their efficacy for reducing hydrogen sulfide.

Besides adding agents to the film, Hausmann et al. [119] show that the selection of the sealant resin may help control odor for certain foods and package forms. An ionomer sealant is compared with a standard PE sealant for the vacuum skin packaging of beef loin. The package structure is (PA-EVOH-PA-Tie-Sealant) [80-μm]. Samples are tested after 28, 42 and 56 days of storage at 4°C. The packages with ionomer sealant show reduced bacteria growth, lower drip (or purge) loss and lower odor while keeping an acceptable meat color. These attributes are associated with the ionomer package preventing the pooling of juices over time where bacteria can grow. A second study with skin packaging of pork roast, beef flank, and beef loin show that the ionomer sealant provides lower drip loss, higher bonding of water to the meat (which correlates with more tender meat) and slightly better color after two weeks.

16.2.3.2 Antimicrobial

The drivers for antimicrobial packaging include improving food safety (controlling deleterious microbes such as listeria); maintaining product color, taste and odor; preventing spoilage; and extending shelf life. Post-packaging processes such as retort sterilization, in-package pasteurization, irradiation and HPP are technologies used to destroy microbes in food products. All of these technologies have shortcomings: thermal treatments may cause undesirable changes to the food taste, irradiation has labeling and cost issues and HPP is just emerging as a sterilization technology. Refrigeration and control of temperature excursions during handling, transportation and storage are also important. Antimicrobial packaging may provide additional options for the food industry.

An ideal antimicrobial active package solution would meet the following critical to quality attributes: [112]

- GRAS (generally accepted as safe) and/or complies with appropriate government food safety, packaging and labeling regulations (e.g., FDA, USDA, EPA, FTC for the US).
- Effective against a broad range of microorganisms.
- Nontoxic, nonallergenic, and harmless to human health.
- Stable over a wide range of temperatures, lighting and environmental conditions.
- Fast acting, even in the presence of organic or biological substances.
- Functional at refrigerated conditions.
- Imparts no color, odor or taste.
- Effective over the life of the packaged product.
- Ideally, extrudable into the packaging.
- Inexpensive.

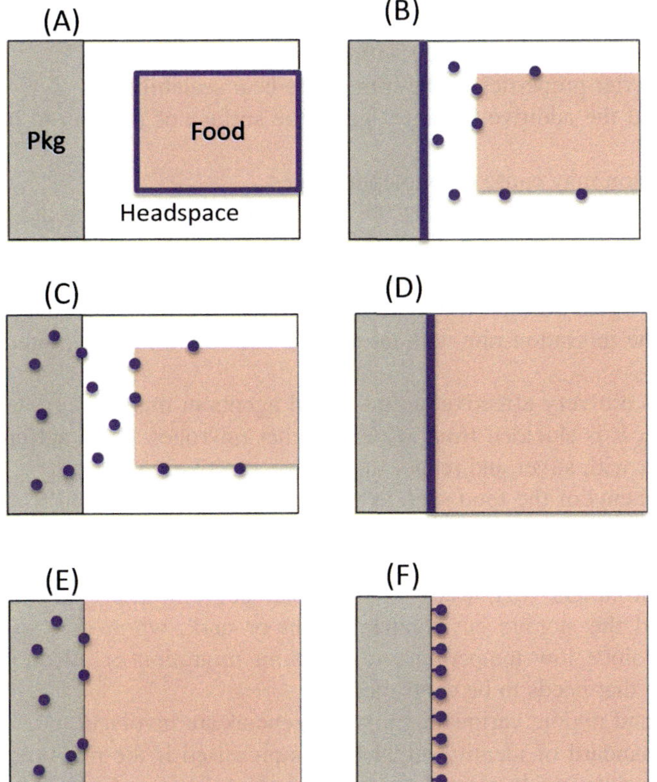

FIGURE 16.7 Strategies for applying antimicrobial technologies: coating or incorporating the active agent onto the food directly (A); adding a volatile active agent to the headspace of the package either by coating the inside of the package with the agent (B) or adding it to the packaging material or sachet insert (C); transfer the active agent from the package to the food by direct contact through a coating on the package (D), diffusion of active agents from the packaging material to the food surface (E) or incorporating an immobile active agent on the surface of the package that is in direct contact with the food (F).

There are three general strategies for using antimicrobials to control microbial growth inside the package, as illustrated in Fig. 16.7. One is to add the antimicrobial agent to the food product itself, as an ingredient or coating; this food additive approach is subject to different regulations from those for packaging. Another is to add the agent to the headspace. Modified atmosphere packaging (MAP) and inserting sachets with volatile active ingredients work in this way. MAP provides an initial environment for controlling microbial growth, but since the headspace atmosphere changes with time, the microbial growth rate may accelerate with passing time. Also, since MAP typically is a low oxygen environment, control of anaerobic microbe growth is critical (see Section 3.1.4 for more information about MAP).

The third strategy is to add the agent to the package material, either by blending with the polymer during extrusion or by applying as a coating. Since antimicrobial additives typically need to contact the food to be effective, the additive either needs to be food grade (regulatory approval as a food ingredient) or be tightly bound to the package surface so that it is not extracted into the food. Most microbial growth occurs on the surface of solid and semisolid food products. There must be a mechanism for the antimicrobial agent to transfer from the package to the surface of the food. One method is to use volatile agents, where the agent can transfer from the film into the headspace, and from there to the food product. If the agent is a nonvolatile migrating agent or is immobile, the food must be in direct contact with the surface of the packaging film. The package form can help provide direct contact; vacuum packaging, for example, provides contact with solid products. Some types of food products, such as ground meat, are not amenable to direct contact strategies.

There are several methods for applying antimicrobial agents as food additives or to the package itself. One is to dip coat the agent onto the food product prior to packaging. However, as the agent diffuses from the surface into the interior of the food product, efficacy may be lost. Another method is to spray the agent as a solution onto the package. The final method is to incorporate the agent into the packaging structure. This includes the use of sachets, coatings, surface attachment and extrusion into the film.

Several review articles are available [110,116,117,120–131] that provide details on specific chemistries and technologies tested over the last twenty or more years. Almost all the studies are academic. Many antimicrobial systems show promise in lab cultures but fail in actual food systems. Antimicrobial packaging has many challenges:

- Extrusion of organic antimicrobials into the film poses several difficulties. The extrusion temperature for most packaging polymers is too high for many organic molecules. Volatile molecules vaporize and are lost. Nonvolatile molecules such as enzymes are prone to thermal degradation. Low MW additives do not mix well with high MW polymers due to their differences in viscosity. Polar additives are incompatible with nonpolar polymers.
- Some additives may change the mechanical, optical or barrier properties of the film or the heat sealability.
- Encapsulation of the additive by the polymer may prevent the additive from getting to the surface of the film to be effective.
- The agent itself or its degradation byproducts from extrusion may cause organoleptic issues.
- Migrating agents must have a rate of migration that matches the rate of loss of the agent at the surface of the product. If the migration rate is too fast, and the agent is too soluble in the product, the agent may initially be effective at reducing the microbial population, but this will subside as the agent diffuses into the core of the product. Similarly, if the migration is too slow, not enough of the agent will get to the product surface to effectively control the microbial population. It is very difficult to control the migration rate without using a controlled release system that adds considerable cost.
- Inorganic agents such as silver, silver salts and zinc salts are very effective antimicrobial agents in the lab. In practice, once a layer of dead cells cover the sliver molecule, it is shielded from affecting other microbes. Also, sulfur-containing amino acids present in some proteins can react with silver and render the surface inactive.
- Chelating agents such as chitosan may react with components of the food such as lipids and be unavailable to react with the microbes.
- The efficacy of antimicrobial agents is influenced by many factors. Antimicrobials that can survive the extrusion or coating process may not be suitable for the food environment (pH, water activity, salt concentration, intrinsic enzymes, antioxidants and other microbes) or withstand the storage environment (light or dark, warm or cold). Refrigerated environments are particularly problematic since low temperature may inhibit migration or efficacy. Finally, the agent may not deactivate the type of microbe that needs to be controlled.
- Regulatory approval is required. Depending on the food and region, various regulatory agencies are involved.
- For natural foods, compliance with label laws and the standard of identity might be compromised if the migratory agent must be listed as an ingredient in the food. One anecdote involves a volatile antimicrobial that was found to be

very effective for a particular cheese product. No trace of the agent could be found on the product itself, but the food company's legal team argued that since the agent was shown to be effective, it must be there. Adding the agent to the food product list of ingredients would have required the product to be relabeled as a different category of food due to labeling conventions. This was unacceptable to the company and commercialization of the system was abandoned.

16.3 Aging

The properties of polymers may change over time due to a number of factors:

- Environmental exposure. Temperature, pressure, humidity and irradiation are discussed in Section 16.1. UV radiation is another factor for some applications, such as agricultural films and heavy duty shipping sacks stored outdoors. Most products using flexible packaging, however, have a short shelf life and UV exposure is not an issue.
- Chemical attack. Oil, grease and chemical resistance are discussed in Chapter 8. Food interactions are described in Section 16.2.
- Physical aging—the subject of this section.

Physical aging is defined as changes in properties that occur over time at constant temperature and zero imposed external stresses [132]. It is generally the result of the movement of molecular chains over long time scales to more energy favorable configurations [132–136]. As the film is fabricated, it is cooled too quickly for the polymer chains to be in their equilibrium state. This results in excess free volume. Chains can move even below Tg, albeit very slowly. Over time they fill the free volume: chains get closer together and interact more. As a result, the polymer may get stiffer and stronger (due to more chemical interactions amongst the polymer chains) but lose ductility. For semicrystalline polymers, chain movement may result in the thickening of crystalline lamellae or the formation of smaller, secondary crystals. This behavior occurs as long as the ambient temperature is above the beta transition temperature of the polymer, which for most polymers is below room temperature (for example, for PET it is −5°C). The change in the polymer is more rapid at elevated temperatures (up to T_g or Tm).

Sepe [133] provides an example of how this affects packaging. PET bottles introduced in the 1970s became brittle when stored in warehouses at 50°C. Impact resistance decreased while strength and modulus increased. Several explanations were considered:

- Loss of MW due to hydrolytic instability (over a long time—similar to what happens at short times in the melt in an extruder);
- UV degradation;
- Chemical exposure;
- Increase in crystallinity.

Each of these was eliminated as possible mechanisms: the conditions were not severe enough to cause hydrolytic or thermal degradation, the bottles were not exposed to sunlight, no chemicals were present and the temperature was not high enough for crystals to form. The loss in ductility, however, fits well with the physical aging hypothesis.

Aging can be delayed by storing the film/part at a cold temperature. Higher polymer MW decreases physical aging, as does cooling the polymer more slowly during film fabrication.

Some polymers are more susceptible to aging than others. Ionomer films, for example, exhibit an increase in modulus over the first week after fabrication. This behavior correlates with the onset and growth of a secondary melting peak in a DSC curve. Upon reheating, the peak disappears, only to reform with storage time. The exact origin of this peak is debated in the literature [137–145], but it is likely a combination of the formation of secondary PE crystals and reorganization within the ionic clusters. The same behavior is seen for acid copolymers and other ethylene copolymers, although to a lesser extent. Conditions that suppress primary crystallization, such as increasing neutralization of the acid groups, give rise to the greater formation of the secondary crystals [137].

PA 6 undergoes secondary crystallization, especially in high humidity environments, that affects stiffness and dimensional stability. This is described in Section 13.5 as it pertains to film curl.

Beyond physical property changes and changes in dimension, adhesion is most affected by aging. Peel strength may age up or down with time. An example showing the change in peel strength with a time of several ethylene-based polymers to aluminum foil is shown in Fig. 16.8. A number of factors may cause peel strength to age up, including:

- The movement of polymer chains or functional groups near the interface to create additional bond sites, particularly polar interactions between the two substrates.

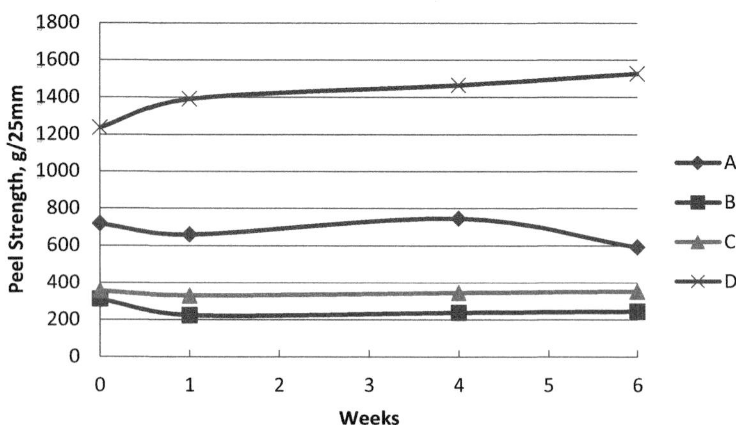

FIGURE 16.8 Example of peel strength as a function of storage time for an extrusion lamination. Structure is 12 μm OPET/19 μm ACR/9-μ Al/19 μm ECP/51 μm LLDPE-film. Peel strength is between the ethylene copolymer (ECP) and Al foil layers. Storage was at 23°C, 50% RH. Samples A–D are different ethylene-based copolymers or homopolymers. *Reproduced with permission from The Dow Chemical Company.*

- Bulk movement of polymer chains near the interface to fill in free volume. Lee and Wool [146–148] describe this phenomenon (see Section 10.3.3). During bonding, the formation of reacted sites at the interface sterically hinders the positioning of the bulk chains near the interface. They postulate that this creates a region of lesser density near the interface, contributing to lower peel strength. Over time, as the chains move to fill in the voids and increase entanglement, the strength near the interface increases.
- Stiffening of the peel arms due to secondary crystallinity or physical aging. This may increase the energy needed to bend the peel arm during the peel test. The actual angle of separation and the amount of elongation may also change. These may act to either increase or decrease the peel strength.
- Changes in viscoelastic dissipation at the peel front due to physical aging. This may affect the peel strength measurement. The local energy dissipation at the peel front is likely linked to changes in crystallinity and free volume. See Section 10.2.3 for a discussion of the importance of bending stiffness and viscoelastic dissipation on peel strength measurement.
- Relaxation of residual stress as the molecules rearrange.

Age down of peel strength may also occur. Some of the factors that may be involved include:

- Migration of competing species to the interface, such as residual acids and processing oils, or in the case of product interactions, acids from the product itself, as described in Section 16.2.1.2.
- An increase in stress due to shrinkage from secondary crystallization. This is thought to be behind the cracking and age-down of bonds between LDPE and metallized films.
- Changes in the chemistry of one or both substrates. Moisture, for example, may change the oxide structure of aluminum foil (see Section 16.1.2).
- Changes in the stress versus strain behavior of the polymer due to physical aging, resulting in a change in failure mode during the peel test.

An example of the last point is described by Hoh et al. [149] They find that the peel strength of thin extrusion coatings of EMAA (12 wt.% MAA) on aluminum foil increases slightly with age. Aging is conducted at room temperature in a standard lab environment (no product present) for 42 days. The peel strength of a thick coating of 100 μm ages up for 28 days and then decreases to less than a quarter of its original value. This is illustrated in Fig. 16.9.

They find that the yield stress and secant modulus of the thick coating increases over the four week time period, as shown in Fig. 16.10. They attribute this to the rearrangement of hydrogen-bonded groups within the polymer, but secondary crystallization may also be involved. They use an analysis by Kaelble [150] to explain the results. Kaelble derives equations for the peel force when pulling a flexible membrane from a rigid substrate. The peel strength is related to the thickness of the film raised to the ¾ power and the modulus to the ¼ power:

$$\text{Peel Strength} \propto X^{3/4}E^{1/4} \tag{16.1}$$

where X = film thickness and E = tensile modulus of the film.

As the thickness of the coating is increased, its yield strength (yield stress times the cross-sectional area) increases. At some critical thickness, it will equal the peel strength. Below the critical thickness, the peel arm yields and

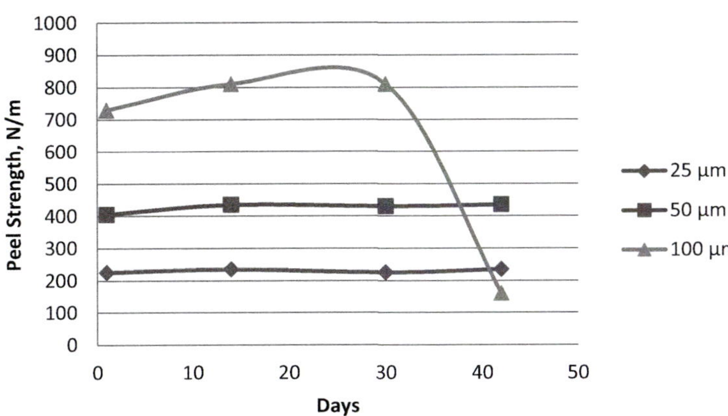

FIGURE 16.9 Peel strength of EMAA to aluminum foil as a function of storage time. The curves represent different coating thicknesses of the EMAA on the foil. *From G. Hoh, S.A. Sadik, J.A. Gates, Aluminum foil adhesion of ethylene/methacrylic acid copolymer extrusion coatings, TAPPI Polymers, Laminations Coat. Conf. (1989) 361–366. Reproduced with permission from The Dow Chemical Company.*

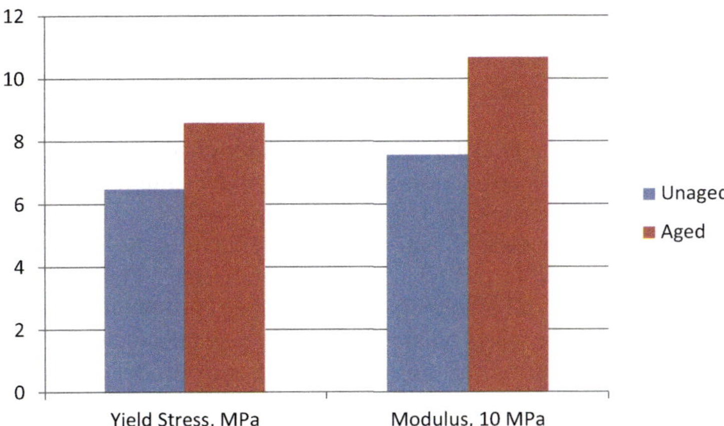

FIGURE 16.10 Aging of yield stress and secant modulus of EMAA film. Modulus is in units of 10 MPa; multiply by 10 to get MPa. *Data from G. Hoh, S.A. Sadik, J.A. Gates, Aluminum foil adhesion of ethylene/methacrylic acid copolymer extrusion coatings, TAPPI Polymers, Laminations Coat. Conf. (1989) 361–366.*

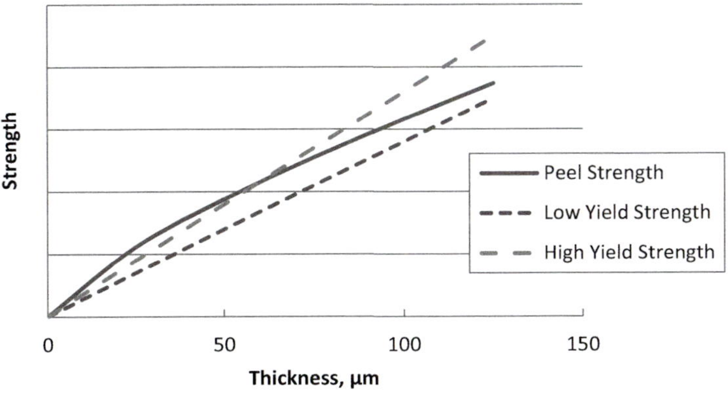

FIGURE 16.11 Effect of yield strength changes on apparent adhesion. When the yield strength of the peel arm exceeds the peel strength, the failure mode changes from yielding to non-yielding. *From G. Hoh, S.A. Sadik, J.A. Gates, Aluminum foil adhesion of ethylene/methacrylic acid copolymer extrusion coatings, TAPPI Polymers, Laminations Coat. Conf. (1989) 361–366. Reproduced with permission from The Dow Chemical Company.*

elongates, distributing the energy of the peel test. Above the critical thickness, the peel arm does not yield and the peel energy is concentrated at the peel front, lowering the overall peel strength.

One way to illustrate this is in Fig. 16.11. Here the peel strength versus thickness curve is shown as the solid curved line ($X^{3/4}$). The yield strength is a linear function of thickness as represented by the straight dashed lines. Initially, the yield strength curve is below the peel strength curve; yielding takes place during the peel test. As the yield stress increases with age, the yield strength curve crosses the peel strength curve. For this system, this occurs around 38–75 μm. Once the thickness crosses the threshold, yielding no longer takes place and the peel strength decreases.

Another way to visualize the effect of the change in yield strength is to plot the stress versus strain curve, as shown in Fig. 16.12. The horizontal dashed line represents the interfacial stress threshold above which failure occurs. If the

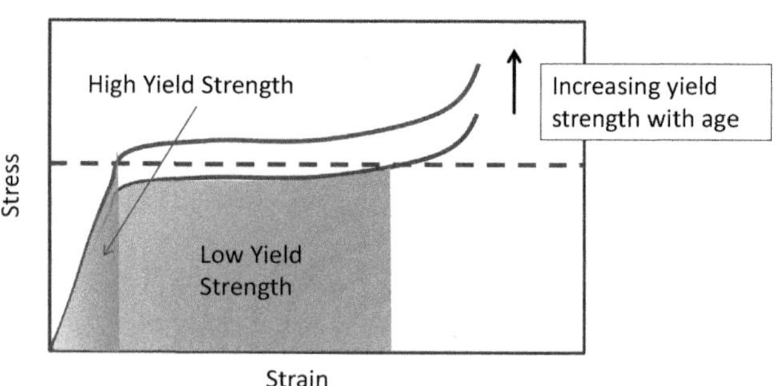

FIGURE 16.12 Energy analysis of the effect of changing yield strength on peel strength. Peel strength is related to the area under the stress versus strain curve.

yield stress is below this threshold, the sample will yield and elongate until the threshold is reached (bottom curve). The peel strength is related to the area under the curve. With aging, the yield stress increases. If the increase takes it above the failure threshold, the peel arm does not elongate prior to failure and the area under the curve is considerably smaller.

16.4 General considerations

The examples in Sections 16.1–16.3 underscore the need to test the packaging structure under conditions that simulate the end-use. Changes in properties due to the environment, sterilization processes, product interactions and physical aging may compromise the integrity of the package. The package may affect the quality of the product, especially the sensory perception of the product due to the loss or gain of migratory species. Often these changes are complex and cannot be easily predicted in advance. Also, simple testing may not be sufficient. For example, simulating scalping with a single food simulation may not capture the effects of co-agents that may swell the polymer. Accelerated testing, especially at elevated temperatures, may also create confounding results—degradation mechanisms or polymer transitions that would not take place at the actual use temperature may produce spurious results.

16.5 Cost

Cost is a critical consideration for the end-user. The driving force behind growth in flexible packaging, and specifically multilayer flexible packaging, is cost savings over other forms of packaging. The total cost of procuring and using the film should be considered, both from a financial and increasingly a sustainability perspective. The sustainability aspects are covered in Chapter 17. Here the financial costs are considered.

The cost of a packaging solution must be weighed against the total value the package provides and the next best competitive alternative. When comparing alternatives, differences in the price of the film, packaging line efficiencies and throughput, and failure rates throughout the value chain (packer, transporter/distrubutor, and retailer or direct delivery for e-commerce) should be considered. These factors are outlined in Table 16.4.

16.5.1 Material costs

The majority of the cost to make a roll of the film comes from raw materials. Judicious selection of raw materials and effective use of structure design allows the material costs to be optimized. Optimized is a better word than minimize. Often using a higher cost material lowers the total cost through better package performance (fewer failures), increased packaging efficiencies or lower roll stock costs (through material substitution, thickness reduction, or greater productivity in the converting process). Premium materials have their place, but not for every application.

The density of the material affects its cost since most polymer resins are sold on a weight basis but the final film is used on an area basis. The conversion from weight to area is called the yield and is inversely proportional to the density and thickness.

$$\text{Yield} = \frac{1}{X\rho} \tag{16.2}$$

where X = film thickness and ρ = density.

TABLE 16.4 Financial cost factors in flexible packaging.

Package material costs	Package converting costs	Packing/brand owner costs
Material selection	Line speed	**Labor and overhead—packer**
Structure design (thicknesses)	Length of runs (transitions)	Labor rate
Price	First quality yield	Number of workers
Quality	Number of processing steps (coex vs lamination)	Machine hour rate
Density/yield	Ability to reuse scrap	Roll stock cost
	Labor rates	Transition losses
	Energy costs	Output (line speed)
	Profit margin	**Package failures**
		Off-machine rewraps
		Off-machine losses (including product)
		Distribution and In-store losses
		Other
		Environmental taxes, fees
		Waste due to package form (e.g., thermoforming round shapes in rectangular sheet)
		Operations inefficiencies (e.g., shape factors, number of packages in a box or shipment)

In English units:

$$\text{Yield}(\text{MSI/lb}) = 27.7/(\text{X}[\text{mils}]\rho[\text{g/cm}^3]) \tag{16.3}$$

where MSI = thousand square inches.

In metric units:

$$\text{Yield}(\text{m}^2/\text{kg}) = 1000/(\text{X}[\mu\text{m}]\rho[\text{g/cm}^3]). \tag{16.4}$$

For example, in English units, a 1 mil (25.4 μm) LLDPE with a density of 0.92 g/cm^3 has a yield of 30.1 MSI/lb (42.8 m^2/kg). Polyolefins enjoy a large share of the flexible packaging market not only because of their excellent properties, but because they have a lower density than most other polymers. An OPP film weighs 35% less than an OPET film of the same thickness; one unit weight of OPP will yield about 35% more area of film.

Quality should not be overlooked as a factor in material costs. A material that is more consistent with fewer defects will run more efficiently in the film converting process, increasing the first quality product.

16.5.2 Package converting costs

A number of factors impact the cost of converting the raw materials and substrates into flexible packaging films and laminates, as described in Table 16.3. There are often several different processes that can be used to implement a packaging solution: lamination, extrusion coating, coextrusion cast film, coextrusion blown film, etc. The lowest cost process may depend on the size of the run and may change during the product's life cycle. Short runs may favor lamination or blown film, whereas large dedicated cast or extrusion coating lines may be used as the product scales up to larger quantities.

A number of technologies, many described elsewhere in this book, have helped improve productivity and reduce cost. Some examples include:

- The development of coextrusion, eliminating lamination steps for some packaging structures.
- The development of more robust metallized films that replace aluminum foil.
- The increase in the number of layers available for coextrusion, allowing the asset to be more flexible for making different structures and increasing the control over individual layer thickness.
- Improved air rings in blown film for helping maintain bubble stability (allowing higher output and better control over film thickness).
- Edge bead reduction dies that reduce edge trim in cast film and extrusion coating.
- Encapsulation dies that allow edge trim of multilayer films to be recycled.
- Higher melt strength resins that help improve bubble stability in blown film and reduce neck-in in cast film/extrusion coating.

16.5.3 Packaging labor and overhead costs

Packaging labor and overhead costs depend on the number of employees operating the line, local labor rates, the allocated cost of the equipment, utilities, and line utilization (packaging rate, downtime, number of shifts per week, etc.). Higher packaging line speeds generally translate into lower cost since the machine time and labor is spread out over more packages. The line speed will depend on what the rate-limiting step is. If the machine is hand-fed, investment in automation or more laborers may help. Often the bottleneck is the sealing process—getting heat to the interface to melt and bond the sealant. Switching to a sealant with a lower seal initiation temperature may help. Alternative methods of sealing, such as ultrasonics, may be considered.

16.5.4 Indirect packaging costs due to package failure

The cost of package failures due to poor sealing, puncture or general abuse through handling and distribution can be significant. The failure rate (often called the leaker rate) typically ranges from less than 1% to several percent. The cost will depend on the cost of the product and volume: leakers may be more costly per item in terms of returns for a fresh meat product than for a bag of potato chips, but because of the sheer volume and scale, small reductions in failure rates are a big driver for snack food companies.

The failures can be categorized into three types. Off-machine rewraps are those that the operator notices during packing. The product is returned to be repackaged and so the loss is only the film itself. Off-machine losses are those that are detected while the product is still at the packer but both the film and product must be disposed of. In-store losses are returns and credits from failures detected at the retailer (or the consumer for e-commerce). These involve loss of both product and film. A fourth category that is difficult to quantify are those packages that prematurely fail after the consumer has purchased the product. Many of these go unreported but can potentially damage brand image and result in loss of repeat sales.

16.5.5 Other costs

Environmental costs are increasingly a factor in evaluating packaging design. The concept of life cycle assessment, which allows quantification of environmental costs, benefits of preventing product loss, end-of-life scenarios and recycling, is discussed in Chapter 17. In terms of financial costs, some countries or regions, such as Europe, have enacted laws imposing taxes or fees on packaging structures to help encourage reducing and recycling of packaging. In some cases, a film with a higher roll stock cost will result in a lower total cost if it has a lower environmental fee.

Other costs include the waste associated with the package form. An example is thermoforming of round shapes from a rectangular sheet. About 50% of the sheet is not used due to simple geometric considerations. If this portion of the sheet cannot be reused in the existing structure, it often must be downgraded, sold at a loss, or disposed of. Package form factor inefficiencies may also be a factor for shipping and handling. For example, an optimum form factor may allow more packages on a pallet for shipping efficiencies.

16.5.6 Modeling financial costs

Models are available that estimate the financial costs of packaging. A simple model is used here to illustrate the benefit of considering more than just the material costs of packaging. The model was originally developed by the Ethylene Copolymer business at DuPont (now part of Dow) in the 1980s as a tool to help packers understand the value premium sealants provide in lowering total cost [151]. It includes the roll stock cost, direct packaging costs (except the transition losses) and indirect packaging costs due to failures.

To demonstrate the utility of cost modeling, a hypothetical cereal liner bag-in-box structure is considered. A flexible packaging structure with an ionomer sealant is compared with one using an EVA sealant. EVA is often used in cereal liner packages for its low seal initiation temperature and ease of formulation for an easy-open experience for the consumer. Ionomers generally cost more but in addition to the performance attributes of EVA they provide outstanding hot tack, oil resistance, and puncture resistance that often translates into higher packaging line speeds and less waste. In this analysis, it is assumed the EVA sealant is the incumbent and its performance is known and fixed. The analysis shows the savings that may occur if the package performance is improved and how sensitive this is to the cost of the roll stock and other cost parameters.

The inputs to the model are provided in Table 16.5. A 10% improvement in line speed and 50% lower failure rate is assumed for the ionomer sealant. The roll stock cost is 6% higher for the film containing the ionomer sealant.

The analysis results are given in Table 16.6 and shown graphically in Fig. 16.13. In this example, switching to the ionomer sealant increases the package material costs, but reduces the overall costs by $140,000/year through savings in labor and overhead and a reduction in waste (including loss of product).

The model includes a sensitivity analysis where each input parameter is varied while the others are held constant. This provides an assessment of how much of a change is needed for that input parameter to match the cost performance of the incumbent film. An example is shown in Fig. 16.14 for the effect of packaging line speed. The EVA-sealant film is assumed to be constant (at 58 packages/minute) since it is the known control. The cost of using the ionomer film at various line speeds is plotted; at low line speed the cost is high due to the greater machine hour time and labor to produce the same number of packages. The cost decreases with increasing line

TABLE 16.5 Cost model input table for a hypothetical cereal liner pouch application.

Case name: cereal liner		
Packing Line Input		
Number of employees	1	
Cost per man-hour (labor rate), US$ per hour	20	
Fixed overhead and equipment costs per hour, US$	20	
Packages produced per week (in thousands)	596	
Weeks per year packing product	50	
Product cost per 1000 packages, US$	500	
Film Specific Input	**Ionomer Sealant**	**EVA Sealant**
Packaging line speed, packs per minute	64	58
Roll stock price per 1000 in^2, US$	0.09	0.0854
Surface area of roll stock used per package, per 1000 in^2	0.65	0.65
Percent off-machine rewraps	1.5	3
Percent off-machine losses	0.5	1
Percent in-store losses	0.5	1

Note: The model compares two film structures with ionomers and EVA sealants, respectively.
Source: Data from B.A. Morris, Optimizing flexible packaging ... with unique tools from DuPont, Present. PackExpo (2011). Reproduced with permission from The Dow Chemical Company.

TABLE 16.6 Cost model output for a hypothetical cereal liner application comparing ionomer and EVA sealant films.

Analysis		
Summary of Costs: US$ per 1000 packages	**Ionomer**	**EVA**
Package materials cost	58.5	55.51
Labor and Overhead cost	10.42	11.49
Waste cost due to off-machine package failures	3.88	7.68
Waste cost due to in-store package failures	2.84	5.67
Total packaging costs with store rejects	75.64	80.35
Total Savings, US$		
Savings per 1000 packages using ionomer	4.71	
Savings per week using ionomer	2810	
Savings per year using ionomer	140500	
Savings per machine hour	18	

Source: Data B.A. Morris, Optimizing flexible packaging ... with unique tools from DuPont, Present. PackExpo (2011). Reproduced with permission from The Dow Chemical Company.

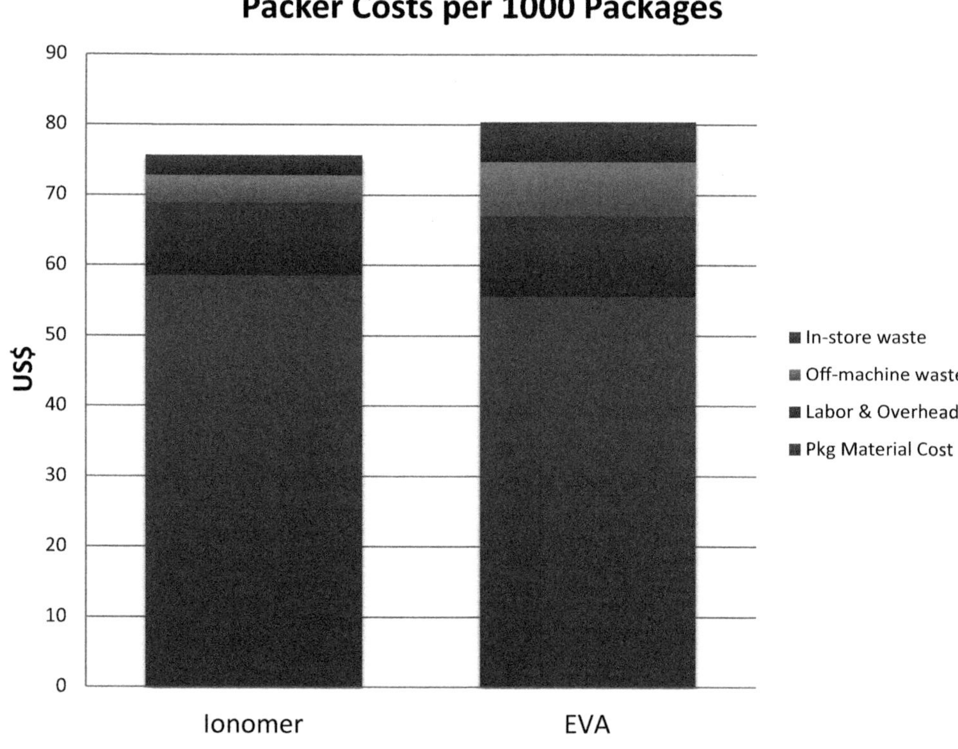

FIGURE 16.13 Graphical output of cost model results for a hypothetical cereal liner film application comparing ionomers and EVA sealants. *Data from B.A. Morris, Optimizing flexible packaging ... with unique tools from DuPont, Present. PackExpo (2011). Reproduced with permission from The Dow Chemical Company.*

speed. Where the two curves intersect is the breakeven point. Keeping all the other inputs constant, the analysis shows that the ionomer film would still be cost-effective at >49 pkg/minute, 16% less than the 58 pkg/minute of the EVA film. The cost-saving due to the improved failure rate allows for this, if needed. More than likely, however, the ionomer sealant will allow higher line speed than the incumbent because of its outstanding sealing performance.

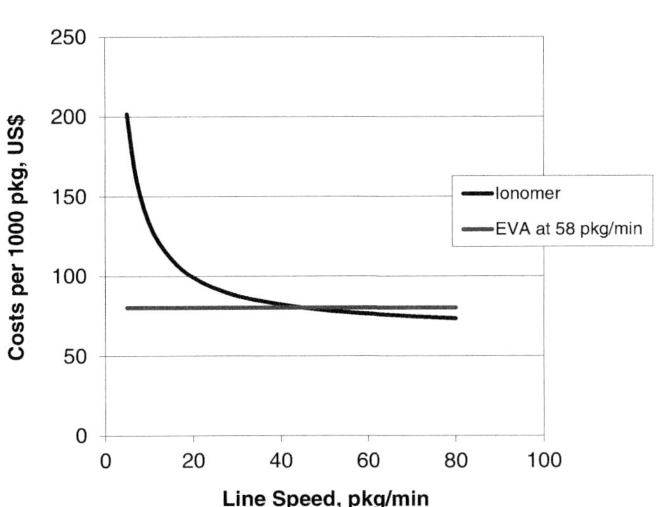

FIGURE 16.14 Break-even analysis for line speed. *Data from B.A. Morris, Optimizing flexible packaging ... with unique tools from DuPont, Present. PackExpo (2011). Reproduced with permission from The Dow Chemical Company.*

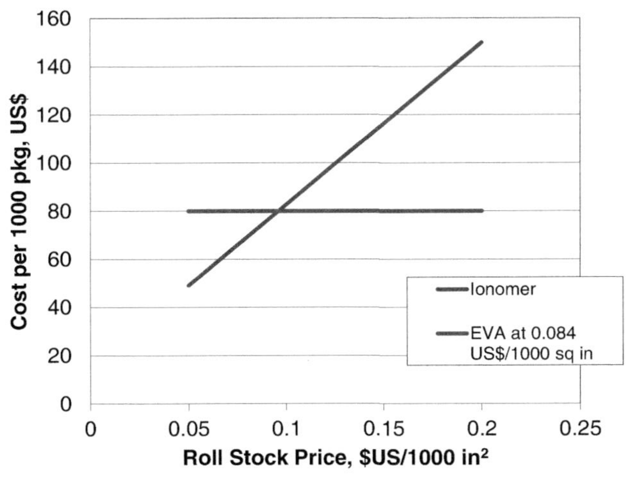

FIGURE 16.15 Break-even analysis for roll stock price. *Data from B.A. Morris, Optimizing flexible packaging ... with unique tools from DuPont, Present. PackExpo (2011). Reproduced with permission from The Dow Chemical Company.*

A similar break-even analysis is shown in Fig. 16.15 for roll stock costs. The result shows that the price of the ionomer film could be 10% higher than the value used in the original analysis (and 17% higher than the EVA film) and still be more cost-effective than the EVA film.

These types of analyses can be very helpful for determining very quickly whether a premium material is worth considering for a given application.

References

[1] T. Osswald, Mechanical behavior of polymers, Underst. Polym. Processing: Process. Gov. Equations, Hanser (2011).

[2] B.C. Tsai, B.J. Jenkins, Effect of retorting on the barrier properties of EVOH, J. Plastic Film. Sheeting 4 (1988) 63–71.

[3] A. Lopez-Rubio, J.M. Lagaron, E. Gimenez, D. Cava, Morphological altercations induced by temperature and humidity in ethylene-vinyl alcohol copolymers, Macromolecules 36 (2003) 9467–9476.

[4] A. Lopez-Rubio, P. Hernandez-Munoz, E. Gimenez, T. Yamamoto, R. Gavara, J.M. Lagaron, Gas barrier changes and morphological alterations induced by retorting in ethylene vinyl alcohol-based food packaging structures, J. Appl. Polym. Sci. 96 (2005) 2192–2202.

[5] K.K. Mokwena, J. Tang, Ethylene vinyl alcohol: a review of barrier properties for packaging shelf stable foods, Crit. Rev. Food Sci. Nutr. 52 (2012) 640–650.

[6] R.J. Hernandez, J.R. Giacin, Factors affecting permeation, sorption and migration processes in package-product systems, in: I.A. Taub, R.P. Singh (Eds.), Food Storage Stability, CRC Press, Boca Raton, FL, 1998, pp. 269–330.

[7] G. Schubert, Adhesion of aluminum foil coatings – stick with it, TAPPI Eur. PLACE. Conf. (2004).

[8] C. Fersti, P. Fersti, High pressure processing: insights on technology and regulatory requirements, Natl Food Lab, NFL White Pap. Ser. 10 (2013) 2–6.

[9] Avure HPP Food Processing, <https://www.jbtc.com/foodtech/products-and-solutions/brands/avure-technologies>, n.d. (accessed 14.06.21).

[10] Hiperbaric USA, HPP Technology. <https://www.hiperbaric.com/en/hpp-technology/>, n.d. (accessed 14.06.21).

[11] U.S. Food and Drug Administration, Kinetics of microbial inactivation for alternative food processing technologies. <https://www.fda.gov/files/food/published/Kinetics-of-Microbial-Inactivation-for-Alternative-Food-Processing-Technologies.pdf>, 2014 (accessed 14.06.21).

[12] D. Salvi, N. Gosavi, M. Karwe, High pressure cold pasteurization, 10.1016/B978-0-08-100596-5.21075-5. <https://www.researchgate.net/publication/311756538_High_Pressure_Cold_Pasteurization>, 2016 (accessed 14.06.21).

[13] E.J. Turek, Preserving foods through high-pressure processing, Food Technol. Mag. 62 (11) (2008)Viewed on 14 June 2021 at. Available from: https://www.ift.org/news-and-publications/food-technology-magazine/issues/2008/november/features/preserving-foods-through-high-pressure-processing.

[14] D.G. Yordanov, G.V. Angelova, High pressure processing for foods," Preserving, Biotechnol. Biotechnological Equip. 24 (3) (2010) 1940–1945. Available from: https://doi.org/10.2478/V10133-010-0057-8Viewed on 14 June 2021 at. Available from: https://www.tandfonline.com/doi/pdf/10.2478/V10133-010-0057-8.

[15] G. Mensitieri, G. Scherillo, S. Iannace, Flexible packaging structures for high pressure treatments, Innovative Food Sci. Emerg. Technol. 17 (2013) 12–21.

[16] M.K. Bull, R.J. Steele, M. Kelly, S.A. Olivier, B. Chapman, Packaging under pressure: effects of high pressure, high temperature processing on the barrier properties of commonly available packaging materials, Innovative Food Sci. Emerg. Technol. 11 (4) (2010) 533–537.

[17] L. Marangoni Júnior, R.M.V. Alves, C.Q. Moreira, M. Cristianini, M. Padula, C.A.R. Anjos, High-pressure processing effects on the barrier properties of flexible packaging materials, J. Food Process. Preservation 44 (11) (2020) e14865.

[18] T. Koutchma, Y. Song, I. Setikaite, P. Juliano, G.V. Barbosa-Cánovas, C.P. Dunne, et al., Packaging evaluation for high-pressure high-temperature sterilization of shelf-stable foods, J. Food Process. Eng. 33 (2010) 1097–1114.

[19] M. Fraldi, A. Cutolo, L. Esposito, G. Perrella, C. Pastore, G. Maria, et al., Delamination onset and design criteria of multilayer flexible packaging under high pressure treatments, Innovative Food Sci. Emerg. Technol. 23 (2014) 39–53.

[20] L. Marangoni Júnior, M. Cristianini, M. Padula, C.A.R. Anjos, Effect of high-pressure processing on characteristics of flexible packaging for foods and beverages, Food Res. Int. 119 (2019) 920–930.

[21] A. Lopez-Rubio, J.M. Lagaron, P. Hernandez-Munoz, E. Almenar, R. Catala, R. Gavara, et al., Effect of high pressure treatments on the properties of EVOH-based food packaging materials, Innovative Food Sci. Emerg. Technol. 6 (2005) 51–58.

[22] C. Caner, R.J. Hernandez, M. Pascall, V.M. Balasubramaniam, B.R. Harte, The effect of high-pressure food processing on the sorption behaviour of selected packaging materials, Packaging Technol. Sci. 17 (3) (2004) 139–153.

[23] A. Seeger, D. Freitag, F. Freidel, G. Luft, Melting point of polymers under high pressure: Part I: Influence of the polymer properties, Thermochim. Acta 424 (2004) 175–181.

[24] L. Sansone, Investigation on the Effect of New Treatments Techniques on Functional and Structural Properties of Polymeric Materials for Food Packaging (PhD thesis), University of Naples Federico II. <http://www.fedoa.unina.it/3302/>, 2008 (accessed 15.07.21).

[25] K.J. Hemmerich, Polymer materials selection for radiation-sterilized products, MDDI (2000)viewed 14 June 2021 at. Available from: http://www.mddionline.com/article/polymer-materials-selection-radiation-sterilized-products.

[26] B.J. Lyons, F.E. Weir, The effect of radiation on the mechanical properties of polymers, in: E. Dole (Ed.), The Radiation Chemistry of Macromolecules, Vol. II, Academic Press, 1973.

[27] G. Hoh, G.M. Cumberbatch, Stability of Surlyn® films to radiation, TAPPI Polymers, Laminations Coat. Conf. (1990) 251–292.

[28] U.S. Code of Federal Regulations, 21 CFR 179.26. <http://www.accessdata.fda.gov/scripts/cdrh/cfdocs/cfcfr/cfrsearch.cfm?fr = 179.26> (accessed 14.06.21).

[29] V. Komolprasert, Packaging for foods treated with ionizing radiation, in: J.H. Han (Ed.), Packaging for Non Thermal Processing of Food, Blackwell Publishing, 2007 viewed 25 Nov 2015 from. Available from: http://www.fda.gov/Food/IngredientsPackagingLabeling/IrradiatedFoodPackaging/ucm081399.htm.

[30] F. Li, Y. Wang, X. Liu, B. Yang, Evaluation of plastic packaging materials used in radiation sterilized medical products and food, Radiat. Phys. Chem. 57 (2000) 435–439.

[31] D.H. Jeon, K.H. Lee, H.J. Park, The effects of irradiation on physicochemical characteristics of PET packaging film, Radiat. Phys. Chem. 71 (2004) 1059–1064.

[32] S. Shang, T. Yang, C. Sandford, M.T.K. Ling, L. Woo, Effect of electron beam irradiation on medical packaging materials, Annu. Technical Conf. - Soc. Plast. Eng. 58th 3 (2000) 2740–2745.

[33] A.E. Goulas, K.A. Riganakos, M.G. Kontominas, Effect of ionizing radiation on physicochemical and mechanical properties of commercial multilayer coextruded flexible plastics packaging materials, Radiat. Phys. Chem. 68 (2003) 865–872.

[34] P.G. Demertzis, R. Franz, F. Welle, The effects of γ-irradiation on compositional changes in plastic packaging films, Packaging Technol. Sci. 12 (3) (1999) 119–130.

[35] H. Barakat, C. Aymes-Chodur, J. Saunier, N. Yagoubi, Effect of electron beam radio sterilization on cyclic olefin copolymers used as pharmaceutical storage materials, Radiat. Phys. Chem. 84 (2013) 223–231.

[36] E. Pasbrig, Influence of radiation sterilisation for flexible packaging material for pharmaceutical and medical products, in: H.R. Skov (ed.), 17th Pharmaceutical and Medical Packaging 2007, Collected Papers of the Annual International Conference, Copenhagen, Denmark, May 8–9, 2007.

[37] G. Pieper, Interaction phenomenon in multilayer flexible packaging, in: P. Ackermann, M. Jagerstad, T. Ohlsson (Eds.), Foods and Packaging Materials – Chemical Interactions, 162, Special Publication – Royal Society of Chemistry, 1995, pp. 95–100.

[38] R.W.G. Van Willige, J.P.H. Linssen, M.B.J. Meinders, H.J. Van der Stege, A.G.J. Voragen, Influence of flavor absorption on oxygen permeation through LDPE, PP, PC and PET plastics food packaging, Food Addit. Contam. 19 (3) (2002) 303–313.

[39] F. Johansson, A. Leufven, Influence of sorbed vegetable oil and relative humidity on the oxygen transmission rate through various polymer packaging films, Packaging Technol. Sci. 7 (6) (1994) 275–281.

[40] V. Korbonits, B.A. Morris, Interaction of abrasion and oil resistance of sealant materials, in: SPE FlexPackCon Confernece, Naples, FL, October 2017.

[41] G. Ortiz, M.T. Tena, Headspace solid-phase microextraction gas chromatography-mass spectrometry method for the identification of cosmetic ingredients causing delamination of packagings, J. Chromatogr. A 1101 (1-2) (2006) 32–37.

[42] M.A. Schroeder, B.R. Harte, J.R. Giacin, R.J. Hernandez, Effect of flavor sorption on adhesive and cohesive bond strengths in laminations, J. Plastic Film. Sheeting 6 (1990) 232–246.

[43] G. Olafsson, M. Jaegerstad, R. Oeste, B. Wesslen, Delamination of polyethylene and aluminum foil layers of laminated packaging material by acetic acid, J. Food Sci. 58 (1993) 215–219.

[44] G. Olafsson, M. Jagerstad, R. Oste, B. Wesslen, T. Hjertberg, Effects of different organic acids on the adhesion between polyethylene film and aluminum foil, Food Chem. 47 (1993) 227–233.

[45] G. Olafsson, I. Hildingsson, Sorption of fatty acids into low density polyethylene and its effect on adhesion with aluminum foil in laminated packaging material, J. Agr. Food Chem. 43 (1995) 306–312.

[46] G. Olafsson, I. Hildingsson, B. Bergenstahl, Transport of oleic and acetic acids from emulsions into low-density polyethylene; effects on adhesion with aluminum foil in laminated packaging, J. Food Sci. 60 (1995) 420–425.

[47] G. Olafsson, I. Hildingsson, Interactions between carboxylic acids and laminated food packaging material, in: P. Ackermann, M. Jagerstad, T. Ohlsson (Eds.), Foods and Packaging Materials – Chemical Interactions, 162, Special Publication – Royal Society of Chemistry, 1995, pp. 101–107.

[48] J. Koch, L. Robinson, K. Figge, Determination of fat release by foods. I. Methods, Fette, Seifen, Anstrichm. 78 (1976) 371.

[49] J. Koch, K. Figge, Determination of fat release by foods. II. Application of method to additional foods, Fette, Seifen, Anstrichm. 80 (1978) 158.

[50] W.D. Bieber, K. Figge, J. Koch, Interaction between plastic packaging materials and foodstuffs with different fat content and different fat release properties, Food Addit. Contam. 2 (1985) 113–124.

[51] R. Bernacki, Adhesive polymers for extrusion coating and lamination,", TAPPI Polymers, Laminations & Coatings Conference Proceedings, TAPPI PRESS, Atlanta, 1985, pp. 311–318.

[52] V. Antonov, A. Soutar, Foil adhesion with copolymers: time in the air gap, TAPPI 1991 Polymers, Laminations & Coatings Conference Proceedings, TAPPI PRESS, Atlanta, 1991, p. 553.

[53] D. Hester, S. Sadik, Y. Trouilhet, Using acid copolymers as coextrusion tie layers, TAPPI Polymers, Laminations Coat. Conf. (1989) 459.

[54] S. Uegdueler, T. De Somer, K.M. Van Geem, M. Roosen, A. Kulawig, R. Leineweber, et al., Towards a better understanding of delamination of multilayer flexible packaging films by carboxylic acids, ChemSusChem 2021 (14) (2021) 1–7.

[55] L.M. Robeson, Environmental stress cracking: a review, Polym. Eng. Sci. 53 (3) (2013) 453–467.

[56] C.M. Hansen, L. Just, Environmental stress cracking in plastics, in: H.R. Skov (ed.), 9th Pharmaceutical and Medical Packaging '99, Collected Papers of the International Conference, Copenhagen, Denmark, May 18–19, 1999.

[57] C.M. Hansen, On predicting environmental stress cracking in polymers, Polym. Degrad. Stab. 77 (1) (2002) 43–53.

[58] A.N. Gent, Hypothetical mechanism of crazing in glassy plastics, J. Mater. Sci. 5 (1970) 925–932.

[59] A. Lustiger, R.L. Markham, Importance of tie molecules in preventing polyethylene fracture under long-term loading conditions, Polymer 24 (1983) 1647–1654.

[60] C.H. Mannheim, J. Miltz, N. Passy, Interaction between aseptically filled citrus products and laminated structures, inJ.H. Hotchkiss (ed.), Food and Packaging Interactions, American Chemical Society Symposium Series, 1988, pp. 68–82.

[61] H. Brown, J. Williams, Packaged product quality and shelf life, in: R. Coles, D. McDowell, M.J. Kirwan (eds.), Food Packaging Technolgoy, CRC Press, Boca Raton, FL, pp. 65–94.

[62] A.L. Perou, J.M. Vergnaud, Contaminant transfer during the coextrusion of thin three-layer food packages with a recycled polymer between two virgin polymer layers, J. Polym. Eng. 17 (5) (1997) 349–361.

[63] A.L. Perou, J.M. Vergnaud, Contaminant transfer during the processing of thick three-layer food packages with a recycled polymer between two virgin polymer layers, Int. J. Numer. Methods Heat. Fluid Flow. 8 (7) (1998) 841–852.

[64] J.M. Vergnaud, Problems encountered for food safety with polymer packages: chemical exchange, recycling, Adv. Colloid Interface Sci. 78 (3) (1998) 267–297.

[65] T. Andersson, B. Wesslen, Degradation of LDPE, LLDPE and HDPE in film extrusion, TAPPI PLACE. Conf. (2004).

[66] T. Andersson, B. Wesslen, J. Sandstrom, Degradation of low density polyethylene during extrusion. I. Volatile compounds in smoke from extruded films, J. Appl. Polym. Sci. 86 (2002) 1580–1586.

[67] T. Andersson, B. Stalbom, B. Wesslen, Degradation of polyethylene during extrusion. II. Degradation of low-density polyethylene, linear low-density polyethylene, and high density polyethylene in film extrusion, J. Appl. Polym. Sci. 91 (2004) 1525–1537.

[68] T. Andersson, T. Nielsen, B. Wesslen, Degradation of low density polyethylene during extrusion. III. Volatile compounds in extruded films creating off-flavor, J. Appl. Polym. Sci. 95 (2005) 847–858.

[69] T. Andersson, M.H.J. Holmgren, T. Nielsen, B. Wesslen, Degradation of low density polyethylene during extrusion. IV. Off-flavor compounds in extruded films of stabilized LDPE, J. Appl. Polym. Sci. 95 (2005) 583–595.

[70] T. Andersson, B. Wesslen, Degradation of low density polyethylene during extrusion. VI. Effects of oxygen content in air gap, J. Appl. Polym. Sci. 96 (2005) 1767–1775.

[71] A. Bravo, J.H. Hotchkiss, T.E. Acree, Identification of odor-active compounds resulting from thermal oxidation of polyethylene, J. Agric. Food Chem. 40 (10) (1992) 1881–1885.

[72] A. Bravo, J.H. Hotchkiss, Identification of volatile compounds resulting from the thermal oxidation of polyethylene, J. Appl. Polym. Sci. 47 (10) (1993) 1741–1748.

[73] F. Gugumus, Re-examination of the thermal oxidation reactions of polymers 2. Thermal oxidation of polyethylene, Polym. Degrad. Stab. 76 (2002) 329–340.

[74] F.B. Whitfield, T.H. Ly Nguyen, K.J. Shaw, J.H. Last, C.R. Tindale, G. Stanley, Contamination of dried fruit by 2,4,6-trichloroanisole and 2,3,4,6-tetrachloroanisole adsorbed from packaging materials, Chem. Ind. (London, U Kingd.) 19 (1985) 661–663.

[75] M.G. Heydanek Jr., G. Woolford, L.C. Baugh, Premiums and coupons as a potential source of objectionable flavor in cereal products, J. Food Sci. 44 (3) (1979) 850–853. 856.

[76] K. Cooksey, Interaction of food and packaging contents, in: C.L. Wilson (Ed.), Intelligent and Active Packaging for Fruits and Vegetables, CRC Press, 2007, pp. 187–199.

[77] S.E. Duncan, J.B. Webster, Sensory impacts of food-packaging interactions, Adv. Food Nutr. Res. 56 (2009) 17–64.

[78] M.G. Sajilata, K. Savitha, R.S. Singhal, V.R. Kanetkar, Scalping of flavors in packaged foods, Compr. Rev. Food Sci. Food Saf. 6 (2007) 17–35.

[79] E.A. Tehrany, S. Desobry, Partition coefficients in food/packaging systems: a review, Food Addit. Contam. 21 (2004) 1186–1202.

[80] J.P.H. Linssen, R.W.G. van Willige, M. Dekker, Packaging-flavor interactions, in: R. Ahvenainen (Ed.), Novel Food Packaging Techniques, CRC Press, 2003, pp. 144–171.

[81] S.C. Fayoux, A.-M. Seuvre, A.J. Voilley, Aroma transfers in and through plastic packagings: orange juice and d-limonene. A review. Part II: overall sorption mechanisms and parameters - a literature survey, Packaging Technol. Sci. 10 (1997) 145–160.

[82] S.C. Fayoux, A.-M. Seuvre, A.J. Voilley, Aroma transfers in and through plastic packagings: orange juice and d-limonene. A review. Part I: orange juice aroma sorption, Packaging Technol. Sci. 10 (1997) 69–82.

[83] A.A. Kadam, T. Karbowiak, A. Voilley, F. Debeaufort, Techniques to measure sorption and migration between small molecules and packaging. A critical review, J. Sci. Food Agriculture 95 (7) (2015) 1395–1407.

[84] J.H. Hotchkiss, Overview on chemical interactions between food and packaging materials, Spec. Publ. - R. Soc. Chem. 162 (1995) 3–11.

[85] G. Pieper, L. Borgudd, P. Acermann, P. Fellers, Absorption of aroma volatiles of orange juice into laminated carton packages did not affect sensory quality, J. Food Sci. 57 (1992) 1408–1411.

[86] O.Y. Kwapong, J.H. Hotchkiss, Comparative sorption of compounds by polyethylene and ionomer food-contact plastics, J. Food Sci. 52 (1987) 761–785.

[87] C.H. Mannheim, J. Miltz, A. Letzter, Interaction between polyethylene laminated cartons and aseptically packed citrus juices, J. Food Sci. 52 (1987) 737–740.

[88] K.S.M. Sheung, S. Min, S.K. Sastry, Dynamic head space analyses of orange juice flavor compounds and their absorption into packaging materials, J. Food Sci. 69 (2004) C549–C556.

[89] R.W.G. van Willige, J.P.H. Linssen, A. Legger-Huysman, A.G.J. Voragen, Influence of flavour absorption by food-packaging materials (low-density polyethylene, polycarbonate and polyethylene terephthalate) on taste perception of a model solution and orange juice, Food Addit. Contam. 20 (2003) 84–91.

[90] V.K. Haugaard, C.J. Weber, B. Danielsen, G. Bertelsen, Quality changes in orange juice packed in materials based on polylactate, Eur. Food Res. Technol. 214 (2002) 423–428.

[91] R. Lebosse, V. Ducruet, A. Feigenbaum, Interactions between reactive aroma compounds from model citrus juice with polypropylene packaging film, J. Agric. Food Chem. 45 (1997) 2836–2842.

[92] R. Gavara, R. Catala, P. Hernandez-Munoz, Study of aroma scalping through thermosealable polymers used in food packaging by inverse gas chromatography, Food Addit. Contam. 14 (1997) 609–616.

[93] P. Hernandez-Munoz, R. Catala, R. Gavara, Effect of sorbed oil on food aroma loss through packaging materials, J. Agric. Food Chem. 47 (1999) 4370–4374.

[94] P. Hernandez-Munoz, R. Catala, R. Gavara, Food aroma partition between packaging materials and fatty food simulants, Food Addit. Contam. 18 (2001) 673–682.

[95] P. Hernandez-Muñoz, R. Catalá, R.J. Hernandez, R. Gavara, Food aroma mass transport in metallocene ethylene-based copolymers for packaging applications, J. Agric. Food Chem. 46 (1998) 5238–5243.

[96] A. Adams, S. van Bloois, B. Otte, E. Caro, D. Mekap, P. Sandkuehler, Flavor scalping by polyethylene sealants, Food Packaging Shelf Life 21 (2019) 100371.

[97] G.W. Halek, Relationship between polymer structure and performance in food packaging applications, in: J.H. Hotchkiss (Ed.), Food and Packaging Interactions, American Chemical Society Symposium Series, 1988, pp. 195–202.

[98] M. Rubino, M.A. Tung, S. Yada, I.J. Britt, Permeation of oxygen, water vapor, and limonene through printed and unprinted biaxially oriented polypropylene films, J. Agric. Food Chem. 49 (6) (2001) 3041–3045.

[99] T. Imai, R.R. Harte, J.R. Giaicin, Partition distribution of aroma volatiles from orange juice into selected polymeric sealant films, J. Food Sci. 55 (1990) 158–161.

[100] G. Colomines, V. Ducruet, C. Courgneau, A. Guinault, S. Domenek, Barrier properties of poly(lactic acid) and its morphological changes induced by aroma compound sorption, Polym. Int. 59 (2010) 818–826.

[101] G. Lopez-Carballo, D. Cava, J.M. Lagaron, R. Catala, R. Gavara, Characterization of the interaction between two food aroma components, α-pinene and ethyl butyrate, and ethylene-vinyl alcohol copolymer (EVOH) packaging films as a function of environmental humidity, J. Agri. Food Chem. 53 (2005) 7212–7216.

[102] F. Johansson, A. Leufven, Food packaging polymer films as aroma vapor barriers at different relative humidities, J. Food Sci. 59 (6) (1994) 1328–1331.

[103] L. Safa, B. Abbes, Experimental and numerical study of sorption/diffusion of esters into polypropylene packaging films, Packaging Technol. Sci. 15 (2002) 55–64.

[104] R.W.G. Van Willige, J.P.H. Linssen, A.G.J. Voragen, Influence of food matrix on absorption of flavour compounds by linear low-density polyethylene: oil and real food products, J. Sci. Food Agric. 80 (12) (2000) 1790–1797.

[105] R.W.G. Van Willige, J.P.H. Linssen, A.G.J. Voragen, Influence of food matrix on absorption of flavour compounds by linear low-density polyethylene: proteins and carbohydrates, J. Sci. Food Agric. 80 (12) (2000) 1779–1789.

[106] R. Van Willige, D. Schoolmeester, A. Van Ooij, J. Linssen, A. Voragen, Influence of storage time and temperature on absorption of flavor compounds from solutions by plastic packaging materials, J. Food Sci. 67 (2002) 2023–2031.

[107] B.P.F. Day, Active packaging, in: R Coles, D McDowell, MJ Kirwan (Eds.), Food Packaging Technology, CRC Press, Boco Ratan, FL, 2003, pp. 282–302.

[108] G.L. Robertson, Active and intelligent packaging, Food Packaging: Principles and Practice, second ed., CRC Press, Boca Raton, FL, 2006. Chap. 14.

[109] A.L. Brody, E.R. Strupinsky, L.R. Kline, Active Packaging for Food Applications, Technomic Publishing Co, Lancaster, PA, 2001.

[110] J.H. Han, Antimicrobial packaging systems, in: S. Ebnesajjad (Ed.), Plastic Films in Food Packaging: Materials, Technology and Applications, Elsevier, Amsterdam, 2013, pp. 151–180.

[111] M.L. Rooney, Introduction to active food packaging technologies, in: S. Ebnesajjad (Ed.), Plastic Films in Food Packaging: Materials, Technology and Applications, Elsevier, Amsterdam, 2013, pp. 127–138.

[112] C. Ebner, Successes and challenges encountered in the development of active packaging for food, SPE FlexPackCon Conf. (2015).

[113] M. Ozdemir, J.D. Floros, Active food packaging technologies, Crit. Rev. Food Sci. Nutr. 44 (2004) 185–193.

[114] M.L. Rooney, History of active packaging, in: CL Wilson (Ed.), Intelligent and Active Packaging for Fruits and Vegetables, CRC Press, 2007, pp. 11–30.

[115] D.A. Pereira de Abreu, J.M. Cruz, P. Paseiro Losada, Active and intelligent packaging for the food industry, Food Rev. Int. 28 (2012) 146–187.

[116] C. Realini, B. Marcos, Active and intelligent packaging systems for a modern society, Meat Sci. 98 (2014) 404–419.

[117] P. Suppakul, J. Miltz, K. Sonneveld, S.W. Bigger, Active packaging technologies with an emphasis on antimicrobial packaging and its applications, J. Food Sci. 68 (2003) 408–420.

[118] K. Cooksey, What's next in active and intelligent food packaging? Packaging World (2014)viewed 24 Nov 2015 from. Available from: http://www.packworld.com/trends-and-issues/smartactive-packaging/whats-next-active-and-intelligent-food-packaging.

[119] K. Hausmann, L. Ziche, H. Schenck, More sustainable meat packages through novel sealant solutions, Eur. TAPPI PLACE. Conf. (2015).

[120] K. Cooksey, Effectiveness of antimicrobial food packaging materials, Food Addit. Contam. 22 (10) (2005) 980–987.

[121] A. Balasubramanian, L.E. Rosenberg, K. Yam, M.L. Chikindas, Antimicrobial packaging: potential vs. reality - a review, J. Appl. Packaging Res. 3 (4) (2009) 193–221.

[122] S. Quintavalla, L. Vicini, Antimicrobial food packaging in meat industry, Meat Sci. 62 (3) (2002) 373–380.

[123] T. Huang, Y. Qian, J. Wei, C. Zhou, Polymeric antimicrobial food packaging and its applications, Polymers 11 (3) (2019) 560.

[124] A. Khaneghah, S. Hashemi, S. Limbo, Antimicrobial agents and packaging systems in antimicrobial active food packaging: an overview of approaches and interactions, Food Bioprod. Process. 111 (2018) 1–19.

[125] P. Appendini, J.H. Hotchkiss, Review of antimicrobial food packaging, Innovative Food Sci. Emerg. Technol. 3 (2) (2002) 113–126.

[126] C.G. Otoni, P.J.P. Espitia, R.J. Avena-Bustillos, T.H. McHugh, Trends in antimicrobial food packaging systems: emitting sachets and absorbent pads, Food Res. Int. 83 (2016) 60–73.

[127] M. Hoseinnejad, S.M. Jafari, I. Katouzian, Inorganic and metal nanoparticles and their antimicrobial activity in food packaging applications, Crit. Rev. Microbiol. 44 (2) (2018) 161–181.

[128] S.A. Sofi, J. Singh, S. Rafiq, U. Ashraf, B.N. Dar, G.A. Nayik, A comprehensive review on antimicrobial packaging and its use in food packaging, Curr. Nutr. Food Sci. 14 (4) (2018) 305–312.

[129] H.M.C. Azeredo, C.G. Otoni, D.S. Correa, O.B.G. Assis, M.R. de Moura, L.H.C. Mattoso, Nanostructured antimicrobials in food packaging-recent advances, Biotechnol. J. 14 (12) (2019) 1900068.

[130] S.-Y. Sung, L.T. Sin, T.-T. Tee, S.-T. Bee, A.R. Rahmat, W.A.W.A. Rahman, et al., Antimicrobial agents for food packaging applications, Trends Food Sci. Technol. 33 (2) (2013) 110–123.

[131] M.J. Galotto, A. Guarda, C. Lopez de Dicastillo, Antimicrobial active polymers in food packaging, in: G. Cirillo, U.G. Spizzirri, F. Iemma (Eds.), Functional Polymers in Food Science, 2015, pp. 323–353.

[132] J.M. Hutchinson, Physical aging of polymers, Prog. Polym. Sci. 20 (4) (1995) 703–760.

[133] M. Sepe, Materials: the mystery of physical aging: part 1, Plast. Technol. (2014)November Issue, viewed 14 June 2021 at. Available from: http://www.ptonline.com/columns/the-mystery-of-physical-aging-part-1.

[134] M. Sepe, Materials: the mystery of physical aging: part 2, Plast. Technol. (2014)December Issue, viewed 14 June 2021 at. Available from: http://www.ptonline.com/columns/the-mystery-of-physical-aging-part-2.

[135] M. Sepe, Materials: the mystery of physical aging: part 3, Plast. Technol. (2015)January Issue, viewed 14 June 2021 at. Available from: http://www.ptonline.com/columns/materials-the-mystery-of-physical-aging-part-3.

[136] L.C.E. Struik, Physical aging of plastics and other glassy materials, Polm. Engr. Sci. 17 (1977) 165–173.

[137] B.A. Morris, J.C. Chen, The stiffness of ionomers: how it is achieved and its importance to flexible packaging applications, SPE ANTEC Conf. (2003).

[138] H. Yoshimizu, Y. Tsujita, Applications of NMR spectroscopy to the structure and ionic aggregates of ionomers, Ann. Rep. NMR Spectrosc. 44 (2001) 1–21.

[139] S. Kutsumizu, K. Tadano, Y. Matsuda, M. Goto, H. Tachino, H. Hara, et al., Investigation of microphase separation and thermal properties of noncrystalline ethylene ionomers. 2. IR, DSC, and dielectric characterization, Macromolecules 33 (2000) 9044–9053.

[140] E. Hirasawa, Y. Yamamoto, K. Tadano, S. Yano, Formation of ionic crystallites and its effect on the modulus of ethylene ionomers, Macromolecules 22 (1989) 2776.

[141] K. Tadano, E. Hirasawa, H. Yamamoto, S. Yano, Order-disorder transition of ionic clusters in ionomers, Macromolecules 22 (1989) 226–233.

[142] R. Goddard, B. Grady, S. Cooper, The room temperature annealing peak in inomers: ionic crystallites or water absorption? Macromolecules 27 (1994) 1710–1719.

[143] C.L. Marx, S.L. Cooper, The crystallinity of ionomers, J. Macromol. Sci. Phys. 9 (1974) 19–33.

[144] Y.-L. Loo, K. Wakabayashi, Y.E. Huang, R.A. Register, B.S. Hsiao, Thin crystal melting produces the low-temperature endotherm in ethylene/methacrylic acid ionomers, Polymer 46 (14) (2005) 5118–5124.

[145] K. Wakabayashi, R.A. Register, Morphological origin of the multistep relaxation behavior in semicrystalline ethylene/methacrylic acid ionomers, Macromolecules 39 (3) (2006) 1079–1086.

[146] I. Lee, R.P. Wool, Optimum polymer solid interface design for adhesion strength, J. Adhes. 75 (2001) 299–324.

[147] I. Lee, R.P. Wool, Polymer adhesion vs substrate receptor group density, Macromolecules 33 (2000) 2680–2687.

[148] L. Gong, A.D. Friend, R.P. Wool, Polymer-solid interfaces: Influence of sticker groups on structure and strength, Macromolecules 31 (1998) 3706–3714.

[149] G. Hoh, S.A. Sadik, J.A. Gates, Aluminum foil adhesion of ethylene/methacrylic acid copolymer extrusion coatings, TAPPI Polymers, Laminations Coat. Conf. (1989) 361–366.

[150] D.H. Kailble, Theory and analysis of peel adhesion: bond stresses and distribution, Trans. Soc. Rheol., IV (1960) 45–73.

[151] B.A. Morris, Optimizing flexible packaging ... with unique tools from DuPont, Present. PackExpo (2011).

Chapter 17

Designing flexible packaging for sustainability

17.1 Introduction

Plastic production has increased by 20-fold from 31 to 311 million tons/year in the last 50 years and is expected to almost quadruple again by 2050 [1,2]. About a third of all plastics go into packaging [3], and 95% of plastic packaging's material value is lost to the economy after one use [2]. Only 14% of all plastic packaging is recovered today by recycling; 72% is not recovered at all with 40% going to landfills and 32% leaking to the environment [1,2]. Even less flexible packaging is recovered by recycling. While flexible packaging provides tremendous benefits by preventing food waste and lowering the overall environmental footprint vs alternative forms of packaging, its end-of-life challenges are driving changes in infrastructure and packaging design. This chapter begins by considering what sustainability means with respect to packaging and how flexible packaging can be part of the solution if end-of-life challenges can be overcome. Quantifying the benefits and challenges of flexible packaging begins with using life cycle analysis (LCA) tools on both the packaging and product; redesigning packaging for ease of recycling does not make sense if it diminishes the package's ability to prevent food waste. Next, a brief review of current recycling technologies is provided, focusing on how challenges with these technologies may impact package design. Finally, concepts for designing flexible packaging for sustainability are described. These include redesigning packaging for ease of recycling, such as converting multimaterial structures to all-polyethylene (PE) or all-polypropylene (PP) structures, or for biodegradability. Specific case studies illustrate difficulties that must be overcome to implement these designs.

This chapter is not meant to be a comprehensive review of sustainability and end-of-life technologies. There are many review articles and book chapters that this chapter draws upon that the reader is referred to for additional information [3–16]. It provides a snapshot of current thinking with the recognition that sustainability is a dynamic driver for packaging innovation. Recycling technology is evolving, and new collecting, sorting and recycling infrastructure will be needed, each of which has its own economic challenges. Many brand owners have set aggressive sustainability goals that involve redesigning packaging to include recycle content, renewable-based materials or be recycled. Government regulations around the circular economy and extended producer responsibility (EPR) programs will create new opportunities for packaging innovation. While specific solutions are still evolving, the principles set forth in this book provide the basis for the reader to be better equipped to meet the challenges ahead.

17.1.1 What is sustainability?

The United Nations Brundtland Report defines sustainability "as meeting the needs of the present without compromising the ability of future generations to meet their own need" [6]. This can mean different things to different people or organizations. Some of the many concerns about current industrial activity that may harm the future environment and human health include:

- Climate change from the result of human activities, specifically CO_2 equivalent gas emissions.
- Air, land and water pollution.
- Water consumption.
- Freshwater eutrophication (degradation of freshwater sources from harmful nutrients).
- Acidification of soil and water (from release of gases such as nitrogen oxides and sulfur oxides).
- Landfill exhaustion.
- Marine litter.
- Fossil fuel depletion.

The Science and Technology of Flexible Packaging. DOI: https://doi.org/10.1016/B978-0-323-85435-1.00018-1

- Food quality and availability/food waste.
- Human health and safety.
- Economic subsistence and equality.

Aspirations for sustainable packaging were laid out by the Sustainable Packaging Coalition in 2011 as [17]:

- "Beneficial, safe and healthy for individuals and communities throughout its life cycle.
- Sourced, manufactured, transported and recycled using renewable energy.
- Optimizing the use of renewable or recycled source materials.
- Manufactured using clean production technologies and best practices.
- Made from materials healthy throughout their life cycle.
- Being physically designed to optimize materials and energy.
- Effectively recovered and utilized in biological and/or industrial closed-loop cycles."

These sustainability goals can be achieved with the help of engineering and design solutions but may involve trade-offs. Flexible packaging offers advantages over rigid and other packaging formats towards many of these goals due to its lower material usage per product weight but has challenges meeting others, particularly with recycling and end-of-life scenarios. It is important not to overly focus on one aspect of sustainability without considering the others. Redesigning a package for better recyclability, for example, at the expense of increasing food waste may result in higher overall damage to the environment (see Section 17.2).

17.1.2 Benefits of flexible packaging

Plastic packaging has advantages over suitable alternatives such as metal, glass, wood and paper due to plastic's lower density and the lesser amount of energy needed to fabricate it into packaging articles. An industry-sponsored study using a life cycle assessment across the entire packaging market shows plastic packaging is advantaged in almost all quantitative measurements, as shown in Table 17.1 [18]. On average, 4.1 tons of alternative materials are needed to replace 1 ton of plastics [19]. For the US market, plastic packaging creates less than half the global warming potential (GWP) [as measured by greenhouse gas (GHG) emissions] than the alternatives. For other categories of environmental impact, the benefits of using plastics are even higher.

Flexible packaging has further advantages, including [10]:

- Lower cost, less material usage, less waste, and lower transportation cost than other formats.
- Lower environmental impact than other formats. This refers to life cycle assessment environmental impact factors and not to end-of-life scenarios.

TABLE 17.1 Environmental impact of plastic packaging compared with suitable alternative materials, expressed as percent.

LCA category	Impact of plastic packaging as percent of alternative materials
Energy demand	52%
Water consumption	17%
Solid waste by weight	20%
Solid waste by volume	42%
Global warming potential (greenhouse gas emissions)	45%
Acidification potential	30%
Eutrophication potential	2%
Smog formation potential	32%
Ozone depletion potential	26%

Source: Data from Franklin Associates, Life cycle impacts of plastic packaging compared to substitutes in the united states and Canada theoretical substitution analysis, viewed online on 22 Dec 2020 at https://plastics.americanchemistry.com/Reports-and-Publications/LCA-of-Plastic-Packaging-Compared-to-Substitutes.pdf, 2018.

TABLE 17.2 Sustainability advantages of flexible packaging vs other packaging formats.

Stage of life cycle	Benefit
Manufacturing/converting	Lightweight/source reduction
Transportation	Form factor (flat, rolls), lighter weight
Transportation/consumer usage	Product protection/breakage reduction (esp. for e-commerce)
Consumer usage	Extended shelf life (barrier packaging)
End of life	Reduced materials to landfill (due to source reduction)
End of life	Energy recovery (for places that have infrastructure)
Overall	Favorable life cycle assessment (fossil carbon, GHG, water usage, etc.) due to source reduction
Overall	Higher product to package ratio

Source: Based on T. Bukowski, M. Richmond, M., A holistic view of the role of flexible packaging in a sustainable world, Flex. Packaging Assoc. Rep., 2018, viewed online on 21 Dec 2020 at https://www.flexpack.org/resources/sustainability-resources#a-holistic-view-of-the-role-of-flexible-packaging-in-a-sustainable-world.

- Highly tunable barrier properties to protect food (lowers food waste).
- Convenience for consumer—easy open, portion control, etc.

Other benefits throughout the flexible packaging life cycle are given in Table 17.2. The European Flexible Packaging Association [20] estimates that converting all packaging for fast-moving consumer goods to flexible packaging would save 70% of materials. Even assuming no recycling of the flexible packaging, this would result in a 33% reduction in GWP. Bukowski and Richmond [6] conduct life cycle assessments (see Section 17.2) on six product categories, comparing flexible packaging formats with rigid packaging alternatives. A cradle-to-grave approach is used with a focus on the primary packaging. Fossil fuel consumption, GHG emissions and water consumption rates are computed. One case compares a stand-up pouch (SUP) with a steel can and a high-density polyethylene (HDPE) cannister for packaging coffee. The weight reduction of the SUP format overwhelms everything for fossil fuel consumption, GHG and water consumption. The steel can has one of the highest recycling rates (71%) and this is used in the analysis. Even assuming zero recycling for the SUP, the amount that ends up in landfills is still one-third that of the steel can. The results give a similar story for the SUP versus the HDPE cannister. Weight reduction is key. The SUP also has the highest product to package ratio; the importance of this is discussed in Section 17.2.

17.1.3 Sustainability challenges for flexible packaging

While flexible packaging generally has favorable environmental impact attributes associated with its production and distribution (GHG, water usage, etc.), it often falls short in three areas

1. Recyclability, which includes collection, sorting and physical or chemical recycling;
2. Leakage into the environment;
3. Use of renewable materials.

The first two are related to end-of-life challenges. Flexible packaging is difficult to mechanically recycle. Its low bulk density is a disincentive to collect; most curbside mixed waste collection programs do not include flexible packaging. Many packaging structures contain multiple layers and materials which are difficult to identify and sort with current technology and infrastructure. The multimaterials may be difficult to separate, or if not separated, their incompatibility can result in poor properties. Printing and food contamination, often 10%–20% of the package weight, is another hurdle. Alternative chemical or feedstock recycling technologies are still in their commercial infancy; these may mitigate the mixed material and contamination issues, but their cost is still out of line with the cost of producing polymers from fossil resources.

The leakage of waste flexible packaging into the environment is visible and attracts negative attention to the industry [21]. The problem may be mostly attributed to poor waste handling infrastructure in a few developing countries but is global in nature since waste from developed countries has been finding its way to these developing countries for processing. The impact is real. While most of the leakage occurs on land, much ends up in the oceans. It is estimated that over 8 million tons/year of plastic waste reaches the ocean each year with an accumulation of over 150 million tons. Plastics generally only slowly degrade; even compostable materials that degrade on land often do not degrade appreciably in the sea. The effects on wildlife and the food chain are twofold. Ingestion of floating or other debris causes damage to sea animals. Over time the package is broken up into small "micro" particles. Microplastics (defined as particles less than 5 microns) have been found throughout the ocean and in the food chain. While the underlying polymer is generally inert, these microparticles may contain additives, absorb toxic chemicals or attract microbes that could be harmful. Research is underway to better understand the potential dangers [22–25].

The use of renewable materials is part of the goal of achieving a circular economy (Section 17.1.4). The production of plastics consumes only about 4%–6% of fossil fuels produced today, and since packaging comprises 34% of plastics consumption, plastic packaging consumes about 1%–2% of fossil fuel production [16]. By 2050, however, plastics' share of the global oil production may increase to 20% and plastics' share of the global carbon budget may increase from 1% today to 15% [2]. These statistics are driving a move away from nonrenewable resources. One approach is to increase recycling, creating a circular economy where resources, whether they be from renewable or nonrenewable sources, are reused/recycled continuously. This reduces the amount of virgin materials needed. A second approach is to substitute fossil-derived plastics with renewably sourced materials in flexible packaging. This may involve increased use of paper-based products, for example, although these materials are often combined with PE or aluminum foil in packaging applications in ways that make them difficult to recycle. This may also involve biobased polymers. Chapter 4 describes currently available biobased polymers for flexible packaging. These include conventional polymers like PE where monomers have been sourced from bio feedstocks rather than fossil sources predominant today. Examples include Braskem's process to produce ethylene from bioethanol derived from sugarcane and Dow's partnership with UPM Biofuels to use byproducts of the papermaking process to feed ethylene crackers [26]. A second approach is to make new polymers from biosources. Examples include polylactic acid (PLA) and thermoplastic starch. Hybrids, such as polyethylene terephthalate (PET) partially derived from biosources, have been commercialized that provide an intermediate point. In general, biosourced polymers are more expensive or have yet to be optimized for flexible packaging, and so are still a niche technology for this market.

The second aspect of renewable packaging is biodegradability or compostability. This describes the end-of-life scenario for plastic and not its composition. Some biodegradable polymers are also biosourced (such as PLA), but not all. Similarly, some fossil-derived polymers are biodegradable. See Table 4.5 for details. Polymers that are both biosourced and biodegradable may provide potential new routes toward flexible packaging sustainability and circularity. This is explored in Section 17.4.3.

17.1.4 Policy frameworks

Two policy frameworks for ensuring the sustainability of packaging are Sustainable Materials Management (SMM) and Circular Economy (CE). SMM focuses on carbon footprint reduction and CE on recycling. Usually, they drive towards the same behaviors. For example, adding postconsumer recycle (PCR) content to packaging satisfies the CE framework by promoting more recycling of resources and SMM by reducing the amount of fossil fuel feedstocks needed (by reducing the amount of virgin polymer) [4,6].

SMM is the primary framework supported by the US Environmental Protection Agency (EPA). It espouses the efficient use, reuse and recycling of materials and the productive use of resources throughout the life cycle. Its focus is on minimizing material usage and carbon impact. Corporate Social Responsibility is a policy network promoting social, ethical and environmental behaviors that are espoused by many major corporations. It drives similar behavior for holding manufactures accountable for efficient use of resources in a way that minimizes their environmental impact.

The CE framework is an approach to sustainability that involves rethinking our linear packaging value chain, as shown in Fig. 17.1. The linear value chain—from producer to consumer to landfill or incineration—results in the single use of plastics. As put forth by the Ellen MacArthur Foundation, a new circular economy that promotes reuse and recycling is a way for society "to promote prosperity, while reducing demands on finite raw materials and minimizing negative externalities." The European Union's Strategy for Plastics embraces CE and has as its objective that by 2030 all plastic packaging is reusable or easily recyclable in a cost-effective manner. China has also adopted CE principles, as have many brand owners around the world who have announced aggressive packaging recycling goals [4]. The process to move toward this new economy involves three aspirational goals [2]:

1. Create an effective after-use-market plastics economy. Without markets, there is no economic incentive for recycling. This involves dramatically increasing the economics, quality and use of recycling, increasing adoption of reusable packaging, and scaling-up industrial compositing for specific applications.

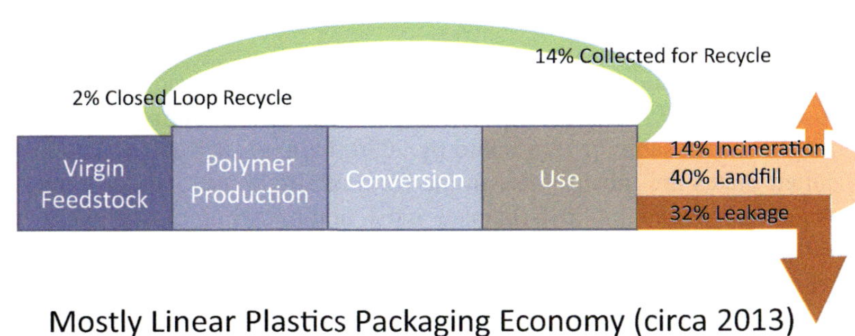

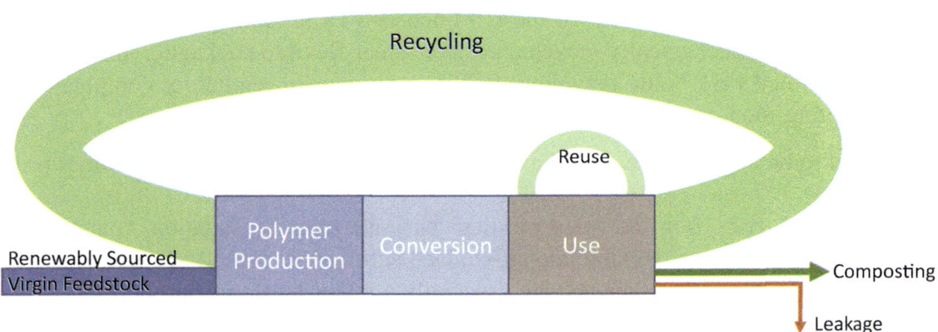

FIGURE 17.1 Linear and circular economies as it pertains to plastics packaging. Top: Linear economy with 2013 end-of-life statistics. Bottom: Circular economy as envisioned by Ellen MacArthur Foundation [2].

2. Drastically reduce leakage of plastics into natural systems. As a first step, improving the collection, storage and reprocessing infrastructure in high-leakage countries is envisioned. Creating the market for recycled plastics (first goal) feeds into this, as does steering investment into the innovation of materials and infrastructure, research into the safety of recycled materials and the development of safe alternatives.
3. Decouple plastics from fossil feedstocks. As recycling becomes more prevalent, the need for virgin plastics is reduced. The development of renewable-based plastics will eventually fill the need for virgin feedstock to make up for the cycle losses.

Recycling can be defined as the process by which industrial or municipal waste is collected, segregated, cleaned and reprocessed to become a new product or a secondary raw material to offset the use of virgin material [3]. There are three aspects that must be met for a package to be considered recyclable [3,4,27]:

1. The material must be capable of being reprocessed. For example, this may exclude certain multimaterial packaging structures if the layers cannot be separated prior to processing.
2. Collection, sorting and processing infrastructure must be in place to handle it. This is currently a limitation for many types of flexible packaging. The Association of Plastic Recyclers stipulates that at least 60% of communities or consumers must have access to collection systems that accepts the item for it to be deemed recyclable. Furthermore, the item must be able to be sorted into a market-ready bale.
3. There must be an end-use application(s) for the reclaimed material.

Currently, there is no widespread system or process to economically recycle multilayer (multimaterial) flexible packaging due to many factors, including [4]:

- The large variety of materials that may be in the structures.
- Large differences in the processing properties of some materials used in the layers.
- Lack of systems to identify multilayer films.
- Absence of system solutions for collecting these films.
- Lack of economically viable systems for separating the various materials.
- Little research on the use of composites made from these recycled materials (properties, processing, applications).

17.1.5 Regulatory drivers and considerations

Government policy and regulatory environment for plastics and flexible packaging is evolving and differs around the world. Some regions, such as Europe, have embraced CE concepts and EPR. The European Commission, for example, has set aggressive goals for plastic packaging to be reusable or recyclable by 2030. Others, such as the US EPA, have guidelines but regulatory policies in the United States are mostly local in the form of single-use bag bans and the like. China temporarily restricted waste imports in 2013 and in 2018 banned most nonindustrial plastic waste imports (National Sword). Prior to this much of the plastic waste in the United States and Europe was exported to China. Many other Asian countries followed suit with their own import bans, particularly since the supply often overwhelmed local infrastructure and ended up contributing to the leakage of plastic waste into the marine environment. The disruption of this supply chain jump-started much needed global discussion and infrastructure build to deal with plastic waste in a more sustainable way.

One of the largest markets for flexible packaging is for food products which are governed by country or regional regulations for food safety. Chapter 4 provides a brief overview of the US Food and Drug Administration (FDA) and European Union regulations. The burden of using mechanically recycled postconsumer flexible packaging in food packaging applications is high. So far, outside limited PET bottle recycling, PCR has not gained regulatory compliance. In the EU, there are several challenges to meeting food contact regulations [4,28]:

- Full traceability. Producers are required to be able to trace materials throughout the value chain. This becomes problematic for PCR where the link is lost.
- Potential misuse during the lifetime. Health risks from recycled plastics can arise from degradation, contaminants from packaged goods, or misuse during the life cycle. Nonintentionally added substances (NIAS) are also a concern for complying with food contact material (FCM) regulations and are especially a major hurtle in Europe.
- Separation of food contact and nonfood contact waste. The initial recycler must keep in mind the intention for the use of the recycling. If for FCM, the recycler must pay attention to traceability and declaration of compliance rules laid out in the FCM framework.
- Recycled resins can be used as FCM in packaging only if they were processed through an authorized recycling process/facility. The regulations describe the authorization process. The focus of the approval process has been on the quality of the input materials, traceability and decontamination efficiency of the process. There are some exemptions: polymers from depolymerized polymers (chemical or feedstock recycling), unused plastic cutoffs and scrap (postindustrial recycling where the materials are traceable and known), and recycled material used behind a functional barrier in a multilayer structure.

In the United States, recycled materials used for food contact must comply with the same regulations as virgin resins. Producers of recycled materials may ask the FDA to review their recycling process. Of particular concern is the "possibility that chemical contaminants in plastic materials intended for recycling may remain in the recycled material and could migrate into the food the material contacts" [29]. This is similar to the concerns for NIAS by the EU regulators. The FDA provides guidelines and testing protocols. Compliance is more clear cut if the original material is FDA compliant for food contact, is kept free of contaminants during the recycling process and is tested to establish suitability for reuse in food packaging [4]. Recyclers may also petition the FDA for a letter of nonobjection; it is up to the recycler to ensure their feedstocks and processes will consistently produce food-grade resins [15]. Like the European Union, the FDA regulations also provide an exception if a functional barrier is between the recycle and food in a multilayer structure. The demonstration that the barrier is "functional" is up to the user of the package.

Given these regulatory hurdles, and the need for high quality, it is not likely there will be widespread use of postconsumer mechanically recycled plastics in flexible packaging for food applications in the near term. Alternative sources (such as postindustrial recycling or secondary packaging waste streams) or polymers made from monomers derived from chemical recycling of PCR may possibly make inroads and are discussed in subsequent sections. The use of recycling in secondary packaging films, such as collation films and pallet wrap, is expected to see growth in the use of recycling.

FDA regulations permit the use of pulp from recycled paper and paperboard to be used in food contact packaging provided the recycled paper does not contain any poisonous or deleterious substances that may migrate into the food (21 CFR Part 176B). To comply with this requirement, pulp mills must demonstrate that their process effectively removes any unwanted substances that may migrate into the food. This may involve restricting the types of feedstock they use, such as paper containing coatings or additives that may contain unwanted substances that cannot be removed in their process. As with plastic packaging, certain coatings may be applied to the paper package to serve as a functional barrier between the paper and food if the paper itself has not been demonstrated to be FDA compliant [15].

Food contact regulations also should be considered for alternative package designs. One technology getting considerable attention is nanocomposites (see Section 8.5.3) and nanomaterials. Thin coatings or blends with polyolefins may provide enhanced barrier performance, allowing single-material solutions that are easier to recycle. Nanomaterials are also being evaluated as compatibilizers for mixed plastic waste streams. Nanoparticles, however, have human health concerns and so far, even if encapsulated in a polymer matrix, they have not gained approval for use in food packaging [16,30].

17.2 Quantifying environmental impact

17.2.1 Life cycle analysis of packaging

The packaging industry is becoming more sophisticated in understanding the nuances of what represents a truly sustainable product. Initially, the industry focused on such factors as biobased content and package weight reduction. While these factors are readily measured and communicated, they are not metrics of actual environmental impact. As such, there is the possibility of a seemingly sustainable solution to have a negative environmental impact relative to the so-called nonsustainable incumbent. LCA is a method to access sustainability; standards have been adopted and results are increasingly becoming available to the public. This section introduces the LCA methodology (following Casarino et al. [31]) with a focus on the package. Section 17.2.2 provides a more holistic analysis considering both the product and package.

An LCA is a detailed assessment of the environmental impact of a product or offering throughout its life cycle. Depending on the scope of the LCA, different environmental impacts can be considered. Typically, LCA focuses on the nonrenewable energy consumed and GHG emitted (also called GWP). Other environmental impacts may also be considered, such as acidification, eutrophication, ozone depletion, smog generation and water use. The nonrenewable energy consumed is net of any energy generated and recycled in the process and is reported as megajoules/unit, typically megajoules per kg of product (MJ/kg). The GHG emissions include emissions from utility generation (e.g., electricity, steam), as well as process emissions of any GHG, and are reported as kilograms of carbon dioxide equivalent/unit, again typically per kg of product (kg CO_2eq/kg).

The scope of the LCA is important. Referring to Fig. 17.2, a Cradle-to-Grave LCA covers raw materials, manufacturing, use, and end-of-life phases. While it is certainly desirable to understand the environmental footprint all the way through the life cycle, the number of value chain participants makes it extremely difficult to obtain reliable data. For this reason, most available LCA data is for so-called Cradle-to-Gate, which includes the environmental footprint of the production of raw materials and all required manufacturing steps. Although the scope of a Cradle-to-Gate LCA is narrower than that of a Cradle-to-Grave, it is still very much a challenge to get reliable and complete data sets so sensible assumptions and approximations are often unavoidable. Also, comparing LCA results from different sources can be challenging since each study may make different assumptions on, for example, heat and electricity generation, which may result in slight differences in the final results. For these reasons, in general, the accuracy of any given LCA should be considered to be no better than $\pm$ 10% when comparing studies not performed within the exact same scope.

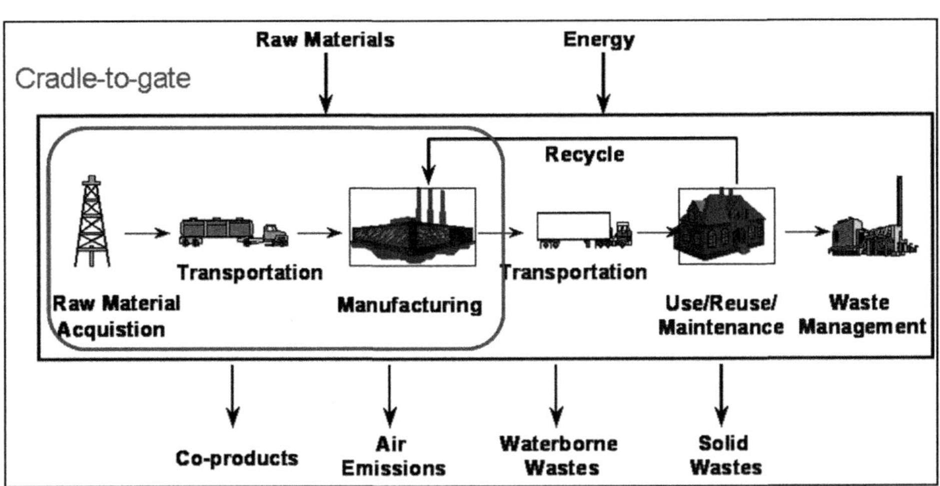

FIGURE 17.2 Schematic of a product life cycle. Cradle-to-grave refers to raw materials through waste management (dark box). Cradle-to-gate refers to raw materials through manufacturing (light box). *Courtesy of E. I. du Pont de Nemours and Company.*

The results of LCA's that consider end-of-life scenarios depend on the assumptions made for the waste source, energy source, quality of the product and efficiency of the treatment options. Energy recovery options from plastics may be favored, for example, when compared to incumbent energy from coal-fired electricity plants. Mechanical recycling comes out on top when the electricity is derived from natural gas [7].

A Cradle-to-Gate LCA for a polymer supplier includes monomers, additives, process chemicals, etc. The resin manufacturer obtains data on the energy and GHG for each raw material—this is often called the embodied or feedstock energy or embodied GHG. Embodied energy and GHG data can be obtained from a variety of sources—public databases, LCA software packages, from the supplier, or depending on the degree of backward integration, from the resin manufacturer's own operations. The LCA also includes manufacturing, which for the polymer supplier may include polymerization, pelletization, compounding, etc.

Key inputs to enable a packaging converter to conduct their own Cradle-to-Gate LCA are the embodied energy and GHG in the materials used to make packaging structures. There are a growing number of publicly available databases providing data for many monomers, ingredients, and resins. Examples include PlasticsEurope Eco-Profiles, NREL US Lifecycle Inventory Database, Ecoinvent, and other LCA software databases. These are generally based on industry averages and/or input from key industry players, and the results are aggregated to conceal any individual organization's proprietary information. As such, these data are generally available only for large volume commodity materials. Data for nonrenewable energy consumption and GHG for several polymers used in packaging are given in Table 17.3.

LCA data is useful to polymer manufacturers to help them prioritize process improvements or product development that may have significant footprint reduction potential. LCAs can help film converters optimize packaging structures by allowing comparison between functional alternatives. For films, this will typically mean a footprint per unit area that gives the same degree of product protection, shelf appeal, handling, abuse resistance, and other properties. While LCAs are typically computed in terms of energy requirements or emissions per unit weight, converting such information into meaningful functional alternatives is not always straightforward. For example, when comparing different package forms, the functional unit should be posed in terms of the unit of product packaged (such as per kg of food product), or even better would be per unit of food waste avoided. Another challenge is that packaging structures can often be put together in more than one way using a variety of converting processes. If the film converting processes are assumed to have similar footprints, then the overall impact of multilayer film structures can be compared by summing the material quantities and footprint/kg of each material that makes up the structure.

Casarino et al. [31] provides an example comparing three resins used to make a clamshell package: amorphous polyester (APET), high-impact polystyrene (HIPS), and PLA. Some of the characteristics of these resins are given in Table 17.4. The traditional approach is to compare these materials based on cost and physical properties. The price and

TABLE 17.3 LCA data for packaging resins and films (cradle-to-gate analysis).

Polymer	Nonrenewable energy consumption, MJ/kg	Greenhouse gas emissions, kg eq. CO2/kg	Source, date of analysis
PET resin	69	2.3	PlasticsEurope, 2013 [32]
PA 6 resin	129	6.7	PlasticsEurope, 2014 [32]
HDPE resin	79	1.8	PlasticsEurope, 2014 [32]
LDPE resin	82	1.9	PlasticsEurope, 2014 [32]
LDPE film	88	2.4	PlasticsEurope, 2005 [32]
LLDPE resin	79	1.8	PlasticsEurope, 2014 [32]
PP resin	77	1.6	PlasticsEurope, 2014 [32]
OPP film	98	3.2	PlasticsEurope, 2005 [32]
APET	80	3.3	PlasticsEurope, 2005 [32]
HIPS	87	3.4	PlasticsEurope, 2005 [32]
PLA	54	1.8	Vink, 2003 [33]

TABLE 17.4 Properties of clamshell resins.

Resin	Cost $/m^3	Tensile Modulus GPa	OPV cc-25μm/m^2-d	MVTR g-25μm/m^2-d
APET	2500	2	45–90	15
HIPS	2300	1.8	4700–9300	155
PLA	2500	3.4	620	325

Source: From C.M. Casarino, B.A. Morris, S.R. Veith, Resin life cycle estimation to help guide sustainability choices, SPE ANTEC Conf., 2009, reproduced with permission from The Dow Chemical Company.

TABLE 17.5 Results of LCA calculation for clamshell packaging.

Resin	Density g/cc	Thickness mm	Weight/Area kg/m^3	NREC MJ/ m^2	Reduction %	GHG kg eq CO2/kg	Reduction %
APET	1.35	0.38	0.51	41	0	1.7	0
HIPS	1.05	0.39	0.41	36	-12%	1.4	-17%
PLA	1.24	0.32	0.40	21	-48%	0.7	-58%

NREC, Nonrenewable energy consumption
Source: From C.M. Casarino, B.A. Morris, S.R. Veith, Resin life cycle estimation to help guide sustainability choices, SPE ANTEC Conf., 2009, reproduced with permission from The Dow Chemical Company.

densities of these resins differ; however, on a unit volume basis, all three materials cost about the same. Because of its lower oxygen and moisture permeability, APET is favored for applications requiring a moderate barrier. HIPS is favored for applications requiring breathability.

Environmental factors can also be considered. Table 17.3 lists the nonrenewable energy consumption and GHG emissions for these materials. This information needs to be put on an equal basis for comparison. The high stiffness of PLA may allow it to be downgauged in this structure while maintaining the overall feel and function of the package. The bending stiffness of a film/sheet scales as the modulus times the thickness cubed (see Chapter 9). For comparison, the APET sheet is selected as the control with a thickness of 0.38 mm (15 mils). The thickness of HIPS and PLA sheets is computed to be 0.39 and 0.32 mm, respectively, to have the same bending stiffness as the APET sheet. A correction for the density of the resins is also made. The combination of the density and thickness allows the conversion of the footprint data from a per kg basis to the functional basis per unit area. Table 17.5 gives the results of the calculations, which show that:

- Using HIPS allows some reduction in energy and GHG emissions over APET because of its lower density.
- PLA gives the greatest reduction in energy and GHG emissions as a result of having an inherently lower footprint and the opportunity for reducing the package thickness (downgauging). Downgauging may also allow for cost reduction.

In the above example, it is assumed the brittleness and use temperature limitations of PLA can be ignored. The analysis highlights the potential of PLA if these problems can be mitigated through the use of additives or other solutions [34–37].

LCA information can be incorporated into packaging design models. An example of this is provided in Chapter 18.

17.2.2 Food waste considerations

Often LCAs are conducted on the package, which drives package design towards reducing its environmental impact by light weighting, enhancing recyclability, use of renewable materials, etc. In the case of food packaging, however, the food product may have an environmental impact that overwhelms that of the package. A new package design with a lower package impact that increases food waste may result in an overall increase in environmental impact. Food production accounts for 30%–35% of climate change impact and 70% of freshwater usage. The United Nations estimates that about one-third of food is lost or wasted through the supply chain, or about 1.3 gigatons of waste (worth about US$1 trillion/year) [38]. A better approach is to consider the food waste in the LCA and treat the product and package holistically.

To illustrate this point, Heller [39] presents several case studies comparing the total environmental impact of changing package formats. For a fresh beef product, a PP tray with LDPE lidding film is switched to a modified atmosphere package (MAP) using a larger PP tray with a barrier lidding film. Retail food waste decreases from 7% to 5%. The LCA model shows that GHG emissions are dominated by the production and processing of beef (96% of the total). The LCA model shows the packaging GHG increases by 29%, the resulting food waste reduction more than compensates. The breakeven point is from 7% to 6.7% retail loss. For mushrooms, switching from bulk packaging (bulk corrugated box plus LDPE bags for retail) to a retail-ready PET tray with PVC overwrap increases the GHG emissions of the packaging by 21%. A reduction in retail food waste from 15.4% to 1% (from actual retail data) overwhelms this impact for a 14% overall reduction in GHG emissions. The reduction in food waste does not always justify a change in packaging, however. In an example with fresh spinach, bulk corrugated boxes with LDPE bags (for every 0.14 kg (5 oz) of product) are replaced with retail size (0.14 kg product) PET clamshell packages. The package weight increases by 50% and retail waste decrease from 5.3% to 1.1%. The LCA model shows that the package GHG emissions increase by 211% and the 80% reduction in food waste is not enough to compensate; total GHG emissions increase.

Denkstatt [40] conducts a study using food loss data from Austrian retailers. For packaging of sirloin steak, they compare a top sealed foamed-PS tray with MAP to a PS/EVA/PE vacuum skin package (VSP). The switch to VSP extends shelf life from 6 to 16 days and reduces retail waste from 12% to 3%. The new package format saves 5 g CO_2e and the reduction in food waste 730 g CO_2e per 330 g of beef. In a second example, two delivery modes are examined for a sliced cheese product: a delicatessen center versus self-serve. For the delicatessen center, 5 kg of bulk cheese is packaged in a barrier shrink bag and after slicing repackaged in wrapping paper at the counter. The self-serve cheese format is presliced at a central location and packaged in an APET tray with a lidding film. Retail waste is 5% and 0.14% for the delicatessen and self-serve formats, respectively. Switching to self-service increases the GHG emissions for the package by 28 g (per 150 g cheese) but the waste reduction reduced GHG emissions by 69 g for a net savings of about 40 g per 150 g cheese. In a third example, the packaging of a yeast bun is changed from a paper bag to a PP film bag. Less dehydration with the PP film reduces food waste from 11% to 0.8%. Both the packaging and waste reduction contribute to a decrease in GHG emissions: 12 and 136 g CO_2e/400 g bun, respectively.

The extent that minimizing food waste may outweigh minimizing the package environmental footprint will depend on the type of food, weight of the food relative to the package and package composition. Representative environmental contributions for various food types are tabulated in Table 17.6. In general, meat, dairy, cereals and processed foods have a higher impact than fruits and vegetables. Comparing these values with Table 17.3, on an equal weight basis these food products have a similar impact as common plastics used in packaging. The weight of a flexible package, however, is often considerably less than the product it is protecting. A useful description of the relative contribution of the food and package is the food-to-packaging environmental ratio (FTP_E), defined by Eq. (17.1).

$$FTP_E = \left[E\left(\frac{Agricultural\ production}{kg\ food}\right) + E\left(\frac{Food\ processing}{kg\ food}\right)\right] / E\left(\frac{Packaging\ materials}{kg\ food}\right) \tag{17.1}$$

where E = environmental impact indicator, such as GHG emissions or cumulative energy demand [41]. A higher value of FTP_E, the greater the food contribution to the environmental impact. Heller et al. [41] compile data on several food categories and package types and compute FTP_{GHG} for about 700 products. The individual values are highly variable within a category, but the medium values are compiled in Table 17.7. Meat, cereal and grains, dairy, fish and seafood have high values of FTP whereas vegetables, fruits, legumes and nuts and beverages have low values. For a subset of these products and packages, they conduct a cradle-to-grave LCA that includes the end-of-life and food waste considerations. They estimate the baseline amount of food waste from USDA data for each food type. They plot the increase in GHG emissions from the package that can be tolerated (while maintaining overall GHG emission constant) assuming the new package resulted in a 10% reduction in food waste. The amount of package GHG that can be tolerated increases with the FTP. It is not a monotonic relationship – it also depends on the baseline amount of food waste for that product category. A similar relationship is found for cumulative energy demand (CED), although the FTP_{CED} is smaller than FTP_{GHG} since there is a lot of energy embodied in the package materials. The results show that food packaging design aimed at reducing food waste can reduce the environmental impact of the product system even if the new packaging itself has a higher impact. For high FTP products, the focus should be on reducing food waste. For low FTP products the focus should be on reducing packaging impact.

TABLE 17.6 Greenhouse gas emissions and nonrenewable energy contributions for various food products.

Food type	GHG emissions (kg CO_2 eq./kg)	Nonrenewable cumulative energy demand (MJ/kg)
Spinach	0.18	0.7
Romaine lettuce (ready to eat)	0.14	10.4
Orange juice (not from concentrate)	0.71	9.0
Tomatoes (chopped)	0.67	9.2
Mushrooms	1.75	25.3
Potatoes	0.20	1.3
Eggs	1.7	12.3
Potato chips	1.98	22.8
Milk	1.05	4.0
Turkey	5.42	29.4
Pork	6.45	22.5
Cheese	6.62	25.2
Beef	14.5	30.0

Source: Data from M.C. Heller, S.E.M. Selke, G.A. Keoleian, Mapping the influence of food waste in food packaging environmental performance assessments. J. Ind. Ecol. 00 (0) (2018), 1e16. https://doi.org/10.1111/jiec.12743.

TABLE 17.7 Food to package ratio for greenhouse gas emissions for various food categories. Mean values from analysis of Heller et al. [41].

Food Type	Medium FTP_{GHG}
Beverages	1.9
Cereals and grains	50
Dairy	19
Fish and seafood	18
Fruit	1.7
Legumes and Nuts	2.2
Meat	50
Vegetables	2

Design features of flexible packaging that help minimize food waste include: [38–42]

- Gas barrier and physical strength to protect food from physical damage and biochemical degradation and extend shelf life at the retailer.
- Easy to reseal to extend shelf life at the consumer.
- Easy to empty completely.
- Portionally sized to avoid leftovers.
- Graphics that provide correct information (content, best before date, etc.).

Some of these may increase the environmental impact of the package or hinder recyclability. For example, adding gas barrier to extend shelf life may lead to multilayer constructions with incompatible and difficult to recycle materials. Packages with smaller portions increases the amount of packaging per unit weight of product.

Williams and Wilström [43] conduct LCA's for the combined product and package for five products: ketchup, bread, milk, cheese and beef. A description of the package formats, the computed energy use for the product and package, and FTP are provided in Table 17.8. They conduct a sensitivity analysis on the package design computing how much of an increase in the energy use for the package can be tolerated if the package results in less food waste. The initial food waste is assumed to be 20%. Fig. 17.3 plots the results. The y-axis represents the maximum increase in environmental impact that can be tolerated from the new package to maintain the same overall environmental impact of the product plus the package. This is expressed as a ratio: E_2/E_1 where E_2 is the environmental impact of the new package

TABLE 17.8 Product, package and computed energy usage for five food products.

Product	Package	Product Energy (MJ/kg food)	Package Energy (MJ/kg food)	FTP_E
Ketchup	Paste: steel barrels, polypropylene, low-density polyethylene, wood pallet Ketchup: polypropylene, low-density polyethylene, corrugated cardboard, wood pallet	11	5.7	1.9
Bread	Plastic bags of low-density polyethylene with plastic clips (polystyrene), and returnable plastic boxes as secondary packaging.	7.8	0.77	10
Milk	One-liter paperboard cartons	4.6	0.64	7.2
Cheese	Plastic packaging of low-density polyethylene and polyamide, and cardboard boxes as secondary packaging	38	0.65	58
Beef	Plastic packaging of plastic film of low-density polyethylene, polyamide and ethylene vinyl alcohol, and corrugated cardboard boxes as secondary packaging	48	3.1	15

Source: Based on analysis of H. Williams, F. Wikström, Environmental impact of packaging and food losses in a life cycle perspective: A comparative analysis of five food items, J. Clean. Prod.,19 (2011) 43–48.

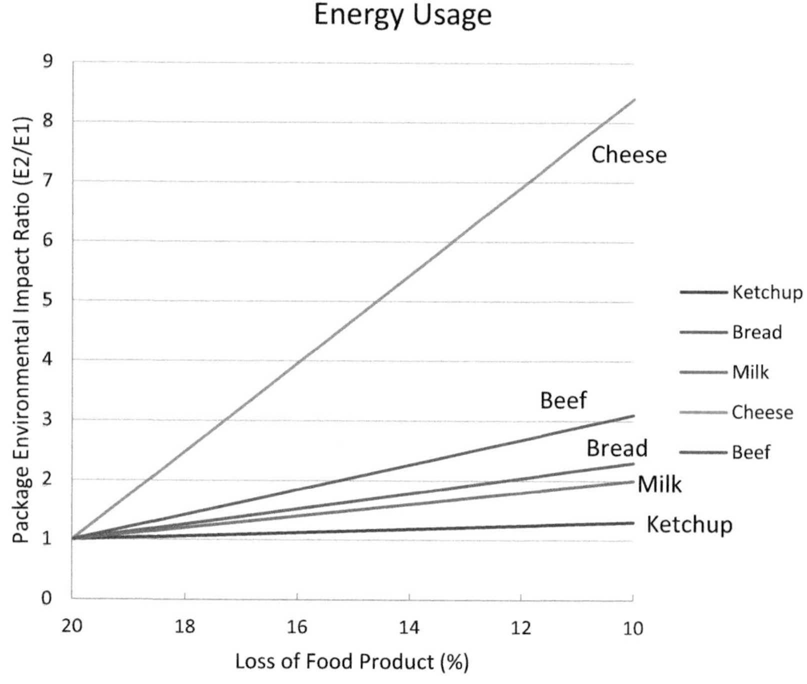

FIGURE 17.3 Effect of food waste reduction on package environmental impact. See text for description of axes. *Based on analysis of H. Williams, F. Wikström, Environmental impact of packaging and food losses in a life cycle perspective: A comparative analysis of five food items, J. Clean. Prod.,19 (2011) 43–48.*

and E_1 the environmental impact of the incumbent. The lines represent the maximum gain in package impact that can be tolerated for a given reduction in food loss. The steeper the slope, the more important reducing food waste is compared with reducing the impact of the package itself. For example, for beef, if the new package can deliver a 50% reduction in food loss (from 20% to 10%), the package can be designed with up to three times the environmental impact. Results shown here are for total energy usage (for the food product and package). Similar results are shown by the authors for other environmental impacts such as GWP.

Pauer et al. [44] analyze six flexible packaging formats for block bacon shown in Table 17.9. Their LCA includes production of the films and their end-of-life, as well as the product. They run several scenarios to account for assumptions that are uncertain (sensitivity analysis). These include recycle rate, transportation distance, type of PE (LDPE or LLDPE) and energy for shrinkage (for shrink bag formats). Recyclability is determined using the RecyClass method (Plastics Recycle Europe), a free online tool (see Section 17.4.2.1). Only package 1b (thermoformed film with ethylene vinyl alcohol [EVOH] as the barrier) is assumed to be recyclable, but with loss of quality. The results show:

- Raw materials used in the package dominate the package environmental impact – there is a strong correlation between packaging weight and impact.
- Freshwater eutrophication trends in the opposite direction due to credits for electricity generation for end of life. (The end-of-life scenario for the unrecycled packaging is assumed in part to be incineration, as is done in Europe. The electricity generation from incineration of the package replaces some electricity generated by coal-fired generation which releases phosphates into groundwater when the coal mining overburden is treated.)
- Increasing the recycle rate of the thermoformed films (1b) reduces GHG emissions: 18% recycle rate results in 7% reduction in GHG; 72% recycle rate gives 27% reduction in GHG. Freshwater eutrophication, however, goes up with the recycle rate (due to lower levels of incineration).
- Changing LLDPE to LDPE does little to the results, except for freshwater eutrophication (up by 40%) which is highly sensitive to energy consumption assumptions.
- The food-to-packaging (FTP) ratio ranges from 23 to 94. On average only 2% of the environmental impact comes from the package – the meat dominates.
- The production of polyamide (PA) creates four times the GHG emissions as PE. This influences the results somewhat.
- Light weighting to unrecyclable package forms (from thermoformed films to shrink bags and vacuum bags) outweighs the effect of increased recycling rate of the recyclable thermoformed structure.
- A change in the transportation distance has little impact on the results.
- The energy for the heat tunnel to shrink the shrink bags has a significant effect on the results.

TABLE 17.9 Six packaging formats for block bacon.

Package Format	Structure	Recycle Rate (%)	Package Weight (g)	Pkg GHG (g CO_2 eq.)	FTP_{GWP}
Thermoformed film – PA/PE (1a)	Nonforming: A-PET/PE/PA/PE–23/50/40/50 (163-μm) Forming: PE/PA/PE–120/90/120 (330-μm)	0	14.0	110	23
Thermoformed film – PO/EVOH (1b)	Nonforming: OPP/PE/EVOH/PE–35/80/5/80 (200-μm) Forming: PE/EVOH/PE 145/10/145–(300-μm)	0, 18, 72 scenarios	13.2	71	44
Vacuum bag (2a)	PA/PE–30/115 (145-μm)	0	8.2	56	44
Vacuum bag (2b)	PA/PE–20/70 (90-μm)	0	5.3	46	69
Shrink bag (3a)	PE/PVdC/PE–36/3/36 (75-μm)	0	4.4	27	94
Shrink bag (3b)	PA/Tie/EVOH/Tie/PE–30/5/5/5/55 (100-μm)	0	5.9	50	48

Source: Based on analysis of E. Pauer, M. Tacker, V. Gabriel, V. Krauter, Sustainability of flexible multilayer packaging: Environmental impacts and recyclability of packaging for bacon in block, Clean. Environ. Syst.1 (2020). https://doi.org/10.1016/j.cesys.2020.100001.

- The study does not consider differences in mechanical and barrier properties of the packaging films. This could have a significant effect on actual preservation of the bacon and food loss, which is not explicitly studied in this analysis.

17.3 End-of-Life Considerations

The general hierarchy for waste management is source reduction, reuse, recycle/composting, energy recovery and disposal (typically landfill) [45]. Much attention has been given to flexible packaging's role in source reduction where it offers considerable advantages over other forms of packaging. Where it continues to be challenged is in how it can be incorporated to a greater extent into the circular economy to avoid landfill, incineration and leakage into the environment.

Flexible packaging normally does not lend itself to reuse, although specific formats find alternative uses. For example, some consumers reuse plastic retail carry-out bags as garbage bags. Studies find that in places where the carry-out bags are banned, purchases of garbage bags increase. Since garbage bags are thicker than the grocery bags, this results in a significant amount of additional plastic use. According to one researcher, about 30% of plastic eliminated by grocery bag bans comes back in the form of thicker garbage bags [46].

Recycling, including energy recovery, is discussed in detail in Sections 17.3.2 and 17.3.3. While some materials used in flexible packaging, such as paper, paperboard, cellophane, polyvinyl alcohol and PLA, will either biodegrade or are compostable, these are often combined with nondegradable polymers, making the system nondegradable. There is opportunity for redesign of packaging structures for compostability, which is discussed in Section 17.4.3. Biodegradability is a subjective term that lacks rigorous definitions or standards. Compostability is a better defined term with specific conditions and appropriate international standards for compliance. Even plastics that are compostable may not degrade much in the open environment or during in-home composting; most require higher temperatures found in industrial composting facilities. There is currently little infrastructure for industrial composting which limits the applications. Most packaging that is truly composted comes from closed-loop collection systems such as amusement parks and sporting events/stadiums where the use and collection of the packaging can be controlled.

Bioplastics are described in Sections 4.2.14, 4.2.15, and 17.1.3 and include biosourced plastics as well as biodegradable plastics. Biosourced plastics are derived, at least partly, from nonfossil carbon sources such as plants or microbes. The monomers that make up the building blocks of these polymers come from natural sources, but the polymer manufacturing process may involve traditional industrial chemical methods. Examples of biosourced polymers include PLA, polyhydroxyalkanoate (PHA), bio PET and bio PE. Biosourced plastics may be compostable or not. Similarly, fossil or petroleum derived polymers can be compostable or not (most are not). Composting a fossil-derived polymer releases fossil-CO_2 into the atmosphere and contributes to GWP, whereas composting a biosourced polymer releases renewable carbon that closes a circle from atmosphere to plant to polymer to atmosphere. Clearly from a climate change perspective, composting a biosourced polymer is preferred.

A similar argument is used by critics of biodegradation additives for PE and other nondegradable polymers. These additives typically act through oxidation catalysts or enzymatic activity to fragment the polymer into microscopic pieces that are more amendable for biological activity. Not only is the carbon released into the atmosphere fossil derived, but the fragments have their own health and safety concerns [47–49].

Bioplastics still represent only a small fraction of materials used in flexible packaging. One concern is the use of foodstuff feedstocks such as corn and sugarcane. Growth will depend on developing routes to nonfood feedstocks. Water, land use and the carbon footprint of agricultural fuel inputs are further concerns. The growth of bioplastics will depend on: [50]

- Future trends in oil and gas prices.
- The ability of bioplastics to provide equal functionality and processing at lower cost than incumbents, or new functionality at the right price (for value).
- The magnitude of green premiums consumers are willing to pay – whether they remain low or increase.
- Societal and government pressures on use of renewable products.

The use of biodegradable packaging may lead to increased leakage of waste into the environment if consumers misunderstand how these materials degrade. Most biodegradable plastics require the specific conditions of composting to break down in a reasonable amount of time. Degradation slows in common soil and marine environments. Degradation in landfills without methane capture may cause more damage to the environment than nondegradable packaging. These factors need to be considered as society transitions to more renewable sourced materials.

Finally, there may be technology on the horizon that will allow more materials to decompose in the environment. Bacterial and plant enzymes have been discovered that break down PET or PE [51–53] and worms that eat PE [54,55]. Should these be successfully scaled up they open up further end-of-life options to possibly deal with flexible packaging waste.

17.3.1 Challenges for recycling flexible packaging

For a package to be considered recyclable there must be systems in place to collect, sort and separate, clean, and reprocess it and there must be a definable market for its use. Each of these considerations have challenges for flexible packaging. Today, very little postconsumer flexible packaging waste is recycled, but considerable research and development is underway to overcome the challenges [4,56].

17.3.1.1 Collection

In-plant scrap, edge trim and the like, is known as postindustrial recycle (PIR). PIR is clean and its composition is well defined, making recycling more attractive, especially for films that have not been printed or laminated. These are often ground into flake and fed back into the film converter's process, typically by blending with virgin resin in one or more layers of a film structure.

Used secondary packaging such as stretch films and pallet wrap films is another category of flexible packaging that is amenable to recycling. This "back of the store" waste stream is generally clean, has a well-known composition of mostly compatible plastics, and is handled by a relatively few industrial companies (retailers, wholesalers, shippers, etc.) that makes collection possible.

An additional developed flexible packaging recycle collection stream is WRAP (Wrap Recycling Action Program), the "all-PE" store drop-off stream that consists of grocery bags and other suitable films returned by the consumer. The "How2Recyle" label teaches consumers which packaging to include. The current end-uses for this collection stream are mostly nonfilm applications such as lumber and railroad ties [57].

Most municipalities ban the inclusion of flexible packaging in curbside mixed recycle programs in the United States and much of Europe. In 2018, only 1% of the 1 billion pounds (454 million kg) of recovered PE-based film in the United States and Canada came from municipal Material Recycling Facilities (MRFs) [58]. MRFs have difficulty handling flexible packaging; it gets entangled in conveyor systems and often ends up in the paper fiber stream. If it were to be collected, its mixed composition is challenging to identify and sort, especially multilayer flexible packaging. Research is underway to improve the situation. Materials Recovery for the Future (MRFF), a consortium of companies and organizations in the value chain, ran a pilot trial at a MFR in Pennsylvania in the United States to better understand the handling issues. They found that they could capture 74% of the flexible plastics packaging (FPP) stream (with an objective of >90%) and keep the amount of cross contamination in the paper and FPP streams below their goals [59].

17.3.1.2 Sorting, separation, and cleaning

Sorting is often done manually, but automation is increasingly used. Methods include magnets for ferrous materials, eddy current for nonferrous metals, ballistic or air separation for separating light (films and paper) from heavy (e.g., rigid packaging) materials, flotation for dense materials, and froth flotation for separating materials by differences in surface tension – useful for plastics with similar density [3,4,7,9,56].

Optical and near infrared (NIR) sensors are also used. Optical sensors can separate colors and articles by shape. NIR can detect PE, PP and mixed polyolefins to separate PET from PE and PP from PE streams. NIR has limitations, however:

- It only detects what is on the surface. If there is a label or nested waste it will not see what is underneath.
- It cannot detect compositions of multilayer structures – it only sees what is on the surface.
- It cannot identify dark or black products.
- It cannot detect thin coatings such as polyvinylidene chloride (PVDC) or distinguish between reverse or surface printing.

Metal detectors are often used in MRFs to prevent metal contamination in the downstream processes; these remove metallized films even though they may be recyclable in their base film recycle stream (e.g., PET). The recycling industry, therefore, would like to see metallized films eliminated from recycle streams [60]. Other challenges include dealing with additives and blends. In general, the greater the purity of the final stream, the more value it has as a recycled material. Each separation, sorting and cleaning step, however, adds cost.

Ragaert et al. [9] provide two examples of separation processes used in commercial operations in Europe. One is a Belgium facility that deals with PMD (plastic bottles and containers, metal packaging, and drink cartons), a system where the consumer fills a plastic bag with bottles and cartons of a limited type. The sorting involves:

- Progressive rotating sieves to remove small and large objects
- Wind sifter to remove loose paper and film
- Magnet to remove ferrous metal
- Eddy current to remove nonferrous metal
- Ballistic separator to remove soft (films) from hard plastic (bottles)
- FT-NIR sensor that sorts plastics into PET and HDPE streams
- Optical sensor that separates colors.

A second example is a facility that sorts mixed packaging waste (Germany and Netherlands allow the consumer to place all plastic waste at the curb). The process has the following steps:

- Crude shredder – reduces waste to fist-sized
- Rotary drum washer to remove rocks, metals and glass (in water)
- Friction washer (2nd washing step)
- 2nd shredder – reduces size to 1–12 mm
- Friction washer (3rd washing)
- Float-sink separation
  - o PET, PVC and PS sink
  - o Polyolefins float
- Drier
- Float fraction is put through a wind surfer where soft particles (mostly films made with LDPE or PP) are separated from hard particles (mostly HDPE and PP from bottles). The soft fraction is compounded into pellets. The hard fraction can be sold as is.

Postconsumer waste typically has a density range, so float-sink separation techniques do not completely separate the waste into mono-streams. Melt filtration may be used to separate nonmelting contaminates from the polymer.

Multimaterial, multilayer flexible packaging is generally considered low-quality waste and is landfilled or incinerated. In addition to the challenge of detecting the composition of individual layers, the components may be incompatible and not suitable for mechanical recycling (plastics stream) or repulping (paper stream). One approach is to selectively dissolve and purify the layers or use solvents, surfactants or other chemicals to physically separate the layers (see Section 17.3.2.2). These technologies are still mostly practiced at small scale, more suited for postindustrial waste where the multilayer structures are better defined, and have uncertain economic viability. Another approach for all-plastic films and laminates may be chemical recycling technologies (see Section 17.3.3). Watermarks, fluorescent markers added to labels, and other technologies are also being evaluated to aid in the identification of materials during sorting. WRAP UK has developed novel organic pigments that look black in visible light but are transparent in near IR to allow NIR sorters to detect the underlying plastic [1].

The cleanliness of the product feeding into the mechanical or chemical recycling process is important. Besides other materials in the recycle stream, printing inks, polymer additives and product waste (such as residual food waste) can be problematic. Surfactants are being investigated to deink polymer films [7]. Various washing technologies such as improved sorting, hot washing, and thermal extraction processes are starting to be used to reduce odor of recycled plastics [1].

17.3.1.3 Reprocessing

There are several methods for recycling plastics waste as illustrated in Fig. 17.4. These methods fall into four main classifications: [61–63]

- *Primary recycling*. This is also known as mechanical recycling, reextrusion, reuse, and closed-loop recycling. This involves putting the waste back into the extrusion process to make the same product, either as a blend with virgin resin or as a separate "regrind" layer. The waste is typically preconsumer (or postindustrial) edge trim, scrap and parts from the film or part making process. It is clean and of known composition. Using it in the converting process

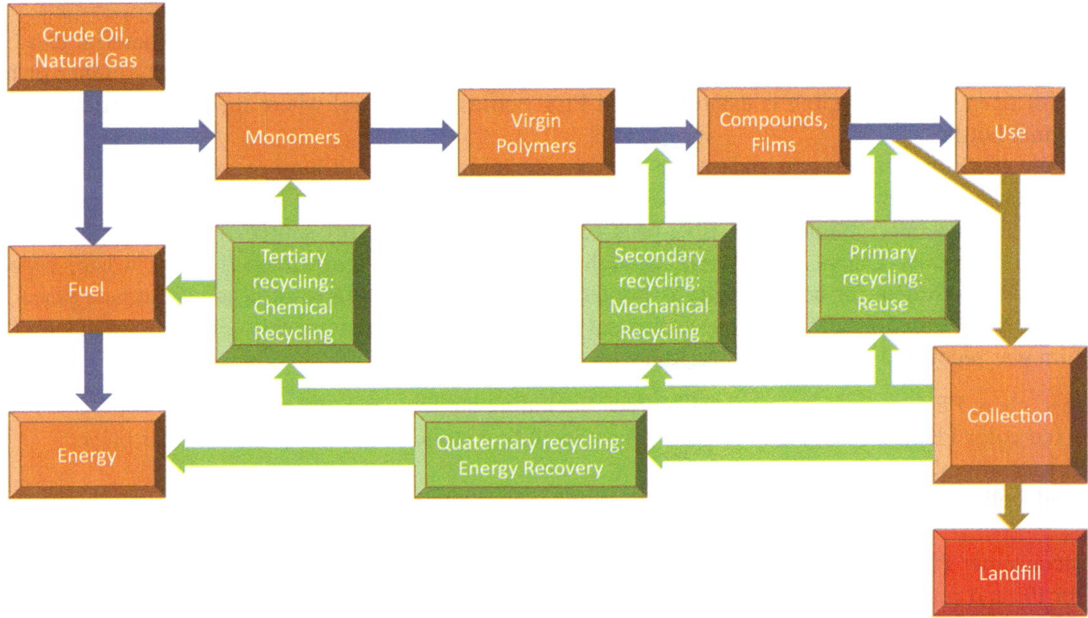

FIGURE 17.4 Categories of recycling. *Based on J. Maris, S. Bourdon, J. Brossard, L. Cauret, L. Fontaine, V. Montembault, Mechanical recycling: compatibilization of mixed thermoplastic wastes, Polym. Degrad. Stab., 147 (2018) 245–266.*

is the most economical and environmentally friendly way to recycle the product. Some well-identified PCR may also follow this path. Solvent dissolution/precipitation and chemically assisted layer separation processes (Section 17.3.2.2) have the potential to upgrade multilayer packaging waste to primary recycling. These processes selectively dissolve and purify the materials, removing coatings, inks and other contamination. Since the polymer is simply dissolved or separated it is never chemically transformed as in chemical or feedstock recycling.

- *Secondary recycling*. This is also known as mechanical recycling, open-loop recycling and downgrading. Like primary recycling, secondary recycling involves mechanically transforming the plastic via extrusion to substitute for virgin polymer in new applications. The difference is that the quality of the end product is not good enough to be used in the same application. The source of the waste may be postconsumer where contaminants, mixed plastics, additives and subsequent heat histories during reextrusion degrade the physical properties. Virgin polymer and various additives (compatibilizers, stabilizers, fillers, etc.) may be incorporated into the recycle to upgrade the performance.
- *Tertiary recycling*. This is also known as chemical or feedstock recycling. It involves chemically breaking down the polymer molecules into small molecules suitable as feedstocks for fuel, new polymers or other chemicals.
- *Quaternary recycling*. This is also known as energy recovery or incineration. It is the most common recovery route for plastics in Europe; in 2018 42.6% of plastics waste was incinerated [33]. While the EU does not consider incineration as recycling, quaternary recycling does recover the high energy content of these materials and is suitable for mixed plastics and other plastics waste that is no longer recoverable by mechanical recycling.

These technologies are explored in greater detail in the following sections. With multimaterial flexible packaging, such as plastic film-aluminum foil laminates, the layers may need to be separated prior to recycling or be limited to quaternary recycling. For example, paper and paperboard are recycled in a repulping process. For polymer-coated paper or paperboard, the requirements for repulpability require greater than 80% fiber yield after filtering a simulated pulp through a slotted screen [64]. Fibers with polymer attached are generally left behind and reduce the yield. The fiber yield can be increased by coating less polymer onto the paper, increasing the thickness of the paper or paperboard relative to the polymer coating, increasing the paper density, or using polymers that are water soluble or release from the fibers when exposed to moisture.

17.3.1.4 End markets

Developing markets for recycled materials is critical to create value and drive supply. Two barriers to using PCR most often cited by companies are "not enough price advantage over virgin resin" and "not enough PCR available that matches our specifications." [58] Both speak to the need for developing outlets for recycled materials.

The terms up-cycling or down-cycling are out of vogue. Better descriptors are closed and open-loop recycling. In closed-loop recycling the material goes back into same application as its original use. Virgin material can be added. In open-loop recycling the material goes back into a different application. These terms do not imply higher or lower value. An example of open-loop up-cycling is using recycled PET bottles for apparel where the value of the apparel is higher than the original bottle.

While some postindustrial recycle ends up in flexible packaging as closed-loop recycling, most flexible packaging PIR and PCR ends up elsewhere. The domestic North America recycle market is mostly for clean, dry PE film or single-resin waste, generally coming from PIR, back-of-store or store drop-off sources [58]. With the cut-off of some export markets, especially China, there is not enough capacity to process all domestic film waste, especially from MRFs. Return on investment for new capacity is limited by the price of virgin resin, low importance placed on recycle content, narrow energy savings and low cost of disposal in the US [65]. Approximately 1 billion pounds (500 billion kg) of plastic film was recovered in 2018, with the following breakdown: [65]

- 47% clear, clean PE film from commercial sources, including stretch wrap and poly bags
- 13% mixed color PE film from commercial sources, including stretch wrap; no postconsumer bags
- 12% PE agricultural film
- 24% retail collected PE mixed color, clean PE film, including stretch wrap and postconsumer bags, sacks, and wraps
- 1% postconsumer PE mixed film collected curbside and sorted at an MRF
- 3% other PE and non-PE films.

The breakdown of uses for this recycle are reported as: 46% lumber and decking, 34% film/sheet, 12% injection molding (pallets, crates, buckets, etc.) and 8% other. In terms of recycle content used in these markets, More Recycling [65] reports that lumber and fencing have the highest at almost 100% recycle, rigid parts such as plumbing and carts range from 25 to 100%, and bottles and bags have the lowest at around 25%.

Some example applications include wood plastic composites (WPC) for decking and other applications, components of polymer composites, and green building construction (polyolefin-based concrete aggregates and composite mortars for lightweight building and asphalt pavement). The use of recycled PE in WPC has advantages over wood and all-plastic lumber. Wood absorbs moisture and decays over time, whereas WPC's are more durable. WPC can also be extruded into shapes. The incorporation of waste wood flour reduces the carbon footprint of an all-plastic part. Using recycled PE reduces the mechanical performance somewhat, but much of this can be regained using compatibilizers, such as maleic anhydride (MAH) grafted PE, to improve the bonding between the polymer and wood flour. The MAH grafted PE compatibilizer also helps reduce water absorption which slows decay [3].

Kosior and Mitchell [1] provide an example of upgrading the recycle of coffee cups in the United Kingdom. PE coated paper cups generally see as single use and as low-value waste end up in landfill. The polymer-coated fibers, however, enhance the properties of polymer composites for molding resins. They are being considered as rainforest timber replacements for cabinetry, pallets and other applications. Where hardwood timber is structurally strong, it has poor durability if exposed to moisture. Plastics have high durability but lack stiffness and warp and buckle. The composite made from recycled coffee cups is stiffer than all-plastic but still durable and can be readily shaped and is less prone to buckling.

PCR is being used as a modifier for asphalt roads. Combined with the appropriate compatibilizer, plastic recycle helps improve the resilience of roads to extend durability [66–70]. This has the potential for using mixed plastics waste streams that are otherwise difficult to recycle.

The use of PCR in primary flexible packaging is limited by food regulatory constraints, quality and supply: the shift towards lighter, highly engineered packaging with multiple layers and materials makes the incorporation of PCR into the same package more difficult. While many major consumer brand companies have announced goals for increasing their use of recycle in packaging, much of this initially is focused on rigid packaging. The use of PCR in flexible packaging remains more challenging for a variety of reasons, including:

- Mismatch of the available recycled PE-type with application needs. Much of the available recycle is high-density PE made from milk cartons whereas many film applications require lower density versions of PE.
- Flow properties of recycle resin that meet film processing needs.
- Loss of mechanical properties.
- Quality and consistency. Film processes and applications generally have high needs around low odor, low number of defects (gel particles, black specs, contamination), high clarity, low color, good thermal stability and no off-gassing that are more demanding than most other processes/applications.

As demand increases for use of recycle resin in flexible packaging, some of the larger PE suppliers have started to get involved. For example, Dow Chemical has announced an association with Avangard Innovative, a PE recycler, and has begun offering products with high recycle content to the market [71,72]. The initial target market is for secondary packaging films such as collation shrink films. By using their knowledge of polymer science and quality assurance expertise, the development of high-quality products will enable further use of recycle in flexible packaging.

17.3.2 Mechanical and advanced physical recycling

Physical recycling refers to mechanical and solvent-based purification processes that transform the waste plastic back into a useful form without changing the polymer chemically. In mechanical recycling this is done through melt processing. In solvent-based purification processes, the waste stream is selectively dissolved to separate out impurities. Physical recycling also includes processes that delaminate layers using solvents, surfactants and other chemical means. These chemically assisted processes do not change the molecular weight or chemistry of the polymer (hence the name physical recycling) [73]. This differs from chemical/feedstock recycling processes where the polymer is chemically broken down into smaller molecular weight units, such as reverting the polymer back to its monomers during depolymerization (chemolysis), or creating liquids or gases that can be refined into basic chemicals or fuels (thermochemical). There is some confusion in the literature. Some references use the term solvolysis to refer to solvent-based processes; here we will use chemolysis to refer to depolymerization processes where the solvent reacts with the polymer and dissolution/precipitation for solvent-assisted physical recycling processes where the polymer is not chemically altered. Some references place dissolution/precipitation processes under the category of chemical recycling. Here they are placed under physical recycling.

17.3.2.1 Mechanical recycling

Mechanical recycling of plastics involves melting and reprocessing the polymer back into a pellet for reuse in a polymer fabrication process. In some cases, the recycle is fed directly into the fabrication extruder, skipping the pelletization step. The feedstock is a sorted, clean, dry, size-reduced (typically shredded or ground) recycle stream. Virgin polymer and various additives may also be added. The reprocessing is typically done in an extruder and may involve melting, devolatilization and steam stripping, mixing in additives, filtering out contaminants with a screen pack, and pelletization. Specialized compounding extruders, such as twin-screw extruders (see Chapters 2 and 6), are typically used.

One of the challenges for mechanical recycling of plastics is controlling polymer degradation and thermo-oxidative stability. The mechanical properties of polyolefins, the primary resins used in flexible packaging, decrease with reprocessing due to chain scission and branching/crosslinking reactions. Fourier Transform Infrared (FTIR) spectroscopy shows little chemical change after recycling PE or PP suggesting just chain scission and crosslinking reactions are taking place. (The polar oxidation byproducts formed at high temperatures and described in Chapter 10 are not prevalant). This results in lower physical properties such as impact strength and % elongation to break. Environmental stress crack resistance (ESCR) of HDPE is shown to decrease by 60% in one study [3]. The viscosity of the polymer changes depending on whether scission or crosslinking is predominant. The melt flow rate of PP, for example, increases since chain scission is typically dominant for this polymer. A slight decrease in melting point and crystallinity (and density) is observed for polyolefins after recycling. Water content/absorption may be higher after recycling due to the lower crystallinity [3]. These changes should be considered when making products with recycle content. As such, mechanical recycling is limited to about seven recycle histories due to thermal degradation [11]. Ways to mitigate the loss of physical properties is further explored in Section 17.4.1.1.

Plastics that have long-time outdoor exposure undergo photo-oxidative degradation. Low molecular weight (MW) volatile compounds can form which are trapped in the solid polymer but are released during melt processing during recycling. These can affect the process (corrosion, etc.) and performance [9].

A second challenge is contamination. As is described in Chapter 6, most polymers are immiscible; when they are blended with one another they form separate phases, which can detract from mechanical and optical performance. Unless the recycle stream is pure, the presence of even small amounts of other polymers can affect the quality of the recycle. With flexible packaging, other contaminants may be present, such as fragments of paper labels, pigments, additives, printing inks, cross-linked adhesives, and product waste (such as food residue). Pigments can act as prodegradants during thermal processing. Slip additives (fatty acid amides) used to control the frictional properties of the film can impart a rancid odor when they degrade. Inks can volatilize in an extruder and cause off-gassing. They also impart dark colors unsuitable for transparent flexible film applications [7]. The shear variety of materials and structures used in flexible packaging make it difficult to develop a single waste management hierarchy for plastic film waste. For many flexible packaging structures chemical recycling and energy recovery may be better options due to these contamination and mixed plastics challenges [7].

These challenges can be mitigated by employing various technologies, including

- Starting with pure recycle streams. This involves either sourcing the waste from known, homogenous outlets such as industrial waste, or employing comprehensive sorting, segregation and cleaning steps in the front end of the waste handling process. For flexible packaging, this may involve improved technologies for identification, sorting, separation of individual layers within multilayer films and laminates, deinking and cleaning. Each of these steps adds cost to the overall process.
- Adding virgin resin. This is often done by the user of the recycle.
- Devolatilizing during extrusion. Extruders with vacuum ports can be utilized to remove volatile components. This can be enhanced by introducing water/steam or solvents to the process.
- Melt filtering. Screens can be introduced at the end of the extruder to filter out contamination and unmelted polymer. Too much contamination, however, will plug the screens and introduce excessive pressure drop.
- Incorporating additives during extrusion. Antioxidants, odor scavengers, impact modifiers, processing aids and other additives can be used to minimize oxidation/degradation, mask odor, and regain physical or flow properties.
- Adding compatibilizers. Compatibilizers can be used to help with mixed polymer systems as well as with some forms of inorganic contamination.

The theory behind improving compatibility of polymer blends is described in Section 6.4. The principles for improving compatibility are the same as for enhancing interfacial adhesion, which are described in Chapter 10. Here the term compatibilizer [74] refers not only to modifiers that stabilize the interface of immiscible polymers, but also "impact modifiers" [75] and "tie chain boosters" [76] that act as toughening agents to enhance properties lost due to the presence of the contaminants. One compatibilizing mechanism is illustrated in Fig. 17.5. Immiscible polymers form dispersed domains in the bulk polymer matrix. The high interfacial energy between the two polymers without a compatibilizer (left) results in course domains with weak interfaces that allow cracks to propagate around them. Compatibilizers modify and stabilize the interface. Nonreactive modifiers offer an intermediate polarity between the polymer components or have a block copolymer structure with phases compatible with each polymer. Reactive compatibilizers have functional groups that reside in one component and react with the other, essentially forming a copolymer at the interface. With the stabilized interface, the crack must propagate through the dispersed phase instead of shortcutting around it. Reducing the interfacial energy not only promotes good adhesion to prevent the dispersed phase from debonding, but also reduces the size of the dispersed domains. Smaller domain size of the dispersed phase contributes to good optics by reducing the haze and increasing the clarity of the films. Additionally, rubber-like compatibilizers residing in the bulk phase can provide toughening through various energy dissipation mechanisms [77–80]. These mechanisms include both dilatational deformation (e.g., cavitation, voiding, and the formation of craze and shear bands) and constant-volume deformation to enhance the energy adsorption capacity of the fabricated articles without creating catastrophic fractures [81].

Table 17.10 provides some chemistries that are used to compatibilize various pairs of polymers likely to be found in flexible packaging recycle streams. PE and PP are the largest volume polymers used in flexible packaging and most

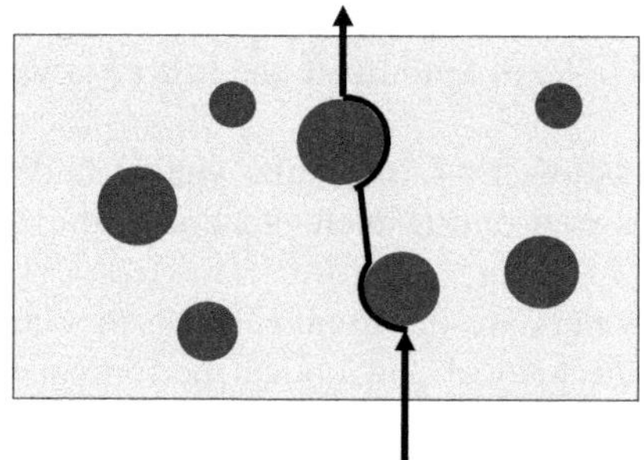

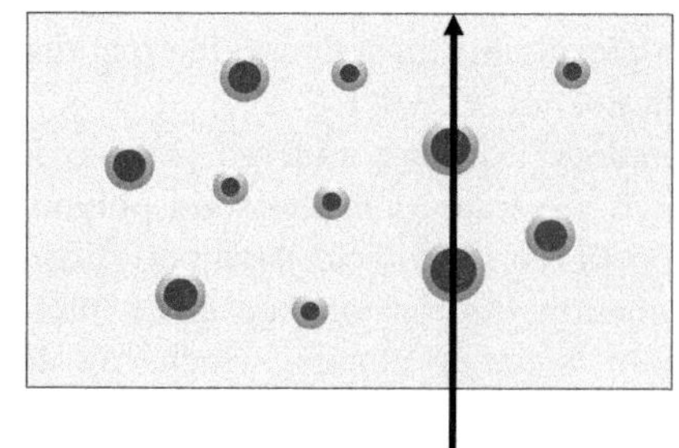

FIGURE 17.5 Mechanism for compatibilization of immiscible polymer blends. *Y. Zeng, D. Abebe, A. Singh, L. Effler, Y. Hu, M. Kapur, et al., Compatibilizing polypropylene (PP) contaminated polyethylene (PE) waste streams in film applications, SPE Int. Polyolefins Conf. (And 2021 ANTEC), 2020.*

TABLE 17.10 Compatibilizers for combinations of polymers.

Combination	Chemistry	Compatibilizer examples
PE/PP	Nonreactive	OBC, POE, EPDM, Nanoparticles
PA/PE	Anhydride + Amine (NH_2) →	PE-g-MAH
PA/PE	Carboxylic Acid + Amine (NH_2) → + H_2O	EAA, Ionomer
PA/PE	Epoxy + Amine (NH_2) →	E/GMA, E/AE/GMA, PE-g-GMA
PA/PP	Anhydride + Amine (NH_2) →	PP-g-MAH, SEBS-g-MAH
PS/PE	Nonreactive (polarity)	EP-g-MAH, EVA
PP/HIPS	Nonreactive (polarity)	SEBS, EVA
PET/PE	Epoxy + Carboxylic Acid →	E/GMA, E/AE/GMA, SEBS-g-GMA
PET/PE	Nonreactive (polarity)	EMA, Ionomer, PE-g-AA, POE-g-MAH, EPDM, Nanoparticles (MMT), SEBS, SEBS-g-MAH
PET/PP	Nonreactive (polarity)	Nanoparticles (MMT), PP-g-MAH
PET/PE/PP	Epoxy + Carboxylic Acid →	E/GMA, E/AE/GMA
PE/EVOH	Anhydride + Hydroxyl (OH) →	PE-g-MAH
PP/EVOH	Anhydride + Hydroxyl (OH) →	PP-g-MAH

PE, polyethylene, *PP*, polypropylene, *PA*, polyamide, *PS*, polystyrene, *HIPS*, high-impact polystyrene, *PET*, polyethylene terephthalate, *OBC*, olefin block copolymer, *POE*, polyolefin elastomer, *EPDM*, ethylene-propylene diene monomer rubber, *PE-g-MAH*, polyethylene grafted with maleic anhydride (MAH), *AA*, acrylic acid, *SEBS*, styrene-ethylene-butadiene-styrene copolymer, *EP*, ethylene-propylene rubber, *EVA*, ethylene acrylate copolymer, *EMA*, ethylene methyl acrylate copolymer, *E/GMA*, ethylene glycidyl methacrylate copolymer; *E/AE/GMA*, ethylene-acrylate ester-glycidyl methacrylate terpolymer, *MMT*, montmorillonite nanoclay.
Source: Based on J. Maris, S. Bourdon, J. Brossard, L. Cauret, L. Fontaine, V. Montembault, Mechanical recycling: compatibilization of mixed thermoplastic wastes, Polym. Degrad. Stab., 147 (2018) 245–266 and K. Hausmann, Compatibilizers, polymeric (recycling of multilayer structures), in J.C. Salamone, (ed.), Polymeric materials encyclopedia, Vol 2, CRC Press, 1996, pp. 1364–1377.

likely to be found together in a recycle stream. Even though both are polyolefins, they are not miscible. PP-PE block copolymers have been found to work well as compatibilizers [80–82]. Zeng et al. [81] evaluate seven different compatibilizer chemistries in a PE-rich blend with PP. Ethylene-octene copolymers are found to be most effective at recovering the toughness of the films (as measured by dart impact) whereas an olefin diblock copolymer (PP-P/E blocks) provides the best optical properties (See Section 17.4.1.1).

For polar-nonpolar polymer compositions, ethylene copolymers are often utilized as compatibilizers. Maleic anhydride (MAH) chemistry is one of the most widely used. It can be grafted onto various polyolefins, ethylene copolymers and polyolefin elastomers. MAH bonds particularly well to PA and EVOH and is used in tie resin compositions (see Chapter 10). Glycidyl methacrylate (GMA) provides an epoxy functional group and is useful for compatibilizing polyolefins with polyester. Ionomers are also used for PA and PET modification [5,83]. Some of the relevant reactions include:

- Anhydride with amine or hydroxyl groups
- Epoxy with carboxylic acid or amine groups
- Carboxylic acid with oxazoline or amine groups
- Isocyanate with hydroxyl or amine groups.

Nanoparticles (NP), such as montmorillonite clay, have been evaluated in the literature to increase compatibility. The hypothesis is that they stabilize the interface between incompatible polymers. They work much like block copolymers. Theory suggests the best NP's for compatibilization have strong interactions with each polymer phase, are spherical, have a size greater than the radius of gyration of the polymer ($>$ 50 nm) and do not self-agglomerate. Most inorganic nanoparticles are chemically treated on their surface with an organic coating, which affects their dispersion in the polymer [5].

Other approaches include the use of free radical crosslinking, where the free radicals are initiated with peroxide, mechanical shear, heat or radiation [5,84], and the use of duel compatibilizers. Examples of the latter include SEBS (styrene-ethylene-butadiene-styrene copolymer) + PE-g-MAH to compatibilize PET and HDPE [5,85] and combining nanoparticles with reactive compatibilizers [5].

The stability and degradation of the blend can differ from the individual polymers and has not been well studied.

17.3.2.2 Dissolution/precipitation and other advanced physical recycling

In advanced physical recycling, chemicals are used to separate layers and remove impurities without chemically changing the polymer. These processes generally involve collection, sorting, shredding and cleaning steps prior to the addition of solvents, surfactants or other chemicals to separate layers or selectively dissolve materials. Standard chemical separation processes such as distillation followed by precipitation are used to purify the polymer.

The basic process technology was developed in the 1990s but it has not gained wide commercial use due to its higher cost than mechanical recycling [86]. New demand for recycling complex plastics waste streams has given it a fresh look. Advanced physical recycling overcomes two deficiencies of mechanical recycling: (1) it removes impurities such as inks, pigments, additives and food residues and (2) it allows for separation of multilayer, multimaterial packaging structures into their pure components. It also has an improved environmental footprint over chemical recycling since it does not involve a repolymerization step.

Several companies have developed or are in the process of scaling up advanced physical recycling technologies. These include:

- Fraunhofer Institute for Process Engineering and Packaging IVV (Germany) has introduced CreaSolv dissolution/precipitation technology. The process has been shown to be effective for PS in scrap electronics, expanded polystyrene (EPS) in insulation and recycling of mixed plastic waste (PS, polyoefins, PET). Unilever has a 1000 ton/year pilot plant for recycling PCR multilayer film packaging (sachets) [73,86–90].
- APK (Germany) Newcyling technology uses a solvent-based process for recycling PIR of PE-PA multilayer packaging waste, recovering pure streams of PE and PA. They have a 10,000 ton/year plant [86,91,92].
- Proctor and Gamble's proprietary PureCycle dissolution process recovers PP from carpet and packaging streams [86,92].
- Solvay (Italy) developed and operated VinyLoop for recovering PVC polymer from flexible PVC waste streams. The process was shuttered in 2018 after 15 years of operation due to new regulations around phthalate plasticizers [92].

- PVC Separation (Australia) have developed a process to delaminate multilayers by swelling the films in low boiling solvents. Exposure to hot water flashes off the solvent and the polymer layers are recovered [86].
- Saperatec (Germany) use an organic solvent and carboxylic acid to accelerate the delamination of multilayer film/sheet [10,86].

Walker et al. [92] note that there is little scientific literature on these processes [93]. They conduct a study to demonstrate the use of solvent-targeted recovery and precipitation (STRAP) to separate and recover the pure components in multilayer plastic films. They use a series of solvent washes guided by thermodynamic calculations of polymer solubility. They begin by using a physical blend of pellets of PE, EVOH and PET but later validate the technique using multilayer film produced by Amcor Flexibles. The film is predominately PET (90%) with some PE and EVOH. They were able to recover essentially the pure components with the physical properties retained. They expect the MW to be unchanged by the process but did not measure it. The residual solvent level after recovery was less than 1000 ppm.

Walker et al. also conduct an economic analysis. The cost of recovering PET decreases with increasing capacity of the facility. At around 3800 ton/year the cost is about equal to that of virgin PET; this is less than half the capacity of APK's Newcycling plant. The separation (such as distillation) steps are major cost drivers: 33% of capital investment cost and 79% of operating costs. The energy use (and CO_2 emissions) is estimated to be 37% less than that for virgin PET.

Vollmer et al. [88] conduct an LCA of several end-of-life scenarios for a PET-PE laminate. When compared with incineration, the dissolution/precipitation process is found to have lower CO_2 emissions than chemical recycling since carbon bonds do not have to be reformed for reuse of the polymers. Versus landfilling, the recycle process is not advantaged in CO_2 emissions since the carbon is sequestered in the landfill, but landfilling is not circular. Their analysis shows dissolution/precipitation to have lower CO_2 emissions than mechanical recycling due to the greater purity of the product that allows for more substitution for virgin materials.

Advanced physical recycling has disadvantages and uncertainties. It can be challenging to find an environmentally benign solvent that is effective and easy to remove. Also it is most useful for well-defined postindustrial recycle since the solvents that are selected depend on a steady composition. This makes it challenging for mixed PCR waste streams; advances in sorting and identification may be needed for the technology to be viable. Cost vs alternatives is another uncertainty. Indeed, while Unilever has started up a CreaSolv pilot plant in Indonesia for recycling multimaterial packaging for sachets, they continue to explore other alternatives such as mono-material structures [90].

17.3.3 Chemical or feedstock recycling

Chemical or feedstock recycling is attracting a lot of interest since it can potentially overcome many of the shortcomings of mechanical recycling such as handling mixed plastic/material waste and product contamination. By breaking the waste plastic down to its molecular constituents and using conventional distillation and other separation processes, the products of chemical recycling can ultimately be recovered as chemical building blocks for polymers and other chemicals, or for use as fuels. Basic technologies are illustrated in Fig. 17.6. Most of the technologies are still under

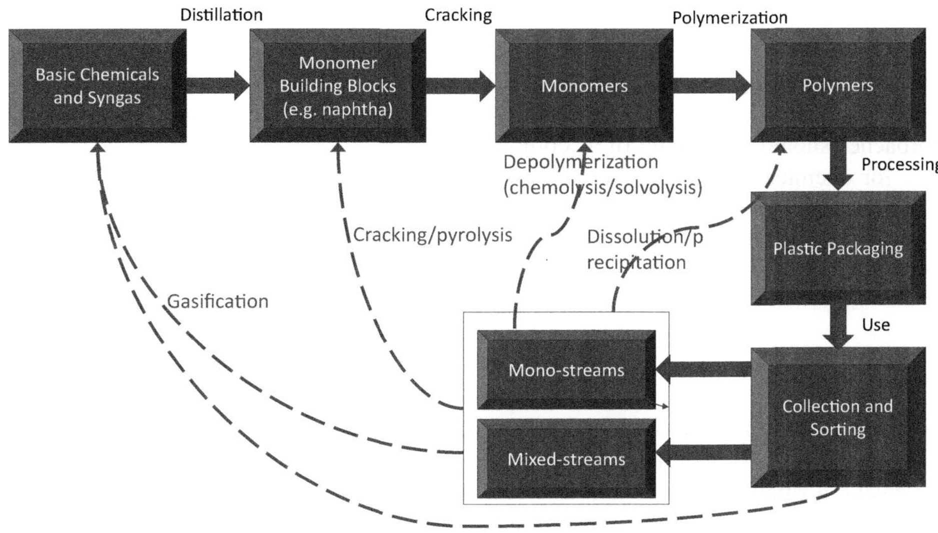

FIGURE 17.6 Feedstock recycling of plastics packaging. *Adapted from G. Bergsma, Chemical recycling and its CO_2 reduction potential. CE Delft, Brussels. https://cedelft.eu/assets/upload/file/Presentaties/2019/Chemical%20recycling%20an%20environmental%20perspective%20CE%20Delft%2020%20february%202019%20brussel.pdf, 2019. Bergsma [94].*

TABLE 17.11 Summary of feedstock recovery technologies.

Process	Advantages	Challenges	Commercial status	Sensitivity to feedstock quality
Chemolysis (solvolysis, depolymerization)	• Generates pure value added products • Operational for PET	• Needs large scale to be economical • Mainly limited to condensation polymers	Commercial for PET	Very high—Requires homogenous polymer waste
Pyrolysis	• Suitable for mixed plastic waste • Simple technology	• Needs large scale to be economical • Low tolerance for PVC contamination • Needs consistent, high-quality feedstock to be economically viable	A few operations in United States, Europe, Japan and India. Renewed interest by several major petrochemical companies	High
Catalytic cracking	• Narrow product output • Favorable economics if less stringent reaction conditions used	• Catalyst deactivation by chlorine and contaminants • Reactor technology not developed	Several commercial processes operating at industrial scale. One of the largest for plastics waste shut down in 2010	High
Hydrocracking	• Quality of naphtha output • Suitable for mixed plastic waste	• Presence of inorganics • High cost of hydrogen • Cannot be used for PVC since it will poison the catalyst	Currently only pilot scale facilities; process used in petrochemical industry	High
Gasification	• Syngas is a valuable intermediate • Cost of air • Well-known technology	• NOx production • Specific drawbacks of using air (formation of nitrogen byproducts); Pure oxygen produces purer product but more expensive • Formation of tar with use with plastics waste	Enerkem in Canada has full-scale commercial steam gasification plant for converting waste into methanol, then ethanol and ethylene. It requires presorting of the plastic waste and very high energy	Medium

Source: Based on information from K. Ragaert, L. Delva, K. Van Geem, Mechanical and chemical recycling of solid plastic waste, Waste Manag., 69 (2017) 24–58 and M. Solis, S. Silveira, Technologies for chemical recycling of household plastics – a technical review and TRL assessment, Waste Manag., 105 (2020) 128–138.

development and little infrastructure currently exists. Some processes still require pure or sorted feedstocks and most require large scales to be economically feasible. Current costs and energy requirements are higher than for mechanical recycling and therefore the two approaches should be thought of as complimentary.

There are two basic technologies for chemical recycling: chemolysis and thermochemical. Chemolysis, also called depolymerization and solvolysis, involves depolymerizing the polymer back to its monomers using chemical means. Thermochemical processes, such as pyrolysis and gasification, involve transforming the plastic into gas and liquid byproducts by thermal or catalytic means. The advantages and disadvantages of specific technologies are summarized in Table 17.11.

17.3.3.1 Chemolysis

Depolymerization by chemolysis is limited to condensation polymers. As described in Section 4.2.1.4, condensation polymers involve a stepwise reaction between two monomers, each with at least two functional groups. These groups undergo a condensation reaction, releasing a small molecule such as water. Examples are PET, PA and polyurethanes. The other class of polymers, addition polymers, undergo a chain reaction between carbon groups during polymerization

– the resulting main chain contains carbon-carbon bonds. Examples are PE, PP, and PS. Depolymerization is easier for condensation polymers since it involves breaking ester or amide bonds in the polymer chain rather than carbon-carbon bonds found in addition polymers. Polycondensation is an equilibrium reaction that can be driven in either direction. During depolymerization, the polymer is typically dissolved in a hydrogen containing compound that reacts with a polar group within the chain of the polymer to cleave off the monomer.

There are several chemolysis depolymerization processes used industrially; the choice depends on the polymer and the desired end product. For polyesters, methanolysis, hydrolysis and glycolysis are the primary methods. Each uses a different reactant and results in a different set of monomer products that can be used for making new polyesters. For PET, these are:

- Methanolysis: PET reacts with methanol to depolymerize into terephthalic acid, dimethyl terephthalate (DMT), bis (hydroxyethyl) terephthalate (BHET), and ethylene glycol (EG).
- Hydrolysis: PET reacts with water to yield EG and DMT.
- Glycolysis: involves transesterification of PET with EG in presence of a catalyst (typically zinc acetate) to form BHET.

Several companies have pilot scale operations for depolymerization of polyester waste including IBM, Ioniqa (with Unilever, Indorama and Coca-Cola), Eastman, and Loop Industries/Suez [10]. Enzymes have been used to catalyze these reactions. For example, the French company Carbios has developed technology for using enzymes to catalyze the hydrolysis of PET [10,95].

Processes to depolymerize polyamides have been developed using hydrolysis chemistry.

One drawback with chemolysis as it relates to flexible packaging waste is that it requires pure polymer feedstocks. Most polyesters and polyamides used in flexible packaging are found in multilayer laminates or coextruded films. Until layer separation technology is well developed (Section 17.3.2.2), chemolysis may not have much impact on flexible packaging, other than the feedstock used to make virgin PET or PA may come from chemical recycling of other waste streams (completing their circle). For example, PET bottles and PA carpets have been chemically depolymerized back to their monomers.

17.3.3.2 Thermochemical processes

17.3.3.2.1 Pyrolysis

Thermochemical processes have potential for dealing with polyolefin and mixed plastic waste. These processes fall under the general categories of pyrolysis and gasification. Pyrolysis involves subjecting the waste to temperature (230°C–800°C) and pressure in the absence of oxygen to form pyrolytic gases that can be condensed into liquids. The resulting gases and liquids may be further refined for fuels and basic chemicals. Pyrolysis of polymers forms higher fractions of olefins than fossil crude oils. If fuel is the desired product, hydrogen may need to be added (in a process called hydrocracking).

Many factors determine the composition of the end-products of pyrolysis: composition of the plastic waste, reactor type, process conditions, and catalyst. Polyolefins can form fractions of olefin gases (ethene, propene, butadiene) that can be used as chemical building blocks. Some polymers, such as PS and poly methyl methacrylate (PMMA), can be depolymerized by pyrolysis into high concentrations of their monomers. The output of pyrolysis will require some level of purification; due to the high cost of separation technologies pyrolysis is often used to produce fuel oils rather than chemicals [7].

While in theory pyrolysis is well suited for handling heterogeneous plastic waste, the required economies of scale and potential for contaminants limit this in practice. Different reactions can occur when mixed plastics are present. As already noted, some polymers such as PTFE, PMMA, PA, PS depolymerize into their monomers. PE and PP tend to fragment into different MW fractions. The presence of different outputs may require totally different separation or refinement steps. Given the large economies of scale needed to be economically feasible, only common polymers, or well-defined mixtures, are being considered for scale-up.

Pyrolysis is sensitive to the presence of small amounts of polyvinyl chloride (PVC) contamination in the feed stream. Hydrogen chloride (HCl) forms which must be removed from the products to prevent corrosion and contamination of the products for their end use. Also, there may be sulfur in the product coming from antioxidants or flame retardants that may be detrimental to the end use. It is very expensive to separate out such additives in the waste stream [9].

In catalytic cracking, a catalyst is used to increase the selectivity of the products of pyrolysis, which may upgrade their value. The catalyst lowers the reactor temperature (from around 450 to 300°C–350°C); increases the reaction rate;

increases the production of isoalkanes, branched and cyclic molecules and aromatics; increases efficiency; and improves the selectivity and quality of the output [95]. A challenge for plastics waste, particularly heterogenous postconsumer waste, is the potential for impurities to deactivate the catalyst. As a result, most of the work to date with this process has been with pure polymer feed streams.

Another variation of pyrolysis technology is hydrocracking. Here addition of hydrogen to the reactor significantly improves quality yield and allows the use of mixed plastic waste streams. The cost of hydrogen, however, is a primary drawback.

Several large petrochemical companies are actively involved with the development of pyrolysis or catalytic cracking processes. The first attempts were in the 1970–80's, resulting in some commercial plants in India (Pyrocat Systems), United States (Agilyx), Europe (Thermofuel), and Japan (Toshiba Corporation). In Europe economic viability has been a struggle. Recent focus is on not just producing fuel but new processes that also produce raw materials for chemicals. Some examples include: [95]

- SABIC is partnering with United Kingdom-based Plastics Energy using mixed plastic waste to produce pyrolysis oil called TACOIL. SABIC is looking at using it as cracker feedstock for PE and PP production.
- BASF is partnering with Recenso GmbH, which makes pyrolysis oil from plastic waste. BASF uses this in its crackers to make ethylene and propylene.
- Neste (a Finland-based petrochemical company) is partnering with ReNew ELP (UK recycling company) and Licella (Australia-based technology company) to develop a catalytic process for plastic waste.
- Agilyx (United States) has developed a pyrolytic process for depolymerizing PS. ReVital Polymers, Pyrowave, and INEOS Styrolution are also developing PS depolymerization technology.
- Renewology (United States based) offers pyrolysis systems for converting plastic waste into petrochemicals such as naphtha [feedstock to plastic (PE) and other chemicals]. Two facilities have been built in North America so far.
- Braskem has announced partnerships for chemical recycling PCR such as grocery bags.
- LyondellBasell has a pilot plant for MoReTec, a catalytic pyrolysis process [96].
- Dow is partnering with Fuenix Ecogy Group (The Netherlands) to supply pyrolysis oils from plastics waste as feedstock to Dow's polymer plant in The Netherlands [97].

Two additional lab-scale pyrolysis technologies offer potential: [11]

- Plasma pyrolysis: Pyrolysis is conducted in the presence of a plasma. This process operates at very high temperature and typically breaks down the polymer to CO and H_2 syngas. The reaction is very fast and does not produce toxic compounds due to running at such high temperature (1730°C–9730°C). This has the potential of dealing with PVC contamination issues. The process is used in other applications but is still in the early R&D stage for plastics waste.
- Microwave assisted pyrolysis: This process involves mixing the waste with a microwave-absorbent dielectric material to generate heat efficiently. It produces very high temperatures to achieve high conversion to products. So far it is only at the lab scale.

17.3.3.2.2 Gasification

Gasification processes involve subjecting the waste to heat (700°C–1100°C) and pressure in the presence of a controlled amount of oxygen, air, oxygen-enriched air or steam to produce syngas. The main components of syngas are hydrogen (H_2) and carbon monoxide (CO), along with smaller amounts of carbon dioxide (CO_2) and hydrocarbons. Syngas can be refined to make fuels or chemicals. Like pyrolysis, gasification avoids the toxic emissions formed during direct incineration of waste plastics [83].

The output depends on several factors, such as composition of the waste feed, the type of reactor, and process conditions. Using pure oxygen creates a higher purity syngas than air, which produces products containing nitrogen. Byproducts are tar and char. The use of gasification with plastics waste dates to the 1970s but will little commercial activity. Enerkem (Canada) uses waste from municipal solid waste and MFR's to produce syngas that is further refined to make methanol and ethanol. New attention is being given to gasification due to the possibility of utilizing heterogenous and contaminated waste and its potential for producing syngas for chemical synthesis as well as for fuel. In many cases metals or minerals in the waste are retained in the ash of a gasification process and therefore, the process lends itself for handling barrier packaging containing aluminum foil and plastics. Corenso operates a commercial facility in Finland using Al/PE laminate waste from liquid packaging where the aluminum is recovered.

Additional gasification technologies that offer potential for plastics waste include: [11]

- Plasma gasification. Plasma is introduced to create very high temperatures (around 3900°C, but up to 15000°C) at very short residence time. This allows for a high tolerance for low feedstock quality and produces a higher purity output (more gas and less tar). Its challenges include high energy and investment cost.
- Pyrolysis with in-line reforming. In a first reactor hydrogen is produced from pyrolysis of the waste, which is fed into a second reactor for a catalyst reforming step. Contaminants remain in the first reactor. So far, this process has only been demonstrated on a lab scale.

Chemical recycling can be complimentary to mechanical recycling, but operations have not yet reached industrial or commercial scale [98]. Cost and complexity are major hurtles. Capital costs are high and large plants are necessary to be economically feasible. Current technology has an unfavorable LCA footprint. For example, Bergsma [94] conducts LCA's for chemolysis and thermochemical processes assuming a footprint in the Netherlands. Depolymerization and chemolysis provide greater climate change benefit (less net CO_2 emissions) than thermochemical recycling but require cleaner/pure feedstock. The benefits are similar to mechanical recycling. Gasification and pyrolysis are more flexible in their feedstocks but have lower climate change benefits (about 50% of mechanical recycling).

One challenge is that the generation of the waste may not be near where the chemical industry infrastructure lies. For example, in the United States the population is concentrated on the east and west coasts, yet the chemical and polymer manufacturing infrastructure lies mostly on the Gulf coast. It would make sense to build large-scale facilities near the chemical plants that can use the output as feedstock. Yet that would require transportation of huge amounts of waste (or of chemicals if the facilities are built near the population centers).

The economics of chemical recycling will depend on the prevailing cost of fossil-derived chemicals and government regulation. The projected cost of naphtha (a feedstock for PE) from chemical recycling was higher than naphtha derived from oil in 2018, except in China where there is existing gas infrastructure and large-scale facilities [98]. One trend in chemical recycling's favor is the rising cost of virgin plastic packaging due to regulations in the European Union and elsewhere. EPR fees and new taxes on packaging make recycling more attractive. Policies that promote chemical recycling, however, may hinder mechanical and advanced physical recycling which may have a lower environmental impact. Chemical recycling processes that preserve the chemical structure (depolymerization and chemolysis) offer the highest benefit. Pyrolysis and gasification, even though they have less benefit, should be considered for where other processes may not be feasible (such as mixed plastics and multilayer, multimaterial flexible packaging waste) due to their benefits over incineration and landfill [94].

17.3.4 Energy recovery

Energy recovery follows recycling in waste management hierarchy [45]. Some 90% of the embedded energy potential in plastics can be recovered. This makes good economic sense and avoids landfilling and leakage to the environment. Incineration is not considered the ideal option by environmentalists since it deviates from the circular economy. Given that most plastics are fossil based, waste to energy recovery through incineration is a source of nonrenewable carbon entering the atmosphere, primarily as CO_2. There are also toxic byproducts of incineration, such as NOx gases and dioxins. Advanced pollution controls are required in regulated economies such as Europe. Countries and regions with the infrastructure, such as Japan and Europe, use this option extensively for flexible packaging waste.

US EPA statistics show of the 35.7 million tons of plastics municipal solid waste generated in 2018, 5.65 million tons were combusted with energy recovery, or 15.8%. Only 8.5% was recycled and 75.7% ended up in landfill [99]. In Europe, 42.6% of postconsumer waste was incinerated with energy recovery, with 32.5% recycled and 24.9% landfilled [32]. Japan uses energy recovery for 74% of its municipal waste [100].

World Bank data shows plastics represent 12% of global solid waste and globally 11.2% of solid waste is treated by incineration and 13.5% is recycled. The amount varies considerably by region, as shown in Table 17.12.

17.4 Packaging design for sustainability

As previously discussed, there are many aspects to sustainability, including minimizing environmental footprint (e.g., GHG emissions, nonrenewable energy consumption, and water usage), reducing product waste (especially food waste) and enabling a more circular economy. Due to its light weight and efficient use of materials, flexible packaging generally has a lower environmental footprint than other forms of packaging. The scientific principles described elsewhere in this book around sealing, abuse resistance, and barrier are useful for minimizing material usage as well as designing

TABLE 17.12 Statistics on regional solid waste treatment [101].

Region	% Recycle	% Incineration	% Landfill and open dump
East Asia & Pacific	9	24	46
Europe & Central Asia	20	17.8	20
Latin America & Caribbean	4.5	<1	43
Middle East & East Africa	9	<1	76
North America	33.3	12	54.3
South Asia	5	<1	79
Sub-Saharan Africa	6.6	<1	92

Source: Data from S. Kaza, L. Yao, P. Bhada-Tata, F. Van Woerden, What a waste 2.0, A global snapshot of solid waste, Management to 2050, World Bank Group, viewed online on 03 Jan 2021 at https://openknowledge.worldbank.org/handle/10986/30317, 2018.

packaging that reduces food waste. These attributes are often most efficiently achieved by layering different materials using adhesives or tie resins. The use of diverse materials in flexible packaging makes recycling more challenging. In this section, designing flexible packaging for enabling sustainable end-of-life options is examined. For many products with high economic value or environmental impact, such as meat, cheese, grains and processed foods, product protection should not be compromised. As described in Section 17.2.2, packaging for products with a high product to package environmental impact ratio should focus on minimizing waste.

Three aspects of design of packaging for sustainability are addressed in this section: the use of mechanically recycled plastics in flexible packaging, designing packaging for ease of mechanical recycling, and design considerations for composting. Design for advanced physical recycling (dissolution/precipitation and delamination technologies) is not covered here; these technologies offer potential for recycling well-defined multimaterial packaging structures, as described in Section 17.3.2.2. Advances in identification and sorting technology may be required to apply these to PCR. Design for chemical recycling is also not specifically described since feedstock recycling technology is still being developed. Some types of chemical recycling have the possibility of handling mixed plastics streams that would enable many multilayer flexible packages to be recycled. Indeed, this may be the best option for some types of packaging. Other feedstock recycling technologies may benefit from using pure material feed streams, requiring the use of delamination and other technologies.

Features for package design that help minimize food waste are described in Section 17.2.2. These include traditional protection properties (barrier and abuse resistance) but also smaller sized packaging to "right-size" the contents to the user consumption patterns, recloseable technology (such as zippers) to reduce spoilage after the package is opened, and proper labeling around shelf life. Often the shelf life requirements are estimated rather than rigorously tested. Underestimation of the true shelf life requirements and not designing enough barrier performance into the package can lead to waste.

17.4.1 Incorporation of postindustrial recycle/postconsumer recycle into flexible packaging

Many consumer goods companies have set aggressive recycle content goals for their packaging. These often include objectives for PCR content in the package and for the package to be recyclable. PCR incorporation has initially been focused on rigid packaging where a separate layer can often be utilized for PCR incorporation without detracting too much from the function or esthetics of the package. Quality needs are often more demanding for flexible packaging, making PCR addition more challenging. Increasingly, consumer goods companies and their suppliers are finding opportunities. For example, Kimberly Clark successfully incorporated up to 20% HDPE PCR into overwrap and bundling films by screening several sources of PCR and engineering the film structure to ensure quality and esthetics are not compromised [102].

The supply of recycle for film applications is currently not enough to meet demand. Most plastics PCR today is collected from rigid packaging: PET bottles and HDPE containers. These materials are not optimized for flexible packaging. As demand increases, other sources of supply will be developed, such as use of back-of-store secondary packaging films or the store drop-off recycle stream.

The most straightforward way to incorporate plastics PCR in flexible packaging is to use virgin materials sourced from feedstock recycling of postconsumer plastic waste. Converting the waste back to its constituent molecules and using chemical refining processes allows the feedstock to be essentially identical to feedstock derived from petrochemicals or other virgin feedstocks. Using mechanically recycled plastic waste has several technical and regulatory challenges described below.

17.4.1.1 Mitigating loss of physical properties of mechanically recycled resins

One key challenge with incorporating plastic PCR or PIR into films is loss of physical properties such as impact resistance and tear strength. Two mechanisms are known to be involved: thermal or oxidative degradation with each subsequent heat history during film converting and incompatibility of mixed plastic wastes.

Several mechanisms have been proposed for the thermal degradation of polyolefins, including chain scission, chain branching (recombination of scissioned components) and crosslinking. These reactions arise from the creation of free radicals through thermal and shear energy imparted during extrusion, presence of residual catalysts, and other factors. The formation of radicals may be accelerated through oxidative reactions to create more radicals. PP degrades primarily through thermally induced chain scission whereas scission, branching and crosslinking occur in PE. Gonzalez-Gonzalez et al. [103] run PP through 20 extrusion cycles, measuring number average molecular weight (Mn), weight average molecular weight (Mw), melt flow rate (MFR) and carbonyl index (by FTIR). The carbonyl index is a measure of amount of oxidative species that form. Extrusion temperature has a strong effect on the flow properties, as shown for MFR in Fig. 17.7. Below 230°C little change in MFR occurs. Between 230°C and 250°C the MFR changes rapidly with extrusion cycle and above 250°C the changes are less so. Similar trends are found for the reduction in Mn and Mw. They find very little formation of carbonyl groups indicating the degradation is occuring primarily through thermally induced chain scission processes.

The degradation of PE involves competition between chain scission and branching or crosslinking mechanisms that often result in complex changes in mechanical and flow properties with extrusion cycles. Oblak et al. [104] subject HDPE (7.5 g/10 min MFI [190°C/2160 g], 0.965 g/cc density) to 100 extrusion cycles in a twin-screw extruder at 240°C. The MFI drops to about 0.1 g/10 minutes after about 30 cycles (Fig. 17.8) and levels off. Dynamic viscosity and elastic modulus (G′) increase substantially between 5 and 20 cycles. Pressure drop in the extrusion process also increases to about 500% of the virgin resin over the first 30 cycles. These results suggest branching and crosslinking mechanisms are predominant in the early cycles, but chain scission becomes more important with later heat histories. Hardness and tensile modulus start decreasing only after about 10 cycles and crystallinity does not change until after 20 cycles (and then decreases). Even after 100 cycles, however, 80% of the mechanical properties are perserved. Pattanakul et al. [105] find significant changes in elongation at break for recycled HDPE milk bottles but conclude that the properties are still useful if blended at the appropriate level with virgin PE.

Santos et al. [106] test the properties of a 90% HDPE, 10% PP blend that simulates a typical PE PCR stream. They subject the blend to three extrusion passes at two processing temperature regimes: 150°C–180°C and 210°C–250°C. The MFI does not change much after three passes at the low temperature settings but decreases in the high temperature regime. The presence of the PP does not appear to have a signficant effect on the processing. The carbonyl index

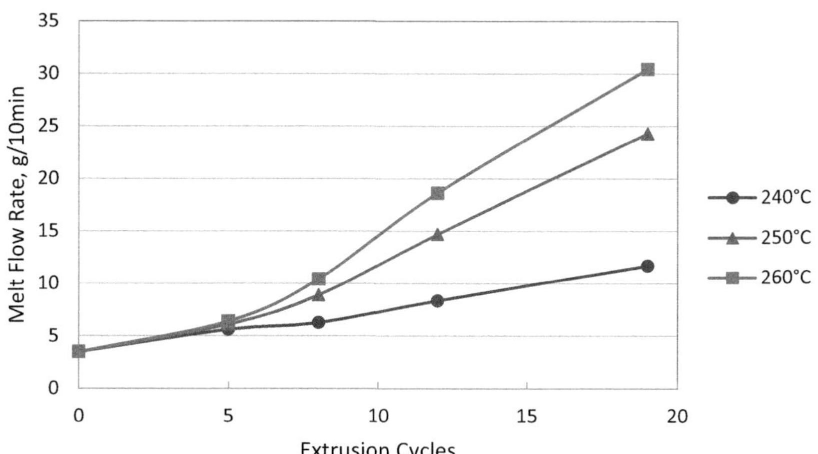

FIGURE 17.7 Effect of extrusion cycles on melt flow rate (230°C, 2160 g) of PP. Curves represent data for three different extrusion temperatures. *Data from V.A. González-González, G. Neira-Velázquez, J. L. Angulo-Sánchez, Polypropylene chain scissions and molecular weight changes in multiple extrusion, Polym. Degrad. Stab., 60 (1998) 33–42.*

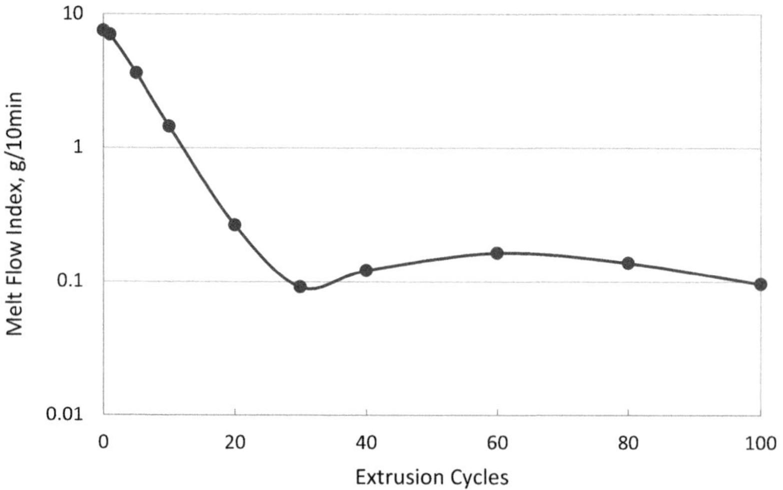

FIGURE 17.8 Effect of extrusion cycles on melt flow index (190°C, 2160 g) of HDPE. *Data from P. Oblak, J. Gonzalez-Gutierra, B. Zupancic, A. Aulova, I. Emri, Processibility and mechanical properties of extensively recycled high density polyethylene, Polym. Degrad. Stab., 114 (2015) 133–145.*

(FTIR) only increased at the high temperature conditions and the MFI could be recovered by adding additional antioxidant additives after the 1st extrusion cycle.

Abad et al. [107] study HDPE and LDPE recycling. The primary degradation mechanism for HDPE is found to be chain scission. For LDPE it is chain branching and crosslinking. FTIR shows an increase in carbonyl groups with each extrusion history. The tensile properties degrade with each pass, but thermal properties and crystallinity are retained for both materials. Adding primary phenolic and secondary phoshite antioxidants are found to be effective for retaining tensile properties of LDPE but not for HDPE.

Dostal et al. [108] subject LDPE to 20 extrusion cycles (single-screw extruder, 30 rpm, 170°C). Degradation is found to proceed by chain scission in the early stages (first 1–2 cycles) followed by recombination and formation of crosslinking and microgel (3–20 cycles). MW results are shown in Fig. 17.9 where the recombination effects are most pronounced in the weight averaged MW. Jin et al. [109] run LDPE through 100 cycles on a twin-screw extruder at 240°C. They find evidence for simultaneous scission and crosslinking of the polymer chains. No change in crystallinity and melting behavior is observed but thermal degradation and gelation occur after several cycles.

The conditions the polymer sees in its service life also impact its stability during extrusion processing. Luzuriaga et al. [110] simulate the effect of thermal or UV exposure on several polymers prior to melt processing by aging samples in an oven at 100°C or in an weatherometer, respectively. LDPE and high-impact polystyrene (HIPS) are more susceptible to thermoxidation than HDPE and PP, whereas HDPE and PP are more susceptible to UV exposure. The PP sample undergoes chain scission during oven aging, but scission is even faster with UV exposure. LDPE chain scissions during thermal exposure, but crosslinks with UV. LDPE can be restabilized with antoxidant additives. For HDPE, the investigators find that scission and crosslinking are competing against one another for both the thermal and UV exposure situations.

As suggested by these studies, the effects of repeated thermal processing depends on the polymer type (and additives that may be present initially), end-use environment and the thermal process. Since many of these factors are unknown for PCR streams, the performance of PCR in subsequent film structures can be variable. Some of the negative effects of repeated thermal processing can be mitigated by incorporating antioxidants. Also, flow and mechanical properties can sometimes be offset by blending in virgin resin. Fig. 17.10 shows an example of recovering impact strength of recycled PP by blending in virgin PP.

The second mechanism for loss of physical properties of PCR is contamination with other plastics. A straightforward example is the effect of mixed polyethyelene types in an all-PE waste stream. The physical properties of PE are highly influenced by the density, comonomer type and amount of long chain branching. These in turn influence the amount of crystallinity, crystalline morphology and tie chain density which influence stiffness, tear strength and toughness (see Chapter 9 for more detail). PCR of unknown origin may vary in its composition and properties. For example, LDPE is often blended with LLDPE to improve bubble stability during blown film converting. Adding LDPE reduces the tear strength and dart impact of the LLDPE film. There are no simple analytical techniques to measure the amount of LDPE in a PCR sample to predict film properties.

Even after sorting, many PCR waste streams are not pure. (One exception is postindustrial recycle where the waste is from a known and controlled source). There will likely be some amount of other plastics mixed in with the waste

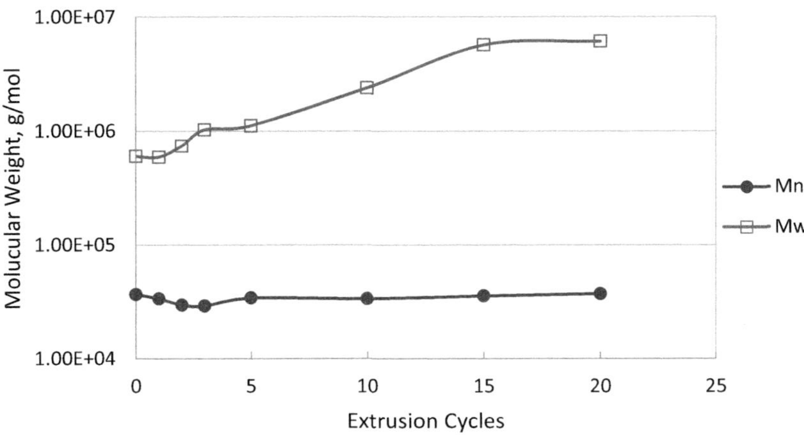

FIGURE 17.9 Effect of extrusion cycles on Mn and Mw of LDPE. *Data from J. Dostál, V. Kašpárková, M. Zatloukal, J. Muras, L. Šimek, Influence of the repeated extrusion on the degradation of polyethylene. Structural changes in low density polyethylene, Eur. Polym. J., 44 (2008), 2652–2658.*

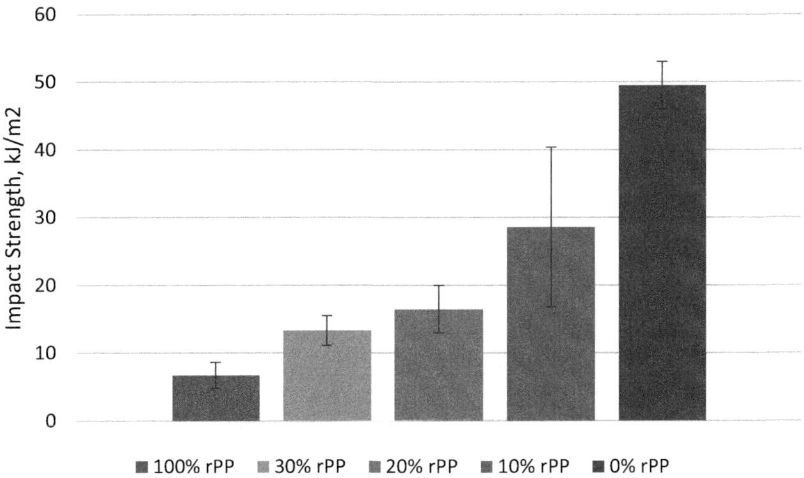

FIGURE 17.10 Effect of recycle PP on impact resistance (Izod) of blends of virgin and recycled PP (rPP) [111]. *Data from L.G. Barbosa, M. Piaia, G.H. Ceni, Analysis of impact and tensile properties of recycled polypropylene, Int. J. Mater. Eng., 7 (2017) 117–120.*

stream. For all-PE streams, generally a small amount of PP is present. Similarly, for all-PP streams, some PE may be present. Even though these are both polyolefins, they are not miscible and will form separate phases when blended. These phases can increase haze, reduce physical performance and in some cases act like defects in a film. Adding virgin resin can often help mitigate these effects but may be limiting. Another approach is to add a compabilitizer, as discussed in Section 17.3.2.1. For example, Zeng, et al. [81] study the effect of various ethylene- and propylene-based copolymers as compatibilizers for LLDPE and LDPE contaminated with homopolymer PP. Model blends with 20% homopolymer PP and 7% compatibilizer (and controls) are made using resins shown in Table 17.13. Seven compatibilizers described in Table 17.14 were screened using the experimental plan summarized in Table 17.15. The compatibilizers range in chemistry from polar ethylene copolymers (ethylene vinyl acetate (EVA) and ethylene methyl acrylate (EMA)), ethylene and propylene-based elastomers, and olefin block copolymers. All are expected to be nonreactive since there are no functional groups on the PE and PP to bond to. The blends are fabricated into 50-μm (2-mil) thick monolayer films on a 60-mm (2.375-inch) diameter blown film line (2.5 BUR, 2-mm (80-mil) die gap). Haze, dart impact (method A), machine direction (MD) Elmendorf tear strength and secant modulus are measured.

Results from the LLDPE film study are shown in Figs. 17.11–17.13. The presence of hPP-1 does not decrease the dart drop resistance of the LLDPE film, but compatibilizers D and E are able to substantially increase it. These are ethylene-octene based copolymers. Compatiblizers E and F (PP-based elastomers) are detrimental. The Elmendorf tear in the machine direction decreases by almost 40% in the presence of 20% hPP-1. The E/O based elastomers (D and E) successfully recover about half of the loss in tear resistance (20% less than the control). The two propylene-based elastomers do not contribute to the recovery of the tear resistance. The addition of hPP-1 to LLDPE increases haze; only compatibilizer C (PP-E block copolymer) brings it back down to that of the LLDPE control. Adding PP to LLDPE substantially increases the modulus with the compatibilizers having only marginal influence (not shown).

TABLE 17.13 Materials in model systems in PE recycle compatibilization study.

Designation	Melt flow rate (g/10 min)	Density (g/cm^3)	Commercial grade	Supplier
LLDPE-1	1[a]	0.920	DOWLEX 2045 G	Dow
LDPE-1	1.85[a]	0.920	AGILITY 1021	Dow
hPP-1	2[b]	0.900	H110–02N	Braskem

[a] *190°C, 2.16 kg.*
[b] *230°C, 2.16 kg.*
Source: Reproduced from Y. Zeng, D. Abebe, A. Singh, L. Effler, Y. Hu, M. Kapur, et al., Compatibilizing polypropylene (PP) contaminated polyethylene (PE) waste streams in film applications, SPE Int. Polyolefins Conf. (And 2021 ANTEC), 2020, with permission from The Dow Chemical Company.

TABLE 17.14 Compatibilizers for PE-PP study.

Compatibilizer	MFR (g/10 min)	Density (g/cm^3)	Comment
A	6.0[a]	0.950	Ethylene vinyl acetate (VA) copolymer, 28 wt% of VA
B	4.0[a]	0.926	Ethylene butyl-acrylate (BA) copolymer, 27 wt% of BA
C	6.5[b]	0.879	Ethylene/propylene (E/P) – iPP diblock copolymer
D	1.0[a]	0.857	Ethylene/octene (E/O) random copolymer
E	1.0[a]	0.866	E/O olefin diblock copolymer (OBC)
F	2.0[b]	0.867	Propylene/Ethylene (P/E) random copolymer
G	2.0[b]	0.876	P/E random copolymer

[a] *190°C, 2.16 kg.*
[b] *230°C, 2.16 kg.*
Source: Reproduced from Y. Zeng, D. Abebe, A. Singh, L. Effler, Y. Hu, M. Kapur, et al., Compatibilizing polypropylene (PP) contaminated polyethylene (PE) waste streams in film applications, SPE Int. Polyolefins Conf. (And 2021 ANTEC), 2020, with permission from The Dow Chemical Company.

TABLE 17.15 Design of PE-PP compatibilization study.

Model composition	Film #	Film description
LLDPE-rich Formulation	#1	100% LLDPE-1
	#2	Control: 80% LLDPE-1 + 20% hPP-1
	#3 to #9	73% LLDPE-1 + 20% hPP-1 + 7% Compatibilizer A to G, respectively
LDPE-rich Formulation	#10	100% LDPE-1
	#11	Control: 80% LDPE-1 + 20% hPP-1
	#12 to #18	73% LDPE-1 + 20% hPP-1 + 7% Compatibilizer A to G, respectively

Source: Reproduced from Y. Zeng, D. Abebe, A. Singh, L. Effler, Y. Hu, M. Kapur, et al., Compatibilizing polypropylene (PP) contaminated polyethylene (PE) waste streams in film applications, SPE Int. Polyolefins Conf. (And 2021 ANTEC), 2020, with permission from The Dow Chemical Company.

The effect of PP contamination in LDPE is more pronounced, as can be seen in Figs. 17.14–17.16. As with the LLDPE films, the ethylene-octene copolymers (D and E) have the greatest impact on improving the dart impact resistance. Elmendorf tear in the MD is significantly reduced by adding hPP and none of the compatibilizers bring it back to close to that of virgin LDPE (which is much lower than that of LLDPE). Compatibilizer D, the ethylene-octene random copolymer elastomer, performs somewhat better than the others. LDPE has much lower haze than LLDPE and the effect

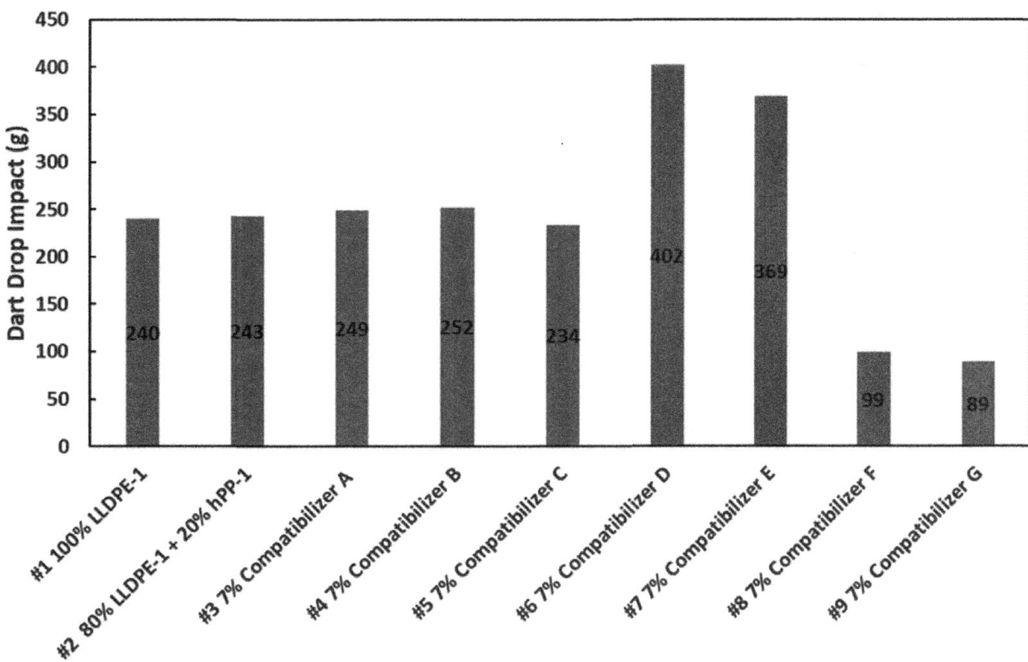

FIGURE 17.11 Effect of compatibilizers on the dart drop impact (Method A) in 20% hPP contaminated LLDPE film. *Reproduced from Y. Zeng, D. Abebe, A. Singh, L. Effler, Y. Hu, M. Kapur, et al., Compatibilizing polypropylene (PP) contaminated polyethylene (PE) waste streams in film applications, SPE Int. Polyolefins Conf. (And 2021 ANTEC), 2020, with permission from The Dow Chemical Company.*

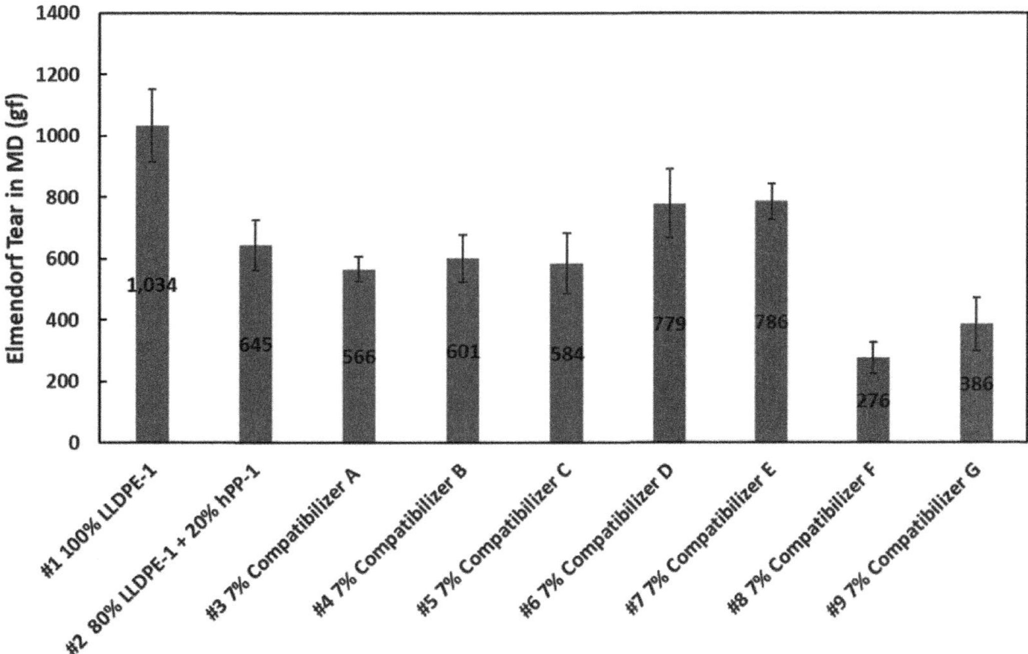

FIGURE 17.12 Effect of compatibilizers on the Elmendorf tear in MD direction in 20% hPP contaminated LLDPE film. *Reproduced from Y. Zeng, D. Abebe, A. Singh, L. Effler, Y. Hu, M. Kapur, et al., Compatibilizing polypropylene (PP) contaminated polyethylene (PE) waste streams in film applications, SPE Int. Polyolefins Conf. (And 2021 ANTEC), 2020, with permission from The Dow Chemical Company.*

of PP contamination is greater. All the contaminated LDPE films have substantially higher haze than the LDPE control with compatibilizer C (PP-E block copolymer) performing better than the others. As with the LLDPE film, adding PP increases the modulus of the film with the compatibilizers having little effect (not shown).

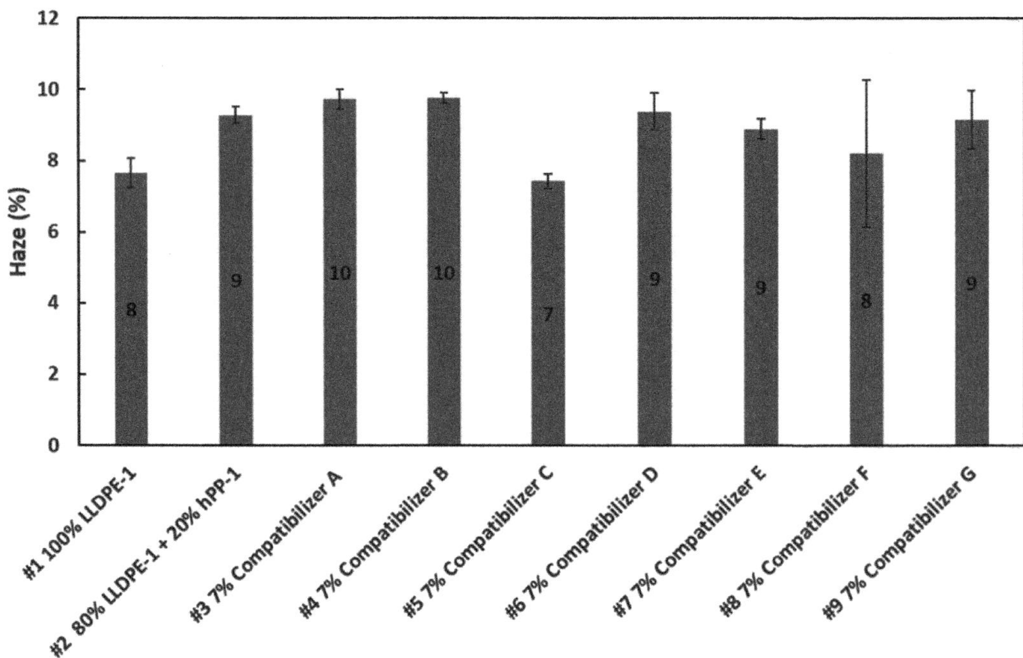

FIGURE 17.13 Effect of compatibilizers on the haze in 20% hPP contaminated LLDPE film. *Reproduced from Y. Zeng, D. Abebe, A. Singh, L. Effler, Y. Hu, M. Kapur, et al., Compatibilizing polypropylene (PP) contaminated polyethylene (PE) waste streams in film applications, SPE Int. Polyolefins Conf. (And 2021 ANTEC), 2020, with permission from The Dow Chemical Company.*

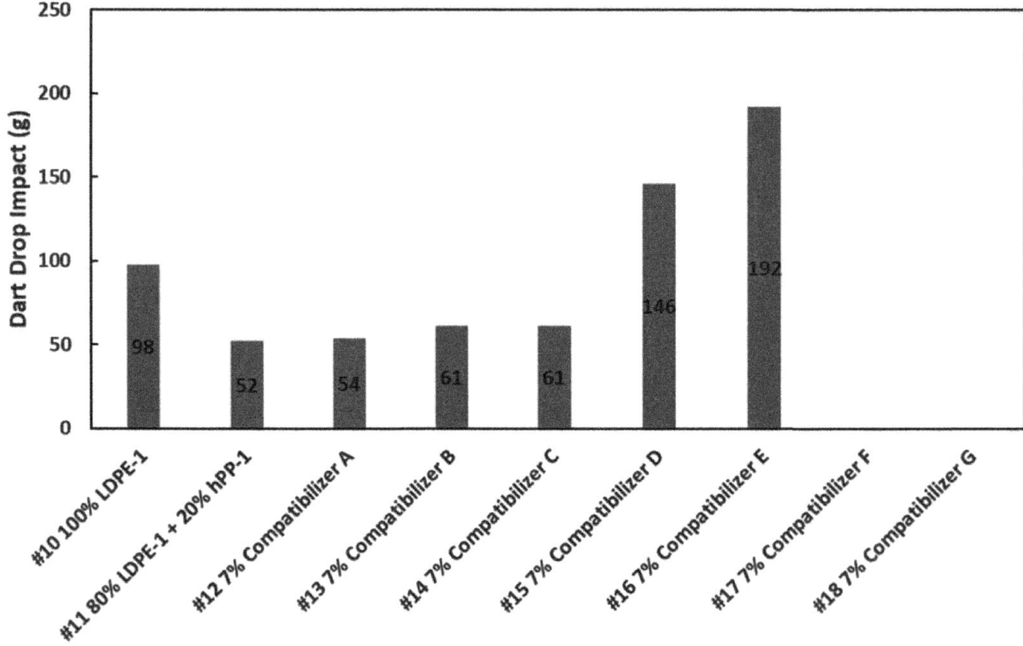

FIGURE 17.14 Effect of compatibilizers on the dart drop impact (Method A) in 20% hPP contaminated LDPE film. *Reproduced from Y. Zeng, D. Abebe, A. Singh, L. Effler, Y. Hu, M. Kapur, et al., Compatibilizing polypropylene (PP) contaminated polyethylene (PE) waste streams in film applications, SPE Int. Polyolefins Conf. (And 2021 ANTEC), 2020, with permission from The Dow Chemical Company.*

These results show that compatibilizers are one approach to mitigate polymer contamination effects. For the PE contaminated with PP, ethylene-octene copolymers are found to be most effective at recovering the toughness of the films whereas an olefin diblock copolymer (PP-P/E blocks) provides the best optical properties. The authors use transmission electron microscopy to understand these results. Both the diblock copolymer and the ethylene-octene copolymers are

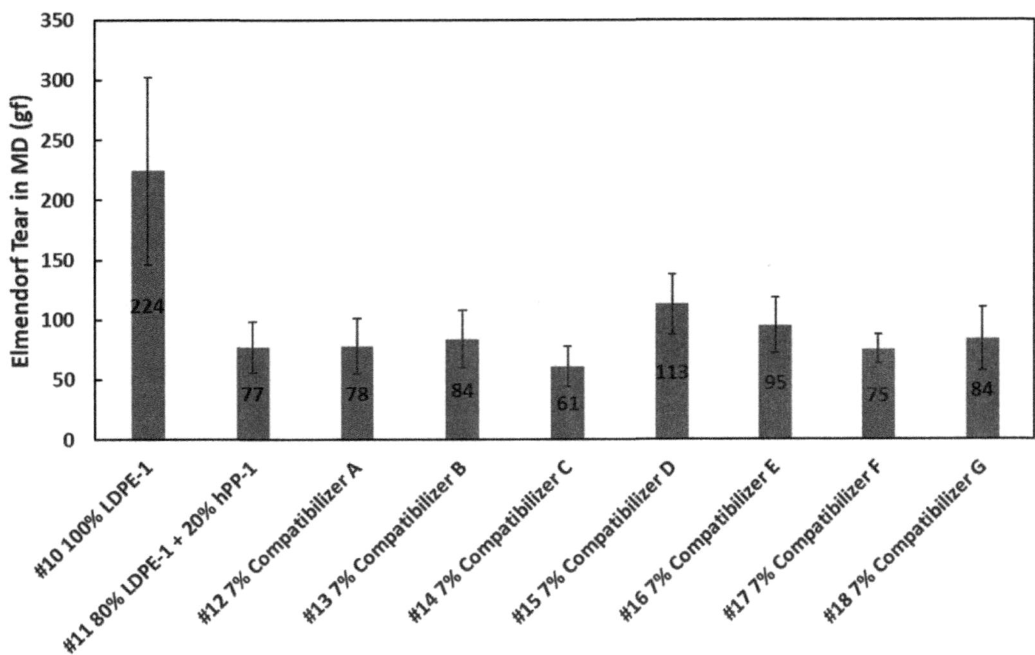

FIGURE 17.15 Effect of compatibilizers on the Elmendorf tear in MD direction in 20% hPP contaminated LDPE film. *Reproduced from Y. Zeng, D. Abebe, A. Singh, L. Effler, Y. Hu, M. Kapur, et al., Compatibilizing polypropylene (PP) contaminated polyethylene (PE) waste streams in film applications, SPE Int. Polyolefins Conf. (And 2021 ANTEC), 2020, with permission from The Dow Chemical Company.*

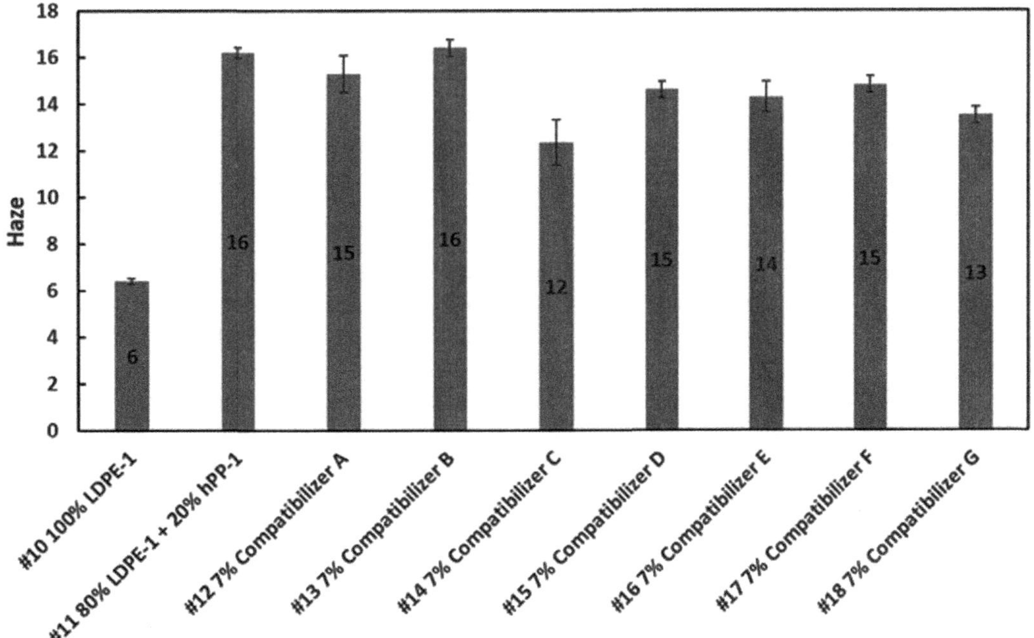

FIGURE 17.16 Effect of compatibilizers on the haze in 20% hPP contaminated LDPE film. *Reproduced from Y. Zeng, D. Abebe, A. Singh, L. Effler, Y. Hu, M. Kapur, et al., Compatibilizing polypropylene (PP) contaminated polyethylene (PE) waste streams in film applications, SPE Int. Polyolefins Conf. (And 2021 ANTEC), 2020, with permission from The Dow Chemical Company.*

found to reduce the domain size of the PP phase, but the ethylene-octene copolymer also toughens the PE matrix. The PP-based elastomers swell the PP phase and coarsen their domain size.

Other polymers may also contaminate the recycle stream. In packaging structures, these may include PET, EVOH, PA and other commonly used resins. Various reactive and nonreactive compatiblizers have been considered to deal with these contaminants, as described in Section 17.3.2.1 and Table 17.10. In addition to being incompatible, they may have substantially higher melting points than the matrix polyolefin resin. For example, PET melts at around 280°C, substantially above the typical processing temperature for PE (around 200°C). The PET may not melt and appear as small particles or defects in a reprocessed film.

Some polymer contaminants in the recycle may also not have good thermal stability and degrade during subsequent reprocessing. Chlorine containing polymers, such as polyvinyl chloride (PVC) or PVDC, create byproducts that are corrosive to extrusion equipment. EVOH is more prone to thermal degradation than polyolefins. Early versions were particularly problematic but suppliers have introduced more stable versions over time. EVOH has been successfully used in PIR for many years, especially in rigid barrier packaging. In some formats, such as blow molded bottles and thermoformed cups, the scrap rate approaches 50% of production. This is reground and fed back into the packaging structure as a separate layer. Virgin polymer, additives and compatibilizers are sometimes added to help enhance the quality of this stream [112]. Dow has introduced a compatibilizer technology that can be added to the package structure so that the film is ready to be recycled without the addition of compatiblizers in the recycle operation (see Section 17.4.2.2.2).

The incorporation of recycled paper fiber into paper or paperboard decreases its strength. Every time paper is recycled, the fibers are shorted by the repulping process. Cell walls and fibrils needed as bonding sites collapse during drying. This reduces bonding and flexibility of the fibers. Additives may be used, such as starches or dry strength resins, to compensate. Blending in virgin fibers also improves the strength. Overall, applications where high strength is needed such as packages bearing considerable weight are not suited for high recycle content paper [15].

17.4.1.2 Other quality factors

Incorporating PCR into packaging films not only affects mechanical performance but also may impact processing, esthetics and organoleptics. Variable composition and changing molecular weight from degradation alters the flow properties of the polymer. As seen in Figs. 17.7 and 17.8, the melt flow rate can change considerably with subsequent heat histories. Chain scission may reduce melt strength and contribute to bubble instability in blown film processes or increased neck-in in cast film and extrusion coating. This affects productivity and consistency of the product (e.g., bubble instability can affect gauge control).

Devolatilization during extrusion is one of the most cited equipment needs for handling PCR according to a major PCR broker [58]. Off-gasing from water and other volatile components or from degradation byproducts can create voids in the film or coating if proper drying or devolatilization is not used.

Film appearance can be degraded through color formation, haze and defects. Off-color may arise from contaminants such as inks from printed substrates, food remnants or degraded polymer. Incompatible polymers or inorganic additives (such as mineral fillers and pigments) can contribute to haze and opacity. Additives such as antioxidants are known to cause yellowing. Film defects may be created by gel from degraded (cross-linked) polymer, black specs from charred polymer, undispersed mineral fillers, and other solid contaminants (such as paper labels, cross-linked adhesives, and foreign particles such as wood scraps). Not only can these be unpleasing visually, they can contribute to film breaks on the line and reduce productivity. Recycle incorporation may also produce bumps in the film or surface roughness that affects print quality [15].

Odor is one of the most challenging issues to determine the root cause and address. It may arise from a number of sources including residual organic waste, polymer degradation or interaction between various additives.

These quality challenges can be difficult to mitigate. Good sorting and cleaning of the waste is critical. Washing, deinking, layer separation, dissolution of adhesives, melt filtration, melt devolatilization are some of the technologies used to address these issues. The cleaner and purer the feedstock, the easier it is to incorporate it into a flexible packaging structure. Various additives may also help mitigate problems. Examples include odor scavengers, blue pigments to mask yellowness, stabilizers such as antioxidants, lubricants and process aids, and compatibilizers.

17.4.1.3 Regulatory considerations

The regulatory hurtles for using mechanical recycled plastics in food packaging are described in Section 17.1.5. Currently, outside of a few specific PET bottle streams, the use of mechanically recycled PCR plastics in food packaging has not gained general regulatory compliance. This often limits the use of PCR to secondary packaging, such as stretch wrap, collation films and the like, or for nonfood applications. Most regulatory bodies do allow the use of PCR incorporated into a layer of the package that is separated from the food by a functional barrier. However, the trend towards mono-material packaging for ease of recycling (see next section) may limit the use of functional barriers in the future.

17.4.2 Design for mechanical recycling

Many consumer goods companies have committed to using packaging that is recyclable. This involves designing the package so it can be collected, sorted and processed through the commercial recycling value chain, and have a marketable end-use. This often entails "simplifying" the package structure or incorporating features that allow it to be easily reprocessed and may involve material substitution. The goal is to replace multimaterial structures with mono-material ones or at least ones that are compatible with a PE, PP or PET recycle stream. For flexible packaging, much of the focus has been on creating structures that can be collected in the "all-PE" recycle stream at store drop-off locations. All-PP structures are also being considered.

17.4.2.1 "Simplification" of packaging structures poses special challenges

Several organizations around the world have developed guidelines for designing flexible packaging for mechanical recycling. In Europe these include CEFLEX, a consortium of European companies across the packaging and recycling value chain, Plastics Recyclers Europe (PRE) with their RecyClass platform and HTP-Cyclos in Germany. The latter two organizations provide certifications. RecyClass offers an online assessment tool [113]. In the United States the Association of Plastics Recycling (APR), a trade association for recyclers, has issued specific protocols and the Sustainable Packaging Coalition offers an online guide [15]. There are similar endeavors in other regions such as Latin America and Asia (e.g., Central Institute of Petrochemicals Engineering & Technology in India (CIPET), China Petroleum and Chemical Industry Federation (CPCIF), Packaging Technology Center in Brazil (CETEA)).

CEFLEX, Recyclass, and APR guidelines are summarized here. Their methodology and rules differ but have many similarities. Methodologies are described in Table 17.16. The CEFLEX guidelines are more comprehensive and focus primarily on polyolefin-based flexible packaging but also include paper and foil. APR covers all plastic packaging but has specific guidelines for PE-based films. Recyclass have initiated protocols on LDPE and PP films, HDPE and PP

TABLE 17.16 Methodology for CEFLEX [13], RecyClass [114], and APR [12] guidelines on packaging design for recycle.

CEFLEX: preferred order for packaging design	RecyClass: six classifications, only A-C considered recyclable	APR: packaging features placed into four categories
1. Flexible PE mono-material packaging. This recycle stream is already collected (e.g., store drop-off of grocery bags and an increasing number of other films with the store drop-off label) 2. Flexible PP mono-material packaging. This is not a collected recycle stream today but CEFLEX anticipates it will be soon 3. Mixed polyolefin (PO) packaging. There is some organized collection of this in Europe in specific countries such as Germany 4. Mixed plastics including flexibles. This should only be used if others are not available and for the highest value products	• Class A: No recyclability issues and the recycled plastic can potentially be used in a closed-loop system for use in the same quality application • Class B: Minor recyclability issues that slightly affect quality of recycle. Majority can still be used in a closed-loop system • Class C: Some recyclability issues that affect the quality of recycle or lead to material loss during recycling. For the former, the recycle may be suitable for a cascade closed-loop system. For the latter, the plastic could potentially feed a closed-loop scheme • Class D: Significant design issues that highly affect recyclablility or result in large material loss. The recycled plastic can only be fed into low-value applications • Class E: Major design issues that jeopardize its recyclability or effect large material loss. The packaging is not considered recyclable and suitable only for incineration with energy recovery • Class F: Not considered recyclable at all due to fundamental design issues or lack of specific infrastructure for collection, sorting and recycling	• Preferred. These are features or materials that enable the highest value end use for the recycled product • Detrimental. Packages with these features are considered recyclable with negative impact. Package designers are encouraged to redesign their packages to avoid these when possible • Renders package non-recyclable. Packages with these features are not recyclable • Unknown. These package designs must be tested. A testing protocol has been established for getting new packages approved

rigids and lately PS, and with this anticipates rules for not yet existing recycling streams (PP and PS), which are in the process of being established in Europe.

Protocols for testing new resins, additives, coatings, layers, adhesives or blends not already covered by the guidelines are available. For example, the APR protocol are has the following steps: [60]

- Obtain samples of a control film and the innovation film.
- Granulate and make 3 states: Pure control, 50/50 blend and pure innovation film.
- Extrude each and measure properties of extruded pellets.
- Make blown film from each and measure specified properties (mechanical properties, physical appearance, etc.). Properties of the 50/50 blend must be no more than a 25% change from the control. There are also conditions on processing (pressure change, etc.).

The Recyclass testing protocol [115] is similar except that an additional 75% control-25% innovation film blend is added to the pelletized states and during film blowing each of the pelletized states is blended with 50% virgin resin.

CEFLEX and RecyClass take a holistic approach and provide guidelines for many aspects of flexible packaging design, as shown in Table 17.17. These go beyond just materials and consider decorative features such as printing and labels and physical features such as density and closures. CEFLEX and RecyClass consider whether the packaging feature is compatible with PE or PP recycle streams. They advise against using packaging features/materials that are incompatible with these streams where possible. A summary of the CEFLEX, RecyClass and APR guidelines is provided in Table 17.18. These guidelines are updated periodically and are posted on the organizations' websites.

The transition from multimaterial flexible packaging structures to mono-material designs is challenging for many applications. As is described throughout this book, multilayer, multimaterial structures have evolved over many years of development to meet the demanding needs of packaging. These include protecting the product (e.g., abuse resistance, strength, barrier, hermetic sealing), providing convenience (e.g., easy-open, recloseability) and conveying information (e.g., printing, labels) in a cost-effective way. The demand for recyclability opens up new opportunities for package design that will surely lead to new innovation.

Many companies and organizations have gotten involved. The Sustainable Packaging Coalition's "How2Recycle" labeling program has helped inform the consumer what structures can be placed into the store drop-off collection bins. Resin suppliers have worked with film converters, brand owners and other supply chain partners to overcome obstacles. Many resin suppliers have developed new polyolefins with improved moisture barrier, heat seal or orientation performance directed at facilitating recyclable flexible packaging [10]. For example, NOVA and Dow have helped develop oxygen barrier structures that can be recyclable in the "all-PE" store drop-off stream [116–120]. One Dow technology involves an improved compatibilizer that allows EVOH film structures to be recycled (see Fig. 17.17). Dow, NOVA,

TABLE 17.17 CEFLEX [13] and RecyClass [14] design for recycle categories.

CEFLEX	RecyClass
Materials • Plastics • Barrier/coatings • Paper • Aluminum • Adhesives • Additivs & Fillers	Main body • Material • Material composition • Colors • Size • Product residues • Barrier • Additives
Decorative Features • Printing inks % lacquers • Pigments • Labels	Attachments • Closure systems • Liners, seals and valves • Other components
Physical Features • Size & shape • Density • Additional features	Decoration • Inks • Labels • Adhesives for labels • Direct printing

TABLE 17.18 Summary of CEFLEX, RecyClass and APR guidelines for designing polyolefin-based flexible packaging for recycle. Classifications for RecyClass are in Table 17.16.

Category	CEFLEX: polyolefin-based packaging	RecyClass: PE film	APR: PE-based packaging
Materials			
Thresholds for mono or laminates with barrier layers or coatings (excluding non-compatible polymers and materials in subsequent sections)	Compatible: >90% PE or PP (or PO for mixed PO stream) Not Compatible: <80% PE or PP Limited compatibility or not compatible structures should be retained if there would be a negative impact on product protection (such as more food waste, which would have an overall negative environmental impact)	Class A-B: PE Class B-C: Multilayer PE/PP with <5% PP Class D-F: Multilayer PE/PP with >5% PP; any other polymer (PET, PVC, etc.) PE content designations: • Class A: > 95% PE • Class B: > 90% PE • Class C: > 70% PE • Class D: > 50% PE • Class E: > 30% PE • Class F: < 30% PE	Preferred: Mono-material preferred (HDPE, LDPE, LLDPE, VLDPE, plastomers). EVA at <5% of total package weight. Ionomers at <20% of package weight. PB-1 at less than 5% of package weight. Blends, coex or laminates with < 80% PE or copolymers require testing
Paper	Not compatible. Small amounts of paper cause defects in plastics recycle due to charring of the fibers. If paper is needed for the package, it should be the majority component so it can be sorted into the paper recycle stream	Not considered	Renders package non-recyclable
Aluminum foil	Not compatible. Aluminum can be separated out. CEFLEX suggests pyrolysis can be used to recover the aluminum	Class D-F: not compatible	Renders package non-recyclable
Barrier layers	Compatible: <5% of structure: AlOx, SiOx, EVOH, PVOH, acrylic; Laminated and printed metallized layers Limited compatibility: Any barrier layer > 5% of structure CEFLEX to study PVDC coatings and PA in 2nd phase study	Class A-B: barrier in polymer matrix; SiOx and AlOx without additional coatings Class B-C: < 5% EVOH; metallized layers without additional coatings; < 15% PA 6/66 with Tm < 192°C and with > = 10% PE-g-MAH tie layers Class D-F: > 5% EVOH; any other PA; PVC, PVDC; any other barrier layer; aluminum	Specific barrier layers are not listed and must be tested; notes that specific formulations have been approved with small amounts of EVOH or PA with compatibilizer. PVDC coatings render packaging non-recyclable
Metallized coatings	Compatible	Class B-C barrier (conditional-limited compatibility)	Recommends against using metallized layers since this often results in the whole package being separated out by metal detectors in the sorting process. Specific metallized films need to be tested
Adhesives	Compatible: < 5% of structure: Polyurethane, acrylic or natural rubber latex adhesives, as well as non-PE- or non-PP based-tie layers	Not specifically discussed except for labels (see below)	Label adhesives need to be tested—must wash off or be compatible with PE

(Continued)

TABLE 17.18 (Continued)

Category	CEFLEX: polyolefin-based packaging	RecyClass: PE film	APR: PE-based packaging
Additives and Fillers	Compatible: Additives and fillers used at minimal levels: thermal stabilizers, UV (ultraviolet) stabilizers, nucleating agents, mineral and polymer cavitating agents, antistatic agents, impact modifiers, chemical blowing agents and tackifiers Not compatible: Fillers in non-PE or non-PP structures that modify the density to be < 1 g/cc (density < 1 g/cc interferes with floatation sorting systems)	Class A-B: additives that do not increase density to >0.97 g/cc Class D-F: additives that increase density to >0.97 g/cc ($CaCO_3$, talc, glass, etc.); Bio/oxo/photodegradable additives	Common additives acceptable when used at minimal levels. Non-PE additives, coatings, layers not listed must be tested
Other polymers	Not compatible: PET layers, PVC layers, non-PE or PP foam layers, biodegradable and compostable materials	See Polyolefin-based packaging and Barrier categories	Non-PE layers not listed must be tested
Other additives	Not compatible: Oxy-degradability additives (not compatible due to negative affect on physical properties of recycle)	Class D-F: Bio/oxo/photodegradable additives	Does not recommend; must be tested
Other materials	Not compatible: substances of very high concern; foamed thermoplastic non-polyolefin elastomers	Class D-F: Metal, aluminum, PVC, PET, PETG, PS, PLA, paper, foams with density < 1 g/cm^3	Renders non-recyclable: PET, PVC, PVDC, paper, aluminum foil, degradable polymers
Recycle content	Preferred. The use of recycle to reduce the amount of virgin material is encouraged. CEFLEX notes that use of recycled materials from polyolefin-based packaging is not approved for food contact applications at this time		Preferred: use of PCR is encouraged up to the maximum amount technically or economically feasible
Decorative features			
Inks & Lacquers	Compatible: Lighter colors, <5% of structure (without PVC binders), surface printing Not compatible: Ink and lacquers with PVC binders	Class A-B: non-toxic (EIPIA guidelines) Class E-F: inks that bleed; toxic or hazardous inks	Direct printing, lightly printed surfaces preferred. Heavily printed surfaces are problematic
Pigments	Compatible: Clear, natural or pale colors Limited compatibility: black dark colors. If black is needed, NIR detectible pigments should be used Not compatible: carbon black (blocks NIR)	For transparent films: Class A-B: unpigmented, transparent Class B-C: light colors; translucent colors Class D-F: dark colors; black; carbon black	Similar to CEFLEX
Labels	Compatible: labels made of the same polymer as the package: mono PE or mono-	Class A-B: PE labels with adhesives that are water	PE labels preferred (for PE packaging). Paper labels are detrimental; metal foil labels

(Continued)

TABLE 17.18 (Continued)

Category	CEFLEX: polyolefin-based packaging	RecyClass: PE film	APR: PE-based packaging
	PP. Ensures correct sorting and high-quality recycle Limited compatibility: labels of different material with <30% of surface area that are easily removed Not compatible: labels not made of the main resin with > 30% of the surface area	soluble or water-releasable at < 60°C Class B-C: PP, paper without fiber loss labels Class D-F: Metallized labels, any other, paper labels with fiberloss; adhesives non-soluble in water or nonreleasable at <60°C	render the package non-recyclable
Physical features			
Size & shape	Material thickness: Minimal viable amount should be used Pack size: Above 20 mm x 20 mm is preferred. Less than 20x20 mm interferes with sorting. Product residue in packaging: the package should be designed to minimize product residue left in the package after use to improve quality and yield of recycle	Class A-B: >A4 or > 50x50 mm once compacted (for sorting) Class B-C: < A4 or between 20x20 mm and 50x50 mm once compacted Class D-F: < 20x20 mm	
Density	Compatible: < 1 g/cc Not compatible: > 1 g/cc (for floating in density separation systems)	Class A-C: < 0.97 g/cc Class D-F: > 0.97 g/cc	
Additional features (spouts, zippers, valves, taps, etc.)	Compatible: features made from same polymer as package Limited compatibility: features made of different material from package but easily separated	Class A-B: PE based Class B-C: PP, PP with Al removable lid Class D-F: Metal, aluminum, PVC, PET, PETG, PS, PLA, foiled paper, non PO or foams with density < 1 g/cm^3	Metal attachments render packaging non-recyclable; non-PE attachments must be tested

Source: Based on CEFLEX [13], RecyClass [14], and APR [12].

ExxonMobil, Sabic, and others have also announced improved PE resins for orientation, a key need for replacing oriented PET film in laminations. Tougher and stiffer grades of polyolefins are positioned to help replace PA in some flexible packaging structures. New barrier coatings are also on the horizon.

More advances will come as there are many challenges. An example is the development of biaxially oriented PE film (BOPE) as is described in Section 11.2.4.1. Biaxally oriented films are generally stronger, clearer and have better barrier properties than their less oriented counterparts. A tenter frame process is often used where the polymer is stretched sequentially in the machine and transverse directions by factors of 4x or more at temperatures below the melting point (see Section 2.4). Because of their increased strength and stiffness, oriented films allow package structures to be lightweighted. Also, many flexible packaging structures are laminates of oriented PET, PA or PP with a PE-sealant film. The oriented film provides a dimensionally stable surface for printing and barrier coatings (such as metallization), abuse resistance, some amount of barrier, and heat resistance during heat sealing. Replacing the oriented film with BOPE would allow reverse printed[1] flexible packaging structures made from all-PE, a key need for recyclability.

1. Reverse printing involves facing the printed surface of the base film towards the inside of the laminate so that the print surface is protected from abuse by the base film.

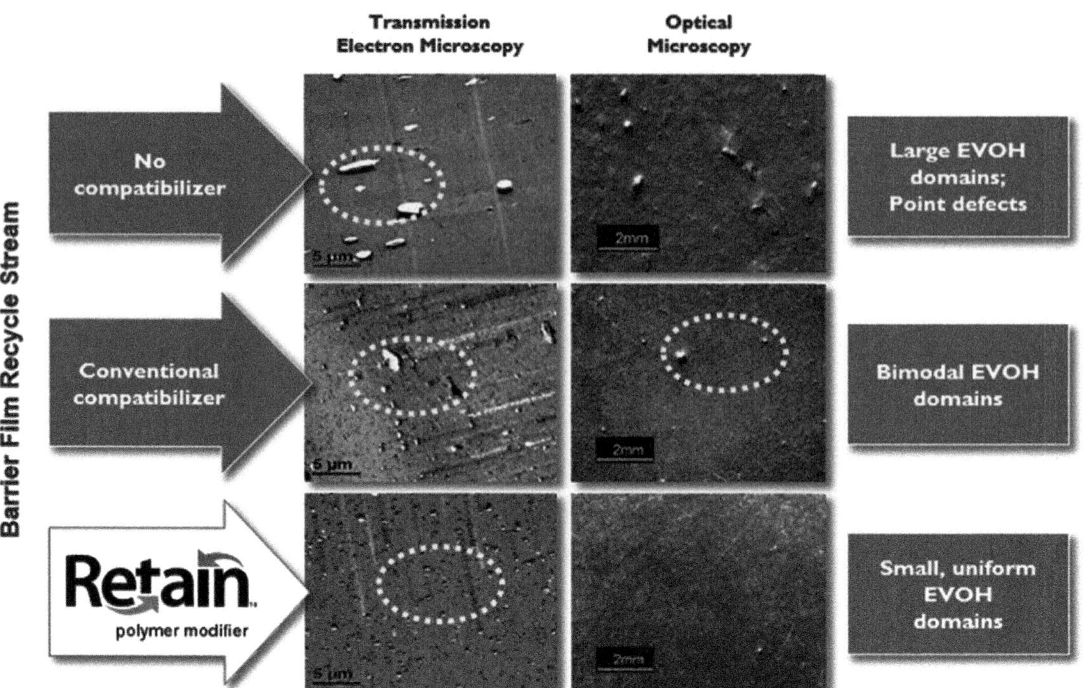

FIGURE 17.17 Effect of compatibilizer on dispersing EVOH in PE recycle stream. *Reproduced with permission from The Dow Chemical Company.*

Conventional PE, however, does not stretch well in commerical tenter frame biaxial orientation processes. Resin suppliers are responding with new PE grades for this purpose. Lin et al. [121] describe a new LLDPE resin for BOPE that overcomes the stretching limitations. They compare the properties of a biaxially oriented film with blown film of a PE with similar density and melt flow index. (The blown film has a thickness of 90 μm, 2.25 times thicker than the BOPE film.) The BOPE film (5x MD, 9x TD stretch ratios) has an 80% increase in MD modulus, 150% increase in TD modulus, 90% increase in dart impact resistance, and 130% increase in puncture resistance over the conventional blown film. The BOPE has 80% lower haze and 48% higher gloss. The BOPE has much lower tear strength in both the machine and transverse directions which may make it ideal for easy-tear (easy-open) packaging applications. To illustrate the performance of the BOPE film, the authors replace the oriented polyamide (OPA) layer in a 3-ply laminate for a liquid detergent pouch with BOPE: OPET/OPA/PE film replaced with OPET/BOPE/PE film. (OPET = biaxially oriented polyethylene terephthalate, OPP = biaxially oriented polypropylene.) Quality tests on the pouch show it passes drop tests.

One should note that many of the property advantages of biaxial orientation of PE are also at least partially achieved with machine direction orientation (MDO). Lawrence et al. [122] compare the properties a 3-layer LLDPE blown film before and after MDO. They report an increase dart impact strength and MD secant modulus, and a reduction in MD tear strength, oxygen and moisture vapor permeability with MDO orientation. Seal strength also increases but with substantially higher seal initiation temperature (which can be mitigated by using a sealant with lower crystallinity).

Despite these new developments and interesting properties, BOPE and MDO PE films are inherently less stiff and dimensionally stable at elevated temperatures than OPET or OPP. Lower stiffness affects film handling, film registration and dimensional stability during converting and packaging operations. The lower melting temperature of PE means lower heat seal temperatures must be used to avoid film distortion during sealing. Stiffness can be built back into the film structure by increasing the thickness of the BOPE layer, or by using the I-beam effect to position other layers into the structure for maximizing bending stiffness (see Section 9.4.2). Higher thickness not only increases material usage, which is counter to many sustainability goals, but slows the conductance of heat through the film during sealing and may slow packaging line speed. Sealants with lower heat seal initiation temperature (HSIT) are required to compensate for the slower heat transfer and the lower seal bar temperature (to avoid distorting the PE surface). Lower HSIT PEs often have greater challenges for preventing blocking and controlling coefficient of friction (COF). (See Section 12.1.) Another approach is to apply a heat resistant coating to the outside of the laminate that contacts the heat seal bar.

This example highlights how solving one issue may create new ones. This is further explored in the case study (17.4.2.2). The scientific principles found throughout this book can be applied to these new challenges to provide solutions.

Most high oxygen barrier flexible packaging involves combining a barrier material with polyolefins in a multilayer structure. Examples of barrier materials include aluminum foil, metallization, SiOx, AlOx, EVOH, PVDC and PA. Some of these materials can be incorporated into an "all-PE" recycle stream whereas others cannot. As described in Table 17.17, aluminum foil is not compatible with the PE stream but can be separated by eddy current techniques and recovered by pyrolysis, although the infrastructure for doing this is still in its infancy. Metallized films often get kicked out of sorting systems by metal detectors and organizations such as APR recommend avoiding metallization. Clear inorganic coatings, such as AlOx and SiOx, eliminate the metal detector issue. These coatings are very thin and so are expected to not be detrimental to the PE recycle stream. Besides their cost, there are still challenges with achieving high barrier performance with these coating technologies (see Section 4.3.2).

PVDC has poor thermal stability in extrusion processes and gives off corrosive byproducts when it degrades. Film structures containing PVDC are deemed unrecyclable by APR. Polyamide is incompatible with PE or PP but anhydride functionalized polyolefins can be used as compatibilizers. Detractors of allowing PA-containing structures in all-PE recycle streams cite its higher melting point (around 240°C for PA6) than typical PE extrusion temperatures. Polyamide copolymers, such as PA6/66 have lower melting points and may be more amenable to recycling.

EVOH has become one of the most compelling options for high barrier "all-PE" recycle-ready packaging. Although incompatible with PE, with tie resins and additional anhydride functionalized polyolefin compatibilizers, it has gained favor with recyclers such as APR. For example, Dow has introduced RETAIN polymer modifier for self-recycle packaging [118–120]. This low molecular weight compatibilizer is added to one or more layers in the multilayer barrier film containing EVOH. When the film is recycled, the compatibilizer is "ready" to do its job, dispersing the EVOH into the PE matrix. As shown in Fig. 17.17, standard anhydride modified polyolefins are higher in molecular weight and not as effective in dispersing the EVOH during recycle, causing haze and other performance deficits.

Alternative barrier coatings for plastic film or paper are being developed. Like inorganic coatings, these thin coatings are expected to meet recycle guidelines. Example technologies include acrylics, clay nanocomposites, nanofibrillated/nanocrystalline cellulose-based coatings, PVOH, polyglycolic acid, and whey proteins [10]. Moisture sensitivity, safety/regulatory compliance, coatability and cost have so far limited their adoption.

17.4.2.2 Case study: designing packaging for polyethylene recycle stream

To put the pieces together, an example of designing stand-up pouch (SUP) structures for the PE recycle stream is provided. This is based on design for recycle methodology developed at The Dow Chemical Company in collaboration with converters, brand owners, and other contributors to the value chain [27]. A real-world adoption is described at the end.

Designing for recycle involves understanding the requirements throughout the packaging and recycle value chains. The converter, fabricator, brand owner, retailer, consumer and recycler all must be considered. Without a collaborative effort the chance of success is slim. The process begins by understanding the package needs. Is this a fresh package design, or replacing an incumbent? What is the product being packaged? What is the package format (SUP, pillow pouch, sachet, etc.) and package machine requirements (film stiffness, COF, etc.)? What are the barrier and shelf life requirements (light, oxygen and moisture barrier)? How strong does the package need to be? Will there be recycle content? What is are the abuse resistance needs? What are the appearance requirements (transparency, gloss, etc.)? How will the package be distributed (traditional or e-commerce)?

The answers to these questions will dictate the package format and materials needs. The capabilities of the converters will also play a role. Some of the key considerations include the film fabrication process (blown, cast, machine direction orientation, extrusion coating) and downstream converting capabilities and requirements (e.g., printing, lamination, pre-made pouch making, zipper or fitment). From this, the package design starts to take shape, with consideration of individual layers and their functions: stiffness and abuse resistance, sealing, adhesion, oxygen barrier, moisture barrier, printing (surface print with clearcoat varnish or reverse print with lamination).

Finally, the means of recycling needs to be to be taken into account. How will the used package be collected (store drop-off, curbside, etc.), what sorting and cleaning technology will be used, will there be contamination (food waste, etc.) and how will it be reprocessed (extrusion, dissolution, chemical, energy recovery)?

For simplicity of discussion, two incumbent SUP formats are considered here: a low oxygen barrier option and a high oxygen barrier option. The evolution of the package design is illustrated in Fig. 17.18.

17.4.2.2.1 Incumbent structures for stand-up pouch

There are many SUP structures in use today, some with three-plies and some with two. A typical low oxygen barrier two-ply SUP has the structure: OPET/ink/adhesive/PE film. The OPET layer provides heat resistance during the heat

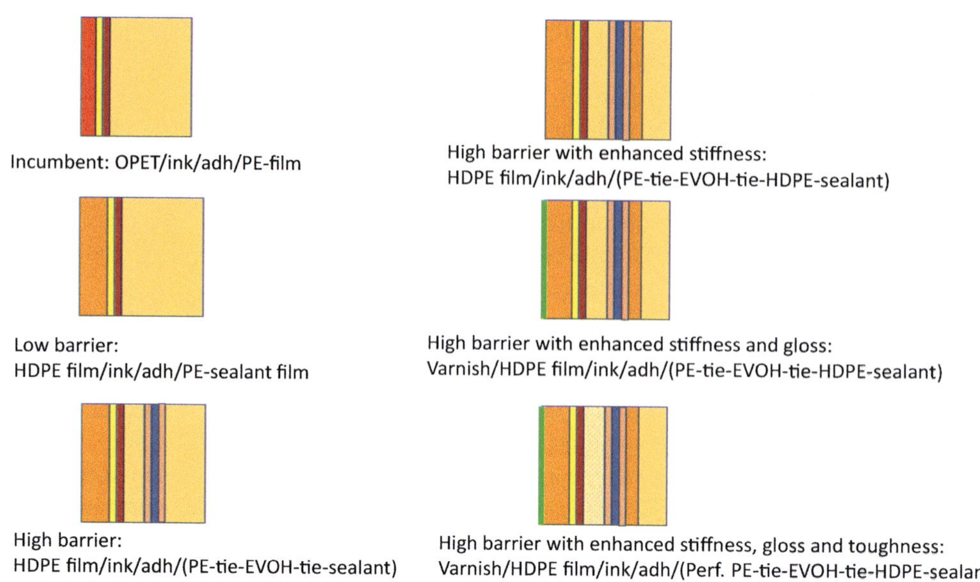

FIGURE 17.18 Conceptual evolution of stand-up pouch design for all-PE recycle stream.

seal process, abuse resistance and a dimensionally stable and smooth surface for printing. The print layer faces the inside of the package so that the PET layer protects the ink from abuse. The PE film is often a multilayer PE film with a LLDPE or HDPE layer for stiffness, strength and moisture barrier and a sealant layer to provide hermetic sealing.

A typical high oxygen barrier two-ply SUP has the structure: OPET/ink/adh/(PE-tie-EVOH-tie-Sealant). Here the PE film has been replaced with a 5-layer coextruded film with EVOH. The EVOH provides oxygen barrier. An alternative structure is a barrier coated OPET film laminated to a PE film.

These structures are not recyclable due to the mixed materials. PET is not compatible with the PE recycle stream. Depending on the types of inks used, they may not be compatible either. EVOH with the proper compatibilization can be recyclable in the PE stream. As described in Section 17.4.2.1, simplifying these structures while meeting all the needs along the value chain is not easy. Examples of ways to overcoming specific challenges are described in the following sections.

17.4.2.2.2 Overcoming limitations with all-polyethylene structure

Sealing The PET layer must be replaced to allow the SUP to be placed in the store drop-off PE recycle stream. The obvious candidate is PE. This may be an oriented PE film, as described in Section 17.4.2.1, with reverse printing: BOPE/ink/adh/PE film. It may also be a conventional blown or cast PE film. To maximize stiffness and thermal resistance, HDPE is favored over other types of PE. Lower density grades may be needed if greater transparency is required.

Another possibility is to use a coextruded PE film and surface print: varnish/ink/(PE-sealant). Here, a varnish is used to protect the print surface. The coextruded film will likely have more than two layers to satisfy other requirements, but two layers are shown here for simplicity. In either scenario the heat resistance of the new structure is not as good as OPET. HDPE melts at 135°C and starts to distort at lower temperatures. BOPE films will distort by around 115°C. This limits the seal bar temperature during sealing. To compensate, the sealant may need to be changed to one with a lower heat seal initiation temperature (HSIT). Generally, premium sealants such as plastomers, ethylene vinyl acetate copolymers and ionomers have the lowest seal initiation temperatures. The selection will need to consider recyclability, handling and cost. The HSIT of plastomers decreases with increasing comonomer content, which reduces crystallinity and increases tackiness. The polymer becomes more prone to blocking in the nip of the blown film process or in the wound roll, and it becomes more challenging to achieve low coefficient of friction (COF) for proper film registration. Resin suppliers are responding with new grades with better performance. Metallocene catalyzed VLDPE and plastomers, such as Dow's AFFINITY product line, have some of the lowest HSIT's available.

Applying a varnish to the outside of the film may have other advantages beyond protecting a print layer. One is to provide some heat resistance during heat sealing to keep the film from sticking to the seal bar and help mitigate distortion of the film. The varnish needs to be compatible with the PE recycle stream.

Oxygen barrier PE alone has high oxygen permeability; for applications requiring high oxygen barrier, a barrier layer is needed. As described in Section 17.4.2.1, EVOH with the appropriate compatibilizer is preferred for recyclability. Some possible structures include:

- HDPE film/print/adh/(PE-tie-EVOH-tie-PE-sealant)
- BOPE/print/adh/(PE-tie-EVOH-tie-PE-sealant)
- Varnish/print/(PE-tie-EVOH-tie-PE-sealant).

Here the PE layers in the coextruded barrier film may be HDPE for maximum moisture barrier and to add stiffness to the laminate (see next section). Dow's RecycleReady technology involves blending RETAIN functional polymer into the PE layers of the coextruded film (see Fig. 17.17). This ensures that the film is recyclable in the PE recycle stream.

Stiffness Maintaining the bending stiffness of the packaging film is important for film handling, registration, and dimensional stability in converting and packaging operations. Stiffness is also important for the function of the package – in this case for the pouch to "stand up." OPET is much stiffer than PE. Table 4.3 provides some typical values for secant modulus, a measure of stiffness. OPET has about four times the stiffness of HDPE blown film and eight times that of current BOPE films on the market, which are lower in density due to limitations on stretching HDPE resins. Machine direction orientation (MDO) will also substantially increase the modulus of PE, but it is still less than OPET of the same thickness [123]. Simply substituting the OPET film in the laminate structure with a PE film of the same thickness will substantially reduce the stiffness of the laminate.

One way build back stiffness is to increase the thickness of the PE film or layer that is replacing the OPET film. Bending stiffness increases with the cube of the thickness. To overcome a 4x reduction in modulus, the thickness needs to be increased by about 60%. For an 8x reduction in modulus, the thickness should be doubled to maintain bending stiffness. Large increases in thickness may increase the cost of the film (although with the density difference between PET and PE, there is a 40% yield advantage with using PE). It also increases the material usage, which may affect its environmental footprint. Increasing the thickness of the film also slows the transfer of heat to the sealant interface, which may slow packaging line speeds. (See Section 7.6 for a discussion on modeling heat transfer through packaging films.)

A second method to build stiffness in multilayer film structures is to arrange layers to take advantage of the I-beam effect. This concept is described in detail in Section 9.4.2. There is an imaginary plane within the film that is in neither compression nor tension when the film is flexed. This is known as the neutral axis. Layers furthest away from the neutral axis have the largest effect on bending stiffness. In the SUP structures, placing a relatively stiff layer as far away from the surface PE layer will be most efficient in gaining stiffness. This is illustrated in Fig. 17.19 where a series of structures are illustrated with their computed bending stiffness factors. The incumbent structure is on the left. Simply replacing the OPET with a BOPE of the same thickness reduces the bending stiffness by 42%. The third structure introduces a surface layer of HDPE that is still insufficient to recover the bending stiffness. The final structure on the right uses two layers of HDPE separated by LLDPE, which provides toughness: HDPE/ink/adh/(LLDPE-HDPE-Sealant). The stiffness is nearly the same as the incumbent structure at the same overall thickness. Incorporating EVOH into barrier structures may also aide in building stiffness since EVOH is stiffer than PE.

Gloss OPET has outstanding gloss and clarity. Replacing it with a HDPE film will lower gloss. One way to increase the gloss is to use a high-gloss coating or varnish. An example is Dow's OPULUX HGT. When applied to surface printed PE films, OPULUX imparts high gloss and surface protection as well as broadens the heat seal temperature

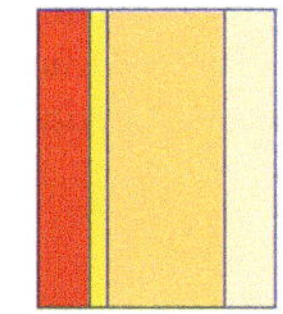

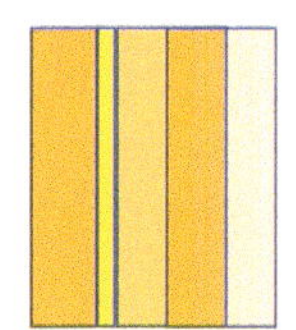

FIGURE 17.19 Use of I-beam principle to build bending stiffness of PE structures.

window by enhancing temperature resistance [124]. As with any coating or layer, the chemistry and thickness must be such that it is compatible with the PE recycle stream.

Abuse resistance Historically, increasing the stiffness of PE film structures meant sacrificing toughness and abuse resistance. Incorporating LLDPE into multilayer structures increases dart impact strength and other abuse resistance properties. Resin suppliers continue to manipulate the molecular architecture of PE to provide a better balance of stiffness and toughness. For example, Dow and ExxonMobil have introduced INNATE and EXCEED XP performance polymers, respectively, that have high stiffness and toughness [123,125]. Material advances such as these give the package designer greater flexibility to meet package requirements with all-PE structures.

Other factors There are other factors that come into play once package structures are changed. For example, standard zippers may no longer seal to the film at the lower temperatures required for all-PE designs. Zipper manufactures have responded with new designs that use lower melt point resins to accommodate this.

The operation of the packaging line may need to be altered to handle less stiff or more tacky films. Operators need to be trained to run at lower seal temperatures and to understand the consequences of running too hot.

17.4.2.2.3 Commercial example

These design ideas were put to practice by Kellogg's for their Bear Naked granola brand [126]. Working closely with resin supplier Dow, film converter Berry Global and many other players in the value chain, they successfully converted their package to a new design suitable for the PE store drop-off recycle stream. Bear Naked is a premium brand, a market leader in healthy foods amongst younger generations that value sustainability, and an innovator in package design. They were the first to package granola in a SUP. Their requirements for the package include:

- Preserving a premium look and feel with a matte surface and clear window to showcase the product.
- Maintaining existing packaging line run rates within acceptable tolerances.
- Meeting quality and sensory requirements with modified atmosphere and seal integrity.

The incumbent package contains an AlOx coated OPET laminated to a PE film: OPET-AlOx/ink/adh/PE film. The solution is illustrated in Fig. 17.20. It incorporates the following features:

- A matted surface varnish that protects the surface printing and provides heat resistance during sealing.
- A coextruded barrier film designed with:
  - o High clarity layer
  - o High stiffness and moisture barrier layer
  - o High oxygen barrier EVOH layer (and tie layers not shown in Fig. 17.20)
  - o Dow RETAIN polymer modifier incorporated into PE layers for recyclability of EVOH
  - o A sealant layer that provides hermetic sealing at low activation temperature and is compatible with the zipper.

It took many iterations and extensive collaboration across the value change to achieve a recyclable design that runs at the target bag-making and filling speeds. Oxygen and moisture barrier performance exceeds that of the incumbent. They were able to maintain the gauge of the film using the I-beam stiffness design principle, judicious selection of materials, and the precision of a 9-layer coextrusion film line. Contact clarity was maintained with the selection of surface resin. The zipper supplier redesigned the zipper for low sealing temperature. Even with the best of film design, the packaging equipment needed optimization to successfully deal with challenges related to channel leakers (zipper crush), linear tear (hole punching), package opening for filling, film extensibility, machine speed and temperature optimization.

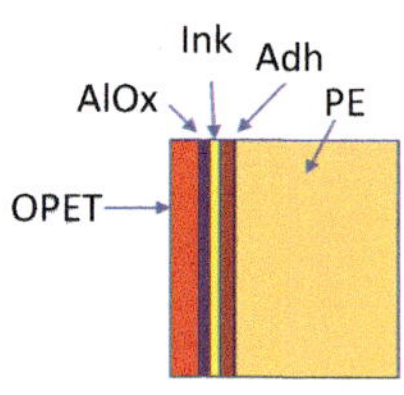

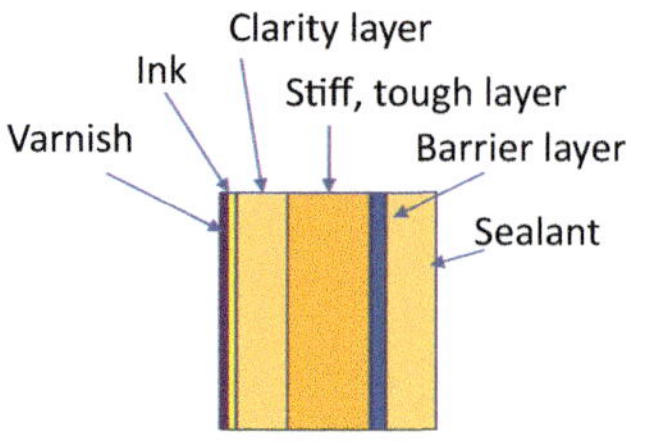

FIGURE 17.20 Example of converting a multimaterial stand-up pouch to a recyclable "all-PE" structure. *Based on S. Moore, P. Wolak, A. Leidolf, Designing for recyclability: a value-chain collaboration, Global Pouch Forum, 2019.*

17.4.3 Designing with biobased materials and for compostability

Besides mechanical and feedstock recycling, composting is another sustainable end-of-life option. The infrastructure for collecting and industrially composting flexible packaging is in its infancy. Most compostable packaging is used in closed-loop scenarios such as amusement parks or sporting events where collection can be closely controlled. Biodegradable materials are not approved by the APR and other recycle organizations for use in the mechanical recycling stream due to their effect on physical properties. Composting does have two significant advantages over mechanical recycling: it can handle food waste contamination as well as multimaterial packaging structures (if all the components are compostable). A disadvantage is that it does not extend the use of the materials beyond a single use.

Bioplastics and biodegradability are described in Section 17.3. From a circular perspective, incorporating biobased products into a package intended for composting is an ideal scenario. Some petroleum-based polymers are compostable but these release nonrenewable carbon into the atmosphere when they decompose. Traditional biobased and compostable materials include paper, paperboard and cellophane. Newer materials that have been introduced include PLA, polyhydroxyalkanoate (PHA), poly(3-hydroxybutyrate-co-3-hydroxyvalerate) (PHBV), and polybutylene succinate (PBS). Each of these has their strengths and weaknesses, as described in Table 17.19. Cost, sensitivity to relative humidity and moisture, raw material variability, narrow processing window, and performance (depending on the application) have limited the use of bioplastics in packaging [14]. Like conventional noncompostable materials, often no single material alone meets the performance and cost requirements of the application. Three approaches are typically considered for optimizing the package design for compostability:

- Blends. For example, BASF's Ecovio [127] product line is a blend of two compostable polymers, polybutylene adipate terephthalate (PBAT) and PLA, used in a variety of applications such as mulch films or paper coatings. Many thermoplastic starches contain polymer and plasticizers (see Section 4.2.14.2.4). Many biobased or compostable polymers are immiscible in one another, however, which limits this approach; such incompatible blends may have high haze and poor cohesive strength. Blends with noncompostable polymers may enhance processability or improve mechanical properties but detract from optical properties and compostability. Sometimes small amounts of polymer additives can be used to strike a balance between performance and compostability. For example, small amounts of acrylic [37] or ethylene copolymers [35] that are not compostable are added to PLA to improve its toughness while retaining optical clarity. Compostability is retained by keeping below the thresholds described in compostability standards such as ASTM D6400 and EN13432.
- Thin coatings. Various coatings have been applied to paper and paperboard to impart moisture or grease barrier and heat sealability. If the coating is thin enough it may meet repulpability and recycling standards even if the coating material itself is not compostable. Coatings may be an option for enhancing the performance of compostable films.
- Multilayer films or laminates. This approach is still relatively unexplored. Much like conventional multilayer films, various materials can be combined to balance mechanical, barrier and sealing properties. Given the narrow processing windows of many biopolymers, attention to ensuring the materials can be processed under the same conditions, have similar flow properties and good interlayer adhesion will be important for coextrusion processes. Suitable tie resins may need to be developed. For laminates, new adhesives that are compostable may be needed.

Cunha et al. [128] find that film blowing of PHVB is difficult due to its poor melt processing and thermal stability. They successfully blow films from a 70/30 blend of PHVB and PBAT and by coextruding the two polymers. The 2-layer coextruded films have better mechanical properties than the blends. Thellen et al. [129] make 3-layer cast film of (PHA-PVOH-PHA). PVOH provides oxygen barrier and PHA moisture barrier. Grafting the PHA with maleic anhydride is found to increase the peel strength of PHA to PVOH by 100%. The oxygen transmission rate (OTR) of the coextruded film is significantly lower than PVOH alone, especially at high relative humidity. Biodegradation studies in a marine environment show that the PHA biodegrades quickly and grafting PHA with MAH only slows this down slightly. PVOH degradation is considerably slower even though it dissolves in the seawater. Dong et al. [130] make three-layer films of (PPC-PVOH-PPC) using a solution casting process. Polypropylene carbonate (PPC) is a biodegradable polymer made from CO_2 and propylene oxide that has poor mechanical properties. The 3-layer films have 500x reduction in oxygen permeation (OPV), 9x increase in modulus and 8x increase in tensile strength compared with a PPC control. The OPV of the 3-layer films is about an order of magnitude higher than a PVOH control. MVTR is unchanged from PPC and is considerably higher than PE films.

Wu et al. [131] note that research into compostable multilayer films is stymied by the poor processability of many biopolymers, the inferior properties compared with their petroleum-based counterparts, and their relatively high cost.

TABLE 17.19 Strengths and weaknesses of several compostable materials.

Material	Strengths	Weaknesses
Paper/paperboard	• Widely used • Inexpensive • Recyclable/repulpable • Many versions from flexible to rigid • Printable • Dimensionally stable	• High moisture permeation • High oxygen permeation • Not sealable • Opaque
Cellophane	• Transparent • Flexible • Strong • Oil resistance • Oxygen barrier	• Poor moisture barrier unless coated • Contribution to pollution (CS2)
Starch	• Widely available feedstock • Can be formulated with other polymers and plasticizers for film applications • Some formulations have good oxygen barrier	• Poor processing/thermal stability • Moisture sensitive—high MVTR • Poor mechanical strength; brittle unless plasticized
Polylactic acid (PLA)	• Transparent • Stiff (high modulus) • Can be made into biaxially oriented films • Versions can be heat sealed	• Brittle • Poor gas barrier
Polyhydroxyalkanoate (PHA), Poly(3-hydroxybutyrate-co-3-hydroxyvalerate) (PHBV), Polyhydroxybutyrate (PHB)	• Good oxygen and moisture barrier • Mechanical properties (similar to PP or PS)	• Brittle • Poor melt strength • Poor thermal stability
Polycaprolactone (PCL)	• Good mechanical properties	• Nonrenewable-based feedstock • High MVTR • High OTR
Polybutylene succinate (PBS)	• Mechanical properties similar to PE	• Partly biobased • Poor flex crack resistance
Polybutylene adipate terephthalate (PBAT)	• Tough • Flexible • PE-like properties	• Petroleum based • Low modulus and strength • High MVTR and OTR

They propose structures with high PHBV with low HV content for good oxygen/moisture barrier, PBS or PBAT with low melting points for sealants and PLA for strength and stiffness.

17.5 Conclusion

Sustainability has increasingly become an important driver for new packaging design and innovation. Concerns around climate change, pollution and depletion of natural resources have propelled plastics and flexible packaging into the spotlight. Consumer goods companies are responding with new goals around reducing packaging, increasing the use of recycled materials, reducing the amount of virgin materials and increasing the circularity of packaging. Designing for sustainability begins with understanding what the goal is—reduce, reuse, recycle, etc. Reducing the amount of packaging materials has driven the growth of flexible packaging over rigid formats for many years. The role flexible packaging plays in minimizing food and other product waste must also be considered; the use of LCA should reflect the impact on product waste since often the product itself has a higher environmental footprint than the package. This can be quantified by the product to package environmental impact ratio. If the ratio is high, any redesign of the package for greater circularity must not result in higher product waste.

Until feedstock or chemical recycling, delamination and dissolution processes, and composting infrastructure is widespread, circularity for flexible packaging will focus on mechanical recycling of plastics (and repulping for paper products). The industry will continue to need to develop the means to collect, sort and clean flexible packaging so that it can be reprocessed in the municipal recycling streams. New markets will also need to be developed since reuse of mechanically recycled materials into some packaging end-uses such as food and medical packaging will likely have limited regulatory compliance. Redesign of flexible packaging for ease of recycling will help enable its reuse in higher quality applications. This will involve moving away from incompatible multimaterial structures towards single material or compatible material structures. Such changes bring new challenges since the performance of incumbent packaging structures has been optimized for cost and performance over many years of package evolution. Overcoming the challenges will involve collaboration across the packaging value chain: new materials from resin, materials and additive suppliers; tweaks to converting and packaging equipment; new packaging designs and formats; advances in collection, sorting, cleaning and reprocessing at municipal recycle centers; and development of new markets. Fortunately, we can apply the learnings from past developments to new designs involving new materials or configurations. Understanding the science behind the technology of flexible packaging, as is the purpose of this book, will aid in this transition.

References

[1] E. Kosior, J. Mitchell, Current industry position on plastic production and recycling, in: T.M. Letcher (Ed.), Plastic Waste and Recycling: Environmental Impact, Societal Issues, Prevention, and Solutions, Academic Press, 2020, pp. 133–162. 2020.

[2] Ellen Macarthur Foundation, The new plastics economy: rethinking the future of plastics & catalysing action, accessed online on 27 June 2020 at https://www.ellenmacarthurfoundation.org/publications/the-new-plastics-economy-rethinking-the-future-of-plastics-catalysing-action, 2017.

[3] D. Jubinville, E. Esmizadeh, S. Saikrishnan, C. Tzoganakis, T. Mekonnen, A comprehensive review of global production and recycling methods of polyolefin (PO) based products and their post recycling applications, Sustain. Mater. Technol. 25 (2020) e00188. Available from: https://doi.org/10.1016/j.susmat.2020.e00188. ISSN 2214–9937.

[4] Michael Niaounakis, Recycling of flexible packaging, Elsevier, 2020.

[5] J. Maris, S. Bourdon, J. Brossard, L. Cauret, L. Fontaine, V. Montembault, Mechanical recycling: compatibilization of mixed thermoplastic wastes, Polym. Degrad. Stab. 147 (2018) 245–266.

[6] T. Bukowski, M. Richmond, A holistic view of the role of flexible packaging in a sustainable world, Flex. Packaging Assoc. Rep. (2018)viewed online on 21 Dec 2020 at. Available from: https://www.flexpack.org/resources/sustainability-resources#a-holistic-view-of-the-role-of-flexible-packaging-in-a-sustainable-world.

[7] O. Horodytska, F.J. Valdés, A. Fullana, Plastic flexible films waste management – a state of art review, Waste Manag. 77 (2018) 413–425.

[8] T.M. Letcher (Ed.), Plastic Waste and Recycling: Environmental Impact, Societal Issues, Prevention, and Solutions, Academic Press, 2020.

[9] K. Ragaert, L. Delva, K. Van Geem, Mechanical and chemical recycling of solid plastic waste, Waste Manag. 69 (2017) 24–58.

[10] T.A. Cooper, Developments and innovations in end-of-life technologies for flexible packaging, SPE FlexPackCon Conf. (2020).

[11] M. Solis, S. Silveira, Technologies for chemical recycling of household plastics – a technical review and TRL assessment, Waste Manag. 105 (2020) 128–138.

[12] Association of Plastics Recyclers, APR Design® Guidance, viewed online on 14 Nov 2021 at https://plasticsrecycling.org/apr-design-guide, 2021.

[13] CEFLEX, “Designing for a circular economy, Recyclability of polyolefin-based flexible packaging,” downloaded 23 Jan 2020 at https://guidelines.ceflex.eu/resources/, 2020.

[14] RecyClass, “Design for recycling guidelines,” viewed online on 13 Nov 2021 at https://recyclass.eu/recyclability/design-for-recycling-guidelines/, 2021.

[15] Sustainable Packaging Coalition, “Design for recycled content guide,” viewed online on 13 Nov 2021 at https://recycledcontent.org, 2020.

[16] F. Licciardello, L. Piergiovanni, Packaging and food sustainability, in: C. Charis Galanakis (Ed.), The Interaction of Food Industry and Environment, Academic Press, 2020, pp. 191–222. ISBN 9780128164495. Available from: https://doi.org/10.1016/B978-0-12-816449-5.00006-0.

[17] T.A. Cooper, Developments in plastic materials and recycling systems for packaging food, beverages and other fast-moving consumer goods, in: N. Farmer (Ed.), Trends in packaging of food, beverages and other fast-moving consumer goods (FMCG), Woodhead Publishing, 2013, pp. 58–107.

[18] Franklin Associates, “Life cycle impacts of plastic packaging compared to substitutes in the united states and Canada theoretical substitution analysis,” viewed online on 22 Dec 2020 at https://plastics.americanchemistry.com/Reports-and-Publications/LCA-of-Plastic-Packaging-Compared-to-Substitutes.pdf, 2018.

[19] American Chemistry Council, Plastics and Sustainability: A Valuation of Environmental Benefits, Costs and Opportunities for Continuous Improvement, AMI Conf. (2017).

[20] Flexible Packaging Europe, Updated IFEU study confirms flexible packaging plays a key role in prevention of packaging waste and mitigation of global warming [Press release]. Düsseldorf, 26 Germany. https://www.flexpack-europe.org/files/FPE/pressreleases/2020/FPE_ifeu_study_update_2020_GB.pdf, 2020.

[21] National Geographic Magazine, We depend on plastic. Now we are drowning in it, National Geographic Society, June 2018.

[22] N.A. Welden, A. Lusher, Microplastics: from origin to impacts, in: T.M. Letcher (Ed.), Plastic Waste and Recycling: Environmental Impact, Societal Issues, Prevention, and Solutions, Academic Press, 2020, pp. 223–249. 2020.

[23] R. Hurley, A. Horton, A. Lusher, L. Nizzetto, Plastic waste in the terrestrial environment, in: T.M. Letcher (Ed.), Plastic Waste and Recycling: Environmental Impact, Societal Issues, Prevention, and Solutions, Academic Press, 2020, pp. 163–193. 2020.

[24] C.R. Cook, R.U. Halden, Ecological and health issues of plastic waste, in: T.M. Letcher (Ed.), Plastic Waste and Recycling: Environmental Impact, Societal Issues, Prevention, and Solutions, Academic Press, 2020, pp. 513–527. 2020.

[25] E. Kosior, I. Crescenzi, Solutions to the plastic waste problem on land and in the oceans, in: T.M. Letcher (Ed.), Plastic Waste and Recycling: Environmental Impact, Societal Issues, Prevention, and Solutions, Academic Press, 2020, pp. 415–446. 2020.

[26] Dow Chemical Company, Dow and UPM partner to produce plastics made with renewable feedstock, Press Release, viewed online on 26 Dec 2020 at https://corporate.dow.com/en-us/news/press-releases/dow-and-upm-partner-to-produce-plastics-made-with-renewable-feedstock.html, 2019.

[27] L. Effler, Commercializing recyclable plastic packaging, piecing the puzzle together, ACS Conf. (2020).

[28] E. De Tandt, C. Demuytere, E. VanAsbroeck, H. Moerman, N. Mys, G. Vyncke, et al., A recycler's perspective on the implications of REACH and food contact material (FCM) regulations for the mechanical recycling of FCM plastics, Waste Manag. 119 (2021) 315–329.

[29] U.S. FDA, Guidance for industry: use of recycled plastics in food packaging (chemistry considerations), viewed online on 27 Dec 2020 at https://www.fda.gov/regulatory-information/search-fda-guidance-documents/guidance-industry-use-recycled-plastics-food-packaging-chemistry-considerations, 2006.

[30] J. Bott, A. Störmer, R. Franz, A model study into the migration potential of nanoparticles from plastics nanocomposites for food contact, Food Packag. Shelf Life 2 (2) (2014) 73e80.

[31] C.M. Casarino, B.A. Morris, S.R. Veith, Resin life cycle estimation to help guide sustainability choices, SPE ANTEC Conf. (2009).

[32] Plastics Europe – Association of Plastics Manufacturers, Plastics – the facts 2020, viewed online on 29 Dec 2020 at https://www.plasticseurope.org/en/resources/market-data, 2020.

[33] E.T.H. Vink, K.R. Rabago, D.A. Glassner, P.R. Gruber, Applications of life cycle assessment to NatureWorks® polylatide (PLA) production, Polym. Degrad. Stab. 80 (2003) 403–419.

[34] NatureWorks, Technology focus report: Toughened PLA, viewed online on 30 Jan 2021 at https://www.natureworksllc.com/~/media/Technical_Resources/Technical_Publications/White_Papers/Toughened-PLA-Technology-Focus-pdf.pdf, 2007.

[35] B.A. Morris, D. Kirschbaum, Modification of PLA for improved processing and properties, Bioplastics Process. Conf. (2007). Charlotte, NC.

[36] J. Uradnisheck, Improved dimensional stability of thermoformed polylactic acid articles, SPE ANTEC Conf. (2009).

[37] No author, Dow introduces impact modifier for polylactic acid, Addit. Polym. 2010 (2) (2010) 2–3.

[38] F. Wikström, K. Verghese, R. Auras, A. Olsson, H. Williams, R. Wever, et al., Packaging strategies that save food: a research agenda for 2030, J. Ind. Ecol. (2018). Available from: https://doi.org/10.1111/jiec.12769.

[39] M. Heller, Demonstration of the environmental interplay between food waste and food packaging via life cycle assessment, LCA XV Conference, Vancouver, BC, 2015.

[40] Denkstatt, How packaging contributes to food waste prevention, denstatt GmbH, viewed online 08 Nov 2021 at https://static.cdn.packhelp.com/wp-content/uploads/2020/02/13101028/How_Packaging_Contributes_to_Food_Waste_Prevention_V2.0-3.pdf, 2017.

[41] M.C. Heller, S.E.M. Selke, G.A. Keoleian, Mapping the influence of food waste in food packaging environmental performance assessments, J. Ind. Ecol. (2018) 1e1600 (0). Available from: https://doi.org/10.1111/jiec.12743.

[42] H. Williams, F. Wikström, T. Otterbring, M. Löfgren, A. Gustafsson, Reasons for household food waste with special attention to packaging, J. Clean. Prod. 24 (2012) 141–148.

[43] H. Williams, F. Wikström, Environmental impact of packaging and food losses in a life cycle perspective: A comparative analysis of five food items, J. Clean. Prod. 19 (2011) 43–48.

[44] E. Pauer, M. Tacker, V. Gabriel, V. Krauter, Sustainability of flexible multilayer packaging: Environmental impacts and recyclability of packaging for bacon in block, Clean. Environ. Syst. 1 (2020). Available from: https://doi.org/10.1016/j.cesys.2020.100001.

[45] U.S. Environmental Protection Agency Sustainable materials management: non-hazardous materials and waste management hierarchy, viewed online on 28 Dec 2020 at https://www.epa.gov/smm/sustainable-materials-management-non-hazardous-materials-and-waste-management-hierarchy, 2017.

[46] G. Rosalsky, Are plastic bag bans garbage? NPR Planet Money, viewed online on 28 Dec 2020 at https://www.npr.org/sections/money/2019/04/09/711181385/are-plastic-bag-bans-garbage, 2019.

[47] T. Clark, Plastics: establishing the path to zero waste, A pragmatic approach to sustainable management of plastic materials, Enso Plastics, viewed 02 Dec 2015 at http://www.ensoplastics.com/download/Plastics_EstablishingthePathtoZeroWaste.pdf, 2015.

[48] Sustainable Packaging Coalition, ND, The SPC Position against Biodegradability Additives for PetroleumBased Plastics, viewed online on 23 May 2020 at https://sustainablepackaging.org/resources/.

[49] P.J. Kershaw, Biodegradable plastics & marine litter: misconceptions, concerns and impacts on marine environments, United Nations Environment Program, viewed 02 Dec 2015 at http://unep.org/gpa/documents/publications/BiodegradablePlastics.pdf, 2015.

[50] T.A. Cooper, Developments and trends in bioplastics technology and markets for flexible, pouch and barrier packaging, SPE ANTEC Conf. (2021).

[51] D. Carrington, Scientists create mutant enzyme that recycles plastic bottles in hours, The Guardian. Accessed from web on April 13, 2020: https://www.theguardian.com/environment/2020/apr/08/scientists-create-mutant-enzyme-that-recycles-plastic-bottles-in-hours, 2020.

[52] V. Tournier, et al., An engineered PET depolymerase to break down and recycle plastic bottles, Nature 580 (2020) 216–219.

[53] Packaging South Asia, 2019, An enzyme based technology for 100% biodegradable polymers/MLP, accessed online April 18, 2020: https://packagingsouthasia.com/type-of-article/industry-news/an-enzyme-based-technology-for-100-biodegradable-polymers-mlp/.

[54] Sample, Ian, Plastic-eating worms could help wage war on waste, The Guardian, accessed from web on April 19, 2020: https://www.theguardian.com/science/2017/apr/24/plastic-munching-worms-could-help-wage-war-on-waste-galleria-mellonella, 2017.

[55] P. Paolo Bombelli, J. Christopher, C.J. Howe, F. Federica Bertocchini, Polyethylene bio-degradation by caterpillars of the wax moth *Galleria mellonella*, Curr. Biol. 27 (2017) R292–R293accessed online on 20 Dec 2020 at. Available from: https://www.cell.com/current-biology/fulltext/S0960-9822(17)30231-2.

[56] A. Feil, T. Pretz, Mechanical recycling of packaging waste, in: T.M. Letcher (Ed.), Plastic Waste and Recycling: Environmental Impact, Societal Issues, Prevention, and Solutions, Academic Press, 2020, pp. 283–319. 2020.

[57] Sustainable Packaging Coalition, Design for recycled content guide, viewed online on 18 Jan 2021 at https://sustainablepackaging.org/resources/, 2019.

[58] More Recycling, End market demand for recycled plastic, Next Generation of Moore Recycling Associates Inc., viewed online on 29 Dec 2020 at https://www.plasticsmarkets.org/jsfcode/srvyfiles/wd_151/endusedemand_report_v8_1.pdf, 2017.

[59] RRS, Flexible packaging recycling in materials recovery facilities pilot, Research Report, Materials Recovery for the Future, viewed online on 29 Dec 2020 at https://www.materialsrecoveryforthefuture.com/research-results/2020-research-results/, 2020.

[60] Association of Plastics Recyclers, Critical guidance protocol for PE film and flexible packaging, viewed online on 18 Jan 2021 at https://plasticsrecycling.org/staging/images/Design-Guidance-Tests/APR-PE-film-critical-guidance-FPE-CG-01.pdf, 2020.

[61] ISO, 2008, “Plastics – Guidelines for the recovery and recycling of plastics waste.” ISO standard 15270:2008.

[62] ASTM, Standard Guide for Waste Reduction, Resource Recovery, and Use of Recycled Polymeric Materials and Products (Withdrawn 2015), ASTM International, West Conshohocken, PA, 2006.

[63] R. Clift, Overview clean technology—the idea and the practice, J. Chem. Tech. Biotechnol. 68 (1999) 347–350.

[64] Fiber Box Association, Voluntary standard for repulping and recycling corrugated fiberboard treated to improve its performance in the presence of water and water vapor, revised August 16, 2013.

[65] More Recycling, 2018 National post-consumer plastic bag & film recycling report, viewed online on 29 Dec 2020 at https://www.plasticsmarkets.org/jsfcontent/FilmReport18_jsf_1.pdf, 2020.

[66] J. McKevitt, 2019, “Dow mixes post-consumer plastic into asphalt roads,” Construction Equipment Guide, viewed on 31 Jan 2021 at https://www.constructionequipmentguide.com/dow-mixes-post-consumer-plastic-into-asphalt-roads/44446.

[67] J. Lombard, 2020, “Use of plastic additives in asphalt mixtures increasing,” Asphalt Contractor, viewed online on 31 Jan 2021 at https://www.forconstructionpros.com/asphalt/article/21196132/use-of-plastics-in-asphalt-mixtures-increasing.

[68] G. White, G. Reid, Recycled waste plastic for extending and modifying asphalt binders, 8th Symposium on Pavement Surface Characteristics: SURF 2018 – Vehicle to Road Connectivity, Brisbane, Queensland, viewed online 31 Jan 2021 at https://www.macrebur.com/pdfs/product/SURF%20-%20Plastic%20Recycling%20for%20Bitumen%20Ver%204.pdf, 2018.

[69] M.A. Dalhat, H.I. Al-Abdul Wahhab, Performance of recycled plastic waste modified asphalt binder in Saudi Arabia, Int. J. Pavement Engr 18 (2015) 349–357.

[70] J. Paben, Study finds benefits to using recycled film in asphalt, Plastics Recycling Update, viewed online 31 Jan 2021 at https://resource-recycling.com/plastics/2020/07/06/study-finds-benefits-to-using-recycled-film-in-asphalt/, 2020.

[71] Dow Chemical, Dow launches new resin made with 70 percent recycled plastic, viewed online on 30 Dec 2020 at https://www.dow.com/en-us/news/new-resin-made-with-70-percent-recycled-plastic.html, 2019.

[72] Dow Chemical, Dow introduces its first recycled plastic resin for shrink film applications in North America, viewed online on 30 Dec 2020 at https://www.dow.com/en-us/news/first-recycled-plastic-resin-shrink-film-north-america.html, 2020.

[73] G. Altnau, The CreaSolv® process is neither a solvolysis nor chemical recycling, viewed online 8 Nov 2021 at https://www.linkedin.com/pulse/creasolv-process-neither-solvolysis-nor-chemical-recycling-altnau, 2020.

[74] C. Chen, J. White, Compatibilizing agents in polymer blends: interfacial tension, phase morphology, and mechanical properties, Polym. Engr Sci. 33 (1993) 923–930.

[75] J. Karger-Kocsis, A. Kalló, A. Szafner, G. Bodor, Z. Sényei, Morphological study on the effect of elastomeric impact modifiers in polypropylene systems, Polymer 20 (1979) 37–43. 1979.

[76] U. Gedde, A. Mattozzi, Polyethylene morphology (review), Adv. Polym. Sci. 169 (2004) 29–73.

[77] J. Munro, G. Wu, J.Z.D. Huang, S. Bawaskar, B. Morris, A review of impact modification technologies for different thermoplastics using ethylene copolymers, ANTEC, *SPE*, San Antonio, TX, 2020.

[78] C. Bucknall, V.L.P. Soares, H.H. Yang, X.C. Zhang, Rubber toughening of plastics: Rubber particle cavitation and its consequences, Macromolecular Symposia, Grenoble, France, 2011.

[79] A. Lazzeri, C.B. Bucknall, Dilatational bands in rubber-toughened polymers, J. Mater. Sci. 24 (1993) 6799–6808.

[80] Y. Lin, V. Yakovleva, H. Chen, A. Hiltner, E. Baer, Comparison of olefin copolymers as compatibilizers for polypropylene and high-density polyethylene, J. Appl. Polym. Sci. 113 (2009) 1945–1952.

[81] Y. Zeng, D. Abebe, A. Singh, L. Effler, Y. Hu, M. Kapur, et al., Compatibilizing polypropylene (PP) contaminated polyethylene (PE) waste streams in film applications, SPE Int. Polyolefins Conf. (2020) (And 2021 ANTEC).

[82] J.M. Eagan, J. Xu, R. Di Girolamo, C.M. Thurber, C.W. Macosko, A.M. LaPointe, et al., Combining polyethylene and polypropylene: enhanced performance with PE/iPP multiblock polymers, Science 355 (2017) 814–816.

[83] K. Hausmann, Compatibilizers, polymeric (recycling of multilayer structures), in: C.Salamone Joseph (Ed.), Polymeric materials encyclopedia, Vol 2, CRC Press, 1996, pp. 1364–1377.

[84] H. Li, Y. Yin, M. Liu, P. Deng, W. Zhang, J. Sun, Improved compatibility of EVOH/LDPE blends by γ-ray irradiation, Adv. Polym. Technol. 28 (3) (2009) 192–198. Available from: https://doi.org/10.1002/adv.20160.

[85] Y. Lei, Q. Wu, C.M. Clemons, W. Guo, Phase structure and properties of poly(ethylene terephthalate)/high-density polyethylene based on recycled materials, J. Appl. Polym. Sci. 113 (2009) 1710–1719.

[86] M. Schlummer, T. Fell, A. Mäurer, G. Altnau, The role of chemistry in plastics recycling, Kunstoffe International, May 2020, viewed 09 Nov 2021 at https://www.ivv.fraunhofer.de/content/dam/ivv/en/documents/Leistungsangebot/Recycling_Environment/The-Role-of-Chemistry-in-Plastics-Recycling.pdf, 2020.

[87] Fraunhofer IVV, Recycling plastics - The CreaSolv® Process, viewed on 08 Dec 2020 at https://www.ivv.fraunhofer.de/en/recycling-environment/recycling-plastics.html#creasolv.

[88] I. Vollmer, M. Jenks, M. Roelands, R. White, T. van Harmelen, P. de Wild, et al., Beyond mechanical recycling: giving new life to plastic waste, Angew. Chem. 59 (2020) 1402–15423.

[89] ZimPackaging, Unilever hails new technology to recycle sachets, viewed online 09 Nov 2021 at https://zimpackaging.co.zw/2020/12/08/unilever-hails-new-technology-to-recycle-sachets/, 2020.

[90] Unilever, n.d., Rethinking plastics packaging, viewed online 09 Nov 2021 at https://www.unilever.com/planet-and-society/waste-free-world/rethinking-plastic-packaging/.

[91] APK, n.d., Newcycling® – the innovative recycling technology from APK AG, viewed online on 09 Nov 2021 at https://www.apk-ag.de/en/.

[92] T. Walker, N. Frelka, Z. Shen, A. Chew, J. Banick, S. Grey, et al., Recycling of multilayer plastic packaging materials by solvent-targeted recovery and precipitation, Sci. Adv. (2020) 6. Available from: https://doi.org/10.1126/sciadv.aba7599.

[93] Y.-B. Zhao, X.-D. Lv, H.-G. Ni, Solvent-based separation and recycling of waste plastics: A review, Chemosphere 209 (2018) 707–720.

[94] G. Bergsma, Chemical recycling and its CO2 reduction potential. CE Delft, Brussels. https://cedelft.eu/assets/upload/file/Presentaties/2019/Chemical%20recycling%20an%20environmental%20perspective%20CE%20Delft%2020%20february%202019%20brussel.pdf, 2019.

[95] M. Pohjakallio, T. Vuorinen, A. Oasmaa, Chemical routes for recycling—dissolving, catalytic, and thermochemical technologies, in: T.M. Letcher (Ed.), Plastic Waste and Recycling: Environmental Impact, Societal Issues, Prevention, and Solutions, Academic Press, 2020, pp. 359–384. 2020.

[96] LyondellBasell, LyondellBasell announces construction of new small-scale molecular recycling facility, viewed online on 2 Jan 2021 at https://www.lyondellbasell.com/en/news-events/products-technology-news/lyondellbasell-announces-construction-of-new-small-scale-molecular-recycling-facility/, 2019.

[97] Dow, Dow and Fuenix enter into a partnership for the production of 100% circular plastic, viewed online on 9 Nov 2021 at https://corporate.dow.com/en-us/news/press-releases/dow-and-fuenix-enter-into-a-partnership-for-the-production-of-10.html, 2019.

[98] M. Morgan, Is chemical recycling a game changer? https://ihsmarkit.com/research-analysis/is-chemical-recycling-a-game-changer.html, 2019.

[99] U.S. EPA, Advancing sustainable materials management: 2018 fact sheet, viewed online on 03 Jan 2021 at https://www.epa.gov/sites/production/files/2020-11/documents/2018_ff_fact_sheet.pdf, 2020.

[100] Organization for Economic Co-Operation and Development, Municipal waste, generation and treatment, OECD.stat, viewed online on 03 Jan 2021 at stats.oecd.org.

[101] S. Kaza, L. Yao, P. Bhada-Tata, F. Van Woerden, What a waste 2.0, A global snapshot of solid waste, Management to 2050, World Bank Group, viewed online on 03 Jan 2021 at https://openknowledge.worldbank.org/handle/10986/30317, 2018.

[102] P.M. Clark, J.H. Wang, Commercial application of post consumer recycled resins in flexible packaging, GPEC 2009, Global Plastics Environmental Conference, Orlando, FL, United States, Feb. 25–27, 2009.

[103] V.A. González-González, G. Neira-Velázquez, J.L. Angulo-Sánchez, Polypropylene chain scissions and molecular weight changes in multiple extrusion, Polym. Degrad. Stab. 60 (1998) 33–42.

[104] P. Oblak, J. Gonzalez-Gutierra, B. Zupancic, A. Aulova, I. Emri, Processibility and mechanical properties of extensively recycled high density polyethylene, Polym. Degrad. Stab. 114 (2015) 133–145.

[105] C. Pattanakul, S. Selke, C. Lai, J. Miltz, Properties of recycled high-density polyethylene from milk bottles, J. Appl. Polym. Sci. 43 (1991) 2147–2150.

[106] A.S.F. Santos, J.A.M. Agnelli, D.W. Trevisan, S. Manrich, Degradation and stabilization of polyolefins from municipal plastic waste during multiple extrusions under different reprocessing conditions, Polym. Degrad. Stab. 77 (2002) 441–447.

[107] M.J. Abad, A. Ares, L. Barral, J. Cano, F.J. Díez, S. García-Garabal, et al., Effects of a mixture of stabilizers on the structure and mechanical properties of polyethylene during reprocessing, J. Appl. Polym. Sci. 92 (2004) 3910–3916.

[108] J. Dostál, V. Kašpárková, M. Zatloukal, J. Muras, L. Šimek, Influence of the repeated extrusion on the degradation of polyethylene. Structural changes in low density polyethylene, Eur. Polym. J. 44 (2008) 2652–2658.

[109] H. Jin, J. Gonzalez-Gutierrez, P. Oblak, B. Zupančič, I. Emri, The effect of extensive mechanical recycling on the properties of low density polyethylene, Polym. Degrad. Stab. 97 (2012) 2262–2272.
[110] S. Luzuriaga, J. Kovářová, I. Fortelný, Degradation of pre-aged polymers exposed to simulated recycling: Properties and thermal stability, Polym. Degrad. Stab. 91 (2006) 1226–1232.
[111] L.G. Barbosa, M. Piaia, G.H. Ceni, Analysis of impact and tensile properties of recycled polypropylene, Int. J. Mater. Eng. 7 (2017) 117–120.
[112] J.M. Torradas, D.P. Dean, R.O. Pimentel, Additives to improve regrind utilization and recycling of high barrier blow molded containers, SPE ANTEC Conf. (2013).
[113] RecyClass, Online tool, viewed online on 13 Nov 2021 at https://recyclass.eu/recyclability/online-tool/, 2022.
[114] RecyClass, Methodology, viewed online 13 Nov 2021 at https://recyclass.eu/recyclability/methodology/, 2021.
[115] RecyClass, Recycability evaluation protocol for PE films, viewed online 13 Nov 2021 at https://recyclass.eu/wp-content/uploads/2021/10/RecyClass-Recyclability-Evaluation-Protocol-for-PE-Films-v.2.0_NEW-DESIGN.pdf, 2021.
[116] NOVA Chemicals, NOVA Chemicals introduces recyclable oxygen barrier film structure design, viewed online on 23 Jan 2021 at https://www.novachem.com/media-center/news-releases/nova-chemicals-introduces-recyclable-oxygen-barrier-film-structure-design/, 2017.
[117] NOVA Chemicals, NOVA Chemicals enters high-density biaxially oriented polyethylene market to expand recyclable packaging options, viewed online 23 Jan 2021 at https://www.novachem.com/media-center/news-releases/nova-chemicals-enters-high-density-biaxially-oriented-polyethylene-market-to-expand-recyclable-packaging-options/, 2020.
[118] Dow, RecycleReady technology for store drop-off recycling, viewed online 23 Jan 2021 at https://www.dow.com/en-us/brand/recycleready.html, 2020.
[119] Dow, Breakthrough Dow Technology enables recyclable flexible plastic packaging, viewed online 23 Jan 2021 at https://corporate.dow.com/en-us/news/press-releases/breakthrough-dow-technology-enables-recyclable-flexible-plastic-packaging.html, 2016.
[120] S. Parkinson, D. Oner-Deliormanli, C. Chirinos, B. Walther, Self-recyclable barrier packaging, U.S. patent 10,300,686 B2, 2019.
[121] Y. Lin, J. Xu, J. Pan, X.-B. Yun, M. Demirors, Biaxially oriented polyethylene (BOPE) films fabricated via tenter frame process and applications thereof2018 SPE ANTEC Conf. (2018).
[122] C. Lawrence, K. Forziatti, J. Garnett, Machine direction orientation and its effects on multilayer sealant film, SPE ANTEC Conf. (2014) 1425–1433.
[123] ExxonMobil, Full PE laminated solutions for sustainable packaging, viewed online 24 Jan 2021 at https://www.exxonmobilchemical.com/en/products/polyethylene/performance-pe-polymers, 2020.
[124] Dow, Dow presents OPULUX™ HGT for high-gloss PE-based packaging at K 2019, viewed online on 24 Jan 2021 at https://www.dow.com/en-us/news/dow-presents-opulux-hgt-for-high-gloss-pe-based-packaging-at-k-2.html, 2019.
[125] Dow, The cast is getting bigger, Introducing the newest INNATE™ precision packaging resins, Dow Chem. Co., accessed online 24 Jan 2021 at https://www.dow.com/content/dam/dcc/documents/en-us/mark-prod-info/500/500-20401-01-introducing-the-newest-innate-precision-packaging-resins.pdf?iframe = true, 2016.
[126] S. Moore, P. Wolak, A. Leidolf, Designing for recyclability: a value-chain collaboration, Global Pouch Forum, 2019.
[127] BASF, "ecovio® – certified compostable polymer with bio-based content," viewed online on 30 Jan 2021 at https://plastics-rubber.basf.com/global/en/performance_polymers/products/ecovio.html.
[128] M. Cunha, B. Fernandes, J.A. Covas, A.A. Vicente, L. Hilliou, Film blowing of PHBV blends and PHBV-based multilayers for the production of biodegradable packages, J. Appl. Polym. Sci. (2015) 133.
[129] C. Thellen, S. Cheney, J.A. Ratto, Melt processing and characterization of polyvinyl alcohol and polyhydroxyalkanoate multilayer films, J. Appl. Polym. Sci. 127 (2012) 2314–2324.
[130] T. Dong, X. Yun, M. Li, W. Sun, Y. Duan, Y. Jin, Biodegradable high oxygen barrier membrane for chilled meat packaging, J. Appl. Polym. Sci. 132 (2015) 242–249.
[131] F. Wu, A. Pal, T. Wang, A. Rodriguez, M. Misra, A.K. Mohanty, Biodegradable multi-layer films for sustainable packaging, Submitted SPE ANTEC Conf. (2021).

Part VII

Structure design and modeling

Chapter 18

Analytical and modeling tools for flexible packaging design and process optimization

This chapter builds upon the proceeding chapters with a discussion of tools that can aid in the design of improved structures for flexible packaging. Chapter 1 introduces the general needs along the value chain for flexible packaging. For a specific application, these needs should be translated to CTQs (critical to quality attributes) that provide the criteria for designing the film structure. The CTQs typically involve sealing, barrier, strength, stiffness, abuse resistance, interlayer adhesion and optical properties described in Part IV while being subjected to end-use conditions and sustainability objectives (Part VI). The challenge to the package engineer is to meet the package CTQs with the optimum combination of materials at minimum cost. As is discussed in Part V, the converting process cannot be ignored; the process not only affects the economics of film production but also film properties. In addition, the combination of some materials may pose special processing difficulties. For example, flow instabilities, curl, adhesion and other factors may make producing a given design a challenge.

Several mathematical, statistical and numerical models have been introduced in previous chapters that aid in the fundamental understanding of process-structure-property relationships. Many of these same models are useful for developing packaging structures. This chapter summarizes the general approach to modeling for structure design, provides examples of models useful for product development and process optimization, and presents a case study of how they may be used in practice.

Often it is helpful to look at how similar products are packaged; or if the goal is to improve upon the existing structure, understand what it is made of and how it is manufactured. The first section of this chapter briefly describes analytical techniques that are helpful for structure identification.

18.1 Identification of packaging structures

There are many reasons for analyzing the structure of an existing commercial packaging film, including:

- Surveying the market to understand what types of materials are used.
- Investigating what structures are used to package similar products.
- Understanding how a competitor is approaching a similar application.

Optical microscopy, Fourier transform infrared (FTIR) spectroscopy and differential scanning calorimetry (DSC) are three invaluable analytical tools that are helpful for identifying the structure of multilayer flexible packaging films.

18.1.1 Microscopy

Microscopy is used to determine the number of layers in a flexible package film and their thicknesses. Generally, optical microscopy is sufficient, although scanning electron microscopy and transmission electron microscopy are sometimes used if very thin layers are involved, such as microlayer technology introduced in Sections 8.5.2, 9.4.7, and 12.2.4. In practice, the resolution limit of an optical microscope is about 0.2 μm [1]. A standard optical microscope with magnification of 4–100x works well for most identification needs. The ability to take a picture and calibrate the dimensions of the image is very helpful. Modern equipment can do this electronically.

The Science and Technology of Flexible Packaging. DOI: https://doi.org/10.1016/B978-0-323-85435-1.00020-X

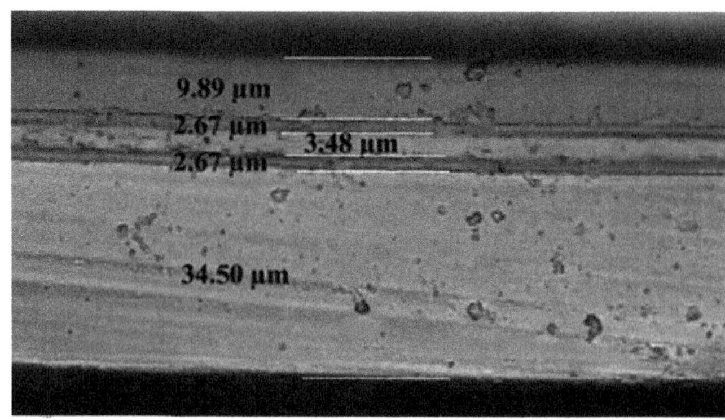

FIGURE 18.1 Cross-sectional view of a packaging film using optical microscope. *Courtesy The Dow Chemical Company.*

The film sample is viewed in the cross-section, as shown in Fig. 18.1. Sample preparation is critical to obtaining a good image without "ghost" layers. These are artifacts that appear as layers that are not really present. One method is to microtome the specimen. A procedure is described by Garcia-Meitin et al. [2]. Cryogenic microtoming of thin sections allows the microscope to be used in the transmission light mode. Differential interference contrast or cross-polarized light can be used to obtain details about interfaces, caulkability and seal failures in coextruded films [2].

Another specimen preparation method is to use a metal frame. The film is sandwiched between the ends of the frame and clamped in place. A razor blade is used to cut a fresh surface of the film flush with the ends of the frame. With practice and replacing the razor blade frequently, a good surface can be obtained for viewing under the microscope using reflected or transmitted light.

Staining can be used to help improve the contrast between layers and make a preliminary identification of the material type. One commonly used stain is tincture of iodine, which stains ethylene vinyl alcohol (EVOH) dark brown and polyamide 6 (PA 6) light brown.

In addition to determining the structure of unknown film samples, microscopy is very useful for quality control of layer thicknesses during coextrusion film production. Microscopy can also be used to establish whether layers contain blends with mineral fillers or with other polymers.

18.1.2 Fourier transform infrared

When infrared radiation (IR) is directed through a film, some of the radiation is absorbed by the material and the rest passes through. In IR spectroscopy, this transmitted radiation is detected and converted into a spectrum that reveals a unique fingerprint of the material. Fourier transforms are a mathematical construct used to convert the signal into a useful spectrum, hence the name "Fourier transform infrared" spectroscopy. FTIR is a powerful technique for identifying the types of polymers in a packaging film. Each polymer type has its own characteristic IR absorption spectrum. Analytical chemists can use FTIR to identify functional groups on a polymer by knowing the spectra of different molecular groups, such as acids or aldehydes. Some of these are tabulated in Table 18.1. Greater detail can be found in reference handbooks [3]. Databases of spectra of standard materials exist that the technician can use as a starting point to determine what materials are present. These libraries, however, may not provide details of specific grades or blends. This takes experience. Many resin suppliers and film converters have built up this experience over time and are good places to consult for help in identifying structures.

There are several different FTIR sampling methods, two of which are often used for film identification [5]

- **Transmission:** This is the simplest method and requires no special accessory. The film is put directly into the IR beam, which passes through the specimen. For multilayer films, however, this creates a spectrum of all the materials present. Some spectra may overlap, making interpretation of the spectrum difficult. Also, the film or laminate must be transparent for the signal to transmit through the specimen.
- **Attenuated total reflectance (ATR):** In this method, an internally reflected IR beam generates an evanescent wave that is contacted with the sample. The wave is attenuated in the regions of the spectrum where the sample absorbs IR radiation. The attenuated wave is recorded by a detector and is converted into an IR spectrum. The evanescent wave decays exponentially with distance from the surface, and so this technique is ideal for measuring the composition of surfaces of films, or for thicker films where strongly absorbing materials create intense peaks in transmission mode. The spacial resolution of ATR is about 150 μm and penetrates a depth of about 2 μm.

TABLE 18.1 IR frequencies of chemical compounds and examples of polymers that contain them [4].

Bond	Compound type	Polymer	Frequency range (cm^{-1})
C–H	Alkanes	PE, PP, ethylene copolymers	2850–2960 1350–1470
C–H	Alkenes	Unsaturated fatty acid slip agents	3020–3080 675–1000
C–H	Aromatic rings	PET, PS	3000–3100 675–870
C–H	Alkynes		3300
C=C	Alkenes	Unsaturated fatty acid slip agents	1640–1680
C≡C	Alkynes		2100–2260
C–C	Aromatic rings	PET, PS	1500, 1600
C–O	Alcohols, ethers, carboxylic acids, esters	EAA, EMAA, EVA, EVOH, PET	1080–1300
C=O	Aldehydes, ketones, carboxylic acids, esters	EAA, EMAA, EVA, EVOH, PET, PA	1690–1760
O–H	Alcohols, phenols	EVOH	3610–3640 3200–3600
N–H	Amines	PA 6	3300–3500
C–N	Amines	PA 6	1180–1360
C≡N	Nitriles		1515–1560
$-NO_2$	Nitro compounds		1345–1385

Source: Data from R.T. Morrison, R.N. Boyd, Organic Chemistry, third ed., Allyn and Bacon, Inc., New York, 1976.

Both ATR and transmission FTIR can be helpful for identifying layers in packaging films. ATR can be used on the outside surfaces and provides the most definitive identification (free of interference from other layers). Sometimes the film can be stretched to expose other layers in the structure which also can be identified with ATR. Modern FTIR equipment can be used in conjunction with microscopy to look at samples in cross-sections. Traversing across the section with multiple scans can help when trying to identify thin layers. Cross-sections often get distorted when they are mounted and thin layers may be below the resolution of the instrument. A single scan will likely be of a combination of layers rather than an isolated spectrum of a single layer. By taking multiple scans across the sample, the changes in the peaks can often provide a clue as to the identity of each layer.

FTIR techniques are very useful for identifying the kind of materials present. They may not be able to differentiate between types or grades of polymer within a polymer family. For example, different categories of polyethylene, such as high-density polyethylene (HDPE) and low-density polyethylene (LDPE) have similar IR spectra. Distinguishing between some types of polyamide or specific grades of EVOH has the same issue. Polymer blends may also be a challenge, especially if the spectra overlap. Sometimes FTIR can identify qualitatively if a type of polymer or functional group is present, but without careful calibration curves generated from known standards, it may not be possible to determine how much is present in a blend or coextrusion. If the film can be delaminated, ATR can be run on the exposed surface to determine the composition of that layer.

18.1.3 Differential scanning calorimetry

DSC is a method that measures changes in the heat capacity of a material. A sample of known mass is heated or cooled and the change in heat capacity is recorded as changes in heat flow. This allows the detection of thermal transitions: glass transition temperature, melting point, and freezing point [6]. These transitions can be very useful for identifying materials in an unknown packaging film. In particular, it may help identify the specific type of polymer within a polymer family: the melting point of LDPE differs from HDPE, for example. Different polyamides have distinctive melting

TABLE 18.2 Analytical techniques that may be useful for identifying the composition of a flexible packaging film.

Method	Uses for flexible packaging identification
Optical and electron microscopy	Number and thickness of layers
FTIR (Fourier transform infrared spectroscopy)	Chemical identity of layers
DSC (differential scanning calorimetry)	Thermal transitions, complements FTIR for identifying layers
DMA (dynamic mechanical analysis)	Thermal transitions
ESCA/XPS and TOF-SIMS (electron spectroscopy for chemical analysis or X-ray photoelectron spectroscopy and time of flight secondary ion mass spectroscopy)	Presence and composition of atoms of the material or additives on or near a surface
LC (liquid chromatography)	Molecular weight distribution, branching
TREF (temperature rise elution fractionation); TGIC (high-temperature thermal gradient interaction chromotagraphy)	Comonomer distribution in PE
NMR (nuclear magnetic resonance)	Chemical structure
Mass Spectroscopy/GC (gas chromatography)	Identity of volatiles
SAXS/WAXS (small angle or wide angle X-ray scattering)	Crystalline structure
Ash	Amount of inorganic filler
ICP/XRF (inductively conductive plasma spectroscopy/X-ray fluorescence spectroscopy)	Identification of metals

points, as do grades of EVOH. Table 16.1 provides thermal transition temperatures for some commonly used polymers in flexible packaging.

For multilayer films, sometimes the melt transition peak for one layer will overlap with another, hiding one or more of the layers (delaminating the layers prior to testing may help). Also, it can be difficult to differentiate between some polymers that have similar melting points, such as specific grades of ethylene vinyl acetate (EVA) and metallocene plastomers. For these reasons, DSC is often used together with FTIR and microscopy to determine the identity of each of the layers. One method alone may not tell the whole story.

18.1.4 Other techniques

In general, optical microscopy, FTIR and DSC will provide a good basis for the identification of a packaging film structure. They are easy to run, the results are straightforward to interpret and the equipment is relatively inexpensive to acquire. There are many other analytical techniques that may be helpful for answering specific questions regarding the composition of a packaging structure. Several of these are given in Table 18.2. Most of these techniques require expensive equipment to run and expertise to interpret the results. Also, many of these techniques work best on a single material—separating the layers may be needed before running them.

18.2 Modeling

Designing new flexible packaging structures can be a time consuming and expensive task. The packaging engineer must take into consideration many factors, including the package strength, ability to withstand abuse during filling and shipping, barrier, sealability, optical properties, and cost (including not only the cost of the materials, but also converting, packaging, waste, and cost to the environment). In addition to these performance and cost factors, the engineer also needs to determine if the proposed package structure can be made by converting equipment at a productive rate.

Often developing a new package structure requires several iterations of producing and testing prototypes. Unless developmental scale equipment is available in the lab, this involves breaking into production to make prototypes, which can incur significant costs and delays as commercial orders typically take priority over development trials. Testing at outside labs can also involve lengthy delays.

Computer modeling and simulation can help minimize iterations and allow the engineer to optimize function and cost on the computer before any film is made. It can also help determine if problems such as poor extrusion, flow instabilities or curl may occur during film fabrication. If so, potential solutions involving material selection, structure and process can be scoped out prior to running on the production line. Process models can also be used to troubleshoot problems with the production of existing structures.

18.2.1 Types of modeling

There are three general types of models that are used in flexible packaging: attribute; empirical/statistical; and fundamental/theoretical.

18.2.1.1 Attribute models

Attribute-based modeling is a way to quantify choices amongst options that have several qualities. One form of attribute modeling is database filtering such as polymer or grade selector tools that allow the user to select property ranges. The tool filters the database based on the user input to show materials that fit the criteria. Examples include general polymer selector tools such as Matweb, Prospector, and others [7–10], grade selector tools on resin producer websites [11–13], and general engineering databases such as Knovel [14]. These tools are more suited for mono-material applications but can be useful even for multilayer film structure design to zero in on possible resin choices.

An example of an attribute model that provides the user with more interpretation is a sealant selector tool developed by DuPont [15–18]. This tool embodies the information in Section 7.11. The user selects the attributes that are most important for their application and ranks their importance on a scale of 1–10. For example, the sealant layer in the package may be called upon to provide low-temperature sealability, be transparent, and provide grease and oil resistance. Some sealant resins may work well for one or more of the attributes but not as well as other options for the other attributes. The model embodies the knowledge of an expert panel that has rated different sealant polymers based on these attributes. The model calculates a numerical score for each resin type based on the importance of the attribute to the user and the performance of that resin based on the expert panel. The output provides a quantitative comparison of the resins as a starting point for the user to consider in the package design. An example illustrating the use of this tool is provided in Section 18.2.3.

Attribute models are a type of expert system. As well as providing selection tools such as the one illustrated above, they are useful for capturing the expert knowledge of an organization, evaluating project ideas, and many other things. A simple version of an attribute model is a cause and effect matrix. This is typically a spreadsheet, as illustrated in Fig. 18.2. Here the products or ideas being evaluated are listed in the first column. Along the top, the attributes that are important are listed. For each attribute, the user rates its importance, on a scale of 1 to 10. These are placed above the attribute in the spreadsheet. An expert panel (which may be the user) rates the performance of each product against each attribute on a scale of 1 to 10. A numerical score is computed by summing up the products of the important factors and attribute ratings. The resulting scores show the relative performance of each product weighted by the importance of the attributes.

18.2.1.2 Empirical and statistical models

Statistical or empirical-based models do not rely on the underlying science behind the property being considered. Instead, they use statistically based tools to develop predictive equations. Often industrial systems are complex with a number of possible variables and phenomena that may influence the outcome of a particular test result. In these situations, theoretical understanding may not exist, or what does exist does not explain all of the variability in the system. Furthermore, complex systems can be very difficult to model theoretically or numerically. The use of statistics allows predictive capability where it may not otherwise be possible, and often with a reduced number of experiments.

One of the most powerful techniques in empirical modeling is the statistical design of experiments (DOEs). In scientific experiments, often all the possible variables are held constant except for one, which is varied systematically to determine the response of the system. If the system has many variables, varying one at a time will require a substantial amount of experiments. DOE uses statistical-based algorithms to economize on the design by simultaneously combining multiple factors. Rather than varying one variable at a time, all the variables are varied and only a few of the possible combinations are used. The results are analyzed to determine which variables and their interactions are statistically important and a statistical-based model is obtained. Many good references are available that provide details into the method [19–27].

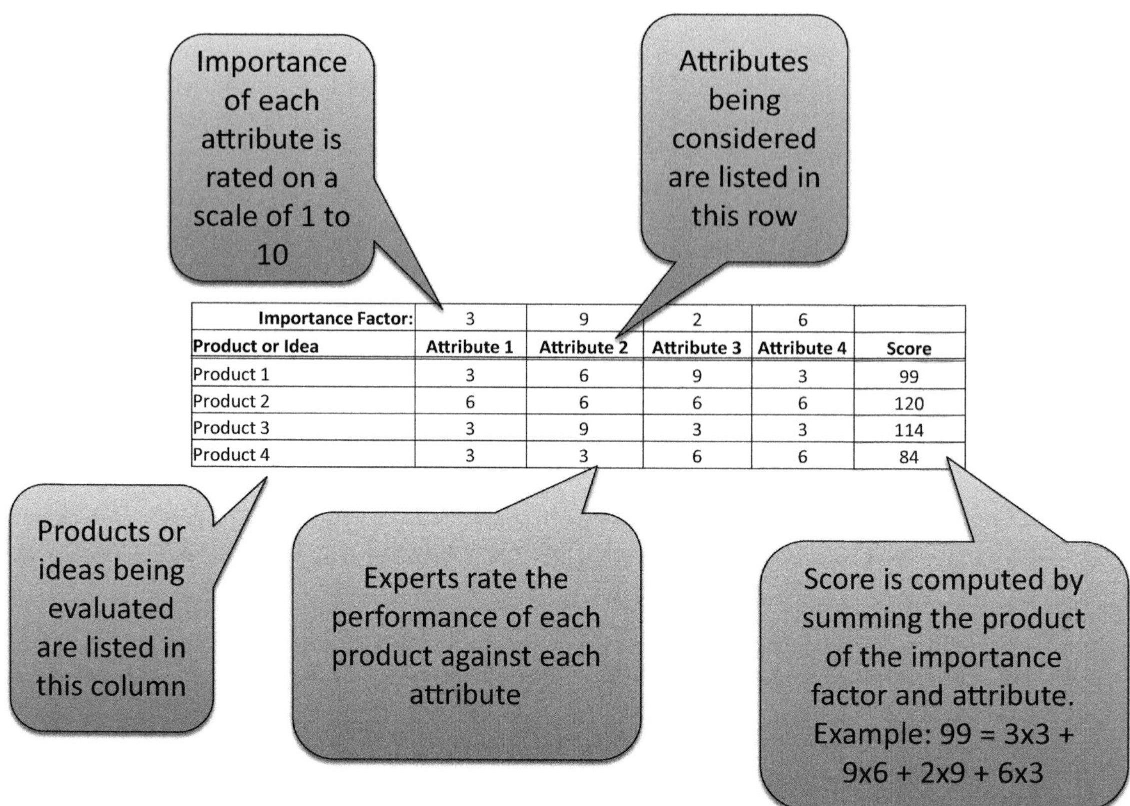

Importance Factor:	3	9	2	6	
Product or Idea	**Attribute 1**	**Attribute 2**	**Attribute 3**	**Attribute 4**	**Score**
Product 1	3	6	9	3	99
Product 2	6	6	6	6	120
Product 3	3	9	3	3	114
Product 4	3	3	6	6	84

FIGURE 18.2 Cause and effect matrix for attribute modeling.

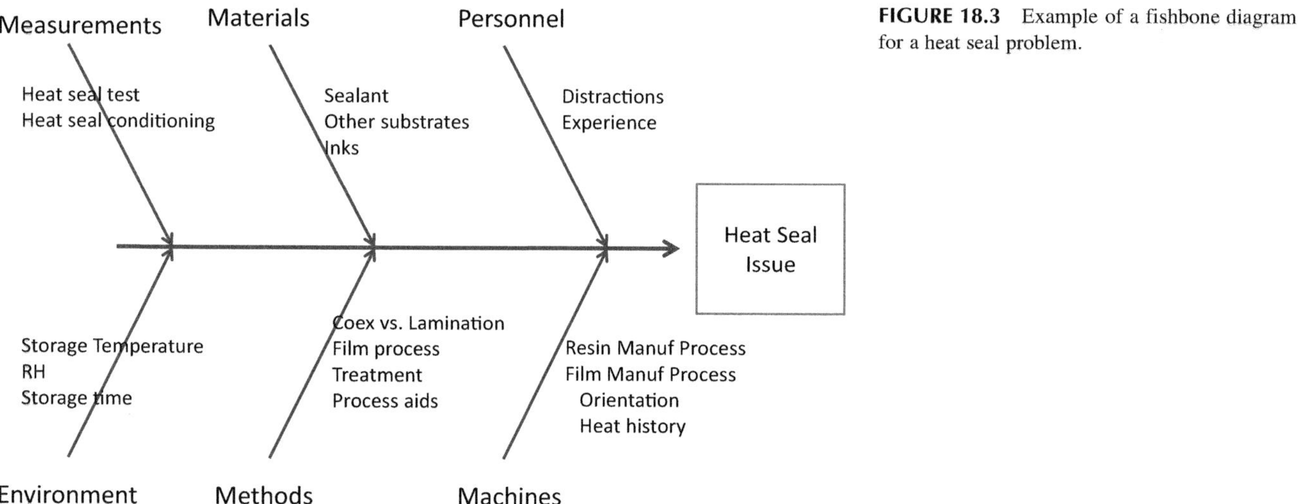

FIGURE 18.3 Example of a fishbone diagram for a heat seal problem.

The first step in designing a DOE is to determine the independent and response variables. The independent variables are those factors that may influence the outcome of the experiment. It is important at the beginning of a project to consider a wide range of possible factors that may affect the final properties. One tool that may be useful is to construct a fishbone diagram, which can help identify possible factors to consider or causes of a problem. An example, given in Fig. 18.3 for a heat seal issue, resulted from a team consideration of possible causes in the general categories that make up the fish bones: personnel, materials, measurements, environment, methods and machines. From this input, the most important factors can be used to conduct the experiments and construct the model.

The response variables are the outcomes, such as tear strength or impact resistance. Multiple response variables may be measured and analyzed. One of the benefits of the analysis is that it can be used to optimize for more than one response, such as maximum tear resistance with minimum haze.

Often there are a number of possible independent variables that may impact the responses. A two-step process is used to decrease the number of experiments. A screening design uses just a few experiments, typically on the order of 10–20 states, to test up to 10 variables for importance. The model that arises from such an experiment is enough to discern the direction of the response (such as increasing film thickness increases impact resistance) but does not provide information on interactions between variables. A second experiment is typically conducted using the three or four most important variables coming out of the screening experiment. Called a response surface design, the second experiment will typically involve more states to determine curvature and interaction effects.

Commercial software is available to design and analyze DOE's, such as JMP [28] and Minitab [29]. A number of designs are possible and the complexity of the design depends on whether the variables are continuous (e.g., thickness or temperature) or categorical (nonnumerical, such as grade names or polymer types), and whether any variables are constrained in certain regions of the design space (some combinations of variables may not be possible, such as high haul-off speed and high film thickness). Statisticians recommend that when conducting the experiment, the order of running the experimental states should be randomized. If some variables are more difficult to change (e.g., extruder temperature), blocking can be used to minimize the number of transitions.

One of the most useful types of DOE is factorial design. A simple 2-level factorial design considers the high and low value of each independent variable. In a full factorial design, each combination of factors is run. The number of runs or states is simply 2^n where n is the number of variables. For example, if there are 3 variables, the number of states is 8. Such an experiment can be conceptualized by the corners of a cube, as shown in Fig. 18.4. Here the range of each variable is arbitrarily set from 0 to 1 (low to high). The responses can be plotted on the corners of the cube to visually see the effects of the variables. In the example in Fig. 18.4, the results show that Variable C has the strongest effect: increasing C increases the response. Variables A and B have weaker effects by themselves, but have some strong interactions. For example, at low A, increasing B increases the response, but at high A, the opposite is true.

For more general designs contour plots are often used to graphically depict the results. Examples of contour plots can be seen in Figs. 10.19 and 10.20.

To complete the DOE replicates are needed for two or more states so that the statistical analysis can calculate the replicate error. Once the replicate error is known, the software computes the level of confidence that one or more of the variables has a statistical impact on the response. In statistical terms, the analysis computes a p-value for each term of the model. If the p-value is less than 0.1, the independent variable is important.

An important consideration when designing a DOE experiment is to choose the range of each variable to be as large as practically possible. This is because statistical models generally work well inside the original design space, but not very well when extrapolating outside the design space. This is one of the fundamental limitations of this approach.

Many DOE experiments are reported in the literature. For example, Borse and Aubee [30] use a full factorial DOE approach to optimize the properties of a three-layer blown film. The structure is (sLLDPE—PE blend—sLLDPE) where sLLDPE is a single site catalyzed LLDPE and the PE core layer is a blend of either butene or hexene LLDPE (Ziegler–Natta) with LDPE. The independent variables are the ratio of the skin to core layers, the type of LLDPE in the core (butene or hexene), the percentage of LDPE in the core layer and the film gauge. The responses include a number of mechanical and optical properties. One set of results is reproduced in Fig. 18.5, which shows the predicted

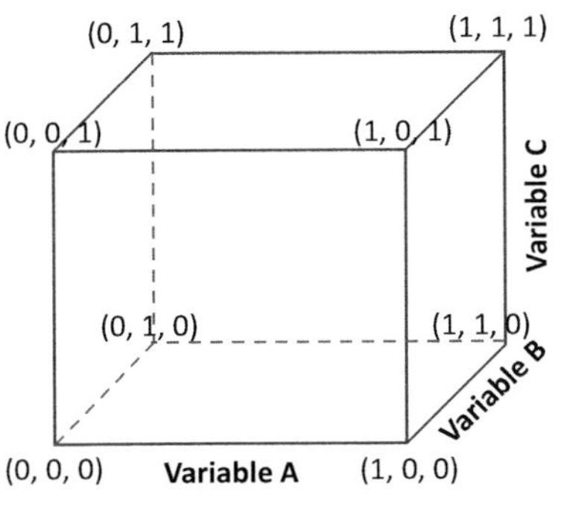

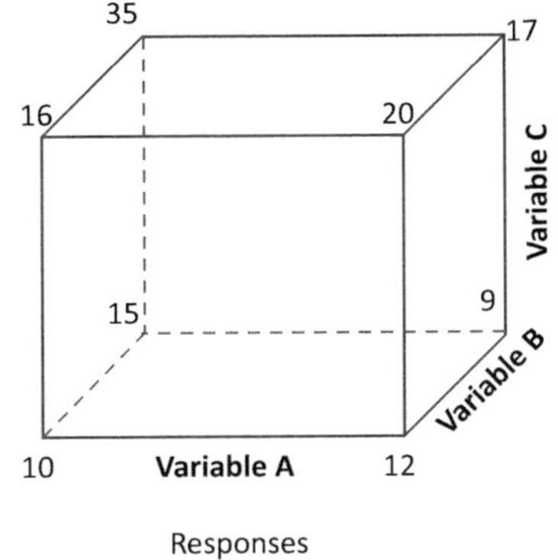

FIGURE 18.4 Graphical representation of a 2-level, 3-variable factorial design.

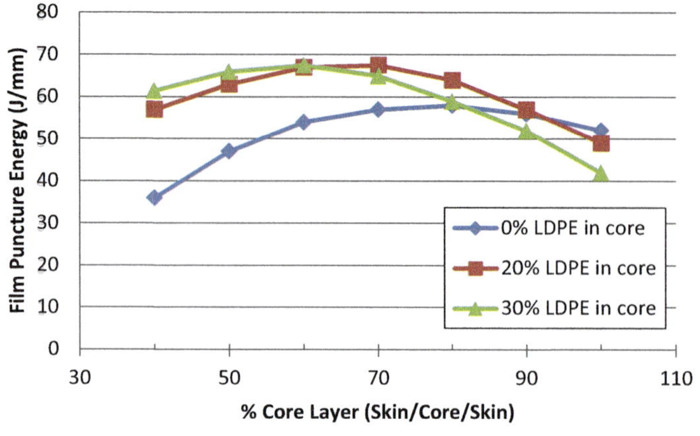

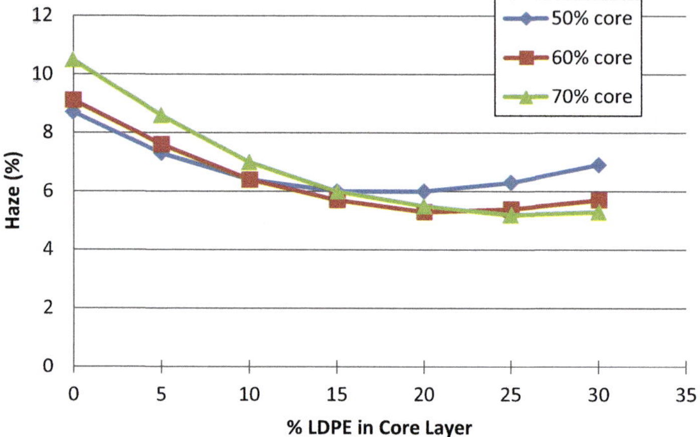

FIGURE 18.5 Effect of layer ratio and % LDPE in the core for a 25-μm coextruded blown film with structure (sLLDPE–bLLDPE + LDPE core–sLLDPE). Top graph is statistical model predictions for film puncture energy and bottom graph is for haze. *Data from N. Borse, N. Aubee, Predicting physical and optical properties of co-extruded blown films using design of experiment based model, SPE ANTEC Conf., (2015) 1584–1588.*

response of varying layer ratio and % LDPE in the core for a 25-μm film with the core layer containing b-LLDPE. The results show that low gloss and high puncture energy can be obtained at 20% LDPE in the core with a layer ratio of 20/60/20.

Several examples of DOE experiments and models are given in this book. One can be found in Section 10.4.1.4 where a response surface design was used to explore the effects of line speed, extrusion temperature, coating thickness and percent of a modifier added to LDPE on the peel strength of LDPE to aluminum foil in extrusion coating. The model predicts the processing conditions needed to obtain a given peel strength at the optimum additive level. Another example can be found in Section 13.5 on curl.

Regression analysis and other techniques used to fit experimental data with a mathematical formula is another type of statistical model. For example, Blom [31] (Section 9.4.3.4) uses graphical and regression techniques to gain insight into the abrasion resistance of plastic films. The work led to the definition of an abrasion coefficient (Eq. 9.12) as a material parameter.

Todd and Podborny [32] use a nonlinear regression model to develop a new grade of HDPE with a good balance of mechanical properties for a barrier film application. The barrier of HDPE film generally increases with higher crystallinity (density) and melt flow index (MFI). Higher MFI decreases film mechanical properties. What was unknown is the effect of molecular weight distribution (MWD). Blown films from 14 different HDPE resins with different densities, MFI and MWD are made and tested for Elmendorf tear strength, dart drop impact resistance and moisture vapor transmission rate (MVTR). A statistical model is fit to the results and used to design a new resin, which when tested, validates the model.

A combination of theoretical and statistical analysis can be used to produce models that better fit the data. Wang et al. [33] look at the effect of LLDPE resin and film processing parameters on dart impact strength. These parameters include MW, MWD, and comonomer distribution, as measured by MFI, stress exponent and TREF (temperature rise elution fractionation); melt relaxation time from rheology measurements, and blown film die gap, film thickness, blow-up ratio, melt temperature and output. Rather than simply use the parameters as is, some are combined or transformed into fundamental factors, such as the strain rate and a corrected Deborah number [see Eq. (15.5)]. Wang et al. find that the fit is better than simply using the raw parameters. The resulting model is highly nonlinear with significant interactions. For example, the dart impact strength depends on the Deborah number raised to the power of 3.75 and the strain rate to the power of 3. The model is fairly robust and is able to predict dart impact strength of resins made on pilot and commercial resin manufacturing lines, as well as resins from a competitor.

Section 15.1.2.2 describes how common extrusion coating process parameters are related to fundamental factors that describe the stress imparted on the melt curtain during coating, such as the stress imparted in the extruder and die, elongation and rate of elongation in the air gap, and the process time. Because of the interactions amongst the parameters, a DOE approach is used and the results are fit to a quadratic statistical model. The results of the analysis (see Figs. 15.16 and 15.17) are used to draw conclusions about the effect of stress on peel strength performance [34].

As shown in Section 15.1.4, the use of DOE models can be used to test hypotheses about the fundamental factors that influence the properties of films and coatings. In this case study, the model developed in Section 10.4.1.4 is used to test potential mechanisms governing the effect of coating thickness on peel strength.

The primary limitation of DOE and other statistical modeling approaches is that they generally cannot be extended beyond the original experimental design space. They are excellent for troubleshooting problems and optimizing processes or film structures, but for totally new structures and processes they may be limiting.

18.2.1.3 Theoretical/fundamental models

Theoretical or fundamental models are derived from scientific or engineering first principles. Examples include the bending stiffness (9.4.2), curl (13.5.3), heat seal (7.6), permeability (8.4) and heat transfer (15.1.2 and 15.1.3) models discussed in this book. These are derived from equations that describe the underlying physics or chemistry behind the phenomenon.

Numerical or computer simulations also fit into this category. Examples include simulations for extrusion, blown film and cast film. Engineering analysis of films and packages is another example, using finite elements, finite difference and finite volume numerical techniques to optimize package design.

Theoretical models are very powerful since they can be used to extend beyond the original design space, provided they accurately capture the fundamental mechanism behind the response. They have their limitations, however. One is that industrial processes are generally complex and noisy. Several phenomena or mechanisms may be influencing the outcome that a simple model cannot accurately project.

Fully representing a physical phenomenon requires complex partial differential equations in three spacial dimensions and a time component. For example, Agassant and Mackley [35] describe 10 equations that must be solved for polymer flow simulation arising from mass, momentum and energy balances. Coupled with this are complex constitutive equations that describe relationships between stress and flow. Solving these systems of equations analytically usually requires many simplifications to make the math tractable. For example, the flow may be assumed to be Newtonian when it actually is highly viscoelastic. The system may be assumed to be isothermal even though temperatures are changing. These simplifications must be carefully evaluated for the system being analyzed. In many cases, simple analytical models can provide tremendous insight into complex processes but may result in a significant error when making quantitative predictions.

The alternative is to solve the equations numerically. Tremendous strides have been made in the last 50 years in the field of numerical simulation, but there still are many limitations. Agassant and Mackley [35] discuss a number of these limitations as they pertain to flow simulation, but the comments apply to modeling mechanical processes as well:

- Finite element, volume or difference techniques are typically used. These involve dividing the space into a mesh of volumes. The accuracy of the simulation can depend on how representative the volumes are, which can be computationally difficult for complex flow situations.
- Nonlinear behavior is difficult to model. For flow simulation, for example, constitutive equations relating the stress tensor to the rate of strain tensor may have many parameters that are challenging to measure with standard lab techniques.
- The most sophisticated constitutive equation may not be the best for a given simulation. It will depend on the type of flow, the available computing power of the facility, and the ability to measure the parameters.
- Direct analysis can take a lot of computational time and may not be practical for parametric studies. The equations can sometimes be simplified through the use of dimensional analysis and other techniques. Often it is not obvious, however, whether the direct or approximate approach is best for a given situation.

- Modeling has become progressively more sophisticated over the years. It is now possible to use commercial software, such as Moldflow or ANSYS Polyflow in the area of polymer flow simulations, to simulate many industrial processes. A trend is toward a "black box" approach where the user simply plugs in the conditions into the model. This may be fine for some situations, but often complex flows require some judgment as to what approximations to make. It is best to develop a good physical understanding of the process, and make appropriate simplifications and realistic rheology assessments before using the black-box approach. Extensional flow remains an area where accurate simulation is uncertain. No universal constitutive equation exists for single-phase polymer melts, much less blends and other multiphase materials. Melt transitions (solid to melt and vice versa) are still a challenge to model accurately. Crystallization is particularly a challenge, and the effect of flow exacerbates this. The future challenge is to model orientation and crystalline morphology development to predict mechanical properties.

When it comes to numerical simulation of the mechanical performance of films and packages, obtaining the model material parameters at the appropriate rate of strain is challenging. The properties of plastics are rate dependent and less understood from a modeling perspective than metals. The lower stiffness and deformation resistance of polymer films, as well as inhomogeneous strain fields, makes testing of polymeric materials difficult [36]. Special test rigs may need to be developed [37]. Another approach, employed by Carbajal et al. [38], is to model the test method itself. To model high-speed impact puncture resistance of multilayer films, they needed high-speed mechanical properties of the individual layers. Using lower speed tensile properties as a starting point, they tweaked the parameters by comparing numerical simulations of the high-speed test rig with experimental results. Their work is explored more fully in the next section.

18.2.2 Examples of models and their use in flexible packaging

Models for flexible packaging can be grouped into two categories, package design and processing/converting. The package design tools help the user select materials and design the film and/or laminate structure to meet the packaging needs and minimize cost. The process tools help the user determine if the structure can be made and troubleshoot in advance potential problems. A summary of the modeling tools described here is provided in Table 18.3. Many of these tools were introduced in earlier chapters and are helpful for understanding the science behind the technology of flexible packaging.

18.2.2.1 Package design

As described throughout this book, most flexible packages have multilayer constructions to achieve the desired functional requirements in a cost-effective manner. Each layer has its role. Often multiple materials are available for each functional role, and some materials may play more than one role. Various modeling tools are available or can be developed which help optimize the structure and select the best materials for a given application.

In many packaging applications controlling the permeation of gases into or out of the package is important. To limit oxygen permeation, barrier polymers such as EVOH, poly vinylidene chloride (PVDC) or polyamide 6 (PA 6), or barrier films, such as aluminum foil or metalized films (e.g., oriented polypropylene (OPP), oriented polyester (OPET) or oriented PE (OPE)), are incorporated into the structure. In modified atmosphere packaging, the headspace is injected

TABLE 18.3 Examples of models for package development.

Package design	Processing/converting
Oxygen and moisture permeability	Extrusion simulation
Sealant selection	Blown/cast film/extrusion coating simulation
Migration/impartation	Flow instability in coextrusion
Sealant performance	Heat transfer in air gap and nip
Tie resin selection	Curl
Physical properties: bending stiffness, other	COF/slip migration
Physical properties: engineering analysis	Adhesion
Cost analysis	
Life cycle assessment	

with a mixture of gases designed to extend the shelf life of the product. It is important that this mixture not permeate out of the package. In other applications, a moisture barrier is important. The permeability of gases through a multilayer package is a function of the inherent permeability and thickness of each of the materials.

Chapter 8 develops the equation for predicting the permeability of a multilayer film structure given the permeability and thickness of each of the layers [see Eq. (8.21)]. This equation can be incorporated into a permeability calculator to help select the best resin and thickness for a given barrier need. Such tools can be found online or can be easily programmed into a spreadsheet. More sophisticated models use finite elements and other techniques to compute the oxygen ingress of complex packages (e.g., bottles) or to predict the composition of modified atmosphere packages over time. The simple approach works well for many applications.

The sealant is the innermost layer of the package that provides a hermetic seal to protect the product. Ethylene copolymers are often used as sealants because of their low melting points. These include low density polyethylene (LDPE), EVA, linear low density polyethylene (LLDPE), metallocene polyethylene (mPE) plastomers, acid copolymer resins (ACRs) and ionomers. The selection of the best sealant for a given application depends on the sealing needs as well as other contributions the sealant makes to the overall performance of the package. Several such attributes are described in Section 7.11. An attribute-based model for selecting the sealant that best meets the criteria is described in Section 18.2.1.1.

Migration of unwanted compounds (monomers, oligomers, additives, solvents, etc.) from the package to the product may affect the sensory perception of a food product (see Section 16.2.2.1) as well as have food regulatory compliance issues. The European Union Commission Regulation No. 10/2011 on the use of plastic materials that come in contact with food has specific migration limits for compounds [39]. The regulation calls for the testing of the migration of these compounds in food simulates according to food type and shelf-life requirements. Such testing often requires sophisticated techniques and can be very expensive. The regulation allows for the use of migration modeling to predict migration under extreme conditions so that the regulations are not exceeded.

A number of migration models are available commercially, including those by Combase [40], AKTS [41], and FABES [42]. These models numerically solve the time-dependent diffusion equation [see Eq. (12.11)] by finite difference or element techniques to estimate the maximum amount of the compound that migrates into the food [43–47]. Similar to the model described in Chapter 12 for slip additive migration, these models use as inputs the partition coefficient of the compound between the contact layer and food medium and the diffusion coefficient of the compound in the polymer. The former is estimated based on the polarity of the compound and medium derived from measurements in octanol and water. The diffusion coefficient is estimated based on generally recognized molecular diffusion models for noninteracting compounds where only the molecular weight of the compound, storage temperature and several parameters specific to the polymer are required [41,45]. This allows for a wide range of migratory species and polymers to be modeled.

Cost is an important consideration, both financial cost as well as cost to the environment, and is covered in detail in Chapters 16 and 17. Down gauging the thickness is one approach to reducing cost, but unless material or process changes are made, the overall stiffness and strength of the package may suffer. Bending stiffness is a complex function of the thickness, modulus and position of each of the layers. A model is described in Section 9.4.2 that has proven very useful for designing structures with less material without sacrificing bending stiffness. This is accomplished through material substitution and position, sometimes in ways that may not be thought of without the use of the model [48]. The model can be combined with information on the material cost, as well as environmental cost from a life cycle analysis (described in Section 17.2.1), to provide overall guidance on cost versus performance [18].

Other physical properties of multilayer film have also been modeled, usually with statistical-based approaches. Examples include abrasion resistance [31], tear-resistance [30,49], energy to break [30], and tensile strength [30]. Carjabal et al. [38] use finite element analysis to model high-speed impact resistance of multilayer films. Their approach is worth highlighting since their combination of experimental and modeling work allowed them to overcome the challenges with material parameter measurement.

As described in Chapter 9, Carjabal et al.'s [38,50] study is motivated by packaging failures during filling of crackers (also known as biscuits or wafers in some parts of the world) on vertical form-fill-seal lines. The drop of the sharp cracker creates pinholes as it impacts the package. Two characteristics of their study are important for the model development. First, they took the time to understand the fundamentals of what happens as the cracker impacts the film. Using high-speed video they record individual drops. They find that (1) the impact is a high-speed event (around 4 m/second), (2) the cracker does not significantly deform during the event, and (3) the size of the pinhole is much smaller than typical laboratory impact tests. Based on this information they design, test and validate a new laboratory test that is described in Section 9.4.3.2. The fundamental study also provides insight into how to model the impact. Because the video shows that the event is only a few thousandths of a second in duration, it is necessary to select software that can deal with short time frames. The lack of deformation of the cracker allows them to represent the cracker in the model as a rigid object, simplifying the calculations.

Second, the test method they develop can also be used to fine tune the material parameters needed for the model. The test, a high-speed reverse impact test, provides information on the work versus displacement during the puncture of the film with a needle probe. Since the model is expected to provide insight into the contribution of individual layer properties, thickness and position within the film structure, each material must be considered explicitly. This requires material parameters at high strain rates. Films of the individual layers are made and the tensile properties at low strain rates are measured. For each individual material, the work versus displacement is measured using the reverse impact tester. Using the low strain rate tensile properties as a starting point, the parameters are tweaked until there is good agreement between the model prediction and the experimental result. With the material parameters for the individual layers in hand, the model can be extended to multilayer structures.

The approach proves to be very successful. A comparison of the model predictions with the test results of the ultimate work to puncture for seven different barrier snack films is shown in Fig. 18.6. The films have the general structure (HDPE–tie–PA 6–tie–sealant) and vary by composition (% PA, thickness, etc.). The model and test results are in excellent agreement and are able to discriminate amongst the samples. The model also provides insight into the dynamics of the impact event, such as the fluttering of the film (Fig. 18.7) that qualitatively agrees with high-speed video images. Further, the model parameters are the same as those used in a second model of a bullet drop test, which is utilized in the initial study to validate the reverse impact puncture test. Bullets are chosen since they have better flying stability than crackers (or screws, which are sometimes used in empirical drop tests). The bullets have similar weight and tip dimensions as the crackers. The model predictions agree well with the measured puncture velocities of the multilayer films. Now that the model is correctly predicting the impact performance of the test films, it can be used to explore design concepts such as replacing or repositioning materials within the structure.

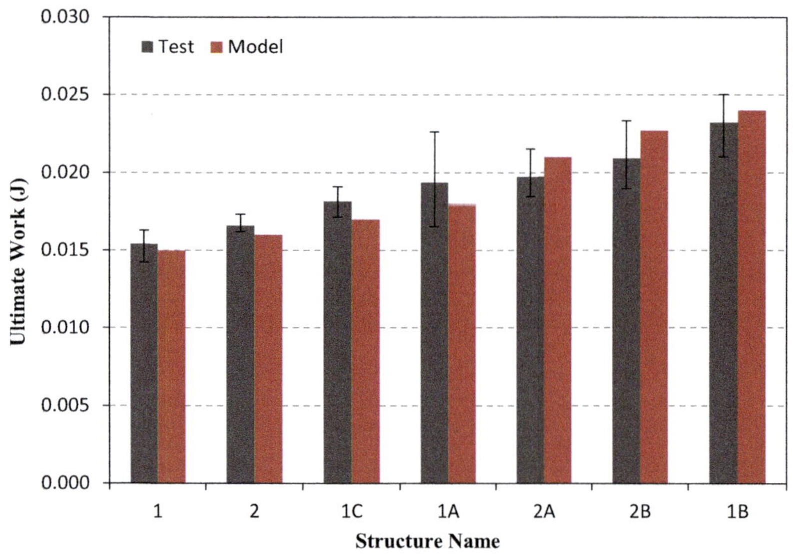

FIGURE 18.6 Comparison of model prediction with test results for ultimate work for failure in high-speed reverse impact puncture test. Data is for seven five-layer barrier blown films. *From L.A. Carbajal, R. Jiao, D.M. Hahm, B.A. Morris, R.R. Kendzirski, Predicting the impact puncture response of multilayer flexible food packages using explicit finite element models, SPE ANTEC Conf., (2016). Reproduced with permission from The Dow Chemical Company.*

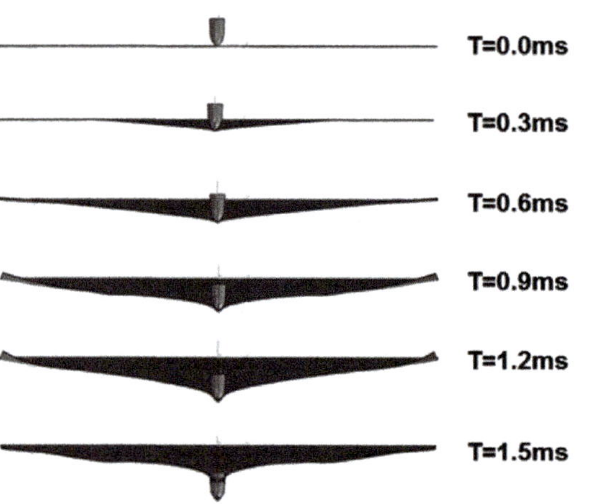

FIGURE 18.7 Typical predicted deformed shapes of during impact of a falling object onto a film. *From L.A. Carbajal, R. Jiao, D.M. Hahm, B.A. Morris, R.R. Kendzirski, Predicting the impact puncture response of multilayer flexible food packages using explicit finite element models, SPE ANTEC Conf., (2016). Reproduced with permission from The Dow Chemical Company.*

Once a candidate's packaging structure has been determined, it is helpful to know how well it will seal. This will determine what packaging line speeds may be obtainable, which affects the total cost to the packer. Morris [51] and Moffitt [52] have developed heat seal models that can help with this assessment; these models are described in Section 7.6.2. They take into account the heat transfer through the film, melting of the sealant, molecular diffusion and strength build-up. As is shown in Section 7.6.2, Morris's model agrees well with experiments on a packaging line.

Cost is an overriding consideration of package design. Total cost to the packer should be considered, including the roll stock costs (which incorporate material and converting costs) and packing costs. The packing costs will depend on the packaging labor and equipment costs (including overhead), packing efficiencies and waste. Using a premium sealant, for example, may increase the roll stock cost but reduce the overall cost by increasing line speed and reducing leakers and waste. Section 16.5.6 describes a simple packaging cost model that incorporates these concepts [15].

An engineering analysis using numerical simulation such as finite element analysis (FEA) can be helpful in some situations. FEA is widely used in rigid packaging but is only beginning to see use in flexible packaging. Yuan and Haynes [37] explain that the flexibility of flexible pouches creates simulation challenges such as structural instabilities, wrinkles and complex contact situations. Accurate characterization of the film for the specific loading situation is critical to success. But once done, it can be very useful for package design.

Morris [18] provides an example where FEA is used for the design of the seal geometry of a dual compartment pouch. The concept for the design is shown in Fig. 18.8. When the consumer squeezes the pouch, the internal seal bursts and the contents mix. The design of this structure requires a special ionomer sealant as well as a seal geometry that focuses the stress in a way that facilitates opening between compartments and not at the pouch edge. An FEA model is used to predict the forces on the seal area of different seal configurations. One such example is shown in Fig. 18.9.

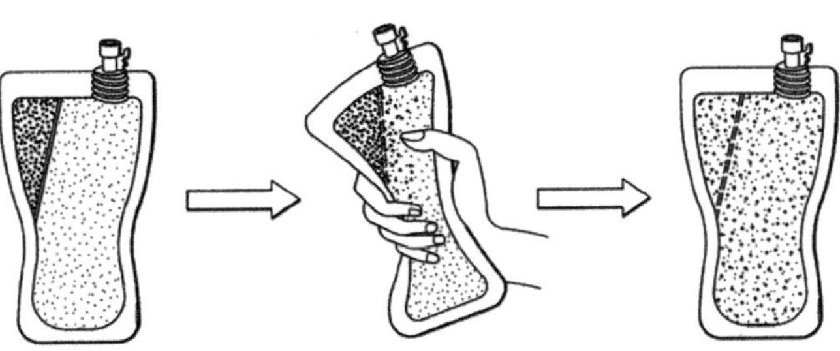

FIGURE 18.8 Dual compartment pouch concept [53]. *Reproduced from R.A. Bourque et al., Multiple compartment pouch and beverage container with frangible seal, US Patent 7306095 B1, 2007.*

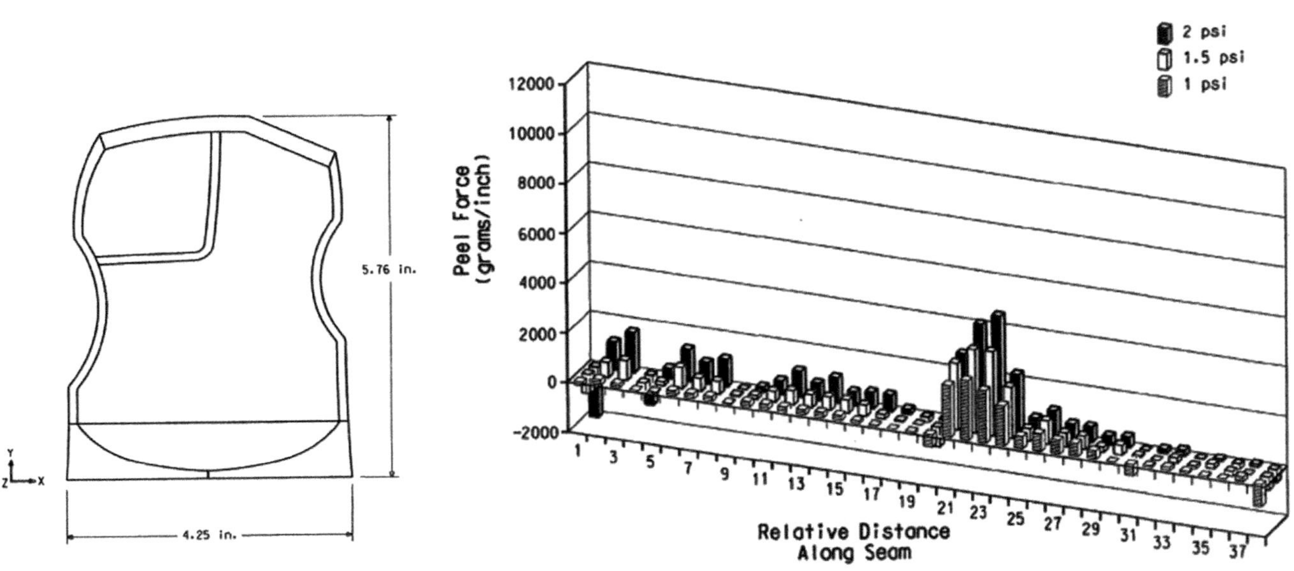

FIGURE 18.9 Prediction of peel force (right) using finite element analysis of a dual compartment pouch (left) at different pressures as a function of position along the interior seam. *Reproduced from R.A. Bourque et al., Multiple compartment pouch and beverage container with frangible seal, US Patent 7306095 B1, 2007.*

Lokhande et al. [54] use FEA to model the deformation behavior during a Gelbo film test, a laboratory flex fatigue test often used to simulate flexing during shipping (see Section 9.4.5). Two cases are analyzed. One uses simple elastic film behavior, which is found to capture the complex mechanics of the Gelbo test. The results show that high stresses develop along the creases during flexing. The locations of the high-stress zones correspond to experimentally observed pinhole locations. The maximum stress increases with increasing stiffness and thickness of the film. In the second case, two LLDPE grades are compared using an elastic-plastic model to approximate the film material behavior. One grade has a low stiffness-to-toughness ratio and the other a high stiffness-to-toughness ratio. The model results predict that lower stiffness-to-toughness ratio LLDPE performs better which is verified with experimental data.

Yuan and Haynes [37] provide six additional case studies that illustrate the utility of applying FEA modeling to pouches:

- **Case 1:** Venting seal bar design for microwavable and retortable pouches. The seal must be strong enough to withstand the temperature and pressure during retort sterilization (121°C, 0.10–0.17 MPa) without debonding, but weak enough to open up after 90 seconds of microwaving for venting at 99°C and 0.027–0.034 MPa. One challenge for the simulation is obtaining a good debonding model, which required measuring peel strength at different angles. This underscores the importance of good experiments to go along with the model development. The model successfully predicts the pressure of the venting as well as the pressure and location of where partial separation occurs in the package.
- **Case 2:** Stand-up pouch filling. An FEA model is developed to simulate a filling operation: folding, followed by opening for filling, stand-up base opening, filling with liquid or solid, and heat sealing. The model is used to determine the optimum location of side suction cups, the magnitude of air pressure and grip locations.
- **Case 3:** Pouch heat sealing. The top of the pouch is initially an oval during sealing; closing and sealing without forming wrinkles is a challenge. An FEA simulation successfully captures the formation of a wrinkle caused by improper transfer grips and is used to design a solution.
- **Case 4:** Retort sterilization. The integrity of the seal can be compromised by the creep behavior of the polymer and hot spots that occur during the cooling portion of the retort cycle. Simulation is used along with a seal failure model to predict where the hot spots will be, accounting for the creep behavior.
- **Case 5:** Pouch drop impact. In this dynamic situation the critical performance attribute is ductility under high strain rate loading and under biaxial stress. The peak strain rate in a 0.9 m (3-ft) drop can be 30 s^{-1}, 100 times higher than in a tensile test. The stress state dependence of the material is needed since the strain rate and stress differ at each location during the impact event. The authors devise custom fixtures to test the materials in the lab at high strain rates and stresses. The comparison between the drop impact test results and the simulation is excellent.
- **Case 6:** Pouch tear-opening. Consumers want a consistent, easy-tear opening experience. Developing a good propagation of tear model is important. The material properties may change due to thermal exposure during retort or hot fill. These changes need to be accounted for in the propagation model. If done properly, the tear performance can be correctly predicted.

18.2.2.2 Processing/converting

The first step in converting polymer to film is extrusion. General three-dimensional computational fluid dynamics simulation software tools from such companies/foundations as ANSYS (Polyflow), ABAQUS, OpenFOAM Foundation (OpenFOAM) and Compuplast (three-dimensional FEM module) can be used to analyze polymer flow in an extruder but these may be too detailed for the general practitioner. While useful for analyzing specific issues, more tailored software simulation tools for single-screw extrusion are commercially available, including those from Compuplast, Polydynamics, and Paderborn University. These programs simulate the solids conveying, melting, and pumping or metering sections of the extruder based on fundamental models from the literature [55,56]. They compute melting, temperature, and pressure profiles along the extruder as well as output rates. They are particularly useful for extrusion engineers for designing screws and troubleshooting extrusion issues [57]. They have limited value for packaging design although may be useful if new materials are being considered.

Combining dissimilar materials in coextrusion can result in flow instabilities as the materials are combined or near the die exit. These instabilities manifest themselves as “wood grain,” “orange peel,” or waves or zig–zag patterns. Sophisticated numerical simulation has been employed to understand the origins of these instabilities, as described in Section 13.4.5. These models often require detailed knowledge of the flow channels and material parameters that may be difficult to measure. A simpler modeling approach is outlined in Section 13.4.5.2. Commercial software is also available to help with these calculations. These tools are helpful for flagging potential problems with matching viscosities and developing strategies for dealing with them, such as changing grades, thicknesses, output or flow channel dimensions. In the design phase of new packaging structures, these tools may help the packaging engineer avoid combinations of materials that may prove difficult to coextrude.

Numerical simulations have been used to better understand flow in dies. A number of studies have been published [58–66] with a representative set of examples provided in the cited references (a complete review is beyond the scope of this book). The function of the die is to distribute the flow so that it is uniform across the width or circumference at the exit. Various methods are used to accomplish this, such as the use of multiple manifolds in flat dies and spirals in annual dies. Rheology, especially viscous encapsulation and elastic effects (see Section 13.4.4.2), may affect the uniformity of the output and film thickness, especially in coextrusion. These models are mostly in the domain of the equipment manufacturers who design the dies, but it is important for the converter and package designer to know that OEM's have tools that can help them optimize dies for a given application.

The region past the die exit, where the polymer is drawn down to its final thickness and quenched, has a substantial impact on film properties. There are several models that have been developed in the literature for the blown film process [67–74]. The goal of these models is to try to predict the temperature, bubble shape, and velocity and stress profiles in the film. The stress near the frost line is particularly important for predicting film properties. These models, however, are still mostly academic in nature. The challenges with modeling blown film are described in Section 14.1.1. Limitations of these numerical simulations include:

- Simplifications that are typically used to make the computations tractable. These may limit the applicability of the model.
- Difficulties in measuring the material parameters needed for simulation. For example, Doufas [57] has created one of the more advanced models that account for many of the physical phenomena that occur during film blowing, including flow-induced crystallization. To obtain the material parameters, however, he resorts to running the material first on the line he is modeling to fit the model to the experimental results.
- Lack of good models for coextrusion.
- Challenges in modeling the cooling part of the process (air flow, heat transfer, etc.) and scaling this up to commercial lines to include both air ring design and inner bubble cooling [75,76].

These models are good for understanding the fundamentals of the process and are continuing to progress. For now, however, they have not reached the point where the package engineer can use them to predict the properties of untested structures.

Similar statements can be made for cast films. Most of the simulation that can be found in the literature uses sophisticated numerical methods and are still in the academic domain [77–96]. A sampling of studies reported in the literature is summarized in Table 18.4 and Barborik and Zatlaukal [96] give a more complete review. Simplifying assumptions regarding the number of dimensions, type of flow (Newtonian or viscoelastic) and temperature change are made to various degrees to make the computation tractable. One-dimensional models fix the width and so provide no information on neck-in or edge bead (see Section 13.4.2). These models can provide information on draw resonance and thickness uniformity. 1.5D models use the one-dimensional framework but impose variable film width in a simplified way by assuming the velocity is a linear function of position from the die exit. These models allow exploration of neck-in. Two-dimensional models use a membrane approximation where the thickness is assumed to be small and the melt curtain behaves as a thin sheet in the state of planar stress. This simplifies the computations. These models can provide information about neck-in and edge bead. Very few investigations model the full three-dimensional case which would be required to capture the effect of modern edge-bead reduction dies.

Heat transfer models like the one developed in Section 15.1.3 have been used to simulate cooling in the nip of extrusion coating [97,98] and sheet lines [99]. They have also been helpful for evaluating new concepts for blown film cooling [100] and annealing of thin sheets and films [101]. The output of the model is particularly helpful for troubleshooting adhesion issues. Commercial software is available for conductive heat transfer modeling, and some specialized programs such as one developed by aiXtrusion [99] for sheet cooling are available. These can be versatile tools for troubleshooting not only adhesion issues but heat sealing and anywhere heat is involved.

During the quenching of a multilayer film, the layers may shrink by different amounts, resulting in film curl. This is described in detail in Section 13.5. A model is developed in Section 13.5.3 based on beam theory that has proven very helpful for troubleshooting curl issues [102].

Models for thermoforming of sheet and film are described in Section 11.1.4. These can be useful for screening new film structures, although obtaining material parameters can be a challenge.

TABLE 18.4 Examples of numerical simulation studies of the cast film process.

References	Model dimensions	Constitutive equation	Comment
[77]	1	Newtonian	Stability analysis
[78]	1	Viscoelastic (generalized UCM)	Draw resonance
[79,80]	1	Viscoelastic (Giesekes)	Isothermal
[81,82]	1	Viscoelastic (Giesekes)	Multiple layers
[83]	1	Viscoelastic (extended Pom Pom)	Isothermal, neck-in
[84]	1	Viscoelastic (extended Pom Pom)	Nonisothermal, neck-in
[85]	1.5	Viscoelastic (extended Pom Pom)	Isothermal, neck-in
[86]	1.5	Viscous (Cross)	Nonisothermal, flow induced crystallization
[87,88]	1.5	Viscoelastic (Leonov)	Isothermal, neck-in
[89]	2	Viscoelastic	Isothermal
[90]	2	Viscoelastic (Maxwell)	Isothermal
[91]	2	Viscoelastic (Giesekes)	Isothermal
[92]	2	Viscoelastic	Nonisothermal
[93]	2	Visoelastic (Giesekes)	Nonisothermal
[94]	3	Newtonian	Isothermal, neck-in
[95]	3	Viscoelastic (PTT)	Isothermal, strain hardening

Section 12.1.4.2 describes models for predicting the level of slip additive accumulating at the surface of a film as a function of time, film thickness and initial slip additive concentration. Presumably, one then could correlate slip surface concentration with coefficient of friction (COF), using empirical relationships. These models have been used primarily to better understand the mechanisms of slip migration, and have limitations for predicting actual COF in packaging films. One of these limitations is the lack of knowledge of how slip additives are partitioned at interfaces between layers in a multilayer film. This knowledge would be required to extend the approach to coextruded films.

Chapter 15 covers a number of fundamental studies on the effect of processing on adhesion, such as the development of a peel strength master curve that relates the material properties of the tie resin and the time scales of the process and polymer relaxation to peel strength (see Section 15.2.3). The master curve approach can be used to predict adhesion performance.

Finally, there are scale-up relationships for duplicating the properties obtained on a lab line on a larger commercial line. Such relationships are discussed in Section 14.1.1 for blown film.

18.2.3 Case study

To illustrate the use of models to speed the development of flexible packaging, a hypothetical package development is considered. The analysis is based on a case study presented by Morris [18] that describes the use of modeling tools developed at DuPont (these tools transferred to Dow when the ethylene copolymer business of DuPont became part of Dow in 2019).

The example package is a barrier bag-in-box package for dry foods such as cereal, crackers, and snacks. The inner liner bag is made and filled on a vertical form-fill-seal line and then placed into a secondary paperboard box. The first step in the design process is to determine the requirements for the package. The brand owner or end-user generally provides these. The converter or designer must translate the stated needs into engineering requirements that can be used in the design process. The requirements for the package considered for this case study are given in Table 18.5.

Because of the barrier requirements, the general structure (HDPE-tie-barrier-tie-sealant) is considered. This is a fairly standard structure used in these types of applications, as is shown in Appendix B. HDPE provides a moisture barrier. Either polyamide or EVOH is typically used as the oxygen barrier layer. The sealant is an EVA, mPE, or ionomer

TABLE 18.5 Requirements of barrier bag-in-box liner bag used in case study.

Package attribute	Specification
Easy open function	Seal strength: between 500 and 1000 g/25 mm
Low heat seal initiation temperature (HSIT)	HSIT <120°C; packaging line speed >60 pkg/min
Moisture barrier	MVTR <0.3 g/day/package
Oxygen barrier	OTR <3 cc O_2/day/package
Puncture resistance	Qualitative for this case study, but may be specified as number of pinholes or package failure rate
Cost effective and easy to run on vertical form fill seal line	Qualitative for this case study, but there would likely be specific cost targets

TABLE 18.6 Description of resins used in the case study.

Resin	Description	MFI (190°C/ 2160g) (g/10 min)	Tm (°C)	Tensile modulus (MPa)	MVTR (38°C/90%RH) (g-mil/100in^2-d)	OPV (25°C/0%RH) (cc-mil/100in^2-d)
EVA	18% VA	2.5	89	70	50	10000
Ionomer 1	Zinc neutralized	5.3	100	240	27	5400
Ionomer 2	Zinc neutralized	0.7	90	480	27	5400
Tie	LLDPE-based	1.1	125	210	19	8500
PA 6	Nylon 6		220	700	200	90
EVOH	32 mol% ethylene	1.6	183	4500	930	0.3
HDPE	0.95 g/cc	0.26	135	770	8	220

formulated for easy opening and must have a low seal initiation temperature to prevent the distortion of the HDPE layer during sealing and to achieve high filling rates. The models and tools will help determine which materials to use, their thicknesses, how cost-effective the structure may be, as well as whether any difficulties in making the structure might be encountered. Table 18.6 describes some of the resins that will be considered in the analysis.

18.2.3.1 Barrier selection

A simple permeability calculator based on Eq. (8.21) can be used to determine what barrier resin to use. An example spreadsheet is shown in Fig. 18.10. The requirements call for a fairly moderate oxygen barrier to prevent spoilage. The oxygen barrier is also often used as a predictor for the aroma barrier, even though aroma involves larger molecules and may not correlate with oxygen permeability (see Section 8.2.4). Highly flavored snacks and cereals may require an aroma barrier to prevent too much of the aroma from permeating through the package and into the grocery aisles. The model results given in Fig. 18.10 show that a 12-μm layer of PA 6 will be sufficient. A similar calculation using EVOH shows that less than 1 μm would be needed to achieve the barrier requirement. Limitations on coextrusion equipment generally constrain the thickness to about 3-μm or higher. Furthermore, films containing PA generally have better impact puncture resistance than EVOH. Therefore, 12-μm PA 6 is chosen as the oxygen barrier layer.

The results show that 50 μm of HDPE will provide enough moisture barrier performance.

Permeabilty Calculator
(Fill in Cells in Blue)

Square inches of film: 130 0.084 square meters
Film Thickness (mils): 4.1 104.1 microns

		% of	Thickness		OPV	MVTR
Layer	Resin	Thickness	mil	microns	cc-mil/100in2-d-atm	g-mil/100in2-d-atm
1	Ionomer 1	25	1.03	26.0	350	1.75
2	tie	7.5	0.31	7.8	550	1.2
3	PA6	12	0.49	12.5	6	13
4	tie	7.5	0.31	7.8	550	1.2
5	HDPE	48	1.97	50.0	220	0.5
6						
7						
8						
9						
Total		100			10.5	0.2

Summary Results:

OTR (cc/100in2-d-atm)	10.5
OTR (cc/m2-d-atm)	163.2
OTR cc/day (100% O2)	13.69
OTR cc/day (air)	2.87

MTR (g/100in2-d-atm)	0.2
MTR (g/m2-d-atm)	3.1
MTR (g/day)	0.26

Calculation:

OTR = 1/(thickness1/OPV1 + thickness2/OPV2 + thickness3/OPV3)
OTR (cc/day) = OTR(cc/100In2-d-atm)*Square inches of film/100

FIGURE 18.10 Permeability calculator showing results for the barrier bag-in-box dry foods package example. *OTR*, Oxygen transmission rate; *MTR*, moisture transmission rate; *OPV*, oxygen permeation value; *MVTR*, moisture vapor transmission rate. *Reproduced from B.A. Morris, Application of modeling to speed flexible packaging development, SPE ANTEC Conf., (2012). Also B.A. Morris, J. Plastic Film & Sheeting, 29 (2013) 249–270 with permission from The Dow Chemical Company.*

TABLE 18.7 Input to the sealant selector tool for the case study.

Attribute	Importance rating (1–10 scale)
Blown film processing	3
Low-seal initiation temperature	10
Broad hot tack temperature range	5
Seal around contamination	3
Caulk gussets; seal through wrinkles	3
Grease and oil resistance	6
Perforation resistance	10
Easy to open seal technology	10

Note: Scale is from 0 to 10 with 10 being most important.
Source: Reproduced from B.A. Morris, Application of modeling to speed flexible packaging development, SPE ANTEC Conf., (2012). Also B.A. Morris, J. Plastic Film & Sheeting, 29 (2013) 249–270 with permission from The Dow Chemical Company.

18.2.3.2 Sealant selection

The attributes the sealant needs to bring to the structure are tabulated in Table 18.7. Each attribute has been assigned an importance rating on a scale of 1–10. These are entered into the sealant selector tool described in Section 18.2.1.1. The results are shown in Fig. 18.11. The functional score for each sealant type has been normalized to a scale of 0–10. The width of each bar is the sum of the important factors multiplied by the matrix score for each attribute. The individual

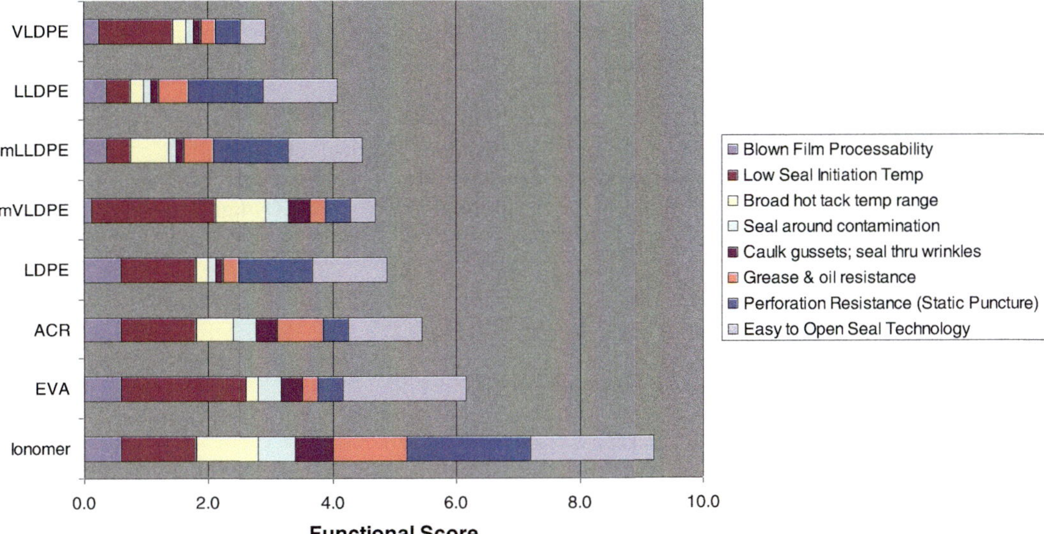

FIGURE 18.11 Sealant selector tool results for the case study example. A higher functional score indicates better predicted performance against the criteria (Table 18.7). *VLDPE*, Very low-density polyethylene; *LLDPE*, linear low-density polyethylene; *mLLDPE*, metallocene linear low-density polyethylene; *mVLDPE*, metallocene very low-density polyethylene plastomer; *LDPE*, low-density polyethylene; *ACR*, acid copolymer resin; *EVA*, ethylene vinyl acetate copolymer. *Reproduced from B.A. Morris, Application of modeling to speed flexible packaging development, SPE ANTEC Conf., (2012) with permission from The Dow Chemical Company.*

contribution from each attribute is highlighted by color and can be used to quickly see why one sealant is scoring better than another. In this example, ionomers and EVA have the highest scores. The ionomer has strong contributions from its ability to be formulated as an easy-open sealant while maintaining a hermetic seal, perforation resistance, grease and oil resistance, and low-seal temperatures.

18.2.3.3 Stiffness and life cycle assessment model

DuPont developed a model that computes bending stiffness [48] and provides a cradle-to-gate life cycle assessment (LCA) [103] of the materials used in the package structure. Table 18.8 shows the typical output from this model. Material cost can also be computed by the model, but it is not shown here. Summary inputs are given in the first six columns, and the computed results are tabulated in the final column. By running several cases, the user can see how changes in materials and thickness can affect package performance (stiffness) and cost (material cost, nonrenewable energy consumption, and greenhouse gas emissions).

Figs. 18.12–18.14 show how varying the sealant type and thickness can affect bending stiffness, nonrenewable energy consumption, and greenhouse gas emissions of the barrier bag-in-box liner film. As is described in Section 9.4.2, the stiffness of the sealant layer often has a disproportionate influence on the overall bending stiffness of a film. Table 18.6 shows that the ionomers have substantially higher stiffness than EVA, as indicated by their tensile moduli. Substituting an ionomer sealant for an EVA sealant increases the film bending stiffness (Fig. 18.12). This may allow the sealant layer to be down gauged (third bar in Fig. 18.12) while maintaining overall package stiffness. A package using an ionomer with a higher modulus can be further down gauged which results in additional reductions in the environmental footprint (Figs. 18.13 and 18.14). Of course, there are practical limits to how thin the sealant layer can be and still maintain good seal functionality (caulking, etc.). As with all the models described here, the packaging concepts will need to be tested under conditions of actual use to validate the results.

18.2.3.4 Packaging costs

Fig. 18.15 shows the output from the packaging cost model described in Section 16.5.6 that considers total packaging costs. The structure with an ionomer sealant is compared with one with an EVA sealant. The packaging material costs

TABLE 18.8 Example output from stiffness/life cycle assessment model.

Example							
Barrier bag in box base case							
Layer description	Thickness (microns)	Modulus (M Pa)	Density (g/cm^3)	Nonrenewable energy consumption (MJ/kg)	Greenhouse gas emissions (eq CO_2/kg)	Model results	
HDPE	50.8	770	0.95	76	1.9	Weight:	0.102 kg/m^2
Tie	7.5	210	0.91	76.9	2.1	Thickness:	104.2 μm
PA 6	13	700	1.2	120	9.1	Stiffness factor:	2.56
Tie	7.5	210	0.91	76.9	2.1	Neutral axis from top:	36.7 μm
EVA	25.4	70	0.95	76.9	2.1	Total material cost:	0.220 $/$m^2$
						Nonrenewable energy consumption:	8.46 MJ/m^2
						Greenhouse gas emissions:	0.313 eq $CO2/m^2$

Source: Reproduced from B.A. Morris, Application of modeling to speed flexible packaging development, SPE ANTEC Conf., (2012). Also B.A. Morris, J. Plastic Film & Sheeting, 29 (2013) 249–270 with permission from The Dow Chemical Company.

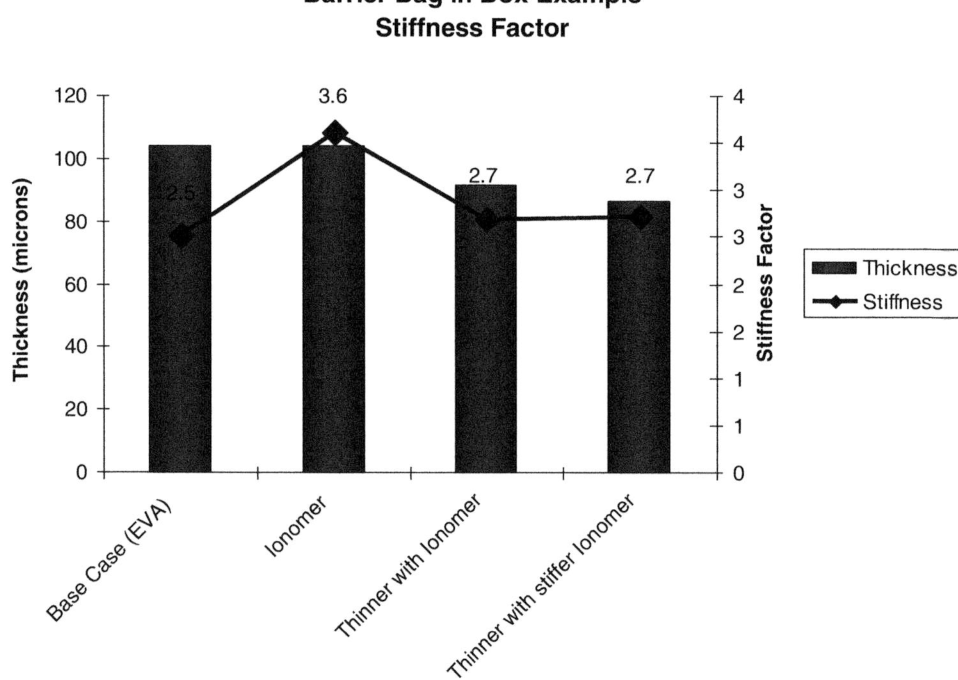

FIGURE 18.12 Cross plot of model output of thickness and bending stiffness (stiffness factor) for barrier bag-in-box film structures. *Reproduced from B.A. Morris, Application of modeling to speed flexible packaging development, SPE ANTEC Conf., (2012) with permission from The Dow Chemical Company.*

are higher for the ionomer structure (bottom/blue bar), but the overall packaging cost is lower (total bar height) due to lower cost of labor, overhead, and waste. Here it is assumed the line speed is 10% faster (64 packages/minute) and leaker rates 50% lower for the ionomer film.

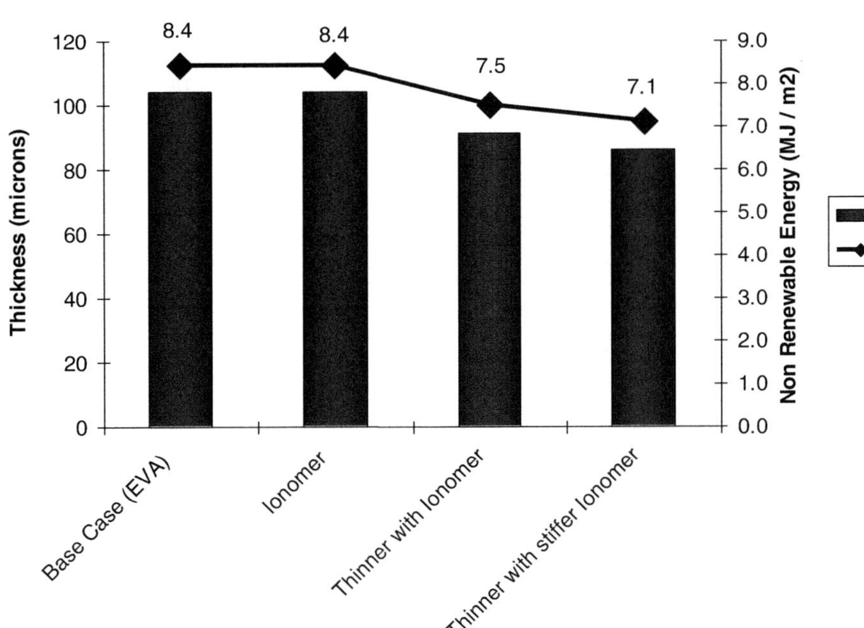

FIGURE 18.13 Cross plot of model output of thickness and nonrenewable energy for barrier bag-in-box film structures. *Reproduced from B.A. Morris, Application of modeling to speed flexible packaging development, SPE ANTEC Conf., (2012) with permission from The Dow Chemical Company.*

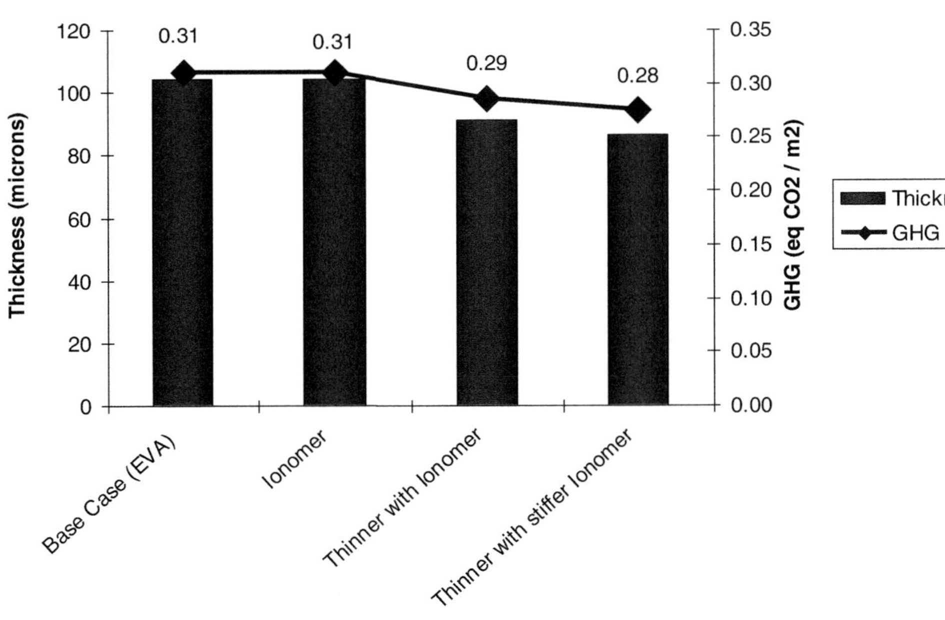

FIGURE 18.14 Cross plot of model output of thickness and greenhouse gas emissions for barrier bag-in-box film structures. *Reproduced from B.A. Morris, Application of modeling to speed flexible packaging development, SPE ANTEC Conf., (2012) with permission from The Dow Chemical Company.*

As is discussed in Section 16.5.6 this tool also provides a sensitivity analysis on each of the key input variables such as line speed, leaker rate and roll stock cost. This example is the same as the one used in Section 16.5.6—the analysis shows that the line speed could be as low as 49 packs/min and the ionomer film would still save cost due to its lower leaker rate.

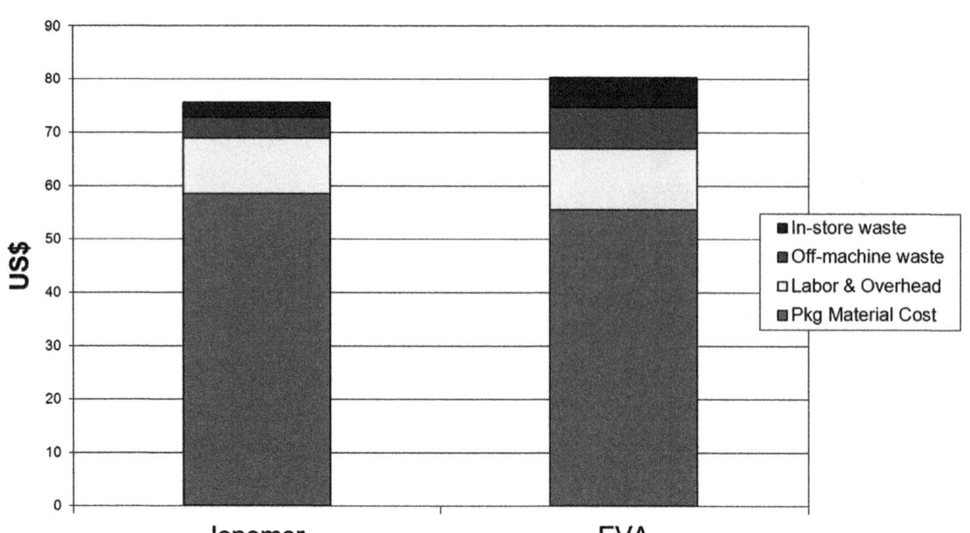

FIGURE 18.15 Example output from the Sealant Value Estimator tool. *Reproduced from B.A. Morris, Application of modeling to speed flexible packaging development, SPE ANTEC Conf., (2012) with permission The Dow Chemical Company.*

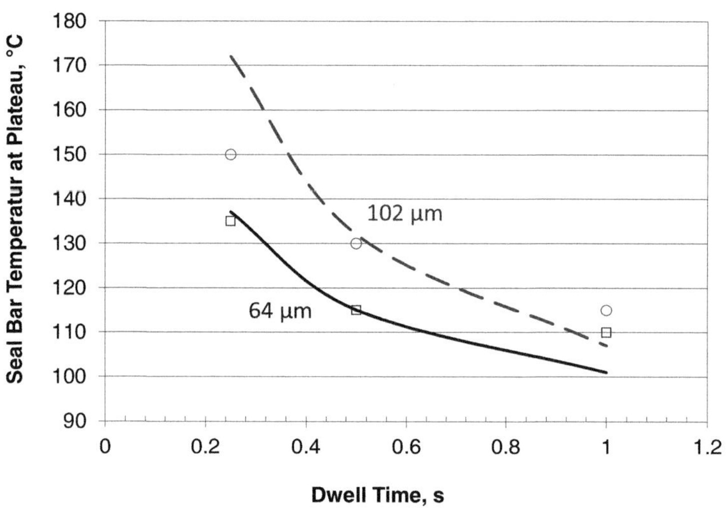

FIGURE 18.16 Heat-seal model (solid curves) and experimental results (data points) for ionomer films. Plateau seal temperature is the minimum temperature at which the maximum heat-seal strength is obtained. Films have structure OPET/ionomer (5), where *OPET* = oriented polyester film. *Reproduced with permission from B.A. Morris, Predicting the heat seal of ionomer films, J. Plastic Film. Sheeting, 18 (2002) 157–167.*

18.2.3.5 Heat seal

Fig. 18.16 shows experimental and model results of plateau seal temperature vs. dwell time for ionomer films from work described in Section 7.6.2 [51]. These curves show the minimum operating temperature for a specified seal dwell time for achieving the ultimate seal strength. For the case study, the snack package structure is roughly 100 μm thick. The model predicts for a sealing temperature of 130°C that the dwell time needs to be around 0.5 seconds to achieve ultimate seal strength. At this dwell time, a typical vertical form fill-seal machine will have a packing speed of 60 packs/minute (assuming that sealing occurs during 50% of the registration cycle). This is consistent with the assumptions used in the total cost model and the requirements specified in Table 18.5. The high sealing temperature may be close to causing distortion of the HDPE layer, which has a peak melting point of around 135°C. The model suggests using a grade of ionomer with a low seal initiation temperature may be advantageous, such as ionomer2 in Table 18.6.

18.2.3.6 Flow instabilities in coextrusion

The flow-matching model from Section 13.4.5.2 is used to estimate whether the designed structure may have flow instability issues during the blown film converting process. The input parameters for the model are given in Table 18.9. The output is shown graphically in Fig. 18.17. The shaded region in the middle is the ideal region for no flow instabilities

TABLE 18.9 Input for the flow-matching model for predicting onset of coextrusion instabilities.

Viscosity matching in coextrusion			
Input parameters			
Total throughput (kg/h):		260	
Width of channel (mm):		1100	
Height of channel, (mm):		1.80	
Layer	**Resin**	**Thickness (μm)**	**Temperature (°C)**
1	Ionomer1	20	200
2	Tie1	7.5	220
3	PA 6	12	250
4	Tie1	7.5	220
5	HDPE	50	200

Source: Reproduced from B.A. Morris, Application of modeling to speed flexible packaging development, SPE ANTEC Conf., (2012). Also B.A. Morris, J. Plastic Film & Sheeting, 29 (2013) 249–270 with permission from The Dow Chemical Company.

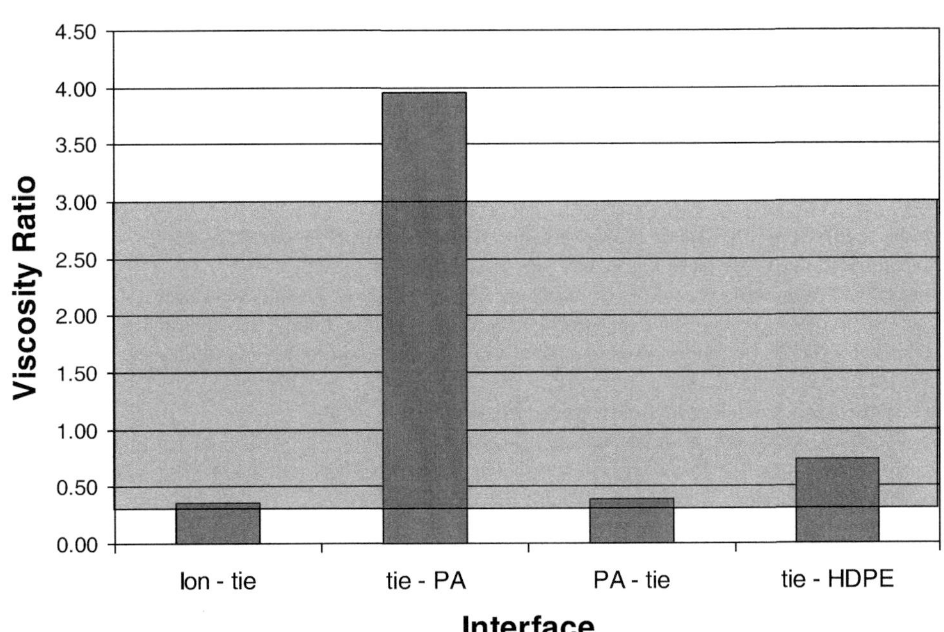

FIGURE 18.17 Graphical representation of output of coextrusion flow-matching model. The viscosity ratio is the ratio of the viscosity of the first resin on the list to the viscosity of the second at the shear stress and temperature at the interface. The shaded region is the desired region for avoiding flow instabilities. *Ion*, Ionomer; *PA*, polyamide 6; *HDPE*, high-density polyethylene.

(where the viscosity ratios fall within the 3:1 to 1:3 criteria). The analysis suggests that the PA viscosity is too low compared to the tie resin; the ratio of the PA viscosity to the first and second tie layers is almost 1:4 and 1:2.6, respectively. A better match can be obtained by reducing the PA extrusion temperature or switching to a higher molecular weight grade of PA (both approaches increase the viscosity of the PA layer). Lowering the viscosity of the tie layer may also be effective but this may cause problems at the HDPE-tie interface. These parameters can be adjusted in the model to determine the best approach without resorting to trial and error on the film line.

Note that the analysis is not symmetrical in that the viscosity ratio of the tie1-PA and PA-tie2 are not the inverse of one another even though the tie is the same composition, thickness and temperature on both sides of the PA. This is due to differences in shear stress (and shear rates) on each side of the PA layer.

18.2.3.7 Curl

Curl can be significant in blown films of this structure. The PA layer freezes first followed by the HDPE layer. As it freezes, the HDPE shrinks due to the more ordered structure (higher density) of the crystal domains. The combination of the shrinkage, stiffness, and thickness of the HDPE layer causes the film to curl toward the HDPE layer. The sealant side of the structure, being thinner and less crystalline, is generally not enough to compensate.

The curl model [102] described in Section 13.5.3 is used to estimate the curvature of the films. The model results suggest films with either an EVA or an ionomer sealant will have a significant curl toward the HDPE. Changing the thickness of the sealant has little impact. Techniques that have been found to reduce curl include increasing the quench rate and adding amorphous PA to PA 6. PA 6 is also known to undergo postquench crystallization, especially if the film is moisture conditioned, which can balance out some of the curl toward the HDPE layer. The model does not capture postquench crystallization.

18.3 Final word

This chapter introduces tools that can be used to help optimize package design and ensure that packaging concepts can be readily made on commercial converting equipment. The case study is intended to illustrate the practical use of such tools. These tools encompass many of the fundamental concepts underlying flexible packaging technology introduced earlier in this book. The use of these and other models is merely the first step in the design process. The utility of using models is that many concepts can be quickly analyzed and screened using a computer before undergoing the expensive and time-consuming process of making the best designs on commercial converting equipment and testing them under conditions of their end-use. Nevertheless, there is no escaping the last step of making and testing the designs. As highlighted throughout this book, there are many aspects of package design, manufacture and use (and end-of-use); no model can take into account all of these. This book attempts to equip the reader with a scientific basis for many of the complexities of flexible packaging technology in a way that will help address future challenges and needs.

References

[1] Nikon, Microscopy U: resolution. <https://www.microscopyu.com/microscopy-basics/resolution>, n.d. (accessed 20.07.20).

[2] E.I. Garcia-Meitin, C.Y. Lai, P. O'Connell, S. Bensason, Development of sealants for flexible packaging using light microscopy, SPE ANTEC Conf. (2015).

[3] R.C. Weast (Ed.), Handbook of Chemistry and Physics, fifty sixth ed., CRC Press, Inc., Cleveland, OH, 1975, pp. F245–F254.

[4] R.T. Morrison, R.N. Boyd, Organic Chemistry, third ed., Allyn and Bacon, Inc, New York, 1976.

[5] Thermo Scientific, FTIR basics. <https://www.thermofisher.com/us/en/home/industrial/spectroscopy-elemental-isotope-analysis/spectroscopy-elemental-isotope-analysis-learning-center/molecular-spectroscopy-information/ftir-information/ftir-basics.html>, n.d. (accessed 20.07.21).

[6] PerkinElmer, Differential scanning calorimetry (DSC), frequently asked questions. <http://www.perkinelmer.com/CMSResources/Images/44-74542GDE_DSCBeginnersGuide.pdf>, 2013 (accessed 21.07.21).

[7] Matweb, Property search. <http://www.matweb.com/search/PropertySearch.aspx>, n.d. (accessed 21.07.21).

[8] Prospector, Property search. <https://www.ulprospector.com/propertysearch>, n.d. (accessed 21.07.21).

[9] Chemnetbase, Polymers: a property database. <https://poly.chemnetbase.com/faces/polymers/PolymerSearch.xhtml>, n.d. (accessed 21.07.21).

[10] National Institute for Material Science, PolyInfo. <https://polymer.nims.go.jp/en/>, n.d. (accessed 21.07.21).

[11] Evonik, Evonik's plastic database. <https://www.plastics-database.com>, n.d. (accessed 21.07.21).

[12] ExxonMobil, Digital product selector. <https://www.exxonmobilchemical.com/en/products/polyethylene/performance-pe-polymers/productselector#/datatable/landing>, n.d. (accessed 21.07.21).

[13] NOVA Chemicals, Product finder. <https://www.novachem.com/product-finder/?pg = 1>, n.d. (accessed 21.07.21).

[14] Elsevier, Knovel. <https://app.knovel.com/kn>, n.d. (accessed 21.07.21).

[15] B.A. Morris, Optimizing flexible packaging ... with unique tools from DuPont, Presentation at PackExpo, 2011.

[16] B.A. Morris, Selecting Sealants for Meat Packaging, American Meat Institute Expo, Chicago, 2011. April 2011.

[17] B.A. Morris, Use of modeling to speed packaging development, in: TAPPI PLACE Conference, Seattle, WA, May 9–11, 2012.

[18] B.A. Morris, Application of modeling to speed flexible packaging development, SPE ANTEC Conf. 29 (2012) 249–270. Also Morris, B. A., 2013, J. Plastic Film & Sheeting.

[19] G.E. Box, W.G. Hunter, J.S. Hunter, W.G. Hunter, Statistics for Experimenters: Design, Innovation, and Discovery, second ed., Wiley, 2005.

[20] D.C. Montgomery, Design and Analysis of Experiments, eight ed., John Wiley and Sons, Inc, 2012.

[21] G.E.P. Box, N.R. Draper, Empirical Model Building and Response Surfaces, John Wiley and Sons, 1987.

[22] J. Cornell, Experiments With Mixtures: Models, and the Analysis of Mixture Data, third ed., John Wiley and Sons, 2002.

[23] R.H. Myers, D.C. Montgomery, C.M. Anderson-Cook, Response Surface Methodology: Process and Product Optimization Using Designed Experiments, third ed., John Wiley and Sons, 2016.

[24] R.L. Mason, R.F. Gunst, J.L. Hess, Statistical Design & Analysis of Experiments with Applications to Engineering and Science, second ed., John Wiley and Sons, 2003.

[25] K. Hinkelmann, O. Kempthorne, Design and Analysis of Experiments, Volume I: Introduction to Experimental Design, second ed., John Wiley and Sons, 2008.

[26] M.J. Anderson, P.J. Whitcomb, DOE simplified, Practical Tools for Effective Experimentation, third ed., CRC Press, New York, 2015.

[27] A. Dean, D. Voss, D. Draguljic, Design and Analysis of Experiments, second ed., Springer, 2017.

[28] SAS, JMP®, data anslysis software. <https://www.jmp.com/en_us/software/data-analysis-software.html>, n.d. (accessed 21.07.21).

[29] Minitab, Minitab. <https://www.minitab.com/en-us/>, n.d. (accessed 06.12.15).

[30] N. Borse, N. Aubee, Predicting physical and optical properties of co-extruded blown films using design of experiment based model, SPE ANTEC Conf. (2015) 1584–1588.

[31] H. Blom, Flexible packaging durability studies, SPE FlexPackCon Conf. (2014).

[32] W.G. Todd, W.R. Podborny, Predictive models helps develop new high-performance HDPE for barrier film application, SPE ANTEC Conf. (2001).

[33] X. Wang, A predictive model for dart impact strengths of solution octene LLDPE film resins, SPE ANTEC Conf. (2003) 1386–1390.

[34] B.A. Morris, The influence of stress on peel strength of acid copolymers to foil, TAPPI Polymers, Laminations Coat. Conf. 21 (1998) 1243–1255. Also Morris, B. A., 2002, J. Reinforced Plastics and Composites.

[35] J.F. Agassant, M.R. Mackley, A personal perspective on the use of modeling simulation for polymer melt processing, Intern. Polym. Proces. XXX (2015) 121–140.

[36] J. Bergstrom, D. Quinn, E. Schmitt, S. Brown, S. Chow, High strain rate testing and modeling of polymers for impact simulations, SPE ANTEC Conf. (2014) 2305–2307.

[37] J.Z. Yuan, C.A. Haynes, Predictive analysis of flexible pouch performance, SPE ANTEC Conf. (2014) 1436–1440.

[38] L.A. Carbajal, R. Jiao, D.M. Hahm, B.A. Morris, R.R. Kendzirski, Predicting the impact puncture response of multilayer flexible food packages using explicit finite element models, SPE ANTEC Conf. (2016).

[39] European Union Commission Regulation No. 10/2011, On plastic materials and articles that come into contact with food., <https://www.fsai.ie/uploadedFiles/Reg10_2011.pdf>, (accessed 21.07.21).

[40] Combase, Combase: a web resource for quantitative and predictive food microbiology. <http://www.combase.cc/>, 2021 (accessed 21.07.21).

[41] AKTS, Specific migration limits software (SML). <http://www.akts.com/sml-diffusion-migration-multilayer-packaging/download-diffusion-prediction-software.html>, n.d. (accessed 21.07.21).

[42] FABES Forschungs-GmbH, Migratest ESP. <http://www.fabes-online.de/software.php?lang = en&mode = migratest>, 2021 (accessed 21.07.21).

[43] B. Roduit, C.H. Borgeat, S. Cavin, C. Fragniere, V. Dudler, Application of finite element analysis (FEA) for the simulation of release of additives from multilayer polymeric packaging structures, Food Addit. Contaminates 22 (2005) 945–955.

[44] M.F. Pocas, J.C. Oliveira, F.A.R. Oliveira, T. Hogg, A critical survey of predictive mathematical models for migration from packaging, Crit. Rev. Food Sci. Nutr. 48 (2008) 913–928.

[45] I.D. Lopez, J.C. Estefan, A. Betancur, Packaging and food interactions modeling: validation for compliance with specific migration regulations, SPE ANTEC Conf. (2015) 1545–1549.

[46] M.G. Sajilata, K. Savitha, R.S. Singhal, V.R. Kanetkar, Scalping of flavors in packaged foods, Compr. Rev. Food Sci. Food Saf. 6 (2007) 17–35.

[47] R. Brandsch, C. Dequatre, P. Mercea, M. Milana, A. Stoermer, X. Trier, et al., Practical guidelines on the application of migration modelling for the estimation of specific migration, in: EUR 27529. Luxembourg (Luxembourg): Publications Office of the European Union; JRC98028. <https://publications.jrc.ec.europa.eu/repository/handle/JRC98028> (accessed 21.07.21).

[48] B.A. Morris, J.D. Vansant, The influence of sealant modulus on the bending stiffness of multilayer films, TAPPI Polymers, Laminations Coat. Conf. (1997).

[49] R.-Y. Wu, L.D. McCarthy, Z.H. Stachurski, Tearing resistance of multi-layer plastic films, Intern. J. Fract. 68 (1994) 141–150.

[50] L.A. Carbajal, R. Jiao, D.M. Hahm, B.A. Morris, R.R. Kendzirski, Impact puncture resistance of multilayer flexible food packages, SPE ANTEC Conf. (2015).

[51] B.A. Morris, Predicting the heat seal of ionomer films, J. Plastic Film. Sheeting 18 (2002) 157–167.

[52] R.D. Moffitt, A mathematical model for the heat sealing of linear, semi-crystalline polymers, SPE ANTEC Conf. (2006) 1027–1032.

[53] R.A. Bourque, et al., Multiple compartment pouch and beverage container with frangible seal, US Patent 7306095 B1, 2007.

[54] A. Lokhande, P. Valavala, P. Thakre, S. Athreya, J. Zawisza, S. Crabtree, et al., Role of polymer film mechanical behavior in liquid packaging application: deformation modeling of Gelbo test, SPE ANTEC Conf. (2017).

[55] Z. Tadmor, C.G. Gogos, Principles of Polymer Processing, John Wiley & Sons, New York, 1982 (1979).

[56] E.E. Agur, J. Vlachopoulos, Numerical simulation of a single-screw plasticating extruder, Polym. Eng. Sci. 22 (1982) 1084–1094.

[57] J. Vleck, J. Perdikoulias, Application of a single screw extrusion simulation towards design, SPE ANTEC Conf. (2002).

[58] S.R. Vaddiraju, M. Kostic, L. Reifneider, A. Pla-Dalmau, V. Rykalin, A. Bross, Extrusion simulation and experimental validation to optimize precision die design, SPE ANTEC Conf. (2004) 76–80.

[59] M. Gupta, Three dimensional simulation of coextrusion in a complex profile die, SPE ANTEC Conf. (2010) 2030–2036.

[60] J. Perdikoulias, J. Svabik, 3D FEM simulation of feed-block profiling for flat die coextrusion, SPE ANTEC Conf. (2002).

[61] P.C. Lee, C.W. Macosko, Die design for layer thickness uniformity in multilayer coextrusion, SPE ANTEC Conf. (2009) 1542–1546.

[62] Y. Sun, M. Gupta, J. Dooley, K.A. Koppi, M.A. Spalding, Experimental and numerical investigation of the elongational viscosity effects in a coat-hanger die, SPE ANTEC Conf. (2005) 81–85.

[63] O. Catherine, Effect of typical melt temperature non-uniformity on flow distribution in flat dies, SPE ANTEC Conf. (2012).

[64] S. Iuliano, A. Svenningsen, The effect of flow channel aspect ratio on layer uniformity in flat extrusion dies, SPE ANTEC Conf. (2016).

[65] S. Kihara, T. Gouda, K. Matsunaga, K. Funatsu, Numerical simulation of three-dimenional viscoelastic flow within dies, Polym. Engr. Sci. 39 (1999) 142–163.

[66] B. Debbaut, J. Dooley, Secondary motions in straight and tapered channels: experiments and three-dimensional finite element simulation with a multimode differential viscoelastic model, J. Rheology 43 (1999) 1525–1545.

[67] J.R.A. Pearson, C.J.S. Petrie, The flow of a tubular film. Part 1. Formal mathematical representation, J. Fluid Mech. 40 (1970) 1–19.

[68] J.R.A. Pearson, C.J.S. Petrie, The flow of a tubular film. Part 2. Interpretation of the model and discussion of solutions, J. Fluid Mech. 42 (1970) 609–625.

[69] B.K. Ashok, G.A. Campbell, Two-phase simulation of tubular film blowing of crystalline polymers, Int. Polym. Process. 7 (1992) 240–247.

[70] A.K. Doufas, A.J. McHugh, Simulation of film blowing including flow-induced crystallization, J. Rheology 45 (2001) 1085–1104.

[71] J.C. Pirkle Jr., R.D. Braatz, A thin-shell two-phase microstructural model for blown film extrusion,", J. Rheology 54 (2010) 471–505. Ashok, B.K. and Campbell, G.A., Int Polym Process, **7**, 240-247 (1992).

[72] A.K. Doufas, A microstructural flow-induced crystallization model for film blowing: validation with experimental data, Rheol. Acta 53 (2014) 269–293.

[73] R. Kolarik, M. Zatloukal, Modeling of nonisothermal film blowing process for non-Newtonian fluids by using variational principles, J. Appl. Polym. Sci. 122 (2011) 2807–2820.

[74] S. Muke, H. Connell, I. Sbarski, S.N. Bhattacharya, Numerical modelling and experimental verification of blown film processing, J. Non-Newtonian Fluid Mech. 116 (1) (2003) 113–138.

[75] V. Sidiropoulos, J. Vlachopoulos, Numerical simulation of blown film cooling, SPE ANTEC Conf. (2000).

[76] M. Jonas, J. Wortberg, T. Uniewski, S. Blume, A prediction model for the numerical optimization of a novel blown film cooling system, SPE ANTEC Conf. (2014) 1194–1199.

[77] Y.L. Yeow, On the stability of extending films: a model for the film casting process, J. Fluid Mech. 66 (1974) 613–622.

[78] N.R. Anturkar, A. Co, Draw resonance in film casting of viscoelastic fluids: a linear stability analysis, J. Non-Newt. Fluid Mech. 28 (1988) 287.

[79] V.R. Iyengar, A. Co, Film casting of a modified Giesekus fluid: a steady-state analysis, J. non-Newtonian Fluid Mech. 48 (1993) 1–20.

[80] V.R. Iyengar, A. Co, Film casting of a modified Giesekus fluid: a steady-state analysis, Chem. Engr. Sci. 51 (1996) 1417–1430.

[81] M.E. Pis-Lopez, A. Co, Multilayer film casting of modified Giesekus fluids Part 1. Steady-state analysis, J. Non-Newtonian Fluid Mech. 66 (1996) 71–93.

[82] M.E. Pis-Lopez, A. Co, Multilayer film casting of modified Giesekus fluids Part 2. Linear stability analysis, J. Non-Newtonian Fluid Mech. 66 (1996) 95–114.

[83] H. Pol, S.S. Thete, P. Doshi, A.K. Lele, Necking in extrusion film casting: the role of molecular architecture, J. Rheology 57 (2013) 559–583.

[84] H. Pol, S. Banik, L.B. Azad, S. Thete, P. Doshi, A. Lele, Nonisothermal analysis of extrusion film casting process using molecular constitutive equations, Rheol. Acta. 53 (2014) 85–101.

[85] K. Chikhalikar, S. Banik, L.B. Azad, K. Jadhav, S. Mahjan, Z. Ahmad, et al., Extrusion film casting of long chain branched polypropylene, Polym. Engr. Sci. 55 (2015) 1977–1987.

[86] G. Lamberti, Flow-induced crystallization during isotactic polypropylene film casting, Polym. Eng. Sci. 51 (2011) 851–861.

[87] M. Zatloukal, T. Barborik, Effect of extensional viscosity, elasticity and die exit stress state on neck-in phenomenon during extrusion film casting: theoretical study, SPE ANTEC Conf. (2016).

[88] T. Barborik, M. Zatloukal, Effect of die exit stress state, Deborah number, uniaxial and planar extensional rheology on the neck-in phenomenon in polymeric flat film production, J. Non-Newtonian Fluid Mech. 255 (2018) 39–56.

[89] B. Debbaut, J.M. Marchal, Viscoelastic effects in film casting, Z. Angew. Math. Phys. (1995) 46. Special Issue, Theoretical, experimental, and numerical contributions to the mechanics of fluids and solids, pp. S679-S698.

[90] D. Silagy, Y. Demay, J.F. Agassant, Stationary and stability analysis of the film casting process, J. Non-Newtonian Fluid Mech. 79 (1998) 563–583.

[91] D. Rajagopalan, Impact of viscoelasticity on gage variation in film casting, J. Rheology 43 (1999) 73–83.

[92] C. Sollogoub, Y. Demay, J.F. Agassant, Nonisothermal viscoelastic numerical model of the cast film process, J. Non-Newtonian Fluid Mech. 138 (2006) 76–86.

[93] C. Cox, J.B. von Oehsen, A finite element model of non-isothermal viscoelastic flow in the cast film process, SPE ANTEC Conf. (2007) 1596–1600.

[94] K. Sakaki, R. Katsumoto, T. Kajiwara, K. Funatsu, Three-dimensional flow simulation of a film-casting process, Polym. Engr. Sci. 36 (1996) 1821–1831.

[95] H. Zheng, W. Yu, C. Zhou, H. Zhang, Three dimensional simulation of viscoelastic polymer melts flow in a cast film process, Fibers Polym. 8 (1) (2007) 50.

[96] T. Barborik, M. Zatloukal, Steady-state modeling of extrusion cast film process, neck-in phenomenon, and related experimental research: a review, Phys. Fluids (2020) 32. published online on June 5, 2020.

[97] Y. Trouilhet, Morris, Prediction of temperature profiles across coating and substrate in the nip, TAPPI Polymers, Laminations Coat. Conf. (1999).

[98] S. Devisme, J.-M. Haudin, J.-F. Agassant, D. Rauline, F. Chopinez, Numerical simulation of extrusion coating: Contribution to the understanding of adhesion mechanisms between grafted polypropylene and aluminum, Intern. Polym. Process. XXII (2007) 90–104.

[99] AiXtrusion, Film and sheet extrusion – simulation of cooling. <https://aixtrusion.de/de/produkte/simulationsprogramm-sheetcoolaix/> (accessed 22.07.21).

[100] C. Hopmann, M. Hennings, Increased throughputs in blow film extrusion by using a contact cooling sleeve, SPE ANTEC Conf. (2016).

[101] W. Huang, M.A. Spalding, M.D. Read, T.A. Hogan, Heat transfer simulation for a continuous annealing process of plastic sheets, SPE ANTEC Conf. (2015).

[102] B.A. Morris, Reducing curl in multilayer blown film: experimental results, model development, and application to a cereal liner film, J. Plastic Film. Sheeting 19 (2003) 31–54.

[103] C.M. Casarino, B.A. Morris, S.R. Veith, Resin life cycle estimation to help guide sustainability choices, SPE ANTEC Conf. (2009).

Appendix A

Writing guide for packaging films and other multilayer structures

SB Marks
The Dow Chemical Company, Wilmington, DE, United States

A wide variety of multilayer structures are used in the flexible and rigid packaging industry, which are put together by an assortment of techniques including extrusion coating, extrusion lamination, adhesive lamination, as well as coextrusion blown film, cast film, cast sheet, and blow molding.

There currently is no standard method for communicating structures in a written format between companies (and for the most part, even within companies). The guidelines described below were derived from formats consolidated by the author during his long career working in international markets. Although no one method is best, we have found this guide to be useful in that it presents a systematic way to not only communicate the components of the structure, but also the technology used to put it together. By following the guidelines, one can better understand whether, for example, a structure is manufactured using coextrusion or by tandem extrusion coating. Such concise writing can improve communication. Therefore we offer these guidelines in the hope that they will increase communication effectiveness and understanding between all persons in our industry. The formatting guidelines allow for use in hand-written communication or for typing using a keyboard or keypad.

A.1 Guide to multilayer structure writing: packaging and industrial applications

A.1.1 Key definitions

Primer: A bonding agent that is applied in a liquid form to a solid surface. It is intended to promote the adhesion of another liquid or molten material that will subsequently be applied on top of the primer. Typically it is dried prior to having the subsequent material applied to it. *(This is usually associated with extrusion coating.)* In some parts of the world, these are referred to as anchor coatings ("*AC*"). Adhesion generally increases after curing time.

Adhesive: A bonding agent that is applied in a liquid or paste form to one or more solid surfaces. It is intended to promote the adhesion of the solid surfaces to one another. The adhesive is typically dried, and then the solid surfaces are brought together with heat and/or pressure to activate the adhesive. *(This usually is associated with water-based, solvent-based, or solventless dry lamination.)* Adhesion generally increases after curing time. *** can be abbreviated as "*ADH*," or "*adh.*"

A.1.2 Key abbreviations

VM—a vacuum metallization process for applying a thin metal layer to a solid surface, for example OPET or OPP films. This can also be applied to paper, printed surfaces, and other materials. The metal is usually aluminum, but can also be other metals such as copper, gold, tin, etc. (If it is not aluminum, please specify in your communication.)
"*vm*" is an alternate abbreviation for "vacuum metallized."
"*met*" is an abbreviation for "metallized."
SiOx—a silicon oxide coating that is typically applied to a film such as OPET. The coating is applied in a vacuum deposition process. The coating is usually transparent.
O.L.—an overlacquer that is applied over ink to typically add scratch protection and gloss.
W.O.—"white opaque," indicates that a material is pigmented with a white colorant to create opacity.
Tie—generically indicates a coextrudable adhesive resin. This is used in structure notations when the resin brand/grade are unknown. If the tie material base resin is known then please indicate:

For example—LLDPE tie, EVA tie, PP tie, etc.
If the brand is known, please indicate, such as Dow Bynel or Mitsui Admer. If the grade number is known, please indicate it as well.

MB—indicates a “masterbatch” or concentrate. This could be for pigments such as TiO_2 (white MB) or for slip and antiblock concentrates.

For example—white MB, slip MB, A/B MB, etc.

A.1.3 Key symbols

“/”—indicates the boundary between a solid interface and a liquid or molten material being applied to it.

For example: OPET/ink/primer/33 u EMAA
*** an oriented polyester film is printed with one or more inks, and dried. The printed surface then has a primer applied to it, dried, then extrusion coated with 33 microns of an EMAA resin.

“()”—a set of parentheses. These indicate that the materials within are being coextruded.
Look for a dash “-” sign between materials. The dash indicates the boundary between two molten materials being coextruded together. They should be enclosed in a set of parentheses, “().” Use space on each side of the dash.

For example: (1 mil LDPE – 2 mil EAA)
*** a two-layer coextruded film of low-density polyethylene and EAA.

“[ ]” –- a set of straight-sided brackets. These indicate that the materials within are being blended. If known, percentages by *weight ratio* for the materials being blended should be indicated. Look for a plus sign “ + ” between the materials. The plus indicates two materials being blended together. They should be enclosed in a set of straight-sided brackets, “[ ].” Use space on each side of the plus sign.

For example:
([75% Nylon 6 + 25% Amorphous PA] – tie–LLDPE)
*** a blend of Nylon 6 and Amorphous Nylon, in the *weight ratios* indicated, is the outer layer of a three-layer coextrusion. The middle layer is an unknown coextrudable adhesive resin, and the inner layer of the coextrusion is an LLDPE.

“//”—indicates the boundary between two solid surfaces, one of which is a film, and the other of which is a film that is often coated with an adhesive. (Typical for adhesive lamination.)

For example: OPET/adhesive//25 u ionomer film
*** an oriented polyester film that has an adhesive applied to it and dried. A film of ionomer is then combined to this typically through a heated nip roll to activate the adhesive.

“*//TL or BL//*”—indicates the boundary between two warm solid layers that are joined together by only heat and pressure. Examples would be:

“*//TL//*” Thermally laminated films. Two films (or one film and one foil), combined typically between rubber rolls, following preheating of one or both films.
“*//BL//*” A “*Blocked Lamination.*” Blown films that are intentionally blocked in the collapsing nip. In addition, if a coextruded film, surround the structure with curved brackets, “{ },” to indicate a combined web of coextrusion. See examples in the following pages.

“ < ” *or* “ > ”—indicates the direction to which some type of surface treatment has been applied, such as flame, corona, ozone, plasma.

For example—Paper < flame / LDPE / foil / LDPE

A.1.4 Materials (generic representations)

A.1.4.1 Webs/films/substrates

OPET—biaxially oriented polyester film. (Occasionally called *BOPET*.)
*** Writing “*OPET*” for a film will prevent confusion with extruded PET.
vm-OPET—vacuum metallized OPET film. Metallization on the left.
*** also known as “met-PET” or “met-OPET”
OPET-vm—vacuum metallized OPET film. Metallization on the right.

*** also known as "PET-met" or "OPET-met"
OPET-SiOx—OPET film with silicon oxide on the right side of the film.
SiOx-OPET—OPET film with silicon oxide on the left side of the film.
OPET-PVdC—oriented polyester film with a PVdC coating on the right.
PVdC-OPET—oriented polyester film with a PVdC coating on the left.
PVdC-OPET-PVdC—oriented polyester film, two side coated with PVdC.

OPP—biaxially oriented polypropylene film, also known as BOPP.
BOPP—biaxially oriented polypropylene film, also known as OPP.
vm-OPP—oriented polypropylene film, with metallization on the left side
OPP-vm—oriented polypropylene film, with metallization on the right side
OPP-A—oriented polypropylene film, with an acrylic coating on the right side
A-OPP—oriented polypropylene film, with an acrylic coating on the left side.
CPP—cast polypropylene film.
vm-CPP—cast polypropylene film with metallization on the left side.
CPP-vm—cast polypropylene film with metallization on the right side.

C-Nylon—cast nylon film, also written as *Cast PA* for cast polyamide film.
O-Nylon—uniaxially oriented nylon film.
BONy—biaxially oriented nylon film.
BOPA—biaxially oriented polyamide (nylon) film.
K-Nylon—biaxially oriented nylon film, with PVdC on one side.
BONy-PVdC—"K-Nylon," but written to specify PVdC on the right side.

Alu—aluminum (aluminium), can also be written as "*AL of alu*."
Foil—typically refers to thin aluminum. Can be used for other metals, so if it is not aluminum, please specify. Examples: tin foil, gold foil, etc.

(In Europe sometimes the word "foil" is used to indicate any type of thin film, and can cause communication issues globally.)

Cello—plain, uncoated cellophane.
K-Cello—two-side PVdC coated cellophane.
NC-Cello—two-side nitrocellulose coated cellophane.

Paper—use for lightweight paper-based webs.

*** There are various types, so please specify when possible:
Kraft, Bleached Kraft, SBS, clay-coated, etc.

Board—use for heavyweight paper-based webs. Various types exist.

Please specify when possible.

Nonwoven—generic for nonwoven webs if the type or brand is not known.

*** Nonwoven HDPE, Nonwoven PP, Nonwoven PET, – please specify
*** use brand name if known; such as DuPont Tyvek, BerryGlobal Reemay, BerryGlobal Typar, Mitsui Tafnel.

Woven Fabric—use for generic fabric weaves, with a "tight weave".

*** woven PP, woven HDPE, woven PP.

Scrim—use for woven mesh type fabrics with an "open weave".

*** Scrim PP, Scrim PET, Scrim HDPE.

A.1.4.2 Thickness or Gauge representations

#—usually means, "pounds per ream" (typically a ream being 3000 square feet).
mils—used for the English measurement of 0.001 or 1/1000 of an inch.

inches—English unit for thick layers, such as in bottles and sheet applications.
u—abbreviation for micron of the Metric measurement, or 0.001 or 1/1000 of a millimeter.
mm—millimeters, metric unit used for thick layers, such as in bottle and sheet applications.
gsm—grams per square meter.
pt.—usually means, "point," which is a unit of thickness for paperboard. 1 point = 1 mil or 0.001".
ga—used to indicate the "gage" or thickness of a film in some systems.

100ga = 1 mil = 25.4 u,
For example—48 ga = 12 u = 0.5 mils.

A.1.4.3 Adhesives and primers

PUR—polyurethane; *PUR adhesive* or *PUR primer*; can be solvent or water-based.
PEI—polyethylene-imine; *PEI primer*, usually a water-based primer.
EAA—ethylene acrylic acid; *EAA primer* (water-based).
EAC—ethylene acrylate; *EAC primer* (water-based).
(*** there are others in use, though less common to our industry, for example, organic titanates).

A.1.4.4 Surface treatments

Corona—electron discharge excitement of atmospheric air over a material surface.
Plasma—electron discharge excitement of a gas other than standard air over the material surface. Examples would include nitrogen, argon, helium, neon, hydrogen, as well as various gas mixtures.
Flame—surface treatment with burning gases. The gas mixture could be an oxidizing flame, a reducing flame, or a stoichiometric flame.
Ozone—molten web oxidation by exposing the polymer to a dry ozone flow created off-line and pumped to the molten web area, through a tube with small outlet holes along its length. When used it is typically seen on an extrusion coating line versus other processes.

A.1.4.5 Extrudable resins

(generic names for when brand and/or grade are unknown.)
HDPE—high density polyethylene.
HMW-HDPE—high molecular weight HDPE.
MDPE—medium density polyethylene.
LDPE—low density polyethylene.
LLDPE—linear low density polyethylene.
VLDPE—very low density (linear) polyethylene; (also known as—VLLDPE).
ULDPE—ultra low density polyethylene (also known as—ULLDPE).
mPE—metallocene polyethylene, generic.
mLMDPE—metallocene linear medium density polyethylene.
mLLDPE—metallocene linear low density polyethylene.
mVLDPE—metallocene very low (linear) density polyethylene, (sometimes referred to as a "plastomer").

PP—polypropylene; (generic indication when minimal information is known).
CoPP—copolymer polypropylene.
HoPP—homopolymer polypropylene.

ACR—acid copolymer resin, generic name for EAA and EMAA resins.
EAA—ethylene acrylic acid copolymer, such as Dow Nucrel, ExxonMobil Escor, SK Primacor.
EMAA—ethylene methacrylic acid copolymer, such as Dow Nucrel.
Ionomer—generic name for ionomeric copolymer resins, such as Dow Surlyn.

Acrylate—generic name for various acrylate copolymers, such as Westlake EMAC, ExxonMobil Optema Dow Elvaloy AC, SK Lotryl.
EBA—ethylene butyl acrylate.
EEA—ethylene ethyl acrylate.

EMA—ethylene methyl acrylate.
EMMA—ethylene methyl methacrylate.
EiBA—ethylene iso-butyl acrylate.
EnBA—ethylene normal-butyl acrylate.
EVA—ethylene vinyl acetate, such as; Celanese Ateva, Dow Elvax, LyondellBasell Ultrathene, ExxonMobil Escorene Ultra *** If % VA is known, please indicate; for example 9% EVA, 15% EVA or 28% EVA.

PS—polystyrene.
EPS—expanded (or foamed) polystyrene.
HIPS—high impact polystyrene.
GPPS—general purpose polystyrene.

PVC—polyvinyl chloride.
PVdC—polyvinylidene chloride; polyvinyl di-chloride. Most commonly seen as a coating on a film, but there are also extrudable grades, such as SK Saran, and Solvay Ixan.

PA—polyamide, *(nylon); for example*,- Advansix Aegis, BASF Ultramid, DSM Akulon, or UBE Nylon.
Nylon—polyamide, when unspecified, quite often is *Nylon 6* in the packaging film industry.
Nylon 6—polymer from caprolactam (also written "*PA 6*").
Nylon 6,6—polymer from adipic acid & hexamethylenediamine.
*** also written as *"PA 6,6" or "PA 6:6" or "PA 66"*.
Nylon 6/6,6—copolymer of 6 and 6,6 types of polyamide;
*** also written as *"PA 6/6,6" or "PA 6:66."*
Amorphous PA—amorphous nylon; such as DuPont Selar PA, or EMS Grivory.
MXD6—crystalline nylon of the MXD6 type; Mitsubishi Nylon-MXD6.
NOTE: *** other polymers exist: Nylon 11, Nylon 12, Nylon 6/ 12, etc.

EVOH—ethylene vinyl alcohol, such as; Kuraray Eval, Nippon-Gohsei Soarnol.
*** if the mole% ethylene is known, it should be indicated, for example 32% EVOH

PAN—polyacrylontrile, sometimes written as "ACN".
SAN—styrene acrylonitrile copolymer.
AN-MA—acrylonitrile methyl acrylate copolymer, such as MSM Poly Anobex; Ineos Barex(discontinued).
ABS—acrylonitrile butadiene styrene copolymer.
LCP—liquid crystal polymer.
COC—cyclic olefin copolymer, such as TAP Topas or Mitsui Apel.
PUR—polyurethane, extrudable type polymer.
SBC—styrene butadiene copolymer, such as Ineos Styrolution K-Resin.
TPS—thermoplastic starch polymer, such as Novamont MaterBi, Kuraray Plantic, others, etc.
PET—extruded polyester (polyethylene terephthalate). Could be monolayer or in a coextrusion.
PET ext.ctg.—extrusion coating of polyester.
APET—amorphous polyester.
CPET—crystalline polyester.
PETG—polyester copolymer with glycol.
PEN—polyethylene napthalate.
PTT—polytrimethylene terephthalate, such as DuPont Sorona.
PLA—polylactic acid.
PGA—polyglycolic acid.
PHA—polyhydroxyalkanote.
PHB—polyhydroxybutyrate.
PBAT—polybutylene adipate terephthalate.
PHBV—poly-hydroxybutyrate-hydroxyvalerate.

Terpolymers
EVACO—terpolymer of ethylene, vinyl acetate, and carbon monoxide.

EnBACO—terpolymer of ethylene, normal-butyl acrylate, and carbon monoxide.
EnBAgMA—terpolymer of ethylene, normal-butyl acrylate, and glacial methacrylate.
EiBAMAA—terpolymer of ethylene, isobutyl acrylate, and methacrylic acid.
EBAMAh—terpolymer of ethylene, butyl acrylate, and maleic anhydride.
EEAMAh—terpolymer of ethylene, ethyl acrylate, and maleic anhydride.
EMAMAh—terpolymer of ethylene, methyl acrylate, and maleic anhydride.
NOTE: there are various other terpolymers in the industry for specialized applications.

Grafted resins
EVA-gMAh—ethylene vinyl acetate with a graft of maleic anhydride (also may be written as EVA-g-MAH).
LLDPE-gMAh—LLDPE with a graft of maleic anhydride.
HDPE-gMAh—HDPE with a graft of maleic anhydride.
PP-gMAh—polypropylene with a graft of maleic anhydride.
NOTE: there are various other grafted resins in the industry for specialized applications.

Others
Peel Seal or Easy Peel—generic name for a peelable sealant resin of unknown brand/grade.
** if brand is known, please indicate, such as Dow Appeel, MDP(Japan) CMPS, Yasuhara Hirodyne, Tosoh Melthene.
H.S. lacquer—generic name for a "heat seal lacquer," usually applied by gravure.
Hot Melt—generic name for a hot melt adhesive applied as a sealant.

A.1.5 Structure examples

****Structures should be written from "outside to inside," with regards to the order of layers, starting from the left.*

OPET / ink / adhesive // CPP
*** an adhcsivc lamination of cast polypropylene film to reverse printed polyester film. An adhesive of unknown type is applied to the printed surface of the polyester film.

OPET / primer / CoPP
*** an extrusion coating of copolymer polypropylene on to an unprinted oriented polyester film. There is a primer of unknown type applied to the OPET.

48ga OPET / ink / PEI primer / 28# LDPE
*** a reverse printed 48 gage OPET film that is primed with a PEI primer, and then extrusion coated with 28 pounds per ream of low density polyethylene.

15 u BONy / ink / PUR adh // 88 u LDPE
*** an adhesive lamination of an LDPE film to a reverse printed biax nylon film. The PUR adhesive is applied to the reverse printed nylon film.

0.5 mils OPET-PVdC / ink / primer / 0.8 mils LDPE / 1.5 mils Peel Seal
*** a 0.5 mils OPET film with PVdC coating is reverse printed on the PVdC side. The printed side is then primed and extrusion coated with 0.8 mils LDPE. A 1.5 mils extrusion coating of a peelable sealant resin is applied onto the LDPE. This is done in either via tandem or a two-pass extrusion coating, not specified.

12 u OPET / primer / 15 u LDPE / 6 u alu / (11 u EMAA – 25 u Peel Seal)
*** a 12-micron OPET film is primed and extrusion laminated with 15 microns of LDPE to 6-micron aluminum foil. The bare aluminum side is then coextrusion coated with a two-layer coating comprising 11 microns of EMAA and 25 microns of a peelable sealant resin.

OPET / primer / LDPE / sealant // TL // leather
*** an oriented polyester film that is primed and extrusion coated with LDPE and then a sealant resin. This structure is then thermally laminated to leather. (The application may be for an emblem or other fabric decoration that requires surface protection.)

Ink / 30 u alu < flame / (EMAA – Peel Seal)
*** a surface printed aluminum foil that is in-line flame treated, and then coextrusion coated with an EMAA copolymer and a peelable sealant resin. (A cup or tray type lidding application.)

OPET / primer / (LDPE - [LDPE + white MB] - LDPE) / foil / EMAA
*** an OPET film which is primed, and then foil is extrusion laminated to it using a three-layer coextrusion, with the center layer containing a white masterbatch blended into LDPE. The foil is then extrusion coated on the other side with EMAA.

20 u OPP / ink / adhesive // 7 u alu / 33 gsm EAA
*** a 20-micron OPP film is reverse printed, then adhesively laminated to 7-micron aluminum foil. The adhesive is applied to the printed surface, not the aluminum. The laminate is then extrusion coated with 33 grams per square meter of EAA.

20 u OPP-PVdC / low temp PUR primer / 35 u 18% EVA
*** PVdC coated OPP is extrusion coated with an 18% EVA.

A special low activation temperature PUR primer is used to allow for direct extrusion coating of an EVA at 235°C onto a primed film. (235°C is the upper limit for EVA extrusion temperature.)

BONy / ink / solventless adh // (LLDPE – LDPE – ionomer)
*** a solventless adhesive lamination of a reverse-printed biaxially oriented nylon film, to a three-layer coextruded film. The adhesive is applied to the printed surface of the nylon film.

OPP < corona / solvent adhesive // (Nylon 6 – tie – LDPE – ionomer)
*** an OPP film that is corona treated in-line, then has a solventless adhesive applied to its surface. This film is then combined with a four-layer coex film, usually with a heated pressure nip.

O.L. / ink / 40# Paper < flame / 0.7 mils LDPE / 0.35 mils alu / 1.8 mils ionomer
*** a surface printed and overlacquered 40 pound paper is flame treated, and extrusion laminated using LDPE to aluminum foil. The other side of the alu is then extrusion coated with the ionomer. You will need to specify if this is done in a two-pass operation on a single station extrusion coating line, or one pass on a tandem extrusion coating line.

O.L. / ink / Paper / EMAA / alu / [ionomer + Slip MB]
*** a surface printed and overlacquered paper is extrusion laminated to alu foil with EMAA. The other side of the alu is then extrusion coated with a blend of ionomer and masterbatch.

OPET / ink /adh // SiOx-OPET / primer / (LDPE – ionomer)
*** a reverse-printed OPET is adhesively laminated to a silicon oxide coated OPET. The adhesive is applied to the printed surface of the outer OPET. Then the plain side of the silicon oxide coated OPET is primed and coextrusion coated with a two-layer coating of LDPE and ionomer.

OPET / ink / primer / met / adhesive // (Nylon - tie - LLDPE)
*** a reverse-printed OPET is primed on the ink surface, then metallized. This film is then adhesively laminated to a three-layer coex film. The adhesive is applied to the metallized surface.

OPET / ink / adhesive // OPET-VM / (LDPE – ionomer)
*** a reverse-printed OPET is adhesively laminated to a metallized OPET. The adhesive is applied to the ink surface. The metallized layer is facing to the inside of the package structure, not the outside. The metallized surface is coextrusion coated with a two-layer coating of LDPE and ionomer.

OPP / ink / primer / LDPE / primer / VM-CPP
*** a reverse-printed OPP that is extrusion laminated using LDPE to a metallized cast polypropylene film. Primers are applied to both the printed OPP, and the metallized CPP.

OPET / ink / primer / LDPE / VM-OPET // adhesive / CPP
*** a reverse-printed OPET that is extrusion laminated using LDPE to a metallized OPET. This is then adhesively laminated to a cast polypropylene film. The adhesive is applied to the CPP film.

OPET / ink / primer / LDPE / VM-OPET / adhesive // CPP
*** a reverse-printed OPET that is extrusion laminated using LDPE to a metallized OPET. This is then adhesively laminated to a cast polypropylene film. The adhesive is applied to the OPET film.

12 u OPET-met / adh // 15 u BONy / adhesive // corona > (20 u LLDPE – 10 u LDPE – 45 u EVA)
*** an OPET that has been metallized, then adhesively laminated to BONy film. The other side of the nylon film is then adhesively laminated to a coex film which has been in-line corona treated (likely re-treated to burn off slip agents in the film).

OPET / ink / primer / LDPE / met-OPET < corona / (tie – LDPE) / LLDPE film
*** a reverse-printed OPET film that is coated with an unknown primer, and then extrusion laminated to metallized OPET film using an LDPE. The plain side of the metallized OPET film is then extrusion laminated to LLDPE film using a special tie resin and LDPE in a coextrusion process.

20 u OPP / ink / primer / 15 u LDPE / 13 u met-OPET < corona / (6 u tie – 6 u LDPE) / 12 u ionomer
*** a reverse-printed OPP that is primed and extrusion laminated to a metallized OPET film. The plain side of the PET film is coextrusion coated with a tie resin and LDPE, and the LDPE is extrusion coated with ionomer.

(LLDPE – W.O. LLDPE – LDPE) film / EAA / alu / EAA / (LDPE – LLDPE) film
*** an extrusion lamination with EAA of a three-layer film to one side of the alu. The outer film is white, but the pigment is only in the middle layer. The other side of the alu has a two-layer film extrusion laminated to it with an EAA. (A toothpaste tube laminate type application.)

Cello / ink / primer / LDPE / alu / EMAA
*** cellophane that is reverse printed, primed, and extrusion laminated to aluminum with LDPE. The other side of the alu is then extrusion coated with EMAA.

K-Cello / ink / primer / LDPE / alu / ionomer
*** a PVdC coated cellophane (two sides coated), that is reverse printed, primed, and extrusion laminated to alu with LDPE. The other side of the alu is then extrusion coated with an ionomer.

O.L. / Ink / 50gsm Paper / LDPE / 7 u Alu / 35 u EAA
Note: (made by tandem extrusion line)
*** 50 gsm Paper that is surface printed and overlacquered. It is extrusion laminated with LDPE to alu. The other side of the alu is extrusion coated with EAA. This is specified as being processed by tandem extrusion lamination instead of two-pass coating.

LDPE / ink / 18 pt. Board / LDPE / alu / (EAA- LDPE)
*** 18 point paperboard that is printed and surface extrusion coated with LDPE, and then extrusion laminated with LDPE to alu. The other side of the alu is then coextrusion coated with a combination of LDPE and EAA.

50 u Blue EAA //TL// alu //TL// 50 u EAA
*** a thermal lamination of blue pigmented EAA film to alu on one side. Another film of clear EAA is thermally laminated to the other side of the alu. (A cable shielding laminate type structure.)

Green EAA / 200 u alu / EAA
*** an extrusion coating of green pigmented EAA is applied to one side of the 200 u alu. A clear EAA is extrusion coated to the other side of the alu. (A cable shielding laminate type structure.)

(LDPE–EAA) //TL// 150 u alu //TL// (EAA–LDPE)
*** 150 u aluminum that has coex films thermally laminated to each side. (A cable shielding type structure.)

{(mLLDPE–LDPE–tie–Nylon–EVOH–Nylon–EVA tie) //BL// (VA tie–Nylon–EVOH–Nylon–tie–LDPE–mLLDPE)}
*** a thirteen layer structure that is made by intentionally blocking a coex blown film in the main collapsing nip. The film from the die is actually seven layers. In the main nip, the bubble is blocked on purpose. The nip rolls may be heated to help this. The tower height is usually very short so that the film is still quite warm entering the nips. The tie resin at the blocking interface is known to be an EVA based tie resin, but the exact grade of brand is not known.

(Nylon 6,6–Tie A–Nylon 6–Tie B–ionomer)
*** a five-layer coextruded film. You will need to specify if this is made by a cast film or blown film process. In a straight coextrusion, this is helpful to know for physical property issues.

(PP–tie A–Nylon 6–EVOH–Nylon 6–tie B–LLDPE)
*** a seven-layer coextruded film. The tie layer resins are not known, but it is known that there are two different grades being used in the one structure. You will need to specify whether it is made by a cast film or blown film process.

(Nylon–tie–Nylon–tie–LLDPE–mPE–ionomer)
*** a seven-layer coextruded film. The tie layers are believed to be the same, so are not labeled differently. An mPE is used to adhere the ionomer to the LLDPE. An upgrade would be to specify the grade or density of the mPE and the grade or ion type of the ionomer. You will need to specify whether it is made by a cast film or blown film process.

(15 u Nylon 6–5 u LLDPE tie–45 u LLDPE)
*** a three-layer coextruded film. The thickness of each layer is known, so is specified in the structure. The tie resin grade is not known, but it is known that it is based on LLDPE. You will need to specify whether it is a cast film or blown film.

(20 u [Nylon 6 + Amorphous PA]–4 u tie–40 u LLDPE)
*** a three-layer coex film. The blend ratio of the outer layer is unknown. The layer thickness for each layer is specified. You will have to specify if this is made by a blown or cast film process.

([80% Nylon 6 + 20% Amorphous PA]–tie–EVOH–tie–LLDPE)
*** a five-layer coextruded film. The outer layer is a blend of 80% Nylon 6 plus 20% amorphous nylon. You will need to specify if the film is made by a blown or cast process.

(850 u HDPE–55 u HDPE tie–55 u Nylon 6)
*** a three-layer coextrusion, which by looking at the thickness, likely is a coex blow molded bottle. This type of layer thickness is often used for barrier bottles, such as for agricultural chemicals, where nylon is often used as the inner layer. Please indicate the application when drafting out structures such as this to prevent confusion, as this could also be a sheeting application.

(725 u HDPE–12 u regrind–55 u HDPE tie–55 u Nylon 6)
*** similar to the above structure for an agchem bottle, except with a layer of regrind material between the outer HDPE and the tie layer.

(225 u PP–PP tie–35 u EVOH–PP tie–175 u PP)
*** a coextrusion that by the thickness is likely a sheet material. The type of structure indicates that it might be used to thermoform into some type of cup or tray. A description of the actual application for the structure should be included in any communication.

Sample structures for practice descriptions:
K-Nylon < corona / adhesive // (mLLDPE–12% EVA–peel seal resin)
BONy-PVdC / adhesive // (mLLDPE–mLLDPE–EMAA)
BONy-PVdC / primer / (LDPE–LDPE–EMAA)
OPET / ink / adh / alu / adhesive // (LLDPE–LLDPE–EMAA)
PET-VM / adhesive // BONy / adhesive // corona > (LLDPE - LDPE - EVA)
OPP / ink / adh // met-OPET / primer / (LDPE–ionomer)
OPET-PVdC /ink / adhesive / (LDPE–EMAA–ionomer)
PVdC-OPET-PVdC / ink / adhesive / LLDPE
Woven PP < corona / (tie–LDPE) / LLDPE film

Woven PP < corona / (tie−CoPP)
OPP / ink < corona / (tie−[LDPE + EMAA] / met-OPET / primer / ozone > (LDPE−EVA)

A.1.6 Trademark references

The following references for Companies and Trademarks are as of December 2020. Please note that companies change names and businesses are sold and acquired, and in the future, the owner of given trademarks may shift. Please consult current literature when wanting to reference a trademark of a material.
Advansix: Aegis
Amcor: EZ Peel, Reseal
Arkema: Pebax, Orgalloy
BASF: Ultramid
BerryGlobal: Reemay, Typar
Celanese Corp: Ateva
ChevronPhillips Chemical: Marlex
Dow Chemical: Dowlex, Elite, Affinity, Agility, Engage, Infuse, Attane, Versify, Adcote, Appeel, Biomax, Booster, Bynel, Conpol, Elvaloy, Elvax, Entira, Fusabond, Nucrel, Surlyn
DSM: Akulon
DuPont-Teijin Films: Mylar, Melinex
DuPont: Hytrel, Selar, Sorona, Tyvek, Zytel
EMS Group: Grivory, Grilon
ExxonMobil Chemical: Escor, Exact, Exceed, Escorene, Optema, Enable, Exxco, VistaMaxx, Vistalon
Henkel: Liofol, Tycel, Loctite
Ineos: Eltex, Rigidex
Ineos Styrolution: K-Resin, Styrofex, Styrolux, Terluran
Kuraray: Eval, Plantic
LyondellBasell Chemical: Ultrathene, Petrothene, Microthene, Plexar, Toppyl, Alathon, Lupolen, Hostalen, Moplen, Clyrell, Adsyl, Adstif, Pro-Fax, StarFlex
Mica Corp: MICA
Michelman Corp: Michem
Mitsubishi Chemical: Modic
Mitsui Chemical: Apel, Admer, Evolue, Tafmer, TPX, Tafnel
Mitsui-Dow Polymers (Japan): CMPS, Evaflex, Hi-Milan
Nippon Gohsei: Soarnol
Nova Chemical: Novapol, Sclair, Surpass
Reliance Industries Ltd: Relene, Repol
SK Chemical: Yuclair, Primacor, Saran, Evatane, Lotader, Lotryl, Lotryl BestPeel, Orevac
Solvay: Ixan, Diofan
Topas Advanced Polymers: Topas
Tosoh: Melthene, Nipoflex,
Toyo Chemical: Oribain, Tocryl
Versalis (Eni SpA): Clearflex, Flexirene, Eraclene
Westlake Chemical: EMAC, EBAC, Mxsten, Tymax, Hifor, Elevate, Epolene
Yasuhara Chemical: Hirodine, Hirotac

Appendix B

Examples of flexible packaging film structures

This appendix is adapted from sections of Butler, T. I. and Morris, B. A., 2010, "PE based multilayer film structures," in Wagner, J. (ed.), *Multilayer flexible packaging, technology and applications for the food, personal care and over-the-counter pharmaceutical industries*, Elsevier, Oxford, pp. 205–230.

Packages may be formed in-line by several techniques or may be supplied to the packer as pre-formed pouches or bags. Packaging may also be created by wrapping or shrinking a basic film around a bundle of goods. In-line package-forming examples include vertical form-fill-seal (VFFS), horizontal form-fill-seal (HFFS) and thermoform-fill-seal (see Chapter 3). In VFFS operations, film from a roll is guided through rollers and then shaped by a forming collar into a tube. The film moves in a vertical direction (down) over a filling tube. A vertical seal is made, forming the film into a continuous tube. As the film continues through the machine, a horizontal seal is made, perpendicular to the film machine direction, forming the bottom end-seal of the bag being formed and the top seal of the previously filled bag. The product is dropped into the partially formed bag, advanced to the seal bars and the next bottom and top end-seal are made. The process may operate in a step-wise or continuous manner. One example of a product normally packaged on VFFS equipment is fresh-cut produce. In HFFS operations, the film moves in a horizontal direction during the packaging step. One application that typically uses HFFS equipment is chunk cheese. In thermoform-fill-seal operations, a bottom web is formed, product is added and the top web, which is normally flat, is sealed to the bottom web. Thermoform-fill-seal packaging is frequently used for bacon and processed meats. Stand-up pouches and other types of packaging may be formed in-line with the filling equipment or maybe fully or partially prefabricated prior to the filling step.

Markets for flexible packaging films have continued to grow in many applications. Polyethylene (PE) and various copolymers account for more than 75% of flexible packaging film. Some major market segments for coextruded films and laminates where PE is a major component include:

1. medical packaging;
2. food packaging;
3. heavy duty shipping bags;
4. stretch wrap;
5. trash bags;
6. condiment sachets (ketchup, salad dressing, etc.);
7. aseptic juice packaging;
8. towlettes;
9. condoms;
10. laminates for toothpaste tubes;
11. stand-up pouches.

The sections that follow provide some of the package requirements for these markets and past structures used to meet them. They are drawn primarily from North America with an emphasis on coextruded films containing PE. By no means are the structures provided exhaustive in what has or could be used for the application. Nor are they meant to predict what structures may be best for the future. They are merely provided as a reference and thought starter. As described throughout this book, there are often many ways to construct a flexible package structure to meet the needs of the application. Overall economics is typically the driving force for selecting one polymer or structure over another. But this depends on the capabilities and assets of the provider, as well as the desires of the packager. Factors such as sustainability may drive different material and structural choices in the future.

B.1 Medical packaging

Consumer and industrial healthcare products using flexible packaging include:

1. Medical disposables;
2. Surgical instruments;
3. Syringes and hypodermic needles;
4. Sutures;
5. Pharmaceuticals;
6. Condoms and towlettes.

A wide variety of structures are used for packaging medical devices such as surgical instruments, syringes and sutures (Table B.1). The structure requirements include:

1. Sterilization capability;
2. Microbial barrier;
3. Linear tear properties;
4. Puncture resistance.

Sterilization methods used for medical device packaging include ethylene oxide gas and radiation. The package is typically a forming web sealed to a nonwoven fabric (like Tyvek) or special paper, which lets the ethylene oxide in and out of the package. Medical device packaging usually does not require oxygen barrier properties so EVOH or other barrier resins are normally not required. All the major converting processes are used: blown film, cast film and extrusion coating/lamination. Coextrusion is increasingly used with three to eleven layers commonly available. The time to qualify new structures in these markets is often long.

The forming webs were historically three-layer EVA and ionomer films, which are still in use today. Heavy gauge films were sometimes produced by combining a three-layer film to form six layers. Newer film structures have seen more layers being used and incorporating polyamide (PA) for forming and puncture resistance.

Sachets for condom and towelette packaging typically use the structure OPET/print/LDPE/Al/ION and variations, with ionomers used as the sealant for its chemical and flex crack resistance. Some types of lidding stock are produced by extrusion coating and/or lamination processes and combine paper, polymers and foil to form multilayer structures.

TABLE B.1 Example medical device packaging structures.

3-layer structure	Layers (%)	Gauge (μm)
(EVA-ION-EVA)	(20/60/20)	50–400
(ION-EAA-EVA)	(30/30/40)	50–400
Paper/LDPE/Al	Lamination	
5-layer structure		
(LLDPE-Tie-PA-Tie-LLDPE)	(40/5/10/5/40)	50–400
(mLLDPE-Tie-PA-Tie-mLLDPE)	(40/5/10/5/40)	50–400
(LLDPE-Tie-PA-Tie-PA)	(70/5/10/5/10)	50–400
LDPE/Paper/LDPE/Foil/LDPE	Lamination	
OPET/adh/(LDPE-EVA-ION)	Lamination	
7-layer structure		
(LLDPE-Tie-PA-EVOH-PA-Tie-LLDPE)	(30/5/10/10/10/5/30)	50–400
9-layer structure		
(PE-Tie-PA-Tie-PE-Tie-PA-Tie-PE)	(20/|5/10/5/20/5/10/5/20)	50–400
(LLDPE-LDPE-Tie-PA-EVOH-PA-Tie-LDPE-LLDPE)	(10/15/10/10/10//10/10/15/10)	50–400

EVA, ethylene vinyl acetate; *EVOH*, ethylene vinyl alcohol; *LLDPE*, linear low-density polyethylene; *LDPE*, low-density polyethylene; *PA*, polyamide; *OPET*, biaxially oriented polyethylene terephthalate.

B.2 Food packaging

B.2.1 Primal meat packaging (shrink bags)

Packaging primal and subprimal meat requires a package that:

1. Provides high shrinkage to fully collapse around irregular shapes.
2. Has excellent optical properties.
3. Shrinks at low temperature to prevent product damage.
4. Imparts good softness and elasticity.
5. Provides excellent oxygen, moisture, odor, and grease barrier protection.
6. Prevents freezer burn.
7. Facilitates using individual cuts by food preparers.
8. Helps reduce purge loss.
9. Extends shelf-life.
10. Offers easy disposal.
11. Has good machinability.
12. Has an oxygen transmission rate (OTR) of less than 15 cc/m^2-day-atm (nonfrozen only).

Shrinkable PVDC barrier films with the sealant layer designed to provide toughness and puncture resistance are often used. These films must be oriented to provide acceptable shrink properties using a double bubble process. PA and EVOH have replaced PVDC in some applications. Table B.2 shows some typical film structures used in shrink film for primal and subprimal meat packaging.

B.2.2 Processed meat packaging

Processed and cook-in meat such as luncheon meat, ham, bologna and salami are packaged in barrier films that are designed to keep oxygen from entering the package. This extends shelf-life and gives the retailer longer product display times. It also allows the consumer to keep the product in their refrigerator, unopened, for some time after purchase. These packages are often printed with eye-catching graphics to increase sales. These films may contain:

1. A barrier polymer.
2. Printing surface, such as PET or PA, also provides thermal resistance during sealing and helps provide abuse resistance during distribution.
3. LLDPE or VLDPE toughness layers.
4. A sealant layer that is typically an EVA, LLDPE, polyolefin plastomer (POP) or ionomer.

The processed meat package comprises a forming film and a backing film (nonforming film). The forming film is thermoformed to the meat product shape. In addition, low oxygen permeability, abuse resistance and seal integrity are critical to maintaining the proper atmosphere inside the package. Optical properties, such as high gloss and high clarity, are important on the backing film where reverse-printed OPET is used to create consumer appeal. Barrier requirements

TABLE B.2 Example shrink bag structures for primal and subprimal meat packaging.

5-layer structure	Layers (%)	Gauge (μm)
(ULDPE-EVA-PVDC-EVA-ULDPE)	(40/5/10/5/40)	50–120
(mLLDPE-EVA-PVDC-EVA-mLLDPE)	(40/5/10/5/40)	50–120
(PA-EVOH-Tie-LLDPE-mLLDPE)	(30/40/10/10/10)	50–120
(PA-EVOH-Tie-LLDPE-ION)	(30/40/10/10/10)	200–300
OPA/PVDC/(LLDPE-EVA-ION)	(20/30/35/5/10)	50–100

Note: /PVDC/ is an adhesive lamination with PVDC. See Appendix A for convention used. *EVA*, ethylene vinyl acetate; *EVOH*, ethylene vinyl alcohol; *LLDPE*, linear low-density polyethylene; *OPA*, oriented polyamide; PA, polyamide; *PVDC*, polyvinylidene chloride; ULDPE, ultralow-density polyethylene, *mLLDPE*, metallocene LLDPE; *ION*, ionomer.

for processed meats range from 3 to 15 cc/m^2-day-atm for OTR and 3–8 g/m^2-day for water vapor transmission rates (WVTRs). Table B.3 shows some typical film structures used in processed meat packaging.

B.2.3 Poultry/fish packaging

Moisture barrier properties are critical for poultry and fish packaging. The package form is normally vacuum packaging with a good sealant polymer such as EVA, ionomer or LLDPE (Table B.4).

TABLE B.3 Example processed meat packaging film structures.

Product structure	Layers (%)	Gauge (μm)
Ground beef		
(PA-Tie-LLDPE)	(20/5/75)	150–200
(PA-EVOH-Tie-LLDPE)	(10/10/5/75)	150–200
(LLDPE-Tie-PA-EVOH-PA-Tie-LLDPE)	(25/10/10/10/10/10/25)	40–150
Barrier overwrap		
(PA-EVOH-Tie-LLDPE-mLLDPE)	(10/10/10/40/30)	150–200
(PA-EVOH-Tie-LDPE-LDPE-mLLDPE)	(15/10/15/10/15/20)	150–200
Chub films		
PA//PVDC/LLDPE	(20/5/75)	150–200
Foodservice portion: steaks/chops/roasts		
Forming web		
(PA-EVOH-Tie-LLDPE)	(15/5/5/75)	150–200
(PA-EVOH-Tie-ION)	(15/5/5/75)	150–200
(PA-Tie-LLDPE)	(20/5/75)	150–200
(PA-ION)	(20/80)	150–200
PA//PVDC/(EVA-ION)	(25/5/10/60)	150–200
(LLDPE-Tie-PA-EVOH-PA-Tie-LLDPE)	(25/10/10/10/10/10/25)	150–200
Nonforming web		
(PA-EVOH-Tie-LLDPE)	(10/10/10/70)	50–80
(PA-EVOH-Tie-ION)	(10/10/10/70)	50–80
OPET//PVDC//LLDPE	(10/5/85)	50–80
OPET//PVDC//ION	(10/5/85)	50–80
(PA-Tie-ION)	(10/10/80)	50–80
(LLDPE-Tie-PA-EVOH-PA-Tie-LLDPE)	(25/10/10/10/10/10/25)	50–80
Skin packaging forming web		
(EVA-Tie-EVOH-Tie-ION)	(35/10/10//10/35)	150–250
Skin packaging nonforming web		
(EVA-Tie-EVOH-Tie-ION)	(35/10/10//10/35)	50–80
Luncheon meat		
Forming web		
(PA-EVOH-Tie-LLDPE)	(15/5/5/75)	150–200
(PA-EVOH-Tie-ION)	(15/5/5/75)	150–200

(Continued)

TABLE B.3 (Continued)

Product structure	Layers (%)	Gauge (μm)
(PA-EVOH-Tie-mLLDPE)	(15/5/5/75)	150–200
(LLDPE-Tie-PA-EVOH-PA-Tie-LLDPE)	(25/10/10/10/10/10/25)	150–200
Nonforming web		
OPET//PVDC//(EVA-LLDPE)	(25/5/10/60)	50–100
OPET//adh//(EVOH-Tie-mLLDPE)	(25/2/5/10/60)	50–100
OPET//adh//(EVOH-Tie-ION)	(25/2/5/10/60)	50–100
(PA-EVOH-Tie-LLDPE)	(25/10/5/25)	50–100
OPA//PVDC//LLDPE	(65/5/25)	50–100
(PA-EVOH-PA-Tie-LLDPE)	(10/10/10/10/60)	50–100
(LLDPE-Tie-PA-EVOH-PA-Tie-LLDPE)	(25/10/10/10/10/10/25)	50–100
Frankfurters		
Forming web		
PA-EVOH-Tie-LLDPE	(15/5/5/75)	150–200
(PA-EVOH-Tie-mLLDPE)	(15/5/5/75)	150–200
(PA-EVOH-Tie-EVA)	(15/5/5/75)	150–200
(PA-EVOH-Tie-ION)	(15/5/5/75)	150–200
(LLDPE-Tie-PA-EVOH-PA-Tie-LLDPE)	(25/10/10/10/10/10/25)	150–200
(PA-Tie-PA-EVOH-PA-Tie-ION)	(15/20/10/5/10/25/15)	150–200
(PA-Tie-PA-EVOH-PA-Tie-LLDPE)	(15/15/10/5/10/25/20)	150–200
(PP-Tie-PA-EVOH-PA-Tie-ION)	(15/20/10/5/10/25/15)	150–200
Nonforming web		
OPA//PVDC/(EVA-LLDE)	(10/5/25/60)	50–100
OPET//adh/(EVOH-Tie-mLLDPE)	(15/10/15/60)	50–100
OPET//PVDC//ION or EVA	(10/5/85)	50–100
(LLDPE-Tie-PA-EVOH-PA-Tie-LLDPE)	(25/10/10/10/10/10/25)	50–100
OPET//adh//(LLDPE-ION)	(25/2/48/25)	50–100
OPET//adh//(LLDPE-Tie-EVOH-Tie-ION)	(15/2/38/10/10/10/15)	50–100
Sausage		
Forming web		
(LLDPE-LDPE-Tie-EVOH-PA-Tie-EVA)	(30/5/10/10/10/5/30)	150–200
(PA-EVOH-Tie-mLLDPE)	(75/5/5/15)	150–200
(PA-EVOH-Tie-ION)	(15/5/5/75)	150–200
(LLDPE-Tie-PA-EVOH-PA-Tie-LLDPE)	(25/10/10/10/10/10/25)	150–200
(PA-Tie-PA-EVOH-PA-Tie-ION)	(15/25/10/5/10/20/15)	150–200
(PA-Tie-PA-EVOH-PA-Tie-LLDPE)	(15/25/10/5/10/15/20)	150–200
(PP-Tie-PA-EVOH-PA-Tie-ION)	(15/25/10/5/10/25/15)	150–200
OPET//adh//(EVOH-Tie-LLDPE)	(15/2/14/10/14/60)	50–100
(PA-EVOH-Tie-mLLDPE)	(15/10/15/60)	50–100

(Continued)

TABLE B.3 (Continued)

Product structure	Layers (%)	Gauge (μm)
(LLDPE-Tie-PA-EVOH-PA-Tie-LLDPE)	(25/10/10/10/10/10/25)	50–100
Shrink bags		
(EVA-PVDC-EVA)	(48/5/47)	50–80
(EVA-Tie-EVOH-Tie-EVA)	(38/10/5/10/37)	50–80
(LLDPE-Tie-PA-EVOH-PA-Tie-LLDPE)	(25/10/10/10/10/10/25)	50–80
Ham		
Forming web		
(PA-EVOH-Tie-LLDPE)	(15/10/15/60)	150–200
(PA-EVOH-Tie-ION)	(15/10/15/60)	150–200
(LLDPE-Tie-PA-EVOH-PA-Tie-LLDPE)	(25/10/10/10/10/10/25)	50–100
Nonforming web		
PA//PVDC//LLDPE	(25/5/70)	50–80
PA//PVDC/ION	(35/5/70)	50–80
OPET//PVDC//LLDPE	(25/5/70)	50–80
OPET//PVDC//ION	(25/5/70)	50–80
(LLDPE-Tie-PA-EVOH-PA-Tie-LLDPE)	(25/10/10/10/10/10/25)	50–100
Shrink bags		
(EVA-PVDC-EVA)	(48/5/47)	50–80
(EVA-Tie-EVOH-Tie-EVA)	(38/10/5/10/37)	50–80
(mLLDPE-Tie-EVOH-Tie-mLLDPE)	(30/10/10/10/30)	50–80
Bacon		
(PA-EVOH-Tie-ION)	(30/10/10/50)	50–100
PA/primer/(PE-ION)	(15/50/35)	50–100
(PA-Tie-PA-EVOH-PA-Tie-ION)	(15/20/10/5/10/20/20)	50–100
Deli meats		
Shrink bags		
(EVA-PVDC-EVA)	(48/5/47)	50–80
(EVA-Tie-EVOH-Tie-EVA)	(38/10/5/10/37)	50–80
(mLLDPE-Tie-EVOH-Tie-mLLDPE)	(30/10/10/10/30)	50–80
Forming web		
(PA-EVOH-Tie-LLDPE)	(10/5/5/80)	200–250
(PA-EVOH-Tie-ION)	(10/5/5/80)	200–250
(LLDPE-Tie-PA-EVOH-PA-Tie-LLDPE)	(25/10/10/10/10/10/25)	50–100
Nonforming web		
LLDPE/adh/(PA-EVOH-Tie-EVA)	(40//10/10/20/30)	50–80
(PA-EVOH-Tie-ION)	(15/10/15/60)	50–80
(LLDPE-Tie-PA-EVOH-PA-Tie-LLDPE)	(25/10/10/10/10/10/25)	50–100

(Continued)

TABLE B.3 (Continued)

Product structure	Layers (%)	Gauge (μm)
Casings		
OPA//LLDPE	(30//70)	50–80
LLDPE//OPA//PVDC	(70//20//10)	50–80

EVA, ethylene vinyl acetate; *EVOH*, ethylene vinyl alcohol; *LDPE*, low-density polyethylene; *LLDPE*, linear low-density polyethylene; *OPA*, oriented polyamide; *OPET*, oriented polyester; *PA*, polyamide; PP, polypropylene; *PVDC*, polyvinylidene chloride; *mLLDPE*, metallocene LLDPE; *ION*, ionomer.

TABLE B.4 Example poultry/fish packaging.

7-layers structure	Layers (%)	Gauge (μm)
(LLDPE-Tie-PA-EVOH-PA-Tie-LLDPE)	(25/10/10/10/10/10/25)	40–150
(LLDPE-Tie-PA-EVOH-PA-Tie-ION)	(25/10/10/10/10/10/25)	40–150

EVOH, ethylene vinyl alcohol; *LLDPE*, linear low-density polyethylene; *PA*, polyamide; *ION*, ionomer.

TABLE B.5 Example cereal packaging structures.

Market structure	Layers (%)	Gauge (μm)
Bag-in-box		
(HDPE-HDPE-EVA + PB)	(45/45/10)	40–60
(HDPE-HDPE-ION)	(45/45/10)	40–60
Bag-in-box (peelable seal)		
(HDPE-HDPE-EVA + ION)	(45/45/10)	40–60
Bag-in-box (barrier bag)		
(HDPE-Tie-EVOH-Tie-EVA + PB)	(60/10/10/10)	40–60
(HDPE-Tie-PA-Tie-EVA + PB)	(60/10/10/10)	40–60

EVA, ethylene vinyl acetate; *EVOH*, ethylene vinyl alcohol; *HDPE*, high-density polyethylene; *PA*, polyamide; *ION*, ionomer; *PB*, polybutene.

B.2.4 Cereal bag-in-box liners

The bag liners for bag-in-box cereal packaging require good moisture barrier properties to provide good taste and freshness protection. High-density PE polymers (HDPEs) are typically used to provide the moisture barrier. Sealant polymers such as EVA, ionomer, plastomers or blends are used for low-temperature seals, form-fill-seal packaging and easy opening seals. Certain products have additional requirements, such as puncture resistance to keep the product from poking through the packaging film and flavor and aroma barrier for highly flavored cereals. Heat seal initiation temperatures of 90°C and below are commonly required. Moisture vapor transmission rates less than or equal to 15 g/m^2-day-atm are often required. Packages requiring aroma or taste barrier properties will contain either PA or EVOH polymers. (Table B.5).

B.2.5 Snack food packaging

Potato chips are often packaged in structures that contain metallized films (Table B.6). Polymer film metallization provides oxygen, moisture and light barrier. The light barrier is to protect the potato chips from ultraviolet radiation that initializes an oxidation mechanism. Seal strength must be optimized to provide a secure package that can be easily opened by the consumer. Seal integrity and consumer appeal are also critical.

TABLE B.6 Examples of snack food packaging films.

Structure	Layers (%)	Gauge (μm)
Potato chips (OTR < 31 cc/m^2-d, MVTR < 0.31 g/m^2-d)		
OPP//met-OPP	(50/50)	20–60
OPP//LDPE//met-OPP	(25/50/25)	20–60
OPP-met//(HDPE-EVA)	(15/70/15)	40–80
Tortilla and corn chips (OTR <31 cc/m^2-d, MVTR <0.5.4 g/m^2-d)		
OPP//LDPE//OPP	(25/50/25)	20–60
OPP//PVDC//OPP	(50/5/45)	20–60
Pretzels (OTR <31 cc/m^2-d, MVTR <7.7 g/m^2-d)		
OPP//LDPE//OPET	(25/50/25)	30–80
OPP//LDPE//OPP	(25/50/25)	30–80
OPP//PVDC/OPP	(45/5/50)	30–60
Meat snacks		
OPET//PVDC//LDPE	(50/5/45)	30–60
(PA-Tie-EVOH-Tie-LLDPE)	(10/10/10/10/60)	40–80

MVTR, moisture vapor transmission rate; *OTR*, oxygen transmission rate; *EVA*, ethylene vinyl acetate; *EVOH*, ethylene vinyl alcohol; *LDPE*, low-density polyethylene; *LLDPE*, linear low-density polyethylene; *OPP*, oriented polypropylene; *OPET*, oriented polyester; *PA*, polyamide; *PVDC*, polyvinylidene chloride.

TABLE B.7 Examples of snack nuts packaging.

Structure	Layers (%)	Gauge (μm)
OPP//LDPE//OPP	(20/60/20)	40–60
OPP//LDPE//Foil//LDPE	(10/20/5/65)	40–60

OPP; oriented polypropylene; *LDPE*, low-density polyethylene; Foil, aluminum foil.

B.2.6 Salty snack packaging

Salty snacks are frequently high in fat content and may require a package that provides an oxygen barrier in order to prevent the fat in the food from going rancid. They may also require grease resistance to keep the package from leaving an oily spot. Salty snacks may be packaged in barrier films containing foil, a metallized polymer film or a barrier polymer such as EVOH or PVDC and are gas flushed with nitrogen to maintain a low oxygen concentration inside the package (Table B.7).

B.2.7 Bakery

Moisture barrier is normally the critical property in bakery applications. Polymers used for moisture barrier include LDPE, LLDPE, HDPE or PP. Typically EVA or plastomers are used for sealability and optics. Applications such as cake mix pouches require aroma, taste and moisture barrier properties. PA is used for taste and aroma barriers. In bread bags, the toughness of LLDPE allows downgauging, while LDPE provides good optics and printability (Table B.8).

B.2.8 Cheese packaging

Most cheese sold in the United States is prepackaged in flexible packaging. The cheese packaging includes:

1. Individually wrapped slices (IWS) of processed cheese;
2. Chunk cheese;
3. Shredded cheese.

TABLE B.8 Examples of bakery packaging.

Structure	Layers (%)	Gauge (μm)
cPP/PP/cPP	(10/80/10)	30–60
PP/EVA	(80/20)	30–60
(HDPE-tie-PA-tie-mPE or EVA)	(65/5/10/5/15)	30–70
(LLDPE-PP-LLDPE)	(10/80/10)	30–60
(PA-EVOH-PA-Tie-LLDPE)	(20/10/20/10/30)	100–160

cPP, cast polypropylene; *PP*, polypropylene; *EVA*, ethylene vinyl acetate; *HDPE*, high-density polyethylene; *mPE*, metallocene plastomer; *PA*, polyamide; *LLDPE*, linear low-density polyethylene; *EVOH*, ethylene vinyl alcohol.

TABLE B.9 Examples of cheese packaging.

Structure	Layers (%)	Gauge (μm)
Natural chuck cheese pouches		
OPET/adh/(EVOH-Tie-LLDPE)	(15/5/10/10/55)	40–60
LLDPE//PVDC//OPP//PVDC//LLDPE	(15/10/75)	50–60
OPET//PVDC//LLDPE	(15/10/75)	45–50
EVA//PVDC	(95/5)	50–60
OPA//PVDC//LLDPE	(15/10/75)	45–50
Vacuum bags for aging		
OPA//PVDC//EVA	(110/5/85)	40–100
(PA-Tie-LLDPE)	(10/10/80)	40–100
(PA-Tie-EVA)	(10/10/80)	40–100
Shredded cheese		
PVDC//PA//LDPE	(5/20/75)	80–100
PVDC//PET//LDPE	(5/20/75)	80–100
(PA-EVOH-Tie-LDPE)	(10/10/10/PA)	80–100
Processed cheese slices		
OPP//EVA	(50/50)	20
PP/EVA	(20/80)	35–40

EVA, ethylene vinyl acetate; *EVOH*, ethylene vinyl alcohol; *LDPE*, low-density polyethylene; *LLDPE*, linear low-density polyethylene; *OPA*, oriented polyamide; *OPET*, oriented polyester; *PA*, polyamide; *PP*, polypropylene; *PVDC*, polyvinylidene chloride; *mLLDPE*, metallocene LLDPE; *ION*, ionomer.

Both chunk cheese and shredded cheese require a substantial oxygen barrier to prevent mold growth and spoilage. Ethylene vinyl alcohol, PVDC or PVOH may provide the oxygen barrier. While EVOH is generally coextruded into the film structure, PVDC or PVOH may be coated on a film via a coating process. Cheese packaging also requires excellent seal integrity and abuse resistance to prevent the controlled atmosphere inside the package from being lost. Cheese packages are often laminations made with reverse printed outer webs containing PET or PA for superior graphical presentation. The sealant layer may be extrusion coated onto the outer layer or adhesively laminated. Sealant layers may consist of EVA, ionomer or polyolefin plastomer. Low heat seal initiation temperature (90°C or below) and good seal through contamination performance are required. Processed cheese typically requires films with OTR of 9–15 cc/m^2-day-atm and WVTR of 15 g/m^2-day (Table B.9).

An acrylic, PVOH coated OPP film is also used in cheese packaging in both extrusion and adhesive laminations. Acrylic is coated on one side and PVOH on the other side.

B.2.9 Milk pouches

LLDPE or LDPE/LLDPE blends provide the sealant in both milk powder and liquid pouches (Table B.10). If an oxygen barrier is required for a long shelf-life, then PA or EVOH layers are included.

B.2.10 Frozen food

Frozen foods are packaged in a variety of packaging types. Examples of frozen foods packaged in flexible packaging include:

1. Frozen fruits;
2. Vegetables;
3. French fries;
4. Individually packed quick-frozen chicken breasts.

Many frozen foods are packaged in surface printed PE films. In recent years in the US, reverse-printed laminations, which may be shaped into stand-up pouches, have grown in use in the frozen food supermarket aisle. These packages have attractive graphics and may incorporate convenience features such as self-venting for microwave cooking.

Key requirements for frozen food packaging are:

1. Low-temperature toughness;
2. Stiffness;
3. High hot tack strength;
4. High seal strength.

Some packages are clear and require good clarity, while others are pigmented and require high gloss. LLDPE, ULDPE, EVA, and plastomer resins are all commonly used in creating frozen food packaging. Stiffness must be adequate for high-speed packaging and packaging films must have tear and puncture strength high enough to prevent package damage during transportation and storage (Table B.11).

TABLE B.10 Examples of flexible packaging structures for milk.

Structure	Layers (%)	Gauge (μm)
(LLDPE-Tie-PA-Tie-LLDPE)	(35/10/10/10/35)	40–70
HDPE/(LDPE + LLDPE)	(40/60)	40–80

LLDPE, linear low-density polyethylene; *PA*, polyamide; *HDPE*, high-density polyethylene; *LDPE*, low-density polyethylene.

TABLE B.11 Examples of frozen food packaging.

Structure	Layers (%)	Gauge (μm)
(EVA-LLDPE-EVA)	(15/70/15)	40–80
OPET//Tie//(LDPE-ION)	(15/5/40/40)	40–80
(mLLDPE-LLDPE-mLLDPE)	(15/70/15)	40–80
(HDPE-MDPE-EVA)	(15/70/15)	40–80

EVA, ethylene vinyl acetate; *LDPE*, low-density polyethylene; *LLDPE*, linear low-density polyethylene; *OPET*, oriented polyester; *mLLDPE*, metallocene LLDPE; ION, ionomer, *MDPE*, medium density polyethylene.

B.2.11 Fresh-cut produce

Key performance requirements for fresh-cut produce packaging include proper oxygen and carbon dioxide permeability, seal integrity, machinability and consumer appeal. Consumer appeal includes both feel and appearance. Feel is generally determined by film thickness and modulus while appearance is governed by the print quality and film optical properties, such as clarity, haze and gloss. In order to extend the shelf-life of the produce being packaged, films must provide the proper oxygen permeability that is matched to the packed produce respiration rate. Cut produce respires after harvesting, consuming oxygen and giving off carbon dioxide. By controlling the permeation of gases through the package, the environment inside the package is controlled, respiration is slowed and shelf-life is extended. The bags must have complete seal integrity in order to prevent the unplanned transfer of gases between the bags and the environment. Bags may contain PP, LLDPE, ULDPE, EVA, or plastomer. Oxygen transmission rate requirements vary widely depending on the produce being packaged, but common items range from about 1500 cc/m^2-day-atm for Caesar salad mixes to 2300–3000 cc/m^2-day-atm for iceberg salad mixes and 3000–5500 cc/100 in^2-day-atm for specialty salad mixes such as baby greens and exotic lettuces. Perforation may be used to obtain high transmission rates for some applications.

B.2.12 Retortable pouches

A growing flexible packaging use is in the replacement of metal cans with retortable pouches. These pouches are typically laminations containing biaxially oriented PA for toughness, aluminum foil for oxygen barrier and a polypropylene sealant film. These pouches may contain items like tuna, pet food and soup. The food items are held at an elevated temperature after packaging, so the packages must remain intact at these temperatures. In addition to temperature resistance, toughness, seal strength and barrier properties are critically important.

B.2.13 Edible oil packaging

The packaging of cooking oil uses PA to provide oxygen barrier properties. Ethylene acrylic acid (EAA) is typically used as the sealant layer. The seal type determines the PA layer location (Table B.12).

B.2.14 Bag-in-box

Laminates with aluminum foil or metallized films are often used in bag-in-box structures for liquid packaging. Coextruded films containing oxygen barrier polymers are replacing some metallized film laminates where flex crack resistance is required. Linear low-density polyethylene (LLDPE) or EVA polymers are used as sealants (Table B.13).

TABLE B.12 Examples of edible oil packaging.

Structure	Layers (%)	Gauge (μm)
(PA-Tie-EAA)	(10/15/75)	50–150
(EAA-Tie-PA-Tie-EAA)	(35/10/10/10/35)	50–150
(EAA-PA-EAA)	(40/20/40)	50–150

EAA, ethylene acrylic acid; *PA*, polyamide.

TABLE B.13 Example bag-in-box packaging structures without aluminum foil or metallized film.

Structure	Layers (%)	Gauge (μm)
(LLDPE-Tie-PA-Tie-LLDPE)	(35/10/10/10/35)	40–80
(EVA-Tie-PA-Tie-EVA)	(35/10/10/10/35)	40–80
(LLDPE-Tie-EVOH-Tie-LLDPE)	(35/10/10/10/35)	40–80

LLDPE, linear low-density polyethylene; *PA*, polyamide; *EVA*, ethylene vinyl acetate; *EVOH*, ethylene vinyl alcohol.

B.3 Industrial/consumer films

B.3.1 Stretch wrap

Stretch film, or stretch/cling film, is used to unitize goods for transportation. A thin film is stretched, either by machine or by hand and wrapped around packages to hold the goods together. The film clings to itself and to the pallet, securing the load. In its most common form, stretch/cling film is applied to a stacked pallet using a power prestretch pallet wrapper in an automated operation. In this operation, the film is stretched, between 100% and 300%, by rollers turning at different speeds and is then applied to a loaded pallet which sits atop a moving turntable. Machine wrap film is typically supplied on rolls that are 20 or 30 inches (51 or 76 cm) wide. Hand wrap film is supplied on smaller rolls. Stretch/cling films may be manufactured by either a cast or blown film process. Most stretch/cling films are coextruded structures containing three to seven layers. LLDPE is the primary component in most stretch films. For specialized applications, coextrusions containing minor PP, EVA, POP, m-LLDPE, or ULDPE layers may be employed (Table B.14 and Section 11.2.4.5). For most stretch film structures, a resin with good inherent cling is used on either one or both surface layers. A tackifier, such as poly-isobutylene, can also be added to the structure to provide the desired cling force. Stretch film is used to unitize entire or partial pallets stacked with products such as resin bags, fertilizer bags, consumer goods and food products during distribution. Most stretch film is removed by the retailer prior to displaying the packaged items for sale. Stretch films must have good:

1. Cling;
2. Stretchability;
3. Load retention;
4. Puncture resistance.

B.3.2 Heavy-duty bags (shipping bags)

Heavy-duty shipping sacks are used to transport items such as:

1. Resin;
2. Salt;
3. Pet food;
4. Fertilizer;
5. Chemicals;
6. Topsoil;
7. Bark mulch;
8. Compressed bales of fiberglass insulation.

TABLE B.14 Examples of stretch cling pallet wrap.

Structure	Layers (%)	Gauge (μm)
Stretch cling film		
(EVA-ULDPE-LLDPE)	(10/80/10)	15–30
(ULDPE-LLDPE-ULDPE)	(10/80/10)	15–30
(ULDPE-LLDPE-mLLDPE-LLDPE-ULDPE)	(10/25/30/25/10)	15–30
One-side cling film		
(mLLDPE-LLDPE-MDPE)	(10/80/10)	15–30
(EMA-LLDPE-PP)	(10/80/10)	15–30
(POP-LLDPE-PP)	(10/80/10)	15–30
(ULDPE-LLDPE-LLDPE-eLLDPE-PP)	(10/20/30/30/10)	10–30

EMA, ethylene methacrylic acid; *EVA*, ethylene vinyl acetate; *LLDPE*, linear low-density polyethylene; *MDPE*, medium density polyethylene; *PP*, polypropylene; *POP*, polyolefin plastomer; *ULDPE*, ultralow-density polyethylene; *mLLDPE*, metallocene LLDPE, *eLLDPE*, enhanced LLDPE.

When filled they weigh 40 pounds (18 kg) or more. Heavy-duty shipping sacks may be supplied as preformed bags or as roll stock, which is formed into bags in a continuous VFFS operation. Special machinery is required to form heavy-duty shipping sacks on VFFS machinery in a high-speed continuous operation. These bags need moderate COF because they must easily pass through the packaging equipment, but stacked bags must not slide off each other. Bags filled with hot products, such as salt, must also withstand the filling temperatures without excessive stretching or dimpling. Film toughness and creep resistance are also important in many heavy-duty shipping sack applications. LLDPE has allowed significant gauge reduction. LDPE is used to reduce creep and improve processability. High density polyethylene (HDPE) and PP are used for stiffness and higher end-use temperature resistance. EVA polymers are used for low-temperature sealability in form-fill-seal applications. Multimaterial laminates are often used but polyolefin coextruded structures also find a use for many applications (Table B.15).

B.3.3 Trash bags

The introduction of LLDPE accelerated using coextrusion in trash (or refuse) bags (Table B.16). Down gauging and using recycled material allows for improved economics and environmental impact. High molecular weight-HDPE polymers are also finding increased usage due to further downgauging opportunities. This is the largest coextruded film market segment.

B.3.4 Grocery sacks (merchandise bags)

High molecular weight-HDPE coextruded with LLDPE provides improved sealability with good downgauging potential (Table B.17). This film is typically made on high stalk HDPE blown film coextrusion lines.

TABLE B.15 Examples of polyethylene-based heavy-duty bag structures.

Structure	Layers (%)	Gauge (μm)
(LLDPE-EPE-LLDPE + LDPE)	(20/60/20)	100–200
(EPE-PP-EPE)	(20/60/20)	100–200

EPE, enhanced polyethylene; *LDPE*, low-density polyethylene; *LLDPE*, linear low-density polyethylene; *PP*, polypropylene.

TABLE B.16 Example trash bag coextrusion structures.

Structure	Layers (%)	Gauge (μm)
(LLDPE-LLDPE)	(50/50)	15–70
(LLDPE-RECYCLE-LLDPE)	(33/34/33)	15–70
(LLDPE—HMW-HDPE—LLDPE)	(10/80/10)	15–25

HMW-HDPE, high molecular weight-high-density polyethylene; *LLDPE*, linear low-density polyethylene.

TABLE B.17 Example structure for grocery bags.

Structure	Layers (%)	Gauge (μm)
(LLDPE—HMW-HDPE—LLDPE)	(10/80/10)	12–20

HMW-HDPE, high molecular weight-high-density polyethylene; *LLDPE*, linear low-density polyethylene.

TABLE B.18 Example structure for high clarity shrink film.

Structure	Layers (%)	Gauge (μm)
(PP-LLDPE-PP)	(25/50/25)	15–25

LLDPE, linear low-density polyethylene; *PP*, polypropylene.

B.3.5 High clarity shrink film (oriented)

Oriented, high clarity shrink film is used to protect and display high-value consumer goods. It is distinguished from regular shrink film by its superior clarity and appearance, as well as increased shrinkage properties and higher stiffness. Goods are packaged by wrapping the film loosely around the goods, sealing the film to make a completely enclosed bag and then shrinking the film with heat in a shrink tunnel or oven. Small holes may be poked in the film before wrapping to allow air to escape while the film is shrinking. As in industrial shrink film, heat causes the polymer molecules to relax, causing the film to return to its original unoriented size and shrink tightly around the packaged goods. Since the polymer molecules in oriented shrink film are much more highly oriented, greater shrinkage may be obtained. Boxed retail goods and stationery products are often wrapped with high clarity shrink film. Ice cream cartons and other food products are also wrapped in high clarity shrink film. Optical properties, seal properties, shrinkage and holding force are key requirements for oriented shrink film. These structures are normally biaxially oriented films of LLDPE and PP (Table B.18).

Index

Note: Page numbers followed by "*f*" and "*t*" refer to figures and tables, respectively.

F

I

J

K

L

M

N

Q

R

S

T

U

V

9780323854351